Analysis of Rotor Systems

The first volume of the second edition of *Rotor Systems* covers the analysis of simple rotors, including bearing and gyroscopic effects, with the help of simple rotor-bearing models. It is biased toward the finite element method analysis for complex rotors, both for torsional and transverse vibrations, including the transfer matrix method. In torsional vibration geared-rotor and branched system is also covered. The torsional and transverse vibration analysis covers the free and forced vibration analyses. Eigenvalue analysis helps in developing Campbell diagram for obtaining critical speeds. Unbalance response analysis helps in finding rotor response with rotor speed.

Key Features:

- Covers rotor analysis from simple single degree of freedom to multi-degree of freedom with lots of solved examples
- Illustrates bearing and seals theory with a perspective on rotor dynamics
- Discusses the finite element method and transfer matrix methods for torsional and transverse vibration analysis
- Includes elemental equations derived from the Euler–Bernoulli beam model, Rayleigh's beam model, and Timoshenko beam model
- Explores the Campbell diagram for forward and backward critical speed analysis, also Bode plot for unbalance response analysis with rotor speed.

This book is aimed at senior undergraduate/graduate students in mechanical engineering, rotor dynamics, rotating machinery/turbomachinery, and related fields.

Analysis of Rotor Systems
Volume I

Second Edition

Rajiv Tiwari

CRC Press
Taylor & Francis Group
Boca Raton London New York

CRC Press is an imprint of the
Taylor & Francis Group, an **informa** business

Designed cover image: Rajiv Tiwari

MATLAB® are trademarks of The MathWorks, Inc. and are used with permission. The MathWorks does not warrant the accuracy of the text or exercises in this book. This book's use or discussion of MATLAB® or Simulink® software or related products does not constitute endorsement or sponsorship by The MathWorks of a particular pedagogical approach or particular use of the MATLAB® and Simulink® software.

Second edition published 2027
by CRC Press
2385 NW Executive Center Drive, Suite 320, Boca Raton FL 33431

and by CRC Press
4 Park Square, Milton Park, Abingdon, Oxon, OX14 4RN

CRC Press is an imprint of Taylor & Francis Group, LLC

First edition published by CRC Press 2018

ISBN: 978-1-041-10355-4 (hbk)
ISBN: 978-1-041-11776-6 (pbk)
ISBN: 978-1-003-66155-9 (ebk)

DOI: 10.1201/9781003661559

Typeset in Times
by codeMantra

To the pioneers of rotating machinery, whose foundational work keeps our modern world spinning safely

Contents

Preface

This book is a second edition of the book *Rotor Systems: Analysis and Identification* published in 2017 by CRC Press as a single volume. Based on feedback received on the first edition of the book, now this book is proposed to have two volumes for comfortable reading. This book materials are the outcome of an elective course on "Rotor Dynamics" offered by me to undergraduate, graduate, and postgraduate students at IIT Guwahati over last 29 years. Moreover, it contains materials of some of the research studies done by me and my graduate students. Hopefully, the content will be useful for classroom teaching and will serve as a reference book for pursuing research and development in the field of rotor dynamics.

In the new edition of this book, few more sections have been added in various chapters. For example, in Chapter 3 of Volume I, the foil bearing basic theory has been introduced. Chapter 2 of Volume II has been overhauled thoroughly and included internal damping in the finite element instability analysis with additional examples. Chapter 8 of Volume II has been added with a few case studies on the experimental testing for vibration control of simple rotor systems for its balancing, control of cracked rotor, control of geared rotor, and control of misaligned rotor train. Also, few more examples have been added now with more detailed steps. Apart from these, in most of the chapters of both volumes, examples have been rechecked for accuracy, and the explanations have been expanded. More multiple choice questions have been added in most of the chapters, and answers to most of the exercise problems are provided. During the course of reading the book for the past 9 years, enumerable number of typos have been detected and are corrected in the present edition. For conciseness of equations, bold fonts are used to represent vectors and matrices (without brackets). Also for better representation of torsional displacement and torque now double headed arrow is used, hope this will remove the confusion of sense of rotation direction.

In rotor-bearing dynamics, modeling and analysis have now reached a developed state. In a wider perspective, this field spans numerous classes, namely the modeling, analysis, instability, control, identification, and condition monitoring in rotor-bearing systems. The finite element method (FEM) and the transfer matrix method (TMM) have been used extensively for modeling and analysis of rotor systems. The FEM has been very popular and applied extensively for identification and control in rotors. Till today, the condition monitoring of rotor-bearing systems based on vibrations is mainly concerned with the feature-based fault detection and diagnostics using advanced machine learning tools, which rely on past history data. As a result of this, the methods available so far are not up to the mark, reliable, and fail-safe up to the expectation of fellow engineers working in the fields due to lack of data. For model-based condition monitoring of the rotor-bearing systems, identification methods for system parameters are under development. For the identification of rotor system parameters, the literature available is not so rich, and a lot of possibilities have been immerged in the literature. The very purpose of this book is to give a basic understanding of the rotor dynamics phenomena with the help of simple rotor models and subsequently the modern analysis methods for real-life rotor systems. This background will be helpful in the identification of rotor-bearing system parameters and their use in futuristic model-based or physics-based condition monitoring leading to the digital models, digital shadows, and digital twins.

The content of this book has been split into two volumes. The first volume mainly concerns with the analysis of rotor systems, and the second volume is devoted to the instability, identification and control of rotors. The first volume contains the analysis of simple rotors, which will give a systematic presentation of some of the rotor dynamic phenomena including gyroscopic effects with the help of simple rotor models. For analyzing rotors used in practice, the free and forced response analyses of multi-DOF rotor system in torsional and transverse vibration have been presented. This book material is biased toward the FEM for the analysis of complex rotors because it is the most practical approach, and it can be applied to very large systems with its easy implementation in computer code and due to development in linear algebra, specially the condensation or reduction schemes. Rotor-bearing transverse vibration analysis has been performed considering the Euler–Bernoulli beam model, Rayleigh's beam model, and Timoshenko beam model for the shaft and the hydrodynamic bearing models for the bearings. In FEM, the Galerkin method has been used extensively because of its advantages over other methods. The dynamic analysis of rotors involves finding the whirl natural frequency, mode shapes, Campbell diagram, critical speeds, unbalance responses, and instability frequency bands. Apart from conventional bearings, i.e., the rolling element, hydrodynamic and foil bearings, and dynamic seals and dampers covered in volune I, this book also touches upon the contemporary active magnetic bearings and associated rotor unbalance response and instability analyses for the rigid and flexible rotors in Volume II.

The second volume deals with instability, identification and active control in rotor systems. Instability in rotors due to fluid-film bearing, internal damping, asymmetrical shaft, seals, steam whirl are covered with simple rotor models. For more complex rotor-bearing system the finite element modelling and associated eigenvalue analysis is used for instability analysis. System parameter identification algorithms are being developed, which are used to characterize critical component parameters with the help of experimental measurements and to identify the fault. Under this category, the unbalance identification and bearing dynamic parameter identification algorithms have been discussed in great detail. Associated with these, various sensors and exciters used in the rotor system and related signal conditioning equipment are described in detail. Measurement of acoustics is also covered. The error analysis is discussed in length to understand the associated measurement error. The signal processing is very important, and it has been covered in depth with special focus on the full spectrum, which is very relevant for rotor response analysis. It also contains some information on the condition monitoring of the rotating machinery. Condition monitoring of simple machine elements and subsystems based on time and frequency domain signals has been presented. It covers unbalance, shaft bow, rubs, shaft cracks, shaft misalignment, rolling element bearings, gears, pumps, coupling, and induction motors. Extensive information is available on feature-based identification of different types of faults in the rotor systems. These methods are being used by industries off late in conjunction with machine learning algorithms. The topic on active control of rotor system has been expanded to include experimentation on the flexible balancing of rotors, geared rotor systems, and misaligned rotor train system. This book compiles some of the available literature in a systematic and lucid form so as to boost research in the developing area, including sustainability, of rotor dynamics in relevant chapters.

As the title suggests, Chapters 1–10 of Volume I and Chapters 1–2 of Volume II are on the dynamic analysis and instability of rotor systems, Chapters 3–7 of Volume II are on the rotor system identification and its condition monitoring, whereas Chapter 8 of Volume II is on active control of rotors. Volume I contains Chapters 1–10, and Volume II contains Chapters 1–8. Chapter 1 of Volume I gives a historical development of the subject through problems faced by fellow engineers in the field in addition to the overall progress that has been made. In terms of analysis, a major portion of the text is devoted to finding the rotor system's natural frequencies (free vibration) and critical speeds (forced vibration) for transverse vibrations (Chapters 2–5, 8–10 of Volume I, and Chapter 2 of Volume II) as well as torsional vibration (Chapters 6 and 7 of Volume I). Chapter 3 of Volume I provides analyses of various bearings, seals, and dampers, as well as how to obtain rotordynamic parameters for them to be used for rotordynamic analysis. Chapters 1 and 2 of Volume II are devoted to the analysis of instabilities in rotor systems due to various sources. Dynamic balancing of rotors is covered in Chapter 3 of Volume II, and Chapter 4 of Volume II provides the experimental methodology of identifying rotordynamic parameters for bearings, seals, and dampers. Chapter 5 of Volume II describes various instruments used for rotordynamic measurement purposes, and Chapter 6 of Volume II describes the processing of measured signals to be used for rotordynamic identification. Monitoring the condition of various rotor elements and systems is covered in Chapter 7 of Volume II. Analysis of rotors integrated with active magnetic bearings is covered in Chapter 8 of Volume II, with an emphasis on dynamics and control of the rigid and flexible rotors. Few laboratory case studies on the control of rotors have been added. Overall, this book contains all the facets of the rotor analysis, instability, identification, and control. The content of this book can be judiciously chosen for a semester course depending upon the focus of the course. For example, Chapters 1–9 of Volume I and Chapters 1–3 of Volume II can be chosen for 40 hours of lectures, except that more details of Chapter 3 can be excluded. However, if the transfer matrix method is to be excluded from Chapters 6 and 8 of Volume I, then selected topics from the remaining chapters can be taken, especially from Chapters 4 to 8 of Volume II.

Throughout this book, materials are supplemented by numerical examples, both by hand calculation and computation extensive methods (MATLAB codes for most of the examples are available in the CRC Press repository). It is expected that with this book material, students will get sufficient exposure and motivation for applying FEM in rotor dynamics and allied areas. Practice engineers and researchers will get sufficient information on measurement and signal analysis. Exercise problems and multiple-choice questions would definitely enlarge the understanding of the subject, and some of them can be taken as term projects due to their effort involvement.

Acknowledgments

I sincerely acknowledge the Ministry of Human Resources and Development, New Delhi, for funding toward the development of the web and video courses under the National Programme on Technology Enhanced Learning on the subject. I would like to express special thanks and gratitude to my teachers (Dr N. S. Vyas, Dr B. P. Singh, Dr K. Gupta, Dr J. S. Rao, Dr V. Arunkumar, Dr A. W. Lees, Dr M. I. Friswell, and Dr R. Markert who are well-known figures in the field of rotor dynamics) and academic collaborators (Dr J. K. Sinha, Dr Fadi Dohnal, Dr S. Jana, Dr A. S. Sekhar, Dr A. Darpe, Dr M. Tiwari, Dr A. A. Khan, and Dr A. Chasalevris). My heartfelt thanks to the help offered by the undergraduate and graduate students, research scholars (Dr M. Karthikeyan, Dr Jagu S. Rao, Dr Sachin Singh, Dr Mohit Lal, Dr D. J. Bordoloi, Dr C. Shravankumar, Dr Sandeep Singh, Mr Dipendra K. Roy, Dr Purushottam Gangsar, Dr Shruti J. Rapur, Dr Siva Srinivas R., Dr Nilakshi Sarmah, Dr Gargi Majumder, Dr Prabhat Kumar, Dr D. Gayen, Dr Gyan Ranjan, Dr Rajeswara Rao Bhyri, Mr Pantha P Das, Mr Atul K Gautam, Ms Madhumita Mohanti, Ms Shivani Raj, Mr Ankit Ranjan, Mr Amit Khare, and Mr Kirti Sharma), and the project, technical, and office staff as well as the faculty at IIT Guwahati. A special thanks to Spinnergy Private Limited, a startup incubated at IIT Guwahati, and its team for utilizing my expertise for development of high capacity and fast response flywheel energy storage system (FESS). I also thank the innumerable students and practicing engineers worldwide who approached me for clarification on their work in this field—this book makes use of some of those discussions. I would also like to show appreciation for the technical support provided during the production of this book by Dr. Gagandeep Singh and Ms. Amrita Sood of CRC Press, Taylor & Francis Books India Pvt. Ltd.; Mr. Matthew Twigg, Senior Project Editor, CRC Press and Taylor & Francis Group USA; and Ms. Vinodhini Kumaran, Production Manager and her capable team members of Codemantra Pvt. Ltd., India. This work is dedicated to my family (Vibha, wife; Antariksh, son; Rimjhim, daughter), for being the anchor that kept me grounded while my mind revolved around the dynamics of rotating machinery.

About the Author

Rajiv Tiwari was born in 1967 at Raipur, Chhattisgarh, India. He graduated with a BE (Mechanical Engineering) in 1988 from Government College of Engineering and Technology, Raipur under Pt. Ravishankar University, Raipur, and an M.Tech. (Mechanical Engineering) in 1991 and a Ph.D. (Mechanical Engineering) in 1997 from the Indian Institute of Technology (IIT) Kanpur, India. He started his career as a lecturer in 1996 at Regional Engineering College, Hamirpur (Himachal Pradesh), India, and worked there for 1 year. In 1997, he joined the Indian Institute of Technology Guwahati as an assistant professor in the Department of Mechanical Engineering. He worked as a research officer at the University of Wales, Swansea, UK, for 1 year in 2001 on deputation. He was elevated to associate professor in 2002 and to professor in 2007 at IIT Guwahati. He was the head of the Center of Educational Technology and Institute Coordinator of the National Programme on Technology Enhanced Learning during 2005–2009, and the National Coordinator of the Quality Improvement Programme (QIP) for engineering college teachers during 2003–2009. He also visited the University of Darmstadt Germany under a DAAD fellowship during May–July 2011. He has been deeply involved in various research areas of rotordynamics, especially *identifying mechanical system parameters* (e.g., the unbalance, misalignment, bearings, seals, gears, coupling and rotor crack dynamic parameters), *diagnosing the faults of machine components* (e.g., bearings, gears, pumps, and induction motors), and *applying active magnetic bearings to monitoring the condition of rotating machinery.* His research areas also include rolling element bearing design and analysis for high-speed applications. He has completed three projects for the Aeronautical Research and Development Board (ARDB), India, on these topics. He has been offering consulting services for the last several years to Indian industries like the Indian Space Research Organisation (ISRO), Trivandrum; Combat Vehicle R&D Establishment (CVRDE), Chennai; Aeronautical Developmenet Agency, Banaglore; NBC Bearings, Jaipur and Tata Bearings, Kharagpur, as well as other local industries in the northeast of India. One of the European power industries, Skoda Power, Czech Republic, has also consulted him on seal dynamic parameter estimates for steam turbine applications. He has authored more than 250 journal and conference papers. He has guided 60 M.Tech. students and 17 Ph.D. students and 8 more students are currently pursuing research projects. He has successfully initiated and organized a national-level symposium on rotor dynamics (NSRD-2003), five short-term courses on rotor dynamics (2004, 2005, 2008, 2015, 2026), and a national workshop on "Use and Deployment of Web and Video Courses for Enriching Engineering Education" (2007) at IIT Guwahati, India. He has jointly organized an "International Conference on Vibration Problems" (ICOVP 2015) at IIT Guwahati. He has developed two web- and video-based freely available online courses under the National Programme on Technology Enhanced Learning, (1) Mechanical Vibration and (2) Rotor Dynamics, and under the MHRD-sponsored virtual lab on the "Mechanical Vibration Virtual Lab."

He has authored a text book on *Rotor Systems: Analysis and Identification* from Taylor & Francis Group, CRC Press, USA and the tutorial book *Simple Rotor Analysis through Tutorial Problems* from Taylor & Francis Group, CRC Press, USA. Prof. Tiwari has been featured among World's Top 2% Scientists List created by Stanford University since 2020. He is Associate Editor to the *Journal of Vibration and Control* (JVC), and *Journal of Vibration Engineering and Technology* (JVET). Also, he is in Worldwide Technical Committee on Rotor Dynamics of International Federation of Machines and Mechanisms (IFToMM). He has been included in The Bearings Sectional Committee, PGD 13, of Bureau of Indian Standards (BIS), which is responsible for the development of the national and international standards on the 'Magnetic Bearings', 'Aerospace Bearings' and 'Railway Bearings'.

1 A Brief History of Rotor Systems and Recent Trends

This chapter presents a brief history of the rotor dynamics field. It reviews the early development of simple rotor models, starting with the Rankine to Jeffcott rotor models and physical interpretations of various kinds of instabilities in rotor-bearing systems. It also reviews the development of analysis methods for the multiple degrees of freedom and continuous systems to allow practicing engineers to apply these methods to real turbomachineries. It also summarizes work on condition monitoring and recent trends in the area of rotor dynamics.

First, however, it would be relevant to examine the importance of this subject. Also, the main difference of this subject as compared to orthodox structural dynamics will be looked into, both in terms of dynamic analysis and condition monitoring (or system identification). The rotating machinery application exists in the domestic, medical, manufacturing, automotive, marine, and aerospace fields. Predicting the dynamic behavior of such rotating machinery (Figure 1.1a) is essential to prolong the life of the machinery along with the comfort and protection of humans. Rotor dynamics covers these analyses, and hence it is imperative that rotordynamicists have a firm understanding of it. Rotor dynamics is different in the following aspects as compared to structural vibrations:

i. Due to the relative motion between machine elements, rotating machineries have intrinsic forces and moments.
ii. For high-speed applications, the gyroscopic effect is very important, which provides a natural frequency as speed-dependent and the associated forward and backward whirl phenomena.
iii. Similar speed dependency of natural frequency occurs due to the presence of bearings and seals.
iv. Rotors can have instability due to various reasons: bearings and seals (due to cross-coupled anisotropic stiffness), asymmetric rotors (such as keyways or slots in rotors and turbine blades), internal (rotating) damping (material damping and friction between two rotating components), steam whirls (due to the high pressure of the steam in turbines), rubs (between rotor/blade/blisk and stator), interaction of aerodynamic forces with blades, thermal asymmetry (the Morton effect), electromechanical interactions, and several other reasons.
v. Due to high absolute motion of machine elements and due to the high relative motion among machine elements, the condition monitoring and maintenance practice also differ as compared to structures.
vi. Faults in rotors can give rise to excessive dynamic loads and even lead to instability (for example, cracks in rotors, and rotor stator rub).
vii. Sensors and instrumentation differ due to the vibration measurement of rotating components with respect to the stator.
viii. Vibration signal processing also differs due to phase measurements with respect to the rotating component reference point during balancing of rotors, especially when acquiring data during run-up and rundown of the machinery and when observing the forward and backward whirls in the full spectrum (see Chapter 6 in Volume II).

These issues make the rotor dynamics subject more demanding compared to structural dynamics.

DOI: 10.1201/9781003661559-1

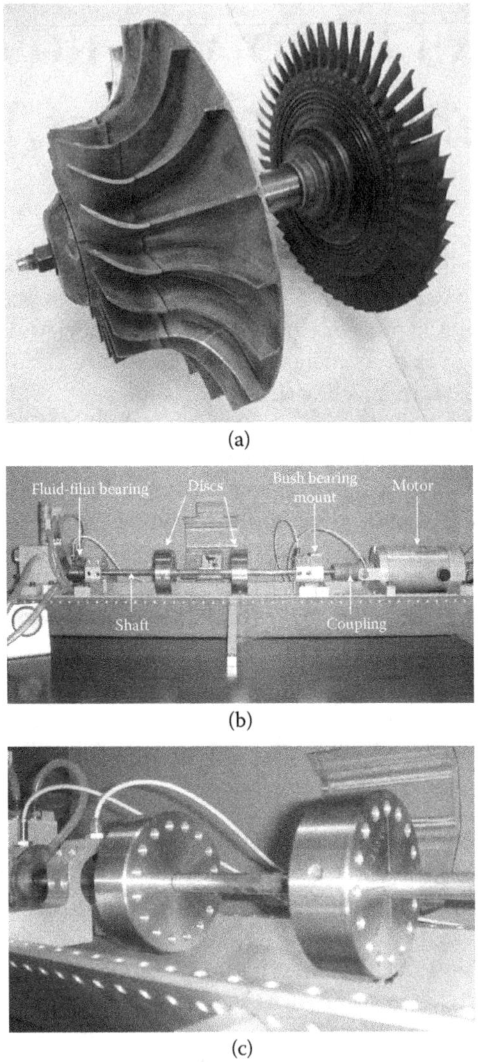

FIGURE 1.1 (a) A typical industrial rotor of a turbo-charger. (b) A typical rotor-bearing laboratory test rig. (c) A close view of a rotor consisting of two disks mounted on a flexible shaft.

Rotating machinery can be categorized based on various factors, like the speed of operation, the power it handles, its size, and so on. Applications with varied operating speeds include 3–4 rpm for cement factory kilns, stone crushers, and escalators; 100–2500 rpm for fans, and helicopter main and tail rotors; 3000 rpm for steam turbine generators; 9000–15,000 rpm for industrial compressors; 20,000 rpm for jet engines for airplanes; 50,000 rpm for cryogenic fuel pumps in rockets; 100,000 rpm for vacuum pumps for centrifuges; and above 100,000 rpm for micromachining applications and turbochargers. Similarly, these rotating machineries have varied power capacities: 0.5–3 W for the medical, micromachining, and household appliances; 2–3 MW for wind turbines, flywheel energy storage system, and train locomotives; 10–190 MW for gasoline pumps, heavy machine tools, and jet engines; 600–1200 MW for steam turbines and fossil fuel power stations; and 2–10 GW for hydraulic turbines and nuclear power plants. Depending on the application, the length of the rotor could be as long as 50 m (for the multistage steam turbine generator), 2 m for jet engines and missiles, 0.5 m for cryogenic pumps and electric motors, and few centimeters for helicopters and machine tool spindles.

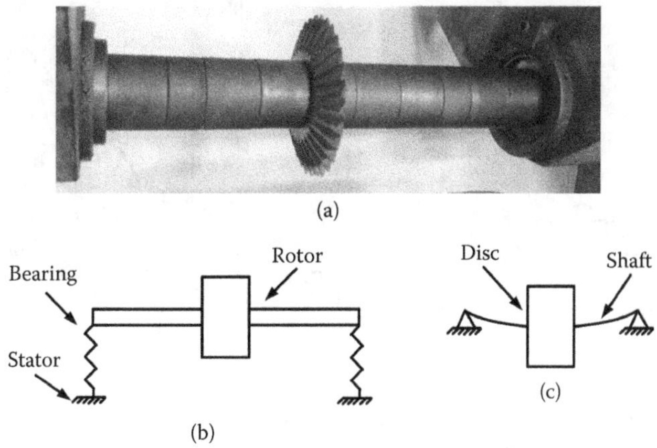

FIGURE 1.2 (a) A rigid rotor mounted on flexible bearings. (b) A flexible rotor mounted on rigid bearings.

A *rotor* is a body suspended through a set of cylindrical hinges or bearings that allow it to rotate freely about an axis fixed in space (Figures 1.1 and 1.2). Rotors can often be represented as a single beam or a series of beam elements with rigid disks and flexible blades. The beam is frequently considered to be flexible. Rigid disks are mounted on the flexible beam either by shrink-fit or other mechanical means. Practically, a rigid disk model represents flywheels, blades, cranks, rotary wings, coupling, disc brakes, impellers, and rolling bearings. Engineering components concerned with rotor dynamics include the rotating components of machines, especially of turbines, generators, motors, compressors, blowers, and the like. The parts of the machine that do not rotate are referred to by the general term *stator*. The machine element that allows relative motion of the rotor relative to the stator is called the *bearing*. Rotors of machines have a great deal of rotational energy and a small amount of vibrational energy while in operation. This is evident from the fact that a relatively small gas turbine propels a huge aircraft. The purpose of rotor dynamics as a subject is to keep the vibrational energy as small as possible. In operation, rotors undergo transverse (lateral or bending), longitudinal (axial), and torsional (or twisting) vibrations, either individually or in combination.

1.1 FROM THE RANKINE TO JEFFCOTT ROTOR MODELS

Rotor dynamics has a remarkable history of development, largely due to the interplay between its theory and its practice (Nelson, 2003). Rotor dynamics have been driven more by practice than by theory. This statement is particularly relevant to the early history of rotor dynamics. Research on industrial rotor dynamics spans more than 15 decades of history.

Rankine (1869) performed the first analysis of a spinning shaft (see Figure 1.3a). He predicted that beyond a certain spin speed, "the shaft is considerably bent and whirls around in this bent form." He defined this certain speed as the *whirling speed* of the shaft. In fact, it can be shown that beyond this whirling speed, the radial deflection of Rankine's model increases without limit, which is not true in actuality. However, Rankine did add the term *whirling* to the rotor dynamics vocabulary. Whirling refers to the movement of the center of the deflected disk (or disks) in a plane perpendicular to the bearing axis (see Figure 1.3b). In general, the frequency of whirl, ν, depends on the stiffness and damping of the rotor, as with the case of free vibration of a system (except for the synchronous whirl in which case it is equal to the unbalanced excitation force frequency, ω, i.e., the spin speed of the rotor). The whirl amplitude is a function of the excitation force's frequency, ω, and its magnitude. A bending critical speed, ω, occurs when the excitation frequency coincides with a (transverse) natural frequency, ω_{nf}, of the rotor system and can lead to excessive vibration amplitudes. Rankine neglected the Coriolis acceleration in his analysis, which led to erroneous conclusions that confused practicing engineers for half a century.

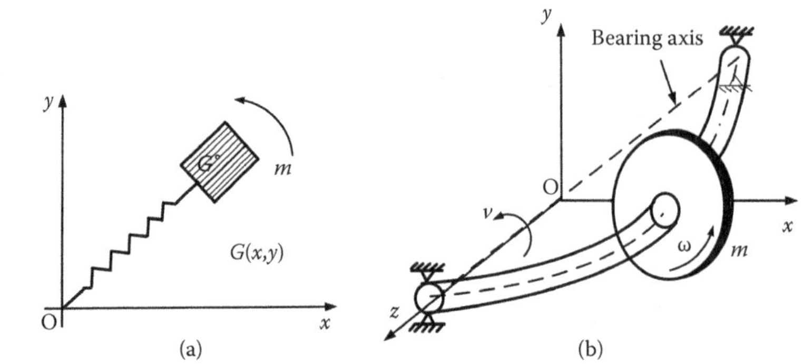

FIGURE 1.3 (a) Rankine rotor model (2 degrees of freedom spring-mass rotor model). (b) A Jeffcott (or Laval or Föppl) rotor model in general motion.

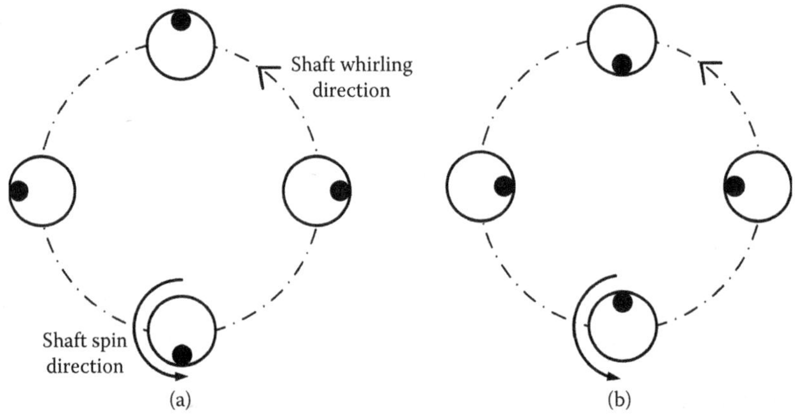

FIGURE 1.4 Synchronous whirls: (a) heavy side flying out (before critical speed). (b) Heavy side flying in (after critical speed).

The turbine built by Parsons (1948) operated at speeds of around 18,000 rpm, which was 50 times faster than the existing engine at that time. In 1883, Swedish engineer de Laval developed a single-stage impulse steam turbine (Figure 1.3b) (named after him) for marine applications and succeeded in operating it at 42,000 rpm. He aimed at the self-centering of the disk above the critical speed (Figure 1.4), a phenomenon that he instinctively recognized. He first used a rigid rotor, but later used a flexible rotor and showed that it was possible to operate above critical speed by operating at a rotational speed about seven times the critical speed (Stodola, 1924).

In order to calculate the critical speeds of cylindrical shafts with several disks and bearings, the general theory of Reynolds (Dunkerley, 1895) was applied. The gyroscopic effect was also considered, together with its dependence on speed (i.e., a Campbell diagram—see Figure 1.5).

A Campbell diagram is designed to show rotor whirl frequency, ν, (natural frequency) with the spin speed, ω, of the shaft. In rotors, the rotor whirl frequency changes with speed due to various reasons (e.g., gyroscopic effects, speed-dependent bearing parameters, etc.). However, this can be used to obtain critical speeds of rotors (intersections of whirl frequency curves with the $\nu=\omega$ line, e.g., at resonance when whirl frequency is equal to the spin speed). Apart from this whirl frequency, the logarithmic decrement is also shown in the Campbell diagram, and the sign of that shows the stable/unstable stage of the rotor. Today more information is being put into these graphs, which are now are called Lee's diagram (Lee, 1993). The Campbell diagram can be drawn from theoretical/numerical analyses as well as through the actual measurement from the machine.

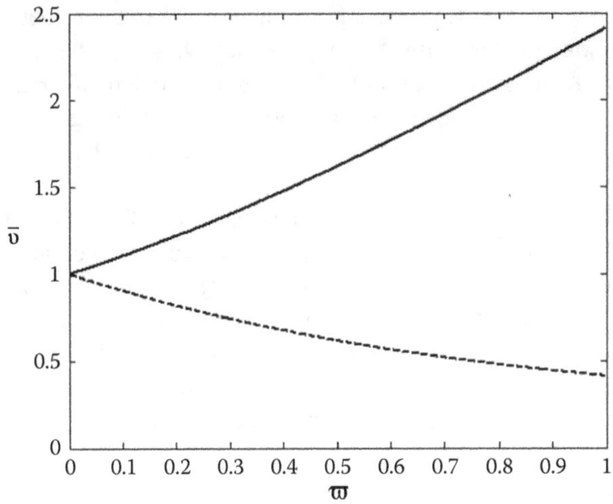

FIGURE 1.5 The non-dimensional natural frequency, $\bar{\nu}$, variation with the non-dimensional spin speed, $\bar{\omega}$, (a typical Campbell diagram).

Dunkerley found through numerous measurements the relationship known today through the work of Southwell, by which the first critical speed can be calculated, even for multi-degree-of-freedom rotor cases. Dunkerley was the first to use the term *critical speed* for the resonance spin speed. Even with general knowledge of critical speeds, the shaft behavior at any general speed was still unclear, but more was learned from the calculation of unbalance vibrations, as given by Föppl (1895). He used an undamped model to show that an unbalanced disk would whirl synchronously with the heavy side (shown as a black spot) flying out (Figure 1.4a) when the rotation was subcritical and with the heavy side flying in (Figure 1.4b) when the rotation was supercritical. Also, the behavior of Laval rotors at high speeds was confirmed by his theory.

Engineers at that time were perplexed by the concepts equating Rankine's whirling speed with Dunkerley's critical speed. This was particularly frustrating because Rankine was far more of an authority than Dunkerley and, as a result, his predictions were widely accepted and were responsible for discouraging the development of high-speed rotors for almost 50 years. It was in England in 1916 that things came to an end. Kerr published experimental evidence that a *second critical speed* existed, and it was obvious to all that a second critical speed could only be attained by the safe traversal of the first critical speed.

The first recorded fundamental theory of rotor dynamics can be found in a classic paper by Jeffcott in 1919. Jeffcott confirmed Föppl's prediction that a stable supercritical solution existed, and he extended Föppl's analysis by including external damping (i.e., damping forces that depend upon only the absolute velocity of the rotor, whereas the internal damping comes from the rate of deformation of the shaft, often called the rotating damping) and showed that the phase of the heavy spot varies continuously as the rotation rate passes through the critical speed (Figure 1.4). We can appreciate Jeffcott's great contributions if we recall that a flexible shaft of negligible mass with a rigid disk at the midspan is called a *Jeffcott rotor* (Figure 1.3b). The bearings are rigidly supported, and viscous damping acts to oppose the absolute motion of the disk. This simplified model is also called the *Laval rotor* and *Föppl rotor*, named after de Laval and Föppl, respectively.

1.2 ROTOR DYNAMICS PHENOMENA STUDIES FROM STODOLA TO LUND

Developments made in rotor dynamics up to the beginning of the twentieth century are detailed in the masterpiece book written by Stodola (1924). Among other things, this book includes the dynamics of an elastic shaft with disks, the dynamics of continuous rotors without considering gyroscopic moment, the secondary resonance phenomenon due to the effect of gravity, the balancing of rotors, and methods of determining the approximate values of critical speeds of rotors with variable cross-sections. He presented a graphical procedure to calculate critical speeds, which was widely used. He showed that these supercritical solutions were stabilized by Coriolis accelerations (which eventually give the gyroscopic moments). The constraint of these accelerations was the defect in Rankine's model.

It is interesting to note that Rankine's model is a sensible one for a rotor whose stiffness in one direction is much greater than its stiffness in the quadrature (perpendicular) direction. Indeed, it is now well known that such a rotor will have regions of divergent instability (see Figure 1.6). It is less well known that Prandtl (1918) was the first to study a Jeffcott rotor with a noncircular cross-section (i.e., elastic asymmetry in the shaft). In Jeffcott's analytical model, the disk did not wobble or precess (see Figure 1.7) due to the disc at mid-span. As a result, the angular velocity vector and the angular momentum vector were collinear and no gyroscopic moments were generated. This restriction was removed by Stodola.

About a decade later, the study of asymmetrical shaft systems and asymmetrical rotor systems began. The former are systems with a directional variation in the shaft stiffness (Figure 1.8a), and the latter are those with a directional difference in rotor inertia (Figure 1.8b). Two-pole generator rotors and propeller rotors are examples of such systems. As these directional differences rotate with the shaft, terms with time-varying coefficients appear in the governing equations. These systems, therefore, fall into the category of parametrically excited systems, in which vibrations depend on the motion itself; however, they may occur in a linear or a nonlinear system. The most characteristic property of an asymmetrical system is the appearance of unstable vibrations in some spin speed ranges. In 1933, Smith produced a pioneering work in the form of simple formulas that predicted the

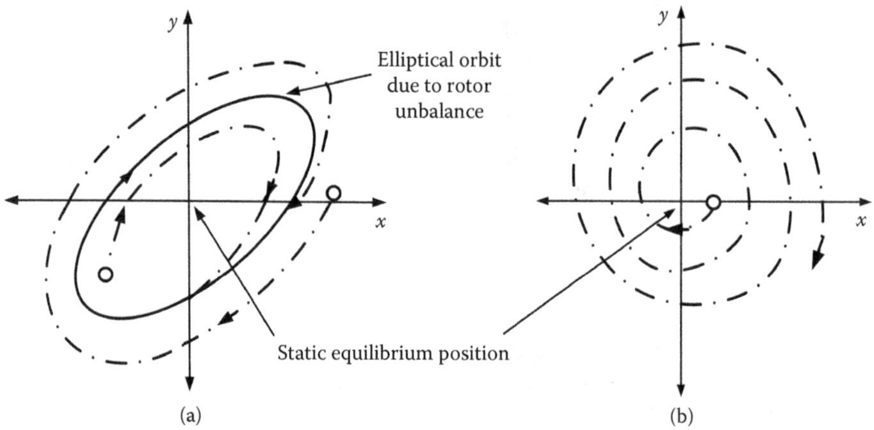

FIGURE 1.6 Journal center path due to perturbation: (a) stable motion and (b) unstable motion.

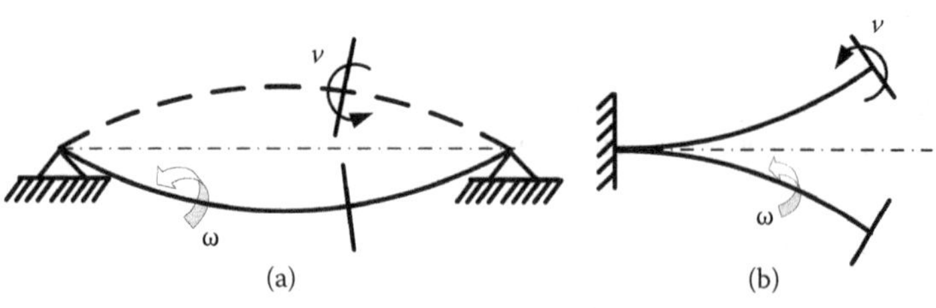

FIGURE 1.7 Wobbling of a disk in a rotor system. (a) A simply supported shaft with a disk near the bearing. (b) A cantilever shaft with a rigid disk at the free end.

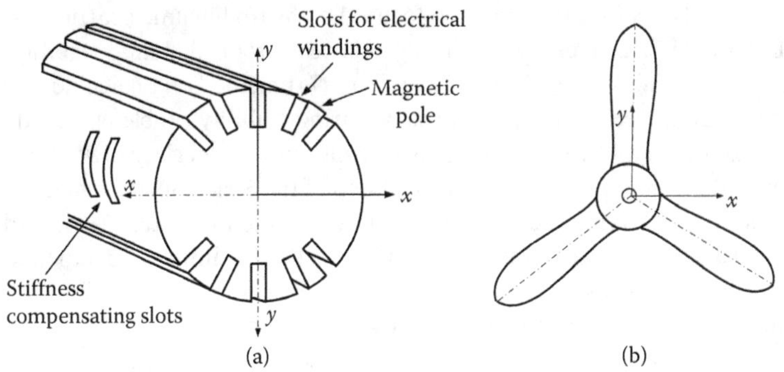

FIGURE 1.8 Asymmetry of the shaft and the rotor: (a) a generator rotor. (b) A three-bladed propeller.

threshold spin speed for the supercritical instability varied with the bearing stiffness and with the ratio of external to internal viscous damping. The formula for damping was obtained independently by Crandall and Brosens (1961) some 30 years later.

In the early 1920s, a supercritical instability in built-up rotors was encountered. Thereafter, it was first shown by Newkirk (1924) and Kimball (1924) to be a manifestation of the rotor's internal damping (i.e., the friction damping between rotor components). Then, Newkirk and Taylor (1925) described an instability caused by the nonlinear action of the oil wedge in a journal bearing, which was named the *oil whip*. Baker (1933) described self-excited vibrations due to contact between the rotor and the stator (i.e., the dry whip). The Soviet scientist Nikolai (1937) examined the stability of transverse and torsional vibrations in a shaft with a disk mounted in the center and the stability of a shaft with a disk attached to the free end. Kapitsa (1939) pointed out that a flexible shaft could become unstable due to friction conditions in its sliding bearings. In the middle of the twentieth century, Hori (1959) succeeded in explaining various fundamental characteristics of the oil whip by investigating the stability of shaft motion and considering pressure forces due to oil films. The mechanism of vibrations due to the steam whirl in turbines was explained by Thomas (1958), and that in compressors was explained by Alford (1965). The vibration of a hollow rotor containing the fluid was the problem of flow-induced vibrations. Instability due to liquids partially filling the interior cavities of rotors was demonstrated by Kollmann (1962), and, in 1967, Ehrich reported that fluid trapped in engine shafts induced asynchronous vibration and changed the shape of resonance curves. Kuipers (1964) and Wolf (1968) independently succeeded in explaining the appearance of an unstable speed range in a postcritical region of a rotor system containing inviscid fluid. In the 1980s, the rotor dynamic effects of seals in fluid-handling machines received a great deal of attention. Rotor destabilization due to seals was predicted and demonstrated in an operational compressor by Jenny (1980).

As rotors became lighter and rotational speeds higher, the occurrence of nonlinear resonances such as subharmonics and superharmonics became a serious problem. Yamamoto (1955, 1957) studied various kinds of nonlinear resonances after he reported on the subharmonic resonance due to ball bearings in 1955. He also investigated *combination resonances*. Tondl (1965) studied nonlinear resonances due to oil films in journal bearings. Ehrich (1966) reported on subharmonic resonances observed in an aircraft gas turbine due to strong nonlinearity produced by the radial clearance of squeeze-film dampers.

The nonstationary phenomena during passage through critical speeds have been studied since Lewis (1932) reported his investigation on the Jeffcott rotor. Nonstationary phenomena that occur are one in a process with a constant acceleration (unlimited driving torque) and another with variable acceleration (limited driving torque). Natanzon (1952) studied shaft vibrations at critical speeds, and Grobov (1953, 1955) investigated in a general form the shaft vibrations resulting from varying rotational speeds. The development of the asymptotic method by Mitropolskii (1965) for nonlinear problems considerably boosted the research on this subject.

Beginning in the early 1960s, most attention focused on hydrodynamic bearings—this was largely stimulated by Lund (1964). Gunter's work (1966) related to rotor dynamic stability problems and, combined with Ruhl and Booker's (1972) and Lund's (1974) methods for calculating damped critical speeds, stimulated a great deal of interest in rotor-bearing stability problems. Lund (1987) gave an overview of the field. In the mid-1970s, rotor dynamic instability experienced with various high-pressure compressors and the high-pressure fuel turbo pump of the Space Shuttle's main engine focused a great deal of attention on the influence of fluid–structure–interaction forces, particularly forces due to the liquid and gas seals, in pumps and turbines. Shaft seals have a similar effect to fluid-film bearings. They influence the critical speeds and can either provide damping or cause instability. Shaft seals have acquired a significant role in their effect on rotor dynamics. Someya (1989) and Tiwari et al. (2004, 2005) compiled extensive numerical and experimental results and presented literature review related to the identification of rotor dynamic parameters of bearings and seals.

Self-excited vibrations, which occur due to nonconservative forces, generally lead to large vibration amplitudes, which may ultimately damage or even destroy rotating machinery (Childs, 1993; Gasch et al., 2002; Tondl, 1965; Yamamoto and Ishida, 2001). Therefore, it is essential during the design stage of a new machine to consider the possibility of self-excitation and take measures against it. A strategy to suppress self-excited vibrations that is based on an antiresonance phenomenon (two neighboring modes having opposite effects) that can occur in parametrically excited systems (Tondl, 1978, 1991, 1998) was described by Ecker and Tondl (2004). The basic idea of parametric stabilization was adopted by introducing a time-dependent variable stiffness located at the bearing mounts. The nonconservative forces were generated through the bearings of the rotor. They showed the cancellation of the self-excited vibrations through the parametric excitation.

Shaw and Balachandran (2008) provided a comprehensive review of nonlinear dynamics of mechanical systems, including the rotating machineries. For rotating machineries, they considered both the ideal and nonideal excitations. In an ideal excitation case, it is assumed that the rotor speed is a specified function of time, which is a classic in the theory of nonstationary problems in dynamics and is extensively covered in the book by Mitropolskii (1965). The problem of passage through resonance of nonideal vibrating systems has received special attention from engineering researchers in recent years, but unfortunately little literature on this subject is available (Balthazar et al., 2003). Generally, nonideal vibrating systems are those for which the power supply is limited. Laval likely was the first one to work with nonideal problems via an experiment. In 1889, he built a single-stage turbine and demonstrated that in the case of rapid passage through the resonance with enough power, the maximum vibration amplitude may be reduced significantly compared with that obtained in the steady-state resonant vibration. Simultaneously, it was also known that sometimes the passage through resonance required more input power than the excitation source had available. The consequence is the so-called *Sommerfeld effect* in which the vibrating system cannot pass the resonance or requires an intensive interaction between the dynamic system and the motor to do it. The worst case is that of a dynamic system constructed for an overcritical operation to become stuck just before resonance conditions are reached. A strong interaction results with fluctuating motor speed and fairly large vibration amplitudes. This phenomenon was studied intensively by Sommerfeld (1927). Balthazar et al. (2003) provided an excellent review on the topic of the limited power source, in which case the system is called the nonideal vibrating system. Yamamoto and Ota (1964), Dimentberg (1961), Crandall and Yeh (1989), and Jei and Lee (1990) reported the curve veering in rotor-bearing systems. In Crandall and Yeh's words (1989), "It is interesting that when the curve for an even rotor mode approaches the curve for an even stator mode, or an odd rotor mode approaches an odd stator mode, the two modes form a coupled system and the curves repel each other avoiding an intersection." This was in reference to the Campbell diagram of the natural frequency of a uniform rotor rotating in a uniform stator, as a function of rotational speed.

Instability from fluid-film bearings and shaft seals arises from the fact that, during the radial displacement of a rotor, a restoring force is produced that has a component at right angles to this displacement. The phenomenon of instability was described in detail by Newkirk (1924), whose interest was in turbomachineries. The cause of this instability, in fact, lay in the oil-film bearings. In the following years, it was established that, in a few cases, internal friction or damping could indeed be a cause of instability.

A synchronous instability due to differential journal surface temperature, called Morton effect or thermally induced instability, may be mostly observed on overhung rotors supported on lightly loaded bearings. A differential journal surface temperature could cause a thermal bow, produce a hot spot chasing for a high spot, and introduce thermal unbalance with a continuous phase shift (de Jongh and Morton, 1994; de Jongh and va der Hoeven, 1998; Berot and Dourlens, 1999; de Jongh, 2008; Tong et al., 2017; Gu, 2018).

The designer must thus be aware of these possibilities. Some of the important phenomena in rotor-bearing systems, their main causes, and investigators' details are summarized in Table 1.1.

1.3 DEVELOPMENT OF ROTOR DYNAMICS ANALYSIS TOOLS

In rotor dynamics, a remarkable number of phenomena can be explained by the dynamics of a single mass Jeffcott rotor model. This model, introduced in 1895 by Föppl, was named after Jeffcott because in 1919 he first explained the science of rotor dynamics in a graphic and illuminating way. Gradually, the Jeffcott rotor model, in its many variations, came closer to the practical needs of rotor dynamicists of those days.

Many practical rotors, especially those designed for steam/gas turbines for power stations (Figures 1.9 and 1.10a) or for aircraft gas turbines (Figure 1.10b), were not suitable for a Jeffcott model. For one thing, the distinction between the disk and the shaft is blurred in the typical aircraft gas turbine (Dimentberg, 1961). In the practical design of rotating machinery, it is necessary to know accurately the natural frequencies, mode shapes, and forced responses to unbalances in complex-shaped rotor systems. The technique for this was supplied by Prohl (1945) in the late 1930s and published in 1945 for the critical speed evaluation of a turbine shaft. It is similar to the method published about the same time by Myklestad (1944) for the natural frequencies of aircraft wings but was developed independently. Together, Prohl's and Myklestad's work led to a broader method, now called the transfer matrix method. This method is particularly useful for multi-rotor-bearing systems and has developed rapidly since the 1960s by the contribution of many researchers such as Lund and Orcutt (1967), Lund (1974), and Rao (1996). The transfer matrix method for rotors remains viable; indeed, it seems still to be the method of choice for most industrial rotor dynamic analyses. Another representative technique used for this purpose is the finite element method (FEM). The name first appeared in the title of a paper by Clough (1960). The first application of the FEM to a rotor system was made by Ruhl and Booker (1972). Then, Nelson and McVaugh (1976) generalized it by considering the rotary inertia, gyroscopic moment, and axial force. It was soon recognized that the large number of nodes necessary to provide accurate stress distribution created dynamic systems too large for economical calculation. Condensation of the number of degrees of freedom by division into the master and slave degrees of freedom was introduced by Guyan (1965). Other dynamic condensation techniques were described by Uhrig (1966), Friswell and Mottershead (1996), Tiwari and Dharmaraju (2006), Prasad and Tiwari (2019), and Baur et al. (2014). A related technique for the dynamic analysis of structures assembled from distinct components or substructures in the component mode synthesis was introduced by Hurty (1960) and applied to rotor dynamics by Glasgow and Nelson (1980), Geradin and Kill (1984), and Crandall and Yeh (1986). Each substructure interacts only through its constraint modes. The next section provides a brief summary of software in use for rotor dynamics analyses.

1.4 SOFTWARE FOR ROTOR DYNAMICS ANALYSIS

World War II can be considered the demarcation between the early stages of rotor dynamics and what might be called modern rotor dynamics. In the 1960s, there was a coalescence of numerical methods applied to structural dynamics and of digital computer capacity that fostered the development of a

TABLE 1.1
Summary of Various Rotor Dynamics Phenomena

S.N.	Phenomena	Caused by	Reported/Interpreted by	Remarks (Theoretical: Th/Experimental: Exp)
1	Whirling	Unbalance	Rankine (1869)	General motion (Th)
2	Self-centering of the rotor	Unbalance	de Laval in 1883 (Meher-Homji, 2000)	Unbalance response (Exp)
3	Synchronous whirling	Unbalance	Föppl (1895)	Resonance (Th)
4	Critical speed	Unbalance	Dunkerley (1895)	Resonance (Th)
5	Sommerfeld effect	Limited drive power	Sommerfeld (1927)	Crossing through resonance (Exp)
6	Second critical speed	Unbalance	Kerr (1916)	Resonance (Exp)
7	Stable supercritical response	Unbalance	Jeffcott (1919)	Unbalance response (Th)
8	Secondary resonance (or gravity critical speed)	Gravity	Stodola (1924)	Unbalance response (Th)
9	Instability	Shaft asymmetry	Prandtl (1918)	Instability analysis (Th)
10	Gyroscopic effect	Rotor wobbling	Stodola (1924)	Free vibrations (Th)
11	Threshold spin speed for instability from rotor	Internal damping	Newkirk (1924), Kimball (1924), Smith (1933), and Crandall and Brosens (1961)	Instability analysis (Th)
12	Threshold spin speed for instability from bearing	Anisotropy of bearing stiffness	Smith (1933)	Instability analysis (Th)
13	Oil whirl	Nonlinear action of the oil wedge in a journal bearing	Newkirk and Taylor (1925)	Instability analysis (Th)
14	Dry whip (self-excited vibration)	Contact between the rotor and the stator	Baker (1933)	Instability analysis (Th)
15	Oil whip	Hydrodynamic bearing	Hori (1959)	Instability analysis (Th)
16	Steam whirl	High-pressure steam injection on turbine blades	Thomas (1958)	Instability analysis (Th)
17	Flow-induced vibrations	Hollow rotor containing fluid	Kollmann (1962), Ehrich (1967, 1992), Kuipers (1964), and Wolf (1968)	Instability analysis (Th)
18	Subharmonics	Combination resonance	Yamamoto (1955, 1957)	Nonlinearity (ball bearing)
19	Nonlinear resonance	Oil films in journal bearings	Tondl (1965)	Nonlinear analysis (Th)
20	Mistuning of turbine blade	Turbine blades	Whitehead (1966)	Linear analysis (Th)
21	Subharmonic resonances	Squeeze-film dampers	Ehrich (1966)	Nonlinear study (Exp)
22	Nonstationary response	Constant/variable accelerations of the rotor	Lewis (1932)	Nonlinear analysis (Th)
23	Shaft vibrations at critical speeds	Varying spin speeds	Natanzon (1952)	Transient analysis (Th)
24	Shaft general motion	Varying spin speeds	Grobov (1953, 1955)	Transient analysis (Th)
25	Damped critical speeds	Hydrodynamic bearings	Ruhl and Booker (1972) and Lund (1974)	Finite element analysis (Th)
26	Instability	Seals	Jenny (1980)	Instability analysis (Exp)
27	Morton's effect	Journal thermal gradients	Keogh and Morton (1994)	Thermal instability analysis (Th and Exp)
28	Curve veering phenomenon	Speed-dependent parameters	Dimentberg (1961)	Eigenvalue analysis (Th) and Campbell diagram (Th/Exp)

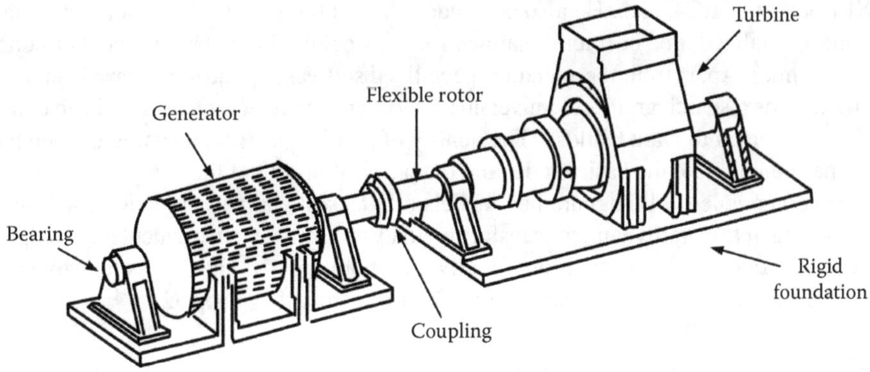

FIGURE 1.9 A rotor-bearing-foundation system.

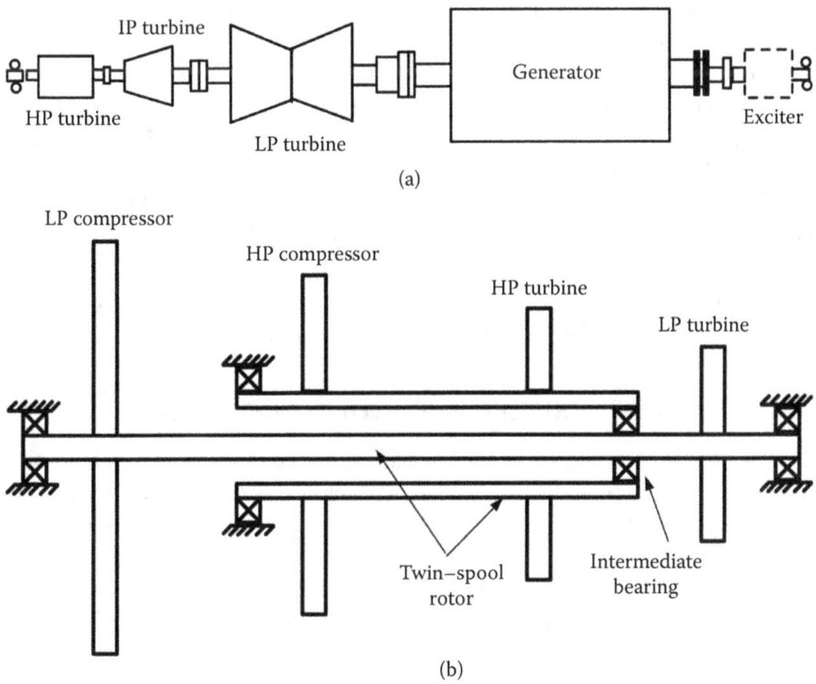

FIGURE 1.10 (a) A typical turbomachinery configuration in a power plant. (b) A schematic of the twin-spool aero-engine.

series of general-purpose computer codes. The initial application of these codes to rotor dynamics was based on the transfer matrix method, but in the 1970s another underlying algorithm, the FEM, became available for the solution of the prevailing beam-based models. Now, in the beginning of the twenty-first century, rotor dynamicists are combining the FEM and solid modeling techniques to generate simulations that accommodate the coupled behavior of flexible disks, flexible shafts, and flexible support structures into a single, massive, multidimensional model. Crandall (1992) gave an overview of the rotor dynamic computer codes (e.g., Abaqus, ANSYS, ARMD, AxSTREAM, CADENSE, ComboRotor, COMSOL, DYNAMICS R4, DynaRot, DyRoBeS, iSTRDYN, MADYN 2000, MSC NASTRAN ROTORDYNAMICS, OpenFoam, RAPPID-RDA, RODYN, ROMAC, ROSS, ROTECH, RSR, SAMCEF, Siemens Simcenter 3D, TURBINE-PAK, VibronRotor, VT-FAST,

XLOG/XLRotor, XLTRC4, etc.). He also concluded that with regard to the quality and quantity of software, the specialized area of rotor dynamics still lags behind in the broader field of nonrotating structural dynamics. Apart from these, many specialized software applications have been developed in-house by various research groups in universities. Modern computer models have been commented on in a quote attributed to Dara Childs: "The quality of predictions from a computer code has more to do with the soundness of the basic model and the physical insight of the analyst.… Superior algorithms or computer codes will not cure bad models or a lack of engineering judgment." Integrating machine learning into rotordynamics transforms analysis from reactive calculation to predictive intelligence. Traditional software tools are rapidly incorporating ML algorithms to automate model tuning, accelerate simulation speeds, and predict failures before they happen.

1.5 DYNAMIC BALANCING OF ROTORS

The most important and fundamental procedure to reduce unfavorable vibrations is to eliminate the geometric unbalance in the rotor. The balancing procedure for a rigid rotor was established relatively early. The arrival of high-speed rotating machines made it necessary to develop a balancing technique for flexible rotors (i.e., the rotor that operates above critical speed). Two representative theories were proposed for flexible rotors. One was the modal balancing method proposed in the 1950s by Federn (1957), and Bishop and Gladwell (1959). The other is the influence coefficient method proposed in the late 1930s by Rathbone (1929), and later by Thearle (1932), and developed mainly in the United States, along with the progress of computers and instruments for vibration measurements (Darlow, 1989; Wowk, 1995). Modern methods based on FEM require a good model of the rotor-bearing foundation systems along with startup/rundown data, and hence the trend is to couple the estimation of unbalance along with the bearing and foundation dynamic parameters (in which the modeling error is large), subject to a minimum number of rundowns (Rieger, 1986; Schneider, 1991; Edwards et al., 1998; Robert, 2003; Tiwari et al., 2004).

1.6 CONDITION MONITORING OF ROTATING MACHINERY

Another area in which a lot of development took place is on the assessment of turbomachinery condition monitoring and failure prognosis technology (Collacott, 1977; Lipovszky et al., 1990; Wowk, 1991; Mitchell, 1993; Davies, 1998; Adams, 2001; Rao, 2007; Lees, 2016). High-performance turbomachines are now extremely important elements of the industry worldwide. The electric power, petrochemical, mining, marine, and aircraft industries are prime examples for which turbomachinery is crucial to business success (Figure 1.10). Condition monitoring involves the continuous or periodic assessment of the condition of a plant or a machinery component during its operation. Basically, condition monitoring is the process of monitoring some parameters from the machinery, such that a significant change in a parameter can give information about the health of the machinery. The acoustics and vibration signals from machines can contain vital information about the internal process and can provide valuable information about a machine's running condition. Acoustic signals are measured in a region of proximity to the external surface of the machine, whereas vibration signals are measured on the surface of the machine. Most acoustic and vibration analysis instruments utilize a fast Fourier transform, which is a special case of the generalized discrete Fourier transform. A spin-off of it is the full spectrum that displays both the forward and backward whirl amplitude with frequency (refer to Chapter 6 of Volume II).

According to Eshleman (1990), over the past several years, instrumentation and monitoring capabilities have increased dramatically, but techniques for fault diagnosis have evolved slowly. The tools are therefore still more advanced than the techniques. Edwards et al. (1998) provided a broad review of the state of the art in fault identification and diagnosis techniques, with particular regard to rotating machinery. Special treatment was given to the areas of mass unbalance, bowed shafts, misalignments, and cracked shafts, these being among the most common rotor dynamic faults.

Vibration response measurements yield a great deal of information concerning any faults within a rotating machine. Cracks in shafts have long been identified as factors limiting the safe and reliable operation of turbomachines. They can sometimes result in catastrophic failure of equipment (rotor bursts) and, more often, in costly process upsets, repairs, and premature scrapping and replacement of equipment. In the past two decades, much research and many resources have gone into developing various online (Figure 1.11) and offline (Figure 1.12) diagnostic techniques (expert systems) to effectively detect faults before they cause serious damage. The expert system uses a probability table such as that compiled by Sohre (1991), or in modern days, artificial intelligence and machine learning techniques (refer to Chapter 7 of Volume II).

With acoustic condition monitoring, a machine running in good condition has a stable noise spectrum. Spectrum changes when the condition changes. Identification of noise sources and comparing their spectrum with that of a stable spectrum of a machine or plant in good condition can prove to be an important tool for condition monitoring. One of the earliest documented applications of acoustic emission technology to rotating machinery monitoring was in the late 1960s. Since then, there has been an explosion in research- and application-based studies covering bearings, pumps,

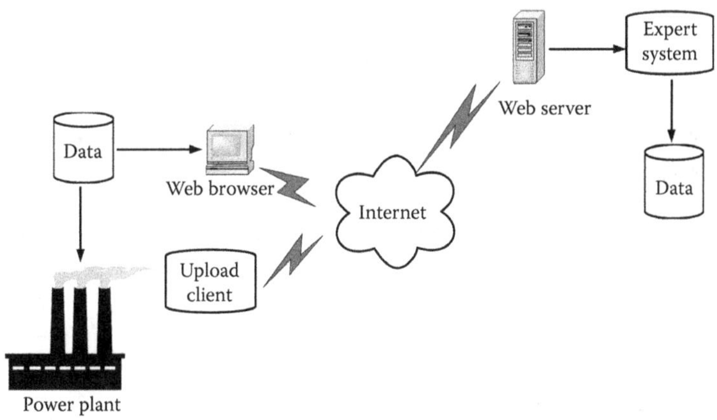

FIGURE 1.11 On-site model of a condition monitoring system.

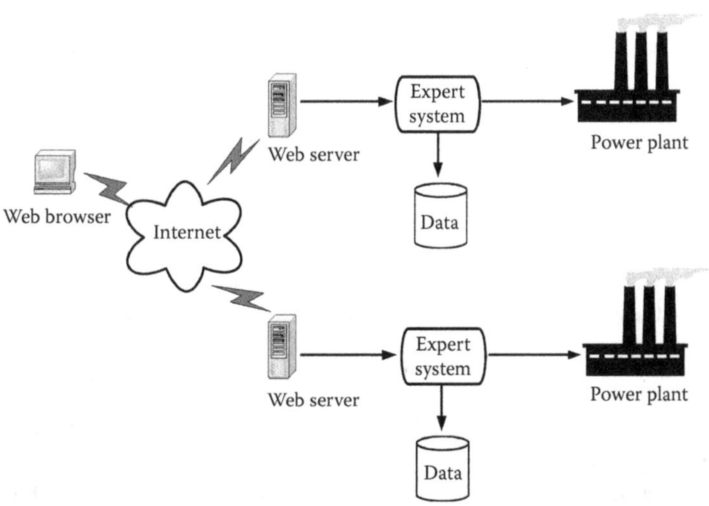

FIGURE 1.12 Off-site model of a condition monitoring system.

motors, gearboxes, engines, and rotating structures (Robert and Ben, 1993; Sato, 1990; Norton, 1989; Hall and Mba, 2004).

1.7 CONFERENCES ON ROTORDYNAMICS

The requirement of close interaction and dissimilation of knowledge in the field of rotor dynamics was recognized as early as the mid-1950s in the form of the American Society of Mechanical Engineers (ASME) Turbo Expo international conference and exposition, and subsequently in the mid-1960s in the form of the ASME Conference on Mechanical Vibration and Noise. A dedicated conference on rotating machinery (quadrennial) took place first in 1976 with the efforts of the Institution of Mechanical Engineers, UK, and subsequently, another event occurred in 1982 by the International Federation for the Promotion of Mechanism and Machine Science on rotor dynamics (quadrennial). Since then, these conferences and others have taken place on a regular basis, as listed in Table 1.2. These conferences are aimed at bringing together both leading and up-and-coming researchers and engineers from all over the world to share their research and case study experiences, and more importantly, to set up the directions for future research in rotor dynamics; they have been able to fulfill these objectives quite well.

TABLE 1.2
Details of Conferences on the Rotor Dynamics Field

S.N.	Name of Conference/ Symposium/Workshop/ Exposition	Year	Frequency	Organizers
1	ASME Turbo Expo	Since 1956	Annual	ASME International Gas Turbine Institute
2	ASME Biennial Conference on Mechanical Vibration and Noise	Since 1967	Biennial	ASME
3	International Conference on Vibrations in Rotating Machinery	Since 1976	Quadrennial	Institution of Mechanical Engineers, UK
4	Workshop on Rotordynamic Instability	1980–2000	Biennial	NASA & Texas A&M Univ.
5	International Conference on Rotordynamics	Since 1982 (To be held in 2030 at IIT Guwahati)	Quadrennial	International Federation for the Promotion of Mechanism and Machine Science
6	International Symposium on Rotating Machinery	Since 1988	Biennial	ISROMAC
7	International Symposium on Stability Control of Rotating Machinery	2001–2007	Biennial	Bently Pressurized Bearing Company
8	International Conference on Vibrations in Rotating Machines	Since 1995	Biennial	German Institutes
9	National Symposium on Rotor Dynamics	2003, 2011, 2014, 2016, 2017, 2019, 2024	Biennial since 2011	Indian Institutes
10	Symposium on Emerging Trends in Rotor Dynamics	1974, 2000	Intermittent	IUTAM
11	International Conference on Maintenance Engineering	2016	Annual	The University of Manchester, UK

1.8 CONCLUDING REMARKS

Research in rotor dynamics is aimed at improving the understanding of rotor dynamic phenomena and improving the performance of rotating machinery. In most rotordynamic systems, the vibratory amplitudes are sufficiently small that the linear analysis of the rotor and stator deformations is satisfactory. In rotor dynamics, the structural modeling is generally adequate (Den Hartog, 1956; Rieger, 1977; Dimarogonas and Paipetis, 1983; Mahrenholtz, 1984; Vance, 1988; Goodwin, 1989; Rao, 1991; Krämer, 1993; Lalanne and Ferraris, 1998; Rao, 2000; Walker, 2003; Chen and Gunter, 2005; Genta, 2005; Muszynska, 2005; Prabhu and Sekhar, 2008; Friswell et al., 2010; Tiwari, 2017, 2024), and most research is centered on fluid–structure interactions: bearings, seals, blade forces, squeeze-film dampers, active support elements, etc. It is here that the nonlinearities are concentrated. Equations of motion of such systems consist of a great many linear equations coupled to a small handful of nonlinear equations (Yamamoto and Ishida, 2001). Several efforts have been made to develop intelligent condition monitoring systems with advanced practicability, sensitivity, reliability, and automation (Han and Song, 2003).

Apart from these, the three-dimensional finite element transient analysis of rotors, analysis of flocks of birds hitting turbine blades, and dynamic analysis of a coupled motor-rotor-bearing system are some of the challenging areas to be covered in future. The most promising area of research for performance improvement is active control (Figures 1.13 and 1.14). The latest topic in rotor dynamics is a study of active magnetic bearings (a *mechatronics* product or, more specifically, *rotortronics*), which support a rotor without contacting it, and active dampers (Schweitzer et al., 2003; Chiba et al., 2005). This study has received considerable attention since Schweitzer reported his work in 1974 and 1975. Now the trend is to perform the condition monitoring of rotating machineries with the help

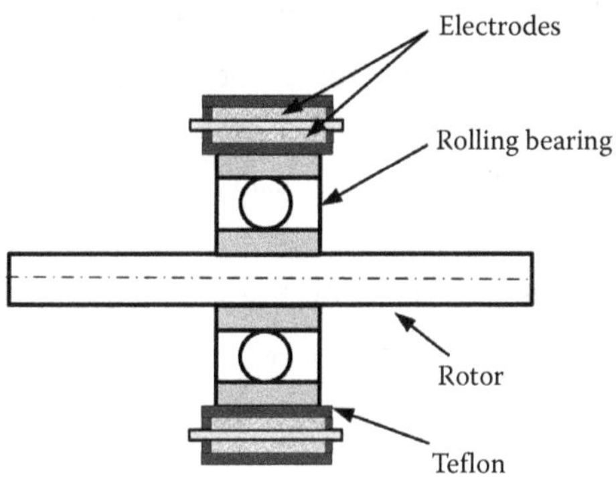

FIGURE 1.13 Smart (active) fluid-film dampers.

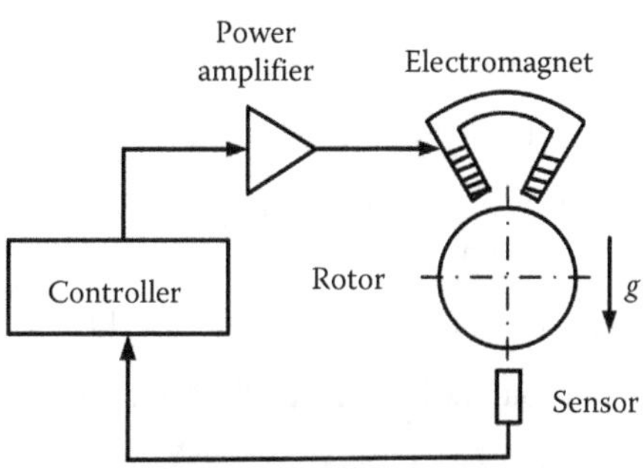

FIGURE 1.14 Operating principle of active magnetic bearings.

of active magnetic bearings (Srinivas et al., 2018; Kumar and Tiwari, 2023, 2024). Interdisciplinary areas like molecular bearings and nano rotor-bearing systems (nano-pump, nano-turbine, etc.) have yet to be explored in terms of fundamental theory and possible applications. Hence, from the present review, it can be concluded that the age of discovery in rotor dynamics lasted to the late 1960s; from the early 1970s to the late 1990s was the age of complex analysis; and the twenty-first century brought the age of the application of smart rotors, which includes identification, control, diagnostics, and prognostics of machines with the help of active control devices integrated into rotor dynamics modeling rather than standalone devices. The present decade is overwhelmed by the application of machine learning in every field, and rotor dynamics is also not untouched. Digital model, digital shadow, and digital twins are now buzzwords, and soon their application will be a reality in rotor dynamics also (Srivastva and Tiwari, 2025). We are now a long way from the approaches of Jeffcott and Prohl—a journey that deserves its own history. It is worth closing this chapter with a quote from Prof. Tondl in one of my email exchanges with him: "Of course, I believe that it is not an easy matter to present a good and complete history of rotor dynamics because there exists an enormous number of publications in different languages." The sustainable design of historic and futuristic rotors marks a major shift from heavy, low-efficiency structures to highly optimized, circular-economy systems. Early rotordynamic and windmill systems focused strictly on survivability and raw mechanical power. In contrast, next-generation futuristic rotors integrate eco-friendly materials, biomimetic aerodynamics, and advanced active control to minimize lifecycle impact.

EXERCISE PROBLEMS

Briefly Describe the Following

Exercise 1.1 Who was the famous person who first analyzed the rotor dynamics problem in 1869 but wrongly predicted that it is impossible to operate industrial rotors at very high speeds?

Exercise 1.2 Who was the first engineer who experimentally demonstrated and reported that it was possible to rotate the rotor safely at very high speeds?

Exercise 1.3 Who clarified theoretically the confusion whether it is possible to rotate the rotor safely above critical speeds?

Exercise 1.4 Who were the first to use the terms *whirling motion* and *critical speed*?

Exercise 1.5 Define the transverse natural frequency and the critical speed of a rotor system (with point masses).

Exercise 1.6 How many critical speeds would there be for a two-disk, massless, flexible shaft rotor system?

Exercise 1.7 Is natural frequency dependent on the spin speed of the rotor? If yes, then under what conditions?

Exercise 1.8 What is the most common cause of a synchronous motion in a rotor system?

Exercise 1.9 In a synchronous whirl of a rotor, what is the whirl frequency?

Exercise 1.10 For a perfectly balanced rotor rotating at a speed, what is the frequency of whirl when it is perturbed from its equilibrium?

Exercise 1.11 In a general motion of a rotor, what is the whirl frequency?

Exercise 1.12 Do bearings and foundations have any effect on the critical speed of a rotor system?

Exercise 1.13 What is the internal and external damping in a rotor system?

Exercise 1.14 How do you distinguish between rigid and flexible rotors?

Exercise 1.15 Is there any difference between the rigid and flexible rotor dynamic balancing?

Exercise 1.16 What is a Campbell diagram?

Exercise 1.17 Explain the inertia asymmetry and shaft asymmetry in a rotor system. What are the effects of these asymmetries on rotor behavior?

Exercise 1.18 What are the different active control mechanisms that can be applied in rotor systems?

Exercise 1.19 Splitting of whirl frequencies occurs due to which factors?

Exercise 1.20 Define instability of a rotor system.

Exercise 1.21 What is the Sommerfeld effect in transient rotor systems?

Exercise 1.22 What is Morton's effect?

Exercise 1.23 What is the curve veering in the Campbell diagram? Why does it happen in rotor-bearing systems?

Exercise 1.24 Multiple Choice Questions

i. In a rotor system, the whirl natural frequency depends on the spin speed of the rotor due to
   A. gyroscopic couple only
   B. fluid-film bearing only
   C. both gyroscopic couple and fluid-film bearing
   D. whirl natural frequency does not depend on the spin speed of the rotor

ii. In a rotor system, the backward whirl refers to
   A. sudden change of spin speed direction
   B. when rotor spin speed and whirling motion have opposite sense of rotation
   C. sudden reduction in spin speed value
   D. when both rotor spin speed and whirling motion have reversal in sense of rotation

iii. In a rotor system, instability may occur due to
   A. seals, bearings, shaft asymmetry, gyroscopic effect
   B. bearings, shaft asymmetry, seals, rubs
   C. rubs, bearings, gyroscopic effect, rotor asymmetry
   D. gyroscopic effect, rotor asymmetry, internal damping, seals

iv. Rotors can have following types of vibration
   A. transverse and axial
   B. torsional and transverse
   C. axial and torsional
   D. transverse, axial, and torsional

v. Which of the following statements is true for a rotor system?
   A. rotational kinetic energy to be maximum
   B. vibration kinetic energy to be maximum
   C. both rotational and vibration kinetic energies to be maximum
   D. potential energy to be maximum

vi. Whirling motion in a rotor system refers to
   A. the movement of the center of the deflected disk or disks. in a plane perpendicular to the bearing axis
   B. spinning motion of a disk about its own axis
   C. tilting motion of a disk about its diameter
   D. torsion motion of a disk relative to the shaft

vii. The whirl natural frequency in a perfectly balanced rotor mainly depends on
   A. effective mass, stiffness and damping
   B. effective stiffness
   C. effective damping only
   D. rotor spin speed only

viii. The whirl frequency in an unbalanced rotor mainly depends on
   A. effective stiffness and damping
   B. effective stiffness only
   C. effective damping only
   D. rotor spin speed only

ix. In a rotor system, the self-centering phenomenon is
   A. rotation of the rotor about its center of gravity
   B. rotation of the rotor about its bearing axis
   C. rotation of the rotor about its geometrical center
   D. rotation of the rotor about its center of percussion

x. A Campbell diagram is
   A. a plot between whirl natural frequency and rotor spin speed
   B. a plot of vibration amplitude with rotor spin speed
   C. a plot of a vibration phase with rotor spin speed
   D. a plot of a rotor orbit

xi. In a Campbell diagram, intersection of curves with $\nu=\omega$ line gives ν whirl natural frequency and ω spin rotor speed)
   A. critical speeds
   B. whirl natural frequencies
   C. the phase of vibration response
   D. the amplitude of vibration response

xii. The critical speed is a
   A. free vibration phenomenon
   B. forced vibration phenomenon
   C. transient vibration phenomenon
   D. unstable vibration phenomenon

xiii. Synchronous whirl condition prevails when
   A. the rotor whirl frequency is the same as rotor spin speed
   B. the rotor system is not operated at the critical speed
   C. the whirl frequency is not the same as spin speed
   D. the shaft has no rigid body mode

xiv. A Jeffcott rotor is
   A. a flexible shaft with a rigid disk at midspan mounted on simply supports
   B. a flexible shaft with an offset rigid disk mounted on simply supports
   C. a rigid shaft with a rigid disk at midspan mounted on flexible bearings
   D. a rigid shaft with an offset rigid disk mounted on flexible bearings

xv. Condition monitoring of rotating machinery is related to
  A. predicting faults in machinery based on measurement of physical parameters from machines
  B. preventive periodic maintenance of machinery
  C. investigating cause of faults in machinery once it has occurred
  D. maintenance of machinery after failure

xvi. Unbalance is defined as
  A. offset of the disk center of gravity from its center of rotation
  B. offset of the disk center of gravity from bearing center
  C. product of disk mass and eccentricity
  D. product of eccentricity and square of spin speed of the rotor

xvii. For a perfectly balanced Jeffcott rotor system, which is rotating at certain speed, ω, if the disk is given a small initial disturbance in transverse direction, the frequency of whirl will be
  A. same as the spin speed, ω
  B. whirl natural frequency
  C. zero
  D. depending on the initial disturbance

ANSWERS TO SELECTED EXERCISE PROBLEMS

1.24

i. C	ii. B	iii. B	iv. D	v. D	vi. A	vii. A
viii. D	ix. A	x. A	xi. A	xii. B	xiii. A	xiv. A
xv. A	xvi. C	xvii. B				

REFERENCES

Adams, M.L. Jr., 2001, *Rotating Machinery Vibration: From Analysis to Troubleshooting*. New York, NY: Marcel Dekker.

Alford, J.S., 1965, Protecting turbomachinery from self-excited rotor whirl, *Transactions of the ASME, Journal of Engineering and Power*, *86*(*2*), 141–148.

Baker, J.G., 1933, Self-induced vibrations, *Journal of Applied Mechanics*, *1*(*1*), 5–12.

Balthazar, J.M., Mook, D.T., Weber, H.I., Brasil, R.M.L.R.F., Fenili, A., Belato, D., and Felix, J.L.P., 2003, An overview on non-ideal vibrations, *Meccanica*, *38*, 613–621.

Baur, U., Benner, P., and Feng, L., 2014, Model order reduction for linear and nonlinear systems: A systemtheoretic perspective, *Archives of Computational Methods in Engineering*, *21*(*4*), 331–358.

Berot, F. and Dourlens, H., 1999, On instability of Overhung Centrifugal Compressors, *Proceedings of ASME Turbo Expo 1999 (99-GT-202), Turbomachinery Technical Conference and Exposition*, Indianapolis, IN, June 7–10, 1999.

Bishop, R.E.D. and Gladwell, G.M.L., 1959, The vibration and balancing of an unbalanced flexible rotor, *Journal of Mechanical Engineering Science*, *1*(*1*), 66–77.

Chen, W.J. and Gunter, E.J., 2005, *Introduction to Dynamics of Rotor-Bearing Systems*. Victoria, BC: Trafford Publishing.

Chiba, A., Fukao, T., Ichikawa, O., Oshima, M., Takemoto, M., and Dorrell, D.G., 2005, *Magnetic Bearings and Bearingless Drives*. Newnes: Elsevier.

Childs, D., 1993, *Turbomachinery Rotordynamics*. New York, NY: Wiley Interscience.

Clough, R.W., 1960, The Finite Element Method in Plane Stress Analysis, *Proceedings of 2nd ASCE Conference on Electronic Computation*, Pittsburgh, Pennsylvania, September 8–9, pp. 345–378.

Collacott, R.A., 1977, *Mechanical Fault Diagnosis and Condition Monitoring*. London: Chapman & Hall.

Crandall, S.H., 1992, Rotordynamic Software. In *Rotating Machinery, Transport Phenomena*, J.H. Kim, W.J. Yang (eds.). Washington, DC: Hemisphere Publishing Corporation, pp. 3–21.

Crandall, S.H. and Brosens, P.J., 1961, On the stability of rotation of a rotor with unsymmetric inertia and stiffness properties, *Journal of Applied Mechanics*, *28*, 567–570.

Crandall, S.H. and Yeh, N.A., 1986, Component Mode Synthesis of Multi-Rotor System. In *Refined Dynamical Theories of Beams, Plates and Shells and their Applications, Proceedings of Euromech Colloquium 219*, I. Elishakoff, H. Irretier (eds.). Berlin: Springer, pp. 44–55.

Crandall, S.H. and Yeh, N.A., 1989, "Automatic generation of component modes for rotordynamic substructures," *ASME, Journal of Vibration, Acoustics, Stress, and Reliability in Design*, *111*, 6–10.

Darlow, M.S., 1989, *Balancing of High-Speed Machinery*. New York, NY: Springer-Verlag.

Davies, A., 1998, *Handbook of Condition Monitoring: Techniques and Methodology*. London: Chapman & Hall.

de Jongh, F.M., 2008, The Synchronous Rotor Instability Phenomenon – Morton Effect, *Proceedings of the 37th Turbomachinery Symposium*, Houston, TX, September 2008, pp. 159–167.

de Jongh, F.M. and Morton, P.G., 1994, The Synchronous Instability of a Compressor Rotor due to Bearing Journal Differential Heating, *Proceedings of ASME Turbo Expo 1994 (94-GT-35), Turbomachinery Technical Conference and Exposition*, The Hague, Netherlands, also published in ASME Transaction, *Journal of Engineering for Gas Turbines and Power*, 1996, *118*, 816–824.

de Jongh, F.M. and va der Hoeven, P., 1998, Application of a Hear Barrier Sleeve of Prevent Synchronous Rotor Instability, *Proceedings of the 27th Turbomachinery Symposium*, Houston, TX, September 1998, pp. 17–26.

Den Hartog, J.P., 1956, *Mechanical Vibration*, 4th ed. New York, NY: McGraw-Hill.

Dimarogonas, A.D. and Paipetis, S.A., 1983, *Analytical Methods in Rotor Dynamics*. London: Applied Science Publications.

Dimentberg, F.M., 1961, *Flexural Vibrations of Rotating Shafts*. London: Butterworths; originally published in Russian (1959).

Dunkerley, S., 1895, On the whirling and vibrations of shafts, *Philosophical Transactions of the Royal Society A*, *185(1)*, 279–360.

Ecker, H. and Tondl, A., 2004, Stabilization of a Rigid Rotor by a Time-Varying Stiffness of the Bearing Mounts, *Eighth International Conference on Vibrations in Rotating Machinery, IMechE*, September 7–9, 2004, University of Wales, Swansea, pp. 45–54 (C623/014/004).

Edwards, S., Lees, A.W., and Friswell, M.I., 1998, Fault diagnosis of rotating machinery, *Shock and Vibration Digest*, *30(1)*, 4–13.

Ehrich, F.F. (ed.), 1992, *Handbook of Rotordynamics*. New York, NY: McGraw-Hill.

Ehrich, F.F., 1966, Subharmonic vibration of rotors in bearing clearance, ASME paper 66-MD-1, *American Society of Mechanical Engineers*, New York.

Ehrich, F.F., 1967, The influence of trapped fluids on high speed rotor vibration, *Transactions of the ASME, Journal of Engineering for Industry*, *91*(*4*), 806–812.

Eshleman, R.L., 1990, *Detection, Diagnosis and Prognosis: An Evaluation of Current Technology*, Proceedings of MFPG 44. Willowbrook, IL: Vibration Institute.

Federn, K., 1957, Grunlagen einer systematischen schwingungsentstörung Wellenelastischer rotoren, VDI Ber. Bd., pp. 9–25.

Föppl, A., 1895, Das Problem der Lavalschen Turbinenwelle, *Der Civillingerenieur*, *4*, 335–342.

Friswell, M.I. and Mottershead, J.E., 1996, *Finite Element Model Updating in Structural Dynamics*. New York, NY: Kluwer Academic Publishers.

Friswell, M.I., Penny, J.E.T., Garvey, S.D., and Lees, A.W., 2010, *Dynamics of Rotating Machineries*. Cambridge: Cambridge University Press.

Gasch, R., Nordmann, R., and Pfützner, H., 2002, *Rotordynamik*, 2nd ed. Berlin: Springer Verlag.

Genta, G., 2005, *Dynamics of Rotating Systems*. New York, NY: Springer.

Geradin, M. and Kill, N., 1984, A new approach to finite element modeling of flexible rotors, *Engineering Computation*, *1*, 52–64.

Glasgow, D.A. and Nelson, H.D., 1980, Stability analysis of rotor-bearing systems using component mode synthesis, *Transactions of the ASME, Journal Mechanical Design*, *102*, 352–359.

Goodwin, M.J., 1989, *Dynamics of Rotor-Bearing Systems*. Sydney: Unwin Hyman.

Grobov, V.A., 1953, O poperechnykh kolebaniyakh vraschchayushchegosya vala pri peremennoi skorosti vrashcheniya, Sbornik Voprosy dinamiki I dinamicheskoi prochnosti, Izdatel'stvo AN Latv. SSR, Vypusk 1. (Transverse Vibrations of a Shaft Rotating with Variable Angular Velocity, *Symposium Problems of Dynamics and Dynamics Stability*, Publishing House of the Latvian SSR, Academy of Sciences, No. 1).

Grobov, V.A., 1955, Poperechnye kolebaniya rotora s raspredelennoi po dline massoi pri peremennoi skorosti vrashcheniya, Izvestiya AN Latv. SSR, Vypusk 5. (Transverse vibrations at variable speed of a rotor carrying an axial distributed mass, *Journal of the Latvian SSR Academy of Sciences*) *5*, 109–118.

Gu, L., 2018, A review of morton effect: From theory to industrial practice, *Tribology Transactions*, *61*(2), 381–391.

Gunter, E., 1966, Dynamic stability of rotor bearing systems, NASA paper No. SP-113, 29.

Guyan, R.J., 1965, Reduction of stiffness and mass matrices, *AIAA Journal*, *3*, 380.

Hall, L.D. and Mba, D., 2004, Diagnosis of continuous rotor–stator rubbing in large scale turbine units using acoustic emissions, *Ultrasonic*, *41(9)*, 765–773.

Han, Y. and Song, Y.H., 2003, Condition monitoring techniques for electrical equipment: A literature survey. *IEEE Transactions on Power Delivery*, *18(1)*, 4–13.

Hori, Y., 1959, A theory of oil whip, *Transactions ASME, Journal of Applied Mechanics*, *26*(*2*), 189–198.

Hurty, W.C., 1960, Vibrations of structural systems by component mode synthesis, *Journal of Engineering Mechanics. Division, Proceedings of the ASCE*, *86*, EM4, 51–69.

Jeffcott, H.H., 1919, The lateral vibration of loaded shafts in neighbourhood of a whirling speed: The effect of want of balance, *Philosophical Magazine*, Series 6, *37*, 304–314.

Jei, Y.-G. and Lee, C.W., 1990, Curve veering in the eigenvalue problem of rotor-bearing systems, *JSME Journal*, *4*(2), 128–135.

Jenny, R., 1980, Labyrinths as a cause of self-excited rotor oscillations in centrifugal compressors, *Sulzer Technical Review*, *4*, 149–156.

Kapitsa, P.L., 1939, Ustoichivost I perekhod cherez kriticheskie oboroty bistro vrash-chayushchikhsya rotorov pri nalichii treniya, *Zhurnal Technicheskoi fiziki, 9, Vypusk 2*. (Stability and transition through critical speeds of high-speed rotors subject to friction, *Journal of Technical Physics*, *9*(*2*), 124–147).

Keogh, P.S. and Morton, P.G., 1994, The dynamic nature of rotor thermal bending due to unsteady lubricant shearing within bearing, *Proceedings of Royal Society of London*, *A*, *445*, 273–290.

Kerr, W., 1916, The whirling speeds of loaded shafts. *Engineering*, *101*, 150, 197, 224 and 245.

Kimball, A.L., 1924, Internal friction theory of shaft whirling, *General Electric Review*, *27*(*4*), 244–251.

Kollmann, F.G., 1962, Experimentelle und theoretische Untersuchen über die Kritischen drezahlen flüssig-keits-gefullter Hohlkörper, *Forschung auf dem Gebiete des Ingerieurwesens*, *28*(*4–5*), 115–123, 147–153.

Krämer, E., 1993, *Dynamics of Rotors and Foundations*. New York, NY: Springer-Verlag.

Kuipers, M., 1964, On the stability of a flexibly mounted rotating cylinder partially filled with liquid, *Applied Scientific Research*, *A13*, 121–137.

Kumar, P and Tiwari, R., 2023, A review: Multiplicative faults and model-based condition monitoring strategies for fault diagnosis in rotary machines. *Journal of the Brazilian Society of Mechanical Sciences and Engineering*, *45*, 282, 1–30.

Kumar, P. and Tiwari, R., 2024, Additive fault diagnosis techniques in rotor systems: A state-of-the-art review. *Sādhanā*, *49* (*3*), 1–49.

Lalanne, M. and Ferraris, G., 1998, *Rotordynamics Prediction in Engineering*. New York, NY: Wiley.

Lee, C.-W., 1993, *Vibration Analysis of Rotors*. London: Kluwer Academic Publishers.

Lees, A.W., 2016, *Vibration Problems in Machines: Diagnosis and Resolution*. Boca Raton, FL: CRC Press.

Lewis, F.M., 1932, Vibrations during acceleration through a critical speed, *Transactions of the ASME*, *54*(*3*), 253–261.

Lipovszky, G., Solyomvari, K., and Varga, G., 1990, *Vibration Testing of Machines and Their Maintenance*. New York, NY: Elsevier.

Lund, J.W., 1964, Spring and damping coefficients for the tilting pad journal bearing, *ASLE Transations*, *7*, 342–352.

Lund, J.W., 1974, Stability and damped critical speed of a flexible rotor in fluid-film bearings, *Transactions of the ASME, Journal of Engineering for Industry*, *96*(*2*), 509–517.

Lund, J.W., 1987, Review of the concept of dynamic coefficients for fluid film journal bearings, *Transactions of the ASME, Journal of Tribology*, *109*, 37–41.

Lund, J.W. and Orcutt, F.K., 1967, Calculation and experiments on the unbalance response of a flexible rotor, *Transactions of the ASME, Journal of Engineering for Industry*, *89*(*4*), 785–795.

Mahrenholtz, O. (ed.), 1984, *Dynamics of Rotors: Stability and System Identification*. New York, NY: International Center for Mechanical Science.

Meher-Homji, C.B., 2000, The Historical Evolution of Turbomachinery, *29th Turbomachinery Symposium*, Texas A & M University, Texas, September 18–21, 2000, pp. 281–331.

Mitchell, J.S., 1993, *An Introduction to Machinery Analysis and Monitoring*. Tulsa, OK: PannWell Books.

Mitropolskii, Y.A., 1965, *Problems of Asymptotic Theory of Nonstationary Vibrations*. Jerusalem: Israel Program for Science Translation.

Muszynska, A., 2005, *Rotordynamics, Series: Dekker Mechanical Engineering*, Vol. *188*. New York, NY: CRC Press.

Myklestad, N.O., 1944, A new method of calculating natural modes of uncoupled bending vibrations, *Journal of Aeronautical Science*, *11*(2), 153–162.

Natanzon, V.Y., 1952, Dvizhenie gibkogo vala na kriticheskoi skotosti, Sbornik Dinamika aviadvigatelei, Oborongiz, No. 8. (Movement of a Flexible Shaft at Critical Speed, *Symposium on Aircraft Engine Dynamics*, Oborongiz, No. 8).

Nelson, F.C. and McVaugh, J.M., 1976, The dynamics of rotor bearing systems using finite elements, *Transactions of the ASME, Journal of Engineering for Industry*, *98(2)*, 593–600.

Nelson, H.D., 2003, A brief history of early rotor dynamics, *Sound and Vibration*, *37(6)*, 8–11.

Newkirk, B.L., 1924, Shaft whipping, *General Electric Review*, *27(3)*, 169–178.

Newkirk, B.L. and Taylor, H.D., 1925, Shaft whirling due to oil action in journal bearings, *General Electric Review*, *28(7)*, 559–568.

Nikolai, E.L., 1937, K teorii gibkogo vala, *Trud Leningr. Ind. Inst.*, No. 6, Razdelenie Fiziko-Matematicheskikh Nauk, Vypusk 3. (The theory of a flexible shaft, *Translations of the Leningrad Industrial Institute*, No. 6, Department of Physico-Mathematical Sciences, No.3).

Norton, M.P., 1989, *Fundamentals of Noise and Vibration Analysis for Engineers*. London: Cambridge University Press.

Parsons, R.H., 1948, *The Steam Turbine and Other Inventions of Sir Charles Parsons, OM*. London: Longman, Green and Company.

Prabhu, B.S. and Sekhar, A.S., 2008, *Dynamic Analysis of Rotating Systems and Applications*. Essex: Multi-Science.

Prandtl, L., 1918, Beitrage zur Frage der Kritischen Drehzahlen, *Dinglers Polytechnisches Journal*, *333*, 179–182.

Prasad, V. and Tiwari, R., 2019, Identification of speed-dependent active magnetic bearing parameters and rotor balancing in high-speed rotor systems. *Journal of Dynamic Systems, Measurement, and Control*, *141*(4), 04101.

Prohl, M.A., 1945, A general method for calculating the critical speeds of flexible rotors, *Journal of Applied Mechanics*, *12(3)*, 142–148.

Rankine, W.J.M., 1869, On the centrifugal force of rotating shaft, *The Engineer*, *27*, 249.

Rao, J.S., 1991, *Turbomachine Blade Vibration*. New Delhi: New Age International.

Rao, J.S., 1996, *Rotor Dynamics*, 3rd ed. New Delhi: New Age.

Rao, J.S., 2000, *Turbine Blades Life Estimation*. Oxford: Alpha Science International Limited.

Rao, J.S., 2007, *Vibratory Condition Monitoring of Machines*. New Delhi: Narosa Publishing House.

Rathbone, T.C., 1929, Turbine vibration and balancing, *ASME Transactions*, *51*, Paper APM-51-23.

Rieger, N.F., 1977, *Vibrations of Rotating Machinery*. Clarendon Hills, IL: The Vibration Institute.

Rieger, N.F., 1986, *Balancing of Rigid and Flexible Rotors*. Washington, DC: Shock and Vibration Information Center.

Robert, B.M., 2003, *Rotating Machinery: Practical Solutions to Unbalance and Misalignment*. New York, NY: CRC Press.

Robert, D.F. and Ben, H.J., 1993, Acoustic signatures: From natural to systems science (A), *The Journal of the Acoustical Society of America*, *93(4)*, 2279.

Ruhl, R.L. and Booker, J.F., 1972, A finite element model for distributed parameter turbogenerator system, *Transaction of the ASME, Journal of Engineering for Industry*, *94(1)*, 126–132.

Sato, I., 1990, Rotating machinery diagnosis with acoustic emission techniques, *Electrical. Engineering Japan*, *110* (2), 115–127.

Schneider, H., 1991, *Balancing Technology*, 4th ed. Darmstadt: Carl Schenck AG.

Schweitzer, G., Stabilization of Self-Excited Rotor Vibrations by an Active Damper, Dynamics of Rotors, *Proceedings of the Symposium*, Lyngby, Denmark, August 12–16, 1974, Springer-Verlag, 1975, pp. 472–493.

Schweitzer, G., Bleuler, H., and Traxler, A., 2003, *Active Magnetic Bearings*. Zurich: VDF Hochschulverlag AGan der ETH.

Shaw, S.W. and Balachandran, B., 2008, A review of the nonlinear dynamics of mechanical systems in year 2008, *The Japan Society of Mechanical Engineers, Journal of System Design and Dynamics*, *2(3)*, 611–640.

Smith, D.M., 1933, The motion of a rotor carried by a flexible shaft in flexible bearings, *Proceedings of the Royal Society London, Series A*, *142*, 92–118.

Sohre, J.S., 1991, Turbomachinery Problems and Their Correction, Standardization and Condition Monitoring, Workshop, Chapter 7, March 11–13, 1991, Houston, Texas.

Someya, T., 1989, *Journal Bearing Hand Book*. Berlin: Springer.

Sommerfeld, A., 1927, Beiträge zum dynamischen ausbau der festigkeitslehe, *Physikal Zeitschr*, *3*, 266, 286.

Srinivas, R.S., Tiwari, R., and Kannababu, C., 2018, Application of active magnetic bearings in flexible rotordynamic systems: A state-of-the-art review. *Mechanical Systems and Signal Processing*, *106*, 537–572.

Srivastava, A. and Tiwari, R., 2025, Development of a digital model of a gear rotor system for fault diagnosis using the finite element method and machine learning, *Journal of Dynamics Monitoring and Diagnostics*. doi:10.37965/jdmd.2025.806

Stodola, A., 1924, *Dampf- und Gasturbinen, 4, Aufl.*, Berlin: Springer. English translation, 1927, *Steam and Gas Turbines*. New York, NY: McGraw-Hill.

Thearle, E.L., 1932, Dynamic balancing of rotating machinery in the field, *ASME Transactions*, Paper APM-56-19.

Thomas, J.J., 1958, Instabile eigenschwingungen von turbinenlaufern, Angefacht durch die spaltstromungen, in stopfbuchsen and beschauflungen, *Bull De l'AIM, 71*, 1039–1063.

Tiwari, R., 2017, *Rotor Systems: Analysis and Identification*. Boca Raton, FL: CRC Press.

Tiwari, R., 2024, *Simple Rotor Analysis through Tutorial Problems*. Boca Raton, FL: CRC Press.

Tiwari, R. and Dharmaraju, N., 2006, Development of a condensation scheme for transverse rotational degrees of freedom elimination in identification of beam crack parameters, *Mechanical Systems and Signal Processing*, *20*(*8*), 2148–2170.

Tiwari, R., Lees, A.W., and Friswell, M.I., 2004, Identification of dynamic bearing parameters: A review, *The Shock & Vibration Digest*, *36*(2), 99–124.

Tiwari, R., Manikandan, S., and Dwivedy, S.K., 2005, A review on experimental estimation of the rotor dynamic parameters of seals, *The Shock & Vibration Digest*, *37*(*3*), 261–284.

Tondl, A., 1965, *Some Problems of Rotor Dynamics*. Prague: Czechoslovak Academy of Sciences.

Tondl, A., 1978, *On the Interaction between Self-Excited and Parametric Vibrations*. Monographs and Memoranda 25. Bechovice: Special Printing of the National Research Institute for Machine Design.

Tondl, A., 1991, *Quenching of Self-Excited Vibrations*. Praha: Academia.

Tondl, A., 1998, To the problem of quenching self-excited vibrations, *Acta Technica CSAV*, *43*, 109–116.

Tong, X., Palazzolo, A., and Suh, J., 2017, A review of the rotordynamic thermally induced synchronous instability (morton) effect, *Applied Mechanics Reviews*, *69*, 1–14.

Uhrig, R., 1966, Reduction of the number of unknown in the displacement method applied to kinetic problems, *Journal of Sound and Vibration*, *4*, 149–155.

Vance, J.M., 1988, *Rotordynamics of Turbomachinery*. New York, NY: John Wiley & Sons.

Walker, D.N., 2003, *Torsional Vibration of Turbomachinery*. New York, NY: McGraw-Hill.

Whitehead, D.S., 1966, Effect of mistuning on the vibration of turbomachine blades induced by Wakes, *Journal of Mechanical Engineering Science*, *8*(*1*), 15.

Wolf, J.A., 1968, Whirl dynamics of a rotor partially filled with liquid, *Transactions of the ASME, Journal of Applied Mechanics*, *35*(*4*), 676–682.

Wowk, V., 1991, *Machinery Vibration: Measurement and Analysis*. New York, NY: McGraw-Hill.

Wowk, V., 1995, *Machinery Vibration: Balancing*. New York, NY: McGraw-Hill.

Yamamoto, T., 1955, On the critical speed of a shaft of sub-harmonic oscillation, *Transaction of the JSME*, *21*(*111*), 853–858 (in Japanese).

Yamamoto, T., 1957, On the vibrations of a rotating shaft, *Memoirs of the Faculty of Engineering, Nagoya University*, *9*(*1*), 25–40.

Yamamoto, T. and Ishida, Y., 2001, *Linear and Nonlinear Rotordynamics*. New York, NY: Wiley & Sons.

Yamamoto, T. and Ota, H., 1964, On the unstable vibrations of a shaft carrying an unsymmetrical rotor, *ASME, Journal of Applied Mechanics*, *31*(*3*), 515–522.

2 Transverse Vibrations of Simple Rotor Systems

In Chapter 1, a brief history and recent trends in the subject of rotor dynamics were provided. The main objective of Chapter 1 was to introduce various phenomena in rotor dynamics so as to have an idea of them before detailed mathematical treatment is provided to comprehensively understand these phenomena. It also briefly introduced various mathematical methodologies, which are used to understand the dynamic behavior of rotor systems, and it summarized the recent and future requirements of the modern high-speed, high-power, and high-reliability rotating machineries.

Rotating machines are extensively used in diverse engineering applications, such as power stations, marine propulsion systems, aircraft engines, helicopters, machine tools, automobiles, household accessories, and futuristic micro- and nano-machines. The design trend of such systems in modern engineering is toward lower weight and operating at supercritical speeds. An accurate prediction of rotor system dynamic characteristics is vitally important in the design of any type of machinery. There have been many studies relating to the field of rotor dynamic systems (Biezeno and Grammel, 1959; Dimentberg, 1961; Tondl, 1965; Rieger, 1977; Dimargonas and Paipetis, 1983; Mahrenholtz, 1984; Vance, 1988; Goodwin, 1989; Childs, 1993; Krämer, 1993; Lee, 1993; Rao, 1996; Lalanne and Ferraris, 1998; Genta, 1999; Admas, 2001; Yamamoto and Ishida, 2001; Robert, 2003; Muszynska, 2005; Genta, 2005; Chen and Gunter, 2005; Prabhu and Sekhar, 2008; Friswell et al., 2010; Vollan and Komzsik, 2025). Of the many published studies, the most extensive portion of the literature on rotor dynamics analysis is concerned with determining natural whirl frequencies, mode shapes, critical speeds, instability thresholds and bands (or regions) of frequencies, and the unbalanced and transient responses. Apart from these analyses, some studies also cover the balancing of rotors, the estimation of bearing dynamic parameters, instrumentation and signal processing, condition monitoring, and nonlinear response analyses. Extensive literatures are available on rolling element bearings and fluid-film bearings and limited literatures on active magnetic bearings.

Understanding the basic phenomena of any dynamic system requires adequate modeling of the system. To start with, transverse vibrations of rotors are considered in the present chapter. The torsional and axial vibrations will be considered in Chapters 6 and 7. The dynamic system can be as simple as a single degree-of-freedom (DOF) system. The rotor is considered as a single mass in the form of a point mass, a rigid disk, or a long rigid cylinder. In a three-dimensional space, a particle and a rigid body can have, at the most, three and six DOFs, respectively. By ignoring the torsional and axial vibration effects, the single mass rotor has, at most, four DOFs. In the present chapter, mathematical treatment is performed of simple rotor models in use over the years by researchers and practicing engineers, for example, the single-DOF undamped and damped model, the two-DOF Rankine's model, the two- and three-DOF Jeffcott rotor models, and four-DOF rotor models derived from the Jeffcott rotor model. Various terminologies are introduced to explain the dynamic behavior of the rotor system, for example, the unbalance; the whirling and wobbling motions; the natural and excitation frequencies; the resonance; the critical speed; the synchronous, antisynchronous, and asynchronous motions; the forward and backward whirls; and the phase. Understanding this present chapter will help in exploring more complex rotor models described in subsequent chapters.

2.1 SINGLE-DOF UNDAMPED ROTOR MODEL

The simplest model of the rotor system can be a single-DOF. Figure 2.1 shows three types of rotor models. In Figure 2.1a, the bearing (support) is assumed to be flexible and the rotor (the shaft and

DOI: 10.1201/9781003661559-2

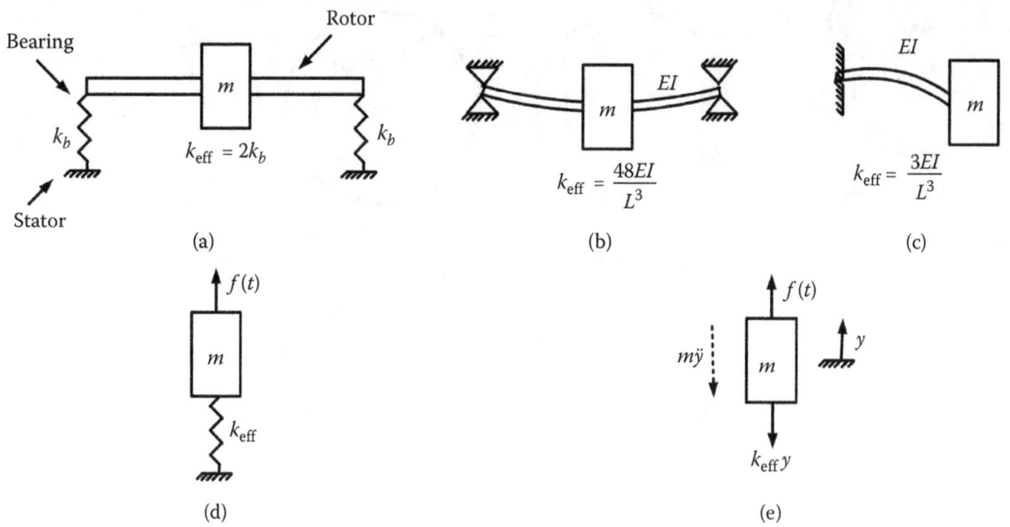

FIGURE 2.1 Simple rotor systems under transverse vibrations and their modeling: (a) a rigid rotor mounted on flexible bearings; (b) a flexible shaft with a rigid disk mounted on rigid bearings; (c) a cantilever beam with a rigid disk; (d) an equivalent single degree of freedom spring-mass system; and (e) the free body diagram of the disk mass.

the disk) as rigid. In Figure 2.1b and c, the bearing is assumed to be rigid (i.e., the simply supported and cantilever, respectively) and the shaft as flexible with the disk as rigid. Here, k_b is the stiffness of each bearing, E is the modulus of elasticity or elastic modulus or Young's modulus of the shaft material, I (which is equal to $\pi d^4/64$, where d is the diameter of the shaft), L is the span length of the shaft, and m is the mass of the disk/rotor.

In all cases, the mass of the rotor is considered to be that of the rigid disk, which is mounted on the massless thin shaft. All the cases can be idealized as a single-DOF for the transverse (lateral or bending) vibration of the rotor as shown in Figure 2.1d, where k_{eff} is the effective stiffness of bearings or the shaft as experienced by the disk mass, and $f(t)$ is the external force on the disk mass (e.g., due to the unbalance). Here only the translatory transverse motion (i.e., a linear motion perpendicular to shaft axis) of the rotor is considered; the simultaneous translatory and rotational (tilting) motions will be considered subsequently. Before considering the free body diagram to derive the governing equation (or equations of motion) and its solution, let us consider first the unbalance force and associated terminologies in more detail.

2.1.1 Unbalance Force Model

If the rotor is perfectly balanced, then theoretically speaking, there will not be any unbalance force as shown in Figure 2.2a, where C and G are the center of rotation (or the geometrical center) and the center of gravity of the rotor, respectively. However, in actual practice, it is impossible to have a perfectly balanced rotor, and various cases are shown in Figure 2.2b through d, where U is the location of an additional unbalance mass (for example, a small screw is attached to the disk). The unbalance may be the result of manufacturing tolerances, operational wear and tear, thermal distortions, repair, etc. The rotor unbalance gives a sinusoidal force at the rotor spin speed. Thus, the unbalance force (in the vertical y-axis direction) is mathematically modeled as

$$f_y(t) = m\omega^2 e \sin \omega t \tag{2.1}$$

where m is the mass of the rotor, ω is the spin speed of the rotor, e is the eccentricity of the rotor (i.e., $e = CG$, see Figure 2.2b), and the product me is normally called the *unbalance*, U. This unbalance

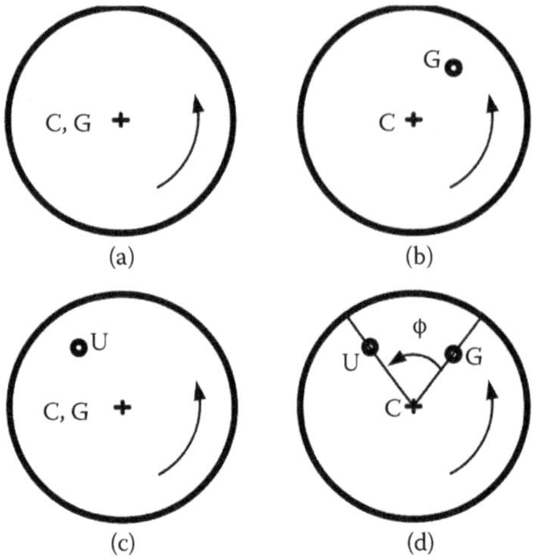

FIGURE 2.2 Different types of unbalances in a single plane. (a) Rotor geometrical center and center of gravity are coincident. (b) Rotor geometrical center and center of gravity are not coincident. (c) An additional unbalance mass location is not coincident with the rotor geometrical center and center of gravity. (d) Rotor geometrical center, center of gravity, and additional unbalance mass location are not coincident.

force changes its direction while the rotor rotates with an angle ωt, so the unbalance force would change sinusoidally, $\sin\omega t$, with respect to the fixed frame of reference (y-axis in the vertical direction). Here, we could have taken in force term $\cos\omega t$ (x-axis in the horizontal direction) or in a more general form $\cos(\omega t-\phi_0)$, with ϕ_0 as an initial *phase* of the unbalance with respect to some convenient reference axis (here the x-axis) at time $t=0$. This unbalance force will come when the rotor is eccentric, that is, the rotor center of rotation and the center of gravity are not coincident. This type of unbalance is called the *residual or inherent unbalance.*

When the rotor is not eccentric (i.e., when C and G are coincident), and when a small unbalance mass, m_i, is attached at a relatively larger radius of $r_i = CU$ as shown in Figure 2.2c, the unbalance force could be written as

$$f_y(t) = m_i\omega^2 r_i \sin\omega t \tag{2.2}$$

This type of unbalance, $m_i r_i$, is called the *trial, correction, or additional unbalance*, and it is often used for the experimental estimation of the residual unbalance using dynamic balancing procedures. For the case when the rotor is eccentric and a small trial unbalance mass is also attached as shown in Figure 2.2d, the total unbalance force would be

$$f_y(t) = m\omega^2 e\sin\omega t + m_i\omega^2 r_i \sin(\omega t + \phi) \tag{2.3}$$

where ϕ is the angular phase difference between the vectors of unbalance forces due to the rotor eccentricity and the trial unbalance mass for the rotation of the disc as shown in Figure 2.2d. For the counterclockwise rotation of the rotor as shown in Figure 2.2d, the trial mass unbalance (CU) is leading by the phase angle of ϕ with respect to the residual mass unbalance (CG). If the unbalance has an initial phase of ϕ_0, the earlier expression would take the following form:

$$f_y(t) = m\omega^2 e\sin(\omega t + \phi_0) + m_i\omega^2 r_i \sin(\omega t + \phi_0 + \phi) \tag{2.4}$$

Figure 2.3 shows positive conventions and variables to define the unbalance location on a rotor system at a particular instant of time, t, in which the initial phase (i.e., ϕ_0) is taken as zero. For a

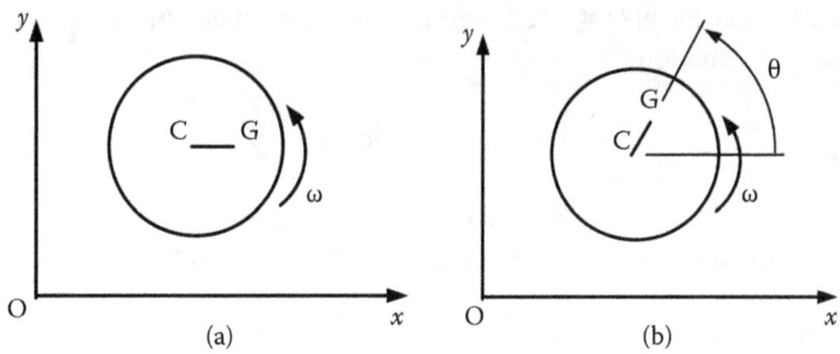

FIGURE 2.3 The unbalance location on a rotor at a particular instant of time: (a) at time $t=0$ and (b) at any time t.

constant angular velocity of the rotor, ω, the location of the unbalance is given as $\theta=\omega t$. In general, $\dot{\theta} \neq \omega$ (the "dot" represents the derivative with respect to the time), that is, when a rotor has some angular acceleration. For example, for a constant angular acceleration $\ddot{\theta} = \alpha$ and for zero initial conditions, we will have $\dot{\theta} = \alpha t$ and $\theta = 0.5\,\alpha t^2$. For the present case $\alpha=0$, that is, $\theta=\omega t$. In general, the angular position θ will have an initial angular position θ_0, and contribution from angular velocity ωt and angular acceleration $0.5\,\alpha t^2$.

2.1.2 Equation of Motion of the Simplest Rotor System

Here, it is assumed that centrifugal force is acting as an external force; hence, the center of gravity could be assumed to be at the center of rotation of disk C itself. Subsequently in Section 2.4, we would see that the centrifugal force actually originates from the inertia force because the center of gravity is offset due to the unbalance. On application of Newton's second law of motion on the free body of the rotor mass as shown in Figure 2.1e, that is, on equating the sum of external forces to the mass of the rotor multiplied by the acceleration at the center of gravity of the rotor mass, we have

$$\sum F_y = m\ddot{y} \quad \text{or} \quad -k_{eff}\,y + m\omega^2 e \sin\omega t = m\ddot{y} \tag{2.5}$$

where k_{eff} is the effective stiffness of the rotor system (see Figure 2.1) and $k_{eff}\,y$ is the restoring spring force that acts opposite to the motion, hence the negative sign. Equation 2.5 is a standard form of equation of motion of a single-DOF spring-mass system, and it could be rearranged as

$$m\ddot{y} + k_{eff}\,y = m\omega^2 e \sin\omega t \tag{2.6}$$

It should be noted that this represents a wide class of rotor systems and the effective stiffness would change depending on the type of rotor system considered. At times even the mass and displacement also will change, but the form of the differential equation will remain the same.

2.1.3 Free Vibration of the Rotor System

From the free vibration, when the external unbalance force is absent (i.e., $e = 0$), it is generally assumed that the rotor mass will have a simple harmonic oscillation, and the free response displacement is expressed as

$$y(t) = Y \sin\omega_{nf} t \tag{2.7}$$

where Y is the amplitude of displacement and ω_{nf} is the frequency of oscillation during the free vibration, which is called the *natural frequency* (herein the subscript *nf* represents the natural frequency) of the rotor system (often in rotor dynamics it is called the natural whirl frequency).

Substituting Equation 2.7 (with $\ddot{y}(t) = -Y\omega_{nf}^2 \sin\omega_{nf}t$) into the homogeneous part (i.e., with $e = 0$) of the equation of motion (Equation 2.6) gives

$$\left(-m\omega_{nf}^2 + k_{eff}\right)Y \sin\omega_{nf}t = 0 \tag{2.8}$$

For a nontrivial solution (i.e., $Y \neq 0$ or $\omega_{nf} \neq 0$) of Equation 2.8, the frequency equation (or natural frequency) of the system can be obtained by equating bracket terms to zero, as

$$\left(-m\omega_{nf}^2 + k_{eff}\right) = 0 \quad \text{or} \quad \omega_{nf} = \pm\sqrt{k_{eff}/m} \quad \text{or} \quad \omega_{nf} = \pm\sqrt{k_{eff}/m} \tag{2.9}$$

where ω_{nf} is the transverse natural frequency of the rotor system because the motion considered is in the transverse direction only. In the earlier equation, the negative sign has no physical meaning because the frequency cannot be a negative quantity.

2.1.4 Forced Response of the Rotor System

The steady-state forced response can be expressed as

$$y(t) = Y\sin(\omega t - \phi) \tag{2.10}$$

where Y is the amplitude of displacement, ω is the spin speed of the rotor, and ϕ is the phase lag of displacement with respect to the unbalance force (normally in the absence of damping, the phase will be zero). On substituting Equation 2.10 (noting $\ddot{y}(t) = -Y\omega^2\sin(\omega t - \phi)$) into Equation 2.6, the steady-state forced response amplitude can be obtained as

$$\left(-m\omega^2 + k_{eff}\right)Y\sin(\omega t - \phi) = me\omega^2 \sin\omega t \tag{2.11}$$

On expanding this equation, we get

$$\left(-m\omega^2 + k_{eff}\right)Y\left(\sin\omega t\cos\phi - \cos\omega t\sin\phi\right) = me\omega^2\sin\omega t \tag{2.12}$$

On separating the sine and cosine terms of ωt on both sides, we have

$$\left(-m\omega^2 + k_{\mathrm{eff}}\right)Y\cos\phi = me\omega^2 \quad \text{and} \quad \left(-m\omega^2 + k_{\mathrm{eff}}\right)Y\left(-\sin\phi\right) = 0 \tag{2.13}$$

On dividing the one over the other, we have

$$\frac{\left(-m\omega^2 + k_{eff}\right)Y\left(-\sin\phi\right)}{\left(-m\omega^2 + k_{eff}\right)Y\cos\phi} = \frac{0}{me\omega^2} \quad \text{or} \quad \tan\phi = 0 \quad \text{or} \quad \phi = 0 \tag{2.14}$$

From the first equation of Equation 2.13, we have

$$\left(-m\omega^2 + k_{eff}\right)Y\cos\phi = me\omega^2 \Rightarrow Y = \frac{me\omega^2}{\left(-m\omega^2 + k_{eff}\right)} \tag{2.15}$$

Herein, $\cos 0 = 1$. It can be written in nondimensional form as

$$\bar{Y} = \frac{Y}{e} = \frac{m\omega^2}{k_{eff} - m\omega^2} = \frac{\omega^2}{\omega_{nf}^2 - \omega^2} = \frac{\bar{\omega}^2}{1 - \bar{\omega}^2} \quad \text{with} \quad \bar{\omega} = \frac{\omega}{\omega_{nf}} \quad \text{and} \quad \omega_{nf} = \sqrt{k_{eff}/m} \tag{2.16}$$

where $\bar{Y}$ is the *nondimensional unbalance response* (ratio of the unbalance response to the eccentricity) and $\bar{\omega}$ is the *frequency ratio* (ratio of the spin speed of the rotor to the natural frequency of the rotor system). The *absolute value* of the nondimensional unbalance response, $\bar{Y}$, is plotted with

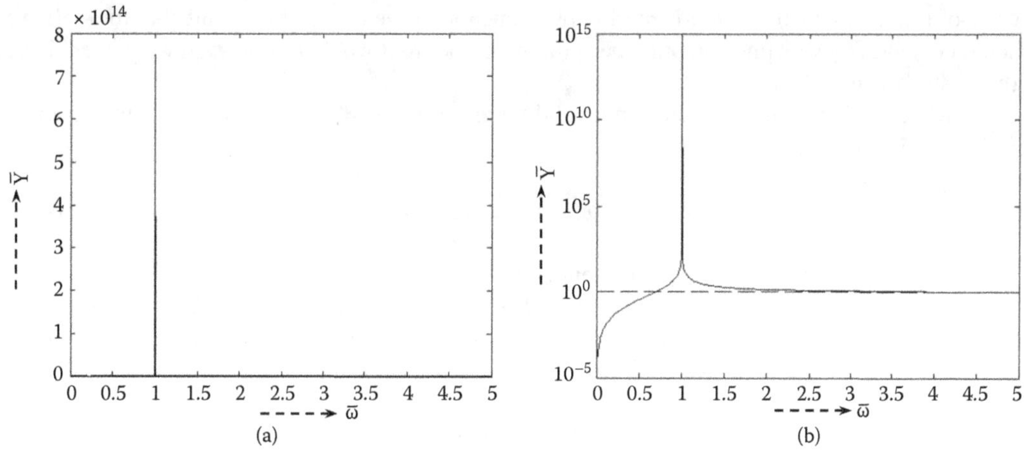

FIGURE 2.4 Variation of the absolute nondimensional unbalance response versus the frequency ratio: (a) linear scale plot; (b) semi-log scale plot.

respect to the frequency ratio, $\bar{\omega}$, as shown in Figure 2.4. From the plot and Equation 2.16, it should be noted that we have an unbounded unbalance response (i.e., it is tending to infinity) when the denominator $\left(1-\bar{\omega}^2\right)$ becomes zero, that is, when the spin speed is

$$\bar{\omega}_{cr} = \frac{\omega_{cr}}{\omega_{nf}} = \pm 1 \quad \text{or} \quad \omega_{cr} = \pm\sqrt{\frac{k_{eff}}{m}} = \pm\omega_{nf} \tag{2.17}$$

This is a *resonance condition*, and the spin speed corresponding to the resonance is defined as the *critical speed*. The subscript *cr* represents the critical speed. For the present case, the critical speed is equal to the transverse natural frequency of the nonrotating rotor system as given by Equation 2.17. It is due to the fact that the unbalance force gives a transverse force, and when its frequency matches the transverse natural frequency of the rotor system, resonance takes place. The undamped, steady-state, forced response amplitude tends to be infinity at the critical speed. The natural frequency and the critical speed concepts have come from the free and forced vibrations, respectively. It should be noted that in rotor dynamics, in general, the natural frequency might not be a constant and might vary with the spin speed of the shaft (e.g., when the gyroscopic couple is considered in the analysis (refer Chapter 5), especially when the spin speed of the rotor is high, and also for the case of speed-dependent bearing dynamic properties (refer Chapter 2 of Volume II), etc.). The ± sign indicates that the rotor will have a critical speed while rotating in either the clockwise or counterclockwise sense (earlier we noted that the natural frequency is always positive; however, the spin speed of the rotor can be described as positive or negative depending upon its sense of rotation). Because the damping is not considered in the analysis, the phase angle, ϕ, remains zero before the critical speed (or 180° after the critical speed; however, its value at the critical speed is 90°), and it will be discussed subsequently.

In Figure 2.4, the response changes its sign (i.e., from positive to negative) after the frequency ratio is equal to unity, which corresponds to the *critical speed* (i.e., the spin speed of the shaft is equal to the transverse natural frequency of the rotor system). This means that the phase difference between the force and the response becomes 180°, which is 0° when the frequency ratio is less than unity in the absence of damping. Both the linear and semi-log plots are shown to have clarity of the response variation, both near to and away from the critical speed. It can be seen that as the frequency ratio increases above unity, the nondimensional response asymptotically approaches unity, which means the unbalance response approaches the eccentricity of the rotor. Physically, this implies that the rotor rotates about its center of gravity at a high frequency ratio and not the actual center of rotation, which is a self-centering phenomenon of the rotor. This means that for $e=0$, we should not have any whirling and only the spinning of the shaft would be observed when there is no other disturbance in the rotor system.

The response from Equation 2.16 is not valid for $\bar{\omega}=1$ (i.e., $\omega=\omega_{cr}$). At this speed, the equation of motion (Equation 2.6) becomes

$$\ddot{y}+\omega_{cr}^2 y=\omega_{cr}^2 e\sin\omega_{cr}t \tag{2.18}$$

Now the amplitude of the rotor displacement is no longer a constant and it depends upon the time (refer to Equation 2.19). Moreover, the phase of the rotor displacement would be 90° with respect to the unbalance force (refer to Equation 2.19 and Section 2.2). Hence, the particular solution for the initial condition $y=0$ and $\dot{y}=0$ can be assumed to have the form

$$y(t)=r(t)\cos\omega_{cr}t \tag{2.19}$$

so that

$$\dot{y}(t)=\dot{r}(t)\cos\omega_{cr}t-\omega_{cr}r(t)\sin\omega_{cr}t \tag{2.20}$$

and

$$\ddot{y}(t)=\ddot{r}(t)\cos\omega_{cr}t-\omega_{cr}\dot{r}(t)\sin\omega_{cr}t-\omega_{cr}\dot{r}(t)\sin\omega_{cr}t-\omega_{cr}^2 r(t)\cos\omega_{cr}t \tag{2.21}$$

On substituting this solution into Equation 2.18, we get

$$\left\{\ddot{r}(t)\cos\omega_{cr}t-2\omega_{cr}\dot{r}(t)\sin\omega_{cr}t-\omega_{cr}^2 r(t)\cos\omega_{cr}t\right\}+\omega_{cr}^2\left\{r(t)\cos\omega_{cr}t\right\}=\omega_{cr}^2 e\sin\omega_{cr}t \tag{2.22}$$

On collecting cosine and sine terms, we get

$$\left\{\ddot{r}(t)\right\}\cos\omega_{cr}t+\left\{-2\omega_{cr}\dot{r}(t)\right\}\sin\omega_{cr}t=\omega_{cr}^2 e\sin\omega_{cr}t \tag{2.23}$$

On equating the cosine and sine terms on either side, we get

$$\ddot{r}(t)=0 \quad \text{and} \quad \dot{r}(t)=-\frac{1}{2}e\omega_{cr}\Rightarrow r(t)=-\frac{1}{2}e\omega_{cr}t \tag{2.24}$$

with zero initial conditions. Then, the solution from Equation 2.19 could be written as

$$y(t)=-\frac{1}{2}e\omega_{cr}t\cos\omega_{cr}t=\frac{1}{2}e\omega_{cr}t\sin\left(\omega_{cr}t-\frac{\pi}{2}\right) \tag{2.25}$$

That means the deflection increases linearly with time and always makes a 90° phase lag with the unbalance force. For the two-plane motion of the rotor, the rotor center trajectory would be a *diverging spiral* in shape with time at the critical speed. However, with damping in the system, the response will be contained with an amplitude of $e/(2\zeta)$ and will not increase with time (refer to Section 2.2, Equation 2.71), where ζ is the damping ratio.

The analysis presented in this section can be applied to the transverse, torsional, and axial vibrations of single-DOF rotors (see Figure 2.5) and accordingly natural frequency can be termed by prefixing respective types of vibrations. For the transverse vibration (refer to Figure 2.5a), we have already discussed. For torsional vibrations (Figure 2.5b, double head arrow represents angular displacement direction), care should be taken that the mass will be replaced by the polar mass moment of inertia, I_p, of the rotor and the stiffness will be replaced by the torsional stiffness (G is the shear

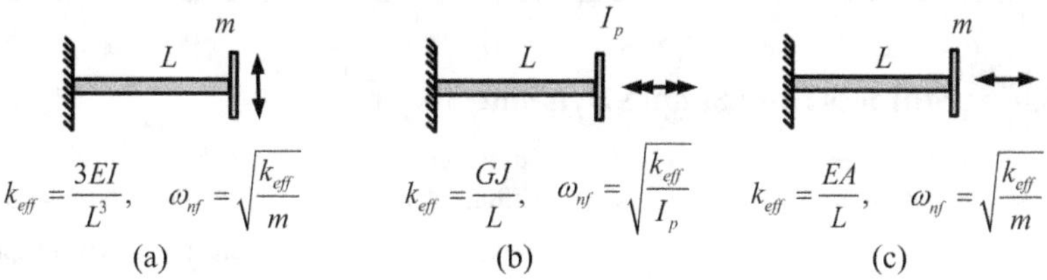

$$k_{eff}=\frac{3EI}{L^3},\quad \omega_{nf}=\sqrt{\frac{k_{eff}}{m}}\qquad k_{eff}=\frac{GJ}{L},\quad \omega_{nf}=\sqrt{\frac{k_{eff}}{I_p}}\qquad k_{eff}=\frac{EA}{L},\quad \omega_{nf}=\sqrt{\frac{k_{eff}}{m}}$$

FIGURE 2.5 A cantilever beam with different modes of vibrations in the (a) transverse direction, (b) torsional direction, and (c) axial (longitudinal) direction.

modulus or modulus of elasticity in shear or modulus of rigidity, l is the length of the shaft, and $J=\pi d^4/32$ where d is the diameter of the shaft). Similarly, for the axial vibration (Figure 2.5c), the mass will remain the same as for transverse vibrations; however, the stiffness will be the axial stiffness (E is the Young's modulus and A is the area of cross-section of the shaft). Table 2.1 summarizes some common single-DOF dynamic systems with the equivalent stiffness. For all such systems, a corresponding natural frequency also could be obtained by attaching inertia (i.e., mass or polar moment of inertia) at load positions. More detailed treatment will be covered through excercise problems and also in subsequent chapters.

2.1.5 Attenuation of Vibrations

The most common cause of vibrations in rotors is the unbalance among the other faults (such as misalignments, bent shaft, cracks, etc.), and the unbalance will always be present in a rotor. However, the unbalance response can be reduced up to a desired level depending upon applications of one or several of the following basic approaches.

1. *Correction at the source*: Balancing the rotor is the most direct approach, because it attacks the problem at the source. However, in practice a rotor cannot be balanced perfectly (Figure 2.6a), and the best achievable state of balance tends to degrade during operation of a rotor (e.g., in the turbomachinery).

   For a rotor system containing a single thin disk of mass m (e.g., propellers of a jet-engine), the single plane balancing would be sufficient. It can be balanced if the $me=m_i r_i$ condition is satisfied, subjected to the condition that the trial mass m_i is kept at radius r_i, diametrically opposite to the disk eccentricity, e, direction.

   However, there are two types of unbalances in rigid rotors. The first is (i) *the static unbalance*, in which the principal axis of the polar mass moment of inertia of the rotor is parallel to the center line of the shaft as shown in Figure 2.6b. The rotor can be balanced by a single plane balancing. The second type is (ii) *the dynamic unbalance* in which the principal axis of the polar mass moment of inertia of the rotor is inclined to the center line of the shaft as shown in Figure 2.6c and d. In Figure 2.6e, both the radial and axial offset of the center of gravity give both the force and the moment (i.e., the dynamic unbalance). For balancing such long, rigid rotors, at least two balancing planes are required. The balancing of rigid and flexible rotors will be covered in more detail in Chapter 3 of Volume II.
2. *Operate the rotor away from the critical speed*: This could be done during the design itself or during operation by providing an extra or auxiliary support. At the design stage, the critical speed can be altered by changing the rotor mass and its distributions and the effective stiffness (for example, by changing dimensions of the shaft, i.e., the shaft diameter and length, or the location and type of bearings). During operation, an auxiliary support can be provided to increase the effective stiffness of the rotor, which in turn increases the critical speed. For the case when the rated operational speed is above the critical speed, the actual rotor critical speed can be safely traversed by this arrangement, by temporarily increasing

TABLE 2.1
Equivalent Spring Stiffness from Simple Dynamic Systems

S.N.	System	Equivalent Stiffness, k_{eff}	Definition of variables
1	Springs in series 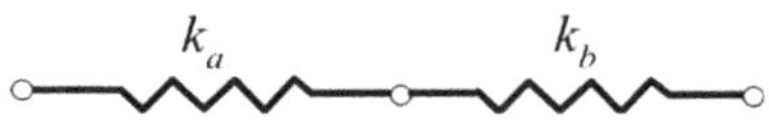	$\frac{k_a k_b}{k_a + k_b}$	k_a and k_b are the stiffness of springs in series. For example, a stepped shaft with cantilever end conditions and a disk at free end executing the torsional or axial vibrations.
2	Springs in parallel 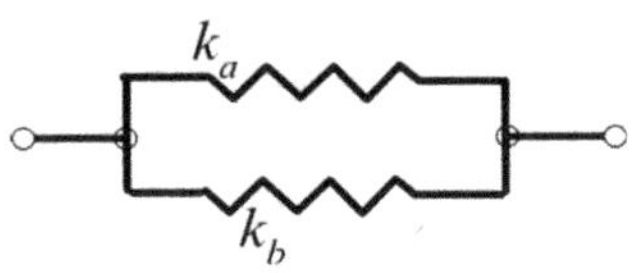	$k_a + k_b$	k_a and k_b are stiffness of springs in parallel. For example, a stepped shaft with fixed-fixed end conditions and a disk at a step executing the torsional or axial vibrations.
3	An inclined spring 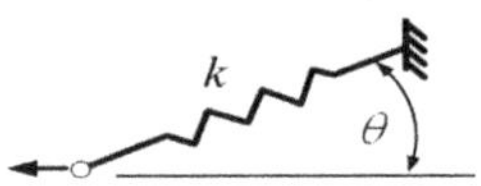	$\frac{k}{\cos^2\theta}$	θ is the angle of spring with force
4	Coil spring 	$\frac{EI}{l}$	E is the Young's modulus. I is the second moment of area of the cross-section. l is the total length.
5	Axial extension of a bar 	$\frac{EA}{l}$	A is the cross-section area.
6	Torsion of a rod 	$\frac{GJ}{l}$	G is the modulus of rigidity. J is the polar moment of area of the cross-section (Double arrow head represents torque load and its direction).
7	Helical spring 	$\frac{Gd^4}{64nR^3}$	R is the mean radius of coil. D is the diameter of wire. N is the number of turns.
8	Fixed-fixed shaft with a load 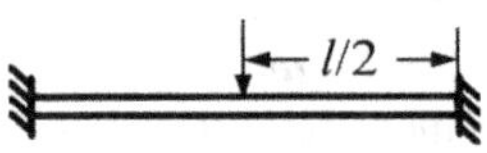	$\frac{192EI}{l^3}$	The vertical downward arrow represents the load position.
9	Simply supported shaft with a load 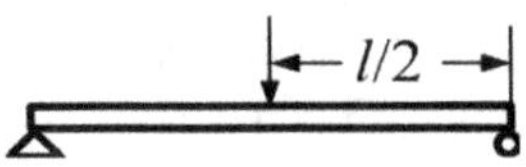	$\frac{48EI}{l^3}$	Stiffness at the load position
10	Simply supported shaft with a load 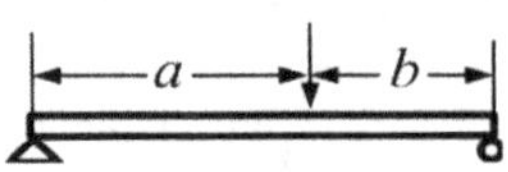	$\frac{3EIl}{a^2b^2}$	Stiffness at the load position

(Continued)

TABLE 2.1 (*Continued*)
Equivalent Spring Stiffness from Simple Dynamic Systems

S.N.	System	Equivalent Stiffness, k_{eff}	Definition of variables
11	Cantilever shaft with a load 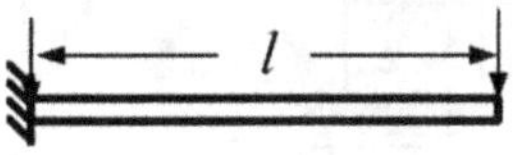	$\frac{3EI}{l^3}$	Stiffness at the load position
12	Fixed-hinged shaft 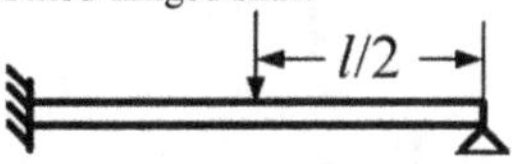	$\frac{768EI}{7l^3}$	Stiffness at the load position
13	Indeterminate shaft-1 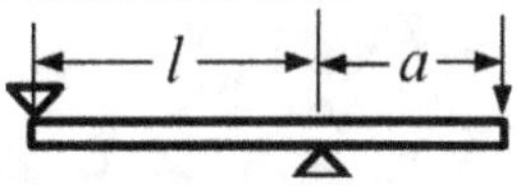	$\frac{3EI}{(l+a)a^2}$	Stiffness at the load position
14	Indeterminate shaft-2 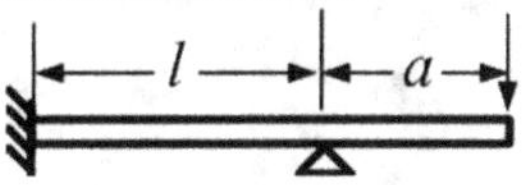	$\frac{24EI}{(3l+8a)a^2}$	Stiffness at the load position

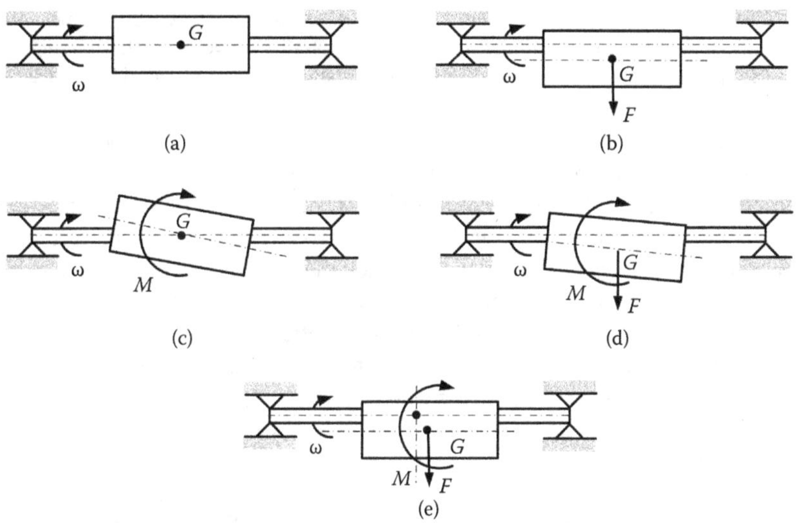

FIGURE 2.6 Classification of unbalances for a rigid rotor: (a) perfectly balanced (no force and moment): (b) static unbalance (pure radial force); (c) dynamic unbalance (pure moment); (d) dynamic unbalance (both force and moment); (e) dynamic unbalance (both force and moment due to radial and axial offset).

the critical speed, and then the auxiliary support can be withdrawn, which brings the critical speed of the rotor again to below the rated operation speed (refer to Example 2.2). Another way to traverse the critical speed is to accelerate the rotor spin speed very quickly to reach the operating speed, which is well above the critical speed of the rotor, because we know that the response at critical speed increases linearly with time. In general, changing the critical speed is useful for machines with a constant or with a narrow range of operational speed (e.g., the turbomachinery in power plants).

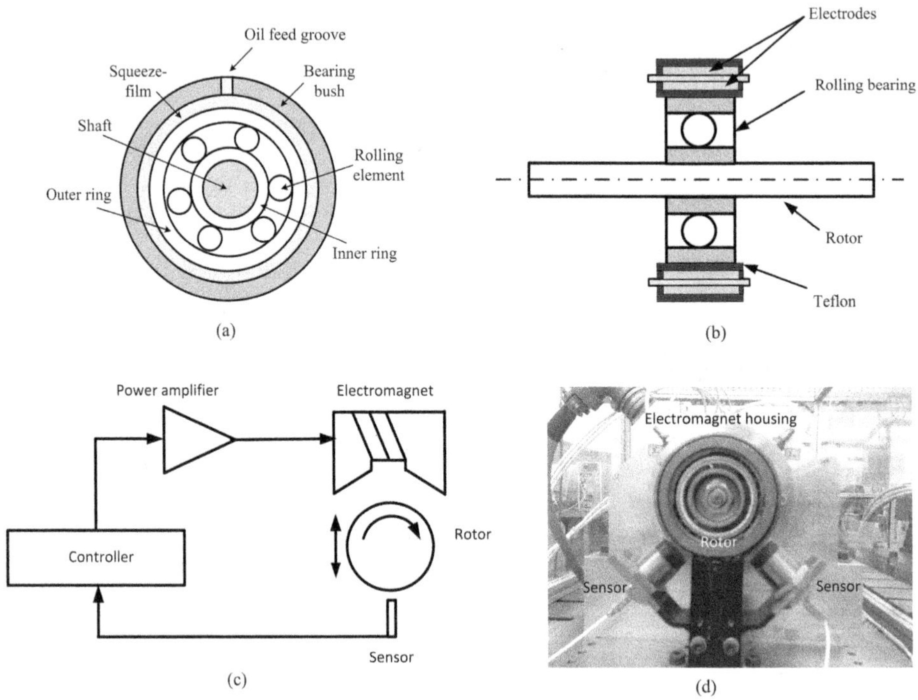

FIGURE 2.7 Passive and active vibration control devices: (a) a passive squeeze-film damper; (b) a schematic diagram of a smart (active) fluid-film damper; (c) basic principles of active magnetic bearings; (d) an active magnetic bearing supporting and controlling the rotor based on its vibrations sensed by displacement probes.

3. *Add damping to the system or the active control of the rotor*: If a critical speed must be traversed slowly or repeatedly, or if machine operation near a critical speed cannot be avoided, then the most effective way to reduce the amplitude of vibration is to add the damping (squeeze-film damper). On the other hand, some other forms of damping (for example, the shaft material or hysteretic and friction between rotating components, i.e., the internal damping) may lead to rotor instability (self-excited vibration).

   Squeeze-film dampers and active bearings are often used to control the dynamics of such rotor systems. Squeeze-film dampers are, in effect, in series with rolling element bearings in which both the bearing outer ring and bearing bush are nonrotating (Figure 2.7a). It is similar to fluid-film bearing in which the ability to provide the damping is retained, but there is no capacity to provide the stiffness (i.e., the load-carrying capacity) as the latter is related to the journal rotation. They are used extensively in applications where it is necessary to eliminate high-amplitude vibrations and to limit the rotor vibration and its effect on the supporting structures of rotor-bearing systems, especially in jet engines.

   The advanced development of electromagnetic bearing technology has enabled the active control of rotor-bearing systems. These bearings are called *active magnetic bearings*. In particular, the electromagnetic suspension of a rotating shaft without the mechanical contact has allowed the development of supercritical shafts in conjunction with modern digital control strategies (Figures 2.7c and d). With the development of *smart fluids* (for example, the electro- and magneto-rheological fluids), new controllable dampers are in the primitive development stage (Figure 2.7b), which similar to squeeze-film dampers. The basic premise of such smart fluids is that their dynamic properties (i.e., the damping and the stiffness) can be controlled by changing the current or magnetic flux in a micro-second. Schematics of typical passive and active (i.e., smart or controllable) squeeze-film dampers and active magnetic bearings are shown in Figure 2.7.

Example 2.1

A rotor has a mass of 10 kg and an operational speed of (100±1) rad/s. What are the bounds of the effective stiffness of the shaft so that the critical speed does not fall within 5% of operating speeds? Assume that there is no damping in the rotor system.

Solution:

The operational speed range is 99–101 rad/s. Now 5% of the lower operational speed would be 99−99×0.05=94.05 rad/s, and 5% of the upper operational speed would be 101+101×0.05=1 06.05 rad/s.

Then, the effective stiffness corresponding to the lower operating speed would be $\omega_{nf}^2 m = 94.05^2 \times 10 = 88.45$ kN/m and the effective stiffness corresponding to the upper operating speed would be $\omega_{nf}^2 m = 106.05^2 \times 10 = 112.5$ kN/m. Hence, the effective stiffness of the shaft should not fall in the range of 88.45–112.5 kN/m. It should be noted that the unit of ω_{nf} is in rad/s when other quantities are in SI units (i.e., m in kg and k in N/m). The angular speed is generally expressed in rpm ($N=60\omega/2\pi$) or in Hz ($f=\omega/2\pi$).

Example 2.2

A rotor system has 100 rad/s as the critical speed and its operating speed is 120 rad/s. If we want to avoid crossing of the critical speed, then what is the enhancement in the support stiffness to be given by an auxiliary support system? To avoid excessive vibration, let us assume that we should have at least 5 rad/s of gap between the operating speed and the critical speed at any instant of time. The rotor has a mass of 10 kg.

Solution:

The initial stiffness of the support is $\omega_{nf1}^2 m = 100^2 \times 10 = 100$ kN/m.

First, we can reach safely a rotor speed of 95 rad/s, which is 5 rad/s lower than the original critical speed of the rotor. Now because we cannot safely increase the rotor speed further, we need to increase the critical speed of the rotor to at least 125 rad/s (ω_{nf2}). This will allow us to reach up to 120 rad/s (i.e., 5 rad/s lower than the new critical speed of 125 rad/s).

The corresponding effective support stiffness should be $\omega_{nf2}^2 m = 125^2 \times 10 = 156.25$ kN/m. Hence, the auxiliary support system should increase the effective stiffness by 56.25 kN/m. Refer to the related Exercise problem 2.20.

Example 2.3

A 2-kg mass of a cantilever rotor caused a static deflection at the free end of 0.5 mm. What are the stiffness and the natural frequency of the system?

Solution:

$$\text{The stiffness of the spring} = k = \frac{\text{Static force}}{\text{Corresponding static deflection}}$$

$$= \frac{mg}{\delta} = \frac{2 \times 9.81}{0.5 \times 10^{-3}} = 39{,}240 \text{ N/m}$$

$$\text{The natural frequency } = \omega_{nf} = \sqrt{\frac{k}{m}} = \sqrt{\frac{mg/\delta}{m}} = \sqrt{\frac{g}{\delta}}$$

$$= \sqrt{\frac{39{,}240}{2}} = 40.07 \text{ rad/s} = 22.29 \text{ Hz}$$

2.2 A SINGLE-DOF DAMPED ROTOR MODEL

In the previous section, the damping was ignored in the rotor model, but now in the present section, its effect will be considered (see Figure 2.8a). The damping in a rotor system can come from bearings, seals, shaft materials, relative motion between loose or shrink-fit components, aerodynamic effects, and rotor-stator rubs. The simplest damping model is the *viscous* (or *proportional*) *external damping*, and the damping force is expressed as

$$f_d(t) = c\dot{y} \tag{2.26}$$

where y is the displacement of the shaft, and c is a constant of proportionality and is called the viscous damping coefficient.

In Figure 2.8a, a rigid rotor mounted on two identical fluid-film bearings is considered a single-DOF system. From the free body diagram (Figure 2.8b) of the rotor, equations of motion could be written as

$$f(t) - ky - c\dot{y} = m\ddot{y} \quad \text{or} \quad m\ddot{y} + c\dot{y} + ky = f(t) \tag{2.27}$$

For the free damped motion (i.e., for a perfectly balanced rotor), Equation 2.27 takes the following form:

$$m\ddot{y} + c\dot{y} + ky = 0 \tag{2.28}$$

Let us assume a solution of the form

$$y = e^{st} \tag{2.29}$$

where s is an unknown parameter (it may be a complex number) and t is the time instant, so that $\dot{y} = se^{st}$ and $\ddot{y} = s^2e^{st}$. On substituting Equation 2.29 in Equation 2.28, we get

$$\left(ms^2 + cs + k\right)e^{st} = 0 \tag{2.30}$$

Hence, from the condition that Equation 2.29 is a solution of Equation 2.28 for all values of t, Equation 2.30 gives the following characteristic equation:

$$s^2 + \frac{c}{m}s + \frac{k}{m} = 0 \tag{2.31}$$

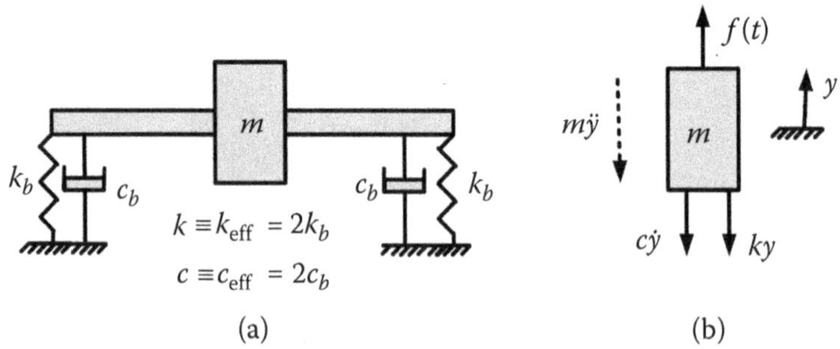

FIGURE 2.8 A rigid rotor mounted on two identical fluid-film bearings. (a) A rigid rotor mounted on flexible bearings. (b) Free body diagram of 1-DOF rotor mass.

which can be solved as

$$s_{1,2} = -\frac{c}{2m} \pm \sqrt{\left(\frac{c}{2m}\right)^2 - \frac{k}{m}} \tag{2.32}$$

Hence, the following form of the general solution is obtained:

$$y(t) = Ae^{s_1 t} + Be^{s_2 t} \tag{2.33}$$

where A and B are constants to be determined from the initial conditions of the problem (e.g., $y(0) = y_0$ and $\dot{y}(0) = v_0$). On substituting Equation 2.32 into Equation 2.33, we get

$$y(t) = e^{-\left(\frac{c}{2m}t\right)}\left[Ae^{\sqrt{\left\{\left(\frac{c}{2m}\right)^2 - \left(\frac{k}{m}\right)\right\}}t} + Be^{-\sqrt{\left\{\left(\frac{c}{2m}\right)^2 - \left(\frac{k}{m}\right)\right\}}t}\right] \tag{2.34}$$

The term outside the square brackets on the right side is an exponentially decaying function for the positive value of c. However, inside the square brackets, the term $\sqrt{\left(\frac{c}{2m}\right)^2 - \frac{k}{m}}$ can have the following three cases:

1. $\left(\frac{c}{2m}\right)^2 < \frac{k}{m}$: Exponents of terms within the square brackets in Equation 2.34 are imaginary numbers $\pm j\sqrt{\frac{k}{m} - \left(\frac{c}{2m}\right)^2}$, which means we can write $e^{\pm j\sqrt{\frac{k}{m} - \left(\frac{c}{2m}\right)^2}t} = \cos\sqrt{\frac{k}{m} - \left(\frac{c}{2m}\right)^2}t \pm j\sin\sqrt{\frac{k}{m} - \left(\frac{c}{2m}\right)^2}t$. Hence, Equation 2.34 can take following form:

$$y(t) = e^{-\left(\frac{c}{2m}\right)t}\left[(A+B)\cos\sqrt{\frac{k}{m} - \left(\frac{c}{2m}\right)^2}t + j(A-B)\sin\sqrt{\frac{k}{m} - \left(\frac{c}{2m}\right)^2}t\right] \tag{2.35}$$

Let $a = (A+B) = Y\cos\phi$ and $b = j(A-B) = Y\sin\phi$; we get

$$y(t) = e^{-\left(\frac{c}{2m}\right)t}\left[Y\cos\phi\cos\sqrt{\frac{k}{m} - \left(\frac{c}{2m}\right)^2}t + Y\sin\phi\sin\sqrt{\frac{k}{m} - \left(\frac{c}{2m}\right)^2}t\right] \tag{2.36}$$

which can alternatively be written as

$$y(t) = Ye^{-\left(\frac{c}{2m}\right)t}\cos\left[\sqrt{\frac{k}{m} - \left(\frac{c}{2m}\right)^2}t - \phi\right] \tag{2.37}$$

with

$$\phi = \tan^{-1}\left(\frac{b}{a}\right); \quad Y = \sqrt{a^2 + b^2}$$

where Y and ϕ are the amplitude and phase, respectively. For the present case, oscillations are possible (with the decaying type), and it is called the *underdamped* system (Figure 2.9). The *damped natural frequency* is given as $\omega_{nfd} = \sqrt{\frac{k}{m} - \left(\frac{c}{2m}\right)^2}$ from Equation 2.37.

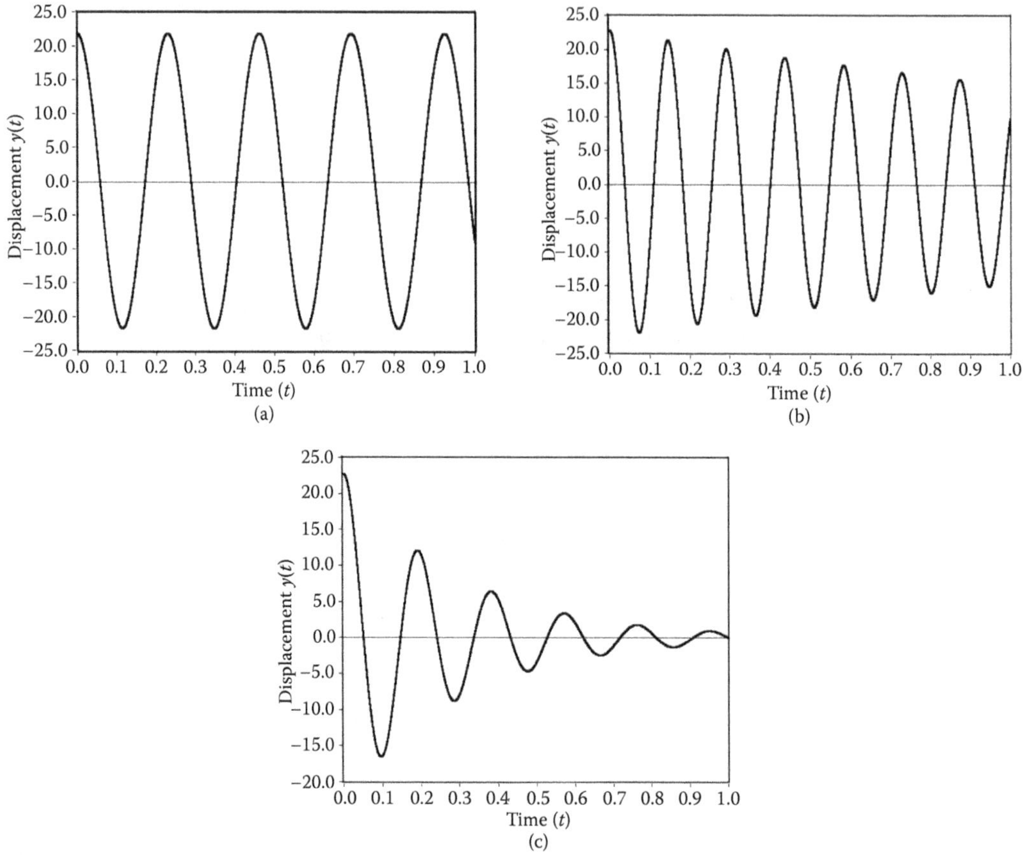

FIGURE 2.9 The response of an (a) undamped system, $\zeta=0$, (b) underdamped system, $\zeta=0.01$, and (c) underdamped system, $\zeta=0.1$.

2. $\left(\dfrac{c}{2m}\right)^2 > \dfrac{k}{m}$: Exponents of terms within the square brackets in Equation 2.34 are real numbers, which means there would not be any harmonic functions. Hence, no oscillation is possible and it is called the *overdamped* system (see Figure 2.10).
3. $\left(\dfrac{c}{2m}\right)^2 = \dfrac{k}{m}$: Exponents of terms within the square brackets in Equation 2.34 are zeros. The damping corresponding to this case is called the *critical damping*, c_c, which is the function of system stiffness and mass and is defined as

$$c_c = 2m\sqrt{k/m} = 2m\omega_{nf} = 2\sqrt{km} \tag{2.38}$$

The viscous damping can be expressed in terms of a nondimensional number, ζ, called the *damping ratio*, as

$$\zeta = c/c_c \tag{2.39}$$

The solution for the critically damped case with two repeated real roots can be expressed as

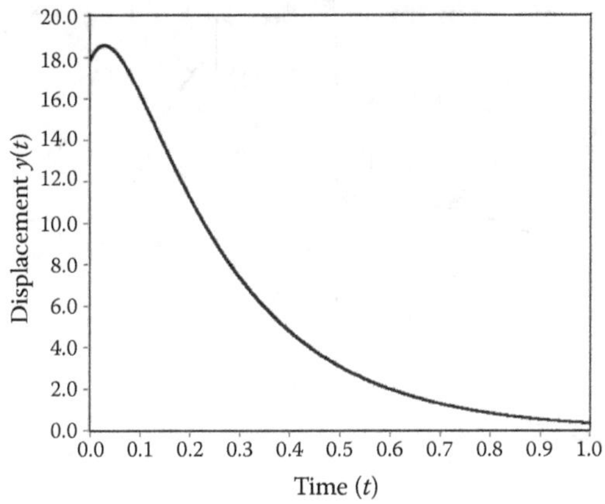

FIGURE 2.10 The response of an overdamped system ($\zeta > 1$).

$$y(t) = e^{-\omega_{nf} t}(A + Bt) \tag{2.40}$$

For this case, either no oscillation (similar to Figure 2.10) or for specific initial conditions a single crossing of the zero-response axis line is possible. Figure 2.11 shows a response of the critically damped system with a single crossing and no crossing of the zero-displacement line.

To summarize, depending upon the value of the damping ratio ζ (let us assume that it is a positive quantity), we can have following cases: (i) $\zeta < 1$: the underdamped condition with the damped natural frequency as $\omega_{nf_d} = \omega_{nf}\sqrt{1-\zeta^2}$; (ii) $\zeta > 1$: the overdamped condition; (iii) $\zeta = 1$: the critical damping; and (iv) $\zeta = 0$: the undamped system. For all the cases, the integration constants A and B are obtained from two initial conditions. Figure 2.12 shows the response of an unstable system with $\zeta < 0$ (i.e., for the negative value of the damping) in which an exponential increase in the amplitude can be seen. More detailed treatment of the present section can be seen in a text on basics of vibrations (Thomson and Dahleh, 1998).

2.3 RANKINE ROTOR MODEL

The single-DOF rotor model has limitations in that it cannot represent the orbital motion of the rotor in two transverse directions, which is the case in actual practice. Rankine (1869) used a two-DOF model to describe the motion of the rotor in two transverse directions as shown in Figure 2.13a. The shape of the orbit produced depends upon the relative amplitude and phase of the motions in two transverse directions (which in turn depend upon the relative difference in stiffness in the two transverse directions). The orbit could be a circular, elliptical, or straight line, inclined to the x- and y-axes, as shown in Figure 2.14. The straight-line motion (Figure 2.14c) could be considered a single-DOF system as described in Section 2.1, whereas the elliptical orbital motion (Figure 2.14b) may occur when the shaft has different stiffness in two orthogonal directions (along with cross-coupled stiffness terms).

The circular orbital motion (Figure 2.14a) may occur for a symmetrical shaft. It can be thought of as a mass attached with a spring, and it revolves about a point. From the free body diagram of the rotor, as shown in Figure 2.13b, for a constant spin speed, the radius of whirling of the rotor center will increase parabolically and is given as $r = F_c/k$, where F_c is the centrifugal force ($= m\omega^2 R$), R is the radius of path of the mass, and r is the extension of the spring of stiffness, k. It also can be physically visualized, as there will not be any resonance condition, as is found in the single-DOF model when the spin speed is increased gradually. This is a serious limitation of the Rankine model. Moreover, this model does not represent the realistic rotating unbalance force.

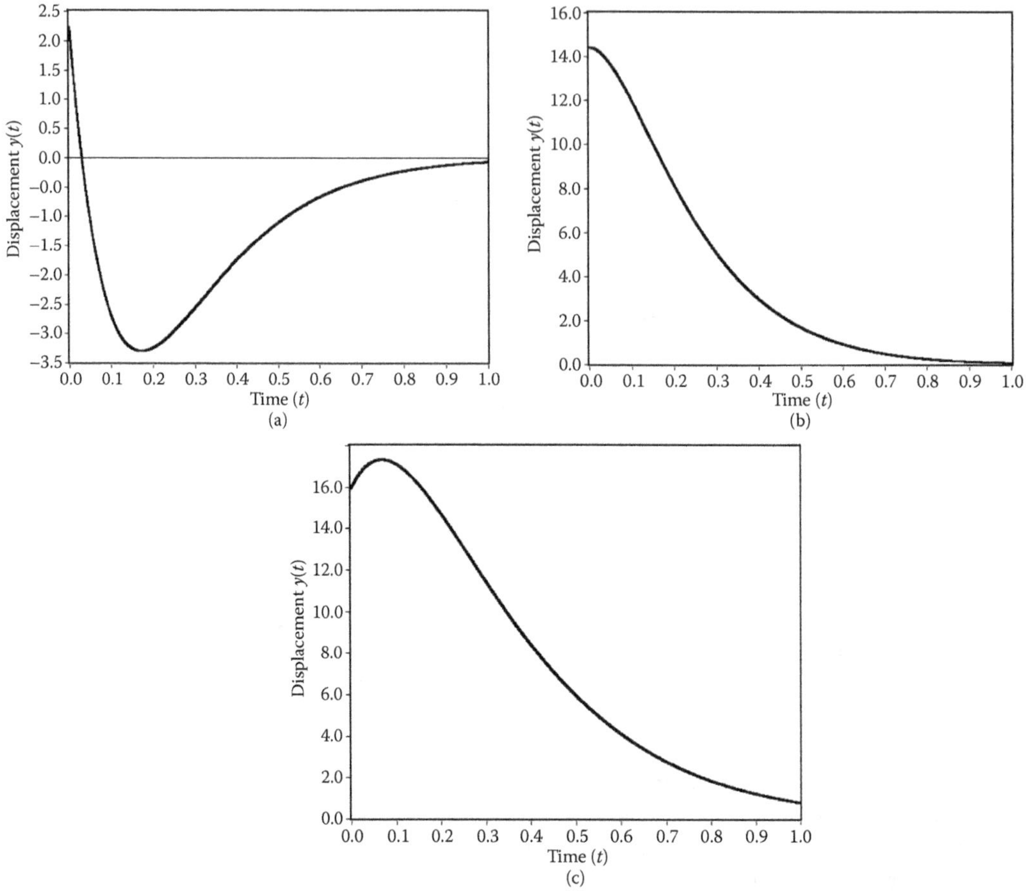

FIGURE 2.11 The response of a critically damped system ($\zeta = 1$) for different initial conditions: (a) $\dot{x}(0) < 0$, (b) $\dot{x}(0) = 0$, and (c) $\dot{x}(0) > 0$.

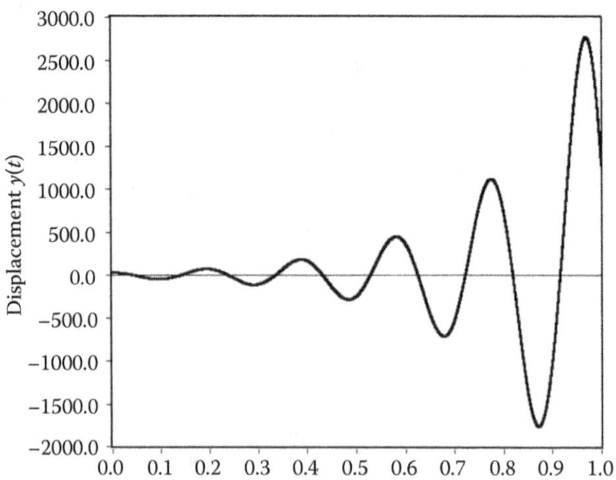

FIGURE 2.12 The response of an unstable system ($\zeta < 0$).

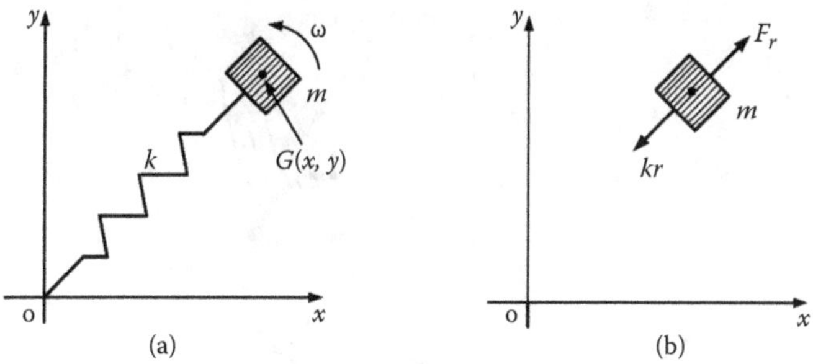

FIGURE 2.13 Rankine rotor model: (a) 2-DOF spring-mass rotor model; (b) free body diagram of the model.

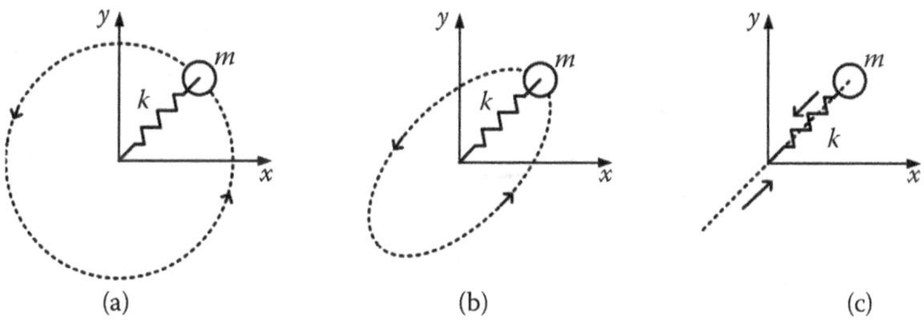

FIGURE 2.14 Orbital motion of the mass: (a) circular motion; (b) elliptical motion; (c) straight-line motion.

2.4 JEFFCOTT ROTOR MODEL

To overcome the limitations of the Rankine model, Jeffcott (1919) proposed a model, and Figure 2.15 shows a typical Jeffcott rotor model (it is also called the Föppl or Laval model). It consists of a simply supported, flexible, massless shaft with a rigid and thin disk mounted at the midspan of the shaft. The disk center of rotation, C, and its center of gravity, G, are offset by a distance, e, which is called the *eccentricity*. The shaft spin speed is ω, and the deformed shaft whirls about the bearing axis with a *whirl frequency*, ν. For the present case, the *synchronous whirl* is assumed (i.e., $\nu = \omega$), which is prevalent in the case of unbalance responses. The synchronous motion also occurs between the earth and the moon, and due to this we see always the same face of the moon from the earth. In the synchronous motion of the shaft, the orbital speed and its own spin speed are equal as shown in Figure 2.16a. The sense of rotation of the shaft spin and the whirling are the same (i.e., counterclockwise for Figure 2.16a) and it is called a *forward synchronous motion*. The black spot on the shaft represents the unbalance location or any other mark on the shaft. The unbalance force, in general, leads to synchronous whirl conditions; hence, this motion is basically a forced response.

Other kinds of whirl motions that may occur in real systems are antisynchronous or backward synchronous (i.e., $\nu = -\omega$, as shown in Figure 2.16b) and asynchronous (i.e., $\nu \neq \omega$). The antisynchronous whirl may occur when there is rubbing between the rotor and the stator; however, it occurs very rarely. For this case, the sense of rotation of the shaft spin and the whirling are opposite. Asynchronous (forward or backward) whirl motion may occur when speeds are high (e.g., when gyroscopic effects are predominant) or when the rotor is asymmetric or when the dynamic properties of the bearing are anisotropic. The asynchronous whirl motion may occur even in the perfectly balanced rotor due to momentarily transverse disturbances. Due to this, it will have the whirl frequency as the natural frequency of the rotor system and not be equal to the spin speed as long as the rotor model is linear. The black mark on the shaft will not be so systematic as in Figure 2.16 and may occupy various positions depending upon the frequency of the whirl.

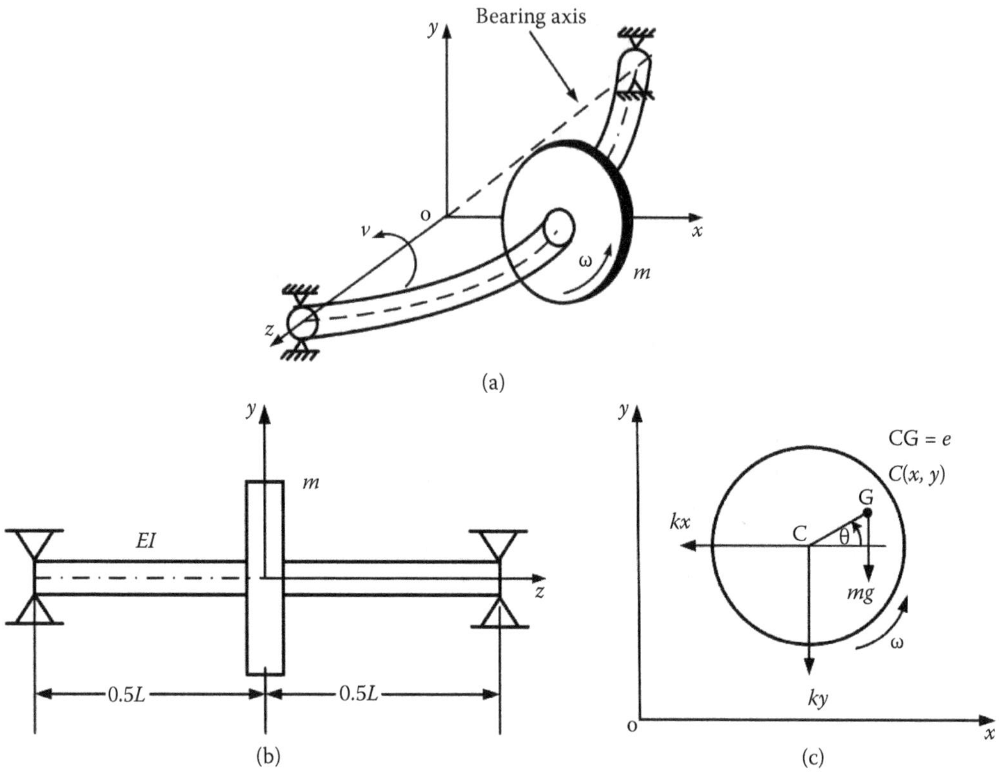

FIGURE 2.15 (a) A Jeffcott rotor model. (b) A Jeffcott rotor model in the y-z plane. (c) Free body diagram of the disk in the x-y plane.

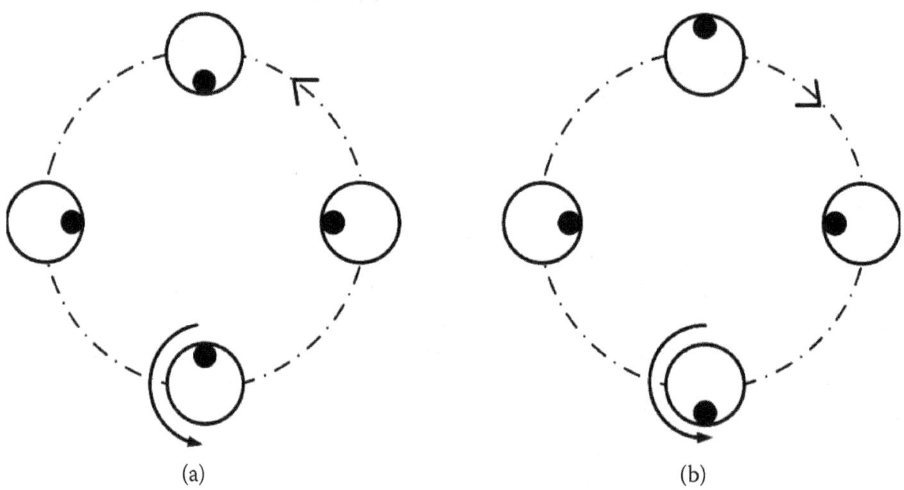

FIGURE 2.16 Whirling motions: (a) synchronous whirl; (b) antisynchronous whirl.

The transverse stiffness, k, of a simply supported shaft is expressed as

$$k = \frac{\text{Transverse load}}{\text{Transverse deflection}} = \frac{P}{PL^3/(48EI)} = \frac{48EI}{L^3} \tag{2.41}$$

where E is the Young's modulus, I is the second moment of area of the shaft cross-section, and L is the span of the shaft. x and y are coordinates to define the position of the center of rotation of the rotor, C (Figure 2.15c). Here o is the static equilibrium point of the disk (for small deflections, it can be approximated as if it is located at the bearing axis as shown in Figure 2.15a). The location of the unbalance is given by θ, which is measured from the x-axis in the counterclockwise direction. Thus, three geometrical coordinates (x, y, θ) are needed to define the position of the Jeffcott rotor (i.e., it has three DOFs with two translational displacements and one rotational displacement). The disk is at the midspan; hence, the tilting of the disk about transverse axes (i.e., x and y) is not there.

From Figure 2.15c, the force and moment balances in the x, y, and θ directions can be written as

$$-kx - c\dot{x} = m\frac{d^2}{dt^2}(x + e\cos\theta) \tag{2.42}$$

$$-ky - c\dot{y} - mg = m\frac{d^2}{dt^2}(y + e\sin\theta) \tag{2.43}$$

and

$$-mge\cos\theta = I_p\ddot{\theta} \tag{2.44}$$

where $(x + e\cos\theta)$ and $(y + e\sin\theta)$ are the position of the center of gravity, G, of the disk; m is the mass; and I_p is the polar mass moment of inertia of the disk. In addition to the restoring force contribution from the shaft, a damping force is considered. The damping force is idealized as *viscous damper*, and it originates primarily from the support and aerodynamic forces at the disk. The material damping of the shaft is not considered, which may lead to the instability in the rotor, which will be considered in detail in Chapters 1 and 2 of Volume II.

Herein, to get Equation 2.44, the moment is taken with respect to the C, the moment can also be taken about G. In that case, moment equation will contain x and y along with θ. But since Equations 2.42 and 2.43 contain θ we need to solve all three equations simultaneously. In Chapter 1 of Volume II, the moment is taken about the G to get equation of motion in θ direction (Example 1.9, Equation 1.44 of Volume II).

For the case $\theta=\omega t$, that is, when the disk is rotating at a constant spin speed, the Jeffcott rotor model reduces to a two-DOF rotor model. Physically, this means that only the transverse vibration is considered and the torsional vibration is ignored. By considering the static equilibrium as the reference for the axis system, the gravity effect could be ignored with the assumption of small static deflection. In the case of large static deflection, when the reference axis is at the bearing axis position, then the gravity force needs to be considered, and it gives a constant deflection along with the time-dependent unbalance response. Such a nonlinear effect of gravity in Equation 2.44 will be considered when we discuss the subcritical speed phenomenon due to gravity in Chapter 1 of Volume II. For the present case, the linear system is considered. Hence, equations of motion in the x and y directions, from Equations 2.42 and 2.43, can be written as

$$m\ddot{x} + c\dot{x} + kx = m\omega^2 e\cos\omega t \tag{2.45}$$

and

$$m\ddot{y} + c\dot{y} + ky = m\omega^2 e\sin\omega t \tag{2.46}$$

In deriving equations of motion, the centrifugal force due to the unbalance is not considered as the external force as it was considered in Section 2.1.2. However, it has been accounted for in the inertia force in the form of the eccentric position of the disk center of gravity as given in Equations 2.42 and 2.43. It should be noted that the equations of motion are uncoupled, and motions can be analyzed independently in two transverse planes (i.e., z-x and y-z planes). Noting Equation 2.9, from the undamped free vibration analyses, it can be seen that because the rotor system is symmetric, the rotor system will have two equal transverse, undamped, natural frequencies in two orthogonal directions, which are given as

$$\omega_{nf1,2} = \sqrt{k/m} \tag{2.47}$$

Because both natural frequencies are the same, a single frequency is observed during the whirling of the rotor in two orthogonal planes. However, for an anisotropic rotor support system, these will be distinct, as we shall observe in Chapter 4. The damped natural frequency can be obtained as described in Section 2.2.

2.4.1 Steady-State Response (Real Domain Method)

The damping does not affect the natural frequency of the system appreciably. However, its effect is more predominant for suppressing the vibration amplitude at the critical speed (i.e., resonance). Steady-state forced responses of Equations 2.45 and 2.46 can be assumed as

$$x = X\cos(\omega t - \phi) \tag{2.48}$$

and

$$y = Y\cos\left[\omega t - (\phi + \pi/2)\right] = Y\sin(\omega t - \phi) \tag{2.49}$$

where X and Y are the steady-state, forced-response amplitude in the x and y directions, respectively; ω is the same as the excitation frequency due to the unbalance force and is equal to the shaft spin speed (synchronous condition is assumed); and ϕ is the phase lag of the displacement, $x(t)$, with respect to the unbalance force. The phase difference between the two orthogonal direction responses for the direction of spinning of the shaft chosen is $\pi/2$ radians. For the direction of shaft whirling shown in Figure 2.15 (i.e., counterclockwise) for the present axis system, the response in the y direction will lag the response in the x direction by $\pi/2$ radians. Hence, the lag of the y direction response with respect to the force will be $(\pi/2+\phi)$. The phase difference between a rotor's response in two orthogonal directions is exactly 90° (or $\pi/2$ radians) for a perfectly isotropic (symmetric) system experiencing synchronous forward whirl due to unbalance, other cases phase difference may not be 90°. On taking the first- and second-time derivatives of the response, $x(t)$, we get

$$\dot{x} = -\omega X\sin(\omega t - \phi) \quad \text{and} \quad \ddot{x} = -\omega^2 X\cos(\omega t - \phi) \tag{2.50}$$

On substituting Equations 2.48 and 2.50 into Equation 2.45, we get

$$-m\omega^2 X\cos(\omega t - \phi) - c\omega X\sin(\omega t - \phi) + kX\cos(\omega t - \phi) = m\omega^2 e\cos\omega t \tag{2.51}$$

which can be expanded as

$$\begin{aligned} &-m\omega^2 X(\cos\omega t\cos\phi + \sin\omega t\sin\phi) - c\omega X(\sin\omega t\cos\phi - \cos\omega t\sin\phi) \\ &+kX(\cos\omega t\cos\phi + \sin\omega t\sin\phi) = m\omega^2 e\cos\omega t \end{aligned} \tag{2.52}$$

On separating the in-phase (i.e., $\cos\omega t$) and quadrature (i.e., $\sin\omega t$) terms of the earlier equation, we get

$$-m\omega^2 X\cos\phi + cX\omega\sin\phi + kX\cos\phi = m\omega^2 e \tag{2.53}$$

and

$$-m\omega^2 X\sin\phi - cX\omega\cos\phi + kX\sin\phi = 0 \tag{2.54}$$

Equation 2.54 can be solved for the phase angle as

$$\tan\phi = \frac{c\omega}{k - m\omega^2} \tag{2.55}$$

which gives

$$\sin\phi = \frac{c\omega}{\sqrt{\left(k - m\omega^2\right)^2 + (c\omega)^2}} \quad \text{and} \quad \cos\phi = \frac{k - m\omega^2}{\sqrt{\left(k - m\omega^2\right)^2 + (c\omega)^2}} \tag{2.56}$$

Substituting Equation 2.56 into Equation 2.53 gives the displacement amplitude as

$$X = \frac{m\omega^2 e}{\sqrt{\left(k - m\omega^2\right)^2 + (c\omega)^2}} \tag{2.57}$$

Similarly, we can obtain the response amplitude in the y direction from Equation 2.46 as

$$Y = \frac{m\omega^2 e}{\sqrt{\left(k - m\omega^2\right)^2 + (c\omega)^2}} \tag{2.58}$$

These amplitudes could be plotted with respect to spin speed of the shaft for overall understanding of the response, which will be seen in the next subsection. From Equations 2.48, 2.49, 2.57, and 2.58, it can be seen that because of the symmetry of the rotor we have $X = Y = R$ and the orbit is circular in nature, that is,

$$x^2 + y^2 = \left\{X\cos(\omega t - \phi)\right\}^2 + \left\{Y\sin(\omega t - \phi)\right\}^2 = R^2 \tag{2.59}$$

with

$$X = Y = R$$

2.4.2 Steady-State Response (Complex Domain Method)

An alternative approach that is very popular in rotor dynamic analyses is to use complex algebra to define the complex whirl radius as

$$r = x + jy \tag{2.60}$$

where $j = \sqrt{-1}$. On multiplying Equation 2.46 by j and adding to Equation 2.45, we get

$$m\ddot{r} + c\dot{r} + kr = me\omega^2 e^{j\omega t} \quad \text{with} \quad e^{j\omega t} = \cos\omega t + j\sin\omega t \tag{2.61}$$

Now the steady-state response can be assumed as

$$r = R\,e^{j(\omega t-\phi)} \tag{2.62}$$

where R is the whirl amplitude (it is a real quantity) and ϕ is the phase lag of response with respect to the unbalance force. Both R and ϕ are constant quantities. On differentiating Equation 2.62 with respect to time, t, we get

$$\dot{r} = j\omega R\,e^{j(\omega t-\phi)} \quad \text{and} \quad \ddot{r} = -\omega^2 R\,e^{j(\omega t-\phi)} \tag{2.63}$$

On substituting Equations 2.62 and 2.63 into Equation 2.61, we get

$$\left\{(k - m\omega^2) + j\omega c\right\} R\,e^{-j\phi} = me\omega^2 \tag{2.64}$$

Equation 2.64 can be written as

$$\left\{(k - m\omega^2) + j\omega c\right\}(R\cos\phi - jR\sin\phi) = me\omega^2 \tag{2.65}$$

On equating the real and imaginary parts from both sides of Equation 2.65, we get

$$\left(k - m\omega^2\right) R\cos\phi + \omega c R\sin\phi = me\omega^2 \tag{2.66}$$

and

$$-(k - m\omega^2) R\sin\phi + \omega c R\cos\phi = 0 \tag{2.67}$$

From Equation 2.67, we get the phase as

$$\tan\phi = \frac{c\omega}{k - m\omega^2} \tag{2.68}$$

On substitution of the phase from Equation 2.68 into 2.66, the whirl amplitude can be written as

$$R = \frac{m\omega^2 e}{\sqrt{(k - m\omega^2)^2 + (c\omega)^2}} \tag{2.69}$$

Equations 2.67 and 2.69 are similar to the previous results, that is, Equations 2.55 and 2.57, respectively. The nondimensional form of Equations 2.68 and 2.69 can be written as

$$\tan\phi = \frac{2\zeta\bar{\omega}}{1-\bar{\omega}^2} \tag{2.70}$$

and

$$\bar{R} = \frac{R}{e} = \frac{\bar{\omega}^2}{\sqrt{(1-\bar{\omega}^2)^2 + (2\zeta\bar{\omega})^2}} \tag{2.71}$$

with

$$\bar{\omega} = \omega/\omega_{nf}; \quad \omega_{nf} = \sqrt{k/m}; \quad \zeta = c/c_c; \quad c_c = 2\sqrt{km} \tag{2.72}$$

where $\bar{R}$ is the whirl amplitude ratio, $\bar{\omega}$ is the frequency ratio, ω_{nf} is the transverse undamped natural frequency of the nonrotating rotor system, ζ is the damping ratio, and c_c is the critical damping factor for which the damping ratio is equal to unity. Figure 2.17 shows the whirl amplitude

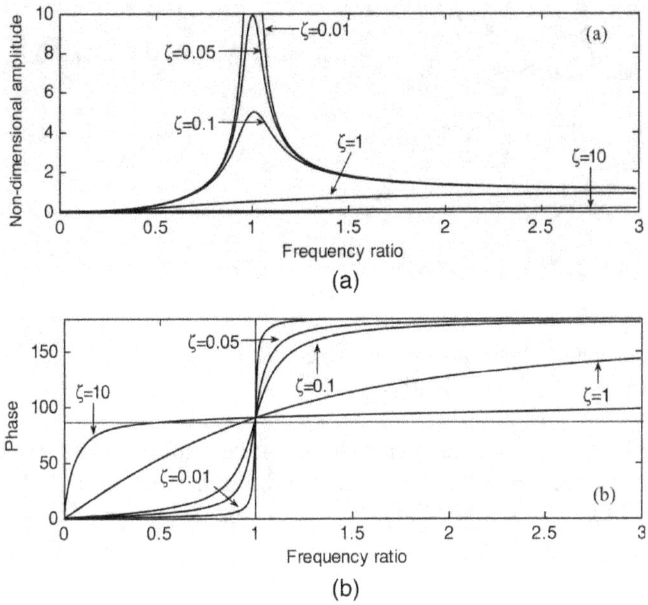

FIGURE 2.17 Plot of (a) nondimensional response and (b) phase versus frequency ratio $\bar{\omega}$.

and the phase variation with the frequency ratio. Figure 2.17a shows that the maximum amplitude (i.e., the location of the critical speed) occurs at $\bar{\omega} = 1$ for the undamped case, albeit at a slightly higher frequency ratio than one (i.e., $\bar{\omega} > 1$), when damping is present in the system (refer Exercise problem 2.1). It should be noted that we have observed previously that damped natural frequency is lesser than the undamped case. It could be observed that the damping is the most important parameter for reducing the whirl amplitude at the critical speed.

It can be seen from Figure 2.17b that for a lightly underdamped system the phase angle changes from 0° to 90° as the spin speed is increased to ω_{nf} (i.e., $\bar{\omega} = 1$) and gradually becomes 180° as the spin speed is increased to a higher frequency ratio. It should be noted that the phase angle is 90° at $\bar{\omega} = 1$ even for the case of various levels of damping in the rotor system. For a highly overdamped system ($\zeta > 10$), the phase angle always remains at 90° before and after $\bar{\omega} = 1$, which might be a physically unrealistic case to attain.

As the spin speed of the rotor crosses the critical speed, the center of gravity of the disk comes inside of the whirl orbit and the rotor tries to rotate about the center of gravity (because the phase changes from 0° to 180° for the undamped case). This can be seen from Figure 2.17a—as the spin speed approaches infinity, the displacement of the shaft tends to equal the disk eccentricity ($\bar{R} = 1$). Because the measurement of amplitude of vibration at the critical speed is difficult, determining the precise critical speed is difficult. To overcome this problem, the measurement of the phase at the critical speed is advantageous (because it remains constant at 90° irrespective of damping in the system). The change in-phase between the force and the response is also shown in Figure 2.18 for three different spin speeds (i.e., below the critical speed, at the critical speed, and above the critical speed). Due to unbalance, we have synchronous whirl and Figure 2.18 is for that. Readers are encouraged to draw such diagrams (i.e., Figure 2.18) when the rotor has antisynchronous whirl due to some reason (refer to Figure 2.16).

Because for the present analysis the synchronous whirl condition is assumed, at a particular spin speed the shaft will not have any flexural vibration and in a particular bent configuration it will whirl (orbiting) about its bearing axis as shown in Figure 2.16a. It can be seen that the black point on the shaft cross-section will have tension during the whirling. However, it can be seen from Figure 2.16b for the antisynchronous whirl that the shaft (the black point on the shaft cross-section) will have

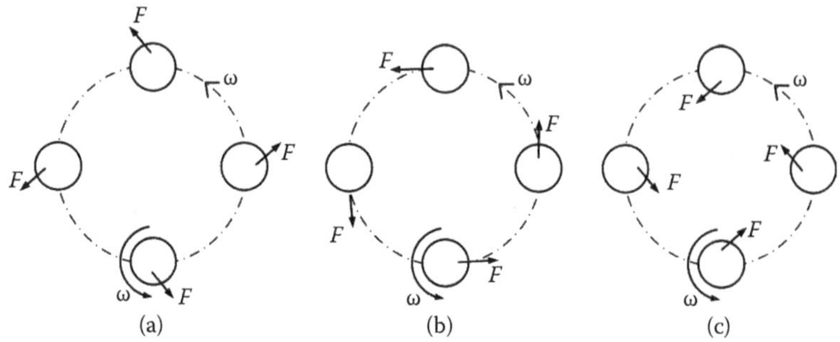

FIGURE 2.18 Orientation of the unbalance force when damping is present in the rotor system: (a) below critical speed; (b) at critical speed; (c) after critical speed.

a reversal of the bending stresses twice per whirling of the shaft. For the asynchronous whirl, the black point on the cross-section of the shaft will take different positions during the whirling of the shaft. These cases will be discussed in Chapter 5 during the discussion of gyroscopic effects.

2.4.3 Steady-State Response (Matrix Method)

By using software (Scilab or MATLAB or Python) that can handle complex matrices, the following matrix procedure may be very helpful for numerical simulation, and it may be extended even for very complex multi-DOF rotor systems. Equations 2.45 and 2.46 can be combined in matrix form as

$$\begin{bmatrix} m & 0 \\ 0 & m \end{bmatrix}\begin{Bmatrix} \ddot{x} \\ \ddot{y} \end{Bmatrix}+\begin{bmatrix} c & 0 \\ 0 & c \end{bmatrix}\begin{Bmatrix} \dot{x} \\ \dot{y} \end{Bmatrix}+\begin{bmatrix} k & 0 \\ 0 & k \end{bmatrix}\begin{Bmatrix} x \\ y \end{Bmatrix}=\begin{Bmatrix} m\omega^2 e\cos\omega t \\ m\omega^2 e\sin\omega t \end{Bmatrix} \tag{2.73}$$

The force vector in Equation 2.73 could be expressed as

$$\begin{Bmatrix} m\omega^2 e\cos\omega t \\ m\omega^2 e\sin\omega t \end{Bmatrix}=\mathrm{Re}\begin{Bmatrix} m\omega^2 e(\cos\omega t+j\sin\omega t) \\ m\omega^2 e(\sin\omega t-j\cos\omega t) \end{Bmatrix}=\mathrm{Re}\begin{Bmatrix} me\omega^2(e^{j\omega t}) \\ me\omega^2(-je^{j\omega t}) \end{Bmatrix}=\mathrm{Re}\left(\begin{Bmatrix} F_{unb_x} \\ F_{unb_y} \end{Bmatrix}e^{j\omega t}\right) \tag{2.74}$$

where

$$F_{unb_x}=me\omega^2 \quad \text{and} \quad F_{unb_y}=-jme\omega^2 \quad \text{so that} \quad F_{unb_y}=-jF_{unb_x} \tag{2.75}$$

where Re(.) represents the real part of the quantity inside the parentheses. F_{unb_x} and F_{unb_y} are the unbalance force components in the x and y directions, respectively. On substituting Equation 2.74 into Equation 2.73, it can be written as (henceforth for brevity the symbol Re(.) will be omitted, because the expression on the left is a real quantity)

$$\begin{bmatrix} m & 0 \\ 0 & m \end{bmatrix}\begin{Bmatrix} \ddot{x} \\ \ddot{y} \end{Bmatrix}+\begin{bmatrix} c & 0 \\ 0 & c \end{bmatrix}\begin{Bmatrix} \dot{x} \\ \dot{y} \end{Bmatrix}+\begin{bmatrix} k & 0 \\ 0 & k \end{bmatrix}\begin{Bmatrix} x \\ y \end{Bmatrix}=\begin{Bmatrix} F_{unb_x} \\ F_{unb_y} \end{Bmatrix}e^{j\omega t} \tag{2.76}$$

The relationship (Equation 2.75) is true for the present axis system along with directions of the whirling (R) and the unbalance force vectors chosen as shown in Figure 2.19a. For this case, F_{unb_y} lags behind F_{unb_x} by 90°. Let us derive this relationship by an alternative way: If $F_{unb_x}=Fe^{j\theta}$, then

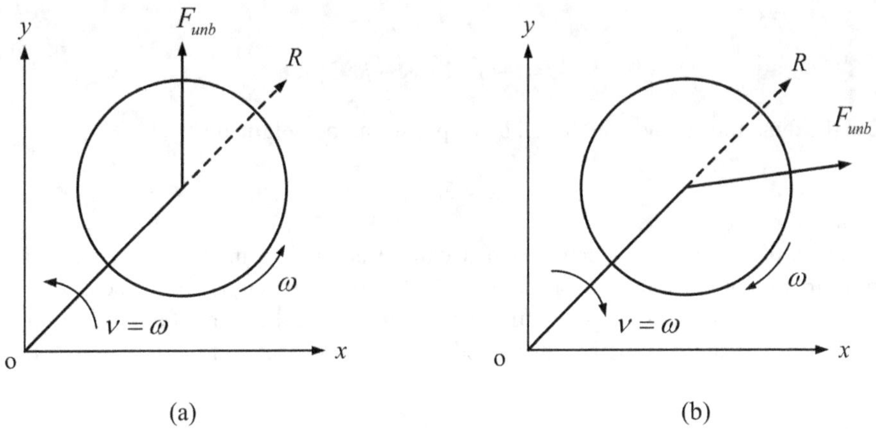

FIGURE 2.19 (a) The direction of whirl is the same as the positive axis direction. (b) The direction of whirl is opposite to the positive axis direction.

$$F_{unb_y} = Fe^{j(\theta-\pi/2)} = Fe^{j\theta}e^{-(\pi/2)j} = Fe^{j\theta}\left\{\cos(-\pi/2)+j\sin(-\pi/2)\right\} = -jFe^{j\theta} = -jF_{unb_x} \quad (2.77)$$

where $\theta=\omega t$. For the direction of whirl (R) as well as spinning of rotor opposite to the positive axis direction as shown in Figure 2.19b (it should be noted that still the motion is synchronous whirl), the following relationship holds:

$$F_{unb_y} = Fe^{j(\theta+\pi/2)} = Fe^{j\theta}e^{(\pi/2)j} = Fe^{j\theta}\left\{\cos(\pi/2)+j\sin(\pi/2)\right\} = jFe^{j\theta} = jF_{unb_x} \quad (2.78)$$

so that

$$F_{unb_y} = jF_{unb_x} \quad (2.79)$$

in which case F_{unb_y} leads F_{unb_x} by 90°. It should be noted in Equation 2.76 that the force vector elements on the right have significance of real parts only, which is quite clear from the expanded form of the force vector in Equation 2.74. Equation 2.76 can be written in more a compact form as

$$\mathbf{M}\ddot{\mathbf{x}}+\mathbf{C}\ddot{\mathbf{x}}+\mathbf{K}\mathbf{x} = \mathbf{F}_{unb}e^{j\omega t} \quad (2.80)$$

$$\text{with } \mathbf{M}=\begin{bmatrix} m & 0 \\ 0 & m \end{bmatrix}, \mathbf{C}=\begin{bmatrix} c & 0 \\ 0 & c \end{bmatrix}, \mathbf{K}=\begin{bmatrix} k & 0 \\ 0 & k \end{bmatrix}, \mathbf{x}=\begin{Bmatrix} x \\ y \end{Bmatrix}, \mathbf{F}_{unb}=\begin{Bmatrix} F_{unb_x} \\ F_{unb_y} \end{Bmatrix}$$

where $\mathbf{M}$ is the mass matrix, $\mathbf{C}$ is the damping matrix, $\mathbf{K}$ is the stiffness matrix, $\mathbf{x}(t)$ is the response vector, and $\mathbf{F}_{unb}$ is the unbalance force complex amplitude vector. The solution can be chosen as

$$\mathbf{x}(t) = \mathbf{X}e^{j\omega t} \quad (2.81)$$

where elements of the vector $\mathbf{X}$ are, in general, complex.

Equation 2.81 can be differentiated to give

$$\dot{\mathbf{x}} = j\omega\mathbf{X}e^{j\omega t} \quad \text{and} \quad \ddot{\mathbf{x}} = -\omega^2\mathbf{X}e^{j\omega t} \quad (2.82)$$

On substituting Equations 2.81 and 2.82 into Equation 2.80, we get

$$\mathbf{Z}\mathbf{X} = \mathbf{F}_{unb} \quad (2.83)$$

with

$$\mathbf{Z} = \left(-\omega^2\mathbf{M} + \mathbf{K} + j\omega\mathbf{C}\right) \tag{2.84}$$

where **Z** is the dynamic stiffness matrix. The response can be obtained as

$$\mathbf{X} = \mathbf{Z}^{-1}\mathbf{F}_{unb} \tag{2.85}$$

This method is quite general in nature and it can be applied to multi-DOF systems also once equations of motion in the standard form are available. To obtain natural frequencies of the system directly, we need to put $|\mathbf{Z}| = 0$ to get the frequency equation in the form of a polynomial. The following example illustrates the method discussed in the present section for a Jeffcott rotor.

Example 2.4

Obtain the unbalance response of a rotor system with the following equations of motion:

$$m\ddot{x} + kx = m\omega^2 e\cos\omega t \quad \text{and} \quad m\ddot{y} + ky = m\omega^2 e\sin\omega t$$

Solution:

Because equations of motion are uncoupled, both equations can be solved independently. The first equation can be written as

$$m\ddot{x} + kx = F_x e^{j\omega t} \quad \text{with} \quad F_x = me\omega^2$$

in which the real part of the term on the right has relevance. The solution can be assumed as

$$x = Xe^{j\omega t}$$

where in general X is a complex quantity. The earlier equation gives

$$\ddot{x} = -\omega^2 Xe^{j\omega t}$$

On substituting assumed solutions into the equation of motion, we get

$$\left(-m\omega^2 X + kX\right) = me\omega^2$$

which gives

$$X = \frac{me\omega^2}{k - m\omega^2}$$

Hence, the response in the x direction becomes

$$x = \frac{me\omega^2}{k - m\omega^2}e^{j\omega t} = \text{Re}\left\{\frac{me\omega^2}{k - m\omega^2}(\cos\omega t + j\sin\omega t)\right\} = \frac{me\omega^2}{k - m\omega^2}\cos\omega t$$

Similarly, the second equation of motion can be written as

$$m\ddot{y} + ky = F_y e^{j\omega t} \quad \text{with} \quad F_y = -jme\omega^2$$

in which the real part of the term on the right has relevance. The solution can be assumed as

$$y = Ye^{j\omega t}$$

On substituting into the equation of motion, we get

$$Y = \frac{F_y}{k - m\omega^2}$$

Hence, the response in the y direction becomes

$$y = \frac{F_y}{k - m\omega^2}e^{j\omega t} = \frac{-jme\omega^2}{k - m\omega^2}e^{j\omega t} = \text{Re}\left\{\frac{me\omega^2}{k - m\omega^2}(-j\cos\omega t + \sin\omega t)\right\} = \frac{me\omega^2}{k - m\omega^2}\sin\omega t$$

Now the same problem is solved in the matrix form. We have

$$\mathbf{M} = \begin{bmatrix} m & 0 \\ 0 & m \end{bmatrix}; \quad \mathbf{K} = \begin{bmatrix} k & 0 \\ 0 & k \end{bmatrix}; \quad \mathbf{C} = \begin{bmatrix} 0 & 0 \\ 0 & 0 \end{bmatrix}; \quad \mathbf{F}_{unb} = \begin{bmatrix} me\omega^2 \\ -jme\omega^2 \end{bmatrix}$$

$$\mathbf{Z} = \left(-\omega^2\mathbf{M} + \mathbf{K}\right) = \begin{bmatrix} k - m\omega^2 & 0 \\ 0 & k - m\omega^2 \end{bmatrix}; \quad \mathbf{Z}^{-1} = \begin{bmatrix} 1/\left(k - m\omega^2\right) & 0 \\ 0 & 1/\left(k - m\omega^2\right) \end{bmatrix}$$

so that

$$\mathbf{X} = \mathbf{Z}^{-1}\mathbf{F}_{unb} = \begin{bmatrix} 1/\left(k - m\omega^2\right) & 0 \\ 0 & 1/\left(k - m\omega^2\right) \end{bmatrix}\begin{Bmatrix} me\omega^2 \\ -jme\omega^2 \end{Bmatrix} = \frac{1}{\left(k - m\omega^2\right)}\begin{Bmatrix} me\omega^2 \\ -jme\omega^2 \end{Bmatrix}$$

The response in the x and y directions can be written as (refer Equation 2.81)

$$\mathbf{x} = \begin{Bmatrix} x \\ y \end{Bmatrix} = \text{Re}\left(\frac{1}{\left(k - m\omega^2\right)}\begin{Bmatrix} me\omega^2 \\ -jme\omega^2 \end{Bmatrix}e^{j\omega t}\right) = \frac{1}{\left(k - m\omega^2\right)}\begin{Bmatrix} me\omega^2\cos\omega t \\ me\omega^2\sin\omega t \end{Bmatrix}$$

It should be noted that for undamped forced vibration, the equation in the frequency domain takes the form $\mathbf{ZX} = \mathbf{0}$, where $\mathbf{0}$ is a null vector of size 2×1. To get the natural frequency of the system, we need to put $|\mathbf{Z}| = 0$, which gives

$$\begin{vmatrix} \left(k - m\omega^2\right) & 0 \\ 0 & \left(k - m\omega^2\right) \end{vmatrix} = 0 \quad \text{or} \quad \left(k - m\omega^2\right)^2 = 0 \quad \text{or} \quad \omega_{nf} = \sqrt{\frac{k}{m}}$$

2.5 A JEFFCOTT ROTOR MODEL WITH AN OFFSET DISK

Figure 2.20a shows a more general case of the Jeffcott rotor when the rigid disk is placed with some offset from the midspan. Using a and b locate the position of the disk in a shaft of length l. The spin speed, ω, of the shaft is considered to be constant. For such rotors, apart from two transverse displacements of the center of the disk (i.e., x and y), the tilting (transverse rotational displacements) of the disk about the x- and y-axis (i.e., φ_x and φ_y) also occurs, and it makes the rotor system a four-DOF system (for three-dimensional visualization of the transverse rotational displacements refer to Figures 5.42 and 5.43). For the present analysis, the rotary inertia of the disk is considered; however, the effect of the gyroscopic moment has been ignored, which otherwise would be present due to the offset of the disk from the midspan of the shaft (see Chapter 5). In Figure 2.20b, points C

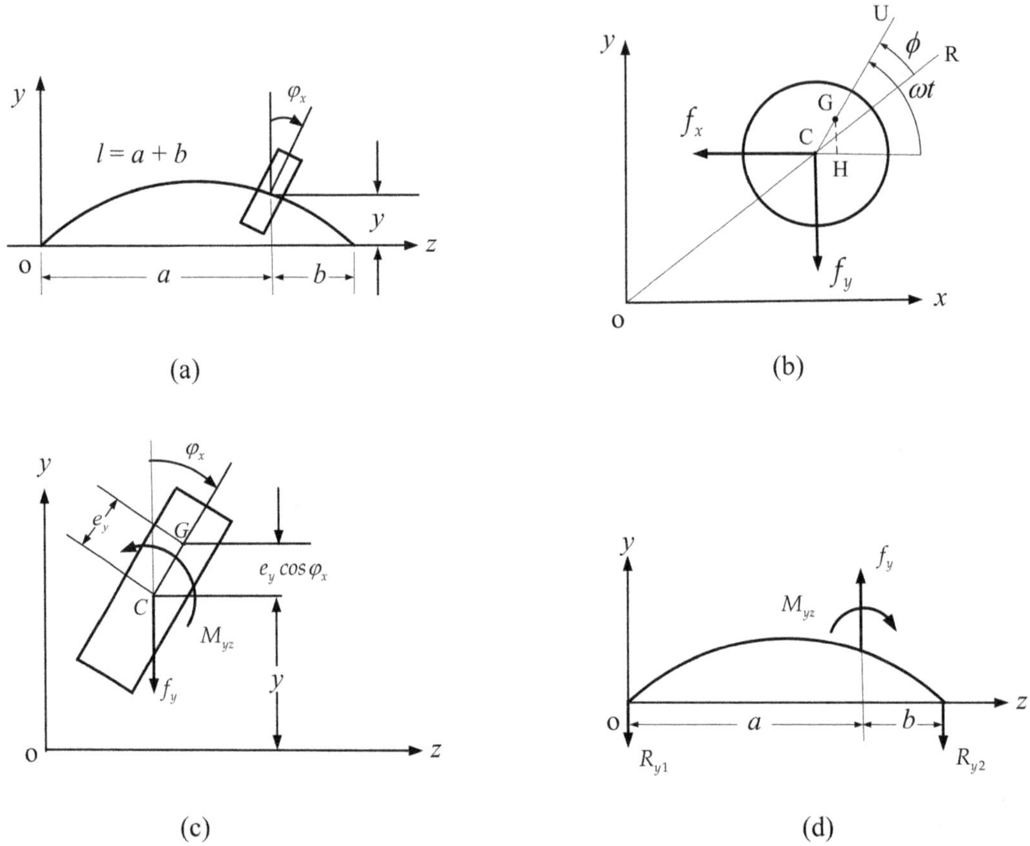

FIGURE 2.20 (a) A Jeffcott rotor with a disk offset from the midspan in the *y-z* plane. (b) Free body diagram of the shaft in the *x-y* plane. (c) Free body diagram of the disk in the *y-z* plane. (d) Free body diagram of the shaft in the *y-z* plane.

and *G* represent the geometrical center and the center of gravity of the disk, respectively. The angle, ϕ, represents the phase between the force and the radial response (for the present case the initial phase of the unbalance is considered as zero).

From Figure 2.20b, we can have the following relations for the eccentricity:

$$e_x = \overline{CH} = e\cos\omega t \quad \text{and} \quad e_y = \overline{GH} = e\sin\omega t \tag{2.86}$$

where e_x and e_y are components of the eccentricity, *e*, at a particular time, *t*, in the *y-z* (Figure 2.20c) and *z-x* planes (Figure 2.21b), respectively.

From Figure 2.20c, equations of motion of the disk in the *y* and φ_x directions can be written as

$$-f_y = m\frac{d^2}{dt^2}\left(y + e_y\cos\varphi_x\right) \Rightarrow -f_y = m\frac{d^2}{dt^2}\left(y + e\sin\omega t\cos\varphi_x\right) \tag{2.87}$$

and

$$-f_y e_y \sin\varphi_x - M_{yz} = I_d\ddot{\varphi}_x \Rightarrow -f_y\left(e\sin\omega t\right)\sin\varphi_x - M_{yz} = I_d\ddot{\varphi}_x \tag{2.88}$$

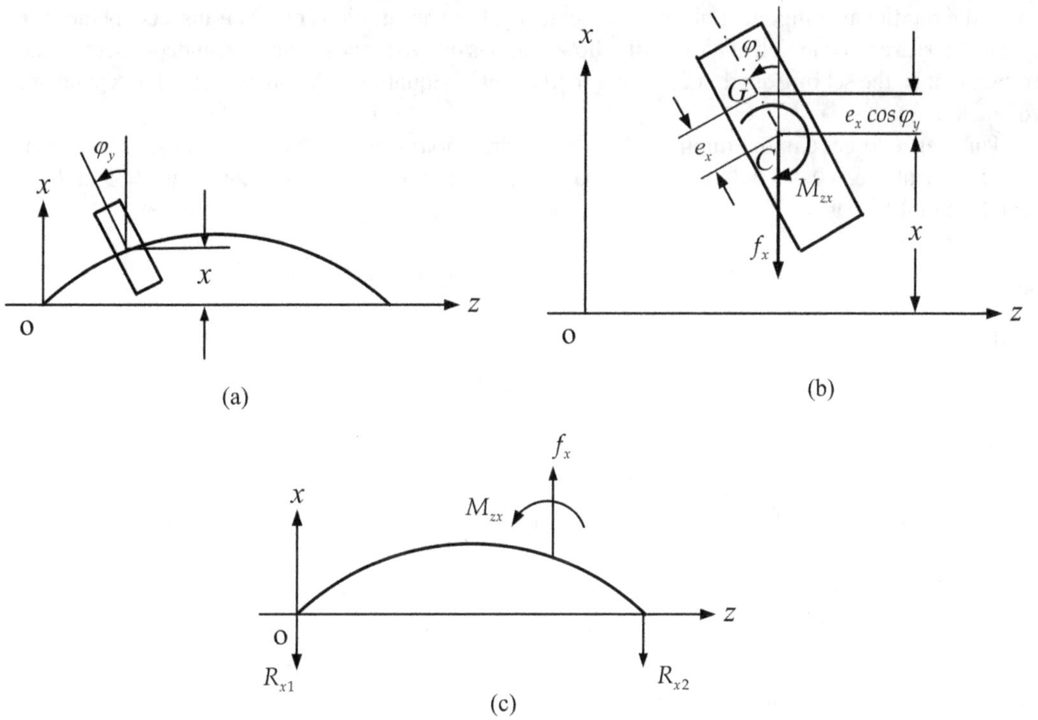

FIGURE 2.21 (a) A Jeffcott rotor with a disk offset from the midspan in the *z-x* plane. (b) Free body diagram of the disk in the *z-x* plane. (c) Free body diagram of the shaft in the *z-x* plane.

which simplifies to

$$m\ddot{y}+f_y=me\frac{d^2}{dt^2}\left(\cos\varphi_x\sin\omega t\right)\quad\text{and}\quad I_d\ddot{\varphi}_x+M_{yz}+e\sin\varphi_x f_y\sin\omega t=0 \tag{2.89}$$

where m is the disk mass, I_d is the diametral mass moment of inertia of the disk, f_y is the reaction force, and M_{yz} is the reaction moment from the shaft on to the disk. It should be noted that the moment is taken about the point G so no unbalance force appears in moment equation. These equations are nonlinearly coupled with the rotational (titling) component of displacement, φ_x.

Figure 2.21a shows the rotor in the *z-x* plane. From Figure 2.21b, we can write equations of motion as

$$-f_x=m\frac{d^2}{dt^2}\left(x+e_x\cos\varphi_y\right)\Rightarrow -f_x=m\frac{d^2}{dt^2}\left(x+e\cos\omega t\cos\varphi_y\right) \tag{2.90}$$

and

$$-f_xe_x\sin\varphi_y-M_{zx}=I_d\ddot{\varphi}_y\Rightarrow -f_x\left(e\cos\omega t\right)\sin\varphi_y-M_{zx}=I_d\ddot{\varphi}_y \tag{2.91}$$

which simplifies to

$$m\ddot{x}+f_x=me\frac{d^2}{dt^2}\left(\cos\varphi_y\cos\omega t\right)\quad\text{and}\quad I_d\ddot{\varphi}_y+M_{zx}+e\sin\varphi_y f_x\cos\omega t=0 \tag{2.92}$$

where I_d is the diametral mass moment of inertia about the *y*-axis, f_x is the reaction force, and M_{zx} is the reaction moment from the shaft onto the disk. These equations are also nonlinearly coupled with the rotational component of displacement, φ_y. However, motions of two transverse planes (i.e., *y-z* and *z-x*) are not coupled and that will allow motions of two planes to analyze independent of each other, that is, the set of Equation 2.89 and another set of Equation 2.92 can be solved independently of each other.

Unbalance forces can be simplified (i.e., by linearization) with the assumption of small rotational displacement (i.e., $\cos\varphi_x$ and $\cos\varphi_y \approx 1$ and $\sin\varphi_x \approx \varphi_x$; $\sin\varphi_y \approx \varphi_y$), and Equations 2.89 and 2.92 can be simplified as

$$m\ddot{y} + f_y = m\omega^2 e \sin\omega t, \quad I_d\ddot{\varphi}_x + M_{yz} + e\varphi_x f_y \sin\omega t = 0 \tag{2.93}$$

and

$$m\ddot{x} + f_x = m\omega^2 e \cos\omega t, \quad I_d\ddot{\varphi}_y + M_{zx} + e\varphi_y f_x \cos\omega t = 0 \tag{2.94}$$

Now the equations are assembled as

$$\begin{bmatrix} m & 0 & 0 & 0 \\ 0 & I_d & 0 & 0 \\ 0 & 0 & m & 0 \\ 0 & 0 & 0 & I_d \end{bmatrix} \begin{Bmatrix} \ddot{y} \\ \ddot{\varphi}_x \\ \ddot{x} \\ \ddot{\varphi}_y \end{Bmatrix} + \begin{Bmatrix} f_y \\ M_{yz} \\ f_x \\ M_{zx} \end{Bmatrix} + \begin{Bmatrix} 0 \\ f_y(e\sin\omega t)\varphi_x \\ 0 \\ f_x(e\cos\omega t)\varphi_y \end{Bmatrix} = \begin{Bmatrix} m\omega^2 e\sin\omega t \\ 0 \\ m\omega^2 e\cos\omega t \\ 0 \end{Bmatrix} \tag{2.95}$$

which can be written in matrix notation as

$$\mathbf{M}\ddot{\mathbf{x}} + \mathbf{R}_L + \mathbf{R}_{NL} = \mathbf{f}_{unb} \tag{2.96}$$

with

$$\mathbf{M} = \begin{bmatrix} m & 0 & 0 & 0 \\ 0 & I_d & 0 & 0 \\ 0 & 0 & m & 0 \\ 0 & 0 & 0 & I_d \end{bmatrix}; \quad \ddot{\mathbf{x}} = \begin{Bmatrix} \ddot{y} \\ \ddot{\varphi}_x \\ \ddot{x} \\ \ddot{\varphi}_y \end{Bmatrix};$$

$$\mathbf{R}_L = \begin{Bmatrix} f_y \\ M_{yz} \\ f_x \\ M_{zx} \end{Bmatrix}; \quad \mathbf{R}_{NL} = \begin{Bmatrix} 0 \\ f_y(e\sin\omega t)\varphi_x \\ 0 \\ f_x(e\cos\omega t)\varphi_y \end{Bmatrix}; \quad \mathbf{f}_{unb} = \begin{Bmatrix} m\omega^2 e\sin\omega t \\ 0 \\ m\omega^2 e\cos\omega t \\ 0 \end{Bmatrix}$$

where **M** represents the mass matrix, $\mathbf{f}_{umb}$ is the unbalance force vector, **x** is the displacement vector, **R** is the reaction force/moment vector, and subscripts *L* and *NL* represent the linear and the nonlinear, respectively. It should be noted that the ordering of the displacement vector can be changed depending upon convenience, and elements of other matrices and vectors will change their positions accordingly. The reaction forces and moments, in *z-x* plane, onto the shaft can be expressed in terms of shaft displacements at the disk location with the help of influence coefficients such as (Timoshenko and Young, 1968)

$$x = \alpha_{11}f_x + \alpha_{12}M_{zx} \quad \text{and} \quad \varphi_y = \alpha_{21}f_x + \alpha_{22}M_{zx} \tag{2.97}$$

where α_{ij} represents the displacement at the ith station due to a unit force at the jth station keeping all other forces at zero. It should be noted that the displacement and force terms are used in a general sense so that displacement can be a translational or a rotational displacement, whereas the force can be a force or a moment. The coupling of the force and the displacement in two orthogonal planes has not been considered because of the symmetry of the shaft. Equation 2.97 can be written in a matrix form as (refer to Appendix 2.1 and Timoshenko and Young (1968))

$$\begin{Bmatrix} x \\ \varphi_y \end{Bmatrix} = \begin{bmatrix} \alpha_{11} & \alpha_{12} \\ \alpha_{21} & \alpha_{22} \end{bmatrix} \begin{Bmatrix} f_x \\ M_{zx} \end{Bmatrix} \tag{2.98}$$

with

$$\begin{bmatrix} \alpha_{11} & \alpha_{12} \\ \alpha_{21} & \alpha_{22} \end{bmatrix} = \begin{bmatrix} \dfrac{a^2b^2}{3EIl} & \dfrac{-\left(3a^2l-2a^3-al^2\right)}{3EIl} \\ \dfrac{ab(b-a)}{3EIl} & \dfrac{-\left(3al-3a^2-l^2\right)}{3EIl} \end{bmatrix}$$

where E is the Young's modulus of elasticity of the shaft material, I is the second moment of area of the shaft cross-section, and length parameters a and b are defined in Figure 2.20a with $l=a+b$. From the simple beam deflection theory, we can get these influence coefficients. Equation 2.98 can be written as

$$\begin{Bmatrix} f_x \\ M_{zx} \end{Bmatrix} = \begin{bmatrix} \alpha_{11} & \alpha_{12} \\ \alpha_{21} & \alpha_{22} \end{bmatrix}^{-1} \begin{Bmatrix} x \\ \varphi_y \end{Bmatrix} = \frac{1}{(\alpha_{11}\alpha_{22}-\alpha_{12}\alpha_{21})} \begin{bmatrix} \alpha_{22} & -\alpha_{12} \\ -\alpha_{21} & \alpha_{11} \end{bmatrix} \begin{Bmatrix} x \\ \varphi_y \end{Bmatrix}$$
$$= \begin{bmatrix} k_{11} & k_{12} \\ k_{21} & k_{22} \end{bmatrix} \begin{Bmatrix} x \\ \varphi_y \end{Bmatrix} \tag{2.99}$$

where k_{ij} is the stiffness coefficient and defined as force at the ith station due to a unit displacement at the jth station keeping all other displacements at zero. Similarly, because the shaft is symmetric about its rotation axis, we can obtain

$$\begin{Bmatrix} f_y \\ M_{yz} \end{Bmatrix} = \begin{bmatrix} k_{11} & k_{12} \\ k_{21} & k_{22} \end{bmatrix} \begin{Bmatrix} y \\ \varphi_x \end{Bmatrix} \tag{2.100}$$

Equations 2.99 and 2.100 can be combined in matrix form as

$$\mathbf{R}_L = \mathbf{K}\mathbf{x} \tag{2.101}$$

with

$$\mathbf{K} = \begin{bmatrix} k_{11} & k_{12} & 0 & 0 \\ k_{21} & k_{22} & 0 & 0 \\ 0 & 0 & k_{11} & k_{12} \\ 0 & 0 & k_{21} & k_{22} \end{bmatrix}; \quad \mathbf{R}_L = \begin{Bmatrix} f_y \\ M_{yz} \\ f_x \\ M_{zx} \end{Bmatrix}; \quad \mathbf{x} = \begin{Bmatrix} y \\ \varphi_x \\ x \\ \varphi_y \end{Bmatrix}$$

Noting Equation 2.101, the nonlinear reaction force vector takes the following form:

$$\mathbf{R}_{NL} = \begin{Bmatrix} 0 \\ f_y e_y \varphi_x \\ 0 \\ f_x e_x \varphi_y \end{Bmatrix} = \begin{Bmatrix} 0 \\ (k_{11}y + k_{12}\varphi_x) e_y \varphi_x \\ 0 \\ (k_{11}x + k_{12}\varphi_y) e_x \varphi_y \end{Bmatrix} \tag{2.102}$$

This equation contains a product of the translational and rotational displacements, which makes the system equations nonlinear. The present analysis considers only linear systems, so contributions from these nonlinear terms can be ignored with the assumption of small displacements. On substituting reaction forces and moments from Equation 2.101 into equations of motion (i.e., Equation 2.96), we get

$$\mathbf{M}\ddot{\mathbf{x}} + \mathbf{K}\mathbf{x} = \mathbf{f}_{unb} \tag{2.103}$$

With

$$\mathbf{M} = \begin{bmatrix} m & 0 & 0 & 0 \\ 0 & I_d & 0 & 0 \\ 0 & 0 & m & 0 \\ 0 & 0 & 0 & I_d \end{bmatrix}; \quad \mathbf{K} = \begin{bmatrix} k_{11} & k_{12} & 0 & 0 \\ k_{21} & k_{22} & 0 & 0 \\ 0 & 0 & k_{11} & k_{12} \\ 0 & 0 & k_{21} & k_{22} \end{bmatrix}; \quad \mathbf{x} = \begin{Bmatrix} y \\ \varphi_x \\ x \\ \varphi_y \end{Bmatrix}; \quad \mathbf{f}_{unb} = \begin{Bmatrix} m\omega^2 e \sin\omega t \\ 0 \\ m\omega^2 e \cos\omega t \\ 0 \end{Bmatrix}$$

2.5.1 Calculation of Natural Frequencies

For obtaining natural frequencies of the system, the determinant of the dynamic stiffness matrix, $\mathbf{Z} = (\mathbf{K} - \omega_{nf}^2 \mathbf{M})$, should be equated to zero and solved for natural frequencies, ω_{nf}, which gives four natural frequencies of the rotor system. It should be noted that because two orthogonal plane motions are uncoupled (i.e., corresponding to y and φ_x, and x and φ_y), equations of motion of each plane could be solved independently. This would decrease the size of the $\mathbf{Z}$ matrix by half, which will be illustrated through Example 2.6. A more general method based on the eigenvalue problem will be discussed in Chapter 7.

2.5.2 Unbalanced Force Responses

The unbalance forcing with frequency, ω, can be written as

$$\mathbf{f}_{unb} = \mathbf{F}_{unb} e^{j\omega t} \quad \text{with} \quad F_{unb_k} = F_{unb_k}^r + jF_{unb_k}^i \quad k = 1, 2, \ldots, N \tag{2.104}$$

where $\mathbf{F}_{unb}$ is the complex unbalance force vector, and it contains the amplitude and the phase information, r and i represent the real and imaginary parts, respectively, k represents the row number in vector $\mathbf{F}_{unb}$, and N is the total DOFs of the system (N=4 for the present case). The response of the system can be written as

$$\mathbf{x} = \mathbf{X}e^{j\omega t} \quad \text{so that} \quad \ddot{\mathbf{x}} = -\omega^2 \mathbf{X} e^{j\omega t} \tag{2.105}$$

where $\mathbf{X}$ is the complex displacement vector and ω is the spin speed of the shaft. On substituting Equations 2.104 and 2.105 into Equation 2.103, we get the unbalance response as

$$\mathbf{X} = \mathbf{Z}^{-1}\mathbf{F}_{umb} \quad \text{with} \quad \mathbf{Z} = \left(\mathbf{K} - \omega^2\mathbf{M}\right) \tag{2.106}$$

where **Z** is the dynamic stiffness matrix. Similar to the force amplitude vector, the response vector will also have complex quantities and can be written as

$$X_k = X_k^r + jK_k^i \quad \text{with} \quad k = 1, 2, \ldots, N \tag{2.107}$$

which will give amplitude and phase information as

$$X_k^{amp} = \sqrt{\left(X_k^r\right)^2 + \left(X_k^i\right)^2} \quad \text{and} \quad X_k^{phase} = \tan^{-1}\left(X_k^i / X_k^k\right) \quad \text{with} \quad k = 1, 2, \ldots, N \tag{2.108}$$

Equation 2.106 is a more general form of the Jeffcott rotor response as that of the disk at the mid-span. However, it is expected to provide four critical speeds corresponding to the four DOFs of the rotor system. Most often it is beneficial to observe the amplitude and the phase of the response rather than the time history. The present method gives the response in the frequency domain. When the damping term is also present, the previous unbalance response procedure can easily handle the additional damping term, and the dynamic stiffness will take the following form:

$$\mathbf{Z} = \left(\mathbf{K} - \omega^2\mathbf{M} + j\omega\mathbf{C}\right) \tag{2.109}$$

where **C** is the damping matrix. It should be noted that **Z** is now a complex matrix and by the numerical simulation critical speeds can be obtained by noticing peaks of responses while varying the spin speed of the shaft. The procedure for obtaining damped natural frequencies for multi-DOF systems will be discussed in Chapter 2 of Volume II. The analysis of the present section is equally valid for other boundary conditions. The only change would be the expressions of influence coefficients corresponding to new boundary conditions (e.g., cantilever, fixed-fixed, free-free, overhang, etc.).

2.5.3 Bearing Reaction Forces

Bearings are, in the present study, assumed to transmit only forces and not moments. Forces transmitted through bearings are those that are related to the deflection of the shaft as shown in Figure 2.22 on the *y*-*z* plane.

On taking moments about ends *L* (left) and *R* (right) of the shaft, we have

$$\sum M_L = 0 \Rightarrow f_y a - M_{yz} - R_{y2} l = 0 \quad \text{or} \quad R_{y2} = \frac{a}{l} f_y - \frac{1}{l} M_{yz} \tag{2.110}$$

and

$$\sum M_R = 0 \Rightarrow R_{y1} l - f_y b - M_{yz} = 0 \quad \text{or} \quad R_{y1} = \frac{b}{l} f_y + \frac{1}{l} M_{yz} \tag{2.111}$$

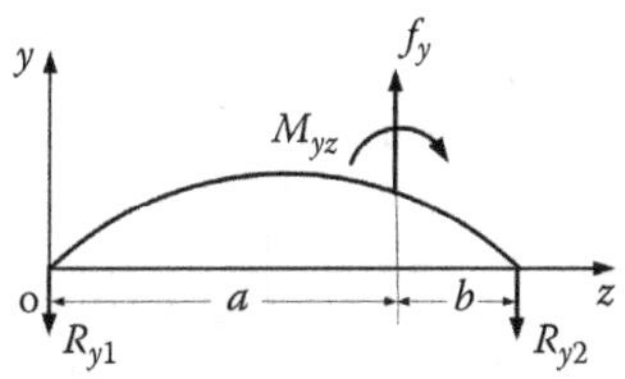

FIGURE 2.22 Bearing reaction forces on the shaft in the *y*-*z* plane.

From these equations, bearing reaction forces at the left and right sides are related to the loading on the shaft, f_y and M_{yz}, from the disk. In matrix form Equations 2.110 and 2.111 can be written as

$$\mathbf{f}_b = \mathbf{D}\mathbf{f}_s \tag{2.112}$$

with

$$\mathbf{f}_b = \mathbf{F}_b e^{j\omega t} = \begin{Bmatrix} R_{y1} \\ R_{y2} \end{Bmatrix}; \quad \mathbf{f}_s = \mathbf{F}_s e^{j\omega t} = \begin{Bmatrix} f_y \\ M_{yz} \end{Bmatrix}; \quad \mathbf{D} = \begin{bmatrix} b/l & 1/l \\ a/l & -1/1 \end{bmatrix}$$

where subscripts b and s represent the bearing and the shaft, respectively. Complex vectors $\mathbf{F}_b$ and $\mathbf{F}_s$ are bearing forces at the shaft ends and shaft reaction forces at the disk, respectively. On using Equations 2.101 and 2.106 into the form of Equation 2.112 for both plane motions (i.e., y-z and z-x), we get

$$\mathbf{F}_b = \mathbf{DKX} = \mathbf{DKZ}^{-1}\mathbf{F}_{unb} = \mathbf{AF}_{unb} \tag{2.113}$$

with

$$\mathbf{A} = \mathbf{DKZ}^{-1}$$

It should be noted that Equation 2.113 has been written for both plane motions (i.e., y-z and z-x); however, they are uncoupled for the present case. Similar to the unbalance force amplitude vector, the bearing force vector will also have complex quantities and can be written as

$$F_{b_k} = F_{b_k}^r + jF_{b_k}^i \qquad k = 1, 2, \ldots, 2n_b \tag{2.114}$$

where n_b is the number of bearings. This will give the amplitude and the phase information as

$$F_{b_k}^{amp} = \sqrt{\left(F_{b_k}^r\right)^2 + \left(F_{b_k}^i\right)^2} \quad \text{and} \quad F_{b_k}^{phase} = \tan^{-1}\left(F_{b_k}^i / F_{b_k}^r\right) \tag{2.115}$$

It should be noted that for the case of no damping, the phase remains zero between a force in one plane and a response in that plane. Now these procedures will be illustrated with simple numerical examples.

Example 2.5

Find the transverse natural frequency of the rotor system shown in Figure 2.23. The disk is rigid and has a mass of 10 kg with negligible diametral mass moment of inertia. Consider the shaft as massless and flexible with $E = 2.1 \times 10^{11}$ N/m². Take one plane motion only.

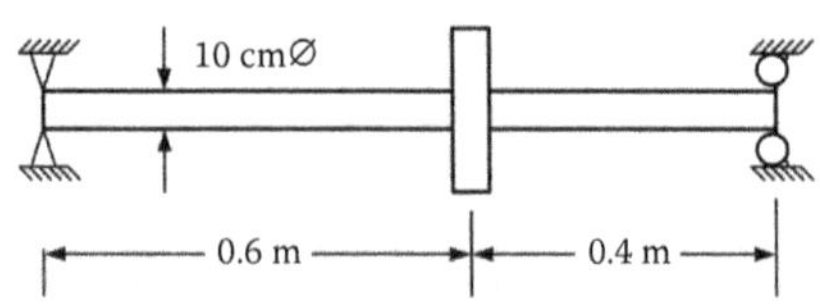

FIGURE 2.23 A simply supported shaft.

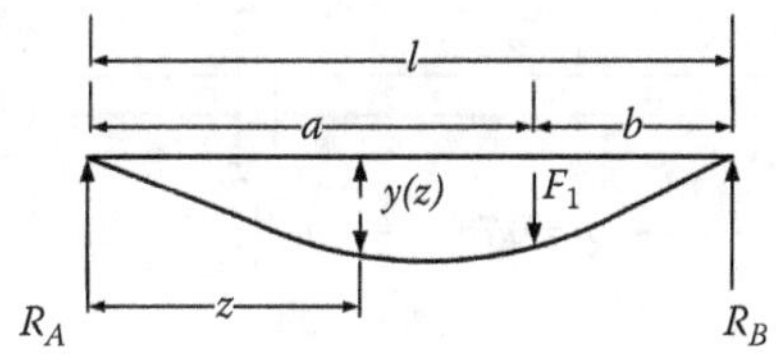

FIGURE 2.24 A shaft in the deflected position.

Solution:

Figure 2.24 shows the deflected position of the shaft. For a simply supported beam, the influence coefficient is defined as (refer to Appendix 2.1 and Chapter 9)

$$\alpha = \frac{y(z)}{F} = \frac{bz\left(l^2 - z^2 - b^2\right)}{6EIl}, \quad (z \le a)$$

For obtaining α_{11} (which is defined as the deflection at station (disk location) 1 for the unit force at station 1), we have $z=0.6$ m, $l=1.0$ m, and $b=0.4$ m. Hence, it can be obtained as

$$\alpha \Big|_{\substack{z=a=0.6 \\ b=0.4}} = \alpha_{11} = \frac{0.4 \times 0.6 \times \left(1^2 - 0.6^2 - 0.4^2\right)}{6 \times 2.1 \times 10^{11} \times \frac{\pi}{64} \times (0.1)^4 \times 1} = 1.863 \times 10^{-8} \text{ m/N}$$

Considering a single plane (*y*-*z*) motion and ignoring the rotational displacement φ_x, the natural frequency can be obtained as (refer to Section 2.5.1)

$$|\mathbf{Z}| = \left|\mathbf{K} - \omega_{nf}^2 \mathbf{M}\right| = \left|\frac{1}{\alpha_{11}} - \omega_{nf}^2 m\right| = 0$$

which gives

$$\omega_{nf} = \sqrt{\frac{1}{m_1 \alpha_{11}}} = \sqrt{\frac{1}{10 \times 1.863 \times 10^{-8}}} = 2316.83 \text{ rad/s} \quad \text{or} \quad 368.92 \text{ Hz}$$

Example 2.6

Obtain transverse natural frequencies of an offset Jeffcott rotor system as shown in Figure 2.25. Assume that the mass of the disk $m=10$ kg, the diametral mass moment of inertia $I_d=0.02$ kg-m^2, and the disk is placed at 0.25 m from the right support. The shaft has a diameter of 10 mm and the total length of the span is 1 m. The shaft is assumed to be massless. Use the influence coefficient method. Take Young's modulus of the shaft material as $E=2.1\times10^{11}$ N/m^2. Ignore the gyroscopic effect and take one plane motion only.

Solution:

Influence coefficients for translational and rotational displacements (y, φ_x) corresponding to a force (f) and a moment (M) acting at the disk are defined as

$$\alpha_{yf} = \frac{a^2b^2}{3EIl} = 1.137 \times 10^{-4} \text{ m/N}; \quad \alpha_{yM} = -\frac{\left(3a^2l - 2a^3 - al^2\right)}{3EIl} = -3.03 \times 10^{-4} \text{ m/N-m}$$

$$\alpha_{\varphi f} = \frac{ab(b-a)}{3EIl} = -3.03 \times 10^{-4} \text{ rad/N}; \quad \alpha_{\varphi M} = -\frac{\left(3al - 3a^2 - l^2\right)}{3EIl} = 1.41 \times 10^{-3} \text{ rad/N-m}$$

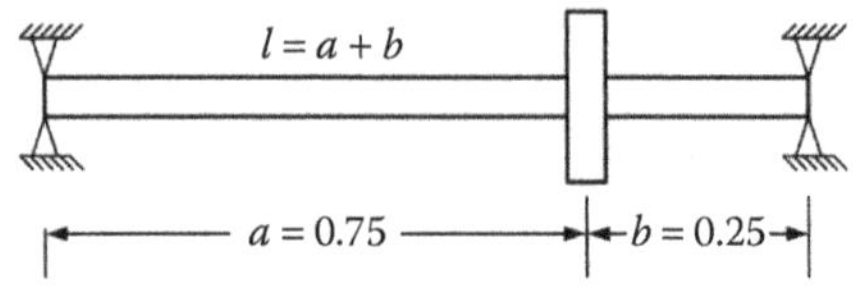

FIGURE 2.25 A Jeffcott rotor with an offset disk.

For the present problem, only a single plane motion is considered. For the free vibration, from Equation 2.103, we get

$$\begin{bmatrix} m & 0 \\ 0 & I_d \end{bmatrix}\begin{Bmatrix} \ddot{y} \\ \ddot{\varphi}_x \end{Bmatrix}+\begin{bmatrix} \alpha_{yf} & \alpha_{yM} \\ \alpha_{\varphi f} & \alpha_{\varphi M} \end{bmatrix}^{-1}\begin{Bmatrix} y \\ \varphi_x \end{Bmatrix}=\begin{Bmatrix} 0 \\ 0 \end{Bmatrix}$$

Because it will execute the simple harmonic motion for the free vibration, we have

$$\left(-\omega_{nf}^2\begin{bmatrix} m & 0 \\ 0 & I_d \end{bmatrix}+\begin{bmatrix} \alpha_{yf} & \alpha_{yM} \\ \alpha_{\varphi f} & \alpha_{\varphi M} \end{bmatrix}^{-1}\right)\begin{Bmatrix} y \\ \varphi_x \end{Bmatrix}=\begin{Bmatrix} 0 \\ 0 \end{Bmatrix} \quad \text{or}$$

$$\left(-\omega_{nf}^2\begin{bmatrix} \alpha_{yf} & \alpha_{yM} \\ \alpha_{\varphi f} & \alpha_{\varphi M} \end{bmatrix}\begin{bmatrix} m & 0 \\ 0 & I_d \end{bmatrix}+\begin{bmatrix} 1 & 0 \\ 0 & 1 \end{bmatrix}\right)\begin{Bmatrix} y \\ \varphi_x \end{Bmatrix}=\begin{Bmatrix} 0 \\ 0 \end{Bmatrix}$$

where ω_{nf} is the natural frequency of the system. The earlier equation is an eigenvalue problem. For the nontrivial solution, we have

$$\begin{vmatrix} m\omega_{nf}^2\alpha_{yf}-1 & I_d\omega_{nf}^2\alpha_{yM} \\ m\omega_{nf}^2\alpha_{\varphi f} & I_d\omega_{nf}^2\alpha_{\varphi M}-1 \end{vmatrix}=0$$

which gives a frequency equation in the form of a polynomial as

$$mI_d\omega_{nf}^4\left(\alpha_{yf}\alpha_{\varphi M}-\alpha_{yM}^2\right)-\omega_{nf}^2\left(\alpha_{yf}m+\alpha_{\varphi M}I_d\right)+1=0$$

On substituting values of the present problem parameters, we get

$$\omega_{nf}^4-8.505\times10^4\omega_{nf}^2+7.3\times10^7=0$$

This can be solved to give two natural frequencies of the system as

$$\omega_{nf_1}=29.4\ \text{rad/s} \quad \text{and} \quad \omega_{nf_2}=290\ \text{rad/s} \quad \text{(Coupled translation and tilting motions)}$$

For the present problem, the translational and rotational displacements in a single plane are coupled. The natural frequencies obtained are system natural frequencies and hence are not as such related to the pure translational or pure rotational motions. If we consider these two motions to be uncoupled, then the corresponding natural frequencies can be obtained as

$$\omega_{nf_1}=\sqrt{\frac{1}{m\alpha_{yf}}}=\sqrt{\frac{1}{10\times1.137\times10^{-4}}}=\sqrt{879.5}=29.65\ \text{rad/s (Pure translation motion of the disc)}$$

and

$$\omega_{nf_2}=\sqrt{\frac{1}{I_d\alpha_{\varphi M}}}=\sqrt{\frac{1}{0.02\times1.4146\times10^{-3}}}=188\ \text{rad/s (Pure tilting motion of the disc)}$$

It can be seen that there is a small difference in the fundamental natural frequency due to the pure translation motion (29.65 rad/s) and the fundamental natural frequency of the coupled system (29.4 rad/s), and a large difference in the natural frequency for the pure tilting motion (188 rad/s) and the second natural frequency of the coupled system (290 rad/s). For the Jeffcott rotor with a disk at the midspan, these equations of motion are uncoupled.

Example 2.7

A disk of mass 13.6 kg and the polar mass moment of inertia 0.02 kg-m² is mounted at the midspan of a shaft with a span length of 0.4064 m. Assume the shaft to be simply supported at the bearings. The rotor is known to have an unbalance of 0.2879 kg-cm. Determine the forces exerted on the bearings at a spin speed of 6000 rpm. The diameter of the steel shaft is 2.54 cm with $E=200$ GNm^{-2}.

Solution:

The following data are available:

$U=me=0.2879$ kg-cm; $m=13.6$ kg; $e=U/m=0.0211$ cm; $\omega=6000$ rpm; $d=2.54$ cm; $E=200\times10^9$ N/m^2

Bearing forces are obtained by considering first the shaft as rigid and then by considering the shaft as flexible. In both cases, bearings are considered rigid in the transverse directions.

i. *For the rigid shaft and rigid bearings* (Figure 2.26):

$$\text{The unbalance force} = me\omega^2 = 0.2879\times10^{-3}\times\left(\frac{2\pi\times6000}{60}\right)^2 = 1136.67\ \text{N}$$

The force at each bearing (amplitude) $=1136.67/2=568.29$ N

The components of the forces in the horizontal and vertical directions are given, respectively, as

$$f_x = 568.29\cos 200\pi t = 568.29\cos 200\pi t\ \text{N}$$

$$f_y = 568.29\sin 200\pi t + mg = (568.29\sin 200\pi t + 133.4)\ \text{N}$$

ii. *For the flexible shaft and rigid bearings* (Method 1) (Figure 2.27):

Because $f_y=ky$, bearing reaction forces can be written as (Figure 2.28b)

$$R_A = R_B = f_y/2 = ky/2$$

The stiffness of the rotor system as shown in Table 2.1 is given as

$$k = \frac{48EI}{l^3} = \frac{48\times(2.0\times10^{11})\frac{\pi}{64}\times0.0254^4}{(0.4064)^3} = 2.92\times10^6\ \text{N/m}$$

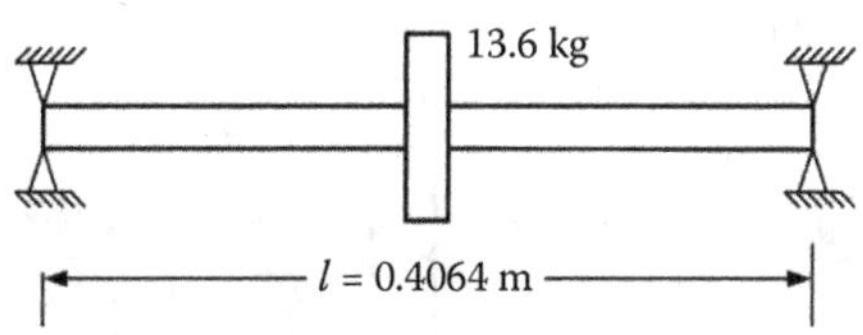

FIGURE 2.26 A Jeffcott rotor.

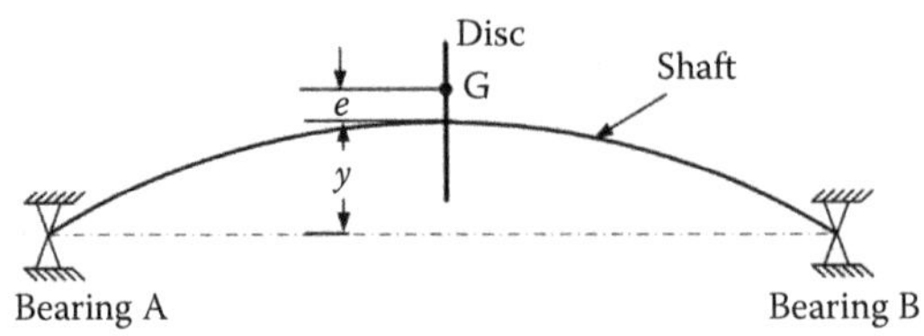

FIGURE 2.27 A flexible rotor system.

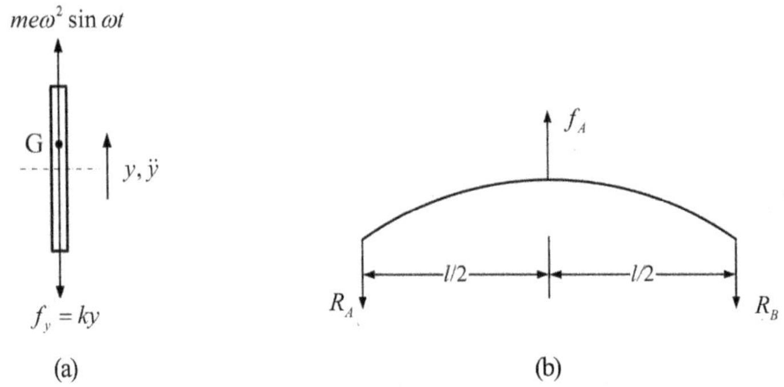

FIGURE 2.28 (a) Free body diagram of the disk. (b) Free body diagram of the shaft.

The equation of motion of the disk, obtained from the free body diagram of the disk (Figure 2.28a), is given as

$$me\omega^2 \sin\omega t - ky = m\ddot{y}$$

For the simple harmonic motion, we have $\ddot{y} = -\omega^2 y$; hence, the previous equation can be written as

$$y = \frac{me\omega^2}{k - m\omega^2} = \frac{\left(0.2879\times10^{-2}\right)\times(200\pi)^2}{2.92\times10^6 - 13.6\times(200\pi)^2} = -4.64\times10^{-4}\ \text{m}$$

The bearing reaction force can be obtained as

$$R_A = ky/2 = 2.92\times10^6\times\left(-4.64\times10^{-4}\right)/2 = -677.6\ \text{N}$$

Components of bearing force in the horizontal and vertical directions can be obtained as, respectively,

$$R_{A_x} = 677.6\cos 200\pi t\ \text{N} \quad \text{and} \quad R_{A_y} = (677.7\sin 200\pi t\ + 133.4)\ \text{N}$$

iii. *For the flexible shaft and rigid bearings* (Method 2): Now the influence coefficient method is used. Bearing forces are given as

$$\begin{Bmatrix} R_A \\ R_B \end{Bmatrix} = \begin{bmatrix} b/l & 1/l \\ a/l & -1/l \end{bmatrix} \begin{Bmatrix} f_y \\ M_{zx} \end{Bmatrix} = \begin{bmatrix} 1/2 & 1/0.4064 \\ 1/2 & -1/0.4064 \end{bmatrix} \begin{Bmatrix} f_y \\ M_{zx} \end{Bmatrix}$$

where the reaction forces from the disk can be expressed as

$$\begin{Bmatrix} f_y \\ M_{zx} \end{Bmatrix} = \begin{bmatrix} k_{11} & k_{12} \\ k_{21} & k_{22} \end{bmatrix} \begin{Bmatrix} y \\ \varphi_x \end{Bmatrix}$$

with

$$K = \begin{bmatrix} \alpha_{11} & \alpha_{12} \\ \alpha_{12} & \alpha_{22} \end{bmatrix}^{-1} = \begin{bmatrix} \alpha_{11}^{-1} & 0 \\ 0 & \alpha_{22}^{-1} \end{bmatrix} = \begin{bmatrix} 48EI/l^3 & 0 \\ 0 & 12EI/l \end{bmatrix} = \begin{bmatrix} 2.92\times10^6 & 0 \\ 0 & 1.21\times10^5 \end{bmatrix}$$

where

$$\alpha_{11} = \frac{l^3}{48EI}; \quad \alpha_{22} = \frac{-\left(3al - 3a^2 - l^2\right)}{3EIl} = \frac{l}{12EI};$$

$$\alpha_{12} = \frac{-\left(3a^2l - 2a^3 - al^2\right)}{3EIl} = 0; \quad \alpha_{21} = \frac{ab(b-a)}{3EIl} = 0$$

Displacement vectors are related to the unbalance force as

$$\begin{Bmatrix} y \\ \varphi_x \end{Bmatrix} = \begin{bmatrix} Z_{11} & Z_{12} \\ Z_{21} & Z_{22} \end{bmatrix}^{-1} \begin{Bmatrix} me\omega^2 \\ 0 \end{Bmatrix}$$

with

$$\mathbf{Z}^{-1} = \begin{bmatrix} k_{11} - m\omega^2 & 0 \\ 0 & k_{22} - I_d\omega^2 \end{bmatrix}^{-1} = \begin{bmatrix} \dfrac{1}{k_{11} - m\omega^2} & 0 \\ 0 & \dfrac{1}{k_{22} - I_d\omega^2} \end{bmatrix} = \begin{bmatrix} -4.08\times10^{-7} & 0 \\ 0 & 8.84\times10^{-6} \end{bmatrix}$$

In view of the previous equations, bearing reaction forces can be written as

$$\begin{Bmatrix} R_A \\ R_B \end{Bmatrix} = \begin{bmatrix} \dfrac{k_{11}(b/l)}{k_{11} - m\omega^2} & \dfrac{k_{22}(1/l)}{k_{22} - I_d\omega^2} \\ \dfrac{k_{11}(a/l)}{k_{11} - m\omega^2} & \dfrac{k_{22}(-1/l)}{k_{22} - I_d\omega^2} \end{bmatrix} \begin{Bmatrix} me\omega^2 \\ 0 \end{Bmatrix} = \begin{Bmatrix} \dfrac{k_{11}(a/l)}{k_{11} - m\omega^2} me\omega^2 \\ \dfrac{k_{11}(a/l)}{k_{11} - m\omega^2} me\omega^2 \end{Bmatrix}$$

From these equations, we have

$$R_A = \frac{2.92\times10^6 \times (1/2)}{2.92\times10^6 - 13.6\times(200\pi)^2} \times \left(0.2879\times10^{-2}\right) \times (200\pi)^2 = -677.6\text{ N}$$

and

$$R_B = -677.6\text{ N}$$

which is the same as that obtained by the previous method. It would be interesting to vary the spin speed and plot the bearing forces with it. It should be noted because the disk is at the midspan, there is no contribution of the diametral mass moment of inertia on bearing reactions. If there had been couple-unbalance present, then the diametral mass moment of inertia would have affected the bearing reactions. As an exercise take the disk location from the left support $a = 0.3l$ and obtain bearing forces for the same.

Example 2.8

Find the transverse natural frequency of a rotor system as shown in Figure 2.29. Consider the shaft as massless and made of steel with a Young's modulus of 2.1×10^{11} N/m^2 and a mass density of 7800 kg/m^3. The disk has 10 kg of mass. The shaft is simply supported at the ends.

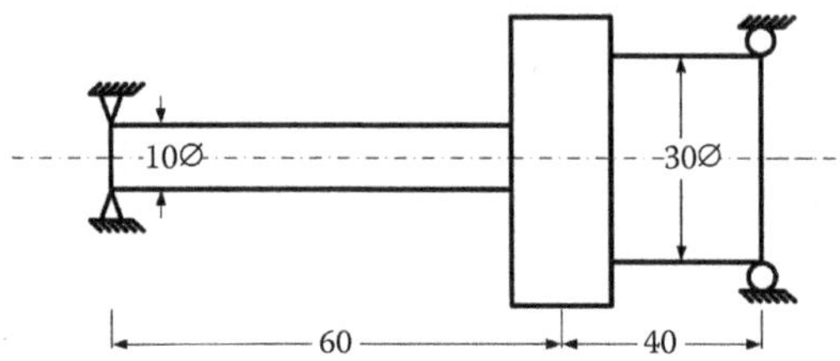

FIGURE 2.29 A stepped shaft with simple supports (all dimensions are in cm).

Solution:

Considering only the translational displacement, first we will obtain the stiffness or the influence coefficient, α_{11} for Figure 2.30 using the energy method. On taking the force and moment balances, we have

$$+\uparrow \Sigma F = 0 \;\Rightarrow F_A + F_B - F = 0 \qquad \text{and} \qquad +\circlearrowright \Sigma M_A = 0 \Rightarrow F_B \times 1 - F \times 0.6 = 0$$

which gives the reaction forces as

$$F_A = 0.4F \quad \text{and} \quad F_B = 0.6F$$

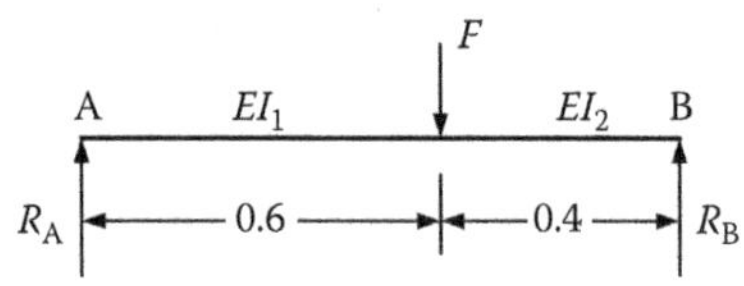

FIGURE 2.30 Free body diagram of the whole shaft (dimensions are in m).

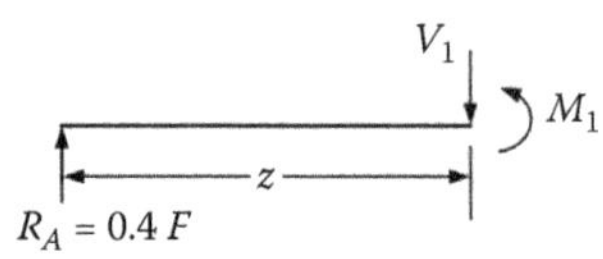

FIGURE 2.31 Free body diagram of the shaft segment for $0 \le z \le 0.6$.

Bending moments are obtained at various segments of the shaft to get the strain energy of the system. On taking the moment balance in the free body diagram as shown in Figure 2.31 of a shaft segment for $0.0 \le z \le 0.6$, we get

$$+\circlearrowright \Sigma M_A = 0 \Rightarrow M_1 - 0.4Fx = 0; \quad \text{or} \quad M_1 = 0.4Fx, \quad 0 \le x \le 0.6$$

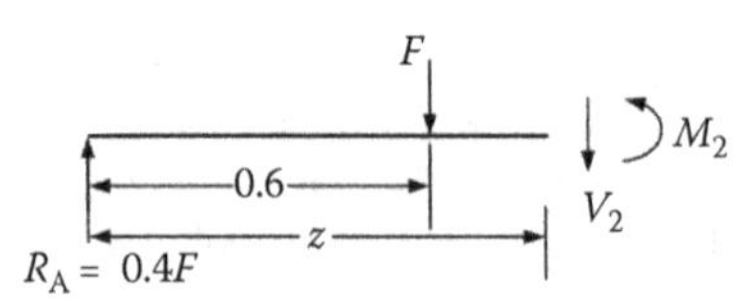

FIGURE 2.32 Free body diagram of the shaft segment for $0.6 \le z \le 1.0$.

On taking the moment balance in the free body diagram as shown in Figure 2.32 of the shaft segment for $0.6 \le z \le 1.0$, we get

$$\circlearrowleft^{+} \sum M_A = 0 \Rightarrow M_2 + F(x-0.6) - 0.4Fx = 0 \quad \text{or} \quad M_2 = 0.6F(1-x); \quad 0.6 \le x \le 1.$$

The strain energy is expressed as

$$U = \int_0^{0.6} \frac{M_1^2 dz}{2EI_1} + \int_{0.6}^{1.0} \frac{M_2^2 dz}{2EI_2}$$

The translational displacement is expressed as

$$\delta = \frac{\partial U}{\partial F} = \int_0^{0.6} \frac{M_1(\partial M_1 / \partial F)dz}{EI_1} + \int_{0.6}^{1.0} \frac{M_2(\partial M_2 / \partial F)dz}{EI_2}$$

On substituting the bending moment expression from equations (a) and (b), we get

$$\delta = \int_0^{0.6} \frac{(-0.4Fz)(-0.4z)dz}{EI_1} + \int_{0.6}^{1} \frac{\{-0.6F(1-z)\}\{-0.6(1-z)\}dz}{EI_2}$$

$$= \left(\frac{0.01152}{EI_1} + \frac{0.00768}{EI_2} \right) F$$

The stiffness is given as

$$k = \frac{F}{\delta} = \left[\frac{0.01152}{EI_1} + \frac{0.00768}{EI_2} \right]^{-1} = 8.45 \times 10^7 \text{ N/m}$$

where

$$E = 2 \times 10^{11} \text{ N/m}^2; \; I_1 = \frac{\pi}{64} \times 0.1^4 = 4.907 \times 10^{-6} \text{ m}^4; \quad I_2 = \frac{\pi}{64} \times 0.3^3 = 3.976 \times 10^{-4} \text{ m}^4$$

which gives the natural frequency as

$$\omega_{nf} = \sqrt{\frac{k}{m}} = \sqrt{\frac{8.45 \times 10^7}{10}} = 2906.81 \text{ rad/s}$$

It should be noted that the tilting motion of the disk has not been considered. For the coupled translational and rotational motions, natural frequencies of the system can be obtained as an exercise by obtaining the corresponding influence coefficients. The purpose of the present exercise is to illustrate the calculation of influence coefficients for uncommon shaft geometries or boundary conditions. In Chapter 8, a more general tabular method for calculating influence coefficients will be presented.

Example 2.9

Obtain the bending natural frequency for the synchronous motion of a rotor as shown in Figure 2.33. The rotor is assumed to be fixed and supported at one end. Take the mass of the disk $m = 1$ kg. The shaft is assumed to be massless, and its length and diameter are 0.2 and 0.01 m, respectively. Take the Young's modulus of the shaft as $E = 2.1 \times 10^{11}$ N/m^2.

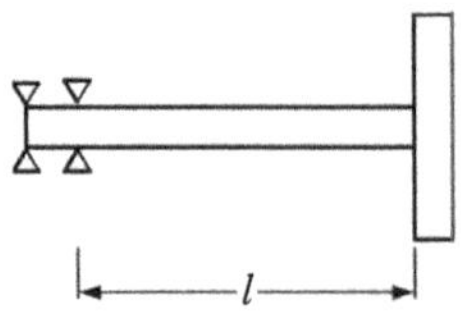

FIGURE 2.33 A cantilever rotor.

Solution:

Let us assume for simplicity that there is no coupling between the translational and rotational motions. Considering only the translational displacement, the transverse stiffness for this case would be (refer to Table 2.1)

$$k_{yf} = \frac{f}{y} = \frac{3EI}{l^3} = \frac{3 \times 2.1 \times 10^{11} \times 4.909 \times 10^{-10}}{0.2^3} = 3.866 \times 10^4 \text{ N/m} \tag{a}$$

with

$$I = \frac{\pi}{64} d^4 = 4.909 \times 10^{-10} \text{ m}^4 \tag{b}$$

Hence, the natural frequency would be

$$\omega_{nf} = \sqrt{\frac{k}{m}} = \sqrt{\frac{3.866 \times 10^4}{1}} = 196.62 \text{ rad/s}$$

2.6 EFFECT OF CROSS-COUPLING STIFFNESS ON CRITICAL SPEEDS

In the present section, an interesting phenomenon will be dealt with in which a critical speed will be shown to be eliminated by suitably choosing system parameters (however, there may be instability regions that we discuss in Chapter 1 of Volume II). For this purpose, the Jeffcott rotor model with a disk offset has been chosen. Now, for a detailed in-depth analysis, a closed-form expression for the response is obtained by defining the following complex displacements:

$$r = x + jy \quad \text{and} \quad \varphi_r = \varphi_y + j\varphi_x \tag{2.116}$$

Equations of motion (Equation 2.103) can be written as

$$m_d \ddot{r} + k_{11} r + k_{12} \varphi_r = m_d e \omega^2 e^{j\omega t} \tag{2.117}$$

and

$$I_d \ddot{\varphi}_r + k_{21} r + k_{22} \varphi_r = 0 \tag{2.118}$$

with

$$I_d = I_{d_x} = I_{d_y}$$

where m_d is the mass of the disk and I_d is the diametral mass moment of inertia. Let the solution be

$$r = R e^{j(\omega t - \phi_r)} \quad \text{and} \quad \varphi_r = \Phi_r e^{j(\omega t - \phi_\varphi)} \tag{2.119}$$

where R and Φ_r are the translational and rotational whirl amplitudes, respectively, and ϕ_r and ϕ_φ are the phase of the translational and rotational whirl amplitudes, respectively (these are all real quantities), so that

$$\ddot{r} = -\omega^2 R e^{j(\omega t - \phi_r)} \quad \text{and} \quad \ddot{\varphi}_r = -\omega^2 \Phi_r e^{j(\omega t - \phi_\varphi)} \tag{2.120}$$

On substituting Equations 2.119 and 2.120 into equations of motion (Equations 2.117 and 2.118), we get

$$\left(k_{11} - m_d\omega^2\right) R e^{-j\phi_r} + k_{12}\Phi_r e^{-j\phi_\varphi} = m_d\omega^2 e \tag{2.121}$$

and

$$k_{21} R e^{-j\phi_r} + \left(k_{22} - I_d\omega^2\right)\Phi_r e^{-j\phi_\varphi} = 0 \tag{2.122}$$

Equation 2.122 can be expressed as

$$\Phi_r e^{-j\phi_\varphi} = \frac{-k_{21}}{\left(k_{22} - I_d\omega^2\right)} R e^{-j\phi_r} \tag{2.123}$$

On substituting Equation 2.123 into Equation 2.121, we get

$$\left\{\frac{\left(k_{11} - m_d\omega^2\right)\left(k_{22} - I_d\omega^2\right) - k_{12}k_{21}}{\left(k_{22} - I_d\omega^2\right)}\right\} R e^{-j\phi_r} = m_d\omega^2 e \tag{2.124}$$

On equating the real and imaginary parts of both sides of Equation 2.124, we get

$$\left\{\frac{\left(k_{11} - m_d\omega^2\right)\left(k_{22} - I_d\omega^2\right) - k_{12}k_{21}}{\left(k_{22} - I_d\omega^2\right)}\right\} R \cos\phi_r = m_d\omega^2 e \tag{2.125}$$

and

$$\left\{\frac{\left(k_{11} - m_d\omega^2\right)\left(k_{22} - I_d\omega^2\right) - k_{12}k_{21}}{\left(k_{22} - I_d\omega^2\right)}\right\} R \sin\phi_r = 0 \tag{2.126}$$

From Equation 2.126, we get

$$\sin\phi_r = 0; \quad \text{i.e.} \quad \phi_r = 0 \tag{2.127}$$

which means there will not be any phase difference between the force and the response (because the damping is zero). On substituting the phase information in Equation 2.126, we get

$$R = \frac{m_d\omega^2 e\left(k_{22} - I_d\omega^2\right)}{\left(k_{11} - m_d\omega^2\right)\left(k_{22} - I_d\omega^2\right) - k_{12}k_{21}} \tag{2.128}$$

which is the whirl amplitude, and the condition of resonance can be obtained by equating the denominator of Equation 2.128 to zero:

$$\left(k_{11} - m_d\omega_{cr}^2\right)\left(k_{22} - I_d\omega_{cr}^2\right) - k_{12}k_{21} = 0 \tag{2.129}$$

where ω_{cr} represents the critical speed. By defining

$$\omega_r^2 = \frac{k_{11}}{m_d}, \quad \omega_\varphi^2 = \frac{k_{22}}{I_d}, \quad \omega_{r\varphi}^2 = \frac{k_{12}}{m_d}, \quad \text{and} \quad \omega_{\varphi r}^2 = \frac{k_{21}}{I_d} \tag{2.130}$$

Equation 2.129 can be written as

$$\omega_{cr}^4 - \left(\omega_r^2 + \omega_\varphi^2\right)\omega_{cr}^2 + \left(\omega_r^2\omega_\varphi^2 - \omega_{r\varphi}^2\omega_{\varphi r}^2\right) = 0 \tag{2.131}$$

The solution of the previous polynomial can be expressed as

$$\omega_{cn_{1,2}}^2 = \frac{1}{2}\left(\omega_r^2 + \omega_\varphi^2\right) \pm \frac{1}{2}\sqrt{\left(\omega_r^2 + \omega_\varphi^2\right)^2 - 4\left(\omega_r^2\omega_\varphi^2 - \omega_{r\varphi}^2\omega_{\varphi r}^2\right)} \tag{2.132}$$

or, more explicitly, as

$$\omega_{cn_{1,2}} = \pm\sqrt{\frac{1}{2}\left(\omega_r^2 + \omega_\varphi^2\right) \pm \frac{1}{2}\sqrt{\left(\omega_r^2 + \omega_\varphi^2\right)^2 - 4\left(\omega_r^2\omega_\varphi^2 - \omega_{r\varphi}^2\omega_{\varphi r}^2\right)}} \tag{2.133}$$

which gives critical speeds of the rotor system (the outermost negative sign has no meaning because frequency cannot be negative; however, the critical speed can be negative in the sense that the rotor can rotate in both directions in some applications). Hence, for the case when the disk is not mounted at the midspan, there are two critical speeds due to coupling of the translational and rotational displacements. The earlier solution (i.e., Equation 2.133) can be more critically analyzed as follows. It can be seen that terms inside the innermost square root are always positive, that is, $\left(\omega_r^2 + \omega_\varphi^2\right)^2 - 4\left(\omega_r^2\omega_\varphi^2 - \omega_{r\varphi}^2\omega_{\varphi r}^2\right) > 0$, because it can be rearranged as

$$\left(\omega_r^2 - \omega_\varphi^2\right)^2 + 4\omega_{r\varphi}^2\omega_{\varphi r}^2 > 0 \tag{2.134}$$

This condition will be always true because all individual terms ω_r, ω_φ, $\omega_{r\varphi}$, and $\omega_{\varphi r}$ are the real quantity. However, if the following condition is valid for terms inside the innermost square root,

$$\left(\omega_r^2\omega_\varphi^2 - \omega_{r\varphi}^2\omega_{\varphi r}^2\right) > 0 \quad \text{or} \quad \omega_r\omega_\varphi > \omega_{r\varphi}\omega_{\varphi r} \quad \text{or} \quad k_{11}k_{22} > k_{12}k_{21} \tag{2.135}$$

then it gives two real critical speeds ($\omega_{cn_{1,2}}$), because Equation 2.133 gives two real roots for both positive and negative signs inside the first square root. However, if the following condition prevails,

$$\left(\omega_r^2\omega_\varphi^2 - \omega_{r\varphi}^2\omega_{\varphi r}^2\right) < 0 \quad \text{or} \quad \omega_r\omega_\varphi < \omega_{r\varphi}\omega_{\varphi r} \quad \text{or} \quad k_{11}k_{22} < k_{12}k_{21} \tag{2.136}$$

then it gives only one real critical speed (corresponding to the positive sign inside the first square root) because the other root will be complex (corresponding to the negative sign inside the first square root). Figure 2.34a and b gives responses for these two cases, respectively. It can be seen that for the first case, two distinct peaks correspond to two critical speeds. For the second case, only one critical speed is observed, and because the system parameters chosen are different, this value is different compared to the previous case. However, there is antiresonance with very low amplitude of vibrations. The following data are taken for the simulation: the disk mass = 1 kg, the unbalance mass eccentricity = 0.0001 m, the diametral mass moment of inertia = 0.03 kgm^2, Case I (Figure 2.34a): $k_{11} = k_{22} = 1000$ N/m, and $k_{12} = k_{21} = 100$ N/m, Case II (Figure 2.34b): $k_{11} = k_{22} = 10$ N/m, and $k_{12} = k_{21} = 100$ N/m. However, generally, the cross-coupled stiffness terms are smaller compared to direct stiffness terms. For the disk at the center of the shaft span, we have $k_{12} = k_{21} = 0$, so Equation 2.128 becomes

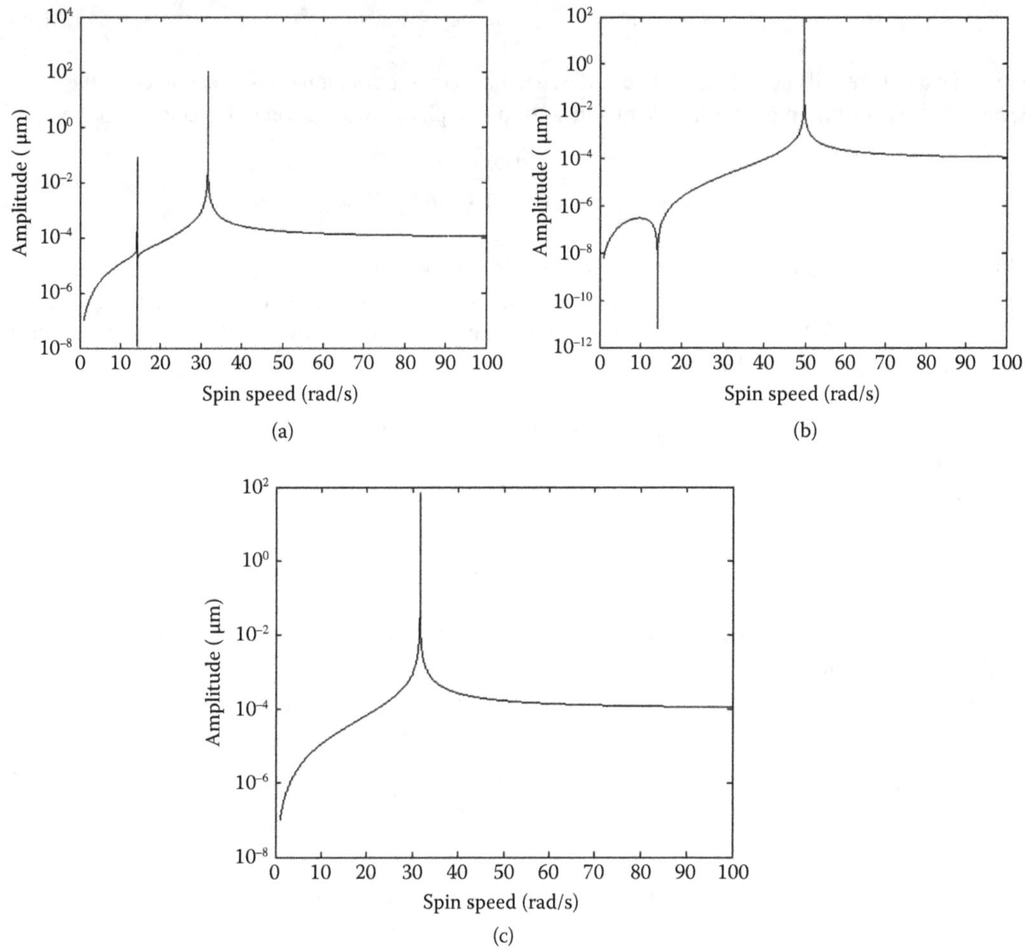

FIGURE 2.34 (a) Variation of the amplitude (R) versus the spin speed (ω) for case I: $k_{11}k_{22} > k_{12}k_{21}$. (b) Variation of the amplitude (R) versus the spin speed (ω) for Case II: $k_{11}k_{22} < k_{12}k_{21}$. (c) Variation of the amplitude (R) versus the spin speed (ω) for $k_{12} = k_{21} = 0$.

$$R = \frac{m_d\omega^2 e\left(k_{22} - I_d\omega^2\right)}{\left(k_{11} - m_d\omega^2\right)\left(k_{22} - I_d\omega^2\right)} = \frac{m_d\omega^2 e}{\left(k_{11} - m_d\omega^2\right)} \tag{2.137}$$

which is the same as discussed in the previous section for the Jeffcott rotor. The response is shown in Figure 2.34c. It can be observed that it has only one critical speed, which may not coincide with the critical speeds obtained by Equation 2.133 in Figure 2.34a and b. However, there will be another critical speed corresponding to rotational displacement, and it is illustrated now.

On substituting Equation 2.128 into Equation 2.123, we get

$$\Phi_r e^{-j\phi_\varphi} = \frac{-k_{21}}{\left(k_{22} - I_d\omega^2\right)} \frac{m_d\omega^2 e\left(k_{22} - I_d\omega^2\right)}{\left[\left(k_{11} - m_d\omega^2\right)\left(k_{22} - I_d\omega^2\right) - k_{12}k_{21}\right]} \tag{2.138}$$

On equating imaginary parts of Equation 2.138, we get

$$\sin\phi_\varphi = 0; \quad \text{i.e.,} \quad \phi_\varphi = 0 \tag{2.139}$$

which means there will not be any phase difference between the rotational displacement and the force because there is no damping in the system. On substituting phase information in Equation 2.138, we get

$$\Phi_r = -\frac{m_d\omega^2 e k_{21}}{\left[\left(k_{11} - m_d\omega^2\right)\left(k_{22} - I_d\omega^2\right) - k_{12}k_{21}\right]} \tag{2.140}$$

which is the whirl amplitude of rotational displacement, and the condition of resonance can be obtained by equating the denominator of Equation 2.140 to zero, which is the same as in Equations 2.128 and 2.133 for the translational displacement. For the disk at the center of the shaft span, we have $k_{12} = k_{21} = 0$. Equation 2.140 becomes

$$\left(k_{11} - m_d\omega^2\right)\left(k_{22} - I_d\omega^2\right) = 0 \tag{2.141}$$

which gives critical speeds as

$$\omega_{cr_1} = \sqrt{\frac{k_{11}}{m_d}} \quad \text{and} \quad \omega_{cr_2} = \sqrt{\frac{k_{22}}{I_d}} \tag{2.142}$$

which is the case when the disk is at the center of the shaft span and the translational and rotational displacements are uncoupled.

For the single plane motion from Equation 2.113, we have

$$\begin{Bmatrix} F_{b_A} \\ F_{b_B} \end{Bmatrix} = \begin{bmatrix} c_{12} & c_{12} \\ c_{21} & c_{22} \end{bmatrix} \begin{Bmatrix} m_d e\omega^2 \\ 0 \end{Bmatrix} = \begin{Bmatrix} m_d e\omega^2 c_{11} \\ m_d e\omega^2 c_{21} \end{Bmatrix}$$
$$= \frac{me\omega^2}{l\Delta} \begin{Bmatrix} (bk_{11} + k_{21})\left(k_{22} - I_d\omega^2\right) - k_{21}(bk_{12} + k_{22}) \\ (ak_{11} - k_{21})\left(k_{22} - I_d\omega^2\right) - k_{21}(ak_{12} - k_{22}) \end{Bmatrix} \tag{2.143}$$

with

$$\Delta = \left(k_{11} - m_d\omega^2\right)\left(k_{22} - I_d\omega^2\right) - k_{12}k_{21}$$

The bearing force amplitude and phase can be obtained from Equation 2.143. Bearing reaction forces will have a similar trend in the variation with spin speed as that of the response, because it has the same denominator, Δ, as that of the response. It can be shown from Equation 2.143 that forces transmitted through bearings also have a maximum at system critical speeds. These forces are dynamic forces and are superimposed on any static loads that may be present, for example, due to gravity loading. In real systems that are designed to operate above their critical speeds, the machine would normally be run through the critical speed very quickly so that very large vibrations and forces associated with the resonance do not have sufficient time to build up. The same is true during the run-down where some form of braking may be employed. If the system is to run at the critical speed and vibrations are allowed to build up, then either the shaft will fracture and a catastrophic failure will result, or there may be sufficient damping in the system to simply limit the vibration and force amplitudes to some very large (albeit tolerable) value. It is common for some of the critical speeds to be suppressed for some combination of bearing parameters, and in Chapter 4, we will see such examples. Similarly, due to the presence of gyroscopic couple also critical speeds are suppressed, as we will see this in Chapter 5 of Volume I and Chapter 1 of Volume II.

2.7 CONCLUDING REMARKS

The present chapter explains various simple rotor models in use to describe some of the important rotor behaviors, especially natural frequencies and critical speeds (i.e., the shaft spin at which the amplitude of rotor is maximum). Basic terminologies generally used to describe the rotor dynamic characteristics are introduced. For a single-DOF system, the natural frequency, and hence the critical speed, decreases by a small amount due to the damping. However, in the Jeffcott rotor model, it is shown that the resonance condition occurs at the critical speed ratio (which is defined in terms of undamped natural frequency) slightly more than one due to the increase in damping in the system (refer to Exercise 2.1 for more detail). Apart from the amplitude of the rotor vibrations, it is shown that the phase between the force and the response is also an important parameter to understand the rotor behavior, especially around the critical speed, where it changes on the order of 180°. The damping is shown to be an important factor in suppressing the rotor vibrations at the resonance. It is shown that the Jeffcott rotor is a very basic model to understand several important phenomena of the rotor system. However, several other phenomena also emanate from the supports, and for this the basic understanding of the support dynamics is very important. The goal of the next chapter will be to find out the dynamic parameters of the rolling element and hydrodynamic bearings, seals, and dampers in isolation to the shaft. This will help in understanding the effect on natural frequencies and some of the instabilities that occur due to support dynamics. Sustainable rotor design focuses on reducing environmental impact across a rotor's lifecycle by optimizing materials, manufacturing, and operational efficiency. Simple rotor models—like those used in small wind turbines, electric motors, or flywheels—benefit significantly from circular economy principles.

EXERCISE PROBLEMS

Exercise 2.1 For a single-DOF damped rotor system, obtain an expression for the frequency ratio ($\bar{\omega} = \omega/\omega_{nf}$) for which the damped response amplitude reaches maximum (i.e., location of the critical speed). Show that it occurs always at frequency ratio of more than one.

Exercise 2.2 Let us define a new frequency ratio in terms of the damped natural frequency, that is, $\bar{\omega}_d = \omega/\omega_{nfd}$ with $\omega_{nfd} = \omega_{nf}\sqrt{1-\zeta^2}$. Obtain an expression for the amplitude ratio (Y/e) and the phase, ϕ, in terms of the new frequency ratio defined. Plot the amplitude ratio and the phase versus the new frequency ratio and discuss the results. Obtain an expression for the frequency ratio ($\bar{\omega}_d = \omega/\omega_{nfd}$) for which the damped response amplitude reaches maximum. In the expression so obtained of frequency ratio, what is the maximum value of damping ratio that is feasible for the underdamped system?

Exercise 2.3 Obtain transverse critical speeds of a cantilever rotor system as shown in Figure 2.35. Take the mass of the disk as m=10 kg and the diametral mass moment of inertia as I_d=0.02 kg-m^2. The shaft diameter, d, is 10 mm, and the total length of the span, l, is 0.2 m. The shaft is assumed to be massless, and its Young's modulus E=2.1×10^{11} N/m^2. Ignore the gyroscopic effect and take one plane motion only. Influence coefficients are given as $\alpha_{yf} = l^3/3EI$, $\alpha_{yM} = \alpha_{\varphi f} = l^2/2EI$, and $\alpha_{\varphi M} = l/EI$, where I is the second moment of area of the shaft cross-section. Subscripts f and M represent force and moment, respectively, and y and φ represent the translational and rotational displacements, respectively.

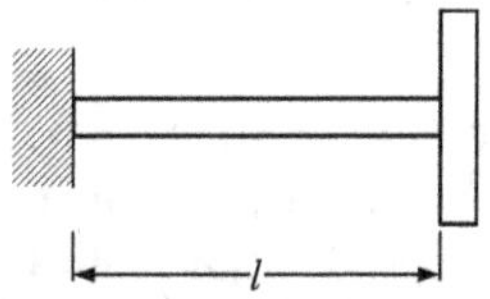

FIGURE 2.35 A cantilever shaft with a disk at the free end.

Exercise 2.4 Obtain the transverse critical speed of a rotor system as shown in Figure 2.36. Assume the mass of the disk $m=5\,\text{kg}$, the diametral mass moment of inertia $I_d=0.02$ kg-m², and the Young's modulus of the shaft $E=2.1\times10^{11}\,\text{N/m}^2$. Take the shaft segment lengths as $a=0.3\,\text{m}$ and $b=0.7\,\text{m}$. The diameter of the shaft is 10 mm. Ignore the gyroscopic effect.

For the present case, influence coefficients are given as $\alpha_{yf}=\dfrac{a^2(a+b)}{3EI}$, $\alpha_{\varphi M}=\dfrac{(3a+b)}{3EI}$, and $\alpha_{\varphi f}=\alpha_{yM}=\dfrac{a(3a+2b)}{6EI}$, where I is the second moment of area of the shaft cross-section. Subscripts f and M represent force and moment, respectively, and y and φ represent the translational and rotational displacements, respectively.

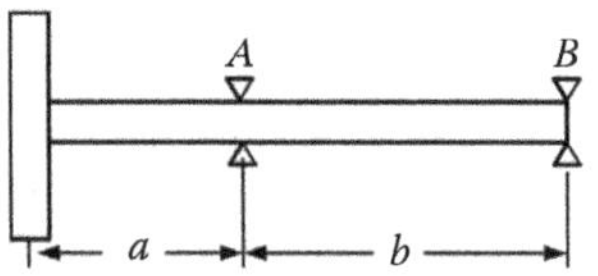

FIGURE 2.36 An overhung rotor system.

Exercise 2.5 Obtain the bearing reaction forces and moments of a cantilever rotor at rotor speeds of (i) $0.5\omega_{nf1}$, (ii) $0.5(\omega_{nf2}+\omega_{nf1})$, and (iii) $1.5\omega_{nf2}$, where ω_{nf1} and ω_{nf2} are the first and second transverse natural frequencies, respectively. Take the mass of the disk to be $m=10\,\text{kg}$, and the diametral mass moment of inertia to be $I_d=0.02$ kg-m². The disk has a residual unbalance of 25 g-cm. The shaft diameter is 10 mm, and the total shaft span is 0.5 m. The shaft is assumed to be massless, and its Young's modulus $E=2.1\times10^{11}\,\text{N/m}^2$. Take one plane motion only.

Influence coefficients are given as $\alpha_{yf}=l^3/3EI$, $\alpha_{yM}=\alpha_{\varphi f}=l^2/2EI$, and $\alpha_{\varphi M}=l/EI$, where I is the second moment of area of the shaft cross-section. Subscripts f and M represent force and moment, respectively, and y and φ represent the translational and rotational displacements, respectively.

Exercise 2.6 Find transverse natural frequencies (or critical speeds) of a cantilever rotor system as shown in Figure 2.37. Consider the shaft to be massless and made of steel with a Young's modulus of $2.1\times10^{11}\,\text{N/m}^2$. A disk is mounted at the free end of the shaft with a mass of 10 kg and a diametral mass moment of inertia of 0.04 kg-m². Do not consider the gyroscopic effect (please note that for this condition only, transverse natural frequencies and critical speeds will be the same). In the diagram, all dimensions are in cm.

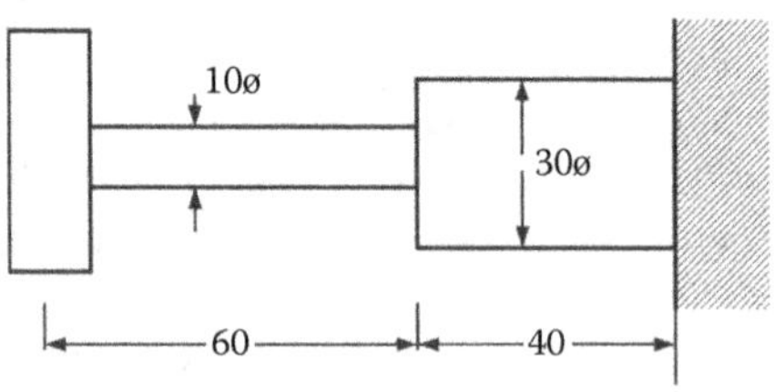

FIGURE 2.37 A stepped shaft with cantilever end conditions.

Exercise 2.7 (a) While the Jeffcott rotor is whirling, with the help of the center of gravity, the center spinning of the disk, and the bearing axis, draw their relative positions in an axial plane when the rotor is (i) below the critical speed, (ii) at critical speed, and (iii) above the critical speed. (b) Define the following terms: natural frequency and critical speed of a rotor and synchronous and asynchronous whirls.

Exercise 2.8 In a design stage of a rotor-bearing system, it has been found that one of the critical speeds is very close to the fixed operating speed of the rotor. List the design modifications a designer can make to overcome this problem.

Exercise 2.9 A cantilever shaft of 1 m length (l) and 30 mm diameter (d) has a disk of 5 kg mass (m) attached at its free end, with a negligibly small diametral mass moment of inertia. The shaft has a through hole parallel to the shaft axis of diameter 3 mm (d_i), which is vertically below the shaft center, with the distance between the centers of the shaft and the hole as 6 mm (e). Consider no cross-coupling in two orthogonal directions as well as between the translational and rotational displacements, and obtain the transverse natural frequencies of the shaft system in two principal planes. Consider the shaft to be massless and Young's modulus $E = 2.1 \times 10^{11}$ N/m^2.

Exercise 2.10 For the Jeffcott rotor, consider pure rotational displacement (tilting) of the disk (without translational displacement) and obtain the transverse natural frequency for the tilting motion. Let EI is the flexural rigidity of the shaft, a and b are the axial location of the disk from the left and right bearings with $l = a + b$, m is the mass of the disk, and r is the radius of the disk.

Exercise 2.11 What are the length and the diameter of a cantilever shaft if the transverse critical speed has to be fixed at 100 Hz (by considering pure translatory motion only) and it has 2 kg of mass at its free end? Because of space limitations, the length of the shaft should not exceed 30 cm. $E = 2.1 \times 10^{11}$ N/m^2.

Exercise 2.12 The transverse critical speed of a rotor system as shown in Figure 2.36 is fixed at 5.98 rad/s. Take the disk as a point mass with $m = 5$ kg. What is the overhung shaft length, a? Take the shaft length as $b = 0.7$ m. The diameter of the shaft is 10 mm. Ignore the gyroscopic effect. $E = 2.1 \times 10^{11}$ N/m^2.

Exercise 2.13 The transverse critical speed of a rotor system as shown in Figure 2.36 is fixed at 5.98 rad/s. Take the disk as a point mass with $m = 5$ kg. What is the diameter of the uniform shaft, d? Take shaft length $a = 0.3$, $b = 0.7$ m. Ignore the gyroscopic effect. $E = 2.1 \times 10^{11}$ N/m^2.

Exercise 2.14 The transverse critical speed of a rotor system as shown in Figure 2.36 is fixed at 5.98 rad/s. Take the disk as a point mass with $m = 5$ kg. What is the diameter of the uniform shaft, d? Take shaft length $2a = b = 0.7$ m. $E = 2.1 \times 10^{11}$ N/m^2. Ignore the gyroscopic effect.

Exercise 2.15 For a Jeffcott rotor with a disk at the midspan, influence coefficients are given as $\alpha_{yf} = l^3/(48EI)$, $\alpha_{yM} = \alpha_{\varphi f} = 0$, and $\alpha_{\varphi M} = l/(12EI)$, where l is the span length and EI is the modulus of rigidity of the shaft. Let m and I_d be the mass and the diametral mass moment of inertia, respectively, of the disk. Obtain the natural frequencies of the rotor system.

Exercise 2.16 Obtain transverse natural frequencies of a rotor system as shown in Figure 2.38. The mass of the disk is $m = 5$ kg, and the diametral mass moment of inertia is $I_d = 0.02$ kg-m^2. Lengths of the shaft are $a = 0.3$ m and $b = 0.7$ m. The diameter of the shaft is 10 mm. Bearing A has the roller support, and Bearing B has the fixed support condition. Ignore the mass of the shaft and gyroscopic effect of the disk. $E = 2.1 \times 10^{11}$ N/m^2.

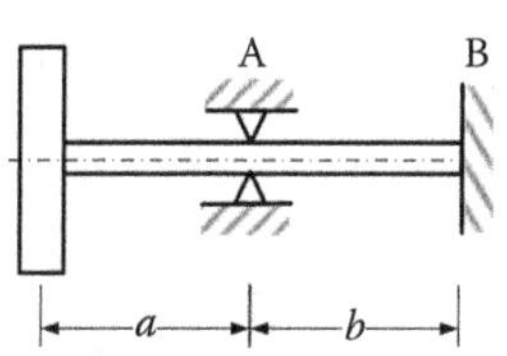

FIGURE 2.38 An overhung rotor system with one end fixed condition.

Exercise 2.17 Consider a rotor system as shown in Figure 2.39 for obtaining the transverse natural frequency. Two flexible *massless* shafts are connected by a coupling (i.e., a pin joint). A thin disk of mass 3 kg and diametral mass moment of inertia 0.05 kg-m^2 is attached to one of the shafts (the shaft on the left), and it is not interfering with the relative motion between the two shafts. The other ends of the shafts have fixed conditions. Take the length of each of the shafts as 0.5 m and the diameter as 0.05 m. Young's modulus $E = 2.1 \times 10^{11}$ N/m^2.

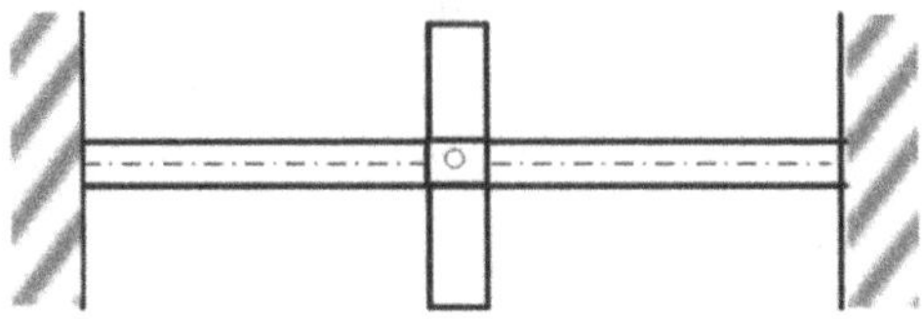

FIGURE 2.39 Two shafts connected by a coupling.

Exercise 2.18 For a Jeffcott rotor, the following energy expressions are given:

$$T = \frac{1}{2} m\dot{x}_G^2 + \frac{1}{2} m\dot{y}_G^2; \quad U = \frac{1}{2} kx^2 + \frac{1}{2} ky^2; \quad \delta W_{nc} = (-c\dot{x})\delta x + (-c\dot{y})\delta y$$

with

$$x_G = x + e\cos\omega t \quad \text{and} \quad y_G = y + e\sin\omega t$$

where T is the kinetic energy, U is the potential energy, δW_{nc} is the nonconservative virtual work done, x and y are the coordinates of the disk geometrical center (i.e., generalized coordinates), x_G and y_G are the coordinates of the disk center of gravity, e is the eccentricity, m is the mass of the disk, k is the stiffness of the shaft, c is the viscous damping in the rotor system, and ω is the spin speed of the rotor. Using Lagrange's equation (refer to Chapter 7), obtain equations of motion of the rotor system.

Exercise 2.19 For a Jeffcott rotor with an offset disk, the following energy expressions are given:

$$T = \frac{1}{2} m\dot{x}_G^2 + \frac{1}{2} m\dot{y}_G^2 + \frac{1}{2} I_d\dot{\varphi}_y^2 + \frac{1}{2} I_d\dot{\varphi}_x^2$$

with

$$x_G = x + e\cos\omega t \quad \text{and} \quad y_G = y + e\sin\omega t$$

and

$$U = \frac{1}{2} k_{11}x^2 + \frac{1}{2} k_{11}y^2 + \frac{1}{2} k_{22}\varphi_x^2 + \frac{1}{2} k_{22}\varphi_y^2 + \frac{1}{2} k_{12}x\varphi_y + \frac{1}{2} k_{21}y\varphi_x$$

with

$$k_{12} = k_{21}$$

where T is the kinetic energy, U is the potential energy, x and y are the translatory coordinates of the disk geometrical center, x_G and y_G are the coordinates of the disk center of gravity, φ_x and φ_y are rotational coordinates of the disk, e is the eccentricity, m is the mass of the disk, k is the stiffness of the shaft, subscripts 1 and 2 represent, respectively, the translational and rotational displacements, and ω is the spin speed of the rotor. Using Lagrange's equation (refer to Chapter 7), obtain equations of motion of the rotor system.

Exercise 2.20 For a Jeffcott rotor to alleviate the crossing of critical speed during run-up, an auxiliary bearing switching technique is applied. The stiffnesses of the rotor without and with auxiliary bearing are k_1 and k_2, respectively. Let the amplitudes of the rotor without and with auxiliary bearing be y_1 and y_2, respectively. Let m be the mass of the rotor and ω the spin speed of the shaft. Obtain the switching frequency, ω_c, at which the system changes from the auxiliary bearing to without auxiliary bearing. Also obtain the maximum nondimensional amplitude of vibration, $|\bar{y}_{max}|=|\bar{y}/e|$. Give the plot of y_1 and y_2 with spin speed and show the switching frequency location.

Exercise 2.21 Obtain the transverse natural frequency of the rotor-bearing system as shown in Figure 2.40. Consider the shaft to be rigid. The bearing on the left is simply supported, and the bearing on the right is having two springs, and each spring has the stiffness k. Take $L=1$ m, $a=0.3$ m, $d=0.15$ m, $k=1$ kN/m, $m=5$ kg, and $I_d=0.02$ kg-m^2. Consider a single plane motion and ignore the gyroscopic couple effect.

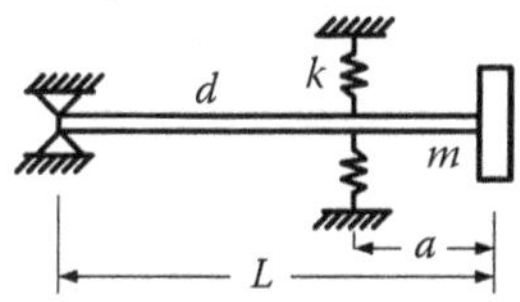

FIGURE 2.40 A rigid shaft supported on a flexible support and a rigid support.

Exercise 2.22 For a Jeffcott rotor with an offset disk, derive the equations of motion using the energy method (e.g., Lagrange's method—refer to Chapter 7) for a single plane. Energy terms are the kinetic energy $T = 0.5\, m\dot{y}^2 + 0.5\, I_d\dot{\varphi}_x^2$, the potential energy $U = 0.5\, k_{11}y^2 + 0.5\, k_{22}\varphi_x^2 + k_{12}y\varphi_x$, and nonconservative virtual work $\delta W_{nc} = \left(me\omega^2 \sin \omega t\right)\delta y$. Here, the stiffness terms have been obtained by inverting the influence coefficient matrix. Here, δ is the variational operator.

Exercise 2.23 *Briefly provide the answer to the following*: In a Jeffcott rotor, the disk has an initial tilt by an angle α in a transverse plane due to improper assembly. During spinning (ω rad/s) of the rotor, the unbalance moment due to this would be what? Let I_d be the diametral mass moment of the disk and I_p the polar mass moment of inertia.

Exercise 2.24 For a single plane motion (let us consider the vertical plane), obtain the whirl natural frequencies in a closed form of the rotor-bearing system shown in Figure 2.41. Consider the shaft to be massless and elastic. Both bearings are identical with no mass, and a thin disk is located at a distance of a from the left end and b from the right end of the shaft (with $a+b=l$). The expression for the deflection of a simple supported shaft is given in Figure 2.42. Do not consider the gyroscopic effect.

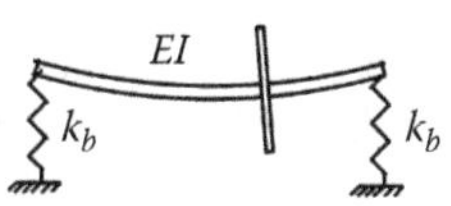

FIGURE 2.41 A flexible shaft on flexible bearings.

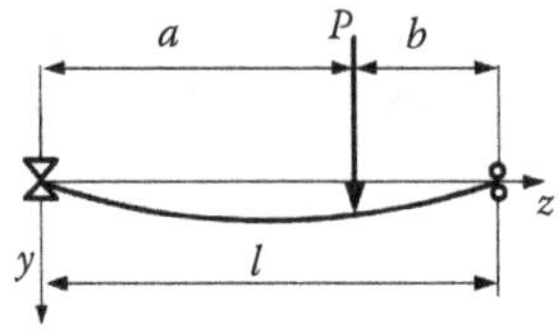

$$y = \frac{Pbz}{6lEI}(l^2 - z^2 - b^2) \text{ for } [0 < z < a]$$

$$y = \frac{Pb}{6lEI}\left[\frac{l}{b}(z-a)^3 + (l^2 - b^2)z - z^3\right] \text{for } [a < z < l]$$

FIGURE 2.42 A simply supported shaft deflection due to an offset load.

Exercise 2.25 Obtain transverse natural frequencies of a rotor system shown in Figure 2.43. It has simply supported conditions, and a thin disk ($m=2$ kg and $I_d=0.01$ kg-m^2) is mounted very close to the support itself so that they have pure tilting motion (transverse rotational displacements). It is assumed that supports do not interfere the motion of the disk. Relations of transverse rotational displacements due to moment at one end are given in Table 2.2 for Figure 2.44. Take $E=2.1\times10^{11}$ N/m^2. The length of the shaft is 0.6 m, and its diameter is 0.01 m. Neglect the mass of the shaft and gyroscopic effect.

FIGURE 2.43 A rotor system with a disk near right support.

TABLE 2.2
A Simply Supported Shaft with a Concentrated Moment *M* at One End Point

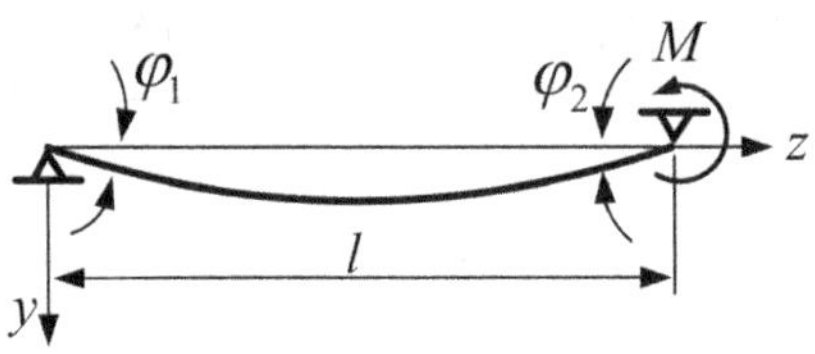

$$\varphi_1 = \frac{Ml}{6EI} \qquad y = \frac{Mlz}{6EI}\left(1 - \frac{z^2}{l^2}\right)$$

$$\varphi_2 = \frac{Ml}{3EI}$$

FIGURE 2.44 A simply supported shaft with a moment at one end.

Exercise 2.26 Two cantilever shafts are connected by a pin joint at free ends and by a flexible coupling having a bending spring of stiffness k_b, which resists the slope of the shaft at ends, as shown in Figure 2.45. Consider single plane motion. In the left cantilever shaft, a rigid disk is mounted at its right extreme end. It is assumed that the disk does not interfere with the shaft in the right side. Obtain the natural frequencies of the rotor system. Take $l=0.3$ m, $d=0.005$ m, $E=2.1\times10^{11}$ N/m^2, $m=3$ kg, $I_p=0.01$ kg-m^2, and $k_b=60$ N-m/rad. Neglect the shaft inertia and ignore the gyroscopic effects of the disk.

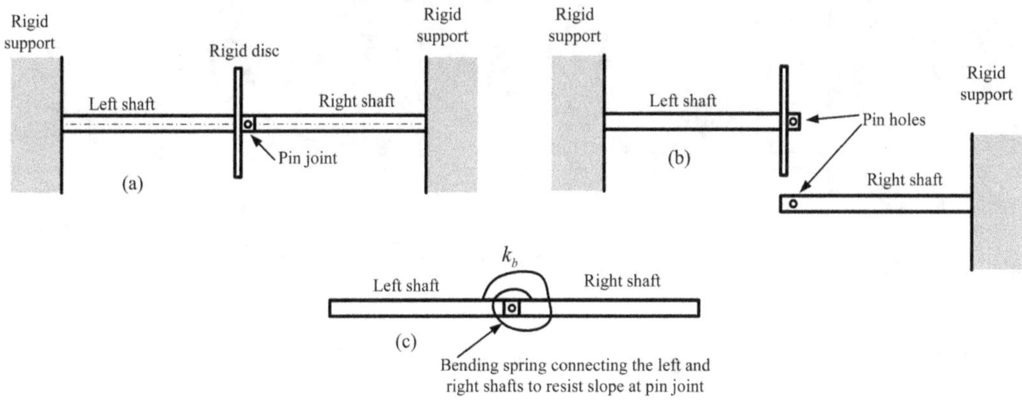

FIGURE 2.45 (a) Two cantilever shafts are connected by a pin joint at free ends and by a flexible coupling, (b) two cantilever shafts in disassembled configuration, and (c) two shaft connected by bending spring.

Exercise 2.27 Obtain transverse natural frequencies of a rotor system for two cases: As shown in Figure 2.46 (with $a=b=0.5l$) and Figure 2.47 (with $a=l$ and $b=0$, it assumed that support is not interfering with the motion of the disk in tilting). Assume that the mass of the thin disk $m=10$ kg, the diametral mass moment of inertia $I_d=0.02$ kg-m². The shaft has a diameter of 10 mm and the total length of the span is 1 m. The shaft is assumed to be massless. Use the influence coefficients provided below respective figures. Take the Young's modulus of the shaft material as $E=2.1\times10^{11}\,\text{N/m}^2$. Ignore the gyroscopic effect and take one plane motion only (Table 2.3).

TABLE 2.3
Influence Coefficients for Simply Supported Shaft

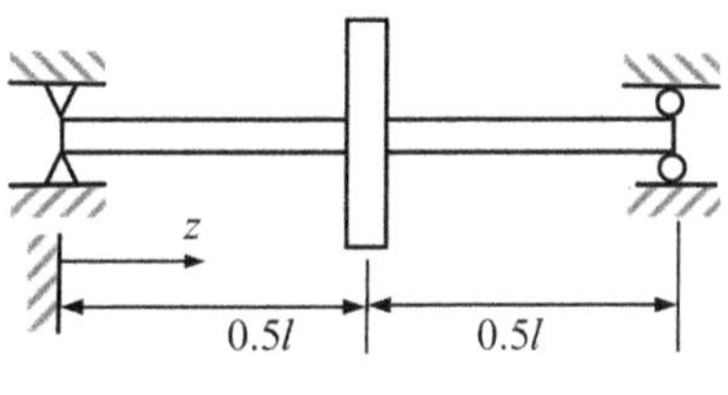

FIGURE 2.46 A simply supported rotor system with a disk at the midspan.

$$\begin{bmatrix} \alpha_{yf} & \alpha_{yM} \\ \alpha_{\varphi f} & \alpha_{\varphi M} \end{bmatrix} = \begin{bmatrix} \dfrac{l^3}{48EI} & 0 \\ 0 & \dfrac{l}{12EI} \end{bmatrix}$$

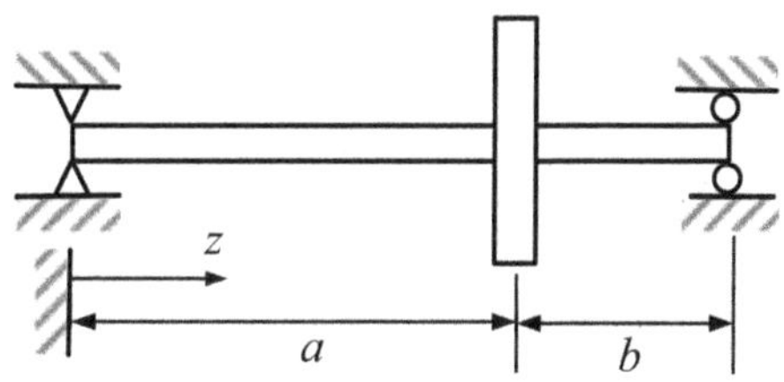

FIGURE 2.47 A simply supported rotor system with disk offset.

$$\begin{bmatrix} \alpha_{yf} & \alpha_{yM} \\ \alpha_{\varphi f} & \alpha_{\varphi M} \end{bmatrix} = \begin{bmatrix} \dfrac{a^2b^2}{3EIl} & \dfrac{-\left(3a^2l-2a^3-al^2\right)}{3EIl} \\ \dfrac{ab(b-a)}{3EIl} & \dfrac{-\left(3al-3a^2-l^2\right)}{3EIl} \end{bmatrix}$$

Exercise 2.28 For a disk-shaft-bearing system shown in Figure 2.48, obtain transverse (bending) natural frequency of system for (i) $k_{shaft}=9$ kN/m, $k_{bearing}=30$ kN/m, (ii) $k_{shaft}=40$ kN/m, $k_{bearing}=30$ kN/m, (iii) $k_{shaft}=\infty$ kN/m, $k_{bearing}=30$ kN/m, and (iv) $k_{shaft}=9$ kN/m, $k_{bearing}=\infty$ kN/m. Interpret your results regarding change in natural frequency for different cases. Take the mass of the disk as 200 kg.

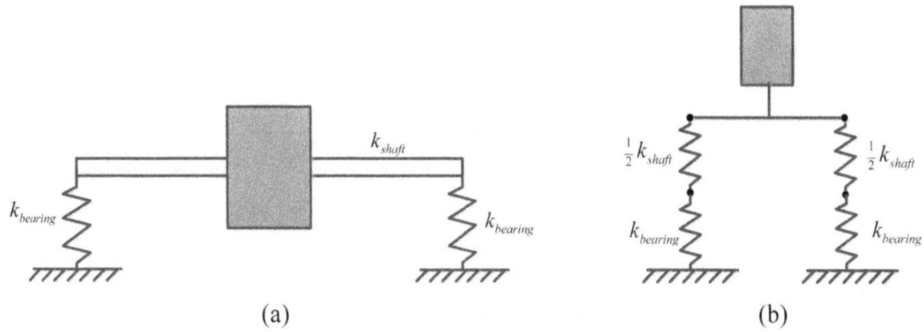

FIGURE 2.48 (a) Disk-shaft-bearing system; (b) dynamic system schematic.

Exercise 2.29 Choose a single correct answer from the multiple choice questions:

i. The critical speed phenomenon of a rotor system is *a(n)*
   A. free vibration
   B. forced vibration
   C. transient vibration
   D. unstable vibration

ii. A rigid body is defined as
   A. a body with no deformation
   B. a body with particles that have fixed distances
   C. both (A) and (B)
   D. a body with large dimensions

iii. A particle has how many degrees of freedom in space?
   A. 1
   B. 2
   C. 3
   D. more than 3

iv. A rigid body has how many degrees of freedom in space?
   A. 1
   B. 3
   C. 6
   D. more than 6

v. A flexible body has how many degrees of freedom in space?
   A. 1
   B. 3
   C. 6
   D. infinite

vi. If three particles have fixed relative distances between them, then it represents a system of
   A. a single particle
   B. a rigid body
   C. a multibody
   D. a flexible body

vii. A system consists of three particles with their relative distances as constant. In space, it has how many degrees of freedom?
A. 1
B. 3
C. 6
D. infinite

viii. A perfectly balanced Jeffcott rotor (i.e., a flexible shaft with a disk at midspan) is rotating at a particular speed. If it is perturbed in the transverse plane from its equilibrium, the frequency of whirl would be equal to
A. the shaft spin speed
B. the transverse natural frequency
C. more than the transverse natural frequency
D. less than the transverse natural frequency

ix. A Jeffcott rotor with an offset disk (i.e., not at the midspan) has a tilt in the transverse plane. The shaft would experience
A. gyroscopic couple
B. an external moment
C. an unbalance force
D. a gyroscopic couple and an external moment

x. The transverse natural frequency of the rotor-bearing system shown in Figure 2.49 would be (Figure 2.49)
A. $\sqrt{\dfrac{k_s + k_b}{m}}$
B. $\sqrt{\dfrac{k_s k_b}{m(k_s + k_b)}}$
C. $\sqrt{\dfrac{(k_s + k_b)}{m k_s k_b}}$
D. $\sqrt{\dfrac{k_b(k_s + k_b)}{m k_s}}$

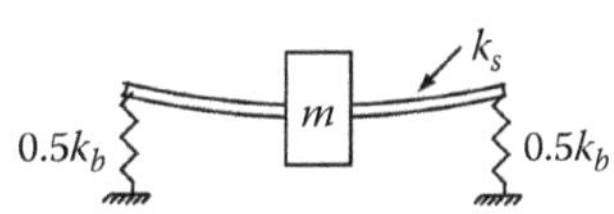

FIGURE 2.49 A symmetric flexible rotor with flexible bearings.

xi. For a Jeffcott rotor operating at supercritical speed (i.e., well above the critical speed), the rotor deflection would be approaching
A. $\dfrac{mg}{k}$
B. infinite
C. zero
D. its eccentricity

xii. For a Jeffcott rotor operating at the critical speed, the rotor response phase with respect to the unbalance force would be approaching
A. 0°
B. 90°
C. 180°
D. some finite value depending upon the damping value

xiii. The transverse critical speed of a rotor system, as shown in Figure 2.36, is to be fixed at 5.98 rad/s. Take the disk as a point mass with m=5 kg. What is the diameter of the uniform shaft, d? Take shaft length $2a=b=0.7$ m. Ignore the gyroscopic effect.
A. 4.25 mm
B. 4.43 mm
C. 5.10 mm
D. 5.22 mm

xiv. For a disk with mass m, eccentricity e, and rotating at ω, the unbalance is defined as
A. $me\omega^2$
B. me
C. $me\omega$
D. $e\omega$

xv. At resonance, the phase between the force and the displacement of a Jeffcott rotor for different levels of damping would
A. vary between 0° and 90°
B. B vary between 90° and 180°
C. 90°
D. 180°

xvi. In a Jeffcott rotor, if the shaft is given a constant axial tensile preload at the ends, then the transverse natural frequency of the rotor would
A. increase
B. decrease
C. remain constant
D. become zero

xvii. For a Jeffcott rotor with the vertical shaft, if a constant axial compressive load is applied on the shaft, the transverse natural frequency of the rotor would
A. remain unchanged
B. increase
C. decrease
D. become zero

xviii. Consider two different rotor systems, the first with a rigid massless shaft and end supported by two identical flexible bearings, each with a stiffness of, k_b, and the second rotor system with simply supported bearing conditions and flexible massless shaft (EI is the flexural rigidity of the shaft and L is its span). Both rotor systems have a disk of mass m, which is symmetrically placed with respect to the ends of the shaft. If both rotor systems have the same whirl frequency, then the following relations would prevail

A. $k_b = \dfrac{48EI}{l^3}$

B. $k_b = \dfrac{12EI}{l^3}$

C. $k_b = \dfrac{24EI}{l^3}$

D. $k_b = \dfrac{96EI}{l^3}$

xix. Two unbalances ($m_{b1}e_1$ and $m_{b2}e_2$) are attached to a Jeffcott rotor disk with an angular phase of ϕ_{12} between them with $m_{b2}e_2$ ahead of $m_{b1}e_1$ with respect to the rotation of the rotor. Initially the phase of $m_{b1}e_1$ is ϕ_0 with the x-axis (i.e., horizontal transverse direction) in the direction of the rotation of the rotor. The unbalance force due to spin speed, ω, of the rotor would be

A. $m_{b1}e_1\omega^2\cos(\omega t+\phi_0)+m_{b2}e_2\omega^2\cos(\omega t+\phi_{12})$

B. $m_{b1}e_1\omega^2\cos(\omega t+\phi_0)+m_{b2}e_2\omega^2\cos(\omega t-\phi_0+\phi_{12})$

C. $m_{b1}e_1\omega^2\cos(\omega t+\phi_0+\phi_{12})+m_{b2}e_2\omega^2\cos(\omega t+\phi_0+\phi_{12})$

D. $m_{b1}e_1\omega^2\cos(\omega t+\phi_0)+m_{b2}e_2\omega^2\cos(\omega t+\phi_0+\phi_{12})$

xx. For a cantilever shaft with a thin disk at the free end, if the transverse and torsional natural frequencies are the same, then the ratio of the length of the shaft to the diameter of the disk would be (take Poisson's ratio as 0.5 for the shaft material)

A. ¾

B. 4/3

C. 2/3

D. 1/3

xxi. For a single-DOF spring-mass-damper rotor model, the equation of motion at the resonance is given as $\ddot{x}+2\zeta\omega_{nf}\dot{x}+\omega_{nf}^2x=e\omega_{nf}^2\cos\omega_{nf}t$. The solution of this equation would have the following form:

A. $x(t)=X(t)\cos\omega_{nf}t$

B. $x(t)=X(t)\sin\omega_{nf}t$

C. $x(t)=X(t)\cos(\omega_{nf}t+\phi)$

D. $x(t)=X(t)\sin(\omega_{nf}t+\phi)$

E. $x(t)=X\cos\omega_{nf}t$

F. $x(t)=X\sin\omega_{nf}t$

G. $x(t)=X\cos(\omega_{nf}t+\phi)$

H. $x(t)=X\sin(\omega_{nf}t+\phi)$

where ζ is the damping ratio, ω_{nf} is the undamped natural frequency, ϕ is the phase, X is time-invariant amplitude, and $X(t)$ is time-variant amplitude.

xxii. In a Jeffcott rotor, the phase of unbalance force with respect to the undamped response below the critical speed approaches

A. 180°

B. 90°

C. 45°

D. 0°

xxiii. In a Jeffcott rotor with a rigid disk having mass and polar mass moment of inertia, when rotating at a constant speed, the torsional vibration may take place due to

A. only gravity
B. gravity with eccentricity
C. only eccentricity
D. none of the above

xxiv. In a Jeffcott rotor, when the speed is very high ($\omega >> \omega_{cr}$), the magnitude of rotor displacement is equal to

A. infinity
B. zero
C. eccentricity of the disk
D. static deflection of the disk

xxv. The elastic coupling in transverse vibration of a circular shaft is

A. coupling of two orthogonal plane motions and no coupling of the linear and angular displacements
B. no coupling of two orthogonal plane motions and coupling of the linear and angular displacements
C. coupling of two orthogonal plane motions and coupling of the linear and angular displacements
D. no coupling of two orthogonal plane motions and no coupling of the linear and angular displacements

xxvi. For the antisynchronous whirl of a Jeffcott rotor when viewed from the bearing center

A. the same face (or mark) of the shaft as that for the synchronous whirl will always be seen
B. the opposite face (or mark) of the shaft as that for the synchronous whirl will always be seen
C. one of the faces (or marks) will be seen twice in one rotation of the shaft
D. a different face (or mark) always will be visible with no fixed pattern

xxvii. The phase between the unbalance force and the response in a Jeffcott rotor at the resonance would be

A. $\pi/2$
B. π
C. 0
D. $-\pi$

xxviii. In a cantilever rotor during maintenance, the mild steel shaft is replaced with an aluminum shaft ($E_{Al}/E_S=1/3$) of the same dimension. The shaft is massless and the disk mass at the free end remains the same. The torsional natural frequency ratio for rotors ($(\omega_{nf\_Al}/(\omega_{nf\_MS})$) would be

A. 3
B. 1/3
C. $\sqrt{3}$
D. $1/\sqrt{3}$

xxix. For the synchronous whirl of a damped Jeffcott rotor model while rotating the rotor above the critical speed, the correct phase representation between the unbalance force (black circle) and disk response would one of the options in Figure 2.50.

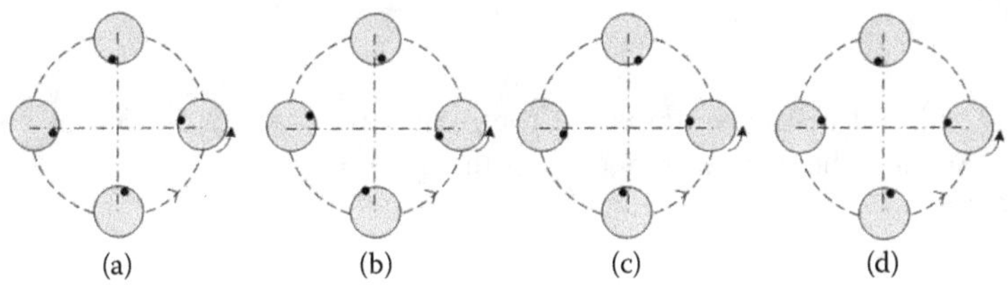

FIGURE 2.50 Shaft whirling and spinning about a bearing center: (a) option A; (b) option B; (c) option C; (d) option D.

xxx. The correct phase between unbalance force and response after the critical speed with damping in the system will have one of the options given in Figure 2.51.

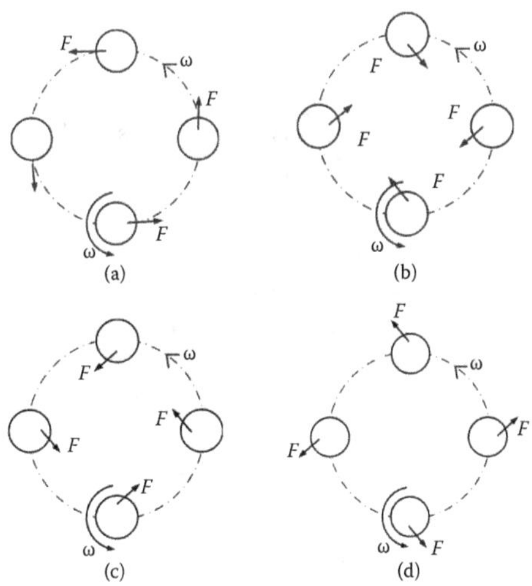

FIGURE 2.51 A shaft orbiting about a bearing center: (a) option A; (b) option B; (c) option C; (d) option D.

xxxi. The critical speed phenomenon of a rotor is a
- A. free vibration
- B. forced vibration
- C. transient vibration
- D. unstable vibration

xxxii. In a Jeffcott rotor (with a disk at the midspan), if the disk center of gravity is at its center of rotation but the disk is tilted by a small angle with respect to the shaft. The critical speed encountered during rotation of the shaft will be

A. $\sqrt{\dfrac{12EI}{I_d L}}$

B. $\sqrt{\dfrac{48EI}{mL^3}}$

C. $\sqrt{\dfrac{3EI}{I_d L}}$

D. $\sqrt{\dfrac{3EI}{mL^3}}$

xxxiii. A 2-kg mass of a simply supported rotor causes a static deflection at a free end of 0.2 cm. What is the natural frequency of the system in rad/s?
A. 221.5
B. 140.1
C. 120.4
D. 220.3

xxxiv. In a Jeffcott rotor, if the rotor rotates at a very high speed (well above the critical speed), the ratio of shaft deflection and disk eccentricity (absolute value) will be approach toward
A. 0
B. 0.5
C. 1
D. 2

xxxv. A long rotor supported on flexible bearings, and has both radial eccentricity and axial offset of its center of gravity, it will produce

xxxvi. The unbalance force in a rotor system gives
A. a synchronous whirl condition of the response
B. an antisynchronous whirl condition of the response
C. an asynchronous whirl condition of the response
D. a multi-harmonic synchronous whirl condition of the response

xxxvii. For a Jeffcott rotor with a positive damping, whether the amplitude ratio (rotor displacement/disk eccentricity) can be infinity for a particular nonzero value of positive damping (in terms of the damping ratio) in unbalanced rotor?
A. 1.414
B. 1
C. –1
D. not feasible

xxxviii. In a Jeffcott rotor with an offset disk, the coupling is present in
A. two orthogonal transverse translational displacements
B. two orthogonal transverse rotational displacements
C. the transverse translational and rotational displacements in same plane
D. the transverse translational and rotational displacements in orthogonal plane

xxxix. The influence coefficient α_{ij} is defined as
A. the displacement at the *i*th station due to a unit force at the *j*th station keeping all other forces at zero.
B. the force at the *i*th station due to a unit displacement at the *j*th station keeping all other displacements at zero.

C. the force at the jth station due to a unit displacement at the ith station keeping all other displacements at zero.
D. the displacement at the jth station due to a unit force at the ith station keeping all other forces at zero.

xl. Obtain the bending natural frequency (in rad/s) for the synchronous motion of a cantilever rotor. Take the mass of the disk $m=1$ kg. The shaft is assumed to be massless, and its length and diameter are 0.3 and 0.01 m, respectively. Take the Young's modulus of the shaft as $E=2.1\times 10^{11}$ N/m^2.
A. 104.4
B. 196.6
C. 121.4
D. 88.3

xli. For a simply supported rotor with a point mass, m, at the midspan. The natural frequency of the system is ω_{nf}. Let L and d be the length and the diameter, respectively, of the shaft, and E be the Young's modulus. The diameter, d, of the shaft expression is given as

A. $\left(\dfrac{4ml^3\omega_{nf}^2}{3E\pi}\right)^{(1/4)}$

B. $\left(\dfrac{64ml^3\omega_{nf}^2}{3E\pi}\right)^{(1/4)}$

C. $\left(\dfrac{4ml^3\omega_{nf}^2}{3E\pi}\right)^{(1/2)}$

D. $\left(\dfrac{64ml^3\omega_{nf}^2}{3E\pi}\right)^{(1/2)}$

APPENDIX 2.1: USEFUL FORMULAS

S.N.	Formulas	Descriptions
1.	For a thin disk (about center of gravity) $I_p = 2I_d = mk^2 = \frac{1}{2}mr^2$ with $k = r/\sqrt{2}$ $I_p = 2I_d = \frac{1}{8}mD^2$	Herein, k is the radius of gyration, r is the radius of the thin disk, D is the disk diameter, m is the mass of the disk, I_d is the diametral mass moment of inertia, and I_p is the polar mass moment of inertia
2.	For a long cylinder (about center of gravity) $I_p = mk^2 = \frac{1}{2}mr^2 = \frac{1}{8}mD^2$ $I_d = \frac{1}{16}mD^2 + \frac{1}{12}ml^2$	D is the diameter of cylinder, m is the mass of the cylinder, l is the length of cylinder, and r is the radius of the cylinder.
3.	$\begin{bmatrix} a & b \\ c & d \end{bmatrix}^{-1} = \dfrac{1}{ad-bc}\begin{bmatrix} d & -b \\ -c & a \end{bmatrix}$	Inverse of 2×2 matrix

(*Continued*)

(Continued)

S.N.	Formulas	Descriptions

4.

$$[M]=\begin{bmatrix} a_r+ja_i & b_r+jb_i \\ c_r+jc_i & d_r+jd_i \end{bmatrix}$$

$$[N]=\begin{bmatrix} \mathrm{Re}[M] & \mathrm{Im}[M] \\ -\mathrm{Im}[M] & \mathrm{Re}[M] \end{bmatrix}=\begin{bmatrix} a_r & b_r & a_i & b_i \\ c_r & d_r & c_i & d_i \\ -a_i & -b_i & a_r & b_r \\ -c_i & -d_i & c_r & d_r \end{bmatrix},$$

$$[P]=[N]^{-1}=\begin{bmatrix} \mathrm{Re}[\mathrm{inv}(M)] & \mathrm{Im}[\mathrm{inv}(M)] \\ -\mathrm{Im}[\mathrm{inv}(M)] & \mathrm{Re}[\mathrm{inv}(M)] \end{bmatrix}=\begin{bmatrix} P_{11} & P_{12} \\ P_{21} & P_{22} \end{bmatrix}$$

$$[M]^{-1}=[P_{11}+jP_{12}]=\left[\mathrm{Re}[\mathrm{inv}(M)]+j\,\mathrm{Im}[\mathrm{inv}(M)]\right]$$

Inverse of a complex matrix (2×2) or $(N\times N)$. Obtain inverse of $[M]_{1\times1}=(a+jb)$ such that $[M][M]^{-1}=(a+jb)(c+jd)=1$.

$$[N]=\begin{bmatrix} a & b \\ -b & a \end{bmatrix},$$

$$[P]=[N]^{-1}=\frac{1}{a^2+b^2}\begin{bmatrix} a & -b \\ b & a \end{bmatrix},$$

$$[M]^{-1}=[P_{11}+jP_{12}]=\left[\frac{a}{a^2+b^2}+j\frac{-b}{a^2+b^2}\right]$$

5.

$$A^{-1}=\frac{1}{|A|}\begin{pmatrix} \begin{vmatrix} a_{22} & a_{23} \\ a_{32} & a_{33} \end{vmatrix} & \begin{vmatrix} a_{13} & a_{12} \\ a_{33} & a_{32} \end{vmatrix} & \begin{vmatrix} a_{12} & a_{13} \\ a_{22} & a_{23} \end{vmatrix} \\ \begin{vmatrix} a_{23} & a_{21} \\ a_{33} & a_{31} \end{vmatrix} & \begin{vmatrix} a_{11} & a_{13} \\ a_{31} & a_{33} \end{vmatrix} & \begin{vmatrix} a_{13} & a_{11} \\ a_{23} & a_{21} \end{vmatrix} \\ \begin{vmatrix} a_{21} & a_{22} \\ a_{31} & a_{32} \end{vmatrix} & \begin{vmatrix} a_{12} & a_{11} \\ a_{32} & a_{31} \end{vmatrix} & \begin{vmatrix} a_{11} & a_{12} \\ a_{21} & a_{22} \end{vmatrix} \end{pmatrix}.$$

It is inverse of a 3×3 matrix

$$A=\begin{bmatrix} a_{11} & a_{12} & a_{13} \\ a_{21} & a_{22} & a_{23} \\ a_{31} & a_{32} & a_{33} \end{bmatrix}$$

6. Influence coefficients for simply supported shaft (refer to Figure 2.52):

$$\begin{bmatrix} \alpha_{yf} & \alpha_{yM} \\ \alpha_{\varphi f} & \alpha_{\varphi M} \end{bmatrix}=\begin{bmatrix} \dfrac{a^2b^2}{3EIl} & \dfrac{-(3a^2{}_l-2a^3-al^2)}{3EIl} \\ \dfrac{ab(b-a)}{3EIl} & \dfrac{-(3al-3a^2-l^2)}{3EIl} \end{bmatrix};$$

$l=a+b$.

Herein, α_{ij} represent the displacement (subscript: y-translational and φ-rotational) at the ith station due to a unit force (subscript: f—force and M—moment) at the jth station keeping all other forces/moments to zero.

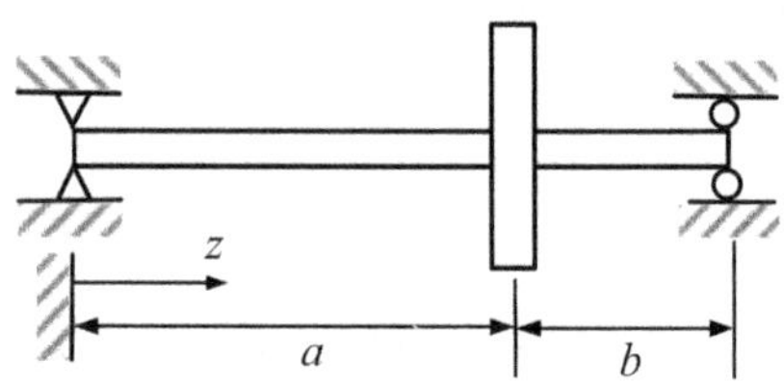

FIGURE 2.52 A simply supported shaft.

7. Influence coefficients for the simply supported shaft (Jeffcott rotor) with $a=b=0.5l$ (refer to Figure 2.53):

$$\begin{bmatrix} \alpha_{yf} & \alpha_{yM} \\ \alpha_{\varphi f} & \alpha_{\varphi M} \end{bmatrix}=\begin{bmatrix} \dfrac{l^3}{3EI} & \dfrac{l^2}{2EI} \\ \dfrac{l^2}{2EI} & \dfrac{l}{EI} \end{bmatrix}$$

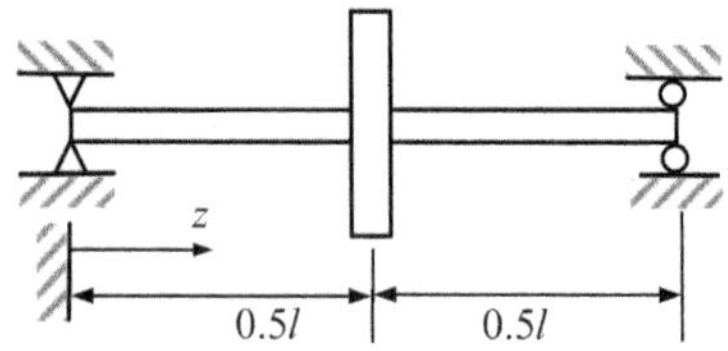

FIGURE 2.53 A simply supported (Jeffcott) rotor system.

8. Influence coefficients for the cantilever shaft (refer to Figure 2.54):

$$\begin{bmatrix} \alpha_{yf} & \alpha_{yM} \\ \alpha_{\varphi f} & \alpha_{\varphi M} \end{bmatrix}=\begin{bmatrix} \dfrac{l^3}{3EI} & \dfrac{l^2}{2EI} \\ \dfrac{l^2}{2EI} & \dfrac{l}{EI} \end{bmatrix}$$

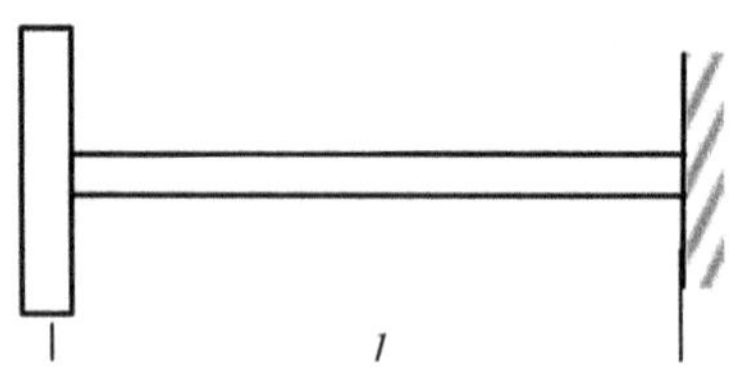

FIGURE 2.54 A cantilever rotor system.

(Continued)

(*Continued*)

S.N.	Formulas	Descriptions

9. Influence coefficients for an overhung shaft with roller supports (refer to Figure 2.55) (tilting considered):

$$\begin{bmatrix} \alpha_{yf} & \alpha_{yM} \\ \alpha_{\varphi f} & \alpha_{\varphi M} \end{bmatrix} = \begin{bmatrix} \dfrac{a^2(a+b)}{3EI} & \dfrac{a(3a+2b)}{6EI} \\ \dfrac{a(3a+2b)}{6EI} & \dfrac{(3a+b)}{3EI} \end{bmatrix}$$

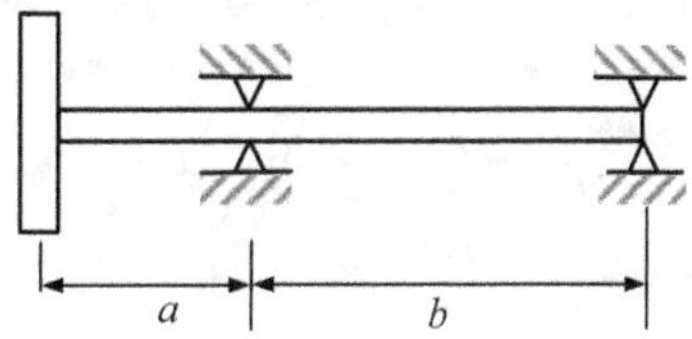

FIGURE 2.55 An overhung shaft with roller supports.

10. Influence coefficients for an overhung shaft with a fixed support (refer to Figure 2.56) (tilting considered):

$$\begin{bmatrix} \alpha_{yf} & \alpha_{yM} \\ \alpha_{\varphi f} & \alpha_{\varphi M} \end{bmatrix} = \begin{bmatrix} \dfrac{a^2(4a+3b)}{12EI} & \dfrac{a(2a+b)}{4EI} \\ \dfrac{a(2a+b)}{4EI} & \dfrac{(4a+b)}{4EI} \end{bmatrix}$$

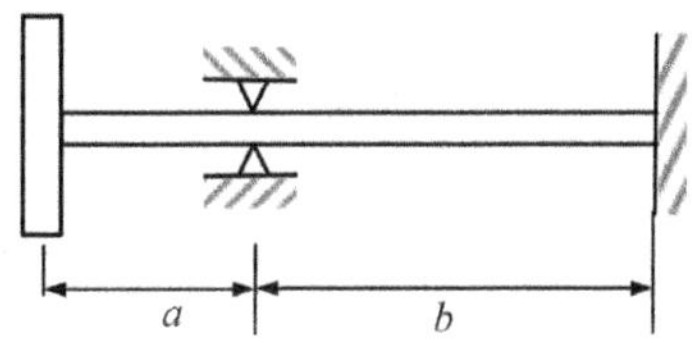

FIGURE 2.56 An overhung rotor system with a fixed support.

11. For a flexible shaft on simply supported bearings (refer to Figure 2.57), due to a point load P at disk the deflection at any point along the length, z, is given by

$$y(z) = \frac{Pbz}{6EIl}(l^2 - z^2 - b^2) \text{ for } 0 < z < a$$

with $\varphi_x = \dfrac{dy}{dz} = \dfrac{Pb}{6EIl}(l^2 - 3z^2 - b^2)$ for $0 < z < a$

$$y(z) = \frac{Pb}{6EIl}\left\{\frac{l}{b}(z-a)^3 + (l^2 - b^2)z - z^3\right\} \text{ for } a < z < l$$

with

$$\varphi_x = \frac{dy}{dz} = \frac{Pb}{6EIl}\left\{\frac{3l}{b}(z-a)^2 + (l^2 - b^2) - 3z^2\right\} \text{ for } a < z < l$$

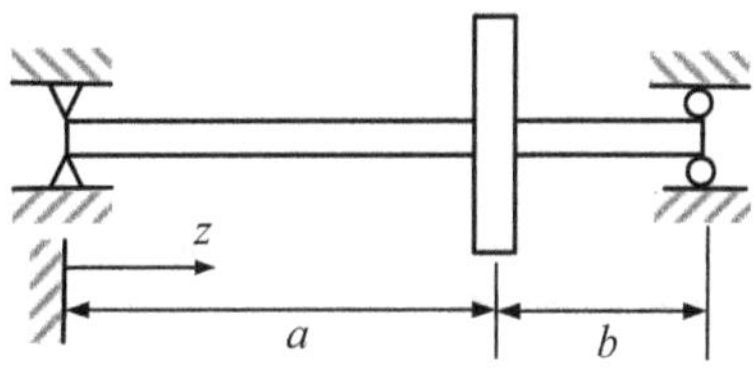

FIGURE 2.57 A simply supported shaft.

12. For fixed-fixed support (refer to Figure 2.58)

$$\begin{bmatrix} \alpha_{yf} & \alpha_{yM} \\ \alpha_{\varphi f} & \alpha_{\varphi M} \end{bmatrix} = \begin{bmatrix} \dfrac{a^3b^3}{3EIl^3} & \dfrac{a^4b^2(b-a)}{2EIl^3} \\ \dfrac{a^4b^2(b-a)}{2EIl^3} & \dfrac{a^2b^2}{3EIl^3} \end{bmatrix}$$

$$\alpha_{yf} = \frac{a^3b^3}{3EIl^3} \text{ with } l = a + b$$

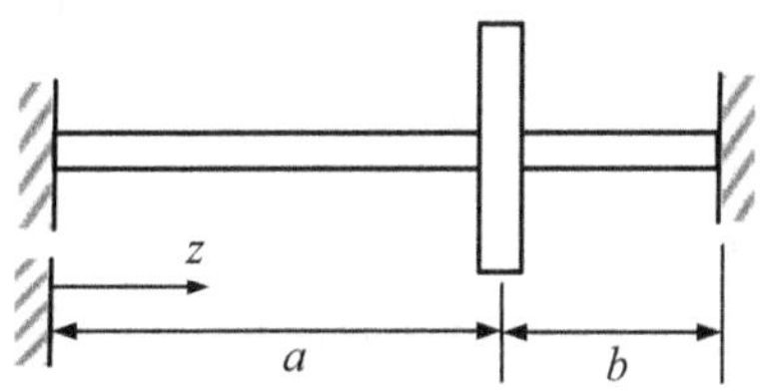

FIGURE 2.58 A fixed-fixed support rotor system.

13. For the fixed-fixed support with load at a midspan (Figure 2.59)

$$\begin{bmatrix} \alpha_{yf} & \alpha_{yM} \\ \alpha_{\varphi f} & \alpha_{\varphi M} \end{bmatrix} = \begin{bmatrix} \dfrac{l^3}{192EI} & 0 \\ 0 & \dfrac{l}{24EI} \end{bmatrix}$$

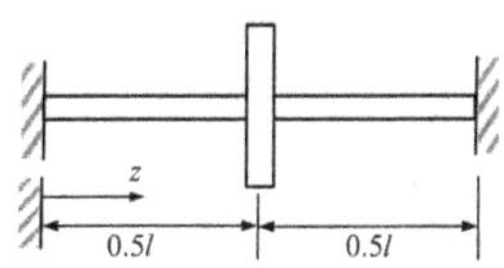

FIGURE 2.59 A fixed-fixed support with load at the midspan.

(*Continued*)

(Continued)

S.N.	Formulas	Descriptions
14.	For a fixed-fixed support, due to a point load P at the disk, the deflection is given as (refer to Figure 2.58) $y=\frac{Pb^2z^2}{6EIl^3}\{3al-z(3a+b)\}$ for $0<z<a$; $\varphi_x=\frac{dy(z)}{dz}=\frac{Pb^2z}{6EIl^3}\{6al-3z(3a+b)\}$ for $0<z<a$. For a fixed-fixed support, due to a point moment M at the disk, the deflection is given as (refer to Figure 2.58) $y=\frac{Mbz^2}{2EIl^2}(2a-b)-\frac{Mabz^3}{EIl^3}$ for $0<z<a$ with $\varphi_x=\frac{dy}{dz}=\frac{Mbz}{EIl^2}(2a-b)-\frac{3Mabz^2}{EIl^3}$ for $0<z<a$. $\begin{bmatrix}\alpha_{yf} & \alpha_{yM}\\ \alpha_{\varphi f} & \alpha_{\varphi M}\end{bmatrix}=\begin{bmatrix}\frac{a^3b^2}{6EIl^3}\{3l-(3a+b)\} & \frac{a^2b^2}{2EIl^3}\{2a^2l-a^2(3a+b)\}\\ \frac{a^2b^2}{2EIl^3}\{2a^2l-a^2(3a+b)\} & -\frac{ab(a^2+b^2-ab)}{EIl^3}\end{bmatrix}$	
15.	For disk locations 1 and 2 (refer to Figure 2.60) (tilting ignored): $\begin{bmatrix}\alpha_{y_1f_1} & \alpha_{y_1f_2}\\ \alpha_{y_2f_1} & \alpha_{y_2f_2}\end{bmatrix}=\begin{bmatrix}\frac{l^3}{8EI} & \frac{l^3}{32EI}\\ \frac{l^3}{32EI} & \frac{l^3}{48EI}\end{bmatrix}$ where l is the length as shown in adjacent figure and 1 and 2 refer to disk location.	Herein, α_{ij} represents the displacement (subscript: y—translational) at the ith station due to a unit force (subscript: f—force) at the jth station keeping all other forces to zero. 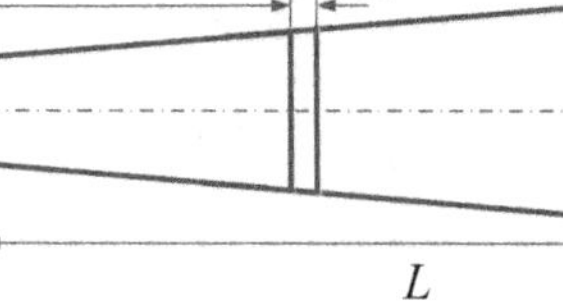 **FIGURE 2.60** An overhung rotor system.
16.	Tapered shaft (refer to Figure 2.61) Diameter: $d(z)=d_1+\frac{d_2-d_1}{L}z$. Second polar moment of area: $J(z)=\frac{\pi d^4(z)}{32}=\frac{\pi}{32}\left(d_1+\frac{d_2-d_1}{L}z\right)^4$. Stiffness: $k_t=\frac{3\pi G(d_2d_1)^3}{32L(d_2^2+d_1d_2+d_1^2)}$. Maximum shear stress at the smaller end ($d=d_1$): $\tau_{max}=\frac{16T}{\pi d_1^3}$	 **FIGURE 2.61** A tapered shaft.

(Continued)

(Continued)

S.N.	Formulas	Descriptions
17.	For a flexible shaft on simply supported bearings (refer to Figure 2.62), the deflection at any point along the length, z, due to a point moment M at the disk is given by	FIGURE 2.62 A simply supported shaft.

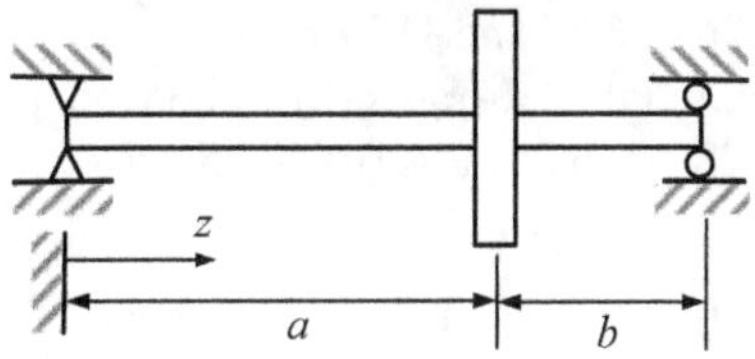

FIGURE 2.62 A simply supported shaft.

$$y(z)=\frac{Mz}{6EIl}(3a^2-6al+2l^2+z^2)\text{ for }0<z<a$$

$$y(z)=\frac{M(l-z)}{6EIl}(2lz-3a^2+z^2)\text{ for }a<z<l$$

with

$$\varphi_x(z)=\frac{dy(z)}{dz}=\frac{M}{6EIl}(3a^2-6al+2l^2+3z^2)\text{ for }0<z<a$$

$$\varphi_x(z)=\frac{M}{6EIl}(3a^2+2l^2-6lz+3z^2)\text{ for }a<z<l$$

ANSWERS TO SELECTED EXERCISE PROBLEMS

2.1 $\omega=\sqrt{1/(1-2\zeta^2)}\omega_{nf}$

2.2 $\zeta<\dfrac{1}{\sqrt{2}}$

2.3 $\omega_{nf_1}=62.17$ rad/s and $\omega_{nf_2}=160.53$ rad/s

2.4 $\omega_{nf_1}=25.47$ rad/s and $\omega_{nf_2}=242.97$ rad/s

2.5 $\omega_{nf_1}=25.47$ rad/s and $\omega_{nf_2}=242.97$ rad/s

2.6 Case I: pure translational motion $\omega_{nf}=1170.60$ rad/s, Case II: pure rotational motion $\omega_{nf}=6526.93$ rad/s

2.7 Not applicable

2.8 Not applicable

2.9 $\omega_{nf_1}=70.55$ rad/s and $\omega_{nf_2}=70.78$ rad/s

2.10 Case I: pure translational motion $\omega_{nf}=\sqrt{\dfrac{48EI}{ml^3}}$, Case II: pure rotational motion $\omega_{nf}=\sqrt{\dfrac{12EI}{I_d l}}$

2.11 $d=0.0288$ m

2.12 $a=1.0067$ m

2.13 $d=4.776$ mm

2.14 $d=5.222$ mm

2.15 $\omega_{nf_1}=\sqrt{1/(\alpha_{yf}m)}=\sqrt{48EI/(ml^3)}$ and $\omega_{nf_2}=\sqrt{1/(\alpha_{\varphi M}I_d)}=\sqrt{12EI/(I_d l)}$

2.16 $\omega_{nf_1}=28.0$ rad/s and $\omega_{nf_2}=245.7$ rad/s

2.17

(a) coupled system $\omega_{nf}=2030.6$ rad/s

(b) coupled system $\omega_{nf_1}=1320.9$ rad/s and $\omega_{nf_2}=3488.2$ rad/s

2.18 $m\ddot{x}+kx+c\dot{x}=me\omega^2\cos\omega t$ and $m\ddot{y}+ky+c\dot{y}=me\omega^2\sin\omega t$

2.19 $m\ddot{x}+k_{11}x+k_{12}\varphi_y=me\omega^2\cos\omega t;\ m\ddot{y}+k_{11}y+k_{12}\varphi_y=me\omega^2\sin\omega t$

$I_d\ddot{\phi}_y+k_{22}\phi_y+k_{12}x=0;\ I_d\ddot{\varphi}_x+k_{22}\varphi_x+k_{12}y=0$

2.20 $\omega_c=\sqrt{\dfrac{k_1+k_2}{2m}},\ \bar{y}_{max}=\left|\dfrac{y}{e}\right|=\dfrac{k_1+k_2}{k_1-k_2}.$

2.21 $\omega_{nf}=14.0$ rad/s

2.22 $\begin{bmatrix} m & 0 \\ 0 & I_d \end{bmatrix}\begin{Bmatrix} \ddot{y} \\ \ddot{\varphi}_x \end{Bmatrix}+\begin{bmatrix} k_{11} & k_{12} \\ k_{21} & k_{22} \end{bmatrix}\begin{Bmatrix} y \\ \varphi_x \end{Bmatrix}=\begin{Bmatrix} me\omega^2\sin\omega t \\ 0 \end{Bmatrix}$

2.23 $M = I_d \omega^2 \alpha$
2.24 Not applicable
2.25 227.03 rad/s
2.26 36.87 and 115.75 rad/s
2.27 (a) 22.24 rad/s (b) 124.35 rad/s
2.28 (i) 6.26 rad/s (ii) 10.95 rad/s (iii) 17.32 rad/s (iv) 6.71 rad/s
2.29

(i) B	(ii) C	(iii) C	(iv) C	(v) D	(vi) B	(vii) C
(viii) B	(ix) D	(x) B	(xi) D	(xii) B	(xiii) D	
(xiv) B	(xv) C	(xvi) A	(xvii) C	(xviii) C	(xix) D	
(xx) A	(xxi) F	(xxii) B	(xxiii) B	(xxiv) C	(xxv) B	
(xxvi) C	(xxvii) A	(xxviii) D	(xxix) A	(xxx) C	(xxxi) B	
(xxxii) A	(xxxiii) A	(xxxiv) C	(xxxv) C	(xxxvi) A	(xxxvii) D	

REFERENCES

Admas, M.L., Jr., 2001, *Rotating Machinery Vibration: From Analysis to Troubleshooting*. New York, NY: Marcel Dekker.

Biezeno, C. and Grammel, R., 1959, *Engineering Dynamics, Vol* III. *of Steam Turbines*. New York, NY: D. Van Nostrand Company.

Chen, W.J. and Gunter, E.J., 2005, *Introduction to Dynamics of Rotor-Bearing Systems*. Victoria, BC: Trafford Publishing.

Childs, D., 1993, *Turbomachinery Rotordynamics: Phenomena, Modeling and Analysis*. New York, NY: Research Studies Publication, A Wiley-Interscience Publication.

Dimentberg, F.M., 1961, *Flexural Vibrations of Rotating Shafts*. London: Butterworths.

Dimargonas, A.D. and Paipetis, S.A., 1983, *Analytical Methods in Rotor Dynamics*. London: Applied Science Publications.

Friswell, M.I., Penny, J.E.T., Garvey, S.D., and Lees, A.W., 2010, *Dynamics of Rotating Machineries*. Cambridge: Cambridge University Press.

Genta, G., 1999, *Vibration of Structures and Machines*, 3rd ed. New York, NY: Springer.

Genta, G., 2005, *Dynamics of Rotating Systems*. New York, NY: Springer.

Goodwin, M.J., 1989, *Dynamics of Rotor-Bearing Systems*. Sydney: Unwin Hyman.

Jeffcott, H.H., 1919, The lateral vibration of loaded shafts in neighbourhood of a whirling speed: The effect of want of balance, *Philosophical Magazine*, Series 6, 37, 304–314.

Krämer, E., 1993, *Dynamics of Rotors and Foundations*. New York, NY: Springer-Verlag.

Lalanne, M. and Ferraris, G., 1998, *Rotordynamics Prediction in Engineering*. New York, NY: Wiley.

Lee, C.-W., 1993, *Vibration Analysis of Rotors*. London: Kluwer Academic Publishers.

Mahrenholtz, O. (eds.), 1984, *Dynamics of Rotors; Stability and System Identification*. New York, NY: International Center for Mechanical Science.

Muszynska, A., 2005, *Rotordynamics (Dekker Mechanical Engineering)*, Vol. 188. Boca Raton, FL: CRC Press.

Prabhu, B.S. and Sekhar, A.S., 2008, *Dynamic Analysis of Rotating Systems and Applications*. Essex: Multi-Science.

Rankine, W.J.M., 1869, On the centrifugal force of rotating shaft, *The Engineer*, 27, 249.

Rao, J.S., 1996, *Rotor Dynamics*, 3rd ed. New Delhi: New Age.

Rieger, N.F., 1977, *Vibrations of Rotating Machinery*. Clarendon Hills, IL: The Vibration Institute.

Robert, B.M., 2003, *Rotating Machinery: Practical Solutions to Unbalance and Misalignment*. New York, NY: CRC Press.

Thomson, W.T. and Dahleh, M.D., 1998, *Theory of Vibration with Applications*, 5th ed. New Delhi: Pearson Education.

Timoshenko, S.P. and Young, D.H., 1968, *Elements of Strength of Materials*, 5th ed. New Delhi: East-West Press.

Tondl, A., 1965, *Some Problems of Rotor Dynamics*. London: Chapman & Hall.

Vance, J.M., 1988, *Rotordynamics of Turbomachinery*. New York, NY: John Wiley & Sons.

Vollan, A. and Komzsik, L., (2025), *Computational Techniques of Rotor Dynamics with the Finite Element Method* (2nd ed.). CRC Press, USA.

Yamamoto, T. and Ishida, Y., 2001, *Linear and Nonlinear Rotordynamics: A Modern Treatment with Applications*. New York, NY: Wiley.

3 Rotordynamic Parameters of Bearings, Seals, and Dampers

In Chapter 2, rotor systems were analyzed for simple cases by considering the shaft to be flexible and the bearing to be rigid. However, in the real world, bearings do provide flexibility to the rotor system. Depending upon the type of bearing or other similar machine elements (e.g., dampers, seals, etc.), these provide the stiffness and/or the damping to the rotor system (or sometimes added-mass or inertia terms). In the present chapter, the procedure to theoretically/numerically obtain these dynamic properties of the support system will be explained. First, rolling element bearings are introduced and classified and the relationship between the load and the deformation is derived from the basic Hertzian contact theory. From the load–deformation relation, then, stiffness of the rolling element bearing is obtained. Hydrodynamic radial bearings are introduced and classified according to the shape of the bore and lubricant groove positions. To obtain the stiffness and damping coefficients of the hydrodynamic radial bearing, a procedure is described using the basic Reynolds equation. For a simple case based on the short bearing assumption, closed-form analytical rotordynamic parameter expressions are presented. Next, dynamic seals are introduced and classified based on various categories. Based on bulk-flow models, governing equations are presented, and for a simple case, the dynamic parameters are obtained and analyzed for various operating parameters. The basics of squeeze-film dampers (SFDs) are introduced and relevant rotordynamic parameters are presented. Finally the foil bearing is introduced regarding its various configurations in use and rotordynamic parameter calculation. Active magnetic bearings are covered in Chapter 8 of Volume II. The main aim of the present chapter is not to provide exhaustive coverage of bearings, seals, and dampers, because each of them is a subject in itself. The basic premise of this chapter is to acquaint readers with the rotordynamic parameters of these machine elements so that, if needed, a detailed analysis could be performed through the available textbooks on these topics. Hydrostatic bearings, gas bearings, etc., have not been dealt with here; however, these do impart similar rotordynamic characteristics to the rotor systems. Chapter 4 of Volume II describes the experimental procedures to obtain rotor dynamic coefficients for all such elements, which are often more reliable and preferred in industries.

All rotating machineries are supported by bearings and often fitted with seals and dampers. The bearings clearly constitute a vital component in any turbomachine, and a good understanding of their dynamic properties is a prerequisite to the prediction of the machine's behavior. The influence of bearings on the performance of rotor-bearing systems has been recognized for many years. One of the earliest attempts to model a journal bearing was reported by Stodola (1925) and Hummel (1926). They represented the fluid film as a simple spring support, but their model was incapable of accounting for the observed finite amplitude of oscillation of a shaft operating at a critical speed. Concurrently, Newkirk (1924) and Newkirk and Taylor (1925) described the phenomenon of bearing-induced instability, which he called oil whip, and it soon occurred to several investigators that the problem of rotor stability could be related to the properties of the rotor dynamic parameters (RDPs), also known as dynamic parameters; bearings/seals force (or moment) coefficients; added-mass, damping, and stiffness coefficients; linearized RDPs; and dynamic impedances.

Seals are mainly used to reduce the leakage of working and lubricating fluids through the interface between the machine's moving parts and the housing. Some leakage is inevitable, and it results in axial fluid velocities through the seal in the direction of the pressure drop. The main difference between seals and bearings is that the former does not support load but the latter does. The present-day requirement of critical sealing applications has a diverse range of operating condition requirements such as (i) cryogenic temperature, (ii) hard vacuum, (iii) ultra-clean systems, (iv) leakage

DOI: 10.1201/9781003661559-3

control of the order of 10^{-12} cc/s, (v) pressures over 100 bar, (vi) temperatures exceeding 800°C, (vii) hard-to-handle liquids and gases, (viii) high-pressure pulsations, and (ix) rotor speeds as high as 10^5rpm. These extreme conditions of seals are challenging tasks in the aviation and aerospace industries. The importance of calculations of RDPs of seals arose in the late 1970s with regard to vibration problems related to the high-pressure oxygen turbo-pump of the Space Shuttle's main engine. Compressors used in many industries also had instability problems within the operating speed range. Seals in high-speed operation of turbomachines can lead to instability (refer to Chapter 1 of Volume II for details). The main factor that governs the instability is the RDPs of the seals. These RDPs are greatly dependent on many physical and mechanical parameters such as lubricant and working fluid temperatures, pressure drop, seal clearances, surface roughness and patterns, rotor speeds, eccentricity, and misalignments.

Designers should know the types of bearings, dampers, and seals that could be used and the performance characteristics associated with each of them. The focus of this chapter will be the operation of bearings, dampers, and seals under the action of dynamic loads; in particular, the RDPs of different types of bearings, dampers, and seals are discussed. These characteristics have a major influence on the overall system dynamics. Bearing types are (i) rolling element bearings, (ii) hydrodynamic bearings, (iii) gas bearings, (iv) squeeze-film bearings/dampers, (v) hydrostatic bearings, (vi) active bearings, and (vii) foil bearings. Dynamic seals can be classified as (i) plain seals, (ii) roughened seals, (iii) contact seals, and (iv) brush seals. Table 3.1 compares some of the criteria of selection between the rolling and journal (or fluid-film) bearings, which find frequent application in rotating machinery.

Developments in bearing technology: A *hybrid bearing* is a combination of two or more bearings as a single bearing unit—for example, a fluid-film journal bearing that combines the physical mechanisms of both hydrostatic and hydrodynamic bearings. Such hybrid bearings are being considered as alternatives to rolling element bearings for future cryogenic turbo-pumps. The hydrostatic characteristics of a hybrid bearing allow it to be used with low-viscosity fluids that could not adequately carry a load with purely hydrodynamic action. Another example is the combination of a hydrodynamic bearing and an active magnetic bearing, which takes advantage of the high load capacity of the former and the active tuning of rotordynamic parameters of the latter. Even the combination of a foil bearing and an active magnetic bearing has been a promising research area, especially for very-high-speed applications.

TABLE 3.1
Criteria for Selection of the Bearing Type

S.N.	Parameter	Rolling Bearing	Journal Bearing
1	Starting friction	Low	High
2	Space	Less axial space	Less radial space
3	Load type	Both radial and axial	Radial or axial
4	Failure time	Gradual	Sudden
5	Clearance	Less	More
6	Displacement of shaft	Less	More
7	Assembly	Simple	Tedious
8	Maintenance	Less	More
9	Replacement	Cheaper	Costly
10	Load-carrying capacity	Moderate	High
11	Electrical insulation	No	Yes
12	Noise	High	Low
13	High overload	For short periods	For moderate periods
14	Speed	Very high	Moderate

A compliant surface *foil bearing* consists of a smooth top foil that provides the bearing surface and a flexible, corrugated foil strip formed by a series of bumps or metal mesh foil that provides a resilient support to this surface. Compared to conventional journal bearings, the advantages offered by the compliant surface foil bearing include its adaptation to shaft misalignment, variations due to tolerance buildups, centrifugal shaft growth, and differential expansion. Apart from the use of the conventional rolling element and fluid-film bearings and seals, which affect the dynamics of the rotor, the squeeze-film and magnetic bearings are often used to control the dynamics of such systems. Squeeze-film bearings are, in effect, fluid-film bearings in which both the journal and bearing are nonrotating. The ability to provide damping is retained, but there is no capacity to provide stiffness, as the latter is related to journal rotation. These are used extensively in applications where it is necessary to eliminate instabilities and to limit rotor vibration and its effect on the supporting structures of rotor-bearing systems, especially in jet engines. In recent years, advanced development of *electromagnetic bearing technology* has enabled the active control of rotor-bearing systems. In particular, the electromagnetic suspension of a rotating shaft without mechanical contact has allowed the development of supercritical shafts in conjunction with modern digital control strategies (refer to Chapter 8 of Volume II for details). With the development of *smart fluids* (for example electro- and magneto-rheological fluids), new controllable bearings are in the primitive development stage.

Now analyses of the rolling element bearings, fluid-film bearings, foil bearings, seals, and dampers will be described one by one.

3.1 ROLLING ELEMENT BEARINGS

Rolling element bearings, or simply rolling bearings, are the most common type of bearing and have around 94% usage in the industry (around 5.5% for hydrodynamic/static fluid-film journal bearings and less than 0.5% for other bearings including active bearings). It requires fewer boundary dimensions and can transmit heavy and variable loads of various forms. They can easily be installed, serviced, and replaced. Figure 3.1 shows a cut section of the ball and roller bearings. The nomenclature and various basic geometries of most simple deep groove ball bearings are shown in Figures 3.2 and 3.3, respectively. d is the bearing bore diameter, D is the bearing outer diameter, D_b is the ball diameter, r_i is the inner raceway curvature radius, r_o is the outer raceway curvature radius, d_i is the inner ring outer diameter, and d_o is the outer ring inner diameter. Various types of rolling element bearings (refer to Figure 3.4) are deep groove ball bearings, self-aligning ball bearings, angular contact ball bearings, split type angular contact ball bearings, cylindrical roller bearings, spherical roller bearings, tapered roller bearings, needle roller bearings, thrust ball/roller bearings, wire race (or slew) bearings, and linear recirculating ball bearings. These bearings are selected on the basis

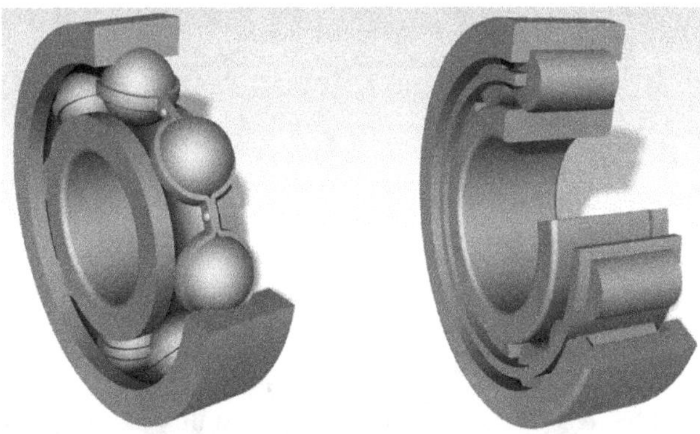

(a) (b)

FIGURE 3.1 A cut section of rolling bearings: (a) ball bearing; (b) roller bearing.

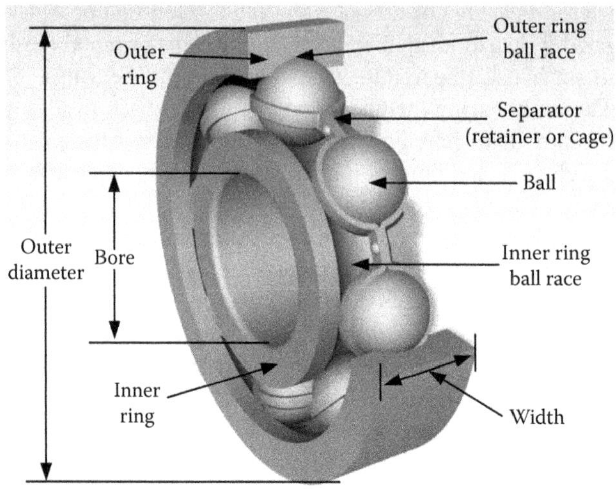

FIGURE 3.2 Nomenclature of a ball bearing.

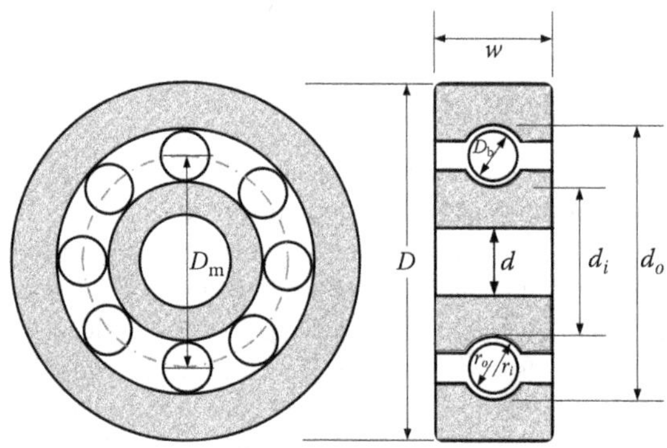

FIGURE 3.3 A line diagram of the radial ball bearing.

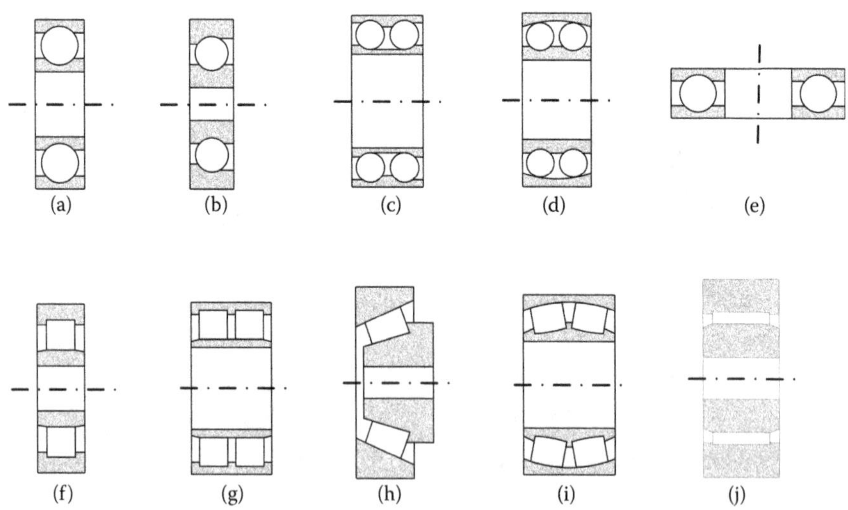

FIGURE 3.4 Types of ball and roller bearings: (a) deep groove; (b) angular contact; (c) double row angular contact; (d) self-aligning; (e) thrust; (f) cylindrical roller; (g) double row cylindrical roller; (h) taper roller; (i) double row spherical roller; and (j) needle roller.

of the magnitude and direction of the loading and speeds. Table 3.2 summarizes the various rolling bearing types available, together with their relative merits, common applications, and friction coefficients.

TABLE 3.2
Types of Rolling Bearings

Bearing Type	Suitable for Load/Speed Type (Friction Coefficient, μ)	Applications and Other Remarks
Deep groove ball bearings	Moderate radial and/or light axial load. Small contact angle, on the order of 10°. Very high speed (0.0015–0.003).[a]	To accommodate higher radial loads, it is available in double row form. To prevent dust and to prevent leakage of grease, they are provided with shields and seals. Used in gearboxes, electric motors, instruments and meters, household electric appliances, internal combustion engine, agriculture machinery, etc.
Self-aligning bearings	Lower radial and axial load-carrying capacity than deep groove ball bearing. Small contact angle. Moderate speeds (0.001–0.003).	Can accommodate large amounts of misalignment because the outer raceway is a part of a sphere whose curvature center is on the axis of the bearing. Used in textile machinery, steel-framed conveyors, agricultural machinery, etc.
Angular contact bearings	Larger axial loads than deep groove ball bearing because of the large contact angle on the order of 15°–35°. Able to carry pure axial or, for a combined case, the axial load at least as large as the radial load. Very high speed and precision conditions (0.0015–0.002).	Available in "matched pairs" to provide accurate axial positioning of shaft and to provide preloading by eliminating the axial clearance. They are used in grinding spindles, headstock spindles, electric motors at high speeds, centrifuges, gyroscopes, jet engines, etc.
Angular contact bearings (duplex pairs)	High load-carrying capacity. To be used when axial load greater than radial load. Moderate speeds (0.0024–0.003).	Available in double row form in face-to-face and back-to-back arrangements. It has a four-point contact. Applications are the same as for angular contact bearings.
Angular contact bearings with (split ring type)	To carry axial loads in both directions and to carry a certain amount of radial loads. Very high speeds (0.0015–0.003).	Used as aircraft gas turbine bearing. Available with a flanged outer ring to facilitate the mounting and locating structure. Available with inner and/or outer ring split. Sufficient lubricating and cooling are extremely important.
Thrust ball bearings	Axial loads only (can be in either direction). The contact angle of the bearings is 90°. Not suitable for high speeds (0.0012).	Axial load must always be present. Can function with a small amount of misalignment when used with a spherical seating washer. Available in cylindrical and needle roller form for very large loads. Used in spindles in machine tools, rotating table of lathes with vertical spindles, etc.
Cylindrical roller bearings	Large radial loads and very light axial loads. Compared to other roller bearings, it has the lowest friction coefficients and is suitable for high-speed conditions (0.001–0.003).	Separate inner and outer rings. Available in "match assembly" form for tight control over internal clearance. Used in medium or large electric motors, axle-boxes of rail vehicles, spindles of machine tools, rolling mills, etc.
Spherical roller bearings	Very high radial loads with shocks and light axial loads. Outer raceway is similar to self-aligning bearings. Used at moderate speeds due to high friction coefficients (0.002–0.003).	Nonseparable. Operate at lower speed than the cylindrical roller. Can accommodate misalignment. Used in rolling mills, paper making machines, construction machinery, crushers, printing machines, etc.

(Continued)

TABLE 3.2 (*Continued*)
Types of Rolling Bearings

Bearing Type	Suitable for Load/Speed Type (Friction Coefficient, μ)	Applications and Other Remarks
Tapered roller bearings	Can support very large radial and axial loads or pure axial loads. Contact angle between 10° and 30°. Moderate speeds (0.002–0.005).	Similar to angular contact ball bearings. Double or four row tapered roller bearings are used for extremely large radial and axial loads. Lower operating speed than angular contact bearings. Provides a very rigid shaft mounting. Used in front and rear wheels of trucks, transmission mechanisms in machine tools, axle-boxes of rail vehicles, rolling mills, lifters, etc.
Needle roller bearings	Can support large radial loads at speeds similar to those of cylindrical roller bearings. Roller diameter less than 5 mm, and the length-to-diameter ratio of the roller is more than 3. Used at low speeds due to roller skewing (0.002).	Very small bearing outer diameter. Can be used without one or both rings to save space provided the seatings are surface treated. Used in gearboxes, universal joints, crankshaft and connecting rods, hydraulic equipment, textile machines, applications where fluctuation of load is high, etc.
Thrust roller bearings	Carry high axial loads. Low speed of operation (cylindrical: 0.004, spherical: 0.003, needle: 0.004).	Available with the cylindrical, needle, tapered, and spherical rollers. Used in vertical electrical motors, propeller shafts, arm cranes, turntables in machine tools, etc.

[a] The number within the parentheses represents typical friction coefficients.

Bearing identification number: Identification numbers of rolling bearings are composed of the *basic numbers* and *the auxiliary codes*. The basic numbers are made up of a *bearing series* symbol and a *dimension series* symbol. For example, in the bearing basic number 6204, the various numbers are as follows: 4 represents diameter series (other options: 8, 9, 0, 1, 2, 3, and 4), 0 represents width series (other options: 1, 2, 3, 4, 5, and 6), and 62 represents bearing series (62 for radial ball bearing, 72 for angular contact ball bearing, 320 for tapered roller bearing, etc.). The auxiliary codes are attached to the basic number to identify *seal or shield codes, race configuration code, clearance code*, and *tolerance class codes* (Harris, 2001).

In rolling element bearings, very high speeds are those speeds with a DN number (this is defined as $d \times N$, where d is the bore diameter in mm and N is the bearing operating speed in rpm) larger than 3 million. For example, $d=30$ mm and $N=100{,}000$ rpm or $d=100$ mm and $N=30{,}000$ rpm or $d=600$ mm and $N=5000$ rpm would give the same DN number of 3 million. High-speed bearings find applications in aerospace and space technologies. One application of low-speed bearings is gyroscopes: They have a very long life on the order of 15 years and require very high-precision bearings. Lubrication for such bearings is quite challenging because they use about one to two drops of lubricant per year. Micro- and nano-pumps may have very useful applications for such a lubrication flow rate. Another application is a rotary kiln: It is used in cement factories and has a rotational speed of 2–3 rpm; however, the size of the bore diameter is on the order of 4 m or so.

Frictional torque and power: The rolling bearing frictional torque, M_t, can be obtained from

$$M_t = \frac{1}{2} fFd \tag{3.1}$$

where f is the coefficient of friction (a general average of about 0.001; see Table 3.2), d is the bore diameter of the bearing in m, and F is the radial or axial load as specified in N. The frictional power loss, P, is given as

$$P = \frac{2\pi M_t N}{60} \tag{3.2}$$

where N is the shaft angular speed in rev/min. The heat generation and temperature distribution at rolling contact is very important because the diametral clearance of the rolling bearing changes with the temperature. A more detailed treatment can be found in Harris (2001) and Kashyap and Tiwari (2006).

Estimation of elastic parameters of bearings involves establishing a relationship between the incident load on the bearing and its resultant deformation. The classical solution for the local stress and deformation of two elastic bodies apparently contacting at a single point was established by Hertz (1896). Hertz's analysis is applied to surface stresses caused by a concentrated force applied perpendicular to the surface. In determining contact deformation versus load, the concentrated load applied normally to the surface alone is considered (i.e., by neglecting the friction forces) for most rolling element bearing applications.

It is possible to determine how the applied load on the bearing is distributed among the balls or rollers (or rolling elements) by considering the equilibrium of rolling elements and the bearing. To do this, it is necessary to develop load–deflection relationships for rolling elements contacting raceways. Most rolling bearing applications involve steady-state rotation of either the inner or outer raceways or both. Rolling element centrifugal forces, gyroscopic moments, and the frictional forces and moments do not significantly influence this load distribution in most general-purpose applications. Theoretical models (Jones, 1946; Palmgren, 1959; Jones, 1960; Ragulskis et al., 1974; Gargiulo, 1980; El-Sayed, 1980; Goodwin, 1989; Stolarski, 1990; Johnson, 1991; Brändlein et al., 1999; Harris, 2001) are available to estimate bearing stiffness under static loading conditions in which the most challenging task is to obtain the load distribution among various rolling elements. In the next section, these analyses will be treated in the most simplistic way.

3.1.1 Linear Radial Stiffness of Rolling Bearings

For a given load, the amount of shaft displacement from its concentric position with respect to the bearing is of interest. This displacement depends upon the elastic deformation of the bearing raceways and of rolling elements themselves. The elastic deformation of these components depends upon the geometry and the material of the bearing and on the bearing internal clearance or preload (i.e., the negative clearance). The bearing internal clearance is an important factor because it determines the size of the stressed area of the rings (i.e., the load zone angle, Ψ_L) as shown in Figure 3.5. The rolling elements falling within the load zone will share the external load and remaining will be unloaded. The lesser the clearance (or the greater the preload), the more rolling elements will share the external applied load from the shaft. This results in lesser load per rolling element and hence lesser maximum contact stress. However, the preload (or negative clearance due to initial compression of rolling elements by shrink fit assembly) may reduce the life of bearings due to high-fatigue stresses in rolling elements, especially in high-temperature environments.

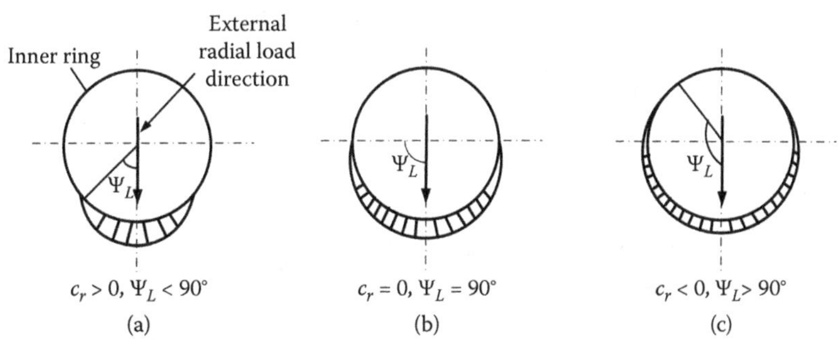

FIGURE 3.5 The load zone in rolling bearings; (a) positive clearance; (b) no clearance; (c) negative clearance (preload).

The Hertzian theory (Hertz, 1896) has been the major method for calculating the contact pressure in a rolling element bearing; it is based on the following assumptions:

- All contact deformation occurs in the elastic range and follows Hooke's law.
- The contact load is perpendicular to the surface; that is, contact surfaces are perfectly smooth and the friction force at surfaces is neglected.
- Sizes of the contact area are much smaller compared with the curvature radii of contact bodies.

For a given load, the size of the contact area (i.e., the elastic deformation between the rolling element and the raceway) determines the magnitude of stresses in the bearing components. In the case of a ball bearing, the contact area between the rolling element and the raceway is zero when no load is applied. This is a point contact scenario. For two bodies in a point contact (with a load the point contact would become an area contact, for example an ellipse for the ball bearing) made of the same material and subjected to a compressive load, F, from the Hertzian contact theory, we have

$$F = K_p \delta_p^{3/2} \quad \text{or} \quad \delta_p = \left(\frac{F}{K_p}\right)^{2/3} \tag{3.3}$$

where the subscript p represents the point contact, δ_p is the contact deformation in a point contact, and K_p is a load–deformation constant for a single point contact, which depends upon the material properties and the geometry of contacting surfaces (Changsen, 1991; Harris, 2001).

In the case of rolling element bearings, the elastic deformation takes place at both the inner raceway and the outer raceway with the rolling element (see Figure 3.6).

The total deformation, δ, at a single rolling element location is given by (noting the form of Equation 3.3 for each of the contact points at the inner and outer raceways)

$$\delta = \delta_{Pi} + \delta_{Po} = \left(\frac{F}{K_{Pi}}\right)^{2/3} + \left(\frac{F}{K_{Po}}\right)^{2/3} = \left(\frac{1}{K_{Pi}^{2/3}} + \frac{1}{K_{Po}^{2/3}}\right) F^{2/3} = \left(\frac{K_{Po}^{2/3} + K_{Pi}^{2/3}}{K_{Pi}^{2/3} K_{Po}^{2/3}}\right) F^{2/3} \tag{3.4}$$

where subscripts i and o represent the inner and outer raceways. Equation 3.4 can be rewritten as

$$F = K_{Pio} \delta^{3/2} \tag{3.5}$$

with

$$K_{Pio} = \frac{K_{Pi} K_{Po}}{\left(K_{Pi}^{2/3} + K_{Po}^{2/3}\right)^{3/2}} \tag{3.6}$$

where K_{Pio} is the load–deformation constant for two-point contacts of a ball with raceways. K_{Pio} (units in N-mm$^{-3/2}$) depends mainly upon the bearing's curvature ratio f as defined in Equation 3.7

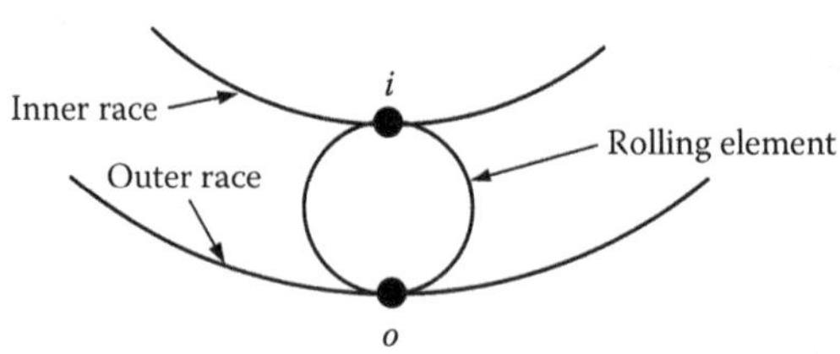

FIGURE 3.6 Contacts of a ball with the inner and outer raceways.

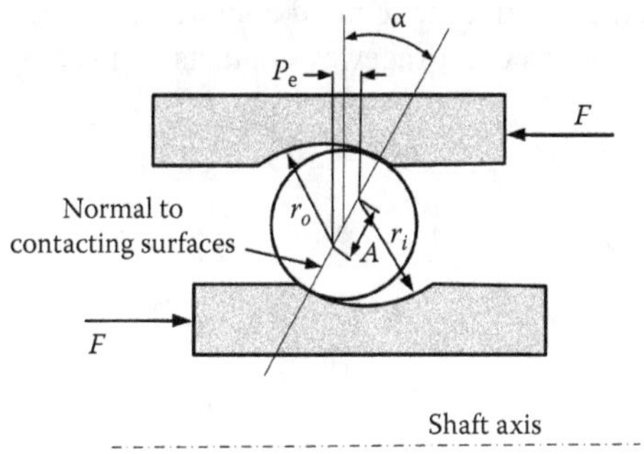

FIGURE 3.7 The contact angle in the ball bearing.

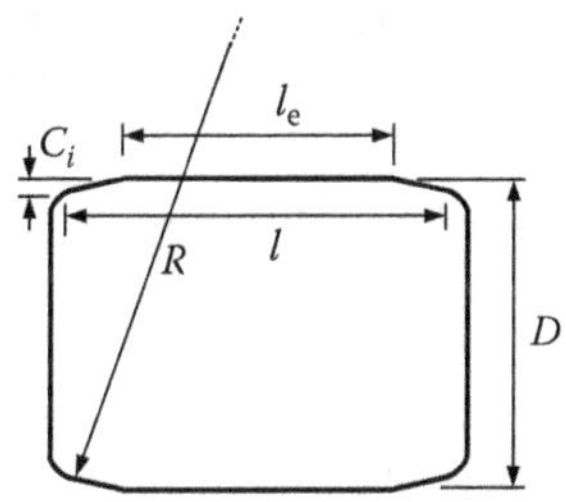

FIGURE 3.8 A crowned roller showing the crown radius and the effective length.

and other terms, that is, D_b, D_m (Figure 3.3), and α, where D_m is the pitch diameter of the bearing and α is the contact angle (refer to Figure 3.7). The curvature ratio is defined as

$$f = \frac{(2r - D_b)}{D_b} \tag{3.7}$$

where r is the radius of curvature of the raceway in mm with subscripts i or o for the inner raceway or outer raceway, respectively.

For detailed accurate calculation of $K_{P_{io}}$, readers are referred to examples in subsequent sections. If the dependence of $K_{P_{io}}$ on the bearing size, D, and contact angle, α, is ignored, then for $f_{inner} \approx f_{outer} < 0.1$ (which is usually the case for deep groove ball bearings) values of $K_{P_{io}}$ may be obtained directly as (Palmgren, 1959)

$$K_{P_{io}} = \frac{34,300}{f^{0.35}} D_b^{1/2} \tag{3.8}$$

For roller bearings, the corresponding approximate load–deformation constant $K_{l_{io}}$ (subscript l represents the line contact of roller) depends only on the effective length of roller, l_e, and it is approximated as (refer to Equation 3.11)

$$K_{l_{io}} = 26,200 l_e^{0.52} \tag{3.9}$$

The effective length of rollers is that which is actually in contact with the raceway—usually this is equal to the actual roller length minus the roller corner radii as shown in Figure 3.8. The relationship between the load–deformation constant, the applied compressive load, and the elastic deformation at a single roller due to the inner and outer raceway contacts is given by

$$F = K_{l_{io}} \delta^{10/9} \tag{3.10}$$

with

$$K_{l_{io}} = \frac{K_{l_i} K_{l_o}}{\left(K_{l_i}^{9/10} + K_{l_o}^{9/10}\right)^{10/9}} \tag{3.11}$$

In roller bearings, there is a *line contact* between the rolling element and bearing raceways at no load conditions. Equations 3.5 and 3.10 indicate relationships of the elastic deformation at a rolling element under the action of a single compressive force applied at the rolling element (i.e., of the ball and roller, respectively). In real bearings, where more than one rolling element is in compression (i.e., all the rolling elements within the load zone), the effect at each element should be incorporated as shown in Figure 3.9.

When the bearing inner ring is displaced from the concentric position by a distance, x_m, with respect to the outer ring center, part of it consists of the radial clearance, c_r. The elastic deformation in the direction of applied radial load will be

$$\delta_0 = x_m - c_r \tag{3.12}$$

and the elastic deformation of any rolling element at an azimuth angle, ψ, as Figure 3.9b, is given by

$$\delta(\psi) = x_m \cos\psi - c_r \tag{3.13}$$

By setting deformation to zero in Equation 3.13, the load zone, ψ_L, can be obtained by

$$\psi_L = \cos^{-1}\left(\frac{c_r}{x_m}\right) \tag{3.14}$$

From the equilibrium of the ring (inner or outer), we have (Figure 3.9)

$$F_r = \sum_{j=1}^{Z} F(\psi_j) \tag{3.15}$$

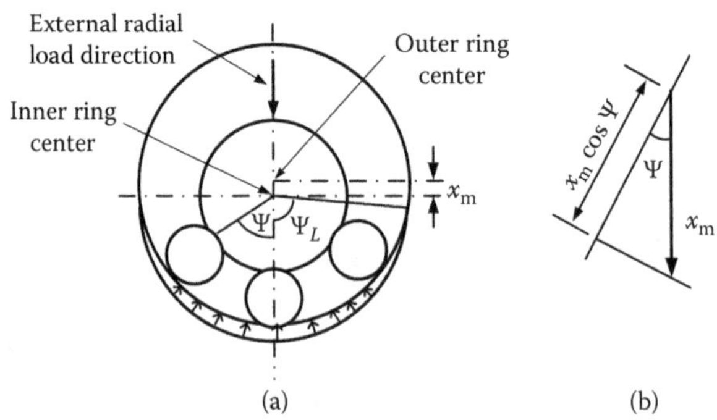

FIGURE 3.9 (a) Displacement of the inner ring with respect to the outer ring. (b) Displacement of a rolling element located at azimuth angle, ψ.

where Z is the number of rolling elements, F_r is the external radial load and F is the load shared by a rolling element at some azimuth angle. For a given bearing with known angular positions of rolling elements (i.e., in general for the analysis, rolling elements are arranged symmetrically to the loading such that one of the rolling elements is just below the radial external load), a displacement of the inner ring x_m is assumed, and thus the resulting elastic deformation, $\delta(\psi)$, at each rolling element is calculated from Equation 3.13. Contact forces at each element may then be evaluated from Equations 3.5 and 3.8. These are added as vectors to give the net radial forces, F_r (i.e., Equation 3.15), applied to the bearing in order to produce the assumed displacement x_m. When there is a difference between the estimated net radial force and applied radial force, a new displacement is chosen and the procedure is repeated until it converges up to the desired accuracy. It can be observed that load distribution calculation is an indeterminate problem. For a single load, we need to obtain multiple support reactions and hence deformation information is used.

If the fraction of the net radial load applied that is transmitted through the rolling element directly in line with the applied load is known, then the resulting inner ring displacement x_m may be calculated directly from Equation 3.5 for the ball bearing as

$$x_m = \delta_0 + c_r \tag{3.16}$$

with

$$\delta_0 = \left(\frac{F_m}{K_{pio}} \right)^{2/3} \tag{3.17}$$

where F_m is the force on the rolling element directly in line with the applied radial load (usually it will share the maximum load) and c_r is the internal clearance. Similarly, for the roller bearing, we have

$$x_m = \left(\frac{F_m}{K_{lio}} \right)^{9/10} + c_r \tag{3.18}$$

Approximate relationships (ignoring the effect of bearing clearance and geometry), respectively, of Equations 3.17 and 3.18 given by Palmgren (1959) are

$$x_m = 4.36 \times 10^{-8} \left(\frac{F_m^2}{D_b} \right)^{1/3} \quad \text{for ball bearings} \tag{3.19}$$

and

$$x_m = 3.06 \times 10^{-10} \left(\frac{F_m^{0.9}}{l_e^{0.8}} \right) \quad \text{for roller bearings} \tag{3.20}$$

where F_m is in N, and D_b and l_e are in m. Before Equations 3.16 through 3.20 can be applied, the maximum compression force, F_m, on a single element must first be determined. The maximum compressive force, F_m, and the net radial force applied, F_r, are approximately related to

$$F_m = \frac{c_1}{Z} F_r \tag{3.21}$$

where Z is the number of rolling elements and c_1 is a number that depends upon the applied load, the number of rolling elements, the load–deformation constant (K_{pio} or K_{lio}), and the bearing clearance. For approximate calculations, we have $c_1 \approx 4.37$ and 4.08, respectively, for the ball and roller bearings.

On substituting Equation 3.21 into Equations 3.16 and 3.18, we get

$$x_m - c_r = \left(\frac{4.37F_r}{ZK_{pio}}\right)^{1/1.08} \text{ for ball bearings} \tag{3.22}$$

and

$$x_m - c_r = \left(\frac{4.08F_r}{ZK_{lio}}\right)^{1/1.08} \text{ for roller bearings} \tag{3.23}$$

Now that the relationship between the applied radial force and the bearing elastic deformation has been established, the bearing stiffness can be calculated. The stiffness of the rolling bearings is highly nonlinear with the displacement of the inner ring with respect to the outer ring. On substituting $F = F_m$ and $\delta = x_m - c_r$ into Equations 3.5 and 3.10, we get

$$F_m = K_{pio}\left(x_m - c_r\right)^{3/2} \text{ for ball bearings} \tag{3.24}$$

and

$$F_m = K_{lio}\left(x_m - c_r\right)^{1.08} \text{ for roller bearings} \tag{3.25}$$

Equations 3.24 and 3.25 give the load–deformation relationship for a single ball, which shares the maximum load (or the ball which is in line with the radial external load). The bearing stiffness $k(x_m)$ can be obtained by differentiating F_r (noting Equation 3.21) with respect to displacement x_m to give

$$k\left(x_m\right) = \frac{dF_r}{dx_m} = \frac{Z}{4.37}\frac{dF_m}{dx_m} = 1.5\frac{Z}{4.37}K_{pio}\left(x_m - c_r\right)^{0.5} \text{ for ball bearings} \tag{3.26}$$

and

$$k\left(x_m\right) = \frac{dF_r}{dx_m} = \frac{Z}{4.08}\frac{dF_m}{dx_m} = 1.08\frac{Z}{4.08}K_{lio}\left(x_m - c_r\right)^{0.08} \text{ for roller bearings} \tag{3.27}$$

The stiffness expressions are functions of x_m, which varies with the steady load applied to the bearing. It should be noted that we effectively linearize the bearing stiffness at the static equilibrium. This is based on the assumption of small oscillations due to the small unbalance. If dynamic forces are large (due to large unbalance), then large amplitudes of vibration are bound to occur at the bearing. In this case, the bearing stiffness nonlinearity plays an important role in machine response calculations. In subsequent sections, the procedures for calculating these stiffnesses are described.

This discussion has made no reference to the effects of movement of rolling elements around raceways on the bearing force–displacement relationship and the stiffness. In fact, although the number and location of rolling elements in the load zone change as the bearing is rotated, the displacement and the stiffness of the bearing are little affected. Hence, the rotation of the bearing need not be allowed for stiffness calculations. For more detail in this regard, readers are referred to Ragulskis et al. (1974). The stiffness characteristics of the rolling element bearing are studied under combined loading, that is, under the action of the radial and axial forces, and the moment; hence, there should be corresponding radial, axial, and angular displacements. All the displacements are coupled, that is, they depend upon the combined effect of all the forces and moments acting on the bearing. The stiffness matrix thus obtained is of the size 5×5. To obtain the displacement, a set of five nonlinear coupled equilibrium equations are solved for a given loading (Lim and Singh, 1990a, b, 1991, 1992, 1994).

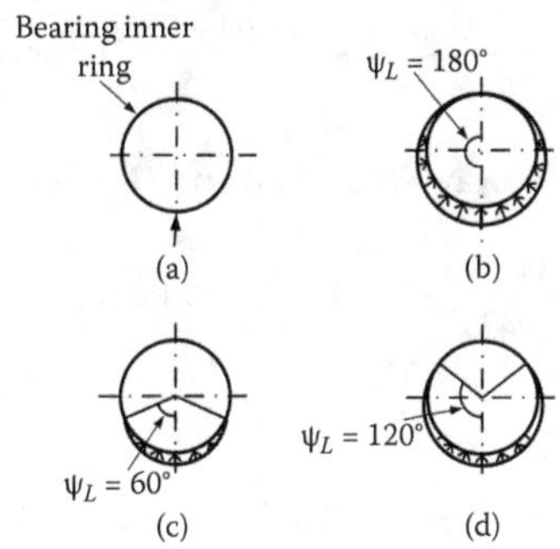

FIGURE 3.10 (a) Case I, load zone for $\psi_L = 0°$. (b) Case II, load zone for $\psi_L = 180°$. (c) Case III, load zone for $\psi_L = 60°$. (d) Case IV, load zone for $\psi_L = 120°$.

Example 3.1

Obtain the load zone angle of a rolling bearing for the following cases: (i) $c_r = x_m$, (ii) $c_r = -x_m$, (iii) $c_r = 0.5x_m$, and (iv) $c_r = -0.5x_m$, where x_m is the displacement of the inner ring from the concentric position and c_r is the radial clearance. Draw load zones for each case.

Solution:

Case I: We have $c_r = x_m$. Thus, the load zone angle $\psi_L = \cos(c_r/x_m) = \cos^{-1}(1) = 0°$. Hence, none of the bearing elements are loaded (Figure 3.10a).

Case II: We have $c_r = -x_m$. Thus, $\psi_L = \cos^{-1}(c_r/x_m) = \cos^{-1}(-1) = 180°$. Hence, all rolling elements will take part in the load sharing (Figure 3.10b) (observe the load zone region).

Case III: We have $c_r = 0.5x_m$. Thus, $\psi_L = \cos^{-1}(c_r / x_m) = \cos^{-1}(0.5) = 60°$. Hence, very few rolling elements will take part in the load sharing (Figure 3.10c).

Case IV: We have $c_r = -0.5x_m$. Thus, $\psi_L = \cos^{-1}(c_r / x_m) = \cos^{-1}(0.5) = 120°$. Hence, a sufficient number of rolling elements will take part in the load sharing (Figure 3.10d).

Example 3.2

The specifications of an SKF 7218 angular contact ball bearing are as follows: inner and outer grove curvature radius = 11.64 mm, inner raceway diameter = 102.92 mm, outer raceway diameter = 147.9 mm, bore diameter = 90.0 mm, outer diameter = 160.0 mm, ball diameter = 22.25 mm, number of balls = 16, and preload = 0.6 μm. Plot the bearing radial stiffness versus the elastic radial deformation. For plotting purpose, take six points, that is, take elastic deformations 0.002, 0.006, 0.01, 0.014, 0.018, and 0.022 mm.

Solution:

From Equation 3.26, for the stiffness of the ball bearing, we have

$$k(x_m) = 1.5\frac{Z}{4.37}K_{Pio}(x_m - c_r)^{0.5}$$

where Z is the number of balls = 16; $(-c_r) = \text{preload} = 0.6\ \mu\text{m} = 0.6 \times 10^{-3}$ mm. From Equation 3.8, we have

$$K_{Pio} = \frac{34{,}300}{f^{0.35}}D_b^{1/2}$$

with (from Equation 3.7)

$$f = \frac{(2r - D_b)}{D_b} = \frac{(2 \times 11.64 - 22.25)}{22.25} = 0.043$$

where r = radius of curvature of raceway = 11.64 mm and D_b = diameter of ball = 22.25 mm. Hence,

$$K_{Pio} = \frac{34{,}300}{(0.043)^{0.35}}(22.25)^{1/2} = 474{,}276.5 \text{ N / mm}^{1.5}$$

The stiffness can be obtained as

$$k(x_m) = 1.5 \times \frac{16}{4.37} \times 474{,}276.5 \times (x_m + 0.0006)^{1/2}$$

$$= 2.6 \times 10^6 \times (x_m + 0.0006)^{1/2} \text{ N / mm}$$

Figure 3.11 shows the variation of the stiffness with deformation due to static loads. The stiffness values for various bearing static deformations are also provided in Table 3.3. So basically, the rolling bearing stiffness changes with the static deformation of the rolling bearing (i.e., relative displacement between the inner and outer races on subtraction of the clearance from it). Shaft is assumed to oscillate about the static equilibrium position with small amplitude to maintain linearity in the stiffness term.

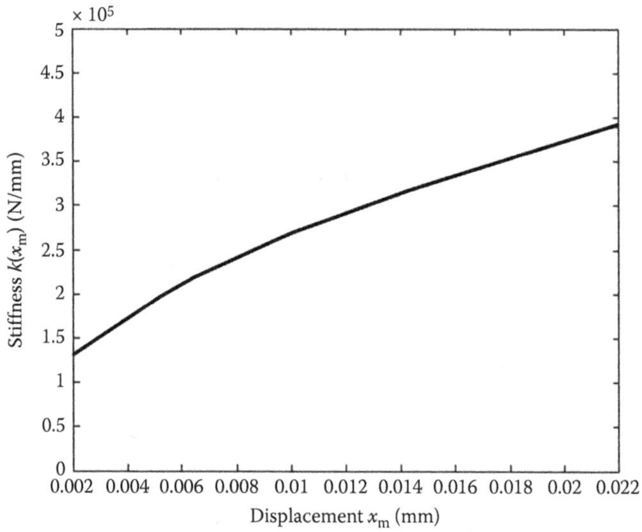

FIGURE 3.11 The stiffness of the bearing.

TABLE 3.3
Bearing Stiffness

x_m (mm)	$k(x_m)$, (N/mm)
0.002	3.52×10^5
0.006	2.12×10^5
0.010	2.36×10^6
0.014	3.36×10^6
0.018	4.35×10^6
0.022	5.39×10^6

3.1.2 Nonlinear Stiffness of Rolling Bearings

Alternatively, a more accurate determination of the stiffness of rolling contact bearings can be performed as follows (Harris, 2001). From the theory of elasticity for two bodies in contact (Timoshenko and Goodier, 1951), we have

$$\delta_p = \delta^* \left[\frac{3F}{2\sum\rho} \left\{ \frac{\left(1-\nu_{\rm I}^2\right)}{E_{\rm I}} + \frac{\left(1-\nu_{\rm II}^2\right)}{E_{\rm II}} \right\} \right]^{2/3} \frac{\sum\rho}{2} \tag{3.28}$$

with

$$\delta^* = \frac{2\mathfrak{R}}{\pi}\left(\frac{\pi}{2\kappa^2\mathfrak{I}}\right)^{1/3}; \quad \mathfrak{I} = \int_0^{\pi/2}\left[1-\left(1-\frac{1}{\kappa^2}\right)^2 \sin^2\phi\right]^{-1/2} d\phi; \quad \mathfrak{R} = \int_0^{\pi/2}\left[1-\left(1-\frac{1}{\kappa^2}\right)^2 \sin^2\phi\right]^{-1/2} d\phi$$

$$a = 0.0236a^*\left(\frac{Q}{\sum\rho}\right)^{1/3}; \quad b = 0.0236b^*\left(\frac{Q}{\sum\rho}\right)^{1/3}; \quad a^* = \left(\frac{2\varepsilon\kappa^2}{\pi}\right)^{1/3}; \quad b^* = \left(\frac{2\varepsilon}{\pi\kappa}\right)^{1/3} \tag{3.29}$$

where δ_p is the deformation at the contact, F refers to the normal force between contacting bodies, subscripts I and II refer to the first and second contacting bodies, respectively, E refers to the Young's modulus, ν refers to the Poisson's ratio, the elliptical eccentricity parameter $\kappa = a/b$, a and b are the semi-major and semi-minor axes of the elliptical contact area for the point contact, respectively, a^* and b^* are the dimensionless semi-major and semi-minor axes, respectively, $\sum\rho$ is the curvature sum of contacting bodies, and δ^* is the dimensionless contact deformation. Equation 3.28 can be rearranged as

$$F = \left[\frac{2^{5/2}}{3\left(\sum\rho\right)^{1/2}} \left[\frac{\left(1-\nu_{\rm I}^2\right)}{E_{\rm I}} + \frac{\left(1-\nu_{\rm II}^2\right)}{E_{\rm II}} \right]^{-1} \delta^{*-3/2} \right] \delta_p^{3/2} \tag{3.30}$$

On comparing Equations 3.30 and 3.3, we have

$$K_p = \left(\frac{2^{5/2}}{3}\right)\left(\frac{\left(1-\nu_{\rm I}^2\right)}{E_{\rm I}} + \frac{\left(1-\nu_{\rm II}^2\right)}{E_{\rm II}}\right)^{-1} \left(\Sigma\rho\right)^{-1/2}\left(\delta^*\right)^{-3/2} \tag{3.31}$$

where p refers to the point contact (either at the inner or outer raceway). For the steel ball and the steel raceway contact, we have

$$K_p = 2.15\times10^5\left(\Sigma\rho\right)^{-1/2}\left(\delta^*\right)^{-3/2} \ \text{N/mm}^{1.5} \tag{3.32}$$

which can be written for the inner and outer ring contacts as

$$K_{p_i} = 2.15\times10^5\left(\Sigma\rho_i\right)^{-1/2}\left(\delta_i^*\right)^{-3/2} \ \text{N/mm}^{1.5} \text{ and } K_{p_o} = 2.15\times10^5\left(\Sigma\rho_o\right)^{-1/2}\left(\delta_o^*\right)^{-3/2} \ \text{N/mm}^{1.5} \tag{3.33}$$

with

$$\Sigma\rho_i = \frac{1}{D_b}\left(4-\frac{1}{f_i}+\frac{2\gamma}{1-\gamma}\right); \quad \Sigma\rho_o = \frac{1}{D_b}\left(4-\frac{1}{f_o}-\frac{2\gamma}{1-\gamma}\right) \tag{3.34}$$

$$\gamma = \frac{D_b \cos\alpha^\circ}{D_m}; \quad f_i = \frac{r_i}{D_b}; \quad f_o = \frac{r_o}{D_b} \quad \text{and} \quad D_m = \frac{1}{2}(d_i + d_o) \approx \frac{1}{2}(D+d) \tag{3.35}$$

where D_b is the ball diameter, D_m is the pitch diameter, α° is the free contact angle (it is zero for deep groove ball bearings), d_i and d_o are the inner and outer ring raceway contact diameters, respectively, and r_i and r_o are the inner and outer raceway groove radii, respectively. We have (for ball bearings)

$$F(\rho)_i = \frac{\dfrac{1}{f_i} + \dfrac{2\gamma}{1-\gamma}}{4 - \dfrac{1}{f_i} + \dfrac{2\gamma}{1-\gamma}} \quad \text{and} \quad F(\rho)_o = \frac{\dfrac{1}{f_o} + \dfrac{2\gamma}{1-\gamma}}{4 - \dfrac{1}{f_o} + \dfrac{2\gamma}{1-\gamma}} \tag{3.36}$$

where $F(\rho)$ is the curvature difference. The plot of δ^* as a function of $F(\rho)$ is shown in Figure 3.12 and listed in Table 3.4 for ball bearings.

To find K_p from Equation 3.33, we need the (i) type of bearing, (ii) inner raceway diameter d_i, (iii) outer raceway diameter d_o, (iv) ball diameter D_b, (v) number of balls Z, (vi) inner groove radius r_i, (vii) outer groove radius d_o, (viii) pitch diameter D_m, (viii) material modulus of elasticity E_{I} and E_{II}, and (ix) material Poisson's ratio ν_{I} and ν_{II} of two contact bodies.

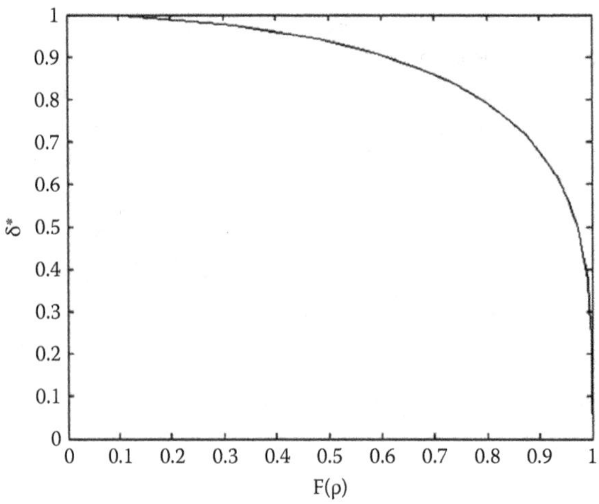

FIGURE 3.12 Variation of δ^* as a function of $F(\rho)$.

TABLE 3.4
Dimensionless Contact Parameters

$F(\rho)$	a^*	b^*	δ^*
0.0000	1.0000	1.0000	1.0000
0.1075	1.0760	0.9318	0.9974
0.3204	1.2623	0.8114	0.9761
0.4795	1.4556	0.7278	0.9429
0.5916	1.6440	0.6687	0.9077
0.6716	1.8258	0.6245	0.8733
0.7332	2.0110	0.5881	0.8394
0.7948	2.2650	0.5480	0.7961

(*Continued*)

TABLE 3.4 (*Continued*)
Dimensionless Contact Parameters

F(ρ)	*a**	*b**	*δ**
0.8350	2.4940	0.5186	0.7602
0.8737	2.8000	0.4863	0.7169
0.9100	3.2330	0.4499	0.6636
0.9366	3.7380	0.4166	0.6112
0.9574	4.3950	0.3830	0.5551
0.9729	5.2670	0.3490	0.4960
0.9838	6.4480	0.3150	0.4352
0.9909	8.0620	0.2814	0.3745
0.9951	10.2220	0.2497	0.3176
0.9973	12.7890	0.2232	0.2705
0.9982	14.8390	0.2072	0.2427
0.9989	17.9740	0.1882	0.2106
0.9995	23.5500	0.1644	0.1717
0.9995	37.3800	0.1305	0.1199
1.0000	∞	0.0000	0.0000

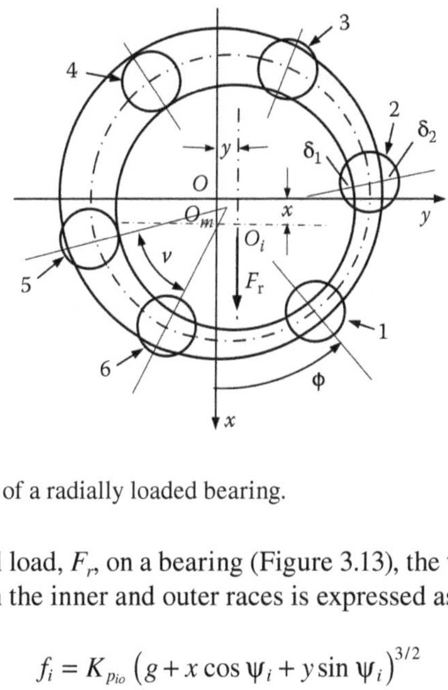

FIGURE 3.13 A line diagram of a radially loaded bearing.

For a given external radial load, F_r, on a bearing (Figure 3.13), the total elastic force at the points of contact of the *i*th ball with the inner and outer races is expressed as (Ragulskis et al., 1974)

$$f_i = K_{pio}\left(g + x\cos\psi_i + y\sin\psi_i\right)^{3/2} \tag{3.37}$$

with

$$\psi_i = \phi + (i-1)v \quad \text{and} \quad i = 1, 2, 3, \ldots, Z$$

where g is the radial preload or negative of radial clearance ($g = -c_r$) between the ball and the races, x and y are the displacements of the moving ring (here inner ring) with respect to the stationary ring (outer ring) in the direction of the radial load and perpendicular to the direction of the radial load, respectively, ψ_i is the angle between the lines of action of the radial load (direction of displacement of the moving ring) and the radius passing through the center of the ith ball, K_{Pio} is a coefficient of proportionality depending on the geometric and material properties of the bearing (the value of K_{Pio}, for the bearing with its specifications, can be estimated by the method suggested by Harris, 2001 and briefly described in this section), v and ϕ are angles (see Figure 3.13) with $v = 2\pi/Z$, and Z is the number of rolling elements. The projection of f_i along the line of action of the applied force is

$$F_i = K_{Pio}\left(g + x\cos\psi_i + y\sin\psi_i\right)^{3/2}\cos\psi_i \tag{3.38}$$

Herein, only the positive value within the bracket need to be considered since the rolling elements will not contribute when it is in the unloaded region. The total elastic force in the direction of the applied force is (i.e., in the x-direction)

$$F_r = \sum_{i=1}^{z} F_i \tag{3.39}$$

where Z is the total number of balls in the bearing. However, the net force in the y direction will be zero. Using the condition of zero net elastic force in the direction perpendicular to the external load, the deformation, y, perpendicular to the radial force line is expressed as (after dropping higher power of y on expanding using McLaurin power series)

$$y = \frac{\sum_{i=1}^{z}\left(g + x\cos\psi_i\right)^{3/2}\sin\psi_i}{1.5\sum_{i=1}^{z}\left(g + x\cos\psi_i\right)^{1/2}\sin^2\psi_i} \tag{3.40}$$

Equations 3.38 and 3.40 are used in Equation 3.39, and the bearing stiffness is determined as a function of the deformation x as

$$k(x) = \frac{\partial F(x)}{\partial x} \tag{3.41}$$

On substituting Equation 3.39 into Equation 3.41, taking into account Equation 3.40, the bearing stiffness is expressed as a function of deformation as

$$k(x) = K_{Pio}\sum_{i=1}^{z}\left\{g + x\cos\psi_i - \left(\frac{A}{1.5B}\right)\sin\psi_i\right\}^{1/2}\left\{\cos\psi_i - \frac{CBZ - 0.5AD}{(1.5B)^2}\sin\psi_i\right\}\cos\psi_i \tag{3.42}$$

where

$$A = \sum_{i=1}^{z}\left(g + x\cos\psi_i\right)^{3/2}\sin\psi_i \qquad B = \sum_{i=1}^{z}\left(g + x\cos\psi_i\right)^{1/2}\sin^2\psi_i$$
$$C = \sum_{i=1}^{z}\left(g + x\cos\psi_i\right)^{1/2}\sin\psi_i\cos\psi_i \qquad D = \sum_{i=1}^{z}\left(g + x\cos\psi_i\right)^{-1/2}\sin^2\psi_i\cos\psi_i \tag{3.43}$$

It can be seen that the bearing stiffness is critically dependent on the preloading, g, of the rolling elements. Although the manufacturer may, at times, provide the preload range, the exact value of the preloading of the bearing rolling elements in the shaft casing assembly, especially during operations that have involved wear and tear, would be difficult to determine. It is also to be noted that the theoretical stiffness calculations are based on formulations that analyze the bearing in isolation of the shaft.

For a special case, $\phi = v/2$, (i.e., when the radial load is passing through the bisector of the two neighboring rolling elements) and $y = 0$ (i.e., the coupling effect has been ignored, which is often the case for rolling bearings), formulas for $k(x)$ are appreciably simplified. For $y=0$, from Equation 3.40, we get

$$\sum_{i=1}^{z}(g + x\cos\psi_i)^{3/2}\sin\psi_i = 0 \equiv A \tag{3.44}$$

Hence, for $y=0$, $A=0$. The stiffness expression (Equation 3.42) can be simplified to

$$k(x) = K_{pio}\sum_{i=1}^{z}(g - x\cos\psi_i)^{0.5}\left\{\cos\eta_i - \frac{C}{1.5B}\sin\psi_i\right\}\cos\psi_i \tag{3.45}$$

with

$$\psi_i = \frac{\pi}{Z}(2i-1); \quad i = 1, 2, \ldots, Z \tag{3.46}$$

$$B = \sum_{i=1}^{z}(g + x\cos\psi_i)^{0.5}\sin^2\psi_i; \quad C = \sum_{i=1}^{z}(g + x\cos\psi_i)^{0.5}\sin\psi_i\cos\psi_i \tag{3.47}$$

The preloading is given as

$$g = -\frac{1}{2}P_d = D_b - \frac{1}{2}(d_o - d_i) \tag{3.48}$$

where P_d is the diametral clearance $(P_d = 2c_r)$. When $x=0$, let $k(x)=k(0)=k_o$. From Equation 3.45, we have

$$k_o = K_{pio}\sum_{i=1}^{z}g^{0.5}\left(\cos\psi_i - \frac{C}{1.5B}\sin\psi_i\right)\cos\psi_i \tag{3.49}$$

$$B = \sum_{i=1}^{z}g^{0.5}\sin^2\psi_i; \quad C = \sum_{i=1}^{z}g^{0.5}\sin\psi_i\cos\psi_i \tag{3.50}$$

We can have different levels of preloading conditions as follows:

Case I: $g>0$ (positive preloading). A simplified mathematical expression of the bearing stiffness is taken as

$$k(x) = a - bx^2 \quad \text{when} \quad |x| \le g \tag{3.51}$$

and

$$k(x) = b_1x^{2/3} \quad \text{when} \quad |x| \ge g \tag{3.52}$$

Let $k(x=g)=k(g)=k_g$. At $x=0$, from Equation 3.51, we get $a=k(0)=k_o$, and at $x=g$, we get $b=\frac{k_o-k_g}{g^2}$ so that the stiffness can be expressed as

$$k(x)=k_o-\frac{k_g-k_o}{g^2}x^2 \quad \text{when} \quad |x|\le g \tag{3.53}$$

Similarly, from Equation 3.52 from information at $x=g$, we get

$$k(x)=k_g g^{-2/3}x^{2/3} \quad \text{when} \quad |x|\ge g \tag{3.54}$$

Case II: $g=0$ (no clearance). A simplified mathematical expression of the bearing stiffness is taken as

$$k(x)=b_2x^{3/2} \tag{3.55}$$

At $x=c$ (c is a constant), we get $b_2=k_c c^{-3/2}$ with $k_c=k(c)$ so that the stiffness can be expressed as

$$k(x)=k_c c^{-3/2}x^{3/2} \tag{3.56}$$

Case III: $g<0$ (preclearance or negative preload). A simplified mathematical expression of the bearing stiffness is taken as

$$\begin{aligned} k(x)&=0 && \text{when} \quad |x|\le g \\ &=b_3x^{2/3} && \text{when} \quad |x|>g \end{aligned} \tag{3.57}$$

For $x=d$ ($d>|g|$, herein d is a constant), we get $b_3=k_d(d)^{-3/2}$ with $k_d=k(d)$ so that the stiffness can be expressed as

$$k(x)=k_d d^{-3/2}x^{3/2} \quad \text{when} \quad x>|g| \tag{3.58}$$

Equations 3.53, 3.54, 3.56, and 3.58 are basically simplified expressions for the bearing stiffness expression given by Equation 3.45. These forms of the bearing stiffness can be used to analyze the nonlinear behavior of rotor-bearing systems (Tiwari and Vyas, 1995). It should be noted that the analysis presented here can be extended to roller bearings by having the load–deflection exponent equal to $n=1.08$. Various geometrical relations for roller bearings would be

$$\Sigma\rho_i=\frac{1}{D_r}\left(\frac{2}{1-\gamma}\right);\quad \Sigma\rho_o=\frac{1}{D_r}\left(\frac{2}{1+\gamma}\right);\quad \gamma=\frac{D_r\cos\alpha^{\circ}}{D_m} \tag{3.59}$$

and

$$F(\rho)_i=F(\rho)_o=1 \tag{3.60}$$

where D_r is the nominal roller diameter.

Example 3.3

Obtain the stiffness of a ball bearing with the following specifications: deep groove ball bearings, bearing no. 6200, d=bore diameter=10 mm, D=outer diameter=30 mm, and width=9 mm.

Solution:

In general, the internal geometric dimensions of the bearings are not provided in the bearing manufacturers' catalogs (e.g., SKF General Catalog). These dimensions can be either measured after disassembling the bearing (these may be different for the same bearing number by different manufacturers) or they can be approximately calculated by the expressions in terms of boundary dimensions of the bearing provided in the bearing catalogs to obtain the stiffness of the bearing. Alternatively, a more advanced optimum design can be attempted (Gupta et al., 2007; Rao and Tiwari, 2007; Tiwari and Chandran, 2015). General proportions for ball bearings, as shown in Figure 3.14, are obtained as follows (PSG, 1982):

The pitch diameter or the mean diameter: $D_{\text{m}} \approx \frac{1}{2}(D+d) = 20\,\text{mm}$.

The diameter of the ball $D_{\text{b}} = 0.3(D-d) = 6\,\text{mm}$.

The radial thickness of rings, $s = 0.15(D-d) = 3\,\text{mm}$.

The number of balls, $Z \approx 2.9\frac{D+d}{D-d} = 2.9\frac{30+10}{30-10} = 5.8$. Hence $Z=6$ can be used.

The inner and outer ring groove radius of curvatures, $r_i = r_o = 0.515\,D_{\text{b}} = 3.09\,\text{mm}$.

For zero radial clearance, the inner and outer ring diameters at the raceway grooves are as follows:

$$d_o = 20+6 = 26\text{ mm};\; d_i = 20-6 = 14\text{ mm}$$

For a ball bearing of the bore diameter 10 mm, from the design data books, we have a radial clearance, c_r, on the order of 2–13 μm. Let us for the present example take the maximum preload; hence $g = -13\times10^{-3}\,\text{mm}$.

Finally, the bearing internal dimensions are obtained as $r_i = r_o = 3.09$ mm, D_b=6 mm, Z=6, d_i=14 mm, d_o=26 mm, D_m=20 mm, and $g = -13 \times 10^{-3}\,\text{mm}$. Based on these dimensions, the following parameters, which will be used to find the bearing stiffness, can be estimated:

$$\gamma = \frac{D_b}{D_{\text{m}}} = \frac{6}{20} = 0.3;\; f_i = \frac{r_i}{D_{\text{b}}} = \frac{3.09}{6} = 0.515;\; f_o = \frac{r_o}{D_{\text{b}}} = \frac{3.09}{6} = 0.515$$

$$\Sigma\rho_i = \frac{1}{D_{\text{b}}}\left(4 - \frac{1}{f_i} + \frac{2\gamma}{1-\gamma}\right) = \frac{1}{6}\left(4 - \frac{1}{0.515} + \frac{2(0.3)}{1-0.3}\right) = 0.4859\text{ mm}^{-1}$$

$$\Sigma\rho_o = \frac{1}{D_{\text{b}}}\left(4 - \frac{1}{f_o} - \frac{2\gamma}{1+\gamma}\right) = \frac{1}{6}\left(4 - \frac{1}{0.515} - \frac{2\times0.3}{1+0.3}\right) = 0.2661\text{ mm}^{-1}$$

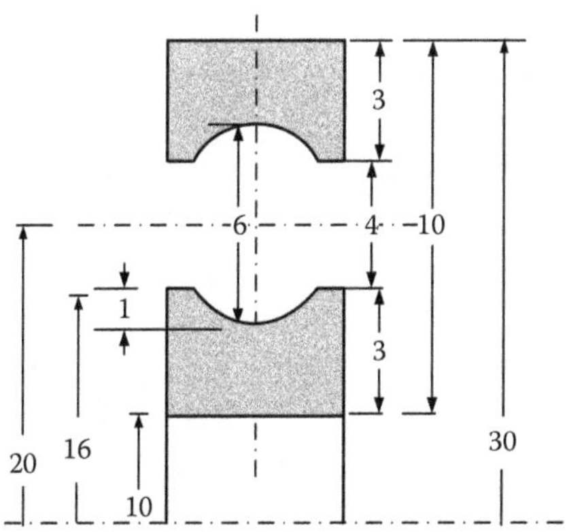

FIGURE 3.14 Bearing dimensions (all dimensions are in mm).

$$F(\rho)_i = \frac{\dfrac{1}{f_i} + \dfrac{2\gamma}{1-\gamma}}{4 - \dfrac{1}{f_i} + \dfrac{2\gamma}{1-\gamma}} = \frac{\dfrac{1}{0.515} + \dfrac{2\times 0.3}{1-0.3}}{4 - \dfrac{1}{0.515} + \dfrac{2\times 0.3}{1-0.3}} = \frac{2.799}{2.915} = 0.9600$$

In relation to the curvature difference of $F(\rho_i) = 0.96$, the nondimensional relative displacement is given as (Figure 3.12, Table 3.4) $\delta_i^* = 0.545$.

$$F(\rho_o) = \frac{\dfrac{1}{f_o} - \dfrac{2\gamma}{1+\gamma}}{4 - \dfrac{1}{f_o} - \dfrac{2\gamma}{1+\gamma}} = \frac{\dfrac{1}{0.515} - \dfrac{2\times 0.3}{1+0.3}}{4 - \dfrac{1}{0.515} - \dfrac{2\times 0.3}{1+0.3}} = \frac{1.4802}{1.5967} = 0.9270$$

In relation to the curvature difference of $F(\rho_i) = 0.96$, the nondimensional relative displacement is given as (Figure 3.12, Table 3.4) $\delta_0^* = 0.63$.

The load–deflection constant at the inner and outer raceway contacts are given as (refer to Equation 3.32)

$$K_{p_i} = 2.15\times 10^5 \left(\Sigma\rho_i\right)^{-1/2} \left(\delta_i^*\right)^{-3/2} = 2.15\times 10^5 \times 0.4859^{-1/2} \times 0.545^{-3/2}$$
$$= 7.67\times 10^5 \text{ N/mm}^{1.5}$$

and

$$K_{p_o} = 2.15\times 10^5 \times 0.2661^{-1/2} \times 0.63^{-3/2} \text{ N/mm}^{1.5}$$
$$= 8.335\times 10^5 \text{ N/mm}^{1.5}$$

The load–deflection factor for a single ball is given as

$$K_{Pio} = \left\{ \frac{1}{\left(1/K_{Pi}\right)^{2/3} + \left(1/K_{Po}\right)^{2/3}} \right\}^{3/2}$$

$$= \left\{ \frac{1}{\left(1/7.67\times 10^5\right)^{0.667} + \left(1/8.335\times 10^5\right)^{0.667}} \right\}^{3/2}$$

$$= 2.8445\times 10^5 \text{ N/mm}^{1.5}$$

The bearing stiffness in a polynomial form can be expressed as

$$k(x) = a + bx^2 \quad \text{with} \quad a = k(0) \quad \text{and} \quad b = \frac{k(g) - k(0)}{g^2}$$

The theoretical bearing stiffness is given as

$$k(x) = K_{Pio} \sum_{i=1}^{z} \left(g - x\cos\psi_i\right)^{0.5} \left(\cos\psi_i - \frac{C}{1.5B}\sin\psi_i\right)\cos\psi_i$$

with $n=3/2$, $Z=6$, $g=0.013$ mm.

$$\psi_i = \frac{\pi}{6}(2i-1),\ i = 1,2,\ldots,6 \quad \text{so that} \quad \psi_i = \frac{\pi}{6}, \frac{3\pi}{6}, \frac{5\pi}{6}, \frac{7\pi}{6}, \frac{9\pi}{6}, \frac{11\pi}{6}$$

$$B=\sum_{i=1}^{z}\left(g+x\cos\psi_i\right)^{0.5}\sin^2\psi_i \quad \text{and}$$

$$C=\sum_{i=1}^{z}\left(g+x\cos\psi_i\right)^{0.5}\sin\psi_i\cos\psi_i$$

At $x=0$, we have

$$B=\sum_{i=1}^{6}0.013^{0.5}\sin^2\psi_i=0.114\left(\sin^2\frac{\pi}{6}+\sin^2\frac{3\pi}{6}+\sin^2\frac{5\pi}{6}+\sin^2\frac{7\pi}{6}+\sin^2\frac{9\pi}{6}+\sin^2\frac{11\pi}{6}\right)$$

$$=0.114\left\{(0.5)^2+1+(0.5)^2+(-0.5)^2+(-1)^2+(-0.5)^2\right\}$$

$$=0.114\times 3=0.342$$

$$C=\sum_{i=1}^{6}0.013^{0.5}\sin\psi_i\cos\psi_{-i}$$

$$=0.114\left\{0.5\times 0.866+0+0.5\times(-0.866)+(-0.5)\times(-0.866)+0+(-0.5)\times 0.866\right\}=0$$

and

$$k(0)=K_{Pio}\sum_{i=1}^{6}0.013^{0.5}\left(\cos\psi_i-0\right)\cos\psi_i$$

$$=2.844\times 10^5\times 0.114\times\left\{0.866^2+0+0.866^2+(-0.866)^2+0+0.866^2\right\}$$

$$=9.73\times 10^4 \quad \text{N/mm}$$

At $x=g$, we have

$$B=\sum_{i=1}^{6}\left(0.013+0.013\cos\psi_i\right)^{0.5}\sin^2\psi_i$$

$$=0.013^{0.5}\times\left\{(1+0.866)^{0.5}\times 0.5^2+(1+0)^{0.5}\times 1.0^2+(1-0.866)^{0.5}\times 0.5^2\right.$$

$$\left.+(1-0.866)^{0.5}\times(-0.5)^2+(1-0)^{0.5}\times(-1)^2+(1+0.866)^{0.5}\times(-0.5)^2\right\}$$

$$=0.114\times(0.3415+1.0+0.0915+0.0915+1.0+0.3415)=0.3267$$

and

$$C=\sum_{i=1}^{6}\left(0.013+0.013\cos\psi_i\right)^{0.5}\sin\psi_i\cos\psi_i$$

$$=0.114\times\begin{Bmatrix}(1+0.866)^{0.5}\times 0.5\times 0.866+0+(1-0.866)^{0.5}\times 0.5\times(-0.866)\\ +(1-0.866)^{0.5}\times(-0.5)\times(-0.866)+0+(1+0.866)^{0.5}(-0.5)\times 0.866\end{Bmatrix}=0$$

$$k(g) = K_{Pio} \sum_{i=1}^{6} \left(0.013 + (0.013)\cos\psi_i\right)^{0.5} \cos^2\psi_i$$

$$= K_{Pio} \times 0.114 \times \begin{Bmatrix} 1.866^{0.5} \times 0.866^2 + 0 + 0.134^{0.5} \times (-0.866)^2 \\ +0.134^{0.5} \times (-0.866)^2 + 0 + 1.866^{0.5} \times 0.866^2 \end{Bmatrix}$$

$$= K_{Pio} \times 0.296 = 8.423 \times 10^4 \text{ N/mm}$$

For $P_d < g$, we have

$$k(x) = a + bx^2$$

with

$$a = k(0) = 9.73 \times 10^4 \text{ N/mm} \quad \text{and} \quad b = \frac{k(g) - k(0)}{g^2} = \frac{8.423 - 9.73}{(0.013)^2} 10^4 = -7.734 \times 10^7 \text{ N/mm}^3$$

Hence,

$$k(x) = a + bx^2 = \left(9.73 \times 10^4 - 7.734 \times 10^7 x^2\right)$$
$$= \left(9.73 - 7734.0\, x^2\right) \times 10^4 \text{ N/mm}$$

However, $k(x)$ is negative for

$$7734x^2 > 9.73 \quad \text{or} \quad x^2 > 1.258 \times 10^{-3} \text{ which gives } x > 0.0354 \text{ mm}$$
$$\text{at } x = 0.1947 \text{ mm} \quad k(x) = -2.83 \times 10^6 \text{ N/mm}$$
$$\text{at } x = g \quad k(x) = 8.423 \times 10^4 \text{ N/mm}$$

It can be seen that the bearing stiffness is critically dependent on the clearance, c_r, in the bearing. Although the manufacturer may at times provide the clearance range depending upon standards of fits and tolerances, the exact value of the clearance of the bearing in the shaft casing assembly, especially during operations that have involved wear and tear, and expansion due to temperature, would be difficult to determine. In Chapter 4 of Volume II, we deal with experimental determinations of the stiffness and damping parameters to avoid such problems.

Example 3.4

Determine the nonlinear stiffness of a bearing with the following specifications: ball bearing type: SKF 6200, number of balls = 6, ball diameter = 6 mm, bore diameter = 10 mm, outer diameter = 30 mm, pitch diameter = 20 mm, inner groove radius = 3.09 mm, outer groove radius = 3.09 mm, allowable preload = 0–2 μm, and rotor mass per bearing = 0.41 kg.

Solution:

The value of K_{Pio} for the bearing with given specifications from Example 3.3 is estimated to be 2.844×10^5 N/mm$^{1.5}$. Let us take a general form of the nonlinear restoring force as

$$f_s = k_L x + k_{NL} G(x) \tag{a}$$

where G can be a nonlinear in x, and k_L and k_{NL} are, respectively, the linear and nonlinear stiffness contribution parameters. For the present problem $G(x)$ is taken as cubic in x so that

$$k(x) = \frac{\partial f_s}{\partial x} = k_L + k_{NL} x^2 \tag{b}$$

Based on the method discussed in the previous section, the stiffness of the bearing is plotted in Figure 3.15 as a function of the radial deformation, x, for various allowable preload values, $-c_r$. The stiffness calculations are based on formulations that analyze the bearing in isolation of the shaft. The expressions for the stiffness in Table 3.5 have been obtained by curve fitting the stiffness values obtained from Equation 3.45 through a quadratic in x (Equation b). Herein, the bearing becomes softer with increasing load. General perception is that the contact area increases with load and would make the bearing stiffer. However, apart from contact area at ball locations, overall load distribution also contributes to the stiffness. The overall load distribution depends upon preload/clearance. For present case, it is for softening type (preloaded) stiffness behavior. However, for the case of clearance, the hardening type stiffness behavior will prevail (as can be seen in Figure 3.11).

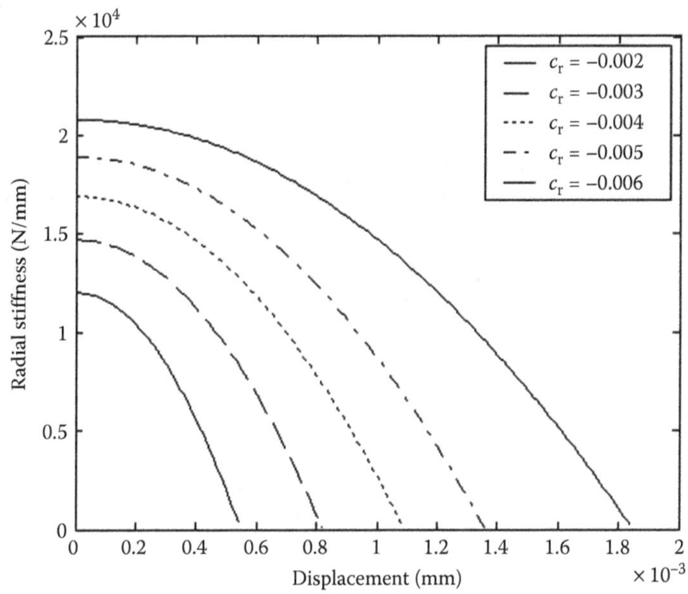

FIGURE 3.15 Stiffness variation with displacement for diverse radial clearances.

TABLE 3.5
Radial Stiffness Parameters of the Ball Bearing

Preload, c_r (mm)	Radial Stiffness, $k(x)$ (N/mm)
0.0002	$1.20\times10^4 - 4.01\times10^{10}\,x^2$
0.0003	$1.47\times10^4 - 2.18\times10^{10}\,x^2$
0.0004	$1.69\times10^4 - 1.42\times10^{10}\,x^2$
0.0005	$1.89\times10^4 - 1.02\times10^{10}\,x^2$
0.0006	$2.08\times10^4 - 0.61\times10^{10}\,x^2$

3.1.3 Linear Stiffness Determination under Combined Loadings

Stiffness characteristics of rolling element bearings are studied under combined loading (see Figure 3.16), that is, under the action of the radial and axial forces, and the moment. Hence, there will be corresponding radial, axial, and angular displacements. All the displacements are assumed to be coupled, that is, they will depend upon the combined effect of all the forces and moments acting on the bearing. The stiffness matrix obtained is of size 5×5. To obtain displacements, a set of five nonlinear equilibrium equations are solved for a given loading condition (Lim and Singh, 1990a, b). Apart from these equilibrium equations, the geometrical and deformation equations would also be needed.

The load–deflection relationship for a single jth rolling element can be expressed as

$$Q_j = K_{p_{io}} \delta_j^{3/2} \tag{3.61}$$

where Q_j is the normal load on the rolling element and $K_{p_{io}}$ is the effective load–deflection constant for the contact of a rolling element with the inner and outer races. The actual contact angle at the jth rolling element position after loading is given as

$$\tan \alpha_j = \frac{A_o \sin \alpha_o + \delta_{z_j}}{A_o \cos \alpha_o + \delta_{r_j}} \quad \text{with} \quad A_o = r_i + r_o - D_b \tag{3.62}$$

Herein, α_o is free contact angle and subscripts in δ: r and z represent deformation in radial and axial directions, respectively. Now the force and moment balance equations can be obtained as follows:

$$F_{b_x} = \sum_{j=1}^{z} Q_j \cos \alpha_j \cos \psi_j \tag{3.63}$$

$$F_{b_y} = \sum_{j=1}^{z} Q_j \cos \alpha_j \sin \psi_j \tag{3.64}$$

$$F_{b_z} = \sum_{j=1}^{z} Q_j \sin \alpha_j \tag{3.65}$$

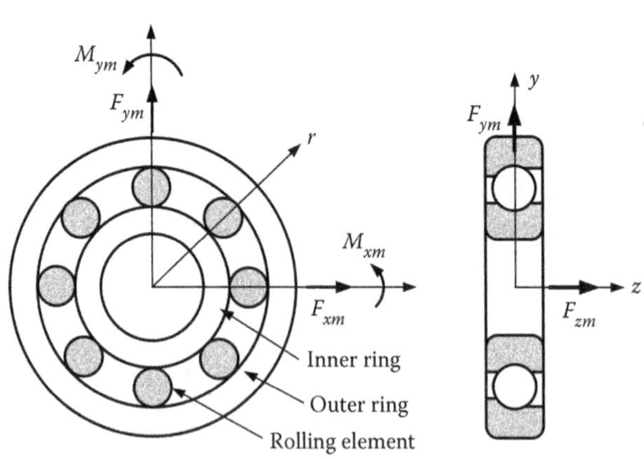

FIGURE 3.16 A bearing with the force and moment loadings.

$$M_{b_x} = \sum_{j=1}^{z} Q_j r_j \sin\alpha_j \sin\psi_j \tag{3.66}$$

$$M_{b_y} = \sum_{j=1}^{z} Q_j r_j \sin\alpha_j \cos\psi_j \tag{3.67}$$

$$M_{b_z} = 0 \tag{3.68}$$

Stiffness coefficients can be obtained by differentiating the force matrix $\mathbf{f}_b$ with respect to the displacement vector $\mathbf{q}_b$:

$$\mathbf{K}_b = \frac{\partial \mathbf{f}_b}{\partial \mathbf{q}_b} \tag{3.69}$$

This will give a 5×5 stiffness matrix. Displacements can be obtained by solving Equations 3.62 through 3.68 along with certain geometrical constraints. Then the stiffness matrix can be obtained by using Equation 3.69. The details can be seen in Lim and Singh (1990a, b).

As an illustration, an analysis of the previous case has been conducted for a deep groove ball bearing with the following parameters: load–deflection exponent $n=3/2$, load–deflection constant $K_{Pio} = 8.5\times10^9$ N/mm$^{1.5}$, number of rolling elements $Z=12$, diametral clearance $P_d=0.00005$ mm, pitch diameter $D_m=19.65$ mm, and $A_o=0.05$ mm.

Table 3.6 summarizes the analysis for various loading conditions. When only an axial load is applied, the axial displacement is found to be significant and other displacements are negligible. Along with the axial load, if some radial load is applied, the radial displacement is found to vary significantly and angular displacements are very small. Now, if, in addition to the axial load, some moment is applied, a considerable amount of the angular displacement is observed. For combined loading, all displacements are found to be significant. These load–deflection relations can be used to obtain the stiffness coefficients by giving small perturbations to these forces (or moments) and estimating the change in the deflections. Now the calculation of stiffness for a simple loading of a pure radial case is considered. However, we will be adding the effect of centrifugal forces in rolling elements.

TABLE 3.6
Force Deformation Relationships

Applied Forces and Moments					Displacements				
F_{bx} (kN)	F_{bym} (kN)	F_{bzm} (kN)	M_{bxm} (kN-mm)	M_{bym} (kN-mm)	δ_{xm} (mm)	δ_{ym} (mm)	δ_{zm} (mm)	β_{xm} (rad)	β_{ym} (rad)
0	0	15	0	0	-9.04×10^{-11}	9.63×10^{-10}	6.61×10^{-02}	1.04×10^{-10}	-2.63×10^{-13}
10	0	15	0	0	4.38×10^{-02}	1.17×10^{-08}	6.64×10^{-02}	-2.28×10^{-09}	6.40×10^{-05}
0	10	15	0	0	1.96×10^{-08}	5.13×10^{-02}	6.13×10^{-02}	-1.07×10^{-03}	1.67×10^{-09}
10	10	15	0	0	4.62×10^{-02}	4.66×10^{-02}	5.92×10^{-02}	-8.83×10^{-04}	8.36×10^{-04}
0	0	15	50	0	3.84×10^{-07}	-7.84×10^{-03}	6.53×10^{-02}	6.01×10^{-04}	8.25×10^{-10}
0	0	15	0	50	8.03×10^{-03}	7.78×10^{-09}	6.53×10^{-02}	-4.13×10^{-10}	6.16×10^{-04}
0	0	15	50	50	7.96×10^{-03}	-7.76×10^{-03}	6.45×10^{-02}	6.06×10^{-04}	6.21×10^{-04}
10	10	15	50	50	5.08×10^{-02}	4.19×10^{-02}	5.80×10^{-02}	-3.43×10^{-04}	1.42×10^{-04}

3.1.4 Radial Stiffness under High-Speed Conditions

In analyses in previous sections, only the static load was applied to the rolling bearings, and the linear stiffness was obtained at the static deformation by placing a small perturbation around it. In high-speed operations, rolling element centrifugal forces can be significantly large compared to the static forces applied to the bearing. Increased loading due to centrifugal force causes larger contact deformation at the outer raceway, which can be compared with the effect of increasing the clearance on displacement characteristics of the rolling element. High speed also affects the lubrication characteristics and thereby the friction in the bearings. This will have an influence on bearing internal speeds, which in turn alters the rolling element internal loading (i.e., centrifugal forces and gyroscopic moments).

Figure 3.17 shows the displacement of a ball bearing due to a radial loading. Under no load condition, centers of raceway groove curvatures are separated by a distance BD_b, that is, $(f_o+f_i-1)D_b$, where $f_o=r_o/D_b$ and $f_i=r_i/D_b$. Distance a is given by $\delta_r \cos\psi_j$, $b=\delta_{oj}$, and $d=(f_i-0.5)D_b+\delta_{ij}$, where δ_r is the radial displacement of the bearing, and δ_{ij} and δ_{oj} are contact deformations at the inner and outer raceways, respectively, of the jth rolling element at azimuth angle ψ_j.

Under an applied static load, the distance between the centers of the groove curvature radii will increase by the amount of contact deformations $(\delta_i+\delta_o)$. The line of action between the centers is collinear, with the line joining the fixed outer raceway groove curvature's center with the initial position of the inner raceway curvature's center. Even when the centrifugal force acts on the ball, because the inner and outer raceway contact angles are similar, the line of action remains collinear. It is assumed that the outer raceway groove curvature's center remains fixed in the space and the inner raceway groove curvature's center moves relative to that fixed center. The distance between the fixed outer raceway groove curvature center and the final position of the ball center at any ball location is given by

$$\Delta_{oj} = (f_o - 0.5)\, D_b + \delta_{oj} \tag{3.70}$$

Similarly, the distance between the final position of the inner raceway groove curvature's center and the final position of the ball's center at any location is given by

$$\Delta_{ij} = (f_i - 0.5)\, D_b + \delta_{ij} \tag{3.71}$$

where δ_{oj} and δ_{ij} are normal contact deformations at the outer and inner raceway contacts, respectively. As evident from Figure 3.17,

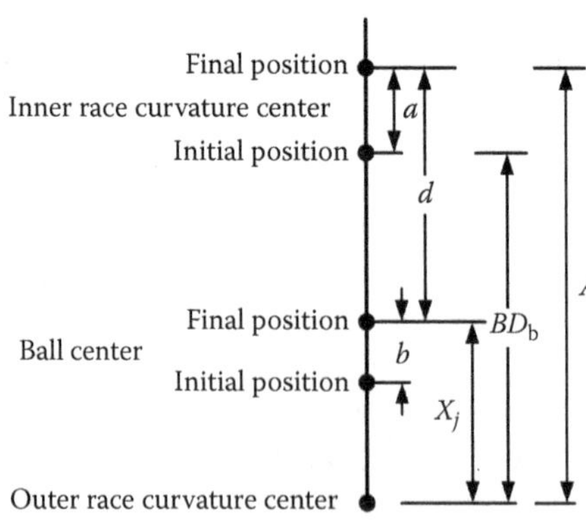

FIGURE 3.17 Displacements in a high-speed deep groove ball bearing for the radial static load.

$$A_j - X_j = (f_i - 0.5)\,D_b + \delta_{ij} \tag{3.72}$$

$$X_j = (f_o - 0.5)\,D_b + \delta_{oj} \tag{3.73}$$

$$K_{p_{oj}}\delta_{oj}^{1.5} - K_{p_{ij}}\delta_{ij}^{1.5} - F_{c_j} = 0 \tag{3.74}$$

Equations 3.72 through 3.74 can be solved simultaneously for X_j, δ_{ij}, and δ_{oj} at each angular location of ball for a given assumed value of δ_r using the Newton–Raphson numerical method. The radial displacement δ_r can be obtained by the following equation, which should be satisfied for the entire bearing:

$$F_r - \sum Q_{ij}\cos\psi_j = 0 \tag{3.75}$$

where F_r is the radial static load and Q_{ij} is the normal load at the inner raceway. The linearized stiffness can be obtained by placing small perturbation about the equilibrium position of the ball bearing and then obtaining the force required to give that perturbation. Let the perturbation given be $\pm(d\delta_r)$ about the mean equilibrium position. The stiffness is defined as follows:

$$K_{rr} = \frac{F_{r_1} - F_{r_2}}{2(d\delta_r)} \tag{3.76}$$

where F_{r_1} is the force on the rolling bearing, when $+(d\delta_r)$ perturbation is placed about the mean position and F_{r_2} is the force on the rolling bearing for a perturbation of $-(d\delta_r)$ about the mean position. For detailed treatment, readers are referred to books by Changsen (1991) and Harris (2001).

3.2 HYDRODYNAMIC FLUID-LUBRICATED JOURNAL BEARINGS

Hydrodynamic fluid-lubricated journal bearings were studied for hundreds of years using empirical methods before there was awareness of the nature of the physical action by which a hydrodynamic lubricating film is formed. Empirical methods still have an important place in the journal bearing design of today. The buildup of pressure in the film was experimentally reported independently by Tower (1883, 1884, 1885) in England and Petroff (1883) in Russia. Tower observed the development of hydrodynamic pressure first in partial journal bearings used in the railway car wheel. The pressure generated is high enough to displace plugs used in the fluid holes. However, he could not explain the physics behind the pressure generation. The first analytical investigation on losses in a hydrodynamic journal bearing was given by Petroff (1883). He applied Newton's law of viscous flow to a rotating concentric journal bearing and expressed the coefficient of fluid friction in terms of lubricant viscosity, speed, and dimensions of journal bearings. This is given by

$$\mu = 2\pi^2\,\frac{\eta N}{q}\left(\frac{R}{c_r}\right) \tag{3.77}$$

with

$$q = \frac{W}{2LR}$$

where μ is the coefficient of friction, η is the coefficient of the absolute viscosity of the lubricant, N is the journal speed (rpm), q is the unit load, R is the journal radius, L is the length of journal, and c_r is the radial clearance. Equation 3.77 is called Petroff's equation. The equation has been found to be a very good approximation for lightly loaded conditions. Three years later Reynolds (1886) explained the buildup of pressure by the viscous action in a convergent film. The physical understanding thus obtained assisted designers in choosing the position of lubricant supply and drain grooves in journal bearings.

For a normal load to be supported by a bearing, positive-pressure profiles must be developed over the bearing length. Figure 3.18 illustrates three ways of developing positive pressure in fluid-lubricated bearings. For a positive pressure to be developed in a hydrodynamic bearing (Figure 3.18a), the lubricant film thickness must be decreasing in the direction of motion. This action can be realized as motion of an inclined palm (e.g., 30° with the direction of motion) over the surface of the water. The palm will feel an upload thrust because of this wedge action between the palm and the water. In a squeeze-film bearing (Figure 3.18b), the squeeze action takes place because the bearing surfaces approach each other with the squeeze velocity. The squeeze mechanism of pressure generation provides a valuable cushioning effect when the bearing surfaces come closer to each other. Positive pressures develop only when the film thickness diminishes. This cushioning (squeeze) action can be felt when the palm is immersed in a bucket of water and then tries to approach the wall (or bottom) of the bucket at a certain velocity. In externally pressurized bearing, sometimes referred to as "a hydrostatic bearing" (Figure 3.18c), the pressure drop across the bearing supports the load. The load-carrying capacity is independent of bearing motion and lubricant viscosity. There is no wear due to surface contact at the starting and the stopping of a rotor as there is with the slider bearing. This action can be felt if a thumb is pressed against the hole of a tap and then gradually the tap is opened. Because of the static pressure of the water, the thumb will be lifted up (pushed away) from the hole of the tap.

There are other types of lubrications (i.e., the elasto-hydrodynamic and boundary lubrications). The elasto-hydrodynamic lubrication is a form of hydrodynamic lubrication where elastic deformation of the lubricated surfaces becomes significant. The maximum pressure is typically between 0.5 and 3 GPa (for hydrodynamic lubrication, it is on the order of 5 MPa); the minimum film thickness normally exceeds 0.1 μm (for hydrodynamic lubrication, it is on the order of 1 μm). Rolling bearings and gears are some of the examples where the elasto-hydrodynamic lubrication exists (Hamrock, 1994). In the boundary lubrication, the contacting solids are not separated by the lubricant, fluid-film effects are negligible, and there is considerable asperity contact in the microscopic level. The contact lubrication mechanism is governed by the physical and chemical properties of thin surface films of molecular proportions. The properties of the bulk lubricant are of minor importance, and friction coefficients are essentially independent of fluid viscosity. Boundary lubrication is usually not planned by the designer, and characteristics are determined by the properties of the solids and the lubricant film at the common interfaces. However, it is better than completely dry friction. The surface films vary in thickness from 1 to 10 nm. Hence, the coefficient of friction for the hydrodynamic

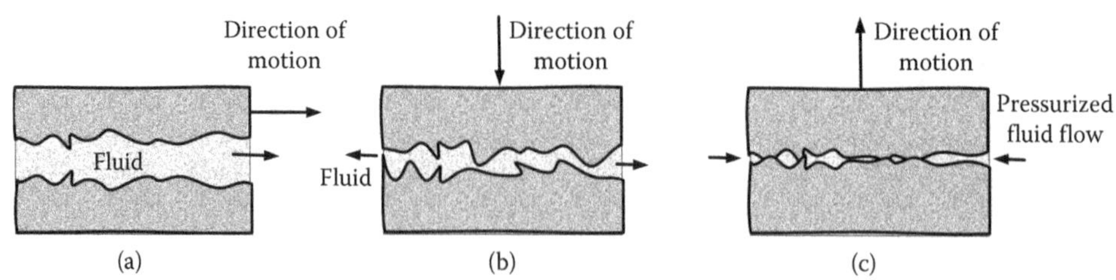

FIGURE 3.18 Various form of fluid-film bearing: (a) hydrodynamic; (b) squeeze film; (c) hydrostatic.

bearing lubrication is on the order of 10^{-3}, while for the elasto-hydrodynamic lubrication it is 10^{-2}, for the boundary lubrication it is 10^{-1}, and for nonlubricated (or dry lubrication) it is 0.7–1.

3.2.1 Types of Hydrodynamic Bearings

A hydrodynamic journal (radial) bearing consists of a bearing bore that is normally circular and inside of which a circular length of shaft (journal) is rotated. The journal diameter is usually 99.8%–99.9% of that of the bore diameter of the bearing, and the clearance space between the two is partially filled by the lubricating fluid. At zero rotational speed, under a steady load, the journal touches the bottom of the clearance space in the bearing bore. When the journal rotates, fluid is dragged along by the journal due to the fluid viscosity, and a thin film of fluid builds up between the journal and bearing bore so that under normal operating conditions there is no surface-to-surface (or metal-to-metal) contact. The aim when designing for steady loads is to ensure that there is sufficient film clearance to take up the effects of the surface asperity and likely changes in load during the machine operation over time.

Most journal bearings operate at steady-state eccentricity ratios (journal eccentricity/radial clearance) of about 0.6–0.7 so as to ensure that the bearing is not prone to self-excited vibrations. A hydrodynamic journal bearing can have a significant role in dynamics of rotors. The fluid film is equivalent to a complicated arrangement of the direct and cross-coupled springs and dampers, and thus influences the machine critical speeds and unbalance response. Moreover, bearing fluid-film forces can cause rotor instability, which results in serious levels of self-excited vibration (e.g., the fluid-film whirl/whip). The most common forms of the bore (solid line) and lubricant inlet ports (dotted line) of journal bearings are provided in Figure 3.19 and compared in Table 3.7.

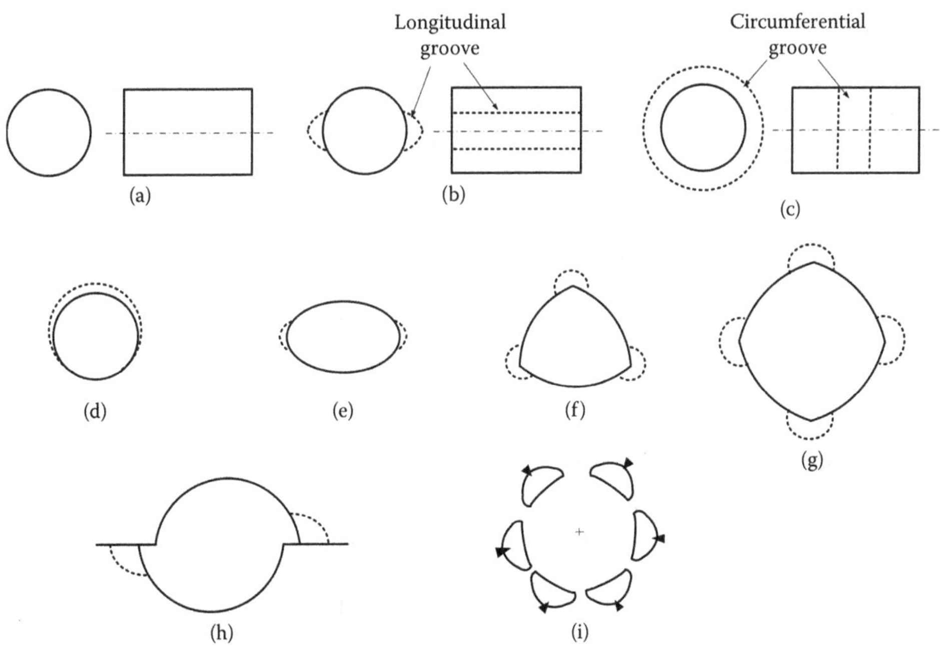

FIGURE 3.19 Various geometric shapes of hydrodynamic bearings: (a) plain cylindrical; (b) longitudinal; (c) partial arc; (d) lemon bore (2-lobe); (e) 3-lobe; (f) 4-lobe; (g) offset halves; (h) tilting pad.

TABLE 3.7
Relative Comparison of Various Forms of Hydrodynamic Radial Bearings

S.N.	Name of the Bearing	Load-Carrying Capacity/Stability	Common Features	Other Special Features
1	Plain cylindrical bearing (PCB)	Carry very heavy loads. Prone to oil whirl.	Simplest to manufacture and the most common in use. Fluid inlet in the form of a hole or axial groove.	Not suitable for loads in the horizontal direction. The limiting space in which the center of the shaft can move is called the clearance circle.
2	Circumferential groove bearings	Carry loads less than half of PCB. Higher frictional losses.	Similar to PCB, except the fluid inlet is in the form of a circumferential groove at the middle.	Used when the direction of load is not fixed.
3	Partial-arc bearings	Carry loads slightly less than PCB. Less friction loss as compared to circumferential groove bearings.	Another variation of PCB with circumferential groove in the upper part only (around 260°–300°).	Not suitable for loads that change directions.
4	Lemon bore bearing	Carry loads comparable to partial-arc bearings. More resistant to oil whirl.	Two partial (circular) arc bearings mounted in opposition to each other.	Stiffness and damping in the horizontal direction are relatively low. The shape of the limiting space of the center of the shaft would be the same as the bearing bore.
5	Three-lobe and four-lobe bearings	Load-carrying capacity is about half that of partial-arc bearings. Greater resistance to fluid-film whirl.	The orientation of lobes (partial arcs) has effects on bearing characteristics.	Relatively large fluid-film stiffness and damping.
6	Offset halves bearings	Support loads that are slightly less than PCB. Very high resistance to oil whirl.	The offset lobes allow machines to rotate in one direction only.	High fluid-film stiffness and damping.
7	Tilting pad bearings	Load-carrying capacity is about half of that of PCB. Perfectly ideal for resisting fluid-film whirl.	Lobes are independently mounted on a fulcrum and thus able to find their own equilibrium alignment position during bearing operation.	It is most costly to manufacture.

3.2.2 Reynolds Equation and Its Basic Assumptions

It is known from fluid mechanics that a necessary condition for the pressure to develop in a thin film of fluid is that the gradient and slope of the velocity profile must vary across the thickness of the film. The analysis of hydrodynamic bearings can be performed theoretically by correctly modeling the bearing fluid film. The assumptions for calculating the performance of the hydrodynamic bearing include the following:

1. Film thickness is small compared with journal dimensions.
2. Journal is cylindrical and the bearing surface is without local distortion.
3. The journal axis is parallel to the bearing axis.
4. Inertia of fluid in film is negligible.

5. Fluid film is unable to sustain sub-atmospheric pressure.
6. Fluid pressure is atmospheric in the supply, in the drain, and at the region where the fluid film is broken or cavitated.
7. There is laminar flow in the bearing fluid film.
8. There is viscous shearing loss in the clearance region outside the pressure field, and this space is partly filled with fluid.
9. No frictional losses from the fluid in the groove or drain spaces adjoining the journal.
10. The fluid is a simple Newtonian liquid with viscosity independent of the shear rate.
11. The viscosity and density of the fluid are constant throughout the bearing.

In real bearings, departures from these assumptions are bound to be there; however, it appears that in bearings for turbomachinery a number of the assumptions are approximately fulfilled, and in many cases, departures do not seriously invalidate the overall results of the conventional theory. A departure is likely to be serious when the film flow is turbulent.

The analysis by which these assumptions are applied in calculating the dynamic characteristics of the hydrodynamic bearings is only outlined here. A detailed treatment can be found in Cameron (1981) and Hamrock (1994). The governing equation that represents the dynamic behavior of a hydrodynamic bearing was first derived by Reynolds (1886) and is

$$\frac{\partial}{\partial s}\left(\frac{\rho h^3}{\mu}\frac{\partial p}{\partial s}\right)+\frac{\partial}{\partial z}\left(\frac{\rho h^3}{\mu}\frac{\partial p}{\partial z}\right)=6U\frac{\partial}{\partial s}(\rho h)+12\frac{\partial}{\partial t}(\rho h) \tag{3.78}$$

where s is the distance around the bearing circumference of the point under consideration measured from some arbitrary reference, z is the position of the point under consideration in the axial direction, $h(s, t)$ is the film clearance, p is the lubricant pressure, U is the tangential velocity of the journal surface, ρ is the density of the lubricant, μ is the dynamic viscosity of the lubricant, and t is the time. The effects of curvature can be ignored for the purposes of evaluating the lubricant pressure variation because the fluid-film thickness is very small compared with the journal diameter. By doing this we reduce the problem to two dimensions, and for the purpose of mathematical modeling, the lubricant film may be considered as through it were unwrapped from around the journal as indicated in Figure 3.20.

In Figure 3.20, θ is the angular position around the bearing measured from the line JB produced, ϕ is the attitude angle, L is the bearing length, B is the bearing center, J is the journal center, R is the journal radius, and $\overline{BJ} = e$ is the eccentricity. In most applications, the lubricant density does not vary substantially throughout the fluid film. Its viscosity might vary; a constant effective viscosity based on the lubricant thermal balance may be used for calculations. With constant lubricant density and viscosity, the factor ρ may be dropped because it is common in all terms, and ρ may be assumed to be outside the differential in Equation 3.78. With the basic assumptions mentioned and for the steady-state operation (i.e., $\partial h/\partial t = 0$), from Equation 3.78 we get

$$\frac{\partial}{\partial s}\left(h^3\frac{\partial p}{\partial s}\right)+\frac{\partial}{\partial z}\left(h^3\frac{\partial p}{\partial z}\right)=6\mu U\frac{\partial h}{\partial s} \tag{3.79}$$

The equation remains applicable to a noncircular bore of the bearing.

3.2.3 Basic Concepts and Assumptions of Fluid-Film-Bearing Models

For a given bearing and rotational speed, the lubrication theory may be used to calculate the reaction forces on the journal from the lubricant film. These forces are functions of displacements of the journal from the bearing's center and of the instantaneous journal center velocities and accelerations. Hence, for small amplitude motions measured from the static equilibrium position

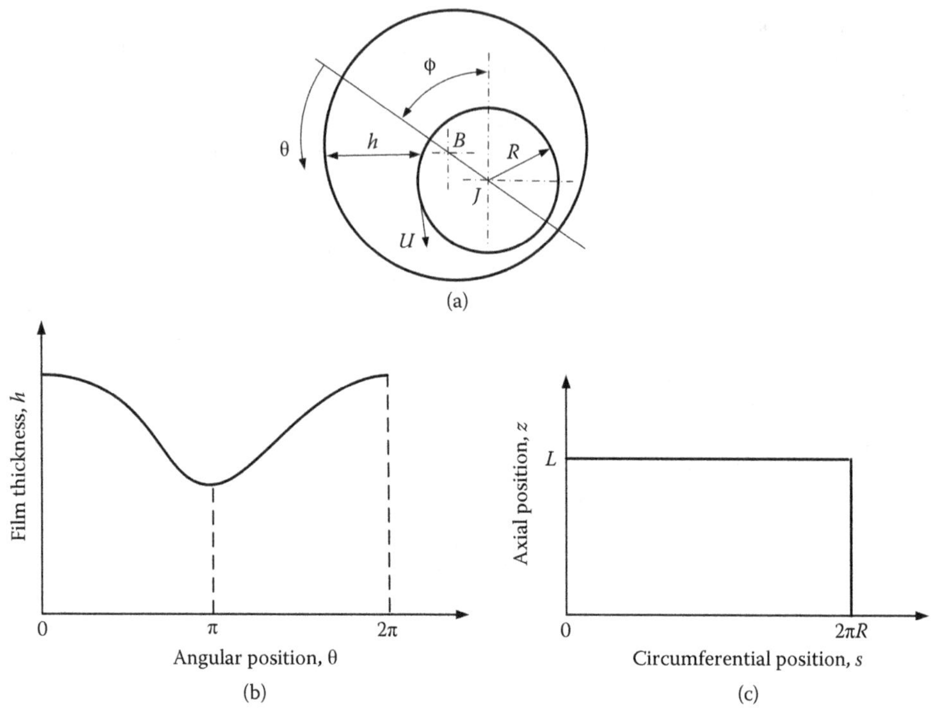

FIGURE 3.20 Analysis of the hydrodynamic bearings: (a) hydrodynamic bearing with floating journal at a static equilibrium; (b) unwrapped diagram of the bearing film thickness; (c) unwrapping of the bearing inner surface.

(see Figure 3.21) of the journal (u_0, v_0), a first-order Taylor series expansion of fluid force functions yields (more detailed steps are provided in Section 3.2.7)

where

$$\begin{aligned} \mathfrak{R}_x &= \mathfrak{R}_{u_o} + k_{xx}x + k_{xy}y + c_{xx}\dot{x} + c_{xy}\dot{y} + m_{xx}\ddot{x} + m_{xy}\ddot{y} \\ \mathfrak{R}_y &= \mathfrak{R}_{v_o} + k_{yx}x + k_{yy}y + c_{yx}\dot{x} + c_{yy}\dot{y} + m_{yx}\ddot{x} + m_{yy}\ddot{y} \end{aligned} \tag{3.80}$$

$$k_{xy} = \left(\frac{\partial \mathfrak{R}_x}{\partial y}\right)_{(u_o v_o)} \qquad c_{xy} = \left(\frac{\partial \mathfrak{R}_x}{\partial \dot{y}}\right)_{(u_o, v_o)} \qquad m_{xy} = \left(\frac{\partial \mathfrak{R}_x}{\partial \ddot{y}}\right)_{(u_o, v_o)}$$

$$\mathfrak{R}_x = f_x - m\ddot{x} \qquad \mathfrak{R}_y = f_y - m\ddot{y}$$

and analogously the remaining bearing dynamic coefficients may be defined. In a matrix form of Equation 3.80, all diagonal terms are called *direct* coefficients and off-diagonal terms are called *cross-coupled* terms. The latter terms arise due to the fluid rotation within the bearing. $\mathfrak{R}$ is the reaction force of the fluid film on the journal, $f(t)$ is the external excitation force on the journal, m is the journal mass, u_0 and v_0 are the static equilibrium position of the journal from the bearing center, x and y are the displacements of the journal from its static equilibrium position, $\dot{x}$ and $\dot{y}$ are the instantaneous journal center velocities, and $\ddot{x}$ and $\ddot{y}$ are the instantaneous journal center accelerations in the horizontal and vertical directions, respectively. The "dot" indicates the time derivatives, and k_{ij}, c_{ij}, and m_{ij} ($i, j = x, y$) are the bearing stiffness, damping, and added-mass (also termed the virtual fluid-film mass or inertia) coefficients, respectively. The indices of the stiffness, damping, and added-mass coefficients have the following significance: The first index gives the direction of

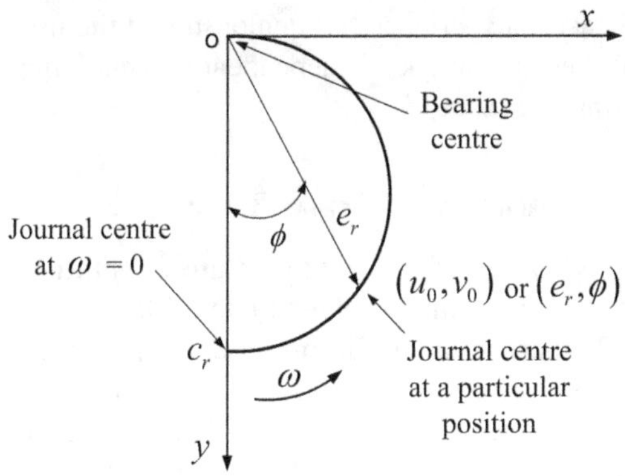

FIGURE 3.21 The equilibrium locus of the journal center of a fluid-film bearing.

loading, which produces the elastic (damping/inertia) force, and the second index gives the direction of the displacement (velocity/acceleration). Because (u_0, v_0) is the equilibrium position, then $\mathfrak{R}_{v0}$ equals the static load, W.

The equilibrium position depends on a unique value of the dimensionless Sommerfeld number, $S = (\mu NDL/W)(R/c_r)^2$, where ρ is the lubricant viscosity, N is the journal rotational speed (in rev/s), D is the bearing bore, R is the journal radius, L is the bearing length, and c_r is the bearing radial clearance. The Sommerfeld number, S, defines the operating conditions (speed, lubricant viscosity, static load, and geometry). The dynamic coefficients are evaluated for a particular static equilibrium position, which is a function of the Sommerfeld number, S, which means that for a given application, they are functions of rotor speed. Moreover, bearing dynamic coefficients could be dependent on the external excitation frequency, Ω.

It should be noted that Equation 3.80 is a complete form of a linearized fluid-film dynamic equation and it contains 12 stiffness, damping, and added-mass coefficients. Consistent with the assumptions inherent in reducing the Navier–Stokes equations to the Reynolds equation, the conventional laminar, thin film lubrication theory ignores the inertia forces in the fluid film (Pinkus and Sternlicht, 1961; Schlichting, 2004). This is theoretically justified for small values of the Reynolds number (on the order of 1). On the other hand, the assumption of laminar flow ceases to be valid when there is a transition to either Taylor vortex flow or to turbulent flow, which, for the fluid-film cylindrical journal bearing, occurs at a Reynolds number of approximately 500–1000. Thus, there is an intermediate range, say, for values of Reynolds number on the order of 100, where added-mass effects may become noticeable (several times more than the mass of the journal itself) without affecting the assumption of laminar flow. The added-mass coefficients represent the mass of the bearing fluid film (Reinhardt and Lund, 1975), but are significant only in exceptional cases, and in most analyses the added-mass coefficients of the bearing film are ignored. The stiffness and damping coefficients can be obtained by a finite difference solution of the perturbed Reynolds equation (Lund and Thomsen, 1978).

From the lubrication theory (ignoring inertia effects), damping coefficients are symmetric but stiffness coefficients are not. Therefore, principal directions do not exist, although this was assumed by Hagg and Sankey (1956) and Duffin and Johnson (1966–1967), and in the experimental determination of the coefficients, it is necessary to obtain two independent sets of amplitude-force measurements. Lund (1987) emphasized the experimental measurement of the bearing coefficients and established more uniform agreement with analytical calculations by considering the influence of thermal and elastic deformations and practical problems of manufacturing and operating tolerances of bearing geometry, clearance, and lubricant viscosity. Although the load–displacement characteristics of a journal bearing are evidently nonlinear, the concept of linear dynamic coefficients is still used for modern rotor dynamic calculations for the unbalance responses, damped natural frequencies, and stability because experience has demonstrated the usefulness of the coefficients. Clearly the applicability of these techniques is dependent on small amplitudes of vibration at the journal (relative to the bearing clearance).

3.2.4 Short and Long Hydrodynamic Radial Bearings

Equation 3.79 describes the variation of lubricant pressure in both the axial (i.e., z) and circumferential (i.e., s) directions. An approximate solution (closed form) may be obtained by using the *short bearing* (Ocvirk, 1952) approximation. In this case, the pressure variation in the circumferential direction (i.e., $\partial p/\partial s$) is assumed to be negligible compared with that in the axial direction (i.e., $\partial p/\partial z$) such that $\partial p/\partial s$ is set to zero (Figure 3.22a). This approximation also enables the closed-form solution of Equation 3.79 to be obtained, provided that appropriate boundary conditions (Figure 3.23) are selected to enable evolution of the constants of integration (Cameron, 1981; Pinkus and Sternlicht, 1961; Rao, 2000).

The conventional boundary condition for the hydrodynamic bearing film is $p=0$, with p measured relative to atmospheric pressure. The position of the boundary where this condition holds is known when the film starts at a fluid-supply groove and when it ends at the side of a bearing or at a fluid-drain groove. An additional assumption is required to determine the position at which the hydrodynamic bearing film starts or terminates when this is not fixed by a supply or drain groove. A widely used assumption takes the hydrodynamic bearing film as starting at $\theta=0$, the position of maximum clearance (Figure 3.20a). This assumption is

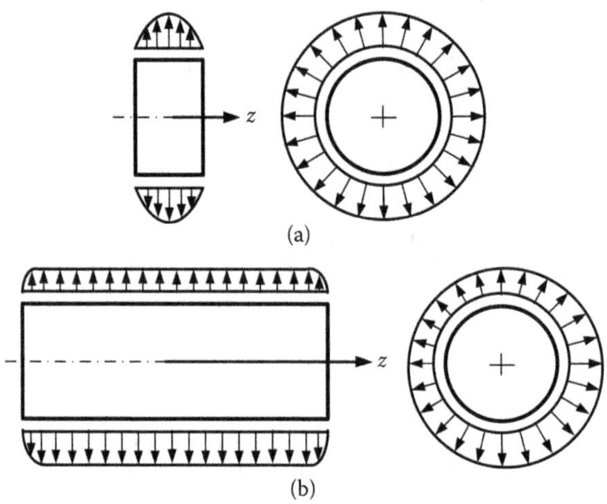

FIGURE 3.22 The short and long bearing approximation: (a) a short bearing; (b) a long bearing.

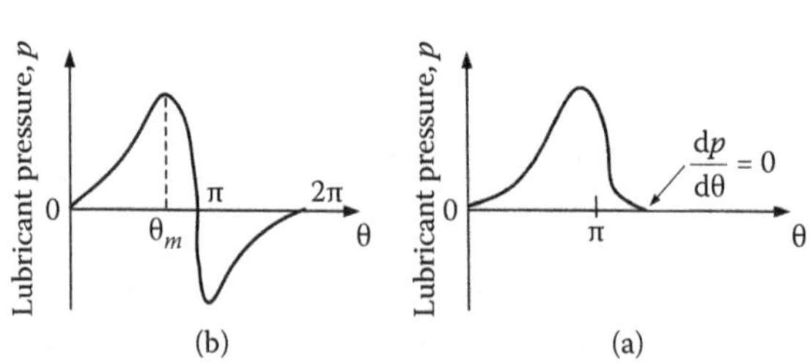

FIGURE 3.23 Boundary conditions used in journal bearing analysis: (a) full ($p=0$ at $\theta=0$ and $\theta=2\pi$) and half (set $p=0$ if computed $p<0$); Sommerfeld conditions; (b) Reynolds condition ($dp/d\theta=0$ when $p=0$).

convenient for calculation, but is not fulfilled in practice unless there is sufficient fluid available to form a continuous film at that position. Calculations show, however, that if the film is assumed to start at a small angle after the position of maximum clearance, the overall characteristics of the bearings are only slightly affected at normal operating attitudes. The additional condition, which is taken in the conventional calculation for defining the position at which the hydrodynamic bearing film terminates within clearance, is expressed as $\partial p / \partial \theta = 0$. This condition is justified by the following argument, which applies when $p=0$ holds at the boundary—if $\partial p / \partial \theta$ had been positive, there could have been a longer film with positive pressure; if $\partial p / \partial \theta$ had been negative, there would have been negative pressure in the film.

The eight linearized stiffness and damping coefficients depend on the steady-state operating conditions of the journal, and in particular upon the angular speed, ω. For the short bearing, the dimensionless bearing stiffness and damping coefficients, $\bar{k}_{ij} = k_{ij}c_r / W$, $\bar{c}_{ij} = c_{ij}c_r\omega / W$, and $i, j = x, y$, as a function of the steady eccentricity ratio ε of the bearing are given as (Smith, 1969; Hamrock, 1994)

$$
\begin{aligned}
\bar{k}_{xy} &= \frac{\pi\left\{\pi^2 - 2\pi^2\varepsilon^2 - \left(16-\pi^2\right)\varepsilon^4\right\}Q(\varepsilon)}{\varepsilon\sqrt{1-\varepsilon^2}} \\
\bar{k}_{yy} &= \frac{4\left\{\pi^2 + \left(32+\pi^2\right)\varepsilon^2 + 2\left(16-\pi^2\right)\varepsilon^4\right\}Q(\varepsilon)}{\left(1-\varepsilon^2\right)} \\
\bar{k}_{yx} &= \frac{-\pi\left\{\pi^2 + \left(32+\pi^2\right)\varepsilon^2 + 2\left(16-\pi^2\right)\varepsilon^4\right\}Q(\varepsilon)}{\varepsilon\sqrt{1-\varepsilon^2}} \\
\bar{k}_{xx} &= 4\left\{2\pi^2 + \left(16-\pi^2\right)\varepsilon^4\right\}Q(\varepsilon) \\
\bar{c}_{xx} &= \frac{2\pi\sqrt{1-\varepsilon^2}\left\{\pi^2 + 2\left(\pi^2-8\right)\varepsilon^2\right\}Q(\varepsilon)}{\varepsilon} \\
\bar{c}_{xy} &= \bar{c}_{yx} = -8\left\{\pi^2 + 2\left(\pi^2-8\right)\varepsilon^2\right\}Q(\varepsilon) \\
\bar{c}_{yy} &= \frac{2\pi\left\{\pi^2 + 2\left(24-\pi^2\right)\varepsilon^2 + \pi^2\varepsilon^4\right\}Q(\varepsilon)}{\varepsilon\sqrt{1-\varepsilon^2}}
\end{aligned}
\tag{3.81}
$$

with

$$
Q(\varepsilon) = \frac{1}{\left\{\pi^2 + \left(16-\pi^2\right)\varepsilon^2\right\}^{3/2}} \quad \text{and} \quad \varepsilon = \frac{e_r}{c_r} \tag{3.82}
$$

where e_r is the journal eccentricity and c_r is the radial clearance. It should be noted that $k_{xy} \neq k_{yx}$, whereas $c_{xy} = c_{yx}$. To determine the stiffness and damping coefficients of a short bearing, the Sommerfeld number or bearing characteristic number

$$
S = \frac{\mu DLN}{W}\left(\frac{R}{c_r}\right)^2 \tag{3.83}
$$

is first determined, where W is the load on the bearing, D is the bearing bore, R is the journal radius, L is the length of the bearing, μ is the viscosity of the lubricant at operating temperature, $\omega=(2\pi N)$ is the angular speed of the journal in rad/s, and N is the number of revolutions per second. We can then determine the eccentricity ratio under steady-state operating conditions by

$$S\left(\frac{L}{D}\right)^2 = \frac{\left(1-\varepsilon^2\right)^2}{\pi\varepsilon\sqrt{\pi^2\left(1-\varepsilon^2\right)+16\varepsilon^2}} \tag{3.84}$$

Variations of the dimensionless stiffness and damping coefficients with the Sommerfeld number for the case of short bearing approximation are shown in Figure 3.24a and b. Variation of the eccentricity ratio with the Sommerfeld number is given in Figure 3.25. An algorithm for finding dimensionless bearing dynamic parameters from the Sommerfeld number is given in Figure 3.26.

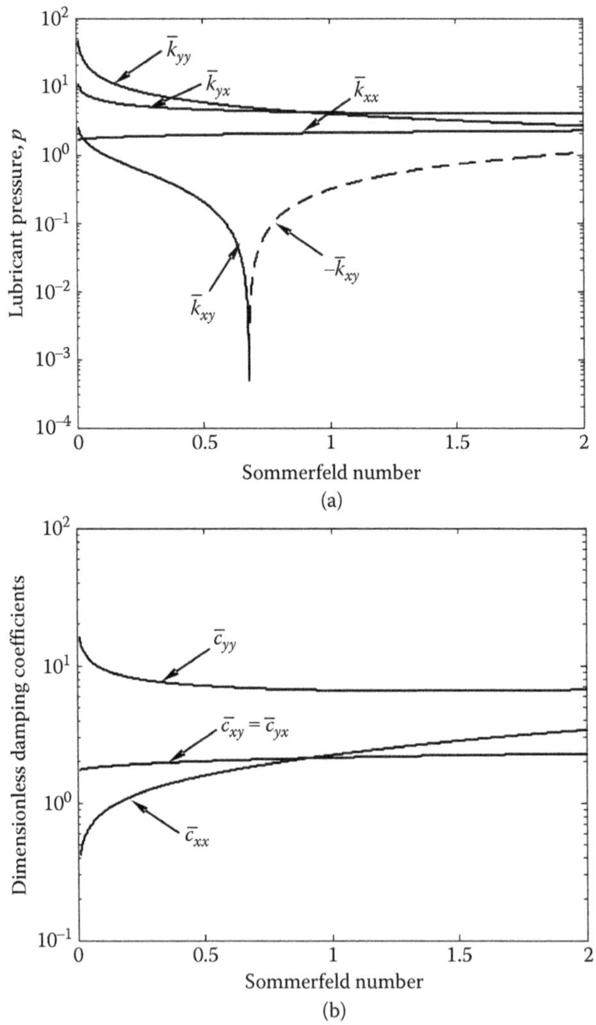

FIGURE 3.24 (a) Variations of dimensionless stiffness coefficients for short bearings. (b) Variations of dimensionless damping coefficients for short bearings.

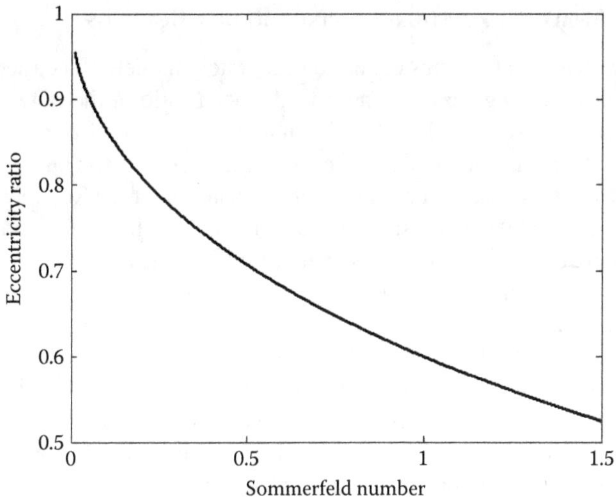

FIGURE 3.25 Variation of the eccentricity ratio with the Sommerfeld number.

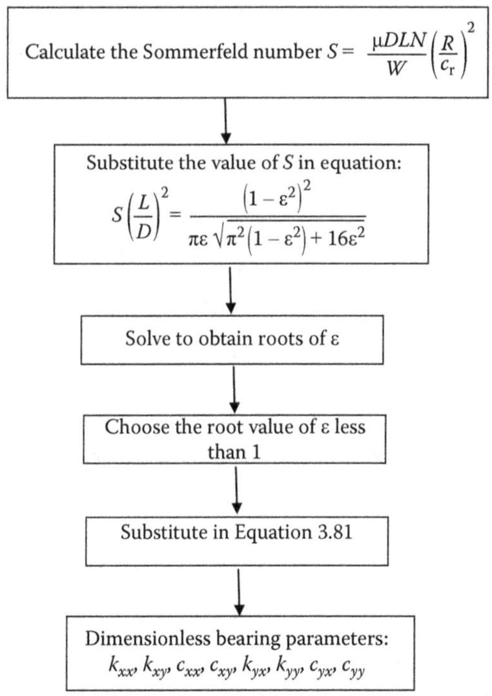

FIGURE 3.26 Flow chart of the procedure for obtaining the bearing dynamic parameters with the short bearing approximation.

For the "long bearing" approximation (Hamrock, 1994), the pressure variation in the axial direction is assumed to be negligible compared with that in the circumferential direction and so $\partial p/\partial z$ is set to zero (Figure 3.22b).

3.2.5 Numerical Analysis Procedure of Finite Radial Bearings

In practice, real hydrodynamic bearings cannot be accurately modeled as either short or long bearing approximation. Most bearings have a length-to-diameter ratio in the range 0.5–1.5. Hence, the approximate solutions of the short or long bearing may be used in preliminary design calculations only. However, the final design studies will often require a more realistic analysis by at least using a two-dimensional Reynolds equation. For a more general solution of the Reynolds equation, which is a partial differential equation, it is necessary to use numerical methods. The most common numerical technique used in hydrodynamic journal bearing analysis is the finite difference method because of the relative ease with which it can be adapted to the most bearing geometries. The problem is reduced to two dimensions by considering the fluid film to be unwrapped from around the journal (Figure 3.20c). The unwrapped fluid film is then divided into a number of sections of a finite size by describing it in terms of mesh nodes, as shown in Figure 3.27a. Here s is the circumferential position and z is the axial position. In between each node described, a number of further points are considered—for example, those surrounding the node (i, j) are shown in Figure 3.27b. Because the Reynolds equation describes the behavior of the lubricant at any location in the fluid film, it can be written for every specific node contained within the finite difference mesh.

Equation 3.79 may be written for the node (i, j) as

$$\frac{h^3_{i,j+\frac{1}{2}}\dfrac{p_{i,j+1}-p_{i,j}}{\Delta s}-h^3_{i,j-\frac{1}{2}}\dfrac{p_{i,j}-p_{i,j-1}}{\Delta z}}{\Delta s}+\frac{h^3_{i+\frac{1}{2},j}\dfrac{p_{i+1,j}-p_{i,j}}{\Delta z}-h^3_{i-\frac{1}{2},j}\dfrac{p_{i,j}-p_{i-1,j}}{\Delta z}}{\Delta z}=6\mu U\frac{h_{i,j+\frac{1}{2}}-h_{i,j-\frac{1}{2}}}{\Delta s} \tag{3.85}$$

which can be rearranged as

$$\begin{aligned}&p_{i,j}\left(-\frac{h^3_{i,j+1/2}}{\Delta s^2}-\frac{h^3_{i,j-1/2}}{\Delta s^2}-\frac{h^3_{i+1/2,j}}{\Delta z^2}-\frac{h^3_{i-1/2,j}}{\Delta z^2}\right)+p_{i,j+1}\left(\frac{h^3_{i,j+1/2}}{\Delta s^2}\right)+p_{i,j-1}\left(\frac{h^3_{i,j-1/2}}{\Delta s^2}\right)\\&+p_{i+1,j}\left(\frac{h^3_{i+1/2,j}}{\Delta z^2}\right)+p_{i-1,j}\left(\frac{h^3_{i-1/2,j}}{\Delta z^2}\right)=6\mu U\frac{h_{i,j+1/2}-h_{i,j-1/2}}{\Delta s}\end{aligned} \tag{3.86}$$

and finally can be simply written as

$$p_{i,j}=\frac{a_6}{a_1}+\frac{a_5}{a_1}p_{i+1,j}+\frac{a_4}{a_1}p_{i-1,j}+\frac{a_3}{a_1}p_{i,j+1}+\frac{a_2}{a_1}p_{i,j-1} \tag{3.87}$$

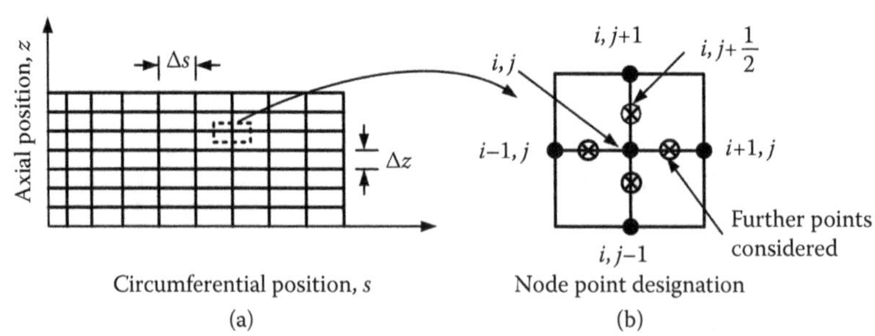

FIGURE 3.27 Finite difference mesh for modeling of the hydrodynamic bearing surface in two dimensions: (a) meshing of the two-dimensional bearing surface: (b) mesh node designation.

with

$$a_1 = -\frac{h^3_{i,j+1/2}}{\Delta s^2} - \frac{h^3_{i,j-1/2}}{\Delta s^2} - \frac{h^3_{i+1/2,j}}{\Delta z^2} - \frac{h^3_{i-1/2,j}}{\Delta z^2}; \quad a_2 = \frac{h^3_{i,j-1/2}}{\Delta s^2}; \quad a_3 = \frac{h^3_{i,j+1/2}}{\Delta s^2}$$

$$a_4 = \frac{h^3_{i-1/2,j}}{\Delta z^2}; \quad a_5 = \frac{h^3_{i+1/2,j}}{\Delta z^2}; \quad a_6 = 6\mu U \frac{h_{i,j+1/2} - h_{i,j-1/2}}{\Delta s}$$

where $h_{i,j+1/2}$ is the film thickness between the (i, j) and $(i, j+1)$ nodes, and a_1, a_2, etc., are constants whose values are known for every node. Equation 3.87 may then be written for every node in the mesh, except for those where the lubricant pressure is already known (for example, at the nodes representing the ends of the bearing (at $z=0$ and $z=L$) where the lubricant pressure must be equal to the ambient pressure). These equations can then be solved simultaneously to get lubrication pressure at all nodes.

On estimating the lubricant pressure for a particular journal attitude angle, ϕ, and eccentricity, e_r, (i.e., then film thickness will be known), the corresponding forces provided by the lubricant on the journal may be evaluated by summing numerically all of the elemental forces associated with each node. For example, each node pressure is considered to contribute to the force on the journal over an area Δs and Δz surrounding it, as shown in the shaded area in Figure 3.28a, when integrating using a trapezium rule.

Thus, if the node is situated at some angle ψ to the horizontal direction as shown in Figure 3.28b, then the lubricant force on the journal in the horizontal direction (from left to right) is

$$F_h = \sum_{j=1}^{m} \sum_{i=1}^{n} p_{ij} \cos \psi_{ij} \Delta s \Delta z \tag{3.88}$$

and that in the vertical direction (upwards) is

$$F_v = \sum_{j=1}^{m} \sum_{i=1}^{n} p_{ij} \sin \psi_{ij} \Delta s \Delta z \tag{3.89}$$

where i is the axial node positions (total of n such nodes) and j is the circumferential node position (total of m such nodes). For the present formulations, lubricant forces acting on the journal should also include the tangential friction forces; however, the friction forces are small when compared with the pressure forces.

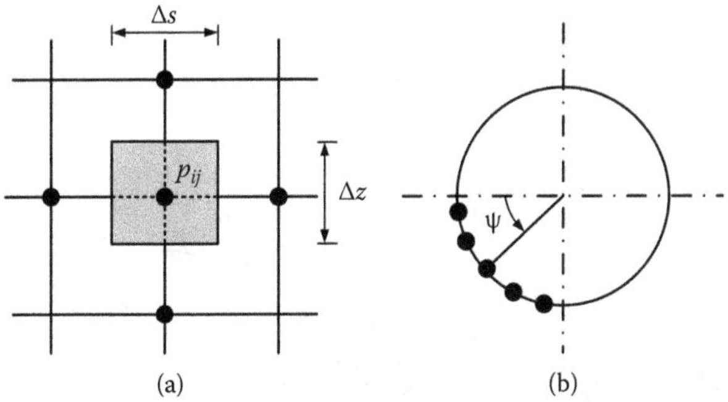

FIGURE 3.28 A closer view of (a) the grid to calculate pressure $p_{i,j}$ and (b) the node around the bearing circumference.

For a bearing designed to carry vertical loads only (for example, the gravity load), the relationship between the eccentricity ratio, ε, and the journal attitude angle, ϕ, may be determined by investigating different values of ϕ for a given value of ε until the value of F_h is found to be zero. For example, once we choose an arbitrary value of (ε, ϕ), then the corresponding film thickness can be obtained (because ε and ϕ determine the position of the shaft with respect to the bearing bore). Using an iterative procedure (or using a simultaneous-linear equation solution procedure), the pressure distribution is obtained by putting negative pressure equal to zero wherever it occurs during the calculation. The resultant forces of the journal are obtained by using Equations 3.88 and 3.89, for $F_h = 0$ and $F_v = W$. If these force conditions are not satisfied, then a different value of ϕ could be chosen until the force conditions are satisfied up to the desired accuracy. When this process is completed, it is found that because the Reynolds equation is a continuous function, the final pressure distribution corresponds to the Reynolds boundary conditions with the constraint of $\partial p / \partial \theta = 0$ at the trailing edge of the lubricant film automatically catered for. This procedure can be repeated for different values of ε to get the relationship between $\varepsilon$and ϕ for a particular bearing under different operating conditions. This trial-and-error method enables a corresponding value of (ε, ϕ), and the Sommerfeld number S to be found (refer to Figure 3.29). A typical steady-state journal locus for plain cylindrical bearings (PCB) and their relationship with the Sommerfeld number are shown in Figure 3.30. For details, readers are referred to Hamrock (1994).

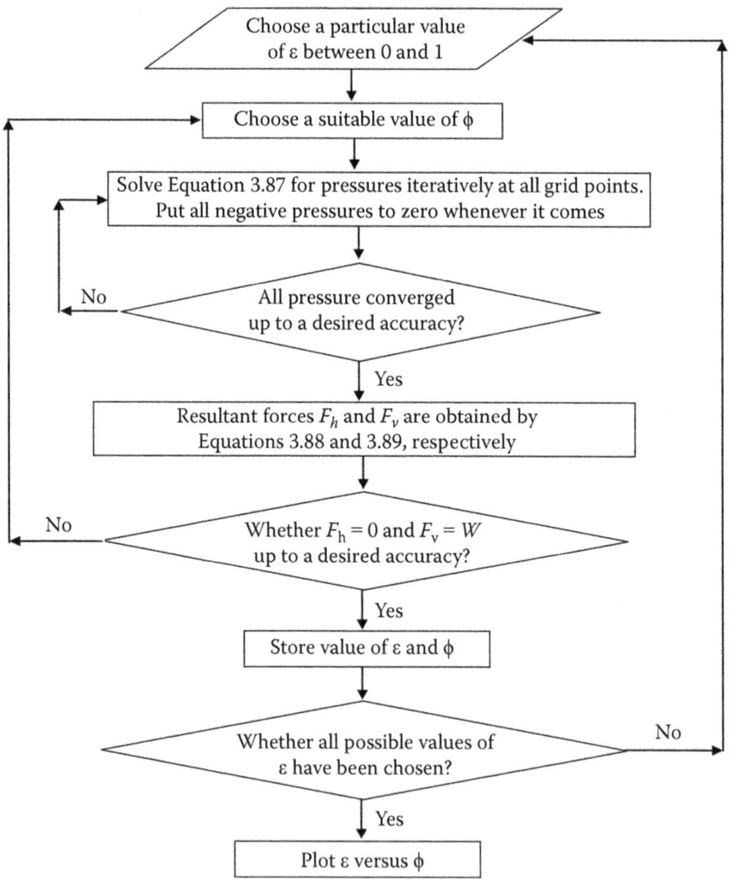

FIGURE 3.29 Flow chart of the solution of the Reynolds equation.

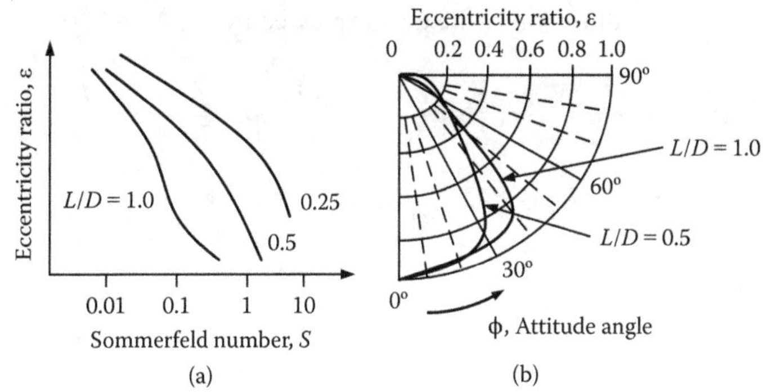

FIGURE 3.30 Typical variation of (a) the eccentricity ratio with the Sommerfeld number and (b) the eccentricity ratio with journal attitude angle for PCB for $L/D = 0.5$ and 1.0.

3.2.6 Rotordynamic Coefficients of Fluid-Film Bearings

The center of a hydrodynamic bearing journal takes a static equilibrium position for a particular speed when it is subjected to a static load (e.g., the weight of the journal). However, when a dynamic load (e.g., unbalance) is applied to the bearing, the journal oscillates about the static equilibrium position. The journal is subjected to the effective stiffness and damping forces associated with this flexibility of the lubricating film. These forces have a significant effect on the system whirl natural frequencies, critical speeds, forced responses, and the system stability. This demands an evaluation of the bearing dynamic characteristics at the design stage itself. This is obtained by providing some small displacements (perturbations) dx and dy of the journal away from its static equilibrium position in the horizontal and vertical directions, respectively. The journal center is also assumed to possess velocities $d\dot{x}$ and $d\dot{y}$ in the horizontal and vertical directions, respectively (whereas under the action of only a steady load, its velocity would be zero). The effects of these perturbations on the state of the journal on the lubricant film forces are then calculated. When there is no dynamic load on the journal (i.e., when $\dot{x}$ and $\dot{y}$ are both zero), the lubricant steady-state film forces in the horizontal and vertical directions could be expressed as functions of the journal positions, x and y, and velocities, $\dot{x}$ and $\dot{y}$, away from the bearing center; that is

$$F_x = f_x(x, y, \dot{x}, \dot{y}) \quad \text{and} \quad F_y = f_y(x, y, \dot{x}, \dot{y}) \tag{3.90}$$

While there are perturbations in these displacements and velocities, the new values of lubricant film forces will be $F_x + dF_x$ and $F_y + dF_y$, respectively. These forces may also be written using Equation 3.90 as a four-variable Taylor series, neglecting higher-order small terms such as

$$\begin{aligned} F_x + dF_x &= f_x(x+dx, y+dy, \dot{x}+d\dot{x}, \dot{y}+d\dot{y}) \\ &= f_x(x, y, \dot{x}, \dot{y}) + dx\frac{\partial F_x}{\partial x} + dy\frac{\partial F_x}{\partial y} + d\dot{x}\frac{\partial F_x}{\partial \dot{x}} + d\dot{y}\frac{\partial F_x}{\partial \dot{y}} \end{aligned} \tag{3.91}$$

and

$$\begin{aligned} F_y + dF_y &= f_y(x+dx, y+dy, \dot{x}+d\dot{x}, \dot{y}+d\dot{y}) \\ &= f_y(x, y, \dot{x}, \dot{y}) + dx\frac{\partial F_y}{\partial x} + dy\frac{\partial F_y}{\partial y} + d\dot{x}\frac{\partial F_y}{\partial \dot{x}} + d\dot{y}\frac{\partial F_y}{\partial \dot{y}} \end{aligned} \tag{3.92}$$

The variations in the lubricant film forces are then expressed by

$$dF_x = (F_x + dF_x) - F_x = dx\frac{\partial F_x}{\partial x} + dy\frac{\partial F_x}{\partial y} + d\dot{x}\frac{\partial F_x}{\partial \dot{x}} + d\dot{y}\frac{\partial F_x}{\partial \dot{y}}$$
$$= k_{xx}dx + k_{xy}dy + c_{xx}d\dot{x} + c_{xy}d\dot{y} \tag{3.93}$$

and

$$dF_y = (F_y + dF_y) - F_y = dx\frac{\partial F_y}{\partial x} + dy\frac{\partial F_y}{\partial y} + d\dot{x}\frac{\partial F_y}{\partial \dot{x}} + d\dot{y}\frac{\partial F_y}{\partial \dot{y}}$$
$$= k_{yx}dx + k_{yy}dy + c_{yx}d\dot{x} + c_{yy}d\dot{y} \tag{3.94}$$

where k_{ij} and c_{ij} are known as the fluid-film stiffness and damping coefficients, respectively, and i and j can take values of x and y, respectively. These bearing dynamic coefficients must be known in order to calculate the overall system dynamic response to an external forcing and its stability. Values of k_{xx} and k_{yx} can be found from the Reynolds equation by calculating the lubricant film forces F_{x1} and F_{y1} when a small displacement dx is imposed on the journal (from its static equilibrium position) with dy, $d\dot{x}$, and $d\dot{y}$ all set to zero. The process is then repeated with a displacement of $-dx$ to calculate F_{x2} and F_{y2}. The stiffness coefficients k_{xx} and k_{yx} are then given by

$$k_{xx} = \frac{F_{x1} - F_{x2}}{2dx} \quad \text{and} \quad k_{yx} = \frac{F_{y1} - F_{y2}}{2dx} \tag{3.95}$$

Remaining stiffness and damping coefficients may be obtained by giving small variations in dy, $d\dot{x}$, and $d\dot{y}$ on the journal one at a time, and finding their effect on the lubricant film forces, as

$$\begin{aligned} k_{xy} &= \frac{F_{x3} - F_{x4}}{2dy}; \quad & k_{yy} &= \frac{F_{y3} - F_{y4}}{2dy}; \quad & c_{xx} &= \frac{F_{x5} - F_{x6}}{2d\dot{x}} \\ c_{yx} &= \frac{F_{y5} - F_{y6}}{2d\dot{x}}; \quad & c_{xy} &= \frac{F_{x7} - F_{x8}}{2d\dot{y}}; \quad & c_{yy} &= \frac{F_{y7} - F_{y8}}{2d\dot{y}} \end{aligned} \tag{3.96}$$

While formulating Equation 3.79 in the finite difference form, the quantity $\partial h / \partial t$ is no longer zero, when $d\dot{x}$ or $d\dot{y}$ are nonzero, and the appropriate value of U is also dependent on $d\dot{x}$ and $d\dot{y}$. Solutions for several types of bearings have been carried out (Lund et al., 1965; Someya, 1989). In addition to the "spring" and "damping" coefficients introduced earlier, some authors have suggested the use of "inertia" coefficient and "moment" coefficients, although in common applications such coefficients have a very small effect on the rotor dynamics. Inertia coefficients can be defined from fluid forces, as follows:

$$F_x(x, y, \dot{x}, \dot{y}, \ddot{x}, \ddot{y}) = 0 \quad \text{and} \quad F_y(x, y, \dot{x}, \dot{y}, \ddot{x}, \ddot{y}) = 0 \tag{3.97}$$

For perturbations in displacements, velocities, and accelerations, we have

$$dF_x = \cdots + \frac{\partial F_x}{\partial \ddot{x}}d\ddot{x} + \frac{\partial F_x}{\partial \ddot{y}}d\ddot{y} = \cdots + m_{xx}d\ddot{x} + m_{xy}d\ddot{y} \tag{3.98}$$

where m_{ij} represents the inertia coefficients. Moment coefficients can be defined from the fluid force, as follows:

$$F_x(x, y, \varphi_y, \varphi_x, \dot{x}, \dot{y}, \dot{\varphi}_y, \dot{\varphi}_x) = 0 \tag{3.99}$$

For perturbations in the linear and angular displacements, velocities, and accelerations, we have

$$dF_x = \cdots + \frac{\partial F_x}{\partial \varphi_y} d\varphi_y + \frac{\partial F_x}{\partial \varphi_x} d\varphi_x + \cdots = \cdots + k_{x\varphi_y} d\varphi_y + k_{x\varphi_x} d\varphi_x + \cdots \quad (3.100)$$

and

$$dM_{zx} = \cdots + \frac{\partial M_{zx}}{\partial \varphi_y} d\varphi_y + \frac{\partial M_{zx}}{\partial \varphi_x} d\varphi_x + \cdots = \cdots + k_{\varphi_y\varphi_y} d\varphi_y + k_{\varphi_y\varphi_x} d\varphi_x + \cdots \quad (3.101)$$

where $k_{\varphi_y\varphi_y}, k_{\varphi_y\varphi_x}, \ldots$ are moment stiffness coefficients. These moment coefficients are predominant only for long journal bearings.

3.3 DYNAMIC SEALS

Dynamic seals are used where sealing takes place between two surfaces with relative movement, viz. the rotary, reciprocating, and oscillating. The main focus of the present section is on rotary seals for rotordynamic applications. This has a wide variety of applications in the high-speed, high-pressure, and cryogenic temperature conditions of aviation and space industries such as in turbine stages, turbo-pumps, compressors, and gearboxes.

3.3.1 Classification of Seals

Seals are broadly classified as liquid and gas seals according to the working fluid used in the system. The most common working fluids are water/steam, air, nitrogen, trifluorobromomethane ($CBrF_3$), liquid oxygen, and liquid hydrogen. In addition, they can be categorized as static and dynamic seals. Static seals are used where the two surfaces do not move relative to one another. Gasket-type seals are static seals (Figure 3.31). Rotary seals can be subdivided into two main categories: clearance seals and contact seals. Clearance seals are circumferential noncontacting seals (Figure 3.32a). In contact seals, the contact is formed by the positive pressure, whereas in the case of clearance seals, they operate with the positive clearance (no rubbing contact). The most commonly used material for dynamic seals (especially for rotary seals) are stainless steel, bronze, aluminum, nickel-based alloys, and polytetrafluoroethylene. Figure 3.32a shows a typical rotary seal with the clearance exaggerated. Rotary seals based on geometry can be classified as follows:

1. *Ungrooved plain seals (or smooth annular seals)*: (i) straight (Figure 3.32a and b), (ii) tapered (Figure 3.32c), and (iii) stepped (Figure 3.32d). In geometry, they are similar to journal bearings but the clearance/radius ratio is as low as two times and as high as ten times (or more) larger to avoid rotor/stator contact.

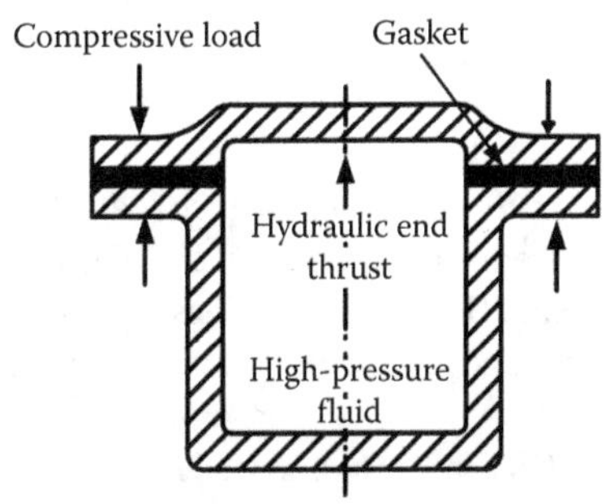

FIGURE 3.31 Static seal (gasket).

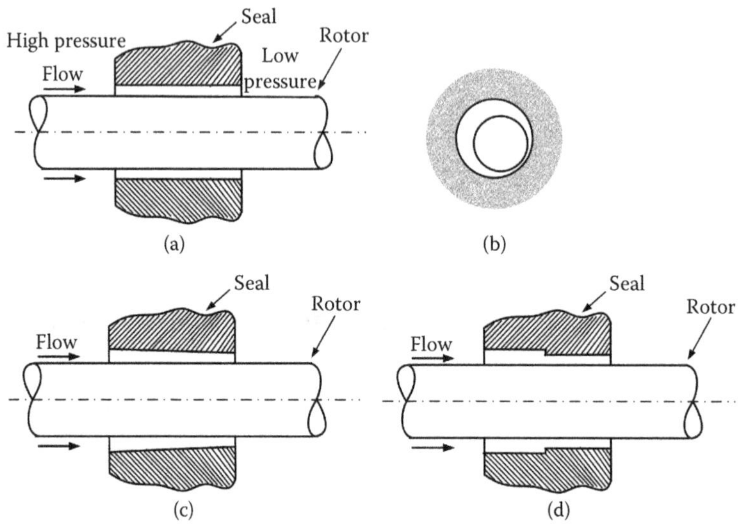

FIGURE 3.32 (a) Rotor-seal assembly. (b) Straight annular seal. (c) Tapered annular seal (converging). (d) Stepped annular seal.

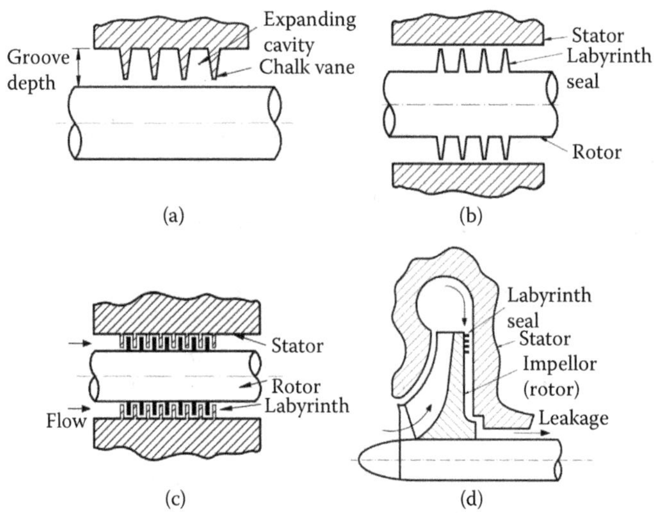

FIGURE 3.33 (a) Labyrinth seal (teeth-on-stator) axial flow type. (b) Labyrinth seal (teeth-on-rotor) axial flow type. (c) Labyrinth seal (teeth-on-stator and teeth-on-rotor) axial flow type. (d) Labyrinth seal radial flow type.

2. *Grooved/roughened surface seals*: (i) porous surface seals, (ii) labyrinth seals (Figure 3.33a through d), (iii) helically grooved/screw seals, (iv) circular holes or triangular pattern seals, and (v) honeycomb pattern seals (Figure 3.34). These seals are used in the centrifugal and axial compressors and pumps and in turbines. The different internal surface patterns of seals are shown in Figure 3.35.
3. *Contact seals*: (i) brush seals (Figure 3.36a), (ii) face seals, and (iii) lip seals (Figure 3.36b). Because of rubbing, these seals are used commonly in low-speed pumps or where the working fluid can act as a coolant. Contact seals provide much lower leakage rates than

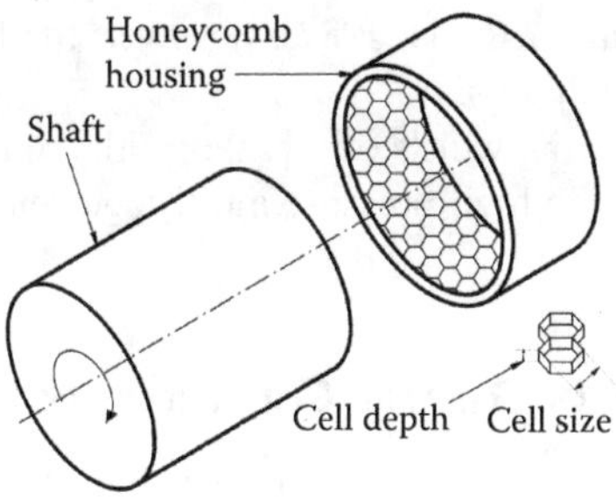

FIGURE 3.34 Honeycomb seal.

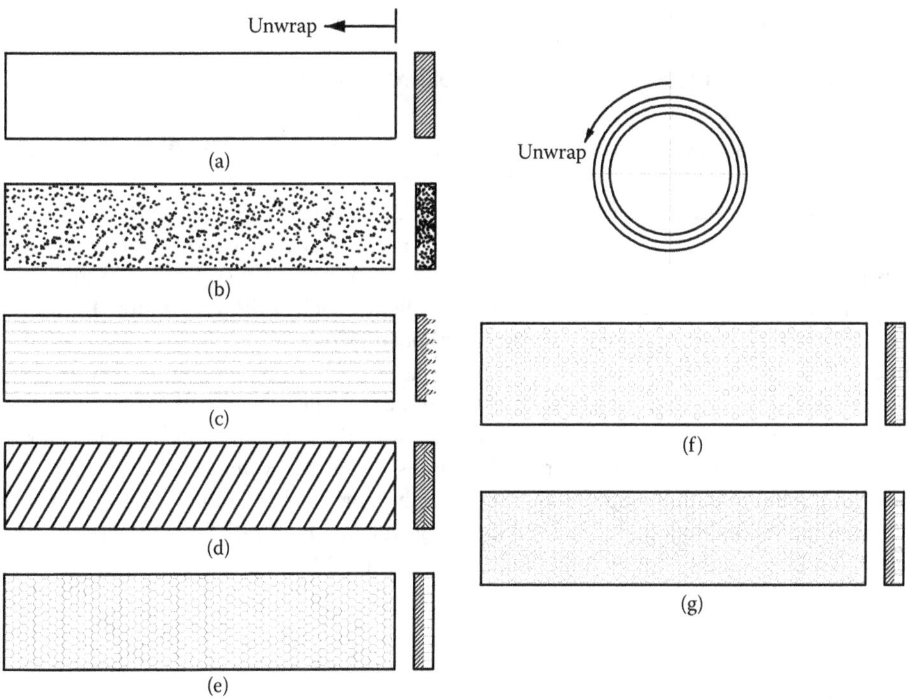

FIGURE 3.35 Different internal surface patterns on seals: (a) plain seal; (b) plain seal with porous material; (c) labyrinth seal; (d) helically grooved seal; (e) honeycomb seal; (f) hole pattern seal; (g) triangular pattern roughness seal.

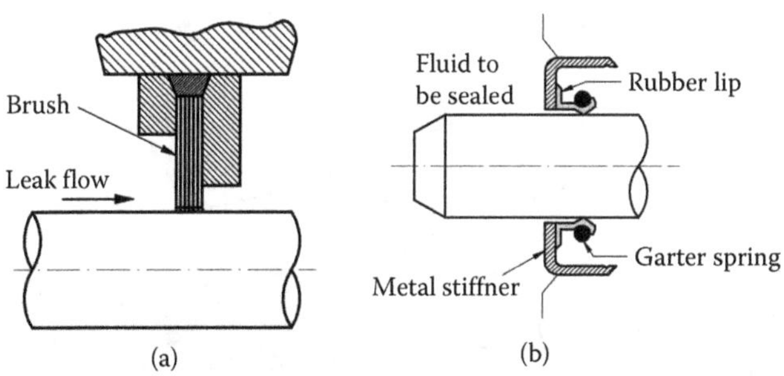

FIGURE 3.36 (a) Brush seal. (b) Lip seal.

either of the noncontact seals (Adams, 1987); however, the latter can operate under very high-speed and pressure conditions.

4. *Floating-ring oil seals*: The ring whirls or vibrates with the rotor in the lubricating oil but does not spin. They are used in high-pressure multistage centrifugal compressors.

3.3.2 Theoretical Estimation of Dynamic Coefficients of Seals

In this section, basic governing equations to obtain the dynamic coefficients of smooth annular turbulent seals (smooth seals) are presented. Dynamic coefficients are calculated from the approximate solution of the bulk-flow theory for the configuration of the test rig. Effects of rotor speeds, seal dimensions, and operating conditions on these dynamic coefficients are also presented and discussed in detail in Example 3.5.

3.3.2.1 Basic Governing Equations and Approximate Solution

In an annular seal, flows are usually turbulent because of the high Reynolds numbers at which they operate. Black and his coworkers (Black, 1969; Black and Jenssen, 1969–1970) were the first to attempt to model and identify the rotor dynamics effects of turbulent annular seals using bulk-flow models (along similar lines to those of Reynolds lubrication equations). Bulk-flow models employ velocity components, $\bar{u}_z(z,\theta)$ and $\bar{u}_\theta(z,\theta)$, that are averaged over the clearance, where $\bar{u}_z$ and $\bar{u}_\theta$ are velocities in the subscript directions, and z and θ are the coordinates as shown in Figure 3.37. Black and Jensen used several heuristic assumptions in their model, such as the assumption that $\bar{u}_\theta = R\omega/2$, where R is the radius of the seal and ω is the rotor speed. Moreover, their governing equations do not reduce to recognizable turbulent lubrication equations. These issues caused Childs (1983b) to publish a revised version of the bulk-flow model, and the present section will focus on this.

The geometry of the seal annulus, which is filled with fluid, is sketched in Figure 3.37 and is described by coordinates of the meridian in the gap as given by $z(s)$ and $R(s)$, $0<s<L$, where the coordinate, s, is measured along that meridian and t is the time. The clearance is denoted by $h(s, \theta, t)$, where the unperturbed value of h is $\delta(s)$. Equations governing the bulk flow are averaged over the clearance. This leads to a continuity of the form in Equation 3.102:

$$\frac{\partial h}{\partial t}+\frac{\partial}{\partial s}(h\bar{u}_s)+\frac{1}{R}\frac{\partial}{\partial\theta}(h\bar{u}_\theta)+\frac{h}{R}\frac{dR}{ds}\bar{u}_s=0 \tag{3.102}$$

where $\bar{u}_s$ and $\bar{u}_\theta$ are velocities averaged over the local clearance.

The axial and circumferential momentum equations are as follows:

$$-\frac{1}{\rho}\frac{\partial P}{\partial s}=\frac{\tau_{ss}}{\rho h}+\frac{\tau_{sr}}{\rho h}-\frac{\bar{u}_\theta^2}{R}\frac{dR}{ds}+\frac{\partial\bar{u}_s}{\partial t}+\frac{\bar{u}_\theta}{R}\frac{\partial\bar{u}_s}{\partial\theta}+\bar{u}_s\frac{\partial\bar{u}_s}{\partial\theta} \tag{3.103}$$

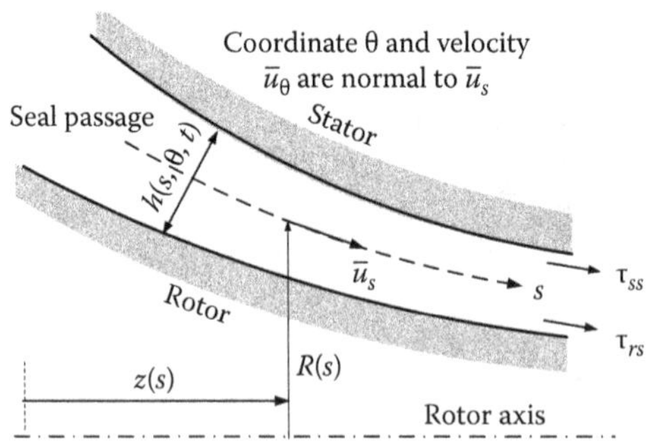

FIGURE 3.37 Fluid-filled annulus between a rotor and a stator for turbulent lubrication analysis.

and

$$-\frac{1}{\rho R}\frac{\partial P}{\partial \theta}=\frac{\tau_{\theta s}}{\rho h}+\frac{\tau_{\theta r}}{\rho h}-\frac{\partial \bar{u}_\theta}{\partial t}+\frac{\bar{u}_\theta}{R}\frac{\partial \bar{u}_\theta}{\partial \theta}+\bar{u}_s\frac{\partial \bar{u}_\theta}{\partial s}+\frac{\bar{u}_\theta \bar{u}_s}{R}\frac{\partial R}{\partial s} \tag{3.104}$$

The approach used by Hirs (1973) is employed to determine the turbulent shear stresses, τ_{ss} and $\tau_{\theta s}$, applied to the stator by the fluid in the s and θ directions, respectively, which take the form

$$\frac{\tau_{ss}}{\rho \bar{u}_s}=\frac{\tau_{\theta s}}{\rho \bar{u}_\theta}=\frac{A_s \bar{u}_s}{2}\left\{1,+,\left(\frac{\bar{u}_\theta}{\bar{u}_s}\right)\right\}^{\frac{m_s+1}{2}}(Re_s)^{m_s} \tag{3.105}$$

and the stresses, τ_{sr} and $\tau_{\theta r}$, applied to the rotor by the fluid in the s and θ directions, respectively, which are obtained as

$$\frac{\tau_{sr}}{\rho \bar{u}_s}=\frac{\tau_{\theta r}}{\rho(\bar{u}_\theta-\omega R)}=\frac{A_r \bar{u}_s}{2}\left\{1+\left(\frac{\bar{u}_\theta-\omega R}{\bar{u}_s}\right)^2\right\}^{\frac{m_\theta+1}{2}}(Re_s)^{m_\theta} \tag{3.106}$$

where the local meridian Reynolds number is given as

$$Re_s=\frac{h\bar{u}_s}{\nu} \tag{3.107}$$

and constants A_s, A_r, m_s, and m_θ are chosen to fit the available data on turbulent shear stresses. Childs (1983a) typically used the following values of these constants:

$$A_s = A_s = 0.066; \quad m_s = m_\theta = -0.25 \tag{3.108}$$

Now the solutions for the governing equation are presented and discussed in detail.

3.3.2.2 Approximate Dynamic Coefficients of Seals

A model of a typical annual (or clearance) seal is shown in Figure 3.32a. The geometrical shape of a clearance seal is similar to that of a hydrodynamic bearing; however, they are different in the following aspects. To avoid contact between a rotor and a stator, the ratio of the clearance to the shaft radius in seals is made two to ten times larger than that of hydrodynamic bearings. The flow in seals is turbulent, and in hydrodynamic bearings it is laminar. Therefore, unlike hydrodynamic bearings, one cannot use the Reynolds equation for analysis of seals.

This section compiles the theoretical and computational analyses performed by various researchers. Lomakin (1958) was the first to propose a theoretical model of a plain seal, which predicted that the axial pressure drop across the seal caused a radial stiffness, which is independent of the shaft rotation. The radial direct stiffness (k_d) is given by

$$k_d = 4.7R\left(\frac{\Delta P}{\lambda}\right)\left(\frac{\lambda L/c_r}{1.5+2\lambda L/c_r}\right)^2 \quad \text{with} \quad \lambda=\frac{0.079}{R_e^{0.25}} \tag{3.109}$$

where ΔP is the pressure drop and R, L, and c_r are the radius, axial length, and radial clearance of the seal, respectively. If the direct stiffness were the only effect of the plain seal, then its effect on critical speeds would be easily and accurately predictable. Black's work (1969, 1971) provided the major initial impetus for the extensive research and the state-of-the-art design information developed on this topic over more than last four decades. Black developed the classical theory for turbulent annular seals, considering the axial fluid flow caused by a pressure drop along the seal, the rotational fluid flow as a consequence of the shaft rotation, and a relative motion of the seal between the rotor and housing. Black (1969, 1971) and Childs (1983a, b) formulated and extended Lomakin's theory to be applicable to the rotor dynamic analysis of centrifugal pumps. Black, Childs, and others have shown, however, that k_d increases with the shaft speed (at constant ΔP) and that the seal also produces the cross-coupled stiffness (k_c), direct and cross-coupled damping (c_d and c_c), and direct inertia coefficients (m_d). Moreover, the pressure drop will vary with the speed in most turbomachineries, and the rotor dynamic effects are quite complex.

Clearances, pressures, and velocities are divided into mean components (subscript 0) that would pertain in the absence of whirl, and small linear perturbations (subscript 1) due to the eccentricity, ε, rotating at the whirl frequency, ω, as

$$\begin{aligned} h &= h_0 + \varepsilon h_1; \qquad P = P_0 + \varepsilon P_1 \\ \bar{u}_s &= \bar{u}_{s0} + \varepsilon \bar{u}_{s1}; \qquad \bar{u}_\theta = \bar{u}_{\theta 0} + \varepsilon \bar{u}_{\theta 1} \end{aligned} \tag{3.110}$$

These expressions are substituted into governing Equations 3.102 through 3.104 to yield a set of equations for mean flow quantities and a second set of equations for perturbation quantities; the terms which are of quadratic or higher order in ε are neglected. The resulting zeroth-order equations define the leakage and the circumferential velocity development and are solved by numerical methods. From the first-order equations, the time and θ dependency are eliminated to obtain the pressure distribution solution. It is then integrated along and around the seal clearance to yield reaction force components. From the rotor dynamic force components, the following rotor dynamic coefficients and constants are obtained (Childs, 1983a, b):

$$\begin{aligned} k_d &= \left\{a_0 - 0.25a_2(\omega T)^2\right\}k^*; \quad k_c = 0.5a_1\omega T k^* \\ c_d &= a_1 c^*; \quad c_c = a_2 \omega T c^*; \quad m_d = a_2 m^* \end{aligned} \tag{3.111}$$

with

$$k^* = \Delta P L \frac{R}{c_r}; \quad c^* = k^* T; \quad m^* = k^* T^2 \tag{3.112}$$

$$T = \frac{L}{V} \tag{3.113}$$

$$a_0 = 2.5AE; \quad a_1 = 2A\left\{\frac{E}{\sigma} + \frac{B}{2}\left(E + \frac{1}{6}\right)\right\}; \quad a_2 = \frac{A}{\sigma}\left(E + \frac{1}{6}\right) \tag{3.114}$$

$$A = \frac{\pi\sigma}{1+\xi+2\sigma}; \quad B = \frac{1+7b^2}{1+4b^2}; \quad E = \frac{1+\xi}{2(1+\xi+B\sigma)} \tag{3.115}$$

$$\sigma = \frac{\lambda L}{c_r}; \quad b = R_a/R_c; \quad R_a = \frac{\rho V c_r}{\mu} \quad \text{and} \quad R_c = \frac{\rho R \Omega c_r}{\mu} \tag{3.116}$$

where k, m, and c are the stiffness, mass, and damping coefficients, respectively; k^*, c^*, and m^* are reference values of corresponding quantities, respectively; a_0, a_1, and a_2 are dimensionless coefficients, respectively; ω is the speed of the rotor; T is the transit time as given in Equation 3.113; L is the length of the seal; V is the average axial stream velocity; ξ is the entrance loss coefficient; ρ is the fluid density; λ is the friction coefficient; R is the radius of the seal; c_r is the radial clearance of the seal; and ΔP is the difference between pressures at the inlet and the exit of the seal. Subscripts d and c represent the direct and cross-coupled terms, respectively. R_a is the Reynolds number for the axial flow, and R_c is the Reynolds number for the circumferential flow for smooth annulus seals. Dimensional coefficients are thus functions of ξ, σ, and b. To determine coefficients a_0, a_1, and a_2 coefficients σ and b are required for the frequently occurring value of $\xi=0.5$. From Childs (1983a), we have

$$\lambda = 0.066R_a^{-1/4}\left[1+\left(\frac{1}{2b}\right)^2\right]^{0.375} \tag{3.117}$$

To calculate λ, the average velocity V is inserted into Equation 3.116. The expression for V can be obtained from the fundamental relationship for the pressure difference,

$$\Delta P = (1+\xi+2\sigma)\frac{\rho}{2}V^2 \tag{3.118}$$

So the average axial stream velocity can be expressed as

$$V = \sqrt{\frac{2\Delta P}{\rho(1+\xi+2\sigma)}} \tag{3.119}$$

Because the desired value of λ is also function of V and thereby σ, it is best obtained iteratively. From σ the dynamic coefficients can be obtained for different speed, ω. Figure 3.38 shows an algorithm for the solution of dynamic coefficients of seals using Equations 3.111 through 3.119.

So, in the preceding section, basic governing equations to obtain dynamic coefficients of the smooth annular turbulent seals (i.e., smooth seals) were explained briefly. Dynamic coefficients are calculated from the bulk-flow theory for a seal dimension, and now effects of the rotor speeds, seal dimensions, and operation conditions on dynamic coefficients of seals are presented and discussed in detail in Example 3.5.

Example 3.5

Obtain plots for a seal's dynamic coefficients with respect to various parameters—for example, the speeds (maximum up to 50,000 rpm), seal dimensions, and pressure differences. The input data for the present numerical example are given in Table 3.8.

Solution:

The seal's dynamic coefficients are dependent on speeds, seal dimensions, and pressure differences. The stiffness (k_d and k_c), damping (c_d and c_c), and mass (m_d) coefficients are presented for various speeds (ω), pressure differences (ΔP), and ratios (L/D).

Figures 3.39 through 3.51 show the variation of the direct and cross-coupled stiffness and damping coefficients and direct inertia coefficients with respect to speeds up to 50,000 rpm, for different values of clearances (0.2 and 0.4 mm), L/D ratios (0.25, 0.50, 0.75, and 1.00), and pressure differences (2, 5, 15, 40, and 80 bars). The effect of these variables on seal dynamic coefficients is discussed in detail next.

EFFECT OF ROTATIONAL SPEEDS AND PRESSURE DIFFERENCES

Direct stiffness coefficients increase with an increase in the pressure difference (Figure 3.39). At low-pressure differences (2 and 5 bars), the direct stiffness coefficient becomes negative as shown in Figure 3.39. The direct stiffness coefficient reaches a maximum nearly at 5000 rpm and then

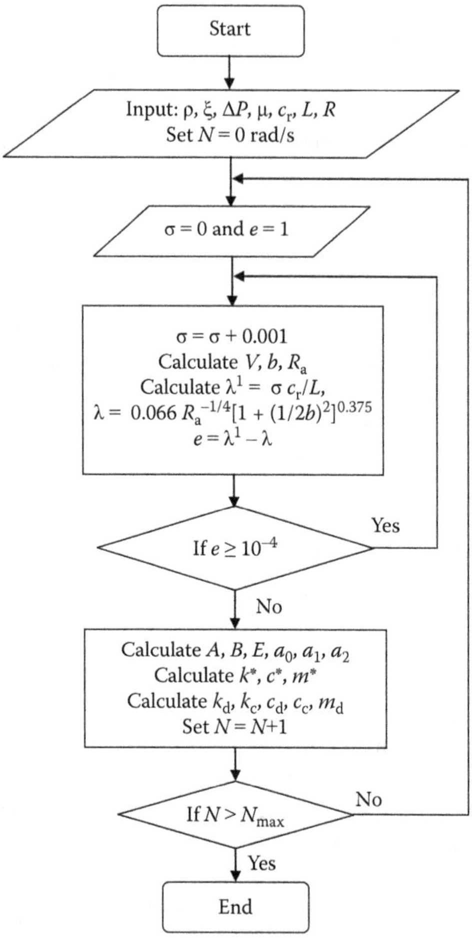

FIGURE 3.38 Flow chart for the theoretical estimation of dynamic coefficients of seals.

TABLE 3.8
Input Data for Numerical Simulation of Dynamic Coefficients of Seals

Parameters	Values
Length of the seal	11, 22, 33, and 44 mm
Radius of the seal	22 mm
Clearance of the seal	0.2 and 0.4 mm
Dynamic viscosity of water at 32°C	0.8×10^{-6} m²/s
Entrance loss coefficient	0.5
Inlet pressure	3, 6, 16, 41, 81 bar
Seal exit pressure	1 bar
Speed of the rotor	1–50,000 rpm

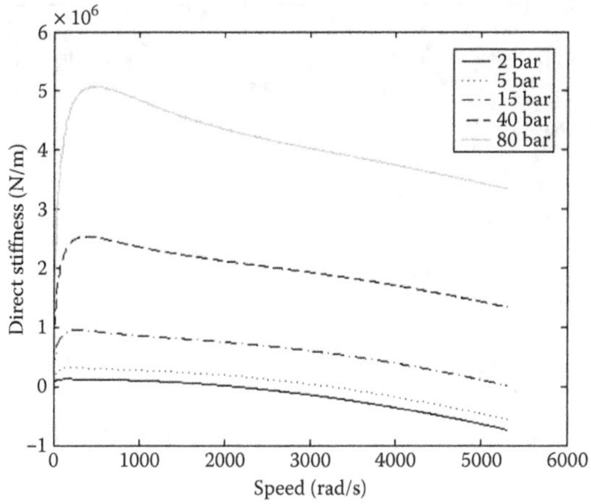

FIGURE 3.39 Direct stiffness coefficients for c_r=0.2 mm, L/D=0.25, and ΔP=2–80 bars.

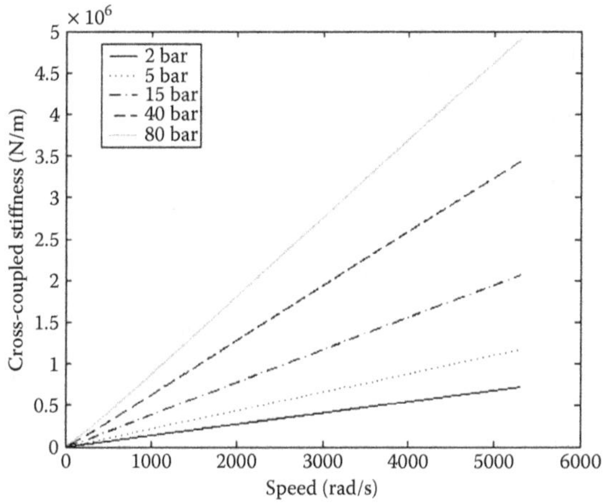

FIGURE 3.40 Cross-coupled stiffness coefficients for c_r=0.2 mm, L/D=0.25, and ΔP=2–80 bars.

slowly declines as shown in Figure 3.39. The cross-coupled stiffness linearly increases with the rotor speed and increases with the pressure difference (Figure 3.40). The direct damping coefficient increases slightly with the speed; however, it increases marginally with the pressure difference (Figure 3.41). The cross-coupled damping increases linearly with the speed, but is insensitive to the pressure difference (Figure 3.42). The direct inertia coefficient increases sharply with the rotor speed and it is almost insensitive to the pressure difference (Figure 3.43).

EFFECT OF L/D RATIOS

The *L/D* ratio has a significant effect on the rotor dynamic coefficients of seals. The direct stiffness increases with the increase in the *L/D* ratio (Figure 3.44). For *L/D*=1.00, after reaching a maximum value nearly at 8000 rpm, it starts declining and becomes negative with an increase in the rotor speed. The cross-coupled stiffness and the direct and cross-coupled damping coefficients increase with the increase in *L/D* ratio as shown in Figures 3.45, 3.46, and 3.47. At *L/D*=0.25, the direct stiffness coefficient always has positive values (Figure 3.46).

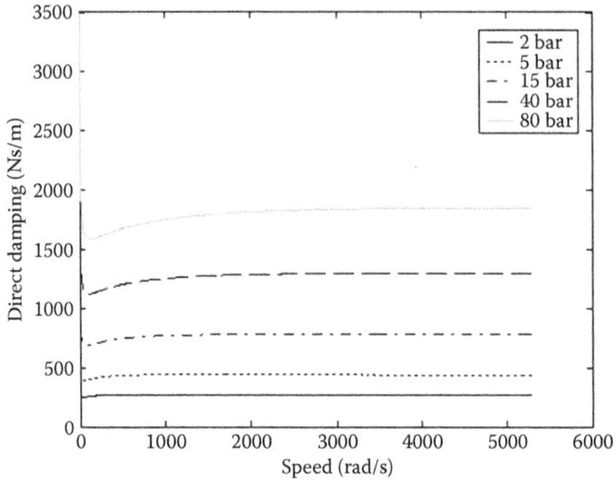

FIGURE 3.41 Direct damping coefficients for c_r=0.2 mm, L/D=0.25, and ΔP=2–80 bars.

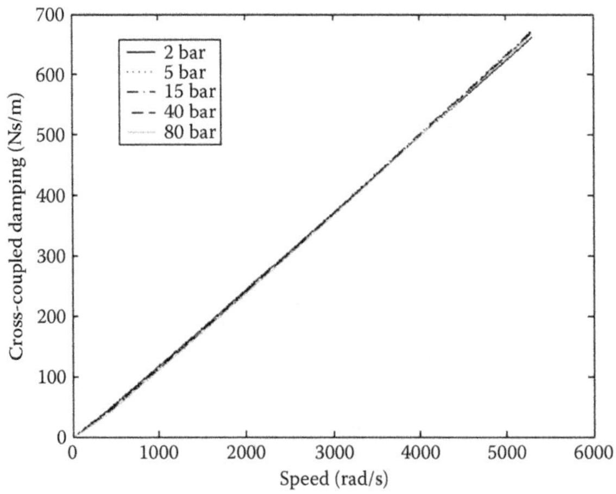

FIGURE 3.42 Cross-coupled damping coefficients for c_r=0.2 mm, L/D=0.25, and ΔP=2–80 bars.

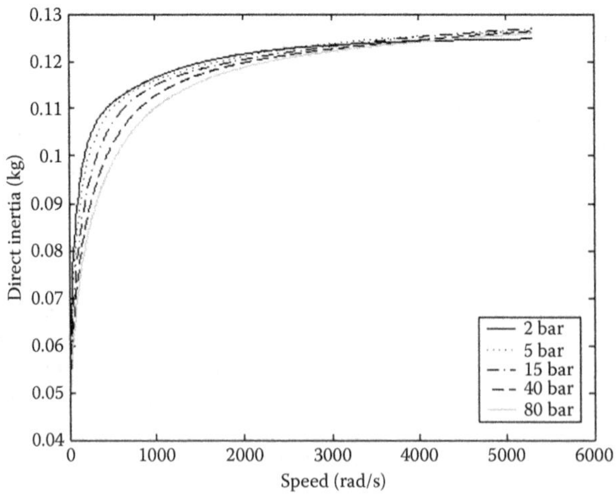

FIGURE 3.43 Direct inertia coefficients for c_r=0.2 mm, L/D=0.25, and ΔP=2–80 bars.

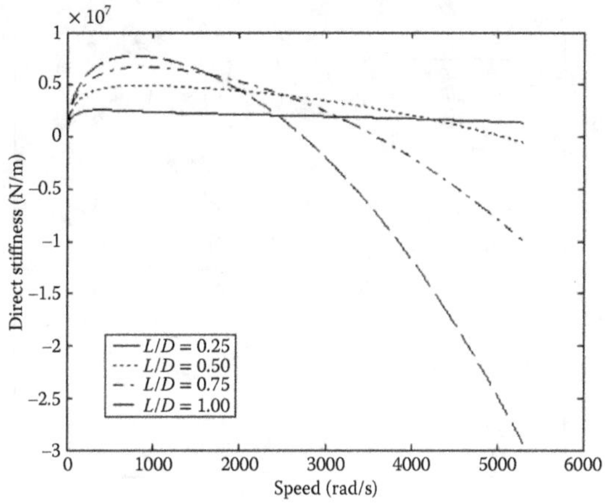

FIGURE 3.44 Direct stiffness coefficients for c_r=0.2 mm, ΔP=40 bars, and L/D=0.25–1.00.

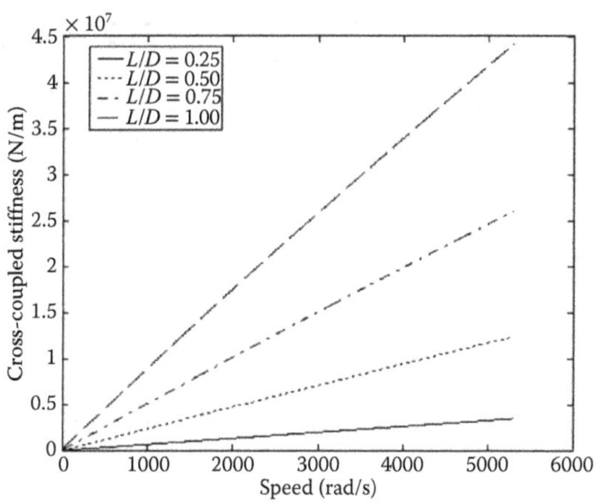

FIGURE 3.45 Cross-coupled stiffness coefficients for c_r=0.2 mm, ΔP=40 bars, and L/D=0.25–1.00.

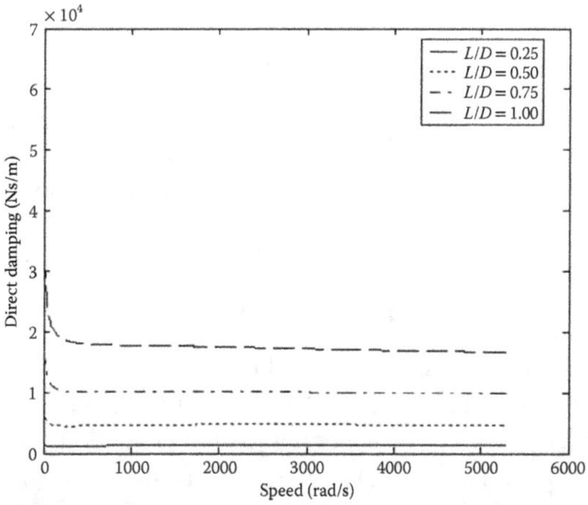

FIGURE 3.46 Direct damping coefficients for c_r=0.2 mm, ΔP=40 bars, and L/D=0.25–1.00.

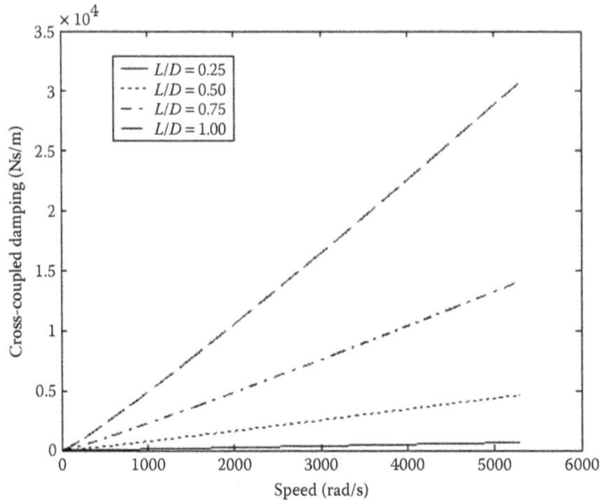

FIGURE 3.47 Cross-coupled damping coefficients for c_r=0.2 mm, ΔP=40 bars, and L/D=0.25–1.00.

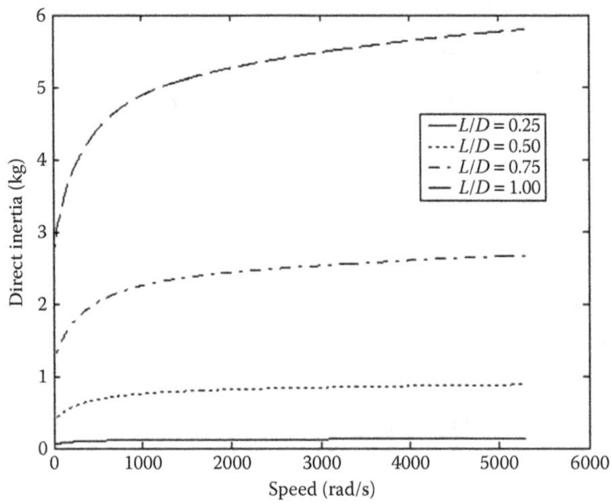

FIGURE 3.48 Direct inertia coefficients for c_r=0.2 mm, ΔP=40 bars, and L/D=0.25–1.00.

EFFECT OF SEAL CLEARANCES

Doubling the clearance shows a huge drop in the direct stiffness and damping coefficients while increasing speeds up to 50,000 rpm. The drop in the cross-coupled stiffness and damping and direct inertia coefficients with an increase in clearance is also significant (Figures 3.48 through 3.51).

3.3.3 General Fluid-Film Dynamic Force Equations

When a rotor vibrates, the reaction force of the fluid in the seal acts upon the rotor. In the case of a small vibration around the equilibrium position, the fluid force can be linearized on the assumption that deflections, Δx and Δy, are small. The general governing equations of fluid-film forces on seals, which have small oscillations relative to the rotor, are given by the following linearized force–displacement model (Childs et al., 1986):

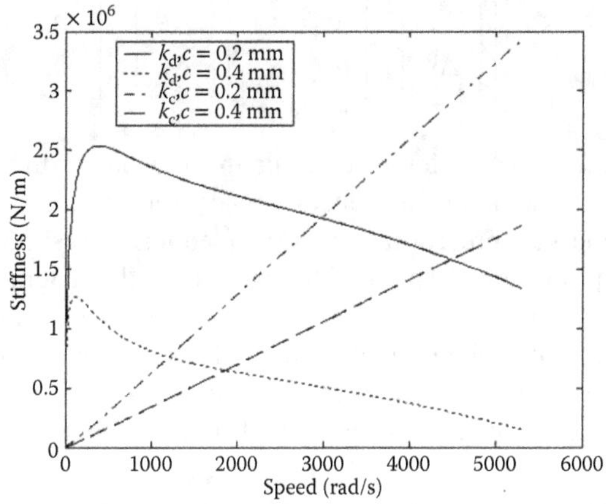

FIGURE 3.49 Direct and cross-coupled stiffness coefficients for $\Delta P = 40$ bars, $L/D = 0.25$, $c_r = 0.2$, and 0.4 mm.

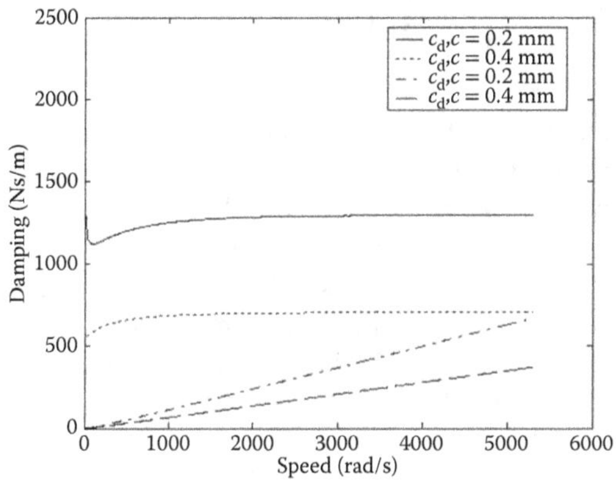

FIGURE 3.50 Direct and cross-coupled damping coefficients for $\Delta P = 40$ bars, $L/D = 0.25$, $c_r = 0.2$, and 0.4 mm.

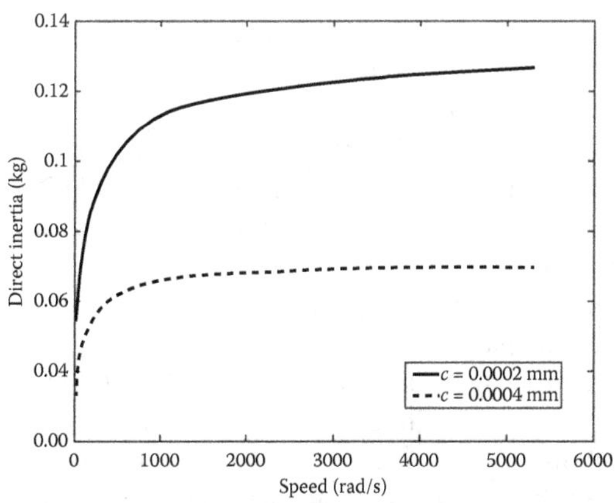

FIGURE 3.51 Direct inertia coefficients for $\Delta P = 40$ bars, $L/D = 0.25$, $c_r = 0.2$, and 0.4 mm.

$$-\begin{Bmatrix} f_x \\ f_y \end{Bmatrix} = \begin{bmatrix} k_{xx} & k_{xy} \\ k_{yx} & k_{yy} \end{bmatrix} \begin{Bmatrix} \Delta x \\ \Delta y \end{Bmatrix} + \begin{bmatrix} c_{xx} & c_{xy} \\ c_{yx} & c_{yy} \end{bmatrix} \begin{Bmatrix} \Delta \dot{x} \\ \Delta \dot{y} \end{Bmatrix} + \begin{bmatrix} m_{xx} & m_{xy} \\ m_{yx} & m_{yy} \end{bmatrix} \begin{Bmatrix} \Delta \ddot{x} \\ \Delta \ddot{y} \end{Bmatrix} \tag{3.120}$$

where f_x and f_y are fluid-film reaction forces on seals in the x and y directions, respectively. k, c, and m represent the stiffness, damping, and added-mass coefficients, subscripts xx and yy represent the direct terms, and xy and yx represent the cross-coupled terms, respectively. These coefficients vary depending on the equilibrium position of the rotor (i.e., the magnitude of the eccentricity), rotational speed, pressure drop, and temperature conditions. The off-diagonal coefficients in Equation 3.120 arise due to fluid rotation within the seal, and unstable vibrations may appear due to these coefficients. Equation 3.120 is applicable to liquid annular seals. But for the gas annular seals, the added-mass terms are negligible. For small motion about a centered position (or with very small eccentricity), the cross-coupled terms are equal and opposite (e.g., $k_{xy}=-k_{yx}=k_c$ and $c_{xy}=-c_{yx}=c_c$) and the diagonal terms are the same (e.g., $k_{xx}=k_{yy}=k_d$ and $c_{xx}=c_{yy}=c_d$) (Childs et al., 1986). Considering these relationships and neglecting the cross-coupled added-mass terms, Equation 3.120 takes the following form:

$$-\begin{Bmatrix} f_x \\ f_y \end{Bmatrix} = \begin{bmatrix} k_d & k_c \\ k_c & k_d \end{bmatrix} \begin{Bmatrix} \Delta x \\ \Delta y \end{Bmatrix} + \begin{bmatrix} c_d & c_c \\ c_c & c_d \end{bmatrix} \begin{Bmatrix} \Delta \dot{x} \\ \Delta \dot{y} \end{Bmatrix} + \begin{bmatrix} m_d & 0 \\ 0 & m_d \end{bmatrix} \begin{Bmatrix} \Delta \ddot{x} \\ \Delta \ddot{y} \end{Bmatrix} \tag{3.121}$$

where subscripts d and c represent the direct and the cross-coupled dynamic coefficients, respectively. The RDPs largely affect the performance of the turbomachineries as they lead to serious synchronous and subsynchronous vibration problems. The whirl frequency ratio, $f=k_c/(c_d\omega)$, is a useful nondimensional parameter for comparing the stability properties of seals. For circular synchronous orbits, it provides a ratio between the destabilizing force component due to k_c and the stabilizing force component due to c_d. In experimental estimation of RDPs of seals, these coefficients of Equations 3.120 and 3.121 are determined with the help of measured vibrations data from a seal test rig (refer to Chapter 4 of Volume II).

More recent textbooks on rotor dynamics include information on rotor dynamic characteristics of rotary seals. Vance (1988), Childs (1993), Krämer (1993), Rao (1996), Adams (2001), and Tiwari et al. (2005) provide good introductory treatments of rotary seal dynamics.

3.4 SQUEEZE-FILM DAMPERS

SFDs can be regarded as a fluid-film cylindrical bearing without the rotation of the journal. The fluid film obtains its shearing action due to the relative translatory motion of the bearing (Figure 3.52a) rather than the squeezing action due to the relative transverse motion of the damper (Figure 3.52b). SFDs usage in attenuating vibrations is widespread in aircraft gas turbine industries as compared to the process industries, where its preferred usage is as a Band-Aid solution without any preplanning. This is because SFDs are more complex to design than any other machine elements. The attenuation of subharmonic rotor instabilities would be the main concern in the designer's mind while designing an SFD because steady-state synchronous vibrations are relatively easier to tackle with modifications to the rotor and the bearing parameters (Zeidan et al., 1996).

SFDss can be used both for rolling bearings and fluid-film bearings by providing an arrangement for the squeezing of the film (Figure 3.52b) between the outer ring (or sleeve) and the housing, and the bearing bush (or sleeve) and the housing. The sleeve is slotted or pinned loosely to prevent its rotation. For rolling bearings, three such arrangements are shown in Figure 3.53, in which based on the seal shape (circular (O) or rectangular (piston) cross-section of seal rings) and its position, the damper classification is made. Rolling bearings are extensively used in aircraft turbine engines, and in such bearings the damping is very low. The SFD is designed in series with rolling bearings to provide enough damping to suppress unstable vibrations of the rotor system.

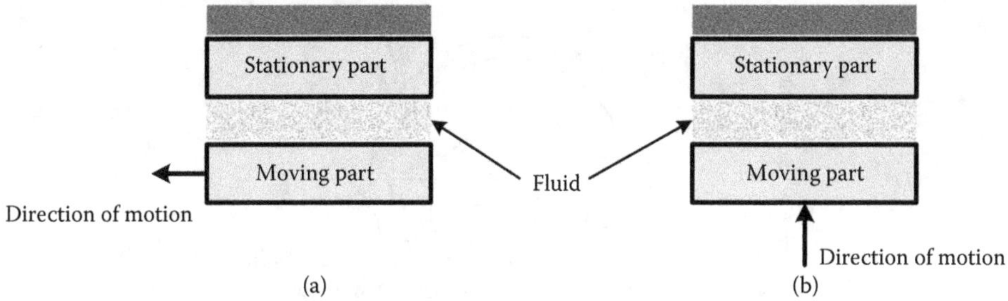

FIGURE 3.52 (a) Shearing of the fluid film. (b) Squeezing of the fluid film.

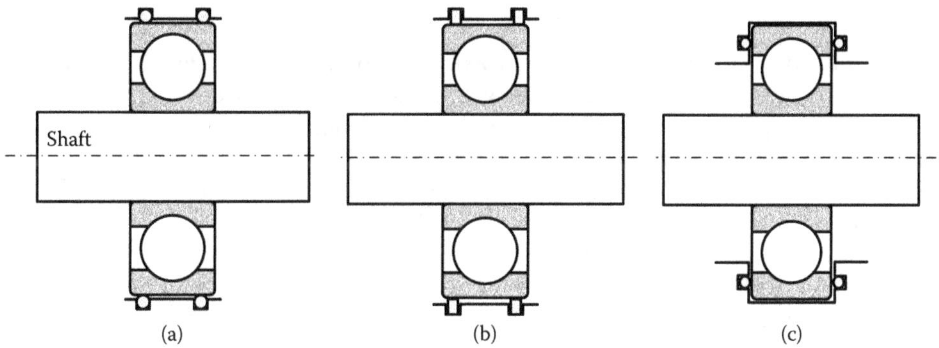

FIGURE 3.53 Three end seal configurations of SFDs: (a) radial O-ring seal; (b) piston-ring seal; (c) side O-ring seal.

The performance of SFDs depends upon several geometrical, physical, and operating parameters. Some of the geometrical parameters are length, the diameter, the radial clearance, the feed and discharge mechanisms, and the types of seal ends. One physical parameter is the viscosity of the lubricant, and some operating parameters are the lubricant supply pressure, the fluid inertia, and dynamic cavitations. Similar to bearings, the incompressible-fluid Reynolds equation is generally used to model the SFDs. However, because of cavitation phenomena in dampers, the correlation between theory and experiment is less as compared to bearings.

3.4.1 Rotordynamic Force Coefficients

Similar to fluid-film bearings, for SFDs, the fluid-film forces are generated by the small amplitude of oscillations about a static equilibrium (off-center) position and are useful for analyzing the stability of rotating machinery. A schematic of squeeze-film geometry and the motion of a journal performing small oscillations at an off-centered static equilibrium position are shown in Figure 3.54, where e_s is the static eccentricity.

Because the damper cannot provide any stiffening effect due to the nonrotating journal, in general, the damper force will be

$$-f_x = c_{xx}\dot{x} + c_{xy}\dot{y} + m_{xx}\ddot{x} + m_{xy}\ddot{y} \tag{3.122}$$

and

$$-f_y = c_{yx}\dot{x} + c_{yy}\dot{y} + m_{yx}\ddot{x} + m_{yy}\ddot{y} \tag{3.123}$$

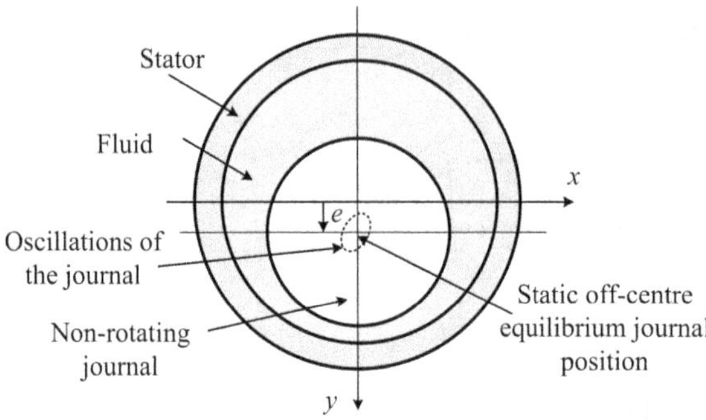

FIGURE 3.54 A schematic illustrating the basic principle of the SFD.

where x and y are displacements in the x- and y-axis directions, respectively; the single and double dots represent the first and second derivatives with respect to time, respectively; c and m are the damping and inertia coefficients, respectively, of the damper fluid film with their subscripts denoting the force and motion directions, respectively; and f is the fluid-film force and its subscripts denote the direction of application. For the short, open-ended SFD, the rotordynamic coefficients are given as (for no cavitations, i.e., the full-film model)

$$c_{xx} = \mu D\left(\frac{L}{c_r}\right)^3 \frac{\pi\left(1+2\varepsilon^2\right)}{2\left(1-\varepsilon^2\right)^{2.5}}, \quad c_{xy} = c_{yx} = 0, \quad c_{yy} = \mu D\left(\frac{L}{c_r}\right)^3 \frac{\pi}{2\left(1-\varepsilon^2\right)^{1.5}}$$

$$m_{xx} = \rho D\left(\frac{L^3}{c_r}\right)\frac{\alpha\pi\left\{1-\left(1-\varepsilon^2\right)^{0.5}\right\}}{12\varepsilon^2\left(1-\varepsilon^2\right)^{0.5}}, \quad m_{xy} = m_{yx} = 0, \quad m_{yy} = \rho D\left(\frac{L^3}{c_r}\right)\frac{\alpha\pi\left\{1-\left(1-\varepsilon^2\right)^{0.5}\right\}}{12\varepsilon^2}$$

with

$$\varepsilon = e_s/c_r \tag{3.124}$$

where α ranges from 1.2 to 1.0 for small to moderately large squeeze-film Reynolds numbers ($R_e = \rho v c_r^2/\mu$); L, D, and c_r represent the damper axial length, diameter, and radial clearance, respectively; ρ and μ are the fluid density and viscosity, respectively; and ε is the eccentricity ratio. Expressions are valid for the iso-viscous and incompressible fluids undergoing an isothermal process within the damper. For a short-length, open-ended SFD, the rotordynamic coefficients are given as (with cavitations, i.e., the half-film model)

$$c_{xx} = \mu D\left(\frac{L}{c_r}\right)^3 \frac{\pi\left\{3\varepsilon+0.5i\left(1+2\varepsilon^2\right)\right\}}{2\left(1-\varepsilon^2\right)^2}, \quad c_{xy} = \mu D\left(\frac{L}{c_r}\right)^3 \frac{\pi}{\left(1-\varepsilon^2\right)^2}, \quad c_{yy} = \mu D\left(\frac{L}{c_r}\right)^3 \frac{\pi}{4\left(1-\varepsilon^2\right)^{1.5}}$$

$$m_{xx} = \rho D\left(\frac{L^3}{c_r}\right)\frac{\alpha\{i-\pi-2\varepsilon\}}{24\varepsilon^2}, \quad m_{xy} = \rho D\left(\frac{L^3}{c_r}\right)\frac{\alpha\left[\ln\{(1-\varepsilon)/(1+\varepsilon)\}-2\varepsilon\right]}{24\varepsilon^2},$$

$$m_{yy} = \rho D\left(\frac{L^3}{c_r}\right)\frac{\alpha\pi\left\{1-\left(1-\varepsilon^2\right)^{0.5}\right\}}{24\varepsilon^2}, \quad c_{yx} = 0, \quad m_{yx} = 0$$

with

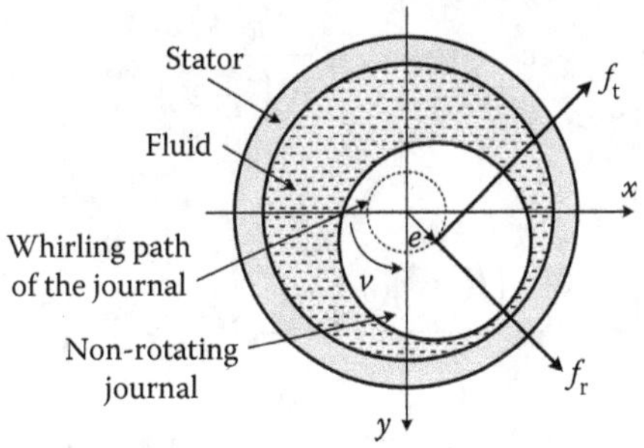

FIGURE 3.55 A schematic illustrating the basic principle of the circular-centered SFD.

$$i = 2\cos(-\varepsilon)/\sqrt{(1-\varepsilon^2)} \tag{3.125}$$

As illustrated in Figure 3.55, for circular-centered journal motions (the journal is the outer ring in rolling bearings, or the bush in fluid-film bearings) with amplitude e and whirl frequency v in a perfectly cylindrical bearing, a damper produces a constant fluid force in a reference frame rotating with angular frequency of ν. The radial and tangential components of the fluid force are equal to

$$f_r = -c_{rt}V_t - m_{rr}a_r \quad \text{and} \quad f_t = -c_{tt}V_t - m_{tr}a_r \tag{3.126}$$

with

$$V_t = e\nu \quad \text{and} \quad a_r = -e\nu^2 \tag{3.127}$$

where V_t and a_r are the tangential velocity and the radial acceleration, respectively, of the journal center, and (c_{rt}, c_{tt}) and (m_{rt}, m_{tt}) are the cross-coupled and direct damping and inertia force coefficients, respectively. It should be noted that these linearized coefficients are defined for small motions only. For the short-length approximation, π-film assumption (i.e., half the circumferential film extent is cavitated), and open-ended SFD model, the force coefficients are given as (Vance, 1988)

$$c_{rt} = \frac{2\mu RL^3\varepsilon}{c_r^3\left(1-\varepsilon^2\right)^2}, \quad c_{tt} = \frac{\pi\mu RL^3}{2c_r^3\left(1-\varepsilon^2\right)^{1.5}}$$

$$m_{rr} = \frac{\rho RL^3}{12c_r}\frac{\left\{1-\left(1-\varepsilon^2\right)^{-0.5}\right\}}{\varepsilon^2}\left\{1-2\left(1-\varepsilon^2\right)^{0.5}\right\}$$

and

$$m_{tr} = -\frac{\rho RL^3}{c_r}\left(\frac{27}{70\varepsilon}\right)\left\{2+\frac{1}{\varepsilon}\ln\left(\frac{1-\varepsilon}{1+\varepsilon}\right)\right\} \tag{3.128}$$

where L, D, and c_r represent the damper axial length, diameter, and radial clearance, respectively; ρ and μ are the fluid density and viscosity, respectively; and ε is the eccentricity ratio. Expressions are valid for the iso-viscous and incompressible fluids undergoing an isothermal process within the damper. For the short-length approximation, 2π-film assumption (i.e., no cavitations), and open-ended SFD model, the force coefficients are given as

$$c_{rt} = 0, \quad c_{tt} = \frac{\pi\mu RL^3}{c_r^3\left(1-\varepsilon^2\right)^{1.5}}$$

and

$$m_{rr} = \frac{\rho RL^3}{6c_r}\frac{\left\{1-\left(1-\varepsilon^2\right)^{-0.5}\right\}}{\varepsilon^2}\left\{1-2\left(1-\varepsilon^2\right)^{0.5}\right\}, \quad m_{tr} = 0 \tag{3.129}$$

It should be noted that cross-coupled coefficients are zero and direct coefficients are doubled as compared to the π-film assumption. A large clearance, low viscosity, and high whirl frequency would lead to appreciable inertia coefficients and may change drastically the rotor-damper dynamic response. Tichy and Modest (1978) and San Andres and Vance (1987) provided fluid inertia coefficients for finite-length dampers. Tao et al. (2000) presented an analysis of SFDs operating in bubbly lubricants. A continuum model was proposed for describing the motion of a bubbly fluid in an open-ended SFD operating with circular-centered journal orbits. Computed predictions for peak-peak dynamic pressures and fluid-film forces were compared with experimental measurements conducted on an SFD test rig operating with a controlled air in oil mixture. Pietra and Adietta (2002) and Adietta and Pietra (2002) gave a comprehensive survey of the SFD investigations in two parts. In the first part, a survey of the construction characteristics and operating features of the SFD as a separate component was reported, as inferred from the technical literature consulted through a number of theoretical approaches and experimental results. The second part of the paper reviewed research work on the dynamic behavior of rotors on supports equipped with SFDs. Because of the complexity in the analysis of SFDs, often practice engineers prefer to find the rotordynamic properties through experiments (Tiwari et al., 2004, 2005) (refer to Chapter 4 of Volume II).

3.5 FOIL BEARING

3.5.1 Basic Terminology, Variants, and Applications

Air (or gas) foil bearing technology has accomplished noteworthy growth throughout the last 50 years. Compliant foil bearings accomplish utmost of the necessities of the challenging oil-free turbomachinery and other applications (not limited to turbo compressors/blowers, turbo alternators, air cycle machines used for aircraft cabin, and turbines drives) by enhancing their reliability by tenfold in contrast to rolling elements bearings. Furthermore, it has features like lower friction and running costs, low noise and vibration, maintenance free, almost zero wear, self-acting, no lubrication, compliant features, and adaptable. Foil bearings can run to extremely high speeds of up to 200 m/s and have extremely low frictional losses, less than half than those of the conventional oil bearings. Compliant foils allow the bearing to accommodate for shaft thermal growth, allowing the bearing to operate in temperatures above 700°C and bump foils provide important damping features to suppress shaft instabilities (Agrawal, 1997).

Foil gas bearings or compliant surface foil gas bearings are a variant of hydrodynamic bearings that use the ambient gas as their operational fluid and thus need no devoted lubrication system. The hydrodynamic fluid-film is developed between the moving shaft surface and a flexible bearing surface characteristically arrange from several layers of sheet metal foils and from which they derive their name (refer to Figure 3.56). Foil bearings are manufactured of one or more compliant surfaces of corrugated sheet metal and one or more layers of top foil surfaces. The compliant surface delivers bearing stiffness and have several configurations, such as tape type (Figure 3.56a), leaf-type (Figure 3.56b), bump-type (Figure 3.56c), wire (metal) mesh type (Figure 3.56d), and hybrid bump and polymer (i.e., visco-elastic) layer type (Figure 3.56e and f) (Park and Sim, 2021). In fact, various combination of all these is also are being tried out to take advantage of hybrid foil bearing (Feng et al., 2016).

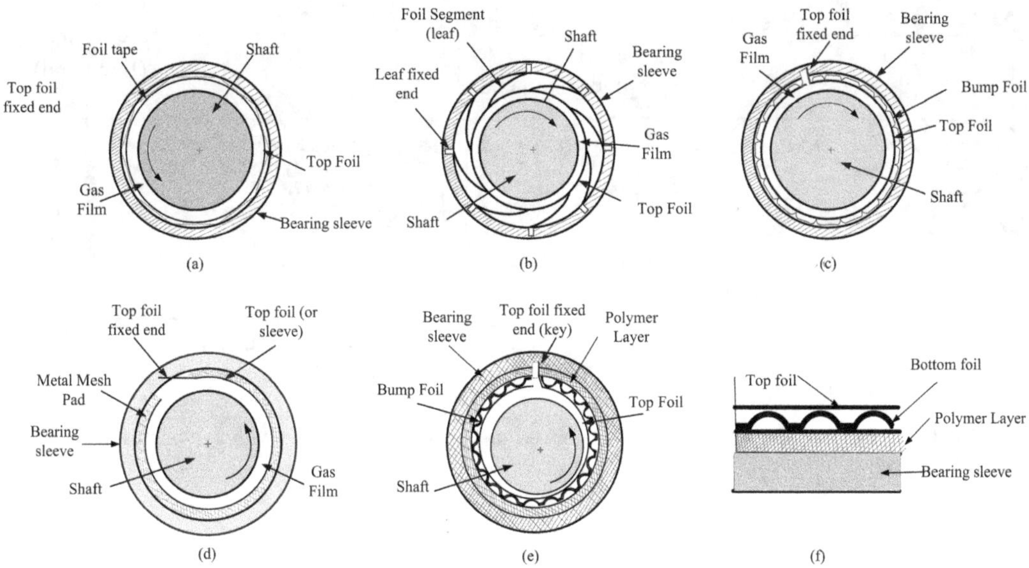

FIGURE 3.56 Various designs of foil bearings: (a) tape type, (b) leaf-type, (c) bump-type, (d) wire (metal) mesh type, (e) hybrid bump and polymer (visco-elastic) layer type, and (f) a close view of hybrid bump and polymer layer type.

In the tape type foil bearing (refer to Figure 3.56a), a single piece of foil wrapped around three or four times. The multi-leaf journal foil bearing is largely made of a number of foil leaves and a top foil (or sleeve) that is utilized to encompass and connect the foil leaves, as depicted in Figure 3.56b. The foil leaf brings together as one overlay on another in sequence toward the circumferential direction, and the foil leaf number is not fixed and is usually 5, 8, 12, etc. depending upon the flexibility requirement. The overly amid foil leaves gives elasticity additionally produces to the converged gas film clearances. Figure 3.56c depicts the common bump type of foil bearing. In bump-type foil bearings, the top foil (or sleeve) is braced by compliant bumps and elastically deforms due to pressure formed by a hydrodynamic gas film. The bearing stiffness adds up that resulting from the deformation of the bumps and also by the hydrodynamic film produced while the shaft revolves. Damping comes in the foil bearing owing to the relative motion amid the bumps and the top foil or among the bumps and the bearing wall and it is usually the Coulomb type damping due to direction contact friction (Ku and Heshmant, 1992a).

The main components of a typical bump foil bearings are as follows (Agrawal, 1997; Rubio and San Andrés, 2003): (i) *Bearing sleeve:* A rigid cylindrical ring backups the bump foils and the top foil. This is generally fitted to the bearing housing or pedestal. (ii) *Top foil:* The top foil is a thin flat metal sheet fixed to the bearing sleeve at one end and free at the other end. It separates the shaft and bump foil, and its smooth surface helps in the development of gas (or air) film due to hydrodynamic action at high speed of the bearing. (iii) *Bump foil:* The bump foil shape comprises of single or multiple strips (five in the diagram) stacked one over another, each with multiple bumps (five in the diagram), aligned axially (refer to Figure 3.57). The end of a strip is welded to the bearing sleeve whereas the other end is unrestricted, as shown in Figure 3.57. Multiple segments (e.g., five) are placed around the bearing sleeve (refer to Figure 3.58) so that it covers the entire circumference of the bearing sleeve. Note that each segment is welded at one end and unrestricted at the other. The foil bearings shown in the diagram have a total of 25 bumps around the bearing sleeve. (iv) *Spot Weld:* The weld line attaches the top foil and the bump foil strips to the bearing sleeve. The spot weld location is characterized by the absence of bumps (i.e., straight part) and it spans roughly 10°–12°. The spot weld restricts the displacement of the top foil and the bump foil at one end

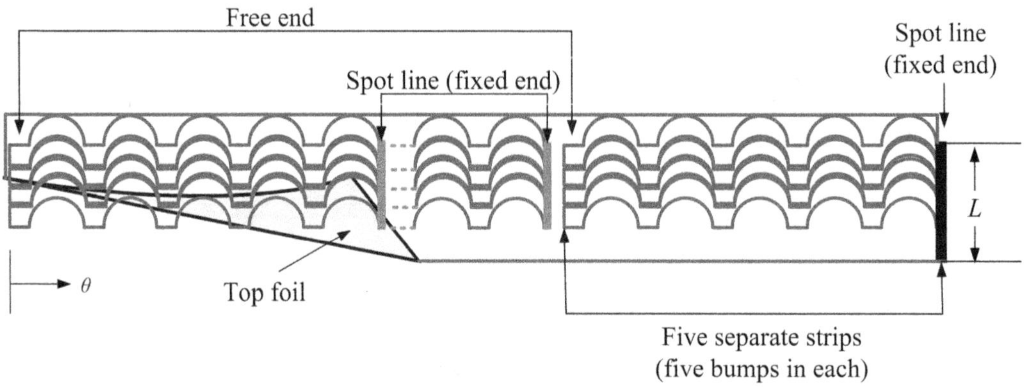

FIGURE 3.57 Detailed view of unwrapped bump strips having five segments with five bumps in each and five strips in each segment overlapped on each other.

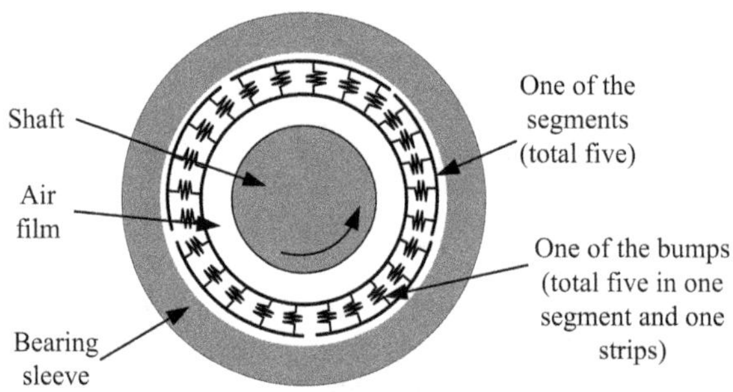

FIGURE 3.58 Equivalent mechanical model of the five segments of bottom foil with each having five bumps in a single strip (refer to Figure 3.57).

(fixed end). It should be noted a lot of possibilities are there to play with to get required performance from such arrangements of various bearing components.

3.5.2 Performance Parameters

Foil gas bearings are a crucial technology in numerous commercial and evolving oil-free turbomachinery systems as mentioned previously. These bearings are nonlinear in nature in terms of rotordynamic properties (i.e., the stiffness and damping properties) and have been tough to analytically model in terms of performance characteristics, such as the load capacity, power loss, and stiffness and damping confidents. Foil gas bearings are self-acting hydrodynamic bearings but experience a brief period of dry sliding contact during low-speed operation (i.e., start-up and shutdown). They are compliant bearings that offer high-speed rotor support while accommodating shaft misalignment and distortion frequently encounter in turbomachinery.

Foil bearings can be arranged as the radial or thrust bearings backing the radial or axial shaft loads, respectively, as shown in Figures 3.59 and 3.60. During action, dynamic rotor motions cause variations in the film pressure and generate momentarily motions in the foils. This generates internal sliding contact inside the bearing structure and contributes a Coulomb damping effect to the system (Ku and Heshmat, 1994). Thus, effectively a foil bearing is both a bearing and a damper. The compliant nature of the bearing also permits it to retort to and take care of thermal and mechanical distortions more effectively than conventional rigid gas bearings. These noticeable characteristics of foil gas bearings have been recognized long ago.

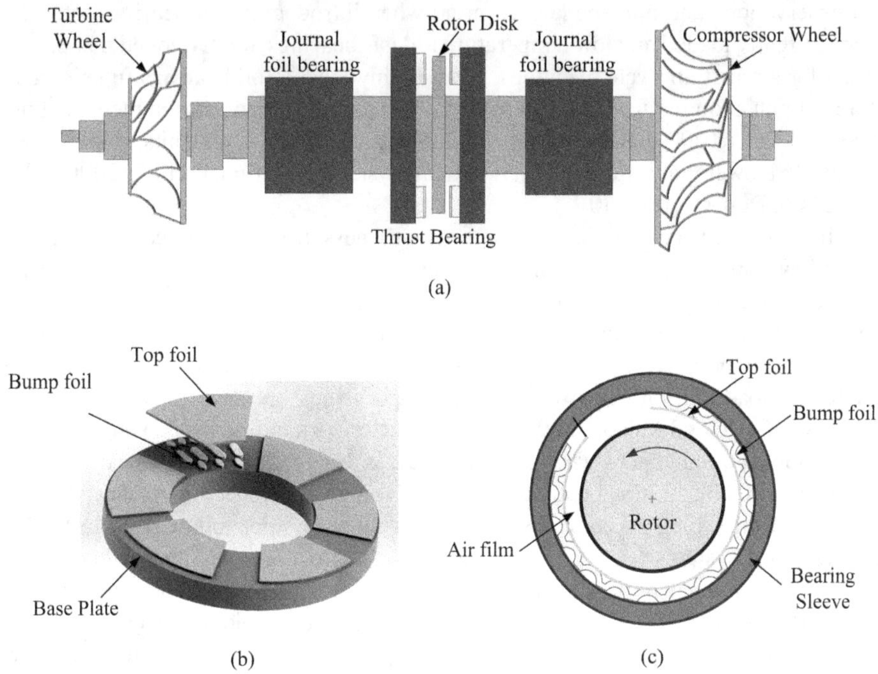

FIGURE 3.59 Turbomachinery supported by the radial and thrust foil bearings: (a) turbine-compression supported by the radial and thrust foil bearings, (b) radial foil bearing, and (c) thrust foil bearing.

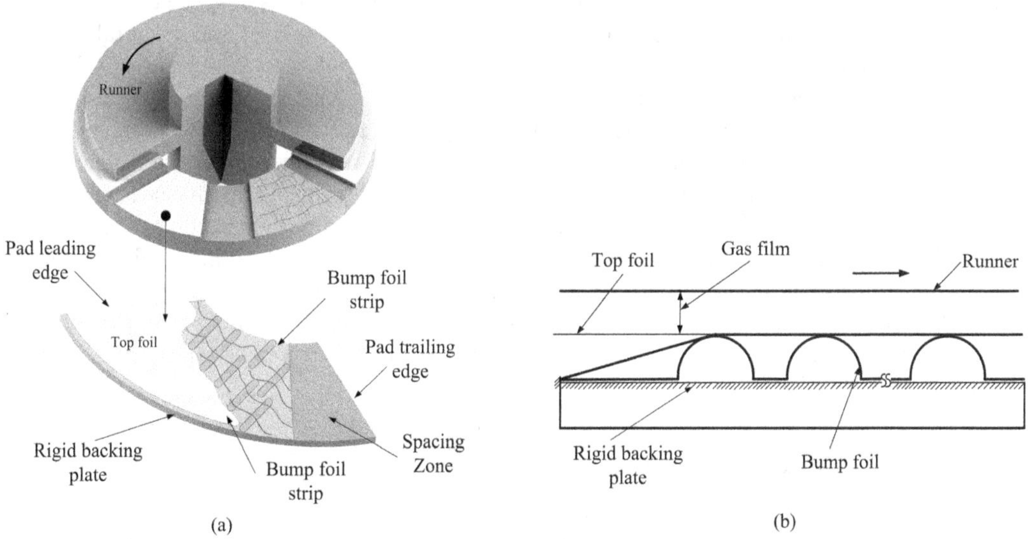

FIGURE 3.60 Basic layout of the thrust bump foil bearing: (a) macro-geometry of thrust foil bearing and (b) a close view of bumps and the gas film.

Gross (1969) reviewed gas bearings and commented that the foil bearings were advanced to minimize instability issues, decrease manufacturing tolerances, and allow adaptation of the bearing to variations in shaft diameter due to centrifugal force or temperature gradients. The fluid-film between the shaft and foil has a greater stiffness than the foil itself. The following five and half decades of research and development in foil bearings has basically reinforced Gross' above mentioned observations and must be kept in mind when introducing foil bearings into innovative rotor systems. Lightly loaded and low temperature foil gas bearings are developed that prevail in the rotor support for aircraft air cycle machines. More highly loaded foil bearings functioning at high temperatures are an evolving technology creating commercial inroads into numerous applications (DellaCorte and Valco, 2003). These comprise aircraft auxiliary power units, microturbines, gas compressors and blowers, and turbochargers (Kolanowski, 2004; Lubell et al., 2006; Klaass and DellaCorte, 2006; DellaCorte, 2010).

Foil bearings deliver very nominal load capacity, stiffness, and damping compared to traditionally supported systems. The stiffness and damping are an order-of-magnitude lower than similar dimension oil-lubricated bearings. However, the friction can be lower especially at high rotational speeds and foil bearings have no inherent DN speed limits, as do have rolling element bearings. Foil bearing load capacity is severely prejudiced by speed also. At high speeds, foil gas bearings display similar or higher load capacity than rolling element bearings but have inadequate capability at low speeds (DellaCorte and Valco, 2000). These characteristics of foil gas bearings command that their effective application happens when a rotor system is designed around the bearing capabilities. This is in contrast to the common practice of first designing the aero-components and determining speeds and loads from which the rotor design and bearing necessities follow (Lynch, 1993).

For rotor support systems (e.g., bearings, dampers, seals), a few key performance parameters are important, i.e., the bearing load capacity, and the bearing stiffness and damping capabilities (DellaCorte and Bruckner, 2010). For traditional rotor support technologies, catalog data, empirical models, and computer-based predictive tools available provide these crucial inputs to a rotor system designer. For a traditional oil-lubricated machine to perform a rotor arrangement design, an engineer merely plans out the rotor system, augments crucial components, like the turbine and compressor wheels, evaluates the shaft loads, and supplements rolling element or hydrodynamic bearings. Then the rotordynamic performance of the layout is evaluated and compared to the anticipated functional range. If the system is not stable, bearing location can be changed, functional speed can be altered, or SFDs can be supplemented. If the bearing loads and associated stress on the structural components or shaft orbits are extreme, then larger bearings can be utilized (Rao, 2000; Tiwari, 2017).

If the bearing performance parameters are well understood and if the general rotor design and layout are comparable to those previously established by practice then only such an iterative analytical design technique is feasible. For foil bearing supported machinery, the development path continues to be less clear. Foil bearings are nonlinear structural elements that include a hydrodynamic lubrication gas film with a compliant elastic foundation. It comprises Coulomb friction that can be slip-stick in nature. Combination of the fluid dynamics with structural mechanics has made modeling of foil gas bearings and rotors that depend on upon them very stimulating and uncertain (DellaCorte and Valco, 2000). Historically, the development of novel machines has frequently been relied on formerly fielded machines or depend on heavily on hardware rigorous experiments (Klaass and DellaCorte, 2006).

3.5.3 Foil Gas Bearings-Solid Lubricants

Foil bearings undergo transitory sliding contact while start-up and shutdown when surface velocity and hydrodynamic gas film pressure is inadequate to endorse a fully established lubricant film. Solid lubricants are desirable to alleviate wear of the foils, diminish friction, and preserve polished surfaces required by gas bearings. Solid lubricants give additional purpose as a sacrificial, non-structural material that permits the growth of compatible surfaces due to the wear-in process (DellaCorte et al., 2002). This characteristic is extremely significant for bearings with non-optimal geometries or unwanted manufacturing discrepancies. Soft solid lubricants are characteristically supplemented to the foil surface, and hard wear-resistant coatings are applied on shaft surfaces. A substitute method applies composite shaft coatings that are concurrently pools for solid lubricants

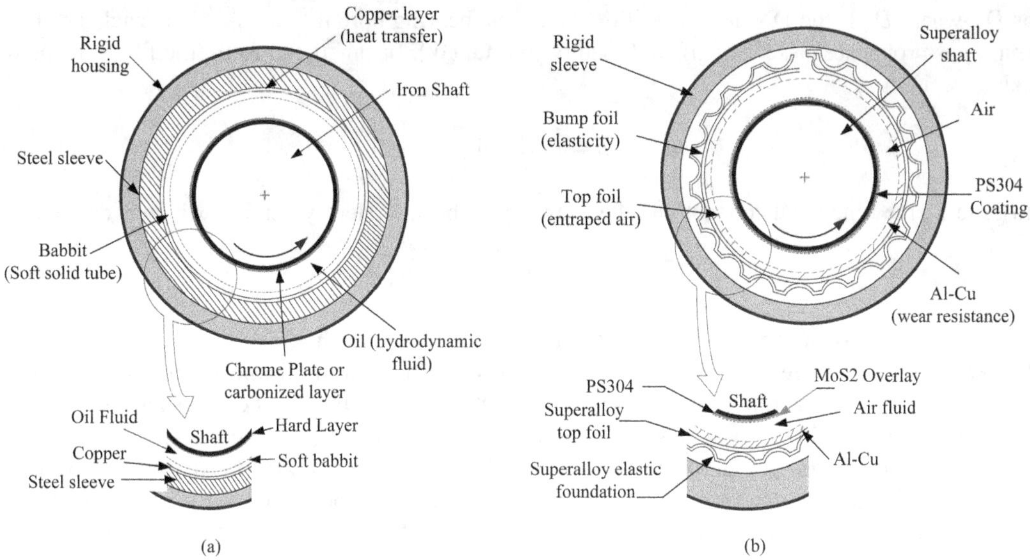

FIGURE 3.61 Detailed components and coating of (a) traditional oil-lubricated bearing and (b) bump-type foil bearing.

and provide wear shield by additionally holding harder segments. Figure 3.61 portrays the system nature of foil and shaft coatings utilized for the solid lubrication of foil gas bearings (DellaCorte and Edmonds, 2009; DellaCorte et al., 2002).

Foil coating advancement and characteristic is suitably recognized in the literature and can be largely categorize based on temperature necessities of a usage. For low temperatures up to 300°C, thin polymer films made from polytetrafluoroethylene, polyimide, and polyamides are conventionally used. These films are frequently mixed with solid lubricants, such as graphite, molybdenum disulfide, and boron nitride, to additionally diminish friction and improve life. For temperatures above 300°C, foil coatings can either be severely hard wear-resistant ceramics, such as aluminum oxide and titanium carbide, or metal-ceramic-solid lubricant composites deposited by spraying, dipping, or electrochemical procedures proceeded by high temperature curing (Bhushan, 1980). Such films are frequently mechanically burnished (polished) to achieve additional unvarying thickness and smooth surface finishes. The polymer coatings produce low friction and long life at a low cost for any fabrication techniques and materials used, and are thus preferred whenever temperatures are not extreme.

For decent solid lubrication to happen, the mating surface of foil must be planned suitably. When functioning against polymer foil coatings, the mating surface is characteristically coated with chrome, either hard chrome or thin dense chrome. Chrome plating provides a smooth and durable texture that prevents galling with the foil in the occasion of a foil coating wear out. When high temperatures are met, either bare or hard coated foils are functioned against ceramic or composite shaft coatings, like the plasma sprayed coatings. Such methods provide reasonable friction levels and the capability to activate foil bearings up to 650°C or higher (Lubell et al., 2006; DellaCorte and Edmonds, 2009).

3.5.4 Film Thickness and Load Capacity

Foil bearings depend upon a thin (typically less than 10 μm) gas film to avoid rubbing contact between the foils and their mating surface (the shaft or thrust runner) (Ruscitto et al., 1978). This film thickness, h, is dependent on the functioning speed (surface velocity, generally represented as $D_{N,}$ where D_N is the DN number with D being the bearing bore in mm and N, journal speed in rpm), gas properties (viscosity, μ), and load of journal (W). In functional form, the film thickness can be written as

$$h = f\left(D_N, W, \mu\right) \tag{3.130}$$

In general, the functional connection reflects in the Stribeck or Hersey number, which is defined as

$$h = f\left(\mu D_N / W\right) \tag{3.131}$$

in which an increase in the speed or viscosity tends to increase the film thickness and an increase in the load leads to a decrease in the film thickness. The operating film thicknesses in practice, however, are constrained by other aspects. It is significant to mention that foil bearings have no factual clearance, unlike rigid gas bearings. At beginning, the foil surface is elastically preloaded on the journal. As the rotation of journal starts, rubbing takes place between the foil surface and journal, which can be slip-stick in nature. As the journal speed increases above the lift-off speed (Bagiński and Żywica, 2018), the hydrodynamic pressure deforms the foil and its elastic supporting structure away from the journal surface leading in a hydrodynamic film thickness. This generally one would compare to a clearance for a rigid bearing, thus creating an eccentricity. Because of these slight differences between foil bearings and conventional gas bearings, it is found that viewing bearing behavior in terms of lubricating film thickness is a balanced approach to analyzing foil bearings.

Presently, however, only a method to estimate foil bearing load capacity exists (Ruscitto et al., 1978). This load capacity estimation technique was developed by examining experimentally measured load capacity data from numerous foil bearing tests to develop a simple linear algebraic relationship between bearing size, design complexity and operating speed, and the resulting load capacity. The ensuing load capacity "rule-of-thumb" (ROT), shown below, has proven effective in guiding new machine development. The load capacity for a foil gas journal bearing is expressed as a function of load capacity coefficient, C_{LC}, related to bearing design, bearing length, L, bearing diameter, D, and shaft speed in thousands of revolutions per minute (N_{krpm}):

$$W_{LC} = C_{LC} L D^2 N_{krpm} \tag{3.132}$$

Early primitive foil bearings exhibit load capacity coefficients, C_{LC}, of about 0.3 (lb/in^3/krpm or 8.1434×10^{-05} N/mm^3/krpm), while more advanced bearings with enhanced structural elastic foundations have coefficients around 1.0 (lb/in^3/krpm or 2.7145×10^{-04} N/mm^3/krpm) (DellaCorte and Bruckner, 2010). Herein, 1 krpm = 10^3 rpm. With this model, one can easily size a foil bearing for a particular application.

Example 3.6

Find out the load capacity of foil bearing using ROT of dimeter 60 mm and length 60 mm. The journal speed is 1,00,000 rpm. Take load capacity coefficients, C_{LC}=2.7145 ´ 10^{-04} N/mm^3/krpm. What will be the load capacity, if the diameter and the length are doubled?

Solution:

A 60 mm diameter bearing bore and operating at 30,000 rpm (N_{krpm} = 30 krpm) will have the *DN* value as

$$DN = D\left(1000 N_{krpm}\right) = 60 \times 30{,}000 = 1.8 \times 10^6$$

which gives million DB of 1.8. The load capacity of foil bearing using ROT from equation 3.132 is given as

$$W_{LC} = C_{LC} L D^2 N_{krpm} = 2.7145 \times 10^{-04} \times 60 \times 60^2 \times 30 = 1758.99 \text{ N}$$

For the diameter and the length are doubled, the load capacity of foil bearing would be

$$W_{LC} = C_{LC} L D^2 N_{krpm} = 2.7145 \times 10^{-04} \times 120 \times 120^2 \times 30 = 14{,}071.82 \text{ N}$$

That means the load capacity of foil bearing would be eight times more with corresponding million DB of 3.6.

Thrust foil bearings attribute to the same fundamental principles as radial foil bearings and can be modeled in a similar fashion (Dykas et al. 2009). Using a similar method, the thrust foil bearing load capacity ROT is as follows:

$$W_{LCT} = C_{LCT} w D_m^2 N_{krpm} \tag{3.133}$$

where the value w signifies the radial pad extent (outer radius—inner radius), the D_m signifies the median bearing diameter, and C_{LCT} is the foil thrust bearing load capacity coefficient {varies from 0.02 lb/in^3/krpm (or 5.4290 ´ 10^{-06} N/mm^3/krpm) to 0.13 lb/in^3/krpm (or 3.5288 ´ 10^{-05} N/mm^3/krpm) or about five to ten times smaller than those for foil radial bearings}. Low thrust bearing load capacity levels may be due to thermal management challenges leading to thermal distortion, and centrifugal distortion due to their size is frequently larger than radial foil bearings.

3.5.5 Stiffness and Damping of Foil Bearing

The ROT is being established for bearing stiffness and damping. Foil gas bearing stiffness and damping are characterized by nonlinearity, frequency, and temperature dependence. Stiffness for foil bearings can be anticipated to be alike for all bearings when normalized for bearing size (projected area) since the gas film properties (hydrodynamic pressures) are identical. Hydrodynamic gas pressures are characteristically 1–5 atm. To operate suitably, a foil bearing structure must be adequately deflected by such low pressures and this demands slightly modest structural stiffness. In practice, this unit stiffness is around 6 MN/m for each square-cm of projected bearing area (DL). Lightly loaded bearings exhibit a reduction in this value of about 50%, and heavily loaded bearings see this value increased by about 50%. Thus, the range one can expect for stiffness is about 3–9 MN/m/cm^2 (2500–7500 lbs/in/in^2) (DellaCorte et al., 2008). Bearings working close to their load capacity limit may deliver static and dynamic stiffness values one to two times higher than this range but can lead to thermal instability and rapid failure (Radil and Zeszotek, 2004).

Damping in foil bearings is mainly due to a sliding friction dissipation mechanism between foil layers and between the sleeve and foils, where contact happens. Since such Coulomb damping is predicated on the low amplitude rubbing motion among compliant surfaces, top foil, and bearing sleeve, lightly loaded components (foils), the damping effect is much lower to fluid-film (oil) dampers. Further, the energy dissipation that results must be managed via conduction or convective cooling to uphold thermal equilibrium. For these reasons, the observed damping action of a foil gas bearing is typically 1–2 orders-of-magnitude lower than that of oil-lubricated sleeve bearings and SFDs but higher than that of a hard-mounted ball bearing (DellaCorte, 2001). Experimental values vary from 100 to 1800 N-s/m/cm^2 (0.5–10 lbs/in/in^2) with the utmost representative values being about 200 N-s/m/cm^2 (DellaCorte, 2010). These values have been normalized by the bearing size (DL) in cm^2. The huge variation of damping values is due to the dearth of consistency in the measurement procedure and the extensive variety of support structures applied by foil bearings (i.e., bump foils, lever springs, wire mesh, visco-elastic layers, etc.).

Exploiting these guidelines, one can layout (design) a rotor system successfully sizing the bearings and defining general performance characteristics. Bearing load capacity margin, lift-off speed, shaft orbits, and critical speed margins can be coarsely evaluated. This type of first-order investigation is the first step for machine development integrated with foil bearing.

3.5.6 Analytical Prediction of Foil Bearing Stiffness

It should be noted that variation in bearing stiffness may be there based on the relative direction of the static load and the fixed end of the top sleeve (refer to Figure 3.62a); however, the variation is expected to be small as it is for the case of rolling bearing. In rolling bearings, the relative positions of rolling elements and the static load may give rise to small variations in stiffness. However, for rolling elements, it will be varying with time during operation. For foil bearings, the since the bumps are fixed to the fixed outer sleeve so for a given configuration the relative position of static load and the bump will be fixed so stiffness may not vary with the rotation of the journal.

The bump stiffnesses (k_1, k_2, ..., k_N), depicted in Figure 3.62, represent the stiffness of an individual bump, where N is the number of bumps. The bump stiffness k_1 is expected to be lesser as compared to bump stiffness k_N, as it is located at the free end. Hence, the softer bump (k_1) is triggered before the harder bump (k_{25}). Thus, at position 1, the bump deformation is more than at other angular locations. For small loads in position 1, the structural stiffness is virtually provided by the stiffness of the softest bumps. Though, when the load surges, the hardest bumps deliver additional stiffness to the bearing, and then the rate of deformation of the bumps leads to decline. The preloading of the bearing removes the unusual deflection characteristic at position 1 (0°) (refer to Figure 3.62a). The bearing preload enlarges the top foil and pressurizes the bumps earlier of the loading procedure initiates. At position 1, the preload adequately pressurizes the bumps at the free end, after which the loading procedure is initiated and the bumps at either ends start acting concurrently. At other positions, it is expected that the deformation is more gradual, with the deformation rate decreasing gradually.

Analytical methods to envisage foil bearing structural stiffness have been advanced throughout the last 45 years. Ku and Heshmant (1992b) advanced a model for envisaging bump foil strip deformation and its equivalent structural stiffness and viscous damping coefficients. This model considers the interaction of nearby bumps and the friction forces present in foil bearings. Iordanoff (1999) advanced a more naïve model and offers the structural stiffness of a single bump for two diverse cases, i.e., one with a fixed end and one with both ends free. The analysis envisages the foil bearing structural stiffness considering the influence of each bump and including the dry friction coefficient and the radial location of the external load. Iordanoff's formulas (1999) are employed for calculating the individual bump's structural stiffness. For the analysis, the bump pitch, p, is presumed to be constant and the interlink among bumps is neglected. The structural stiffness for a one end welded bump (k_W) and a free end bump (k_F) is given by Iordanoff (1999), which are summarized as follows (refer to Figure 3.63):

The Welded Bump: It is presumed to be fixed at one end and simply touching at the other end. Between bump and plate Coulomb friction forces are considered. The stiffness expression is specified by (Iordanoff, 1999)

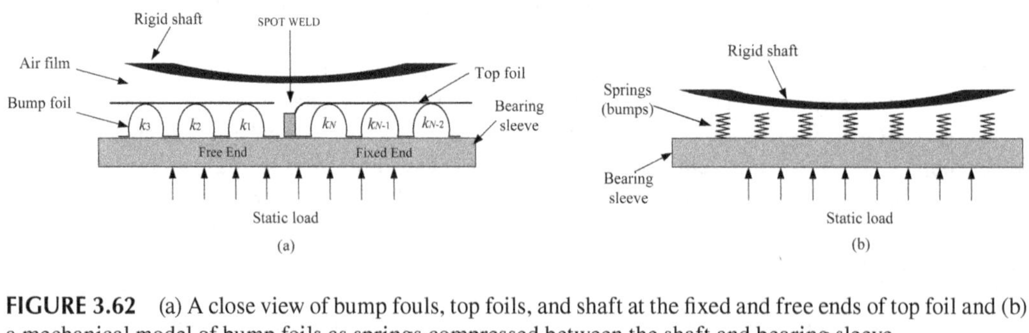

FIGURE 3.62 (a) A close view of bump fouls, top foils, and shaft at the fixed and free ends of top foil and (b) a mechanical model of bump foils as springs compressed between the shaft and bearing sleeve.

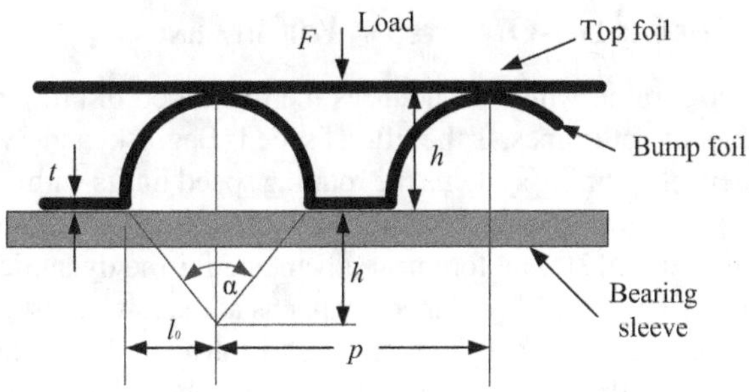

FIGURE 3.63 Macro-geometry of a bump for analytical calculation of bump stiffness in foil bearing.

$$k_W = \frac{F}{\delta} = \frac{Et^3 \sin^3(0.5\alpha)}{12 l_o^3 p^2 J(\alpha, \mu_f)(1 - \nu_p^2) L} \tag{3.134}$$

with

$$J(\alpha, \mu_f) = \{(\cos 0.5\alpha - 1)(\sin 0.5\alpha + \mu_f \cos 0.5\alpha) + 0.25\mu_f(1 - \cos\alpha)\} + 0.25(1 - y_a)(\alpha - \sin\alpha)$$

and

$$y_a(\alpha, \mu_f) = \frac{-(0.75 \sin\alpha - 0.25\alpha - \sin 0.5\alpha)}{a \sin 0.5\alpha(\sin 0.5\alpha + \mu_f \cos 0.5\alpha) - 2\mu_f \sin^2 0.5\alpha + 0.5\alpha - 0.5 \sin\alpha}$$

The Free Bump: It is presumed to be simply applied at both ends. Between the bump and plate Coulomb friction forces are considered. The stiffness expression is specified by (Iordanoff, 1999)

$$k_F = \frac{F}{\delta} = \frac{Et^3 \sin^3(0.5\alpha)}{6 l_o^3 p^2 I(\alpha, \mu_f)(1 - \nu_p^2) L} \tag{3.135}$$

with

$$I(\alpha, \mu_f) = 05\alpha\{A^2 + 0.5(1 + \mu_f^2)\} - 0.25(1 - \mu_f^2)\sin\alpha - 0.5\mu_f(\cos\alpha - 1)$$
$$- aA(1 - \cos 0.5\alpha + \mu_f \sin 0.5\alpha)$$

and

$$A(\alpha, \mu_f) = \sin 0.5\alpha + \mu_f \cos 0.5\alpha$$

where E and ν are the material elastic modulus and Poisson ratio, respectively, pL is the projected area of the top foil (refer to Figure 3.63), $J(\alpha, \mu_f)$ and $I(\alpha, \mu_f)$ are a function of α and the dry coefficient μ_f, p is the bump pitch, t is the foil thickness, h is the bump height, δ is the bump deflection, $2l_o$ represents the bump length, F is the applied load, and α is the bump arc angle (refer to Figure 3.63).

3.5.7 Determination of the Lift-Off Speed in Foil Bearing

An important operative constraint, which pronounces the likelihood of utilizing gas foil bearings in rotor systems, for example microturbines, is the lift-off speed (Bagiński and Żywica, 2018). This constraint conveys to engineers of spinning systems the rotating speed limits within which the bearing has a load capacity and can operate for a prolonged duration without wear of the top foil surface (Kim and Park, 2009; Kim and Zimbru, 2012). The foremost advantage for the dynamics and the usage of the intended rotating system is so that the lift-off speed happens at the lowermost rotating speed (Paouris et al., 2013). This behavior is construed by investigators in diverse directions. The utmost general declaration is, nevertheless, that the lift-off speed is the instant when, during the runup and run-down, dry friction transitions into fluid friction. One of the approaches for finding the rotating speed at which this condition happens is determining the frictional moment on a test stand with single foil bearing (Norsworthy, 2014; Radil et al., 2002). Subsequently, throughout the acceleration of the rotor, the frictional moment increases (surfaces of the journal and top foil rub against one another), and later, upon attainment of the speed at which the lubricating film is developed, it drops quickly to the lowest level (Andres and Kim, 2008; San Andres and Norsworthy, 2015).

In the literature, it can also be found that the lift-off speed happens when the amplitude of absolute vibrations during increasing rotating speed reduces to the amplitude of relative vibrations (Manoj et al., 2016). Based on the experimentation performed, they resolved that the temperature measurement can also be utilized to evaluate the lift-off speed in a foil bearing. The temperature variation is related not only to the presence of a lubricating film nevertheless, but also to the form of the vibration trace. They suggested that the temperature should be measured at two points (or at several points around the circumference) to get values independent from the momentary location of the journal. It was also suggested that this type of testing should be repeated for a better-chosen friction pair. The bearing should be loaded with a sleeve that will load it uniformly. Rotating speed control should permit for an extended function at the speed at which the temperature starts to drop.

Air foil bearings have amalgamated interest in the market for utilization in high-speed machinery, such as air cycle machines, turbo blowers, and microgas turbines, with the foremost thrust being oil-free applications. There continue to be robust benefits in air foil bearings, along with progress and patents filing, initially from the US and Korea, and now also from Japan, China, and India. Several of the patents or novelty are rather similar in objective and are focused on refining the consistency of the "bump" foil feature and/or load capacity. Price saving of air foil bearings has continuously been a significant thrust, either by shortening the construction procedure or the quantity of segments.

3.6 SUSTAINABILITY IN BEARINGS, SEALS, AND DAMPERS

Sustainability commonly refers to schemes, characteristics, and events to preserve a specific object or supply. Sustainable engineering emphases producing enduring, competent products that can be appreciated by future peers. Sustainability in mechanical bearings, dampers, and seals comprises selecting the precise bearing for the application, decreasing upkeep and interruption costs, and utilizing innovative bearing know-hows. Some of themes to produce, utilize, and discard mechanical bearings (as well as dampers and seals) in more sustainable ways are as follows: (i) Choice of the precise bearing: Select the finest bearing for the application. (ii) Advance incessantly: Utilize the state-of-the-art design characteristics and endlessly advance. (iii) Decrease upkeep and stoppage expenses: Utilize keys to decrease upkeep and interruption charges. (iv) Practice progressive bearing technologies: Utilize forward-thinking bearing technologies, like precision raceway honing, coatings, best material selection, optimum process planning, and cloud-based condition monitoring. (v) Augment source exploitation: Improve the usage of bearing material, water, and energy. (vi) Encourage waste diminishing: Promote waste decreasing, and appliances and process that use less energy. (vii) Practice salvaging systems: Utilize reprocessing systems.

3.7 CONCLUDING REMARKS

To summarize, this chapter considered rotordynamics aspects of rolling bearings, hydrodynamic bearings, foil bearings, seals, and SFDs. The main aim is to obtain the stiffness, damping, and inertia effects of these supports on the rotor. For rolling element bearings, a comparative study of several kinds of bearings is provided, and then load–deflection relations are obtained for simple to more general cases. The procedure to calculate the linear and nonlinear stiffness is described in detail. For fluid-film hydrodynamic bearings, the basic Reynolds equation is described and the procedure to calculate the linearized stiffness and damping coefficients for short and finite bearings is provided. These dynamic parameters are found to be dependent on rotor speed. Various kinds of rotary seals are described with their application area. The governing equations of seal dynamics based on bulk fluid theory are described, and the damping and stiffness coefficients are obtained and discussed in detail for a simple case. SFDs are introduced and their rotordynamic parameters are summarized. Various types of foil bearings are covered and basic performance characteristics are described. These dynamic parameters of bearings, seals, and dampers have been obtained at present in isolation of the shaft dynamics. However, in subsequent chapters, we will integrate the dynamics of the rotor and bearing together so that the free and forced response behavior of the rotor can be predicted accurately. This includes the calculation of whirl natural frequencies, critical speeds, and instability of the rotor-bearing system (refer to Chapter 4 of Volume I, and Chapters 1 and 2 of Volume II). One more important bearing is the active magnetic bearing; its basic principles and rotordynamic coefficients are described in Chapter 8 of Volume II.

EXERCISE PROBLEMS

Exercise 3.1 List the parameters on which the stiffness of rolling element bearings depends.

Exercise 3.2 List the parameters on which the stiffness and damping coefficients of fluid-film hydrodynamic bearings and seals depend.

Exercise 3.3 The specification of the ball and roller bearings are given in Table 3.9. Plot the variation of the bearing radial stiffness versus the elastic radial deformation. Show the variation for different radial clearances. Take increments in the load of 100 N and the radial clearance as 0.005 mm for plotting the variation of the radial stiffness.

TABLE 3.9
Geometrical and Load Parameters of the Bearing

		Bearing Parameters	
S.N.	Geometry Type	Ball	Roller
1	Basic bore diameter (mm)	40	20
2	Basic outside diameter (mm)	68	47
3	Inner groove radii (mm)	4.95	—
4	Outer groove radii (mm)	4.96	—
5	Ball/roller diameter (mm)	9.53	6.50
6	Roller length (mm)	—	6.50
7	Radial clearance range (mm)	0.005–0.020	0.005–0.020
8	Radial loads (N)	300–1000	300–1000

Exercise 3.4 For Exercise 3.3, obtain the variation of stiffness of the ball and roller bearings for different orientations of the rolling elements (i.e., for different values of ϕ). Discuss the results obtained.

Exercise 3.5 Obtain the plots of the nondimensional stiffness and damping coefficients versus the Sommerfeld number (between 0.01 and 10) for the short bearing assumption of a hydrodynamic bearing.

Exercise 3.6 Choose a single answer from the multiple choice questions.

i. A high-speed hydrodynamic bearing is modeled by linearized coefficients; these coefficients are
   A. stiffness and damping
   B. virtual or added mass
   C. both (A) and (B)
   D. none of the above

ii. The hydrodynamic fluid-film bearing is governed by the
   A. Bernoulli equation
   B. Reynolds equation
   C. Navier–Stokes equation
   D. Maxwell equation

iii. For variable speed and load conditions, the most appropriate bearing would be a
   A. SFD
   B. fluid-film bearing
   C. gas bearing
   D. rolling element bearing

iv. For the attenuation of rotor vibrations at the resonance, the machine element used is a
   A. seal
   B. fluid-film bearing
   C. rolling element bearing
   D. squeeze-film bearing

v. For the short bearing approximation in the fluid-film bearing analysis
   A. the radial and circumferential pressure gradients are negligible
   B. the radial and axial pressure gradients are negligible
   C. the radial pressure gradient is negligible
   D. the circumferential and axial pressure gradients are negligible

vi. In a rotor system, the cross-coupling of two orthogonal transverse motions due to the spring and damping bearing forces are predominant in the case of the
   A. bush or sleeve bearing
   B. journal bearing
   C. rolling bearing
   D. none of the bearings

vii. Due to flexible bearings, the rotor critical speeds as compared to the rigid supports are expected to
   A. increase
   B. decrease
   C. have no effect
   D. increase or decrease

viii. In fluid-film bearings for the short bearing approximate solution of the Reynolds equation
   A. the pressure variation along the circumferential direction is neglected
   B. the pressure variation along the radial direction is neglected

C. the pressure variation along the axial direction is neglected
D. the pressure variation in any direction is not neglected

ix. For a deep groove ball bearing, the radial clearance is 20 μm and a radial load is applied such that the inner ring center is displaced with respect to the outer ring center by 40 μm in the direction of radial load; *the load zone* would be
A. 30°
B. 60°
C. 90°
D. 120°

x. The contact angle in rolling bearings is related to the
A. static capacity of bearing
B. dynamic capacity of bearing
C. radial load capacity of bearing
D. axial load capacity of bearing

xi. For a long bearing approximation in the fluid-film hydrodynamic bearing
A. only the radial pressure gradients is negligible
B. the radial and circumferential pressure gradients are negligible
C. the radial and axial pressure gradients are negligible
D. the axial and circumferential pressure gradients are negligible

xii. The most stable rotor operation is expected in a
A. cylindrical journal bearing
B. four-lobe journal bearing
C. offset halves journal bearing
D. titling-pad journal bearing

xiii. In hydrodynamic bearings, the pressure in fluid is generated mainly due to
A. external pressure
B. motion of journal
C. squeezing action of fluid
D. heating

xiv. In rolling bearing dry contact, the load–deformation relation is governed by
A. Hooke's law
B. Newton's law
C. Hertz's law
D. Reynolds equation

xv. For high-speed applications, seals provide the following to the rotor system (herein the added mass is the fluid mass)
A. damping and added mass
B. stiffness and damping
C. stiffness and added mass
D. stiffness, damping, and added mass

xvi. The journal center position in a fluid-film bearing for a particular speed and without unbalance force occupies
A. a fixed position
B. a variable position

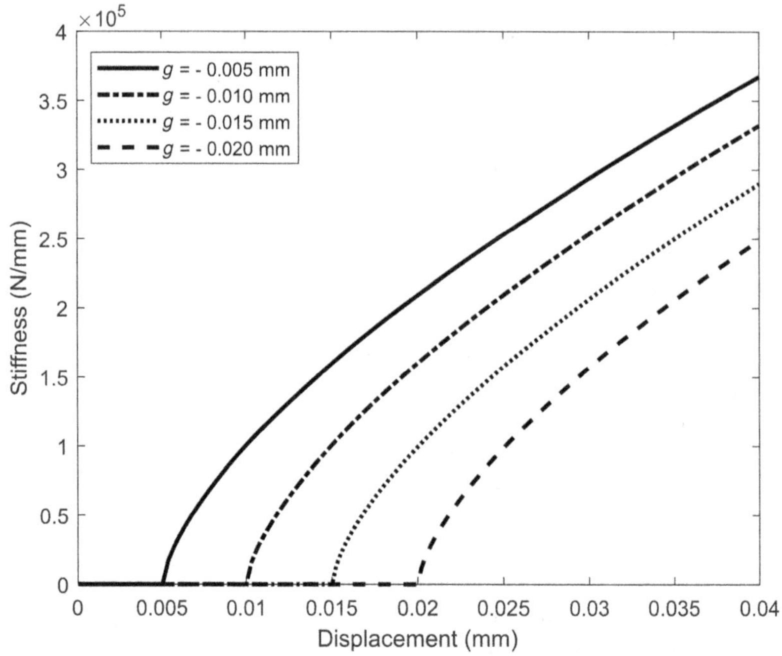

FIGURE 3.64 Variation of stiffness of ball bearing with displacement for different clearances (or negative preloads).

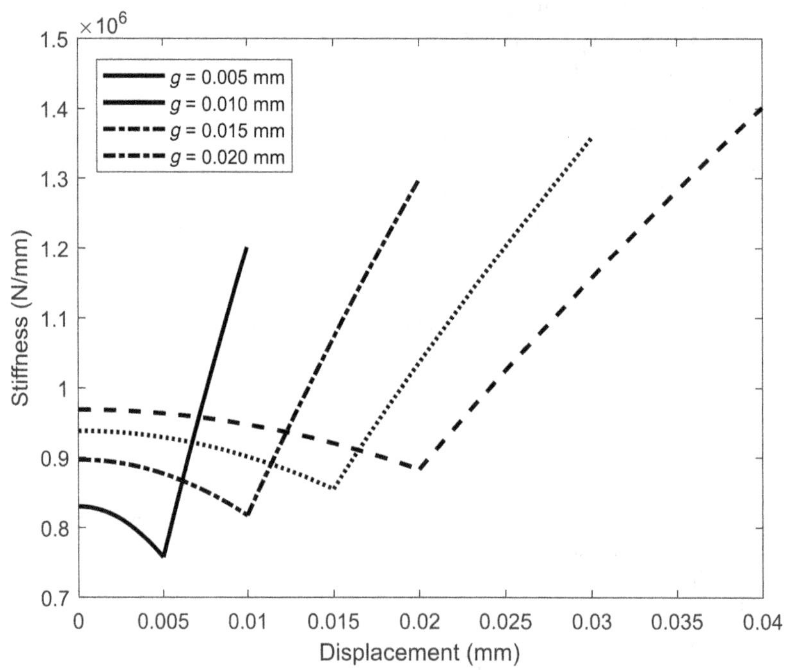

FIGURE 3.65 Variation of stiffness of roller bearing with displacement for different preloads ($x > g$).

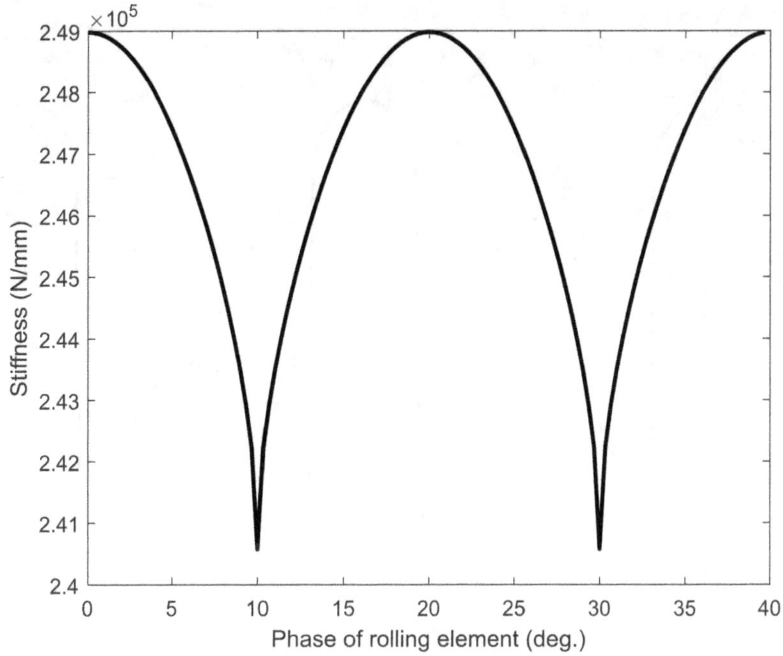

FIGURE 3.66 Variation of stiffness at $x=g$ with different phases of the rolling (ball) element ($g=-0.02$ mm).

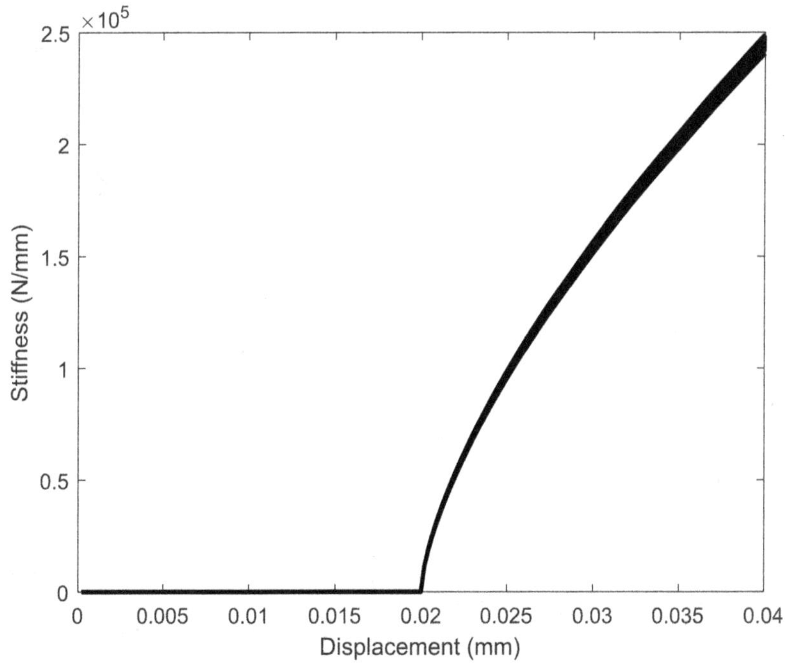

FIGURE 3.67 Variation of stiffness with displacement of ball bearing for different phases ($g=-0.02$ mm, phase 0–40°: two rotations of balls).

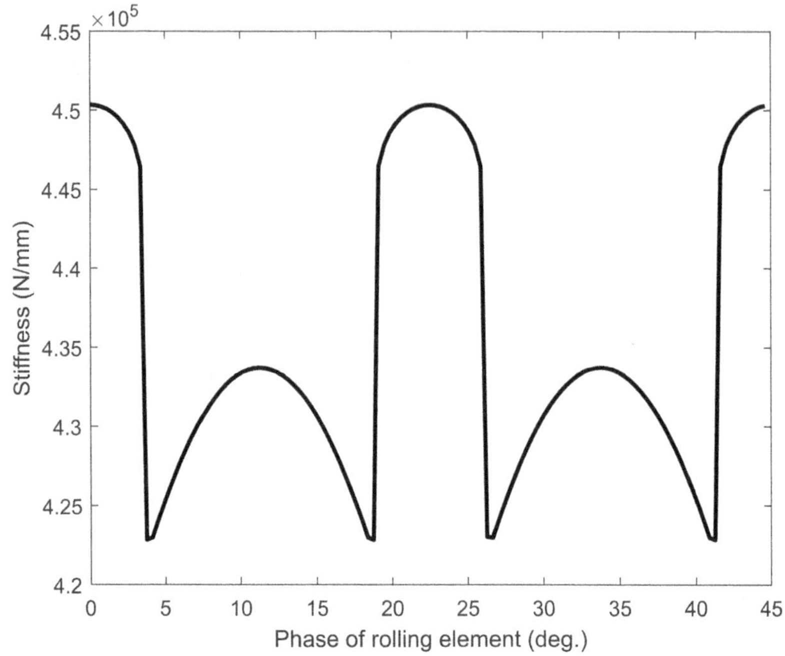

FIGURE 3.68 Variation of stiffness at $x=g$ with different phases of the rolling (roller) element ($g=-0.02$ mm).

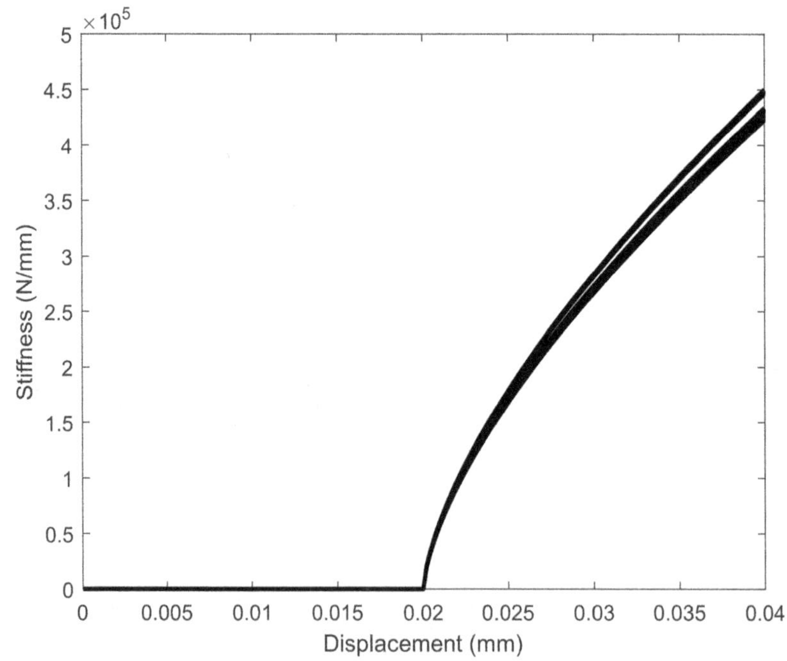

FIGURE 3.69 Variation of stiffness with displacement of roller bearing for different phases ($g=-0.02$ mm, phase 0–40°: two rotations of rollers).

C. the center of the bearing bore
D. a point vertically downward from the center of the bearing bore

xvii. Bearing dynamic coefficients are obtained based on linearization at the
A. bearing center
B. shaft center
C. static equilibrium position
D. dynamic equilibrium position

ANSWERS TO SELECTED EXERCISE PROBLEMS

3.1 Not applicable
3.2 Not applicable
3.3 Refer to Figures 3.64 and 3.65
3.4 Refer to Figures 3.66 to 3.69
3.5 Not applicable
3.6

(i) C	(ii) B	(iii) D	(iv) D	(v) A	(vi) B
(vii) B	(viii) A	(ix) B	(x) D	(xi) C	(xii) D
(xiii) B	(xiv) C	(xv) D	(xvi) A	(xvii) C	

REFERENCES

Adams, M.L., 1987, *Rotating Machinery Vibration: From Analysis to Troubleshooting*. New York, NY: Marcel Dekker

Adams, M.L. Jr., 2001, *Rotating Machinery Vibration*. New York, NY: Marcel Dekker.

Adietta, G. and Pietra, L.D., 2002, The squeeze film damper over four decades of investigations. Part II: Rotordynamic analyses with rigid and flexible rotors, *The Shock & Vibration Digest*, *34*(2), 97–126.

Agrawal, G.L., 1997, Foil Air/Gas Bearing Technology: An Overview, ASME 1997 *International Gas Turbine & Aeroengine Congress & Exhibition*, Orlando, FL, June 2–5, Paper 97-GT-347.

Bagiński, P. and Żywica, G., 2018, Determination of the lift-off speed in foil bearings using various measurement methods. *Mechanics and Mechanical Engineering*, *22*(2), 415–424.

Bhushan, B., 1980, High temperature self-lubricating coatings for air lubricated foil bearings for automotive gas turbine engine. NASA CR-159848, April 1980.

Black, H., 1969, Hydraulic forces in annular pressure on the vibration of centrifugal pump rotors, *Journal of Mechanical Engineering Science*, *11*(2), 206–213.

Black, H.F., 1971, Effects of high pressure ring seals on the pump rotor vibrations. *ASME* No. 71-WA/FF-38.

Black, H.F., and Jenssen, D.N., 1969–1970, Dynamic hybrid properties of annular pressure seals, *Proceedings of Institute of Mechanical Engineering*, *184*(*3N*), 92–100.

Brändlein, J., Eschmann, P., Hasbargen, I., and Weigand, K., 1999, *Ball and Roller Bearings, Theory, Design and Application*, 3rd ed. New York, NY: John Wiley & Sons.

Cameron, A., 1981, *The Principles of Lubrication*. New York, NY: John Wiley & Sons.

Changsen, W., 1991, *Analysis of Rolling Element Bearings*. London: Mechanical Engineering Publications.

Childs, D., Nelson, C., Nicks, C., Scharrer, J., Elrod, D., and Hale, K., 1986, Theory versus experiment for the rotordynamic coefficients of annular gas seals: Part I-test facility and apparatus, *ASME Journal of Tribology*, *108*, 426–432.

Childs, D.W., 1983a, Dynamic analysis of turbulent annual seals based on Hirs lubrication equations, *ASME Journal of Lubrication Technology*, *105*(*3*), 429–436.

Childs, D.W., 1983b, Finite length solutions for rotordynamic coefficients of turbulent annular seals, *ASME Journal of Lubrication Technology*, *105*(*3*), 437–445.

Childs, D.W., 1993, *Turbomachinery Rotordynamics: Phenomena, Modeling and Analysis*. New York, NY: John Wiley & Sons.

DellaCorte, C., 2001, Oil-free shaft support system rotordynamics: Past, present, and future challenges and opportunities, NASA/TM—2011-217003, 1–13.

DellaCorte, C., 2010, Stiffness and Damping Coefficient Estimation of Compliant Surface Gas Bearings for Oil-Free Turbomachinery. NASA/TM—2010–216924, *International Joint Tribology Conference cosponsored by the STLE and ASME*, San Francisco, CA, October 17–20, 2010, IJTC2010–41232.

DellaCorte, C., and Bruckner, R.J., 2010, Remaining Technical Challenges and Future Plans for Oil-Free Turbomachinery, *Proceedings of 2010 ASME-IGTI Turbo Expo*, Glasgow, UK, June 14–18, 2010, GT2010–22086.

DellaCorte, C. and Edmonds, B.J., 2009, NASA PS400: A new high temperature solid lubricant coating for high temperature wear applications, NASA/TM—2009–215678, August 2009.

DellaCorte, C., Radil, K.C., Bruckner, and Howard, S.A., 2008, Design, fabrication, and performance of open source generation I and II compliant hydrodynamic gas foil bearings, NASA/TM-007–214691, ARL-TR-4102, and *STLE Tribology Transactions*, *51*, 254–264.

DellaCorte, C. and Valco, M.J., 2000, Load capacity estimation of foil air journal bearings for oil-free turbomachinery applications, *Tribology Transactions*, *43(4)*, 795–801.

DellaCorte, C. and Valco, M.J., 2003, Oil-Free Turbomachinery Technology for Regional Jet, Rotorcraft and Supersonic Business Jet Propulsion Engines, *Proceedings of the 2003 ISABE Conference*, Cleveland, Ohio, August 31–September 5, 2003, AIAA/ISABE 2003–1182.

DellaCorte, C., Zaldana, A., and Radil, K., 2002, A systems approach to the solid lubrication of foil air bearings for oil-free turbomachinery. *ASME Journal of Tribology*, *126*, 200–207, January 2004. Also, NASA/TM-2002-211482, ARL-TR-2867, October 2002.

Duffin, S. and Johnson, B.T., 1966–1967, Some experimental and theoretical studies of journal bearings for large turbine-generator sets, *Proceedings of Institute of Mechanical Engineering*, *181(3B)*, 89–97.

Dykas, B., Bruckner, R.J., DellaCorte, C., Edmonds, B.J., and Prahl, J.M., 2009, Design, fabrication and performance of foil gas thrust bearings for micro-turbomachinery applications," NASA/TM-2008–215–062, ASME Paper GT2008–50377, and *ASME Journal of Engineering for Gas Turbines and Power*, *131*, 012301-1 to 012301-8.

El-Sayed, H.R., 1980, Stiffness of deep-groove ball bearing, *Wear*, *63*, 89–94.

Feng, K., Zhao, X., Huo, C., and Zhang, Z., 2016, Analysis of novel hybrid bump-metal mesh foil bearings. *Tribology International*, *103*, 529–539.

Gargiulo, E.P., 1980, A simple way to estimate bearing stiffness, *Machine Design*, *52*, 107–110.

Goodwin, M.J., 1989, *Dynamics of Rotor-Bearing Systems*. Sydney: Unwin Hyman.

Gross,' W.A., 1969, A review of developments in externally pressurized gas bearing technology since 1959, *Journal of Lubrication Technology*, *91*, 163.

Gupta, S., Tiwari, R., and Nair, S.B., 2007, Multi-objective design optimization of rolling bearings using genetic algorithms, *Mechanism and Machine Theory*, *42*, 1418–1443.

Hagg, A.C. and Sankey, G.O., 1956, Some dynamic properties of oil-film journal bearings with reference to the unbalance vibration of rotors, *ASME Journal of Applied Mechanics*, *78(2)*, 302–305.

Hamrock, B.J., 1994, *Fundamentals of Fluid Film Lubrication*. New York, NY: McGraw-Hill International Editions, Mechanical Engineering Series.

Harris, T.A., 2001, *Rolling Bearing Analysis*. New York, NY: Wiley.

Hertz, H., 1896, *On the Contact of Rigid Elastic Solids and on Hardness*, Miscellaneous Papers, 163–183. London: Macmillan.

Hirs, G.G., 1973, A bulk-flow theory for turbulence in lubricant films, *ASME Journal of Lubrication Technology*, *95(2)*, 137–146.

Hummel, C., 1926, *Kritische Drehzahlen als Folge der Nachgiebigkeit des Schmiermittels im Lager*. Doctoral thesis, Eidgenössischen Technischen Hochschule in Zürich (Federal Technical University in Zurich). Osterwieck-Harz (present-day Germany): A.W. Zickfeldt, 1926.

Iordanoff, I., 1999, Analysis of an aerodynamic complaint foil thrust bearing: Method for a rapid design. *ASME Journal of Tribology*, *121*, 816–822.

Johnson, T.L., 1991, *Contact Mechanics*, 2nd ed. New York, NY: McGraw-Hill.

Jones, A.B., 1946, *Analysis of Stresses and Deflections*. Briston: New Departure Engineering Data.

Jones, A.B., 1960, A general theory for elastically constrained ball and radial roller bearings under arbitrary load and speed conditions, *Transactions of ASME, Journal of Basic Engineering*, *82(2)*, 309–320.

Kashyap, R. and Tiwari, R., 2006, Prediction of Heat Generations and Temperature Distributions at Critical Contact Zones of High-Speed Rolling Bearings, *Proceedings of 18th National & 7th ISHMT-ASME Heat and Mass Transfer Conference*, IIT Guwahati, January 4–6, 2006.

Kim, D. and Park, S., 2009, Hydrostatic air foil bearings: Analytical and experimental investigation, *Tribology International*, *42(3)*, 413–425.

Kim, D. and Zimbru, G., 2012, Start-stop characteristics and thermal behavior of a large hybrid airfoil bearing for aero-propulsion applications, *Journal of Engineering for Gas Turbines and Power*, *134(3)*, 032502.

Klaass, R.F. and DellaCorte, C., 2006, The Quest for Oil-Free Gas Turbine Engines, *Proceedings of the Power Systems Conference*, New Orleans, Louisiana, November 7–9, 2006, SAE 2006-01-3055.

Kolanowski, B.F., 2004, *Guide to Microturbines*. New York, NY: Marcel Dekker, Ch. 1, pp. 1–15.

Krämer, E., 1993, *Dynamics of Rotors and Foundations*. New York, NY: Springer-Verlag.

Ku, C.-P. and Heshmant, H., 1992a, Complaint foil bearing structural stiffness analysis Part I: Theoretical model - Including strip and variable bump foil geometry, *ASME Journal of Tribology*, *114*(2), 394–400.

Ku, C.P.R. and Heshmat, H., 1994, Structural stiffness and coulomb damping in compliant foil journal bearings: Parametric studies, *STLE Transactions*, *37*(*1*), 455–462.

Ku, R. and Heshmant, H., 1992b, Compliant foil bearing structural stiffness analysis, *ASME Journal of Tribology*, *113*(*1*), 364–369.

Lim, T.C. and Singh, R., 1990a, Vibration transmission through rolling element bearings, Part I: Bearing stiffness formulation, *Journal of Sound and Vibration*, *139*(2), 179–199.

Lim, T.C. and Singh, R., 1990b, Vibration transmission through rolling element bearings, Part II: System studies, *Journal of Sound and Vibration*, *139*(2), 201–225.

Lim, T.C. and Singh, R., 1991, Vibration transmission through rolling element bearings, Part III: Geared rotor system studies, *Journal of Sound and Vibration*, *151*(*1*), 31–54.

Lim, T.C. and Singh, R., 1992, Vibration transmission through rolling element bearings, Part IV: Statistical energy analysis, *Journal of Sound and Vibration*, *153*(*1*), 37–50.

Lim, T.C. and Singh, R., 1994, Vibration transmission through rolling element bearings, Part V: Effect of distributed contact load on roller bearing stiffness matrix, *Journal of Sound and Vibration*, *169*(*4*), 547–553.

Lomakin, A.A., 1958, Calculation of critical speed and securing of dynamic stability of the rotor of hydraulic high pressure machines with reference to forces arising in the seal gaps (in Russian), *Energomashinostroenie*, *4*(*4*), 1–5.

Lubell, D., DellaCorte, C., and Stanford, M.K., 2006, Test Evolution and Oil-Free Engine Experience of a High Temperature Foil Air Bearing Coating, *Proceedings of GT2006: ASME Turbo Expo 2006*, Barcelona, Spain, May 8–11, 2006, GT2006–90572.

Lund, J.W., 1987, Review of the concept of dynamic stiffness coefficients for fluid film journal bearings, *ASME Journal of Tribology*, *109*(*1*), 37–41.

Lund, J.W., Arwas, E.B., Cheng, H.S., Ng, C.W., Pan, C.H.T., and Sternlicht, B., 1965, *Rotor-Bearing Dynamics Design Technology, Part III: Design Handbook for Fluid-Film Type Bearings*, Technical Report AFAPL-TR-65–45, Wright-Patterson Air Force Base, p. 229. Latham, NY: Mechanical Technology, Inc.

Lund, J.W. and Thomsen, K.K., 1978, A Calculation Method and Data for the Dynamic Coefficients of Oil-Lubricated Journal Bearings. In *Topics in Fluid Film Bearing and Rotor Bearing System Design and Optimization*, S.M. Rodhe, P.E. Allaire, and D.J. Maday (eds.). New York, NY: American Society of Mechanical Engineers, pp. 1–28.

Lynch, T., 1993, Smart bearing boosts air turbine efficiency, Design News, November 1993, pp. 101–103.

Manoj, I.V., Srihari, P.V., Kulkarni, S.S., Kumar, K.S., and Bharatish, A., 2016, Assessment of thermal effects on the levitation speed of bump foil bearings made of low-cost spring steel, *Measurement: Journal of the International Measurement Confederation*, *92*, 453–463.

Newkirk, B.L., 1924, Shaft whipping, *General Electric Review*, *24*, 169.

Newkirk, B.L. and Taylor, H.D., 1925, Shaft whipping due to oil action in journal bearing, *General Electric Review*, *28*(*8*), 559–568.

Norsworthy, J.D., 2014, *Measurement of drag torque, lift off speed, and identification of frequency dependent stiffness and damping coefficients of a shimmed bump-type foil bearing*. M.S. thesis, Texas A&M University, College Station, TX, 2014.

Ocvirk, F.W., 1952, *Short Bearing Approximation for Full Journal Bearings*, NACA Technical Note 2808. Washington, DC: National Advisory Committee for Aeronautics.

Palmgren, A., 1959, *Ball and Roller Bearing Engineering*, 3rd ed. Philadelphia, PA: SKF Industries Inc.

Paouris, L.I., Bompos, D.A., and Nikolakopoulos, P.G., 2013, Simulation of static performance of air foil bearings using coupled finite element and computational fluid dynamics techniques, *Journal of Engineering for Gas Turbines and Power*, *136*, 022503.

Park, J. and Sim, K., 2021, Rotordynamic analysis of gas foil-polymer bearings based on a structural elasticity model of polymer layer along with static-load deflection tests. *Applied Science*, *11*, 1789.

Petroff, N.P., 1883, Friction in machines and the effect of the lubricant, *Inzhenerno Zhurnal St. Petersburg*, *1–4*, 71–140, 227–279, 377–436, 535–564.

Pietra, L.D. and Adietta, G., 2002, The squeeze film damper over four decades of investigations. Part I: Characteristics and operating features, *The Shock & Vibration Digest*, *34*(*1*), 3–26.

Pinkus, O. and Sternlicht, B., 1961, *Theory of Hydrodynamic Lubrication*. New York, NY: McGraw-Hill.

PSG, 1982, Faculty of Mechanical Engineering, PSG College of Technology, Design Data. Coimbatore: DPV Printers.

Radil, K., Howard, S., and Dykas, B., 2002, The role of radial clearance on the performance of foil air bearings, *Tribology Transactions*, *45(4)*, 485–490.

Radil, K.C. and Zeszotek, M., 2004, An experimental investigation into the temperature profile of a compliant foil air bearing. NASA/TM—2004–213100 and ARL-TR-3200, May 2004.

Ragulskis, K.M., Jurkauskas, A.Y., Atstupenas, V.V., Vitkute, A.Y., and Kulvec, A.P., 1974, *Vibration of Bearings*. Vilnyus: Mintis Publishers.

Rao, B.R. and Tiwari, R., 2007, Optimum design of rolling element bearings using genetic algorithms, *Mechanism and Machine Theory*, *42*(2), 233–250.

Rao, J.S., 1996, *Rotor Dynamics*, 3rd ed. New Delhi: New Age.

Rao, J.S., 2000, *Vibratory Condition Monitoring of Machines*. New Delhi: Narosa Publishing House.

Reinhardt, E. and Lund, J.W., 1975, The influence of fluid inertia on the dynamic properties of journal bearings, *ASME Journal of Lubrication Technology*, *97(2)*, 159–167.

Reynolds, O., 1886, On the theory of lubrication and its application to Mr. Beauchamp tower's experiments, including an experimental determination of the viscosity of olive oil, *Philosophical Transactions of the Royal Society, A*, *177*, 157–234.

Rubio, D. and San Andrés, L., 2003, Bump-type foil bearing structural stiffness: Experiments and predictions. TRC Project: Identification of Structural Stiffness and Damping in Foil Bearings, Research Progress Report to the Turbomachinery Research Consortium, TRC-B&C-5-03, May 2003.

Ruscitto, D., McCormick, J., and Gray, S., 1978, Hydrodynamic air lubricated compliant surface bearing for an automotive gas turbine engine: I-Journal bearing performance, NASA CR–135368, 1978.

San Andres, L. and Kim, T.H., 2008, Forced nonlinear response of gas foil bearing supported rotors, *Tribology International*, *41(4)*, 704–715.

San Andres, L. and Norsworthy, J., 2015, Structural and rotordynamic force coefficients of a shimmed bump foil bearing: An assessment of a simple engineering practice, *Journal of Engineering for Gas Turbines and Power*, *138(1)*, 012505.

San Andres, L. and Vance, J.M., 1987, Effect of fluid inertia on finite length squeeze-film dampers, *ASLE Transactions*, *30(3)*, 384–393.

Schlichting, H.T., 2004, *Boundary-Layer Theory*, 8th ed. New York, NY: Springer.

Smith, D.M., 1969, *Journal Bearings in Turbomachinery*. London: Chapman and Hall.

Someya, T. (ed.), 1989, *Journal Bearing Data Book*, London: Springer-Verlag.

Stodola, A., 1925, Kritische Wellenstörung infolge der Nachgiebigkeit des Ölpolslers im Lager (Critical shaft perturbations as a result of the elasticity of the oil cushion in the bearings), *Schweizerische Bauzeitung*, *85(21)*, 265–266.

Stolarski, T.A., 1990, *Tribology in Machine Design*. Oxford: Heinemann Newnes.

Tao, L., Diaz, S., San Andres, L., and Rajagopla, K.R., 2000, Analysis of squeeze film dampers operating with bubbly lubricants, *ASME Journal of Tribology*, *122(1)*, 205–210.

Tichy, J.A. and Modest, M., 1978, Squeeze-film in arbitrary shaped journal bearings subject to oscillations, *ASME Journal of Lubrication Technology*, *100*, 323–330.

Timoshenko, S. and Goodier, J., 1951, *Theory of Elasticity*, 2nd ed. New York, NY: McGraw-Hills.

Tiwari, R., 2017, *Rotor Systems: Analysis and Identification*. Boca Raton, FL: CRC Press.

Tiwari, R. and Chandran, R.M.P., 2015, Multitude of objectives based optimum designs of cylindrical roller bearings using evolutionary algorithms, *ASME Journal of Tribology*, *137*, 041504-1–041504-12.

Tiwari, R., Lees, A.W., and Friswell, M.I., 2004, Identification of dynamic bearing parameters: A review, *The Shock & Vibration Digest*, *36*(2), 99–124.

Tiwari, R., Manikandan, S., and Dwivedy, S.K., 2005, A review on the experimental estimation of the rotor dynamic parameters of seals, *The Shock & Vibration Digest*, *37(3)*, 261–284.

Tiwari, R. and Vyas, N.S., 1995, Estimation of non-linear stiffness parameters of rolling element bearings from random response of rotor-bearing systems, *Journal of Sound and Vibration*, *187*(2), 229–239.

Tower, B., 1883, First Report on Friction Experiments (Friction of Lubricated Bearing), *Proceedings of the Institution of Mechanical Engineers*, November 1, 632–659, 29–35. London.

Tower, B., 1884, Adjourned Discussion for First Report on Friction Experiments, *Proceedings of the Institution of Mechanical Engineers*, January 24–25, 29–35. London.

Tower, B., 1885, Second Report on Friction Experiments (Experiments on the Oil Pressure in a Bearing), *Proceedings of the Institution of Mechanical Engineers*, January 30, 58–70. London.

Vance, J.M., 1988, *Rotordynamics of Turbomachinery*. New York, NY: John Wiley & Sons.

Zeidan, F.Y., San Andres, L.A., and Vance, J.M., 1996, Design and Application of Squeeze Film Dampers in Rotating Machinery, *Proceedings of the Twenty-Fifth Turbomachinery Symposium*, September 17–19, Houston, TX, *2518*, 169–188. College Station: Texas A&M University.

4 Transverse Vibrations of Simple Rotor-Bearing-Foundation Systems

In Chapter 2, we studied the dynamic behaviors of rotors with a rigid disk and a flexible massless shaft. These simple rotor models have an advantage in that the mathematical modeling is simple, and it predicts some of the vital phenomena with relative ease. However, in actuality, as the previous chapter has demonstrated, the supports of rotors, that is, bearings as well as the foundation, are flexible and have a considerable amount of damping. Consequently, they play a vital role in predicting the dynamic behavior of rotor systems. In the present chapter, we will incorporate bearing dynamic parameters in mathematical models of rotor systems. The previous chapter showed that a bearing normally has eight linearized dynamic parameters (four for the stiffness and four for the damping, with the direct and cross-coupled terms). To start with, first we will consider a long rigid rotor mounted on flexible anisotropic bearings (without damping and without cross-coupled stiffness terms). Next, in the long rigid rotor system, we will incorporate a more general bearing model with eight linearized bearing dynamic coefficients. Subsequently, along with the flexibility of the bearing, the shaft flexibility with rigid disks is considered. Finally, the flexibility of the shaft, bearings, and foundations has been included for predicting the dynamic behavior of a simple rotor system and the forces transmitted through the supports. Wherever equations are large, the matrix and vector forms are preferred. In fact, the aim of this chapter is to demonstrate that for complex rotor-bearing-foundation systems, conventional modeling and analysis procedures are difficult to apply. A more systematic approach is required, like the transfer matrix method (refer to Chapters 6 and 8) and the finite element (FE) method (refer to Chapters 7, 9 and 10).

4.1 A SYMMETRICAL LONG RIGID SHAFT ON FLEXIBLE ANISOTROPIC UNDAMPED BEARINGS

When bearings are far more flexible than the shaft in rotor systems, then the bearings will have the highest effect on the motion of the rotor. These rotors may be considered rigid. In the present analysis, it is assumed that the shaft has no flexibility, and the bearings are assumed to behave as linear springs with the stiffness k_x and k_y in the horizontal and vertical directions, respectively (refer to Figure 4.1). The center of gravity of the rotor mass, m, is offset from the geometrical center by distances e and e_z as shown in Figure 4.1, respectively, in the radial and axial directions. x and y are translational displacements of the rotor (the geometrical center) in the horizontal and vertical directions, respectively, whereas φ_y and φ_x are rotational (tilt) displacements of the rotor (the geometrical center line) in the z-x and y-z planes, respectively. Figure 4.1b shows the positive convention for rotational displacements. For the present case, it is assumed that there is no coupling among various displacements, that is, x, y, φ_y, and φ_x. Hence, free body diagrams (Figure 4.2) and equations of motion have been obtained by giving such displacements independent of each other. It is assumed that the translational and angular displacements are small and the direction of the unbalance force and moment does not change with the tilting of the shaft (i.e., due to φ_y and φ_x). In fact, without this assumption, we will get equations of motion with force containing unknown rotational

DOI: 10.1201/9781003661559-4

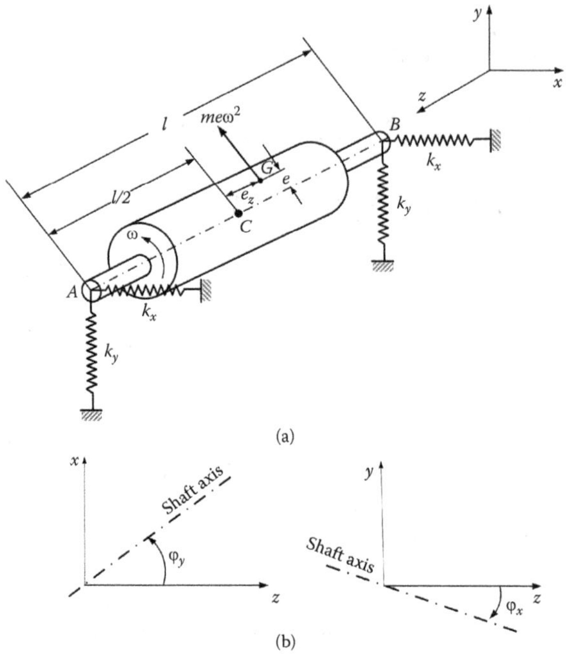

FIGURE 4.1 (a) A rigid rotor mounted on flexible bearings. (b) Positive conventions of rotational displacements.

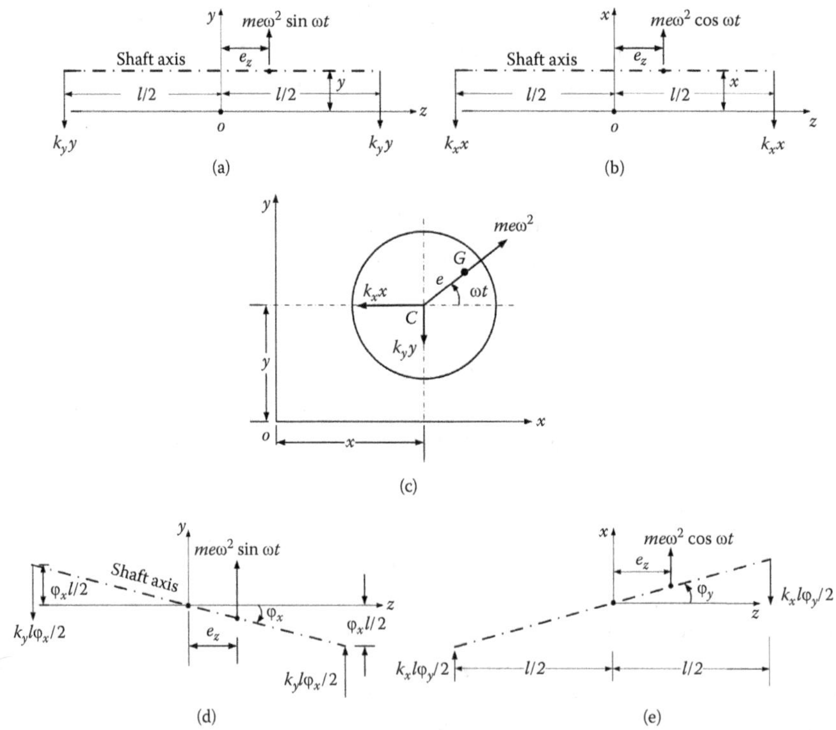

FIGURE 4.2 (a) A free body diagram of the rotor in the y-z plane for a pure translational motion. (b) A free body diagram of the rotor in the z-x plane for a pure translational motion. (c) A free body diagram of the rotor in the x-y plane. (d) A free body diagram of the rotor in the y-z plane for the pure rotational motion. (e) A free body diagram of the rotor in the z-x plane for the pure rotational motion.

displacements (i.e., φ_y and φ_x) and equations would be parametrically excited systems (i.e., differential equations with time-dependent coefficients), which is beyond the scope of the present text.

From Figures 4.2a–c, equations of motion in the x and y directions are

$$me\omega^2 \cos\omega t - 2k_x x = m\ddot{x} \quad \text{and} \quad me\omega^2 \sin\omega t - 2k_y y = m\ddot{y} \tag{4.1}$$

Here, m is the mass of the rotor, e is the radial eccentricity, and ω is the spin speed of the shaft. From Figure 4.2d and e, the equations of motion in the φ_x and φ_y directions are

$$-me\omega^2 e_z \sin\omega t - 0.5k_y l^2 \varphi_x = I_d \ddot{\varphi}_x \quad \text{and} \quad me\omega^2 e_z \cos\omega t - 0.5k_x l^2 \varphi_y = I_d \ddot{\varphi}_y \tag{4.2}$$

Here, I_d is the diametral mass moment of inertia of the rotor, l is the bearing span, and e_z is the axial eccentricity. The effect of a rigid rotor with a small eccentricity in the axial direction, e_z, is the same as a disk mounted on a massless shaft with an offset by the same small axial eccentricity, e_z. However, if the disk is offset large, then the translational and rotational motions will be coupled. It will give time-dependent moment terms multiplied by angular displacement terms (refer to Section 2.5). For simplicity of analysis, we often ignore this term.

For the simple harmonic vibration, we can write

$$\ddot{x} = -\omega^2 x, \quad \ddot{y} = -\omega^2 y, \quad \ddot{\varphi}_x = -\omega^2 \varphi_x \quad \text{and} \quad \ddot{\varphi}_y = -\omega^2 \varphi_y \tag{4.3}$$

On substituting Equation 4.3 into Equations 4.1 and 4.2, unbalance responses can be expressed as

$$x = \frac{me\omega^2}{2k_x - m\omega^2}\cos\omega t = X\cos\omega t; \quad y = \frac{me\omega^2}{2k_y - m\omega^2}\sin\omega t = Y\sin\omega t \tag{4.4}$$

and

$$\varphi_x = \frac{-me\omega^2 e_z}{0.5k_y l^2 - \omega^2 I_d}\sin\omega t = -\Phi_x \sin\omega t \quad \text{and} \quad \varphi_y = \frac{me\omega^2 e_z}{0.5k_x l^2 - \omega^2 I_d}\cos\omega t = \Phi_y \cos\omega t \tag{4.5}$$

where X and Y are translational displacement amplitudes and Φ_x and Φ_y are rotational displacement amplitudes. From the denominators of these amplitudes, it can be seen that the system has four critical speeds: two for translational displacements in the x and y directions, and two for rotational displacements in the x-z and y-z planes. On equating these denominators to zero to have a resonance condition, then critical speeds can be obtained as

$$\omega_{cr1} = \sqrt{\frac{2k_x}{m}}; \quad \omega_{cr2} = \sqrt{\frac{2k_y}{m}}; \quad \omega_{cr3} = \sqrt{\frac{0.5k_x l^2}{I_d}}; \quad \text{and} \quad \omega_{cr4} = \sqrt{\frac{0.5k_y l^2}{I_d}} \tag{4.6}$$

From Equation 4.4, upon squaring expressions for x and y and adding them, this gives

$$\frac{x^2}{X^2} + \frac{y^2}{Y^2} = 1 \tag{4.7}$$

This is an equation of the ellipse. Hence, for the pure transverse translational motion, each particle of the rotor will have an elliptical path, orbit, or trajectory (Figure 4.3a and c). Such a motion would be possible with $e \neq 0$ and $e_z = 0$. Similarly, from Equation 4.5, we get

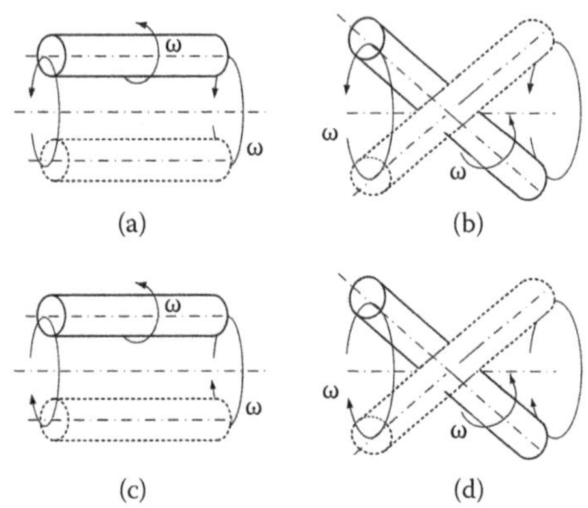

FIGURE 4.3 Mode shapes for a rigid rotor mounted on flexible bearings: (a) synchronous translational whirl; (b) synchronous conical whirl; (c) antisynchronous translational whirl; (d) antisynchronous conical whirl.

$$\frac{\varphi_x^2}{\Phi_x^2} + \frac{\varphi_y^2}{\Phi_y^2} = 1 \tag{4.8}$$

Hence, for the purely rotational (tilting) motion in the transverse planes, each particle of the rotor will have an elliptical path (Figure 4.3b and d). Such a motion would be possible with $e_z \neq 0$ and $e = 0$. Such pure rotational motion in practice could only be possible if a long rotor is supported at the midspan by self-aligning rolling bearings and at the ends by fluid-film bearings. It would be interesting to observe the whirl direction (i.e., clockwise or counterclockwise [CCW]) with respect to the spin speed direction. In general, all modes may be excited simultaneously. It should be noted that a synchronous whirl condition has been assumed (i.e., $\nu = \omega$, where ν is the whirl frequency and ω is the spin speed of the rotor). Let us first consider the translational displacement only, that is, Equation 4.4, and assume that $k_x < k_y$ (i.e., $\omega_{cr_1} < \omega_{cr_2}$).

Case I: When the rotor operates below the first critical speed, that is, $\omega < \omega_{cr_1}$, from Equation 4.4, both X and Y are positive. From Figure 4.4a–c, we can observe that the location of points P_1 and P_2 corresponding to two-time instances t_1 and t_2 (with $t_2 > t_1$) is such that it moves in the CCW direction. Thus, the whirl direction is CCW, which is the same as the spin direction according to the positive convention. Hence, the rotor whirls in the same direction as the spin of the shaft, as shown in Figure 4.3a. This type of whirl is called the *forward (or in-) synchronous whirl* ($\nu = \omega$).

Case II: For $\omega_{cr_1} < \omega < \omega_{cr_2}$, from Equation 4.4, the displacement amplitude X becomes negative, indicating that the horizontal displacement of the rotor is always in the opposite direction to that in Case I. In Figure 4.4d–f, the movement of P_1 and P_2 could be observed on the orbit (clockwise), which is now opposite to the spin direction (CCW). It follows from this that the rotor will whirl in the opposite direction to the spinning of the shaft as shown in Figure 4.3c. This type of whirl is called the *backward (or anti-) synchronous whirl* ($\nu = -\omega$).

Case III: At higher shaft speeds $\omega > \omega_{cr_2}$, both X and Y are negative, which leads to both displacements being in the opposite direction to that in Case I. Figure 4.4g–4.4i reveals that the rotor whirls once more in the same direction as the spin of the shaft (CCW); however, the phase may be different. In all three cases, if we observe the motion of a point P on the orbit at time instants $t = t_1$ and t_2 the point will move from P_1 to P_2 as shown in Figure 4.4c, f, and i, which clearly shows the direction of whirl with respect to the spin speed.

Equation 4.8 is related to the rotational motion of the rotor and is also an equation of the ellipse. This means that there is an elliptical orbital trajectory of the rotor due to the rotational motion of the rotor, as shown in Figure 4.4b and d. This rotor motion is caused by the unbalance couple $me\omega^2 e_z$ acting on the rotor. For $\omega < \omega_{cr_3}$ and $\omega > \omega_{cr_4}$, the synchronous whirl persists. A reversal of the direction of the orbit occurs for the rotor spin speed between two critical speeds associated with

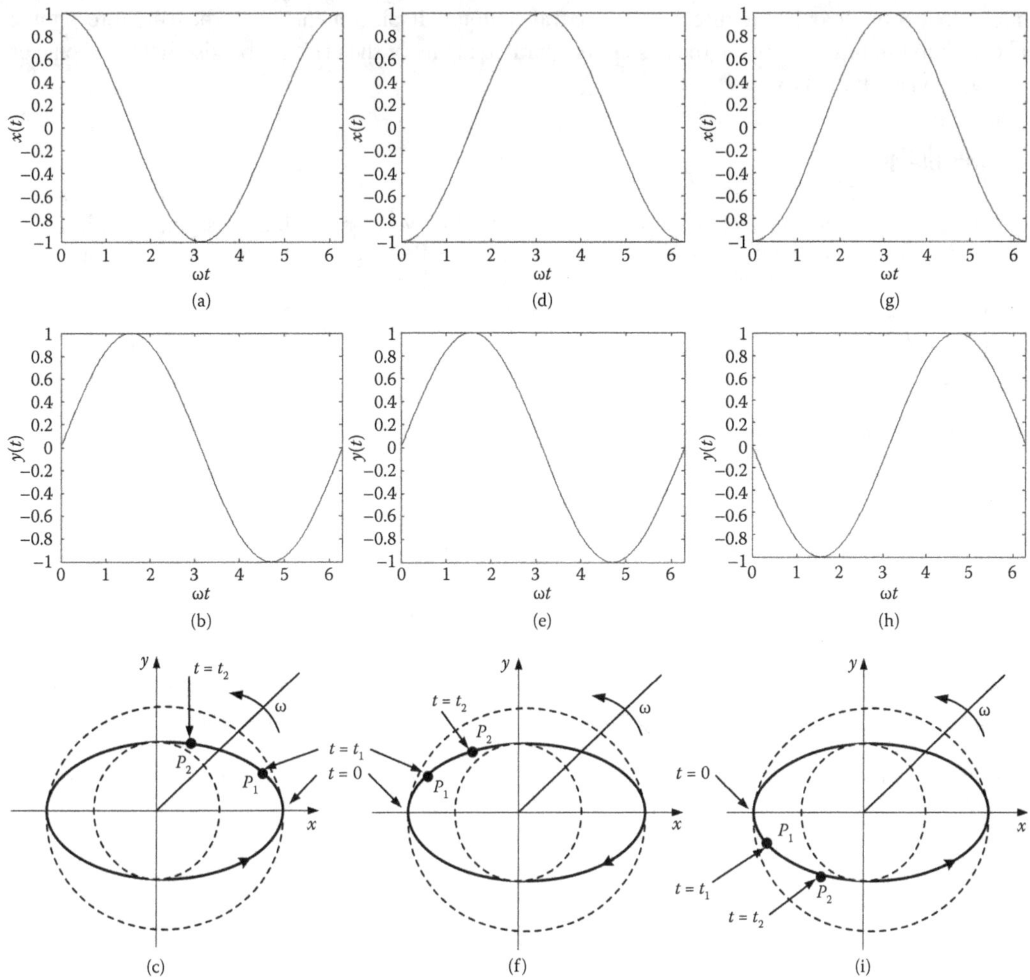

FIGURE 4.4 Whirl directions with respect to the shaft spin frequency. (i) Synchronous whirl, $\omega < \omega_{cr1}$: (a) plot of x versus ωt; (b) plot of y versus ωt; (c) plot of x-y (the rotor center orbit) (at $t=0$, $x=+x_{max}$ and $y=0$). (ii) Antisynchronous whirl, $\omega_{cr1} < \omega < \omega_{cr2}$: (d) plot of x versus ωt; (e) plot of y versus ωt; (f) plot of x-y (the rotor center orbit) (at $t=0$, $x=-x_{max}$ and $y=0$). (iii) Synchronous whirl, $\omega > \omega_{cr2}$: (g) plot of x versus ωt; (h) plot of y versus ωt; (i) plot of x-y (the rotor center orbit) (at $t=0$, $x=-x_{max}$ and $y=0$).

the rotational motion of the rotor (i.e., $\omega_{cr3} < \omega < \omega_{cr4}$) as shown in Figure 4.4f. In general, because both the force and the moment are acting, the motion of the rotor will be a combination of both the translational and rotational displacements, and only at or near to the corresponding critical speeds will such whirling be distinguishable.

The amplitude of forces transmitted to the bearing is now different in the horizontal and vertical directions, as well as at each end of the rotor. The force transmitted through the bearing is given by the product of spring stiffness and rotor deflection at the bearing, which is

$$f_x = k_x\left(x \pm 0.5l\varphi_y\right) \quad \text{and} \quad f_y = k_y\left(y \pm 0.5l\varphi_x\right) \tag{4.9}$$

in the horizontal and vertical directions, respectively. In Equation 4.9, the + sign refers to the rotational displacement of the rotor, which causes its end to deflect in the same direction to the translational displacement, and the − sign refers to the rotational displacement of the rotor, which causes its end to deflect in the opposite direction to the translational displacement. These bearing forces must take on maximum values when the system is operated at any of the critical speeds, that is, whenever x, y, φ_x, and φ_y are maximum.

Example 4.1

A long rigid symmetric rotor is supported at the ends by two identical bearings. The rotor has a 0.2 m diameter, 1 m length, and 7800 kg/m^3 mass density. Bearing dynamic parameters are as follows: $k_{xx}=k_{yy}=1$ kN/mm with the rest of the linearized stiffness and damping coefficients equal to zero. Considering the gyroscopic effect to be negligible, obtain the transverse critical speeds of the system.

Solution:

We have the following data:

$$k = 1\times 10^6 \text{ N/m}, \quad m = \rho\pi r^2 l = 245.04 \text{ kg}$$

and

$$I_d = \frac{1}{16}md^2 + \frac{1}{12}ml^2 = 21.0330 \text{ kg-m}^2$$

Because cross-coupled stiffness coefficients in the x and y directions are zero and no gyroscopic effect is considered, a single-plane motion can be considered. For the present analysis, the coupling is not considered between the translational and rotational displacements. The stiffnesses in the x and y directions are the same; hence, critical speeds corresponding to the translational motion can be written as

$$\omega_{cr1,2} = \sqrt{\frac{2k}{m}} = \sqrt{\frac{2\times 1\times 10^6}{245.04}} = 90.34 \text{ rad/s}$$

Similarly critical speeds corresponding to the titling motion can be written as

$$\omega_{cr3,4} = \sqrt{\frac{kl^2}{2I_d}} = \sqrt{\frac{1\times 10^6\times 1^2}{2\times 21.0330}} = 154.18 \text{ rad/s}$$

Example 4.2

Find the transverse natural frequencies of a system as shown in Figure 4.5. The mass and the diametral mass moment of inertia of the rotor are 2.51 kg and 0.00504 kg-m^2, respectively. The total span of the shaft between the bearings is 508 mm, and the shaft is treated as rigid. (i) Bearings have equal flexibility in the horizontal and vertical directions $k_x=k_y=k=175$ N/m. (ii) Solve the same problem as part (i) except that the bearings have different vertical and horizontal flexibilities: $k_{hoz}=175$ N/m and $k_{ver}=350$ N/m for each of the bearings.

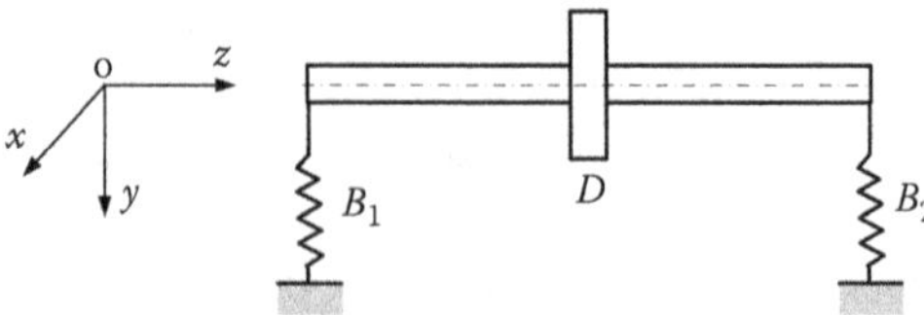

FIGURE 4.5 A rigid rotor on flexible supports.

Solution:

a. Considering a single-plane (e.g., the horizontal) motion with the assumption of uncoupled translational and rotational motions; equations of motion for free vibrations can be written as

$$m\ddot{x} + 2k_x x = 0 \quad \text{and} \quad I_d\ddot{\varphi}_y + 0.5k_x l^2\varphi_y = 0$$

where x is the translational displacement of the center of gravity of the shaft and φ_y is the rotational displacement of the shaft center line with respect to the z-axis in the z-x plane. The mass and mass moment of inertia of the disk are given as $m = 2.51$ kg and $I_d = 0.00504$ kg-m^2. We have $k_x = k = 175$ N/m and $l = 0.508$ m. Hence, natural frequencies (or critical speeds) can be written as

$$\omega_{nf1} = \sqrt{\frac{2k}{m}} = 11.81 \text{ rad/s} \quad \text{and} \quad \omega_{nf2} = \sqrt{\frac{kl^2}{2I_d}} = 66.93 \text{ rad/s}$$

b. With different stiffness properties in the horizontal and vertical directions, the four natural frequencies are given as

$$\omega_{nf1} = \sqrt{\frac{2k_{hoz}}{m}} = 11.81 \text{ rad/s}; \quad \omega_{nf2} = \sqrt{\frac{2k_{ver}}{m}} = 16.70 \text{ rad/s}$$

$$\omega_{nf3} = \sqrt{\frac{k_{hoz}l^2}{2I_d}} = 66.93 \text{ rad/s}; \quad \omega_{nf4} = \sqrt{\frac{k_{ver}l^2}{2I_d}} = 94.66 \text{ rad/s}$$

Example 4.3

Find the critical speeds of a rotor system as shown in Figure 4.6. The bearing stiffness properties are $k_{x_A} = 1.1$ kN/mm, $k_{y_A} = 1.8$ kN/mm, $k_{x_B} = 3.1$ kN/mm, and $k_{y_B} = 3.8$ kN/mm. The rotor has $m = 10$ kg and $I_d = 0.1$ kg-m^2. The rotor length $l = 1$ m.

Solution:

Consider two plane (e.g., the horizontal and the vertical) motions with the assumption of uncoupled translational and rotational motions and equations of motion in the x and y directions with the radial unbalance, me, given as

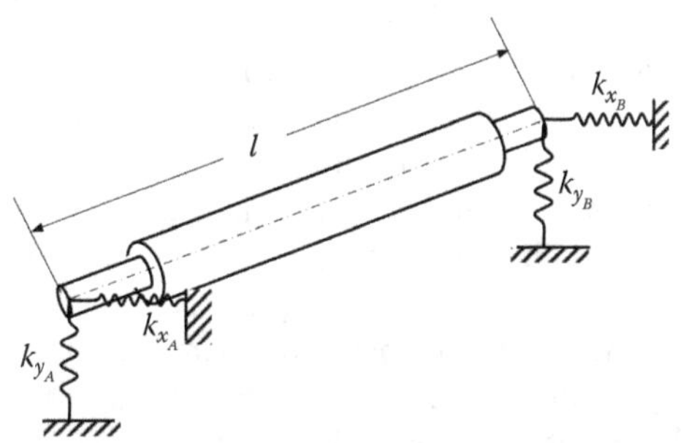

FIGURE 4.6 A long rigid rotor on flexible bearings.

$$me\omega^2\cos\omega t - k_{x_A}x - k_{x_B}x = m\ddot{x} \quad \text{and} \quad me\omega^2\sin\omega t - k_{y_A}y - k_{y_B}y = m\ddot{y} \tag{a}$$

where x and y are the translational displacements of the shaft center. Bearings at ends A and B are not the same; however, it is assumed that the difference in bearing dynamic parameters is small and the shaft remains horizontal in the static equilibrium position. Equations of motion in the φ_y and φ_x directions with the moment unbalance, mee_z , are

$$me\omega^2 e_z\cos\omega t - 0.5^2 k_{x_A} l^2\varphi_y - 0.5^2 k_{x_B} l^2\varphi_y = I_d\ddot{\varphi}_y$$

and

$$me\omega^2 e_z\sin\omega t - 0.5^2 k_{y_A} l^2\varphi_x - 0.5^2 k_{y_B} l^2\varphi_x = I_d\ddot{\varphi}_x \tag{b}$$

where φ_y and φ_x are rotational displacements of the shaft center line. The steady-state unbalance (forced) vibration responses from Equations (a) and (b) can be obtained as

$$x = \frac{me\omega^2}{k_{x_A} + k_{x_B} - m\omega^2}\cos\omega t; \quad y = \frac{me\omega^2}{k_{y_A} + k_{y_B} - m\omega^2}\sin\omega t \tag{c}$$

$$\varphi_y = \frac{me\omega^2 e_z}{0.5^2\left(k_{x_A} + k_{x_B}\right)l^2 - \omega^2 I_d}\cos\omega t; \quad \varphi_x = \frac{me\omega^2 e_z}{0.5^2\left(k_{y_A} + k_{y_B}\right)l^2 - \omega^2 I_d}\sin\omega t \tag{d}$$

On equating the denominator terms of responses in Equations (c) and (d) to zero, critical speeds of the rotor system can be obtained as

$$\omega_{cr1} = \sqrt{\frac{k_{x_A} + k_{x_B}}{m}} = \sqrt{\frac{(1.1+3.1)\times 10^6}{10}} = 648.10 \text{ rad/s}$$

$$\omega_{cr2} = \sqrt{\frac{k_{y_A} + k_{y_B}}{m}} = \sqrt{\frac{(1.8+3.8)\times 10^6}{10}} = 748.33 \text{ rad/s}$$

$$\omega_{cr3} = \sqrt{\frac{0.5^2\left(k_{x_A} + k_{x_B}\right)l^2}{I_d}} = \sqrt{\frac{0.5^2(1.1+3.1)\times 10^6\times 1^2}{0.1}} = 3240.37 \text{ rad/s}$$

$$\omega_{cr4} = \sqrt{\frac{0.5^2\left(k_{y_A} + k_{y_B}\right)l^2}{I_d}} = \sqrt{\frac{0.5^2(1.8+3.8)\times 10^6\times 1^2}{0.1}} = 3741.66 \text{ rad/s}$$

4.2 A SYMMETRICAL LONG RIGID SHAFT ON ANISOTROPIC BEARINGS

Fluid-film bearings impart the damping as well as stiffness forces to the rotor system. The shaft motion in the horizontal direction is coupled with the motion in the vertical direction in the presence of fluid-film bearings. That means a vertical transverse force may produce both the vertical and horizontal translational displacements; similarly, a horizontal transverse force may produce both the vertical and horizontal translational displacements. The same is valid for rotational displacements due to transverse moments. However, the coupling between the translational (i.e., x and y) and rotational (i.e., φ_y and φ_x) motion has not been considered. That means a symmetric force on a rigid rotor may not produce rotational displacement, and similarly, a symmetric moment may not produce a translational displacement. As discussed in Chapter 3, in most applications, the properties of such bearings are described in terms of the eight linearized stiffness and damping coefficients.

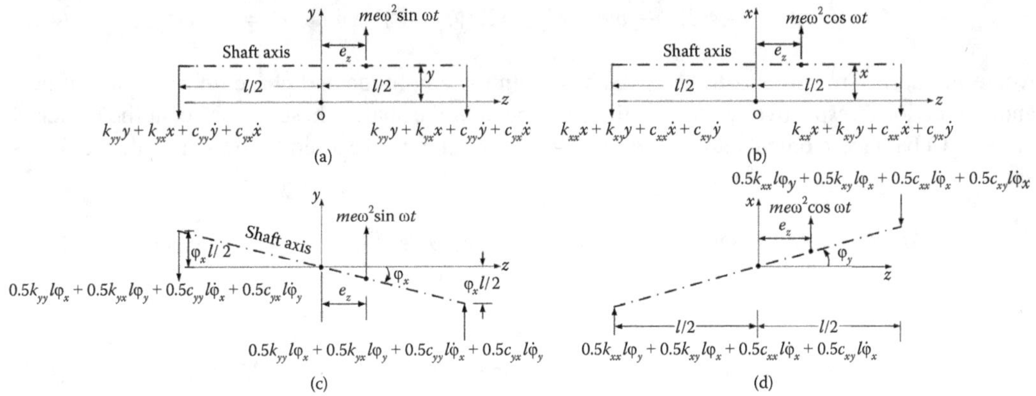

FIGURE 4.7 (a) A free body diagram of the shaft for the pure translatory motion in the y-z plane. (b) A free body diagram of the shaft for the pure translatory motion in the z-x plane. (c) A free body diagram of the shaft for the pure tilting motion in the y-z plane. (d) A free body diagram of the shaft for the pure tilting motion in the z-x plane.

Both bearings are assumed to be identical for the present case. A symmetrical long rigid shaft on anisotropic bearings will be similar to Figure 4.1 with additional damping and cross-coupled force terms at the shaft ends in free body diagrams (Figure 4.7).

Equations of motion for the rotor can be written from the free body diagram (Figure 4.7) as

$$f_x(t) - 2k_{xx}x - 2k_{xy}y - 2c_{xx}\dot{x} - 2c_{xy}\dot{y} = m\ddot{x} \tag{4.10}$$

$$f_y(t) - 2k_{yx}x - 2k_{yy}y - 2c_{yx}\dot{x} - 2c_{yy}\dot{y} = m\ddot{y} \tag{4.11}$$

$$M_{xz}(t) - k_{xx}\left(0.5l^2\varphi_y\right) - k_{xy}\left(0.5l^2\varphi_x\right) - c_{xx}\left(0.5l^2\dot{\varphi}_y\right) - c_{xy}\left(0.5l^2\dot{\varphi}_x\right) = I_d\ddot{\varphi}_y \tag{4.12}$$

$$M_{yz}(t) - k_{yx}\left(0.5l^2\varphi_y\right) - k_{yy}\left(0.5l^2\varphi_x\right) - c_{yx}\left(0.5l^2\dot{\varphi}_y\right) - c_{yy}\left(0.5l^2\dot{\varphi}_x\right) = I_d\ddot{\varphi}_x \tag{4.13}$$

in the x, y, φ_y, and φ_x directions, respectively. Here, f and M represent the external force and moment (e.g., due to an unbalance), respectively. k_{ij} and c_{ij} $(i, j=x, y)$ are the eight linearized bearing stiffness and damping coefficients, respectively.

4.2.1 Unbalance Response

The center of gravity is located at a radial eccentricity of e and an axial eccentricity, e_z, from the rotor geometrical center (Figure 4.2c), and these are assumed to be small quantities for the present case. Unbalance forces in the horizontal and vertical directions can then be written as

$$f_x = me\omega^2\cos\omega t = \text{Re}\left(me\omega^2 e^{j\omega t}\right) = \text{Re}\left(F_x e^{j\omega t}\right) \quad \text{with} \quad F_x = me\omega^2 \tag{4.14}$$

and

$$f_y = me\omega^2 \sin \omega t = \text{Re}\left(-jme\omega^2 e^{j\omega t}\right) = \text{Re}\left(F_y e^{j\omega t}\right) \quad \text{with} \quad F_y = -jme\omega^2 \tag{4.15}$$

where F_x and F_y are complex forces (which contain the amplitude and phase information) in the x and y directions, respectively, and ω is the spin speed of the shaft. These forces act at the center of gravity. Unbalance moments about the rotor geometrical center caused by these unbalance forces are

$$M_{zx} = me\omega^2 e_z \cos \omega t = \text{Re}\left(me\omega^2 e_z e^{j\omega t}\right) = \text{Re}\left(\bar{M}_{xz} e^{j\omega t}\right) \quad \text{with} \quad \bar{M}_{zx} = me\omega^2 e_z \tag{4.16}$$

and

$$M_{yz} = -me\omega^2 e_z \sin \omega t = \text{Re}\left(jme\omega^2 e_z e^{j\omega t}\right) = \text{Re}\left(\bar{M}_{yz} e^{j\omega t}\right) \quad \text{with} \quad \bar{M}_{yz} = jme\omega^2 e_z \tag{4.17}$$

where $\bar{M}_{xz}$ and $\bar{M}_{yz}$ are complex moments (which contain the amplitude and phase information) about the y and x axes, respectively. Then unbalance responses could be assumed to be

$$x = Xe^{j\omega t}; \quad y = Ye^{j\omega t}; \quad \varphi_y = \Phi_y e^{j\omega t}; \quad \varphi_x = \Phi_x e^{j\omega t} \tag{4.18}$$

where X, Y, Φ_y, and Φ_x are complex displacements. Equations 4.10–4.13 can be written as

$$\mathbf{M}\ddot{\mathbf{x}} + \mathbf{C}\dot{\mathbf{x}} + \mathbf{K}\mathbf{x} = \mathbf{f}(t) \tag{4.19}$$

with

$$\mathbf{M} = \begin{bmatrix} m & 0 & 0 & 0 \\ 0 & m & 0 & 0 \\ 0 & 0 & I_d & 0 \\ 0 & 0 & 0 & I_d \end{bmatrix}; \quad \mathbf{C} = \begin{bmatrix} 2c_{xx} & 2c_{xy} & 0 & 0 \\ 2c_{yx} & 2c_{yy} & 0 & 0 \\ 0 & 0 & 0.5l^2 c_{xx} & 0.5l^2 c_{xy} \\ 0 & 0 & 0.5l^2 c_{yx} & 0.5l^2 c_{yy} \end{bmatrix};$$

$$\mathbf{K} = \begin{bmatrix} 2k_{xx} & 2k_{xy} & 0 & 0 \\ 2k_{yx} & 2k_{yy} & 0 & 0 \\ 0 & 0 & 0.5l^2 k_{xx} & 0.5l^2 k_{xy} \\ 0 & 0 & 0.5l^2 k_{yx} & 0.5l^2 k_{yy} \end{bmatrix}; \quad \mathbf{x}(t) = \begin{Bmatrix} x \\ y \\ \varphi_y \\ \varphi_x \end{Bmatrix}; \quad \mathbf{f}(t) = \begin{Bmatrix} f_x \\ f_y \\ M_{zx} \\ M_{yz} \end{Bmatrix} = \mathrm{F}e^{j\omega t}$$

The response takes the following form:

$$\mathbf{x} = \mathbf{X}e^{j\omega t}; \quad \text{so that} \quad \dot{\mathbf{x}} = j\omega \mathbf{X}e^{j\omega t} \quad \text{and} \quad \ddot{\mathbf{x}} = -\omega^2 \mathbf{X}e^{j\omega t} \tag{4.20}$$

where the dot above x represents the time derivative (single dot: single derivative and double dots: double derivative with respect to time). On substituting Equations 4.14 through 4.17 and 4.20 into equations of motion (Equation 4.19), we get

$$\left(-\omega^2 \mathbf{M} + j\omega \mathbf{C} + \mathbf{K}\right)\mathbf{X} = \mathbf{F} \tag{4.21}$$

with

$$\mathbf{X} = \begin{Bmatrix} X \\ Y \\ \Phi_y \\ \Phi_x \end{Bmatrix}; \quad \mathbf{F} = \begin{Bmatrix} F_x \\ F_y \\ \bar{M}_{zx} \\ \bar{M}_{yz} \end{Bmatrix}$$

which can be written as

$$\mathbf{DX} = \mathbf{F} \quad \text{with} \quad \mathbf{D} = \left(\mathbf{K} - \omega^2\mathbf{M} + j\omega\mathbf{C}\right) \tag{4.22}$$

The unbalance responses can be obtained by

$$\mathbf{X} = \mathbf{D}^{-1}\mathbf{F} \tag{4.23}$$

The displacement amplitudes of the rotor are given by

$$X = \sqrt{X_r^2 + X_i^2}, \quad Y = \sqrt{Y_r^2 + Y_i^2}, \quad \Phi_y = \sqrt{\Phi_{y_r}^2 + \Phi_{y_i}^2}, \quad \Phi_x = \sqrt{\Phi_{x_r}^2 + \Phi_{x_i}^2} \tag{4.24}$$

and the corresponding phase lag will be given by

$$\alpha = \tan^{-1}\left(\frac{X_i}{X_r}\right), \quad \beta = \tan^{-1}\left(\frac{Y_i}{Y_r}\right), \quad \gamma = \tan^{-1}\left(\frac{\Phi_{y_i}}{\Phi_{y_r}}\right), \quad \delta = \tan^{-1}\left(\frac{\Phi_{x_i}}{\Phi_{x_r}}\right) \tag{4.25}$$

where subscripts r and i represent the real and imaginary parts, respectively. The resulting shaft whirl orbit can be plotted using Equations 4.18 and 4.23 (i.e., $x = Xe^{j\omega t}$ and $y = Ye^{j\omega t}$), and in general, for a stable rotor-bearing system, the orbit will take the form of an inclined ellipse as shown in Figure 4.8 due to cross-coupled stiffness terms.

The form of the orbit is still elliptical, as in the previous section; however, the major and minor axes are no longer along the x and y directions, respectively. A typical force vector is also shown on the diagram, and it precedes the displacement vector—in the presence of damping, it will be less than 180°. However, during the crossing of critical speeds, it may change the phase to 90°, irrespective of the damping value.

The undamped free vibration analysis could be performed using Equation 4.21 by dropping the force vector **F** and the damping matrix **C** as discussed in Chapter 2. In the present case, the coupling is considered between the vertical and horizontal planes, and no coupling is considered between the translational and rotational motions. It can be observed that Equations 4.10 and 4.11 (without damping) are similar in form to those discussed in Chapter 2, that is, the case when we considered

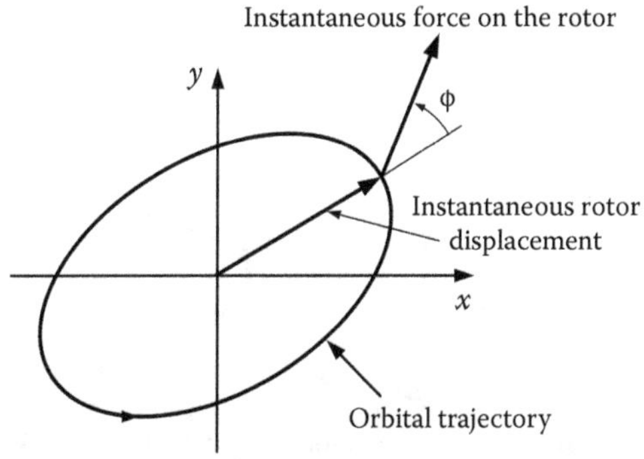

FIGURE 4.8 A rotor whirl orbit.

no coupling between the vertical and horizontal planes and the coupling was considered between the translational and rotational motions (i.e., a Jeffcott rotor with an offset disk). Mathematically, we can write $2k_{xx} = k_{11}, 2k_{xy} = k_{12}, \ldots$ and the rest of the analysis and interpretations will be similar to those discussed in Chapter 3. Similar analyses can be performed by considering Equations 4.12 and 4.13 unbalanced for the moment. The following examples will illustrate this analysis.

Example 4.4

Obtain the transverse natural frequencies of a rotor-bearing system as shown in Figure 4.9 *for a pure tilting motion of the shaft*. Consider the shaft to be rigid and the whole mass of the shaft to be concentrated at its midspan. The shaft is of 1 m of span and the diameter is 0.05 m with a mass density of 7800 kg/m³. The shaft is supported at the ends by identical bearings. Consider the motion in both the vertical and horizontal planes. Take the following bearing properties: for both bearings *A* and *B*, k_{xx}=200 MN/m, k_{yy}=150 MN/m, k_{xy}=15 MN/m, k_{yx}=10 MN/m.

Solution

The following data are given for the present problem:

$$m = \pi r^2 l \rho = 15.3075\,\text{kg}$$

$$I_d = \frac{1}{12} m\left(3r^2 + l^2\right) = 1.2780\ \text{kg-m}^2$$

Equations of motion would be obtained by considering the free body diagram of the rotor as shown in Figure 4.10. A pure tilting of the shaft is considered with φ_x and φ_y as the rotational displacements in the *y-z* (Figure 4.10a) and *z-x* (Figure 4.10b) planes, respectively. Let/be the total length of the rotor.

FIGURE 4.9 A rigid rotor mounted on two identical bearings.

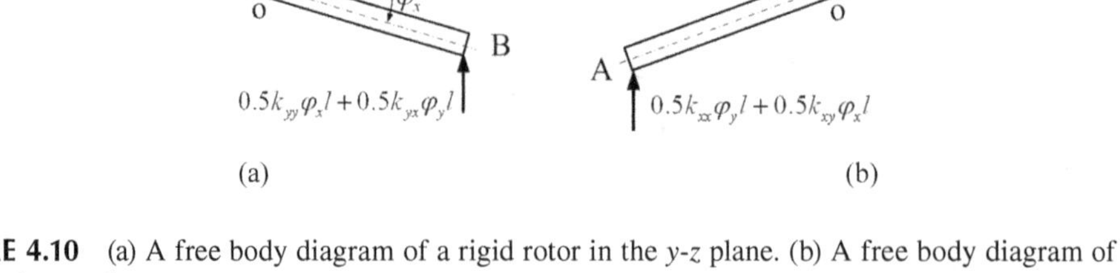

FIGURE 4.10 (a) A free body diagram of a rigid rotor in the *y-z* plane. (b) A free body diagram of a rigid rotor in the *z-x* plane.

Equations of motion in the φ_y and φ_x directions (Figure 4.10), respectively, can be written as

$$-(0.5k_{xx}l\varphi_y + 0.5k_{xy}l\varphi_x)l = I_d\ddot{\varphi}_y \tag{a}$$

and

$$-(0.5k_{yy}l\varphi_x + 0.5k_{yx}l\ddot{\varphi}_y)l = I_d\ddot{\varphi}_x \tag{b}$$

In the matrix form, these equations can be written as

$$\begin{bmatrix} 2I_d & 0 \\ 0 & 2I_d \end{bmatrix}\begin{Bmatrix} \ddot{\varphi}_y \\ \ddot{\varphi}_x \end{Bmatrix} + \begin{bmatrix} k_{xx}l^2 & k_{xy}l^2 \\ k_{yx}l^2 & k_{yy}l^2 \end{bmatrix}\begin{Bmatrix} \varphi_y \\ \varphi_x \end{Bmatrix} = \begin{Bmatrix} 0 \\ 0 \end{Bmatrix} \tag{c}$$

For the free vibration, it takes the form

$$\left(-\omega_{nf}^2\begin{bmatrix} 2I_d & 0 \\ 0 & 2I_d \end{bmatrix} + \begin{bmatrix} k_{xx}l^2 & k_{xy}l^2 \\ k_{yx}l^2 & k_{yy}l^2 \end{bmatrix}\right)\begin{Bmatrix} \varphi_y \\ \varphi_x \end{Bmatrix} = \begin{Bmatrix} 0 \\ 0 \end{Bmatrix} \tag{d}$$

which gives the frequency equation as

$$4I_d^2\omega_{nf}^4 - 2I_dl^2(k_{xx} + k_{yy})\omega_{nf}^2 + (k_{xx}k_{yy} - k_{xy}k_{yx})l^4 = 0 \tag{e}$$

or

$$\omega_{nf}^4 - \frac{l^2(k_{xx} + k_{yy})}{2I_d}\omega_{nf}^2 + \frac{(k_{xx}k_{yy} - k_{xy}k_{yx})l^4}{4I_d^2} = 0 \tag{f}$$

For numerical values of the present problem, the frequency equation becomes

$$4\times(1.2780)^2\,\omega_{nf}^4 - 2\times1.2780\times1^2\times(200+150)\times10^6\omega_{nf}^2 + \{200\times150 - 15\times10\}\times10^6\times1^4 = 0$$

or

$$6.5333\omega_{nf}^4 - 8.9461\times10^8\omega_{nf}^2 + 2.9850\times10^{16} = 0 \tag{g}$$

which gives $\omega_{nf_1} = 7587.8$ rad/s and $\omega_{nf_2} = 8908.2$ rad/s.

Example 4.5

Obtain transverse critical speeds of a long rigid rotor supported on two identical fluid-film bearings at the ends that has a 2 m shaft span, 5 kg rotor mass, and 0.1 kg-m² diametral mass moment of inertia of the rotor. The dynamic properties of both bearings are $k_{xx}=2.0\times10^4$ N/m, $k_{yy}=8.8\times10^4$ N/m, $k_{xy}=1.0\times10^3$ N/m, $k_{yx}=1.5\times10^3$ N/m, $c_{xx}=1.0$ N-s/m, $c_{yy}=1.0$ N-s/m, $c_{xy}=1.0\times10^{-1}$ N-s/m, and $c_{yx}=1.0\times10^{-1}$ N-s/m. Obtain the unbalance responses (the amplitude and the phase) with the spin speed of shaft at the shaft center when the radial eccentricity of 0.1 mm and axial eccentricity of 1 mm are present in the rotor, and from these plots locate critical speeds.

Solution:

By the method described in the previous section, unbalance responses are obtained for the given data. Figure 4.11 shows the unbalance responses both for the translational (left side) and rotational (right side) displacements. Both the amplitude and phase are plotted. It can be observed that in each of the plots of the translational and rotational displacements, two peaks appear, and they correspond to critical speeds of the rotor-bearing system. Because the translational and rotational

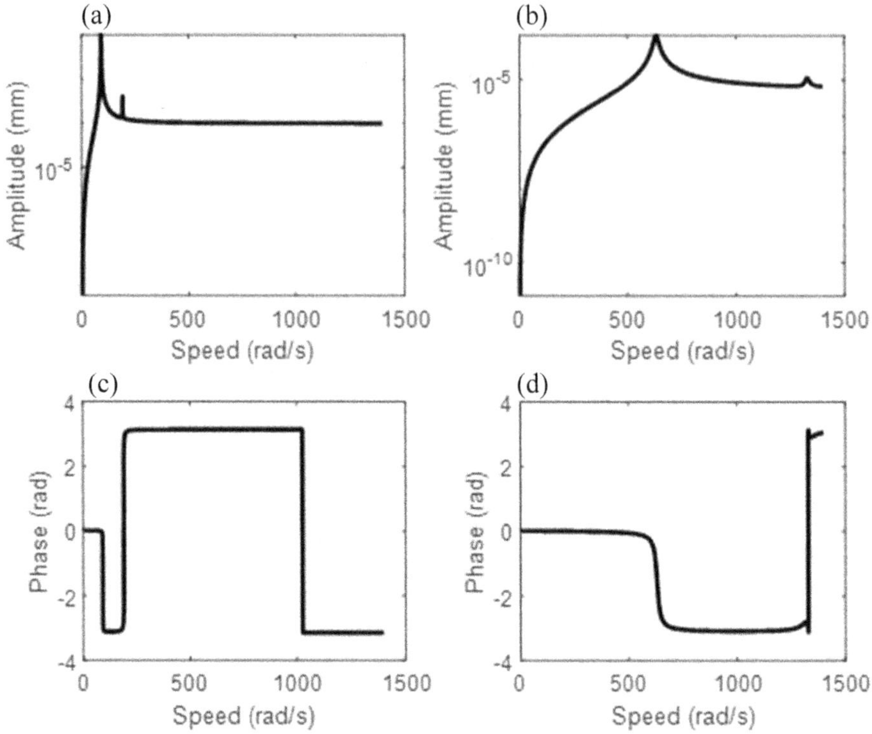

FIGURE 4.11 The amplitude and phase variation with respect to spin speeds: (a) the translational displacement amplitude; (b) the rotational displacement amplitude; (c) the translational displacement phase; (d) the rotational displacement phase.

displacements are uncoupled for the present case, the corresponding critical speeds appear in their respective plots only. The four (undamped) critical speeds corresponding to transverse translational modes are 89.39 and 187.64 rad/s, and those corresponding to the transverse rotational modes are 632.11 and 1326.80 rad/s. The change in phase angles at critical speeds can be seen in each of the plots.

4.2.2 Bearing Forces

Forces transmitted through bearings are due to translational deformations of the bearing lubricant film. These bearing forces would not include rotor inertia terms. In general, bearing forces will lag behind the unbalance force such that the bearing horizontal and vertical force components at both ends A and B of the machine can be represented as

$$ {}_Af_{bx} = {}_Ak_{xx}x + k_{xy}y + {}_Ac_{xx}\dot{x} + {}_Ac_{xy}\dot{y} + \frac{1}{2}\,{}_Ak_{xx}l\varphi_y + \frac{1}{2}\,{}_Ak_{xy}l\varphi_x + \frac{1}{2}\,{}_Ac_{xx}l\dot{\varphi}_y + \frac{1}{2}\,{}_Ac_{xy}l\dot{\varphi}_x \tag{4.26} $$

$$ \begin{aligned} {}_Af_{by} = {}_Ak_{yx}x + {}_Ak_{yy}y + {}_Ac_{yx}\dot{x}c_{yy}\dot{y} + \frac{1}{2}\,{}_Ak_{yx}l\varphi_y \\ + \frac{1}{2}\,{}_Ak_{yy}l\varphi_x + \frac{1}{2}\,{}_Ac_{yx}l\dot{\varphi}_y + \frac{1}{2}\,{}_Ac_{yy}l\dot{\varphi}_x \end{aligned} \tag{4.27} $$

$$
\begin{aligned}
{}_Bf_{bx} &= {}_Bk_{xx}x + {}_Bk_{xy}y + {}_Bc_{xx}\dot{x} + {}_Bc_{xy}\dot{y} + \frac{1}{2}\,{}_Bk_{xx}l\varphi_y \\
&\quad + \frac{1}{2}\,{}_Bk_{xy}l\varphi_x + {}_Bc_{xx}l\dot{\varphi}_y + \frac{1}{2}\,{}_Bc_{xy}l\dot{\varphi}_x
\end{aligned} \tag{4.28}
$$

$$
\begin{aligned}
{}_Bf_{by} &= {}_Bk_{yx}x + k_{yy}y + {}_Bc_{yx}\dot{x} + {}_Bc_{yy}\dot{y} + \frac{1}{2}\,{}_Bk_{yx}l\varphi_y \\
&\quad + \frac{1}{2}\,{}_Bk_{yy}l\varphi_x + \frac{1}{2}\,{}_Bc_{yx}l\dot{\varphi}_y + \frac{1}{2}\,{}_Bc_{yy}l\dot{\varphi}_x
\end{aligned} \tag{4.29}
$$

where ${}_Ak_{xx} = k_{xx}, k_{xx} = k_{xx}$, etc. Hence, Equations 4.26–4.29 can be written in the matrix form as

$$
\mathbf{f}_b = \mathbf{C}_b\dot{\mathbf{x}} + \mathbf{K}_b\mathbf{x} \tag{4.30}
$$

with

$$
\mathbf{f}_b = \begin{Bmatrix} f_{bx} \\ {}_Af_{by} \\ {}_Bf_{bx} \\ {}_Bf_{by} \end{Bmatrix}; \quad \dot{\mathbf{x}} = \begin{Bmatrix} \dot{x} \\ \dot{y} \\ \dot{\varphi}_y \\ \dot{\varphi}_x \end{Bmatrix}; \quad \mathbf{x} = \begin{Bmatrix} x \\ y \\ \varphi_y \\ \varphi_x \end{Bmatrix}
$$

$$
\mathbf{C}_b = \begin{bmatrix} {}_Ac_{xx} & {}_Ac_{xy} & 0.5\,{}_Ac_{xx}l & 0.5\,{}_Ac_{xy}l \\ {}_Ac_{yx} & {}_Ac_{yy} & 0.5\,{}_Ac_{yx}l & 0.5\,{}_Ac_{yy}l \\ {}_Bc_{xx} & {}_Bc_{xy} & 0.5\,{}_Bc_{xx}l & 0.5\,{}_Bc_{xy}l \\ {}_Bc_{yx} & {}_Bc_{yy} & 0.5\,{}_Bc_{yx}l & 0.5\,{}_Bc_{yy}l \end{bmatrix};
$$

$$
\mathbf{K}_b = \begin{bmatrix} {}_Ak_{xx} & {}_Ak_{xy} & 0.5\,{}_Ak_{xx}l & 0.5\,{}_Ak_{xy}l \\ {}_Ak_{yx} & {}_Ak_{yy} & 0.5\,{}_Ak_{yx}l & 0.5\,{}_Ak_{yy}l \\ {}_Bk_{xx} & {}_Bk_{xy} & 0.5\,{}_Bk_{xx}l & 0.5\,{}_Bk_{xy}l \\ {}_Bk_{yx} & {}_Bk_{yy} & 0.5\,{}_Bk_{yx}l & 0.5\,{}_Bk_{yy}l \end{bmatrix}
$$

For the unbalance, excitation with a frequency ω, rotor responses, and bearing forces can be expressed as

$$
\mathbf{x} = \mathbf{X}e^{j\omega t}; \quad \dot{\mathbf{x}} = j\omega\mathbf{X}e^{j\omega t}; \quad \mathbf{f}_b = \mathbf{F}_b e^{j\omega t} \tag{4.31}
$$

Here, $\mathbf{F}_b$ is the complex bearing force vector, and $\mathbf{X}$ is the complex displacement vector. On substituting Equation 4.31 into Equation 4.33, we get

$$
\mathbf{F}_b = \left(\mathbf{K}_b + j\omega\mathbf{C}_b\right)\mathbf{X} \tag{4.32}
$$

This can be used to evaluate the bearing forces. The amplitudes of forces transmitted through the bearings are then given by

$$\begin{aligned} {}_AF_{bx} &= \sqrt{{}_AF^2_{bx_r} + {}_AF^2_{bx_i}}; \quad F_{by} = \sqrt{{}_AF^2_{by_r} + {}_AF^2_{by_i}}; \\ {}_BF_{bx} &= \sqrt{{}_BF^2_{bx_r} + {}_BF^2_{bx_i}}; \quad F_{by} = \sqrt{{}_BF^2_{by_r} + {}_BF^2_{by_i}} \end{aligned} \tag{4.33}$$

and the corresponding phase angles are given by

$$\varepsilon = \tan^{-1}\left(\frac{{}_AF_{bx_i}}{{}_AF_{bx_r}}\right); \quad \varsigma = \tan^{-1}\left(\frac{{}_AF_{by_i}}{{}_AF_{by_r}}\right); \quad \eta = \tan^{-1}\left(\frac{{}_BF_{bx_i}}{{}_BF_{bx_r}}\right); \quad \lambda = \tan^{-1}\left(\frac{{}_BF_{by_i}}{{}_BF_{by_r}}\right) \tag{4.34}$$

where subscripts r and i represent the real and imaginary quantities, respectively, of a complex quantity. It should be noted that bearing forces will be maximum whenever a rotor-bearing system is rotating at or near critical speeds.

4.3 A SYMMETRICAL FLEXIBLE SHAFT ON ANISOTROPIC BEARINGS

For the present case, both the shaft and bearings are flexible, as shown in Figure 4.12. All generalized coordinates are with respect to a fixed frame of reference, and they are absolute displacements. The motion in two orthogonal planes will be considered simultaneously. The analysis allows for finding different instantaneous displacements of the shaft at the disk (offset from the midspan of the shaft) and at the bearings. The system will behave in a similar manner to that described in the previous section, except that the flexibility of the shaft will increase the overall flexibility of the support system as experienced by the rigid disk. An equivalent set of system stiffness and damping coefficients is first evaluated, which allows for the flexibility of the shaft in addition to that of bearings, and is used in place of the bearing coefficients in the previous section analysis. The total deflection of the disk is the vector sum of the deflection of the disk relative to the shaft ends, plus that of the shaft ends in the bearings. For the disk, we observe the displacement of its geometrical center.

The deflection of the shaft ends in the bearings is related to the force transmitted through the bearings by the bearing stiffness and damping coefficients as

$$f_{bx} = k_{xx}x_b + k_{xy}y_b + c_{xx}\dot{x}_b + c_{xy}\dot{y}_b \quad \text{and} \quad f_{by} = k_{yx}x_b + k_{yy}y_b + c_{yx}\dot{x}_b + c_{yy}\dot{y}_b \tag{4.35}$$

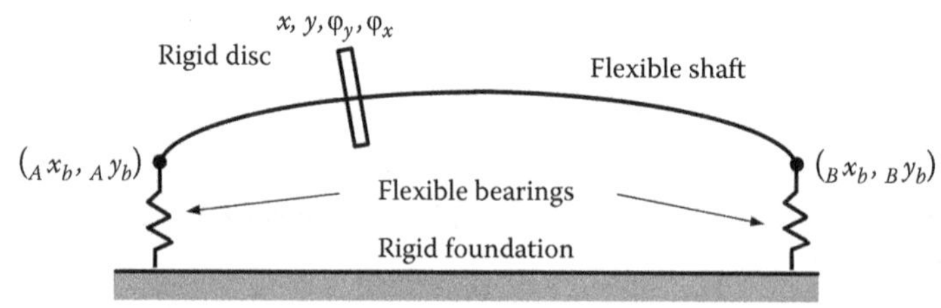

FIGURE 4.12 A flexible shaft with an offset disk on flexible bearings.

where x_b and y_b are instantaneous displacements of the shaft ends relative to the bearings in the horizontal and vertical directions, respectively, and they take the following form:

$$x_b = X_b e^{j\omega t} \quad \text{and} \quad y_b = Y_b e^{j\omega t} \tag{4.36}$$

where X_b and Y_b are complex displacements in the x and y directions, respectively. Equation 4.36 can be differentiated once with respect to time to give

$$\dot{x}_b = j\omega X_b e^{j\omega t} \quad \text{and} \quad \dot{y}_b = j\omega Y_b e^{j\omega t} \tag{4.37}$$

It should be noted that bearings are modeled as a point connection with the shaft and only translational displacements are considered, because they support mainly radial (transverse) loads. Bearing forces have the following form:

$$f_{bx} = F_{bx} e^{j\omega t} \quad \text{and} \quad f_{by} = F_{by} e^{j\omega t} \tag{4.38}$$

where F_{bx} and F_{by} are complex forces in the x and y directions, respectively. On substituting in equations of motion (Equation 4.35), we get

$$F_{bx} = k_{xx}X_b + k_{xy}Y_b + j\omega c_{xx}X_b + j\omega c_{xy}Y_b \tag{4.39}$$

and

$$F_{by} = k_{yx}X_b + k_{yy}Y_b + j\omega c_{yx}X_b + j\omega c_{yy}Y_b \tag{4.40}$$

which can be written in a matrix form for both bearings A and B as

$$\mathbf{F}_b = \mathbf{K}\mathbf{X}_b \tag{4.41}$$

with

$$\mathbf{F}_b = \begin{Bmatrix} {}_AF_{bx} \\ {}_AF_{by} \\ {}_BF_{bx} \\ {}_BF_{by} \end{Bmatrix};$$

$$\mathbf{K} = \begin{bmatrix} {}_A(k_{xx} + j\omega c_{xx}) & {}_A(k_{xy} + j\omega c_{xy}) & 0 & 0 \\ {}_A(k_{yx} + j\omega c_{yx}) & {}_A(k_{yy} + j\omega c_{yy}) & 0 & 0 \\ 0 & 0 & {}_B(k_{xx} + j\omega c_{xx}) & {}_B(k_{xy} + j\omega c_{xy}) \\ 0 & 0 & {}_B(k_{yx} + j\omega c_{yx}) & {}_B(k_{yy} + j\omega c_{yy}) \end{bmatrix}$$

and

$$\mathbf{X}_b = \begin{bmatrix} {}_AX_b & {}_AY_b & {}_BX_b & {}_BY_b \end{bmatrix}^T$$

The magnitude of reaction forces transmitted by bearings can also be evaluated in terms of forces applied to the shaft by the disk (refer to Figure 4.13).

From Figure 4.13, the moment balance will be

$$\sum M_B = 0 \Rightarrow f_{by}l = f_y(l-a) + M_{yz} \quad \text{or} \quad {}_Af_{by} = f_y(1-a/l) + M_{yz}(1/l) \tag{4.42}$$

and

$$\sum M_A = 0 \Rightarrow {}_Af_{by}l = f_ya - M_{yz} \quad \text{or} \quad {}_Bf_{by} = f_y(a/l) - M_{yz}(1/l) \tag{4.43}$$

Similarly, forces in the horizontal direction may be written as

$${}_Af_{bx} = f_x(1-a/l) - M_{zx}(1/l) \tag{4.44}$$

and

$${}_Bf_{bx} = f_x(a/l) + M_{zx}(1/l) \tag{4.45}$$

Equations 4.42 through 4.45 can be combined in a matrix form as

$$\mathbf{f}_b = \mathbf{A}\mathbf{f}_s \tag{4.46}$$

with

$$\mathbf{f}_b = \begin{Bmatrix} {}_Af_{bx} \\ {}_Af_{by} \\ {}_Bf_{bx} \\ {}_Bf_{by} \end{Bmatrix}; \quad \mathbf{f}_s = \begin{Bmatrix} f_x \\ f_y \\ M_{zx} \\ M_{yz} \end{Bmatrix}; \quad \mathbf{A} = \begin{bmatrix} (1-a/l) & 0 & -1/l & 0 \\ 0 & (1-a/l) & 0 & 1/l \\ a/l & 0 & 1/l & 0 \\ 0 & a/l & 0 & -1/l \end{bmatrix}$$

For the unbalance excitation, we have

$$\mathbf{f}_b = \mathbf{F}_b e^{j\omega t} \quad \text{and} \quad \mathbf{f}_s = \mathbf{F}_s e^{j\omega t} \tag{4.47}$$

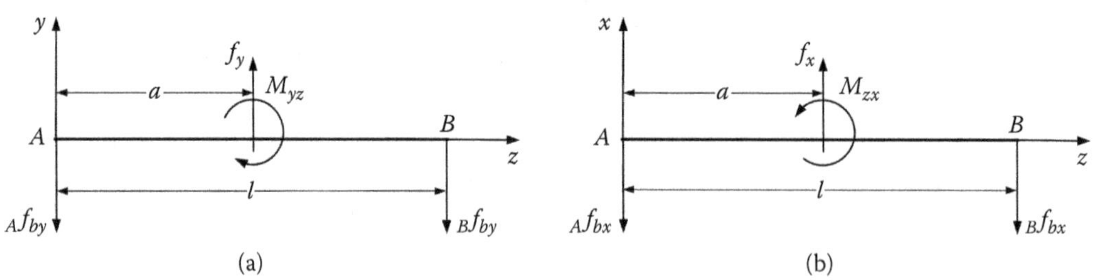

FIGURE 4.13 A free body diagram of the shaft: (a) *y*-*z* plane; (b) *z*-*x* plane.

where subscript b refers to the bearing and s refers to the shaft. On substituting Equation 4.47 into Equation 4.46, we get

$$\mathbf{F}_b = \mathbf{A}\mathbf{F}_s \tag{4.48}$$

In Equation 4.48, the bearing forces are related to the reaction forces and moments on the shaft by the disk. On equating Equations 4.41 and 4.48, we get

$$\mathbf{K}\mathbf{X}_b = \mathbf{A}\mathbf{F}_s \quad \text{or} \quad \mathbf{X}_b = \mathbf{K}^{-1}\mathbf{A}\mathbf{F}_s \tag{4.49}$$

Equation 4.49 relates the shaft end deflections to the reaction forces and moments on the shaft by the disk. The deflection at the location of the disk due to movement of the shaft ends can be obtained as follows. Consider the shaft to be rigid for some instant, and let us denote shaft end deflections in the horizontal direction to be ${}_Ax_b$ and ${}_Bx_b$ at ends A and B, respectively, as shown in Figure 4.14. These displacements are assumed to be small.

The translational displacement in the x direction can be written as (Figure 4.14a)

$$x = {}_Ax_b + \frac{\left({}_Bx_b - {}_Ax_b\right)}{l}a = \left(1 - \frac{a}{l}\right){}_Ax_b + \left(\frac{a}{l}\right){}_Bx_b \tag{4.50}$$

The rotational displacement of the shaft in the x-z plane will be (Figure 4.14a)

$$\varphi_y = \frac{\left({}_Bx_b - {}_Ax_b\right)}{l} = \left(-\frac{1}{l}\right){}_Ax_b + \left(\frac{1}{l}\right){}_Bx_b \tag{4.51}$$

Similarly, for the translational and rotational displacements in the y direction and in the y-z plane, respectively, we have (refer to Figure 4.14b)

$$y = {}_Ay_b - \frac{\left({}_Ay_b - {}_By_b\right)}{l}a = \left(\frac{l-a}{l}\right){}_Ay_b + \left(\frac{a}{l}\right){}_By_b \tag{4.52}$$

and

$$\varphi_x = \frac{\left({}_Ay_b - {}_By_b\right)}{l} = \left(\frac{1}{l}\right){}_Ay_b + \left(-\frac{1}{l}\right){}_By_b \tag{4.53}$$

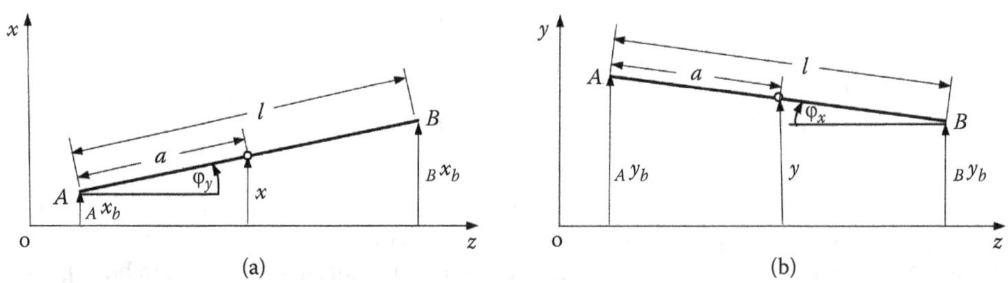

FIGURE 4.14 Rigid body movement of the shaft in the (a) z-x plane and (b) y-z plane.

Equations 4.50–4.53 can be combined in a matrix form as

$$\mathbf{u}_{s1} = \mathbf{B}\mathbf{x}_b \tag{4.54}$$

with

$$\mathbf{u}_{s1} = \begin{Bmatrix} x \\ y \\ \varphi_y \\ \varphi_x \end{Bmatrix}_{s1}; \quad \mathbf{x}_b = \begin{Bmatrix} {}_Ax_b \\ {}_Ay_b \\ {}_Bx_b \\ {}_By_b \end{Bmatrix}; \quad \mathbf{B} = \begin{bmatrix} (1-a/l) & 0 & a/l & 0 \\ 0 & (1-a/l) & 0 & (a/l) \\ -1/l & 0 & 1/l & 0 \\ 0 & 1/l & 0 & -1/l \end{bmatrix}$$

where subscript s_1 indicates that these displacements are due to the rigid body motion of the shaft. For the unbalance excitation (or for the free vibration analysis), shaft displacements at the bearing locations and at the disk center vary sinusoidally such that

$$\mathbf{u}_{s1} = \mathbf{U}_{s1}e^{j\omega t} \quad \text{and} \quad \mathbf{x}_b = \mathbf{X}_b e^{j\omega t} \tag{4.55}$$

where ω is the spin speed (or the natural frequency in the case of free vibrations). On substituting Equation 4.55 into Equation 4.54, we have

$$\mathbf{U}_{s1} = \mathbf{B}\mathbf{X}_b \tag{4.56}$$

On substituting Equation 4.49 into Equation 4.56, we get

$$\mathbf{U}_{s1} = \mathbf{B}\mathbf{K}^{-1}\mathbf{A}\mathbf{F}_s = \mathbf{C}\mathbf{F}_s \tag{4.57}$$

which gives the deflection of the disk due to the shaft reaction force and moment from the disk, when the shaft is rigid. Equation 4.57 will give deflections of the disk that are caused by only the movement of the shaft ends (rigid body movement) on flexible bearings. In order to obtain the net rotor deflection under a given load, we have to add the deflection due to the deformation of the shaft with respect to bearing locations also in Equation 4.57. The deflection associated with flexure of the shaft alone has already been calculated in Chapter 2, which can be combined in a matrix form as

$$\mathbf{u}_{s2} = \alpha \mathbf{f}_s \tag{4.58}$$

with

$$\mathbf{u}_{s2} = \begin{Bmatrix} x \\ y \\ \varphi_y \\ \varphi_x \end{Bmatrix}_{s2}; \quad \mathbf{f}_s = \begin{Bmatrix} f_x \\ f_y \\ M_{zx} \\ M_{yz} \end{Bmatrix}; \quad \alpha = \begin{bmatrix} \alpha_{11} & 0 & \alpha_{12} & 0 \\ 0 & \alpha_{11} & 0 & \alpha_{12} \\ \alpha_{21} & 0 & \alpha_{22} & 0 \\ 0 & \alpha_{21} & 0 & \alpha_{22} \end{bmatrix}$$

where subscript s_2 indicates that these displacements are due to the pure deformation of the shaft without any rigid body motion. For the unbalance excitation (or for the free vibration analysis), shaft reaction forces at the disk location and disk displacements vary sinusoidally and can be expressed as

$$\mathbf{u}_{s2} = \mathbf{U}_{s2}e^{j\omega t} \quad \text{and} \quad \mathbf{f}_s = \mathbf{F}_s e^{j\omega t} \tag{4.59}$$

On substituting Equation 4.59 into Equation 4.58, we get

$$\mathbf{U}_{s2} = \alpha \mathbf{F}_s \tag{4.60}$$

which is the deflection of the disk due to the flexure of the shaft alone, without considering the bearing flexibility. The net deflection of the disk that is caused by the deflection of bearings plus that due to the flexure of the shaft is then given by

$$\mathbf{U}_s = \mathbf{U}_{s1} + \mathbf{U}_{s2} = (\mathbf{C} + \alpha)\mathbf{F}_s = \mathbf{D}\mathbf{F}_s \tag{4.61}$$

where $\mathbf{U}_s$ contains absolute displacements of the shaft at the location of the disk. Equation 4.61 describes displacements of the shaft at the disk under the action of sinusoidal forces and moments applied at the disk (hence the matrix **D** is similar to the influence coefficient matrix). Equation 4.61 can be written as

$$\mathbf{F}_s = \mathbf{D}^{-1}\mathbf{U}_s = \mathbf{E}\mathbf{U}_s \tag{4.62}$$

where the matrix **E** is similar to the stiffness matrix (it is the equivalent stiffness of the shaft and bearings experienced at the disk location). Equations of motion of the disk can be written in the x direction and on the z-x plane (see Figure 4.15a) as

$$-f_x = m\frac{d^2}{dt^2}(x + e\cos\omega t) \quad \text{or} \quad me\omega^2\cos\omega t - f_x = m\ddot{x} \tag{4.63}$$

and

$$-M_{zx} = I_d\ddot{\varphi}_y \tag{4.64}$$

Similarly, equations of motion in the y direction and on the y-z plane (see Figure 4.15b) can be written as

$$-f_y = m\frac{d^2}{dt^2}(y + e\sin\omega t) \quad \text{or} \quad me\omega^2\sin\omega t - f_y = m\ddot{y} \tag{4.65}$$

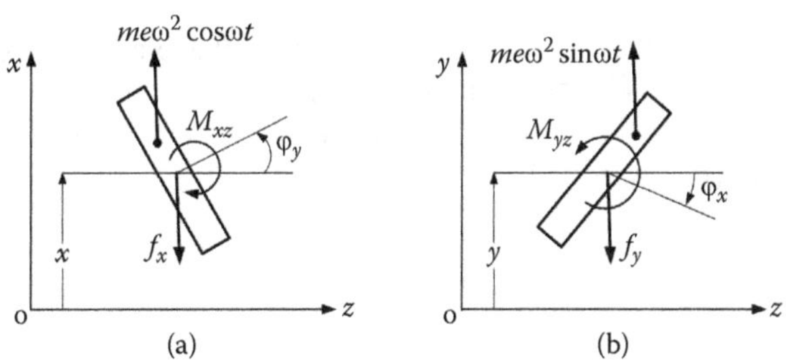

FIGURE 4.15 Free body diagram of the disk (a) in the z-x plane and (b) in the y-z plane.

and

$$-M_{yz} = I_d \ddot{\varphi}_x \tag{4.66}$$

Equations of motion (Equations 4.63 through 4.66) of the disk can be written in a matrix form as

$$\mathbf{M}\ddot{\mathbf{u}} + \mathbf{f}_s = \mathbf{f}_{unb} \tag{4.67}$$

with

$$\mathbf{M} = \begin{bmatrix} m & 0 & 0 & 0 \\ 0 & m & 0 & 0 \\ 0 & 0 & I_d & 0 \\ 0 & 0 & 0 & I_d \end{bmatrix}; \quad \mathbf{u} = \begin{Bmatrix} x \\ y \\ \varphi_y \\ \varphi_x \end{Bmatrix}; \quad \mathbf{f}_s = \begin{Bmatrix} f_x \\ f_y \\ M_{xz} \\ M_{yz} \end{Bmatrix}; \quad \mathbf{f}_{unb} = \begin{Bmatrix} me\omega^2 \\ -jme\omega^2 \\ 0 \\ 0 \end{Bmatrix} e^{j\omega t} = \mathbf{F}_{unb} e^{j\omega t}$$

Noting $\mathbf{f}_s = \mathbf{F}_s e^{j\omega t}$ and $\mathbf{u} = \mathbf{U}_s e^{j\omega t}$, equations of motion take the following form:

$$-\omega^2 \mathbf{M}\mathbf{U}_s + \mathbf{F}_s = \mathbf{F}_{unb} \tag{4.68}$$

Noting Equation 4.62, Equation 4.68 becomes

$$-\omega^2 \mathbf{M}\mathbf{U}_s + \mathbf{E}\mathbf{U}_s = \mathbf{F}_{unb} \tag{4.69}$$

which gives

$$\mathbf{U}_s = \mathbf{H}\mathbf{F}_{unb} \tag{4.70}$$

with

$$\mathbf{H} = \left(-\omega^2 \mathbf{M} + \mathbf{E}\right)^{-1}$$

where $\mathbf{H}^{-1}$ is the equivalent dynamic stiffness matrix, as experienced by the disk, of the shaft and the bearing system. Once the response of the disk has been obtained from the previous equation for a given unbalance force, the loading applied to the shaft by the disk can be obtained with Equation 4.62. Then from Equation 4.49, we can get the shaft end deflections $\mathbf{X}_b$ at each bearing, which is substituted in Equation 4.41 to get bearing forces $\mathbf{F}_b$. Alternately, bearing forces can be used directly from Equation 4.48. Displacements and forces have a complex form; the amplitude and the phase information can be extracted from the real and imaginary parts. Amplitudes will be the modulus of complex numbers, and phase angles of all these displacements can be evaluated by calculating the arctangent of the ratio of the imaginary to the real components as given by Equations 4.33 and 4.34.

Example 4.6

(a) Obtain transverse critical speeds of a rotor-bearing system as shown in Figure 4.16. Consider the shaft to be rigid, and the whole mass of the shaft is assumed to be concentrated at its mid-span. The shaft has a span of 1 m, and the diameter is 0.05 m with a mass density of 7800 kg/m^3 and Young's modulus of 2.1×10^{11} N/m^2. The shaft is supported at the ends by flexible bearings.

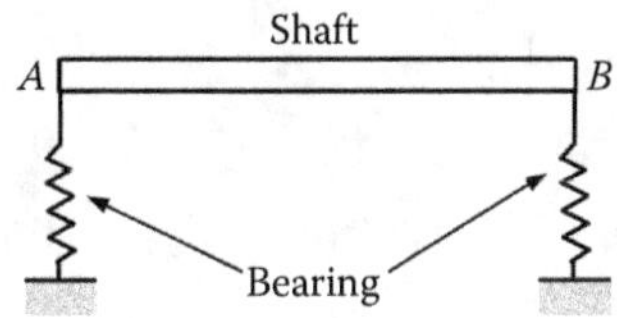

FIGURE 4.16 A flexible rotor mounted on flexible bearings.

Consider the motion in both the vertical and horizontal planes. Take the following bearing properties: for bearing A: $k_{xx}=200$ MN/m, $k_{yy}=150$ MN/m, $k_{xy}=15$ MN/m, $k_{yx}=10$ MN/m, $c_{xx}=200$ N-s/m, $c_{yy}=150$ N-s/m, $c_{xy}=14$ N-s/m, $c_{yx}=21$ N-s/m; and for bearing B: $k_{xx}=240$ MN/m, $k_{yy}=170$ MN/m, $k_{xy}=12$ MN/m, $k_{yx}=16$ MN/m, $c_{xx}=210$ N-s/m, $c_{yy}=160$ N-s/m, $c_{yx}=13$ N-s/m, $c_{yy}=18$ N-s/m. Use a numerical simulation to get the unbalance response to cross-check the critical speeds for an assumed unbalance. The axial eccentricity is 1 mm. (b) For the previous problem, consider the shaft as flexible and attach a rigid disk of 2 kg on the shaft at a distance of 0.6 m from end A. Obtain the transverse critical speeds of the system by attaching an unbalance on the disk. Take 40 g-mm unbalance in the disk at 130° from a shaft reference point.

Solution

a. *Rigid shaft*: First, we would consider the shaft as rigid in the present case, and this assumption would be relaxed in the next case (b). Equations of motion in the frequency domain for the present case can be written as

$$-\omega^2\mathbf{M}\bar{\eta}+j\omega\mathbf{C}\bar{\eta}+\mathbf{K}\bar{\eta}=\mathbf{F} \tag{a}$$

with

$$\mathbf{M}=\begin{bmatrix} m^s & 0 & 0 & 0 \\ 0 & m^s & 0 & 0 \\ 0 & 0 & I_d^s & 0 \\ 0 & 0 & 0 & I_d^s \end{bmatrix}=\begin{bmatrix} 15.30 & 0 & 0 & 0 \\ 0 & 15.30 & 0 & 0 \\ 0 & 0 & 1.27 & 0 \\ 0 & 0 & 0 & 1.27 \end{bmatrix}$$

$$m^s=\pi r^2 l\rho=15.3\text{ kg};\quad I_d^s=\frac{1}{12}m^s\left(3r^2+l^2\right)=1.27\text{ kg-m}^2$$

$$\mathbf{K}=\begin{bmatrix} (k_{xxA}+k_{xxB}) & (k_{xyA}+k_{xyB}) & 0 & 0 \\ (k_{yxA}+k_{yxB}) & (k_{yyA}+k_{yyB}) & 0 & 0 \\ 0 & 0 & 0.25l^2(k_{xxA}+k_{xxB}) & 0.25l^2(k_{xyA}+k_{xyB}) \\ 0 & 0 & 0.25l^2(k_{yxA}+k_{yxB}) & 0.25l^2(k_{yyA}+k_{yyB}) \end{bmatrix}$$

$$=10^6\begin{bmatrix} 440 & 27 & 0 & 0 \\ 26 & 320 & 0 & 0 \\ 0 & 0 & 110 & 6.75 \\ 0 & 0 & 6.50 & 80 \end{bmatrix}$$

$$\mathbf{C}=\begin{bmatrix}(c_{xxA}+c_{xxB}) & (c_{xyA}+c_{xyB}) & 0 & 0\\ (c_{yxA}+c_{yxB}) & (c_{yyA}+c_{yyB}) & 0 & 0\\ 0 & 0 & 0.25l^2(c_{xxA}+c_{xxB}) & 0.25l^2(c_{xyA}+c_{xyB})\\ 0 & 0 & 0.25l^2(c_{yxA}+c_{yxB}) & 0.25l^2(c_{yyA}+c_{yyB})\end{bmatrix}$$

$$=\begin{bmatrix}410 & 27 & 0 & 0\\ 39 & 310 & 0 & 0\\ 0 & 0 & 102.5 & 6.75\\ 0 & 0 & 9.75 & 77.5\end{bmatrix}$$

and

$$\mathbf{F}=\begin{Bmatrix} me\omega^2\\ -jme\omega^2\\ me\omega^2 e_z\\ -jme\omega^2 e_z\end{Bmatrix};\quad \bar{\eta}=\begin{Bmatrix} X\\ Y\\ \varphi_y\\ \varphi_x\end{Bmatrix}$$

Now first by considering the undamped free vibration of the system, natural frequencies of the system can be obtained from the eigenvalue problem as

$$(\mathbf{K}-\omega^2\mathbf{M})\bar{\eta}=\mathbf{0} \tag{b}$$

Transverse natural frequencies of the system obtained from this eigenvalue problem are

$$\omega_{nf_1}=4530.9\text{ rad/s};\quad \omega_{nf_2}=5393.9\text{ rad/s};\quad \omega_{nf_3}=7840.4\text{ rad/s};\quad \omega_{nf_4}=9333.8\text{ rad/s}$$

Now the unbalance response is calculated by the expression

$$\bar{\eta}=\mathbf{D}^{-1}\mathbf{F}\quad\text{where}\quad \mathbf{D}=(\mathbf{K}+j\omega\mathbf{C}-\omega^2\mathbf{M}) \tag{c}$$

Figure 4.17 shows the unbalance response plots, and it can be seen that at all noted natural frequencies, the resonance condition prevails both in the amplitude and the phase. Because transverse translational (x and y) and rotational (φ_x and φ_y) motions are uncoupled in two planes, the first two peaks are present in translational displacement plots (Figure 4.17a and b), and the other two are present in rotational displacement plots (Figure 4.17c and d).

b. *Flexible shaft*: Now the shaft is considered to be flexible, and a disk is attached at a distance of $a=0.6$ m as shown in Figure 4.18.

Now first considering the undamped free vibration, the system natural frequencies can be calculated by the eigenvalue problem of the form

$$(\mathbf{E}-\omega^2\mathbf{M})\bar{\eta}=\mathbf{0} \tag{d}$$

with

$$\mathbf{E}=\mathbf{D}^{-1};\quad \mathbf{D}=\mathbf{C}+\alpha;\quad \mathbf{C}=\mathbf{B}\mathbf{K}^{-1}\mathbf{A}$$

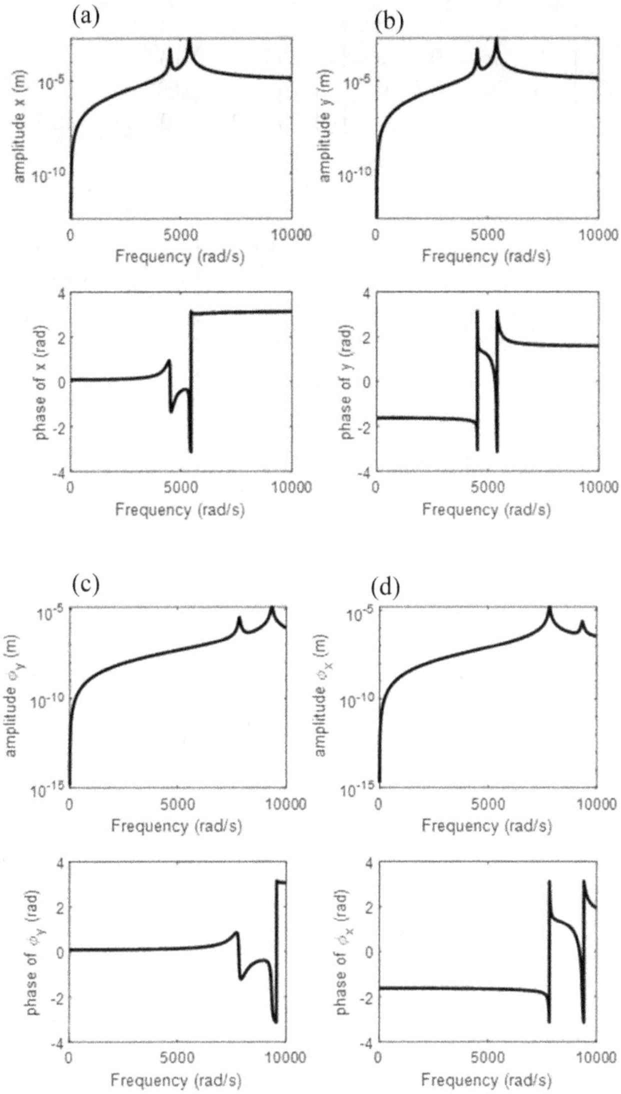

FIGURE 4.17 Amplitude and phase variation with respect to spin speed (a) horizontal displacement x_1, (b) vertical displacement y_1, (c) angular displacement φ_y, and (d) angular displacement φ_x.

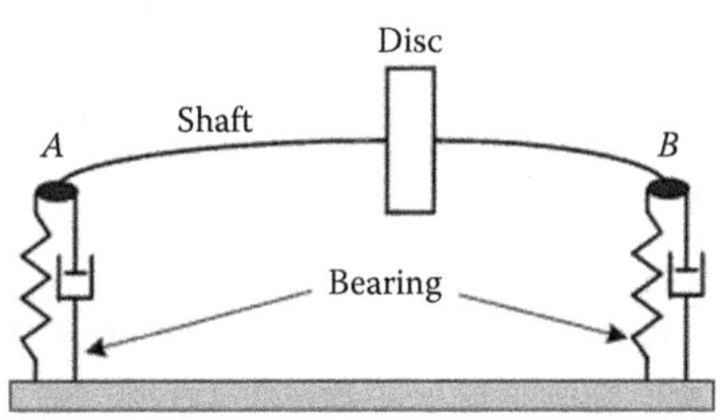

FIGURE 4.18 Flexible shaft and flexible bearing system.

$$\mathbf{M}=\begin{bmatrix} m & 0 & 0 & 0 \\ 0 & m & 0 & 0 \\ 0 & 0 & I_d & 0 \\ 0 & 0 & 0 & I_d \end{bmatrix}=\begin{bmatrix} 2 & 0 & 0 & 0 \\ 0 & 2 & 0 & 0 \\ 0 & 0 & 0.0025 & 0 \\ 0 & 0 & 0 & 0.0025 \end{bmatrix}$$

$$\begin{bmatrix} \alpha_{11} & \alpha_{12} \\ \alpha_{21} & \alpha_{22} \end{bmatrix}=\begin{bmatrix} \dfrac{a^2b^2}{3EIl} & \dfrac{-\left(3a^2l-2a^3-al^2\right)}{3EIl} \\ \dfrac{ab(b-a)}{3EIl} & \dfrac{-\left(3al-3a^2-l^2\right)}{3EIl} \end{bmatrix}=10^{-5}\begin{bmatrix} 0.0298 & -0.0248 \\ -0.0248 & 0.1449 \end{bmatrix}$$

$$\boldsymbol{\alpha}=\begin{bmatrix} \alpha_{11} & 0 & \alpha_{12} & 0 \\ 0 & \alpha_{11} & 0 & \alpha_{12} \\ \alpha_{21} & 0 & \alpha_{22} & 0 \\ 0 & \alpha_{21} & 0 & \alpha_{22} \end{bmatrix}=10^{-5}\begin{bmatrix} 0.0298 & 0 & -0.0248 & 0 \\ 0 & 0.0298 & 0 & -0.0248 \\ -0.0248 & 0 & 0.1449 & 0 \\ 0 & -0.0248 & 0 & 0.1449 \end{bmatrix}$$

$$\mathbf{A}=\begin{bmatrix} (1-a/l) & 0 & -1/l & 0 \\ 0 & (1-a/l) & 0 & 1/l \\ a/l & 0 & 1/l & 0 \\ 0 & a/l & 0 & -1/l \end{bmatrix}=\begin{bmatrix} 0.4 & 0 & -1 & 0 \\ 0 & 0.4 & 0 & 1 \\ 0.6 & 0 & 1 & 0 \\ 0 & 0.6 & 0 & -1 \end{bmatrix}$$

$$l=a+b=1\text{ m}$$

$$\mathbf{B}=\begin{bmatrix} (1-a/l) & 0 & a/l & 0 \\ 0 & (1-a/l) & 0 & a/l \\ -1/l & 0 & 1/l & 0 \\ 0 & 1/l & 0 & -1/l \end{bmatrix}=\begin{bmatrix} 0.4 & 0 & 0.6 & 0 \\ 0 & 0.4 & 0 & 0.6 \\ -1 & 0 & 1 & 0 \\ 0 & 1 & 0 & -1 \end{bmatrix}$$

$$\mathbf{K}=\begin{bmatrix} {}_A\left(k_{xx}+j\omega c_{xx}\right) & {}_A\left(k_{xy}+j\omega c_{xy}\right) & 0 & 0 \\ {}_A\left(k_{yx}+j\omega c_{yx}\right) & {}_A\left(k_{yy}+j\omega c_{yy}\right) & 0 & 0 \\ 0 & 0 & {}_B\left(k_{xx}+j\omega c_{xx}\right) & {}_B\left(k_{xy}+j\omega c_{xy}\right) \\ 0 & 0 & {}_B\left(k_{yx}+j\omega c_{yx}\right) & {}_B\left(k_{yy}+j\omega c_{yy}\right) \end{bmatrix}$$

Herein, details of matrix in numerical form are not provided for brevity. The undamped transverse natural frequencies of the system obtained from this eigenvalue problem (Equation (d)) are

$$\omega_{nf_1} = 1287.8 \text{ rad/s}; \quad \omega_{nf_2} = 1289.8 \text{ rad/s}; \quad \omega_{nf_3} = 17{,}858.5 \text{ rad/s}; \quad \omega_{nf_4} = 17{,}873.8 \text{ rad/s}$$

Herein, first two natural frequencies are close by and similarly, last two natural frequencies are close by.

The unbalance response is given as

$$\eta = \mathbf{HF} \tag{e}$$

with

$$\mathbf{H} = \left(-\omega^2\mathbf{M} + \mathbf{E}\right)^{-1}; \quad \mathbf{E} = \mathbf{D}^{-1}; \quad \mathbf{D} = \mathbf{C} + \alpha; \quad \mathbf{C} = \mathbf{BK}^{-1}\mathbf{A}$$

and

$$\mathbf{U} = \begin{Bmatrix} X \\ Y \\ \Phi_y \\ \Phi_x \end{Bmatrix}; \quad \mathbf{F} = \begin{Bmatrix} me\omega^2 \\ -jme\omega^2 \\ 0 \\ 0 \end{Bmatrix}$$

From the unbalance response shown in Figure 4.19a–d, it can be seen that at the noted natural frequencies, the critical speeds appear in the form of amplitude peaks and phase changes.

4.4 A FLEXIBLE ROTOR ON FLEXIBLE BEARINGS AND FOUNDATIONS

In some rotating machines, for example, turbines, generators, stone crushers, etc., the rotor-bearing systems themselves may be mounted on flexible foundations (Figure 4.20), which may in turn influence the motion of the disk masses. In the present section, a simple model of the foundation is considered by ignoring cross-coupled terms of the stiffness and the damping. For more detailed treatment on foundation effects, readers can refer to the book by Krämer (1993).

The net displacement of the disk is given by the vector sum of (i) the disk displacement relative to the shaft ends, (ii) that of shaft ends relative to the bearing, and (iii) that of the bearing relative to the space. The theoretical analysis of the disk, shaft, and bearing unbalance responses, and that of the force transmissibility of such a system, can be carried out in a similar manner to that described in the previous section. Additional governing equations related to the foundation are derived, and how to relate them with governing equations of the disk and bearings is detailed here.

The relationship between forces transmitted through bearings and displacements of shaft ends is governed by the bearing stiffness and damping coefficients. The form of the governing equation is given by Equation 4.41, which is

$$\mathbf{f}_b = \mathbf{KX}_b \tag{4.71}$$

In the frequency domain, it takes the form

$$\mathbf{F}_b = \mathbf{KX}_b \quad \text{or} \quad \mathbf{X}_b = \mathbf{K}^{-1}\mathbf{F}_b \tag{4.72}$$

where $\mathbf{X}_b$ is the shaft end displacement relative to the bearing. Displacements of bearings with respect to foundations and forces transmitted through bearings are shown in Figure 4.21.

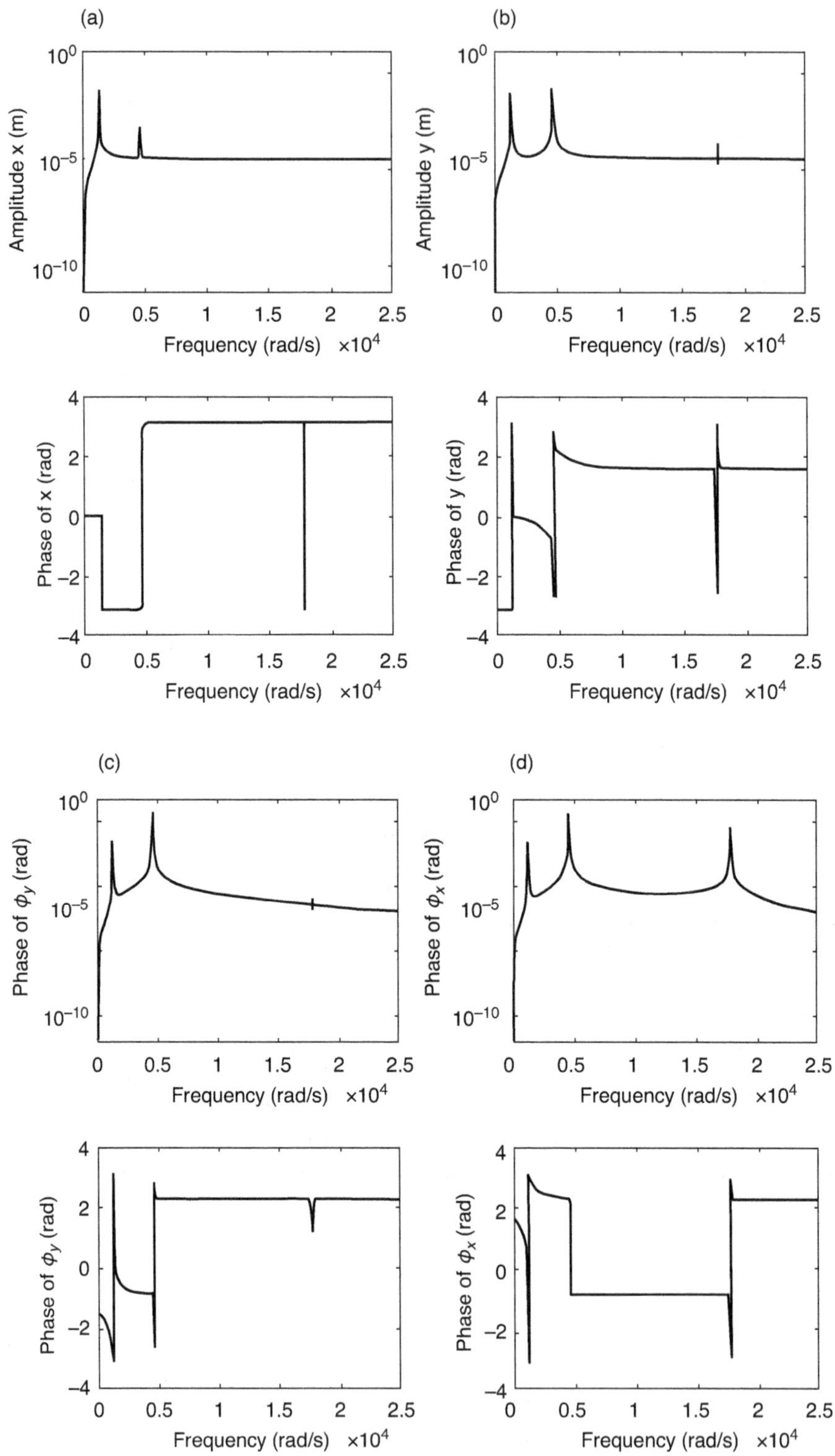

FIGURE 4.19 Amplitude and phase variation with respect to spin speed: (a) horizontal displacement x_1; (b) vertical displacement y_1; (c) angular displacement φ_{x1}; (d) angular displacement φ_{y1}.

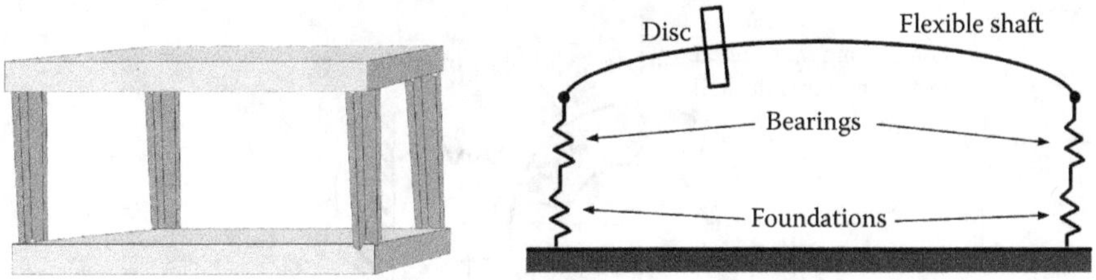

FIGURE 4.20 (a) A typical foundation with channel columns on top of which the machine mounts. (b) A flexible rotor-bearing foundation system.

The bearing will respond in the horizontal direction for an external force f_{bx}, which is governed by the following equation:

$$f_{bx} - k_{fx}x_f - c_{fx}\dot{x}_f = m_b\ddot{x}_f \tag{4.73}$$

where x_f is the horizontal displacement of the bearing, with respect to foundation. m_b is the bearing, mass of one bearing, and k_{fx}, c_{fx}, k_{fy}, and c_{fy} are the foundation stiffness and damping coefficients. Similarly, the response of the bearing in the vertical direction to a force f_{by} is given as

$$f_{by} - k_{fy}y_f - c_{fy}\dot{y}_f = m_b\ddot{y}_f \tag{4.74}$$

where y_f is the vertical displacement of the bearing with respect to foundation. The displacement of the bearing will take the form

$$x_f(t) = X_f e^{j\omega t} \quad \text{and} \quad y_f(t) = Y_f e^{j\omega t} \tag{4.75}$$

where X_f and Y_f are complex displacements in the x and y directions, respectively. On substituting Equation 4.75 in equations of motion (Equations 4.73 and 4.74), and on combining in this a matrix form (for bearing A), we obtain

$${}_A\mathbf{D}\,{}_A\mathbf{X}_f = {}_A\mathbf{F}_b \tag{4.76}$$

with

$${}_A\mathbf{D} = \left(\begin{bmatrix} {}_Ak_{f_x} & 0 \\ 0 & k_{f_y} \end{bmatrix} - \omega^2\begin{bmatrix} m_b & 0 \\ 0 & m_b \end{bmatrix} + j\omega\begin{bmatrix} c_{f_x} & 0 \\ 0 & c_{f_y} \end{bmatrix}\right)$$

$${}_A\mathbf{X}_f = \begin{Bmatrix} X_f \\ Y_f \end{Bmatrix}_A \quad \text{and} \quad {}_A\mathbf{F}_b = \begin{Bmatrix} F_{b_x} \\ F_{b_y} \end{Bmatrix}_A$$

For both bearings A and B, equations of form, Equation 4.76, can be combined as

$$\mathbf{F}_b = \mathbf{D}\mathbf{X}_f \tag{4.77}$$

with

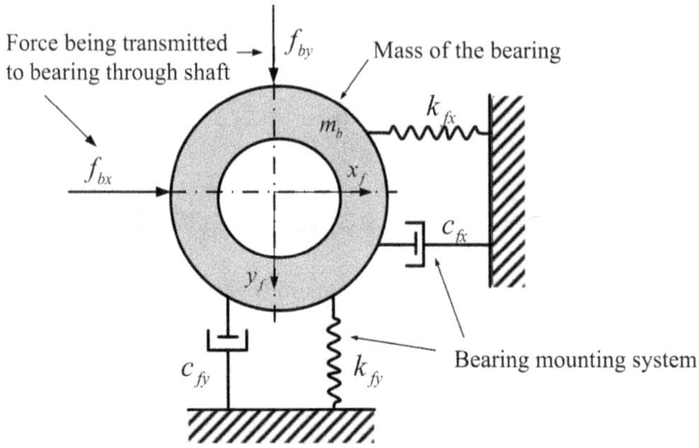

FIGURE 4.21 A bearing block mounted on a foundation.

$$\mathbf{F}_b = \begin{Bmatrix} {}_A\mathbf{F}_b \\ {}_B\mathbf{F}_b \end{Bmatrix}; \quad \mathbf{D} = \begin{bmatrix} {}_A\mathbf{D} & \mathbf{0} \\ \mathbf{0} & {}_B\mathbf{D} \end{bmatrix}; \quad \mathbf{X}_f = \begin{Bmatrix} {}_A\mathbf{X}_f \\ {}_B\mathbf{X}_f \end{Bmatrix}$$

which gives the relative displacements between the bearings and foundations as

$$\mathbf{X}_f = \mathbf{D}^{-1}\mathbf{F}_b \tag{4.78}$$

The total displacement of the shaft ends under the action of an applied force $\mathbf{F}_b$ is given by summing the individual displacements $\mathbf{X}_b$ (by Equation 4.72) and $\mathbf{X}_f$ (by Equation 4.78), that is,

$$\mathbf{W} = \mathbf{X}_b + \mathbf{X}_f = \left(\mathbf{K}^{-1} + \mathbf{D}^{-1}\right)\mathbf{F}_b = \boldsymbol{\alpha}'\mathbf{F}_b \tag{4.79}$$

with

$$\boldsymbol{\alpha}' = \mathbf{K}^{-1} + \mathbf{D}^{-1}$$

where $\boldsymbol{\alpha}'$ is a system-equivalent dynamic receptance matrix describing the overall shaft support characteristics and it allows for flexibilities of both bearings and foundations. The study of the disk motion may now proceed in the same manner as described in the previous section, except the equivalent dynamic stiffness matrix $(\boldsymbol{\alpha}')^{-1}$ should be substituted for $\mathbf{K}$ in Equation 4.41. Once the disk displacement vector $\mathbf{U}_s$ is known, it is possible to substitute back and obtain $\mathbf{F}_s$, $\mathbf{X}_b$, and $\mathbf{F}_b$. Forces transmitted to foundations are given as

$$f_{fx} = k_{fx}x_f + c_{fx}\dot{x}_f \quad \text{and} \quad f_{fy} = k_{fy}y_f + c_{fy}\dot{y}_f \tag{4.80}$$

For the unbalance excitation, we have

$$f_{fx} = F_{fx}e^{j\omega t} \quad \text{and} \quad f_{fy} = F_{fy}e^{j\omega t} \tag{4.81}$$

On substituting Equation 4.81 into Equation 4.80, we get

$$\left\{\begin{array}{c} F_{f_x} \\ F_{f_y} \end{array}\right\}=\left(\left[\begin{array}{cc} k_{f_x} & 0 \\ 0 & k_{f_y} \end{array}\right]+j\omega\left[\begin{array}{cc} c_{f_x} & 0 \\ 0 & c_{f_y} \end{array}\right]\right)\left\{\begin{array}{c} X_f \\ Y_f \end{array}\right\} \tag{4.82}$$

Forces transmitted through foundations will not be the same as forces transmitted through bearings. Bearing masses (i.e., inertia forces) will absorb some force toward acceleration. If bearing masses are negligible, then bearings and foundations will transmit the same amount of force, although there may be some phase lag due to the presence of damping. The amplitude and the phase of forces transmitted through the foundations can be obtained from F_{fx1}, F_{fx2}, F_{fy1}, and F_{fy2} using the procedure described in the previous sections.

More detailed study on the foundation effects is beyond the scope of the present book (Tiwari, 2024); however, various studies have incorporated foundation effects in the analysis of a rotor-bearing system, and some of them are summarized here. Smith (1933) investigated the Jeffcott rotor with internal damping to include a massless, damped, and flexible support system. Lund (1965) and Gunter (1967) showed that damped and flexible supports may improve the stability of high-speed rotors. Also, Lund and Sternlicht (1962), Dworski (1964), and Gunter (1970) demonstrated that a significant reduction in the transmitted force could be achieved by a proper design of a bearing support system. Kirk and Gunter (1972) analyzed the steady-state and transient responses of the Jeffcott rotor for elastic bearings mounted on the damped and flexible supports. Gasch (1976) dealt with the flexible rotating shaft of a large turbo-rotor using FE analysis. He introduced foundation dynamics into the rotor equations via receptance matrices, which were obtained from modal testing and modal analysis. Vance et al. (1987) provided comparison of results from computer-based predictions and experimental measurements on a rotor-bearing test apparatus. They modeled the rotor-bearing system to include foundation impedance effects by using the transfer matrix method. Stephenson and Rouch (1992) utilized the FE method to analyze rotor-bearing foundation systems. They provided a procedure using modal analysis techniques, which could be applied in measured frequency response functions to include the dynamic effects of the foundation structure. Kang et al. (2000) studied the foundation effects on the dynamic characteristics of rotor-bearing systems. The modeling and analysis of rotor-bearing-foundation systems based on the FE method were discussed. A substructure procedure that included the foundation effects in equations of motion and the application of the dynamic solver of a commercial package were addressed.

A good model of the rotor and a reasonably accurate model of fluid journal bearings may be constructed using the FE method or any other reliable method. Indeed, a number of FE-based software programs are available for such modeling. However, a reliable FE model for the foundation is extremely difficult to construct due to a number of practical difficulties (Lees and Simpson, 1983). Experimental modal analysis (Ewins, 2000) is a possible solution, but this requires that the rotor be removed from the foundation, which is not practical for an existing power station. With these difficulties, it is unlikely that the techniques of FE model updating (Friswell and Mottershead, 1995) could be used, and the direct estimation of the foundation model from measured responses at the bearing pedestals from the machine run-down data has been accepted as a viable alternative technique (Lees, 1988; Smart et al., 2000). The estimation technique assumes that the state of unbalance is known from balancing runs, either by the difference in the response from two run-downs, or by the estimated unbalance from a single run-down (Edwards et al., 2000; Sinha et al., 2002).

4.5 A TURBINE-COUPLING–GENERATOR ROTOR ON FLEXIBLE BEARINGS

A simple analytical model of the turbo-generator system (Figure 4.22) is considered by Lal and Tiwari (2012), which is composed of two rigid rotors (m is the mass and I_d is the diametral mass moment of inertia of the rotor), with residual unbalances ($F=me$ is the unbalance and e is the eccentricity of the rotor) in each rotor, connected by a flexible coupling (composed of the translational,

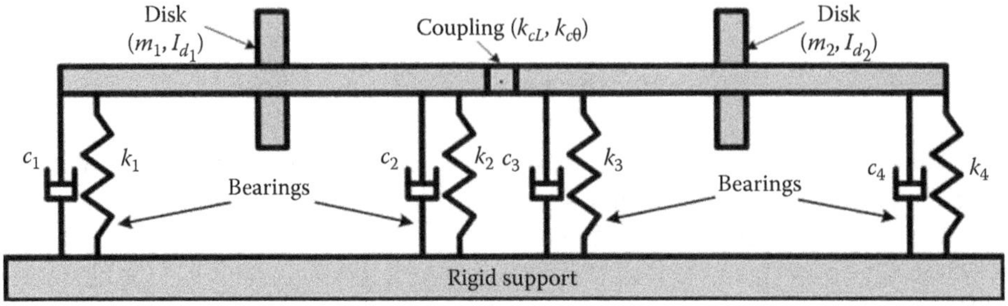

FIGURE 4.22 Rotor-bearing coupling system.

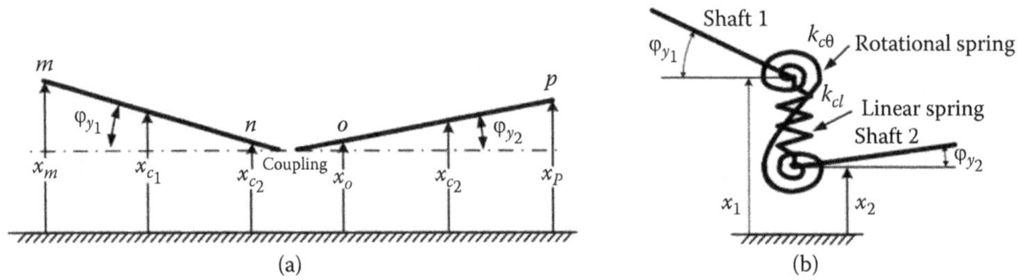

FIGURE 4.23 (a) Schematic diagram of the deflected rotor in the *z-x* plane. (b) Schematic diagram of the coupling model.

k_{cL}, and torsional, $k_{c\theta}$, stiffness). Rotors are mounted on flexible bearings with damping, c, and stiffness, k, properties. The schematic diagram of the turbo-generator model in a deflected position in the vertical plane (x-z) is shown in Figure 4.23a, in which (x, φ_y) are the translational and rotational displacements, and the schematic diagram of the coupling is shown in Figure 4.23b. Again, detailed treatment of equations of motion and critical speed analysis is beyond the scope of this book; however, a typical response at a coupling location would appear as shown in Figure 4.24 for the rotor parameters given in the publication by Lal and Tiwari (2014). For coupling impelling readers can refer to Srinivas et al. (2019, 2021).

The emphasis of the present chapter is to analytically analyze a rotor-bearing-foundation system. The methodology becomes cumbersome for even simple rotor systems, and every system requires individual analysis. This demands a systematic methodology that can handle complex rotor-bearing foundation systems. In subsequent chapters, two representative numerical-based methods—the transfer matrix and FE methods—will be discussed in Chapters 6 through 10 and also in Chapter 2 of Volume II. These methods do not require different formulations for individual rotor systems; however, in a more algorithmic way governing equations can be obtained. Nonetheless, the analysis of the present chapter will be beneficial to check the validity of such numerical-based analysis.

CONCLUDING REMARKS

In the present chapter, we dealt mainly with dynamic responses (critical speeds and unbalance responses) of a single mass rotor with flexible supports. Dynamic parameters of supports not only provide the stiffness and damping forces to the rotor, but also provide asymmetry in these dynamic parameters in two orthogonal directions. The translatory and conical whirl motions are obtained for the long rigid rotor supported on anisotropic bearings. It is found that the orbit of the shaft center not only becomes elliptical, but its major axis becomes inclined to both orthogonal axes. The forward and backward whirls are observed on mounting the rigid rotor on anisotropic bearings. The flexibility of the foundation resulted in an increase in the effective flexibility experienced

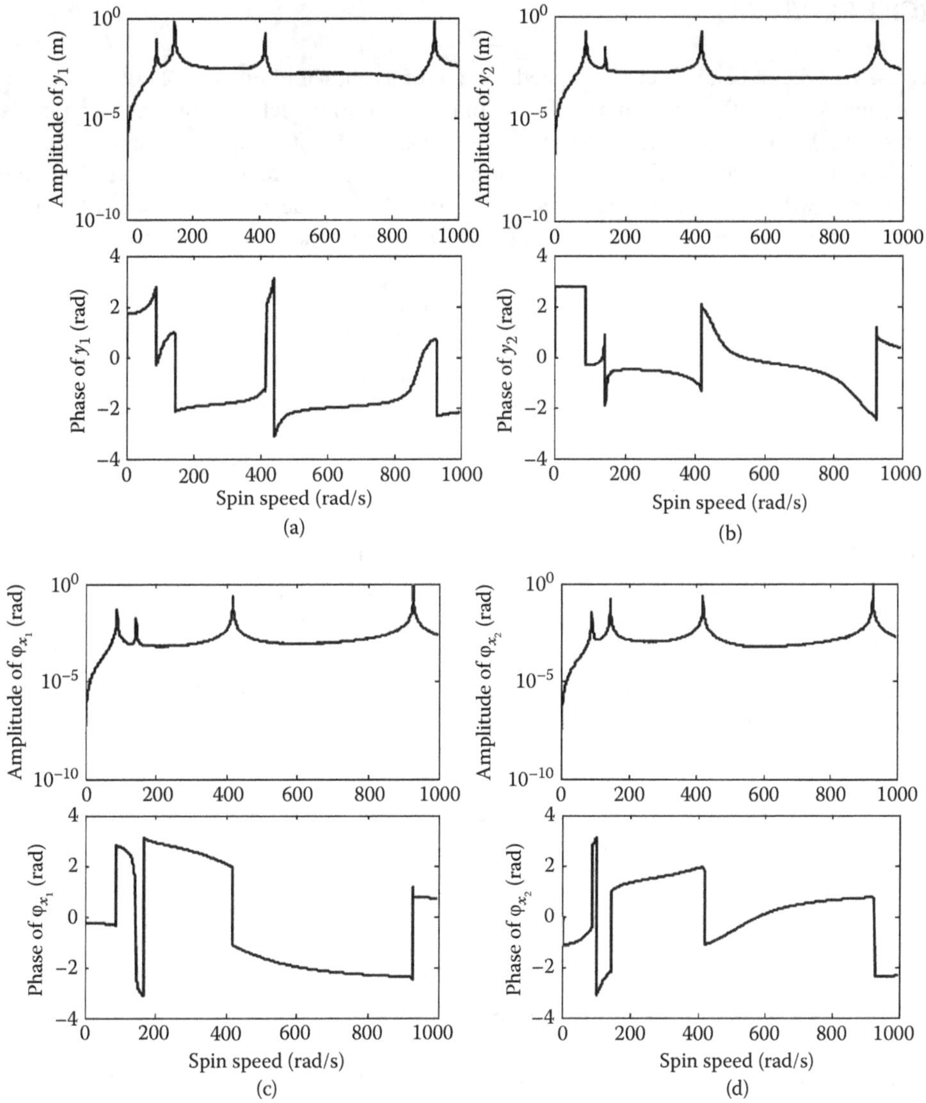

FIGURE 4.24 Amplitude and phase variation with respect to spin speed; (a) vertical translational displacement y_1; (b) vertical translational displacement y_2; (c) rotational displacement φ_{x1}; (d) rotational displacement φ_{x2}.

by the rotor system, which is expected to decrease the critical speeds. Overall complexity of the dynamic analysis procedure becomes cumbersome while considering bearings and foundations even with a single mass rotor. More systematic methods for dynamically analyzing multimass rotors are required. In subsequent chapters, while considering the torsional and transverse vibrations of multi-degree-of-freedom (DOF) rotor systems, two representative methods called the transfer matrix method and the FE method will be dealt with in detail. In the next chapter, we will consider a single mass rotor only; however, a very important effect of gyroscopic moments is included in that chapter. In the end the sustainable design of a rotor-bearing system focuses on minimizing energy consumption, maximizing operational lifespan, reducing environmental contamination, and ensuring safe structural integrity.

EXERCISE PROBLEMS

Exercise 4.1 Obtain bending critical speeds of a rotor as shown in Figure 4.25. It consists of a massless rigid shaft (with 1 m span and 0.7 m from the disk to the left bearing) and a rigid disk (5 kg mass and 0.1 kg-m^2 diametral mass moment of inertia), and it is supported on two identical flexible bearings (1 kN/m of stiffness for each bearing). Consider motion in the vertical plane only. Is there any difference in critical speeds when the disk is placed at the center of the rotor? If not, then justify the same, and if yes, then obtain the same. Ignore the gyroscopic couple.

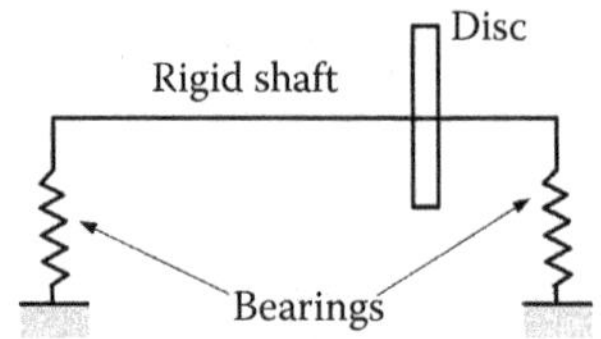

FIGURE 4.25 A rigid rotor supported on flexible bearings.

Exercise 4.2 Consider a long rigid rotor, R, supported on two identical bearings, B_1 and B_2, as shown in Figure 4.26. The direct stiffness coefficients of both bearings in the horizontal and vertical directions are equal (i.e., k). Ignore the direct damping and the cross-coupled stiffness and damping coefficients of both bearings. The mass of the rotor is m, the span of the rotor is l, and the diametral mass moment of inertia is I_d. Derive equations of motion for the system and obtain natural frequencies of whirl. Ignore the gyroscopic effect.

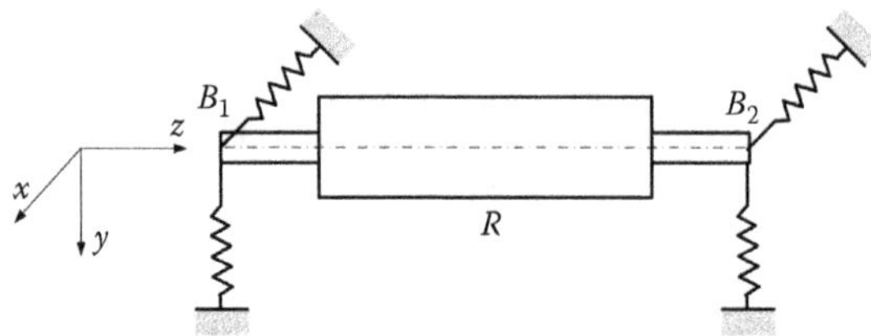

FIGURE 4.26 A long rigid rotor supported on flexible bearings.

Exercise 4.3 Find critical speeds of the rotor-bearing system shown in Figure 4.27. The shaft is rigid and massless. The mass of the disk is m_d = 1 kg with negligible diametral mass moment of inertia. Bearings B_1 and B_2 are identical bearings and have the following properties: k_{xx}=1.1 kN/m, k_{yy}=1.8 kN/m, k_{xy}=0.2 kN/m, and k_{yx}=0.1 kN/m. Let B_1D=75 mm and DB_2=50 mm.

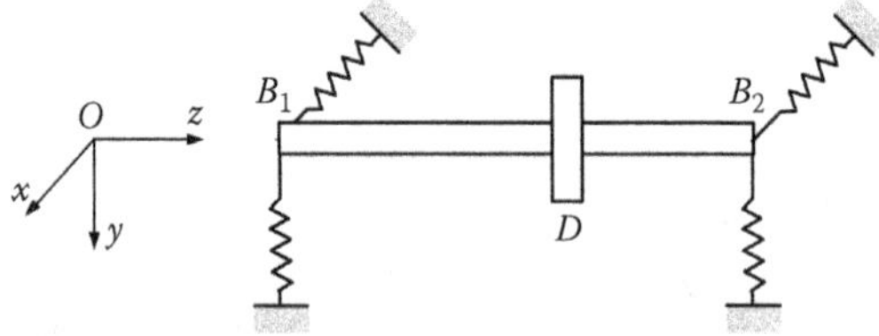

FIGURE 4.27 A rigid rotor on flexible bearings.

Exercise 4.4 For Exercise 4.3, take 25 g-mm of unbalance in the disk at a phase of 38° from a shaft reference point. Plot the disk response amplitude and phase versus the spin speed of the rotor to show all critical speeds. Plot the variation of bearing forces with the spin speed of the rotor.

Exercise 4.5 For Example 4.4, obtain critical speeds of the rotor-bearing-foundation system when the foundation (which is in series with bearing) has the following dynamic characteristics: $k_{f_x} = k_{f_y} = 100$ MN/m and $c_{f_x} = c_{f_y} = 50$ kN-s/m. Let the mass of each bearing be 2 kg. Plot the unbalance response amplitude and phase of the shaft end and the bearing at B_1 with respect to the spin speed of shaft to show all critical speeds of the system. Take 25 g-mm of unbalance on the disk at a phase of 38° from a shaft reference point. Plot the variation of the bearing and foundation forces at end B_1 with the spin speed. Choose a suitable range to cover all critical spin speeds.

Exercise 4.6 Consider a simple rigid rotor flexible bearing system as shown in Figure 4.28. The rotor is supported on two different flexible bearings. In the figure, L_1 and L_2 are the distances of bearings 1 and 2 from the center of gravity of the rotor with $L = L_1 + L_2$, R_1 and R_2 are distances of balancing planes (i.e., rigid disks) from the center of gravity of the rotor, and u is the unbalance. Consider linearized eight bearing dynamic parameters for each of the bearings based on short bearing approximation (refer to Chapter 3).

Let m be the mass of the rotor. I_t is the transverse mass moment of inertia of the rotor about an axis passing through the center of gravity, I_p is the polar mass moment of inertia of the rotor, k and c are, respectively, the stiffness and damping parameters, $f_x(t)$ and $f_y(t)$ are, respectively, the force in the horizontal and vertical directions, u is the unbalance, φ is the phase, x and y are translational displacements in the horizontal and vertical directions, respectively, t is the time, and subscripts 1 and 2 represent the right and left-hand sides from the midspan of the rotor, respectively. Obtain equations of motion of the rotor-bearing system in terms of translational displacements (four in number, i.e., x_1, y_1, x_2, y_2) at two bearings. The motivation behind obtaining the equations of motion in terms of bearing response is that, in real-life terms, often these responses can only be accessible to the practicing engineers.

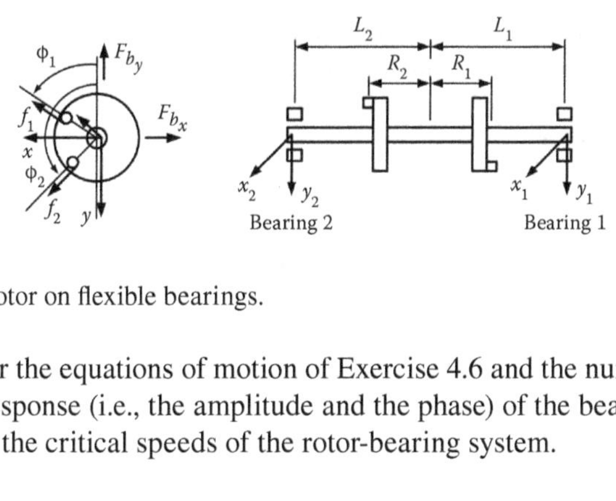

FIGURE 4.28 A rigid rotor on flexible bearings.

Exercise 4.7 Consider the equations of motion of Exercise 4.6 and the numerical data given in Table 4.1. Obtain the response (i.e., the amplitude and the phase) of the bearings with respect to the rotor speed and list the critical speeds of the rotor-bearing system.

TABLE 4.1
Details of the Rotor Model for the Numerical Example

Property	Numerical Value
Rotor	
Rotor shaft diameter	10 mm
Rotational speed, ω	100 Hz
Mass, m	4 kg
Length of the rotor, L	0.425 m
Distance of bearings from the center of the rotor	0.2125 m

(Continued)

TABLE 4.1 (*Continued*)
Details of the Rotor Model for the Numerical Example

Property	Numerical Value
Distance of disks from the center of the rotor	0.130 m
Transverse mass moment of inertia, I_d	0.0786 kg-m^2
Rigid disks	
Inner diameter	10 mm
Outer diameter	74 mm
Thickness	25 mm
Bearings	
Diameter	25.4 mm
Length to diameter ratio	1
Radial clearance, c_r of Bearing 2	0.075 mm
Kinetic viscosity	20.11 cS (centi-Stokes)
Temperature of lubricant	40°C
Specific gravity of the lubricant	0.87

Exercise 4.8 Consider a rigid rotor mounted on two bearings at the ends, as shown in Figure 4.29, that has a varying cross-section along the longitudinal axis (e.g., a tapered rotor). For this case, the center of gravity, G, of the rotor will be offset from the midspan of the rotor, C. It is assumed that the rotor is perfectly balanced (i.e., it has no external radial force and corresponding external moment). Let m be the mass, I_d be the diametral mass moment of inertia of the rotor about the center of gravity, k_A and k_B be the stiffnesses of bearings A and B, respectively, and l be the length of the rotor. Obtain governing equations of motion for the following six sets of chosen generalized coordinates for a single-plane motion of the rotor. Discuss forms of the mass and stiffness matrices in regards to the diagonal and off-diagonal terms (i.e., the mass and stiffness couplings) for a linearized system.

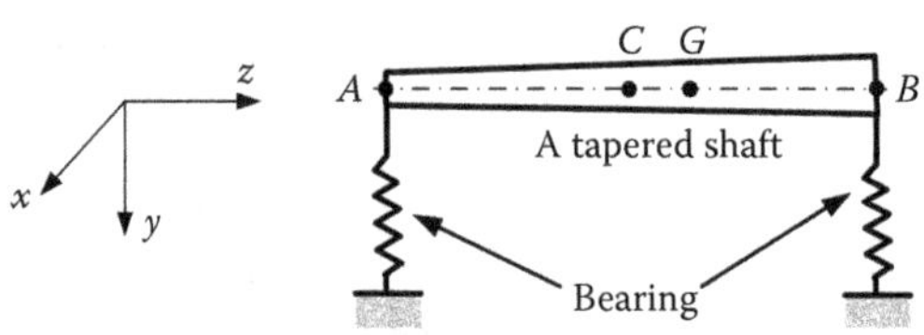

FIGURE 4.29 An axially asymmetric shaft mounted on flexible dissimilar bearings.

1. If we choose generalized coordinates as $\left(x_G, \varphi_y\right)$, where the translational displacement of the center of gravity is x_G, and tilting of the rotor from the horizontal (i.e., z-axis) is φ_y.
2. If we choose generalized coordinates as $\left(x_E, \varphi_y\right)$, where the translational displacement of a point on the rotor where if a transverse force is applied then it produces pure translation of the rotor (i.e., $k_A l_{AE} = k_B l_{BE}$) is x_E, and tilting of the rotor remains the same as for the first case.
3. If we choose generalized coordinates as $\left(x_A, \varphi_y\right)$, where the translational displacement of the extreme left end of the rotor is x_A, and tilting of the rotor remains the same as for the first case.
4. If we choose generalized coordinates as (x_A, x_B), where the translational displacements of the extreme left and right ends of the rotor are x_A and x_B, respectively.

5. If we choose generalized coordinates as (x_C, φ_y), where the translational displacement of the midspan is x_C, and tilting of the rotor from the horizontal (i.e., z-axis) is φ_y.
6. If we choose generalized coordinates as (x_E, x_G), where displacements have similar meanings as defined previously.

Exercise 4.9 For a perfectly balanced rigid rotor mounted on flexible bearings as shown in Figure 4.29, the following data are given: $m=10\,\text{kg}$, $I_d=0.015$ kg-m^2, $l=1\,\text{m}$, $l_{AG}=0.6\,\text{m}$, $k_A=120$ kN/m, and $k_B=140$ kN/m. Consider one-plane motion with two DOFs and coupling in the generalized coordinates. Obtain the transverse natural frequencies and mode shapes of the rotor-bearing system.

Exercise 4.10 Obtain transverse natural frequencies of a rotor-bearing system as shown in Figure 4.30 *for pure translatory motion of the shaft*. Consider the shaft to be rigid, and the whole mass of the shaft is assumed to be concentrated at its midspan. The shaft has a span of 1 m and the diameter is 0.05 m with a mass density of 7800 kg/m^3. The shaft is supported at the ends by flexible bearings. Consider the motion in both the vertical and horizontal planes. Use the following bearing properties: for both bearings A and B, $k_{xx}=200$ MN/m, $k_{yy}=150$ MN/m, $k_{xy}=15$ MN/m, and $k_{yx}=10$ MN/m.

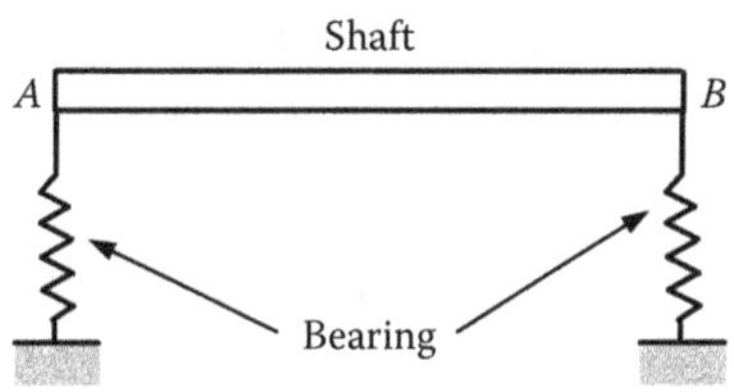

FIGURE 4.30 A rigid rotor mounted on two bearings.

Exercise 4.11 Obtain transverse critical speeds of a rotor-bearing system as shown in Figure 4.30, *for pure tilting motion of the shaft*. Consider the shaft to be rigid and the whole mass of the shaft is assumed to be concentrated at its midspan. The shaft has a span of 1 m and the diameter is 0.05 m with a mass density of 7800 kg/m^3. The shaft is supported at the ends by flexible bearings. *Consider the motion in both the vertical and horizontal planes.* Use the following bearing properties: for both bearings A and B, $k_{xx}=200$ MN/m, $k_{yy}=150$ MN/m, $k_{xy}=1.5$ MN/m, and $k_{yx}=0.5$ MN/m.

Exercise 4.12 Obtain equations of motion (put them in expanded matrix form) for transverse vibration of a rotor-bearing coupling system as shown in Figure 4.31. Two identical rigid rotors are connected by a coupling (a pin joint or a universal joint) and are supported on four bearings (modeled as springs, each of which has a stiffness of k) as shown in the figure. Choose appropriately generalized coordinates to define the motion of the system. In the figure, L is the distance between the bearings and a is the overhang portion of each rotor. Consider a single-plane motion of the system.

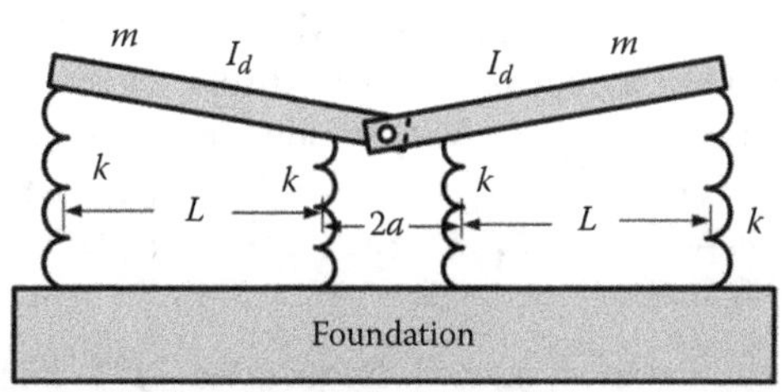

FIGURE 4.31 A rotor-bearing-coupling system model.

Exercise 4.13 For a symmetrical long rigid shaft on flexible anisotropic bearings, the following energy expressions are given:

$$T = \frac{1}{2}m\dot{x}^2 + \frac{1}{2}m\dot{y}^2 + \frac{1}{2}I_d\dot{\varphi}_x^2 + \frac{1}{2}I_d\dot{\varphi}_y^2$$

$$U = \frac{1}{2}(2k_x)x^2 + \frac{1}{2}(2k_y)y^2 + \frac{1}{2}(k_x l^2/2)\varphi_y^2 + \frac{1}{2}(k_y l^2/2)\varphi_x^2$$

$$\delta W_{nc} = \left(me\omega^2 \cos\omega t\right)\delta x + \left(me\omega^2 \sin\omega t\right)\delta y + \left(me\omega^2 e_z \cos\omega t\right)\delta\varphi_y + \left(me\omega^2 e_z \sin\omega t\right)\delta\varphi_x$$

where T is the kinetic energy, U is the potential energy, δW_{nc} is the nonconservative virtual work done, x and y are the coordinates of the rotor geometrical center (i.e., generalized coordinates), φ_x and φ_y are rotational coordinates of the rotor (i.e., generalized coordinates), e is the radial eccentricity, e_z is the axial eccentricity, m is the mass of the rotor, k_x and k_y are the stiffness of each bearing, l is the length of the rotor, I_d is the diametral mass moment of inertia of the rotor, and ω is the spin speed of the rotor. Using Lagrange's equation (refer to Chapter 7), obtain equations of motion of the rotor system.

Exercise 4.14 For a symmetrical long rigid shaft on anisotropic bearings, the following energy expressions are given:

$$T = \frac{1}{2}m\dot{x}^2 + \frac{1}{2}m\dot{y}^2 + \frac{1}{2}I_d\dot{\varphi}_x^2 + \frac{1}{2}I_d\dot{\varphi}_y^2$$

$$U = \frac{1}{2}(2k_{xx})x^2 + \frac{1}{2}(2k_{yy})y^2 + \frac{1}{2}(k_{xx}l^2/2)\varphi_y^2 + \frac{1}{2}(k_{yy}l^2/2)\varphi_x^2$$

$$\delta W_{nc} = \left(me\omega^2 \cos\omega t - 2k_{xy}y - 2c_{xx}\dot{x} - 2c_{xy}\dot{y}\right)\delta x + \left(me\omega^2 \sin\omega t - 2k_{yx}x - 2c_{yy}\dot{y} - 2c_{yx}\dot{x}\right)\delta y$$
$$+ \left(me\omega^2 e_z \cos\omega t - 0.5l^2 k_{xy}\varphi_x - 0.5l^2 c_{xx}\dot{\varphi}_y - 0.5l^2 c_{xy}\dot{\varphi}_x\right)\delta\varphi_y$$
$$+ \left(me\omega^2 e_z \sin\omega t - 0.5l^2 k_{yx}\varphi_y - 0.5l^2 c_{yy}\dot{\varphi}_x - 0.5l^2 c_{yx}\dot{\varphi}_y\right)\delta\varphi_x$$

where T is the kinetic energy, U is the potential energy, δW_{nc} is the nonconservative virtual work done, x and y are the coordinates of the rotor geometrical center (i.e., generalized coordinates), φ_x and φ_y are rotational coordinates of the rotor (i.e., generalized coordinates), e is the radial eccentricity, e_z is the axial eccentricity, m is the mass of the rotor, k_{xx} and k_{yy} are the direct stiffness of each bearing, k_{xy} and k_{yx} are the cross-coupled stiffness of each bearing ($k_{xy} \neq k_{yx}$), c_{xx} and c_{yy} are the direct damping of each bearing, c_{xy} and c_{yx} are the cross-coupled damping of each bearing ($c_{xy} \neq c_{yx}$), l is the length of the rotor, I_d is the diametral mass moment of inertia of the rotor, and ω is the spin speed of the rotor. Using Lagrange's equation (refer to Chapter 7), obtain equations of motion of the rotor system.

Exercise 4.15 Let $x(t) = \sin(mt)$ and $y(t) = \sin(nt + \phi)$, where $m = 1, 2, 3, \ldots$, $n = 1, 2, 3, 4, \ldots$, phase $\phi = (0 - 360)$. $x(t)$ and $y(t)$ could be thought as transverse displacements of a rotor. Plot (i) $x(t)$ versus t, (ii) $y(t)$ versus t, and (iii) $x(t)$ versus $y(t)$. Vary various variables and see the plots (Lissajous figures) and interpret them. For example, for $\phi = 0$, (i) $m = 1$, $n = 1, 2, \ldots, 5$, (ii) $m = 2$, $n = 1, 2, \ldots, 5$, (iii) $m = 3$, $n = 1, 2, \ldots, 5$, (iv) $m = 4$, $n = 1, 2, \ldots, 5$, and (v) $m = 5$, $n = 1, 2, \ldots, 5$. Similarly for $\phi = 0$, $\pi/4$, $\pi/2$, $3\pi/4$, π, $5\pi/4$, $3\pi/2$, $7\pi/4$, 2π.

Exercise 4.16 Obtain the transverse natural frequencies of a long, rigid, tapered rotor mounted on dissimilar springs. The length of the rotor is 1 m, and the diameters of two ends are 0.10 m (LHS) and 0.15 m (RHS), respectively. The rotor is supported on springs at the ends (L: left and R: right) of the rotor with $k_L = 1$ kN/m and $k_R = 1.2$ kN/m. Consider single-plane motion only. The density of the rotor material is 7800 kg/m^3 and Young's modulus of $E = 2.1 \times 10^{11}$ N/m^2. Give the steps for the formulations and state the valid assumptions that are made.

Exercise 4.17 Briefly distinguish (i) whirl frequency and critical speed and (ii) translatory and conical whirls.

Exercise 4.18 Obtain equations of motion of a rotor-bearing system shown in Figure 4.32 in explicit form. Assume single-plane motion with both translational and tilting motions of the disk; however, ignore the gyroscopic effect. Consider the shaft to be massless and flexible with EI as the flexural rigidity. Let m be the mass of the disk, and I_d be the diametral mass moment of inertia of the disk.

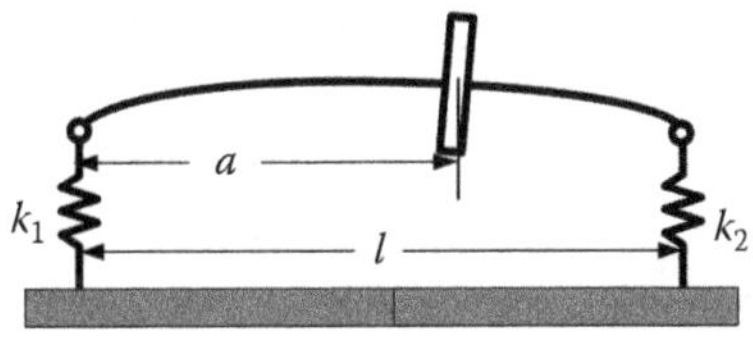

FIGURE 4.32 A flexible rotor on flexible bearings.

Exercise 4.19 Obtain the transverse natural frequencies of a rotor-bearing system as shown in Figure 4.33. Consider the shaft to be rigid. Both bearings have identical torsional springs (of stiffness k_b) to resist tilting (or transverse rotational motion) and identical linear springs (of stiffness k) to resist translational motions. Let $a = 0.7$ m, $b = 0.3$ m, $k_b = 100$ kN-m/rad, and $k = 1$ kN/m. The disk has $m = 5$ kg and $I_d = 0.02$ kg-m^2. Consider only the single-plane motion and *ignore the gyroscopic couple effect.*

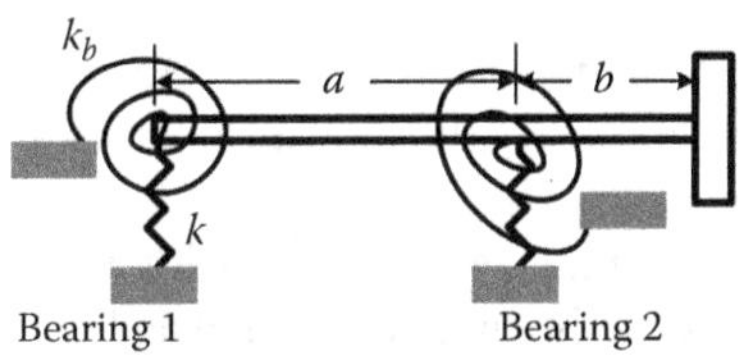

FIGURE 4.33 A rigid rotor mounted on flexible supports.

Exercise 4.20 Derive for transverse vibrations the equations of motion (in a single-plane motion only) of a rotor-coupling system (Figure 4.34) that consists of two identical Jeffcott rotors connected by a coil spring (stiffness, k_φ, in the plane shown). The coil spring resists the relative slopes of the shafts near the bearings. Let the shaft have a flexural rigidity of EI, the length of each shaft is l, and each disk has mass m (= P/g). Ignore the effect of the overhung part of the shaft at the coupling, and consider the masses to be symmetrically placed in each rotor. Figure 4.35 gives the basic relations of the load and deflections (linear and slope). State assumptions in deriving equations of motion. Obtain the natural frequencies of the rotor system. Take $l = 0.6$ m, $d = 0.01$ m, $\rho = 7800$ kg/m^3, $E = 2.1 \times 10^{11}$ N/m^2, and disk mass, $m = 2$ kg.

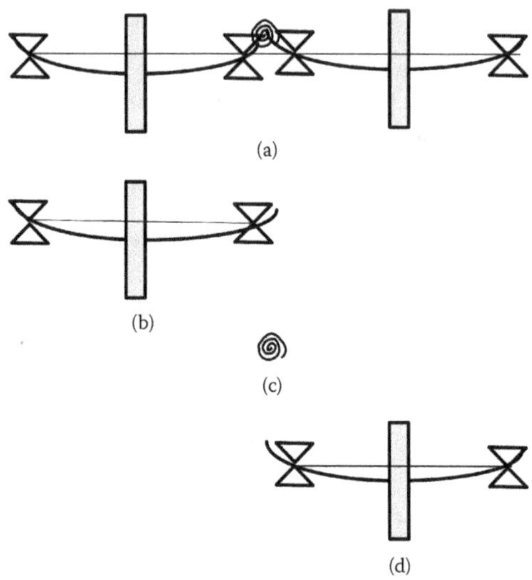

FIGURE 4.34 A rotor-coupling system; (a) two Jeffcott rotors connected by a coupling; (b) rotor 1; (c) the coupling (coil spring); (d) rotor 2.

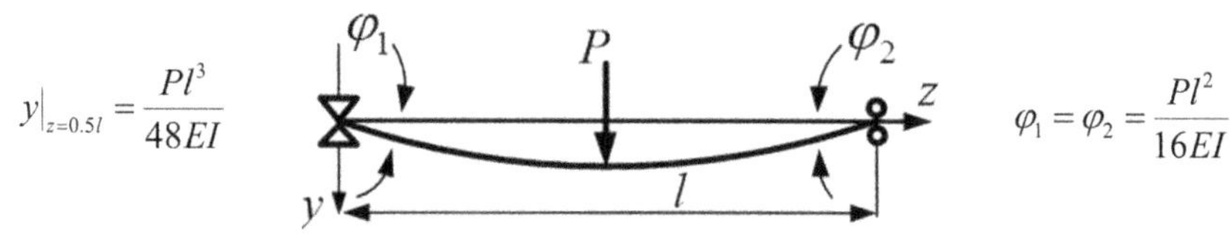

FIGURE 4.35 A simply supported shaft with a concentrated load P at the midspan.

Exercise 4.21 Obtain equations of motion of a two-spool rotor system (two concentric shafts connected by a bearing at annulus space with a stiffness of k_i) modeled as shown in Figure 4.36. Derive equations of motion and write in matrix form in terms of the chosen generalized coordinates. Shaft 1 is supported on two bearings at ends that are connected with the rigid foundation. Shaft 2 is supported by bearings, one connected with the rigid foundation and the other to Shaft 1 at its midspan by a spring of stiffness k_i. Consider one-plane motion of the shafts. Treat the shafts as uniform rigid bars with the transverse translational and rotational motions only. The spring stiffness of the bearing between the two shafts has a stiffness of k_i and the other support (bearing) springs have a stiffness of k. Masses of the shafts are m_1 and m_2, and the diametral mass moments of inertia are I_{d1} and I_{d2}. The lengths of the shafts are l_1 and l_2 with $l_1 = 2\,l_2$, and the radii are r_1 and r_2.

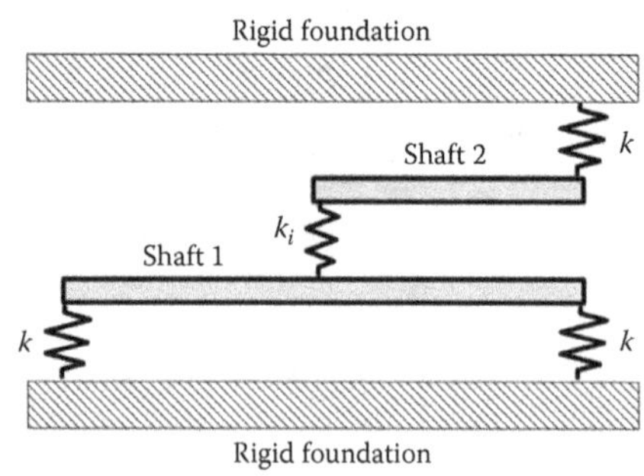

FIGURE 4.36 A two-spool rotor model.

Exercise 4.22 Obtain the equations of motion in matrix form of the system as shown in Figure 4.37. Consider the intermediate bearing is at the midspan of the shaft. The shaft is rigid and massless. *Consider single-plane motion*. The disk has $m=2\,\text{kg}$ and $I_d=0.01$ kg-m^2. The total length of the shaft is 1.0 m, and the disk is located at 0.25 m from the left bearing. All three bearings are identical and have the following properties: $k_{xx}=2.1$ MN/m, $k_{yy}=1.5$ MN/m, $k_{xy}=0.0$ MN/m, $k_{yx=}00$ MN/m, $c_{xx}=200$ kN-s/m, $c_{yy}=200$ kN-s/m, $c_{xy}=0$ kN-s/m, and $c_{yx}=0$ kN-s/m. Obtain critical speeds of the rotor-bearing system.

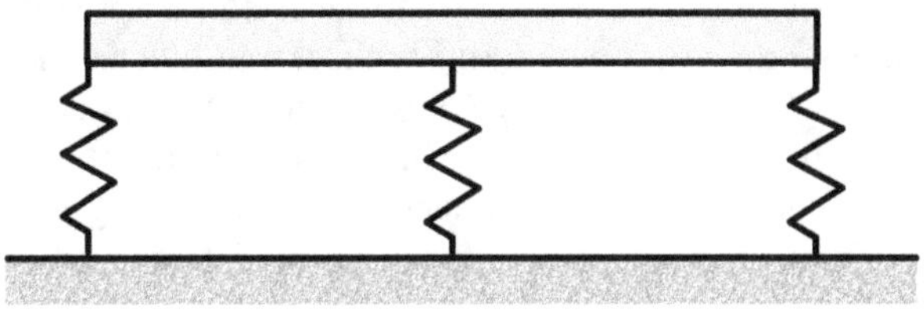

FIGURE 4.37 A rigid shaft with a disk supported by three flexible bearings.

Exercise 4.23 Choose a single correct answer from the multiple choice questions:

i. A rigid long rotor supported on flexible anisotropic bearings can have how many transverse natural frequencies?
   A. 1
   B. 2
   C. 3
   D. 4
   E. more than 4

ii. A rigid long rotor supported on flexible anisotropic bearings can have a reversal of the orbit direction as the spin speed of the rotor is increased.
   A. true
   B. false

iii. A symmetric rigid rotor mounted on fluid-film bearings would have coupling of motions in
   A. only translational displacements (x, y)
   B. only rotational displacements (φ_x, φ_y)
   C. between the translational and rotational displacements (x and φ_y) and/or (y and φ_x)
   D. both (a) and (b)

iv. A flexible rotor (e.g., a Jeffcott rotor with an offset disk) mounted on rigid bearings would have coupling of motions in
   A. only translational displacements (x, y)
   B. only rotational displacements (φ_x, φ_y)
   C. between the translational and rotational displacements (x and φ_y) and (y and φ_x)
   D. both (A) and (B)

v. A flexible rotor (e.g., a Jeffcott rotor with a disk at the midspan) mounted on rigid bearings would have coupling of motions in
   A. only translational displacements (x, y)
   B. only rotational displacements (φ_x, φ_y)
   C. between the translational and rotational displacements (x and φ_y) and/or (y and φ_x)
   D. none of the displacement would be coupled

vi. A flexible rotor (e.g., a Jeffcott rotor with an offset disk) mounted on fluid-film bearings would have coupling of motions in
   A. only translational displacements (x, y)
   B. only rotational displacements (φ_x, φ_y)
   C. between the translational and rotational displacements (x and φ_y) and/or (y and φ_x)
   D. all translational and rotational displacements would be coupled

vii. Obtain the transverse natural frequency of the rotor system shown in Figure 4.38. Consider the shaft to be rigid. It is assumed that it oscillates (processes) about its center of gravity while whirling (i.e., pure tilting without translational motion). The stiffness of each bearing is k_b, and the distance between the bearings is l. Let us assume the center of gravity lies from the right bearing at one quarter of l. The diametral mass moment of inertia of the rotor is I_d. Dimensions of the bearing span and the overhang portion are not required since it is assumed to oscillate about CG. Ignore the gyroscopic effects.

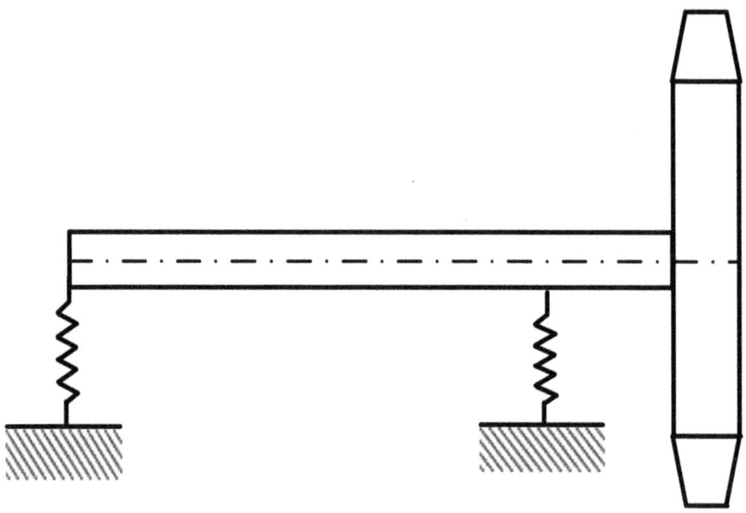

FIGURE 4.38 Rigid rotor mounted on flexible bearings.

   A. $\sqrt{\dfrac{5l^2k_b}{8I_d}}$

   B. $\sqrt{\dfrac{8l^2k_b}{5I_d}}$

   C. $\sqrt{\dfrac{l^2k_b}{2I_d}}$

   D. $\sqrt{\dfrac{2l^2k_b}{I_d}}$

viii. A long rigid rotor with mass m is mounted on symmetrical anisotropic flexible bearings with k_x and k_y as the effective stiffness in two orthogonal directions (with $k_y > k_x$). For a rotor speed of $\sqrt{k_x/m} < \omega < \sqrt{k_y/m}$, the direction of pure translational whirl due to the unbalance would be
   A. synchronous

B. antisynchronous
C. asynchronous (forward)
D. asynchronous (backward)

ix. A rotor mounted on anisotropic bearings has an elliptical orbit, with the major axis tilted with respect to the horizontal or vertical axis. This tilt of the orbit is mainly due to
A. the difference in the direct stiffness coefficients in the two orthogonal directions
B. the cross-coupled damping coefficients
C. the cross-coupled stiffness coefficients
D. the difference in the direct damping coefficients in the two orthogonal directions

x. A rigid symmetrical rotor mounted on isotropic bearings can have how many transverse critical speeds?
A. 1
B. 2
C. 3
D. 4

xi. For a rigid rotor supported on flexible bearings, when it is undergoing conical whirl, then two ends of the shaft would have
A. the same direction of whirl
B. opposite directions of whirl
C. a direction dependent upon whether it is synchronous or antisynchronous whirl
D. a direction dependent upon the asynchronous whirl condition only

xii. The whirling of the shaft shown in Figure 4.39 (the black dot represents the center of gravity of the disk) represents the case of

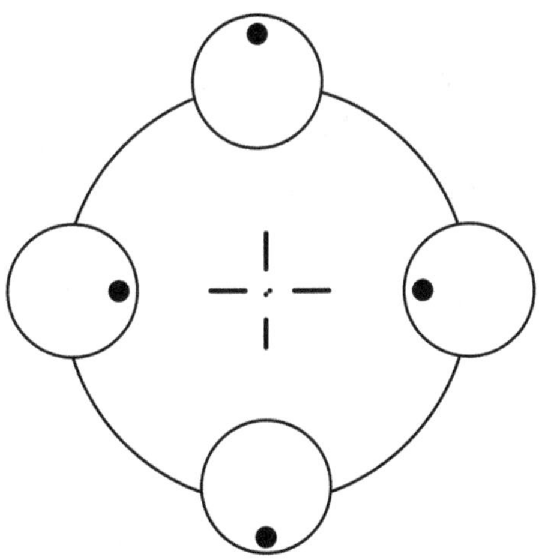

FIGURE 4.39 A shaft in the orbital motion.

A. forward asynchronous whirl
B. synchronous whirl
C. backward asynchronous whirl
D. antisynchronous whirl

xiii. A rigid long rotor mounted on flexible anisotropic springs (with no cross-coupled terms and no damping) would have how many frequency bands in which it has synchronous backward whirls?
A. 0
B. 1
C. 2
D. 3

xiv. A rigid tapered shaft rotating about its geometrical axis (consider the center of the rigid shaft and its center of gravity at the same location, i.e., the center of gravity axis and geometrical axis to be collinear and ignore the gravity effect) will have
A. unbalance force
B. unbalance moment
C. both unbalance force and moment
D. no unbalance force or moment

xv. In the conical whirl of a rigid long rotor mounted on anisotropic bearings, if the rotor is rotating below the first conical whirl critical speed, then the two ends of the rotor will have
A. synchronous whirl
B. asynchronous whirl
C. antisynchronous whirl
D. none of the above

xvi. For the rotor system shown in Figure 4.40, each sphere has a mass m and a polar mass moment of inertia I_p and is connected by a massless rigid shaft of length 2 l. The bearing supports are identical and have a stiffness of k. The natural frequency of the system in pure conical whirl would be

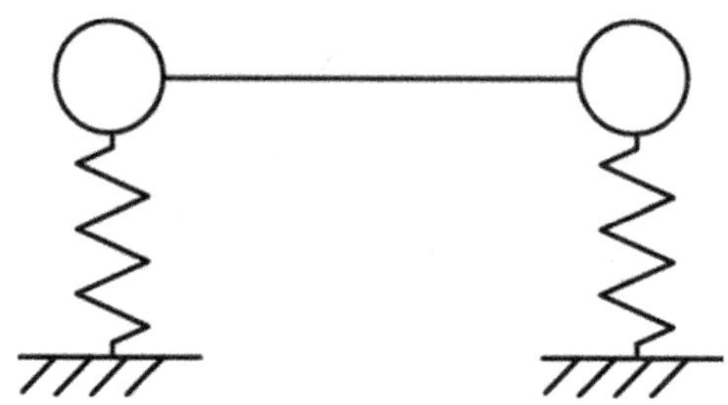

FIGURE 4.40 A rigid rotor supported on flexible bearings.

A. $\sqrt{\dfrac{kl^2}{I_p + ml^2}}$

B. $\sqrt{\dfrac{2kl^2}{I_p + 2ml^2}}$

C. $\sqrt{\dfrac{kl^2}{I_p + 2ml^2}}$

D. $\sqrt{\dfrac{kl^2}{2I_p + 2ml^2}}$

xvii. For a simple rotor-bearing system with a disk at the center of the shaft, the left portion of the shaft stiffness (i.e., half of the shaft transverse stiffness) is assumed to be connected to the left bearing stiffness in series, and similarly in the right half. Then these stiffnesses (left and right sides) so obtained are connected in parallel to the disk. The equivalent stiffness the disk experiences from the shaft as well as the bearing will be

A. $k_{eq} = \dfrac{2}{2/k_{shaft} + 1/k_{bearing}}$

B. $k_{eq} = \dfrac{1}{2/k_{shaft} + 1/k_{bearing}}$

C. $k_{eq} = \dfrac{2}{1/k_{shaft} + 1/k_{bearing}}$

D. $k_{eq} = \dfrac{1}{1/k_{shaft} + 1/k_{bearing}}$

xviii. A rigid rotor is mounted at the ends by springs of stiffness $k = 1$ kN/m each; the rotor has a length of 1 m and a diameter of 3 cm. The shaft material density is 7800 kg/m^3. The pure rotational transverse mode natural frequency (in rad/s) of the rotor-bearing system would be
A. 24
B. 27
C. 31
D. 33

xix. For a long rigid rotor mounted on anisotropic undamped bearings, the path of a point on the shaft during whirling would be
A. elliptical
B. circular
C. spiral
D. a straight line

xx. A flexible rotor mounted on a flexible foundation would have critical speeds that are what compared to a rigid foundation?
A. less
B. more
C. the same
D. more or less depending upon the flexibility of the rotor and foundation

xxi. The pure conical whirl built in a long rigid rotor supported by flexible bearings is due to
A. pure radial unbalance
B. pure axial unbalance
C. both radial and axial unbalances
D. neither radial nor axial unbalance

xxii. Because of the bearings in a rotor system, its critical speed would
A. increase
B. decrease
C. remain the same
D. either increase or decrease

xxiii. Indicate the antisynchronous conical whirl in Figure 4.41.

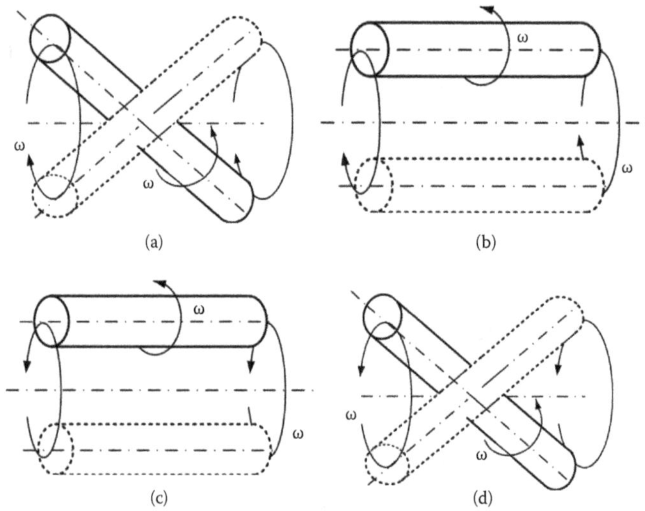

FIGURE 4.41 Whirling of a rigid rotor mounted on flexible bearings.

A. option A
B. option B
C. option C
D. option D

xxiv. A flexible long rotor supported on flexible anisotropic bearings can have how many transverse natural frequencies?
A. 1
B. 2
C. 3
D. 4
E. infinite

xxv. A rigid long rotor supported on flexible anisotropic bearings can have a reversal of the orbit direction as the spin speed of the rotor is increased.
A. true
B. false

xxvi. In an unbalanced rigid long rotor supported on anisotropic flexible bearings with k_x and k_y stiffness of each bearing, no other bearing coefficients are present, and the gyroscopic effect is ignored. For a pure translatory motion, the whirl direction between two critical speeds of the system will be
A. backward but non-synchronous
B. forward but non-synchronous
C. synchronous
D. antisynchronous

xxvii. For a rigid long rotor supported on anisotropic flexible bearings, how many critical speeds related to transverse vibration will be present?
A. 1
B. 2

C. 3
D. 4

xxviii. For a long uniform rigid rotor supported on bearings, each has eight linearized stiffness and damping coefficients. The coupling will be present in
A. all four DOFs
B. only in translatory motion
C. only in rotational motion
D. between translatory motions and between rotational motions

xxix. For a long non-unform rigid rotor supported on bearings, each has eight linearized stiffness and damping coefficients. The coupling will be present in
A. all four DOFs
B. only in translatory motion
C. only in rotational motion
D. between translatory motions and between rotational motions

xxx. For an unbalanced non-uniform rotor with anisotropic bearings, the shape of the shaft center orbit will be (with the y-axis in the vertical direction and the x-axis in the horizontal direction)
A. elliptical with major and minor axes along x and y axes
B. circle
C. elliptical with major and minor axes inclined with x and y axes
D. non-elliptical

xxxi. For a symmetrical long rigid rotor supported by linear and angular springs, the natural frequency in pure tilting will be (with k and k_a are the linear and angular (tilt) direct stiffness, respectively. L is the length of the rigid rotor, m is the mass, and I_d is the diametral mass moment of inertia of the rotor about the center of gravity)

A. $\sqrt{\dfrac{\left(0.5kL^2+2k_a\right)}{I_d}}$

B. $\sqrt{\dfrac{\left(0.5kL^2+k_a\right)}{I_d}}$

C. $\sqrt{\dfrac{\left(kL^2+2k_a\right)}{I_d}}$

D. $\sqrt{\dfrac{\left(0.5kL^2+k_a\right)}{I_d+mL^2}}$

xxxii. For a long symmetrical rotor of the mass, m, diametral mass moment of inertia, I_d, and length L, is supported on anisotropic bearings with each having a stiffness of k_h and k_v in the horizontal and vertical directions, respectively, with $k_v > k_h$. The system second natural frequency is the same as the third natural frequency; the condition for this to happen will be (it is to be noted that always natural frequencies are arranged in ascending order)

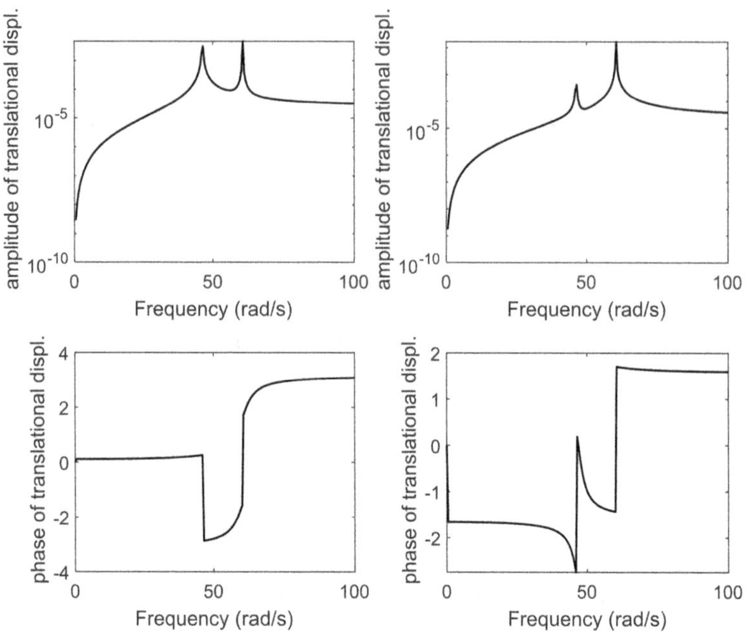

FIGURE 4.42 Displacement amplitude and phase versus rotor spin speed plots: (a) horizontal direction and (b) vertical direction.

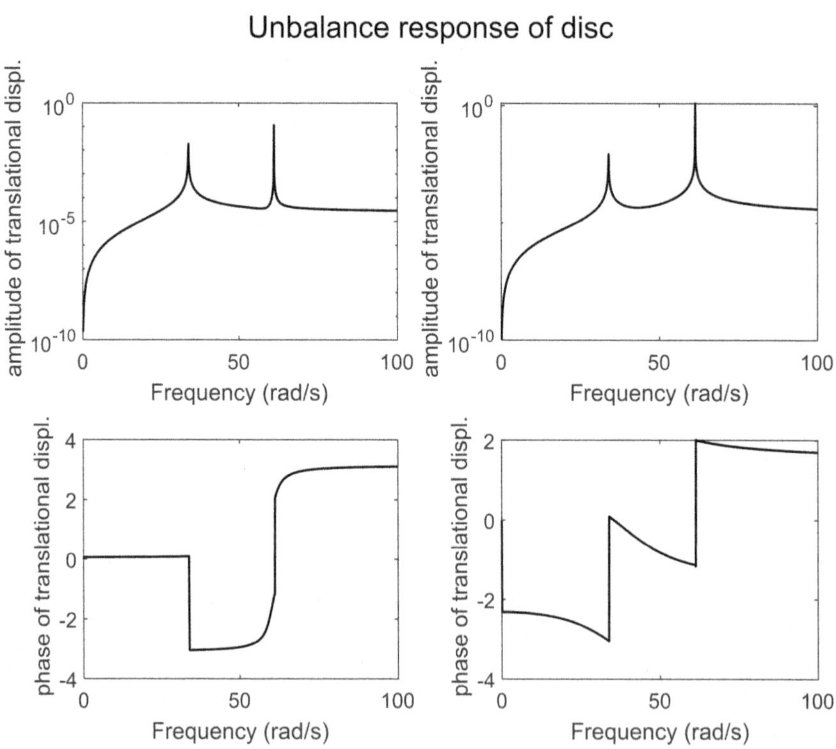

FIGURE 4.43 Translational displacement amplitude and phase versus rotor spin speed plots: (a) horizontal direction and (b) vertical direction.

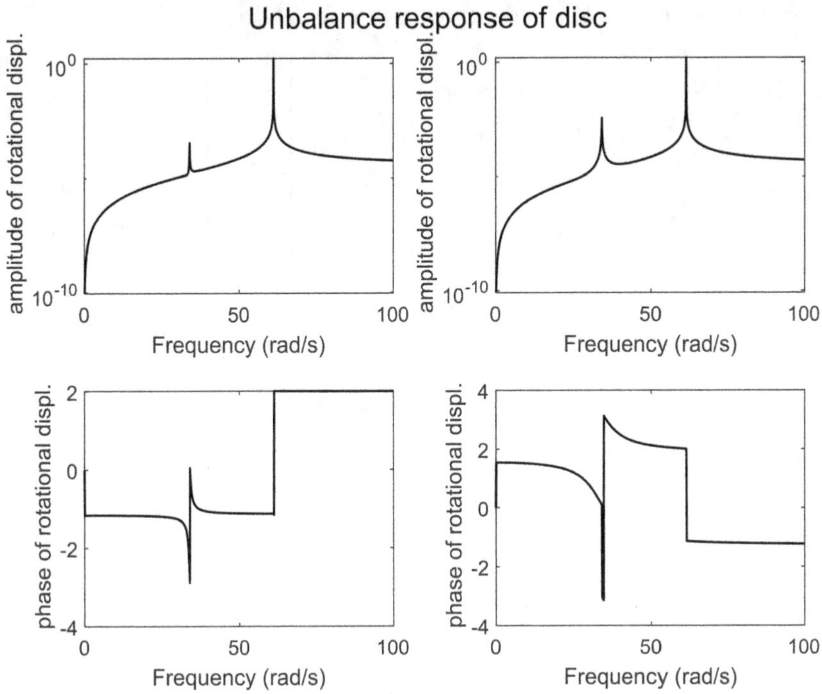

FIGURE 4.44 Rotational displacement amplitude and phase versus rotor spin speed plots: (a) horizontal direction and (b) vertical direction.

A. $\dfrac{k_v}{k_h} = \dfrac{mL^2}{4I_d}$

B. $\dfrac{k_v}{k_h} = \dfrac{mL^2}{2I_d}$

C. $\dfrac{k_v}{k_h} = \dfrac{mL^2}{I_d}$

D. $\dfrac{k_v}{k_h} = \dfrac{mL^2}{8I_d}$

ANSWERS TO SELECTED EXERCISE PROBLEMS

4.1 18.48 rad/s and 76.54 rad/s
4.2 90.34 rad/s and 154.18 rad/s
4.3 46.31 rad/s and 60.46 rad/s
4.4 46.31 rad/s and 60.46 rad/s. Refer Figure 4.42.
4.5 Refer Figures 4.43 and 4.44.
4.6 Refer to Section 18.8 and Example 18.3
The equation of motion of the two offset disks rigid rotor-flexible bearings can be written as

$$\mathbf{M}\ddot{\mathbf{q}}(t) + (\mathbf{C} - \omega\mathbf{G})\dot{\mathbf{q}}(t) + \mathbf{K}\mathbf{q}(t) = \mathbf{f}_{unb}$$

where the mass matrix **M** and the gyroscopic matrix **G** are

$$\mathbf{M}=\begin{bmatrix} \left(m\bar{l}_2^2+i_t\right) & 0 & \left(m\bar{l}_1\bar{l}_2-i_t\right) & 0 \\ 0 & \left(m\bar{l}_2^2+i_t\right) & 0 & \left(m\bar{l}_1\bar{l}_2-i_t\right) \\ \left(m\bar{l}_1\bar{l}_2-i_t\right) & 0 & \left(m\bar{l}_1^2+i_t\right) & 0 \\ 0 & \left(m\bar{l}_1\bar{l}_2-i_t\right) & 0 & \left(m\bar{l}_1^2+i_t\right) \end{bmatrix};\quad \mathbf{G}=\begin{bmatrix} 0 & i_p & 0 & -i_p \\ -i_p & 0 & i_p & 0 \\ 0 & -i_p & 0 & i_p \\ i_p & 0 & -i_p & 0 \end{bmatrix}$$

$$i_t=\frac{I_t}{L^2};\quad i_P=\left(\frac{I_P}{L^2}\right);\quad \bar{l}_1=\frac{L_1}{L};\quad \bar{l}_2=\frac{L_2}{L}$$

The stiffness and damping matrices of bearings are

$$\mathbf{K}=\begin{bmatrix} k_{xx1} & k_{xy1} & 0 & 0 \\ k_{yx1} & k_{yy1} & 0 & 0 \\ 0 & 0 & k_{xx2} & k_{xy2} \\ 0 & 0 & k_{yx2} & k_{yy2} \end{bmatrix};\quad \mathbf{C}=\begin{bmatrix} c_{xx1} & c_{xy1} & 0 & 0 \\ c_{yx1} & c_{yy1} & 0 & 0 \\ 0 & 0 & c_{xx2} & c_{xy2} \\ 0 & 0 & c_{yx2} & c_{yy2} \end{bmatrix}$$

The unbalance force vector is

$$\mathbf{f}_{unb}=\begin{Bmatrix} u_1\omega^2\cos\left(\omega t+\phi_1\right)\left(\bar{l}_2+\bar{r}_1\right)+u_2\omega^2\cos\left(\omega t+\phi_2\right)\left(\bar{l}_2-\bar{r}_2\right) \\ u_1\omega^2\sin\left(\omega t+\phi_1\right)\left(\bar{l}_2+\bar{r}_1\right)+u_2\omega^2\sin\left(\omega t+\phi_2\right)\left(\bar{l}_2-\bar{r}_2\right) \\ u_1\omega^2\cos\left(\omega t+\phi_1\right)\left(\bar{l}_1-\bar{r}_1\right)+u_2\omega^2\cos\left(\omega t+\phi_2\right)\left(\bar{l}_1+\bar{r}_2\right) \\ u_1\omega^2\sin\left(\omega t+\phi_1\right)\left(\bar{l}_1-\bar{r}_1\right)+u_2\omega^2\sin\left(\omega t+\phi_2\right)\left(\bar{l}_1+\bar{r}_2\right) \end{Bmatrix}$$

$$\bar{r}_1=\frac{R_1}{L};\quad \bar{r}_2=\frac{R_2}{L}$$

The displacement vector is

$$\mathbf{q}(t)=\left\{ x_1 \;\; y_1 \;\; x_2 \;\; y_2 \right\}^T$$

4.7 Refer to Chapter 3 and Exercise 4.6

4.8

i. $$\begin{pmatrix} m & 0 \\ 0 & I_G \end{pmatrix}\begin{Bmatrix} \ddot{x}_G \\ \ddot{\varphi}_z \end{Bmatrix}+\begin{bmatrix} 2k & -2ke \\ -2ke & \dfrac{l^2}{2}+2e^2 \end{bmatrix}\begin{Bmatrix} x_G \\ \varphi_z \end{Bmatrix}=\begin{Bmatrix} 0 \\ 0 \end{Bmatrix}$$

ii. $$\begin{pmatrix} m & 0 \\ 0 & I_G \end{pmatrix}\begin{Bmatrix} \ddot{x}_G \\ \ddot{\varphi}_z \end{Bmatrix}+\begin{bmatrix} k_A+k_B & -k_Aa-k_Ba \\ -k_Aa-k_Ba & k_A(l_1+a^2)^2+k_B(l_2+a^2)^2 \end{bmatrix}\begin{Bmatrix} x_G \\ \varphi_z \end{Bmatrix}=\begin{Bmatrix} 0 \\ 0 \end{Bmatrix}$$

iii. $$\begin{pmatrix} m & 0 \\ 0 & I_G \end{pmatrix}\begin{Bmatrix} \ddot{x}_A \\ \ddot{\varphi}_z \end{Bmatrix}+\begin{bmatrix} 2k & -2ke \\ -2ke & \dfrac{l^2}{2}+2e^2 \end{bmatrix}\begin{Bmatrix} x_A \\ \varphi_z \end{Bmatrix}=\begin{Bmatrix} 0 \\ 0 \end{Bmatrix}$$

iv. $$\begin{bmatrix} m-\left(\frac{1}{2}+e\right)\left(\frac{1}{l}\right) & m\left(\frac{1}{2}+e\right)\left(\frac{1}{l}\right) \\ m\left(\frac{1}{2}+e\right)\left(\frac{1}{l}\right) & m-\left(\frac{1}{2}+e\right)\left(\frac{1}{l}\right) \end{bmatrix} \begin{Bmatrix} \ddot{x}_A \\ \ddot{x}_B \end{Bmatrix} = \begin{Bmatrix} 0 \\ 0 \end{Bmatrix}$$

v. $$\begin{bmatrix} m & 0 \\ 0 & I+me^2 \end{bmatrix} \begin{Bmatrix} \ddot{x}_C \\ \ddot{\varphi}_z \end{Bmatrix} + \begin{bmatrix} k_A+k_B & -\frac{l}{2}(k_A-k_B) \\ -\frac{1}{2}(k_A-k_B) & \left(\frac{l}{2}\right)^2(k_A+k_B) \end{bmatrix} \begin{Bmatrix} x_C \\ \varphi_z \end{Bmatrix} = \begin{Bmatrix} 0 \\ 0 \end{Bmatrix}$$

4.9 153.6 rad/s and 2091.3 rad/s
4.10 5146.7 rad/s and 4383.8 rad/s
4.11 7658.3 rad/s and 8843.8 rad/s
4.12

$$\left[(m_1+m_2)\,\ddot{y}\right]+4ky-k(l+2a)(\theta_1+\theta_2)=0$$

$$(I_d\ddot{\theta}_1)-ky(l+2a)+k((l+a)^2+a^2)=\theta_1=0$$

$$(I_d\ddot{\theta}_2)-ky(l+2a)+k((l+a)^2+a^2)=\theta_2=0$$

4.13 $$m\ddot{x}+2k_x x=me\omega^2\cos\omega t$$

$$m\ddot{y}+2k_y x=me\omega^2\sin\omega t$$

$$I_d\ddot{\phi}_x+0.5k_x l^2\phi_x=me\omega^2 e_z\sin\omega t$$

$$I_d\ddot{\phi}_y+0.5k_y l^2\phi_y=me\omega^2 e_z\cos\omega t$$

4.14 $$m\ddot{x}+2k_{xx}x+2k_{xy}y+2c_{xx}\dot{x}+2c_{xy}\dot{y}=me\omega^2\cos\omega t$$

$$m\ddot{y}+2k_{yy}y+2k_{yx}x+2c_{yy}\dot{y}+2c_{yx}\dot{x}=me\omega^2\sin\omega t$$

$$I_d\ddot{\varphi}_x+\frac{k_{yy}l^2}{2}\varphi_x+\frac{k_{xy}l^2}{2}\varphi_x+\frac{c_{xx}l^2}{2}\dot{\varphi}_y+\frac{c_{xy}l^2}{2}\dot{\varphi}_x=me\omega^2 e_z\cos\omega t$$

$$I_d\ddot{\varphi}_y+\frac{k_{xx}l^2}{2}\varphi_y+\frac{k_{yx}l^2}{2}\varphi_y+\frac{c_{yy}l^2}{2}\dot{\varphi}_x+\frac{c_{yx}l^2}{2}\dot{\varphi}_y=me\omega^2 e_z\sin\omega t$$

4.15 Refer Figure 4.45.
4.16 14.1375 rad/s and 190.7494 rad/s
4.17 Not applicable
4.18 Not applicable
4.19 20.0 rad/s and 3170.9 rad/s

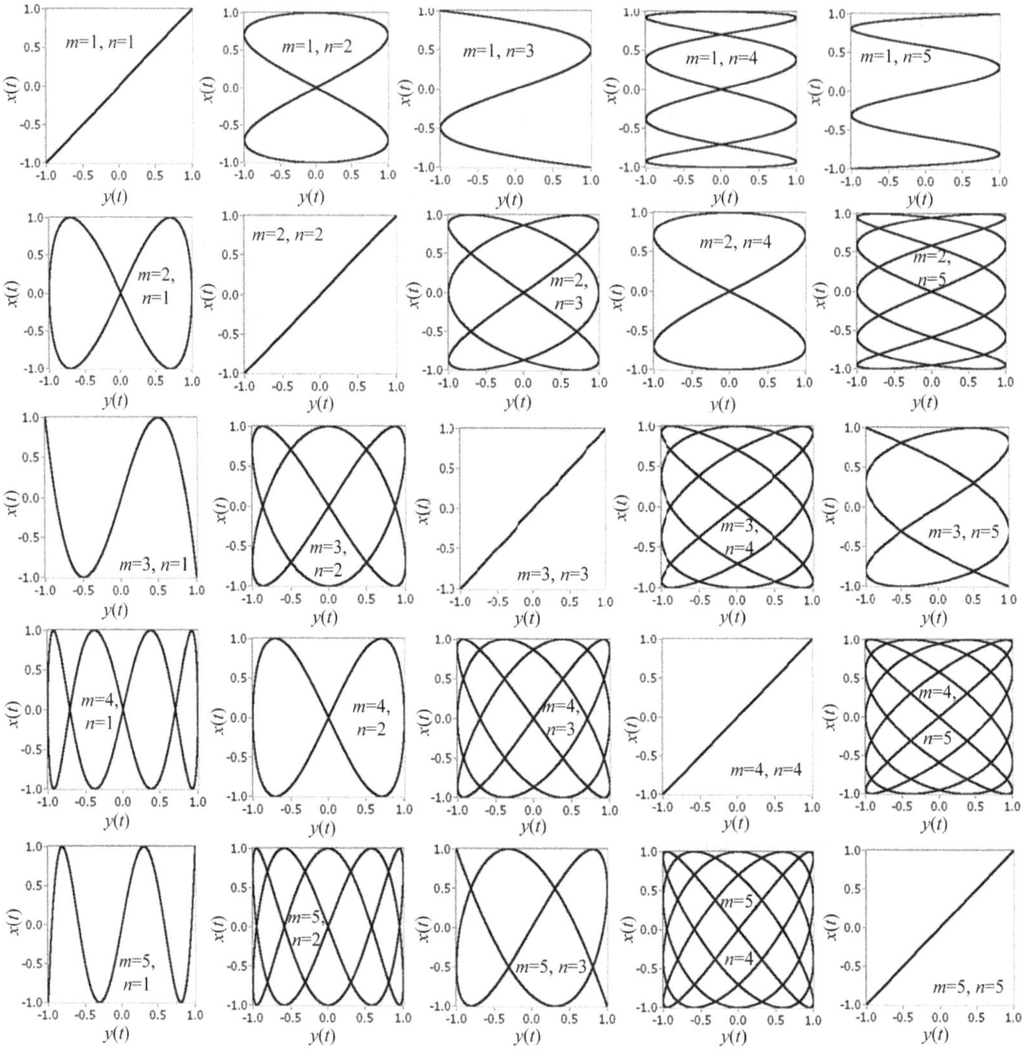

FIGURE 4.45 Lissajous plots for various combinations of coefficients m and n mentioned at the respective top of the plots.

4.20 $$\begin{pmatrix} m & 0 \\ 0 & m \end{pmatrix}\begin{pmatrix} \ddot{y}_1 \\ \ddot{y}_2 \end{pmatrix} + \begin{pmatrix} \dfrac{48EI}{l^3} + \dfrac{k_\varphi}{3a} & \dfrac{-k_\varphi}{3a} \\ \dfrac{-k_\varphi}{3l-3a} & \dfrac{48EI}{l^3} + \dfrac{k_\varphi}{3l-3a} \end{pmatrix}\begin{pmatrix} y_1 \\ y_2 \end{pmatrix} = \begin{pmatrix} 0 \\ 0 \end{pmatrix}$$

4.21 $$\begin{pmatrix} m_1 & 0 & 0 & 0 \\ 0 & I_{d_1} & 0 & 0 \\ 0 & 0 & m_2 & 0 \\ 0 & 0 & 0 & I_{d_2} \end{pmatrix}\begin{Bmatrix} \ddot{y}_1 \\ \ddot{\varphi}_1 \\ \ddot{y}_2 \\ \ddot{\varphi}_2 \end{Bmatrix} + \begin{bmatrix} 2k+k_i & 0 & -k_i & k_il_2 \\ 0 & 2k_il_1^2 & 0 & 0 \\ -k_i & 0 & k+k & (k-k_i)l_2 \\ k_il_2 & 0 & (k-k_i)l_2 & (k+k_i)l_2^2 \end{bmatrix}\begin{Bmatrix} y_1 \\ \varphi_1 \\ y_2 \\ \varphi_2 \end{Bmatrix} = \mathbf{0}$$

4.22 9060.4 rad/s and 64711.0 rad/s

4.23

(i) D	(ii) A	(iii) D	(iv) C	(v) D	(vi) D
(vii) A	(viii) B	(ix) C	(x) D	(xi) C	(xii) D
(xiii) C	(xiv) D	(xv) A	(xvi) A	(xvii) A	(xviii) D
(xix) A	(xx) A	(xxi) B	(xxii) B	(xxiii) A	(xxiv) E
(xxv) A	(xxvi) D	(xxvii) D	(xxviii) D	(xxix) A	(xxx) C

REFERENCES

Dworski, J., 1964, High speed rotor suspension formed by fully floating hydrodynamic radial and thrust bearings, *Journal of Engineering for Power*, *86*, 149–160.

Edwards, S., Lees, A.W., and Friswell, M.I., 2000, Experimental identification of excitation and support parameters of a flexible rotor-bearing-foundation system from a single run-down, *Journal of Sound and Vibration*, *232*(*5*), 963–992.

Ewins, D.J., 2000, *Modal Testing: Theory, Practice and Application*, 2nd ed. Hertfordshire, UK: Research Studies Press Ltd.

Friswell, M.I. and Mottershead, J.E., 1995, *Finite Element Model Updating in Structural Dynamics*. Dodrecht: Kluwer Academic Publishers.

Gasch, R., 1976, Vibration of large turborotors in fluid-film bearing on an elastic foundation, *Journal of Sound and Vibration*, *47*, 53–73.

Gunter, E.J., 1967, The influence of internal friction on the stability of high speed rotors, *Journal of Engineering for Industry*, *89*, 683–688.

Gunter, E.J., 1970, Influence of flexibly mounted rolling element bearing on rotor response, Part I - linear analysis, *Journal of Lubrication Technology*, *92*, 59–75.

Kang, Y., Chang, Y.-P., Tsai, J.-W., Mu, L.-H., and Chang, Y.-F., 2000, An investigation in stiffness effects on dynamics of rotor-bearing-foundation systems, *Journal of Sound and Vibration*, *231*(*2*), 343–374.

Kirk, R.G. and Gunter, E.J., 1972, The effect of support flexibility and damping on the synchronous response of a single-mass flexible rotor, *ASME Journal of Engineering for Industries*, *94*, 221–232.

Krämer, E., 1993, *Dynamics of Rotors and Foundations*. New York, NY: Springer-Verlag.

Lal, M., and Tiwari, R., 2012, Multi-fault identification in simple rotor-bearing-coupling systems based on forced response measurements, *Mechanism and Machine Theory*, *51*, 87–109.

Lees, A.W., 1988, The Least Squares Method Applied to Identified Rotor/Foundation Parameters. *IMechE Conference on Vibrations in Rotating Machinery*, Paper C306/88, UK, September 13–15, pp. 209–216.

Lees, A.W. and Simpson, I.C., 1983, The Dynamics of Turbo-Alternator Foundations. *IMechE Conference on Vibrations in Rotating Machinery*, Paper C6/83, London, UK, September 7–9, pp. 37–44.

Lund, J.W., 1965, The stability of an elastic rotor in journal bearings with flexible, damped supports, *Journal of Applied Mechanics*, *32*, 911–920.

Lund, J.W., and Sternlicht, B., 1962, Rotor-bearing dynamics with emphasis on attenuation, *Journal of Basic Engineering*, *84*, 491–502.

Sinha, J.K., Lees, A.W., and Friswell, M.I., 2002, The identification of the unbalance and the foundation model of a flexible rotating machine from a single run down, *Mechanical Systems and Signal Processing*, *16*(*2–3*), 255–271.

Smart, M.G., Friswell, M.I., and Lees, A.W., 2000, Estimating turbogenerator foundation parameters—model selection and regularisation, *Proceedings of the Royal Society of London, Series A: Mathematical, Physical and Engineering Sciences*, *456*, 1583–1607.

Smith, D.M., 1933, The motion of a rotor carried by a flexible shaft in flexible bearing, *Proceedings of the Royal Society, Series A*, *142*, 92.

Srinivas, S.R, Tiwari, R., Babum, C.K., 2021, Modeling, analysis, and identification of parallel and angular misalignments in a coupled rotor-bearing-active magnetic bearing system. *Journal of Dynamic Systems, Measurement, and Control*, *143*(1), 011007.

Srinivas, S.R, Tiwari, R., Kannababu, C., 2019, Model based analysis and identification of multiple fault parameters in coupled rotor systems with offset discs in the presence of angular misalignment and integrated with an active magnetic bearing. *Journal of Sound and Vibration*, *450*, 109–140.

Stephenson, R.W. and Rouch, K.E., 1992, Generating matrices of the foundation structure of a rotor system from test data, *Journal of Sound and Vibration*, *154*, 467–484.

Tiwari, R., 2024, *Simple Rotor Analysis through Tutorial Problems*. Boca Raton, FL: CRC Press.

Vance, J.M., Murphy, B.T., and Tripp, H.A., 1987, Critical speeds of turbomachinery: Computer predictions vs. experimental measurements-Part II: Effect of tilt-pad bearing and foundation dynamics, *ASME Journal of Vibration Acoustics, Stress, Reliability in Design*, *109*, 8–14.

5 Transverse Vibrations of Simple Rotor Systems with Gyroscopic Effects

In Chapter 4, we considered rotor-bearing systems for a single-mass rotor with different levels of complexity in restoring and dissipating forces of supports. We analyzed the rotor system for the transverse (bending), translatory, and rotary motions by considering the respective inertias. However, we ignored an important dynamic behavior of the rotor system, the *gyroscopic effect*, which especially predominates in high-speed heavy rotors. In the present chapter, we shall still be dealing with a single-mass (in the form of a thin disk or long cylinder) rotor system with a slender massless shaft and will come back to the assumption of rigid bearings in transverse directions unless otherwise stated. However, now we shall include the effect of gyroscopic effects and will explore the motion of the rotor for the synchronous as well as the asynchronous whirl. For the present case, we shall analyze the rotor system using two different approaches: first, the *quasi-static analysis* (which gives better physical insight into the effect of gyroscopic effects; however, it can be applied ideally to simple systems only), and second, the *dynamic analysis* (which can be easily extended to multi-DOF systems). An important aspect, which we will observe from the present chapter, is that because of the gyroscopic effect, the whirl natural frequency is dependent on the rotor spin speed. Another interesting phenomenon that can be observed is that the rotor can have forward and backward whirling motions, and the splitting of whirl frequencies takes place. Moreover, for the present case, the distinction between the rotor spin speed, the whirl natural frequency, and the critical speed will be made clearer through a Campbell diagram (Campbell, 1924).

It should be observed that the whirling of the shaft takes place either due to a momentary disturbance of a perfectly balanced rotor or due to the unbalance of a spinning rotor. The former is the free vibration, and the latter is the forced vibration. In the former case, it would whirl at one of its natural frequencies in forward or backward whirl, and in the latter case, at a frequency equal to the spin speed of the rotor due to unbalance (synchronous whirl) or at any other excitation frequency (forward or backward whirl) due to a variety of other fault conditions.

When a relatively large disk (or a rotor with a large polar mass moment of inertia) spins at a very high speed about its longitudinal (polar) axis, then it has a large angular momentum, and the rotors carry a tremendous amount of the rotary power. However, if it has precession (slow or fast) about its transverse (diametral) axes, which is due to the flexibility of the shaft and/or bearings, then it develops a change in the angular momentum due to the change in its direction. This leads to an inertia moment called the *gyroscopic moment*. Basically, the gyroscopic moment develops due to the Coriolis acceleration component, and it will be illustrated presently.

Figure 5.1a and b shows the motion of a disk in a simply supported shaft (i.e., a Jeffcott rotor) when the disk is at the midspan and at an offset position, respectively. For the former case, no gyroscopic effect would be present, because during whirling of the disk (ν is the whirling frequency and ω is the spin speed of the disk), no precession motion takes place about its diameter. For the latter case, the precession of the disk (i.e., transverse angular displacement, φ_x) about its diametral axes takes place along with the spinning about the polar axis. This leads to the gyroscopic moment on the disk, and it depends upon the spin speed, the precession velocity, and the polar mass moment of inertia of the rotor. Rotor systems with a point mass (with a negligible polar mass moment of inertia) and a thin disk (with an appreciable polar mass moment of inertia and the same mass as the point mass)

DOI: 10.1201/9781003661559-5

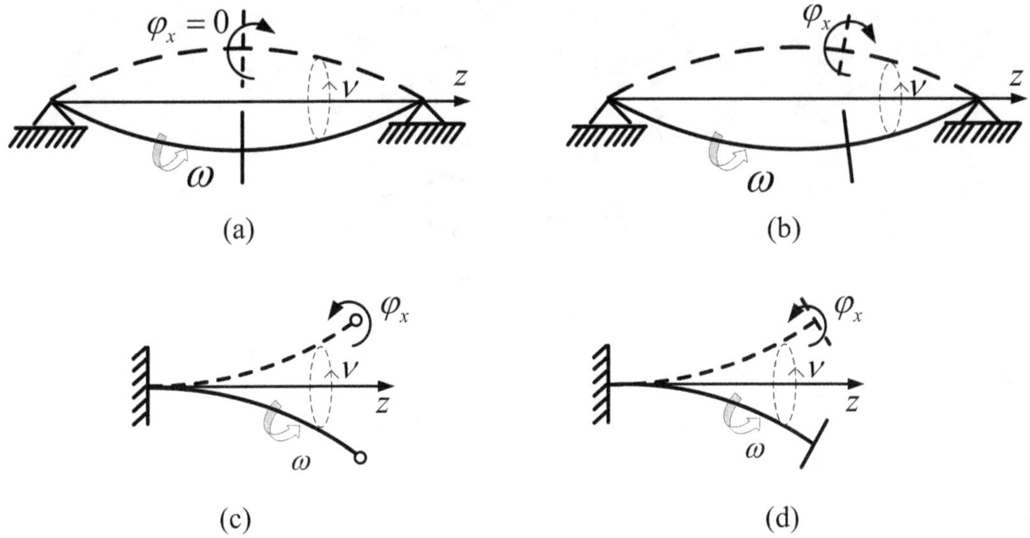

FIGURE 5.1 The spinning and precession (whirling) motions of the disk: (a) a simply supported shaft with a disk at the midspan; (b) a simply supported shaft with a disk away from the midspan; (c) a cantilever shaft with a point disk at the free end; (d) a cantilever shaft with a rigid, thin disk at the free end.

are shown in Figure 5.1c and d, respectively. The critical speeds of these rotor systems will not be the same. This is due to the fact that centrifugal forces of particles of the disk do not lie in one plane during motion and thus form a moment that straightens the shaft, which will be discussed subsequently in more detail. However, in the case of a point mass, the centrifugal forces would not produce any such moment.

Gyroscopic effects and gyroscopes have been a traditional subject, and lots of interesting books have been written on this topic (Gary, 1918; Deimel, 1929; Ferry, 1932; Davidson, 1946; Arnold and Maunder, 1961). However, our main focus would be on the gyroscopic effects on the dynamics of rotors. Let us get acquainted with the basic concepts of gyroscopic moment and related principles and physical properties.

5.1 ANGULAR MOMENTUM

In Figure 5.2, a link *OA* rotates about the *z*-axis in the *x*-*y* plane, and *OB* is the position it takes after an infinitesimal time interval, *dt*. Let $\Delta\theta$ be the infinitesimal angular (rotational) displacement of *OA*; it is the angular displacement vector along the *z*-axis. Similarly, the angular velocity, angular acceleration, and angular momentum are also vector quantities. A change in the magnitude and direction of the angular velocity results in the angular acceleration. The gyroscopic moment can be understood using the principle of angular momentum.

Let a particle of mass *m* be moving with a velocity *v*. From Figure 5.3, the *linear momentum, L*, is defined as

$$L = mv \tag{5.1}$$

The direction and sense of the linear momentum are the same as the linear velocity.

Now, referring to Figure 5.4a in which a point mass *m* revolves in a circle with a radius *r* in a plane, the *angular momentum, H*, is defined as the moment of linear momentum:

$$H = (mv)r = \left(mr^2\right)\omega = I_p\omega \tag{5.2}$$

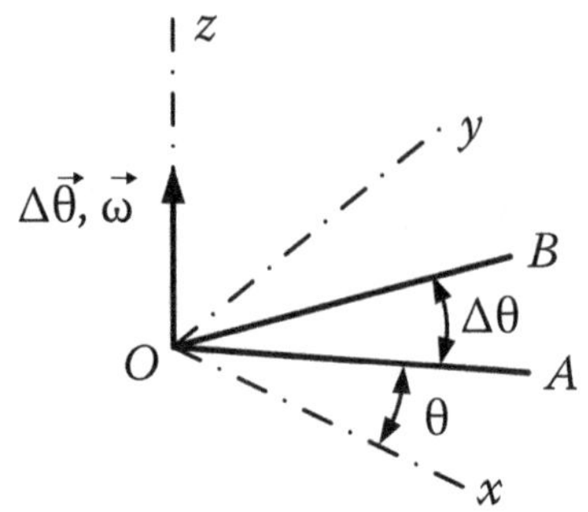

FIGURE 5.2 An angular displacement vector.

FIGURE 5.3 A particle in linear motion.

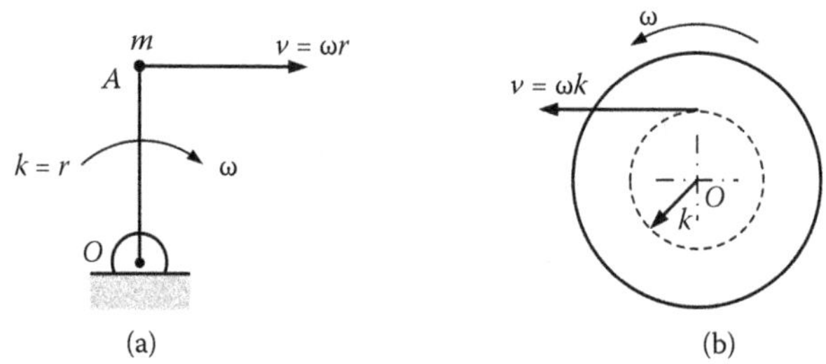

FIGURE 5.4 Angular momentum: (a) a point mass in rotation; (b) a flywheel in rotation.

with

$$v = \omega r \quad \text{and} \quad I_p = mr^2$$

where I_p is the polar mass of inertia of a particle of mass m about its axis of rotation O, and ω is the angular velocity of link OA. The direction of the angular momentum will be the same as the angular velocity.

Referring to Figure 5.4b, in which a flywheel of mass m and of radius of gyration k rotates with an angular speed of ω, the angular momentum of the flywheel is given as

$$H = (mv)k = \left(mk^2\right)\omega = I_p\omega \quad \text{with} \quad v = \omega k \quad \text{and} \quad I_p = mk^2 \tag{5.3}$$

The *radius of gyration* is an imaginary radius at which all the mass of the flywheel is assumed to be concentrated so as to have the same angular momentum as that of the actual flywheel. Now, in the following section, we will investigate the gyroscopic moment in a disk and a propeller of a jet engine.

5.2 GYROSCOPIC MOMENTS IN ROTATING SYSTEMS

In the present section, the concept of the gyroscopic moment will be introduced with the help of simple rotating systems (e.g., disks and propeller blades). You might have studied these basic concepts to some extent in the subject of dynamics of machinery (Wrigley et al., 1969; Bevan, 1984; Mabie and Reinholts, 1987; Rao and Dukkipati, 1995).

5.2.1 Motion of a Rotor Mounted on Two Bearings

Let us assume that a rotor with a flexible massless shaft carries a disk that rotates with angular speed ω and is constrained to move in a vertical (single) plane during whirling with a circular frequency, ν. It is assumed that the constraint does not provide any friction forces during the motion of the shaft in that plane. The rotor is spinning with a constant angular velocity, ω; the angular momentum, H, will be given by $I_p\omega$ along OC as shown in Figure 5.5a. Let x-y-z be the rectangular coordinate system (see Figure 5.5a), where Oz is the spin axis, Ox is the precession axis, and Oy is the gyroscopic moment axis.

Let the disk (or the spin axis) precession angle be $\Delta\varphi$ from the z-axis in the y-z plane as shown in Figure 5.5a. The angular momentum will change from H (i.e., OC) to H' (i.e., OD), which can be written as (see the triangle ΔOCD, which is in the y-z plane)

$$\vec{H'} = \vec{H} + \Delta\vec{H} \tag{5.4}$$

where ΔH is the change in angular momentum (i.e., CD). It is due to a change in the direction of H. From ΔOCD, we have

$$CD = (OC)\Delta\varphi, \quad \Delta H = H(\Delta\varphi), \quad \Delta H = I_p\omega\Delta\varphi \tag{5.5}$$

with

$$H = I_p\omega$$

Noting this expression, now the rate of change of the angular momentum can be written as

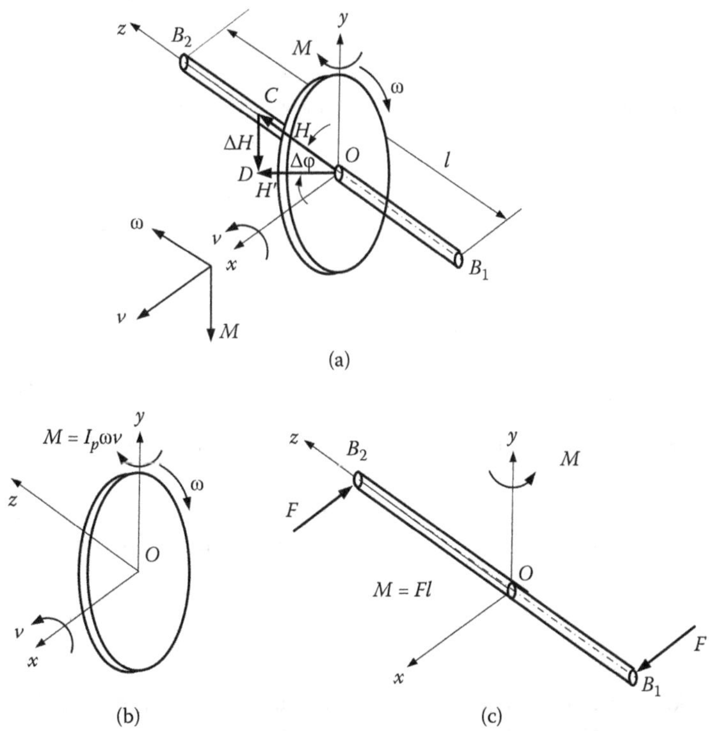

FIGURE 5.5 A rotor mounted on two bearings with a single-plane motion in the y-z plane: (a) a rotor with precession about the x plane; (b) the active gyroscopic couple on the disk; (c) a free body diagram of the shaft.

$$M = \frac{dH}{dt} = \lim_{\Delta t \to 0} I_p \omega \frac{\Delta \varphi}{\Delta t} = I_p \omega \nu = H\nu \tag{5.6}$$

where ν is the uniform angular velocity of precession (or whirl frequency), and M is the gyroscopic moment. The gyroscopic moment will have the same sense and direction as ΔH, that is, $\overrightarrow{CD}$ (in the negative y-axis direction). In the vector form, the equation can be written as

$$\vec{M} = \vec{\nu} \times \vec{H} = I_p\left(\vec{\nu} \times \vec{\omega}\right) \tag{5.7}$$

From the right-hand screw rule, we will get the direction of gyroscopic moment, that is, along the negative y-axis (the clockwise [CW] direction when seen from above; see Figure 5.5a). This is the moment exerted on the disk by the shaft in the negative y-axis direction (by action), and the reaction moment on the shaft by the disk will be opposite (i.e., in the positive y-axis direction). Whenever an axis of rotation or spin axis changes its direction about another orthogonal axis, then a gyroscopic moment acts about the third orthogonal axis. This is an active moment acting on the disk, which means the disk will experience this moment as an inertia moment. The support will also experience the same active moment. In other words, if a moment, M, is applied to a spinning disk (as shown in Figure 5.5a), then precession will take place in an attempt to align the angular momentum vector, $I_p\omega$, with the applied moment vector (i.e., H will try to align itself toward the negative y-axis direction).

If we consider the free body diagram of the disk as shown in Figure 5.5b, the reaction from the shaft onto the disk hub will be the active moment (i.e., in the negative y-axis direction). Hence, the shaft will experience a reactive moment from the disk hub in the opposite direction to the active moment (i.e., in the positive y-axis direction). Let Fl be a couple on the shaft from bearings; then its direction will be as shown in Figure 5.5c (i.e., CW looking from the top). A reactive moment will be experienced by the bearings through the shaft in the opposite direction as the active gyroscopic moment onto the disk (i.e., a couple due to $-F$ forces).

5.2.2 Gyroscopic Moments through the Coriolis Component of Acceleration

Figure 5.6 shows a disk that is spinning with an angular velocity, ω, (in the negative z-axis direction) and a precession angular velocity, ν. Let z and y be the spin and precession axes, respectively. It is assumed here that the disk has a precession only about a single axis (i.e., in the negative y-axis direction). An infinitesimal mass, dm, at point P has coordinates (r, θ) in polar coordinates or (x, y) in rectangular coordinates. From Figure 5.6a, it can be seen that the velocity of point P, that is, ωr, will be perpendicular to OP. The velocity component along the x-axis will be $\omega r \sin\theta = \omega y$ and that along the y-axis is $\omega r \cos\theta = \omega x$ (sign is taken care of in diagrams by the arrow directions; hence, it is not mentioned with magnitudes of velocity vectors). Particle P is in motion along the x-axis (i.e., parallel to the x-axis) and simultaneously it is rotating about the y-axis as shown in Figure 5.6c. Hence, a *Coriolis component* of acceleration, that is, $2\vec{\nu} \times (\vec{\omega} y)$, acts along the positive z-axis direction (Figure 5.6a and b).

Similarly for particle P' (diagonally opposite to particle P), the Coriolis acceleration component will be $2\vec{\nu} \times (\vec{\omega} y)$, and it acts along the negative z-axis direction as shown in Figure 5.6a and b. The force due to the Coriolis component of acceleration of the particle P is given as

$$F = dm\left(2\omega y \nu\right) \tag{5.8}$$

Let us first consider the moment about the x-axis (Figure 5.6b), and due to the particle P, it is $dm\left(2\omega y^2 \nu\right)$; hence, the total moment about the x-axis will be

$$M_{xx} = \int 2\omega\nu y^2 dm = 2\omega\nu I_{xx} = I_p \omega \nu \tag{5.9}$$

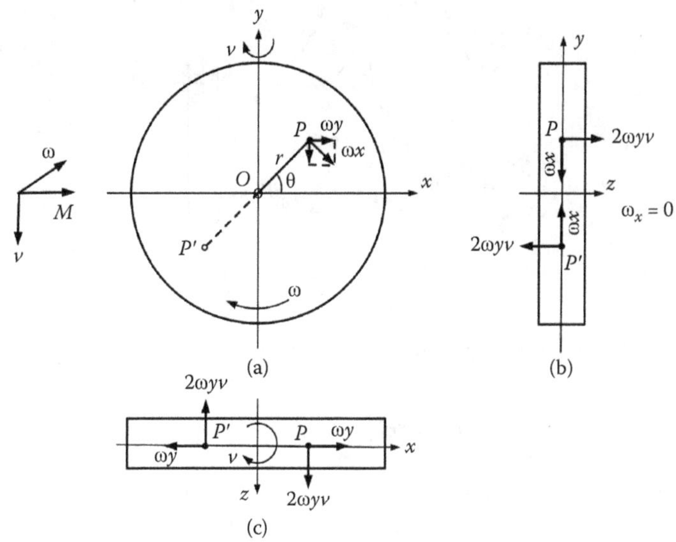

FIGURE 5.6 The gyroscopic moment on a rotating disk in the (a) *x-y* plane, (b) *y-z* plane, and (c) *z-x* plane.

with

$$I_{xx} = \int y^2 dm = \frac{1}{2} I_p \quad \text{(for thin disc)} \tag{5.10}$$

where I_p (or I_{zz}) is the polar moment of inertia of the disk (for the thin disk, we have $I_p = 2I_d$, where I_d is the diametral mass moment of inertia). The component of the gyroscopic moment, M_{xx}, acts along the positive *x*-axis direction (Figure 5.6a and b). Now consider the moment about the *y*-axis (Figure 5.6c), and due to the particle *P*, it is $dm(2\omega xy\nu)$; hence, the total moment about the *y*-axis will be

$$M_{yy} = \int 2\omega\nu xy dm = 2\omega\nu I_{xy} \tag{5.11}$$

with

$$I_{xy} = \int xy dm = 0 \quad \text{(for symmetric disc)} \tag{5.12}$$

It should be noted that if the disk was not symmetric, then the product mass moment of inertia, $I_{xy} \neq 0$, so we would get M_{xx} and M_{yy} both nonzero. Hence, accelerating forces arising out of these Coriolis acceleration components due to the motion of particles in the *x direction* produce a net moment (or a couple), $M = I_p(\vec{\nu} \times \vec{\omega})$, along the positive *x*-axis direction only. There is no Coriolis component of acceleration when we analyze the motion of the particle in the *y* direction because it has no precession about the *x*-axis, $\omega_x = 0$ (i.e., because the single-plane precession is assumed) as shown in Figure 5.6b. It should be noted that the outcome (magnitude and direction) is the same as in the previous section.

5.2.3 Gyroscopic Moments in a Rotating Thin Blade

In Figure 5.7, we have the *z*-axis as the axis of spin and the *y*-axis as the axis of precession. Let ξ and η be the two orthogonal principal axes of the thin rod, with ξ making an angle of θ with the *x*-axis. Due to the Coriolis component of acceleration, the force at a point, *P*, of the mass *dm* is given as

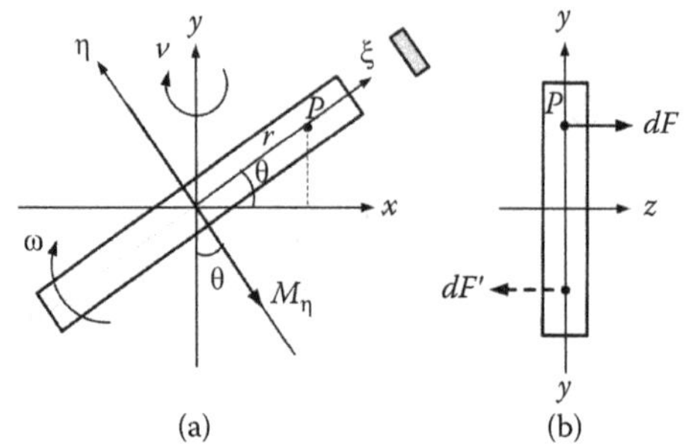

FIGURE 5.7 A thin rod rotating about its centroid axis: (a) x-y plane; (b) y-z plane.

$$dF = dm\{2\nu(\omega r \sin\theta)\} \tag{5.13}$$

which acts along the spin axis (positive z-axis direction in Figure 5.7b). The moment due to this force about the η-axis is given as (with the moment arm of r)

$$dM_\eta = dFr = dm\left(2\omega\nu r^2 \sin\theta\right) \tag{5.14}$$

The total moment of all particles above and below the η-axis is given as

$$M_\eta = \int (2\omega\nu\sin\theta)r^2 dm = 2\omega\nu\sin\theta\int r^2 dm = 2\omega\nu\sin\theta I_\eta \tag{5.15}$$

with

$$I_\eta = \int r^2 dm \tag{5.16}$$

From the parallel axis theorem, we have

$$I_p = I_\xi + I_\eta \approx I_\eta \tag{5.17}$$

where $I_p = I_{zz}$ is the polar moment of inertia. Because the rod is thin, we have $I_\xi \approx 0$. In view of Equation 5.17, Equation 5.15 reduces to

$$M_\eta = 2I_p\omega\nu\sin\theta \tag{5.18}$$

which is along the negative η-axis direction as shown in Figure 5.7. Taking the component of moment M_η along the x- and y-axes as

$$M_{xx} = 2I_p\omega\nu\sin^2\theta = I_p\omega\nu(1-\cos 2\theta) \tag{5.19}$$

and

$$M_{yy} = 2I_p\omega\nu\sin\theta\cos\theta = I_p\omega\nu\sin 2\theta \tag{5.20}$$

there are two gyroscopic moments, respectively, about the x- and y-axes. This is because of the asymmetric body of revolution, that is, $I_\eta \neq I_\xi$. From Equations 5.19 and 5.20, it can be seen that M_{xx} and M_{yy} vary with θ, that is, M_{xx} varies from 0 to $2I_p\omega\nu$, and M_{yy} varies from $-I_p\omega\nu$ to $I_p\omega\nu$. This analysis is applicable to the two-bladed propeller or the airscrew (Figure 5.8) and can be extended to a multibladed propeller (e.g., Figure 5.9).

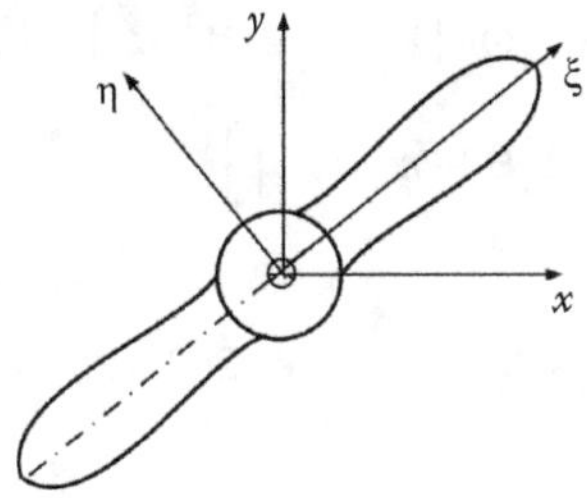

FIGURE 5.8 A two-bladed propeller.

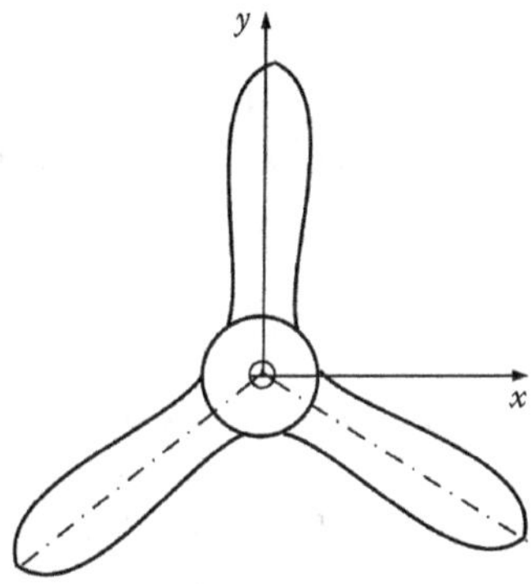

FIGURE 5.9 A three-bladed propeller.

5.2.4 Gyroscopic Moments in a Multibladed Propeller

Let n be the number of blades ($n \geq 3$) and $\alpha = 2\pi/n$ is an equally spaced angle between two blades. Let us consider one of the blades (designated as 1), which is inclined to an angle θ with the x-axis as shown in Figure 5.10. Let the moment of inertia of each blade about the η-axis (i.e., perpendicular to the blade) be equal to I_η, which in turn is equal to the polar mass moment of the inertia of blade 1 alone, that is, I_{p_1} because $I_{p_1} = I_\xi + I_\eta \approx I_\eta$. The total polar moment of inertia of the airscrew about the axis of rotation (z-axis) is $I_p = nI_{p_1}$.

The total moment about the x-axis of blade 1 is given as

$$M_{x_1} = 2\omega\nu\int y^2 dm = 2\omega\nu I_{p_1}\sin^2\theta = I_{p_1}\omega\nu(1-\cos 2\theta) \quad (5.21)$$

The location of other blades is given by the phase angle $(n-1)\alpha$ (the phase angle of various blades with respect to one of the reference blades would be $(n-1)\alpha$, where α is the phase difference between the two blades). Noting Equation 5.21, on summing up the moments due to all n blades, we have

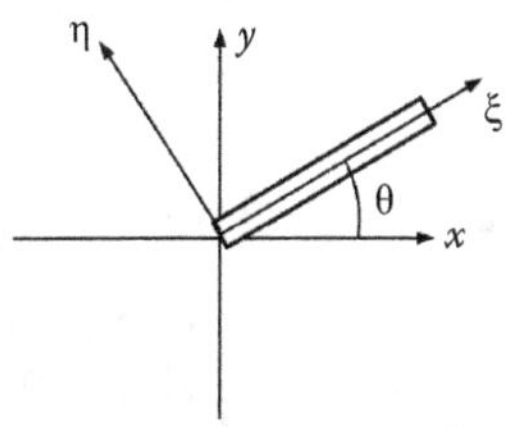

FIGURE 5.10 One of the blade positions.

$$M_x = I_{p1}\omega\nu(1-\cos 2\theta)+I_{p1}\omega\nu\{1-\cos 2(\theta+\alpha)\}+I_{p1}\omega\nu\{1-\cos 2(\theta+2\alpha)\} + \cdots + I_{p1}\omega\nu\left[1-\cos 2\{\theta+(n-1)\alpha\}\right] \tag{5.22}$$

or

$$M_x = I_{p1}\omega\nu\left[n-\{\cos 2\theta+\cos 2(\theta+\alpha)+\cos 2(\theta+2\alpha)+\cdots+\cos 2\{\theta+(n-1)\alpha\}\}\right] \tag{5.23}$$

which can be simplified as

$$M_x = I_{p1}\omega\nu\left[n-\frac{\cos 2\{\theta+0.5(n-1)\alpha\}\sin n\alpha}{\sin\alpha}\right] \tag{5.24}$$

Because $\alpha = 2\pi/n$ for $n > 2$, we have $\alpha = 2\pi/3, 2\pi/4, \ldots$ and $\sin\alpha \neq 0$ for all these values. Moreover, because $n\alpha = 2\pi$ for all values of n, we have $\sin n\alpha = \sin 2\pi = 0$. Hence, for all values of $n > 2$, from Equation 5.24, we can write

$$M_x = nI_{p1}\omega\nu \quad \text{or} \quad M_x = I_p\omega\nu \tag{5.25}$$

For $n = 2$, we have $\sin\alpha = \sin\pi = 0$ and $\sin n\alpha = \sin 2\pi = 0$; hence, from Equation 5.24, we get

$$M_x = I_{p1}\omega\nu(1-\cos 2\theta) \tag{5.26}$$

The gyroscopic moment about the y-axis for a blade, as shown in Figure 5.10, which makes an angle θ with the x-axis, is

$$M_y = I_{p1}\omega\nu\sin 2\theta \tag{5.27}$$

The total moment about the y-axis for n blades with phase angles of $(n-1)\alpha$ for $n = 1, 2, \ldots$, is

$$M_{yy} = I_{p1}\omega\nu\left[\sin 2\theta+\sin 2(\theta+\alpha)+\cdots+\sin\{2(\theta+(n-1)\alpha)\}\right] \tag{5.28}$$

The sine series in Equation 5.28 will be zero for all values $n > 2$; hence

$$M_y = 0 \quad \text{for} \quad n > 2 \tag{5.29}$$

From Equations 5.25 and 5.29, we can conclude that for a multibladed propeller with three or more blades, it resembles a plane disk with a polar mass moment of inertia $I_p = nI_{p1}$ about the axis of rotation.

The objective of this section was to provide an understanding of the gyroscopic moment in rotating components, especially its direction of application. Now we shall deal with the effects of gyroscopic moment and the procedure for obtaining natural frequencies and critical speeds in simple single-mass rotor systems with synchronous and asynchronous whirls (Savet, 1961; Tondl, 1965; Den Hartog, 1984; Shravankumar and Tiwari, 2011).

5.3 SYNCHRONOUS MOTION OF ROTORS

We are considering the case of a perfectly balanced rotor; however, it is assumed to be whirling at its critical speed (synchronous motion, i.e., $\nu=\omega$) in a slightly deflected position. In the presence of a gyroscopic couple, we expect the whirl frequency, ν, to be different. However, in the present analysis, the whirl frequency, ν, of the shaft is assumed to be the same as the spin speed of the shaft, ω.

Hence, the aim of this section is to obtain the synchronous critical speed (because $\nu=\omega$ is assumed). In other words, the aim is to obtain ω where its value is equal to ν, or the speed (or condition) at which the synchronous whirl would take place. This implies that a particular point of the disk, which is outside during whirling, will always be outside and the inside point will always remain inside; this motion is called the *synchronous motion*. The shaft fibers in tension always remain in tension while whirling, and similarly, the compression fibers always remain in compression. Thus, any individual point of the disk moves in a circle in a plane perpendicular to the undistorted center line of the shaft. We will consider two representative cases of cantilever rotors: first, a thin disk at the free end, and second, a long stick at the free end.

5.3.1 A Cantilever Rotor with a Thin Disk

In the present section, a thin disk attached to a flexible massless shaft at its free end is considered (Figure 5.11). In Figure 5.11b, let points C and G be the disk geometrical center and the disk center of gravity, respectively. Because there is no unbalance in the rotor, points C and G are coincident. Let δ and φ_x be the translational and rotational displacements of the disk at its geometrical center, C, respectively. In fact, the rotor is analyzed for one-plane motion (in the y-z plane) with the translational and rotational displacements in the y-axis and x-axis directions, respectively, considered similar to the case in the previous section. The centrifugal force of a mass element dm at point P, with coordinate (r, θ) in the x-y plane (Figure 5.11b) with respect to point B, is $\omega^2 r_1\, dm$ and is directed away from point B (a point on the bearing axis as shown in Figure 5.11b) with $BP=r_1$. It can be considered as two force vectors: one when dm is assumed to be rotated about the disk center C (along CP) and a second force when C itself is rotating about B (along BC) with $PC=r$ and $BC=\delta$. The motion of the disk is considered in two parts: first, the pure spinning, and second, the pure whirling. Two similar triangles, BCP and EFP, can be constructed to represent these three forces (Figure 5.11b). The component in the vertical direction is $\omega^2\delta dm$ and is directed vertically downward (this is due to the pure whirling). When these component forces are added together, it will give a force and no moment (Figure 5.12a and b), whereas the component in the radial direction is $\omega^2 r dm$ and is directed away from the disk center C. When these component forces are added together, they will give zero force, and the moment will be nonzero (Figure 5.12c and d). These will be illustrated now. Note that components of forces in the *z direction* due to tilting of the disk in the x-y plane give forces that will be small and nonlinear. The force $\omega^2\delta dm$ for various masses is added together and gives (Figure 5.12a and b)

$$F_y = \int \omega^2 \delta dm = m\omega^2\delta \tag{5.30}$$

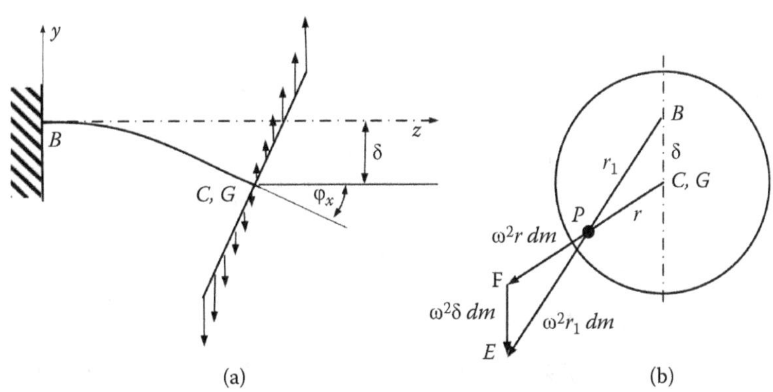

FIGURE 5.11 A cantilever rotor with centrifugal forces on the rigid disk: (a) a cantilever rotor; (b) centrifugal force on a disk particle.

where m is the total mass of the disk. This force acts downward at C on the shaft. The aim is to obtain the critical speed of the system $(\omega_{cr} = \omega = \nu)$.

Force components, $\omega^2 r dm$, all radiate from the center of the disk C and in the plane of the disk. Hence, such forces will balance each other because of the symmetry; however, these will provide a net moment, and it will be clear when we take their components in the x and y directions. From Figure 5.12c and d, the y-component of the force, $\omega^2 r dm$, is $\omega^2 r\sin\theta dm = \omega^2 y dm$ (because $y = r\sin\theta$) and the moment arm of this elemental force is $y\varphi_x$. Thus, the moment of the centrifugal force of a small particle dm is

$$dM_{yz} = \left(\omega^2 y dm\right) y\varphi_x = \omega^2 y^2 \varphi_x dm \tag{5.31}$$

and it will act on the shaft in the negative x-axis direction (counterclockwise [CCW] direction in Figure 5.12c). The total moment M_{yz} of centrifugal forces is

$$M_{yz} = \int \omega^2 y^2 \varphi_x dm = \omega^2 \varphi_x \int y^2 dm = \omega^2 \varphi_x I_d \tag{5.32}$$

where I_d is the diametral mass moment of inertia of the disk. Hence, the y direction component forces balance themselves but produce the moment, M_{yz}. Note that the same could also be obtained by considering the angular momentum. The x-component forces $\omega^2 r\cos\theta dm = \omega^2 x dm$ (because $x = r\cos\theta$) will balance themselves and produce no moment, because these forces are on the plane of the disk (Figure 5.12c and d) due to the symmetry of the disk.

Thus, in totality, the end of the shaft is subjected to a force, $m\omega^2\delta$, and to a moment, $I_d\omega^2\varphi_x$; under the influence of this, it assumes a translational deflection δ and a rotational deflection φ_x. This can happen only at a certain speed ω, that is, at the critical speed. Thus, the calculation of the

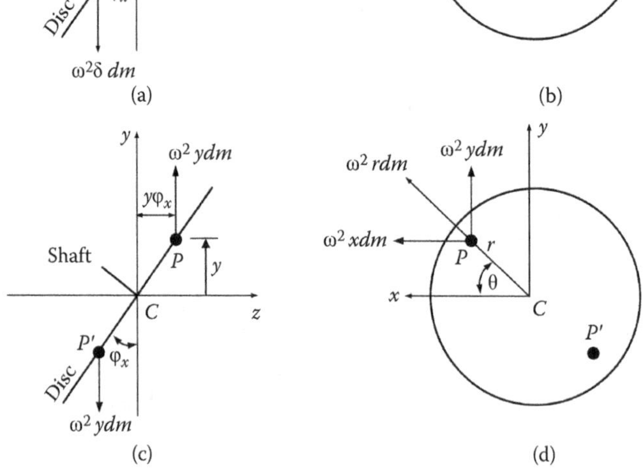

FIGURE 5.12 The centrifugal force on a particle of the disk (a and b) due to pure whirling motion and (c and d) due to pure spinning motion.

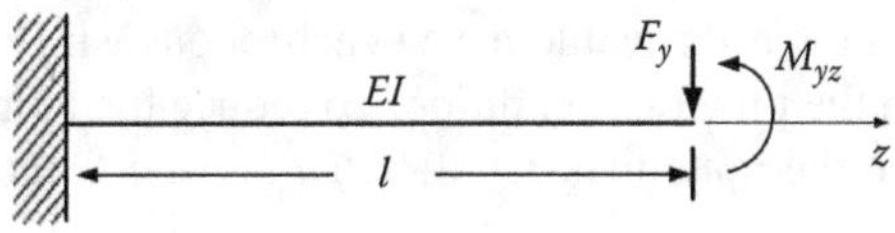

FIGURE 5.13 A cantilever beam with loadings at the free end.

critical speed is reduced to a *quasi-static problem* (in which the dynamic forces are considered time independent). Now the objective is to find at which value of ω the shaft will deflect δ and φ_x under the influence of $F_y = m\omega^2\delta$ and $M_{yz} = \omega^2 I_d \varphi_x$. The direction of the force, F_y, and the moment, M_{yz}, on the shaft is the same as that obtained in Equations 5.8 and 5.32, respectively.

The translational and rotational displacements of the free end of the cantilever (i.e., the fixed-free end conditions) beam, as shown in Figure 5.13, will be (Timoshenko and Young, 1968)

$$\delta = \alpha_{yf}F_y + \alpha_{yM}M_{yz} = \frac{F_y l^3}{3EI} - \frac{M_{yz}l^2}{2EI} = \frac{\left(m\omega^2\delta\right)l^3}{3EI} - \frac{\left(I_d\omega^2\varphi_x\right)l^2}{2EI} \tag{5.33}$$

and

$$\varphi_x = \alpha_{\varphi f}F_y + \alpha_{\varphi M}M_{yz} = \frac{F_y l^2}{2EI} - \frac{M_{yz}l}{EI} = \frac{\left(m\omega^2\delta\right)l^2}{2EI} - \frac{\left(I_d\omega^2\varphi_x\right)l}{EI} \tag{5.34}$$

where α_{ij} (with $i = y, \varphi$ and $j = f, M$) are influence coefficients (refer to Appendix 2.1). It should be noted here that these relations can also be developed for other boundary conditions, such as simply supported, fixed-fixed, and fixed-hinged. This requires calculating the influence coefficients by using the deflection theory of strength of materials. Now these relations can be used to find critical speeds for the fixed-free boundary condition. Equations 5.33 and 5.34 can be rearranged as

$$\left(m\omega^2\frac{l^3}{3EI} - 1\right)\delta + \left(-I_d\omega^2\frac{l^2}{2EI}\right)\varphi_x = 0 \tag{5.35}$$

and

$$\left(-m\omega^2\frac{l^2}{2EI}\right)\delta + \left(I_d\omega^2\frac{l}{EI} + 1\right)\varphi_x = 0 \tag{5.36}$$

This homogeneous set of equations can have a nontrivial solution for δ and φ_x only when the determinant vanishes:

$$\begin{vmatrix} \left(m\omega^2\dfrac{l^3}{3EI} - 1\right) & \left(-I_d\omega^2\dfrac{l^2}{2EI}\right) \\ \left(-m\omega^2\dfrac{l^2}{2EI}\right) & \left(I_d\omega^2\dfrac{l}{EI} + 1\right) \end{vmatrix} = 0 \tag{5.37}$$

which gives the frequency equation (now replacing ω with ω_{cr}) as

$$\omega_{cr}^4 + \omega_{cr}^2\left\{\frac{12EI}{mI_d l^3}\left(\frac{ml^2}{3} - I_d\right)\right\} - \frac{12E^2I^2}{mI_d l^4} = 0 \tag{5.38}$$

This is the condition at which the rotor would have synchronous whirling, and it will take place at critical speeds obtained from the previous condition. To reduce the number of parameters involved in the earlier equation, define the nondimensional *critical speed function*, $\bar{\omega}_{cr}$, and the *disk mass effect*, μ, as

$$\bar{\omega}_{cr} = \omega_{cr}\sqrt{\frac{ml^3}{EI}} \quad \text{and} \quad \mu = \frac{I_d}{ml^2} \tag{5.39}$$

where ω_{cr} is the *synchronous critical speed.* Equation 5.38 can be written as

$$\left(\bar{\omega}_{cr}\sqrt{\frac{EI}{ml^3}}\right)^4 + \left(\bar{\omega}_{cr}\sqrt{\frac{EI}{ml^3}}\right)^2 \left\{\frac{12EI}{mI_d l^3}\left(\frac{ml^2}{3} - I_d\right)\right\} - \frac{12E^2I^2}{mI_d l^4} = 0 \tag{5.40}$$

which will simplify to

$$\bar{\omega}_{cr}^4 + \bar{\omega}_{cr}^2 \left\{\frac{12}{I_d}\left(\frac{ml^2}{3} - I_d\right)\right\} - \frac{12ml^2}{I_d} = 0 \tag{5.41}$$

On using the disk parameter expression, we get

$$\bar{\omega}_{cr}^4 + \bar{\omega}_{cr}^2\left(\frac{4}{\mu} - 12\right) - \frac{12}{\mu} = 0 \tag{5.42}$$

the solution of which is

$$\bar{\omega}_{cr}^2 = \left(6 - \frac{2}{\mu}\right) \pm \sqrt{\left(6 - \frac{2}{\mu}\right)^2 + \frac{12}{\mu}} \tag{5.43}$$

The positive sign will give a positive value for $\bar{\omega}_{cr}^2$ or a real value for $\bar{\omega}_{cr}$, and the negative sign will give a complex value of the critical speed, which has no physical significance. Hence, we will have only a single critical speed for a particular value of μ, so the negative sign may be ignored. The plot of $\bar{\omega}_{cr}^2$ versus μ is given in Figure 5.14.

It should be noted that the critical speed of the rotor increases with the disk mass effect, μ. That means the effective stiffness of the rotor system increases due to the thin disk as compared to a point-mass disk. This can be seen from Figure 5.11a, where the effect of centrifugal forces is to resist the tilting of the disk, thereby increasing the effective stiffness of the rotor system. Overall, for the synchronous whirl condition ($\nu=+\omega$) due to the gyroscopic effect, the (forward) critical speed of the system increases. It will be shown that for the antisynchronous whirl condition ($\nu=-\omega$) due to gyroscopic effect, the (backward) critical speed of the system decreases. Two limiting cases of Figure 5.14 are discussed as follows:

Case 1: The disk as a point mass: For the disk effect $\mu=0$ (i.e., the concentrated mass of the disk) from Equation 5.42, we have

$$4\bar{\omega}_{cr}^2 - 12 = 0 \Rightarrow \bar{\omega}_{cr}^2 = 3 \tag{5.44}$$

Noting Equation 5.39, the previous equation gives

$$\left(\omega_{cr}\sqrt{\frac{ml^3}{EI}}\right)^2 = 3 \Rightarrow \omega_{cr} = \sqrt{\frac{3EI}{ml^3}} = \sqrt{\frac{k_{eq}}{m}} = \omega_{nf} \quad \text{with} \quad k_{eq} = \frac{3EI}{l^3} \tag{5.45}$$

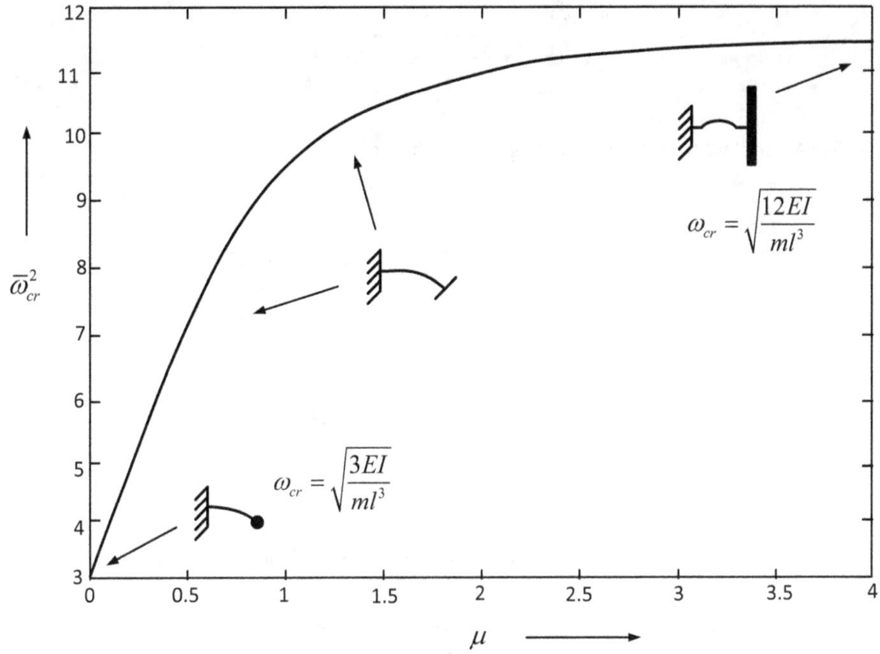

FIGURE 5.14 Variation of the critical speed function with the disk effect.

where ω_{nf} is the transverse natural frequency of the rotor system in the nonrotating condition. It gives the synchronous critical speed of a disk having a point mass for the cantilever beam case. For this case, in fact, the gyroscopic effect is not present, and the critical speed is the same as the transverse natural frequency of the cantilever beam with a point mass at the free end.

Case 2: A disk with an infinite diameter: For $\mu \rightarrow \infty$ (i.e., a disk for which all the mass is concentrated at a relatively large radius so that $I_d \rightarrow \infty$), no finite rotational displacement φ_x is possible, because it would require an infinite torque, which the shaft cannot furnish. The disk remains parallel to itself, and the shaft is much stiffer than without the disk effect (i.e., $\mu = 0$). From Equation 5.42 for $\mu \rightarrow \infty$, we get

$$\bar{\omega}_{cr}^2 \left(\bar{\omega}_{cr}^2 - 12\right) = 0 \text{ since } \bar{\omega}_{cr}^2 \neq 0 \text{ hence } \bar{\omega}_{cr}^2 = 12 \Rightarrow \omega_{cr} = \sqrt{\frac{12EI}{ml^3}} = 2\sqrt{\frac{3EI}{ml^3}} \tag{5.46}$$

or

$$\omega_{cr} = \sqrt{\frac{k_{eff}}{m}} \quad \text{with} \quad k_{eff} = \frac{12EI}{l^3} \tag{5.47}$$

The critical speed for the present case is two times more as compared to Case I. It should be noted that for the present case one of the solutions $\bar{\omega}_{cr}^2 = 0$ is not considered feasible. However, when $\mu \rightarrow -\infty$, it is a feasible solution and the natural frequency of the system would be zero. This particular issue will be considered in the next section by replacing the thin disk with a long stick, which requires a separate analytical treatment; it has been analyzed for pure rotational motion only for a better visualization.

5.3.2 A Cantilever Rotor with a Long Stick

For the present case, the disk at the free end of a cantilever rotor has a considerable length, as shown in Figure 5.15. The couple of centrifugal forces for this case is such that it tries to push the rotor

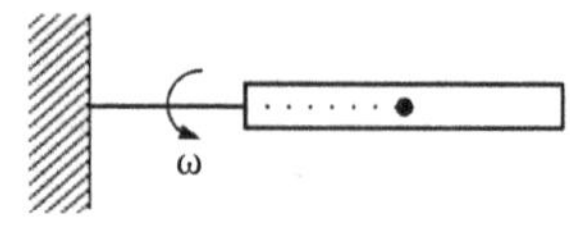

FIGURE 5.15 A cantilevered rotor with a long stick at the free end.

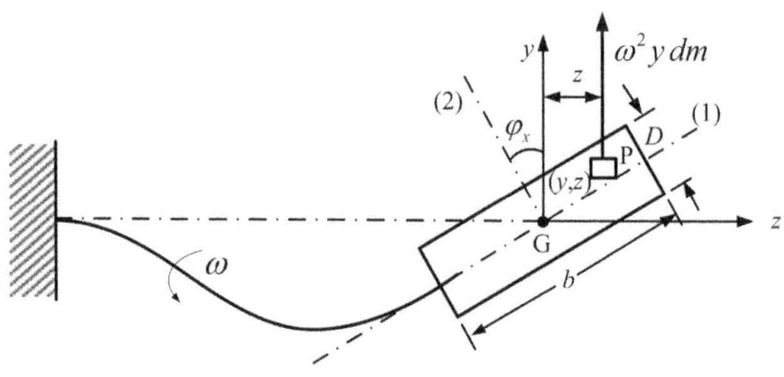

FIGURE 5.16 Centrifugal forces in a cantilever rotor with a long stick at the free end.

away from the static equilibrium position angularly, as shown in Figure 5.16. For the present case also, the synchronous whirl condition is assumed. For the thin disk case, the couple of centrifugal forces tries to bring it back to the static equilibrium position angularly. In Figure 5.16, the principal axis directions are (1) and (2) along the longitudinal and transverse directions. The coordinate of a point mass dm is (y, z).

For the present case, it is assumed that no unbalance is present in the rotor (i.e., the rotor geometrical center C and the center of gravity G are coincident) and the center of gravity G of the body is positioned on the axis of rotation x (i.e., no linear displacement, $\delta=0$). Hence, there is no net centrifugal force, $m\omega^2\delta$ (where m is the mass of the long stick), and only a moment is present. The force on a particle is $\omega^2 y dm$, and its moment arm about the x-axis is z (Figure 5.16), so that the moment is

$$dM_{yz} = \omega^2 yzdm \tag{5.48}$$

For the thin disk, we have the moment arm, $z = y\varphi_x$ (refer to Figure 5.12c), so Equation 5.31 reduces to $dM_{yz} = \omega^2 y^2 \varphi_x dm$. Now for the present case (i.e., the long stick case), for the whole body, we have

$$M_{yz} = \omega^2 \int yzdm \tag{5.49}$$

Let the mass moment of inertia about these principal axes be I_1 and I_2, which are the polar and diametral mass moments of inertia, respectively. This set of axes is at an angle φ_x with respect to the y-z axes, as shown in Figure 5.16. The product mass moment of inertia (it is analogous to the shear stress in terms of the strength of materials) about the y-z plane is defined as

$$\int yzdm = \frac{I_1 - I_2}{2}\sin 2\varphi_x \approx (I_1 - I_2)\varphi_x \tag{5.50}$$

Here, the rotational deformation, φ_x, is assumed to be small. For the thin disk case, $I_1 = I_p = 2I_d$ and $I_2 = I_d$ so that Equations 5.49 and 5.50 give $M_{yz} = \omega^2 I_d \varphi_x$, which is the same as Equation 5.32. However, for the present case of a long stick of diameter D and length b (Figure 5.16), we have

$$I_1 = \frac{mD^2}{8} \quad \text{and} \quad I_2 = \frac{mD^2}{16} + \frac{mb^2}{12} \tag{5.51}$$

TABLE 5.1
Gyroscopic Effects for Different Geometries of the Disk/Stick

Relation between b and D	Sign of I_d	Gyroscopic Moment Sign
b=0.87D	0	0
b>0.87D (long stick)	Negative	Negative
b<0.87D (thin disk)	Positive	Positive

On substituting Equations 5.50 and 5.51 into Equation 5.49, we get

$$M_{yz} = \omega^2 (I_1 - I_2)\varphi_x = \left(\frac{mD^2}{16} - \frac{mb^2}{12}\right)\omega^2 \varphi_x \tag{5.52}$$

For the moment of centrifugal forces to be zero, from the previous equation, we have

$$\frac{mD^2}{16} = \frac{mb^2}{12} \quad \Rightarrow b = \frac{\sqrt{3}}{2}D \quad \text{or} \quad b = 0.866D \tag{5.53}$$

and it becomes negative for $b>0.866D$ (i.e., for the long stick) and positive for $b<0.866D$ (i.e., for the thin disk).

Table 5.1 summarizes the gyroscopic moment with its sign for different ratios of b and D. Equation 5.52 can be written as

$$M_{yz} = I_d \omega^2 \varphi_x \quad \text{with} \quad I_d = \left(\frac{mD^2}{16} - \frac{mb^2}{12}\right) \tag{5.54}$$

It is worth mentioning here that until now we have considered the pure tilting motion of the long stick. It can be found that the net resultant force will be the same for the long stick and the thin disk when we consider both the translational and rotational motions simultaneously. Thus, the critical speed analysis of the previous section will be valid for the present case also when we consider both the translational and rotational motions simultaneously. Hence, Figure 5.14 of the previous case will still be applicable for a range of thin disks. However, the plot from the previous equation will represent both the long stick and thin disk cases, as shown in Figure 5.17, because the form of the net moment in Equations 5.32 and 5.54 is identical. Hence, from Equation 5.43, we have

$$\bar{\omega}_{cr}^2 = \left(6 - \frac{2}{\mu}\right) \pm \sqrt{\left(6 - \frac{2}{\mu}\right)^2 + \frac{12}{\mu}} \tag{5.55}$$

with

$$\mu = \frac{I_d}{ml^2}, \quad \text{for the long stick} \quad I_d = \left(\frac{mD^2}{16} - \frac{mb^2}{12}\right)$$

and

$$\text{For the thin disk} \left(D = r/2 \text{ and } b = t\right), \; I_d = \frac{1}{4}mr^2 + \frac{1}{12}mt^2 \approx \frac{1}{4}mr^2 \; \text{since} \; t^2 \approx 0 \tag{5.56}$$

The condition for which the square root term in Equation 5.55 remains always positive is (because now the disk effect μ can be a negative quantity)

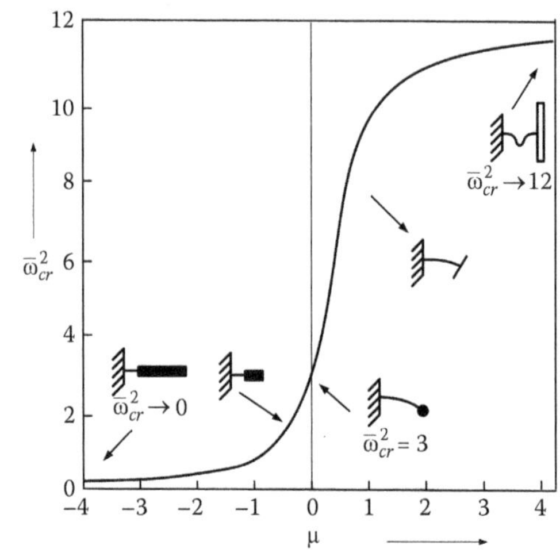

FIGURE 5.17 Variation of the critical speed function with the disk effect.

$$\left(6-\frac{2}{\mu}\right)^2+\frac{12}{\mu}>0 \quad \text{or} \quad (3\mu-1)^2+3\mu>0 \tag{5.57}$$

From this equation, it can be observed that both terms are always positive for a *positive value* of μ, because the first term is a square term and the second term is positive. For a *negative value* of μ, the first term in the earlier inequality is always positive, and it is always greater than the second term. Hence, for all real values of μ, the earlier inequality is true (for imaginary values of μ, we may violate the inequality; however, it is not a feasible case in real systems). This means we will always get a real root from Equation 5.55 when we consider a positive sign in front of the square root. For $\mu \to 0$ again Equation 5.45 would be valid. For $\mu \to \pm\infty$ from Equation 5.55, the critical speeds are $\bar{\omega}_{cr}^2 \to 12$ and $\bar{\omega}_{cr}^2 \to 0$, respectively, for the positive and negative signs in front of the square root. For a very small amount of disturbance, the long stick can have a large angular displacement, so the effective stiffness of the system becomes zero, and hence the natural frequency is shown in Figure 5.17. This analysis is valid for the synchronous whirl and for the antisynchronous whirl—the curve in Figure 5.17 would be just the mirror image with respect to the vertical axis at $\mu=0$.

For the long stick case, it is assumed that the shaft extends to the center of the cylinder without interference. If the shaft is attached to the end of the cylinder, the elastic-influence coefficients are modified. The phenomenon described in the present section is generally referred to as a gyroscopic effect. Now, the calculation of critical speeds will be demonstrated with some numerical examples.

Example 5.1

Obtain the transverse critical speed for the synchronous motion of a cantilever rotor as shown in Figure 5.18. Take the mass of the thin disk, m, as 1 kg and the radius, r, as 3 cm. The shaft is assumed to be massless, and its length and diameter are 0.2 and 0.01 m, respectively. Let the Young's modulus of the shaft material be $E=2.1\times10^{11}$ N/m^2.

Solution:

Case I: For disk effect $\mu=0$ (i.e., the concentrated mass of the disk) from Equation 5.45, we have

$$\omega_{cr}=\sqrt{\frac{3EI}{ml^3}}=\sqrt{\frac{3\times2.1\times10^{11}\times4.909\times10^{-10}}{1\times0.2^3}}=196.61 \text{ rad/s}$$

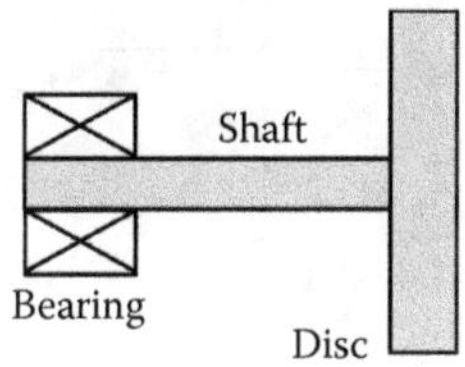

FIGURE 5.18 A cantilever shaft with a rigid disk at the free end.

with

$$I = \frac{\pi}{64}(0.01)^4 = 4.909\times10^{-10}\ \text{m}^4;\quad d = 0.01\ \text{m};\quad m = 1\ \text{kg};\quad l = 0.2\ \text{m} \tag{a}$$

Case II: Considering the disk to be rigid (i.e., $\mu \neq 0$), from Equation 5.43, we have

$$\bar{\omega}_{cr} = \sqrt{\left(6-\frac{2}{\mu}\right)+\sqrt{\left(6-\frac{2}{\mu}\right)^2+\frac{12}{\mu}}} = \sqrt{\left(6-\frac{2}{0.0056}\right)+\sqrt{\left(6-\frac{2}{0.0056}\right)^2+\frac{12}{0.0056}}} = 1.7431 \tag{b}$$

with

$$\mu = \frac{I_d}{ml^2} = \frac{\frac{1}{4}mr^2}{ml^2} = \frac{r^2}{4l^2} = \frac{0.03^2}{4\times0.2^2} = 0.0056 \tag{c}$$

Now, from Equation 5.39, we have

$$\bar{\omega}_{cr} = \omega_{cr}\sqrt{\frac{ml^3}{EI}} = \omega_{cr}\sqrt{\frac{1\times0.2^3}{2.1\times10^{11}\times4.909\times10^{-10}}} = 0.00881\omega_{cr} \tag{d}$$

On substituting the value of $\bar{\omega}_{cr}$ from Equation (b) into (d), the critical speed is given by

$$\omega_{cr} = \frac{\bar{\omega}_{cr}}{0.00881} = \frac{1.7431}{0.00881} = 197.86\ \text{rad/s}$$

It should be noted that compared *d* to Case I, the critical speed is greater, which is expected due to the increase in the effective stiffness in forward whirl while considering the diametral mass moment of inertia of the disk. For the present case $I_d = 0.25\,mr^2 = 0.25\times1\times0.03^2 = 2.25\times10^{-4}$ kg-m^2, which is a great deal less, and that is why the increase in the critical speed is marginal. Readers can check the change in the critical speed, for example, when the radius of the disk equals 6 cm. The effect would be far more predominant in the limiting case, when the disk is very large, that is, $\mu\to\infty$, and the critical speed can be calculated from Equation 5.46 as

$$\omega_{cr} = \sqrt{\frac{12EI}{ml^3}} = \sqrt{\frac{12\times2.1\times10^{11}\times4.909\times10^{-10}}{1\times0.2^3}} = 393.22\ \text{rad/s}$$

Example 5.2

Obtain the transverse critical speed for the synchronous motion of a rotor, as shown in Figure 5.19. The shaft is assumed to have cantilever end conditions. Let the dimensions of the cylinder (stick) be (i) D=0.2 m, b=0.0041 m; (ii) D=0.0547 m, b=0.1093 m; (iii) D=0.0547 m, b=0.0547 m; and (iv) D=0.0361 m, b=0.1649 m. Parameters D and b are the diameter and the length of the cylinder, respectively. The shaft is assumed to be massless, and its length l and diameter d are 0.2

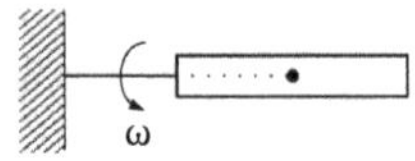

FIGURE 5.19 A rotor with a long stick.

and 0.02 m, respectively. Let the Young's modulus of the shaft material be 2.1×10^{11} N/m² and the density of the cylinder material be 7800 kg/m³.

Solution:

For the long stick, the critical speed is given as

$$\omega_{cr} = \bar{\omega}_{cr}\sqrt{\frac{EI}{ml^3}} \tag{a}$$

with

$$\bar{\omega}_{cr} = \sqrt{\left(6-\frac{2}{\mu}\right)+\sqrt{\left(6-\frac{2}{\mu}\right)^2+\frac{12}{\mu}}};\quad \mu=\frac{I_d}{ml^2};\quad \text{and}\quad I_d=\left(\frac{mD^2}{16}-\frac{mb^2}{12}\right) \tag{b}$$

This equation is valid for all values of μ except for $\mu = 0$ (i.e., the point mass). For $\mu = 0$, we have the relation given by Equation 5.45 given as

$$\omega_{cr}=\sqrt{\frac{3EI}{ml^3}}\quad \text{so that}\quad \bar{\omega}_{cr}=\frac{\omega_{cr}}{\sqrt{\frac{EI}{ml^3}}}=\frac{\sqrt{\frac{3EI}{ml^3}}}{\sqrt{\frac{EI}{ml^3}}}=\sqrt{3}=1.732 \tag{c}$$

with

$$I=\frac{\pi}{64}d^4=\frac{\pi}{64}0.02^4=7.854\times10^{-9}\ \text{m}^4;\quad EI=1649.34\ \text{N-m}^2,\quad m=1\ \text{kg},\quad l=0.2\ \text{m}$$

Hence, for $\mu = 0$ (thin disk), we have

$$\sqrt{\frac{EI}{ml^3}}=\sqrt{\frac{1649.34}{1\times0.2^3}}=454.06\quad \text{and}\quad \omega_{cr}=\sqrt{\frac{3EI}{ml^3}}=\sqrt{\frac{3\times1649.34}{1\times0.2^3}}=786.45\ \text{rad/s} \tag{d}$$

Table 5.2 summarizes the calculation of rotor parameters and critical speeds from the earlier equations. It should be noted that the choice of D and b is such that the mass of the disk remains the same for all cases. For the point mass, parameters D and b are not defined. On comparing the cases of a point mass and a thin disk, an increasing trend in the critical speed is observed. For a short stick (with $D=b$), the disk parameter, μ, becomes negative with a very low value. This leads to a very high value for the critical speed. However, for the long stick again, the trend is toward a decrease in the critical speed, where the disk parameter, μ, remains negative with a relatively high value.

5.4 ASYNCHRONOUS ROTATIONAL MOTION OF A ROTOR SYSTEM

In this section, we will first consider the asynchronous whirling motion of the spinning rotor. Consider a rotor that is suspended practically at its center of gravity by very flexible torsional springs (which resist the transverse tilt motion of the motor), as shown in Figure 5.20. This will enable us to analyze the whirl frequency corresponding to the rotational displacement of the rotor on its whirling frequencies, without complicating with the general motion in which both the translational and

TABLE 5.2
Summary of Rotor Parameters and Synchronous Critical Speeds

Type of Disk	D (m)	b (m)	m (kg)	I_d (kg-m²)	$\mu = \frac{I_d}{ml^2}$	$\bar{\omega}_{cr}$	$\sqrt{\frac{EI}{ml^3}}$ (rad/s)	ω_{cr} (rad/s)
Thin disk	0.2000	0.0041	1	0.0025	0.0625	1.86	454.06	845.00
Point mass	–	–	1	0.0000	0.0000	1.73[a]	454.06	786.43
Short stick	0.0547	0.0547	1	-6.23×10^{-5}	−0.0016	50.89	454.06	23 107.11
Long stick	0.0316	0.1649	1	-2.2×10^{-3}	−0.0550	9.06	454.06	4113.78

[a] From Equation 5.41.

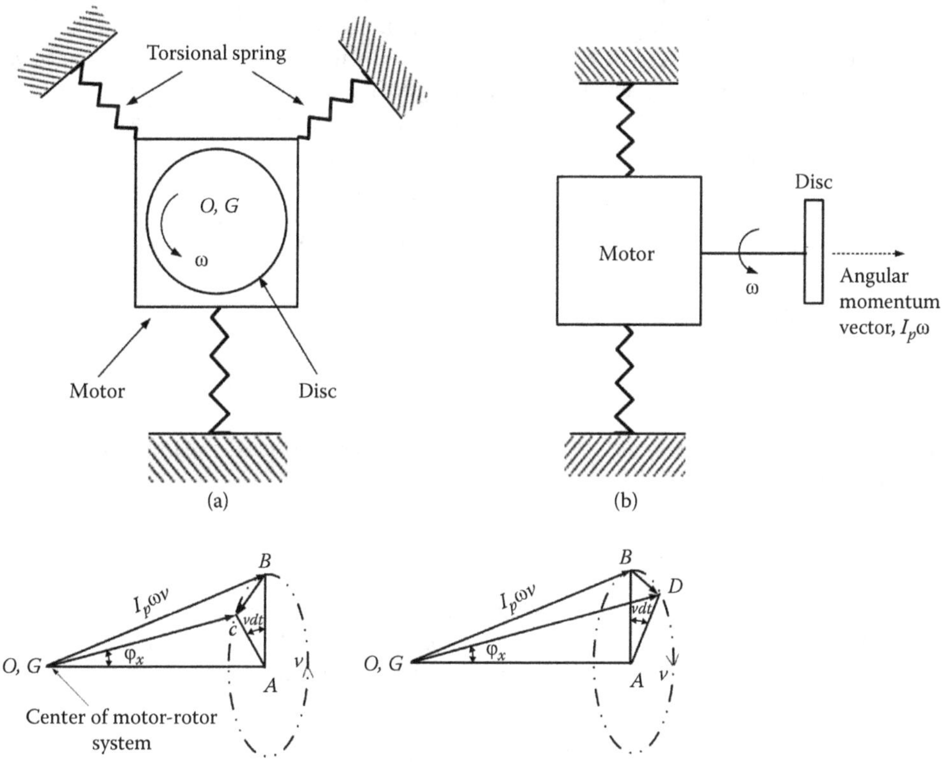

FIGURE 5.20 A motor and a rotor system supported on torsional springs: (a) a motor on torsional springs; (b) a rotor system with motor suspensions; (c) a conical forward whirl of the rotor; (d) a conical backward whirl of the rotor.

rotational displacements take place simultaneously. The shaft and the disk are assumed to be rigid (i.e., their flexibility has been ignored). Such a general motion is quite complicated and will be considered in subsequent sections.

The aim is to calculate the natural frequencies of modes of motion for which the center of gravity O of the whole suspended system (including the motor and the rotor) remains at rest, and the shaft whirls about O in a cone of angle $2\varphi_x$. Let the effective torsional stiffness of the spring support be k_t. The disk on the motor shaft rotates very fast, and because the springs on which the motor is mounted are flexible, the whirling takes place at a slower rate, ν, than the shaft spin speed, ω. It is assumed that the sense of rotation of both ν and ω is the same (herein, the motion is considered as a forward whirl, and it is not a synchronous whirl). The angular momentum, H, of the whole rotor system is given as

$$H = I_p\omega \tag{5.58}$$

where I_p is the polar mass moment of inertia of the rotor (i.e., rotating parts of the motor, the shaft, and the disk). In the case when the whirl is in the same direction as the rotation (i.e., the forward whirl), the time rate of change of angular momentum will be directed from B to C (i.e., out of the plane of the paper in Figure 5.20c). This is equal to the moment applied by the motor frame to the disk. The reaction, that is, the moment acting on the motor frame from the disk is pointing into the paper (in Figure 5.20c) and therefore tends to make φ_x smaller. This acts in addition to the existing spring torque, $k_t\varphi_x$. Hence, it is seen that the whirl in the direction of rotation (i.e., the forward whirl) makes the natural frequency higher. In the same manner, it can be reasoned that for the whirl opposite to the direction of rotation (i.e., the backward whirl), the whirl frequency is made lower by the gyroscopic effect due to the reversal of moment (Figure 5.20d). This is due to the reactive torque moment trying to make φ_x large (Figure 5.20d). To calculate the magnitude of the gyroscopic effect for the forward whirl, we have

$$\frac{d(I_p\omega)}{I_p\omega} = \frac{BC}{OB} = \frac{BC}{AB}\frac{AB}{OB} = (\nu dt)\varphi_x = \nu\varphi_x dt \tag{5.59}$$

Equation 5.59 can be rearranged as

$$\frac{d}{dt}(I_p\omega) = \nu\varphi_x I_p\omega \tag{5.60}$$

Equation 5.60 gives the gyroscopic moment. The elastic moment due to the spring k_t is equal to $k_t\varphi_x$ and the total moment is equal to

$$(k_t \pm I_p\omega\nu)\varphi_x \tag{5.61}$$

where the positive sign for a whirl is in the same sense as the rotation and the negative sign for a whirl is in the opposite sense. It should be noted that k_t will also restrict the transverse angular displacement, φ_x. In Equation 5.61, the term in parentheses is the equivalent spring constant, and hence the natural whirl frequency will be (for the asynchronous whirl)

$$\nu^2 = \frac{\text{Equivalent stiffness}}{\text{Diametral mass moment of inertia}} = \frac{k_t \pm I_p\omega\nu}{I_d} \quad \text{or} \quad \nu^2 \mp \frac{I_p\omega}{I_d}\nu - \frac{k_t}{I_d} = 0 \tag{5.62}$$

where I_d is the diametral mass of whole motor assembly (including frame and disk). The solution of Equation 5.62 can be given as

$$\nu = \mp\frac{I_p\omega}{2I_d} \pm \sqrt{\left(\frac{I_p\omega}{2I_d}\right)^2 + \frac{k_t}{I_d}} \tag{5.63}$$

From Equation 5.63, it can be observed that the natural frequency of the rotor system depends upon the spin speed of the rotor, ω. For the $\pm$ sign before the square root, only the positive sign needs to be retained because the negative sign gives two values of ω_{nf}, which are both negative and equal and opposite to the two positive roots obtained with the positive sign before the square root. This is due

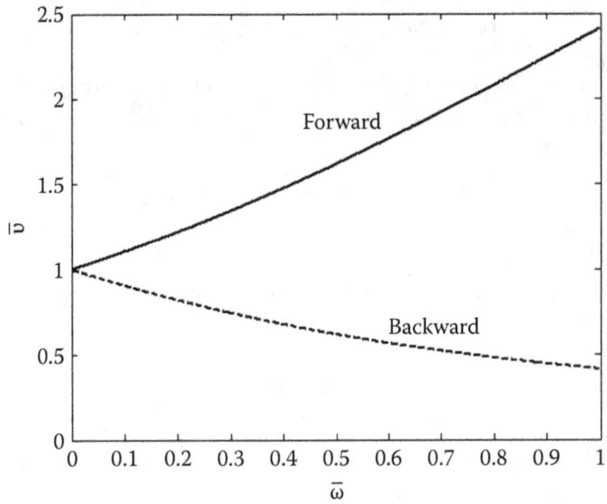

FIGURE 5.21 The natural frequency variation with the spin speed (Campbell diagram).

to the fact that natural frequencies remain the same for the spinning of the rotor in either direction (CW or CCW). Let us define the nondimensional terms of the *frequency ratio* as $\bar{\nu} = \nu/\omega_{nf_{non\text{-}rot}}$, where $\omega_{nf_{non\text{-}rot}} = \sqrt{k_t/I_d}$ is the nonrotating shaft's natural frequency, that is, without the gyroscopic effect, and the *spin ratio* $\bar{\omega} = \left(0.5I_P\omega/\sqrt{k_tI_d}\right)$. Equation 5.63 takes the following form:

$$\bar{\nu} = \pm\bar{\omega} + \sqrt{\bar{\omega}^2 + 1} \quad \text{with} \quad \bar{\nu} = \nu\sqrt{\frac{I_d}{k_t}} \quad \text{and} \quad \bar{\omega} = \frac{0.5I_P\omega}{\sqrt{k_tI_d}} \tag{5.64}$$

Figure 5.21 shows the variation of the nondimensional natural frequency, $\bar{\nu}$, with the nondimensional speed, $\bar{\omega}$, which is governed by Equation 5.64. It can be seen that the natural frequency splits into two frequencies on account of the gyroscopic effect: (i) a slow one whereby the whirl is opposed to the rotor spin (i.e., the backward whirl) and (ii) a fast one where the whirl and spin directions are the same (i.e., the forward whirl). The whirl frequency corresponding to the backward whirl $(\nu = -\omega)$ is drawn by a dashed line, indicating that it should have been in the fourth quadrant. However, for brevity, the practice is to fold the fourth quadrant onto the first quadrant. It should be noted that we will have two critical speeds: one corresponding to the forward whirl and another to a backward whirl. These critical speeds can be obtained by the condition that whenever the spin speed is equal to the natural frequency (i.e., the synchronous whirl case), we will have critical speeds. The present formulation will be valid for a variety of rotor systems supported on flexible supports. These will be illustrated in the following numerical example.

Example 5.3

A long rigid symmetric rotor is supported at the ends by two identical bearings. Let the shaft have a diameter of 0.2 m, a length of 1 m, and a material mass density equal to 7800 kg/m³. The bearing has dynamic parameters as follows: $k_{xx}=k_{yy}=k=1$ kN/mm with the cross-coupled stiffness and all damping terms equal to zero. By considering the pure tilting motion and the gyroscopic effect, obtain the whirl natural frequencies of the system if the rotor is rotating at $N=10{,}000$ rpm. Obtain also the forward and backward critical speeds of the rotor-bearing system. Compare the critical speeds of the rotor without considering the gyroscopic effect.

Solution:

For the circular cylinder, the polar moment of inertia, I_p, and the diameter moment of inertia, I_d, are given as

$$I_p = \frac{1}{2}mr^2 \quad \text{and} \quad I_d = \frac{1}{12}m\left(3r^2 + l^2\right)$$

where m is the mass of the cylinder, r is the radius of the cylinder, and l is the length of the cylinder. For the pure tilting motion of the rotor, we have the following rotor properties:

$$I_p = \frac{1}{2}mr^2 = \frac{1}{2}\times 245.04\times(0.1)^2 = 1.2252 \text{ kgm}^2$$

$$I_d = \frac{1}{12}m\left(3r^2 + l^2\right) = \frac{1}{12}\times 245.04\left\{3\times(0.10)^2 + 1^2\right\} = 21.0326 \text{ kgm}^2$$

$$\omega = \frac{2\pi N}{60} = \frac{2\pi\times 10{,}000}{60} = 1047.198 \text{ rad/s}$$

The effective torsional stiffness due to the bearings on the rotor can be obtained by considering Figure 5.22. A long rotor is supported at the ends by two identical bearings; consider the pure tilting of the rotor by an angle φ_x, which gives $0.5l\varphi_x$ compression of the left bearing, and the same amount of extension of the right bearing. This produces reaction forces at the bearings with a magnitude of $0.5kl\varphi_x$ and the direction shown in Figure 5.22. The moment on the rotor due to these bearing forces would be $0.5kl^2\varphi_x$. Hence, the effective torsional stiffness would be $0.5kl^2$.

From Equation 5.63, we have whirl frequencies for the pure translational rotational motion, as

$$\nu_{3,4} = \pm\frac{I_p\omega}{2I_d} + \sqrt{\left(\frac{I_p\omega}{2I_d}\right)^2 + \frac{k_{eff}}{I_d}} \qquad \text{(a)}$$

with

$$k_{eff} = \frac{1}{2}kl^2 = 0.5\times 1\times 10^6\times 1^2 = 5\times 10^5 \text{ N/m}$$

with a positive sign for the forward whirl and a negative sign for the backward whirl. On substituting values in Equation (a), we get

$$\nu_{3,4} = \left\{\frac{1.2252}{2\times 21.0326}\times 1047.198 \pm \sqrt{\frac{5\times 10^5}{21.0326} + \left(\frac{1.2252}{2\times 21.0326}\times 1047.198\right)^2}\right\} = \{\pm 30.50 + 157.17\}$$

which gives

$$\nu_3 = 187.67 \text{ rad/s (forward whirl)} \quad \text{and} \quad \nu_4 = 126.67 \text{ rad/s (backward whirl)}.$$

It should be noted that these natural whirl frequencies change with the spin speed of the rotor. To obtain the forward and backward critical speeds, respectively, the following conditions, that is, $\omega = \nu$ (synchronous condition) and $\omega = -\nu$ (antisynchronous condition), need to be used. On substituting these conditions in Equation (a), one at a time, we get the following expression for critical speeds:

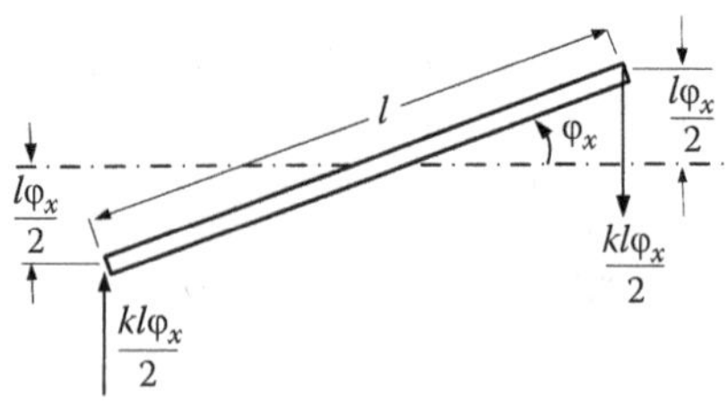

FIGURE 5.22 Free body diagram of the long rigid rotor supported on bearings.

$$\omega_{cr}^{F,B} = \sqrt{\frac{k_t}{(I_d \mp I_p)}} \tag{b}$$

where the negative sign is for the forward critical speed, ω_{cr}^{F}, and the positive sign is for the backward critical speed, ω_{cr}^{B}. It should be noted from Equation (b) that for $I_d < I_p$, the term inside the square bracket will be negative, and there would not be any forward critical speed for the pure transverse rotational motion case. For the present case, we have $I_d > I_p$ and the corresponding critical speeds are

$$\omega_{cr}^{F} = \sqrt{\frac{5\times10^5}{(21.0326-1.2252)}} = 158.88 \text{ rad/s}$$

and

$$\omega_{cr}^{B} = \sqrt{\frac{5\times10^5}{(21.0326+1.2252)}} = 149.88 \text{ rad/s}$$

Hence, the rotor is operating well above the critical speeds, because the operating speed is 1047.2 rad/s (10,000 rpm). That means if we consider a perfectly balanced rotor that is rotating at 1047.2 rad/s and if we perturb the rotor angularly in forward whirl, it will have a whirl frequency equal to 187.67 rad/s. Moreover, if we perturb the rotor angularly in backward whirl, it will have a whirl frequency equal to 126.67 rad/s. While the rotor is coasting up during the passage of critical speeds, large oscillations are expected.

When no gyroscopic effect is considered, ν_3 and ν_4 will be equal, and from Equation (a) for $\bar{\omega}_{cr}^2$ we get

$$\nu_{3,4} = \pm\sqrt{\frac{k_{eff}}{I_d}} = \sqrt{\frac{5\times10^5}{21.0326}} = 154.184 \text{ rad/s}$$

which is between the forward and backward whirl frequencies (187.67, 126.67) rad/s obtained by considering the gyroscopic effect; this whirl frequency is now independent of the rotor spin speed.

If we consider the pure translational motion, then the critical speed will not experience any gyroscopic moments; hence, the corresponding critical speeds can be obtained as

$$\omega_{cr_1} = \omega_{cr_2} = \sqrt{\frac{2k}{m}} = \sqrt{\frac{2\times1\times10^6}{245.04}} = 90.34 \text{ rad/s}$$

with

$$m = \rho\frac{\pi}{4}d^2 l = 7800\times\frac{\pi}{4}(0.2)^2\times1 = 245.04 \text{ kg}$$

where d is the diameter of the cylinder. In this case, the natural frequency does not change with the spin speed of the rotor and remains equal to the critical speeds (90.34 rad/s) obtained earlier. An alternative solution to this problem will be presented in a subsequent section based on dynamic analysis by considering the formulation of governing differential equations of the rotor. When the translational and rotational motions are coupled, then the aforementioned analyses are not valid; a subsequent section deals with such analyses.

5.5 ASYNCHRONOUS GENERAL MOTION OF ROTOR SYSTEMS

In the critical speed analysis of the cantilever rotor in Section 5.3, the rotor was whirling (the whirling is defined as a circular motion of the deflected shaft center line about its undeflected position with small amplitude) and spinning at the same angular speed and in the same direction (i.e., the synchronous whirl case). So, in that analysis, we concentrated more on determining the critical speed rather than obtaining variation of whirl frequency with the spin speed of the shaft for a general motion. In fact, we attempted a special case for which the spin speed was equal to whirl frequency by assuming the operation of the rotor near the critical speed. Cases have been observed where the whirling and the spinning occur at different frequencies and sometimes in opposite directions, as described in the previous section (Section 5.4), in which the analysis was for only the pure rotational motion. The aim of the present section is to calculate natural whirling frequencies, ν, of a rotor (i.e., a shaft with a single disk on it) at any speed of rotation, ω, in the most general manner as shown in Figure 5.23, where ω is the shaft spin speed about the deflected centerline and ν is the shaft whirling frequency about undeformed position *OA*. Because a general motion of the disk is difficult to visualize, the following two cases (i.e., the pure whirling: Case I in Figure 5.24 and the pure spinning: Case II in Figure 5.25) have been considered so that is relatively easier to visualize, and the final motion will be the superposition of two cases (i.e., Cases I and II).

Case I: $\nu=\nu_0$ with $\omega=0$ (pure whirling motion)

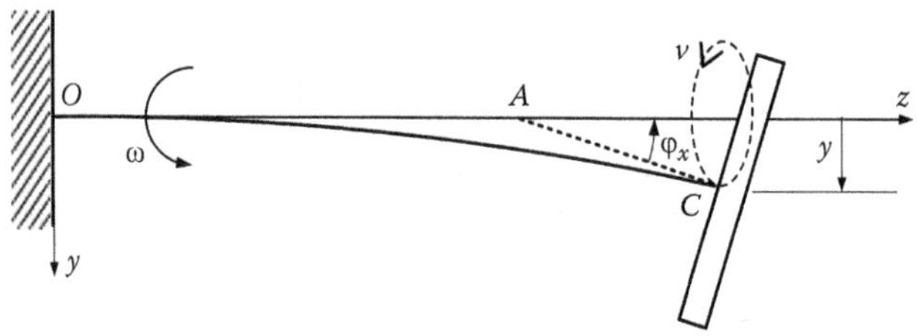

FIGURE 5.23 A cantilever rotor with a general motion of the spinning and whirling.

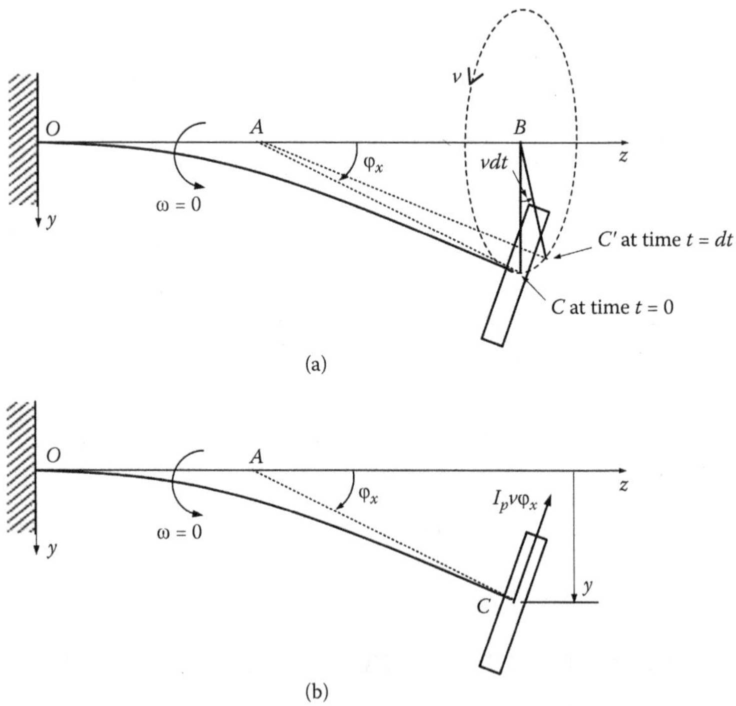

FIGURE 5.24 A pure whirling motion of the rotor (Case I): (a) the pure whirling motion; (b) the angular momentum due to the pure whirling motion.

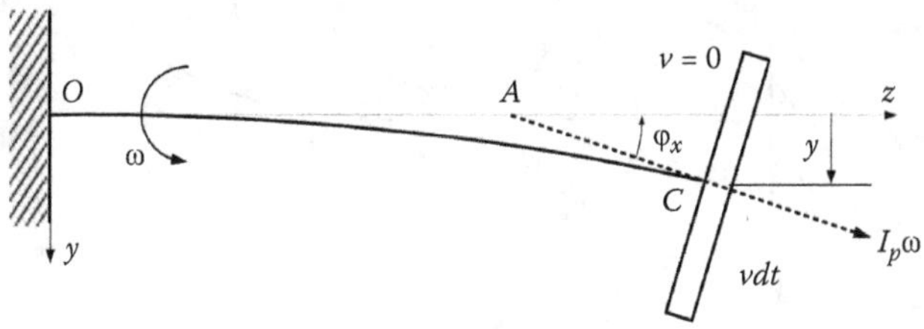

FIGURE 5.25 The angular momentum of the cantilever rotor with a pure spinning motion (Case II).

The shaft does not spin about its deflected centerline, but the deflected centerline *OC* whirls with ν_0 about the bearing center line *OA* (see Figure 5.24).

Case II: $\nu=0$ with $\omega=\omega_0$ (pure spinning motion)

The shaft is in the deflected position *OC*, and it spins with ω_0 about the deflected centerline *OC* (see Figure 5.25).

With these combined ν and ω motions (*Cases I* and *II*), our first aim would be to obtain the angular momentum. For *Case II* when it does not whirl but only spins, the angular momentum is equal to $I_p\omega$ (along *AC* as shown in Figure 5.25), where I_p is the polar moment of inertia of the disk.

For *Case I* when there is no spinning $\omega=0$, but only a whirling, ν, the disk wobbles in the space (about its diameter) and it is difficult to visualize its (wobbling) angular speed. The visualization can be made easier by noting that at the point *C*, the shaft is always perpendicular to the disk, so that we can study the motion of a shaft segment near C instead of the disk (refer to Figure 5.24). The line *CA* is tangential to the shaft at point *C*. The small piece *ds* of the shafting at the disk moves with the line *AC*, describing a cone with point *A* as an apex as shown in Figure 5.24a. The velocity of point *C* for a whirling in the CCW direction, as seen from the right, is perpendicular and into the plane of the paper and its value is νy, where *y* is the translational displacement of the disk center and ν is the whirl frequency.

The line AC lies in the plane of the paper at time $t=0$ (see Figure 5.24a), but at time *dt* later, point *C* is behind the plane of paper by

$$CC' = (\nu dt)\,y = (\nu y)\,dt \tag{5.65}$$

The angle between two positions of line *AC* (i.e., *AC* and *AC′* in Figure 5.24a) is

$$\frac{CC'}{AC} = \nu\left(\frac{y}{AC}\right)dt = \nu\varphi_x dt \quad \text{with} \quad \varphi_x = \frac{y}{AC} \tag{5.66}$$

where φ_x is the rotational transverse displacement of the disk, which is considered to be small. From Equation 5.66, the angle of rotation of *AC* in time *dt* is equal to $\nu\varphi_x dt$. Hence, the angular speed of *AC* (and of the disk) is equal to $\nu\varphi_x$. The disk rotates about a diameter in the plane of the paper and perpendicular to *AC* at *C* so that the appropriate moment of inertia is I_d ($=\frac{1}{2}I_p$ for the thin disk). Hence, the angular momentum vector of the disk due to the pure whirl is $I_d\nu\varphi_x$ and is shown in Figure 5.24b. The direction as shown in the figure of the angular momentum can be obtained by considering the tilting of the disk when point *C* is moving to *C′* (refer to Figures 5.24a and 5.26b). During this motion, the disk will try to tilt such that the left side of the face would be visible to the observer. It will be clearer when the disk center occupies a position along the line *OB* and the disk is inside the plane of the paper (Figure 5.26b) fully. At this instant, it tilts about its diameter; hence, the observer will see motion of the disk tilting in the CW direction when looking at the disk from the bottom along the diagonal. The disk center will occupy the same position on line *OB*, when it is outside the plane of the paper fully, and in this case, the observer will be able to see the right side of the face (Figure 5.26d). The total angular momentum is the vector sum of $I_p\omega$ and $I_d\nu\varphi_x$ (see Figure 5.27), which have been obtained from Cases I and II, respectively.

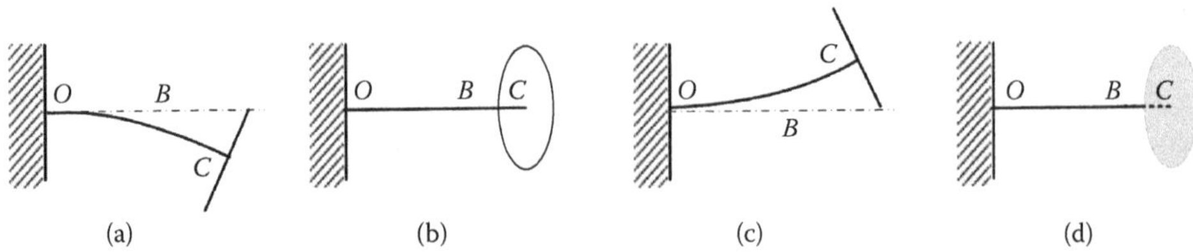

FIGURE 5.26 Disk tilting during pure whirling (CCW direction when looking from right) of the disk when (a) in the downmost position, (b) into the plane of the paper, (c) in the upmost position, and (d) outside the plane of the paper.

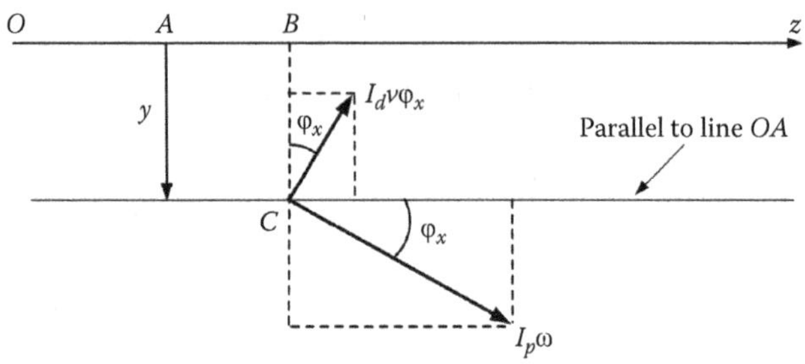

FIGURE 5.27 Angular momentums due to the whirling and spinning of a rotor.

Now our aim is to calculate the rate of change of the angular momentum vector, which has been obtained for Cases I and II, individually. For this purpose, we resolve the vector into components parallel (the direction from O to A as positive) and perpendicular (the direction from B to C as positive) to line OA as shown in Figure 5.27, which are given, respectively, as

$$I_p \omega \cos\varphi_x + I_d \nu \varphi_x \sin\varphi_x \approx I_d\left(2\omega + \nu\varphi_x^2\right) \quad \| OA \tag{5.67}$$

and

$$I_p \omega \sin\varphi_x - I_d \nu \varphi_x \cos\varphi_x \approx I_d \varphi_x (2\omega - \nu) \quad \perp OA \tag{5.68}$$

where $I_p \approx 2I_d$ for the thin disk and for a small rotational displacement, φ_x, of the disk such that $\cos\varphi_x \approx 1$ and $\sin\varphi_x \approx \varphi_x$.

Components parallel to line OA rotate around line OA in a circle with a radius y and keep the magnitude and the direction of the angular momentum constant during the process so that its rate of change is zero, as shown in Figure 5.28a.

Angular momentum components perpendicular to line OA create a vector along the direction of line BC and rotate in a circle with the center as point B, as shown in Figure 5.28b. Let time $t=0$ so this vector lies in the plane of the paper, and at time dt, this vector moves inside the plane of the paper at an angle νdt (see Figure 5.28b). The increment in the vector is CC', which is directed perpendicular to the paper and into it, with the magnitude equal to the length of the vector itself, $I_d \varphi_x (2\omega - \nu)$, multiplied by an angle of νdt. It is given as

$$I_d \varphi_x (2\omega - \nu) \nu dt \tag{5.69}$$

The rate of change of the angular momentum with time (i.e., the moment) is then given as

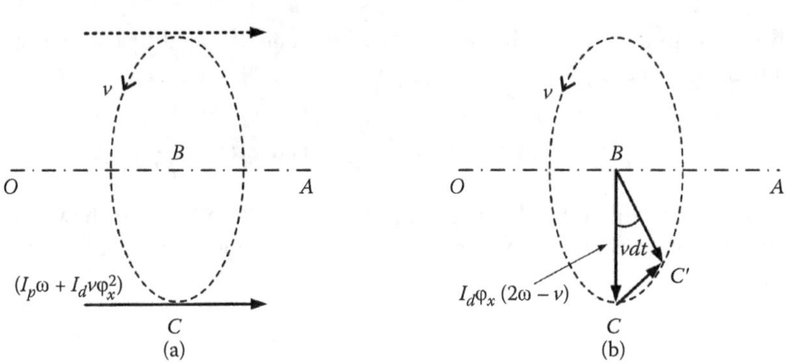

FIGURE 5.28 Angular momentum components: (a) parallel; (b) perpendicular to the undeflected position of the shaft OA.

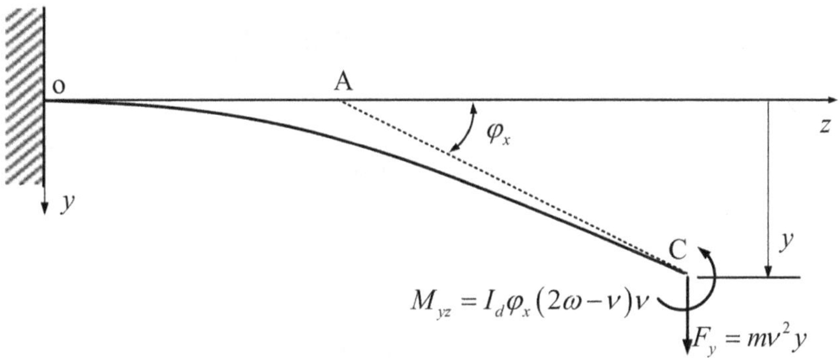

FIGURE 5.29 The inertia force and moment acting from the disk on the shaft caused by the shaft rotation, ω, and the shaft whirling, ν.

$$I_d\varphi_x(2\omega-\nu)\nu \tag{5.70}$$

For the synchronous case $\nu=\omega$, we have from the previous equation the rate of change of angular momentum (or gyroscopic moment) as $I_d\omega^2\varphi_x$, which is the same as Equation 5.32. Earlier, it was obtained by using the concept of the centrifugal force. Now the expression is an active moment; in other words, this is the moment exerted on the disk by the shaft (i.e., by action). The reaction moment exerted by the disk on the shaft is equal and opposite, that is, a vector directed out of the paper and perpendicular to it at *C*. For the antisynchronous case $\nu = -\omega$, from Equation 5.70, the rate of change of angular momentum (or gyroscopic moment) is obtained as $3I_d\omega^2\varphi_x$, which is three times more than the synchronous case. In addition to this moment, there is a centrifugal force, $m\omega^2 y$, acting on the disk from Case I, as shown in Figure 5.29.

Influence coefficients of the shaft can be defined as follows: α_{11} is the deflection y at the disk from a unit force; α_{21} is the angle φ_x at the disk from a unit force or is the deflection y at the disk from a unit moment, that is, $\alpha_{12}=\alpha_{21}$ (by Maxwell's reciprocal theorem of strength of materials); and α_{22} is the angle φ_x at disk from a unit moment (in all cases, remaining forces or moments would be zero). For example, for the cantilever beam (refer to Appendix 8.A and note the sign conventions adopted for the positive shear force and bending moment) with a concentrated load F and moment M as shown in Figure 5.29, we have

$$\alpha_{11} = \frac{l^3}{3EI}, \quad \alpha_{12} = \alpha_{21} = \frac{l^2}{2EI} \quad \text{and} \quad \alpha_{22} = \frac{l}{EI} \tag{5.71}$$

It should be remembered that other boundary conditions can also be used; however, we have to obtain the relevant influence coefficients. The transverse translational and rotational deflections can be expressed as

$$y = F_y\alpha_{11} + M_{yz}\alpha_{12} \quad \text{and} \quad \varphi_x = F_y\alpha_{12} + M_{yz}\alpha_{22} \tag{5.72}$$

It should be noted that the sign convention of M_{yz} used for obtaining the influence coefficient is CW; however, the reactive moment from the disk to the shaft is CCW—hence, the negative sign in the moment term in the following expression. On substituting the force, F_y, and the moment, M_{yz}, from Figure 5.29, we get

$$y = \alpha_{11}m\nu^2 y + \alpha_{12}\left\{-I_d\varphi_x\nu(2\omega - \nu)\right\} \tag{5.73}$$

and

$$\varphi_x = \alpha_{12}m\nu^2 y + \alpha_{22}\left\{-I_d\varphi_x\nu(2\omega - \nu)\right\} \tag{5.74}$$

which can be rearranged in a matrix form as

$$\begin{bmatrix} 1-\alpha_{11}m\nu^2 & I_d\alpha_{12}\nu(2\omega-\nu) \\ -\alpha_{12}m\nu^2 & 1+\alpha_{22}I_d\nu(2\omega-\nu) \end{bmatrix} \begin{Bmatrix} y \\ \varphi_x \end{Bmatrix} = 0 \tag{5.75}$$

Equation 5.75 is a homogeneous equation in y and φ_x; on putting the determinant of the matrix equal to zero, we get the frequency equation as

$$\begin{aligned} &\nu^4\left(-m\alpha_{11}\alpha_{22}I_d + m\alpha_{12}^2 I_d\right) + \nu^3\left(2m\alpha_{11}\alpha_{22}I_d\omega - 2m\alpha_{12}^2 I_d\omega\right) \\ &+\nu^2\left(\alpha_{22}I_d + m\alpha_{11}\right) + \nu\left(-\alpha_{22}I_d 2\omega\right) - 1 = 0 \end{aligned} \tag{5.76}$$

Equation 5.76 contains seven system parameters: ω, ν, m, I_d, α_{11}, α_{12}, and α_{22}, which makes a good understanding of the solution very difficult. It should be kept in mind that the previous frequency equation represents a large class of single-mass rotor systems, and depending upon boundary conditions (e.g., the simply supported, cantilever, fixed-fixed, overhang, etc.), only the influence coefficient expressions (i.e., Equation 5.71) would change. It is worthwhile to diminish the number of parameters as much as possible by nondimensionalizing them. We introduce four new variables:

$$\text{The dimensionless frequency}: \bar{\nu} = \nu\sqrt{\alpha_{11}m}$$

$$\text{The disc effect: } \mu = \frac{I_d\alpha_{22}}{m\alpha_{11}}$$

$$\text{The elastic coupling}: \bar{\alpha} = \frac{\alpha_{12}^2}{\alpha_{11}\alpha_{22}}$$

and

$$\text{The dimensionless speed}: \bar{\omega} = \omega\sqrt{m\alpha_{11}} \tag{5.77}$$

With these new four nondimensional variables, Equation 5.76 can be simplified in the following steps:

$$\nu^4 (mI_d\alpha_{11}\alpha_{22})\left(-1+\frac{\alpha_{12}^2}{\alpha_{11}\alpha_{22}}\right)+\nu^3 (2mI_d\alpha_{11}\alpha_{22}\omega)\left(1-\frac{\alpha_{12}^2}{\alpha_{11}\alpha_{22}}\right)$$

$$+\nu^2 (m\alpha_{11})\left(\frac{\alpha_{22}I_d}{m\alpha_{11}}+1\right)+\nu(-\alpha_{22}I_d 2\omega)-1=0$$

or

$$\left(\frac{\bar{\nu}}{\sqrt{\alpha_{11}m}}\right)^4 (mI_d\alpha_{11}\alpha_{22})(-1+\bar{\alpha})+\left(\frac{\bar{\nu}}{\sqrt{\alpha_{11}m}}\right)^3 (2mI_d\alpha_{11}\alpha_{22}\omega)(1-\bar{\alpha})$$

$$+\left(\frac{\bar{\nu}}{\sqrt{\alpha_{11}m}}\right)^2 (m\alpha_{11})(\mu+1)+\left(\frac{\bar{\nu}}{\sqrt{\alpha_{11}m}}\right)(-\alpha_{22}I_d 2\omega)-1=0$$

or

$$\bar{\nu}^4\left(\frac{I_d\alpha_{22}}{m\alpha_{11}}\right)(-1+\bar{\alpha})+\bar{\nu}^3\left(\frac{2I_d\alpha_{22}\omega}{\sqrt{\alpha_{11}m}}\right)(1-\bar{\alpha})+\bar{\nu}^2(\mu+1)+\bar{\nu}\left(\frac{-\alpha_{22}I_d 2\omega}{\sqrt{\alpha_{11}m}}\right)-1=0$$

or

$$\bar{\nu}^4(\mu)(-1+\bar{\alpha})+\bar{\nu}^3\left(\frac{2I_d\alpha_{22}}{\alpha_{11}m}\right)(\omega\sqrt{\alpha_{11}m})(1-\bar{\alpha})+\bar{\nu}^2(\mu+1)+\bar{\nu}\left(\frac{-\alpha_{22}I_d 2}{\alpha_{11}m}\right)(\omega\sqrt{\alpha_{11}m})-1=0$$

or

$$\bar{\nu}^4(\mu)(\bar{\alpha}-1)-\bar{\nu}^3(2\mu)(\bar{\omega})(\bar{\alpha}-1)+\bar{\nu}^2(\mu+1)-\bar{\nu}(2\mu)(\bar{\omega})-1=0$$

or

$$\bar{\nu}^4-2\bar{\omega}\bar{\nu}^3+\frac{\mu+1}{\mu(\bar{\alpha}-1)}\bar{\nu}^2-\frac{2\bar{\omega}}{\bar{\alpha}-1}\bar{\nu}-\frac{1}{\mu(\bar{\alpha}-1)}=0 \tag{5.78}$$

Equation 5.78 is a fourth-degree polynomial in $\bar{\nu}$, so for a given $\bar{\alpha}$, carrying a given disk μ, and rotating at certain speed $\bar{\omega}$, there will be four natural frequencies of the whirl. Now, different cases will be considered to physically interpret them.

Case I: For a point mass of the disk, that is, $I_d=0$ or $\mu=0$, multiply Equation 5.78 by μ and on substituting $\mu=0$, we get

$$\frac{0+1}{\bar{\alpha}-1}\bar{\nu}^2-\frac{1}{\bar{\alpha}-I}=0 \quad \text{or} \quad \bar{\nu}=\pm 1 \quad \text{or} \quad \nu=\frac{\pm 1}{\sqrt{\alpha_{11}m}} \tag{5.79}$$

Because $\alpha_{11}=1/k$ in the usual natural notation, we have the following expression for the natural frequency of the whirl:

$$\nu=\pm\sqrt{\frac{k}{m}} \tag{5.80}$$

This result is very much familiar.

Case II: $\bar{\alpha} = 0$, that is, no elastic coupling between the translational and rotational displacements. For this case, a force causes the translational deflection only, without any rotational deflection φ_x, whereas a moment causes the rotational deflection φ_x only, without any translational deflection y. For a shaft with simple supports and a disk at the midspan, such a case is possible, as shown in Figure 5.30.

For $\bar{\alpha} = 0$, Equation 5.78 reduces to

$$\bar{\nu}^4 - 2\bar{\omega}\bar{\nu}^3 - \frac{\mu+1}{\mu}\bar{\nu}^2 + 2\bar{\omega}\bar{\nu} + \frac{1}{\mu} = 0 \tag{5.81}$$

Equation 5.81 can be simplified in the following steps:

$$\bar{\nu}^4 - 2\bar{\omega}\bar{\nu}^3 + 2\bar{\omega}\bar{\nu} - \frac{\mu+1}{\mu}\bar{\nu}^2 + \frac{1}{\mu} = 0 \quad \text{or} \quad \bar{\nu}^4 - 2\bar{\omega}\bar{\nu}\left(\bar{\nu}^2 - 1\right) - \bar{\nu}^2 - \frac{1}{\mu}\bar{\nu}^2 + \frac{1}{\mu} = 0$$

or

$$\bar{\nu}^4 - 2\bar{\omega}\bar{\nu}\left(\bar{\nu}^2 - 1\right) - \bar{\nu}^2 - \frac{1}{\mu}\left(\bar{\nu}^2 - 1\right) = 0 \quad \text{or} \quad \bar{\nu}^2\left(\bar{\nu}^2 - 1\right) - 2\bar{\omega}\bar{\nu}\left(\bar{\nu}^2 - 1\right) - \frac{1}{\mu}\left(\bar{\nu}^2 - 1\right) = 0$$

or

$$\left(\bar{\nu}^2 - 1\right)\left(\bar{\nu}^2 - 2\bar{\omega}\bar{\nu} - \frac{1}{\mu}\right) = 0 \quad \text{or} \quad (\bar{\nu}+1)(\bar{\nu}-1)\left(\bar{\nu}^2 - 2\bar{\omega}\bar{\nu} - \frac{1}{\mu}\right) = 0 \tag{5.82}$$

Equation 5.82 gives whirl frequencies as

$$\bar{\nu}_1 = +1; \quad \bar{\nu}_2 = -1 \tag{5.83}$$

and

$$\bar{\nu}_{3,4} = \bar{\omega} \pm \sqrt{\bar{\omega}^2 + \frac{1}{\mu}} \tag{5.84}$$

The nondimensional frequency $\bar{\nu}$ is plotted against the speed $\bar{\omega}$ for a numerical value of $\mu = 1$, as shown in Figure 5.31. For $\bar{\omega} = 0$ (in fact, for any other values) from the first and second solutions, that is, Equation 5.83, we have $\bar{\nu} = \pm 1$, which is the natural frequency related to the up and down motions of the disk without any tilting. For $\bar{\omega} = 0$ (but not for $\bar{\omega} \neq 0$) from the third and fourth solutions, that is, Equation 5.84, we have $\bar{\nu} = \pm\sqrt{1/\mu}$ and because $\mu = 1$, $\bar{\nu} = \pm 1$. This is the natural frequency related to the precessional motion of the disk, that is, wobbling without up and down motions. From Equation 5.84, it can be seen that for the present case, the second forward critical speed (for $\bar{\nu} = \bar{\omega} = \bar{\omega}_{cr2}^F$) would not be feasible because, on substitution of this condition, it gives $\bar{\omega}_{cr2}^F = \sqrt{\frac{-1}{\mu}}$, and only the second backward critical speed (for $\bar{\nu} = -\bar{\omega} = \bar{\omega}_{cr2}^B$) would be $\bar{\omega}_{cr2}^B = \sqrt{\frac{1}{3\mu}}$

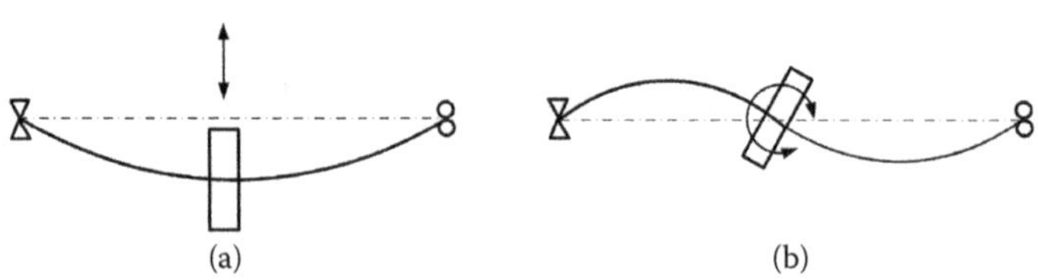

FIGURE 5.30 A Jeffcott rotor with the disk motion as (a) the pure translation and (b) the pure rotational motion.

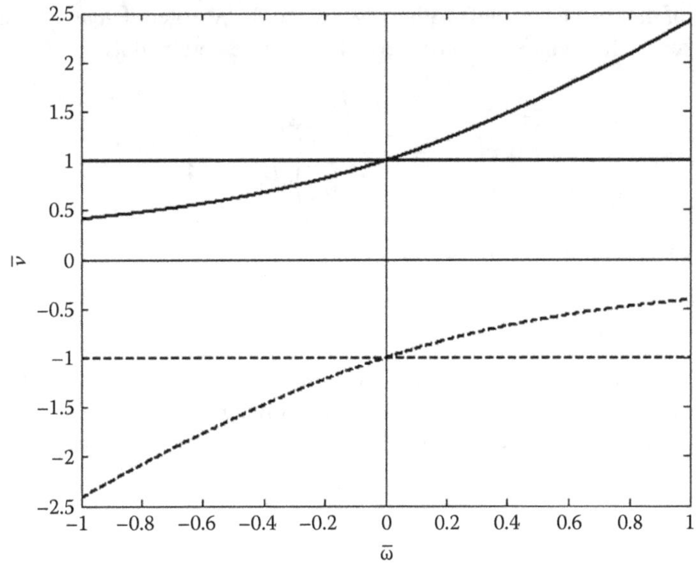

FIGURE 5.31 The variation of the whirl frequency versus the spin speed (for $\bar{\alpha} = 0, \mu = 1$).

(which can be obtained by substitution of the present condition). That means the gyroscopic couple reduces the number of forward critical speeds (Tondl, 1965; Shravankumar and Tiwari, 2011; and refer to Chapter 1 of Volume II).

The disk effect μ is so chosen or dimensioned such that the natural frequency of the translational up and down motions is the same as the natural frequency of the wobbling (tilting) without up and down motions at zero speed, that is, $\bar{\omega} = 0$ (see Figure 5.31). It should be noted that these kinds of uncoupling of the translational and rotational motions are a special case with $\bar{\alpha} = 0$ and in general such uncoupling of motions is not feasible, and for such systems, all natural frequencies—both the translational and rotational motions—would be present simultaneously with different relative magnitudes and directions. From Equation 5.77, we have

$$\mu = \frac{I_d \alpha_{22}}{m\alpha_{11}} = 1 \quad \Rightarrow \frac{1}{\sqrt{m\alpha_{11}}} = \frac{1}{\sqrt{I_d \alpha_{22}}} \quad \Rightarrow \omega_{nf_{linear}} = \omega_{nf_{wobbling}} \tag{5.85}$$

Figure 5.31 shows that there are four natural frequencies at any given speed. For a positive value of $\bar{\omega}$, it could be seen that, corresponding to the translational up and down motion, the natural whirl frequency remains constant, whereas for the tilting motion corresponding to the forward whirl, it increases, and for the backward whirl, it decreases. This is due to the fact that the gyroscopic couple is affected only by the tilting (transverse rotational) motion, and there is no coupling of the transverse translational and rotational motions for the present case (i.e., $\bar{\alpha} = 0$). For the negative value of $\bar{\omega}$, the frequency curve is symmetric with respect to the $\bar{\omega} = 0$ line, which means the natural frequency does not depend upon the direction of rotation of the symmetric rotor. For the zero speed, the four natural frequencies are reduced to two, $\bar{\nu} = \pm 1$, which is really one frequency only because $\bar{\nu} = +1$ corresponds to the forward whirl and $\bar{\nu} = -1$ corresponds to the backward whirl, and the nonrotating shaft cannot distinguish the forward and backward whirls.

Case III: Now, the frequency, Equation 5.78, is plotted, which is for a general motion when the disk has both translational and rotational motions.

Taking a case of the cantilever shaft with a disk (it should be noted that any other boundary conditions can be taken at this stage), the elastic coupling can be obtained as

$$\bar{\alpha} = \frac{\alpha_{12}^2}{\alpha_{11}\alpha_{22}} = \frac{\left(l^2/2EI\right)^2}{\left(l^3/3EI\right)\left(l/EI\right)} = \frac{3}{4}$$

For $\bar{\alpha} = 3/4$ and $\mu = 1$, Equation 5.78 becomes

$$\bar{\nu}^4 - 2\bar{\omega}\bar{\nu}^3 - 8\bar{\nu}^2 + 8\bar{\omega}\bar{\nu} + 4 = 0 \tag{5.86}$$

Equation 5.86 is a fourth-degree polynomial in $\bar{\nu}$ with four roots $\bar{\nu} = f(\bar{\omega})$. It can be solved in two ways: (i) take a value of $\bar{\omega}$ and solve the fourth-degree polynomial equation in $\bar{\nu}$, or (ii) take a value of $\bar{\nu}$ and solve a linear equation in $\bar{\omega}$; that is, Equation 5.86 can be rearranged as

$$\bar{\omega} = \frac{\bar{\nu}^4 - 8\bar{\nu}^2 + 4}{2\bar{\nu}^3 - 8\bar{\nu}} \tag{5.87}$$

As an example, for $\bar{\nu} = 1$ we get $\bar{\omega} = 0.5$, etc., and it will result in Figure 5.32. Similar plots can be obtained for the other values of μ and $\bar{\alpha}$ (i.e., other disks and other shafts) with different boundary conditions. It is seen in Figure 5.32 that for $\bar{\omega} = 0$ there are only two nondimensional whirl natural frequencies, $\bar{\nu} = \pm 0.74$ and $\bar{\nu} = \pm 2.73$, corresponding to two DOFs (i.e., y and φ_x) of a nonrotating disk. When the disk rotates, all four natural frequencies are different. Because in the present case, the elastic coupling terms are not zero, the translational and rotational displacements will take place simultaneously for each of the whirl natural frequencies. The curves are symmetrical about the vertical $\bar{\nu}$-axis, which means that corresponding to $+\bar{\omega}$ and $-\bar{\omega}$ the same values of $\bar{\nu}$ occur—in other words, four natural frequencies remain the same for the CW or CCW rotation of the disk due to the symmetry of the rotor. The line $\bar{\nu} = \bar{\omega}$ intersects the frequency curve at point A, which is called a *forward (synchronous) critical speed* (refer to Figure 5.32).

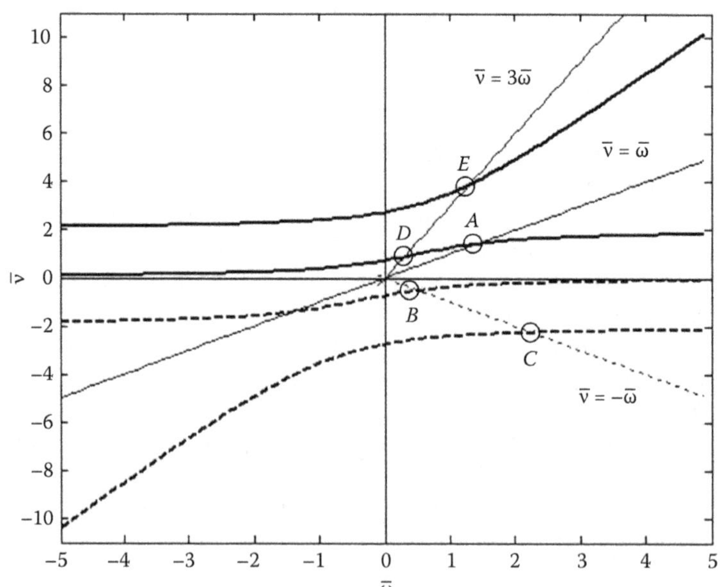

FIGURE 5.32 The variation of the whirl frequency versus the spin speed (Campbell diagram).

At this point, the whirl natural frequency is the same as the spin speed of the rotor; it should be remembered that for the present case, the whirl natural frequency changes with the speed of the rotor due to gyroscopic effects. This kind of motion is obviously excited by the unbalance, which gives the excitation in synchrony with the rotor speed. It is a resonance phenomenon, and the vibration amplitude at this critical speed is proportional to the amount of unbalance. For the present case, it should be noted that the $\bar{\nu} = \bar{\omega}$ line does not intersect the second frequency curve if we extend the line beyond point *A*. So, there is no feasible second forward synchronous critical speed (Tondl, 1965; Shravankumar and Tiwari, 2011, and also refer to Chapter 1 of Volume II).

In Figure 5.32, frequency curves intersect the line $\bar{\nu} = -\bar{\omega}$ at points *B* and *C*, which represent *backward (antisynchronous) critical speeds*. For this case, the natural frequency is the same as the rotor spin speed; however, the sense is in the opposite direction. Such backward critical speeds are very rare in practice; however, they do occur when a rotor-to-stator contact takes place, which gives excitation opposite to the rotation of the shaft due to the friction. It should be noted that in a general case, the maximum of four critical speeds (1*F*, 1*B*, 2*F*, and 2*B*) might occur, two each for the forward and backward whirls; however, for the present case, there are three (1*F*, *1B*, and 2*B*). Apart from these synchronous critical whirls, if the rotor has excitation, which is, for example, thrice the spin speed, then the corresponding critical speeds can be obtained by intersecting a line $\bar{\nu} = 3\bar{\omega}$ with the frequency curve (i.e., at points *D* and *E*). However, the occurrence of such critical speeds is very rare because to excite such critical speeds physically, the excitation mechanism should be present, and in most of the cases, they are not dangerous (i.e., the amplitude of vibration is low). The coordinates of the various points in Figure 5.32 are *A*(1.414, 1.414), *B*(0.0513, –0.513), *C*(2.252, –2.252), *D*(0.2916, 0.8882), and *E*(1.3, 3.9002). Such a plot is called a Campbell diagram (Campbell, 1924).

The summary of the quasi-static analysis of a single-mass system with gyroscopic effects is presented as follows.

Case I: For $\nu = \omega$, that is, the synchronous whirl, a single forward critical speed is obtained for three cases as shown in Figure 5.33 based on the centrifugal force concept.

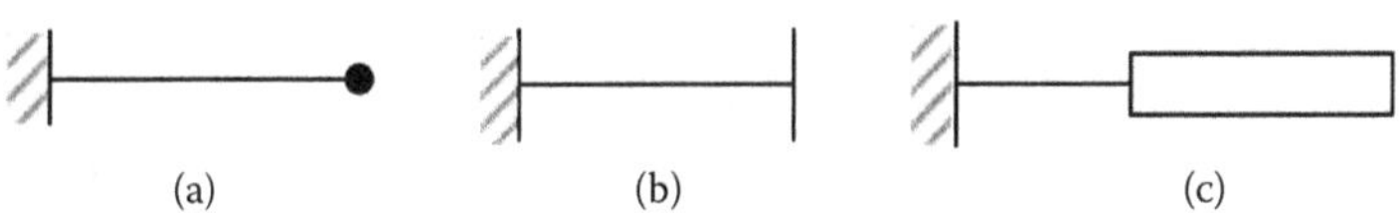

FIGURE 5.33 Cantilever rotors with a (a) point mass, (b) rigid thin disk, and (c) long stick.

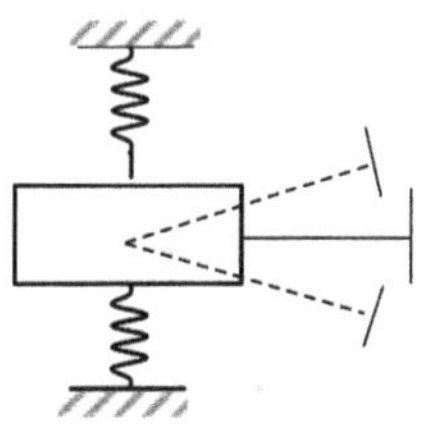

FIGURE 5.34 A flexibly mounted motor carrying a rotor.

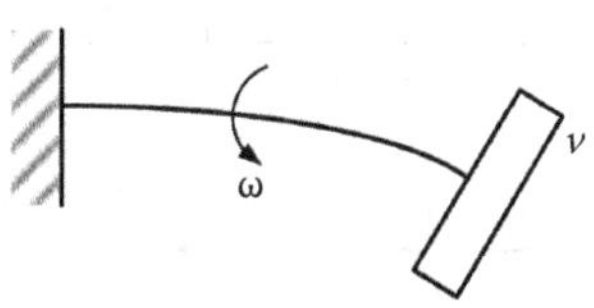

FIGURE 5.35 A cantilever rotor.

Case II: For a rotor mounted on flexible torsional springs so as to have pure rotational displacement, as shown in Figure 5.34, it has two kinds of whirls: fast whirl (*forward*) in the same sense as ω, and slow whirl (*backward*) in the opposite sense as ω. This is analyzed based on the change in the angular momentum principle.

Case III: For a general motion of a cantilever rotor as shown in Figure 5.35, it has four whirl natural frequencies, two forward and two backward. The first forward critical speed is the same as for Case I due to the unbalance in the rotor. This is analyzed based on the change in the angular momentum principle.

For every natural frequency of a rotor-bearing system, we have a backward and a forward natural frequency. If one wants to write the first natural frequency of the system, should I report the average of backward and forward as my first natural frequency? The answer to this is as follows: Generally, force is from unbalance, which gives forward critical speed resonance. So generally, forward critical speed is given when you have gyroscopic effect (or speed-dependent bearing parameters). Backward critical speed is rare. Otherwise, it would be better to provide both *F* and *B*, but the average has no physical significance.

Example 5.4

Obtain the transverse forward and backward synchronous critical speeds for a general motion of a rotor system, as shown in Figure 5.36. Take the mass of the disk as $m=10$ kg, the diametral mass moment of inertia as $I_d=0.02$ kg-m², and the polar mass moment of inertia as $I_p=0.04$ kg-m². The disk is placed at $b=0.25$ m from the right support. The shaft has a diameter of 10 mm, a total span length of 1 m, and Young's modulus of 2.1×10^{11} N/m². The shaft is assumed to be massless, and gyroscopic effects are included. Draw the Campbell diagram and show the forward and backward synchronous critical speeds.

Solution:

For the rotor system shown in Figure 5.36, the influence coefficients are defined as

$$\begin{Bmatrix} y \\ \varphi_x \end{Bmatrix} = \begin{bmatrix} \alpha_{11} & \alpha_{12} \\ \alpha_{21} & \alpha_{22} \end{bmatrix} \begin{Bmatrix} f \\ M \end{Bmatrix}$$

with

$$\alpha_{11} = \frac{a^2b^2}{3EIl} = 1.137\times10^{-4} \text{ m/N}; \quad \alpha_{12} = \frac{\left(3a^2l - 2a^3 - al^2\right)}{3EIl} = 3.031\times10^{-4} \text{ m/N-m}$$

and

$$\alpha_{21} = \frac{ab(b-a)}{3EIl} = 3.031\times10^{-4} \text{ rad/N}; \quad \alpha_{22} = -\frac{\left(3al - 3a^2 - l^2\right)}{3EIl} = 1.4146\times10^{-3} \text{ rad/N-m}$$

From Equation 5.78, the frequency equation is given by

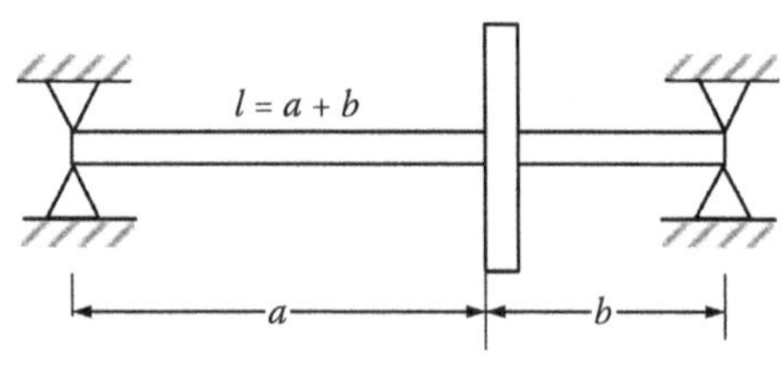

FIGURE 5.36 A rotor system.

$$\bar{\nu}^4 - 2\bar{\omega}\bar{\nu}^3 + \frac{\mu+1}{\mu(\bar{\alpha}-1)}\bar{\nu}^2 - \frac{2\bar{\omega}}{\bar{\alpha}-1}\bar{\nu} - \frac{1}{\mu(\bar{\alpha}-1)} = 0 \tag{a}$$

We have

$$\mu = \frac{I_d\alpha_{22}}{m\alpha_{11}} = \frac{0.02\times 1.4146\times 10^{-3}}{10\times 1.137\times 10^{-4}} = 0.0249$$

and

$$\bar{\alpha} = \frac{\alpha_{12}^2}{\alpha_{11}\alpha_{22}} = \frac{\left(3.031\times 10^{-4}\right)^2}{1.137\times 10^{-4}\times 1.4146\times 10^{-3}} = 0.5712$$

On substituting values of μ and $\bar{\alpha}$ in Equation (a), we get

$$\bar{\nu}^4 - 2\bar{\omega}\bar{\nu}^3 - 95.99\bar{\nu}^2 + 4.664\bar{\omega}\bar{\nu} + 93.658 = 0 \tag{b}$$

For the forward whirl, we have $\bar{\nu} = +\bar{\omega}$; hence, Equation (b) reduces to

$$\bar{\omega}^4 + 91.33\bar{\omega}^2 - 93.66 = 0 \tag{c}$$

which can be solved as

$$\bar{\omega}^2 = \frac{-91.33 \pm \sqrt{91.33^2 + 4\times 93.66}}{2} = \frac{-91.33 \pm 93.36}{2} \Rightarrow \bar{\omega}_{1,2}^2 = 1.015 \quad \text{and} \quad -92.35 \tag{d}$$

A negative value is not a feasible solution, so taking the positive value only, we get

$$\bar{\omega} = \omega_{cn}^F\sqrt{\alpha_{11}m} = \omega_{cn}^F\sqrt{1.137\times 10^{-4}\times 10} = 1.0075 \quad \Rightarrow \omega_{cn}^F = 29.88 \text{ rad/s}$$

Hence, only a single forward critical speed is possible for the present case. For the backward whirl, we have $\bar{\nu} = -\bar{\omega}$; hence from Equation (b), we get

$$\bar{\omega}^4 - 33.55\bar{\omega}^2 + 31.22 = 0 \tag{e}$$

which can be solved to give two positive roots, as

$$\bar{\omega}_{1,2}^2 = 0.96 \quad \text{and} \quad 32.59 \quad \Rightarrow \bar{\omega}_{1,2} = \omega_{cn,2}^B\sqrt{\alpha_{11}m} = 0.98 \quad \text{and} \quad 5.71 \tag{f}$$

Hence, two possible backward critical speeds are

$$\omega_{cn}^B = 0.98\times 29.66 = 28.47 \text{ rad/s} \quad \text{and} \quad \omega_{cr_2}^B = 5.71\times 29.66 = 169.34 \text{ rad/s}$$

Figure 5.37 shows the corresponding Campbell diagram, which can be drawn using Equation (b) of the following form:

$$\bar{\omega} = \frac{\bar{\nu}^4 - 95.99\bar{\nu}^2 + 93.658}{2\bar{\nu}^3 - 4.664\bar{\nu}} \tag{g}$$

In Figure 5.37, forward whirl natural frequencies are drawn with solid lines and backward whirl frequencies by dashed lines. The locations of critical speeds are *A*(1.007, 1.007), *B*(0.9787, −0.9787), and *C*(5.7091, −5.7091). It can be seen that coordinates $(\bar{\nu},\bar{\omega})$ of three intersection points between whirl frequency curves and a line $\bar{\nu} = \bar{\omega}$ match the critical speed ratios obtained in Equations (d) and (f), and there is no intersection corresponding to the negative root in Equation (d).

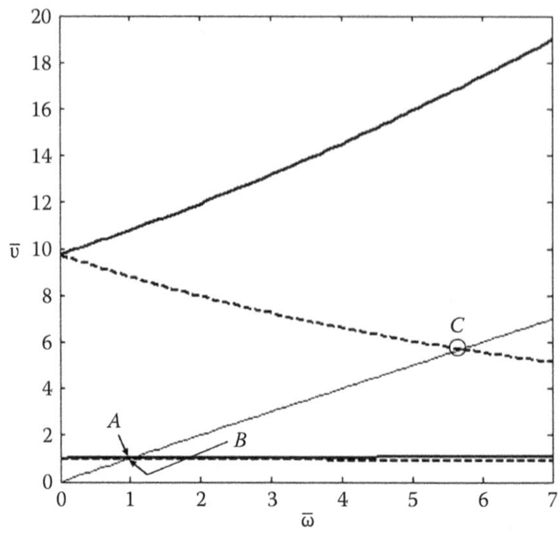

FIGURE 5.37 The Campbell diagram.

Example 5.5

Obtain the forward and backward synchronous transverse critical speeds for a general motion of a rotor as shown in Figure 5.38. The rotor is assumed to be fixed and supported at one end. Take the mass of the disk as $m=2$ kg, the polar mass moment of inertia as $I_p=0.01$ kg-m², and the diametral mass moment of inertia as $I_d=0.005$ kg-m². The shaft is assumed to be massless, and its length and diameter are 0.2 and 0.1 m, respectively. Assume the Young's modulus of the shaft $E=2.1\times10^{11}$ N/m². Draw the Campbell diagram and show the forward and backward synchronous transverse critical speeds.

For the cantilever beam, the influence coefficients are

$$\begin{Bmatrix} y \\ \phi_x \end{Bmatrix} = \begin{bmatrix} \alpha_{11} & \alpha_{12} \\ \alpha_{21} & \alpha_{22} \end{bmatrix} \begin{Bmatrix} f \\ M \end{Bmatrix}$$

with

$$\alpha_{11} = \frac{l^3}{3EI}; \quad \alpha_{12} = \alpha_{21} = \frac{l^2}{2EI}; \quad \alpha_{22} = \frac{l}{EI}$$

Solution:

The frequency equation is

$$\bar{\nu}^4 - 2\bar{\omega}\bar{\nu}^3 + \frac{\mu+1}{\mu(\bar{\alpha}-1)}\bar{\nu}^2 - 2\bar{\omega}\frac{\bar{\nu}}{(\bar{\alpha}-1)} - \frac{1}{\mu(\bar{\alpha}-1)} = 0 \tag{a}$$

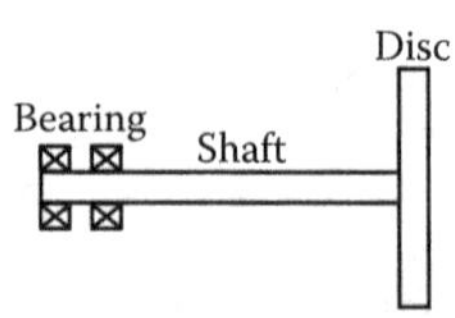

FIGURE 5.38 A cantilever rotor.

From the given parameters, we have

$$\mu = \frac{I_d \alpha_{22}}{m\alpha_{11}} = 0.1875 \quad \text{and} \quad \bar{\alpha} = \frac{\alpha_{12}^2}{\alpha_{22}\alpha_{11}} = 0.75 \tag{b}$$

On substituting Equation (b) into Equation (a), we get

$$\bar{\nu}^4 - 2\bar{\omega}\bar{\nu}^3 - 21.1458\bar{\nu}^2 + 8\bar{\omega}\bar{\nu} + 21.33 = 0 \tag{c}$$

For the forward whirl, we have $\bar{\nu} = +\bar{\omega}$; hence, Equation (c) becomes

$$\bar{\omega}^4 + 13.1458\bar{\omega}^2 - 21.33 = 0 \tag{d}$$

On solving Equation (d), we will get the roots of the equation as

$$\bar{\omega}_{1,2}^2 = \frac{-13.1458 \pm \sqrt{13.1458^2 + 4 \times 21.33}}{2} = \frac{-13.1458 \pm 16.62}{2} \Rightarrow \bar{\omega}_{1,2}^2 = 1.46 \quad \text{and} \quad -21.47 \tag{e}$$

The negative value is not a feasible solution, so on taking the positive value only, we get

$$\bar{\omega}_1 = 1.2 = \bar{\omega}_{cr1}^F \sqrt{\alpha_{11} m} \quad \text{or} \quad \bar{\omega}_{cr1}^F = \frac{1.2}{\sqrt{2.586 \times 10^{-9} \times 2}} = 16{,}744 \text{ rad / s} \tag{f}$$

Hence, only a single forward critical speed is possible.

For the backward whirl, we have $\bar{\nu} = -\bar{\omega}$; on substituting in Equation (c), we get

$$\bar{\omega}^4 - 9.8193\bar{\omega}^2 + 7.11 = 0 \tag{g}$$

which gives

$$\bar{\omega}_{1,2}^2 = \frac{+9.819 \pm \sqrt{9.819^2 + 4 \times 7.11}}{2} = 0.6678 \quad \text{and} \quad 10.24 \tag{h}$$

or

$$\bar{\omega}_1 = 0.8173 = \bar{\omega}_{cr1}^B \sqrt{\alpha_{11} m} \quad \text{and} \quad \bar{\omega}_2 = 3.2 = \bar{\omega}_{cr2}^B \sqrt{\alpha_{11} m} \tag{i}$$

Hence, the possible backward critical speeds are

$$\omega_{cr1}^B = \frac{0.8173}{\sqrt{2.568 \times 10^{-9} \times 2}} = 11{,}396 \text{ rad/s and } \omega_{cr2}^B = \frac{3.2}{\sqrt{2.568 \times 10^{-9} \times 2}} = 44{,}652 \text{ rad/s} \tag{j}$$

To summarize, the forward and backward critical speeds are given as

$$\omega_{cr1}^F = 14{,}906 \text{ rad/s}, \ \omega_{cr1}^B = 44{,}652 \text{ rad/s}, \text{ and } \omega_{cr2}^B = 11{,}396 \text{ rad/s} \tag{k}$$

Campbell diagram: We can draw the Campbell diagram from Equation (c) by expressing it as

$$\bar{\omega} = \frac{\bar{\nu}^4 - 25.33\bar{\nu}^2 + 21.33}{2\bar{\nu}^3 - 8\bar{\nu}} \tag{l}$$

In Figure 5.39, on the abscissa we have taken $\bar{\omega}$ and on the ordinate we have taken $\bar{\nu}$. Forward whirl natural frequencies are drawn with solid lines, and backward whirl frequencies are drawn

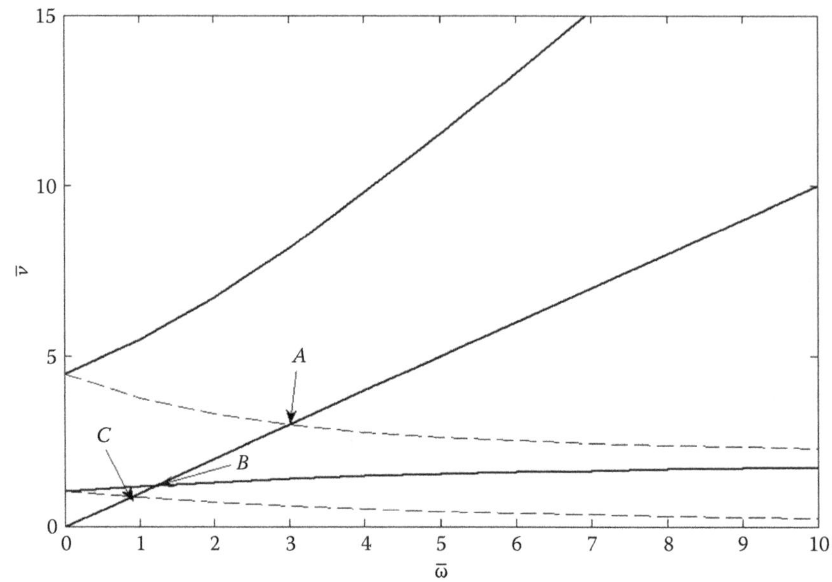

FIGURE 5.39 The Campbell diagram.

with dotted lines. From the diagram, it can be seen that the coordinates of intersection points $A(2.97, 2.97)$, $B(1.21, 1.21)$, and $C(0.84, 0.84)$ between natural whirl curves and a line $\bar{\omega} = \bar{\nu}$ match with the critical speed ratio obtained in Equations (e) and (i). There is no intersection corresponding to the negative root from Equation (e).

5.6 GYROSCOPIC EFFECTS BY THE DYNAMICS APPROACH

In the present section, the gyroscopic effect of a disk on the dynamics of a rotor will be studied. Figure 5.40a shows a Jeffcott rotor with an offset disk. However, it should be noted that for the other boundary conditions, the present analysis also would be valid. First, equations of motion from the free body diagram of the disk are obtained. Free body diagrams of the disk in two orthogonal planes are shown in Figure 5.40b and c. In the z-x plane, the translational and rotational (tilting) displacements are x and φ_y, respectively. Both the translational and rotational (tilting) displacements of the disk are chosen in the positive axis direction. Similarly, in the y-z plane, the translational and rotational displacements are y and φ_x, respectively.

Free body diagrams also show the angular momentum $I_p\omega$ and the change in the angular momentum, $\Delta(I_p\omega)$. For example, in the z-x plane, the change in the angular momentum is in the positive x direction, which gives rise to a CW gyroscopic moment, M_{gx}, onto the disk in the y-z plane as shown in Figure 5.40c. In the y-z plane, the x-axis is into the plane of the paper (it is shown by a symbol ⊗, which represents the tail of the arrow of the x-axis). Similarly, in the y-z plane, the change in the angular momentum is in the negative y direction, which gives rise to a CW gyroscopic moment, M_{gy}, onto the disk in the z-x plane as shown in Figure 5.40b. In the z-x plane, the y-axis is out of the plane of the paper (it is shown by a symbol ⊙, which represents the head or the tip of the arrow of the y-axis). Rotary inertias ($I_d\ddot{\varphi}_y$ and $I_d\ddot{\varphi}_x$) are also shown in both planes, the direction of which is opposite to the rotational displacement. Because the rotary inertia moments are due to inertia, these are shown by dotted arcs.

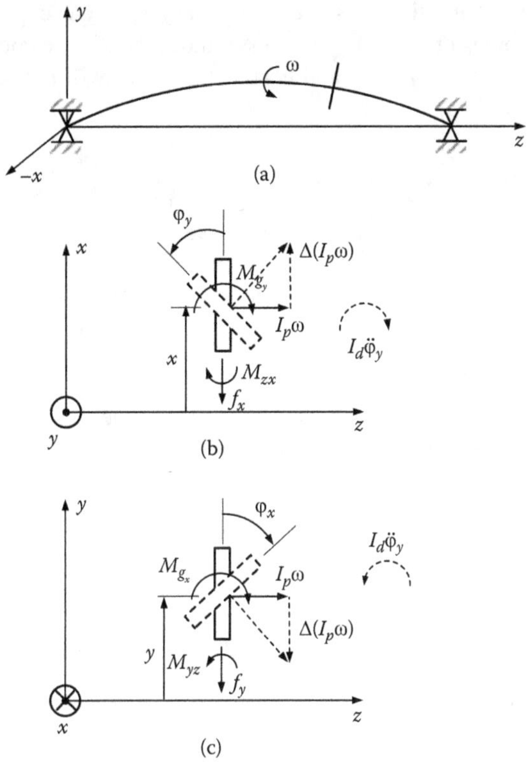

FIGURE 5.40 A Jeffcott rotor with an offset disk from the midspan: (a) a rotor system with an offset disk; (b) a free body diagram of the disk in the z-x plane; (c) a free body diagram of the disk in the y-z plane.

Let f_x and f_y be restoring forces, and M_{yz} and M_{zx} be restoring moments (the signs of which are opposite to the respective translational and rotational displacements). These resulting forces and moments are due to the shaft elasticity. From Figure 5.40, equations of motion can be written as

$$-f_x = m\ddot{x}; \quad -f_y = m\ddot{y} \tag{5.88}$$

$$-M_{zx} - M_{g_y} = I_d\ddot{\varphi}_y; \quad -M_{yz} + M_{g_x} = I_d\ddot{\varphi}_x \tag{5.89}$$

The direction of change in the angular momentum gives the direction of moment to be applied to the disk by the shaft to produce the precession such that the angular momentum tries to align with the external moment. Gyroscopic moments are expressed as

$$M_{g_y} = -I_P\omega\dot{\varphi}_x \quad \text{and} \quad M_{g_x} = -I_P\omega\dot{\varphi}_y \tag{5.90}$$

Because these moments are inertia moments, the negative sign has been added in each of the gyroscopic moments in the previous expressions. On considering the gyroscopic moment also as the inertia moment, equations of motion in Newtonian form can be written as

$$m\ddot{x} = -f_x; \quad m\ddot{y} = -f_y \tag{5.91}$$

$$I_d\ddot{\varphi}_y - I_P\omega\dot{\varphi}_x = -M_{zx}; \quad I_d\ddot{\varphi}_x + I_P\omega\dot{\varphi}_y = -M_{yz} \tag{5.92}$$

where the terms on the left are inertia terms due to the rotary and gyroscopic motions, and the terms on the right are external forces or moments. For a symmetric shaft, the coupling of motions due to elastic forces and moments in the vertical and horizontal planes will not be there; hence, from the elementary strength of materials, we have

$$\begin{Bmatrix} f_x \\ f_y \\ M_{zx} \\ M_{yz} \end{Bmatrix} = \begin{bmatrix} \alpha_{ll} & 0 & \alpha_{la} & 0 \\ & \alpha_{ll} & 0 & \alpha_{la} \\ & & \alpha_{aa} & 0 \\ sym & & & \alpha_{aa} \end{bmatrix}^{-1} \begin{Bmatrix} x \\ y \\ \varphi_y \\ \varphi_x \end{Bmatrix} = \begin{bmatrix} k_{ll} & 0 & k_{la} & 0 \\ & k_{ll} & 0 & k_{la} \\ & & k_{aa} & 0 \\ sym & & & k_{aa} \end{bmatrix} \begin{Bmatrix} x \\ y \\ \varphi_y \\ \varphi_x \end{Bmatrix} \tag{5.93}$$

where subscripts l and a represent the linear (translational) and the angular (rotational), respectively. On substituting Equation 5.93 into Equations 5.91 and 5.92, we get

$$m\ddot{x} + k_{ll}x + k_{la}\varphi_y = 0; \quad m\ddot{y} + k_{ll}y + k_{la}\varphi_x = 0 \tag{5.94}$$

and

$$I_d\ddot{\varphi}_y - I_p\omega\dot{\varphi}_x + k_{la}x + k_{aa}\varphi_y = 0; \quad I_d\ddot{\varphi}_x + I_p\omega\dot{\varphi}_y + k_{la}y + k_{aa}\varphi_x = 0 \tag{5.95}$$

These equations can be combined in the standard matrix form as

$$\mathbf{M}\ddot{\mathbf{x}} - \omega\mathbf{G}\dot{\mathbf{x}} + \mathbf{K}\mathbf{x} = \mathbf{0} \tag{5.96}$$

with

$$\mathbf{M} = \begin{bmatrix} m & 0 & 0 & 0 \\ & m & 0 & 0 \\ & & I_d & 0 \\ sym & & & I_d \end{bmatrix}; \quad \mathbf{G} = \begin{bmatrix} 0 & 0 & 0 & 0 \\ 0 & 0 & 0 & 0 \\ 0 & 0 & 0 & I_p \\ 0 & 0 & -I_p & 0 \end{bmatrix} \tag{5.97}$$

and

$$\mathbf{K} = \begin{bmatrix} k_{ll} & 0 & k_{la} & 0 \\ & k_{ll} & 0 & k_{la} \\ & & k_{aa} & 0 \\ sym & & & k_{aa} \end{bmatrix}; \quad \mathbf{x} = \begin{Bmatrix} x \\ y \\ \varphi_y \\ \varphi_x \end{Bmatrix} \tag{5.98}$$

where **M** is the mass matrix, **K** is the stiffness matrix, and **G** is the gyroscopic matrix, which is a skew-symmetric matrix. Let the general solution be

$$\mathbf{x} = \mathbf{X}e^{\nu t} \tag{5.99}$$

so that

$$\dot{\mathbf{x}} = \nu\mathbf{X}e^{\nu t} \quad \text{and} \quad \ddot{\mathbf{x}} = \nu^2\mathbf{X}e^{\nu t} \tag{5.100}$$

where ν is the asynchronous whirl frequency (eigenvalue). Hence, the equation gives

$$\left(\nu^2\mathbf{M}-\omega\nu\mathbf{G}+\mathbf{K}\right)\mathbf{X}=\mathbf{0} \tag{5.101}$$

With the direct approach, because the previous equation is a homogeneous equation, for the nontrivial solution, the determinant should be zero. Hence, we get

$$\left|\nu^2\mathbf{M}-\omega\nu\mathbf{G}+\mathbf{K}\right|=0 \quad \text{or} \quad \left|\nu^2\mathbf{K}^{-1}\mathbf{M}-\omega\nu\mathbf{K}^{-1}\mathbf{G}+\mathbf{I}\right|=0 \tag{5.102}$$

or

$$\left|\nu^2\alpha\mathbf{M}-\omega\nu\alpha\mathbf{G}+\mathbf{I}\right|=0 \quad \text{with } \alpha=\mathbf{K}^{-1} \tag{5.103}$$

On substituting matrices, we get

$$\left| \nu^2\begin{bmatrix} \alpha_{ll} & 0 & \alpha_{la} & 0 \\ & \alpha_{ll} & 0 & \alpha_{la} \\ & & \alpha_{aa} & 0 \\ sym & & & \alpha_{aa} \end{bmatrix}\begin{bmatrix} m & 0 & 0 & 0 \\ & m & 0 & 0 \\ & & I_d & 0 \\ sym & & & I_d \end{bmatrix} - \right.$$

$$\left. \omega\nu\begin{bmatrix} \alpha_{ll} & 0 & \alpha_{la} & 0 \\ & \alpha_{ll} & 0 & \alpha_{la} \\ & & \alpha_{aa} & 0 \\ sym & & & \alpha_{aa} \end{bmatrix}\begin{bmatrix} 0 & 0 & 0 & 0 \\ 0 & 0 & 0 & 0 \\ 0 & 0 & 0 & I_p \\ 0 & 0 & -I_p & 0 \end{bmatrix} + \begin{bmatrix} 1 & 0 & 0 & 0 \\ & 1 & 0 & 0 \\ & & 1 & 0 \\ sym & & & 1 \end{bmatrix} \right| = 0$$

This can be simplified as

$$\left| \nu^2\begin{bmatrix} m\alpha_{ll} & 0 & I_d\alpha_{la} & 0 \\ 0 & m\alpha_{ll} & 0 & I_d\alpha_{la} \\ m\alpha_{la} & 0 & I_d\alpha_{aa} & 0 \\ 0 & m\alpha_{la} & 0 & I_d\alpha_{aa} \end{bmatrix} - \omega\nu\begin{bmatrix} 0 & 0 & 0 & I_p\alpha_{la} \\ 0 & 0 & -I_p\alpha_{la} & 0 \\ 0 & 0 & 0 & I_p\alpha_{aa} \\ 0 & 0 & -I_p\alpha_{aa} & 0 \end{bmatrix} + \begin{bmatrix} 1 & 0 & 0 & 0 \\ 0 & 1 & 0 & 0 \\ 0 & 0 & 1 & 0 \\ 0 & 0 & 0 & 1 \end{bmatrix} \right| = 0$$

And finally, it takes the following form:

$$\begin{vmatrix} \nu^2 m\alpha_{ll}+1 & 0 & \nu^2 I_d\alpha_{la} & -\omega\nu I_p\alpha_{la} \\ 0 & \nu^2 m\alpha_{ll}+1 & \omega\nu I_p\alpha_{la} & \nu^2 I_d\alpha_{la} \\ \nu^2 m\alpha_{la} & 0 & \nu^2 I_d\alpha_{aa}+1 & -\omega\nu I_p\alpha_{aa} \\ 0 & \nu^2 m\alpha_{la} & \omega\nu I_p\alpha_{aa} & \nu^2 I_d\alpha_{aa}+1 \end{vmatrix} = 0$$

This determinant will give a polynomial of the eighth degree, and the roots can be obtained by root-searching numerical methods. This would give the same natural frequencies as Equation 5.77, which was treated in only one plane with the help of quasi-static analysis. From the present analysis, we will get two sets of similar natural frequencies with different signs.

In the matrix form, the eigenvalue problem for the synchronous whirl (the forward and backward whirls) can be obtained as follows. For the forward synchronous whirl, $\nu=\omega=\omega_{cr}^F$,

$$\left[\left(\omega_{cr}^F\right)^2(\mathbf{M}-\mathbf{G})+\mathbf{K}\right]\mathbf{X}=\mathbf{0} \tag{5.104}$$

which can be written as

$$\left[\left(\omega_{cr}^{F}\right)^{2}\mathbf{M}_{eff}+\mathbf{K}\right]\mathbf{X}=\mathbf{0} \tag{5.105}$$

with

$$\mathbf{M}_{eff}=\mathbf{M}-\mathbf{G}$$

which is a standard eigenvalue problem, and on taking the determinant of the matrix in Equation 5.105, it will give the critical speeds. From this, two critical speeds $\omega_{cr1,2}^{F}$ can be obtained (one of which is expected to be infeasible). For the backward synchronous whirl, $\nu=-\omega=-\omega_{cr}^{B}$, we have

$$\mathbf{M}_{eff}=\mathbf{M}+\mathbf{G} \tag{5.106}$$

where the eigenvalue problem will be of the same form as Equation 5.105. From the determinant of the eigenvalue problem matrix, it would give another two critical speeds $\omega_{cr3,4}^{B}$. It should be noted from Equations 5.105 and 5.106 that the effective mass for the forward whirl is less than the original mass matrix, and for the backward whirl, it is more. Hence, the net effect would be to increase the whirl frequency for the forward whirl and decrease the whirl frequency for the backward whirl. This particular trend is observed in previous sections also while analyzing with the quasi-static method.

5.7 ANALYSIS OF GYROSCOPIC EFFECTS WITH ENERGY METHODS

Governing differential equations can also be obtained by considering Hamilton's principle or Lagrange's equations (see Chapter 7 for details about the method). The kinetic energy without gyroscopic moment of the disk, as shown in Figure 5.40, is given as

$$T_{non-gyro}=\frac{1}{2}m\left(\dot{x}^{2}+\dot{y}^{2}\right)+\frac{1}{2}I_{d}\left(\dot{\varphi}_{x}^{2}+\dot{\varphi}_{y}^{2}\right)+\frac{1}{2}I_{p}\omega^{2} \tag{5.107}$$

In Figure 5.40c, for the *y-z* plane, the gyroscopic moment, M_{g_x}, is in the same direction as the rotational displacement, φ_x, in that plane. Hence, it gives rise to the additional kinetic energy as

$$(T)_{yz}=M_{g_x}\varphi_x=\left(-\frac{1}{2}I_p\omega\dot{\varphi}_y\right)\varphi_x=-\frac{1}{2}I_p\omega\dot{\varphi}_y\varphi_x \quad \text{since} \quad M_{g_x}=-\frac{1}{2}I_p\omega\dot{\varphi}_y \tag{5.108}$$

Similarly, in Figure 5.40b, for the *z-x* plane, the gyroscopic moment, M_{g_y}, is in the opposite direction as the rotational displacement, φ_y. Hence, it gives rise to the additional kinetic energy as

$$(T)_{zx}=-M_{g_y}\varphi_y=-\left(-\frac{1}{2}I_p\omega\dot{\varphi}_x\right)\varphi_y=\frac{1}{2}I_p\omega\dot{\varphi}_x\varphi_y \quad \text{since} \quad M_{g_y}=-\frac{1}{2}I_p\omega\dot{\varphi}_x \tag{5.109}$$

Hence, the total kinetic energy of the disk is given as

$$T=\frac{1}{2}m\left(\dot{x}^{2}+\dot{y}^{2}\right)+\frac{1}{2}I_{d}\left(\dot{\varphi}_{x}^{2}+\dot{\varphi}_{y}^{2}\right)+\frac{1}{2}I_{p}\omega^{2}+\frac{1}{2}I_p\omega\left(\dot{\varphi}_x\varphi_y-\dot{\varphi}_y\varphi_x\right) \tag{5.110}$$

Or the gyroscopic term could also be written as (because for the conservation of angular momentum, we have $\frac{d}{dt}\left(\varphi_x\varphi_y\right)=0$, that is, $\dot{\varphi}_x\varphi_y=-\varphi_x\dot{\varphi}_y$)

$$T=\frac{1}{2}m\left(\dot{x}^{2}+\dot{y}^{2}\right)+\frac{1}{2}I_{d}\left(\dot{\varphi}_{x}^{2}+\dot{\varphi}_{y}^{2}\right)+\frac{1}{2}I_{p}\omega^{2}-\frac{1}{2}I_p\omega\left(2\dot{\varphi}_y\varphi_x\right) \tag{5.111}$$

or

$$T=\frac{1}{2}m\left(\dot{x}^2+\dot{y}^2\right)+\frac{1}{2}I_d\left(\dot{\varphi}_x^2+\dot{\varphi}_y^2\right)+\frac{1}{2}I_p\omega^2+\frac{1}{2}I_p\omega\left(2\dot{\varphi}_x\varphi_y\right) \tag{5.112}$$

Alternatively, the kinetic energy can be obtained as follows by considering instantaneous angular velocities about the ξ-η-ζ axis system that is fixed with the disk and rotated with it. Let $\dot{\varphi}_\xi$, $\dot{\varphi}_\eta$, and $\dot{\varphi}_\zeta$ be instantaneous angular velocities of the disk about the axes ξ, η, and ζ, respectively. The kinetic energy can be written as

$$T=\frac{1}{2}m\left(\dot{x}^2+\dot{y}^2\right)+\frac{1}{2}I_d\left(\dot{\varphi}_\xi^2+\dot{\varphi}_\eta^2\right)+\frac{1}{2}I_p\dot{\varphi}_\zeta^2 \tag{5.113}$$

with

$$\begin{Bmatrix}\dot{\varphi}_\xi\\ \dot{\varphi}_\eta\\ \dot{\varphi}_\zeta\end{Bmatrix}=\begin{Bmatrix}0\\0\\ \omega\end{Bmatrix}+\left[\mathbf{R}(\omega t)\right]_z\begin{Bmatrix}\dot{\varphi}_x\\0\\0\end{Bmatrix}+\left[\mathbf{R}(\omega t)\right]_z\left[\mathbf{R}(\varphi_x)\right]_x\begin{Bmatrix}0\\ \dot{\varphi}_y\\0\end{Bmatrix}$$

$$\left[\mathbf{R}(\varphi_x)\right]_x=\begin{bmatrix}1&0&0\\0&\cos\varphi_x&\sin\varphi_x\\0&-\sin\varphi_x&\cos\varphi_x\end{bmatrix}\quad\text{and}\quad\left[\mathbf{R}(\omega t)\right]_z=\begin{bmatrix}\cos\omega t&\sin\omega t&0\\-\sin\omega t&\cos\omega t&0\\0&0&1\end{bmatrix}$$

$$\dot{\varphi}_\xi=\dot{\varphi}_x\cos\omega t+\dot{\varphi}_y\sin\omega t\cos\varphi_x,\quad \dot{\varphi}_\eta=-\dot{\varphi}_x\sin\omega t+\dot{\varphi}_y\cos\omega t\cos\varphi_x$$

and

$$\dot{\varphi}_\zeta=\omega-\dot{\varphi}_y\sin\varphi_x$$

where $\dot{\varphi}_x$, $\dot{\varphi}_y$, and ω are instantaneous angular velocities of the disk about axes *x*, *y*, and *z*, respectively, that are fixed in space. It should be noted that the order of rotation is first about the *y*-axis by φ_y, then about the new *x*-axis by φ_x, and then finally about the new *z*-axis by ω. The rotational transformation matrix is required when an angular velocity vector is rotated about an axis orthogonal to itself. For example, the angular velocity vector corresponding to spin speed (i.e., the new *z*-axis) would not require any transformation because it is rotated, in the last step, about its own axis. The angular velocity vector corresponding to the new *x*-axis would require one rotational transformation because it is rotated by the ωt angle during the last rotation. Similarly, the angular velocity vector corresponding to the *y*-axis would require two rotational transformations because it is rotated by the ωt angle during the last rotation and by the φ_x angle in the second-to-last rotation. It should be noted that the order of rotation can be chosen arbitrarily, and it would not affect the final equations of motion. Table 5.3 summarizes the angular velocities due to the sequence of rotations discussed earlier.

For small rotational displacements, we can write

$$\dot{\varphi}_\xi\approx\dot{\varphi}_x\cos\omega t+\dot{\varphi}_y\sin\omega t,\quad \dot{\varphi}_\eta\approx-\dot{\varphi}_x\sin\omega t+\dot{\varphi}_y\cos\omega t \tag{5.114}$$

and

$$\dot{\varphi}_\zeta\approx\omega-\dot{\varphi}_y\varphi_x \tag{5.115}$$

so that

TABLE 5.3
Summary of Angular Velocity Vectors for a Sequence of Rotations about the *y*, *x*, and *z* Axes

S.N.	Rotation	Angular Velocity Vector Along		
		z direction	*x* direction	*y* direction
1	Initially	$\begin{Bmatrix} 0 \\ 0 \\ 0 \end{Bmatrix}_z$	$\begin{Bmatrix} 0 \\ 0 \\ 0 \end{Bmatrix}_x$	$\begin{Bmatrix} 0 \\ 0 \\ 0 \end{Bmatrix}_y$
2	About the y-axis	$\begin{Bmatrix} 0 \\ 0 \\ 0 \end{Bmatrix}_z$	$\begin{Bmatrix} 0 \\ 0 \\ 0 \end{Bmatrix}_x$	$\begin{Bmatrix} 0 \\ \dot{\varphi}_y \\ 0 \end{Bmatrix}_y$
3	About the x-axis	$\mathbf{R}_x \begin{Bmatrix} 0 \\ 0 \\ 0 \end{Bmatrix}_z = \begin{Bmatrix} 0 \\ 0 \\ 0 \end{Bmatrix}_z$	$\begin{Bmatrix} \dot{\varphi}_x \\ 0 \\ 0 \end{Bmatrix}_x$	$\mathbf{R}_x \begin{Bmatrix} 0 \\ \dot{\varphi}_y \\ 0 \end{Bmatrix}_y$
4	About the z-axis	$\begin{Bmatrix} 0 \\ 0 \\ \dot{\varphi}_z \end{Bmatrix}_z = \begin{Bmatrix} 0 \\ 0 \\ \omega \end{Bmatrix}_z$	$\mathbf{R}_z \begin{Bmatrix} \dot{\varphi}_x \\ 0 \\ 0 \end{Bmatrix}_x$	$\mathbf{R}_z\mathbf{R}_x \begin{Bmatrix} 0 \\ \dot{\varphi}_y \\ 0 \end{Bmatrix}_y$

$$\dot{\varphi}_\xi^2 + \dot{\varphi}_\eta^2 \approx \dot{\varphi}_x^2 + \dot{\varphi}_y^2 \tag{5.116}$$

and

$$\dot{\varphi}_\xi^2 \approx \omega^2 - 2\omega\dot{\varphi}_y\varphi_x \tag{5.117}$$

In these expressions, higher-order terms have been neglected; hence, the kinetic energy takes the following form:

$$T = \frac{1}{2}m\left(\dot{x}^2 + \dot{y}^2\right) + \frac{1}{2}I_d\left(\dot{\varphi}_x^2 + \dot{\varphi}_y^2\right) + \frac{1}{2}I_p\omega^2 - \frac{1}{2}I_p\omega\left(2\dot{\varphi}_y\varphi_x\right) \tag{5.118}$$

Readers can check that by taking the rotation order first about the x-axis, then about the y-axis, and then about the z-axis, the following expression of kinetic energy results:

$$T = \frac{1}{2}m\left(\dot{x}^2 + \dot{y}^2\right) + \frac{1}{2}I_d\left(\dot{\varphi}_x^2 + \dot{\varphi}_y^2\right) + \frac{1}{2}I_p\omega^2 + \frac{1}{2}I_p\omega\left(2\dot{\varphi}_x\varphi_y\right) \tag{5.119}$$

However, both forms of kinetic energy would produce the same equations of motion. Now, the strain energy stored in the shaft is given by

$$U = \frac{1}{2}k_{ll}x^2 + \frac{1}{2}k_{ul}y^2 + \frac{1}{2}k_{aa}\varphi_x^2 + \frac{1}{2}k_{aa}\varphi_y^2 + \frac{1}{2}k_{la}x\varphi_y + \frac{1}{2}k_{la}y\varphi_x + \frac{1}{2}k_{la}\varphi_y x + \frac{1}{2}k_{la}\varphi_x y \tag{5.120}$$

Lagrange's equation (refer to Chapter 7) could be used to obtain equations of motion as follows:

$$\frac{d}{dt}\left(\frac{\partial T}{\partial \dot{\eta}_i}\right) - \frac{\partial T}{\partial \eta_i} + \frac{\partial U}{\partial \eta_i} = 0, \quad i = 1, 2, 3, 4 \tag{5.121}$$

with

$$\eta_1 = x, \quad \eta_2 = y, \quad \eta_3 = \varphi_y, \quad \eta_4 = \varphi_x \tag{5.122}$$

which gives

$$\frac{d}{dt}\left(\frac{\partial T}{\partial \dot{x}}\right) - \frac{\partial T}{\partial x} + \frac{\partial U}{\partial x} = 0, \quad \frac{d}{dt}(m\dot{x}) - 0 + \left(k_{ll}x + k_{l\varphi}\varphi_y\right) = 0 \tag{5.123}$$

$$\frac{d}{dt}\left(\frac{\partial T}{\partial \dot{y}}\right) - \frac{\partial T}{\partial y} + \frac{\partial U}{\partial y} = 0, \quad \frac{d}{dt}(m\dot{y}) - 0 + \left(k_{ll}y + k_{l\varphi}\varphi_x\right) = 0 \tag{5.124}$$

$$\frac{d}{dt}\left(\frac{\partial T}{\partial \dot{\varphi}_x}\right) - \frac{\partial T}{\partial \varphi_x} + \frac{\partial U}{\partial \varphi_x} = 0, \quad \frac{d}{dt}(I_d\dot{\varphi}_x) - (-I_p\omega\dot{\varphi}_y) + \left(k_{\varphi\varphi}\varphi_x + k_{l\varphi}y\right) = 0 \tag{5.125}$$

and

$$\frac{d}{dt}\left(\frac{\partial T}{\partial \dot{\varphi}_y}\right) - \frac{\partial T}{\partial \varphi_y} + \frac{\partial U}{\partial \varphi_y} = 0, \quad \frac{d}{dt}\left(I_d\dot{\varphi}_y - I_p\omega\varphi_x\right) - 0 + \left(k_{\varphi\varphi}\varphi_y + k_{l\varphi}x\right) = 0 \tag{5.126}$$

On simplification, we get equations of motion, as

$$m\ddot{x} + k_{ll}x + k_{al}\varphi_y = 0; \quad m\ddot{y} + k_{ll}y + k_{al}\varphi_x = 0 \tag{5.127}$$

$$I_d\ddot{\varphi}_x + I_p\omega\dot{\varphi}_y + k_{aa}\varphi_x + k_{al}y = 0; \quad I_d\ddot{\varphi}_y - I_p\omega\dot{\varphi}_x + k_{al}x + k_{aa}\varphi_y = 0 \tag{5.128}$$

which are exactly the same as those obtained earlier. Alternatively, the extended Hamilton principle (refer to Chapter 7) could be used to obtain equations of motion from the kinetic and strain energies, and is expressed as

$$\int_{t_1}^{t_2} (\delta L + \delta W_{nc})\, dt = 0 \tag{5.129}$$

with

$$L = T - U \tag{5.130}$$

where L is the Lagrangian function and δW_{nc} is the nonconservative virtual work (for the present case, $\delta W_{nc} = 0$). On substituting Equations 5.110 and 5.120 into Equation 5.129, we get

$$\begin{aligned} &\int_{t_1}^{t_2} \Big[m\left(\dot{x}\delta\dot{x} + \dot{y}\delta\dot{y}\right) + I_d\left(\dot{\varphi}_x\delta\dot{\varphi}_x + \dot{\varphi}_y\delta\dot{\varphi}_y\right) - k_{ll}x\delta x - k_{ll}y\delta y \\ &-k_{aa}\varphi_x\delta\varphi_x - k_{aa}\varphi_y\delta\varphi_y - k_{la}\left(\varphi_x\delta y + y\delta\varphi_x\right) - k_{la}\left(\varphi_y\delta x + x\delta\varphi_y\right)\Big] dt \\ &+\frac{1}{2}I_p\omega\int_{t_1}^{t_2} \left[\delta\dot{\varphi}_x\varphi_y + \dot{\varphi}_x\delta\varphi_y - \delta\dot{\varphi}_y\varphi_x - \dot{\varphi}_y\delta\varphi_x\right] dt = 0 \end{aligned} \tag{5.131}$$

On performing integration by parts with respect to time of the first four and last four terms of Equation 5.131, which contain variations of time derivatives of generalized coordinates, we have

$$\left[m\dot{x}\delta x+m\dot{y}\delta y+I_d\dot{\varphi}_x\delta\varphi_x+I_d\dot{\varphi}_y\delta\varphi_y\right]_{t_1}^{t_2}-\int_{t_1}^{t_2}\left[m\left\{\ddot{x}\delta x+\ddot{y}\delta y\right\}+I_d\left\{\ddot{\varphi}_x\delta\varphi_x+\ddot{\varphi}_y\delta\varphi_y\right\}\right]dt$$

$$-\int_{t_1}^{t_2}\left[k_{ll}x\delta x+k_{ll}y\delta y+k_{aa}\varphi_x\delta\varphi_x+k_{aa}\varphi_y\delta\varphi_y+k_{al}\left(\varphi_y\delta x+x\delta\varphi_y+\varphi_x\delta y+y\delta\varphi_x\right)\right]dt \tag{5.132}$$

$$+\frac{1}{2}I_p\omega\left[\varphi_y\delta\varphi_x\Big|_{t_1}^{t_2}-\int_{t_1}^{t_2}\dot{\varphi}_y\delta\varphi_x dt+\int_{t_1}^{t_2}\dot{\varphi}_x\delta\varphi_y dt-\varphi_x\delta\varphi_y\Big|_{t_1}^{t_2}+\int_{t_1}^{t_2}\dot{\varphi}_x\delta\varphi_y dt-\int_{t_1}^{t_2}\dot{\varphi}_y\delta\varphi_x dt\right]=0$$

Because the variation of variables is defined to be zero at t_1 and t_2, all terms with time-dependent limits will vanish. Equation 5.132 can be simplified as

$$\int_{t_1}^{t_2}\Big[\left\{m\ddot{x}+k_{ll}x+k_{al}\varphi_y\right\}\delta x+\left\{m\ddot{y}+k_{ll}y+k_{la}\varphi_x\right\}\delta y \tag{5.133}$$

$$+\left\{I_d\ddot{\varphi}_x+I_p\omega\dot{\varphi}_y+k_{aa}\varphi_x+k_{al}y\right\}\delta\varphi_x+\left\{I_d\ddot{\varphi}_y-I_p\omega\dot{\varphi}_x+k_{aa}\varphi_y+k_{al}x\right\}\delta\varphi_y\Big]dt=0$$

Because δx, δy, $\delta\varphi_x$, and $\delta\varphi_y$ are arbitrary variations, we can equate their coefficients in Equation 5.133 to be equal to zero as

$$m\ddot{x}+k_{ll}x+k_{al}\varphi_y=0;\quad m\ddot{y}+k_{ll}y+k_{al}\varphi_x=0 \tag{5.134}$$

and

$$I_d\ddot{\varphi}_x+I_p\omega\dot{\varphi}_y+k_{aa}\varphi_x+k_{al}y=0;\quad I_d\ddot{\varphi}_y-I_p\omega\dot{\varphi}_x+k_{al}x+k_{aa}\varphi_y=0 \tag{5.135}$$

which is the same as Equation 5.95, obtained by Newton's second law. In matrix form, the earlier equations can be written as

$$\begin{bmatrix} m & 0 & 0 & 0 \\ 0 & m & 0 & 0 \\ 0 & 0 & I_d & 0 \\ 0 & 0 & 0 & I_d \end{bmatrix}\begin{Bmatrix} \ddot{x} \\ \ddot{y} \\ \ddot{\varphi}_y \\ \ddot{\varphi}_x \end{Bmatrix}-\omega\begin{bmatrix} 0 & 0 & 0 & 0 \\ 0 & 0 & 0 & 0 \\ 0 & 0 & 0 & I_p \\ 0 & 0 & -I_p & 0 \end{bmatrix}\begin{Bmatrix} \dot{x} \\ \dot{y} \\ \dot{\varphi}_y \\ \dot{\varphi}_x \end{Bmatrix}+$$

$$\begin{bmatrix} k_{ll} & 0 & k_{al} & 0 \\ k_{al} & 0 & k_{aa} & 0 \\ 0 & k_{ll} & 0 & k_{al} \\ 0 & k_{al} & 0 & k_{aa} \end{bmatrix}\begin{Bmatrix} x \\ y \\ \varphi_y \\ \varphi_x \end{Bmatrix}=\begin{Bmatrix} 0 \\ 0 \\ 0 \\ 0 \end{Bmatrix} \tag{5.136}$$

Sometimes the stacking of the vector is different, and for the previous equation, we can have the following form:

$$\begin{bmatrix} m & 0 & 0 & 0 \\ 0 & I_d & 0 & 0 \\ 0 & 0 & m & 0 \\ 0 & 0 & 0 & I_d \end{bmatrix} \begin{Bmatrix} \ddot{x} \\ \ddot{\varphi}_y \\ \ddot{y} \\ \ddot{\varphi}_x \end{Bmatrix} - \omega \begin{bmatrix} 0 & 0 & 0 & 0 \\ 0 & 0 & 0 & I_p \\ 0 & 0 & 0 & 0 \\ 0 & -I_p & 0 & 0 \end{bmatrix} \begin{Bmatrix} \dot{x} \\ \dot{\varphi}_y \\ \dot{y} \\ \dot{\varphi}_x \end{Bmatrix} + \begin{bmatrix} k_{ll} & k_{al} & 0 & 0 \\ k_{al} & k_{aa} & 0 & 0 \\ 0 & 0 & k_{ll} & k_{al} \\ 0 & 0 & k_{al} & k_{aa} \end{bmatrix} \begin{Bmatrix} x \\ \varphi_y \\ y \\ \varphi_x \end{Bmatrix} = \begin{Bmatrix} 0 \\ 0 \\ 0 \\ 0 \end{Bmatrix} \tag{5.137}$$

Now, through simple examples, we determine the whirl frequencies, and critical speeds will be illustrated.

Example 5.6

A long rigid symmetric rotor is supported at the ends by two identical bearings. The shaft has a diameter of 0.2 m, a length of 1 m, and a material mass density equal to 7800 kg/m³. The bearing dynamic characteristics are as follows: $k_{xx}=k_{yy}=k=1$ kN/mm with other stiffness and damping terms equal to zero. Consider the gyroscopic effect to obtain the whirl natural frequencies of the system if the rotor is rotating at 10,000 rpm. Consider the uncoupling of the transverse translational and rotational motions.

Solution:

Figure 5.41 shows a typical long rotor mounted on identical (isotropic) bearings. Noting Equations 5.134 and 5.135, the force and moment balance equations for the present case are

$$m\ddot{x} = \sum f_x; \quad m\ddot{y} = \sum f_y \tag{a}$$

and

$$I_d\ddot{\varphi}_x + I_p\omega\dot{\varphi}_y = \sum M_{yz}; \quad I_d\ddot{\varphi}_y - I_p\omega\dot{\varphi}_x = \sum M_{zx} \tag{b}$$

where the right side of each of these equations represents the summation of external forces or moments, which are different compared to the previous case. Noting Figure 5.41, Equations (a) and (b) can be written as

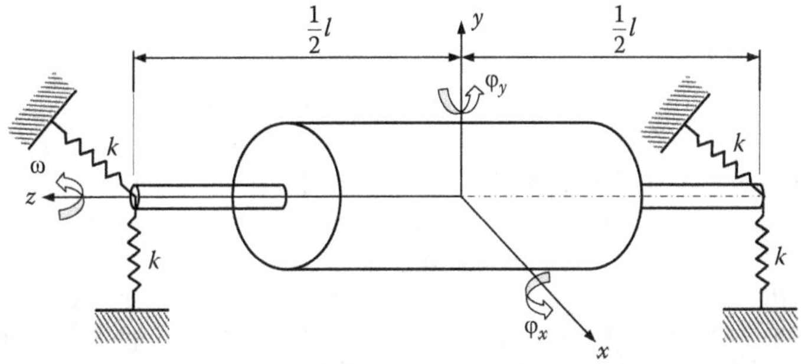

FIGURE 5.41 A long symmetrical rotor mounted on identical flexible bearings.

$$m\ddot{x} = -2kx; \quad m\ddot{y} = -2ky \tag{c}$$

and

$$I_d\ddot{\varphi}_x + I_p\omega\dot{\varphi}_y = -k(0.5\varphi_x l)l; \quad I_d\ddot{\varphi}_y - I_p\omega\dot{\varphi}_x = -k(0.5\varphi_y l)l \tag{d}$$

Let us assume the solutions of Equations (c) and (d) are

$$x = Xe^{\nu t}; \quad y = Ye^{\nu t}; \quad \varphi_x = \Phi_x e^{\nu t}; \quad \varphi_y = \Phi_y e^{\nu t} \tag{e}$$

where X, Y, Φ_x, and Φ_y are complex displacements, which contain the amplitude and phase information, and ν is the frequency of the whirl.

On substituting Equation (e) into Equations (c) and (d), we get

$$\begin{bmatrix} m\nu^2+2k & 0 & 0 & 0 \\ 0 & m\nu^2+2k & 0 & 0 \\ 0 & 0 & I_d\nu^2+0.5kl^2 & I_p\omega\nu \\ 0 & 0 & -I_p\omega\nu & I_d\nu^2+0.5kl^2 \end{bmatrix} \begin{Bmatrix} X \\ Y \\ \Phi_x \\ \Phi_y \end{Bmatrix} = \begin{Bmatrix} 0 \\ 0 \\ 0 \\ 0 \end{Bmatrix} \tag{f}$$

For a nontrivial solution, by placing the determinant of the matrix in Equation (f) equal to zero, we get

$$(m\nu^2+2k)^2\left\{(I_d\nu^2+0.5kl^2)^2+(I_p\omega\nu)^2\right\}=0$$

and on solving the obtained frequency equation, the following pure imaginary solutions are obtained:

$$\nu_i = \pm j\omega_{nfi} \quad \text{where} \quad i = 1,2,3,4 \tag{g}$$

with

$$\omega_{nf1} = \omega_{nf2} = \sqrt{\frac{2k}{m}} \tag{h}$$

and

$$\omega_{nf3,4} = \left[\frac{I_p}{2I_d}\omega \pm \sqrt{\frac{kl^2}{2I_d}+\left(\frac{I_p}{2I_d}\omega\right)^2}\right] \tag{i}$$

For the present case, there is no coupling of motions in the translational and rotational displacements, and because of this, natural frequencies for the translational motion, that is, Equation (h), are not affected by the gyroscopic moment. Equation (i) gives whirl natural frequencies for pure tilting motion of the rotor. Because of the decoupled translational and rotational motions, the whirl natural frequency expressions are simpler compared to the coupled case, as discussed in Sections 5.6 and 5.7. It should be noted that it gives identical expressions as that in Equation 5.64 for asynchronous rotational motion through the quasi-static approach. However, for the present case, the approach is through equations of motion. Numerical values of natural whirl frequencies will be exactly the same as in Example 5.3, which was solved by the quasi-static approach.

5.8 PURE TRANSVERSE ROTATIONAL VIBRATIONS OF A JEFFCOTT ROTOR MODEL WITH MOMENT UNBALANCE

Let us consider a Jeffcott rotor under pure tilting motion (2-DOF model) without translational motion of the disk (i.e., the center of the gravity remains at the bearing axis). In Figure 5.42, let Oz be the bearing axis, and OB be the axis about which the disk is rotating, and let its constant angle with the z-axis be ϕ. Let OA be the axis of the principal polar moment of inertia, and its angle with the z-axis be $\phi_1(t)$; I_p is the polar mass moment of inertia, and I_d is the diametral mass moment of inertia.

At $t=0$, let OA and OB be in the plane of z-x; then at $t \neq 0$, we will have (Figure 5.43b)

$$\phi_y(t) = \phi\cos\omega t \quad \text{and} \quad \phi_x(t) = \phi\sin\omega t \tag{5.138}$$

where ω is the spin speed of the disk, ϕ is the angle between the disk's principal polar moment of inertia axis and the spin axis due to improper mounting of the disk, and ϕ_y and ϕ_x are angles made by projections of OB with the z-axis on the z-x and y-z planes, respectively, at some time instance, t. So that at $t=0$, we have $\phi_y(0) = \phi$ and $\phi_x(0) = 0$.

Referring to the previous analysis of the disk with gyroscopic effect (refer to Section 5.6), we have

$$I_d\ddot{\phi}_{1y} - I_p\omega\dot{\phi}_{1x} = \sum M_{zx} = -\alpha_\varphi\varphi_y \quad \text{and} \quad I_d\ddot{\phi}_{1x} + I_p\omega\dot{\phi}_{1y} = \sum M_{yz} = -\alpha_\varphi\varphi_x \tag{5.139}$$

with

$$\phi_{1y}(t) = \varphi_y(t) + \phi_y(t) = \varphi_y(t) + \phi\cos\omega t \text{ and } \phi_{1x}(t) = \varphi_x(t) + \phi_x(t) = \varphi_x(t) + \phi\sin\omega t \tag{5.140}$$

where α_φ is the influence coefficient corresponding to the pure tilting of the disk at the shaft mid-span and ϕ_{1x} and ϕ_{1y} are angles made by projections of OA (the axis of the principal polar moment of inertia of the disk) with the z-axis on the z-x and y-z planes, respectively (refer to Figure 5.43a).

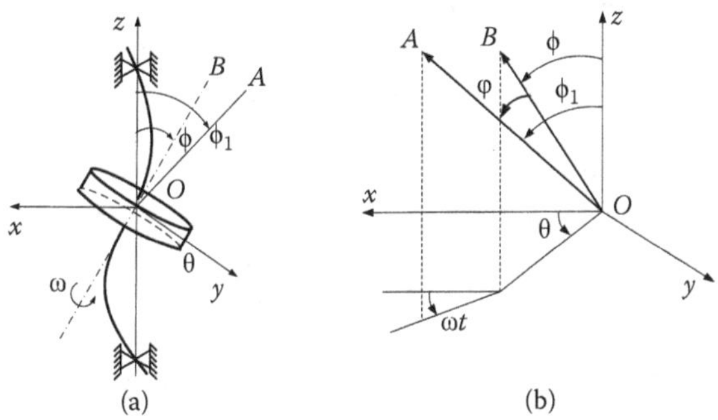

FIGURE 5.42 (a) A Jeffcott rotor with pure tilting motion. (b) Angular positions of various rotor axes.

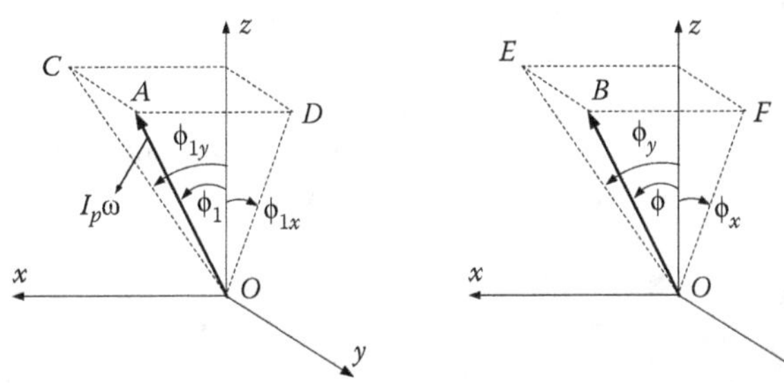

FIGURE 5.43 (a) Disk principal polar moment of inertia axis vector. (b) Disk spin axis vector.

On substituting Equation 5.140 into Equation 5.139, we get

$$I_d\ddot{\varphi}_y - I_p\omega\dot{\varphi}_x + \alpha_\varphi\varphi_y = \left(I_d - I_p\right)\phi\omega^2\cos\omega t \tag{5.141}$$

and

$$I_d\ddot{\varphi}_x - I_p\omega\dot{\varphi}_y + \alpha_\varphi\varphi_x = \left(I_d - I_p\right)\phi\omega^2\sin\omega t \tag{5.142}$$

Hence, from these equations, external moments due to moment unbalances are

$$M_y = \left(I_d - I_p\right)\phi\omega^2\cos\omega t \quad \text{and} \quad M_x = \left(I_d - I_p\right)\phi\omega^2\sin\omega t \tag{5.143}$$

Here, the moment unbalance is $\left(I_d - I_p\right)\phi$, which is analogous to the radial unbalance, me. The former is related to the initial permanent tilt (or angular offset) of the disk, and the latter is related to the radial offset of the disk. Hence, a Jeffcott rotor with an initial tilt in the disk can have a gyroscopic effect near its second critical speed (usually it will be at a very high speed as compared to the first one), which will be excited by the couple unbalance (Figure 5.30b). It will be interesting to analyze whether, for the couple unbalances, we also have a self-correction effect as it is for the radial unbalance after crossing the corresponding critical speed.

5.9 CONCLUDING REMARKS

The present chapter starts with the basic concepts of angular momentum and the gyroscopic moment in simple spinning rotors. We used quasi-static analysis to begin with to study the whirling motion of a single-mass rotor with gyroscopic effects. Initially, synchronous whirl motion is studied for rotor systems with the thin disk and the long stick. It is found that the thin disk has the effect of increasing the critical speed, and the long stick has the effect of decreasing the critical speed. The pure tilting motion has been considered for asynchronous whirl—it is found that the whirl natural frequency depends upon the spin speed. The whirl natural frequency is split into two, and the parts are distinct at high spin speeds of rotors; consequently, the forward and backward critical speeds are defined. For the general motion of the simple rotor, four whirl natural frequencies are obtained; consequently, critical speeds (forward and backward) are extracted from the Campbell diagram. Subsequently, the dynamic analysis of the general motion of the single-mass rotor with gyroscopic effects is undertaken by first obtaining differential equations of motion and then formulating the eigenvalue problem for the synchronous and antisynchronous whirl, respectively, to obtain the forward and backward critical speeds. Governing equations of motion are also derived using a dynamic method for pure rotational motions, considering gyroscopic effects. The present chapter will help in interpreting the results obtained by numerical methods (e.g., transfer matrix method (TMM) and finite element method (FEM)) in subsequent chapters (in Chapters 8 and 10, respectively) with gyroscopic effects. Also in Chapter 2 of Volume II along with speed-dependent bearing dynamic parameters. The sustainable design of a rotor with gyroscopic effects focuses on maximizing energy efficiency and extending component lifespan while managing the complex forces that couple rotational and translational vibration. By integrating advanced rotordynamics with eco-friendly engineering, designers can optimize weight distribution, mitigate structural fatigue, and harness gyroscopic self-centering properties to reduce wear.

EXERCISE PROBLEMS

Exercise 5.1 Obtain the forward and backward synchronous transverse critical speeds for a general motion of a rotor. The rotor is assumed to have cantilever end conditions. Let the mass of the thin disk m be 5 kg and its radius be 15 cm. The shaft is assumed to be massless, and its length and diameter are 0.2 and 0.02 m, respectively. Assume the Young's modulus of the shaft. Obtain the first two forward and backward synchronous transverse critical speeds by drawing the Campbell diagram. Consider the gyroscopic effects. Assume the Young's modulus of the shaft, $E=2.1\times10^{11}\,\text{N/m}^2$.

Exercise 5.2 Consider a rotor system as shown in Figure 5.44. The mass of the thin disk, m, is 5 kg, and the diametral mass moment of inertia, I_d, is 0.02 kg-m². The shaft lengths are $a=0.3$ m and $b=0.7$ m. The diameter of the shaft is 10 mm. Obtain the first two forward and backward synchronous transverse critical speeds by drawing the Campbell diagram. Consider the gyroscopic effects.

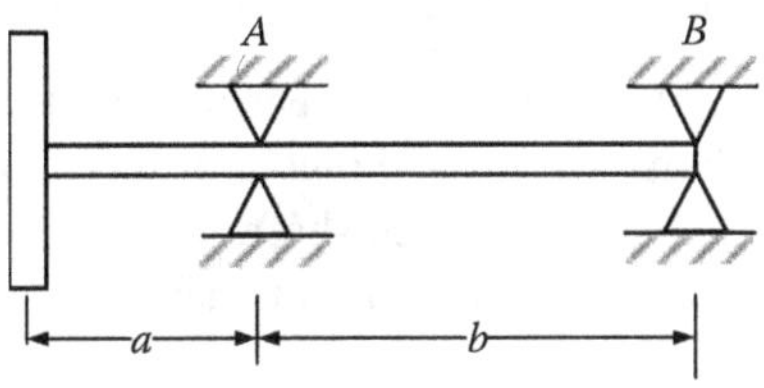

FIGURE 5.44 An overhung rotor system.

Exercise 5.3 Consider a rotor system as shown in Figure 5.45 (in the figure, all dimensions are in cm). Consider the shaft to be massless and made of steel with Young's modulus $E=2.1\times10^{11}\,\text{N/m}^2$. The thin disk has a mass of 10 kg and a radius of 15 cm. The shaft is simply supported at the ends. Obtain the first two forward and backward synchronous transverse critical speeds by drawing the Campbell diagram. Consider the gyroscopic effects.

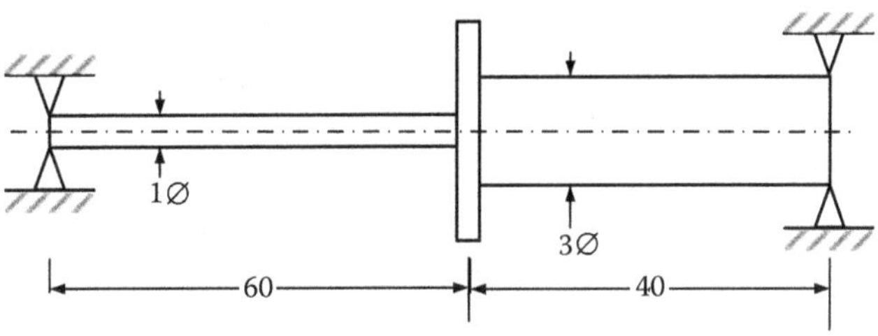

FIGURE 5.45 A stepped shaft with a thin disk and simply supported at the ends.

Exercise 5.4 Formulate the standard eigenvalue problem for the following equations of motion:

$$\mathbf{M}\ddot{\mathbf{x}}+(\mathbf{C}-\omega\mathbf{G})\dot{\mathbf{x}}+\mathbf{K}\mathbf{x}=\mathbf{f}$$

where **M**, **C**, **G**, and **K** are the mass, damping, gyroscopic, and stiffness matrices, respectively; **x** and **f** are the response and force vectors, respectively; and ω is the spin speed of the rotor. Discuss the characteristics of eigenvalues and interpret them physically for the present case.

Exercise 5.5 Find the transverse critical speeds of a rotor system shown in Figure 5.46. The shaft is massless, and Young's modulus $E=2.1\times10^{11}\,\text{N/m}^2$. The thin disk properties about its principal axes are $I_p=0.02$ kg-m² and $I_d=0.01$ kg-m², and the tilt of the disk from the vertical axis is $\phi_0=2°$. Discuss whether gyroscopic effects will be present or not. Obtain critical speeds accordingly. Obtain plots of transverse rotational displacements with spin speed due to angular unbalance.

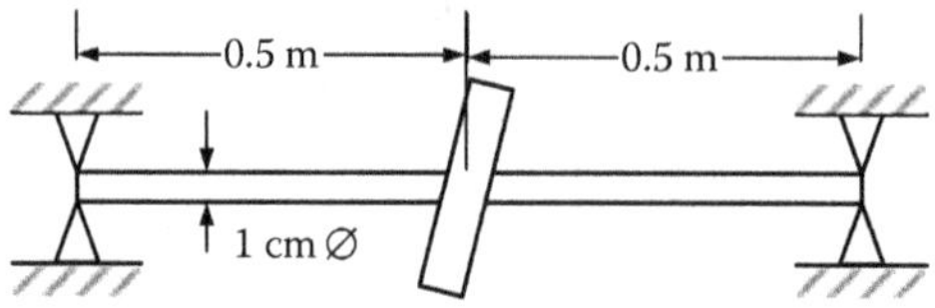

FIGURE 5.46 A simply supported rotor with a tilted thin disk.

Exercise 5.6 A shaft of modulus of rigidity EI and total length, l, is simply supported at its ends. At a quarter-length between the end bearings, the shaft carries a thin disk of mass m and of diameter moment of inertia I_d. Find (i) the nonrotational natural frequency and (ii) the transverse whirl natural frequency. Assume m = 1 kg, l = 1 m, d = 0.01 m, and I_d = 9/112 kg-m^2.

Exercise 5.7 A shaft of total length, l, on the end bearings carries a disk at the quarter-length point. The disk has mass m and diametral mass moment of inertia I_d, and the shaft modulus of rigidity is EI. (i) Derive equations for the whirling shaft, where the whirling frequency is equal to the spin speed, (ii) make the frequency equation dimensionless in terms of the critical speed function $\bar{\omega}_{cr}^2 = \dfrac{ml^2\omega^2}{EI}$ and the disk effect $\mu = \dfrac{I_d}{ml^2}$, and (iii) find the whirling speed for the following three cases: (i) $\mu = 0$, (ii) $\mu \to \infty$, and (iii) $\mu = 1/12$.

Exercise 5.8 Obtain the forward and backward transverse critical speeds of the rotor system shown in Figure 5.47. Assume the shaft is rigid. It is assumed that it oscillates (processes) about its center of gravity while whirling (i.e., pure tilting without translational motion). The effective torsional stiffness of the bearings is $k_t = 2000$ N-m / rad, and the polar and diametral mass moment of inertia of the rotor are 0.03 and 0.2 kg-m^2, respectively. Consider the gyroscopic effects.

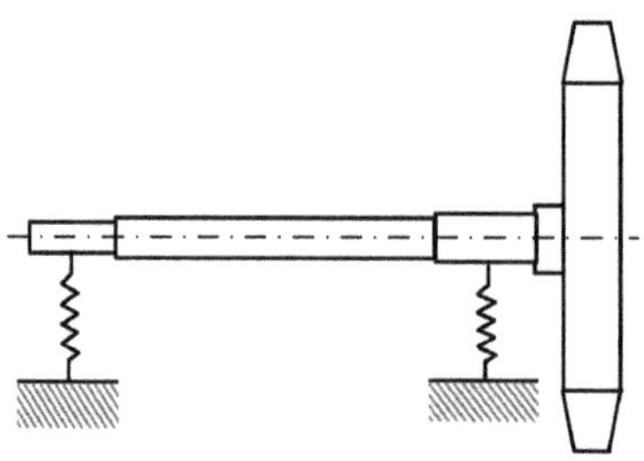

FIGURE 5.47 An overhung rotor supported on flexible bearings.

Exercise 5.9 A massless shaft of total length, l, between bearings (simple supports) carries at its center a disk of diametral mass moment of inertia I_d. The disk is keyed on at a small angle φ_0. When rotating at constant angular speed, ω, the centrifugal forces tend to diminish the angle, φ_0, to a new value, $(\varphi_0 - \varphi)$. Find the ratio, $\varphi(\omega) / \varphi_0$, as a function of the spin speed, ω.

Exercise 5.10 Consider a Jeffcott rotor, a simply supported shaft with a central disk. Consider only the tilting mode of vibration with no translational displacement. (i) Find the natural frequency of this motion for a nonrotating disk. (ii) Find the natural frequency (or frequencies) ω_{nf} for the case of a disk rotating with angular speed ω. (iii) Plot the variation of the obtained natural frequency with respect to the angular speed of the rotor.

Exercise 5.11 Obtain the transverse critical speed for the synchronous motion of a rotor as shown in Figure 5.48 by considering the gyroscopic moment effects. The shaft is assumed to be simply supported at both ends, and the disk has a diameter of D and a thickness of t. It is assumed

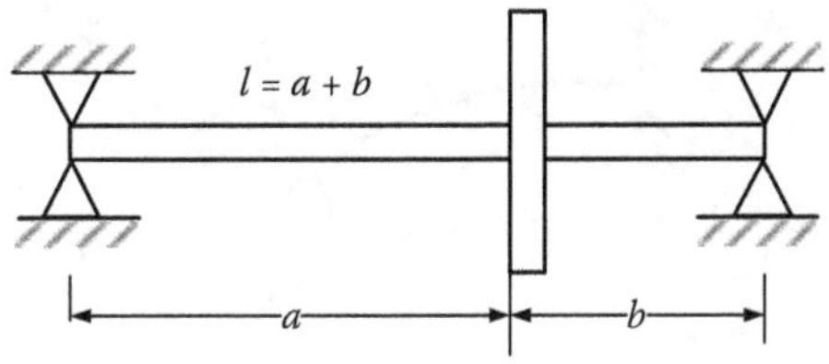

FIGURE 5.48 A Jeffcott rotor system with an offset disk.

that even when the thickness of the disk is large, the disk is attached to the shaft at the location shown without interfering with the motion of the shaft. Consider dimensions of the disk for the following four cases: (i) $D=0.2$ m, $t=0.0082$ m; (ii) $D=0.0721$ m, $t=0.0628$ m; (iii) $D=0.0689$ m, $t=0.0689$ m; and (iv) $D=0.0547$ m, $t=0.1093$ m. The shaft is assumed to be massless, and its length and diameter are 1 and 0.01 m, respectively, with $a=0.75$ m. Let the Young's modulus of the shaft material be 2.1×10^{11} N/m^2 and the density of the disk material be 7800 kg/m^3.

For the simply supported shaft, the influence coefficients are defined as

$$\begin{Bmatrix} y \\ \varphi_x \end{Bmatrix} = \begin{bmatrix} \alpha_{11} & \alpha_{12} \\ \alpha_{21} & \alpha_{22} \end{bmatrix} \begin{Bmatrix} f \\ M \end{Bmatrix}$$

with

$$\alpha_{11} = \frac{a^2b^2}{3EIl};\quad \alpha_{12} = -\frac{\left(3a^2l-2a^3-al^2\right)}{3EIl};\quad \alpha_{21} = \frac{ab(b-a)}{3EIl};\quad \alpha_{22} = -\frac{\left(3al-3a^2-l^2\right)}{3EIl}$$

Exercise 5.12 A shaft of modulus of rigidity EI, length l, and negligible mass is simply supported at both ends. At one end, it carries a thin disk of mass m and diametral mass moment of inertia I_d. This disk is very close to the bearing: Assume it to be at the bearing so that it can tilt, but not displace its center of gravity. (i) For the case of no rotation, find the natural frequency. (ii) Derive the general frequency equation in terms of a rotational speed, ω, and the whirl frequency, ν; make this equation dimensionless in terms of $\bar{\omega}$ and $\bar{\nu}$ -functions,

$$\bar{\omega} = \frac{\omega}{\sqrt{EI/(I_d l)}} \quad \text{and} \quad \bar{\nu} = \frac{\nu}{\sqrt{EI/(I_d l)}}$$

Solve for the nondimensional whirl frequency $\bar{\nu}$ in terms of the nondimensional spin frequency ω; (iii) does the critical speed exist for this system?

Exercise 5.13 For the synchronous backward whirl condition of a cantilever rotor system, derive the frequency equation in terms of the critical speed function $\bar{\omega}_{cr}^2 = \dfrac{ml^3\omega^2}{EI}$, and the disk effect $\mu = \dfrac{I_d}{ml^2}$. Show a plot of $\bar{\omega}_{cr}^2$ versus μ and show limiting cases (i) $\mu = 0$ and (ii) $\mu \to \infty$.

Exercise 5.14 Obtain the transverse critical speed for the forward synchronous motion of a cantilever shaft with a sphere (of mass m and radius r) at the free end (Figure 5.49). It is assumed that the shaft free end is connected to the center of the sphere and does not interfere with the motion of the sphere at any other point of the sphere. The modulus of rigidity of the shaft is EI, and its length is l. The spin speed is ω, and whirl frequency is ν. Discuss the limiting cases (i) when the radius is zero and (ii) when it is infinity. Consider the gyroscopic effects. For the sphere, we have $I_d = I_p = 2mr^2/5$.

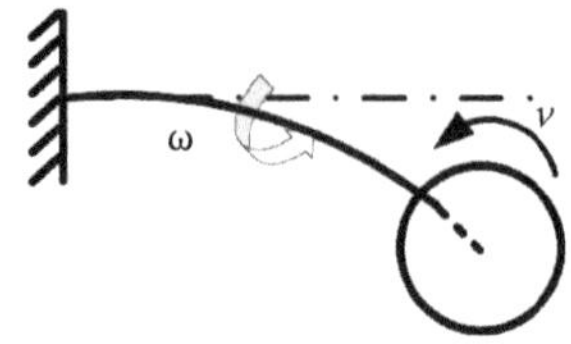

FIGURE 5.49 A cantilever shaft with a sphere at the free end.

Exercise 5.15 Consider an industrial fan that rotates with a very high speed, and its overall body has a slow precession about its center of gravity due to flexible supports. Consider the gyroscopic effect and obtain the forward and backward critical speeds of the fan rotor system. The effective torsional stiffness of the support is 1 MN/m, and the effective diametral and polar mass moments of inertia of the fan are 1 and 0.1 kg-m^2, respectively.

Exercise 5.16 Obtain the frequency equation by using the dynamic method for a general motion (i.e., asynchronous motion) of a single-disk rotor system, and compare the same with the frequency equation obtained using the quasi-static method. Use software that can handle a symbolic form of expressions (e.g., Mathematica or MATLAB or Python) if required to obtain the determinant of a matrix in the symbolic form. Obtain closed-form expressions of critical speeds for the forward and backward whirl conditions by the frequency equation so obtained.

Exercise 5.17 Derive the critical speed expression of a long stick rotor for the backward synchronous whirl based on the centrifugal force concept with the help of neat diagrams, and discuss the same for various disk effect parameters with the help of plots.

Exercise 5.18 Obtain transverse natural frequencies of a rotor system as shown in Figure 5.50. The mass of the thin disk is $m=5$ kg, and the diametral mass moment of inertia is $I_d=0.02$ kg-m^2. The shaft lengths are $a=0.3$ m and $b=0.7$ m. The diameter of the shaft is 10 mm. Bearing A has a roller support, and bearing B has a fixed support condition. Ignore the mass of the shaft; however, consider the gyroscopic effect of the disk. Let $E=2.1 \times 10^{11}$ N/m^2. Compare and discuss the results with the case without a gyroscopic couple.

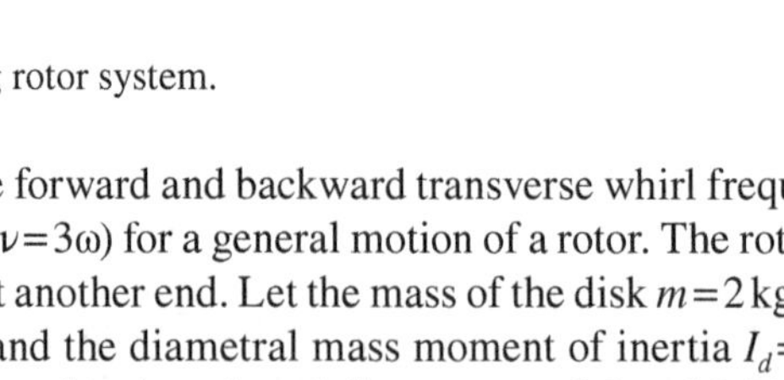

FIGURE 5.50 An overhang rotor system.

Exercise 5.19 Obtain the forward and backward transverse whirl frequencies corresponding to thrice the rotor speed (i.e., $\nu=3\omega$) for a general motion of a rotor. The rotor is assumed to be fixed at one end and supported at another end. Let the mass of the disk $m=2$ kg, the polar mass moment of inertia $I_p=0.01$ kg-m^2, and the diametral mass moment of inertia $I_d=0.005$ kg-m^2. The shaft is assumed to be massless, and its length and diameter are 0.2 and 0.01 m, respectively. Assume the Young's modulus of the shaft $E=2.1 \times 10^{11}$ N/m^2.

Exercise 5.20 For the synchronous motion of a cantilever rotor, derive the expression of critical speed in terms of the disk mass effect, μ. Consider the z-x plane motion instead of y-z as considered in the present chapter. Take positive conventions with respect to displacement, force, moment, etc.

Exercise 5.21 For the synchronous motion of a cantilever rotor, derive the expression of critical speed in terms of the disk mass effect, μ, using angular momentum principles. Compare the result with those achieved using the quasi-static method in the present chapter.

Exercise 5.22 Let the forward synchronous transverse critical speeds for a general motion of a rotor, be fixed at 1700 rad/s due to some practical constraint. The rotor is assumed to be fixed at one end and free at the other end (i.e., the cantilever end conditions). Consider gyroscopic effects in the analysis. Let the mass of the disk $m=2$ kg, the polar mass moment of inertia $I_p=0.01$ kg-m^2, and the diametral mass moment of inertia $I_d=0.005$ kg-m^2. The shaft is assumed to be massless, and its length is 0.2 m. Assume the Young's modulus of the shaft $E=2.1\times10^{11}$ N/m^2. Obtain the suitable diameter of the shaft.

For the cantilever beam, influence coefficients are related as

$$\begin{Bmatrix} y \\ \varphi_x \end{Bmatrix} = \begin{bmatrix} \alpha_{11} & \alpha_{12} \\ \alpha_{21} & \alpha_{22} \end{bmatrix} \begin{Bmatrix} f \\ M \end{Bmatrix} \quad \text{with} \quad \alpha_{11} = \frac{l^3}{3EI}; \quad \alpha_{12} = \alpha_{21} = \frac{l^2}{2EI}; \quad \alpha_{22} = \frac{l}{EI}$$

Exercise 5.23 A rotor is suspended through linear springs (linear stiffness, $k=1$ kN/m) to a rigid ceiling and through torsional springs (torsional stiffness, $k_t=0.1$ kN-m/rad) to rigid supports as shown in Figure 5.51. Both the linear and torsional springs resist the tilting motion of the rotor. The rotor consists of two thin rigid disks connected by a massless thin rigid rod. The distance between two linear springs is $l_1=0.8$ m, and that of torsional springs is $l=1.0$ m (the torsional springs are attached to the shaft close to the disks). Let the rotor spin about its shaft axis with $\omega=100$ rad/s. The diametral mass moment of inertia and the polar mass moment of inertia of the whole rotor system are $I_d=0.8$ kg-m^2 and $I_p=0.1$ kg-m^2, respectively. For the pure tilting motion about its center of gravity, what are the whirl frequencies and critical speeds of the rotor system? Assume asynchronous whirl conditions due to the gyroscopic couple.

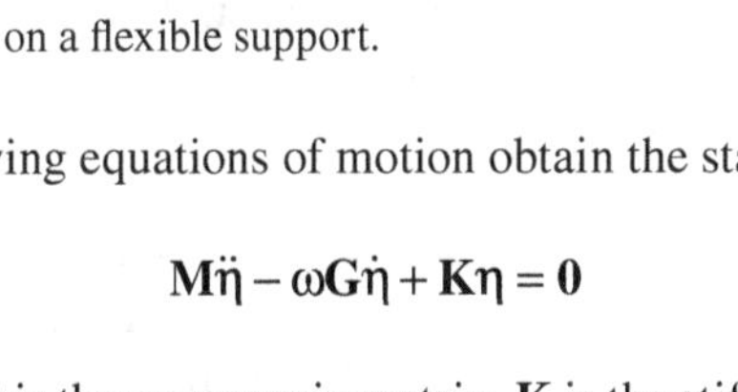

FIGURE 5.51 A rotor mounted on a flexible support.

Exercise 5.24 For the following equations of motion obtain the standard eigenvalue problem:

$$\mathbf{M}\ddot{\boldsymbol{\eta}} - \omega\mathbf{G}\dot{\boldsymbol{\eta}} + \mathbf{K}\boldsymbol{\eta} = \mathbf{0}$$

where **M** is the mass matrix, **G** is the gyroscopic matrix, **K** is the stiffness matrix, **0** is a null vector and $\boldsymbol{\eta}$ is the displacement vector. Discuss the characteristics of eigenvalues for the stability of the system (refer to Chapter 2 of Volume II).

Exercise 5.25 Let the antisynchronous transverse critical speeds for a general motion of a rotor, to be fixed at 1500 rad/s due to some practical constraint. The rotor is assumed to be fixed and supported at one end. *Consider gyroscopic effects in the analysis.* Take the mass of the disk $m=2$ kg, the polar mass moment of inertia $I_p=0.01$ kg-m^2, and the diametral mass moment of inertia $I_d=0.005$ kg-m^2. The shaft is assumed to be massless, and its length is 0.2 m. Assume the Young's modulus of the shaft $E=2.1\times10^{11}$ N/m^2. Obtain the suitable diameter of the shaft.

For the cantilever beam, the influence coefficients are related as

$$\begin{Bmatrix} y \\ \varphi_x \end{Bmatrix} = \begin{bmatrix} \alpha_{11} & \alpha_{12} \\ \alpha_{21} & \alpha_{22} \end{bmatrix} \begin{Bmatrix} f \\ M \end{Bmatrix} \quad \text{with} \quad \alpha_{11} = \frac{l^3}{3EI}; \quad \alpha_{12} = \alpha_{21} = \frac{l^2}{2EI}; \quad \alpha_{22} = \frac{l}{EI}$$

Exercise 5.26 A long symmetrical rigid rotor is supported at both ends by two identical bearings. The shaft has a diameter of 0.2 m and a length of 1 m, with the mass density of the shaft material 7800 kg/m³. Consider each bearing has properties $k_{xx}=k_{yy}=k$ with other stiffness and damping properties ignored. For the pure tilting motion of the shaft, while considering the gyroscopic effects, the first forward critical speed ($\nu=\omega$) is to be fixed at 100 rad/s. Obtain the bearing stiffness, k.

Exercise 5.27 A massless flexible shaft of length 0.6 m and diameter 2 cm is simply supported at the ends. A thin disk of mass 10 kg, polar moment of inertia 0.2 kg-m², and diametral moment of inertia 0.1 kg-m² is at the midspan of the shaft. Consider the gyroscopic effects. Obtain the system natural frequencies when the shaft rotates at 6000 rpm. Obtain all critical speeds of the rotor system, and if the disk has (i) only radial eccentricity e, (ii) only pure initial tilt φ, and (iii) both e and φ, then which of the critical speeds would be encountered in each case by the rotor system? Let $E=2.1\times10^{11}$ N/m².

Exercise 5.28 Briefly describe the following:

i. Synchronous forward whirl and asynchronous forward whirl.
ii. Natural whirl frequency and critical speed in the case of rotor systems with gyroscopic effects.
iii. A sphere is spinning about a diametral axis with ω rad/s and precessing about another diametral axis perpendicular to the spinning axis with ν rad/s. Let I_p be the polar mass moment of inertia of the sphere. What is the value of the gyroscopic moment about the third orthogonal axis?
iv. A sphere is supported at both ends of its diameter by two identical springs. The transverse stiffness of the springs at either end of the diameter is k, the mass of the sphere is m, and its polar mass moment of inertia is I_p. What is the natural frequency(ies) of the system?
v. A cantilever rotor with a massless flexible shaft has a thin heavy disk rigidly fixed at the free end of the shaft (Figure 5.52). On the disk is mounted a massless free disk with the help of a rolling bearing, and with this arrangement, the free disk can rotate (Ω) independent of the heavy disk (ω) by a separate drive arrangement mounted on the shaft itself. The free disk can be rotated independently CW or CCW at the speed of the same magnitude as the shaft speed. Also, on the free disk, a provision is there to keep an unbalance mass. With this arrangement, keeping the shaft speed sense of rotation the same (e.g., CCW), the free disk with an unbalance is rotated first in the CW and then in the CCW sense of rotation. Comment on the critical speed encountered by the rotor system if the speed is increased

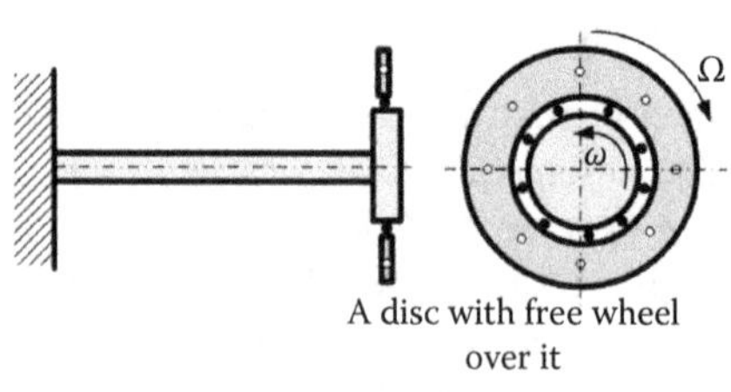

FIGURE 5.52 A cantilever rotor with a free disk at the end.

gradually from zero to beyond the critical speeds of the rotor system. Will the critical speeds observed be the same for the CW and CCW rotation of the free disk? Explain the proper physical reasoning for this.

Exercise 5.29 Obtain the forward and backward transverse critical speeds corresponding to twice the rotor speed ($\nu=2\times\omega$) for a general motion of a rotor. Let the mass of the disk $m=2$ kg, the polar mass moment of inertia $I_p=0.01$ kg-m^2, and the diametral mass moment of inertia $I_d=0.005$ kg-m^2. The shaft is assumed to be massless, and its length and diameter are 0.2 and 0.01 m, respectively. Assume the Young's modulus of the shaft $E=2.1\times10^{11}$ N/m^2.

Exercise 5.30 Obtain transverse critical speeds of the rotor-bearing system as shown in Figure 5.53. Consider the shaft to be rigid. Both bearings have torsional springs (of stiffness k_b) to resist tilting (or transverse rotational motion) and linear springs (of stiffness k) to resist translational motions. Both disks have identical inertia properties and locations from the bearings. Let $a=0.7$ m, $b=0.3$ m, $k_b=100$ kN-m/rad, and $k=10$ kN/m. Each disk has $m=5$ kg and $I_d=0.02$ kg-m^2. Consider the gyroscopic couple effect and pure rotational motion only.

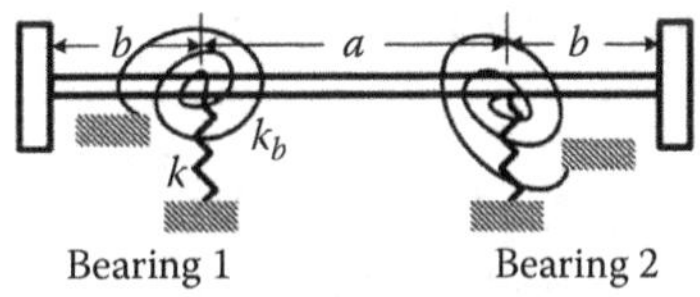

FIGURE 5.53 A rigid rotor mounted on flexible bearings.

Exercise 5.31 Obtain all critical speeds of the rotor system shown in Figures 5.56 and 5.57 by considering the gyroscopic effect. Let $a=0.6$ m and $b=0.4$. Let the shaft diameter $d=0.015$ m and $E=2.1\times10^{11}$ N/m^2. The offset disk has a mass of 2 kg and a diametral mass moment of inertia of 0.01 kg-m^2.

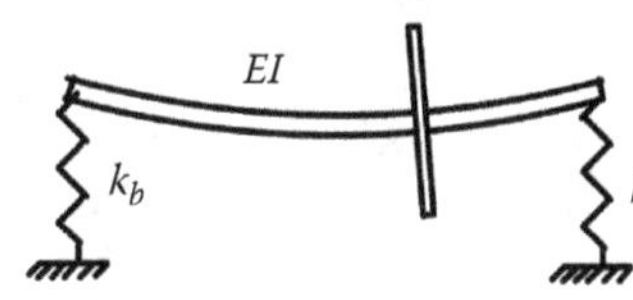

FIGURE 5.54 A flexible shaft on flexible bearings.

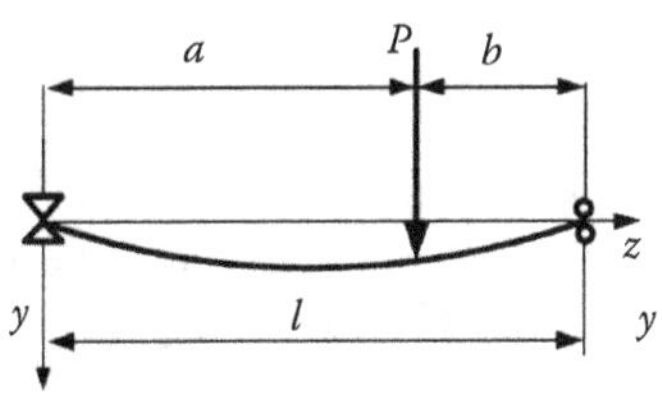

$$y=\frac{Pbz}{6lEI}(l^2-z^2-b^2)\ \text{for}\ [0<z<a]$$

$$y=\frac{Pb}{6lEI}\left[\frac{l}{b}(z-a)^3+(l^2-b^2)z-z^3\right]\text{for}\ [a<z<l]$$

FIGURE 5.55 A simply supported shaft with a concentrated load P at any point.

Exercise 5.32 A rotor, as shown in Figure 5.56, consists of a long stick (cylindrical in shape with diameter D_s and length b_s; it has a mass of m_s) and a thin disk (radius r_d; it has a mass of m_d) mounted at the free end of a cantilever shaft. The shaft is assumed to be connected at the center of gravity of the cylinder–disk system (it is an idealization of the blisk—blade over disk, see Figure 5.57—in turbines, there support conditions may be different) and the shaft is in no way disturbing the motion of the blisk. The shaft flexural rigidity is EI, and its length is l. Consider the pure tilting motion of the blisk, and obtain critical speed expressions for the forward and backward whirls of the rotor system shown in Figure 5.56. If D_s=3.3166D_d, then obtain the critical speeds (with D_d=2r_d).

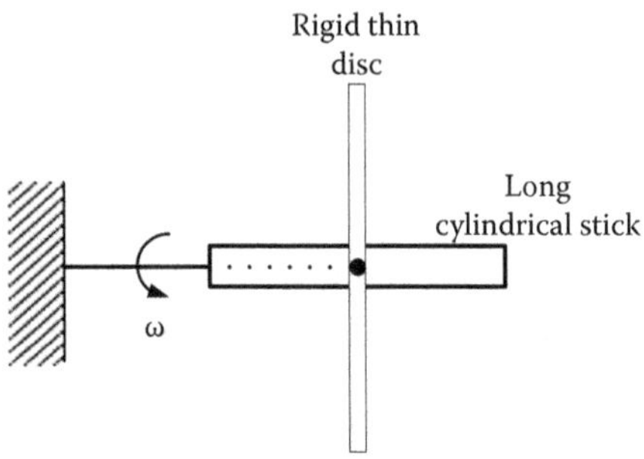

FIGURE 5.56 A cantilever shaft with a rigid thin disk and a long cylindrical stick.

A cantilever shaft with a rigid thin disk and a long cylindrical stick.

FIGURE 5.57 A blade-disk (blisk) of a turbine.

Exercise 5.33 Obtain transverse critical speeds of the rotor-bearing system as shown in Figure 5.58. Consider the shaft to be rigid and massless. The bearing on the left is simply supported, and that on the right is supported by two springs, and each spring has the stiffness k. Let L=1 m, a=0.3 m, k=1 kN/m, m=5 kg, and I_d=0.02 kg-m^2. Consider the gyroscopic couple effect.

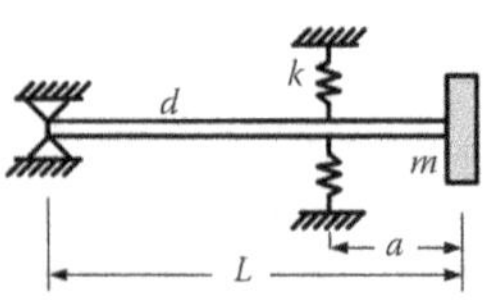

FIGURE 5.58 A rotor system flexibly supported at one end and simply supported at the other end.

Exercise 5.34 A motor drives a rotor with a rigid shaft and a disk at the end. The motor itself is mounted on a torsional spring that resists its motion in the transverse direction with effective torsional stiffness, $k=2$ kN-m/rad. It is assumed that the motor has no rotational motion about the rotor axis, and it has rotational motion in the transverse direction about its center of gravity. The polar mass moment of inertia of the rotor is 0.02 kg-m², and the diametral mass moment of inertia of the rotor and motor as a whole is 0.05 kg-m². Consider gyroscopic effects. Obtain critical speeds of the rotor system. If the speed of the rotor is 5000 rpm, obtain the transverse natural frequencies of the rotor system. Compare the critical speeds of the rotor without considering the gyroscopic effect.

Exercise 5.35 For an antisynchronous whirl of a cantilever shaft with a thin disk at the free end, obtain the corresponding critical speed expression (so-called frequency equation). Express obtained expression in terms of nondimensional disk effect, $\mu=I_d/ml^2$, and critical speed function $\bar{\omega}_{cr}=\omega_{cr}\sqrt{ml^3/(EI)}$. Plot the variation of $\bar{\omega}_{cr}^2$ with the disk effect. Let I_d be the diametral mass moment of inertia of the disk, m be the mass of the disk, l be the length of the shaft, E is the Young's modulus of the shaft, and I is the second moment of area of the shaft. Draw neat diagrams and give detailed steps involved from the very fundamentals.

Exercise 5.36 Obtain transverse natural frequencies of a rotor system shown in Figures 5.61 and 5.62. It has simply supported conditions, and a thin disk ($m=2$ kg and $I_d=0.01$ kg-m²) is mounted very close to the support itself so that they have pure tilting motion (transverse rotational displacements). It is assumed that supports do not interfere with the motion of the disk. The relations of transverse rotational displacements due to the moment at one end are given in Table 5.4. Take $E=2.1\times10^{11}$ N/m². The length of the shaft is 0.6 m, and its diameter is 0.01 m. Neglect the mass of the shaft, but consider the gyroscopic effect due to the disk.

FIGURE 5.59 A rotor system with a disk near right support.

TABLE 5.4
A Simply Supported Shaft with a Concentrated Moment M at One End Point

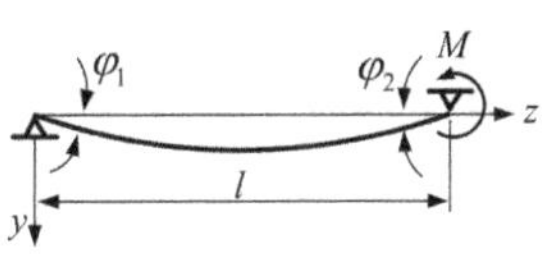

FIGURE 5.60 A simply supported shaft with a moment at one end.

$$\varphi_1=\frac{Ml}{6EI}\qquad \varphi_2=\frac{Ml}{3EI}\qquad y=\frac{Mlz}{6EI}\left(1-\frac{z^2}{l^2}\right)$$

Exercise 5.37 A long symmetrical rigid rotor is supported at both ends by two identical bearings. The shaft has a diameter of 0.2 m and a length of 1 m, with the mass density of the shaft material as 7800 kg/m³. Consider each bearing has properties as $k_{xx}=k_{yy}=k$, with other stiffness and damping properties ignored. For pure titling motion of the shaft while considering the gyroscopic effects, the forward critical speed is to be fixed at 100 rad/s. Obtain the bearing stiffness, k.

Exercise 5.38 Obtain transverse critical speeds of the rotor-bearing system as shown in Figure 5.61. Consider the gyroscopic couple effect. Let the single shaft be rigid and massless. The shaft in the middle is roller supported along with a coil spring of stiffness k_t (it resists angular tilt of the shaft; however, the spring is not shown in the figure). Roller support prevents transverse translational motion; however, it allows transverse rotational motion. The left and right bearings are represented by four linear springs as shown symmetrically placed between two disks, and each spring has the stiffness k. Let $L=1$ m, $a=0.2$ m, $b=0.3$ m, $k=1$ kN/m, $k_t=1$ kN-m/rad, $m_1=m_4=5$ kg, $m_2=m_3=3$ kg, $I_{d1}=I_{d4}=0.015$ kg-m^2, and $I_{d2}=I_{d3}=0.01$ kg-m^2.

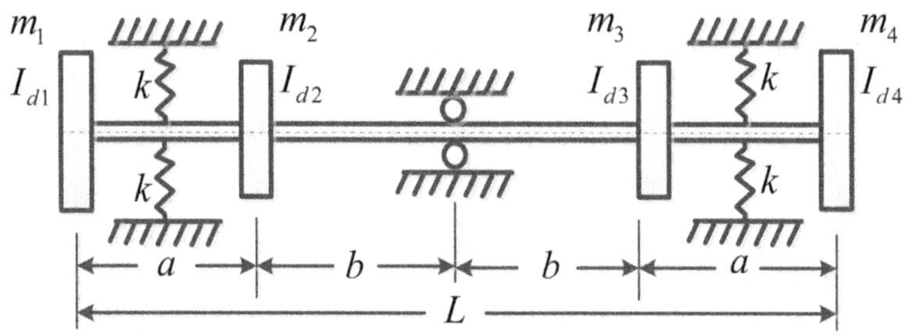

FIGURE 5.61 A rotor-bearing system.

Exercise 5.39 The fundamental (first) transverse critical speed of a rotor system, is fixed at 5.98 rad/s. Take the disk mass $m=5$ kg and diametral mass moment of inertia $I_d=0.15$ kg=m^2. What is the diameter of the uniform shaft, d? Take shaft length $a=0.3$ and $b=0.7$ m. Ignore the gyroscopic effect. $E=2.1\times10^{11}$ N/m^2.

For the present case, influence coefficients are given as $\alpha_{yf}=\dfrac{a^2(a+b)}{3EI}$, $\alpha_{\varphi M}=\dfrac{(3a+b)}{3EI}$, and $\alpha_{\varphi f}=\alpha_{yM}=\dfrac{a(3a+2b)}{6EI}$, where I is the second moment of area of the shaft cross-section. Subscripts f and M represent force and moment, respectively, and y and φ represent the translational and rotational displacements.

Exercise 5.40 Choose a single correct answer for the multiple choice questions.

i. In a rotor system, the gyroscopic moment can give rise to instability.
   A. true
   B. false

ii. A cantilever rotor with a single disk and massless shaft, in a general motion, can have how many transverse natural frequencies?
   A. 1
   B. 2
   C. 3
   D. 4
   E. more than 4

iii. Because of the gyroscopic moment, natural frequencies depend upon the spin speed of the rotor.
   A. true
   B. false

iv. In a general motion of a rotor, the whirl frequency and the spinning frequency (speed) are always the same.
   A. true
   B. false

v. Because of the gyroscopic moment, the forward whirl natural frequency of a rotor
   A. increases only
   B. decreases only
   C. either increases or decreases
   D. remains constant
   E. increases and decreases simultaneously

vi. For a rotor system with gyroscopic effects and without damping, eigenvalues would be purely imaginary.
   A. true
   B. false

vii. For a rotor system with gyroscopic effects and with damping, eigenvalues would be pure imaginary.
   A. true
   B. false

viii. Because of the gyroscopic moment in the forward whirl motion, the effective stiffness of the shaft (or experienced by the disk) increases.
   A. true
   B. false

ix. Because of the gyroscopic moment in the backward whirl motion, the effective stiffness of the shaft (or experienced by the disk) increases.
   A. true
   B. false

x. Because of the gyroscopic moment in a rotor system with underdamping, the system will never be unstable.
   A. true
   B. false

xi. A Campbell diagram is a diagram of
   A. the amplitude versus the rotor spin frequency
   B. the phase versus the rotor spin frequency
   C. both amplitude and phase versus the rotor spin frequency
   D. the whirl natural frequency versus the rotor spin frequency

xii. Because of the gyroscopic effect, a rotor (with a thin disk) backward natural whirl frequency
   A. remains the same
   B. decreases
   C. increases
   D. has no definite trend

xiii. For a cantilever rotor with a long stick (with diameter D and length b) for the pure tilting motion, the gyroscopic couple will be absent if
   A. $b<D$
   B. $b=D$
   C. $b=\sqrt{3}D/2$
   D. $b=1.866D$

xiv. Due to gyroscopic effects
   A. frequencies of the forward and backward whirls become faster compared to the whirling frequency without gyroscopic effects
   B. frequencies of the forward and backward whirls become slower compared to the whirling frequency without gyroscopic effect
   C. frequencies of the forward and backward whirls become slower and faster, respectively, compared to the whirling frequency without gyroscopic effect
   D. frequencies of the forward and backward whirls become faster and slower, respectively, compared to the whirling frequency without gyroscopic effect

xv. For a general motion of a rotor with gyroscopic effects, for a special case when the elastic coupling, α, is absent, the square of the backward critical speed corresponding to the pure rotational motion would be (where the disk effect $\mu = I_d \alpha_{22} / (m\alpha_{11})$):
   A. $1/\mu$
   B. $1/\mu^2$
   C. $1/(2\mu)^2$
   D. $1/(3\mu)$

xvi. For a long rigid rotor mounted on anisotropic elastic bearings with no cross-coupling, the following mode of whirl may be assumed for analysis:
   A. purely translation and purely conical
   B. purely translation only
   C. purely conical only
   D. combination of translation and conical

xvii. While considering the gyroscopic effect in asynchronous pure rotational motion of a rotor mounted on some spring (flexible) support, if the support spring breaks suddenly, then the instantaneous whirl frequencies would be
   A. $\nu = \pm \dfrac{I_d \omega}{I_p}$
   B. $\nu = 0$ and $\nu = \pm \dfrac{I_p \omega}{I_d}$
   C. $\nu = 0$ and $\nu = \pm \dfrac{I_d}{I_p}$
   D. $\nu = \pm \dfrac{I_p}{I_d}$

xviii. For the rotor system shown in Figure 5.62, if the gyroscopic effect is also considered, then the first forward natural frequency, as compared to that without the gyroscopic effect, would

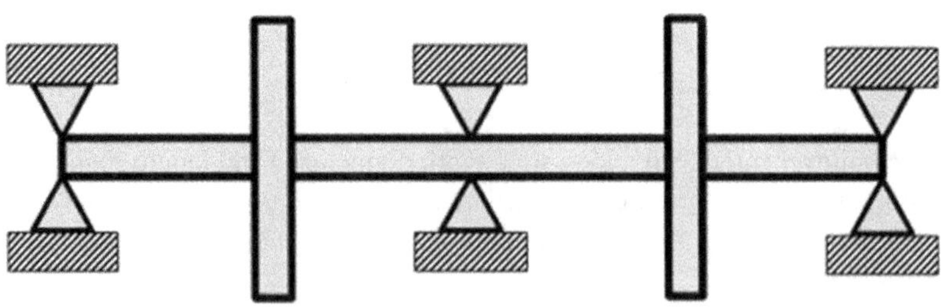

FIGURE 5.62 A multi-support rotor system.

A. remain the same
B. increase
C. decrease
D. increase or decrease

xix. For general motion in a single-disk rotor model with a massless elastic shaft, due to the gyroscopic effect, the total number of transverse forward critical speeds corresponding to 2 × of the rotor speed is
A. 1
B. 2
C. 3
D. 4

xx. For a long rotor with the increase in the absolute value of disk effect, the synchronous whirl frequency
A. remains the same
B. increases
C. decreases
D. increases then decreases

xxi. Which option does not lead to instability?
A. gyroscopic effect
B. asymmetrical shaft
C. seals
D. material damping

xxii. The gyroscopic couple gives
A. coupling of two orthogonal plane motions in translational displacements
B. coupling of two orthogonal plane motions in rotational displacements
C. coupling of the translational and rotational displacements in the same plane
D. coupling of the translational and rotational displacements in two orthogonal planes

xxiii. A cantilever beam has a length of 0.5 m and a diameter of 1 cm with a thin disk of mass 3 kg at the free end. Let Young's modulus $E=2.1\times10^{11}\,\text{N/m}^2$. If the disk diametral mass moment of inertia is such that it tends to infinity, then the critical speed (rad/s) of the rotor system (while considering the gyroscopic effect) will be
A. 57.43
B. 16.6
C. 49.7
D. 99.5

xxiv. For a cantilever massless elastic shaft with a thin balanced rigid disk at the free end undergoing general asynchronous motion (both translational and rotational transverse motion with gyroscopic effect included), if the whirl frequency $\nu = 1/\sqrt{m\alpha_{11}}$ and then the corresponding rotor speed, ω, would be
A. $1/\sqrt{m\alpha_{11}}$
B. $0.5/\sqrt{m\alpha_{11}}$
C. $2/\sqrt{m\alpha_{11}}$
D. $1/\sqrt{m\alpha_{22}}$

xxv. A single-disk cantilever rotor will have four distinct natural frequencies
A. when the disk is a point mass
B. when the disk has appreciable gyroscopic effects
C. when the disk (stick) is long with negligible gyroscopic effects
D. when the disk is thin with negligible gyroscopic effects

xxvi. For a synchronous and pure tilting whirl of a long rotor (stick) for $b=0.866\,D$ (where D is the diameter and b is the length of the stick), which effect will be absent?
A. whirl frequency
B. critical speed
C. gyroscopic effect
D. whirling

xxvii. A cantilever shaft carries a spherical (instead of a point mass) mass as a tip mass. Which effect will be absent?
A. whirl frequency
B. whirling
C. critical speed
D. gyroscopic effect

xxviii. A cantilever massless shaft with a tip mass (thin disk), on considering the gyroscopic effect, would have a whirl natural frequency equal to
A. 1
B. 2
C. 3
D. 4

xxix. The Campbell diagram is a
A. forced response plot
B. free vibration response plot
C. natural whirl frequency plot
D. transient vibration response plot

xxx. In a cantilever rotor, if $\alpha_{12}=y/M_y$ and $\alpha_{21}=\varphi_x/F_x$, then
A. $\alpha_{12}/\alpha_{21}=1$
B. $\alpha_{12}\alpha_{21}=1$
C. $\alpha_{12}/\alpha_{21}=-1$
D. $\alpha_{12}\alpha_{21}=-1$

xxxi. The effective stiffness of a rotor system during the backward whirl due to the gyroscopic effect has a tendency to
A. increase
B. remain the same
C. decrease
D. either increase or decrease

xxxii. Because of the gyroscopic couple, the natural frequencies depend upon the spin speed of the shaft.
A. true
B. false

xxxiii. In a general case of a perfectly balanced rotor, the whirl frequency and the spinning frequency (speed) are always the same.
A. true
B. false

xxxiv. For a long stick supported at the free end of a flexible cantilever shaft, the length-to-diameter ratio of the long stick is equal to $\sqrt{3}/2$. The gyroscopic couple on the shaft, due to the long stick, will
A. try to prevent the angular displacement of the shaft
B. try to help the angular displacement of the shaft
C. have no effect on the angular displacement of the shaft
D. none of the above

xxxv. For the rotor system shown in Figure 5.63 (consider polar moments of inertia of end disks only), which effect would be absent?

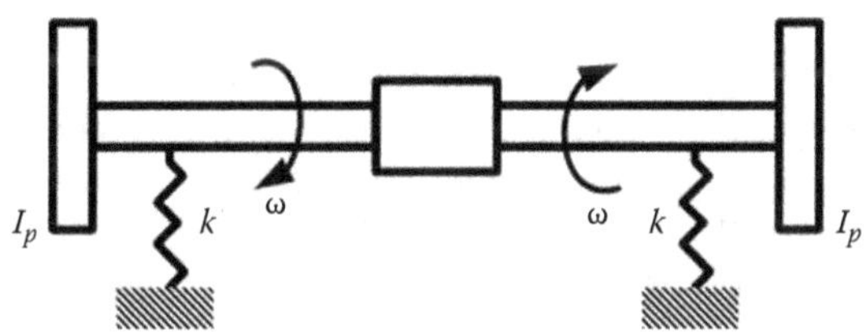

FIGURE 5.63 A rigid symmetric rotor mounted on identical flexible bearings.

A. whirl frequency
B. whirling
C. critical speed
D. gyroscopic effect

xxxvi. The whirl natural frequency of a rotor system is the frequency of oscillation of
A. a non-spinning rotor during free vibration
B. a spinning rotor during free vibration
C. a non-spinning rotor during forced vibration
D. a spinning rotor during forced vibration

xxxvii. The whirl frequency of a rotor system due to unbalance will be equal to
A. the spin speed of the shaft
B. the whirl natural frequency of the spinning rotor system
C. the natural frequency of the non-spinning rotor system
D. the critical speed of the rotor system

xxxviii. In a balanced rotor system, a free wheel has an unbalance and is mounted on the shaft. The free wheel is a disk, which is not fixed to the shaft directly. It is mounted on the shaft through a rolling bearing so that it can rotate freely, independent of shaft rotation. If the free wheel is rotated by a separate drive in the opposite sense of rotation of the shaft, on increasing the speed of the free wheel, it will excite at which critical speed?

A. forward critical speed
B. backward critical speed
C. both forward and backward critical speeds
D. neither forward nor backward critical speed

xxxix. A Campbell diagram is a plot between
A. the whirl natural frequency versus the spin speed of a rotor
B. the critical speed versus the spin speed of a rotor
C. the whirl natural frequency versus the critical speed of a rotor
D. the amplitude of the rotor displacement versus the spin speed of a rotor

xl. The gyroscopic effect occurs due to
A. centrifugal force
B. Coriolis component of acceleration
C. unbalance
D. shear force

xli. In a Jeffcott rotor (with a disk at the midspan), if the disk has no radial unbalance but it is tilted by some angle in a transverse plane, then during spinning, the critical speed excited will be (consider no gyroscopic effect; E is Young's modulus, I is the second moment of area, m is the mass of the disk, I_d is the diametral mass moment of inertia of the disk, and l is the length of the shaft)

A. $\sqrt{\dfrac{12EI}{I_d l}}$

B. $\sqrt{\dfrac{48EI}{I_d l}}$

C. $\sqrt{\dfrac{12EI}{ml^3}}$

D. $\sqrt{\dfrac{48EI}{ml^3}}$

xlii. In a Jeffcott rotor (with a disk at the midspan), if the disk has pure tilting motion in a transverse plane during spinning, then the critical speed will be (consider gyroscopic effect with $\mu = \dfrac{I_d \alpha_{22}}{m\alpha_{11}}$ is the disk effect)

A. $\dfrac{1}{\sqrt{3\mu}}$

B. $\dfrac{1}{2\sqrt{3\mu}}$

C. $\dfrac{1}{\sqrt{2\mu}}$

D. $\dfrac{1}{2\sqrt{\mu}}$

xliii. In a Jeffcott rotor (with a disk at the midspan), if the disk has a pure tilting motion in a transverse plane during spinning, then the feasible critical speed will be
A. both forward and backward critical speeds
B. backward critical speed

C. forward critical speed
D. neither forward nor backward critical speeds

xliv. For a general motion of a single-disk rotor, the gyroscopic moment is given as (where I_d is the diametral mass moment of inertia, I_p is the polar mass moment of inertia, φ_x is the angular tilt of the disk, ω is the rotor spin speed, and ν is the whirl natural frequency)

A. $I_d\varphi_x(2\omega-\nu)\nu$

B. $I_p\varphi_x(2\omega-\nu)\nu$

C. $I_d\varphi_x(\omega-2\nu)\omega$

D. $I_p\varphi_x(\omega-2\nu)\omega$

xlv. A spinning rotor (the diametral mass moment of inertia of the rotor is three times its polar mass moment of inertia with $I_p=1$ kg-m^2) in a pure tilting motion and supported on a flexible support of a torsional stiffness of 20 kN-m/rad will have a forward transverse critical speed (in rad/s) of (consider the gyroscopic effect)
A. 100.00
B. 200.00
C. 81.65
D. 70.71

xlvi. A spinning rotor (the diametral mass moment of inertia is two times its polar mass moment of inertia with $I_p=1$ kg-m^2) in a pure tilting motion and supported on a flexible support of the torsional stiffness of 20 kN-m/rad will have a transverse critical speed (in rad/s) of (ignore the gyroscopic effect)
A. 100.00
B. 200.00
C. 81.65
D. 70.71

ANSWERS TO SELECTED EXERCISE PROBLEMS

5.1 Case I: without gyroscopic effect 303.45 rad/s and 1255.20 rad/s
Case II: with gyroscopic effect 413.40 rad/s ($F1$), 241.78 rad/s ($B1$) and 909.56 rad/s ($B2$)
5.2 Case I: without gyroscopic effect 25.47 rad/s and 242.97 rad/s
Case II: with gyroscopic effect 27.01 rad/s ($F1$), 24.11 rad/s ($B1$) and 148.17 rad/s ($B2$)
5.3 Case I: without gyroscopic effect 26.82 rad/s and 122.26 rad/s
Case II: with gyroscopic effect 33.33 rad/s ($F1$), 22.50 rad/s ($B1$) and 84.16 rad/s ($B2$)
5.4 Refer to Chapter 2 of Volume II
5.5 $\omega_{cr}=\sqrt{\dfrac{12EI}{(I_p+I_d)l}}$
5.6 Case I: without gyroscopic effect 70.66 rad/s and 147.37 rad/s
Case II: with gyroscopic effect 84.76 rad/s ($F1$), 54.43 rad/s ($B1$) and 110.47 rad/s ($B2$)
5.7 Not applicable
5.8 108.47 rad/s (F) and 93.25 rad/s (B)
5.9 Not applicable
5.10 Refer to Case II in Page 255 and Figure 5.30
5.11 (i) 66.75 rad/s (ii) 66.47 rad/s (iii) 66.42 rad/s (iv) 66.54 rad/s
5.12 (i) $\omega_{nf}=\sqrt{\dfrac{3EI}{I_d l}}$ (ii)–(iii) $\omega_{cr}^{B}=\sqrt{\dfrac{EI}{I_d l}}$

5.13 $\omega_{cr}^{B} = \sqrt{\dfrac{1}{3I_d\alpha_{22}}}$

5.14 $\omega_{cr} = \sqrt{\dfrac{3EI}{ml}}$

5.15 1054.1 rad/s (F) and 953.46 rad/s (B)
5.16 Not applicable
5.17 Not applicable
5.18 Case I: without gyroscopic effect 25.47 rad/s and 242.97 rad/s
Case II: with gyroscopic effect 27.01 rad/s ($F1$), 24.11 rad/s ($B1$) and 148.17 rad/s ($B2$)
5.19 With gyroscopic effect 45.28 rad/s ($F1$), 41.45 rad/s ($B1$), 379.45 rad/s ($F2$) and 185.37 rad/s ($B2$)
5.20 Refer to Equation 5.43
5.21 Not applicable
5.22 d=0.03374 m.
5.23 Natural whirl frequencies 32.50 rad/s (F) and 20.00 rad/s (B)
Critical speeds 27.26 rad/s (F) and 24.04 rad/s (B)
5.24 (refer to Chapter 2 of Volume II)
5.25 d=0.03615 m
5.26 3.9639×10^5 N/m
5.27 Case I: without gyroscopic effect 191.45 and 428.09 rad/s
Case II: with gyroscopic effect 191.45 rad/s ($F1$), 191.45 rad/s ($B1$) and 247.16 rad/s ($B2$)
5.28 Not applicable
5.29 $\omega_{cr} = 69.51$ rad/s
5.30 $\omega_{cr}^{F} = 32.61$ and $\omega_{cr}^{B} = 32.00$ rad/s
5.31 $\omega_{cr1F} = 116.81$ rad/s; ω_{cr1B}=115.88 rad/s; ω_{cr2B}=468.65 rad/s
5.32 Not applicable
5.33 $\omega_{crF} = 14.03$ rad/s and $\omega_{crB} = 13.92$ rad/s
5.34 $\omega_{cr}^{F} = 258.20$ and $\omega_{cr}^{B} = 169.03$ rad/s
5.35 Refer to Figure 5.64

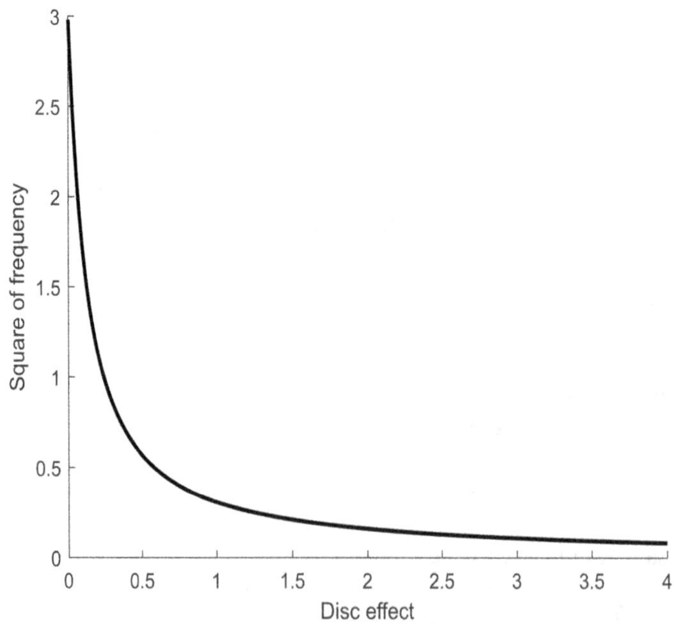

FIGURE 5.64 Variation of the critical speed function with the disk effect.

5.36 No critical speed for forward whirl due to thin disk $I_p > I_d$ and $\omega_{cr}^{B} = 131.07$ rad/s.
5.37 $k = 396.15$ kN/m
5.38 $\omega_{cr}^{F} = 23.42$ and $\omega_{cr}^{B} = 22.67$ rad/s
5.39 $I = 5.2472 \times 10^{-11}\,\text{m}^4$; $d = 5.7179$ mm
5.40

(i) B	(ii) D	(iii) A	(iv) B	(v) A	(vi) A
(vii) B	(viii) A	(ix) B	(x) A	(xi) D	(xii) B
(xiii) C	(xiv) D	(xv) D	(xvi) A	(xvii) B	(xviii) A
(xix) B	(xx) C	(xxi) A	(xxii) B	(xxiii) A	(xxiv) B
(xxv) B	(xxvi) C	(xxvii) D	(xxviii) D	(xxix) C	(xxx) A
(xxxi) C	(xxxii) A	(xxxiii) B	(xxxiv) C	(xxxv) D	(xxxvi) B
(xxxvii) A	(xxxviii) B	(xxxix) A	(xl) B	(xli) A	(xlii) A

REFERENCES

Arnold, R.N. and Maunder, L., 1961, *Gyrodynamics*. New York, NY: Academic Press.
Bevan, T., 1984, *The Theory of Machines*, 3rd ed. New Delhi: CBS Publishers & Distributors.
Campbell, W.E., 1924, The protection of steam turbine disk wheels from axial vibration, *Transactions of the ASME*, *46*, 31–160.
Davidson, M. (ed.), 1946, *The Gyroscope and Its Applications*. London: Hutchinson's Scientific and Technical Publications.
Deimel, R., 1929, *Mechanics of the Gyroscope*. New York, NY: Macmillan.
Den Hartog, J.P., 1984, *Mechanical Vibrations*. New York, NY: Dover Publications.
Ferry, E.S., 1932, *Applied Gyrodynamics*. New York, NY: John Wiley & Sons.
Gary, A., 1918, *A Treatise on Gyrostatics and Rotational Motion*. London: Macmillan.
Mabie, H.H. and Reinholts, C.F., 1987, *Mechanisms and Dynamics of Machinery*, 4th ed. New York, NY: John Wiley & Sons.
Rao, J.S. and Dukkipati, R.V., 1995, *Mechanism and Machine Theory*, 2nd ed. New Delhi: New Age International (P) Limited, Publishers.
Savet, P.H., 1961, *Gyroscopes, Theory and Design*. New York, NY: McGraw-Hill.
Shravankumar, C. and Tiwari, R., 2011, Gyroscopic effects on a cantilever rotor system: A comparative analysis, *National Symposium on Rotor Dynamics (NSRD-2011)*, IIT Madras, India, December 19–21, 2011.
Timoshenko, S.P. and Young, D.H., 1968, *Elements of Strength of Materials*, 5th ed. New Delhi: Affiliated East-West Press Private Limited.
Tondl, A., 1965, *Some Problems of Rotor Dynamics*. Prague: Czechoslovak Academy of Sciences.
Wrigley, W., Hollister, W.M., and Denhard, W.G., 1969, *Gyroscopic Theory, Design, and Instrumentation*. Cambridge: The MIT Press.

6 Torsional Vibrations of Rotor Systems by the Direct Analytical and Transfer Matrix Methods

In previous chapters, we mainly studied transverse vibrations of simple rotor-bearing systems. It was pointed out that transverse vibrations are very common in rotor systems due to residual imbalance, which is the most inherent fault in a rotor. We studied the speed-independent bearing dynamic parameters that affected the behavior of rotors. The effects of gyroscopic couples on natural whirl frequencies and critical speeds have also been investigated in detail. In the present chapter, we extend the analysis of simple rotors to torsional vibrations. We analyze the torsional vibrations of the single-disk, two-disk, and three-disk rotor systems with the conventional Newton's second law of motion as well as energy methods. We also analyze stepped shafts, simple geared rotor systems, and branched geared rotor systems. For the multi-degree-of-freedom (DOF) rotor system, a general procedure of the transfer matrix method (TMM) is discussed for both undamped and damped cases. Advantages and disadvantages of the TMM are outlined. In reciprocating engines, large variations in torque take place periodically. This leads to torsional resonances, and to analyze the free and forced vibrations of these systems, we convert them to an equivalent multi-DOF rotor system. The present chapter paves the road for the TMM to be extended for the analysis of transverse vibrations of multi-DOF rotor systems in Chapter 8.

The study of torsional vibration of rotors is very important, especially in applications where high-power transmission and high speed are present (Figure 6.1). Torsional vibrations are predominant whenever there are large disks on relatively thin shafts (e.g., the flywheel of a punch press). Torsional vibrations may occur due to the following: (i) inertia forces of reciprocating mechanisms (e.g., due to pistons in internal combustion engines); (ii) impulsive loads occurring during a normal machine cycle (e.g., during operations of a punch press); (iii) shock loads applied to electrical machinery (such as a generator line fault followed by fault removal and automatic closure); (iv) torques related to gear mesh frequencies, and the turbine blade and compressor fan passing frequencies; (v) a rotor rubs with the stator; (vi) during angular acceleration/deceleration of engine or flywheel energy storage systems; and (vii) shaft misalignment. For machines having massive rotors and flexible shafts (where system natural frequencies of torsional vibrations may be close to, or within, the source frequency range during normal operation), torsional vibrations constitute a potential design problem area.

Torsional vibration, the twisting of rotating shafts, can develop various failures in machinery. Common faults comprise shaft cracking or breakage, particularly in crankshafts, and failures of couplings, gears, the flange of flywheel, and other components associated with the shaft. These failures frequently happen at a 45° angle to the shaft axis. Torsional vibration problems can be difficult to detect before failure because they are always evident as observable vibrations until a critical point is reached. In such cases, designers should ensure the accurate prediction of machine torsional frequencies, and frequencies of any torsional load fluctuations should not coincide with torsional natural frequencies. Hence, the determination of torsional natural frequencies of the rotor system is very important (Tuplin, 1966; Walker, 2003), and in the present chapter, we deal with it in great detail.

DOI: 10.1201/9781003661559-6

FIGURE 6.1 A heavy duty gearbox.

6.1 A SIMPLE TORSIONAL ROTOR SYSTEM WITH A SINGLE DISK

Consider a rotor system like that shown in Figure 6.2a. The shaft is massless, and it provides torsional stiffness. The disk is thin and rigid (i.e., it has no flexibility). However, it has a large polar mass moment of inertia. If an initial disturbance is given to the non-spinning disk in the pure torsional mode (about its longitudinal or polar axis) and the disk is allowed to oscillate on its own, then free torsional vibrations are generated. It is assumed that there is no coupling of the torsional motion and the transverse or axial motion. Figure 6.3 shows that the rotor is spinning with a nominal speed of ω and generating torsional vibrations, $\varphi_z(t)$; due to this, it has an actual speed of $(\omega + \dot{\varphi}_z(t))$.

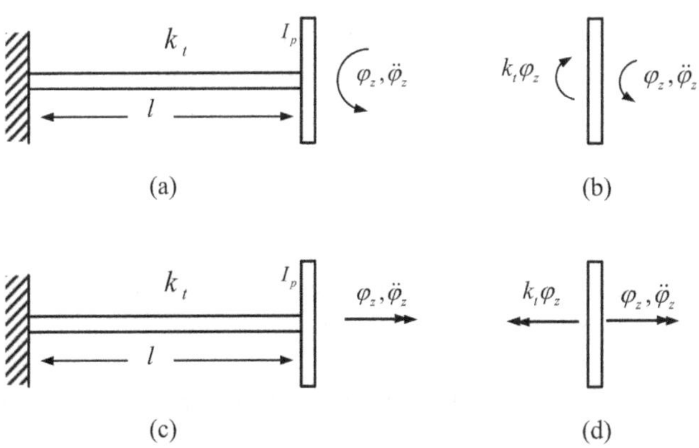

FIGURE 6.2 (a) A single-disk cantilever rotor system. (b) A free body diagram of the disk. (c) A single-disk cantilever rotor system with double arrow representation. (d) A free body diagram of the disk with double arrow representation.

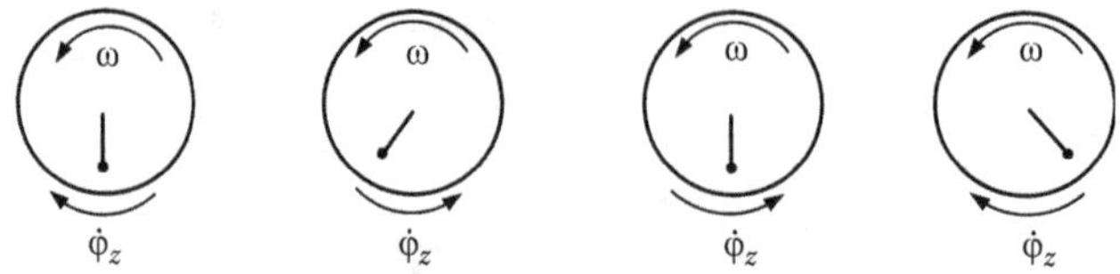

FIGURE 6.3 Torsional vibrations of a spinning rotor as observed with a stroboscope tuned at ω as the flashing frequency.

It should be noted that the spinning speed, ω, remains constant; however, the angular velocity due to the torsion has a varying direction over a period. For a multidisk case, different disks will have different angular velocities (refer to Section 6.2). In actual practice, if we tune the flashing frequency of a *stroboscope* (a speed/frequency measuring instrument based on a light source with an adjustable *flashing frequency*, refer to Chapter 5 of Volume II) to the nominal speed of a rotor, then free torsional oscillations are observed. For the present case and in most of our analysis, it is assumed that the torsional natural frequency does not depend upon the spin speed of the rotor. Hence, the natural frequency of the non-spinning rotor is the same as at any other speed. For the free torsional oscillation, the motion is simple harmonic motion (SHM) with a unique frequency, which is called the *torsional natural frequency* of the rotor system.

From the theory of torsion of the shaft (Timoshenko and Young, 1968), we have

$$k_t = \frac{T}{\varphi_z} = \frac{GJ}{l} \quad \text{with} \quad J = \frac{\pi}{32} d^4 \tag{6.1}$$

where k_t is the torsional stiffness of the shaft, J is the polar second moment of area of the shaft cross-section, l is the length of the shaft, d is the diameter of the shaft, and φ_z is the angular displacement of the disk (the counterclockwise direction is assumed as the positive direction when the observer is looking at the rotor from the right-hand side (RHS) to the left-hand side (LHS); often angular displacement and the torque are indicated by a line with a double arrow with a positive sign toward the left direction and a negative sign toward the right direction, refer to Figure 6.2. It is similar to the right hand rule: curled fingers representing the angular displacement (or torque) direction and thumb representing the angular displacement (or torque) vector). From the free body diagram of the disk as shown in Figure 6.2b, we have

$$\sum \text{External torque on the disc} = I_p \ddot{\varphi}_z(t) \;\Rightarrow\; -k_t \varphi_z(t) = I_p \ddot{\varphi}_z(t) \tag{6.2}$$

where Σ represents the summation operator, I_p is the polar mass moment of inertia of the disk, and two dots at the top of a variable represent the double derivative with respect to time. Equation 6.2 is the equation of motion of the disk for free torsional vibrations. The free (or natural) torsional vibration has SHM, which gives the disk displacement, as

$$\varphi_z(t) = \Phi_z \sin \omega_{\text{nf}} t \quad \text{so that} \quad \ddot{\varphi}_z(t) = -\omega_{\text{nf}}^2 \Phi_z \sin \omega_{\text{nf}} t = -\omega_{\text{nf}}^2 \varphi_z(t) \tag{6.3}$$

where Φ_z is the amplitude of the torsional vibration and ω_{nf} is the torsional natural frequency. On substituting Equation 6.3 into Equation 6.2, we get

$$-k_t \varphi_z = I_p\left(-\omega_{nf}^2 \varphi_z\right) \quad \text{or} \quad \varphi_z\left(\omega_{nf}^2 I_p - k_t\right) = 0 \tag{6.4}$$

Because $\varphi_z \neq 0$, it gives

$$\omega_{nf} = \pm\sqrt{\frac{k_t}{I_p}} = \pm\sqrt{\frac{GJ}{lI_p}} \tag{6.5}$$

which is similar to the case of a single-DOF spring-mass system, where the polar mass moment of inertia and the torsional stiffness replace the mass and the spring stiffness, respectively. Only the positive sign is considered because the natural frequency cannot be a negative quantity. It should be noted that the equation of motion can also be derived by Lagrange's equation (refer to Chapter 7) by considering the potential and kinetic energies of the system ($U = \frac{1}{2}k_t\varphi_z^2$ and $T = \frac{1}{2}I_p\dot{\varphi}_z^2$).

Example 6.1

Obtain the torsional natural frequency of a cantilever rotor system as shown in Figure 6.4. The end B_1 of the shaft has fixed end conditions. The shaft diameter is 10 mm, and the length of the span is 0.2 m. The disk D_1 is thin and has a mass of 10 kg and a polar mass moment of inertia equal to 0.02 kg-m². Neglect the mass of the shaft. Compare the torsional natural frequency so obtained with the transverse natural frequency of the same rotor system.

Solution:

For the present problem, the torsional stiffness of the shaft can be obtained as

$$k_t = \frac{GJ}{l} = \frac{0.8 \times 10^{11} \times \frac{\pi}{32}(0.01)^4}{0.2} = 392.7 \text{ Nm/rad} \quad \text{with} \quad J = \frac{\pi}{32}d^4$$

Hence, the torsional frequency is given as

$$\omega_{nf} = \sqrt{k_t / I_p} = \sqrt{\frac{392.7}{0.02}} = 140.12 \text{ rad/s} = 22.3 \text{ Hz}$$

Hence, if the rotor has a cyclic torque variation with a period of 1/22.3 s, then the rotor might undergo resonance under torsional vibrations. For a comparison with the transverse natural frequency, the bending stiffness is given as

$$k_b = \frac{3EI}{l^3} = \frac{3 \times 2.1 \times 10^{11} \times \frac{\pi}{64}(0.01)^4}{0.2^3} = 3.87 \times 10^4 \text{ N/m} \quad \text{with} \quad I = \frac{\pi}{64}d^4$$

Hence, the transverse natural frequency of the same rotor system is given as

$$\omega_{nf} = \sqrt{k_b / m} = \sqrt{\frac{3.87 \times 10^4}{10}} = 62.21 \text{ rad/s} = 9.9 \text{ Hz}$$

If the same rotor has a small amount of imbalance and if the rotor is spinning at approximately 9.9 Hz, then the rotor might undergo resonance under transverse vibrations. For the present case, the transverse natural frequency is much lower than the torsional natural frequency.

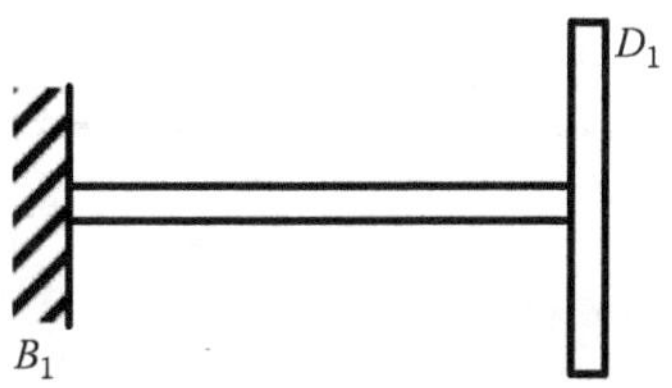

FIGURE 6.4 A cantilever rotor system.

6.2 A TWO-DISK TORSIONAL ROTOR SYSTEM

A two-disk torsional system is shown in Figure 6.5. In this case, the entire rotor is free to rotate about the center line of the shaft, as the shaft is mounted on frictionless bearings. Hence, it has free–free end conditions, the application of which can be found in an aircraft when it is flying, and the whole structure has torsional vibrations due to aerodynamic forces; of course, the present model will be too simplistic.

Let φ_{z1} and φ_{z2} be angular displacements of thin disks 1 and 2, respectively. For both angular displacements, the counterclockwise direction is chosen as the positive direction when looking at the disk from the RHS to LHS. Here again, we want to stress that there are disk angular displacements about its polar axis. Let I_{p1} and I_{p2} be the polar mass moments of inertia of disks 1 and 2, respectively. From the free body diagram of disks shown in Figure 6.6, we have

$$\sum \text{External torques} = I_{p1}\ddot{\varphi}_{z1} \Rightarrow -\left(\varphi_{z1} - \varphi_{z2}\right)k_t = I_{p1}\ddot{\varphi}_{z1} \tag{6.6}$$

and

$$\sum \text{External torques} = I_{p2}\ddot{\varphi}_{z2} \Rightarrow -\left(\varphi_{z1} - \varphi_{z2}\right)k_t = I_{p2}\ddot{\varphi}_{z2} \tag{6.7}$$

where k_t is the torsional stiffness of the shaft, and let $\left(\varphi_{z1} - \varphi_{z2}\right)$ be the relative twist of the shaft ends. If we take the free body diagram of the shaft also, then it will have reaction torques at the ends in opposite directions, as shown for disks. However, finally, if we join disks to the shaft, it is expected that reaction torques cancel. The above expressions give the following equations of motion:

$$I_{p1}\ddot{\varphi}_{z1} + k_t\varphi_{z1} - k_t\varphi_{z2} = 0 \quad \text{and} \quad I_{p2}\ddot{\varphi}_{z2} + k_t\varphi_{z2} - k_t\varphi_{z1} = 0 \tag{6.8}$$

For free vibrations, we have SHM as $\varphi_{z1} = \Phi_{z1}e^{j\omega_{nf}t}$ and $\varphi_{z2} = \Phi_{z2}e^{j\omega_{nf}t}$, where Φ is the amplitude and ω_{nf} is the natural frequency, so the solution is

$$\ddot{\varphi}_{z1} = -\omega_{nf}^2\varphi_{z1} \quad \text{and} \quad \ddot{\varphi}_{z2} = -\omega_{nf}^2\varphi_{z2} \tag{6.9}$$

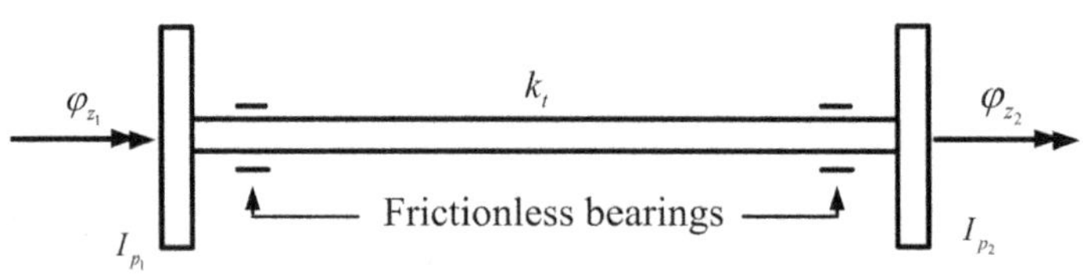

FIGURE 6.5 A two-disk torsional system.

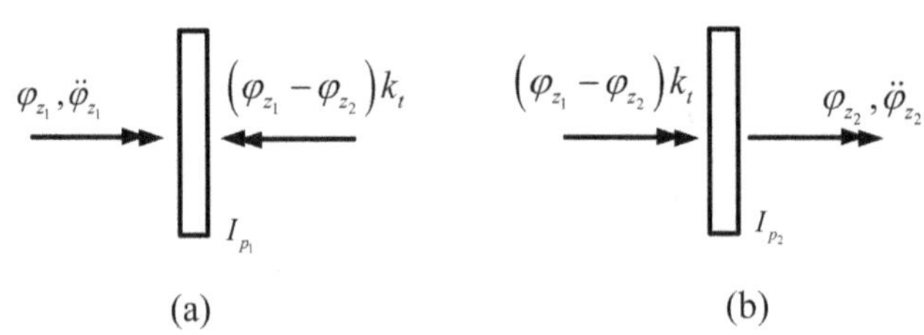

FIGURE 6.6 Free body diagrams of (a) disk 1 and (b) disk 2.

It should be noted that both disks are expected to have the same frequency of oscillation (i.e., the synchronous motion) during the normal mode, when it is oscillating with one of its natural frequencies. Substituting Equation 6.9 into Equation 6.8, we get

$$-I_{p1}\omega_{nf}^2\varphi_{z1}+k_t\varphi_{z_1}-k_t\varphi_{z_2}=0 \quad \text{and} \quad -I_{p2}\omega_{nf}^2\varphi_{z2}+k_t\varphi_{z2}-k_t\varphi_{z1}=0 \tag{6.10}$$

Equation 6.10 can be assembled in matrix form in the frequency domain as

$$\mathbf{D\Phi}_z=\mathbf{0} \tag{6.11}$$

with

$$\mathbf{D}=\begin{bmatrix} k_t-I_{p1}\omega_{nf}^2 & -k_t \\ -k_t & k_t-I_{p2}\omega_{nf}^2 \end{bmatrix};\quad \mathbf{\Phi}_z=\begin{Bmatrix}\Phi_{z1}\\ \Phi_{z2}\end{Bmatrix};\quad \mathbf{0}=\begin{Bmatrix}0\\0\end{Bmatrix}$$

The nontrivial solution of Equation 6.11 is obtained by taking the determinant of the matrix **D** equal to zero:

$$|\mathbf{D}|=0$$

which gives

$$\left(k_t-I_{p1}\omega_n^2\right)\left(k_t-I_{p2}\omega_n^2\right)-k_t^2=0 \quad \text{or} \quad I_{p1}I_{p2}\omega_{nf}^4-\left(I_{p1}+I_{p2}\right)k_t\omega_{nf}^2=0 \tag{6.12}$$

The roots of Equation 6.12 are given as

$$\omega_{nf1}=0 \quad \text{and} \quad \omega_{nf2}=\sqrt{\frac{\left(I_{p1}+I_{p2}\right)k_t}{I_{p1}I_{p2}}} \tag{6.13}$$

Hence, the system has two torsional natural frequencies, and one of them is zero. From Equation 6.11, corresponding to the first natural frequency for $\omega_{nf1}=0$, we get

$$\Phi_{z1}=\Phi_{z2} \tag{6.14}$$

From Equation 6.14, it can be concluded that the first root of Equation 6.12 represents the case when both disks simply roll together in phase with each other, as shown in Figure 6.7. The representation of the relative angular displacement of two disks in this form is called the *mode shape*. The mode shape shown in Figure 6.7 is called the *rigid body mode*, which is of little practical significance because no stresses develop in the shaft during the motion. However, it may be of interest in control engineering. This mode occurs whenever the system has free–free boundary conditions (e.g., an airplane during flight).

Now, from the first set of Equation 6.11, for $\omega_{nf}=\omega_{nf2}$, we get

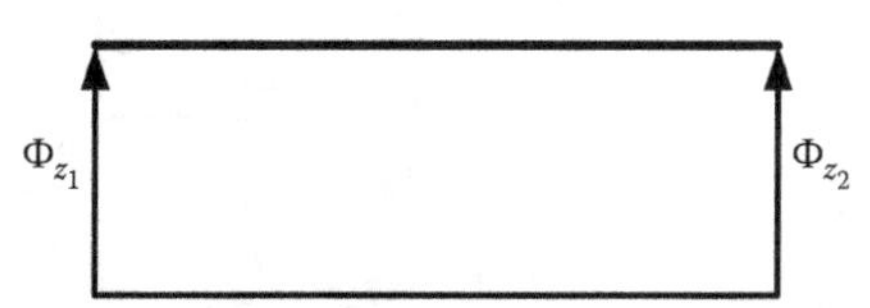

FIGURE 6.7 The first mode shape.

$$\left(k_t - I_{p1}\omega_{nf2}^2\right)\Phi_{z1} - k_t\Phi_{z2} = 0 \quad \text{or} \quad \left[k_t - I_{p1}\frac{I_{p1} + I_{p2}}{I_{p1}I_{p2}}k_t\right]\Phi_{z1} - k_t\Phi_{z2} = 0 \tag{6.15}$$

which gives the relative amplitudes of two disks as

$$\frac{\Phi_{z1}}{\Phi_{z2}} = -\frac{I_{p2}}{I_{p1}} \tag{6.16}$$

The second mode shape from Equation 6.16 represents the case when both masses oscillate in antiphase with each other (i.e., the direction of rotation of one disk will always be the opposite of the other). Both disks reach their extreme angular positions simultaneously, and both reach the static equilibrium (untwisted) position also simultaneously. It should be noted that both disks have the same frequency of oscillation (i.e., *the time period*) but different angular amplitudes. A large disk has a lesser amplitude of oscillations compared to the smaller disk. Figure 6.8a shows the mode shape of the two-rotor system. From two similar triangles in Figure 6.8a, we have

$$\frac{\Phi_{z1}}{l_1} = \frac{\Phi_{z2}}{l_2} \quad \Rightarrow \quad \frac{\Phi_{z1}}{\Phi_{z2}} = \frac{l_1}{l_2} \tag{6.17}$$

where l_1 and l_2 are the node positions from disks 1 and 2, respectively (Figure 6.8a). Because both masses are always vibrating in the opposite direction, there must be a point on the shaft where the torsional displacement is not taking place, i.e., where the angular displacement amplitude is zero. This point is called a *node*, and its location is at present an unknown. The location of the node may be established by treating each end of the real system as a separate single-disk cantilever system, as shown in Figure 6.8b and c. The node is treated as the point where the shaft is rigidly fixed. Hence, basically we will have two single-DOF cantilever rotor systems in torsion (Figure 6.8b and c) instead of one two-DOF free–free rotor system (Figure 6.5). Because the value of the torsional natural frequency, ω_{nf2}, is known from Equation 6.13 (the frequency of oscillation of each of the

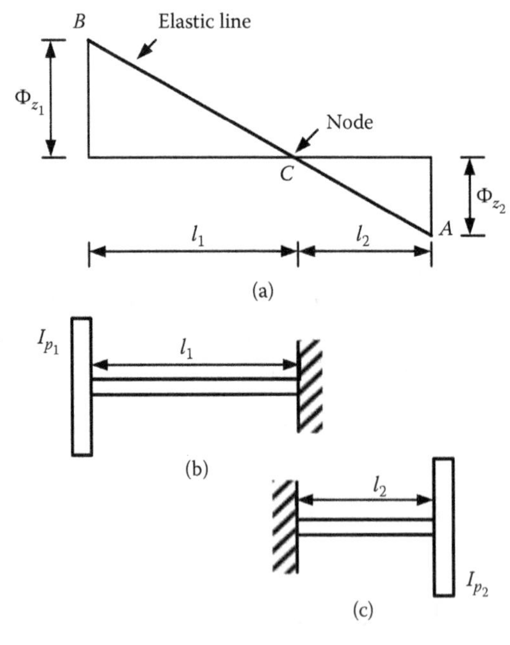

FIGURE 6.8 (a) The second mode. (b) Equivalent system 1. (c) Equivalent system 2.

single-disk cantilever systems must be the same and is equal to ω_{nf_2} because they belong to the same original system), we write

$$\omega_{nf_2}^2 = \frac{k_{t_1}}{I_{p1}} = \frac{k_{t_2}}{I_{p2}} \tag{6.18}$$

where ω_{nf_2} is defined by Equation 6.13, k_{t_2} and k_{t_1} are the torsional stiffnesses of two single-DOF cantilever rotor systems, which can be obtained from Equation 6.18, as

$$k_{t_1} = \omega_{nf_2}^2 I_{p1} \quad \text{and} \quad k_{t_2} = \omega_{nf_2}^2 I_{p2} \tag{6.19}$$

Lengths l_1 and l_2 of these two single-DOF rotor systems can be obtained as

$$l_1 = \frac{GJ}{k_{t_1}} \quad \text{and} \quad l_2 = \frac{GJ}{k_{t_2}} \quad \text{with} \quad l_1 + l_2 = l \tag{6.20}$$

which gives the node position. It should be noted that the shear stress would be maximum at the node point, being a fixed end of cantilever rotor systems, and it is a likely position of failure in the torsional vibration.

Alternatively, from Equations 6.16 and 6.17, we have (ignoring the sign)

$$\frac{l_1}{l_2} = \frac{I_{p2}}{I_{p1}} \quad \text{with} \quad l_1 + l_2 = l \tag{6.21}$$

On solving the above equations, we obtain

$$l_1 = \frac{I_{p2}}{I_{p1} + I_{p2}} l \quad \text{and} \quad l_2 = \frac{I_{p1}}{I_{p1} + I_{p2}} l \tag{6.22}$$

6.2.1 Alternative Method (Indirect Method)

From the flexible mode of the two-disk rotor, it is clear that there should be a node between the two disks on the shaft, from which we can write

$$l_1 + l_2 = l \tag{6.23}$$

From the information that in this flexible mode, the two-disk free–free rotor system could be treated as two single-disk cantilever rotors, and both rotor systems would have the same natural frequency as the original rotor system, we write

$$\omega_{nf_2}^2 = \frac{k_{t_1}}{I_{p1}} = \frac{k_{t_2}}{I_{p2}} \Rightarrow \frac{GJ}{l_1 I_{p1}} = \frac{GJ}{l_2 I_{p2}} \quad \text{or} \quad l_1 = \frac{I_{p2}}{I_{p1}} l_2 \tag{6.24}$$

Equations 6.23 and 6.24 could be solved to get

$$l_1 = \frac{I_{p2}}{I_{p1} + I_{p2}} l, \quad \text{so that} \quad l_2 = \frac{I_{p1}}{I_{p1} + I_{p2}} l \tag{6.25}$$

The earlier expressions give the node location, and now with Equation 6.24, the torsional natural frequency of the two-disk free–free rotor system can be obtained, which will be the same as either of the single-disk cantilever rotor systems.

Equations of motion can also be derived by Lagrange's equation (refer to Chapter 7) by considering the potential and kinetic energies of the system $U = \frac{1}{2}k_t\left(\varphi_{z2} - \varphi_{z1}\right)^2$ and $T = \frac{1}{2}I_{P1}\dot{\varphi}_{z1}^2 + \frac{1}{2}I_{P2}\dot{\varphi}_{z2}^2$.

Example 6.2

Determine torsional natural frequencies and mode shapes for a rotor system mounted on frictionless supports as shown in Figure 6.9. Neglect the mass of the shaft and assume the disks are thin and rigid. The shaft has 1 m length, 0.015 m diameter, and 0.8×10^{11} N/m modulus of rigidity (or the shear modulus). The disks have polar mass moment of inertia $I_{P1} = 0.01$ kg-m^2 and $I_{P2} = 0.015$ kg-m^2.

Solution:

The stiffness of the shaft can be obtained as

$$k_t = \frac{GJ}{l} = \frac{0.8 \times 10^{11} \times \pi (0.015)^4 / 32}{1.0} = 397.61 \text{ Nm/rad} \quad \text{with} \quad J = \frac{\pi}{32} d^4$$

The natural frequency is given as

$$\omega_{nf1} = 0 \quad \text{and} \quad \omega_{nf2} = \sqrt{\frac{\left(I_{P1} + I_{P2}\right)k_t}{I_{P1} I_{P2}}} = \sqrt{\frac{(0.01 + 0.015) \times 397.61}{0.01 \times 0.015}} = 257.43 \text{ rad/s}$$

The relative displacements would be

$$\frac{\Phi_{z1}}{\Phi_{z2}} = -\frac{I_{P2}}{I_{P1}} = -\frac{0.015}{0.01} = -1.5$$

which means disk 1 would have 1.5 times the angular displacement amplitude compared to disk 2, in the opposite direction. The node position can be obtained as

$$\frac{l_1}{l_2} = \frac{I_{P2}}{I_{P1}} = \frac{0.015}{0.01} = 1.5 \quad \text{and} \quad l_1 + l_2 = 1$$

Hence, we get the node location as $l_1 = 0.6$ m (i.e., 0.6 m from disk 1, refer to Figure 6.8a). It can be verified that two equivalent single-mass cantilever rotors have the same natural frequency as

$$k_{t1} = \frac{GJ}{l_1} = \frac{0.8 \times 10^{11} \times \pi (0.015)^4 / 32}{0.6} = 662.68 \text{ Nm/rad}$$

and

$$k_{t2} = \frac{GJ}{l_2} = \frac{0.8 \times 10^{11} \times \pi (0.015)^4 / 32}{0.4} = 994.03 \text{ Nm/rad}$$

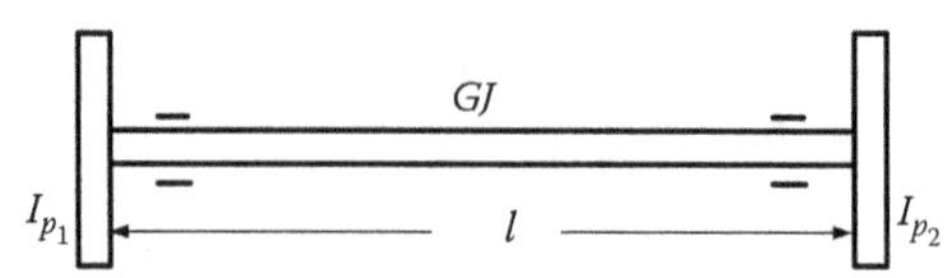

FIGURE 6.9 A two-disk rotor system.

so that

$$\omega_{nf2}^{(1)} = \sqrt{\frac{k_{t1}}{I_{p1}}} = \sqrt{\frac{662.68}{0.01}} = 257.43 \text{ rad/s}$$

and

$$\omega_{nf2}^{(2)} = \sqrt{\frac{k_{t2}}{I_{p2}}} = \sqrt{\frac{994.03}{0.015}} = 257.43 \text{ rad/s}$$

6.3 A TWO-DISK TORSIONAL ROTOR SYSTEM WITH A STEPPED SHAFT

Figure 6.10a shows a stepped shaft with two large disks at the ends, with I_p as the polar mass moment of inertia (subscripts 1 and 2 represent the LHS and RHS disks, respectively). It is assumed that the rotor has free–free boundary conditions and the polar mass moment of inertia of the shaft is negligible compared with that of the two large disks at either end of the shaft. In such cases, the actual shaft should be replaced by an unstepped (uniform) equivalent shaft for the purpose of analysis as shown in Figure 6.10b. The equivalent shaft diameter may be the same as the smallest diameter of the real shaft (or any other diameter). The equivalent shaft must have the same torsional stiffness as the real shaft.

Because the torsional stiffnesses corresponding to different shaft segments are connected in series, the equivalent torsional stiffness can be written as

$$\frac{1}{k_{t_e}} = \frac{1}{k_{t_1}} + \frac{1}{k_{t_2}} + \frac{1}{k_{t_3}} \tag{6.26}$$

where k_t is the torsional stiffness, subscripts 1, 2, 3 represent the shaft segment number, and the subscript e represents the equivalent. Nothing Equation 6.1, Equation 6.26 becomes

$$\frac{l_e}{J_e} = \frac{l_1}{J_1} + \frac{l_2}{J_2} + \frac{l_3}{J_3} \tag{6.27}$$

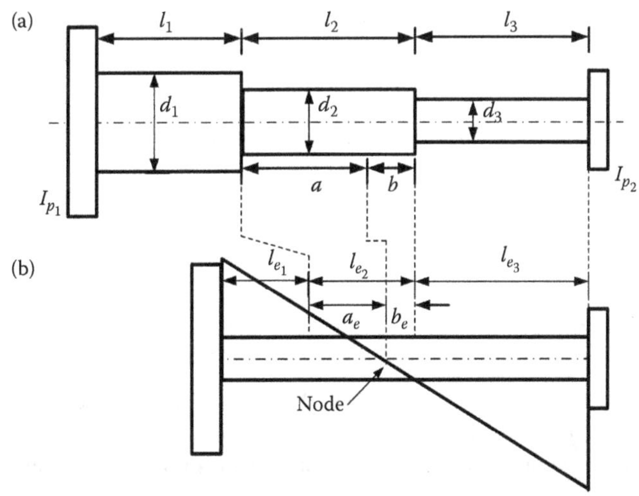

FIGURE 6.10 Two disks with (a) a stepped shaft and (b) an equivalent uniform shaft.

where l represents the length of the shaft segment and J is the polar moment of inertia of the shaft cross-sectional area. The earlier equation can be written as

$$l_e = l_{e_1} + l_{e_2} + l_{e_3} \tag{6.28}$$

with

$$l_{e_1} = l_1 J_e / J_1; \quad l_{e_2} = l_2 J_e / J_2; \quad l_{e_3} = l_3 J_e / J_3$$

where $l_{e_1}, l_{e_2}, l_{e_3}$ are the equivalent lengths of shaft segments having the equivalent shaft diameter d_3, and l_e is the total equivalent length of the uniform equivalent shaft as shown in Figure 6.10b. Let us assume that the node position in the equivalent shaft system comes out in the second shaft segment from the analysis of the two-disk uniform shaft rotor system with free–free boundary conditions in the previous section (in the case where the node position is in the first shaft segment or in the third shaft segment, then the analysis would remain the same, and parameter values and the sign so obtained would take care of the actual position of the node). Noting Equations 6.18 and 6.20, the node location in the equivalent shaft from Figure 6.10b can be obtained as

$$l_{e_1} + a_e = \frac{GJ_e}{\omega_{nf2}^2 I_{p1}} \quad \text{and} \quad l_{e_3} + b_e = \frac{GJ_e}{\omega_{nf2}^2 I_{p2}} \tag{6.29}$$

with

$$\omega_{nf2} = \sqrt{\frac{\left(I_{p1} + I_{p2}\right) k_{t_e}}{I_{p1} I_{p2}}} \quad \text{and} \quad k_{t_e} = \frac{1}{l_1/(GJ_1) + l_2/(GJ_2) + l_3/(GJ_3)}$$

From Equation 6.29, the node position (i.e., a_e or b_e in Figure 6.10b) can be obtained; the corresponding node location in the real shaft system can be obtained as explained here. From Equation 6.28, we have

$$l_{e_2} = l_2 \frac{J_e}{J_2}; \quad J_e = \frac{\pi}{32} d_e^4 = \frac{\pi}{32} d_3^4; \quad J_2 = \frac{\pi}{32} d_2^4 \tag{6.30}$$

Because Equation 6.30 is for the shaft segment in which the node is assumed to be present, we can write

$$a_e = a \frac{J_e}{J_2} \quad \text{and} \quad b_e = b \frac{J_e}{J_2} \quad \text{or} \quad a = a_e \frac{J_2}{J_e} \quad \text{and} \quad b = b_e \frac{J_2}{J_e} \tag{6.31}$$

where a or b represents the node position in the real system (Figure 6.10a). Equation 6.31 can be combined as

$$\frac{a}{b} = \frac{a_e}{b_e} \tag{6.32}$$

So once a_e or b_e is obtained from Equation 6.29, the location of the node in the actual shaft can be obtained from Equation 6.32. Hence, the final location of the node on the shaft in the real system is given in the same proportion as in the shaft of the equivalent system in which the node occurs.

Example 6.3

Consider a stepped shaft with two disks as shown in Figure 6.11. Assume the following shaft dimensions: $l_1=0.5$ m, $l_2=0.3$ m, $l_3=0.2$ m, $d_1=0.015$ m, $d_2=0.012$ m, and $d_3=0.01$ m. Take the modulus of rigidity of the shaft as 0.8×10^{11} N/m. Disks have polar mass moments of inertia $I_{p1}=0.015$ kg-m^2 and $I_{p2}=0.01$ kg-m^2. Obtain the torsional natural frequencies, mode shapes, and location of the node for free–free end conditions. Neglect the inertia of the shaft.

Solution:

Let us represent shaft segments toward the left, middle, and right sides as 1, 2, and 3, respectively. For the present problem, the shaft has the following data:

$$J_1 = \frac{\pi d_1^4}{32} = \frac{\pi \times 0.015^4}{32} = 4.97\times10^{-9}\ \text{m}^4$$

$$J_2 = 2.036\times10^{-9}\ \text{m}^4$$

$$J_3 = 0.982\times10^{-9}\ \text{m}^4$$

$$k_{t_1} = \frac{GJ_1}{l_1} = \frac{0.8\times10^{11}\times4.97\times10^{-9}}{0.5} = 795.20\ \text{Nm/rad}$$

$$k_{t_2} = 542.93\ \text{Nm/rad}$$

$$k_{t_3} = 392.80\ \text{Nm/rad}$$

For the stepped shaft, the first step would be to obtain the equivalent length with respect to reference shaft 3, which has a diameter of 0.01 m, as

$$l_e = \frac{l_1}{J_1}J_e + \frac{l_2}{J_2}J_e + \frac{l_3}{J_3}J_e = \frac{0.5}{4.97}0.982 + \frac{0.3}{2.036}0.982 + \frac{0.2}{0.982}0.982$$
$$= 0.0988 + 0.1447 + 0.2 = 0.4435\ \text{m}$$

Hence, $l_{e_1}=0.0987$ m and $l_{e_2}=0.1447$ m. Now, the equivalent stiffness can be calculated as

$$k_{t_e} = \frac{GJ_e}{l_e} = \frac{0.8\times10^{11}\times0.982\times10^{-9}}{0.4435} = 177.14\ \text{Nm/rad}$$

The torsional natural frequency of the rotor system can be calculated as

$$\omega_{nf2} = \sqrt{\frac{(I_{p1}+I_{p2})k_{t_e}}{I_{p1}I_{p2}}} = \sqrt{\frac{(0.015+0.01)\times177.14}{0.015\times0.01}} = 171.82\ \text{rad/s}$$

Relative displacements of the rotor system would be

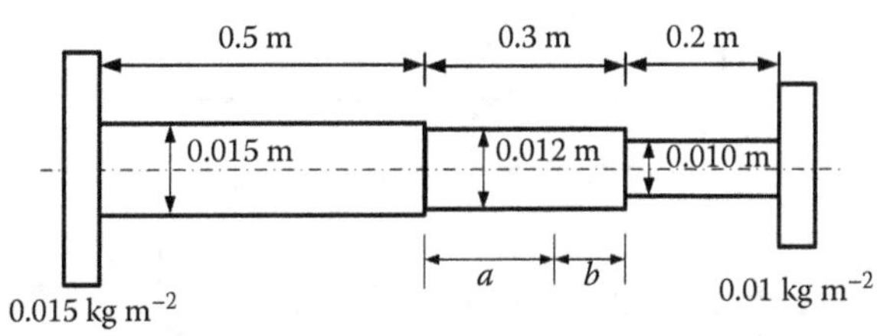

FIGURE 6.11 A stepped shaft with two disks.

$$\frac{\Phi_{z1}}{\Phi_{z2}} = -\frac{I_{P2}}{I_{P1}} = -\frac{0.01}{0.015} = -0.667$$

which means disk 1 would have 0.667 times the angular displacement amplitude compared with disk 2, in the opposite direction. It is interesting that relative displacements remain the same irrespective of shaft characteristics (i.e., stepped, uniform, and tapered) and their stiffness. However, the node position depends upon the shaft characteristics and its stiffness, and can be obtained as for the equivalent shaft as

$$\frac{l_{ne1}}{l_{ne2}} = \frac{I_{P2}}{I_{P1}} = \frac{0.01}{0.015} = 0.667 \quad \text{and} \quad l_{ne1} + l_{ne2} = l_e = 0.4435 \text{ m}$$

Hence, we get the node location as $l_{ne2} = 0.266$ m (i.e., 0.266 m from disk 2 in the equivalent system, see Figure 6.12). Hence, we have $l_{ne1} = 0.1775$ m. This means that the node is in the second (middle) shaft segment. The location in the actual rotor system would be

$$\frac{a}{b} = \frac{a_e}{b_e} = \frac{0.0787}{0.066} = 1.1924 \quad \text{and} \quad a + b = 0.3 \text{ m}$$

with

$$a_e = l_{ne1} - l_{e1} = 0.1775 - 0.0988 = 0.0787 \text{ m}$$

and

$$b_e = l_{ne2} - l_{e3} = 0.266 - 0.2 = 0.066 \text{ m}$$

Hence, we have the position of the node in the actual system, from the earlier equations, as $b = 0.137$ m and $a = 0.163$ m (see Figure 6.11 for locations of a and b).

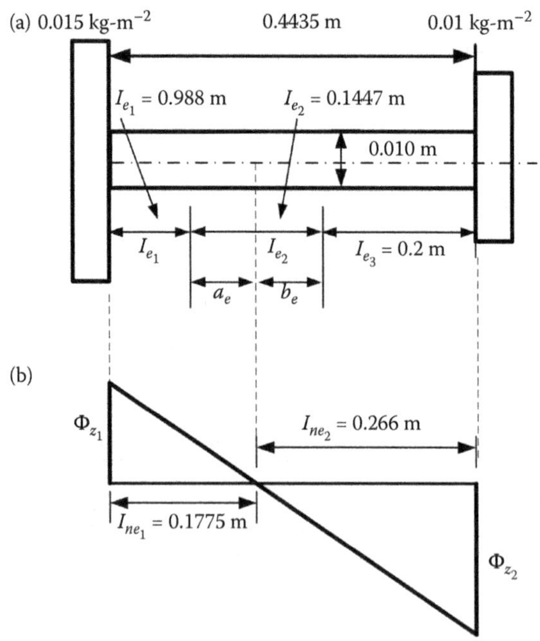

FIGURE 6.12 (a) Equivalent system and (b) its mode shape and node position.

6.4 THREE-DISK TORSIONAL ROTOR SYSTEM

Now the analysis from the previous section is extended for the three-disk rotor system having free–free boundary conditions. Two different approaches are applied for free vibration analysis to get the torsional natural frequencies and the corresponding mode shapes.

6.4.1 Direct Approach

A three-disk rotor system is shown in Figure 6.13. It is assumed that there is no friction at the supports, and the boundary conditions are those of the free–free case. The method using Newton's second law (or Lagrange's equations), with the help of free body diagrams (Figure 6.14), may be applied to analyze the three-mass (or more) rotor system. This method was already demonstrated for the two-mass rotor system in the previous section.

From free body diagrams of individual disks, three equations of motion for free vibrations can be obtained, and in matrix form, we obtain the following form:

$$\begin{bmatrix} I_{P1} & 0 & 0 \\ 0 & I_{P2} & 0 \\ 0 & 0 & I_{P3} \end{bmatrix} \begin{Bmatrix} \ddot{\varphi}_{z1} \\ \ddot{\varphi}_{z2} \\ \ddot{\varphi}_{z3} \end{Bmatrix} + \begin{bmatrix} k_{t1} & -k_{t1} & 0 \\ -k_{t1} & (k_{t1}+k_{t2}) & -k_{t2} \\ 0 & -k_{t2} & k_{t2} \end{bmatrix} \begin{Bmatrix} \varphi_{z1} \\ \varphi_{z2} \\ \varphi_{z3} \end{Bmatrix} = \begin{Bmatrix} 0 \\ 0 \\ 0 \end{Bmatrix} \tag{6.33}$$

Alternatively, equations of motion can be obtained from the potential and kinetic energies of the system and can be written as

$$U = \frac{1}{2}k_{t1}\left(\varphi_{z2} - \varphi_{z1}\right)^2 + \frac{1}{2}k_{t2}\left(\varphi_{z3} - \varphi_{z2}\right)^2 \tag{6.34}$$

and

$$T = \frac{1}{2}I_{P1}\dot{\varphi}_{z1}^2 + \frac{1}{2}I_{P2}\dot{\varphi}_{z2}^2 + \frac{1}{2}I_{P3}\dot{\varphi}_{z3}^2 \tag{6.35}$$

FIGURE 6.13 A three-disk torsional system.

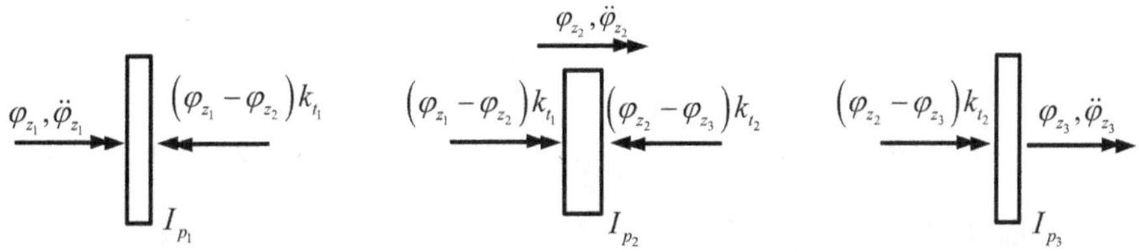

FIGURE 6.14 Free body diagrams of disks: (a) disk 1, (b) disk 2, and (c) disk 3.

For free vibrations, which have SHM, Equation 6.33 takes the following form:

$$\left(-\omega_{nf}^2\begin{bmatrix} I_{p1} & 0 & 0 \\ 0 & I_{p2} & 0 \\ 0 & 0 & I_{p3}\end{bmatrix}+\begin{bmatrix} k_{t1} & -k_{t1} & 0 \\ -k_{t1} & (k_{t1}+k_{t2}) & -k_{t2} \\ 0 & -k_{t2} & k_{t2}\end{bmatrix}\right)\begin{Bmatrix}\varphi_{z1}\\ \varphi_{z2}\\ \varphi_{z3}\end{Bmatrix}=\begin{Bmatrix}0\\0\\0\end{Bmatrix} \tag{6.36}$$

where ω_{nf} is the torsional natural frequency of the rotor system. For finding natural frequencies, two methods can be adopted: (i) obtaining characteristic (or frequency) equations and (ii) formulating an eigenvalue problem.

6.4.1.1 Characteristic (or Frequency) Equations

On equating the determinant of the matrix in Equation 6.36 to zero for a nontrivial solution, we get a characteristic equation of the following form:

$$\omega_{nf}^2\left\{\omega_{nf}^4-\left(k_{t1}\frac{I_{p1}+I_{p2}}{I_{p1}I_{p2}}+k_{t2}\frac{I_{p2}+I_{p3}}{I_{p2}I_{p3}}\right)\omega_{nf}^2+\frac{k_{t1}k_{t2}\left(I_{p1}+I_{p2}+I_{p3}\right)}{I_{p1}I_{p2}I_{p3}}\right\}=0$$

which gives torsional natural frequencies as

$$\omega_{nf1}=0$$

and

$$\omega_{nf2,3}^2=\frac{1}{2}\left(k_{t1}\frac{I_{p1}+I_{p2}}{I_{p1}I_{p2}}+k_{t2}\frac{I_{p2}+I_{p3}}{I_{p2}I_{p3}}\right)\pm\sqrt{\frac{1}{4}\left(k_{t1}\frac{I_{p1}+I_{p2}}{I_{p1}I_{p2}}+k_{t2}\frac{I_{p2}+I_{p3}}{I_{p2}I_{p3}}\right)^2-\left(\frac{k_{t1}k_{t2}\left(I_{p1}+I_{p2}+I_{p3}\right)}{I_{p1}I_{p2}I_{p3}}\right)} \tag{6.37}$$

Mode shapes can be estimated by substituting natural frequencies so obtained, one by one, into Equations 6.36 and obtaining relative amplitudes with the help of any two equations (out of three equations), as

$$\left(k_{t1}-\omega_{nf}^2I_{p1}\right)\varphi_{z1}-k_{t1}\varphi_{z2}=0\Rightarrow\frac{\varphi_{z2}}{\varphi_{z1}}=\frac{\left(k_{t1}-\omega_{nf}^2I_{p1}\right)}{k_{t1}} \tag{6.38}$$

and

$$-k_{t1}\varphi_{z1}+\left\{\left(k_{t1}+k_{t2}\right)-\omega_{nf}^2I_{p2}\right\}\varphi_{z2}-k_{t2}\varphi_{z3}=0 \tag{6.39}$$

On substituting Equation 6.38 in Equation 6.39, we get

$$-k_{t1}\varphi_{z1}+\left\{\left(k_{t1}+k_{t2}\right)-\omega_{nf}^2I_{p2}\right\}\left\{\left(k_{t1}-\omega_{nf}^2I_{p1}\right)\varphi_{z1}/k_{t1}\right\}-k_{t2}\varphi_{z3}=0 \tag{6.40}$$

which can be simplified to

$$\frac{\varphi_{z3}}{\varphi_{z1}}=\frac{\left(I_{p1}I_{p2}\right)\omega_{nf}^4-\left\{\left(I_{p1}+I_{p2}\right)k_{t1}+I_{p1}k_{t2}\right\}\omega_{nf}^2+\left(k_{t1}k_{t2}\right)}{k_{t1}k_{t2}} \tag{6.41}$$

It should be noted that from Equations 6.38 and 6.41 for $\omega_{nf1}=0$, we have $\varphi_{z2}/\varphi_{z1}=\varphi_{z3}/\varphi_{z1}=1$ (or $\varphi_{z1}=\varphi_{z2}=\varphi_{z3}$), which belong to the rigid body mode. Similarly, for the other two natural frequencies (i.e., ω_{nf2} and ω_{nf3}) relative amplitudes of disks can be obtained by substituting these natural frequencies (Equation 6.41 does not contain I_{p3} explicitly, but ω_{nf}^2 contains I_{p3} also for the second and third modes) one by one in Equations 6.38 and 6.41. For rotor systems with more than three disks, expressions become cumbersome to handle by the direct method, and generally, numerical methods would be essential to solve the frequency equation (i.e., the polynomial of higher degree) or eigenvalue problem.

6.4.1.2 An Eigenvalue Problem

A more general method of obtaining natural frequencies and mode shapes is to formulate an eigenvalue problem that can relatively easily be solved by computer routines. Eigenvalues of the eigenvalue problem of Equation 6.36 give natural frequencies, and eigenvectors represent mode shapes. Equation 6.36 can be written as

$$\left(-\omega_{nf}^2\mathbf{M}+\mathbf{K}\right)\mathbf{\Phi}=\mathbf{0} \tag{6.42}$$

with

$$\mathbf{M}=\begin{bmatrix} I_{p1} & 0 & 0 \\ 0 & I_{p2} & 0 \\ 0 & 0 & I_{p3} \end{bmatrix};\quad \mathbf{K}=\begin{bmatrix} k_{t1} & -k_{t1} & 0 \\ -k_{t1} & \left(k_{t1}+k_{t2}\right) & -k_{t2} \\ 0 & -k_{t2} & k_{t2} \end{bmatrix};\quad \mathbf{\Phi}=\begin{Bmatrix} \varphi_{z1} \\ \varphi_{z2} \\ \varphi_{z3} \end{Bmatrix}$$

On multiplying both sides by the inverse of the mass matrix in Equation 6.42, we get a standard eigenvalue problem of the following form:

$$\left(-\omega_{nf}^2\mathbf{I}+\mathbf{D}\right)\mathbf{\Phi}=\mathbf{0} \tag{6.43}$$

with

$$\mathbf{D}=\mathbf{M}^{-1}\mathbf{K}$$

where $\mathbf{I}$ is the identity matrix. The eigenvalue and eigenvector of the matrix $\mathbf{D}$ can be obtained conveniently by hand calculations for a matrix size up to 3×3; however, for the larger size matrix for multi-DOF rotor systems, any standard software (e.g., MATLAB, SCILAB, Python, etc.) could be used. The square root of eigenvalues gives the natural frequencies and corresponding eigenvectors as mode shapes (i.e., relative amplitudes). These methods will now be illustrated through a numerical example.

Example 6.4

Obtain torsional natural frequencies of a turbine-coupling-generator rotor system as shown in Figure 6.15. The rotor is assumed to be supported on frictionless bearings, and the shafts are connected by a rigid coupling. The polar mass moments of inertia of the turbine, coupling, and generator are $I_{p1}=25$ kg-m^2, $I_{p2}=5$ kg-m^2, and $I_{p3}=50$ kg-m^2, respectively; these are assumed to be thin disks. Take the modulus of rigidity of the shaft as $G=0.8\times10^{11}$ N/m^2. Assume that the shaft diameter is uniform throughout and is equal to 0.2 m, and the length of the shafts between the bearing-turbine-coupling-generator-bearing is 1 m each, so that the total span is 4 m. Consider the shaft as massless.

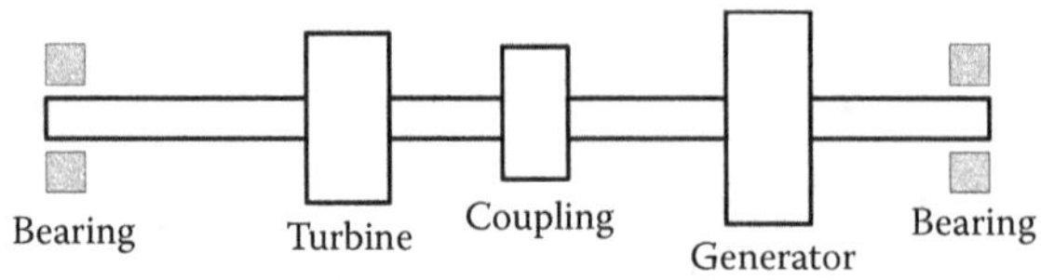

FIGURE 6.15 A turbine-coupling-generator set.

Solution:

It should be noted that for free–free end conditions, both ends of shaft segments (i.e., between bearing and turbine, and generator and bearing) do not have relative torsional displacements due to the absence of polar moments of inertia. Hence, only shaft segments between the turbine and the coupling (let us take it as shaft 1), and between the coupling and the generator (shaft 2) have the torsional stiffness effect. Hence, we have the following data:

$$l_1 = l_2 = l = 1 \text{ m}, \quad J_1 = J_2 = J = \frac{\pi}{32} d^4 = \frac{\pi}{32} 0.2^4 = 1.5708 \times 10^{-4} \text{ m}^4 \tag{a}$$

and

$$k_{t_1} = k_{t_2} = \frac{GJ}{l} = \frac{\left(0.8 \times 10^{11}\right)\left(1.5708 \times 10^{-4}\right)}{1} = 1.257 \times 10^7 \text{ N-rad/m}^2 \tag{b}$$

The natural frequencies of the three-disk rotor system are given as Equation 6.37

$$\omega_{nf_1} = 0 \tag{c}$$

and

$$\omega_{nf2,3}^2 = \frac{1}{2}\left(k_{t_1} \frac{I_{p1} + I_{p2}}{I_{p1} I_{p2}} + k_{t_2} \frac{I_{p2} + I_{p3}}{I_{p2} I_{p3}}\right) \pm \sqrt{\frac{1}{4}\left(k_{t_1} \frac{I_{p1} + I_{p2}}{I_{p1} I_{p2}} + k_{t_2} \frac{I_{p2} + I_{p3}}{I_{p2} I_{p3}}\right)^2 - \left(\frac{k_{t_1} k_{t_2}\left(I_{p1} + I_{p2} + I_{p3}\right)}{I_{p1} I_{p2} I_{p3}}\right)} \tag{d}$$

On substituting the values of various parameters of the present problem in the earlier equation, we get

$$\omega_{nf_1} = 0 \text{ rad/s}; \quad \omega_{nf_2} = 611.56 \text{ rad/s}; \quad \omega_{nf_3} = 2325.55 \text{ rad/s};$$

Natural frequencies are generally arranged in ascending order. The mode shape (relative angular displacements of various disks) can be obtained as summarized in Table 6.1 (refer to Equations 6.38 and 6.41). Figure 6.16 shows mode shapes with node locations; *T*, *C*, and *G* represent the locations of the turbine, coupling, and generator, respectively. It can be observed that the second mode has one node position and the third mode has two node positions. For the third mode, the relative angular displacement at the coupling location is very high due to its relatively lower inertia as compared to the turbine and generator.

Node locations in the second and third modes can be obtained as follows:

Second mode: Only a single node (Figure 6.16b) is present between the coupling and the generator. Hence, from the node position to the generator, a single-DOF rotor system can be assumed, with the length of the shaft $l_{gn}^{(1)}$ (superscript corresponding to the single-node mode and subscript *gn* represents from the generator to the node location) and polar mass moment of inertia I_{p3}.

$$\omega_{nf_2}^2 = \frac{k_{t_{gn}}}{I_{p3}}, \quad \Rightarrow l_{gn}^{(1)} = \frac{GJ}{I_{p3}\omega_{nf_2}^2} = \frac{\left(0.8 \times 10^{11}\right)\left(1.5708 \times 10^{-4}\right)}{50 \times 611.56^2} = 0.6723 \text{ m}$$

TABLE 6.1
Relative Angular Displacements of Various Disks

Relative Displacement	$\omega_{nf1}=0$ rad/s	$\omega_{nf2}=611.56$ rad/s	$\omega_{nf3}=2325.55$ rad/s
$\dfrac{\varphi_{z2}}{\varphi_{z1}}=\dfrac{k_{t1}-\omega_{nf}^2 I_{p1}}{k_{t1}}$	1	0.2563	−9.7600
$\dfrac{\varphi_{z3}}{\varphi_{z1}}=\dfrac{(I_{p1}I_{p2})\omega_{nf}^4-\{(I_{p1}+I_{p2})k_{t1}+I_{p1}k_{t2}\}\omega_{nf}^2+(k_{t1}k_{t2})}{k_{t1}k_{t2}}$	1	−0.5256	0.4754

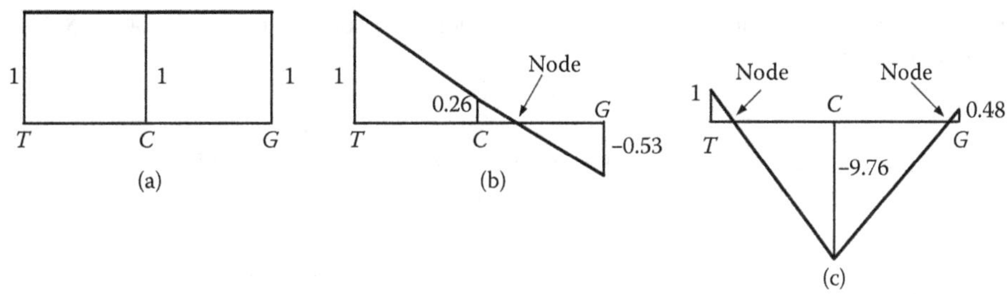

FIGURE 6.16 Three mode shapes corresponding to three torsional natural frequencies: (a) for ω_{nf1}; (b) for ω_{nf2}; (c) for ω_{nf3}.

Third mode: Two nodes are present (Figure 6.16c); thus, this leads to two single-DOF cantilever rotor systems: (i) from the first node to the turbine and (ii) from the second node to the generator. Hence, node locations are obtained as

$$l_{tn}^{(2)}=\frac{GJ}{I_{p1}\omega_{nf3}^2}=\frac{(0.8\times10^{11})(1.5708\times10^{-4})}{25\times2325.55^2}=0.0930\text{ m}$$

and

$$l_{gn}^{(2)}=\frac{GJ}{I_{p3}\omega_{nf3}^2}=\frac{(0.8\times10^{11})(1.5708\times10^{-4})}{50\times2325.55^2}=0.0465\text{ m}$$

where the superscript in the length represents two-node mode and subscript *tn* represents from the turbine to the nearest node.

Now, using the eigenvalue problem solution procedure, the earlier problem is solved again. This demonstrates how powerful this procedure is, especially for multi-DOF systems. The mass and stiffness matrices can be given as

$$\mathbf{M}=\begin{bmatrix} I_{p1} & 0 & 0\\ 0 & I_{p2} & 0\\ 0 & 0 & I_{p3}\end{bmatrix}=\begin{bmatrix} 25 & 0 & 0\\ 0 & 5 & 0\\ 0 & 0 & 50\end{bmatrix}\text{kg-m}^2$$

and

$$\mathbf{K}=\begin{bmatrix} k_{t1} & -k_{t1} & 0\\ -k_{t1} & (k_{t1}+k_{t2}) & -k_{t2}\\ 0 & -k_{t2} & k_{t2}\end{bmatrix}=\begin{bmatrix} 1.257 & -1.257 & 0\\ -1.257 & 2.514 & -1.257\\ 0 & -1.257 & 1.257\end{bmatrix}\times10^7\text{ N-rad/m}$$

Hence, for the eigenvalue problem, the matrix becomes

$$\mathbf{D} = \mathbf{M}^{-1}\mathbf{K}$$

$$= \begin{bmatrix} 25 & 0 & 0 \\ 0 & 5 & 0 \\ 0 & 0 & 50 \end{bmatrix}^{-1} \begin{bmatrix} 1.257 & -1.257 & 0 \\ -1.257 & 2.514 & -1.257 \\ 0 & -1.257 & 1.257 \end{bmatrix} \times 10^7$$

$$= \begin{bmatrix} 0.5028 & -0.5028 & 0 \\ -2.5140 & 5.0280 & -2.5140 \\ 0 & -0.2514 & 0.2514 \end{bmatrix} \times 10^6$$

Eigenvalues and eigenvectors are given as (by the eigenvalue solution command of the MATLAB software to the above matrix)

$$\lambda = \begin{Bmatrix} 5.4082 \\ 0.3740 \\ 0 \end{Bmatrix} \times 10^6 \quad \text{and} \quad \eta = \begin{bmatrix} -0.1018 & -0.8632 & 0.5774 \\ 0.9936 & -0.2212 & 0.5774 \\ -0.0484 & 0.4537 & 0.5774 \end{bmatrix}$$

where the columns of matrix η represent the mode shapes corresponding to eigenvalues given in rows of the vector λ. Hence, natural frequencies are obtained as

$$\omega_{nf} = \begin{Bmatrix} \omega_{nf3} \\ \omega_{nf2} \\ \omega_{nf1} \end{Bmatrix} = \sqrt{\lambda} = \begin{Bmatrix} 2325.55 \\ 611.56 \\ 0 \end{Bmatrix} \text{rad/s}$$

These are exactly the same as obtained earlier. The mode shape can be normalized as (in each column the elements are divided by the corresponding first row element, e.g., 0.9936/(−0.018)=−9.76, −0.0484/(−0.018)=0.48, −0.2212/(−0.8632)=0.26, etc.)

$$\mathbf{X} = \begin{bmatrix} \begin{Bmatrix} \varphi_{xT} \\ \varphi_{xC} \\ \varphi_{xG} \end{Bmatrix}_{\omega_{nf3}} & \begin{Bmatrix} \varphi_{xT} \\ \varphi_{xC} \\ \varphi_{xG} \end{Bmatrix}_{\omega_{nf2}} & \begin{Bmatrix} \varphi_{xT} \\ \varphi_{xC} \\ \varphi_{xG} \end{Bmatrix}_{\omega_{nf1}} \end{bmatrix} = \begin{bmatrix} 1 & 1 & 1 \\ -9.76 & 0.26 & 1 \\ 0.48 & -0.53 & 1 \end{bmatrix}$$

where *T*, *C*, and *G* represent the turbine, compressor, and generator, respectively. These mode shapes are also exactly the same as in Figure 6.16.

6.4.2 Indirect Approach

From the previous method, it is clear that for a particular natural frequency a unique mode shape exists and vice versa. In the present method, the information regarding the possible mode shapes is utilized to get the corresponding natural frequencies. In this case, when the shaft has steps, then the first action is to reduce the actual shaft to an equivalent shaft of a uniform diameter as shown in Figure 6.17a. Here, l_1 and l_2 are the lengths of shaft segments between disks 1 and 2, and between disks 2 and 3, respectively; and I_{pi} is the polar mass moment of inertia of a disk (with $i=1, 2, 3$).

For the three-disk rotor system, three natural frequencies are expected and the correspondingly three natural (or normal) modes of vibrations are possible. For free–free boundary conditions, one of the modes (corresponding to a torsional natural frequency equal to zero) would be the rigid body mode, in which all the disks have the same motion. Apart from the rigid body mode, there will be two flexible natural modes of vibration, in which all disks reach their extreme positions at the same instant of time and all pass through their equilibrium position at the same instant of time. There will be a distinct torsional natural frequency for each of these two normal modes.

In the first flexible mode (corresponding to the first nonzero natural frequency), there will be a single *node* (a point where there will not be any angular displacement) either between disks 1 and 2 or between disks 2 and 3 (see Figure 6.17b). It depends upon the relative polar mass moment of inertia of disks and the stiffness of shaft segments. However, the oscillations of the outside disks 1 and 3 are in the antiphase because of the single node of vibration mode. Disk 2 has the same or opposite phase with disk 1 (or disk 3); this depends upon the node position. It is assumed in Figure 6.17b that the node lies between disks 2 and 3.

However, in the second flexible mode, there are two nodes, one between disks 1 and 2 and the other between disks 2 and 3 (as shown in Figure 6.17c). Oscillations of outside disks (1 and 3) are now in phase, whereas the inside disk has an opposite phase with respect to both disks 1 and 3.

Two-node mode: For two-node vibration (see Figure 6.17c), let $l_1^{(2)}$ be the distance of one node from disk 1 (the superscript represents the two-node mode of vibration), and $l_3^{(2)}$ be the distance of the other node from disk 3. For this case, the three-disk free–free rotor system could be considered as three single-DOF rotor systems {i.e., two cantilever rotor systems with disks 1 (rotor system A) and 3 (rotor system C), and one fixed–fixed rotor system with disk 2 (rotor system C)}. Then, the torsional natural frequency of the single-DOF cantilever system (rotor system A), with disk 1 and shaft length $l_1^{(2)}$, is given as

$$\omega_{nfA}^{(2)} = \sqrt{\frac{k_{t1}^{(2)}}{I_{p1}}} = \sqrt{\frac{GJ}{l_1^{(2)}} \frac{1}{I_{p1}}} \tag{6.44}$$

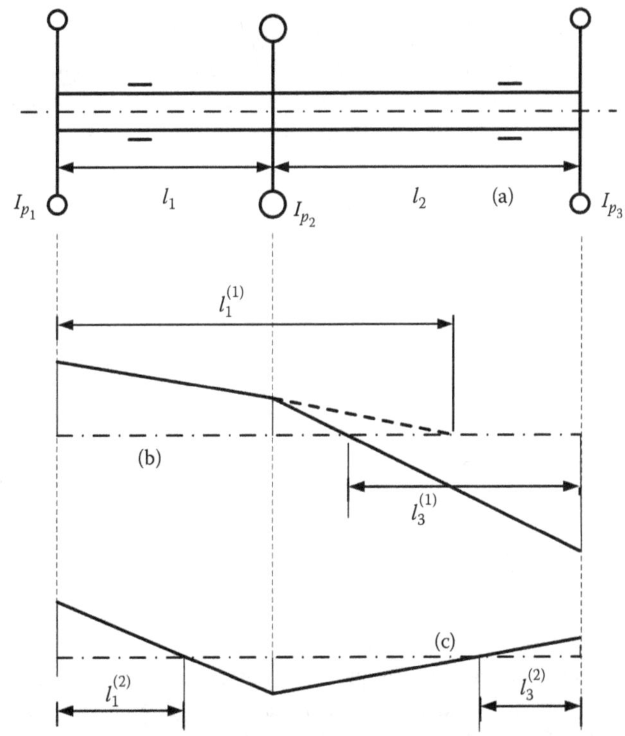

FIGURE 6.17 A three-disk rotor system with two possible flexible modes of vibration: (a) three-disk rotor system; (b) mode shape with a node; (c) mode shape with two nodes.

Similarly, the torsional natural frequency of another single-DOF cantilever system (rotor system C) with disk 3 and shaft length $l_3^{(2)}$ is given as (Figure 6.17c)

$$\omega_{nfC}^{(2)} = \sqrt{\frac{GJ}{l_3^{(2)}} \frac{1}{I_{P3}}} \tag{6.45}$$

For the single-DOF fixed–fixed rotor system (rotor system B), with disk 2 and shaft lengths at either side of the disk as $\left(l_1 - l_1^{(2)}\right)$ and $\left(l_2 - l_3^{(2)}\right)$, the torsional natural frequency is (Figure 6.17c)

$$\omega_{nfB}^{(2)} = \sqrt{\frac{k_{t2}^{(2)}}{I_{P2}}} \tag{6.46}$$

where

$$k_{t2}^{(2)} = \frac{GJ}{l_1 - l_1^{(2)}} + \frac{GJ}{l_2 - l_3^{(2)}} = GJ \frac{l_1 + l_2 - l_1^{(2)} - l_3^{(2)}}{\left(l_1 - l_1^{(2)}\right)\left(l_2 - l_3^{(2)}\right)} \tag{6.47}$$

where k_{t_2} is the torsional stiffness of a rotor system with fixed–fixed end conditions. The torsional stiffnesses of two shaft segments are connected in parallel. On substituting Equation 6.47 into Equation 6.46, we get

$$\omega_{nfB}^{(2)} = \sqrt{\frac{GJ}{I_{P2}} \frac{\left(l_1 + l_2 - l_1^{(2)} - l_3^{(2)}\right)}{\left(l_1 - l_1^{(2)}\right)\left(l_2 - l_3^{(2)}\right)}} \tag{6.48}$$

For a particular mode, all frequencies $\omega_{nfA}^{(2)}$, $\omega_{nfB}^{(2)}$, and $\omega_{nfC}^{(2)}$ must be equal (superscript represents the two-node mode) as they all belong to the same original system. This leads to two independent equations to be solved for $l_1^{(2)}$ and $l_3^{(2)}$. Once we know these node positions, we are able to get the natural frequency of the two-node (or one-node) mode. On equating Equations 6.44 and 6.45, we get

$$l_1^{(2)} I_{P1} = l_3^{(2)} I_{P3} \quad \text{or} \quad l_1^{(2)} = \frac{I_{P3}}{I_{P1}} l_3^{(2)} \tag{6.49}$$

Similarly, on equating Equations 6.44 and 6.48, we get

$$\frac{1}{l_1^{(2)} I_{P1}} = \frac{1}{I_{P2}} \frac{\left(l_1 + l_2 - l_1^{(2)} - l_3^{(2)}\right)}{\left(l_1 - l_1^{(2)}\right)\left(l_2 - l_3^{(2)}\right)} \tag{6.50}$$

Equation 6.49 can be used to eliminate $l_1^{(2)}$ from Equation 6.50, and it gets simplified to

$$\left\{\frac{I_{P2} I_{P3}}{I_{P1}} + \frac{I_{P3}^2}{I_{P1}} + I_{P3}\right\}\left(l_3^{(2)}\right)^2 - \left\{\frac{I_{P3} I_{P2} l_2}{I_{P1}} + I_{P2} l_1 + I_{P3}\left(l_1 + l_2\right)\right\} l_3^{(2)} + \left\{I_{P2} l_1 l_2\right\} = 0 \tag{6.51}$$

The two roots of $l_3^{(2)}$ from this quadratic equation give positions of nodes for the one-node and two-node vibration modes. The actual frequencies corresponding to these flexible modes are obtained by substituting the two values of $l_3^{(2)}$ in Equation 6.45. From Equation 6.49, two values of $l_1^{(2)}$ could be obtained corresponding to two values of $l_3^{(2)}$. Note that only one of these two values of $l_1^{(2)}$ may give the position of a real node (in Figure 6.17c), whereas the other gives the point at which the elastic line between disks 1 and 2, when produced, cuts the axis of the shaft (as shown in

Figure 6.17b by the dotted line). Physically, this means that disks of the two-disk rotor system (in Figure 16.7b disks which are left of the node position will form a two-disk cantilever rotor system) will have angular displacement proportional to their axial location along the shaft from the intersection point. For the single-node mode of vibration, the three-disk rotor system could be considered as a two cantilever rotor system (at either side of the node location, which is chosen between disks 2 and 3) as shown in Figure 6.17b, one with a single disk (disk 3) and the other with two disks (disks 1 and 2). However, this will depend upon the node location, e.g., in the case where the single node is between disk 1 and 2, disk 1 forms a single-DOF cantilever rotor, whereas disks 2 and 3 would form a two-DOF cantilever rotor system.

For the present case, we started with a two-node mode shape; however, a similar exercise could be done with a single-node mode shape. In this case, we would have two equivalent rotor systems, with one as a single-DOF cantilever rotor system and the other as a two-DOF cantilever rotor system. It is left to the reader to explore the same. The present method can be extended for other boundary conditions (fixed–free, fixed–fixed, etc.) and for more disks; however, the complexity of handling a higher degree of polynomials is cumbersome. The present method is now illustrated through an example.

Example 6.5

Solve Example 6.4 by the indirect method described in the previous section.

Solution:

From Equation 6.51, we have

$$\left[\frac{I_{p2}I_{p3}}{I_{p1}}+\frac{I_{p3}^2}{I_{p1}}+I_{p3}\right]\left(l_3^{(2)}\right)^2-\left[\frac{I_{p3}I_{p2}l_2}{I_{p1}}+I_{p2}l_1+I_{p3}\left(l_1+l_2\right)\right]l_3^{(2)}+I_{p2}l_1l_2=0$$

On substituting values of physical parameters of the present problem (Figure 6.18a), we get

$$\left[\frac{5\times 50}{25}+\frac{50^2}{25}+50\right]\left(l_3^{(2)}\right)^2-\left[\frac{5\times 50\times 1}{25}+5\times 1+50\times 2\right]l_3^{(2)}+5\times 1\times 1=0$$

or

$$160\left(l_3^{(2)}\right)^2-115l_3^{(2)}+5=0$$

which gives two values corresponding to two modes (i.e., the one- and two-node modes, Figure 6.18b and d) as

$$l_3^{(2)}=0.6723\text{ m}\quad\text{and}\quad 0.04648\text{ m}$$

Two possible values of $l_1^{(2)}$ can be obtained from Equation 6.49, as

$$l_1^{(2)}=l_3^{(2)}I_{p3}/I_{p1}$$

which gives two values corresponding to two modes (i.e., the one- and two-node modes) as

$$l_1^{(2)}=1.3446\quad\text{and}\quad 0.09297\text{ m}$$

Hence, we have two solutions:

$$\left(l_1^{(2)},l_3^{(2)}\right)=(0.09297,0.04648)\text{ m}\quad\text{and}\quad\left(l_1^{(2)},l_3^{(2)}\right)=(1.3446,0.6723)\text{ m}$$

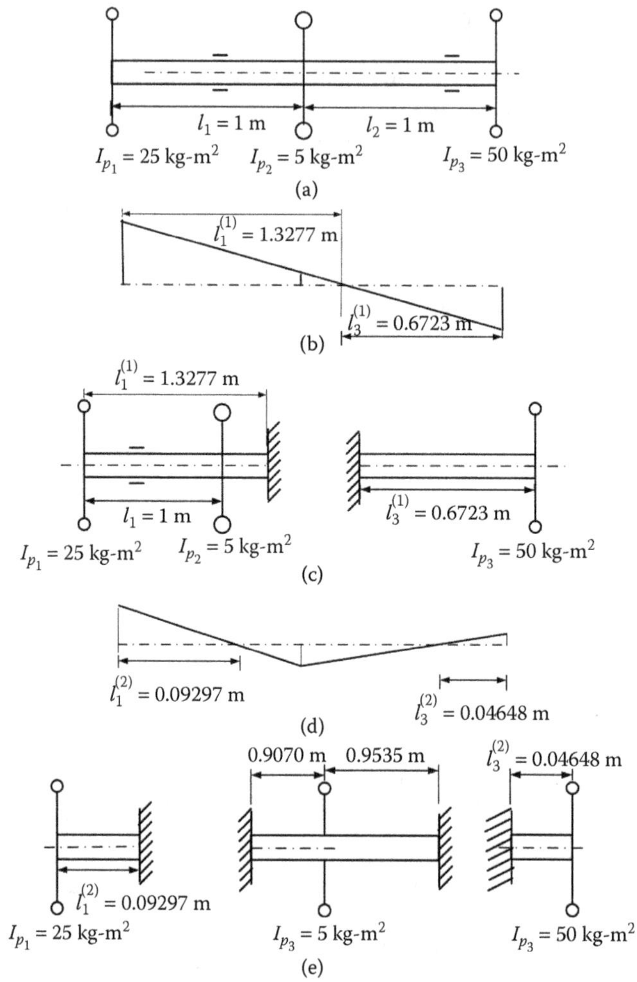

FIGURE 6.18 (a) A three-mass rotor system. (b) Single-node mode shape. (c) An equivalent two system, two-mass and one-mass cantilever rotor system. (d) Two-node mode shape. (e) An equivalent three system, single-mass two cantilevers and one fixed–fixed end condition rotor system.

It is clear that two nodes are possible at $\left(l_1^{(2)}, l_3^{(2)}\right) = (0.09297, 0.04648)$ m (Figure 6.18d and e). Although the single node is possible at $\left(l_1^{(1)}, l_2^{(1)}\right) = (1.3446, 0.6723)$ m, the second one is a feasible solution at $l_3^{(1)} = 0.6723$ m (Figure 6.18b and c), as discussed in the theory part in Section 6.4.2. The first one is not a feasible solution. It should be noted that mode shapes in Figure 6.18b are not to scale; however, qualitative comparison can be made with the previous example. Quantitatively also it can be observed that they are exactly the same.

Now, the natural frequency corresponding to the two-node mode can be obtained as

$$\omega_{nf3}^{(2)} = \sqrt{\frac{GJ}{l_3^{(2)}}\frac{1}{I_{p3}}} = \sqrt{\frac{\left(0.8\times10^{11}\right)\times\left(1.5708\times10^{-4}\right)}{0.04648}\frac{1}{50}} = 2325.34 \text{ rad/s}$$

with

$$J = \frac{\pi}{32}d^4 = \frac{\pi}{32}0.2^4 = 1.5708\times10^{-4} \text{ m}^4$$

The natural frequency corresponding to the single-node mode can be obtained as

$$\omega_{nf2}^{(1)} = \sqrt{\frac{GJ}{l_3^{(1)}}\frac{1}{I_{P3}}} = \sqrt{\frac{(0.8\times10^{11})\times(1.5708\times10^{-4})}{0.6723}\frac{1}{50}} = 611.42 \text{ rad/s}$$

It should be noted that these natural frequencies and the node positions are exactly the same as obtained in Example 6.4.

6.5 TRANSFER MATRIX METHODS

When there are more than three disks in the rotor system or when the mass of the shaft itself may be significant (i.e., continuous systems, which have infinite DOFs) so that more lumped masses are considered, then the analysis described in previous sections (i.e., the single-, two-, or three-disk rotor systems) becomes complicated and it is infeasible to model such systems. Such rotor systems are called multi-DOF systems. Alternative popular methods for the analysis of multi-DOF torsional rotor systems are the TMM (or Myklestad and Prohl method) and the finite element method (FEM). In the present section, we consider TMM with the discrete system approach in detail, and in the next chapter, we consider the FEM along with the continuous system approach.

A typical multidisk rotor system, supported on frictionless supports, is shown in Figure 6.19. The longitudinal axis is taken as the z-axis, about which disks have angular displacements, $\varphi_z(t)$. For the present analysis, disks are considered as thin and rigid, and located at a point, and the shaft is treated as flexible and massless. The number of disks is n, the *station number* is designated from 0 to $(n + 1)$, and hence the system has $(n + 2)$ total number of stations as shown in Figure 6.19. The free body diagram of a representative shaft and a disk is shown in Figure 6.20. At a particular station in the torsional rotor system, we have two state variables: the angular twist, $\varphi_z(t)$, and the torque, $T(t)$. To get familiarize with the double head arrows to represent the torsional displacements and torques direction, in Figures 6.19 and 6.20 both conventions are shown. Now, in subsequent sections, we develop the relationship among these state variables between two neighboring stations in terms of physical properties of the disk and the shaft, and this then can be used to obtain governing equations of motion of the whole rotor system.

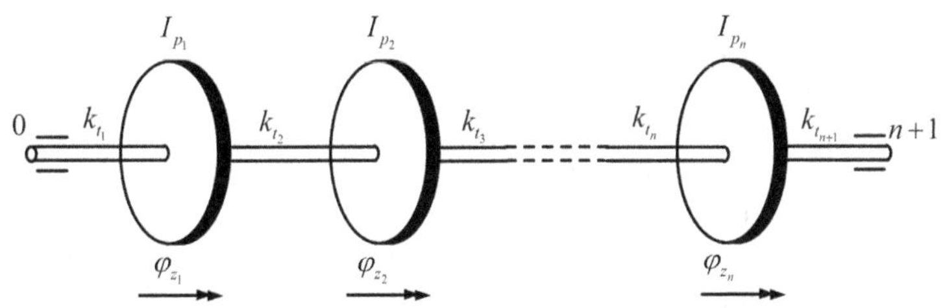

FIGURE 6.19 A multidisk rotor system.

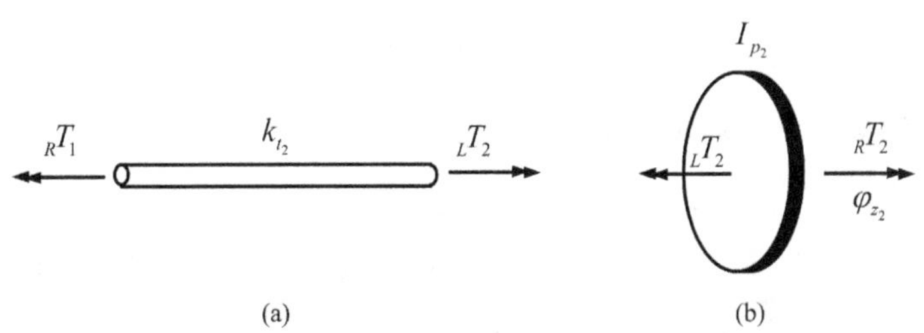

FIGURE 6.20 (a) A free body diagram of shaft section 2. (b) A free body diagram of disk 2.

6.5.1 Point Matrix

In this subsection, we develop a relationship between state variables at either end (i.e., the right and left sides) of a thin disk. The aim would be to obtain the state vector at the RHS in terms of the state vector in the LHS. One could relate the other way around and expect that the relations would be different; however, we would stick to this convention.

Because angular displacements on either side of the rotor are equal

$$ {}_R\varphi_{z2} = {}_L\varphi_{z2} \tag{6.52} $$

The equation of motion for the disk 2 is given by (see Figure 6.20b)

$$ {}_RT_2 - {}_LT_2 = I_{p2}\ddot{\varphi}_{z2} \tag{6.53} $$

where I_p is the polar mass moment of inertia of the disk, and back subscripts R and L represent the *right* and the *left* of a disk, respectively. For free vibrations, the angular oscillation of the disk is given by

$$ \varphi_{z2} = \Phi_{z2}\sin\omega_{nf}t \quad \text{so that} \quad \ddot{\varphi}_{z2} = -\omega_{nf}^2\Phi_{z2}\sin\omega_{nf}t = -\omega_{nf}^2\varphi_{z2} \tag{6.54} $$

where Φ_z is the amplitude of angular displacement and ω_{nf} is the torsional natural frequency. On substituting Equation 6.54 into Equation 6.53, we get

$$ {}_RT_2 - {}_LT_2 = -\omega_{nf}^2 I_{p2}\varphi_{z2} \quad \text{or} \quad {}_RT_2 = \left(-\omega_{nf}^2 I_{p2}\right){}_L\varphi_{z2} + {}_LT_2 \tag{6.55} $$

Note that a back-subscript L (corresponds to the LHS of a disk) has been added to the rotational displacement in lieu of Equation 6.52. Now Equations 6.52 and 6.55 can be combined as

$$ {}_R\mathbf{S}_2 = \mathbf{P}_2\,{}_L\mathbf{S}_2 \tag{6.56} $$

with

$$ \mathbf{P}_2 = \begin{bmatrix} 1 & 0 \\ -\omega_{nf}^2 I_{p2} & 1 \end{bmatrix}; \quad \mathbf{S} = \begin{Bmatrix} \varphi_z \\ T \end{Bmatrix} $$

where $\mathbf{S}_2$ is the *state vector* corresponding to station 2 (toward the left of the disk or right of the disk depending upon whether the back-subscript is L or R, respectively), and $\mathbf{P}_2$ is the *point matrix* for disk 2. Hence, in general the point matrix relates a state vector left of a disk to a state vector right of the disk. When an external torque, $T_E(t)$, is applied to the disk (e.g., the disk is a gear element or a pulley driven by a belt) in the direction of the chosen positive angular displacement direction, then the torque balance equation would be

$$ {}_RT_2 = \left(-\omega_{nf}^2 I_{p2}\right){}_L\varphi_{z2} + {}_LT_2 - T_{E2} \tag{6.57} $$

Then Equation 6.56 will be modified as

$$ {}_R\mathbf{S}_2 = \mathbf{P}_2\,{}_L\mathbf{S}_2 + \mathbf{T}_{E2} \tag{6.58} $$

with

$$ \mathbf{T}_{E2} = \begin{Bmatrix} 0 \\ -T_{E2} \end{Bmatrix} $$

It is more convenient to write Equation 6.58 in the following form:

$$ {}_R\mathbf{S}_2^* = \mathbf{P}_2^* \, {}_L\mathbf{S}_2^* \tag{6.59}$$

with

$$\mathbf{P}_2^* = \begin{bmatrix} 1 & 0 & 0 \\ -\omega_{nf}^2 I_p & 1 & -T_{E2} \\ 0 & 0 & 1 \end{bmatrix}, \quad \mathbf{S}^* = \begin{Bmatrix} \varphi_z \\ T \\ 1 \end{Bmatrix}$$

where $\mathbf{P}^*$ and $\mathbf{S}^*$ are called the *modified point matrix* and the *modified state vector*, respectively. It should be noted the third equation of expression (Equation 6.59) is an identity equation, and it helps in including the external torque in the modified point matrix, which has certain advantages and it will be clear in Section 6.5.3.

6.5.2 Field Matrix

In this subsection, we develop a relationship between state variables at two ends of a shaft segment (i.e., the state vector on the RHS is related to the state vector on the LHS and using similar steps as followed for the point matrix). For shaft element 2, as shown in Figure 6.20a, the angle of twist is related to its torsional stiffness, k_t, and to the torque, $T(t)$, which is transmitted through it, as

$$ {}_L\varphi_{z2} - {}_R\varphi_{z1} = \frac{{}_RT_1}{k_{t2}} \quad \text{or} \quad {}_L\varphi_{z2} = {}_R\varphi_{z1} + \frac{{}_RT_1}{k_{t2}} \tag{6.60}$$

Because the torque is the same at either end of the shaft, we have

$$ {}_LT_2 = {}_RT_1 \tag{6.61}$$

On combining Equations 6.60 and 6.61 in matrix form, we get

$$ {}_L\mathbf{S}_2 = \mathbf{F}_2 \, {}_R\mathbf{S}_1 \tag{6.62}$$

with

$$\mathbf{F}_2 = \begin{bmatrix} 1 & 1/k_{t2} \\ 0 & 1 \end{bmatrix}; \quad \mathbf{S} = \begin{Bmatrix} \varphi_z \\ T \end{Bmatrix}$$

where $\mathbf{F}_2$ is the *field matrix* for shaft element 2. Hence, in general, the field matrix relates a state vector that is at the LH end of a shaft segment to the RH end of the shaft segment. It should be noted that Equation 6.62 is also valid for a torsional spring (e.g., a flexible coupling between two shaft segments), which has k_t as the torsional stiffness; however, such a spring has negligible axial length compared with the shaft length. Ideally, such torsional springs can be considered a *point spring* (similar to a *point mass*). A flexible coupling between a motor and a shaft or between a turbine and a generator could be modeled by such torsional springs.

Equation 6.62 can be modified to consider an external torque in the rotor system (it is assumed here that the external torque is applied at disk locations only and not at the intermediate position of the shaft), as

$$ {}_L\mathbf{S}_2^* = \mathbf{F}_2^* \, {}_R\mathbf{S}_1^* \tag{6.63}$$

with

$$\mathbf{F}_2^* = \begin{bmatrix} 1 & 1/k_{t2} & 0 \\ 0 & 1 & 0 \\ 0 & 0 & 1 \end{bmatrix}; \quad \mathbf{S}^* = \begin{Bmatrix} \varphi_z \\ T \\ 1 \end{Bmatrix}$$

where $\mathbf{F}^*$ is the *modified field matrix,* where the external torque will not appear because of the assumption stated that external torques are considered only on disks.

6.5.3 Transfer Matrix

On substituting Equation 6.62 into Equation 6.56, we get

$$_R\mathbf{S}_2 = \mathbf{U}_2\, {}_R\mathbf{S}_1 \tag{6.64}$$

with

$$\mathbf{U}_2 = \mathbf{P}_2\mathbf{F}_2 = \begin{bmatrix} 1 & 1/k_{t2} \\ -\omega_{nf}^2 I_{p2} & 1 - \dfrac{\omega_{nf}^2 I_{p2}}{k_{t2}} \end{bmatrix}$$

where $\mathbf{U}_2$ is the transfer matrix, which relates the state vector at the right of station 2 to the state vector at the right of station 1, when the external torque is absent (i.e., for the free torsional vibration). Along the same lines, we can write

$$_R\mathbf{S}_1 = \mathbf{U}_1\mathbf{S}_0$$

$$_R\mathbf{S}_2 = \mathbf{U}_2\, {}_R\mathbf{S}_1 = \mathbf{U}_2\mathbf{U}_1\mathbf{S}_0$$

$$_R\mathbf{S}_3 = \mathbf{U}_3\, {}_R\mathbf{S}_2 = \mathbf{U}_3\mathbf{U}_2\mathbf{U}_1\mathbf{S}_0$$

$$_R\mathbf{S}_n = \mathbf{U}_n\, {}_R\mathbf{S}_{n-1} = \mathbf{U}_n\mathbf{U}_{n-1}\cdots\mathbf{U}_1\mathbf{S}_0$$

$$_R\mathbf{S}_{n+1} = \mathbf{U}_{n+1}\, {}_R\mathbf{S}_n = \mathbf{U}_{n+1}\mathbf{U}_n\cdots\mathbf{U}_1\mathbf{S}_0 = \mathbf{T}\mathbf{S}_0$$

with

$$\mathbf{T}(\omega_{nf}) = \mathbf{U}_{n+1}\mathbf{U}_n\cdots\mathbf{U}_1 = \begin{bmatrix} t_{11}(\omega_{nf}) & t_{12}(\omega_{nf}) \\ t_{21}(\omega_{nf}) & t_{22}(\omega_{nf}) \end{bmatrix} \tag{6.65}$$

where $\mathbf{S}_0$ is the state vector at the 0th station (i.e., for the present case the leftmost station of the rotor system), $_R\mathbf{S}_{n+1}$ is the state vector at the $(n+1)$th station (i.e., for the present case the rightmost station of the rotor system), and $\mathbf{T}$ is the *overall system transfer matrix* and its size is 2×2. Hence, it relates the state vector at the far left to the state vector at the far right of the shaft. When the external torque, T_E, is also present, the modified point and field matrices are simply considered as

$$_R\mathbf{S}_2^* = \mathbf{U}_2^*\, {}_R\mathbf{S}_1^* \tag{6.66}$$

with

$$\mathbf{U}_2^* = \mathbf{P}_2^*\mathbf{F}_2^* = \begin{bmatrix} 1 & 0 & 0 \\ -\omega_{nf}^2 I_{p2} & 1 & -T_{E2} \\ 0 & 0 & 1 \end{bmatrix} \begin{bmatrix} 1 & 1/k_{t_2} & 0 \\ 0 & 1 & 0 \\ 0 & 0 & 1 \end{bmatrix} = \begin{bmatrix} 1 & 1/k_{t_2} & 0 \\ -\omega_{nf}^2 I_{p2} & \left(-\omega_{nf}^2 I_{p2}/k_{t_2} + 1\right) & -T_{E2} \\ 0 & 0 & 1 \end{bmatrix}$$

where $\mathbf{U}_2$ is the *modified transfer matrix*, and the overall transfer matrix $\mathbf{T}^*$ is

$$_R\mathbf{S}_{n+1}^* = \mathbf{T}^*(\omega)\ _R\mathbf{S}_0^* \tag{6.67}$$

with

$$\mathbf{T}^*(\omega) = \mathbf{U}_{n+1}^*\mathbf{U}_n^*\cdots\mathbf{U}_2^*\mathbf{U}_1^* = \begin{bmatrix} t_{11}^*(\omega) & t_{12}^*(\omega) & 0 \\ t_{21}^*(\omega) & t_{22}^*(\omega) & t_{23}^*(\omega) \\ 0 & 0 & 1 \end{bmatrix}$$

It should be noted that the size of the overall system transfer matrix remains the same as that of the field matrix or the point matrix, i.e., (2×2) for free vibrations; when the external torque is also considered, then the size becomes (3×3). The overall transformation for free vibrations can be written as

$$_R\begin{Bmatrix} \varphi_z \\ T \end{Bmatrix}_{n+1} = \begin{bmatrix} t_{11}(\omega_{nf}) & t_{12}(\omega_{nf}) \\ t_{21}(\omega_{nf}) & t_{22}(\omega_{nf}) \end{bmatrix} \begin{Bmatrix} \varphi_z \\ T \end{Bmatrix}_0 \tag{6.68}$$

The overall transfer matrix elements are a function of the torsional natural frequency, ω_{nf}, of the rotor system (or the excitation frequency, ω, for the case when the external torque is present, see Equation 6.67).

6.5.4 Application of Boundary Conditions

Now different boundary conditions are considered to illustrate the application of boundary conditions in the overall transfer matrix equation for obtaining natural frequencies and mode shapes of the system (or unbalance responses). In all cases, the number of disks is kept equal to n, and depending upon boundary conditions and the location of disks, the station numbers may change.

6.5.4.1 Free–Free Boundary Conditions

For free–free boundary conditions (Figure 6.19), at each end of the rotor system, the torque transmitted through the shaft is zero, hence

$$_RT_{n+1} = T_0 = 0 \tag{6.69}$$

On using Equation 6.69 in Equation 6.68, the second equation gives

$$t_{21}(\omega_{nf})\ _R\varphi_{z0} = 0 \tag{6.70}$$

Because $_R\varphi_{z0} \neq 0$ for a general case, from Equation 6.70, we get

$$t_{21}\left(\omega_{nf}\right)=0 \tag{6.71}$$

which is satisfied for ω_{nfi}, $i=1,2,\ldots,N$, where N is the number of degrees of freedom of the system (for the present case, it will be equal to the number of disks, n, in the system) and these are system torsional natural frequencies. Equation 6.71 is called the *frequency equation*, and it has the form of a polynomial in terms of the torsional natural frequency, ω_{nf}. For higher degree polynomials, these roots, ω_{nfi}, may be found by any of root searching techniques (e.g., incremental method, bisection method, and Newton–Raphson method; refer to Press et al., 1998).

Root searching method of finding natural frequencies: Briefly, the basic concept of the root searching method is described here, for finding torsional natural frequencies of the rotor system from the frequency equation and for the sake of completeness. Let us define

$$f\left(\omega_{nf}\right)=t_{21}\left(\omega_{nf}\right) \tag{6.72}$$

where $f()$ is a function of ω_{nf}. In the constant increment root searching method, the function, $f(\omega_{nf})$, could be plotted with respect to ω_{nf}. Wherever the curve crosses the zero line, there is a root at that value of ω_{nf}.

Alternatively, if ω_{nf} is the initial guess of the natural frequency, then this will not satisfy Equation 6.72, and hence, $f\left(\omega_{nf}\right)\neq 0$. Then, let the next guess value of the natural frequency be $\left(\omega_{nf}+\Delta\omega_{nf}\right)$ by which the solution is expected to improve. Hence, by using the Taylor series expansion, we have

$$f\left(\omega_{nf}+\Delta\omega_{nf}\right)=f\left(\omega_{nf}\right)+\frac{\partial f}{\partial\omega_{nf}}\Delta\omega_{nf}+\frac{1}{2}\frac{\partial^2 f}{\partial\omega_{nf}^2}\left(\Delta\omega_{nf}\right)^2+\cdots \tag{6.73}$$

where $\Delta\omega_{nf}$ is the increment in the initial guess value of ω_{nf}. On neglecting the second and higher-order terms, we get

$$\Delta\omega_{nf}=\frac{f\left(\omega_{nf}+\Delta\omega_{nf}\right)-f\left(\omega_{nf}\right)}{\partial f\left(\omega_{nf}\right)/\partial\omega_{nf}} \tag{6.74}$$

A flow chart of the overall solution algorithm is shown in Figure 6.21. By this procedure, finding the subsequent increment is called the Newton–Raphson method, which gives faster convergence compared with the case of the constant increment method or bisection root searching method. In the flow chart, ε is a small parameter chosen depending upon the function value minimized and the accuracy of the solution desired. When using such a numerical analysis for finding the natural frequencies, there is no need to multiply various point and field matrices in variable form to get the

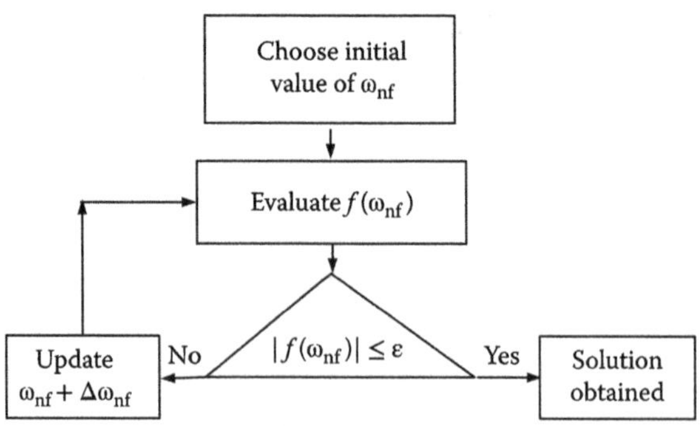

FIGURE 6.21 A flow chart of an algorithm for finding roots of a function.

overall transfer matrix; instead, it can be done in numerical form, which is much easier to perform. Some suitable algorithms are implemented to obtain all natural frequencies of the system in a single run of the computer code up to the desired accuracy.

Mode shapes: Relative angular twists can be determined for each value of ω_{nfi} (where $i = 1, 2, \ldots, n$). From the first set of Equation 6.68 keeping note of boundary conditions in Equations 6.69, we have

$$_R\varphi_{z_{n+1}} = t_{11}\left(\omega_{nfi}\right) {}_R\varphi_{z0} \tag{6.75}$$

Mode shape is nothing but relative angular displacement among various disks. On taking $_R\varphi_{z0} = 1$ as a reference value for obtaining the mode shape, we get

$$_R\varphi_{z_{n+1}} = t_{11}\left(\omega_{nfi}\right) \tag{6.76}$$

Equation 6.76 gives $_R\varphi_{z_{n+1}}$ for a particular value of the natural frequency, ω_{nfi}, and by using Equation 6.65, relative displacements of all other stations can be obtained. The mode shape can be plotted with the station number as the abscissa and the angular displacement as the ordinate. A similar process can be repeated to obtain mode shapes corresponding to all other values of natural frequencies one at a time. In general, for each natural frequency, there is a corresponding distinctive mode shape. This is illustrated subsequently through numerical examples.

6.5.4.2 Fixed–Free Boundary Conditions

For fixed–free boundary conditions (Figure 6.22a), at the fixed end (let this be the 0th station), the angular displacement is zero and at the free end (i.e., at the nth station) the torque is zero, hence

$$_R\varphi_{z0} = 0 \quad \text{and} \quad {}_RT_n = 0 \tag{6.77}$$

On using Equation 6.77 in Equation 6.68, the second set of equations gives

$$t_{22}\left(\omega_{nf}\right) {}_RT_0 = 0 \tag{6.78}$$

Because $_RT_0 \neq 0$ for a general case, hence from Equation 6.78 we have the frequency equation as

$$t_{22}\left(\omega_{nf}\right) = 0 \tag{6.79}$$

It should be noted that for the case when the free end is at the 0th station (i.e., at the extreme left as shown in Figure 6.22b), the fixed end is at the nth station (i.e., at the extreme right). Boundary conditions are $_LT_0 = 0$ and $_R\varphi_{zn} = 0$, and the frequency equation from Equation 6.68 is (it is assumed that the free end and intermediate stations have disks)

$$t_{11}\left(\omega_{nf}\right) = 0 \tag{6.80}$$

It should be noted that the transformation matrix (point or field or overall) has been developed to transfer the state vector from the left to right side and that it has to be maintained strictly. The

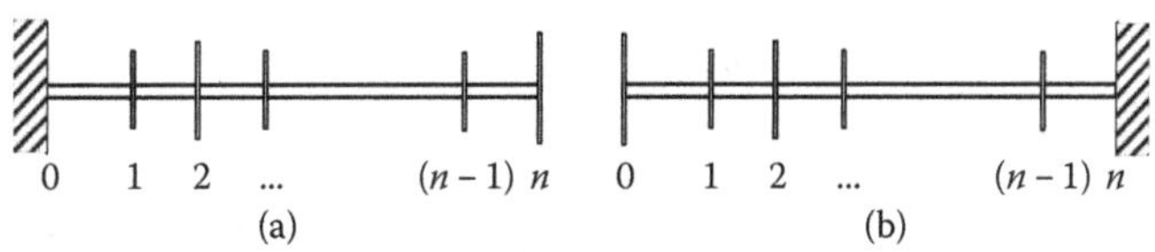

FIGURE 6.22 A multi-DOF rotor system with fixed–free boundary conditions: (a) fixed end toward the LHS and (b) fixed end toward the RHS.

transformation matrix would be different for a transformation from right to left direction; in fact, it affects the signs of some of the terms because we need to invert the matrix. It is illustrated subsequently.

6.5.4.3 Fixed–Fixed Boundary Conditions

For fixed–fixed boundary conditions (Figure 6.23), at both fixed ends {at 0th and $(n+1)$th stations}, the angular displacements are zero, hence

$$\varphi_{z0} = 0 \quad \text{and} \quad \varphi_{z_{n+1}} = 0 \tag{6.81}$$

On using Equation 6.81 into Equation 6.68, the second equation gives

$$t_{12}\left(\omega_{nf}\right) {}_RT_0 = 0 \tag{6.82}$$

Because ${}_RT_0 \neq 0$ for the general case, from Equation 6.82, we have the frequency equation as

$$t_{12}\left(\omega_{nf}\right) = 0 \tag{6.83}$$

Table 6.2 summarizes the frequency equations and equations for the state vector calculation for all the cases discussed earlier.

In the above cases, we have considered intermediate supports as frictionless, and no friction of disks with the medium in which these disks are oscillating. In actual practice, we have supports and disks with friction and this produces some frictional (damping) torque on the shaft or disks. Although the rotor is rotating at a certain constant spin speed, frictions in these supports and disks give a constant damping torque. However, the torque on the shaft and disks is a function of the

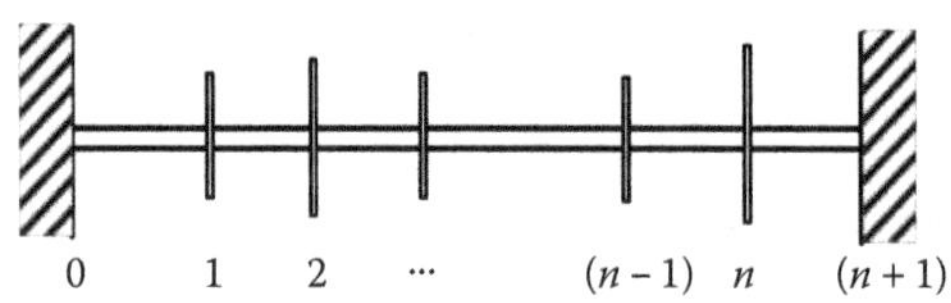

FIGURE 6.23 A rotor system with fixed–fixed boundary conditions.

TABLE 6.2
Equations for the Calculation of Natural Frequencies and Mode Shapes for Some Simple Boundary Conditions

S.N	Boundary Conditions	Station Numbers[a]	Equations to Get Natural Frequencies	Equations to Get Mode Shapes (i.e., Remaining All State Vector Components at the Last Station)
1.	Free–free	0, $n+1$: free ends	$t_{21}(\omega_{nf})=0$	${}_R\varphi_{z_{n+1}} = t_{11}\left(\omega_{nfi}\right)\varphi_{z0}$
2.	Cantilever (fixed–free)	0: fixed end, n: free end, 0: free end, n: fixed end	$t_{22}(\omega_{nf})=0$ $t_{11}(\omega_{nf})=0$	${}_R\varphi_{z_n} = t_{12}\left(\omega_{nfi}\right)T_{z0}$ ${}_RT_{z_n} = t_{21}\left(\omega_{nfi}\right)\varphi_{z0}$
3.	Fixed–fixed	0, $n+1$: fixed ends	$t_{12}(\omega_{nf})=0$	${}_RT_{z_{n+1}} = t_{22}\left(\omega_{nfi}\right)T_{z0}$

[a] The station numbering is done for the present formulation from the extreme left toward the extreme right in the multi-DOF rotor system.

spin speed of the rotor. Due to relative twisting of the shaft ends during torsional oscillations, the hysteretic damping due to shaft material is present, which is a function of the rate of change of relative angular displacements. Overall effects of these frictions are much less on the torsional natural frequencies of rotor systems, and for initial estimates of system dynamic characteristics, it can be ignored. Torsional oscillations of the rotor with torsional dampers are considered subsequently.

A word of caution regarding the numbering of stations: For the present formation, we stick to the numbering scheme where the 0th station is assigned to the extreme LHS of the station, and subsequent station numbers (i.e., 1, 2, …) are given to the station encountered toward the right (Figure 6.22). When numbering stations from the extreme RHS (as 0th) and increasing toward the LHS (i.e., 1, 2, …), the following point and field matrices should be used (which are slightly different compared to Equations 6.56 and 6.62),

$$\overleftarrow{\mathbf{P}} = \begin{bmatrix} 1 & 0 \\ \omega_{nf}^2 I_p & 1 \end{bmatrix} \quad \text{and} \quad \overleftarrow{\mathbf{F}} = \begin{bmatrix} 1 & -1/k \\ 0 & 1 \end{bmatrix} \tag{6.84}$$

where $\overleftarrow{\mathbf{P}}$ and $\overleftarrow{\mathbf{F}}$ are the point and field matrices when the transformation of the state vector is performed from the right to the left (i.e., in the direction of the arrow at the top of P and F). For example, Equations 6.56 and 6.62 can be written as

$${}_L\mathbf{S}_2 = \overleftarrow{\mathbf{P}}_2\, {}_R\mathbf{S}_2 \quad \text{and} \quad {}_R\mathbf{S}_1 = \overleftarrow{\mathbf{F}}_2\, {}_L\mathbf{S}_2 \tag{6.85}$$

with

$$\overleftarrow{\mathbf{P}}_2 = \mathbf{P}_2^{-1} \quad \text{and} \quad \overleftarrow{\mathbf{F}}_2 = \mathbf{F}_2^{-1}$$

It should be noted that these point and field matrices are in fact the inverse of the previous matrices. To avoid this confusion in the present text, the station number is consistently assigned from the left end to the right end, and the transformation of the state vector also follows the same sequence (i.e., from the left to the right). To illustrate the TMM, now several simple numerical problems are given here.

Example 6.6

Obtain torsional natural frequencies of a rotor system as shown in Figure 6.24 by using the TMM. Assume the shaft is massless. Check the result obtained with the closed form solution available. Let $G = 0.8 \times 10^{11}\,\text{N/m}^2$.

Solution:

We have the following properties of the rotor system:

$$G = 0.8 \times 10^{11}\ \text{N/m}^2; \quad l = 0.6\ \text{m}; \quad J = \frac{\pi}{32}(0.1)^4 = 9.82 \times 10^{-6}\ \text{m}^4$$

The torsional stiffness of the shaft is given as

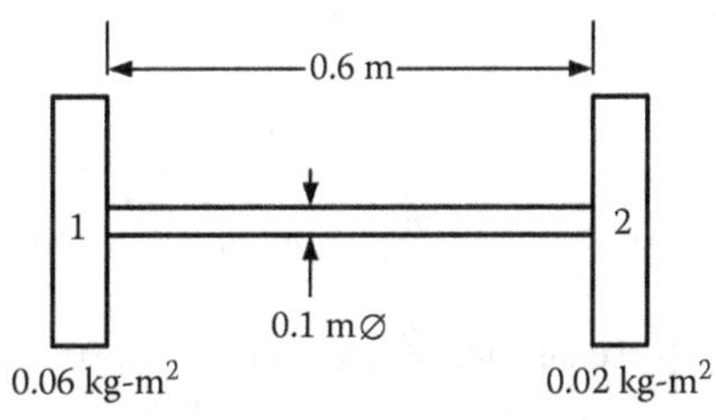

FIGURE 6.24 Two-disk rotor system.

$$k_t = \frac{GJ}{l} = \frac{0.8\times 10^{11}\times 9.82\times 10^{-6}}{0.6} = 1.31\times 10^{6}\ \text{Nm/rad}$$

Analytical method: Natural frequencies in the closed form are given as

$$\omega_{nf_1} = 0; \quad \text{and} \quad \omega_{nf_2} = \sqrt{\frac{(I_{P1}+I_{P2})k_t}{I_{P1}I_{P2}}} = \sqrt{\frac{(0.06+0.02)1.31\times 10^{6}}{0.06\times 0.02}}$$

$$= 9345.23\ \text{rad/s}$$

Mode shapes (relative amplitudes) are given as for

$$\omega_{nf_1} = 0, \quad \frac{\Phi_{z2}}{\Phi_{z_1}} = 1$$

and for

$$\omega_{nf_2} = 9345.23\ \text{rad/s}, \quad \frac{\Phi_{z2}}{\Phi_{z_1}} = -\frac{I_{P1}}{I_{P2}} = -3$$

TMM: Let the station numbers be 1 and 2 as shown in Figure 6.24. State vectors can be related between stations 1 and 2 as

$$_R\mathbf{S}_1 = \mathbf{P}_1\, {}_L\mathbf{S}_1$$
$$_R\mathbf{S}_2 = \mathbf{P}_2\mathbf{F}_2\, {}_R\mathbf{S}_1 = \mathbf{P}_2\mathbf{F}_2\mathbf{P}_1\, {}_L\mathbf{S}_1$$

The overall transformation of the state vectors between 1 and 2 is given as

$$_R\begin{Bmatrix}\varphi_z\\ T\end{Bmatrix}_2 = \begin{bmatrix}1 & 0\\ -\omega_{nf}^2 I_{P2} & 1\end{bmatrix}\begin{bmatrix}1 & 1/k_t\\ 0 & 1\end{bmatrix}\begin{bmatrix}1 & 0\\ -\omega_{nf}^2 I_{P1} & 1\end{bmatrix}{}_L\begin{Bmatrix}\varphi_z\\ T\end{Bmatrix}_1$$

$$= \begin{bmatrix}1 & 1/k_t\\ -\omega_{nf}^2 I_{P2} & \left(1-\omega_{nf}^2 I_{P2}/k_t\right)\end{bmatrix}\begin{bmatrix}1 & 0\\ -\omega_{nf}^2 I_{P_1} & 1\end{bmatrix}{}_L\begin{Bmatrix}\varphi_z\\ T\end{Bmatrix}_1$$

$$= \begin{bmatrix}1-\omega_{nf}^2 I_{P1}/k_t & 1/k_t\\ \left\{-\omega_{nf}^2 I_{P_2} - \omega_{nf}^2 I_{P_1}\left(1-\omega_{nf}^2 I_{P_2}/k_t\right)\right\} & \left(1-\omega_{nf}^2 I_{P_2}/k_t\right)\end{bmatrix}{}_L\begin{Bmatrix}\varphi_z\\ T\end{Bmatrix}_1$$

On substituting values of various rotor parameters, we get

$$_R\begin{Bmatrix}\varphi_z\\ T\end{Bmatrix}_2 = \begin{bmatrix}\left(1-4.58\times 10^{-8}\omega_{nf}^2\right) & 7.64\times 10^{-7}\\ \left(-0.08\omega_{nf}^2 + 9.16\times 10^{-10}\omega_{nf}^4\right) & \left(1-1.53\times 10^{-8}\omega_{nf}^2\right)\end{bmatrix}{}_L\begin{Bmatrix}\varphi_z\\ T\end{Bmatrix}_1 \qquad \text{(a)}$$

Because the ends of the rotor are free–free type, the following boundary conditions are applied:

$$_LT_1 = {}_RT_2 = 0 \qquad \text{(b)}$$

On application of boundary conditions (b) in Equation (a), we get the following condition:

$$\left(-0.08\omega_{nf}^2 + 9.16\times 10^{-10}\omega_{nf}^4\right)_L \varphi_{z_1} = 0$$

which gives, for the nontrivial solution, the following frequency equation:

$$\omega_{nf}^2\left(9.16\times10^{-10}\omega_{nf}^2-0.08\right)=0$$

This gives the natural frequencies as

$$\omega_{nf_1}=0 \quad \text{and} \quad \omega_{nf_2}=9345.23 \text{ rad/s}$$

which are exactly the same as those obtained by the closed form solution. Mode shapes can be obtained by substituting these natural frequencies, one at a time, into the first (or the second) expression of Equation (a), as

$$\frac{\Phi_{z2}}{\Phi_{z1}}=\left(1-4.58\times10^{-8}\omega_{nf}^2\right)\big|_{\omega_{nf1}=0}=1, \text{ rigid body mode}$$

and

$$\frac{\Phi_{z2}}{\Phi_{z1}}=\left(1-4.58\times10^{-8}\omega_{nf}^2\right)\big|_{\omega_{nf2}=9345.23}=-3, \text{ antiphase mode}$$

which are also exactly the same as obtained by closed form solutions. Hence, as such there is no assumption involved in the TMM compared with the analytical method. Moreover, inertia of the shaft is ignored in both analyses.

Example 6.7

Obtain torsional natural frequency for a cantilever shaft carrying a disk and a spring at the free end as shown in Figure 6.25. The disk has a polar mass moment of inertia of 0.02 kg-m². The shaft has 0.4 m length and 0.015 m diameter. The spring has torsional stiffness of $k_{t_2}=100$ N-m/rad. Let $G=0.8\times10^{11}$ N/m² for the shaft.

Solution:

Let the fixed end have station number 0, the shaft free end have station number 1, and the spring at the other end fixed to the fixed support have station number 2 (Figure 6.25). The transformation of the state vector from station 0 to station 1 can be written as

$$
\begin{aligned}
{}_R\mathbf{S}_1=\mathbf{P}_1\mathbf{F}_1\mathbf{S}_0 &= \begin{bmatrix} 1 & 0 \\ -\omega_{nf}^2 I_{p1} & 1 \end{bmatrix}_1 \begin{bmatrix} 1 & \dfrac{l}{GJ} \\ 0 & 1 \end{bmatrix}_1 \begin{Bmatrix} \varphi_z \\ T \end{Bmatrix}_0 \\
&= \begin{bmatrix} 1 & \dfrac{l}{GJ} \\ -\omega_{nf}^2 I_{p1} & 1-\dfrac{\omega_{nf}^2 I_{p1} l}{GJ} \end{bmatrix} \begin{Bmatrix} \varphi_z \\ T \end{Bmatrix}_0
\end{aligned} \tag{a}
$$

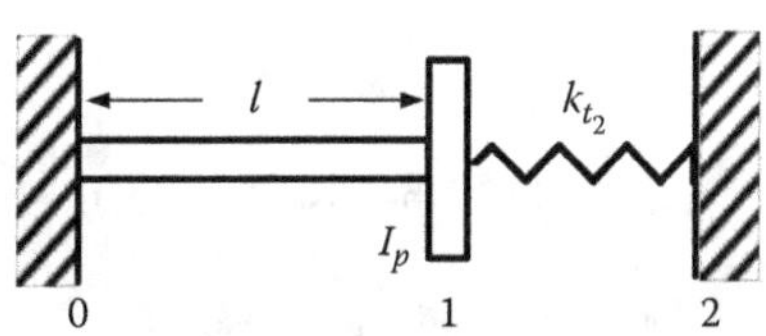

FIGURE 6.25 A cantilever rotor with a spring at the free end.

The spring at the free end can be thought as an equivalent shaft segment with the same stiffness as that of the spring. The overall transfer matrix for such an idealization between stations 0 and 2 is

$$\mathbf{S}_2 = \mathbf{F}_2\mathbf{P}_1\mathbf{F}_1\mathbf{S}_0 = \begin{bmatrix} 1 & \dfrac{1}{k_{t_2}} \\ 0 & 1 \end{bmatrix} \begin{bmatrix} 1 & \dfrac{l}{GJ} \\ -\omega_{nf}^2 I_{p1} & 1-\dfrac{\omega_{nf}^2 I_{p1} l}{GJ} \end{bmatrix} \begin{Bmatrix} \varphi_z \\ T \end{Bmatrix}_0$$

$$= \begin{bmatrix} 1-\dfrac{\omega_{nf}^2 I_{p1}}{k_{t_2}} & \dfrac{l}{GJ}+\dfrac{1}{k_{t_2}}\left(1-\dfrac{\omega_{nf}^2 I_{p1} l}{GJ}\right) \\ -\omega_{nf}^2 I_{p1} & 1-\dfrac{\omega_{nf}^2 I_{p1} l}{GJ} \end{bmatrix} \begin{Bmatrix} \varphi_z \\ T \end{Bmatrix}_0 \tag{b}$$

Boundary conditions for the present case are

$$\varphi_{z0} = \varphi_{z2} = 0 \tag{c}$$

On applying boundary conditions to Equation (b), from the first equation, we get

$$\left\{\frac{l}{GJ}+\frac{1}{k_{t_2}}\left(1-\frac{\omega_{nf}^2 I_{p1} l}{GJ}\right)\right\} T_0 = 0 \tag{d}$$

Because torque T_0 cannot be zero, we get the natural frequency from Equation (d) as

$$\omega_{nf} = \sqrt{\frac{k_{t_2}+(GJ/l)}{I_{p1}}} \text{ rad/s} \tag{e}$$

with

$$J = \frac{\pi}{32} 0.015^4 = 4.97\times 10^{-9}\ \text{m}^4, \quad I_{p1} = 0.02\ \text{kg-m}^2,$$

$$\frac{GJ}{l} = 994.02\ \text{Nm/rad}, \quad k_{t_2} = 100\ \text{Nm/rad}$$

Hence, we have the natural frequency as

$$\omega_{nf} = 233.88\ \text{rad/s}$$

From Equation (e), it can be observed that the effect of the spring at the free end is to increase the effective stiffness of the system (i.e., springs connected in parallel with the equivalent stiffness of $k_{t_2}+(GJ/l)$, where GJ/l is the stiffness of the shaft).

Alternatively, the spring can be included as a boundary condition as follows. In this case, the transformation of Equation (a) is valid. The equilibrium equation at the right of the free end would be

$$_RT_1 + k_{t_2}\varphi_{z1} = 0 \tag{f}$$

where $_RT_1$ is the reaction torque at the right of the disk. Hence, the boundary conditions are

$$\varphi_{z0} = 0 \quad \text{and} \quad {}_RT_1 = -k_{t_2}\varphi_{z1} \tag{g}$$

On application of boundary conditions (g) in Equation (a), we get

$$_R\begin{Bmatrix} \varphi_z \\ -k_{t_2}\varphi_z \end{Bmatrix}_1 = \begin{bmatrix} 1 & \dfrac{l}{GJ} \\ -\omega_{nf}^2 I_{p_1} & 1-\dfrac{\omega_{nf}^2 I_{p_1} l}{GJ} \end{bmatrix} \begin{Bmatrix} 0 \\ T \end{Bmatrix}_0 \tag{h}$$

Equation (h) can be split as follows:

$$_R\varphi_{z_1} = \frac{l}{GJ}T_0 \quad \text{and} \quad -k_{t_2\,R}\varphi_{z_1} = \left(1-\frac{\omega_{nf}^2 I_{p_1} l}{GJ}\right)T_0 \tag{i}$$

which gives an eigenvalue problem of the following form:

$$\begin{bmatrix} 1 & \dfrac{-l}{GJ} \\ k_{t_2} & 1-\dfrac{\omega_{nf}^2 I_{p_1} l}{GJ} \end{bmatrix} \begin{Bmatrix} \varphi_{z_1} \\ T_0 \end{Bmatrix} = \begin{Bmatrix} 0 \\ 0 \end{Bmatrix} \tag{j}$$

For the nontrivial solution, on taking the determinant of the earlier matrix, we obtain a natural frequency exactly the same as in Equation (e).

Example 6.8

Obtain the torsional frequency response at the disk and the support torque at the fixed end of the shaft of the rotor system shown in Figure 6.26. An external sinusoidal torque of amplitude $T_E = 10$ Nm is applied with a single frequency, ω. Identify the torsional critical speed of the system from the response so obtained. The disk has a polar mass moment of inertia of 0.02 kg-m². The shaft has 0.4 m length and 0.015 m diameter. Take $G = 0.8 \times 10^{11}$ N/m².

Solution:

Let the station number of the fixed end be 0 and that of the free end be 1. The transformation of the state vector can be written as

$$_R\mathbf{S}_1^* = \mathbf{P}_1^*\,\mathbf{F}_1^*\,\mathbf{S}_0^* = \begin{bmatrix} 1 & 0 & 0 \\ -\omega^2 I_{p_1} & 1 & -T_{E_1} \\ 0 & 0 & 1 \end{bmatrix} \begin{bmatrix} 1 & 1/k_{t_1} & 0 \\ 0 & 1 & 0 \\ 0 & 0 & 1 \end{bmatrix} \begin{Bmatrix} \varphi_z \\ T \\ 1 \end{Bmatrix}_0$$

$$= \begin{bmatrix} 1 & 1/k_{t_1} & 0 \\ -\omega^2 I_{p_1} & 1-\dfrac{\omega^2 I_{p_1}}{k_{t_1}} & -T_{E_1} \\ 0 & 0 & 1 \end{bmatrix} \begin{Bmatrix} \varphi_z \\ T \\ 1 \end{Bmatrix}_0 \tag{a}$$

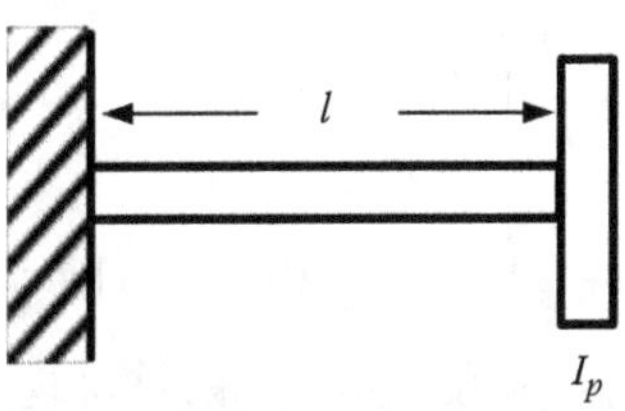

FIGURE 6.26 A shaft with cantilevered end conditions.

where ω is the external excitation frequency and T_E is the external torque amplitude. The boundary conditions of the present problem are

$$\varphi_{z0} = 0 \quad \text{and} \quad {}_R T_1 = 0 \tag{b}$$

On application of boundary conditions (b) in Equation (a), we get

$$_R\begin{Bmatrix} \varphi_z \\ 0 \\ 1 \end{Bmatrix}_1 = \begin{bmatrix} 1 & 1/k_{t_1} & 0 \\ -\omega^2 I_{p_1} & 1-\dfrac{\omega^2 I_{p_1}}{k_{t_1}} & -T_{E_1} \\ 0 & 0 & 1 \end{bmatrix} \begin{Bmatrix} 0 \\ T \\ 1 \end{Bmatrix}_0 \tag{c}$$

which gives the following equations:

$$_R\varphi_{z_1} = \frac{T_0}{k_{t_1}} \quad \text{and} \quad 0 = \left(1-\frac{\omega^2 I_{p_1}}{k_{t_1}}\right)T_0 - T_{E_1} \quad \text{or} \quad T_0 = \frac{T_{E_1}}{\left(1-\dfrac{\omega^2 I_{p_1}}{k_{t_1}}\right)} \tag{d}$$

From the above, the frequency response at station 1 takes the following form:

$$_R\varphi_{z_1} = \frac{T_{E_1}}{\left(1-\dfrac{\omega^2 I_{p_1}}{k_{t_1}}\right)} = \frac{T_{E_1}}{\left(1-\dfrac{\omega^2}{\omega_{nf}^2}\right)} \quad \text{with} \quad \omega_{nf} = \sqrt{k_t/I_p} \tag{e}$$

For the present problem, T_{E_1}=10 Nm, J=4.97×10^{-9} m^4, I_{p_1}=0.02 kg-m^2, and $k_{t_1} = \dfrac{GJ}{l} = 994.02$ Nm/rad, and hence $\omega_{nf} = 222.94$ rad/s.

Hence, from Equations (e) and (d), we have

$$_R\varphi_{z_1} = \frac{10}{\left(1-\dfrac{\omega^2}{4.97\times10^4}\right)} \text{ rad} \quad \text{and} \quad T_0 = \frac{10\times994.02}{\left(1-\dfrac{\omega^2}{4.97\times10^4}\right)} \tag{f}$$

Equation (f) can be used to plot the response amplitudes of the disk and of the reactive torque at the fixed support with respect to the excitation frequency, ω. However, it can be seen from the denominator that the resonance takes place when it becomes zero, i.e., $\omega = \omega_{nf} = 222.94$ rad/s, which is the condition of torsional critical speed, $\omega = \omega_{cr} = 222.94$ rad/s.

Example 6.9

Find the torsional natural frequencies and mode shapes of the rotor system shown in Figure 6.27. B is a fixed end, and D_1 and D_2 are rigid disks. The shaft is made of steel with a modulus of rigidity G=0.8×10^{11} N/m^2 and a uniform diameter d=10 mm. The shaft lengths are BD_1=50 mm and D_1D_2=75 mm. The polar mass moments of inertia of disks are $I_{p_1} = 0.08$ kg-m^2 and $I_{p_2} = 0.2$ kg-m^2. Consider the shaft as massless and apply (a) the analytical method and (b) the TMM.

Solution:

The torsional free vibration is analyzed by the classical analytical method and the TMM to compare the results.

i. *Analytical method*: From the free body diagrams of disks shown in Figure 6.28, equations of motion for free vibrations can be written as

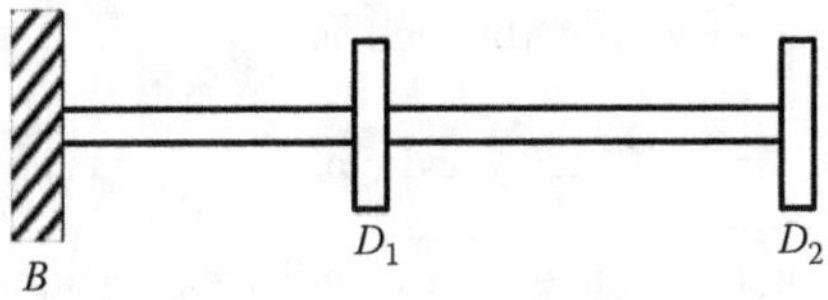

FIGURE 6.27 Two-rotor system.

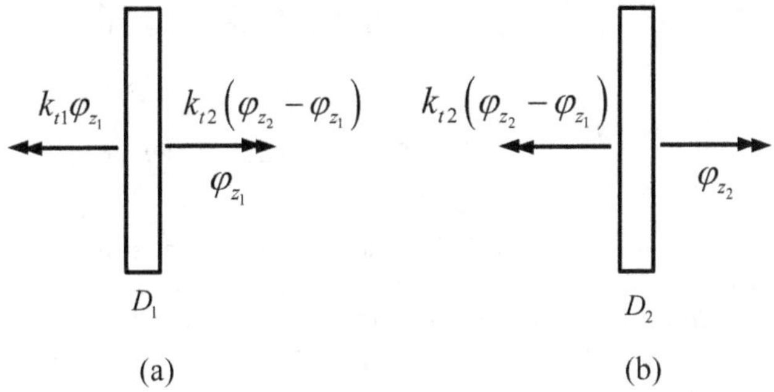

FIGURE 6.28 Free body diagrams of disks: (a) free body diagram of disk 1; (b) free body diagram of disk 2.

$$I_{p1}\ddot{\varphi}_{z1}+k_{t1}\varphi_{z1}+k_{t2}(\varphi_{z1}-\varphi_{z2})=0 \quad \text{and} \quad I_{p2}\ddot{\varphi}_{z2}+k_{t2}(\varphi_{z2}-\varphi_{z1})=0 \tag{a}$$

where k_1 and k_2 are the torsional stiffnesses of shaft segments BD_1 and D_1D_2, respectively. Equations of motion are homogeneous second-order differential equations. In free vibrations, disks generate the SHM.

For the SHM, we have $\ddot{\varphi}_z=-\omega_{nf}^2\varphi_z=-\omega_{nf}^2\Phi_z\sin\omega_{nf}t$; hence, equations of motion take the form

$$\begin{bmatrix} k_{t1}+k_{t2}-I_{p1}\omega_{nf}^2 & -k_{t2} \\ -k_{t2} & k_{t2}-I_{p2}\omega_{nf}^2 \end{bmatrix}\begin{Bmatrix}\Phi_{z1}\\ \Phi_{z2}\end{Bmatrix}=\begin{Bmatrix}0\\0\end{Bmatrix} \tag{b}$$

On taking the determinant of the above matrix, we obtain the frequency equation as

$$I_{p1}I_{p2}\omega_{nf}^4-(I_{p1}k_{t2}+I_{p2}k_{t1}+I_{p2}k_{t2})\omega_{nf}^2+k_{t1}k_{t2}=0 \tag{c}$$

which can be solved for ω_{nf}^2 as

$$\omega_{nf}^2=\frac{I_{p1}k_{t2}+I_{p2}k_{t1}+I_{p2}k_{t2}\pm\sqrt{(I_{p1}k_{t2}+I_{p2}k_{t1}+I_{p2}k_{t2})^2-4k_{t1}k_{t2}I_{p1}I_{p2}}}{2I_{p1}I_{p2}} \tag{d}$$

For the present problem, we have the following properties:

$$J_1=\frac{\pi}{32}d^4=\frac{\pi}{32}(0.01)^4=9.82\times10^{-4}\ \text{m}^4=J_2$$

$$k_{t1}=\frac{GJ_1}{l_1}=1570.79\ \text{Nm/rad} \quad \text{and} \quad k_{t2}=\frac{GJ_2}{l_2}=1047.19\ \text{Nm/rad}$$

$$I_{p1}=0.08\ \text{kgm}^2 \quad \text{and} \quad I_{p2}=0.2\ \text{kgm}^2$$

From Equation (d), natural frequencies are obtained as

$$\omega_{nf_1} = 54.17 \text{ rad/s} \quad \text{and} \quad \omega_{nf_2} = 187.15 \text{ rad/s}$$

The relative amplitude ratio can be obtained from the first expression of Equation (b) as

$$\frac{\Phi_{z1}}{\Phi_{z2}} = \frac{k_{t_2}}{k_{t_1} + k_{t_2} - I_{p1}\omega_{nf}^2} = 0.4394 \text{ for } \omega_{nf_1} \quad \text{and} \quad -5.689 \text{ for } \omega_{nf_2} \tag{e}$$

Alternatively, the relative amplitude ratio can be obtained from the second expression of Equation (b), as

$$\frac{\Phi_{z1}}{\Phi_{z2}} = \frac{k_{t_2} - I_{p2}\omega_{nf}^2}{k_{t_2}} = 0.4394 \text{ for } \omega_{nf_1} \quad \text{and} \quad -5.689 \text{ for } \omega_{nf_2} \tag{f}$$

As expected, this gives the same result as in Equation (e). Mode shapes are shown in Figure 6.29, in which the first one is the in-phase mode and the second is the antiphase mode. Practically, the antiphase mode is difficult to excite (it has more resistive torque) compared with the in-phase mode. Also in the second mode the lighter disk 1 has more angular displacement as compared to the heavier disk 2. The natural frequency value of the former mode is more than the latter.

For Figure 6.30, state vectors between 0th and 2nd stations can be related as (order of multiplication of matrices is very important here)

$${}_R\mathbf{S}_2 = \mathbf{P}_2\mathbf{F}_2\mathbf{P}_1\mathbf{F}_1\mathbf{S}_0 \tag{g}$$

State vectors at neighboring stations (i.e., 1 and 2, and 0 and 1) can be related as

$$_R\begin{Bmatrix} \varphi_z \\ T \end{Bmatrix}_2 = \begin{bmatrix} 1 & 1/k_{t2} \\ -\omega_{nf}^2 I_{p2} & \dfrac{-\omega_{nf}^2 I_{p2}}{k_{t2}} + 1 \end{bmatrix}_R \begin{Bmatrix} \varphi_z \\ T \end{Bmatrix}_1$$

$$\text{and} \quad _R\begin{Bmatrix} \varphi_z \\ T \end{Bmatrix}_1 = \begin{bmatrix} 1 & 1/k_{t1} \\ -\omega_{nf}^2 I_{p1} & \dfrac{-\omega_{nf}^2 I_{p1}}{k_{t1}} + 1 \end{bmatrix}_R \begin{Bmatrix} \varphi_z \\ T \end{Bmatrix}_0 \tag{h}$$

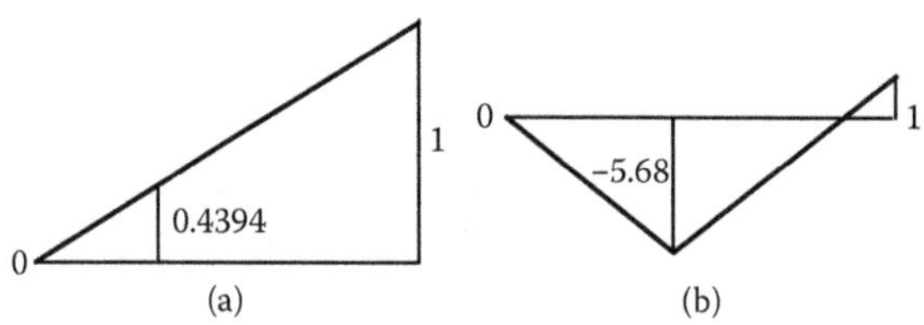

FIGURE 6.29 Mode shapes (a) for ω_{nf_1} and (b) for ω_{nf_2}.

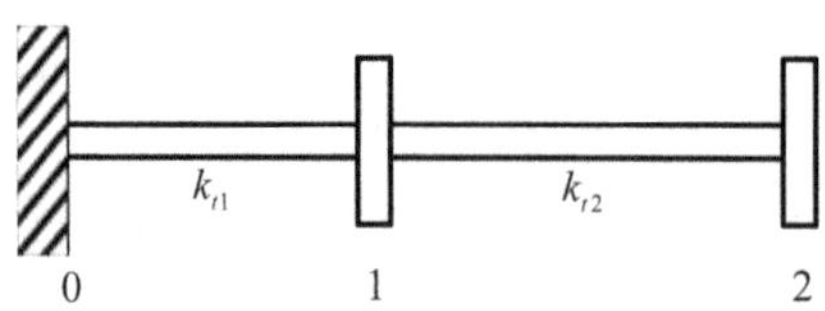

FIGURE 6.30 Two-disk rotor system with numbering of stations.

which can be combined to give

$$
{}_R\begin{Bmatrix} \varphi_z \\ T \end{Bmatrix}_2 = \begin{bmatrix} \left(1-\dfrac{\omega_{nf}^2 I_{p1}}{k_{t2}}\right) & \left\{\dfrac{1}{k_{t2}}\left(1-\dfrac{\omega_{nf}^2 I_{p1}}{k_{t1}}\right)+\dfrac{1}{k_{t1}}\right\} \\ \left\{-\omega_{nf}^2 I_{p2}-\omega_{nf}^2 I_{p1}\left(\dfrac{-\omega_{nf}^2 I_{p2}}{k_{t2}}+1\right)\right\} & \left\{\dfrac{\omega_{nf}^4 I_{p1} I_{p2}}{k_{t1}k_{t2}}-\omega_{nf}^2\left(\dfrac{I_{p1}}{k_{t1}}+\dfrac{I_{p2}}{k_{t2}}+\dfrac{I_{p2}}{k_{t1}}\right)+1\right\} \end{bmatrix} \begin{Bmatrix} \varphi_z \\ T \end{Bmatrix}_0 \tag{i}
$$

Boundary conditions are at station 0, $\varphi_{z0}=0$, and at the right of station 2, ${}_RT_2=0$. On application of boundary conditions in Equation (i), the second equation gives the frequency equation as

$$
t_{22}(\omega_{nf}) = \left\{\frac{\omega_{nf}^4 I_{p1} I_{p2}}{k_{t1}k_{t2}}-\omega_{nf}^2\left(\frac{I_{p1}}{k_{t1}}+\frac{I_{p2}}{k_{t2}}+\frac{I_{p2}}{k_{t1}}\right)+1\right\}=0
$$

which can be simplified as

$$
I_{p1}I_{p2}\omega_{nf}^4-\left(I_{p1}k_{t2}+I_{p2}k_{t1}+I_{p2}k_{t2}\right)\omega_{nf}^2+k_{t1}k_{t2}=0
$$

It should be noted that this is the same as was obtained by the analytical method in Equation (c). Hence, natural frequencies by the TMM will also be given by Equation (d). For obtaining mode shapes from Equations (h) and (i), we have

$$
{}_R\varphi_{z2}=t_{12}T_0;\quad {}_R\varphi_{z2}={}_R\varphi_{z1}+\frac{T_{R1}}{k_{t2}};\quad {}_R\varphi_{z1}=\frac{T_0}{k_{t1}} \tag{j}
$$

From Equation (j), we have

$$
\frac{{}_R\varphi_{z1}}{{}_R\varphi_{z2}}=\frac{{}_R\Phi_{z1}}{{}_R\Phi_{z2}}=\frac{1}{k_{t1}t_{12}}=\frac{k_2}{k_{t1}+k_{t2}-I_{p1}\omega_{nf}^2} \tag{k}
$$

which is again the same as Equation (e). Because mode shapes are relative angular displacements of various disks in the rotor system, on assuming one of the angular displacements as unity (i.e., $\varphi_{z2}=1$), we can get the torque acting at various sections of the shaft from Equation (j), as

$$
T_0=\frac{1}{t_{12}}=\frac{k_{t2}}{\omega_{nf}^2\left(k_{t2}I_{p1}+k_{t2}I_{p2}-I_{p1}I_{p2}\omega_{nf}^2\right)} \tag{l}
$$

and

$$
{}_RT_1=k_{t2}\left({}_R\varphi_{z2}-{}_R\varphi_{z1}\right)=k_{t2}\left(1-\frac{T_0}{k_{t1}}\right)=k_{t2}\left\{1-\frac{k_{t2}}{\omega_{nf}^2 k_{t1}\left(k_{t2}I_{p1}+k_{t2}I_{p2}-I_{p1}I_{p2}\omega_{nf}^2\right)}\right\} \tag{m}
$$

It should be noted that these torques would be produced for a unit angular displacement at disk 2 (i.e., $\varphi_{z2}=1$).

Example 6.10

Find torsional natural frequencies and mode shapes of the rotor system shown in Figure 6.31. B_1 and B_2 are frictionless bearings, which provide free–free end conditions; and D_1, D_2, D_3, and D_4 are rigid disks. The shaft is made of steel with a modulus of rigidity $G=0.8\times10^{11}\,\text{N/m}^2$ and uniform diameter $d=20\,\text{mm}$. Various shaft lengths are as follows: $B_1D_1=150\,\text{mm}$, $D_1D_2=50\,\text{mm}$, $D_2D_3=50\,\text{mm}$, $D_3D_4=50\,\text{mm}$, and $D_4B_2=150\,\text{mm}$. The mass of the disks are $m_1=4\,\text{kg}$, $m_2=5\,\text{kg}$, $m_3=6\,\text{kg}$, and $m_4=7\,\text{kg}$. Consider the shaft as massless. Consider disks as thin and take the diameter of the disks as $d_1=8\text{ cm}$, $d_2=10\text{ cm}$, $d_3=12\text{ cm}$, and $d_4=14\text{ cm}$.

Solution:

The disks have the following data $m_1=4\,\text{kg}$, $m_2=5\,\text{kg}$, $m_3=6\,\text{kg}$, $m_4=7\,\text{kg}$, $d_1=0.08\text{ m}$, $d_2=0.1\text{ m}$, $d_3=0.12\text{ m}$, $d_4=0.14\text{ m}$, $I_{p1}=0.5\,m_1r_1^2=0.5\times4\times0.04^2=0.0032\text{ kg-m}^2$, $I_{p2}=0.5\times5\times0.05^2=0.00625\text{ kg-m}^2$, $I_{p3}=0.5\times6\times0.06^2=0.0108\text{ kg-m}^2$, and $I_{p4}=0.5\times7\times0.07^2=0.01715\text{ kg-m}^2$.

The shaft has $GJ=1256.64\text{ N-m}^2$. Shaft segments at the ends do not contribute to the free vibration for the present case because their inertias are ignored. Station numbers 1, 2, 3, and 4 are given at disk locations, and the lengths of shafts between two neighboring disks are $l_1=50\,\text{mm}$, $l_2=50\,\text{mm}$, and $l_3=50\,\text{mm}$.

Now, the overall transformation of the state vectors can be written as

$$ {}_R\mathbf{S}_4=\mathbf{T}\,{}_L\mathbf{S}_1 \tag{a}$$

with

$$\mathbf{T}=\mathbf{P}_4\mathbf{F}_3\mathbf{P}_3\mathbf{F}_2\mathbf{P}_2\mathbf{F}_1\mathbf{P}_1 \tag{b}$$

$$\mathbf{P}_i=\begin{bmatrix}1 & 0\\ -\omega_{nf}^2 I_{Pi} & 1\end{bmatrix};\quad \mathbf{F}_i=\begin{bmatrix}1 & 1/k_{t_i}\\ 0 & 1\end{bmatrix};\quad \mathbf{S}=\begin{Bmatrix}\varphi_z\\ T\end{Bmatrix} \tag{c}$$

From Table 6.2, for free–free boundary conditions, the frequency equation is

$$f(\omega_{nf})=t_{2,1}(\omega_{nf})=0 \tag{d}$$

On solving the roots of the above function by the root searching method, we obtain the following natural frequencies (refer to Figure 6.32):

$$\omega_{nf_1}=0\text{ rad/s},\quad \omega_{nf_2}=1374\text{ rad/s}$$
$$\omega_{nf_3}=2453\text{ rad/s},\quad \omega_{nf_4}=3756\text{ rad/s}$$

From Table 6.2, the eigenvector can be obtained from the following equation:

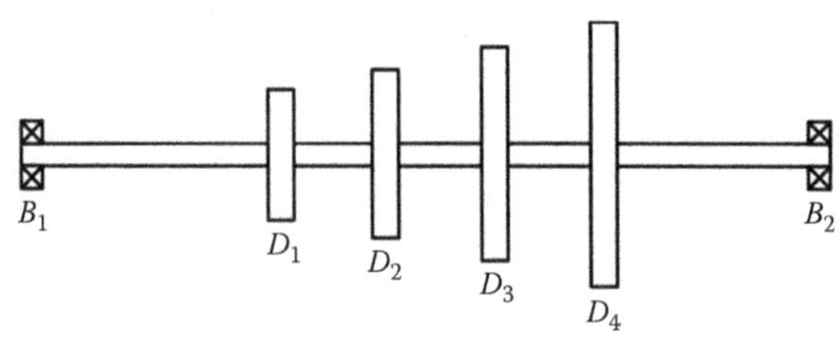

FIGURE 6.31 A multidisk rotor system.

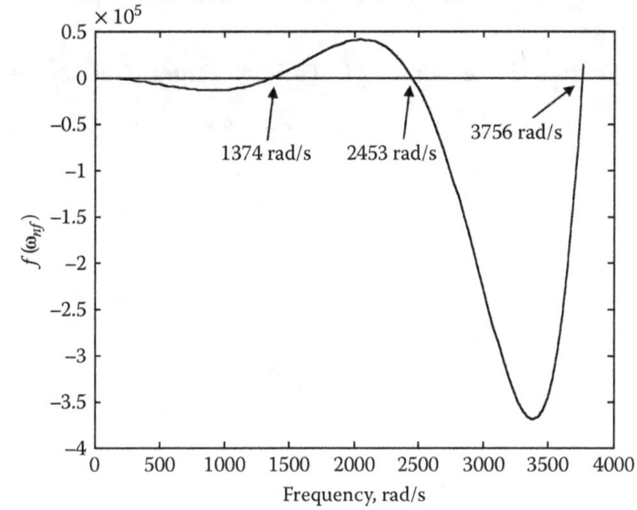

FIGURE 6.32 Variation of function defined in Equation (d) with the frequency.

$$ {}_R\varphi_{z4} = t_{11}\left(\omega_{nfi}\right)\varphi_{z1} \tag{e}$$

Now on choosing $\varphi_{z1} = 1$ as the reference value, let us obtain the state vectors corresponding to the second mode (i.e., $\omega_{nfi} = \omega_{nf2} = 1374$ rad/s). From Equation (e) and noting the boundary conditions, we get the state vector at the first and fourth stations as

$$ {}_L\mathbf{S}_1 = \begin{Bmatrix} 1 \\ 0 \end{Bmatrix} \quad \text{and} \quad {}_R\mathbf{S}_4 = \begin{Bmatrix} t_{11}\left(\omega_{nf2}\right) \\ 0 \end{Bmatrix} = \begin{Bmatrix} -0.59 \\ 0 \end{Bmatrix} $$

At other stations, the state vectors can be obtained as

$$ {}_R\mathbf{S}_1 = \mathbf{P}_1\, {}_L\mathbf{S}_1 = \begin{Bmatrix} 1 \\ -6272 \end{Bmatrix}; \quad {}_L\mathbf{S}_2 = \mathbf{F}_1\, {}_R\mathbf{S}_1 = \begin{Bmatrix} 0.75 \\ -6272 \end{Bmatrix} $$

$$ {}_R\mathbf{S}_2 = \mathbf{P}_2\, {}_L\mathbf{S}_2 = \begin{Bmatrix} 0.75 \\ -15{,}465 \end{Bmatrix}; \quad {}_L\mathbf{S}_3 = \mathbf{F}_2\, {}_R\mathbf{S}_2 = \begin{Bmatrix} 0.14 \\ -15465 \end{Bmatrix} $$

and

$$ {}_R\mathbf{S}_3 = \mathbf{P}_3\, {}_L\mathbf{S}_3 = \begin{Bmatrix} 0.14 \\ -18{,}325 \end{Bmatrix}; \quad {}_L\{\mathbf{S}\}_4 = \mathbf{F}_3\, {}_R\mathbf{S}_3 = \begin{Bmatrix} -0.59 \\ -18{,}325 \end{Bmatrix} $$

Hence, the mode shape (the relative angular displacement) can be drawn for the second mode. Along the same lines, other state vectors corresponding to the remaining natural frequencies can be obtained to get the related mode shapes. All the state vectors are summarized in Table 6.3.

6.6 SIMPLE GEARED ROTOR SYSTEMS

In actual practice, it is rare that a rotor system has a single shaft (with either uniform or stepped cross-sections) with multiple disks as we analyzed in previous sections. In some machines, the shaft may not be continuous from one end of the machine to the other, but may have a gearbox installed

TABLE 6.3
State Vectors Corresponding to Various Natural Frequencies

	$\omega_{nf_2} = 1374$ rad/s	$\omega_{nf_3} = 2453$ rad/s	$\omega_{nf_3} = 3756$ rad/s
${}_L\mathbf{S}_1$	$\begin{Bmatrix} 1 \\ 0 \end{Bmatrix}$	$\begin{Bmatrix} 1 \\ 0 \end{Bmatrix}$	$\begin{Bmatrix} 1 \\ 0 \end{Bmatrix}$
${}_R\mathbf{S}_1$	$\begin{Bmatrix} 1 \\ -6272 \end{Bmatrix}$	$\begin{Bmatrix} 1 \\ -19{,}208 \end{Bmatrix}$	$\begin{Bmatrix} 1 \\ -45{,}000 \end{Bmatrix}$
${}_L\mathbf{S}_2$	$\begin{Bmatrix} 0.75 \\ -6272 \end{Bmatrix}$	$\begin{Bmatrix} 0.24 \\ -19{,}208 \end{Bmatrix}$	$\begin{Bmatrix} -0.79 \\ -45{,}000 \end{Bmatrix}$
${}_R\mathbf{S}_2$	$\begin{Bmatrix} 0.75 \\ -15{,}465 \end{Bmatrix}$	$\begin{Bmatrix} 0.24 \\ -28{,}052 \end{Bmatrix}$	$\begin{Bmatrix} -0.79 \\ -24{,}477 \end{Bmatrix}$
${}_L\mathbf{S}_3$	$\begin{Bmatrix} 0.14 \\ -15{,}465 \end{Bmatrix}$	$\begin{Bmatrix} -0.88 \\ -29{,}022 \end{Bmatrix}$	$\begin{Bmatrix} 0.18 \\ 24{,}477 \end{Bmatrix}$
${}_R\mathbf{S}_3$	$\begin{Bmatrix} 0.14 \\ -18{,}325 \end{Bmatrix}$	$\begin{Bmatrix} -0.88 \\ -29{,}022 \end{Bmatrix}$	$\begin{Bmatrix} 0.18 \\ -3377 \end{Bmatrix}$
${}_L\mathbf{S}_4$	$\begin{Bmatrix} -0.59 \\ -18{,}325 \end{Bmatrix}$	$\begin{Bmatrix} 0.27 \\ 29{,}022 \end{Bmatrix}$	$\begin{Bmatrix} 0.05 \\ -3377 \end{Bmatrix}$
${}_R\mathbf{S}_4$	$\begin{Bmatrix} -0.59 \\ -5751 \end{Bmatrix}$	$\begin{Bmatrix} 0.27 \\ 11{,}237 \end{Bmatrix}$	$\begin{Bmatrix} 0.05 \\ -10{,}822 \end{Bmatrix}$

at one or more locations (Kahraman, 2001), for example, in a stone crusher unit between the motor and the crusher (refer to Figure 6.1), in an automobile between the engine and wheels (or propeller shaft), in rolling mills between the main motor and the main shaft, and then between various stages of shafts. Hence, shafts will have different angular velocities as shown in Figure 6.33a. For the purpose of analysis, the geared rotor system should be reduced to a system with a continuous shaft so that it is analyzed for torsional vibrations by methods as described in the preceding sections (i.e., a two-, three-, or multidisk rotor system).

It is assumed that gears and shafts have negligible polar mass moment of inertia compared with disks in the geared rotor system. This assumption could be relaxed with the addition of one more disk in the equivalent rotor system, for example, in Figure 6.33a, from a two-disk rotor system to a three-disk rotor system due to a gear pair. In next Chapter 7, while using the FEM, these issues will be addressed in more detail. In the actual system, shown in Figure 6.33a, k_{t_2} is the torsional stiffness of the shaft between gear 2 and disk 2, and I_{p_2} is the polar mass moment of inertia of disk 2. Let the equivalent system as shown in Figure 6.33b have shaft torsional stiffness, k_{t_e}, and disk polar mass moment of inertia, I_{p_e}. The strain and kinetic energy values must be the same in both the real and dynamically equivalent systems for the equivalent theoretical model to be valid.

Equivalent stiffness: Let the disk with the polar mass moment of inertia, I_{p_2}, be imagined to be held rigidly in both the real (Figure 6.33a) and equivalent (Figure 6.33c) systems, while pinion shaft 1 is rotated through an angle of φ_{z_1} at the input to the gearbox (i.e., at the pinion). Shaft 2 is rotated through an angle $\varphi_{z_2} = \varphi_{z_1}/n$ at gear 2, where n is the *gear ratio*. It is the ratio of the angular speed of the driving gear (pinion) to that of the driven gear, i.e.,

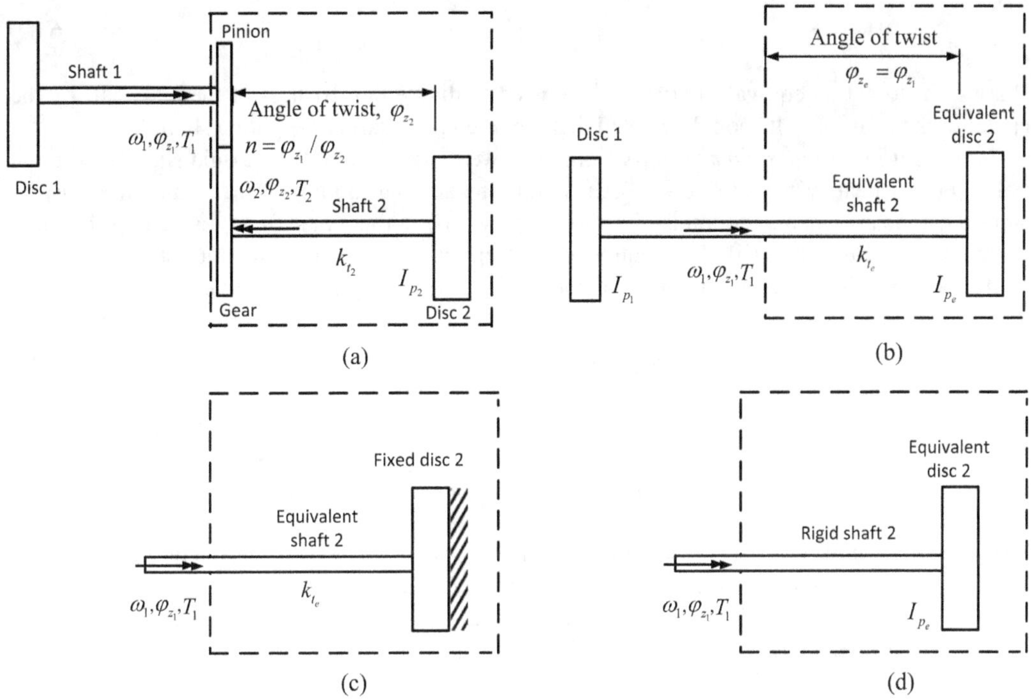

FIGURE 6.33 (a) Actual geared rotor system, (b) an equivalent system without a geared system, (c) the equivalent system with disk 2 fixed, and (d) the equivalent system with shaft 2 rigid.

$$n = \frac{\text{Input shaft speed}}{\text{Output shaft speed}} = \frac{\omega_1}{\omega_2} = \frac{\varphi_{z_1}(t)}{\varphi_{z_2}(t)} = \frac{N_{t_2}}{N_{t_1}} \tag{6.86}$$

where ω is the spin speed of the gear and N_t is the number of teeth of the gear. That means to get the speed of the output shaft (or gear), we need to divide the speed of the input shaft (or pinion) by the gear ratio. The *speed ratio*, *train value*, and *kinematic coefficient* are other terms used for the gear ratio; however, these are an inverse of the gear ratio, i.e., the ratio of the angular speed of the driven gear to that of the driving gear. In the present text, consistently we will be using the gear ratio only. The sense of rotation of the gear can be taken care of two ways, either by observing the sense of rotation in the diagram of the gear train itself or by attaching a positive/negative sign to indicate the same/opposite sense of rotation, and we will follow the former convention. Hence, the strain energy stored in shaft 2 of the actual system, for a twist of φ_{z_1} at the input to the gearbox, can be written as

$$U_r = \frac{1}{2}k_{t_2}\varphi_{z_2}^2 = \frac{1}{2}k_{t_2}\left(\frac{\varphi_{z_1}}{n}\right)^2 \tag{6.87}$$

where U_e is the strain energy in the real system. Now on applying the same angular displacement at the input of the gearbox to the equivalent rotor system (Figure 6.33c), it results in the strain energy stored in the equivalent shaft and can be expressed as

$$U_e = \frac{1}{2}k_{t_e}\varphi_{z_e}^2 = \frac{1}{2}k_{t_e}\varphi_{z_1}^2 \tag{6.88}$$

where U_e is the strain energy in the equivalent system, and we have $\varphi_{z_e} = \varphi_{z_1}$. On equating Equations 6.89 and 6.90, it gives the equivalent stiffness as

$$k_{t_e} = \frac{k_{t2}}{n^2} \tag{6.89}$$

That means to get an equivalent stiffness, we need to divide the stiffness of the real shaft by the square of the gear ratio. It should be noted that the reference shaft is the pinion shaft.

Equivalent polar mass moment of inertia: Now we consider shaft 2 as a rigid shaft in both the real and equivalent systems (Figure 6.33d), so that the angular motion of gear 2 and disk 2 is the same. That means when a motion is given to the pinion (i) in the real system, disk 2 gets the same motion as the gear 2, and (ii) in the equivalent system, disk 2 gets the same motion as the pinion itself. The kinetic energies of both the real and equivalent rotor systems must also be equated:

$$T_r = \frac{1}{2} I_{p2} \bar{\omega}_2^2 \quad \text{and} \quad T_e = \frac{1}{2} I_{pe} \bar{\omega}_e^2 \tag{6.90}$$

with

$$\bar{\omega}_2 = \omega_2 + \dot{\varphi}_{z2} \quad \text{and} \quad \bar{\omega}_e = \omega_1 + \dot{\varphi}_{z_e} = \omega_1 + \dot{\varphi}_{z1}$$

where T_r and T_e are kinetic energies in the real and equivalent systems, respectively, and $\bar{\omega}_2$ and $\bar{\omega}_e$ are angular frequencies of disk 2 of the real $\left(I_{p2}\right)$ and equivalent $\left(I_{p_e}\right)$ systems, respectively. These angular velocities contain the nominal angular speed, ω, and torsional angular velocities, $\dot{\varphi}_z$, as given in Equation 6.90. Now Equation 6.90 can be equated and is written as

$$\frac{1}{2} I_{p2} \left(\omega_2 + \dot{\varphi}_{z2}\right)^2 = \frac{1}{2} I_{p_e} \left(\omega_1 + \dot{\varphi}_{z_e}\right)^2 \tag{6.91}$$

where $\dot{\varphi}_{z2}$ and $\dot{\varphi}_{z_e}$ are the angular velocities due to the torsion vibration in the actual and equivalent systems, respectively. It can be seen from Figure 6.33d that $\varphi_{z_e} = \varphi_{z1}$, and ω_1 and ω_2 are the angular frequencies of shafts 1 and 2, respectively. We have the following basic relations:

$$\varphi_{z1} = \frac{T_1}{k_{t_e}} = \varphi_{z_e} \text{ and } \varphi_{z2} = \frac{T_2}{k_{t2}} = \frac{nT_1}{k_{t2}} \tag{6.92}$$

where T_1 and T_2 are torques at gears 1 and 2, respectively, in the actual system. Noting Equation 6.92, Equation 6.91 can be written as

$$\frac{1}{2} I_{p2} \left\{ \frac{\omega_1}{n} + \frac{d}{dt}\left(\frac{nT_1}{k_{t2}}\right) \right\}^2 = \frac{1}{2} I_{p_e} \left\{ \omega_1 + \frac{d}{dt}\left(\frac{T_1}{k_{t_e}}\right) \right\}^2 \tag{6.93}$$

On substituting relations of angular velocities from Equation 6.89 in Equation 6.93, we get

$$\begin{aligned} \frac{1}{2} I_{p2} \left\{ \frac{\omega_1}{n} + \frac{d}{dt}\left(\frac{nT_1}{k_{t2}}\right) \right\}^2 &= \frac{1}{2} I_{p_e} \left\{ \omega_1 + \frac{d}{dt}\left(\frac{n^2 T_1}{k_{t2}}\right) \right\}^2 \Rightarrow \frac{I_{p2}}{n^2} \left\{ \omega_1 + \frac{d}{dt}\left(\frac{n^2 T_1}{k_{t2}}\right) \right\}^2 \\ &= I_{p_e} \left\{ \omega_1 + \frac{d}{dt}\left(\frac{n^2 T_1}{k_{t2}}\right) \right\}^2 \end{aligned} \tag{6.94}$$

which simplifies to

$$I_{p_e} = \frac{I_{p2}}{n^2} \tag{6.95}$$

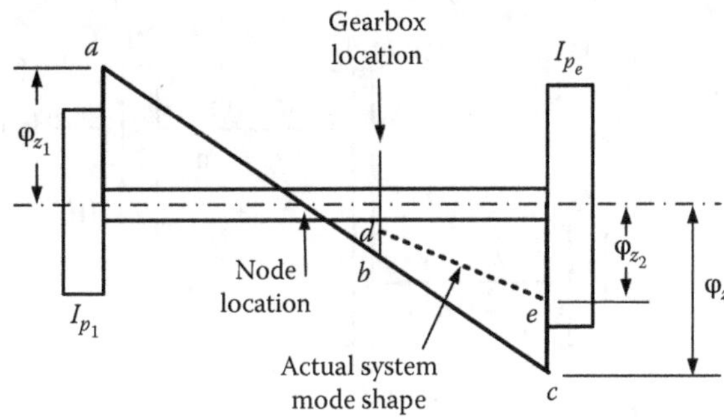

FIGURE 6.34 The elastic line in the equivalent and original systems.

where k_{te} and I_{p_e} are, respectively, the equivalent shaft stiffness and the equivalent polar mass moment of inertia of the geared system referred to as the "reference shaft" speed, i.e., shaft 1. The general rule, for forming the equivalent system for the purpose of analysis, is to divide all shaft stiffnesses and rotor polar mass moment of inertia of the gear system by n^2 (where n is the gear ratio). It can be seen that a gear system is reduced to a two-disk rotor system (Figure 6.34) for which we have already discussed the torsional free vibration analysis. When the analysis is completed, it should be remembered that the elastic line of the mode shape of the equivalent system (i.e., the line *abc* in Figure 6.34) is modified for the real system by dividing the displacement amplitudes of the equivalent shaft by the gear ratio n as shown in Figure 6.34 by the line *abde*. It should be noted that the angular displacements shown in Figure 6.34 are now that of disks 1 and 2 in the real system.

Now through numerical examples, the torsional analysis of the geared rotor system is illustrated.

Example 6.11

For a simple geared system as shown in Figure 6.35, find the torsional natural frequencies and mode shapes. Find also the location of the node point on the shaft (i.e., the location of the point where the angular twist during torsional vibrations is zero). Shaft "A" has a diameter of 5 cm and length 0.75 m, and shaft "B" has diameter 4 cm and length 1.0 m. Take the modulus of rigidity of the shaft *as* $G = 0.8 \times 10^{11}$ N/m², and the polar mass moments of inertia of disks as $I_{PA} = 24$ Nm^2 and $I_{PB} = 10$ Nm^2. Neglect the inertia of gears.

Solution:

On taking shaft "B" as the input shaft (or the reference shaft) as shown in Figure 6.36, the gear ratio can be defined as

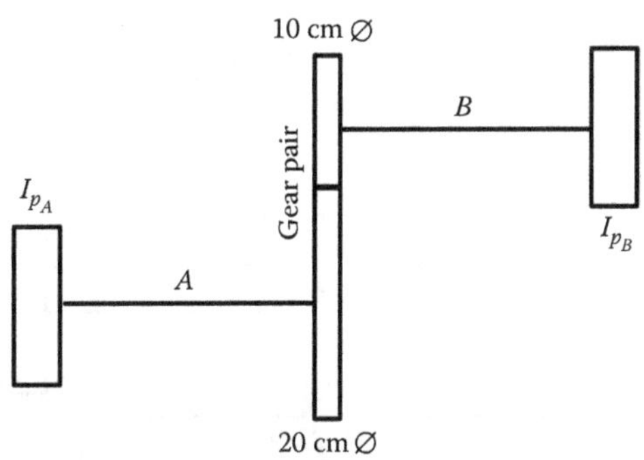

FIGURE 6.35 A two-disk geared rotor system.

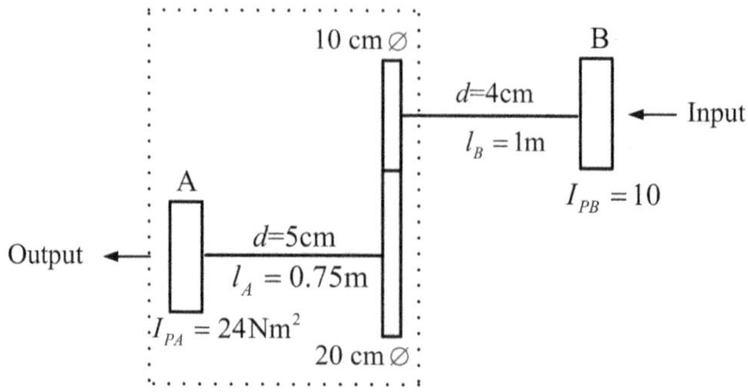

FIGURE 6.36 A geared rotor system.

$$\text{Gear ratio} = n = \frac{\omega_B}{\omega_A} = \frac{D_A}{D_B} = \frac{20}{10} = 2$$

where D is the nominal diameter of the gear. It should be noted that in case shaft "A" is input shaft, the gear ratio will change to ½ and all calculation of the present case needs to be redone accordingly, but the natural frequencies will remain the same.

The polar moment of inertia of the shaft cross-section and the torsional stiffness of the shaft can be obtained as

$$J_A = \frac{\pi}{32} d_A^4 = 6.136 \times 10^{-7}\ \text{m}^4; \quad J_B = \frac{\pi}{32} d_B^4 = 2.51 \times 10^{-7}\ \text{m}^4$$

and

$$k_{tA} = \frac{GJ_A}{l_A} = \frac{0.8 \times 10^{11} \times 6.136 \times 10^{-7}}{0.75} = 6.545 \times 10^4\ \text{Nm/rad}; \quad k_{tB} = 2.011 \times 10^4\ \text{Nm/rad}$$

On treating shaft B as a reference shaft and replacing an equivalent shaft system of shaft "A" (i.e., the same diameter as that of reference shaft "B"), the system becomes as shown in Figure 6.37. In the equivalent system, the shaft system "A" has the following torsional stiffness and polar mass moment of inertia properties:

$$k_{tA_e} = \frac{k_{tA}}{n^2} = \frac{6.545 \times 10^4}{2^2} = 1.6362 \times 10^4\ \text{Nm/rad} \quad \text{and} \quad I_{P_e} = \frac{I_{PA}}{n^2} = \frac{24}{2^2} = 6\ \text{Nm}^2$$

which gives the equivalent length of shaft "A" as (note that now its diameter is that of the reference shaft "B")

$$l_{A_e} = \frac{GJ_B}{k_{tA_e}} = \frac{0.8 \times 10^{11} \times 2.513 \times 10^{-7}}{1.6362 \times 10^4} = 1.229\ \text{m}$$

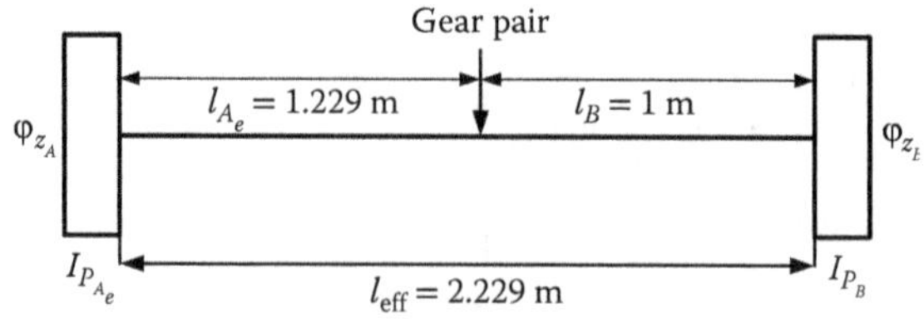

FIGURE 6.37 An equivalent single-shaft geared rotor system with shaft "B" as the reference shaft.

Hence, the equivalent full shaft length is given as $l_e = l_{A_e} + l_B = 1.229 + 1 = 2.229$ m.

The equivalent stiffness of the full shaft (shaft segments are connected in series) is given as (Figure 6.37):

$$\frac{1}{k_{te}} = \frac{1}{k_{tA_e}} + \frac{1}{k_{tB}} = \frac{1}{2.011\times10^4} + \frac{1}{1.6362\times10^4} = 1.1085\times10^{-4} \text{ rad/Nm}$$

which gives $k_{te} = 9021.2$ Nm/rad. The torsional natural frequency corresponding to the flexible mode of the equivalent two-disk rotor system, as shown in Figure 6.37, is given as

$$\omega_{nf2} = \sqrt{\frac{(I_{PA_e} + I_{PB})k_{te}}{(I_{PA_e} I_{PB})}} = \sqrt{\frac{(6+10)\times 9021.2/9.81}{(6\times10)/9.81^2}} = 153.62 \text{ rad/s}$$

The node location can be obtained from Figure 6.38 as

$$\frac{\Phi_{zAe}}{l_{n1}} = \frac{\Phi_{zB}}{l_{n2}}$$

which can be written by noting Equations 6.17 and 6.21, as

$$\frac{l_{n1}}{l_{n2}} = \frac{\Phi_{zAe}}{\Phi_{zB}} = -\frac{I_{PB}}{I_{PAe}} = -\frac{10}{6} = -1.667$$

The negative sign indicates that both disks are at either end of the node location. The absolute location of the node position is given as

$$l_{n_1} = 1.667 l_{n_2}$$

Also from Figure 6.38, we have

$$l_{n_1} + l_{n_2} = 2.2288 \quad \text{which gives} \quad l_{n_2} = 0.8358 \text{ m}$$

Hence, the node is on shaft "B" at 0.8356 m from disk B. Alternatively, from a similar triangle of the mode shape (Figure 6.38), we have

$$\frac{l_{n2}}{2.2288 - l_{n_2}} = \frac{\Phi_{zB}}{\Phi_{zA_e}} = \frac{1}{1.667} \Rightarrow l_{n_2} = 0.8358 \text{ m}$$

Let $\Phi_{zB} = 1$ rad be a reference value, then $\Phi_{zA_e} = -1.667$ rad, and hence we have

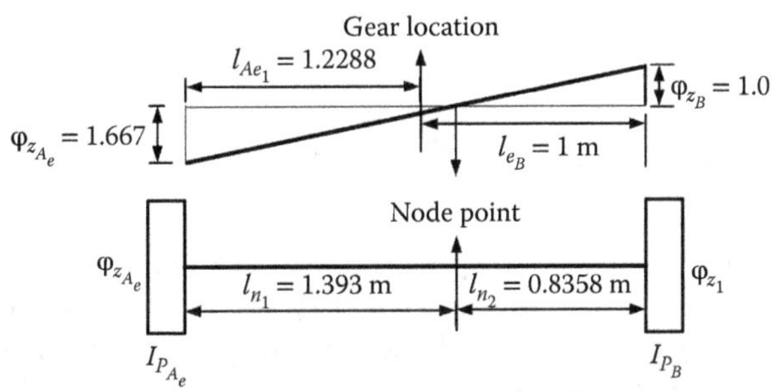

FIGURE 6.38 The mode shape and node point location in the equivalent system.

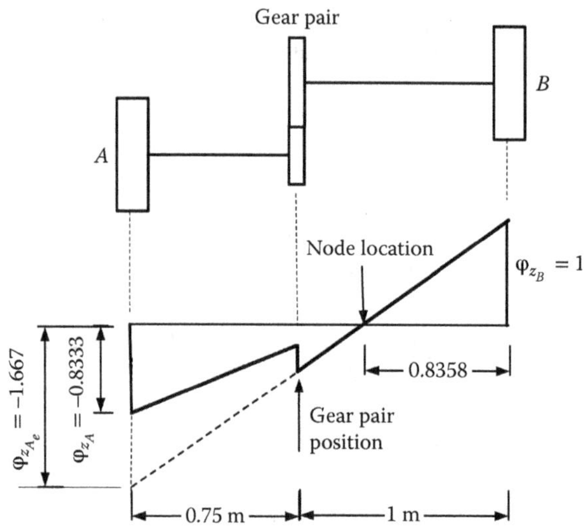

FIGURE 6.39 The mode shape and the node location in the actual geared system.

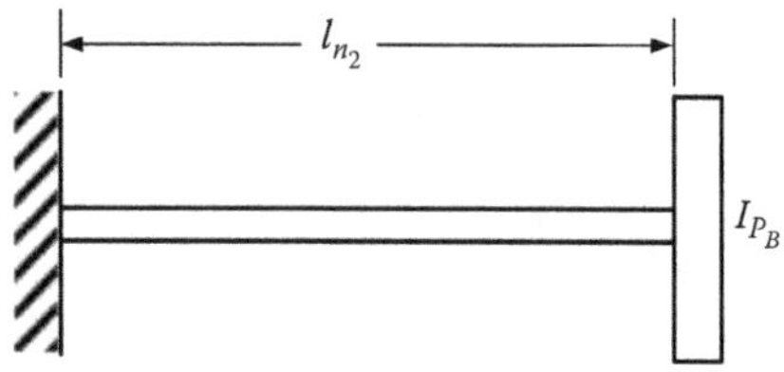

FIGURE 6.40 A single-DOF system.

$$\Phi_{z_A} = \frac{\Phi_{zA_e}}{n} = -0.8333 \text{ rad}$$

The mode shape and the node location in the actual system are shown in Figure 6.39. Using Figures 6.38 and 6.39 the gear and pinion angular displacement amplitudes can be obtained using similar triangles geometry concepts.

An alternative way to obtain the natural frequency is to consider the equivalent two-disk rotor system (Figure 6.37) as two single-DOF systems (one such system is shown in Figure 6.40). The stiffness and polar mass moment of inertia properties of the system are given as

$$k_{tl_{n2}} = \frac{GJ_B}{l_{n2}} = \frac{0.8\times10^{11}\times2.513}{0.8358} = 2.435\times10^4 \text{ Nm/rad} \quad \text{and} \quad I_{PB} = \frac{10}{9.81} \text{ kgm}^2$$

This gives the natural frequency as

$$\omega_{nf2} = \sqrt{\frac{k_{tl_{n2}}}{I_{PB}}} = \sqrt{\frac{2.435\times10^4\times9.81}{10}} = 153.62 \text{ rad/s}$$

which is the same as obtained earlier. The whole free vibration torsional analysis can be done by taking the speed of shaft "A" as the reference and converting shaft "B" by an equivalent system (Figure 6.41). For completeness, some of the basic steps are given as follows.

The gear ratio for the present case will be

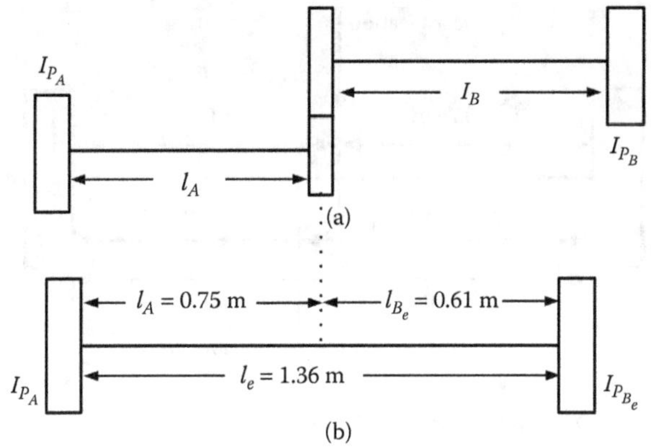

FIGURE 6.41 (a) Actual and (b) equivalent geared rotor systems.

$$n = \frac{\omega_A}{\omega_B} = \frac{D_B}{D_A} = \frac{10}{20} = 0.5$$

It is assumed that the equivalent shaft (i.e., shaft "B") has the same diameter as that of the reference shaft (i.e., shaft "A"). The equivalent polar mass moment of inertia and the torsional stiffness can be written as

$$I_{pB_e} = \frac{I_{pB}}{n^2} = 40 \text{ Nm}^2 \quad \text{and} \quad k_{tB_e} = \frac{k_{tB}}{n^2} = \frac{2.011\times 10^4}{(0.5)^2} = 8.044\times 10^4 \text{ Nm/rad}$$

which gives the equivalent length as

$$k_{tB_e} = \frac{GJ_A}{l_{B_e}} = 8.044\times 10^4 \text{ Nm/rad} \;\Rightarrow\; l_{B_e} = \frac{0.8\times 10^{11}\times 6.136\times 10^{-7}}{8.044\times 10^4} = 0.610 \text{ m}$$

The total equivalent length and the equivalent torsional stiffness would be

$$l_e = l_A + l_{B_e} = 0.75 + 0.61 = 1.36 \text{ m}$$

and

$$k_{t_e} = \frac{GJ_A}{l_e} = \frac{0.8\times 10^{11}\times 6.136\times 10^{-7}}{1.360} = 3.61\times 10^4 \text{ Nm/rad}$$

Alternatively, the effective stiffness can be obtained as

$$\frac{1}{k_{t_e}} = \frac{1}{k_{tA}} + \frac{1}{k_{tB_e}} \Rightarrow$$

$$k_{t_e} = \frac{k_{tA}k_{tB_e}}{k_{tA}+k_{tB_e}} = \frac{6.545\times 10^4\times 8.044\times 10^4}{6.545\times 10^4 + 8.044\times 10^4} = 3.61\times 10^4 \text{ Nm/rad}$$

The natural frequencies of the two-mass rotor system (refer to Figure 6.42) are given as

$$\omega_{nf_1} = 0 \quad \text{and} \quad \omega_{nf_2} = \sqrt{\frac{(I_{pA}+I_{pB_e})}{I_{pA}I_{pB_e}}k_{te}} = \sqrt{9.81\times\frac{(24+40)}{24\times 40}\times 3.61\times 10^4} = 153.62 \text{ rad/s}$$

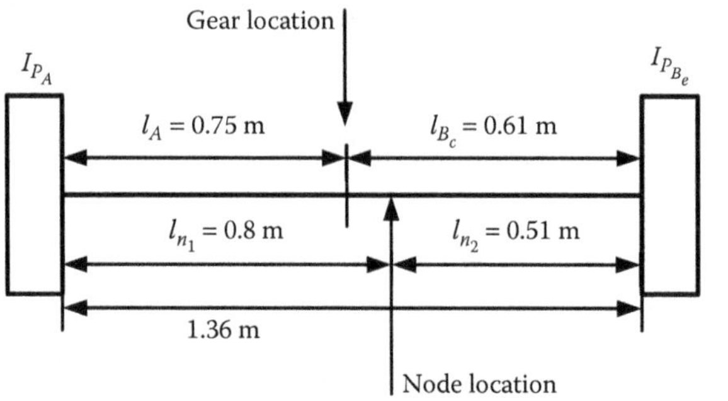

FIGURE 6.42 Equivalent two-mass rotor geared system.

A factor of 9.81 is used because I_{PA} is in Nm² and we need it in kg-m².
The node location can be obtained as

$$\frac{l_{n1}}{l_{n2}} = -\frac{I_{PB_e}}{I_{PA}} = -\frac{10}{6} = -1.667 \quad \text{and} \quad l_{n1} + l_{n2} = l_A + l_{B_e} = 1.36 \text{ m}$$

which gives

$$\left(1.667 l_{n1}\right) + l_{n2} = 1.36 \Rightarrow l_{n1} = 0.85 \text{ m} \quad \text{and} \quad l_{n2} = 0.51 \text{ m}$$

The stiffness of the shaft length equal to l_{n2} will be (equivalent stiffness corresponding to shaft "A" speed)

$$k_{tl_{n2}} = \frac{GJ_A}{l_{n2}}$$

The shaft stiffness corresponding to shaft "B" speed can be defined in two ways, i.e.,

$$k_{tB_2} = \frac{GJ_B}{l_2} \quad \text{and} \quad k_{tB_2} = n^2 k_{tl_{n2}} = n^2 \frac{GJ_A}{l_{n2}}$$

On equating the above equations, the location of the node in the actual system from disk B can be obtained as

$$l_2 = n^2 l_{n2} \frac{J_B}{J_A} = 0.84 \text{ m}$$

which is the same as that obtained using the previous method.

6.7 TMM FOR BRANCHED GEARED ROTOR SYSTEMS

For rolling mills, textile machineries, marine vessel power transmission shafts, and machine tool drives, there may be many rotor inertias in the system and the gearbox may be a branch point where more than two shafts are attached. In such cases, where there are more than two shafts attached as shown in Figure 6.43 to the gearbox, the system is said to be *branched*. It has three branches A, B, and C and each branch has multiple disks, e.g., $(p + 1)$, $(q + 1)$, and $(r + 1)$ number of disks (including gears) in branches A, B, and C, respectively. Such a system cannot be converted to a single-shaft system as we could do to the two shaft geared system discussed in the previous section, because now the system contains several disks; hence, it is a multi-DOF system. Thus, the analysis of the branched system is now done by a more general procedure, i.e., the TMM.

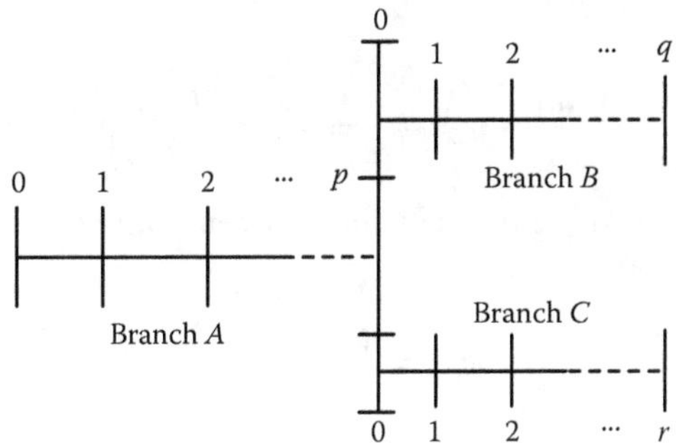

FIGURE 6.43 A branched multi-DOF rotor system.

For the branched system shown in Figure 6.43, state vectors for different branches can be written as

$$ {}_R\mathbf{S}_{pA} = \mathbf{A}\mathbf{S}_{0A};\quad {}_R\mathbf{S}_{qB} = \mathbf{B}\ \mathbf{S}_{0B};\quad {}_R\mathbf{S}_{rC} = \mathbf{C}\ \mathbf{S}_{0C} \tag{6.96} $$

where **A**, **B**, and **C** are the overall transfer matrices for branches A, B, and C, respectively. Now boundary conditions of each branch and at the junction point will be used to connect these individual branch equations to get the frequency equation.

Branch A: For branch A, we take $\varphi_{z0_A} = 1$ as the reference value for the angular displacement, and since the left-hand end of branch A is the free end, for free vibrations, we have $T_{0A} = 0$. It should be pointed out that it is not mandatory to take some reference value for φ_{z0_A}, one can use a variable form. This will finally lead to a homogeneous equation in terms of φ_{z0_A}. The equation for branch A takes the following form:

$$ \begin{Bmatrix} \varphi_z \\ T \end{Bmatrix}_{pA} = \begin{bmatrix} a_{11} & a_{12} \\ a_{21} & a_{22} \end{bmatrix} \begin{Bmatrix} 1 \\ 0 \end{Bmatrix} \tag{6.97} $$

which can be expanded as

$$ T_{z_{pA}} = a_{21} \quad \text{and} \quad \varphi_{z_{pA}} = a_{11} \tag{6.98} $$

Branch B: At the branch point, between shafts A and B, we have

$$ \varphi_{z0_B} = \frac{\varphi_{z_{pA}}}{n_{AB}} = \frac{a_{11}}{n_{AB}} \tag{6.99} $$

where n_{AB} is the gear ratio between shafts A and B. For branch B, $T_{qB} = 0$ because the right-hand end of the branch is free. For branch B from Equation 6.96, and noting the condition described by Equation 6.99, we have

$$ \begin{Bmatrix} \varphi_z \\ 0 \end{Bmatrix}_{qB} = \begin{bmatrix} b_{11} & b_{12} \\ b_{21} & b_{22} \end{bmatrix} \begin{Bmatrix} a_{11}/n_{AB} \\ T \end{Bmatrix}_{0B} \tag{6.100} $$

Equation 6.100 can be expanded as

$$ \varphi_{z_{qB}} = b_{11}\frac{a_{11}}{n_{AB}} + b_{12}T_{0B} \tag{6.101} $$

and

$$0 = b_{21}\frac{a_{11}}{n_{AB}} + b_{22}T_{0B} \quad \Rightarrow T_{0B} = -\left(\frac{b_{21}}{b_{22}}\right)\left(\frac{a_{11}}{n_{AB}}\right) \tag{6.102}$$

Branch C and junction point: At the branch point between shafts A and C, we have the following condition (noting Equation 6.98):

$$\varphi_{z0C} = \frac{\varphi_{zpA}}{n_{AC}} = \frac{a_{11}}{n_{AC}} \tag{6.103}$$

where n_{AC} is the gear ratio between shafts A and C. Another condition at the branch (among shafts A, B, and C) to be satisfied regarding the work done by the torques transmitted at various branches (assuming negligible friction during transmissions) is given as

$$\frac{1}{2}T_{pA}\varphi_{znA} = \frac{1}{2}T_{0B}\varphi_{z0B} + \frac{1}{2}T_{0C}\varphi_{z0C} \Rightarrow T_{pA} = \frac{T_{0B}}{\varphi_{zpA}/\varphi_{z0B}} + \frac{T_{0C}}{\varphi_{zpA}/\varphi_{z0C}} \Rightarrow T_{pA} = \frac{T_{0B}}{n_{AB}} + \frac{T_{0C}}{n_{AC}} \tag{6.104}$$

On substituting Equation 6.102 into Equation 6.104, we can write

$$T_{0C} = n_{AC}\left[T_{pA} - \frac{T_{0B}}{n_{AB}}\right] = n_{AC}\left[T_{nA} + \frac{b_{21}}{b_{22}}\frac{a_{11}}{n_{AB}^2}\right]$$

On substituting Equation 6.98 in the above equation, we get

$$T_{0C} = n_{AC}\left[a_{21} + \frac{b_{21}}{b_{22}}\frac{a_{11}}{n_{AB}^2}\right] \tag{6.105}$$

Substituting Equations 6.104 and 6.105 into Equation 6.96 for branch C, we get

$$\begin{Bmatrix} \varphi_z \\ 0 \end{Bmatrix}_{rc} = \begin{bmatrix} c_{11} & c_{12} \\ c_{21} & c_{22} \end{bmatrix} \begin{Bmatrix} a_{11}/n_{AC} \\ n_{AC}a_{21} + \dfrac{b_{21}a_{11}n_{AC}}{b_{22}n_{AB}^2} \end{Bmatrix} \tag{6.106}$$

where $T_{rc} = 0$ is the boundary condition describing the free right-hand end of branch C.

The frequency equation: From Equation 6.106, the second equation will give the frequency equation as

$$a_{11}b_{22}c_{21}n_{AB}^2 + a_{11}b_{21}c_{22}n_{AC}^2 + a_{21}b_{22}c_{22}n_{AB}^2n_{AC}^2 = 0 \tag{6.107}$$

where the a's, b's, and c's are a function of the natural frequency, ω_{nf}. The roots of the above equation are system torsional natural frequencies. We expect $(p+q+r+1)$ to be such natural frequencies.

Mode shapes: Angular displacements at the beginning and end of various branches can be summarized as

$$\varphi_{z0A} = 1; \quad \varphi_{zpA} = a_{11} \tag{6.108}$$

$$\varphi_{z0B} = \frac{a_{11}}{n_{AB}}; \quad \varphi_{zqB} = \left(b_{11} - \frac{b_{12}b_{21}}{b_{22}}\right)\frac{a_{11}}{n_{AB}} \tag{6.109}$$

$$\varphi_{z0C}=\frac{a_{11}}{n_{AC}};\quad \varphi_{zrC}=\left(c_{11}-\frac{c_{12}c_{21}}{c_{22}}\right)\frac{a_{11}}{n_{AC}} \tag{6.110}$$

On substituting one of the values of the torsional natural frequencies obtained from Equation 6.107 into Equations 6.108 through 6.110, angular displacements at the beginning and end of various branches can be obtained. Then, these may be substituted back into transfer matrices for each branch considered (i.e., Equation 6.96), whereupon the state vector at each station may be evaluated. The plot of angular displacements against shaft positions then indicates the system mode shapes corresponding to the chosen natural frequency. For other natural frequencies, similar steps have to be performed to get the corresponding mode shapes.

Using this method, there will not be any change in the elastic line (mode shape) due to the gear ratio because these have now already been allowed for in the analysis. Moreover, for the present case, we have not gone for the equivalent system at all. For the case when the system can be converted to an equivalent single shaft, the equivalent system approach has the advantage. It should be noted that for the present case, the DOF of the rotor system would be $(p+q+r+1)$. The total number of disks (including gears) is $(p+q+r+3)$; however, at the junction, the DOF of two gears (e.g., at the 0th station of branches B and C) is related to the third (at the pth station of branch A); hence, we would have two DOF less compared with the number of disks in the system. Now, through numerical examples, the procedure will be illustrated.

Example 6.12

For a geared system as shown in Figure 6.44, find the torsional natural frequencies. Shaft "A" has 5 cm diameter and 0.75 m length, and shaft "B" has 4 cm diameter and 1.0 m length. Take the modulus of rigidity of the shaft G equal to 0.8×10^{11} N/m². The polar mass moments of inertia of the disks and gears are $I_{p_A}=24$ N-m², $I_{p_B}=10$ N-m², $I_{p_{gA}}=5$ N-m², and $I_{p_{gB}}=3$ N-m².

Solution:

This problem can now be solved using the TMM for the illustration of the method to a geared rotor system. The pinion and gear have appreciable polar mass moments of inertia. Let us denote the station number of the disk on branch A as 0, the pinion as 1, the station number of the disk on branch B as 2, and the gear as 3 (Figure 6.44).

The state transformation equation for branch A can be written as

$$_R\mathbf{S}_1=\mathbf{A}_L\mathbf{S}_0 \tag{a}$$

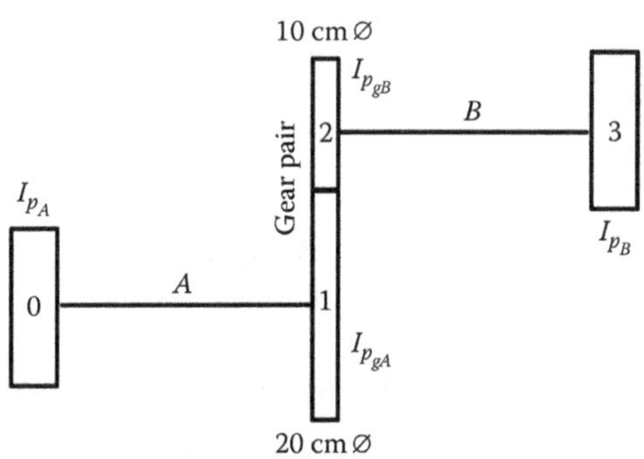

FIGURE 6.44 Two disks with a geared system.

with

$$\mathbf{A}=\mathbf{P}_1\mathbf{F}_1\mathbf{P}_0=\begin{bmatrix}1 & 1/k_{t_A}\\ -\omega_{nf}^2 I_{pgA} & 1-\dfrac{\omega_{nf}^2 I_{pgA}}{k_{t_A}}\end{bmatrix}_1\begin{bmatrix}1 & 0\\ -\omega_{nf}^2 I_{pA} & 1\end{bmatrix}_0$$

$$=\begin{bmatrix}\left(1-\dfrac{\omega_{nf}^2 I_{pA}}{k_{t_A}}\right) & \dfrac{1}{k_{t_A}}\\ \left\{\left(-\omega_{nf}^2 I_{pgA}\right)+\left(-\omega_{nf}^2 I_{pA}\right)\left(1-\dfrac{\omega_{nf}^2 I_{pgA}}{k_{t_A}}\right)\right\} & \left(1-\dfrac{\omega_{nf}^2 I_{pgA}}{k_{t_A}}\right)\end{bmatrix} \tag{b}$$

where k_{t_A} is the stiffness of shaft A. Similarly, the state transformation equation for branch B can be written as

$$_R\mathbf{S}_3=\mathbf{B}\,_L\mathbf{S}_2 \tag{c}$$

with

$$\mathbf{B}=\mathbf{P}_3\mathbf{F}_3\mathbf{P}_2=\begin{bmatrix}\left(1-\dfrac{\omega_{nf}^2 I_{pgB}}{k_{t_B}}\right) & \dfrac{1}{k_{t_B}}\\ \left\{\left(-\omega_{nf}^2 I_{pB}\right)+\left(-\omega_{nf}^2 I_{pgB}\right)\left(1-\dfrac{\omega_{nf}^2 I_{pB}}{k_{t_B}}\right)\right\} & \left(1-\dfrac{\omega_{nf}^2 I_{pB}}{k_{t_B}}\right)\end{bmatrix} \tag{d}$$

where k_{t_B} is the stiffness of shaft B. At the gear pair, the following conditions hold:

$$\varphi_{x2}=-\frac{\varphi_{x1}}{n}\quad\text{and}\quad T_2=-nT_1 \tag{e}$$

where φ_x is the angular displacement. Equation (e) can be combined as

$$_L\begin{Bmatrix}\varphi_x\\ T\end{Bmatrix}_2=-\begin{bmatrix}1/n & 0\\ 0 & n\end{bmatrix}\,_R\begin{Bmatrix}\varphi_x\\ T\end{Bmatrix}_1 \tag{f}$$

or

$$_L\mathbf{S}_2=\mathbf{n}\,_R\mathbf{S}_1 \tag{g}$$

with *the gear ratio transformation matrix* is given as

$$\mathbf{n}=-\begin{bmatrix}1/n & 0\\ 0 & n\end{bmatrix} \tag{h}$$

The negative sign can be ignored if the direction of rotation and torque is shown in the figure itself. Noting Equation (g), Equation (c) can be written as

$$_R\mathbf{S}_3=\mathbf{B}\mathbf{n}\,_R\mathbf{S}_1 \tag{i}$$

Noting Equation (a), Equation (i) can be expressed as

$$_R\mathbf{S}_3=\mathbf{B}\mathbf{n}\mathbf{A}\,_L\mathbf{S}_0=\mathbf{T}\,_L\mathbf{S}_0 \tag{j}$$

with the overall transformation matrix given as

$$\mathbf{T} = \mathbf{BnA} = \begin{bmatrix} b_{11} & b_{12} \\ b_{21} & b_{22} \end{bmatrix} \begin{bmatrix} 1/n & 0 \\ 0 & n \end{bmatrix} \begin{bmatrix} a_{11} & a_{12} \\ a_{21} & a_{22} \end{bmatrix}$$

$$= \begin{bmatrix} \dfrac{a_{11}b_{11}}{n} + na_{21}b_{12} & \dfrac{a_{12}b_{11}}{n} + na_{22}b_{12} \\ \dfrac{a_{11}b_{21}}{n} + na_{21}b_{22} & \dfrac{a_{12}b_{21}}{n} + na_{22}b_{22} \end{bmatrix} \tag{k}$$

The overall transformation matrix can also be written as

$$\mathbf{T} = \mathbf{BnA} = \mathbf{P}_3\mathbf{F}_3\mathbf{P}_2\mathbf{n}\mathbf{P}_1\mathbf{F}_1\mathbf{P}_0 \tag{l}$$

The boundary conditions of the problem are $T_0 = T_3 = 0$, since both ends are free. Equation (j) can be written in expanded form, as

$$_R\begin{Bmatrix} \varphi_x \\ T \end{Bmatrix}_3 = \begin{bmatrix} t_{11} & t_{12} \\ t_{21} & t_{22} \end{bmatrix} {}_L\begin{Bmatrix} \varphi_x \\ T \end{Bmatrix}_0 \quad \text{or} \quad {}_R\begin{Bmatrix} \varphi_x \\ 0 \end{Bmatrix}_0 = \begin{bmatrix} t_{11} & t_{12} \\ t_{21} & t_{22} \end{bmatrix} {}_L\begin{Bmatrix} \varphi_x \\ 0 \end{Bmatrix}_0 \tag{m}$$

which gives the frequency equation as

$$t_{21}\left(\omega_{nf}\right) = 0 \tag{n}$$

From Equation (k), the frequency equation comes out to be

$$a_{11}b_{21} + n^2 a_{21}b_{22} = 0 \tag{o}$$

Noting Equations (b) and (d), in the expanded form, Equation (o) can be written

$$\left(1 - \frac{\omega_{nf}^2 I_{P_A}}{k_{t_A}}\right)\left\{\left(-\omega_{nf}^2 I_{P_B}\right) + \left(-\omega_{nf}^2 I_{P_B}\right)\left(1 - \frac{\omega_{nf}^2 I_{P_B}}{k_{t_B}}\right)\right\}$$
$$+ n^2\left\{\left(-\omega_{nf}^2 I_{P_{gA}}\right) + \left(-\omega_{nf}^2 I_{P_A}\right)\left(1 - \frac{\omega_{nf}^2 I_{P_{gA}}}{k_{t_A}}\right)\right\}\left(1 - \frac{\omega_{nf}^2 I_{P_{gB}}}{k_{t_B}}\right) = 0 \tag{p}$$

For the numerical values of the present problem (please note that mass moments of inertia are taken in kg-m^2), Equation (p) reduces to

$$\omega_{nf}^2\left\{\omega_{nf}^4 - \left(1.68 \times 10^5\right)\omega_{nf}^2 + \left(3.003 \times 10^9\right)\right\} = 0 \tag{q}$$

From frequency Equation (q), the following natural frequencies are obtained:

$$\omega_{nf_1} = 0; \quad \omega_{nf_2} = 142.12 \text{ rad/s}; \quad \omega_{nf_3} = 382.78 \text{ rad/s}$$

In the present case, it should be noted that for the four-disk (polar mass moment of inertia) rotor system, only three natural frequencies are obtained. This is due to the fact that the gear pair is treated as a single polar mass moment of inertia due to the gear ratio. So effectively for the present problem, only three generalized coordinates are sufficient to describe the motion.

For comparison with Example 6.11, let us put the polar mass moment of inertia of the pinion and the gear to zero in Equation (p), then we get

$$\omega_{nf}^2\left\{I_{PA}I_{PB}\left(n^2k_{t_A}+k_{t_B}\right)\omega_{nf}^2-\left(n^2I_{PA}+I_{PB}\right)k_{t_A}k_{t_B}\right\}=0$$

which gives

$$\omega_{nf_1}=0$$

and

$$\omega_{nf_2}^2=\frac{\left(I_{PA}+I_{PB}/n^2\right)\left(n^2k_{t_A}k_{t_B}\right)}{\left\{I_{PA}\left(I_{PB}/n^2\right)\right\}\left(n^2k_{t_A}+k_{t_B}\right)}=\frac{\left(I_{PA}+I_{PB_e}\right)\left\{n^2k_{t_A}k_{t_B}/\left(n^2k_{t_A}+k_{t_B}\right)\right\}}{\left(I_{PA}I_{PB_e}\right)}$$

The second natural frequency can be simplified as

$$\omega_{nf_2}=\sqrt{\frac{\left(I_{PA}+I_{PB_e}\right)k_{t_e}}{\left(I_{PA}I_{PB_e}\right)}} \tag{q}$$

with

$$k_t=n^2k_{t_A}k_{t_B}/\left(n^2k_{t_A}+k_{t_B}\right);\quad I_{PB_e}=I_{PB}/n^2$$

It should be noted that Equation (q) is exactly the same as in the previous example (i.e., for an equivalent two-mass rotor system) for which $\omega_{nf_2}=153.62$ rad/s. For the present example problem, natural frequencies are expected to be reduced because of the increase in the polar mass moment of inertia of the system; moreover, there is an increase in DOFs of the system to three from the two of Example 6.11.

Example 6.13

Obtain torsional natural frequencies and mode shapes of a branched system as shown in Figure 6.45. The polar mass moments of inertia of rotors are $I_{PA}=0.01$ kg-m², $I_{PE}=0.005$ kg-m², $I_{PF}=0.006$ kg-m², and $I_{PB}=I_{PC}=I_{PD}=0$. The gear ratios are $n_{BC}=3$ and $n_{BD}=4$. The shaft lengths are $l_{AB}=l_{CE}=l_{DF}=0.25$ m, and the diameters are $d_{AB}=0.03$ m, $d_{CE}=0.02$ m, and $d_{DF}=0.02$ m. Take the modulus of rigidity of the shaft as $G=0.8\times10^{11}$ N/m².

Solution:

The branched system has the following numerical data:

$$I_{PA}=0.01\text{ kg-m}^2;\quad I_{PE}=0.005\text{ kg-m}^2;\quad I_{PF}=0.006\text{ kg-m}^2$$

$$J_{AB}=\frac{\pi}{32}d_{AB}^4=\frac{\pi}{32}0.03^4=7.95\times10^{-8}\text{ m}^4;\quad J_{CE}=J_{DF}=\frac{\pi}{32}0.02^4=1.57\times10^{-8}\text{ m}^4$$

$$k_{t_{AB}}=\frac{GJ_{AB}}{l_{AB}}=2.55\times10^4\text{ N-m/rad};\quad k_{t_{CE}}=k_{t_{DF}}=0.50\times10^4\text{ N-m/rad}$$

For branch *AB*, state vectors at stations are related as

$${}_R\mathbf{S}_{n_{AB}}=\mathbf{A}\mathbf{S}_{0_{AB}}$$

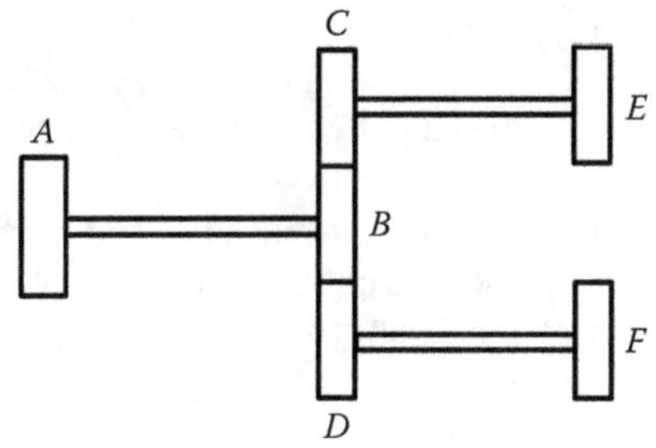

FIGURE 6.45 A branched rotor system.

with

$$\mathbf{A} = \mathbf{F}_{AB}\mathbf{P}_A = \begin{bmatrix} 1 & 3.93\times10^{-5} \\ 0 & 1 \end{bmatrix}\begin{bmatrix} 1 & 0 \\ -0.01\omega_{nf}^2 & 1 \end{bmatrix} = \begin{bmatrix} 1-3.93\times10^{-7}\omega_{nf}^2 & 3.93\times10^{-5} \\ -0.01\omega_{nf}^2 & 1 \end{bmatrix}$$

For branch *CE*, state vectors at stations are related as

$$_R\mathbf{S}_{nCE} = \mathbf{C}\mathbf{S}_{0CE}$$

with

$$\mathbf{C} = \mathbf{P}_E\mathbf{F}_{CE} = \begin{bmatrix} 1 & 0 \\ -0.005\omega_{nf}^2 & 1 \end{bmatrix}\begin{bmatrix} 1 & 2.0\times10^{-4} \\ 0 & 1 \end{bmatrix} = \begin{bmatrix} 1 & 2.0\times10^{-4} \\ -0.005\omega_{nf}^2 & 1-1.0\times10^{-6}\omega_{nf}^2 \end{bmatrix}$$

Similarly, for branch *DF*, we have

$$_R\mathbf{S}_{nDF} = \mathbf{D}\mathbf{S}_{0DF}$$

with

$$\mathbf{D} = \mathbf{P}_F\mathbf{F}_{DF} = \begin{bmatrix} 1 & 0 \\ -0.006\omega_{nf}^2 & 1 \end{bmatrix}\begin{bmatrix} 1 & 2.0\times10^{-4} \\ 0 & 1 \end{bmatrix} = \begin{bmatrix} 1 & 2.0\times10^{-4} \\ -0.006\omega_{nf}^2 & 1-1.2\times10^{-6}\omega_{nf}^2 \end{bmatrix}$$

From Equation 6.107, the frequency equation can be written as

$$a_{11}c_{22}d_{21}n_{BC}^2 + a_{11}c_{21}d_{22}n_{BD}^2 + a_{21}c_{22}d_{22}n_{BC}^2n_{BD}^2 = 0$$

On substitution various terms, we get

$$\begin{aligned}&\left(1-3.93\times10^{-7}\omega_{nf}^2\right)\left(1-1.0\times10^{-6}\omega_{nf}^2\right)\left(-0.006\omega_{nf}^2\right)\times9\\&+\left(1-3.93\times10^{-7}\omega_{nf}^2\right)\left(-0.005\omega_{nf}^2\right)\left(1-1.2\times10^{-6}\omega_{nf}^2\right)\times16\\&+\left(-0.01\omega_{nf}^2\right)\left(1-1.0\times10^{-6}\omega_{nf}^2\right)\left(1-1.2\times10^{-6}\omega_{nf}^2\right)\times9\times16=0\end{aligned}$$

which can be simplified to

$$\omega_{nf}^2\left(1.7685\times10^{-12}\omega_{nf}^4 - 3.3532\times10^{-10}\omega_{nf}^2 + 1.5740\right) = 0$$

from which natural frequencies are given as

$$\omega_{nf_1} = 0; \quad \omega_{nf_2} = 924.4 \text{ rad/s}, \quad \text{and} \quad \omega_{nf_3} = 1020.6 \text{ rad/s}$$

It can be seen that the rigid body mode exists because the ends of the gear train are free. Mode shapes for each of these natural frequencies can be obtained as follows.

For $\omega_{nf_1} = 0$ with $\varphi_{z0AB} = 1$ as a reference value, angular displacements at various disk locations can be written as (refer to Section 6.7)

$$\varphi_{z0AB} = 1; \quad \varphi_{znAB} = a_{11} = 1$$

$$\varphi_{z0CE} = -\frac{a_{11}}{n_{BC}} = -\frac{1}{3} = -0.33; \quad \varphi_{znCE} = \left(c_{11} - \frac{c_{12}c_{21}}{c_{22}}\right)\varphi_{z0CE} = -0.33$$

$$\varphi_{z0DF} = -\frac{a_{11}}{n_{BD}} = -\frac{1}{4} = -0.25; \quad \varphi_{znDF} = \left(d_{11} - \frac{d_{12}d_{21}}{d_{22}}\right)\varphi_{z0DF} = -0.25$$

Figure 6.46 shows the mode shape. Similarly, for other natural frequencies, displacements can be obtained to get mode shapes like those in Figures 6.47 and 6.48.

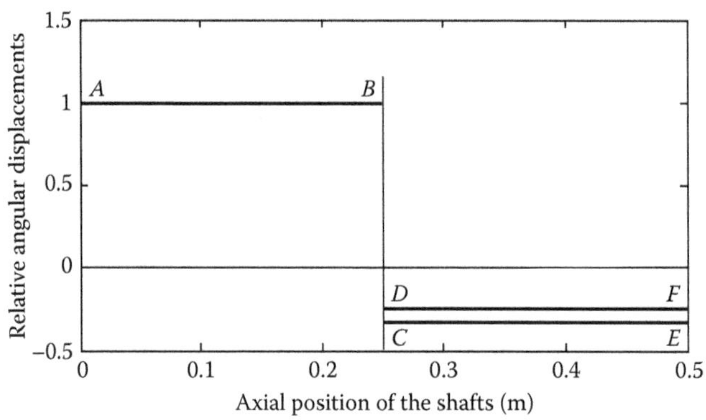

FIGURE 6.46 Mode shape of the branched system for $\omega_{nf_1} = 0$.

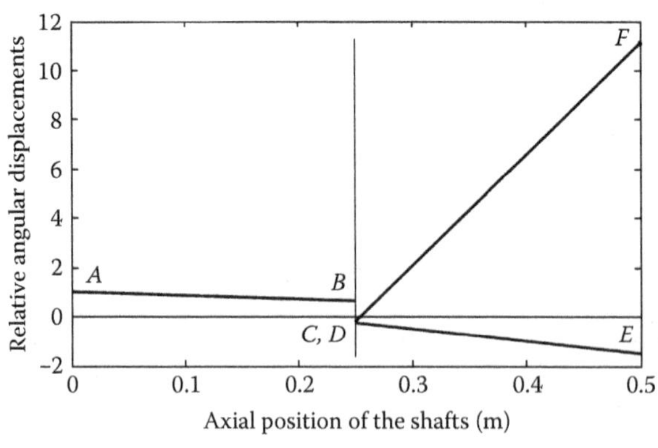

FIGURE 6.47 Mode shape of the branched system for $\omega_{nf_2} = 924.4$ rad/s.

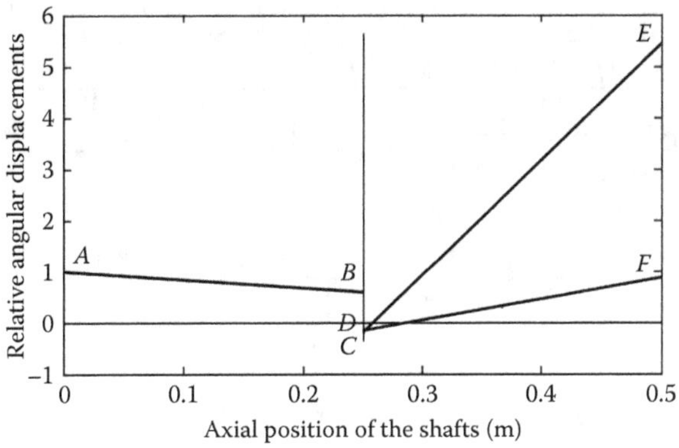

FIGURE 6.48 Mode shape of the branched system for $\omega_{nf_3} = 1020.6$ rad/s.

6.8 TMM FOR DAMPED TORSIONAL VIBRATIONS

In any real rotor systems, damping is always present. The torsional damping comes from several sources, e.g., the shaft material, bearings, couplings, torsional vibration dampers, aerodynamic damping at disks, rubbing of the rotor with the stator, loose components mounted on the shaft, etc. The shaft material or hysteretic damping comes due to intermolecular interaction in the shaft material, which results in an increase in the temperature of the shaft material. Couplings provide a high level of torsional damping. Working fluid provides torsional damping, for example, by water in the impeller of centrifugal pumps, by steam in the blades of turbines, etc. The torsional vibration damper is an external device that may be used between two shaft sections as shown in Figure 6.49. It develops a damping torque, which depends on of the angular velocity of one shaft relative to the other. These types of damping can be considered proportional to the relative angular velocity at either end of the torsional damper to which the shaft is connected and it is represented as c_s. The disk aerodynamic (or rubbing) damping, c_d, comes due to interaction of the disk with the working fluid (like the steam, the gas, the air, etc.), the lubricant, and the coolant, which results in dissipation of the energy in the form of heat. This type of damping is proportional to the angular velocity of the disk itself.

Torsional dampers can be used as a means of attenuating (decreasing) system vibrations and to tune system resonant frequencies to suit particular operating conditions. The damping in the system introduces phase lag between the system displacement and torque.

Point matrix: Here, the aim would be to relate the state vector at the RHS of the disk with the state vector at the LHS of it. Figure 6.50 shows a general arrangement of a torsional multi-DOF rotor system with the disk and the shaft damping. From the free body diagram of the nth disk (Figure 6.51), the following governing equations can be written as

$$ {}_L\varphi_{z_n} = {}_R\varphi_{z_n} \tag{6.111} $$

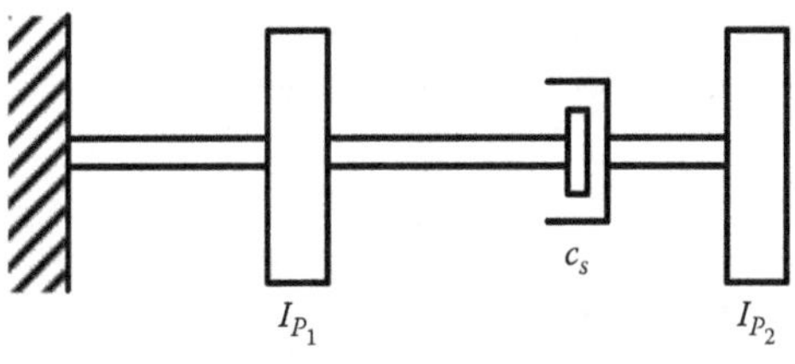

FIGURE 6.49 A schematic of a torsional vibration damper.

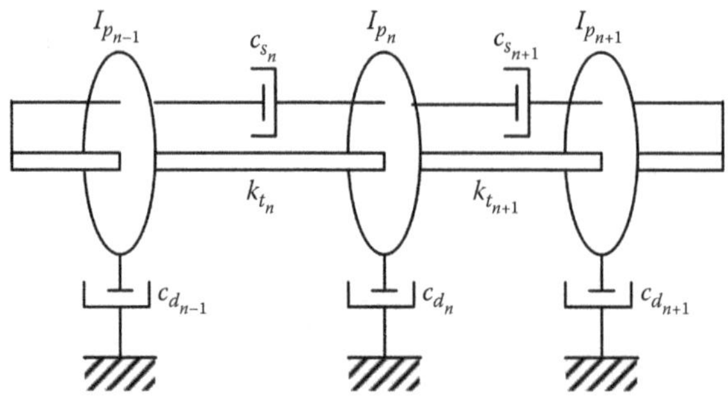

FIGURE 6.50 A general arrangement of a multi-DOF rotor system with torsional dampers.

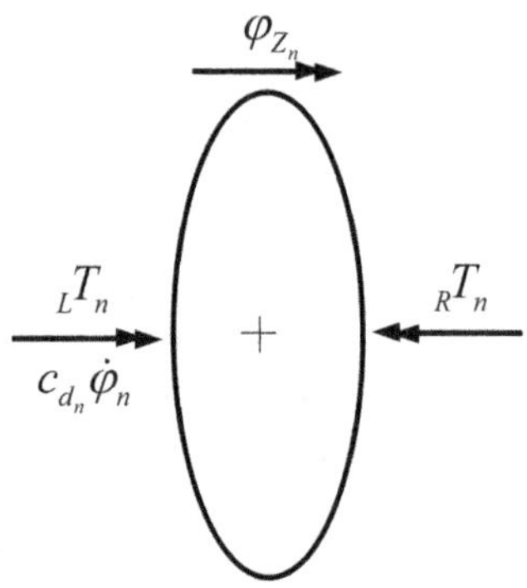

FIGURE 6.51 A free body diagram of the nth rotor

and

$$ {}_RT_n - {}_LT_n - c_{d_n}\, {}_L\dot{\varphi}_{z_n} = I_{p_n}\, {}_R\ddot{\varphi}_{z_n} \tag{6.112}$$

For the free vibration, torques ${}_RT_n$ and ${}_LT_n$ may be written in the form

$$ T_n = \overline{T}_n e^{j\omega_{nf}t} \tag{6.113}$$

where $\overline{T}_n$ is the complex amplitude of the torque at the nth disk (which contains both the amplitude and phase information), and ω_{nf} is the torsional natural frequency of the system. The angular displacement takes the form

$$ \varphi_{z_n} = \Phi_{z_n} e^{j\omega_{nf}t} \tag{6.114}$$

where Φ_{z_n} is the complex amplitude of angular displacement at the nth disk (which also contains both the amplitude and phase information). Differentiating Equations 6.113 and 6.114 with respect to time and substituting in Equations 6.111 and 6.112 leads to

$$ {}_R\Phi_{z_n} = {}_L\Phi_{z_n} $$

and

$$ {}_R\overline{T}_n = \left(-\omega_{nf}^2 I_p + j\omega_{nf} c_{d_n}\right)_L \Phi_{z_n} + {}_R\overline{T}_n + {}_L\overline{T}_n $$

or in the matrix form,

$$ {}_R\begin{Bmatrix} \Phi \\ \bar{T} \end{Bmatrix}_n = \begin{bmatrix} 1 & 0 \\ \left(-\omega_{nf}^2 I_p + j\omega_{nf} c_d\right) & 1 \end{bmatrix}_L \begin{Bmatrix} \Phi \\ \bar{T} \end{Bmatrix}_n \tag{6.115} $$

which can be simplified as

$$ {}_R\mathbf{S}_n = \mathbf{P}_n \, {}_L\mathbf{S}_n \tag{6.116} $$

with

$$ \mathbf{P}_n = \begin{bmatrix} 1 & 0 \\ -\omega_{nf}^2 I_p + j\omega_{nf} c_d & 1 \end{bmatrix}_n; \quad {}_R\mathbf{S}_n = {}_R\begin{Bmatrix} \Phi \\ \bar{T} \end{Bmatrix}_n; \quad {}_L\mathbf{S}_n = {}_L\begin{Bmatrix} \Phi \\ \bar{T} \end{Bmatrix}_n $$

where $\mathbf{P}_n$ is a *point matrix* and $\mathbf{S}_n$ is a state vector. It should be noted that now the point matrix is complex in nature due to the damping term. In fact, these can be written in terms of real quantities by separating the real and imaginary parts of both sides of Equations 6.115 and reassembling them in a point matrix of size 4×4. We need to express the torsional displacement and torque as $\Phi = \Phi^r + j\Phi^i$ and $T = T^r + jT^i$.

Field matrix: Here the aim would be to relate the state vector at the RHS of a shaft segment to the state vector at the LHS of it. The characteristics of the shaft segment at station n (Figure 6.52) are represented in the equation describing the torque applied to the shaft at the location of rotor n, as

$$ {}_L T_n = k_{tn}\left({}_L\varphi_{z_n} - {}_R\varphi_{z_{n-1}}\right) + c_{s_n}\left({}_L\dot{\varphi}_{z_n} - {}_R\dot{\varphi}_{z_{n-1}}\right) \tag{6.117} $$

The torque transmitted through the shaft is the same at both ends, i.e.,

$$ {}_L T_n = {}_R T_{n-1} \tag{6.118} $$

Substituting Equations 6.113 and 6.114 into Equations 6.117 and 6.118, we get

$$ {}_R\bar{T}_{n-1} = k_{tn}\left({}_L\Phi_{zn} - {}_R\Phi_{zn-1}\right) + j\omega_{nf} c_{s_n}\left({}_L\Phi_{zn} - {}_R\Phi_{zn-1}\right) \tag{6.119} $$

and

$$ {}_L\bar{T}_n = {}_R\bar{T}_{n-1} \tag{6.120} $$

The above equations can be rearranged as

$$ \left(k_{tn} + j\omega_{nf} c_{s_n}\right) {}_L\Phi_{z_n} = \left(k_{tn} + j\omega_{nf} c_{s_n}\right) {}_R\Phi_{z_{n-1}} + {}_R\bar{T}_{n-1} $$

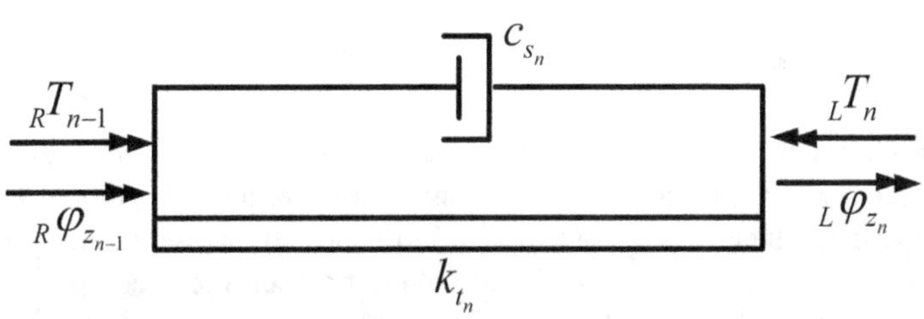

FIGURE 6.52 A free body diagram of the nth shaft segment.

and

$$ {}_L\bar{T}_n = {}_R\bar{T}_{n-1} $$

Combining the above equations in matrix form, we get

$$ \begin{bmatrix} k_t + j\omega_{nf}c_s & 0 \\ 0 & 1 \end{bmatrix}_n {}_L\begin{Bmatrix} \Phi \\ \bar{T} \end{Bmatrix}_n = \begin{bmatrix} k_t + j\omega_{nf}c_s & 1 \\ 0 & 1 \end{bmatrix}_n {}_R\begin{Bmatrix} \Phi \\ \bar{T} \end{Bmatrix}_{n-1} \tag{6.121} $$

which can be written as

$$ \mathbf{L}_n\, {}_L\mathbf{S}_n = \mathbf{M}_n\, {}_R\mathbf{S}_{n-1} $$

with

$$ \mathbf{L}_n = \begin{bmatrix} k_t + j\omega_{nf}c_s & 0 \\ 0 & 1 \end{bmatrix}_n; \quad \mathbf{M}_n = \begin{bmatrix} k_t + j\omega_{nf}c_s & 1 \\ 0 & 1 \end{bmatrix}_n; \quad \mathbf{S} = \begin{Bmatrix} \Phi \\ \bar{T} \end{Bmatrix} $$

which can be simplified as

$$ {}_L\mathbf{S}_n = \mathbf{L}_n^{-1}\,\mathbf{M}_n\, {}_R\mathbf{S}_{n-1} = \mathbf{F}_n\, {}_R\mathbf{S}_{n-1} \tag{6.122} $$

with

$$ \mathbf{F}_n = \mathbf{L}_n^{-1}\,\mathbf{M}_n = \begin{bmatrix} \dfrac{1}{k_t + j\omega_{nf}c_s} & 0 \\ 0 & 1 \end{bmatrix} \begin{bmatrix} k_t + j\omega_{nf}c_s & 1 \\ 0 & 1 \end{bmatrix} = \begin{bmatrix} 1 & \dfrac{1}{k_t + j\omega_{nf}c_s} \\ 0 & 1 \end{bmatrix} $$

where $\mathbf{F}_n$ is a field matrix at station n. From Equations 6.116 and 6.122, we get

$$ {}_R\mathbf{S}_n = \mathbf{P}_n\, {}_L\mathbf{S}_n = \mathbf{P}_n\mathbf{F}_n\, {}_R\mathbf{S}_{n-1} = \mathbf{U}_n\, {}_R\mathbf{S}_{n-1} \tag{6.123} $$

with

$$ \mathbf{U}_n = \mathbf{P}_n\mathbf{F}_n = \begin{bmatrix} 1 & 0 \\ \left(-\omega_{nf}^2 I_p + j\omega_{nf}c_d\right) & 1 \end{bmatrix}_n \begin{bmatrix} 1 & \dfrac{1}{k_t + j\omega_{nf}c_s} \\ 0 & 1 \end{bmatrix}_n $$

$$ = \begin{bmatrix} 1 & \dfrac{1}{k_t + j\omega_{nf}c_s} \\ \left(-\omega_{nf}^2 I_p + j\omega_{nf}c_d\right) & \left(1 + \dfrac{-\omega_{nf}^2 I_p + j\omega_{nf}c_d}{k_t + j\omega_{nf}c_s}\right) \end{bmatrix}_n $$

where $\mathbf{U}_n$ is a transfer matrix between stations n and $(n-1)$. Once we have the point and field matrices, the remaining analysis is the same for obtaining natural frequencies, mode shapes, and forced responses. The only difference is that now we need to handle the complex numbers. On the other hand, if the matrices need to be retained as real quantities, then these complex equations can be split into real and imaginary parts and then again assembled to get real matrices. Such analysis with damped multi-DOF could also be performed in a relatively simpler way with FEM and this will be discussed subsequently.

6.9 MODELING OF RECIPROCATING MACHINE SYSTEMS

One of the simplest analyses that can be carried out is by lumping the inertias of rotating and reciprocating parts at discrete points on the main shaft. The problem then reduces to a forced torsional vibration study of a multi-DOF rotor system subjected to varying torques at different cylinder points. The crankshaft and the other drive or driven shafts are generally flexible in torsion, but have low polar moments of inertia, unlike in the case of some large turbines or compressors. In contrast, parts mounted on the shafting, like the torsional dampers, flywheels, and generators, are rigid and have very high polar moments of inertia. The system containing a crankshaft, coupling, generator, auxiliary drive shaft, other driven shafts like pumps, and mounted parts can then be reduced to a simple system with a series of rigid rotors (representing the inertias) connected by the massless flexible shafts shown in Figure 6.53a. Now simple procedures will be described (i) to reduce reciprocating inertias to equivalent rotating inertias, (ii) to represent an uneven crank shaft geometry (refer to Figure 6.53b) to an equivalent uniform shaft system (refer to Figure 6.53c), and (iii) to convert a periodic torque variation to its components.

Up to now we have considered various machines that have components with pure rotary motions. The advantages of such rotating machineries are that they do not have variable polar moment inertias. Another class of machineries that have the possibility of torsional vibrations is reciprocating machines. A multicylinder reciprocating machine contains many reciprocating and rotating parts

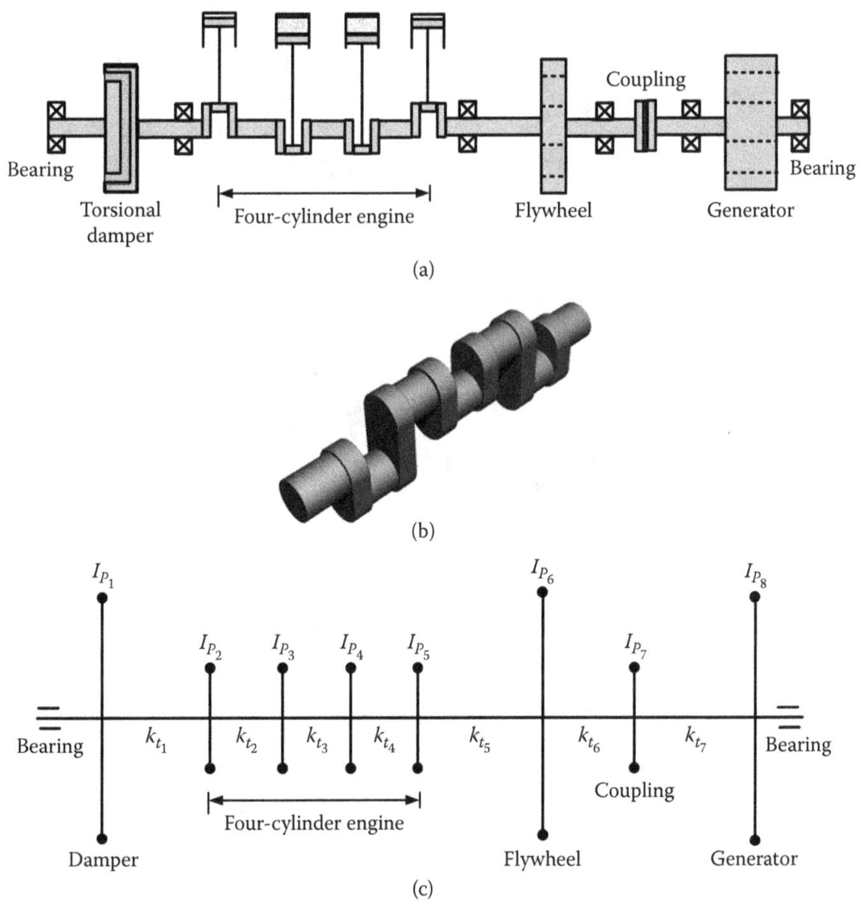

FIGURE 6.53 (a) A typical reciprocating engine installation. (b) A typical four-cylinder engine crankshaft. (c) An equivalent rotor model with *N*-disk.

such as pistons, connecting rods, crankshafts, flywheels, torsional dampers, and couplings. The system is so complicated that it is difficult, if not impossible, to undertake an exact analysis of its torsional vibration characteristics. The actual system is characterized by the presence of unpredictable effects like variable inertia, internal damping, fluid-film bearing forces, misalignments in the transmission units, uneven firing order, unbalance, etc. (Wilson, 1956, 1963, 1965; Bremer, 1979; Hestermann and Stone, 1992, 1995; Rao, 1996; Walker, 2003).

6.9.1 Equivalent Polar Mass Moment of Inertia

In this subsection, the procedure will be described to convert various inertias of reciprocating engine components to equivalent rotary inertias.

6.9.1.1 Due to Revolving and Reciprocating Masses of the Connecting Rod

The inertia of a connecting rod can be obtained by considering a two-mass equivalent dynamic system with one mass at the piston, m_{cn}^{rec} (small end of the connecting rod), and the other mass at the crank pin, m_{cn}^{rev} (big end of the connecting rod) (Figure 6.54). With some approximation (for more

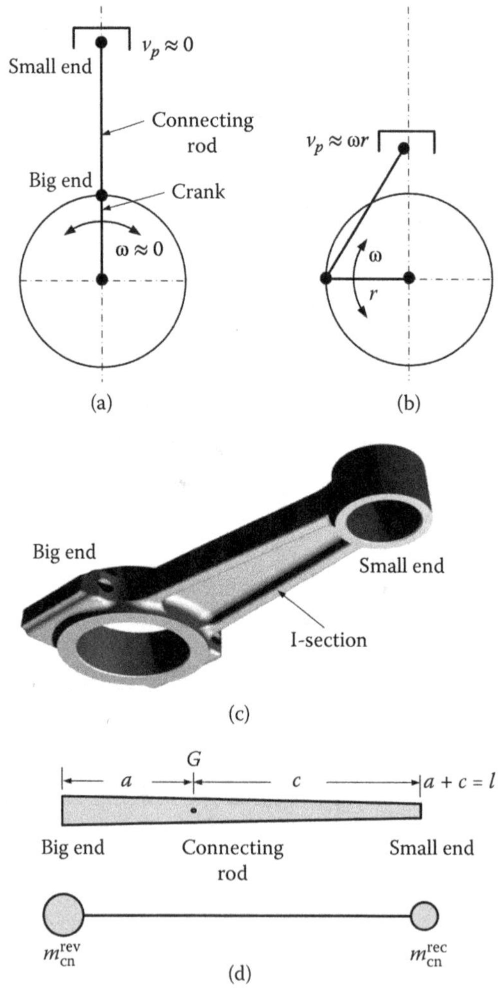

FIGURE 6.54 (a and b) An equivalent polar mass moment of inertia of the piston at two positions. (c) A typical connecting rod. (d) Dynamically equivalent two-mass system of the connecting rod.

accuracy, refer to Bevan, 1984), the mass of the connecting rod m_{cn} can be considered as a two-mass system, one at the piston of magnitude $m_{cn}^{rec} = m_{cn}a/l$, and another at the crank radius of magnitude $m_{cn}^{rev} = m_{cn}c/l$, where l is the length of the connecting rod, c is the distance from the piston pin to the center of gravity of the connecting rod, and a is the distance from the crank pin to the center of gravity of the connecting rod (i.e., $l=a+c$). It contributes to both the total revolving mass, m_{rev}, and the total reciprocating mass, m_{rec}, by a small amount.

6.9.1.2 Due to Revolving Masses on the Crank

Determination of a polar mass moment of inertia is a straightforward matter for rotating parts; however, it is not quite so simple in the case of reciprocating parts. Let us imagine a crankshaft with a polar mass moment of inertia, $m_{rev}r^2$, where m_{rev} is the total revolving mass at the crank radius r (it is also called the crank through). It includes all the revolving parts of the crank m_{cr}, and only the revolving part of connecting rod m_{cn}^{rev} (when its two-mass equivalent dynamic system is considered).

6.9.1.3 Due to Reciprocating Masses of the Piston

Consider the piston shown in two different positions in Figure 6.54a and b and let us assume that the crank is not revolving; however, it is executing small torsional oscillations about the mean position shown in the diagrams (Figure 6.54).

In the first case (Figure 6.54a), there is no motion for the piston, with small oscillations of the crank, and hence, the equivalent polar mass moment of inertia of the piston is zero. However, in the second case (Figure 6.54b), the piston has practically the same acceleration as that of the crank pin and the equivalent polar moment inertia is $m_{rec}r^2$, where m_{rec} is the mass of the reciprocating parts. It includes all the mass of the piston m_p and only the reciprocating part of the connecting rod m_{cn}^{rec} (Figure 6.54c) when it is converted to a two-mass equivalent dynamic system (Figure 6.54d). Hence, the total polar mass moment of inertia varies from $m_{rev}r^2$ to $(m_{rev}r^2+m_{rec}r^2)$ when the crankshaft is rotating.

We consider as an approximation for the system an average inertia (*rms* value could also be taken, i.e., 0.707 times instead of 0.5) that is given by

$$I_p = m_{rev}r^2 + 0.5m_{rec}r^2 \tag{6.124}$$

with

$$m_{rev} = m_{cr} + m_{cn}^{rev} \quad \text{and} \quad m_{rec} = m_p + m_{cn}^{rec} \tag{6.125}$$

6.9.2 Equivalent Torsional Stiffness of Crankshafts

In determining the torsional stiffness of shafts that connects the rotors, the main difficulty arises from the crank webs. Consider a crankshaft as an equivalent uniform diameter shaft having the same torsional stiffness as the original one, as shown in Figure 6.55. Though this idealization is physically possible, the calculations involved are extremely difficult. This is because the crank webs are subjected to bending and the crank pin to twisting when the main shaft is subjected to twisting. Moreover, the beam formula, if used, is not very accurate because of the short stubs involved rather than the long beams usually considered. Further torques applied at the free end also give rise to sidewise displacement, i.e., coupled bending–torsion vibration exists, which is prevented in the machine. For high-speed lightweight engines, crank webs are no longer rectangular blocks but have a continuously varying geometry, and the application of conventional theory becomes extremely difficult. Because of these uncertainties in the analytical calculation to estimate the torsional stiffness of crank throws, several experiments have been carried out on a number of crankshafts of large

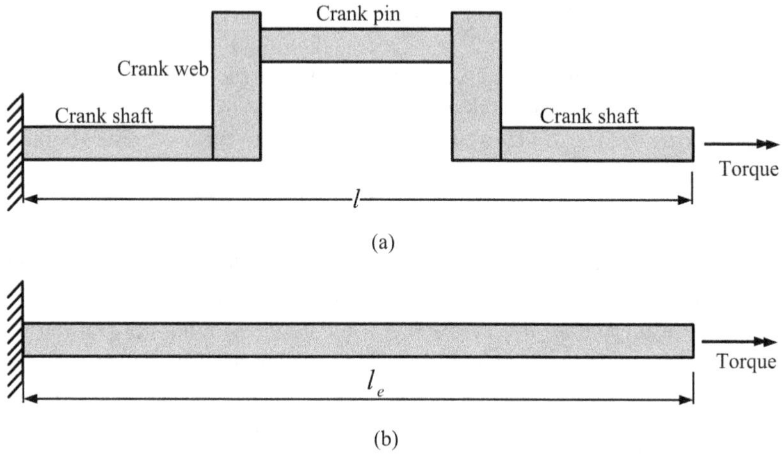

FIGURE 6.55 (a) Actual crankshaft. (b) The equivalent length of the crankshaft.

slow speed engines, which have shown that the equivalent length l_e is nearly equal to the actual length, if the diameter of the crank pin is equal to the diameter of the main shaft.

In general, the procedure that is applied to reduce the reciprocating machine system to a mathematical model is to use a basic diameter, which corresponds to the journal diameter of the crankshaft. The torsional stiffness is all calculated based on the basic diameter, irrespective of their actual diameter. For the end rotors (i.e., the generator rotor), compute the stiffness of the shaft from the coupling up to the point of rigidity. In the case, where one part of the system is connected to the other part through gears, or other transmission units, it is convenient to reduce all the inertias and stiffness to one reference speed.

The FEM (commercial finite element packages) can be used to obtain the equivalent stiffness of the crankshafts. Once the mathematical model is developed, it can be used for the free and forced vibration for illustrating the critical speed calculations and forced vibration responses as described in the next section. As such detailed torsional vibration of multicylinder reciprocating engine is beyond the scope of the present text, interested readers can refer to review paper by Hesterman and Stone (1992) and another recent paper by Mendes, Meirelles, and Zampieri (2008).

6.9.3 Torque Variations in Reciprocating Machinery

Torsional oscillations in the crankshaft and in the shafting of driven machinery are the vibration phenomenon of practical importance in the design of reciprocating engines. The average torque delivered by a cylinder in a reciprocating machine is a small fraction of the maximum torque, which occurs during the firing period. Even though the torque is periodic, the fact that it fluctuates so violently within the period constitutes one of the inherent disadvantages of a reciprocating machine, from the dynamics point of view, as compared with a turbine where the torque is practically uniform. It is possible to express the torque of a reciprocating engine into its harmonic components of several orders of the engine speed and these harmonic components can excite the engine-driven installations into forced torsional vibrations. The crankshaft torque is periodic or repetitive in nature (over a cycle), so we can express torque as a sum of harmonics by Fourier analysis:

$$T(\theta) = C_0 + A_1 \sin\theta + A_2 \sin 2\theta + \cdots + A_n \sin n\theta + \cdots + B_1 \cos\theta + B_2 \cos 2\theta + \cdots + B_n \cos n\theta + \cdots \quad (6.126)$$

With the knowledge of $T(\theta)$ for various values of θ_j, $j=1, 2, \ldots, m$, constants $C_0, A_1, A_2, \ldots A_n$ can be obtained for $m \geq (2n+1)$ using a least-squares fit as follows:

$$
\begin{aligned}
T(\theta_1) &= C_0 + A_1 \sin\theta_1 + A_2 \sin 2\theta_1 + \cdots + A_n \sin n\theta_1 + B_1 \cos\theta_1 + B_2 \cos 2\theta_1 + \cdots + B_n \cos n\theta_1 \\
T(\theta_2) &= C_0 + A_1 \sin\theta_2 + A_2 \sin 2\theta_2 + \cdots + A_n \sin n\theta_2 + B_1 \cos\theta_2 + B_2 \cos 2\theta_2 + \cdots + B_n \cos n\theta_2 \\
&\vdots \\
T(\theta_m) &= C_0 + A_1 \sin\theta_m + A_2 \sin 2\theta_m + \cdots + A_n \sin n\theta_m + B_1 \cos\theta_m + B_2 \cos 2\theta_m + \cdots + B_n \cos n\theta_m
\end{aligned}
\tag{6.127}
$$

These equations can be combined in matrix form as

$$
\mathbf{T}_{m\times 1} = \mathbf{A}_{m\times(2n\times 1)}\mathbf{C}_{(2n\times 1)\times 1} \tag{6.128}
$$

with

$$
\mathbf{T}^T_{1\times m} = \begin{bmatrix} T(\theta_1) & T(\theta_2) & \cdots & T(\theta_m) \end{bmatrix}
$$

$$
\mathbf{C}^T_{1\times(2n\times 1)} = \begin{bmatrix} C_0 & A_1 & A_2 & \cdots & A_n & B_1 & B_2 & \cdots & B_n \end{bmatrix}
$$

$$
\mathbf{A}_{m\times(2n\times 1)} = \begin{bmatrix}
1 & \sin\theta_1 & \sin 2\theta_1 & \cdots & \sin n\theta_1 & \cos\theta_1 & \cos 2\theta_1 & \cdots & \cos n\theta_1 \\
1 & \sin\theta_2 & \sin 2\theta_2 & \cdots & \sin n\theta_2 & \cos\theta_2 & \cos 2\theta_2 & \cdots & \cos n\theta_2 \\
\vdots & \vdots & \vdots & \ddots & \vdots & \vdots & \vdots & \ddots & \vdots \\
1 & \sin\theta_m & \sin 2\theta_m & \cdots & \sin n\theta_m & \cos\theta_m & \cos 2\theta_m & \cdots & \cos n\theta_m
\end{bmatrix}
$$

This can be solved as

$$
\mathbf{C}_{(2n\times 1)\times 1} = \left[\left\{ \left(\mathbf{A}^T_{(2n\times 1)\times m}\, \mathbf{A}_{m\times(2n\times 1)} \right)^{-1} \right\}_{(2n\times 1)\times(2n\times 1)} \mathbf{A}^T_{(2n\times 1)\times m} \right]_{(2n\times 1)\times m} \mathbf{T}_{m\times 1} \tag{6.129}
$$

For all practical purposes, the first few harmonics give a sufficient result. This is very useful in the analysis of the torsional vibration of an engine crankshaft.

The engine and the driven unit, such as a generator or a pump, are normally connected by a flexible coupling. Thus, the driver-driven unit has fairly low torsional natural frequencies. These frequencies fall in the speed range of the unit and its harmonics of different orders. It is a commonly known fact that failures can occur in reciprocating machine installations, when the running speed of the engine is at or near a dominant torsional critical speed of the system. High dynamic stresses can occur in the main shafting of such engine installations, and to avoid these conditions, it is essential that the torsional vibration characteristics of the entire installation be analyzed before the unit is put into operation. Any analysis of torsional vibration characteristics of the reciprocating machinery should finally predict the maximum dynamic stresses or torque developed in the shafting and couplings of the system, as accurately as possible, so that they can be compared with the permissible values, to check the safety of installation.

Example 6.14

A marine reciprocating engine, flywheel, and propeller are approximately equivalent to the following three-rotor system. The engine has a crank 50 cm long and a connecting rod 250 cm long. The engine revolving parts are equivalent to 50 kg at the crank radius, and the piston and pin masses are 41 kg. The connecting rod mass is 52 kg, and its center of gravity is 26 cm from the crank pin center. The mass of the flywheel is 200 kg with a radius of gyration of 25 cm. The propeller has a polar mass moment of inertia of 6 kg-m^2. The shaft between the engine masses and the flywheel is 38 cm in diameter and 5.3 m long and that between the flywheel and the propeller is 36 cm in diameter and 1.5 m long. Find the torsional natural frequencies of the rotor system.

Solution:

The main aim of the present solution procedure is to first find the equivalent rotating mass of the reciprocating engine; once this has been done, then the problem reduces to obtaining torsional natural frequencies of a three-rotor system as shown in Figure 6.56. The three revolving masses correspond to the reciprocating engine, flywheel, and propeller.

The equivalent rotating part of the engine can be obtained as follows. We have the following engine data:

Crank radius, $r=0.5$ m, mass of crank revolving parts, $m_{cr}=50$ kg at the crank radius

Length of the connecting rod $l=2.5$ m, mass of the connecting rod, $m_{cn}=52$ kg

Distance of the center of gravity of the connecting rod from the crank pin, $a=0.26$ m

Distance of the center of gravity of the connecting rod from the piston pin, $c=l-a=2.24$ m

Piston and pin masses, $m_p=41$ kg

Hence, the equivalent (approximate) reciprocating and revolving masses of the connecting rod are

$$m_{cn}^{rec}=m_{cn}a/l=52\times0.26/2.5=5.41\text{ kg};\quad m_{cn}^{rev}=m_{cn}c/l=52\times2.24/2.5=46.59\text{ kg}$$

The equivalent revolving and reciprocating masses are given as

$$m_{rev}=m_{cr}+m_{cn}^{rev}=50+46.59=96.59\text{ kg}$$

and

$$m_{rec}=m_p+m_{cn}^{rec}=41+5.41=46.41\text{ kg}$$

Hence, the equivalent polar mass moment of inertia of the engine is obtained as

$$I_{p_s}=(m_{rev}+0.5m_{rec})r^2=(96.59+0.5\times46.41)\times0.5^2=29.95\text{ kg-m}^2$$

Now, the polar mass moment of inertia of the flywheel $I_{p_f}=mr_f^2=200\times(0.25)^2=12.5\text{ kg-m}^2$. For the propeller, the polar mass moment of inertia is given as $I_{p_p}=6\text{ kg-m}^2$.

The torsional stiffnesses of shaft segments 1 and 2 are given as

$$k_{t_1}=\frac{G_1J_1}{l_1}=\frac{78.9\times10^9\times\pi\times0.38^4}{32\times53}=3\times10^7\text{ N-m/rad}$$

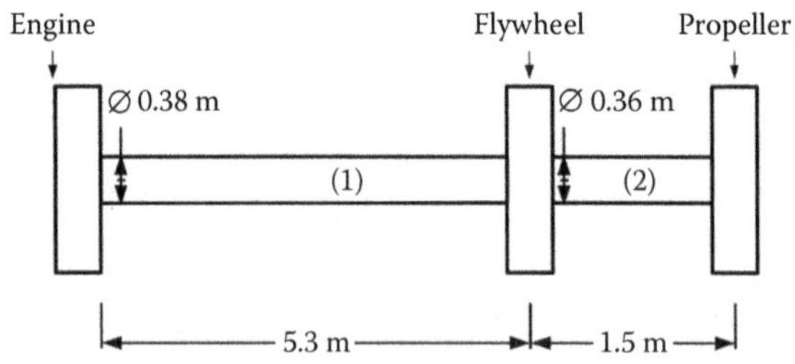

FIGURE 6.56 A three-disk model of the engine, flywheel, and propeller.

and

$$k_{t_2} = \frac{G_2 J_2}{l_2} = \frac{78.9\times10^9\times\pi\times0.36^4}{32\times11.5} = 8.67\times10^7 \text{ N-m/rad}$$

The natural frequencies of the three-disk rotor system (with $I_{P_e} = I_{p1}$, $I_{P_f} = I_{p2}$, and $I_{P_p} = I_{p3}$) are given as (Equation 6.37)

$$\omega_{nf_1} = 0$$

and

$$\omega_{nf2,3}^2 = \frac{1}{2}\left(k_{t_1}\frac{I_{p1}+I_{p2}}{I_{p1}I_{p2}} + k_{t_2}\frac{I_{p2}+I_{p3}}{I_{p2}I_{p3}}\right) \pm \sqrt{\frac{1}{4}\left(k_{t_1}\frac{I_{p1}+I_{p2}}{I_{p1}I_{p2}} + k_{t_2}\frac{I_{p2}+I_{p3}}{I_{p2}I_{p3}}\right)^2 - \left(\frac{k_{t_1}k_{t_2}\left(I_{p1}+I_{p2}+I_{p3}\right)}{I_{p1}I_{p2}I_{p3}}\right)}$$

which gives

$$\omega_{nf1} = 0, \quad \omega_{nf2} = 1.58\times10^3 \text{ rad/s}, \quad \omega_{nf3} = 4.72\times10^3 \text{ rad/s}$$

The finding of mode shapes and the position of the nodes is left to the reader as a practice problem.

6.10 CONCLUDING REMARKS

To summarize, now we have a clear idea about torsional natural frequencies and mode shapes for simple rotor systems. We have obtained torsional natural frequencies and mode shapes using Newton's second law of motion and using the systematic TMM. The TMM is found to be quite versatile and easy to apply, especially for multi-DOF rotor systems. The TMM is also developed for rotor systems with damping in the disk due to aerodynamic forces and in the shaft due to material damping (both the damping models are taken as equivalent viscous damping). Apart from these simple rotor systems, we considered the simple and branched geared rotor systems for obtaining torsional natural frequencies. For multi-DOF geared and branched systems, the TMM is applied because of its simplicity in the application. At the end, the procedure for obtaining an equivalent single-shaft rotor system from multicylinder reciprocating engines is briefly discussed. For simplicity of analysis in the TMM, the point and field matrices have been derived based on lumped-mass analysis (instead of distributed mass and stiffness analysis or continuous system analysis). In subsequent chapters, the continuous system approach and FEM will be discussed in detail. The sustainable design of a rotor for torsional vibration minimizes material waste, extends operational life, and optimizes energy efficiency by tuning the rotor's natural frequencies away from destructive resonance zones. Sustainable design shifts the focus from purely adding mass (which increases carbon footprint and energy consumption) to geometric optimization, advanced materials, and smart damping.

EXERCISE PROBLEMS

Exercise 6.1 Find the torsional natural frequencies and mode shapes of the two-disk rotor system shown in Figure 6.57 by using the TMM. B_1 and B_2 are frictionless bearings, and D_1 and D_2 are rigid disks. The shaft is made of steel with a modulus of rigidity $G = 0.8\times10^{11}$ N/m^2 and a uniform diameter $d = 10$ mm. The various shaft lengths are as follows: $B_1D_1 = 50$ mm, $D_1D_2 = 75$ mm, and $D_2B_2 = 50$ mm. The polar mass moments of inertia of the disks are $I_{p1} = 0.0008$ kg-m^2 and $I_{p2} = 0.002$ kg-m^2. Consider the shaft as massless.

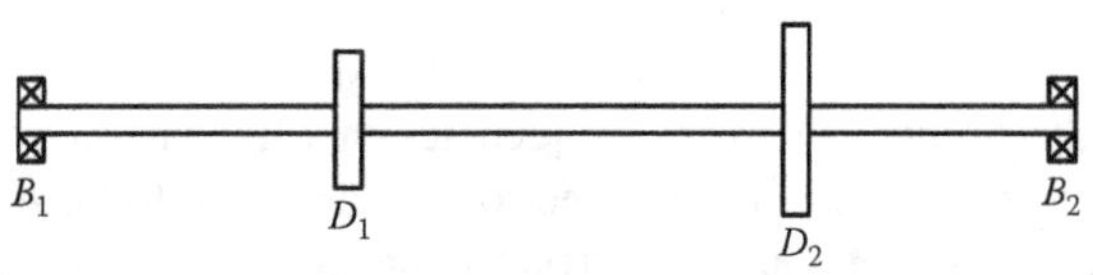

FIGURE 6.57 Two-rotor system.

Exercise 6.2 Obtain the torsional natural frequency of a cantilever rotor system as shown in Figure 6.58. The end B_1 of the shaft has a fixed end condition. The shaft diameter is 10 mm, and the total length of the shaft is 0.2 m. The polar mass moment of inertia of the disk is equal to 0.02 kg-m^2. Neglect the inertia of the shaft. Use the TMM.

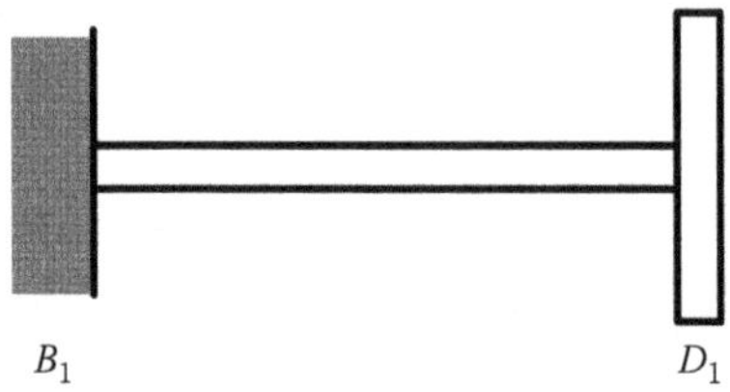

FIGURE 6.58 A cantilever shaft with a rigid disk.

Exercise 6.3 Find the torsional natural frequencies and mode shapes of a rotor system as shown in Figure 6.59 by using the TMM. B_1 and B_2 are fixed supports, and D_1 and D_2 are rigid disks. The shaft is made of steel with a modulus of rigidity of $G = 0.8 \times 10^{11}$ N/m^2 and has a uniform diameter of $d = 10$ mm. The different shaft lengths are as follows: $B_1D_1 = 50$ mm, $D_1D_2 = 75$ mm, and $D_2B_2 = 50$ mm. The polar mass moments of inertia of the disks are $I_{p1} = 0.08$ kg-m^2 and $I_{p2} = 0.2$ kg-m^2. Consider the shaft as massless.

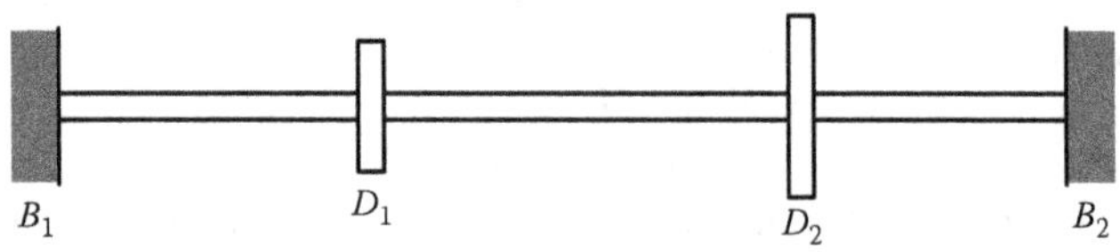

FIGURE 6.59 Two-disk system with fixed supports.

Exercise 6.4 Find the torsional natural frequencies and draw the corresponding mode shapes of the rotor system shown in Figure 6.60. B_1 is a fixed support (with zero angular displacement about the shaft axis), and B_2 and B_3 are simply supported (with nonzero angular displacements). The shaft is made of steel with $G = 0.8 \times 10^{11}$ N/m^2 and uniform diameter $d = 10$ mm. Various shaft lengths are as follows: $B_1D_1 = 50$ mm, $D_1B_2 = 50$ mm, $B_2D_2 = 25$ mm, $D_2B_3 = 25$ mm, and $B_3D_3 = 30$ mm. The polar mass moments of inertia of the disks are $I_{p1} = 0.002$ kg-m^2, $I_{p2} = 0.001$ kg-m^2, and $I_{p3} = 0.008$ kg-m^2. Use the TMM. Give all the detailed steps involved in obtaining the final system of equations and application of boundary conditions. Consider the shaft as massless and the disks as lumped masses.

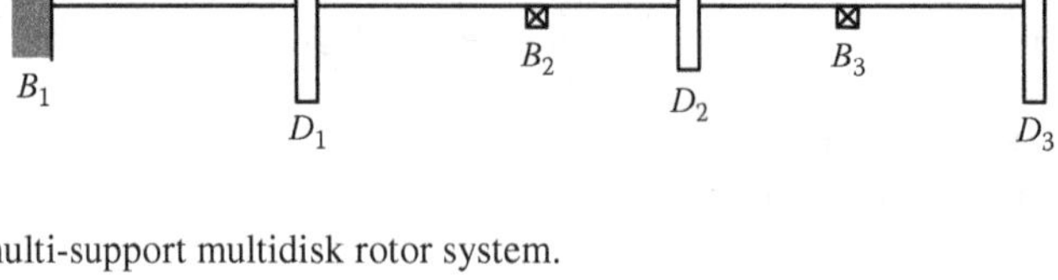

FIGURE 6.60 A multi-support multidisk rotor system.

Exercise 6.5 Obtain the torsional natural frequencies of a turbine-coupling-generator rotor system as shown in Figure 6.61 by the TMM. The rotor is assumed to be supported on frictionless bearings. The polar mass moments of inertia of the turbine, coupling, and generator are

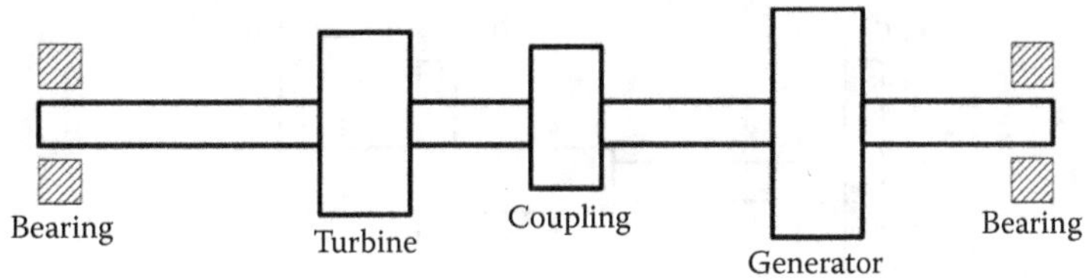

FIGURE 6.61 A turbine-coupling-generator set.

$I_{PT} = 25$ kg-m^2, $I_{PC} = 5$ kg-m^2, and $I_{PG} = 50$ kg-m^2, respectively. Take the modulus of rigidity of the shaft as $G = 0.8 \times 10^{11}$ N/m^2. Assume the shaft diameter throughout is equal to 0.2 m, and the lengths of shafts between the bearing-turbine-coupling-generator-bearing are 1 m each so that the total span is 4 m. The coupling also gives a point torsional flexibility (inverse of torsional stiffness) equivalent to five times that of a shaft with 1 m length and 0.2 m diameter. Consider the shaft as massless.

Coupling can be modeled as two flanges with a flexible element in between. Coupling polar moment of inertia can be split into two parts. So, stiffness of coupling will give a field matrix and two flanges will give point matrices. This case is different as compared to support stiffness (Exercise 6.6) wherein two ends of spring have the same angular twist, whereas in coupling two ends have different angular twists.

Exercise 6.6 Obtain the torsional natural frequency of the overhung rotor system as shown in Figure 6.62. Take the polar mass moment of inertia of the disk as $I_p = 0.04$ kg-m^2. The massless shaft has the following properties: The lengths are $a = 0.3$ m and $b = 0.7$ m, the uniform diameter is 10 mm, and the modulus of rigidity $G = 0.8 \times 10^{11}$ N/m^2. Bearing "A" is flexible and provides a torsional restoring torque with a torsional stiffness equal to 5% of the torsional stiffness of the shaft segment having length a. Consider bearing B as a fixed bearing. Use both the direct method and TMM.

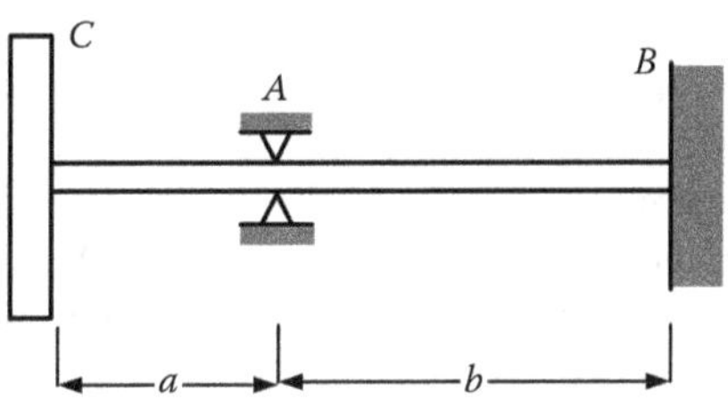

FIGURE 6.62 An overhung rotor system with an intermediate support.

Exercise 6.7 A small electric motor drives another through a long coil spring (wire diameter d, coil diameter D, and n turns). The two motor rotors have inertias I_{p1} and I_{p2}. Calculate the torsional natural frequencies of the setup. Assume the ends of the spring are "built in" to the shafts. The data for the present problem are as follows: $I_{p1} = 0.012$ kg-m^2; $I_{p2} = 0.02$ kg-m^2; $n = 10$; $D = 0.075$ m; $d = 0.01$ m; $E = 2.1 \times 10^{11}$ N/m^2.

Exercise 6.8 For a rotor system with a stepped circular shaft as shown in Figure 6.63, obtain the torsional natural frequencies, mode shapes, and node positions. Consider the free–free end conditions. Neglect the polar mass moment of inertia of the shaft and take $G = 0.8 \times 10^{11}$ N/m^2. Use the direct method, the indirect method (based on the node location information of the mode shapes), and the TMM.

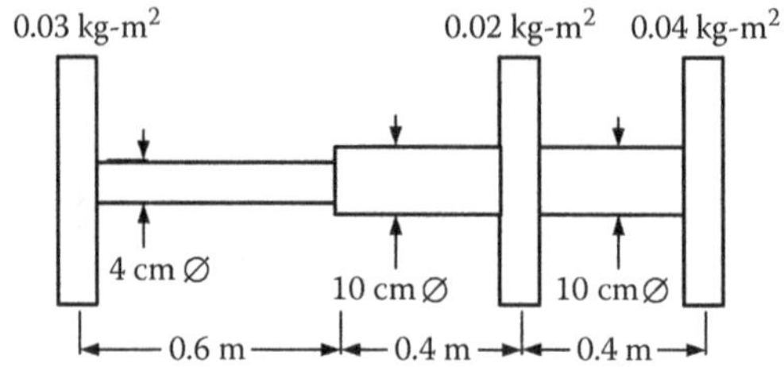

FIGURE 6.63 Three-disk rotor system with a stepped shaft.

Exercise 6.9 For a geared system as shown in Figure 6.64, find the torsional natural frequencies and mode shapes. Shafts "A'" and "B" are supported on frictionless bearings. Find also the location of the node points on the shaft (if any). Shaft "A" has 1.5 cm diameter and 0.3 m length, and shaft "B" has 1 cm diameter and 0.4 m length. Let the modulus of rigidity of shaft G be equal to 0.8×10^{11} N/m^2; the polar mass moments of inertia of disks and gears are $I_{PA} = 0.1$ Nm^2, $I_{PB} = 0.08$ Nm^2, $I_{PgA} = 0.003$ Nm^2, and $I_{PgB} = 0.002$ Nm^2. Use (i) the equivalent system approach and (ii) the TMM.

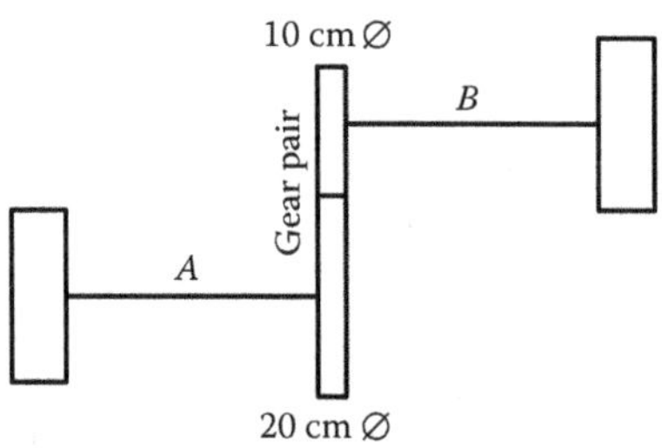

FIGURE 6.64 A geared rotor system.

Exercise 6.10 Obtain the torsional natural frequencies and mode shapes of an epicyclic gear train as shown in Figure 6.65. Consider free–free support conditions. Find also the location of node points on the shaft. The gear mounted on shaft "B" is a planetary gear, and the gear on shaft "A" is a sun gear. Consider the polar mass moments of inertia of the shaft, the arm, and the gears as negligible. Shaft "A" has a 5 cm diameter and 0.75 m length, and shaft "B" has a 4 cm diameter and 1.0 m length. The angular speeds of shaft A and the arm are 300 and 100 rpm, respectively. Let the modulus of rigidity of the shaft G be equal to 0.8×10^{11} N/m^2; the polar mass moments of inertia of the disks are $I_{PA} = 24$ Nm^2 and $I_{PB} = 10$ Nm^2. State the assumptions made in the analysis.

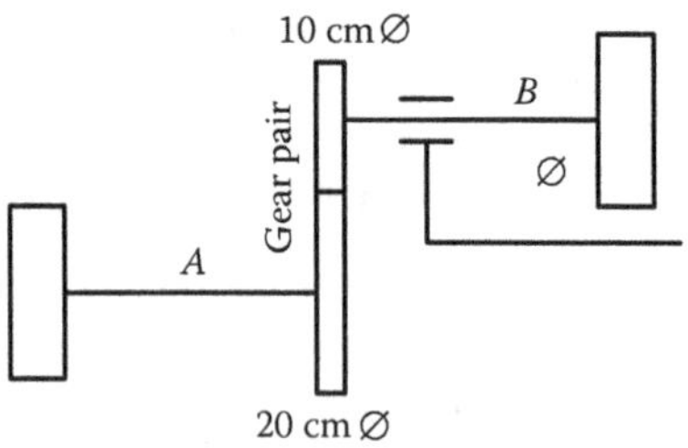

FIGURE 6.65 An epicyclic geared system.

Exercise 6.11 For a gear train as shown in Figure 6.66, obtain the torsional natural frequencies and the location of the nodes. It is assumed that all shafts are mounted on frictionless bearings. The shaft dimensions are as follows: (i) motor shaft: 0.20 m length and 0.015 m diameter, (ii) flywheel shaft: 0.2 m length and 0.01 m diameter, and (iii) intermediate shaft: 0.4 m length and 0.012 m diameter. The polar mass moments of inertia of the motor and flywheel are $I_{p_m} = 0.01$ kg-m^2 and $I_{p_f} = 0.04$ kg-m^2, respectively. The gear ratio of the first and the second gear pairs are 3 and 4, respectively. Neglect inertias of gears and the mass of shafts. Assume free–free support conditions because all shafts are mounted on frictionless bearings. Take $G=0.8\times10^{11}$ N/m^2. Use the TMM.

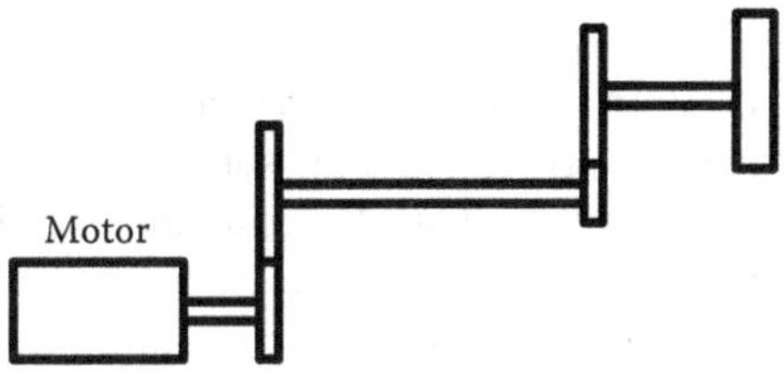

FIGURE 6.66 A motor connected to a flywheel by a two-stage gearbox.

Exercise 6.12 A stepped shaft consists of three segments with lengths 40, 30, and 40 cm and the corresponding diameters d cm, 13 cm, and d cm, where d is an unknown to be obtained. The shaft has two flywheels ($I_{p_1} = 11$ kg-m^2 and $I_{p_2} = 11$ kg-m^2, with the radius of gyration of both flywheels equal to 0.5 m) at the ends, and the shaft is supported on two frictionless rolling bearings 20 cm away from either end. The operating speed of the shaft is 1500 rpm, and due to rotation of the shaft, it has external torque impulses such that it has a period corresponding to a quarter of the shaft rotation. Obtain the diameter, d, such that the torsional critical speed may be 20% above the external torque frequency (fundamental harmonics). Obtain the transverse natural frequency of the rotor system so designed based on the dynamics of the rotor in torsion. Neglect the mass of the shaft and take $G=0.8\times10^{11}$ N/m^2, and $E=2.1\times10^{11}$ N/m^2.

Exercise 6.13 A motor has a rotating mass of polar mass moment of inertia 5.8 kg-m^2, which is connected to one end of a shaft of 6 cm diameter that is 0.30 m long. At the other end, a pinion is attached, with an effective polar mass moment of inertia of 2.20 kg-m^2. The pinion is connected to a gear with a gear ratio of 4 and polar mass moment of inertia of 0.70 kg-m^2, which drives a pump. There is no shaft at the pump end. The measured torsional vibration frequency of the rotor system is 60 Hz. Find the effective polar mass moments of inertia of the pump impeller and entrained water. Take $G=0.8\times10^{11}$ N/m^2.

Exercise 6.14 A cantilever shaft of 1 m length, l, and 30 mm diameter, d, has a thin disk of 5 kg mass, m, attached at its free end, with a disk radius of 5 cm. The shaft has a through hole parallel to the shaft axis of diameter 3 mm, d_i, which is vertically below the shaft center, with a distance between the centers of the shaft and the hole of 6 mm (e). Consider no warping of the shaft plane; obtain the torsional natural frequencies of the shaft system. Consider the shaft as massless and assume the modulus of rigidity $G=0.8\times10^{11}$ N/m^2.

Exercise 6.15 Find the torsional natural frequencies and mode shapes of the rotor system shown in Figure 6.67. B is a fixed bearing that provides a fixed support condition, and D_1, D_2, D_3, and D_4 are rigid disks. The shaft is made of steel with a modulus of rigidity $G=0.8\times10^{11}$ N/m^2 and uniform diameter $d=20$ mm. The various shaft lengths are as follows: $D_1D_2=50$ mm, $D_2D_3=50$ mm, $D_3D_4=50$ mm, and $D_4B_2=150$ mm. The masses of the disks are $m_1=4$ kg, $m_2=5$ kg, $m_3=6$ kg, and $m_4=7$ kg. Consider the shaft as massless. Consider the disks as thin and take the diameter of disks as $d_1 = 12$ cm, $d_2 = 6$ cm, $d_3 = 12$ cm, and $d_4 = 14$ cm.

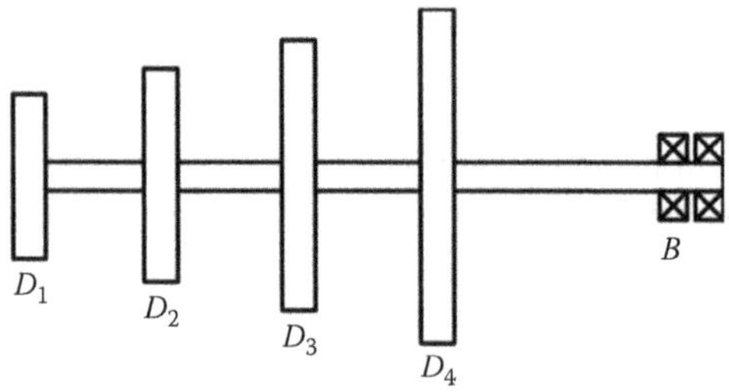

FIGURE 6.67 A four-disk cantilever rotor.

Exercise 6.16 Find the torsional natural frequencies and mode shapes of the rotor system shown in Figure 6.68. B is a fixed bearing that provides a fixed support end condition, and D_1, D_2, D_3, D_4, and D_5 are rigid disks. The shaft is made of steel with a modulus of rigidity $G=0.8\times10^{11}$ N/m² and uniform diameter $d=20$ mm. The various shaft lengths are as follows: $D_1D_2=50$ mm, $D_2D_3=50$ mm, $D_3D_4=50$ mm, $D_4D_5=50$ mm, and $D_5B_2=50$ mm. The masses of the disks are $m_1=4$ kg, $m_2=5$ kg, $m_3=6$ kg, $m_4=7$ kg, and $m_5=8$ kg. Consider the shaft as massless. Consider the disks as thin and take the diameter of the disks as $d_1=12$ cm, $d_2=6$ cm, and $d_3=12$ cm, $d_4=14$ cm, and $d_5=16$ cm.

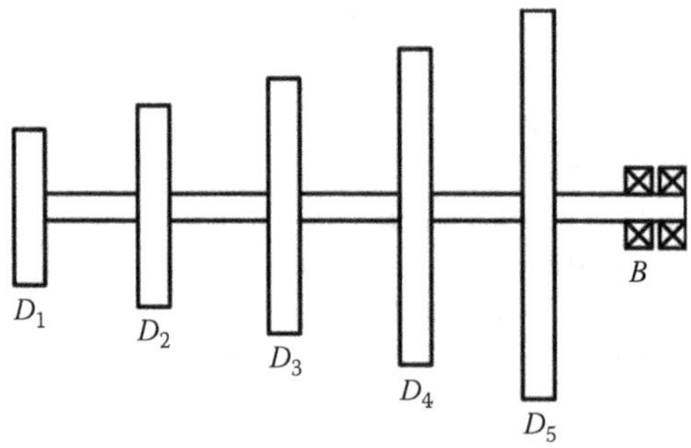

FIGURE 6.68 A five-disk cantilever rotor.

Exercise 6.17 Find the torsional natural frequencies and mode shapes of the rotor system shown in Figure 6.69. B_1 and B_2 are bearings, which provide free–free end conditions, and D_1, D_2, D_3, D_4, and D_5 are rigid disks. The shaft is made of steel with a modulus of rigidity $G=0.8\times10^{11}$ N/m² and uniform diameter $d=20$ mm. The various shaft lengths are as follows: $B_1D_1=150$ mm, $D_1D_2=50$ mm, $D_2D_3=50$ mm, $D_3D_4=50$ mm, $D_4D_5=50$ mm, and $D_5B_2=150$ mm. The masses of the disks are $m_1=4$ kg, $m_2=5$ kg, $m_3=6$ kg, $m_4=7$ kg, and $m_5=8$ kg. Consider the shaft as massless. Consider the disks as thin and take the diameter of the disks as $d_1=8$ cm, $d_2=10$ cm, $d_3=12$ cm, $d_4=14$ cm, and $d_5=16$ cm.

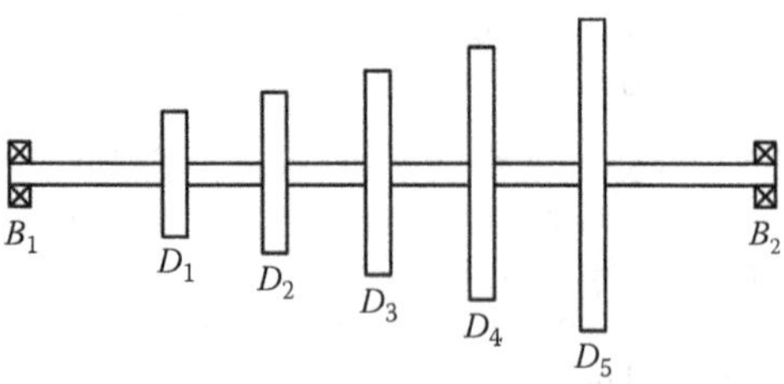

FIGURE 6.69 A five-disk rotor system with simply supported end conditions.

Exercise 6.18 Find the torsional natural frequencies of a cantilever rotor system as shown in Figure 6.70. Consider the shaft as massless and assume it is made of steel with a modulus of rigidity of 0.8×10^{11} N/m^2. A disk is mounted at the free end of the shaft with a polar mass moment of inertia of 0.01 kg-m^2. In the diagram, all dimensions are in cm. Use the TMM.

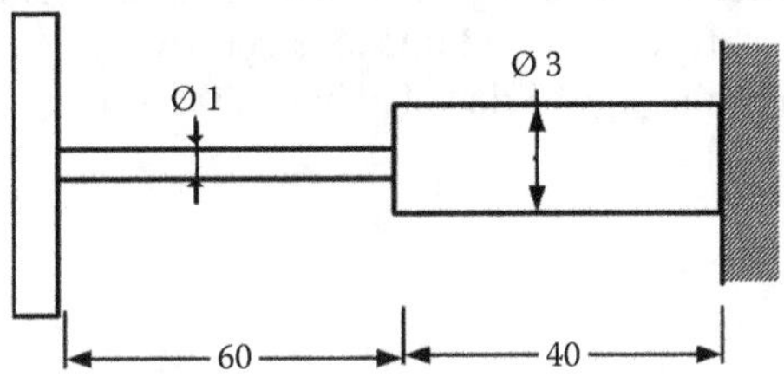

FIGURE 6.70 A cantilever rotor with a stepped shaft.

Exercise 6.19 Obtain the torsional natural frequencies of the geared rotor system shown in Figure 6.71. Consider that the LHS of shaft A and the RHS of shaft B have rigid supports. Assume the gears are thin disks with masses of 3 and 2 kg for gears of diameters 20 and 10 cm, respectively. Shafts A and B are, respectively, 2 and 1.5 cm in diameter, respectively, with a length of 40 cm each. Neglect the inertia of the shafts. The gear tooth pair provides an effective torsional stiffness of 1 kN-m/rad between gears. Take $G = 0.8 \times 10^{11}$ N/m^2. Use the TMM.

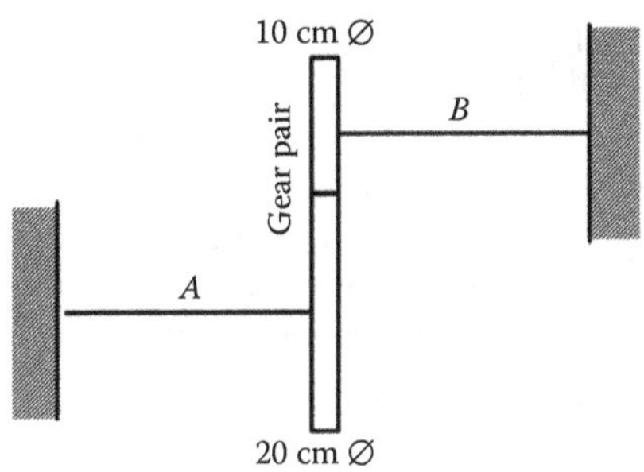

FIGURE 6.71 A gear pair.

Exercise 6.20 Obtain the torsional natural frequencies and mode shapes of the branched system shown in Figure 6.72. The polar mass moments of inertia of the rotors are $I_{PA} = I_{PF} = I_{PI} = 0.01$ kg-m^2, $I_{PB} = I_{PE} = I_{PH} = 0.005$ kg-m^2, and $I_{PC} = I_{PD} = I_{PG} = 0.003$ kg-m^2. The gear ratios are $n_{CD} = 3$ and $n_{CG} = 4$. The shaft lengths are $l_{AB} = l_{BC} = l_{DE} = l_{EF} = l_{GH} = l_{HI} = 0.25$ m, and the diameters are $d_{AC} = 0.03$ m, $d_{DF} = 0.02$ m, and $d_{GI} = 0.02$ m. Take the modulus of rigidity of the shaft as $G = 0.8 \times 10^{11}$ N/m^2. Use the TMM.

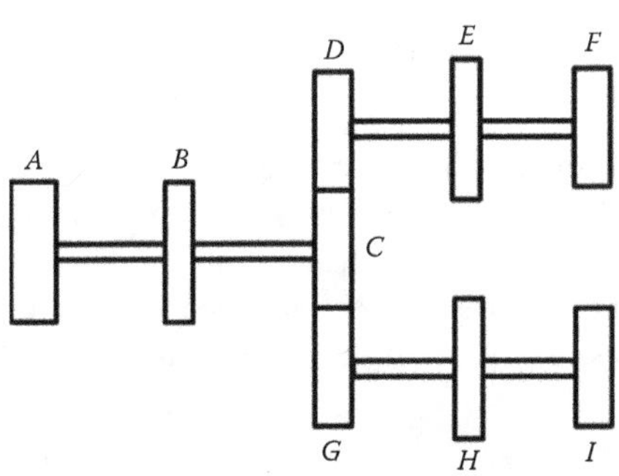

FIGURE 6.72 A branched rotor system.

Exercise 6.21 For the turbine-generator rotor system shown in Figure 6.73, obtain the torsional natural frequencies and mode shapes. Plot the mode shapes and animate them. Take $I_{p1} = 0.90$ kg-m^2, $I_{p2} = 0.88$ kg-m^2, $I_{p3} = 0.86$ kg-m^2, $I_{p4} = 0.84$ kg-m^2, $I_{p5} = 1$ kg-m^2, $I_{p6} = I_{p7} = 0.1$ kg-m^2, and $I_{p8} = I_{p9} = I_{p10} = I_{p11} = I_{p12} = I_{p13} = 0.8$ kg-m^2. Each shaft segment between the disks, between the disk and bearing, and between the bearing and coupling has a length of 0.5 m. The diameters of the turbine and generator shafts are 0.1 and 0.15 m, respectively. Take $G = 0.8 \times 10^{11}$ N/m^2. The coupling has a torsional stiffness that is 30% that of a shaft of 0.5 m length and 0.1 m diameter. Take the bearings as frictionless.

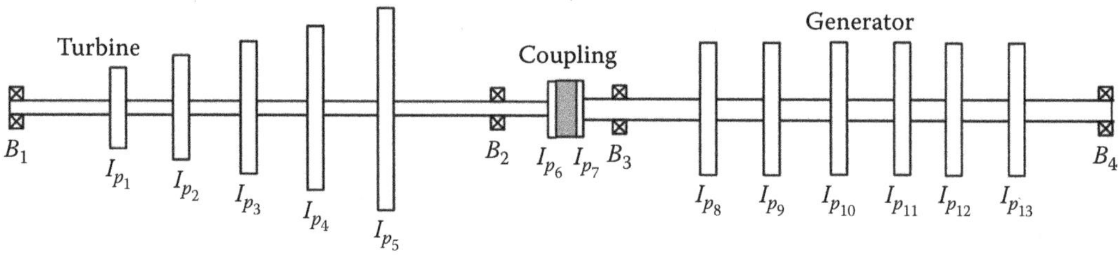

FIGURE 6.73 A turbine-generator model.

Exercise 6.22 For a motor-coupling-flywheel rotor system as shown in Figure 6.74, use TMM to obtain the torsional natural frequency of the rotor system with the help of first obtaining the frequency equation. The rotor of the motor has a polar mass moment of inertia of 0.01 kg-m^2; consider it rigid. The coupling has two flanges; the polar mass moment of inertia of each of them is 0.005 kg-m^2. In between the two flanges, there is a torsional spring with a stiffness of 100 kNm/rad. The shaft between the motor and the coupling has a length of 10 cm with a diameter of 1 cm and that between the coupling and flywheel also has a length of 10 cm with a diameter of 1 cm. The polar mass moment of inertia of the flywheel is 0.05 kg-m^2. The shaft length between the flywheel and bearing is 5 cm with a diameter of 1 cm. Neglect the mass of the shaft and treat the bearings as frictionless. Take $G = 0.8 \times 10^{11}$ N/m^2.

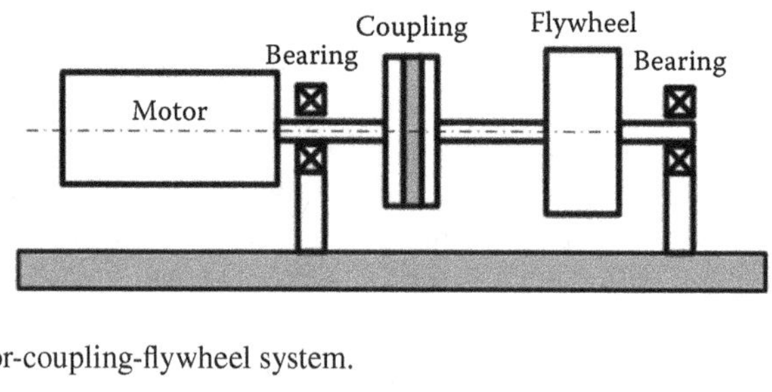

FIGURE 6.74 A motor-coupling-flywheel system.

Exercise 6.23 Obtain the torsional natural frequencies of a bevel-geared rotor system as shown in Figure 6.75 by the TMM. Both shafts between the flywheels and the gears have a length of 20 cm and are 1.5 cm in diameter. Flywheels 1 and 2 are free to rotate and have a polar mass moment of inertia as shown in the figure. Neglect the polar mass moments of inertia of the gears and treat bearings as frictionless. Assume for the shafts that $G = 0.8 \times 10^{11}$ N/m^2.

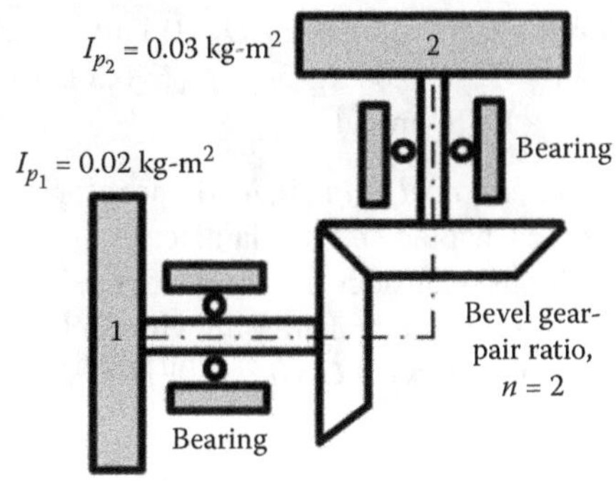

FIGURE 6.75 Flywheels connected by a bevel gear pair.

Exercise 6.24 Obtain the torsional natural frequency of the rotor system shown in Figure 6.76. For all shafts k_t = 100 kN-m/rad. Use the TMM or FEM. Neglect the polar moment of inertia of the gears and consider the bearings as frictionless.

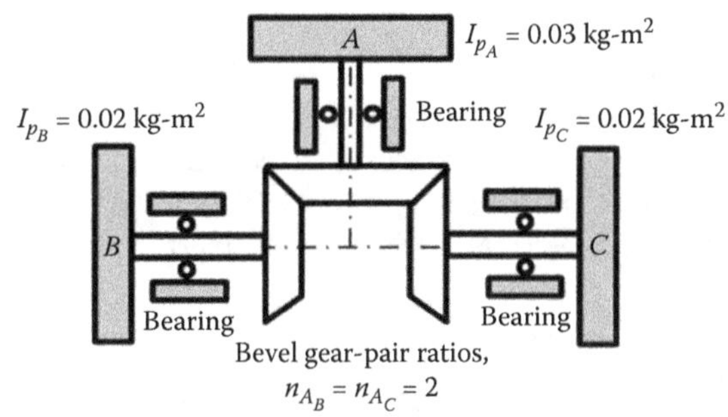

FIGURE 6.76 Three flywheels connected by bevel gears.

Exercise 6.25 Obtain the governing equations and frequency equation of a geared rotor system as shown in Figure 6.77 for torsional vibrations. Two flywheels with polar moments of inertia of I_{p1} and I_{p4} are mounted on two different shafts with stiffnesses k_{t1} and k_{t2}, respectively. The gear pair teeth provide flexibility during their contact, and it is represented as a linear spring as shown in the figure. Let us consider k_g as the stiffness of the spring. Let the diameter of the gears be d_1 and d_2.

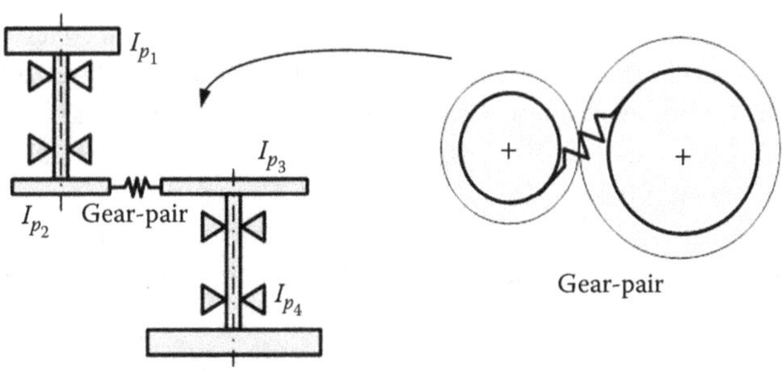

FIGURE 6.77 A gear flywheel system with contact flexibility.

Assume rotor data as follows: gear diameter, D_3=0.1 m; gear diameters, D_4=0.3 m; $I_{p1} = 0.1$ kg-m^2; $I_{p2} = 0.02$ kg-m^2; $I_{p3} = 0.04$ kg-m^2; $I_{p4} = 0.3$ kg-m^2; $k_{t1} = 1\times10^4$ N-m/rad; $k_{t2} = 1.5\times10^4$ N-m/rad; and $k_g = 1\times10^6$ N-m/rad.

Exercise 6.26 Obtain the *torsional natural frequencies* of the rotor system shown in Figure 6.78. Use only the TMM or FEM (refer to Chapter 7). The diameter of the shaft is 1 cm and its length is 90 cm with fixed–fixed end conditions. Consider the shaft as massless. Disks are symmetrically placed (distance between two disks is 30 cm) and are thin with each having mass 3 kg and diameter 10 cm. Take for the shaft material property $G=0.8\times10^{11}$ N/m^2.

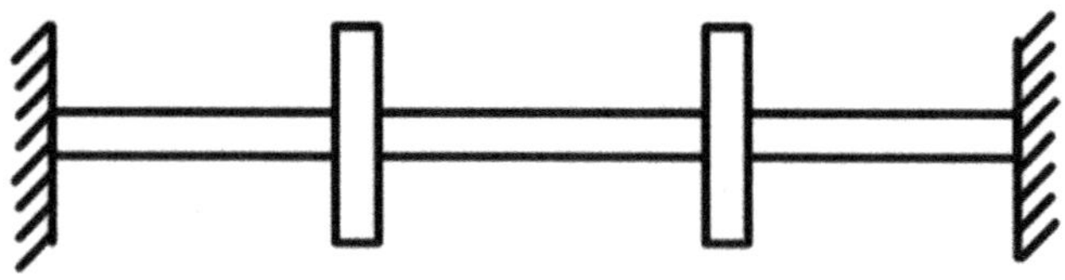

FIGURE 6.78 Two-disk fixed end rotor system.

Exercise 6.27 Obtain the torsional natural frequencies of the rotor system shown in Figure 6.79. The geared rotor consists of two identical flywheels (I_p=0.01 kg-m^2) placed symmetrically on the shaft with each shaft segment length (e.g., between the fixed support to the flywheel, and between the flywheel and the gear) being 0.3 m and a diameter of 0.015 m. The gear pair has a gear ratio of 2. Use the TMM only. Take $G=0.8\times10^{11}$ N/m^2. Neglect the inertia of the shaft and gears.

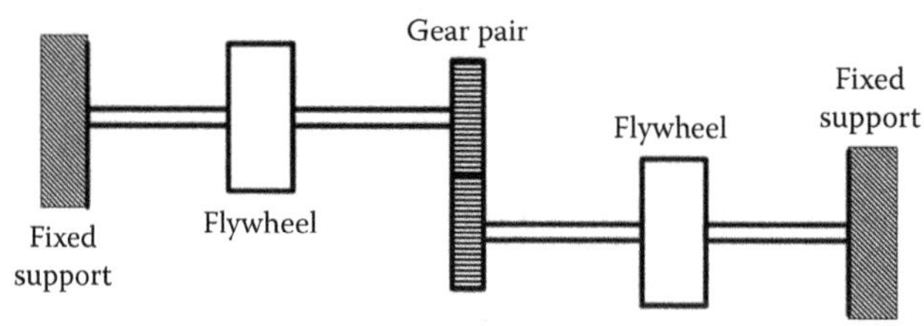

FIGURE 6.79 Two flywheels connected by a gear pair.

Exercise 6.28 Obtain the torsional natural frequencies of the rotor system shown in Figure 6.79 by using the FEM (Chapter 7). Take $G=0.8\times10^{11}$ N/m^2. Neglect the inertia of the shaft and gears.

Exercise 6.29 Obtain the torsional natural frequencies of the rotor system shown in Figure 6.80. Consider a central disk of polar mass moment of inertia 0.01 kg-m^2 and a concentric ring of polar mass moment of inertia 0.02 kg-m^2. These are assumed to be the rigid body and are connected by torsional springs. The effective torsional stiffness of these springs is 1 kN-m/rad. The central disk is attached to a flexible massless shaft with a length of 0.25 m and diameter of 0.02 m, and the other end is fixed to the ground. Take for the shaft $G=0.8\times10^{11}$ N/m^2. It is assumed that the ring has a motion in the horizontal plane and remains concentric to the central disk during motion. This can be considered as a crude model of a helicopter main propeller hub arrangement for torsional vibration analysis.

Exercise 6.30 A centrifugal compressor is gear driven by an electric motor, with a gear reduction of 7.5–1. The motor armature, moment of inertia 22.76 kg-m^2, is connected to the pinion, moment of inertia 0.14 kg-m^2, through a shaft 88.9 mm in diameter and 0.813 m long. The compressor shaft is equivalent to a shaft 152.4 mm in diameter, and 1.372 m long, with a gear wheel,

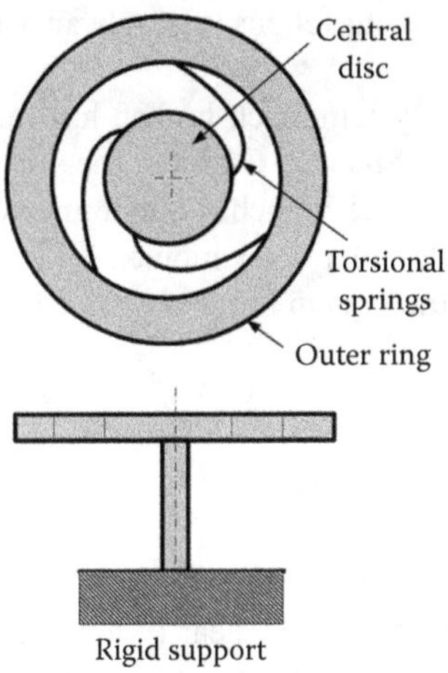

FIGURE 6.80 A cantilever rotor with flexible disk connection.

moment of inertia 303.41 kg-m^2, at one end and a rotor of moment of inertia 16.86 kg-m^2 at the other end. Find the torsional natural frequencies of the system and corresponding mode shapes. Show the node locations in each case. $G=0.8\times10^{11}$ N/m^2.

Exercise 6.31 Two flywheels (each of them has polar mass moment of inertia of I_p) are connected by a splined coupling having torsional stiffness of k_{tc}. In between the coupling and flywheels, each of the shaft segments has a torsional stiffness of k_{ts}. What is the torsional natural frequency of the system?

Exercise 6.32 Obtain *torsional natural frequencies* of the geared system shown in Figure 6.81 using the TMM/FEM (refer to Chapter 7). It has fixed support conditions at left end, and the pinion and the gear are mounted in the right end with a gear ratio of 3 and their polar moments of inertia are 0.012 and 0.01 kg-m^2, respectively. Take for shaft $G=0.8\times10^{11}$ N/m^2 and $\rho=7800$ kg/m^3. The length of the both shafts is 0.6 m, and their diameter is 0.01 m.

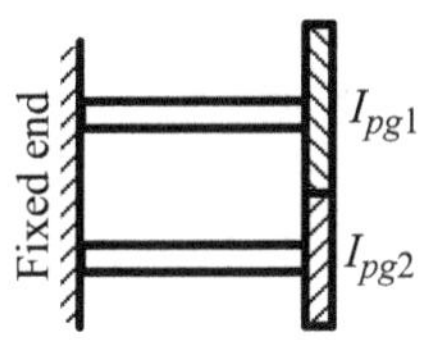

FIGURE 6.81 Geared system with fixed ends.

Exercise 6.33 Obtain *torsional natural frequencies* of the system shown in Figure 6.82 using the TMM/FEM (refer to Chapter 7). The polar moment of inertia of the disk is 0.01 kg-m^2, and the stiffness of the torsional spring is equal to half that of the cantilever shaft. The length of the shaft is 0.6 m, and its diameter is 0.01 m. Take for shaft $G=0.8\times10^{11}$ N/m^2 and $\rho=7800$ kg/m^3.

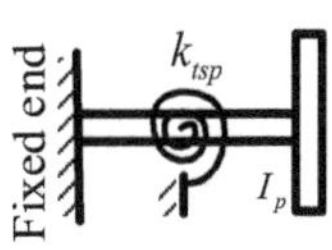

FIGURE 6.82 A cantilever rotor system with a torsional spring.

Exercise 6.34 Multiple choice questions, choose a single answer:

i. In a three-branched rotor system, each branch has three flywheels having an appreciable polar mass moment of inertia. Consider the polar mass moment of inertia of the shaft as negligible. These three branches are connected together by three gears with each having an appreciable polar mass moment of inertia. What number of torsional natural frequencies does the system have?
A. 10
B. 12
C. 9
D. 11

ii. Between two shafts, the motion is transmitted by a gear pair with an idler in between. The gear ratio between the first shaft gear (pinion) and idler is n_1, and the gear ratio between the idler and gear in the second shaft is n_2. The input and output shafts have stiffness k_{t_1} and k_{t_2}, respectively. The idler shaft stiffness is ignored. Then, the equivalent stiffness of uniform shaft with respect to the input shaft would be

A. $\dfrac{n_1^2 n_2^2 k_{t_1} k_{t_2}}{n_1^2 n_2^2 k_{t_1} + k_{t_2}}$

B. $\dfrac{k_{t_1} k_{t_2}}{n_1^2 n_2^2 k_{t_1} + k_{t_2}}$

C. $\dfrac{k_{t_1} k_{t_2}}{k_{t_1} + n_1^2 n_2^2 k_{t_2}}$

D. $\dfrac{n_1^2 n_2^2 k_{t_1} k_{t_2}}{n_1^2 k_{t_1} + n_2^2 k_{t_2}}$

iii. The ratio of the transverse to torsional natural frequencies of a cantilever rotor would be

A. $\sqrt{\dfrac{3Gd^2}{16El^2}}$

B. $\sqrt{\dfrac{3Gd^2}{8El^2}}$

C. $\sqrt{\dfrac{3Ed^2}{8Gl^2}}$

D. $\sqrt{\dfrac{3Ed^2}{16Gl^2}}$

where E is the Young's modulus, G is the modulus of rigidity, l is the length of the shaft, and d is the diameter of the shaft.

iv. The ratio of the axial to torsional natural frequencies of a cantilever rotor would be

A. $\sqrt{\dfrac{4Gd^2}{Er^2}}$

B. $\sqrt{\dfrac{Gd^2}{2Er^2}}$

C. $\sqrt{\dfrac{Gd^2}{4Er^2}}$

D. $\sqrt{\dfrac{2Gd^2}{Er^2}}$

where d is the shaft diameter, r is the disk radius, E is the Young's modulus, and G is the modulus of rigidity of the shaft.

v. A cantilever shaft of torsional stiffness k_t is under torsional vibrations. Let l be the length of the shaft. It has a disk at the free end with polar mass moment of inertia I_p. If the shaft itself has polar mass moment of inertia of I_{ps} per unit length, then the fundamental torsional natural frequency would approximately be

A. $\sqrt{k_t/I_p}$

B. $\sqrt{k_t/(I_p + lI_{ps})}$

C. $\sqrt{k_t/(I_p + 0.5lI_{ps})}$

D. $\sqrt{k_t/(I_p + 0.3lI_{ps})}$

vi. Consider the rotor system shown in Figure 6.83 with symmetrical disks of polar moment of inertia I_p and uniform shaft with GJ as the torsional rigidity property. With the help of antiphase mode shape information and considering the equivalent system concept, the corresponding torsional natural frequency would be

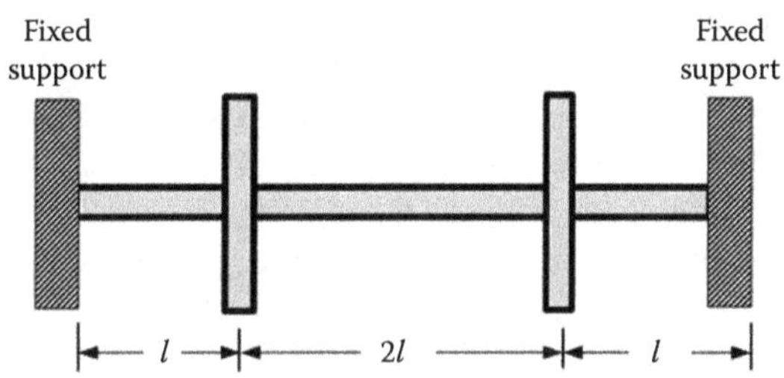

FIGURE 6.83 A cantilever rotor with flexible disk connection.

A. $\sqrt{GJ/(2lI_p)}$

B. $\sqrt{GJ/(3lI_p)}$

C. $\sqrt{3GJ/(lI_p)}$

D. $\sqrt{2GJ/(lI_p)}$

vii. In the equivalent rotor system of a geared system with n as the gear ratio and I_p as the polar moment of inertia of the rotor, the equivalent polar mass moment of inertia of the rotor is given as

A. nI_p

B. n^2I_p

C. I_p/n
D. I_p/n^2

viii. In the torsional vibration analysis of a reciprocating engine where the piston mass is m_p, what would be its approximate contribution toward an equivalent rotating system at the crank pin location?
A. 0
B. $\frac{1}{2}m_p$
C. m_p
D. $2m_p$

ix. In the torsional vibration analysis of a reciprocating engine, the connecting rod mass is m_c with its center of gravity located at ⅓ its length from the crank pin. What would be its approximate contribution toward an equivalent rotating system at the crank pin location?
A. $\frac{1}{3}m_c$
B. $\frac{1}{2}m_c$
C. $\frac{2}{3}m_c$
D. $2m_c$

x. In a gear system when analyzing using the equivalent revolving mass concept for torsional vibration, for converting the mode shape of the equivalent system to the actual system, we need to
A. divide the angular displacements by gear ratio n
B. multiply the angular displacements by n
C. divide the angular displacements n^2
D. multiply the angular displacements by n^2

xi. In a gear system, an idler (polar moment of inertia I_i) is placed between a pinion (polar moment of inertia I_p) and a gear (polar moment of inertia I_g) with a gear ratio between the pinion and idler of n_1 and between the idler and the gear of n_2. What is the equivalent polar moment of inertia of all three gears (i.e., pinion, idler, and gear) with respect to the pinion shaft speed?

A. $I_p + \frac{I_i}{n_1^2} + \frac{I_g}{n_2^2}$

B. $I_p + \frac{I_i}{n_1^2} + \frac{I_g}{n_1 n_2}$

C. $I_p + \frac{I_i}{n_1^2} + \frac{n_1^2 I_g}{n_2^2}$

D. $I_p + \frac{I_i}{n_1^2} + \frac{I_g}{n_1^2 n_2^2}$

xii. For a two-DOF rotor system, which is freely mounted at the ends for a torsional flexible mode, the ratio of torsional natural frequencies of two single-DOF rotor systems situated at either end of the node point would have a value of
A. 1
B. 0
C. 0.5
D. 2

xiii. For a cantilever rotor with a thin disk at the free end, when the torsional and transverse natural frequencies of the system are the same, we get a condition that is
A. independent of the length of the shaft
B. independent of the diameter of the shaft
C. independent of the length and diameter of the shaft
D. dependent of the length and diameter of the shaft

xiv. In an experiment involving a cantilever rotor, the shear stress, τ, is measured using strain gauges at the shaft radius, r. If the length of the shaft is l and E and ν are Young's modulus and Poisson's ratio, respectively, then the angular displacement would be
A. $2\tau(\nu+1)l/(Er)$
B. $2\tau(\nu-1)l/(Er)$
C. $2E(\nu+1)l/(\tau r)$
D. $2E(\nu-1)l/(\nu r)$

xv. If two cantilever beams of the same material with a disk at the free end have the same torsional natural frequencies, then which of the following relations would hold?
A. $\frac{l_1}{d_1}=\frac{l_2}{d_2}$
B. $\frac{\sqrt{l_1}}{d_1}=\frac{\sqrt{l_2}}{d_2}$
C. $\frac{l_1}{d_1^4}=\frac{l_2}{d_2^4}$
D. $l_1 d_1^4 = l_2 d_2^4$
where l and d are the length and diameter of the shaft, respectively, and subscripts 1 and 2 refer to the two different rotor systems.

xvi. A rotor system consists of rigid disks fixed on flexible shafts with free end conditions; node locations will depend
A. only on the polar moments of inertia of disks
B. only on the stiffnesses of the shaft
C. both on the polar moments of inertia of disks and on the stiffnesses of the shaft
D. only on the locations of disks on the shaft

xvii. For a two-disk rotor system having polar moments of inertia $I_{p1} = 2$ kg-m^2 and $I_{p2} = 4$ kg-m^2 undergoing free torsional vibrations (with shaft torsional stiffness, k_t= 120 kN-m/rad), the flexible mode torsional natural frequency (in rad/s) would be
A. $100\sqrt{2}$
B. $100\sqrt{3}$
C. $100\sqrt{6}$
D. $100\sqrt{9}$

xviii. Two shafts are connected by a gear pair (with a gear ratio of 2); each shaft carries a flywheel at the other end (the input shaft has a flywheel of $I_{p1} = 1$ kg-m^2 and torsional stiffness of $k_{t1} = 10$ kN-m/rad, and the output shaft has a flywheel of $I_{p2} = 4$ kg-m^2 and torsional stiffness of $k_{t2} = 40$ kN-m/rad). The flexible mode torsional natural frequency (in rad/s) of the gear system would be
A. 80
B. 90

C. 100
D. 120

xix. In a two-disk free–free rotor system, the node position for torsional vibration is obtained and then a very fine transverse cut equal to the radius of the shaft is made on the shaft by a saw at the node position. What would be its effect?
A. since it is a node position the shaft torsional natural frequency would not change
B. the rotor torsional natural frequency would change
C. the node position would remain the same
D. the node position would remain the same but the rotor torsional natural frequency would change

xx. In the torsional vibration of a rotor at the node position the stresses are
A. maximum
B. minimum
C. zero
D. optimum

xxi. In a three-DOF rotor system with free–free boundary conditions, stresses at a node position corresponding to torsional vibrations of two nodes as compared to one node will be
A. more
B. less
C. equal
D. both zero

xxii. In modeling of a reciprocating engine to an equivalent rotating mass system, the following statement is true:
A. only the equivalent revolving mass of the crank and the connecting rod is considered
B. only the equivalent revolving mass of the crank and the piston is considered
C. only the equivalent revolving mass of the connecting rod and the piston is considered
D. only the equivalent revolving mass of the crack, the connecting rod, and the piston is considered

xxiii. In an equivalent dynamic system (which is used to simplify analysis) as compared to the original conservative dynamic system, the following qualities should be the same:
A. elastic and dissipative energies
B. kinetic and elastic energies
C. kinetic and dissipative energies
D. kinetic energy and virtual work

xxiv. In a gear pair while considering the flexibility of contacting teeth, the angular displacements of two meshing gears are
A. related to the gear ratio
B. related to the square of the gear ratio
C. independent
D. related to the inverse of the gear ratio

xxv. On increasing the gear ratio of a geared rotor system, the torsional natural frequency of the system would
   A. increase
   B. decrease
   C. remain the same
   D. either increase or decrease

xxvi. In a two-stage gearbox with n as the gear ratio in each stage, the equivalent polar moment of inertia of a disk mounted at the last stage with reference to the input shaft speed would be the actual polar moment of inertia times
   A. $1/n$
   B. $1/n^2$
   C. $1/n^3$
   D. $1/n^4$

xxvii. A three-disk mass rotor system is mounted on frictionless bearings for free–free torsional vibration. The shaft is massless. If suddenly one of the disk masses got disengaged from the shaft, how many torsional natural frequencies would the remaining system have?
   A. 0
   B. 1
   C. 2
   D. 3

xxviii. A reciprocating engine has crank revolving mass 4 kg at a radius of 5 cm, and a connecting rod with a mass of 6 kg and length of 30 cm with the center of gravity at one-third of its length from the crack pin; the piston has a mass of 5 kg. What would be the total equivalent revolving mass of the reciprocating engine for the analysis of torsional vibration?
   A. 4 kg
   B. 8 kg
   C. 10.5 kg
   D. 13 kg

xxix. A main shaft of a multicylinder reciprocating engine is analyzed for torsional vibration
   A. by treating the main shaft in isolation
   B. by treating twisting of the crack alone
   C. by treating piston oscillation inside the cylinder
   D. by considering an equivalent of the reciprocating mass as the rotating mass at the main shaft

xxx. The torsional vibration occurs in a rotor system having
   A. a lower polar mass moment of inertia and shaft stiffness
   B. a lower polar mass moment of inertia and higher shaft stiffness
   C. a higher polar mass moment of inertia and shaft stiffness
   D. a higher polar mass moment of inertia and lower shaft torsional stiffness

xxxi. A reciprocating engine shaft gets an excitation torque of
   A. single frequency
   B. double frequencies

C. multiple frequencies
D. continuous frequencies

xxxii. A multidisk rotor system, undergoing torsional vibration and having a nominal angular speed, will have actual angular speeds with respect to the nominal angular speed of different disks
A. the same irrespective of disk positions and modes of vibration
B. lower for all disks irrespective of disk positions and modes of vibration
C. higher for all disks irrespective of disk positions and modes of vibration
D. lower or higher depending upon the disk position and modes of vibration

xxxiii. Two different cantilever rotor systems have shafts (with the stiffnesses k_{t_1} and k_{t_2}) and disks at free ends (with the polar mass moments of inertia I_{p_1} and I_{p_2}) have the same torsional natural frequency. If these two cantilever rotors are connected back-to-back together to form a two-disk single rotor system with free–free end conditions (the disks are at the free ends), then the new rotor system will have flexible mode torsional natural frequency as

A. $\sqrt{\dfrac{k_{t_1}}{I_{p_1}}}$

B. $\sqrt{\dfrac{k_{t_1}+k_{t_2}}{I_{p_1}+I_{p_2}}}$

C. $\sqrt{\dfrac{k_{t_1}k_{t_2}}{(k_{t_1}+k_{t_2})(I_{p_1}+I_{p_2})}}$

D. $\sqrt{\dfrac{(k_{t_1}+k_{t_2})(I_{p_1}+I_{p_2})}{I_{p_1}I_{p_2}}}$

E. $\sqrt{\dfrac{k_{t_1}k_{t_2}}{I_{p_1}I_{p_2}}}$

xxxiv. Two different cantilever rotor systems have shafts (with stiffnesses k_{t_1} and k_{t_2}) and disks at free ends (with polar mass moment of inertia I_{p_1} and I_{p_2}) that have the same torsional natural frequency. If these two cantilever rotors are connected together back-to-back to form a two-disk single-shaft rotor system with free–free end conditions (with disks at free ends), then the new rotor system will have, for the flexible torsional mode, the ratio of angular displacement of two disks (disk 1 and disk 2) as

A. $-\dfrac{I_{p_2}}{I_{p_1}}$

B. $-\dfrac{I_{p_1}}{I_{p_2}}$

C. $-\dfrac{I_{p_2}}{I_{p_1}+I_{p_2}}$

D. $-\dfrac{I_{p_1}}{I_{p_1}+I_{p_2}}$

xxxv. In a two-disk-rotor system under torsional vibration, if the massless shaft has a tapered geometry with the shaft end diameters as d_1 and d_2 (with $d_2 > d_1$), and length L, then the torsional natural frequency of the free–free two-disk rotor system (with I_{p1} and I_{p2} as polar mass moments of inertia) will be

A. $\sqrt{\dfrac{3\pi G(d_2 d_1)^3 (I_{p1}+I_{p2})}{32L(d_2^2+d_1 d_2+d_1^2) I_{p1} I_{p2}}}$

B. $\sqrt{\dfrac{3\pi G(d_2 d_1)^4 (I_{p1}+I_{p2})}{32L(d_2^4-d_1^4) I_{p1} I_{p2}}}$

C. $\sqrt{\dfrac{3\pi G(d_2 d_1)^3 (I_{p1}+I_{p2})}{32L(d_2^2-d_1^2) I_{p1} I_{p2}}}$

D. $\sqrt{\dfrac{3\pi G(d_2^2-d_1^2)(d_2 d_1)^3 (I_{p1}+I_{p2})}{32L(d_2^4-d_1^4) I_{p1} I_{p2}}}$

xxxvi. In a two-disk rotor system (of polar mass moment of inertia I_{p1} and I_{p2}), the shafts of stiffness k_{t1} and k_{t2} are connected by a coupling of stiffness k_{tc}. The outer ends of the shafts have disks and assuming free–free end conditions, the torsional natural frequency of the whole system would be

A. $\sqrt{\dfrac{k_{t1} k_{t2} k_{tc} (I_{p1}+I_{p2})}{(k_{t1} k_{t2}+k_{tc} k_{t1}+k_{t2} k_{tc}) I_{p1} I_{p2}}}$

B. $\sqrt{\dfrac{(k_{t1}+k_{t2}+k_{tc})(I_{p1}+I_{p2})}{I_{p1} I_{p2}}}$

C. $\sqrt{\dfrac{\left(\dfrac{k_{t1} k_{t2}}{k_{t1}+k_{t2}}+k_{tc}\right)(I_{p1}+I_{p2})}{I_{p1} I_{p2}}}$

D. $\sqrt{\dfrac{k_{t1} k_{t2} k_{tc}}{(k_{t1} k_{t2}+k_c k_{t1}+k_{t2} k_c)(I_{p1}+I_{p2})}}$

xxxvii. In a multidisk rotor system for torsional vibration (consider disks are thin), if a node occurs at one of the disks in a particular mode of vibration, the angular displacement of the disk will be

A. zero
B. it will be any other value other than zero
C. a disk location cannot be a node
D. it will be only possible with rigid body mode

xxxviii. In an N-disk rotor system with free–free end conditions under torsional vibration, the maximum number of nodes in the system can be

A. N
B. $N+1$
C. $N-1$
D. N^2

xxxix. Consider a geared rotor system, with two end flywheels, connected by a gear pair through two shafts. If both flywheels and gears have polar mass moment of inertia, then the number of torsional natural frequencies of the rotor system will be

A. 4
B. 3
C. 2
D. it will depend on the gear ratio

xl. In a two-stage compound geared rotor system, two flywheels of polar mass moments of inertia I_{p1} and I_{p2} are connected through two flexible shafts via a two-stage gear pair. The shaft of the intermediate stage has a negligible shaft length (i.e., the gear of the first stage and the pinion of the second stage are very close to each other). The first stage shaft has a stiffness of k_{t1}, and the last stage shaft has a stiffness of k_{t2}. The gears have negligible polar mass moment of inertia. Let us assume that the gear ratios of the first and second stages are n_1 and n_2, respectively. Considering free–free support conditions, the torsional natural frequency of the geared system for the flexible mode will be

A. $\sqrt{\dfrac{k_{t1}k_{t2}\left(n_1^2n_2^2I_{p1}+I_{p2}\right)}{\left(n_1^2n_2^2k_{t1}+k_{t2}\right)I_{p1}I_{p2}}}$

B. $\sqrt{\dfrac{k_{t1}k_{t2}\left\{\left(n_1^2+n_2^2\right)I_{p1}+I_{p2}\right\}}{\left\{\left(n_1^2+n_2^2\right)k_{t1}+k_{t2}\right\}I_{p1}I_{p2}}}$

C. $\sqrt{\dfrac{k_{t1}k_{t2}\left(n_1^2I_{p1}+n_2^2I_{p2}\right)}{\left(n_1^2k_{t1}+n_2^2k_{t2}\right)I_{p1}I_{p2}}}$

D. $\sqrt{\dfrac{n_1^2n_2^2k_{t1}k_{t2}\left(n_1^2n_2^2I_{p1}+I_{p2}\right)}{\left(n_1^2n_2^2k_{t1}+k_{t2}\right)I_{p1}I_{p2}}}$

xli. In the TMM, the point matrix is derived for a rigid thin disk. If the point matrix is to be derived for a rigid cylinder (instead of a rigid thin disk), the form of the point matrix will

A. remain the same and also the field matrix
B. remain the same but not the field matrix
C. not remain the same as well as the field matrix
D. not remain the same but the field matrix will remain the same

xlii. In the TMM, for fixed–fixed end conditions of a rotor system from the overall transfer function matrix, the following component will be zero to get the frequency equation of the system

A. t_{11}
B. t_{12}
C. t_{21}
D. t_{22}

xliii. In a cantilever beam with a fixed support at the right end (station 1) and a disk at the left end (station 0), the overall transfer matrix will be

A. $\mathbf{F}_1\mathbf{P}_0$
B. $\mathbf{P}_0\mathbf{F}_1$
C. $\mathbf{P}_0$
D. $\mathbf{F}_1$

xliv. For an intermediate bearing in a rotor system under torsional vibration, if its torsional stiffness is k_{tb}, the transfer matrix for the intermediate bearing will be

A. $\begin{bmatrix} 1 & 0 \\ k_{tb} & 1 \end{bmatrix}$

B. $\begin{bmatrix} 1 & 0 \\ 1/k_{tb} & 1 \end{bmatrix}$

C. $\begin{bmatrix} 1 & 1/k_{tb} \\ 0 & 1 \end{bmatrix}$

D. $\begin{bmatrix} 1 & k_{tb} \\ 0 & 1 \end{bmatrix}$

xlv. A gear pair has a gear ratio, n, and the transfer matrix related to the gear pair will be

A. $\mathbf{n} = -\begin{bmatrix} 1/n & 0 \\ 0 & n \end{bmatrix}$

B. $\mathbf{n} = -\begin{bmatrix} 1/n & 0 \\ 0 & 1/n \end{bmatrix}$

C. $\mathbf{n} = -\begin{bmatrix} 1/n^2 & 0 \\ 0 & n^2 \end{bmatrix}$

D. $\mathbf{n} = -\begin{bmatrix} 1/n^2 & 0 \\ 0 & n^2 \end{bmatrix}$

xlvi. For a connecting rod in a reciprocating engine, while considering its two masses' dynamic equivalence, the following statement is true
A. One mass becomes reciprocating mass and other pure rotating
B. Both masses become reciprocating masses
C. Both masses become pure rotating masses
D. Both masses remain oscillatory masses

xlvii. For a connecting rod in a reciprocating engine, while considering its two masses' dynamic equivalence, the following statement is true
A. One mass becomes reciprocating mass and other pure rotating
B. Both masses become reciprocating masses
C. Both masses become pure rotating masses
D. Both masses remain oscillatory masses

ANSWERS TO SELECTED EXERCISE PROBLEMS

6.1 $\omega_{nf_1} = 0$ rad/s and $\omega_{nf_2} = \sqrt{\dfrac{k_{t2}\left(I_{p1} + I_{p2}\right)}{I_{p1} I_{p2}}} = 1353.74$ rad/s

6.2 $\omega_{nf} = \sqrt{k_{t0} / I_{p1}} = 140.13$ rad/s

6.3 $\omega_{nf_1} = 100.33$ rad/s and $\omega_{nf_2} = 189.08$ rad/s

6.4 224.41 rad/s, 1000.77 rad/s and 1706.11 rad/s

6.5 0.00 rad/s, 320.77 rad/s, 2323.41 rad/s and 2704.91 rad/s

6.6 46.03 rad/s

6.7 $\omega_{nf1} = 0$ rad/s and $\omega_{nf2} = \sqrt{\dfrac{k_{t2}\left(I_{p1} + I_{p2}\right)}{I_{p1} I_{p2}}} = 76.38$ rad/s

6.8 0.00 rad/s, 1273.41 rad/s and 12180.45 rad/s

6.9 0.00 rad/s, 166.63 rad/s and 1994.67 rad/s

6.10 0.00 rad/s, 146.16 rad/s and 6804.64 rad/s

6.11 0.00 rad/s and 97.49 rad/s

6.12 0.0334 m

6.13 29.018 kg-m2

6.14 1008.90 rad/s and 1007.20 rad/s

6.15 441.65 rad/s, 1323.95 rad/s, 2259.90 rad/s and 5038.25 rad/s

6.16 491.19 rad/s, 1150.85 rad/s, 1709.51 rad/s, 2346.71 rad/s and 5038.29 rad/s

6.17 0.00 rad/s, 980.32 rad/s, 1764.64 rad/s, 2505.27 rad/s and 3756.59 rad/s

6.18 113.94 rad/s

6.19 478.74 rad/s and 918.98 rad/s

6.20 0.00 rad/s, 469.00 rad/s, 493.44 rad/s, 1500.26 rad/s, 1511.84 rad/s, 2117.80 rad/s and 3936.12 rad/s

6.21 0.00 rad/s, 302.68 rad/s, 863.56 rad/s, 1546.88 rad/s, 1647.83 rad/s, 2159.03 rad/s, 2560.42 rad/s, 3087.03 rad/s, 3585.87 rad/s, 4423.81 rad/s, 5419.79 rad/s, 6072.29 rad/s and 7149.01 rad/s

6.22 0.00 rad/s, 202.28 rad/s, 458.50 rad/s and 6336.98 rad/s

6.23 0.00 rad/s and 269.99 rad/s

6.24 0.00 rad/s, 2108.19 rad/s and 2236.07 rad/s

6.25 0.00 rad/s, 188.13 rad/s, 749.91 rad/s and 1055.81 rad/s

6.26 264.22 rad/s and 457.65 rad/s

6.27 364.06 rad/s and 514.85 rad/s

6.28 364.06 rad/s and 514.85 rad/s

6.29 202.72 rad/s and 782.02 rad/s

6.30 0.00 rad/s, 347.30 rad/s and 456.064 rad/s

6.31 $\omega_{nf1} = 0,\ \omega_{nf2} = \sqrt{\dfrac{k_{ts}k_{tc}\left(I_{p1} + I_{p2}\right)}{I_{p1} I_{p2}\left(k_{ts} + 2k_{tc}\right)}}$

6.32 105.32 rad/s

6.33 (without shaft inertia): 120.6 rad/s; (with shaft inertia): 121.0 rad/s and 19612.0 rad/s.

6.34

(i) A	(ii) B	(iii) D	(iv) C	(v) D	(vi) D
(vii) D	(viii) B	(ix) C	(x) A	(xi) D	(xii) A
(xiii) B	(xiv) A	(xv) C	(xvi) C	(xvii) D	(xviii) C
(xix) B	(xx) A	(xxi) A	(xxii) D	(xxiii) B	(xxiv) C
(xxv) D	(xxvi) D	(xxvii) C	(xxviii) B	(xxix) D	(xxx) D
(xxxi) C	(xxxii) D	(xxxiii) A	(xxxiv) A	(xxxv) A	(xxxvi) A
(xxxvii) A	(xxxviii) C	(xxxix) B	(xl) A	(xli) A	(xlii) B

REFERENCES

Bevan, T., 1984, *The Theory of Machines*, 3rd ed. New Delhi: CBS Publishers & Distributors.

Bremer, R.C. Jr., 1979, A Practical Treatise on Engine Crankshaft Torsional Vibration Control. SAE/SP-79/445, *Proceedings of West Coast International Meeting*, August 6–9, Portland, OR, pp. 1–38.

Hestermann, D.C. and Stone, B.J., 1992, The Causes of Torsional Resonance in Reciprocating Engines and Pumps at Other Than Integer Multiples of the Fundamental Excitation Frequency. *IMechE International Conference on Vibration and Rotating Machinery, C432/041*, September 7–10, London: Mechanical Engineering Publications, pp. 517–521.

Hestermann, D.C. and Stone, B.J., 1995, Secondary inertia effects in the torsional vibration of reciprocating engines: A literature review. *Proceedings of the Institution of Mechanical Engineering*, *209*, 11–15.

Kahraman, A., 2001, Free torsional vibration characteristics of compound planetary gear sets. *Mechanism and Machine Theory*, *36*, 953–971.

Mendes, S., Meirelles, P.S., and Zampieri, D.E., 2008, Analysis of torsional vibration in internal combustion engines: Modelling and experimental validation. *Proceedings of the Institution of Mechanical Engineers, Part K: Journal of Multi-body Dynamics*, *222*(*2*), 155–178.

Press, W.H., Teukolsky, S.A., Vetterling, W.T., and Flannery, B.P., 1998, *Numerical Recipes in C: The Art of Scientific Computing*, 2nd ed. New York, NY: Cambridge University Press.

Rao, J.S., 1996, *Rotor Dynamics*, 3rd ed. New Delhi: New Age.

Timoshenko, S.P. and Young, D.H., 1968, *Elements of Strength of Materials*, 5th ed. New Delhi: Affiliated East-West Press.

Tuplin, W.A., 1966, *Torsional Vibrations*. London: Pitman.

Walker, D.N., 2003, *Torsional Vibration of Turbomachinery*. New York, NY: McGraw-Hill.

Wilson, W.K., 1956, *Practical Solution of Torsional Vibration Problems*, Vol. *1*. London: Chapman & Hall.

Wilson, W.K., 1963, *Practical Solution of Torsional Vibration Problems*, Vol. *2*. New York: John Wiley.

Wilson, W.K., 1965, *Practical Solution of Torsional Vibration Problems*, Vol. *3*. London: Chapman & Hall.

7 Torsional Vibrations of Rotor Systems by the Continuous System and Finite Element Methods

In Chapter 6, we considered torsional vibrations of rotor systems using Newton's second law of motion and the transfer matrix method. The transfer matrix method especially proved its applicability for analyzing even large rotor systems in a more systematic (algorithmic) fashion. The advantage of the method is that the size of matrices does not increase with the degrees of freedom (DOFs) of the system. The only disadvantage of the method is that natural frequencies have to be obtained for large systems using root searching numerical techniques. Moreover, it has some other shortcomings that we observe, especially when we analyze transverse vibrations in Chapter 8. There, the method requires special treatment when rotor systems have multiple supports and when the shaft is treated as continuous (i.e., when the mass and the stiffness of the shaft are distributed throughout its span). In the present chapter, we first analyze torsional vibrations using the analytical approach by treating the shaft as continuous, which has infinite DOFs. For this case, the governing equation becomes a partial differential equation (i.e., identical to the *wave equation*). For simple boundary conditions, governing equations can be solved in closed form. However, for complicated rotor-support systems, the continuous rotor system is very difficult to analyze by analytical methods, and often approximate methods are used.

For analyzing complex systems by an approximate method, we have chosen the finite element method (FEM) among various methods available. The FEM is well known for its versatility and flexibility to analyze complex rotor and support systems. The basic steps involved in the solution of torsional vibrations by the FEM are presented. A variety of cases (simple to multidisc systems, simple gear rotor systems, branched-gear rotor systems, and reciprocating engines) are illustrated through simple numerical examples. The only limitation of the FEM compared with the transfer matrix method (TMM) is that with the increase of DOFs of the system the size of matrices to be handled increases. Large matrices require more computer memory, have a higher computational cost, and involve an inherent ill-conditioning problem (i.e., possibility of becoming nearly singular matrices). Again, it should be noted that all the analyses of torsional vibration are equally valid for axial vibrations with the replacement of torsional rotational displacement, $\varphi_z(t)$, with axial translational displacement, $u_x(t)$, and the torque, $T_z(t)$, with the axial force, $f_x(t)$. In fact, the present chapter paves the way for analyzing more complex transverse vibration problems in rotor-bearing systems to be discussed in Chapters 8, 9 and 10; and also in Chapter 2 of Volume II.

7.1 TORSIONAL VIBRATIONS OF CONTINUOUS SHAFT SYSTEMS

Until now, the flexible shaft was considered massless (Figure 7.1a), and a long, heavy shaft was treated as rigid (Figure 7.1b) for the transverse and torsional vibrations. However, there are cases where the mass (i.e., the mass and the diametral and polar mass moments of inertia) and the stiffness (e.g., the transverse, the longitudinal, or the torsional) are both distributed along the shaft, and such flexible heavy rotors (Figure 7.1c) fall under the category of *continuous systems*. Figure 7.1 is for transverse vibrations because it is easy to visualize the motion; however, the shaft models are still valid for other types of vibrations (i.e., the torsional and axial vibrations).

DOI: 10.1201/9781003661559-7

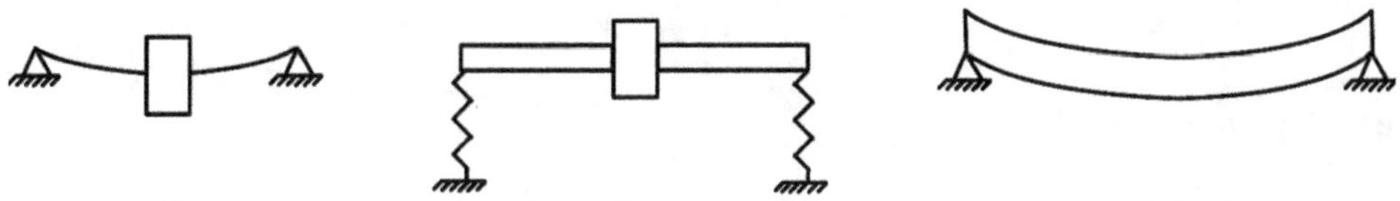

FIGURE 7.1 Different shaft models of a rotor system: (a) a flexible shaft, (b) a rigid shaft, and (c) a continuous shaft.

In continuous systems, equations of motion take the form of a partial differential equation with spatial (e.g., z in the present one-dimensional model; however, for the general solid model, it can have x, y, and z) and temporal (i.e., t, time) derivatives. Equations of motion of a continuous rotor system can be derived by (i) the force and the moment balance of a differential element and (ii) *the extended Hamilton's principle.* Although the force and the moment balance are a convenient approach for simple rotor systems, Hamilton's principle is applied for complex rotor systems. The approach requires the consideration of the potential and kinetic energies of the system. Hamilton's principle is briefly discussed here; for more detailed information, readers are referred to excellent texts by Meirovitch (1986), Rao (1992), and Thomson and Dahleh (1998).

7.1.1 Hamilton's Principle

The *extended Hamilton's principle* is stated as an integral equation in which the energy is integrated over an interval of time. Mathematically, the principle can be stated as

$$\int_{t_1}^{t_2} (\delta T + \delta W)\, dt = 0 \tag{7.1}$$

where δ is the *variational operator* (similar to the commonly used *differential operator* in most mathematical operations), δT is the variation of the kinetic energy, and δW is the virtual work that includes contributions from both the *conservative* (e.g., inertia and restoring) and *nonconservative* (e.g., dissipative internal and external) forces and can be expressed as

$$\delta W = \delta W_c + \delta W_{nc} \tag{7.2}$$

where subscripts c and nc represent the conservative (e.g., the strain (potential) energy in elastic bodies and gravitational potential energy) and nonconservative (the energy due to damping forces and work due to external forces) parts. The extended Hamilton's principle is a quite general method and can be used to derive governing dynamic equations for a system of particles, for a system of rigid bodies, and for continuous (distributed parameter) systems. In deriving *Hamilton's principle* (which is valid for conservative dynamic systems only as compared with the extended Hamilton's principle, which is valid for nonconservative systems as well), it is assumed that virtual displacements must be reversible. It has the restriction that the constraint forces must not perform any work, which means that the principle cannot be used for systems with friction forces. The virtual work performed by the conservative forces can be expressed as the negative of the variation of the potential energy, i.e.,

$$\delta W_c = -\delta U \tag{7.3}$$

where U is the *potential energy* of the system. From Equation 7.2, we get

$$\delta W = -\delta U + \delta W_{nc} \tag{7.4}$$

Defining the *Lagrangian*, *L*, as

$$L = T - U \quad \text{so that} \quad \delta L = \delta T - \delta U \tag{7.5}$$

We now have

$$\delta T + \delta W = \delta T + (-\delta U + \delta W_{nc}) = (\delta T - \delta U) + \delta W_{nc} = \delta L + \delta W_{nc} \tag{7.6}$$

Noting the earlier expression, Equation 7.1 takes the following form:

$$\int_{t_1}^{t_2} (\delta L + \delta W_{nc})\, dt = 0 \quad \text{with} \quad \delta L = \delta T - \delta U \tag{7.7}$$

In a special case, when nonconservative forces are zero, Equation 7.7 gives

$$\int_{t_1}^{t_2} \delta L dt = 0 \tag{7.8}$$

which is Hamilton's principle for conservative systems. The physical interpretation of Equation 7.8 is that out of all possible paths (e.g., paths 1 and 2) of motion of a system during an interval of time from t_1 to t_2 (Figure 7.2), the actual path will be that for which the integral has a *stationary value.*

It can be shown that the stationary value is, in fact, the minimum value of the integral; however, the derivation of it is beyond the scope of the present book. Hamilton's principle can yield governing differential equations as well as boundary conditions. Variations of field variables are small quantities in the spatial domain only, and these need to satisfy the constraints of the problem. However, in the temporal domain, the variations are not defined. For illustration of the method, a very simple, but representative case of a single-DOF spring-mass-damper rotor system is considered through an example.

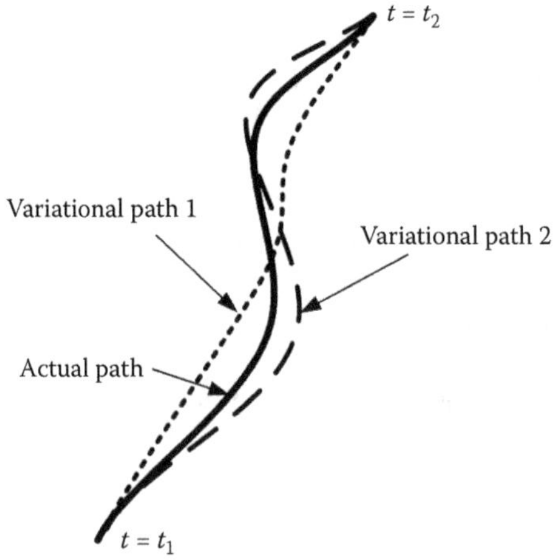

FIGURE 7.2 The actual and variational paths of the motion.

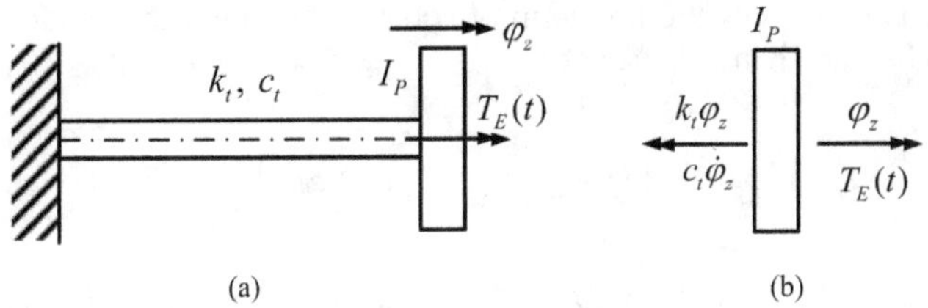

FIGURE 7.3 (a) A single-DOF spring-mass-damper rotor system. (b) A free body diagram.

Example 7.1

A single-DOF spring-mass-damper rotor system (Figure 7.3) is subjected to an external torque, $T_e(t)$; obtain the equation of motion using the extended Hamilton's principle. The shaft torsional stiffness and damping are represented by k_t and c_t, respectively, and the polar mass moment of inertia of the disk is I_p.

Solution:

To illustrate the extended Hamilton's principle, a simple single-DOF spring-mass-damper rotor system is considered; subsequently, this will help in the formulation of governing dynamic equations for continuous systems.

For the present rotor system, the kinetic energy, T, and strain energy, U, can be written as

$$T = \frac{1}{2} I_p \dot{\varphi}_z^2 \quad \text{and} \quad U = -W_c = \frac{1}{2} k_t \varphi_z^2$$

so that

$$\delta T = I_p \dot{\varphi}_z \delta(\dot{\varphi}_z) \quad \text{and} \quad \delta U = -\delta W_c = k_t \varphi_z \delta(\varphi_z) \tag{a}$$

It should be noted that the variation operator is similar in operation to the differential operator. Now, the nonconservative virtual work done is given as (Figure 7.3b)

$$\delta W_{nc} = T_E(t)\delta\varphi_z - (c_t \dot{\varphi}_z)\delta\varphi_z \tag{b}$$

where $T_E(t)$ is an external torque and $c_t\dot{\varphi}_z$ is the torque due to damping. The virtual work done by the external torque is positive; however, the virtual work due to the damping force is negative. The extended Hamilton's principle is written as

$$\int_{t_1}^{t_2} (\delta T + \delta W_c + \delta W_{nc}) \mathrm{d}t = 0 \tag{c}$$

or

$$\int_{t_1}^{t_2} (\delta T - \delta U + \delta W_{nc}) dt = 0 \tag{d}$$

On substituting Equations (a) and (b) into Equation (d), we get

$$\int_{t_1}^{t_2} \left\{ I_p \dot{\varphi}_z (\delta\dot{\varphi}_z) - k_t \varphi_z (\delta\varphi_z) + T_E(t)\delta\varphi_z - (c_t\dot{\varphi}_z)\delta\varphi_z \right\} dt = 0 \tag{e}$$

On integrating by parts of only the first term of Equation (e) in which variation and differential operators need to be interchanged, we get

$$I_p\dot{\varphi}_z(\delta\varphi_z)\Big|_{t_1}^{t_2} - \int_{t_1}^{t_2}\{I_p\ddot{\varphi}_z + k_t\varphi_z - T_E(t) + c_t\dot{\varphi}_z\}(\delta\varphi_z)dt = 0 \tag{f}$$

The first term of Equation (f) will vanish because the variation is defined to be zero at two instants of time t_1 and t_2. The second term gives (because the variation $\delta\varphi_z$ is arbitrary and it may not be zero)

$$I_p\ddot{\varphi}_z + c_t\dot{\varphi}_z + k_t\varphi_z = T_E(t) \tag{g}$$

which is a standard equation of motion of the single-DOF spring-mass-damper rotor system in torsional vibration. For continuous systems, we would see that, along with the equation of motion, boundary conditions also come automatically from derivations. Now, in the subsequent subsection, Lagrange's equation will be introduced, which can also be used to derive equations of motion of discrete systems.

7.1.2 Lagrange's Equation

Hamilton's principle is stated as an integral equation, where the total energy is integrated over a time interval. On the other hand, Lagrange's equation involves differential equations, in which one considers the energies of the system instantaneously in time. Hamilton's principle can be used to derive Lagrange's equation in a set of generalized coordinates, η_i. Lagrange's equation can be written as

$$\frac{d}{dt}\left(\frac{\partial T}{\partial \dot{\eta}_i}\right) - \frac{\partial T}{\partial \eta_i} = Q_i(t), \quad i = 1,2,\ldots,N \tag{7.9}$$

with

$$Q_i = Q_{c_i} + Q_{nc_i} \tag{7.10}$$

where $Q(t)$ is the torque and subscripts c and nc represent the conservative and the nonconservative, respectively. Equation 7.9 is Lagrange's equations of motion in their most general form. The conservative generalized torques have the following form:

$$Q_{c_i} = -\frac{\partial U}{\partial \eta_i}; \quad i = 1,2,\ldots,N \tag{7.11}$$

where U is the potential energy. On substituting Equation 7.11 into Equation 7.9, we get

$$\frac{d}{dt}\left(\frac{\partial T}{\partial \dot{\eta}_i}\right) - \frac{\partial T}{\partial \eta_i} + \frac{\partial U}{\partial \eta_i} = Q_{nc_i}(t); \quad i = 1,2,\ldots,N \tag{7.12}$$

Because the potential energy does not depend upon the velocity, Equation 7.12 can be written as

$$\frac{d}{dt}\left(\frac{\partial L}{\partial \dot{\eta}_i}\right) - \frac{\partial L}{\partial \eta_i} = Q_{nc_i}(t); \quad i = 1,2,\ldots,N \tag{7.13}$$

where $L = (T - U)$ is the Lagrangian. Lagrange's equations are mainly used for discrete systems, including the rigid body. However, for continuous systems, Hamilton's principle is preferred over

Lagrange's equations because, in the formal case, boundary conditions come from the formulation itself.

Example 7.2

For a single-DOF spring-mass-damper rotor system subjected to an external torque, obtain the equation of motion using Lagrange's equation.

Solution:

From the previous example, we have

$$L = T - U = \frac{1}{2} I_p \dot{\varphi}_z^2 - \frac{1}{2} k_t \varphi_z^2; \quad \delta W_{nc} = Q_{nc} \delta\varphi_z \quad \text{with} \quad Q_{nc_i}(t) = T_E(t) + (-c_t \dot{\varphi}_z) \tag{a}$$

From Equation (a), we have

$$\frac{d}{dt}\left(\frac{\partial L}{\partial \dot{\varphi}_z}\right) = \frac{d}{dt}(I_p \dot{\varphi}_z) = I_p \ddot{\varphi}_z \quad \text{and} \quad \frac{\partial L}{\partial \varphi_z} = k_t \varphi_z \tag{b}$$

On substituting Equations (a) and (b) into Lagrange's equation (Equation 7.13), we get

$$I_p \ddot{\varphi}_z + k_t \varphi_z = T_E(t) - c_t \dot{\varphi}_z$$

or

$$I_p \ddot{\varphi}_z + c_t \dot{\varphi}_z + k_t \varphi_z = T_E(t) \tag{c}$$

which is a standard equation of motion in torsional vibration. This example elucidates the application of Lagrange's equation in the simplest form.

7.1.3 Governing Differential Equations

Governing equations for torsional vibrations of continuous shafts are much simpler compared with transverse vibrations. It should be noted that the former is identical to axial vibrations of rods (by replacing the torsional rotational displacement with the axial translational displacement, the polar mass moment of inertia with the mass, and the second polar moment of area, J, with the cross-sectional area, A). To obtain governing differential equations, the variational principle (i.e., Hamilton's principle) is used, which requires the system's *potential energy functional.* The potential energy functional is obtained by considering the displacement field of a point P in the shaft cross-section when the external torque is applied onto the shaft, as shown in Figure 7.4. Let $T(z, t)$ be the distributed external torque applied at some part of the span of the rod, and $T_0(t)$ be the concentrated external torque applied at z_0. Let $\varphi_z(z,t)$ be the rotational displacement of any plane at location z about the z-axis, and L be the total length of the rod. Consider a point P (Figure 7.4) in the cross-section

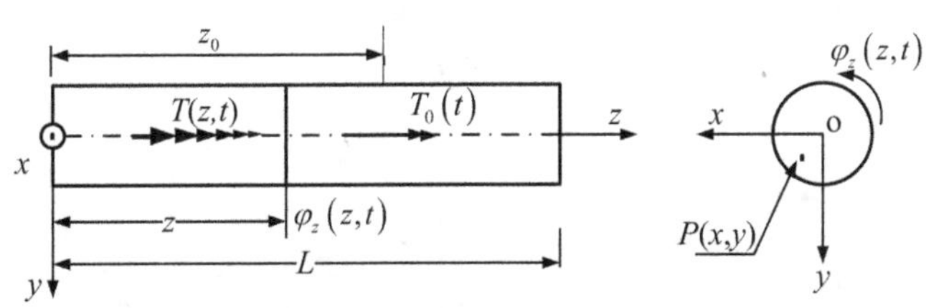

FIGURE 7.4 A continuous rod under time-dependent external torques.

at an axial position z from the origin of the coordinate system x–y–z. From Figure 7.5, we have the following geometric relations:

$$\sin\theta = \frac{y}{r} \quad \text{and} \quad \cos\theta = \frac{x}{r} \tag{7.14}$$

where (x, y, z) is the Cartesian coordinate of the point P, whereas (r, θ, z) is the cylindrical coordinate of the same point.

Derivation of displacement, strain, and stress fields: Due to pure twisting of a plane, the point P moves to P_1 (Figure 7.5a), so that $PP_1 = r\varphi_z$ for a small rotational displacement, φ_z. The displacement field of the point P, located on the cross-section at an axial distance of z from the origin, can be expressed as

$$\begin{aligned} u_x &= PQ = -(PP_1)\sin\theta = -r\varphi_z\sin\theta = -y\varphi_z(z,t) \\ u_y &= P_1Q = (PP_1)\cos\theta = r\varphi_z\cos\theta = x\varphi_z(z,t) \quad \text{and} \quad u_z = 0 \end{aligned} \tag{7.15}$$

where $u_i(x, y, z)$ is the translational displacement of the point with $i=x, y, z$. The strain field is given by

$$\begin{aligned} &\varepsilon_{xx} = u_{x,x} = 0; \quad \varepsilon_{yy} = u_{y,y} = 0; \quad \varepsilon_{zz} = u_{z,z} = 0; \quad \varepsilon_{xy} = \left(u_{x,y} + u_{y,x}\right) = \left(-\varphi_z + \varphi_z\right) = 0 \\ &\varepsilon_{yz} = \left(u_{y,z} + u_{z,y}\right) = \left(x\varphi_{z,z} - 0\right) = x\varphi_{z,z} \quad \text{and} \quad \varepsilon_{zx} = \left(u_{z,x} + u_{x,z}\right) = \left(0 - y\varphi_{z,z}\right) = -y\varphi_{z,z} \end{aligned} \tag{7.16}$$

where $u_{x,x} = \dfrac{\partial u_x}{\partial x}$, $u_{x,y} = \dfrac{\partial u_x}{\partial y}$, etc. The stress field is given by

$$\begin{aligned} &\sigma_{xx} = E\varepsilon_{xx} = 0; \quad \sigma_{yy} = \sigma_{zz} = 0; \quad \tau_{xx} = G\varepsilon_{xy} = 0 \\ &\tau_{yz} = G\varepsilon_{yz} = Gx\varphi_{z,z} \quad \text{and} \quad \tau_{zx} = G\varepsilon_{zx} = -Gy\varphi_{z,z} \end{aligned} \tag{7.17}$$

Derivation of various energy terms: The strain energy is given by

$$\begin{aligned} U &= \frac{1}{2}\int_V \left[\sigma_{xx}\varepsilon_{xx} + \sigma_{yy}\varepsilon_{yy} + \sigma_{zz}\varepsilon_{zz} + \tau_{xy}\varepsilon_{xy} + \tau_{yz}\varepsilon_{yz} + \tau_{zx}\varepsilon_{zx}\right]dV \\ &= \frac{1}{2}\int_0^L\int_A \left[Gy^2\varphi_{z,z}^2 + Gx^2\varphi_{z,z}^2\right]dA\,dz = \frac{1}{2}\int_0^L GJ\varphi_{z,z}^2\,dz \quad \text{with} \quad J = \int_A \left(x^2 + y^2\right)dA \end{aligned} \tag{7.18}$$

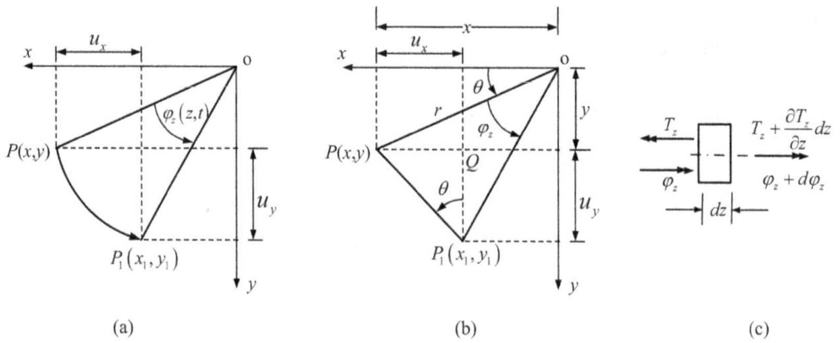

FIGURE 7.5 (a) Displacement of a point on the cross-section and (b) its geometric details. (c) Free body diagram of an infinitesimal shaft segment.

where J is the polar moment of inertia of the shaft cross-section, V is the volume, and A is the area of the shaft cross-section. Velocity components, noting Equation 7.15, can be written as

$$\dot{u}_x = -y\dot{\varphi}_z(z,t); \quad \dot{u}_y = x\dot{\varphi}_z(z,t); \quad \dot{u}_z = 0 \tag{7.19}$$

where $\dot{u}_x = u_{x,t} = \dfrac{\partial u_x}{\partial t}$, etc. The kinetic energy is given by

$$T = \frac{1}{2}\int_0^L\int_A \left[\rho\left\{\dot{u}_x^2 + \dot{u}_y^2 + \dot{u}_z^2\right\}\right] dA dz = \frac{1}{2}\int_0^L \rho J\dot{\varphi}_z^2 dz \tag{7.20}$$

where ρ is the density of the shaft material. The external work done (i.e., the nonconservative energy) for a distributed external torque, $T(z, t)$, and for a concentrated torque, $T_0(t)$, as shown in Figure 7.4, is given by

$$W_{nc} = \int_0^L \left[T(z,t) + T_0(t)\delta^*(z - z_0)\right]\varphi_z dz \tag{7.21}$$

where δ^* is the *Direct delta function*, which is defined as

$$\int_{-\infty}^{+\infty} \delta^*(z - z_0) dz = 1 \text{ for } z = z_0 \text{ and } = 0 \quad \text{for} \quad z \neq z_0 \tag{7.22}$$

Derivation of equations of motion and boundary conditions: On substituting various energy terms from Equations 7.18, 7.20, and 7.21 in the extended Hamilton's principle (Equation 7.7), we obtain

$$\begin{aligned}&\int_{t_1}^{t_2} (\delta T - \delta U + \delta W_{nc}) dt \\ &= \int_{t_1}^{t_2}\int_0^L \left[\rho J\dot{\varphi}_z\delta\dot{\varphi}_z - GJ\varphi_{z,z}\delta\varphi_{z,z} + T(z,t)\delta\varphi_z + T_0(t)\delta^*(z - z_0)\delta\varphi_z\right] dz dt = 0\end{aligned} \tag{7.23}$$

which can be written, after changing the order of differentiation and variation operators of some of the terms, in the following form:

$$\begin{aligned}&\int_0^L\int_{t_1}^{t_2} \left[\rho J\dot{\varphi}_z \frac{\partial}{\partial t}(\delta\varphi_z)\right] dt dz - \int_{t_1}^{t_2}\int_0^L \left[GJ\varphi_{z,z}\frac{\partial}{\partial z}(\delta\varphi_z)\right] dz dt \\ &+ \int_{t_1}^{t_2}\int_0^L \left[T(z,t) + T_0(t)\delta^*(z - z_0)\right]\delta\varphi_z dz dt = 0\end{aligned} \tag{7.24}$$

Here, the aim would be to reduce the order of differentiation in $\delta\varphi_z$. On performing integration by parts of the first and second terms with respect to t and z, respectively, Equation 7.24 gives

$$\begin{aligned}&\int_0^L \rho J\dot{\varphi}_z(\delta\varphi_z)0\Big|_{t_1}^{t_2} dz - \int_0^L\int_{t_1}^{t_2} \left[\rho J\ddot{\varphi}_z(\delta\varphi_z)\right] dt dz - \int_{t_1}^{t_2} GJ\varphi_{z,z}(\delta\varphi_z)\Big|_0^L dt \\ &+ \int_{t_1}^{t_2}\int_0^L \left[GJ\varphi_{z,zz}(\delta\varphi_z)\right] dz dt + \int_{t_1}^{t_2}\int_0^L \left[T(z,t) + T_0(t)\delta^*(z - z_0)\right]\delta\varphi_z dz dt = 0\end{aligned} \tag{7.25}$$

The first term is zero because the variation of the rotational displacement is defined to be zero at two time instants, t_1 and t_2. Equation 7.25 can be written in a compact form as

$$\int_{t_1}^{t_2}\int_{0}^{L}\left[\left\{\rho J\ddot{\varphi}_z(\delta\varphi_z)-GJ\varphi_{z,zz}-T(z,t)-T_0(t)\delta^*(z-z_0)\right\}\delta\varphi_z dz+GJ\varphi_{z,z}(\delta\varphi_z)\Big|_0^L\right]dt=0 \quad (7.26)$$

Because $\delta\varphi_z$, $\delta\varphi_z|_0$, and $\delta\varphi_z|_L$ are variations of the rotational displacement and they are arbitrary (however, at the boundary, these variations may take zero values so these have been retained in boundary conditions), we set their respective coefficients equal to zero, as

$$\rho J\ddot{\varphi}_z-GJ\varphi_{z,zz}-T(z,t)-T_0(t)\delta^*(z-z_0)=0 \quad (7.27)$$

and

$$GJ\varphi_{z,z}(\delta\varphi_z)\Big|_0^L=0 \quad (7.28)$$

Equations 7.27 and 7.28 represent the equation of motion and boundary conditions, respectively, for torsional vibrations of a continuous shaft. It should be noted that $GJ\varphi_{z,z}$ represents the internal reaction torque, T_R, and φ_z is the rotational displacement of the shaft about its longitudinal axis. Accordingly, the boundary conditions can have two components: (i) geometrical or essential boundary conditions (i.e., displacement is specified at a location, z) and (ii) natural boundary conditions (torque is specified at a location, z). Equation 7.27 is similar to the *wave equation* when external torques are absent and can be solved in closed form by using the standard *separation of variables method* (Kreyszig, 2006); this is described subsequently for completeness. The solution involves the satisfaction of both the geometrical and natural boundary conditions of the problem.

7.1.3.1 Alternative Method for Deriving Equation of Motions

On considering the free body diagram of an infinitesimal shaft segment, without any external excitation torque as shown in Figure 7.5c, on application of Newton's second law of motion, we can simply write

$$-T_z+\left(T_z+\frac{\partial T_z}{\partial z}dz\right)=\rho J\ddot{\varphi}_z dz \quad \text{with} \quad T_z=GJ\frac{\partial\varphi_z}{\partial z} \quad (7.29)$$

which gives

$$\frac{\partial T_z}{\partial z}=\rho J\ddot{\varphi}_z \quad \text{or} \quad \rho J\ddot{\varphi}_z-\frac{\partial}{\partial z}\left(GJ\frac{\partial\varphi_z}{\partial z}\right)=0 \quad (7.30)$$

which is the same as Equation 7.27 for free vibrations. The external torque could also be considered in the element to get the equation of motion for forced torsional vibrations.

Torsional natural frequencies and mode shapes for a cantilever rod: Let us obtain the torsional natural frequencies (or eigenvalues) and mode shapes (or eigenfunctions) for a continuous shaft with cantilever boundary conditions as shown in Figure 7.6.

Let the torsional rigidity of a circular shaft be GJ and the mass density of the shaft material be ρ. At the fixed end of the shaft, no rotational displacement takes place, and at the free end of the shaft, the reactive torque is zero. Hence, boundary conditions for the present case are (noting Equation 7.28)

$$\varphi_z(z,t)\Big|_{z=0}=0 \quad (7.31)$$

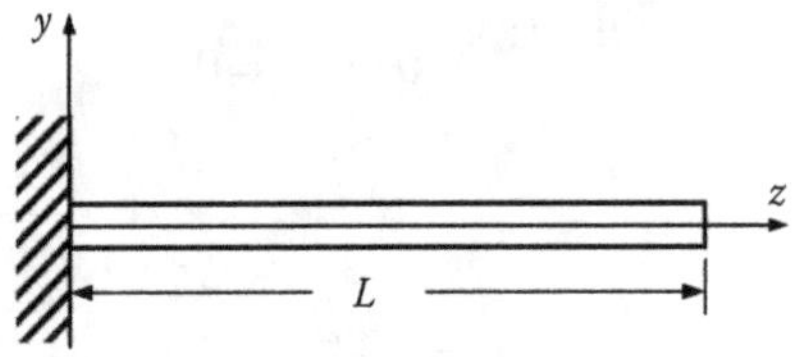

FIGURE 7.6 A shaft with cantilever boundary conditions.

and

$$GJ\frac{\partial\varphi_z(z,t)}{\partial z}\bigg|_{z=L}=0 \tag{7.32}$$

Let us assume the solution of governing Equation 7.30 is in the following form (which ensures that, irrespective of time, the ratio of displacements at various spatial positions remains the same; it is the condition of synchronous motion):

$$\varphi_z(z,t)=\chi(z)\eta(t) \tag{7.33}$$

where $\chi(z)$ is the eigenfunction (it is identical to the eigenvector or mode shape in the discrete rotor system), and for the free vibration, the function $\eta(t)$ is a harmonic function with frequency equal to the natural frequency, ω_{nf}, of the system and has the following form:

$$\eta(t)=A\cos\omega_{nf}t+B\sin\omega_{nf}t \tag{7.34}$$

A and *B* are constants to be determined from initial conditions of the problem, and depending upon that, the free vibration response would change. In view of Equation 7.34, on taking double derivatives with respect to time, *t*, of Equation 7.33, we get

$$\ddot{\varphi}_z(z,t)=-\omega_{nf}^2\chi(z)\eta(t)=-\omega_{nf}^2\varphi_z(z,t) \tag{7.35}$$

where the dot $(\dot{\ })$ represents the single derivative with respect to time. Now, on taking double derivatives with respect to *z* of Equation 7.33, we get

$$\varphi_z''(z,t)=\chi''(z)\eta(t) \tag{7.36}$$

where the double prime (″) represents the double derivative with respect to the spatial coordinate, *z*. On substituting Equations 7.35 and 7.36 into the equation of motion (Equation 7.27), and for free vibrations taking external torque terms equal to zero, we get

$$\left\{\rho\omega_{nf}^2\chi(z)+G\chi''(z)\right\}\eta(t)=0 \tag{7.37}$$

In Equation 7.37, because in general the function $\eta(t)$ cannot be zero, we have

$$G\frac{d^2\chi(z)}{dz^2}+\omega_{nf}^2\rho\chi(z)=0 \tag{7.38}$$

The above equation has to be satisfied throughout the domain $0<z<L$, and at boundaries, we need to satisfy Equations 7.31 and 7.32. Equation 7.38 can be written as

$$\frac{d^2\chi(z)}{dz^2}+\alpha^2\chi(z)=0 \tag{7.39}$$

with

$$\alpha^2=\frac{\rho\omega_{nf}^2}{G} \tag{7.40}$$

which has a solution of the following form:

$$\chi(z)=C\cos\alpha z+D\sin\alpha z \tag{7.41}$$

where C and D are constants to be determined from the boundary conditions of the problem that change depending upon the eigenfunction, $\chi(z)$. On taking the first derivative with respect to z of Equation 7.41, we get

$$\chi'(z)=-C\alpha\sin\alpha z+D\alpha\cos\alpha z \tag{7.42}$$

On applying boundary conditions 7.31 and 7.32, respectively, in Equations 7.41 and 7.42 and noting Equation 7.33, we get

$$\varphi_z(z,t)\big|_{z=0}=\chi(z)\eta(t)\big|_{z=0}=\chi(0)\eta(t)=0\Rightarrow\chi(0)=0\Rightarrow C=0 \tag{7.43}$$

and

$$GJ\frac{\partial\varphi_z(z,t)}{\partial z}\bigg|_{z=L}=GJ\varphi_z'(z,t)\big|_{z=L}=GJ\chi'(L)\eta(t)=0\Rightarrow\chi'(L)=0\Rightarrow D\alpha\cos\alpha L=0 \tag{7.44}$$

where D and α cannot be zero in the general case in Equation 7.44; otherwise, the entire system response is zero. Hence, the *torsional frequency equation* is obtained as

$$\cos\alpha L=0 \tag{7.45}$$

with the solution as

$$\alpha_i L=\frac{\pi i}{2},\quad i=1,3,5,\ldots \tag{7.46}$$

On substituting Equation 7.40 into Equation 7.46, we get torsional natural frequencies as

$$\omega_{nfi}=\frac{i\pi}{2L}\sqrt{\frac{G}{\rho}},\quad i=1,3,5,\ldots \tag{7.47}$$

On substituting Equations 7.46 and 7.43 into Equation 7.41, mode shapes (eigenfunctions) are given by

$$\chi_i(z)=D\sin\frac{\pi i}{2L}z,\quad i=1,3,5,\ldots \tag{7.48}$$

The first three modes are plotted in Figure 7.7. We observe that the second and third modes have one and two nodes (a zero rotational displacement), respectively. On substituting Equations 7.34, 7.47, and 7.48 into Equation 7.33, the torsional free vibration response of the cantilever rod is obtained as

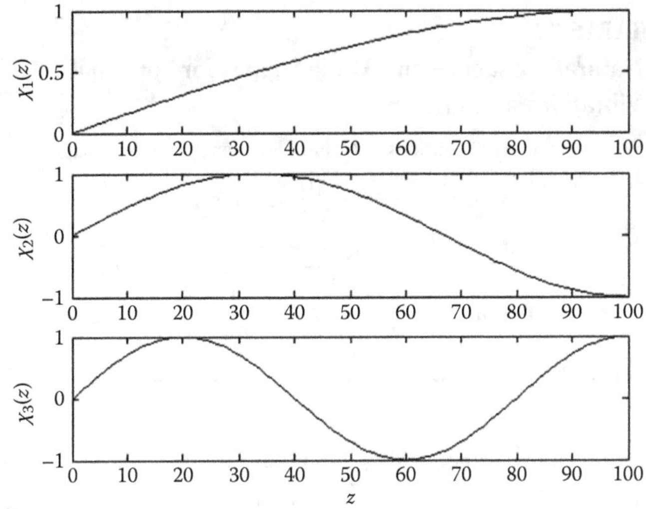

FIGURE 7.7 Torsional mode shapes for the first three modes of a cantilever rod.

$$\varphi_z(z,t) = \sum_{i=1,3,\ldots}^{\infty} \sin\frac{i\pi z}{2L}\left(A_i \cos\frac{i\pi}{2L}\sqrt{\frac{G}{\rho}}t + B_i \sin\frac{i\pi}{2L}\sqrt{\frac{G}{\rho}}t\right) \tag{7.49}$$

where A_i and B_i are constants, which are determined by the initial conditions of the problem. It should be noted that the actual response is the summation of all the modes of the system, and the contributions of each mode (i.e., constants A_i and B_i) depend upon initial conditions. It is interesting to note that if we give initial conditions such that they try to excite the shaft to vibrate closer to its first mode shape, then the main contribution in the response would be from the fundamental natural frequency. Similarly, for higher mode shapes, this will be valid. For example, for zero initial conditions, i.e., $\varphi_z(z,0) = \varphi_0(z)$ and $\dot{\varphi}_z(z,0) = \dot{\varphi}_0(z)$, from Equation 7.49, we get

$$A_i = \frac{2}{L}\int_0^L \varphi_0(z)\sin\frac{i\pi z}{2L}dz \tag{7.50}$$

and

$$B_i = \frac{2}{L}\int_0^L \dot{\varphi}_0(z)\sin\frac{i\pi z}{2L}dz \tag{7.51}$$

which are obtained by the use of the *orthogonality of mode shapes* (Thomson and Dahleh, 1998), i.e.,

$$\int_0^L \rho A \chi_i(z)\chi_j(z)dz \neq 0 \quad \text{for} \quad i = j \tag{7.52}$$

and

$$\int_0^L \rho A \chi_i(z)\chi_j(z)dz = 0 \quad \text{for} \quad i \neq j \tag{7.53}$$

TABLE 7.1
Natural Frequency and Mode Shapes for Torsional Vibrations of Rods

S.N.	Boundary Conditions	Natural Frequency	Mode Shape
1	Fixed–free	$\frac{i\pi}{2L}\sqrt{\frac{G}{\rho}}, \quad i = 1,3,5,\ldots$	$\sin\frac{\pi i}{2L}z$
2	Fixed–fixed	$\frac{i\pi}{L}\sqrt{\frac{G}{\rho}}, \quad i = 1,2,3,\ldots$	$\sin\frac{\pi i}{L}z$
3	Free–free	$\frac{i\pi}{L}\sqrt{\frac{G}{\rho}}, \quad i = 0,1,2,\ldots$	$\cos\frac{\pi i}{L}z$

It can be checked that the above orthogonality conditions are valid for mode shapes of the present problem, i.e., $\chi_i(z) = \sin\frac{\pi i}{2L}z$ and $\chi_j(z) = \sin\frac{\pi j}{2L}z$. Table 7.1 summarizes natural frequencies and mode shapes for some common types of boundary conditions.

Axial vibrations: A detailed treatment of axial (longitudinal) vibrations will not be performed here. There is a one-to-one analogy that can be made between the torsional and axial vibrations by replacing φ_z with u_z, J with A, and G with E. Here, u_z is the axial displacement (i.e., each of the particles of the shaft would have displacement in the z-axis direction only), A is the area of the shaft cross-section, and E is the Young's modulus of the shaft material. It should be noted that for axial vibrations, the strain and kinetic energies can be written as

$$U = \frac{1}{2}\int_0^L EAu_{z,z}^2 dz \quad \text{and} \quad T = \frac{1}{2}\int_0^L \rho A\dot{u}_z^2 dz \tag{7.54}$$

Hence, the equations of motion for free vibrations would be

$$\frac{\partial^2 u_z(z,t)}{\partial t^2} = \alpha\frac{\partial^2 u_z(z,t)}{\partial z^2} \quad \text{with} \quad \alpha = \frac{E}{\rho} \tag{7.55}$$

Hence, all the analyses described in this section for torsional free vibrations are equally valid for the longitudinal vibrations as well, and now the natural frequency is called the *longitudinal natural frequency*.

Now through a simple example, a torsional free vibration analysis is performed for a special case when the boundary condition of the problem depends upon the eigenvalue (i.e., the natural frequency).

Example 7.3

Consider a circular shaft (Figure 7.8) that is rigidly fixed at one end and carries a rigid disk of polar mass moment of inertia, I_p, at the free end. The torsional rigidity of the shaft is GJ, and the mass density of the shaft material is ρ. Obtain the torsional natural frequencies and mode shapes for the rotor system using the continuous approach.

Solution:

The equation of motion of the continuous shaft remains the same as Equation 7.27. However, now the boundary conditions are (refer to Figure 7.8)

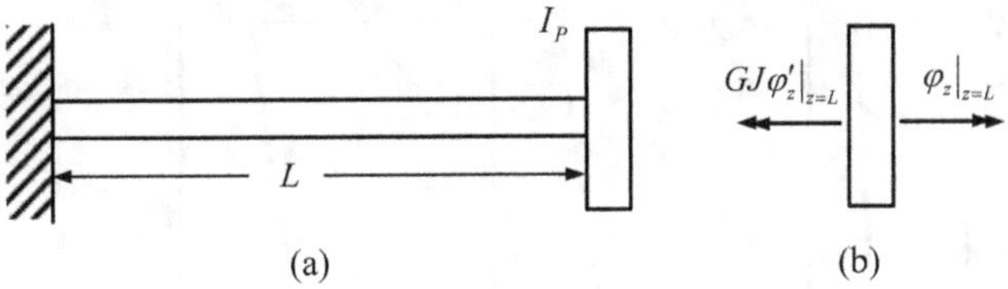

FIGURE 7.8 (a) A cantilever shaft with a disk at the free end. (b) A free body diagram of the disk.

$$\varphi_z(z,t)|_{z=0} = 0 \tag{a}$$

and

$$-GJ\frac{\partial \varphi_z(z,t)}{\partial z}\bigg|_{z=L} = I_p \frac{\partial^2 \varphi_z(z,t)}{\partial z^2}\bigg|_{z=L} \Rightarrow GJ\chi'(z)\big|_{z=L} = \omega_{nf}^2 I_p \chi(z)\big|_{z=L} \tag{b}$$

Hence, the reactive torque on the shaft at the free end is balanced by the inertia of the disk; this can be obtained by using Newton's second law of motion from the free diagram of the disk as shown in Figure 7.8b. From Equation 7.41, we have a general mode shape of the following form:

$$\chi(z) = C\cos\alpha z + D\sin\alpha z \tag{c}$$

with

$$\alpha^2 = \frac{\rho\omega_{nf}^2}{G} \tag{d}$$

On taking the first derivative with respect to z of Equation (d), we get

$$\chi'(z) = -C\alpha\sin\alpha z + D\alpha\cos\alpha z \tag{e}$$

On applying the boundary conditions of Equations (a) and (b), respectively, in Equations (c) and (e), we get

$$\chi(0) = 0 = C \Rightarrow \quad C = 0 \tag{f}$$

and

$$GJ(D\alpha\cos\alpha L) = \omega_{nf}^2 I_p (D\sin\alpha L) \Rightarrow D\left(GJ\alpha\cos\alpha L - \omega_{nf}^2 I_p \sin\alpha L\right) = 0 \tag{g}$$

where D cannot be zero for a general case, otherwise the whole response would be zero. Hence, from Equation (g), noting Equation (d), the frequency equation is obtained as

$$\tan\alpha L = \frac{GJ\alpha}{\omega_{nf}^2 I_p} = \frac{GJ}{\omega_{nf}^2 I_p}\alpha^2\frac{1}{\alpha} = \frac{GJ}{\omega_{nf}^2 I_p}\frac{\rho\omega_{nf}^2}{GJ}\frac{1}{\alpha} = \frac{\rho GL}{I_p}\frac{1}{\alpha L} \tag{h}$$

which is a transcendental equation to be solved for αL by numerical methods (root searching methods, Newton–Raphson method, etc.) and the first three roots are given as

$$\alpha_1 L = 0.8605, \quad \alpha_2 L = 3.4256, \quad \text{and} \quad \alpha_3 L = 6.4373 \tag{i}$$

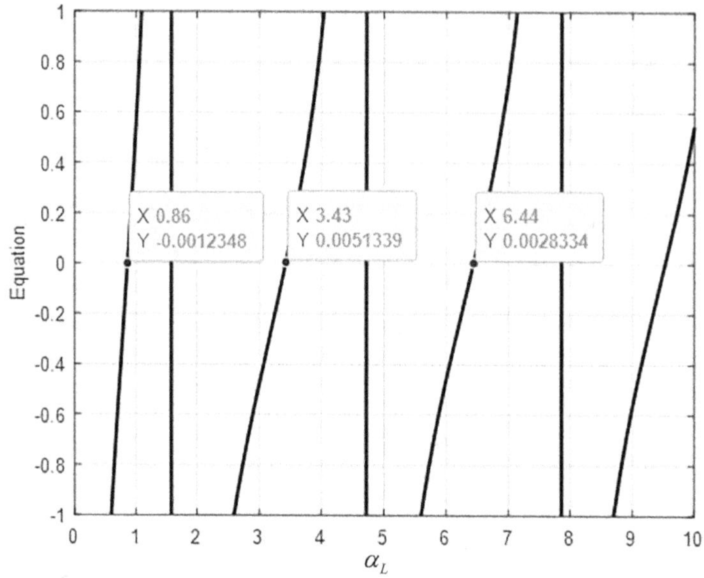

FIGURE 7.9 Plot show the roots of Equation (h).

Figure 7.9 shows the plot of Equation (h) in functional form with respect to $\alpha_i L$ for $\frac{\rho GL}{I_p} = 1$. It can be seen that curves have intersections at the zero line as per Equation (i). The vertical lines are the locations where the function has a discontinuity so those intersections are not valid solutions.

Hence, from Equation (d), the natural frequencies can be expressed as

$$\omega_{nfi} = \alpha_i L \sqrt{\frac{G}{\rho L^2}}, \quad i = 1,2,3,\ldots \tag{j}$$

where values of $\alpha_i L$ for the first three modes are given by Equation (i). The mode shape is expressed as $\chi_i(z) = D\sin\alpha_i z$.

7.2 APPLICATIONS OF FEMs

The method described in the previous section for the analysis of the continuous system having infinite DOFs is feasible for only simple boundary conditions. For analyses of practical rotors, the continuous system model is discretized into a finite-DOF system as an approximation. One such discretization method is called the *FEM*. One of the popular methods of developing finite element equations from governing equations in the differential form is the *weighted-residual method* (the *Galerkin method*, as a special case, will be considered here; other methods are the least-squares error method, collocation least-squares error method, collocation method, and Petrov–Galerkin method). However, by using the *Rayleigh–Ritz method*, the functional (various energies of the system) can be used directly to derive the finite element formulation of the system without deriving the governing equations in the differential form. For the present case, both methods are presented, and it is shown that they give the same finite element formulation. However, the former has the advantage that the boundary conditions come from the finite element formulations themselves. The FEM is briefly introduced in the present section; however, for more detailed information on FEMs, readers are referred to the excellent texts by Bathe (1982), Hughes (1986); Zienkiewicz and Taylor (1989), Reddy (1993), Cook et al. (2002), Huebner et al. (2001), and Dixit (2009).

Figure 7.10 shows a cantilever rod as a complete system (Figure 7.10a), and when it is divided into several segments (e.g., six equal segments in Figure 7.10b; however, it is not necessary to have

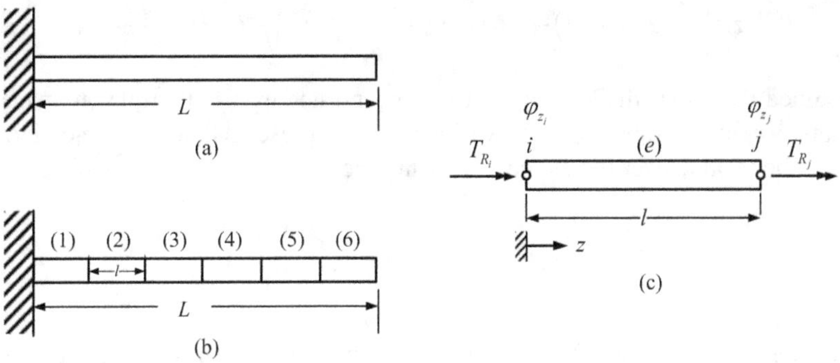

FIGURE 7.10 A cantilever rod: (a) actual system; (b) a discretization of the rod into several elements; (c) a free body diagram of an element.

all segments of the same size and it can be any convenient number), these are called *elements.* The total length of the rod is L, and the element length is l. For each element, for the present case, two nodes are present and a free body diagram of one such element under torsional loadings is shown in Figure 7.10c. An axis system is attached to the element, and this is called *the local coordinate system.* Apart from this, we can have a *global coordinate system* for the whole rod. The element has two node numbers i and j. The directions of *nodal rotational displacements* ($\varphi_{z_i}(t)$ and $\varphi_{z_j}(t)$) and *nodal reaction torques* ($T_{R_i}(t)$ and $T_{R_j}(t)$) at both ends are taken with counterclockwise as positive (while looking from the free end along the polar axis, i.e., the positive z-axis direction). Hence, for the element shown in Figure 7.10c, the DOF is 2, i.e., $\varphi_{z_i}(t)$ and $\varphi_{z_j}(t)$.

In the FEM, we seek the solution of field variables at nodal points (i.e., at nodes i and j) and express an approximate solution of the field variable, $\varphi_z^{(e)}(z,t)$, within an element in terms of nodal variables as a *polynomial*, which is in general defined as

$$\varphi_z^{(e)}(z,t) = a + bz + cz^2 + \cdots = \lfloor N_1(z) \quad N_2(z) \quad \cdots \quad N_r(z) \rfloor \begin{Bmatrix} \varphi_{z1}(t) \\ \varphi_{z2}(t) \\ \vdots \\ \varphi_{zr}(t) \end{Bmatrix} = \mathbf{N}(z)\boldsymbol{\varphi}_z^{(ne)}(t) \tag{7.56}$$

where $N_i(z)$, $i = 1, 2, \ldots, r$, are called *shape functions* (or *approximating functions* or *interpolation functions*) and they are a function of spatial coordinates, z, only; r is the number of DOFs on an element, and $\varphi_{z_i}(t)$ is the *field variable* (i.e., the rotational displacement) value at the ith node. Superscripts (e) and (ne) represent the *element* and the *node of the element*, respectively. The row and column vectors are represented by $\lfloor \quad \rfloor$ and $\{ \quad \}$, respectively. Equation 7.56 represents a general form of the approximate solution of the equations of motion within an element (Equation 7.27), where a, b, c, …, are unknown constants to be determined from boundary conditions (mainly the geometrical boundary conditions and not the natural boundary conditions; that is why it gives an approximate solution) of an element (however, not from the system); this is illustrated subsequently.

7.2.1 Galerkin Method

On substituting the assumed approximate solution (which satisfies only geometrical boundary conditions) from Equation 7.56 into the equation of motion (Equation 7.27), it gives some *residue*, $R^{(e)}$, and can be expressed as

$$R^{(e)}(z,t) = \rho J\ddot{\varphi}_z^{(e)}(z,t) - GJ\varphi_{z,zz}^{(e)}(z,t) - T^{(e)}(z,t) - T_0^{(e)}(t)\delta^*\left(z - z_0^{(e)}\right) \tag{7.57}$$

Here, it is assumed that $z_0^{(e)}$ is the location of the concentrated external torque on an element in the local coordinate system and subscript (e) in various terms represents that these belong to an element. On minimizing the residue over the element by using the Galerkin method, we have

$$\int_0^l N_i R^{(e)} dz = 0; \quad i = 1, 2, \ldots, r \tag{7.58}$$

where l is the element length and $N_i(z)$ is the shape function. On substituting Equation 7.57 into Equation 7.58, we get

$$\int_0^l N_i \left\{\rho J\ddot{\varphi}_z^{(e)} - GJ\varphi_{z,zz}^{(e)} - T^{(e)}(z,t) - T_0^{(e)}(t)\delta^*(z - z_0)\right\} dz = 0; \quad i = 1, 2, \ldots, r \tag{7.59}$$

At this stage, a choice of the interpolation function could be done; however, because the highest derivative of the field variable with respect to the spatial variable is two, we have to choose an interpolation function of at least quadratic form so as to satisfy the *completeness condition* (i.e., the field variable value should not vanish). The completeness requirement can be *weakened* by performing integration by parts with respect to z of the second term in Equation 7.59, so as to give

$$\int_0^t \rho JN_i\ddot{\varphi}_z^{(e)} dz - GJN_i\,\varphi_{z,z}^{(e)}\Big|_0^l + \int_0^t GJN_{i,z}\varphi_{z,z}^{(e)} dz - \int_0^t N_i \left\{T^{(e)}(z,t) + T_0^{(e)}(t)\delta^*(z - z_0)\right\} dz = 0,$$
$$i = 1, 2, \ldots, r \tag{7.60}$$

Interpolation function: When choosing the interpolation function, we need to satisfy two conditions, i.e., the *completeness and compatibility requirements*. The compatibility requirement is to be satisfied at the interelement boundaries, whereas the completeness requirement is to be satisfied inside of the element. In integral Equation 7.60, the highest derivative of the field variable with respect to the spatial parameter, z, is now reduced to 1. Hence, for the interpolation function, the *completeness* requirement is up to the first derivative with respect to the spatial variable, z, i.e., of φ_z and $\varphi_{z,z}$. Moreover, the *compatibility* requirement will be one order less than the highest differentiation with respect to z in the integral only, i.e., up to the field variable, φ_z, only. Hence, the interpolation function of a polynomial with linear form is sufficient and can be expressed as

$$\varphi_z^{(e)}(z,t) = a + bz \tag{7.61}$$

where a and b are constants to be determined by field variables specified at boundaries of the element (i.e., at nodes i and j) as shown in Figure 7.11. It is to be noted that for the present case, only two boundary conditions are present to obtain that many unknowns (i.e., a and b). The z is the local coordinate system. At any general element (e), the following geometric boundary conditions exist (Figure 7.11):

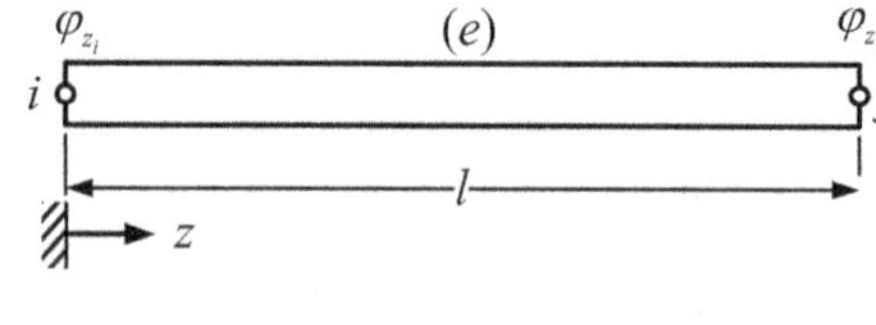

FIGURE 7.11 A shaft element with field variables at boundaries.

$$\text{At } i\text{th node:} \quad z = 0, \quad \varphi_z^{(e)}(0,t) = \varphi_{z_i}(t) \tag{7.62}$$

and

$$\text{At } j\text{th node:} \quad z = l, \quad \varphi_z^{(e)}(l,t) = \varphi_{z_j}(t) \tag{7.63}$$

On application of Equations 7.62 and 7.63 into Equation 7.61, the constants of the interpolation function can be obtained as

$$a = \varphi_{z_i}(t) \quad \text{and} \quad b = \frac{\varphi_{z_j}(t) - \varphi_{z_i}(t)}{l} \tag{7.64}$$

Hence, Equation 7.61 takes the following form:

$$\varphi_z^{(e)}(z,t) = \varphi_{z_i}(t) + \frac{\varphi_{z_j}(t) - \varphi_{z_i}(t)}{l} z \quad \text{or} \quad \varphi_z^{(e)}(z,t) = \left(1 - \frac{z}{l}\right)\varphi_{z_i}(t) + \left(\frac{z}{l}\right)\varphi_{z_j}(t) \tag{7.65}$$

In view of Equation 7.65, a shape function in the form of Equation 7.56 can be written as

$$\varphi_z^{(e)}(z,t) = \mathbf{N}(z)\,\varphi_z^{(ne)} \tag{7.66}$$

with

$$\mathbf{N}(Z) = \lfloor N_i(Z) \quad N_j(Z) \rfloor; \quad \varphi_z^{(ne)} = \begin{Bmatrix} \varphi_{z_i} \\ \varphi_{z_i} \end{Bmatrix}; \quad N_i = \left(1 - \frac{z}{l}\right); \quad N_j = \frac{z}{l} \tag{7.67}$$

where N_i and N_j are interpolation functions corresponding to the *i*th and *j*th nodes, respectively. These have characteristics that they have a unit value at the node they correspond, and have value zero at other nodes (e.g., $N_i=1$ at $z=0$, $N_i=0$ at $z=l$; and $N_j=0$ at $z=0$, $N_j=1$ at $z=l$).

7.2.1.1 Checking Compatibility Requirements

In Figure 7.12, the *j*th node is the common node of the (*e*)th and (*e* + 1)th elements. For the (*e*)th element, we can write Equation 7.66 as

$$\varphi_z^{(ne)}(z,t) = \begin{bmatrix} \left(1 - \frac{z}{l}\right) & \frac{z}{l} \end{bmatrix} \begin{Bmatrix} \varphi_{z_i}(t) \\ \varphi_{z_j}(t) \end{Bmatrix} \tag{7.68}$$

Similarly, for the (*e*+ 1)th element, we can write

$$\varphi_z^{(e+1)}(z,t) = \begin{bmatrix} \left(1 - \frac{z}{l}\right) & \frac{z}{l} \end{bmatrix} \begin{Bmatrix} \varphi_{z_j}(t) \\ \varphi_{zk}(t) \end{Bmatrix} \tag{7.69}$$

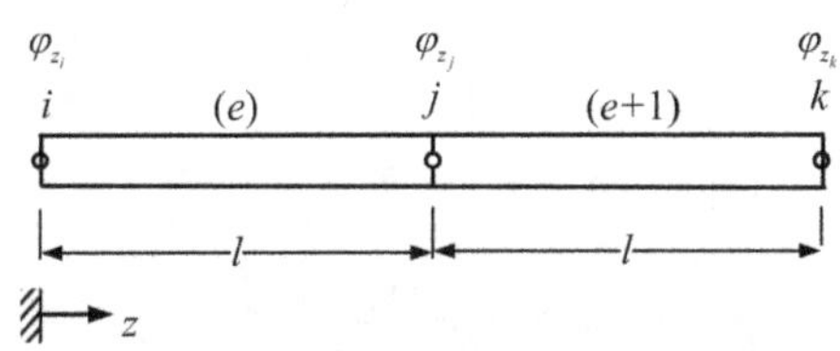

FIGURE 7.12 Two neighboring elements with field variables at boundaries.

We want to check the compatibility requirements at the common node, i.e., at the jth node. Hence, we have the following conditions:

For $z=l$ in the (e)th element, from Equation 7.68, we get

$$\varphi_z^{(e)}(z,t)\Big|_{z=l} = \begin{bmatrix} 0 & 1 \end{bmatrix} \begin{Bmatrix} \varphi_{z_i}(t) \\ \varphi_{z_j}(t) \end{Bmatrix} = \varphi_{z_j}(t) \tag{7.70}$$

For $z=0$ in the $(e+1)$th element, from Equation 7.69, we get

$$\varphi_z^{(e+1)}(z,t)\Big|_{z=0} = \begin{bmatrix} 1 & 0 \end{bmatrix} \begin{Bmatrix} \varphi_{z_j}(t) \\ \varphi_{zk}(t) \end{Bmatrix} = \varphi_{z_j}(t) \tag{7.71}$$

Hence, by using interpolations of two different elements, nodal values of the field variable at the common node are the same, as can be seen from Equations 7.70 and 7.71. This ensures the compatibility of the field variable, $\varphi_z(z,t)$, between two elements. Because in the present analysis, all other elements have similar interpolation functions, the above compatibility is ensured for other common nodes as well. Now, to verify whether the present interpolation function gives compatibility of the higher order also, on taking the first derivative of Equation 7.68 with respect to z, we get

$$\varphi_{z,z}^{(e)}(z,t) = \mathbf{N}_{,z}(z)\varphi_z^{(ne)}(t) \tag{7.72}$$

with

$$\mathbf{N}_{,z}(z) = \begin{bmatrix} N_{i,z}(z) & N_{j,z}(z) \end{bmatrix}; \quad \varphi_z^{(ne)}(t) = \begin{Bmatrix} \varphi_{z_i}(t) \\ \varphi_{z_j}(t) \end{Bmatrix} \tag{7.73}$$

$$N_{i,z} = \frac{-1}{l}; \quad N_{j,z} = \frac{1}{l} \tag{7.74}$$

For the (e)th element, we can write Equation 7.72 as

$$\varphi_{z,z}^{(e)}(z,t) = \begin{bmatrix} \left(\frac{-1}{l}\right) & \left(\frac{1}{l}\right) \end{bmatrix} \begin{Bmatrix} \varphi_{z_i}(t) \\ \varphi_{z_j}(t) \end{Bmatrix} \tag{7.75}$$

Similarly, for the $(e+1)$th element, we can write

$$\varphi_{z,z}^{(e+1)}(z,t) = \begin{bmatrix} \left(\frac{-1}{l}\right) & \left(\frac{1}{l}\right) \end{bmatrix} \begin{Bmatrix} \varphi_{z_j}(t) \\ \varphi_{zk}(t) \end{Bmatrix} \tag{7.76}$$

For $z=l$ in the (e)th element, Equation 7.75 gives

$$\varphi_{z,z}^{(e)}\Big|_{x=1} = \frac{\varphi_{z_j} - \varphi_{z_i}}{l} \tag{7.77}$$

For $z=0$ in the $(e+1)$th element, Equation 7.76 gives

$$\varphi_{z,z}^{(e+1)}\Big|_{x=0} = \frac{\varphi_{zk} - \varphi_{z_j}}{l} \tag{7.78}$$

As can be seen from Equations 7.77 and 7.78, the nodal values of the first spatial derivative of the field variable at a common node of two neighboring elements are not the same. Thus, the linear interpolation function chosen does not give compatibility of the first spatial derivative of the field variable, which is also not needed for the present case.

7.2.1.2 Checking Completeness Requirements

The completeness condition ensures convergence as the size of the elements is decreased, and in the limit, it should converge to the exact value for elements of infinitesimal size. The completeness requirement of the interpolation function, for the present problem, is that it should not vanish up to the first spatial derivative. Because the interpolation function is a linear polynomial, it satisfies the completeness condition.

7.2.1.3 Finite Element Formulations

Now, Equation 7.60 can be written by combining all equations for $i=1, 2, \ldots, r$ in vector form, as

$$\int_0^l \rho J\mathbf{N}(z)\ddot{\varphi}_z^{(e)}dz + \int_0^l GJ\mathbf{N}_{,z}\varphi_{z,z}^{(e)}dz = GJ\mathbf{N}(z)\varphi_{z,z}^{(e)}\Big|_0^l + \int_0^l \mathbf{N}(z)\left[T^{(e)}(z,t)+T_0^{(e)}(t)\delta^*\left(z-z_0^{(e)}\right)\right]dz \tag{7.79}$$

From the assumed solution of the form of Equation 7.66, we have

$$\ddot{\varphi}_z^{(e)}(z,t)=\mathbf{N}(z)\,\ddot{\varphi}_z^{(ne)}(t) \text{ and } \varphi_{z,z}^{(e)}(z,t)=\mathbf{N}_{,z}(z)\,\varphi_z^{(ne)}(t) \tag{7.80}$$

On substituting the above equations into Equation 7.79, we get

$$\begin{aligned}&\int_0^l \rho J\mathbf{N}\mathbf{N}^T dz\ddot{\varphi}_z^{(ne)}(t)+\int_0^l GJ\mathbf{N}_{,z}\,\mathbf{N}_{,z}^T dz\varphi_z^{(ne)}(t)\\ &= GJ\begin{Bmatrix} N_i \\ N_j \end{Bmatrix}\varphi_z^{(e)}(z,t)\Big|_0^l + \int_0^l \mathbf{N}\left[T^{(e)}(z,t)+T_0^{(e)}(t)\delta^*\left(z-z_0^{(e)}\right)\right]dz\end{aligned} \tag{7.81}$$

From Equation 7.67, it should be noted that N_i gives values of 1 and 0 at nodes i and j, respectively, and N_j gives values of 0 and 1 at nodes i and j, respectively. On substituting these values in the first term of the right-hand side of Equation 7.81, we get

$$\begin{aligned}&\int_0^l \rho J\mathbf{N}\mathbf{N}^T dz\ddot{\varphi}_z^{(ne)}(t)+\int_0^l GJ\mathbf{N}_{,z}\,\mathbf{N}_{,z}^T dz\varphi_z^{(ne)}(t)\\ &= \begin{Bmatrix} 0-GJ\varphi_{,z}^{(e)}\big|_{z=0} \\ GJ\varphi_{,z}^{(e)}\big|_{z=l}-0 \end{Bmatrix} + \int_0^l \mathbf{N}\left[T^{(e)}(z,t)+T_0^{(e)}(t)\delta^*(z-z_0)\right]dz\end{aligned} \tag{7.82}$$

The above equation can be simplified to a standard form of equations of motion of a rod element, as

$$\mathbf{M}^{(e)}\ddot{\varphi}_z^{(ne)}(t)+\mathbf{K}^{(e)}\varphi_z^{(ne)}(t)=\mathbf{T}_R^{(ne)}(t)+\mathbf{T}_E^{(ne)}(t) \tag{7.83}$$

with

$$\mathbf{M}^{(e)}=\int_0^l \rho J\mathbf{N}\mathbf{N}^T dz=\int_0^l \rho J\begin{bmatrix} N_i^2 & N_iN_j \\ N_jN_i & N_j^2 \end{bmatrix}dz=\frac{\rho Jl}{6}\begin{bmatrix} 2 & 1 \\ 1 & 2 \end{bmatrix} \tag{7.84}$$

$$\mathbf{K}^{(e)} = \int_0^l GJ\mathbf{N}_{,z}\ \mathbf{N}_{,z}^T dz = \int_0^l GJ\begin{bmatrix} N_{i,z}^2 & N_{i,z}N_{j,z} \\ N_{j,z}N_{i,z} & N_{j,z}^2 \end{bmatrix} dz = \frac{GJ}{l}\begin{bmatrix} 1 & -1 \\ -1 & 1 \end{bmatrix} \tag{7.85}$$

$$\mathbf{T}_R^{(ne)}(t) = \begin{Bmatrix} -T_{R_i}^{(ne)} \\ T_{R_j}^{(ne)} \end{Bmatrix} = \begin{Bmatrix} -GJ\varphi_{z,z}^{(e)}\Big|_{z=0} \\ GJ\varphi_{z,z}^{(e)}\Big|_{z=l} \end{Bmatrix}; \quad \mathbf{T}_E^{(ne)}(t) = \int_0^l \mathbf{N}\left[T^{(e)}(z,t) + T_0^{(e)}(t)\delta^*\left(z - z_0^{(e)}\right)\right]dz \tag{7.86}$$

where $\mathbf{M}^{(e)}$ is the *element mass matrix*, $\mathbf{K}^{(e)}$ is the *element stiffness matrix*, $\mathbf{T}_R^{(ne)}$ is the *nodal reaction torque vector* of the element (because from the strength of materials for the pure torsion theory we have $T/GJ = \varphi_z/l$, hence $T = GJ\varphi_{z,z}^{(e)}$), and $\mathbf{T}_E^{(ne)}$ is the *nodal external torque vector* of the element. Matrices $\mathbf{M}^{(e)}$ and $\mathbf{K}^{(e)}$ are called the *consistent mass and stiffness matrices*, respectively. It should be noted that the governing equation of the finite element is an ordinary differential equation, compared with the governing equation of the continuous system in which it is a partial differential equation. The former form of the equation is easier to solve compared with the latter case.

The mass (or the polar mass moment of inertia) of an element is sometimes lumped at its nodes based on some physical reasoning, and this simplifies the analysis drastically because the mass matrix becomes diagonal in nature. Such a mass matrix is called the *lumped mass matrix*. If there is a rigid disk in the rotor system, then generally we put a node at the disk and the mass (or the polar mass moment of inertia) of the disk is considered to be lumped at that node; the lumped mass appears in the diagonal of the mass matrix corresponding to that nodal variable at which it is lumped. The nodal external torque vector is the equivalent torque acting at nodes corresponding to the actual loading inside the element. For the present case, two types of external loadings, i.e., the distributed and concentrated torques, have been considered. The distributed loading can have different forms (e.g., the linear, quadric, cubic, etc.) and may act inside the element. The concentrated torque may act at nodes directly (more often a node is chosen in such locations) or if it is acting inside the element, then its equivalence has to be obtained at element nodes as the nodal external torque vector. This is discussed in more detail while treating the finite element analysis of transverse vibrations. Now, to have a basic understanding of the Rayleigh–Ritz method, the above finite element formulation is repeated in the following section.

7.2.2 Rayleigh–Ritz Method

The Rayleigh–Ritz method can also be used to develop the finite element formulation from the *functional* (i.e., the conservative and nonconservative energies of the system, i.e., from Equation 7.23 without taking the variational operator). On substituting the approximate solution of Equation 7.56 into the functional, we get the *functional* for the element, as

$$\Pi^{(e)} = \int_{t_1}^{t_2}\int_0^l \left\{\frac{1}{2}\rho J\dot{\varphi}_z^{(e)2} - \frac{1}{2}GJ\varphi_{z,z}^{(e)2} + T^{(e)}\varphi_z^{(e)} + T_0^{(e)}\delta^*(z - z_0)\varphi_z^{(e)}\right\}dzdt \tag{7.87}$$

The minimization of the functional for an element, $\Pi^{(e)}$, is equivalent to

$$\frac{\partial \Pi^{(e)}}{\partial \varphi_{z_i}} = 0, \quad i = 1,2,\ldots,r \tag{7.88}$$

Hence, for a typical ith node, on substituting Equation 7.87 into Equation 7.88, we get

$$\frac{\partial \Pi^{(e)}}{\partial \varphi_{zi}} = 0 = \int_{t_1}^{t_2}\int_{0}^{l}\left[\rho J\dot{\varphi}_z^{(e)}\frac{\partial \dot{\varphi}_z^{(e)}}{\partial \varphi_{zi}} - GJ\varphi_{z,z}^{(e)}\frac{\partial \varphi_{z,z}^{(e)}}{\partial \varphi_{zi}} + \left\{T^{(e)} + T_0^{(e)}\delta^*(z-z_0)\right\}\frac{\partial \varphi_z^{(e)}}{\partial \varphi_{zi}}\right]dzdt \tag{7.89}$$

$$i = 1,2,\ldots,r$$

On performing integration by parts with respect to time of the first term on the left-hand side of Equation 7.89, we get

$$\int_0^l\left[\rho J\dot{\varphi}_z^{(e)}\frac{\partial \varphi_z^{(e)}}{\partial \varphi_{zi}}\bigg|_{t_1}^{t_2}\right]dz - \int_{t_1}^{t_2}\int_{0}^{l}\left[\rho J\ddot{\varphi}_z^{(e)}\frac{\partial \varphi_z^{(e)}}{\partial \varphi_{zi}} + GJ\varphi_{z,z}^{(e)}\frac{\partial \varphi_{z,z}^{(e)}}{\partial \varphi_{zi}} + \left\{T^{(e)} + T_0^{(e)}\delta^*(z-z_0)\right\}\frac{\partial \varphi_z^{(e)}}{\partial \varphi_{zi}}\right]dzdt = 0$$

$$i = 1,2,\ldots,r \tag{7.90}$$

The first term is zero at the t_1 and t_2 time instances. Now, by noting Equation 7.56, we have the following derivatives:

$$\begin{aligned}\frac{\partial \varphi_z^{(e)}}{\partial \varphi_{zi}} &= \frac{\partial}{\partial \varphi_{zi}}\left\{\mathbf{N}^T(z)\boldsymbol{\varphi}_z^{(ne)}(t)\right\}\\ &= \frac{\partial}{\partial \varphi_{zi}}\left\{N_1\varphi_{z1}^{(ne)} + N_2\varphi_{z2}^{(ne)} + \cdots + N_i\varphi_{zi}^{(ne)} + \cdots + N_r\varphi_{zr}^{(ne)}\right\} = N_i\end{aligned} \tag{7.91}$$

$$\begin{aligned}\varphi_{z,z}^{(e)} = \frac{\partial \varphi_z^{(e)}}{\partial z} &= \frac{\partial}{\partial z}\left\{\mathbf{N}^T(z)\boldsymbol{\varphi}_z^{(ne)}(t)\right\} = \left\{N_{1,z}\varphi_{z1}^{(ne)}(t) + N_{2,z}\varphi_{z2}^{(ne)}(t) + \cdots + N_{r,z}\varphi_{zr}^{(ne)}(t)\right\}\\ &= \mathbf{N}_{,z}^T(z)\boldsymbol{\varphi}_z^{(ne)}(t)\end{aligned} \tag{7.92}$$

$$\begin{aligned}\frac{\partial \varphi_{z,z}^{(e)}}{\partial \varphi_{zi}} &= \frac{\partial}{\partial \varphi_{zi}}\left\{\mathbf{N}_{,z}^T(z)\boldsymbol{\varphi}_z^{(ne)}(t)\right\}\\ &= \frac{\partial}{\partial \varphi_{zi}}\left\{N_{1,z}\varphi_{z1}^{(ne)} + N_{2,z}\varphi_{z2}^{(ne)} + \cdots + N_{i,z}\varphi_{zi}^{(ne)} + \cdots + N_{r,z}\varphi_{zr}^{(ne)}\right\} = N_{i,z}\end{aligned} \tag{7.93}$$

and

$$\begin{aligned}\frac{\partial \varphi_z^{(e)}}{\partial t} &= \frac{\partial}{\partial t}\left\{\mathbf{N}^T(z)\boldsymbol{\varphi}_z^{(ne)}(t)\right\} = \left\{N_1\dot{\varphi}_{z1}^{(ne)}(t) + N_2\dot{\varphi}_{z2}^{(ne)}(t) + \cdots + N_r\dot{\varphi}_{zr}^{(ne)}(t)\right\}\\ &= \mathbf{N}^T(z)\dot{\boldsymbol{\varphi}}_z^{(ne)}(t)\end{aligned} \tag{7.94}$$

On substituting Equations 7.56 and 7.91 through 7.94 into Equation 7.90, we get

$$\int_{t_1}^{t_2}\int_0^l\left[\rho JN_i\mathbf{N}^T\ddot{\boldsymbol{\varphi}}_z^{(ne)} + GJ\frac{dN_i}{dz}\mathbf{N}_{,z}^T(z)\boldsymbol{\varphi}_z^{(ne)}(t) - N_i\left\{T^{(e)} + T_0^{(e)}\delta^*(z-z_0)\right\}\right]dzdt = 0 \tag{7.95}$$

Now, the above equations can be combined as

$$\int_{t_1}^{t_2}\left[\int_0^l \rho J\mathbf{N}\mathbf{N}^T dz\ddot{\boldsymbol{\varphi}}_z^{(ne)} + \int_0^l GJ\mathbf{N}_{,z}\mathbf{N}_{,z}^T dz\boldsymbol{\varphi}_z^{(ne)}(t) - \int_0^l\left(T^{(e)} + T_0^{(e)}\delta^*(z-z_0)\right)\mathbf{N}dz\right]dt = 0 \tag{7.96}$$

which gives the standard finite element formulation of the equations of motion as

$$\mathbf{M}^{(e)}\ddot{\boldsymbol{\varphi}}_z^{(ne)} + \mathbf{K}^{(e)}\boldsymbol{\varphi}_z^{(ne)} = \mathbf{T}_E^{(ne)} \tag{7.97}$$

where $\mathbf{M}^{(e)}$ is the element mass matrix, $\mathbf{K}^{(e)}$ is the element stiffness matrix, and $\mathbf{T}_E^{(ne)}$ is the nodal external torque vector as defined earlier. Matrices $\mathbf{M}^{(e)}$ and $\mathbf{K}^{(e)}$ here also are the *consistent* mass and stiffness matrices, respectively. It should be noted that the nodal reaction torque vector, $\mathbf{T}_E^{(ne)}$, does not appear automatically from the formulation, and it has to be added separately.

This element equation is of the general form for the torsional vibration of a rod. It can be used to obtain the element equation for all other elements depending upon the geometrical and physical parameters of elements. However, the question now remains as to how to bring these equations together to represent the system as a whole.

7.2.3 Assembled System Equations

Once the elemental equation of the form in Equation 7.83 has been obtained for all elements, the next step is to obtain the system equation by *assembling* all elements of the system. This is illustrated for a simple cantilever shaft, which for simplicity is divided into only two equal length elements as shown in Figure 7.13 (as mentioned earlier, the element size need not be same and more elements is expected to give better accuracy in the results). By noting Equations 7.83 through 7.86 and that the first element we have $i=1$ and $j=2$, the finite element equation for the first element can be written as

$$\frac{\rho J l}{6}\begin{bmatrix} 2 & 1 \\ 1 & 2 \end{bmatrix}\begin{Bmatrix} \ddot{\varphi}_{z1} \\ \ddot{\varphi}_{z2} \end{Bmatrix} + \frac{GJ}{l}\begin{bmatrix} 1 & -1 \\ -1 & 1 \end{bmatrix}\begin{Bmatrix} \varphi_{z1} \\ \varphi_{z2} \end{Bmatrix} = \begin{Bmatrix} -T_{R_1} \\ T_{R_2} \end{Bmatrix} + \begin{Bmatrix} T_{E_1}^{(1)} \\ T_{E_2}^{(1)} \end{Bmatrix} \tag{7.98}$$

where T_{R_1} and T_{R_2} are the reaction torques node 1 and node 2, respectively. The negative sign represents that the torque is in a negative direction relative to the positive sign convention. $T_{E_1}^{(1)}$ and $T_{E_2}^{(1)}$ represent equivalent external torques, respectively, at nodes 1 and 2 of element (1) due to actual external loadings on element (1). When a concentrated load (torque) is acting at one of the nodes,

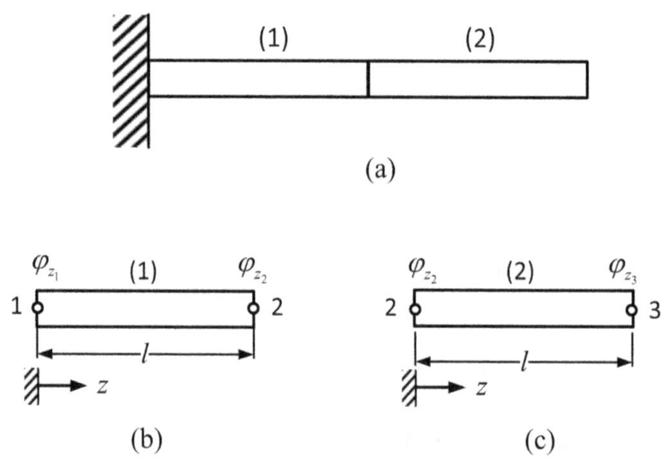

FIGURE 7.13 (a) A cantilever shaft discretized into two elements: (b) element 1; (c) element 2.

say 2, which is a common node for elements (1) and (2), then that torque can be considered to be acting at node 2 of element (1) or node 1 of element (2) and not in both of them (if both of them, then half of the torque in each common node of elements).

Similarly, for the second element, we have $i=2$ and $j=3$, hence in view of Equation 7.83, the finite element equation for the second element can be written as

$$\frac{\rho J l}{6}\begin{bmatrix} 2 & 1 \\ 1 & 2 \end{bmatrix}\begin{Bmatrix} \ddot{\varphi}_{z2} \\ \ddot{\varphi}_{z3} \end{Bmatrix}+\frac{GJ}{l}\begin{bmatrix} 1 & -1 \\ -1 & 1 \end{bmatrix}\begin{Bmatrix} \varphi_{z2} \\ \varphi_{z3} \end{Bmatrix}=\begin{Bmatrix} -T_{R2} \\ T_{R3} \end{Bmatrix}+\begin{Bmatrix} T_{E2}^{(1)} \\ T_{E3}^{(1)} \end{Bmatrix} \tag{7.99}$$

Because the shaft is discretized into elements of equal lengths, the stiffness and mass matrices for elements 1 and 2 are the same. However, the element can be of different lengths, cross-sections, and materials; accordingly, l, J, ρ, and G would be changed in the elemental equation. Now Equations 7.98 and 7.99 can be assembled in the following steps:

Element (1)

$$\frac{\rho J l}{6}\begin{bmatrix} 2 & 1 & 0 \\ 1 & 2 & 0 \\ 0 & 0 & 0 \end{bmatrix}\begin{Bmatrix} \ddot{\varphi}_{z1} \\ \ddot{\varphi}_{z2} \\ \ddot{\varphi}_{z3} \end{Bmatrix}+\frac{GJ}{l}\begin{bmatrix} 1 & -1 & 0 \\ -1 & 1 & 0 \\ 0 & 0 & 0 \end{bmatrix}\begin{Bmatrix} \varphi_{z1} \\ \varphi_{z2} \\ \varphi_{z3} \end{Bmatrix}=\begin{Bmatrix} -T_{R1} \\ T_{R2} \\ 0 \end{Bmatrix}+\begin{Bmatrix} T_{E1}^{(1)} \\ T_{E2}^{(1)} \\ 0 \end{Bmatrix}$$

Element (2)

$$\frac{\rho J l}{6}\begin{bmatrix} 0 & 0 & 0 \\ 0 & 2 & 1 \\ 0 & 1 & 2 \end{bmatrix}\begin{Bmatrix} \ddot{\varphi}_{z1} \\ \ddot{\varphi}_{z2} \\ \ddot{\varphi}_{z3} \end{Bmatrix}+\frac{GJ}{l}\begin{bmatrix} 0 & 0 & 0 \\ 0 & 1 & -1 \\ 0 & -1 & 1 \end{bmatrix}\begin{Bmatrix} \varphi_{z1} \\ \varphi_{z2} \\ \varphi_{z3} \end{Bmatrix}=\begin{Bmatrix} 0 \\ -T_{R2} \\ T_{R3} \end{Bmatrix}+\begin{Bmatrix} 0 \\ T_{E2}^{(2)} \\ T_{E3}^{(2)} \end{Bmatrix}$$

On adding above two equations, the assembled equation takes the following form:

$$\frac{\rho J l}{6}\begin{bmatrix} 2 & 1 & 0 \\ 1 & 2+2 & 1 \\ 0 & 1 & 2 \end{bmatrix}\begin{Bmatrix} \ddot{\varphi}_{z1} \\ \ddot{\varphi}_{z2} \\ \ddot{\varphi}_{z3} \end{Bmatrix}+\frac{GJ}{l}\begin{bmatrix} 1 & -1 & 0 \\ -1 & 1+1 & -1 \\ 0 & -1 & 1 \end{bmatrix}\begin{Bmatrix} \varphi_{z1} \\ \varphi_{z2} \\ \varphi_{z3} \end{Bmatrix}$$

$$=\begin{Bmatrix} -T_{R1} \\ T_{R2}-T_{R2} \\ T_{R3} \end{Bmatrix}+\begin{Bmatrix} T_{E1}^{(1)} \\ T_{E2}^{(1)}+T_{E2}^{(2)} \\ T_{E3}^{(2)} \end{Bmatrix} \tag{7.100}$$

For the clarity of the assembly procedure, the matrix elements are kept in expanded form. For example, in the assembled mass and stiffness matrices, the contribution from the corresponding matrices of the first element is in the first two rows and columns, whereas, from the second element, it is in the second and third rows and columns. Similarly, for torque vectors, the contribution from the first element is in the first two rows of assembled vectors, and the contribution from the second element is in the last two rows. Now the form of the assembled governing equation is clear if we divide the rod in three elements or more. For three elements, for the sake of illustration, the assembled governing equation would be

$$\frac{\rho J l}{6}\begin{bmatrix} 2 & 1 & 0 & 0 \\ 1 & 4 & 1 & 0 \\ 0 & 1 & 4 & 1 \\ 0 & 0 & 1 & 2 \end{bmatrix}\begin{Bmatrix} \ddot{\varphi}_{z1} \\ \ddot{\varphi}_{z2} \\ \ddot{\varphi}_{z3} \\ \ddot{\varphi}_{z4} \end{Bmatrix} + \frac{GJ}{l}\begin{bmatrix} 1 & -1 & 0 & 0 \\ -1 & 2 & -1 & 0 \\ 0 & -1 & 2 & -1 \\ 0 & 0 & -1 & 1 \end{bmatrix}\begin{Bmatrix} \varphi_{z1} \\ \varphi_{z2} \\ \varphi_{z3} \\ \varphi_{z4} \end{Bmatrix}$$

$$= \begin{Bmatrix} -T_{R1} \\ T_{R2} - T_{R2} \\ T_{R3} - T_{R3} \\ T_{R4} \end{Bmatrix} + \begin{Bmatrix} T_{E1}^{(1)} \\ T_{E2}^{(1)} + T_{E2}^{(2)} \\ T_{E3}^{(2)} + T_{E3}^{(3)} \\ T_{E4}^{(3)} \end{Bmatrix}$$

with

$$T_R = GJ\varphi_{z,z}$$

where the internal torque, T_R, symbol has been used. For details, readers are encouraged to refer to basic books on FEMs. Until now, we have not considered boundary conditions of the problem, e.g., for the present case, the boundary conditions of a cantilever rod.

7.2.4 Application of Boundary Conditions

Again, coming to the present illustration of a two-element problem (Figure 7.13), because node 1 is fixed and node 3 is free, the following boundary conditions are specified by the problem:

$$\varphi_{z1} = \ddot{\varphi}_{z1} = 0 \quad \text{and} \quad T_{R3} = 0 \;\Rightarrow\; GJ\varphi_{z,z}\big|_3 = 0 \tag{7.101}$$

On applying the above boundary conditions into Equation 7.100, we get

$$\frac{\rho J l}{6}\begin{bmatrix} 2 & 1 & 0 \\ 1 & 2+2 & 1 \\ 0 & 1 & 2 \end{bmatrix}\begin{Bmatrix} 0 \\ \ddot{\varphi}_{z2} \\ \ddot{\varphi}_{z3} \end{Bmatrix} + \frac{GJ}{l}\begin{bmatrix} 1 & -1 & 0 \\ -1 & 1+1 & -1 \\ 0 & -1 & 1 \end{bmatrix}\begin{Bmatrix} 0 \\ \varphi_{z2} \\ \varphi_{z3} \end{Bmatrix}$$

$$= \begin{Bmatrix} -T_{R1} \\ 0 \\ 0 \end{Bmatrix} + \begin{Bmatrix} T_{E1}^{(1)} \\ T_{E2}^{(1)} + T_{E2}^{(2)} \\ T_{E3}^{(2)} \end{Bmatrix} \tag{7.102}$$

Equation 7.102 is basically three equations and the unknowns are rotational twists φ_{z2} and φ_{z3} (and their second derivatives with respect to time) and the reaction torque at node 1, T_{R1}. The first equation contains the reaction torque, T_{R1}, along with the rotational twist φ_{z2} (and its time derivative) as unknowns. The last two equations contain rotational twists φ_{z2} and φ_{z3} (and their time derivatives) as unknown; however, no reaction torques are unknown. These can be solved if external torques are specified for forced vibrations (or zero for free torsional vibrations). Once rotational twists (and their time derivatives) are obtained from the last two equations, they can be used in the first equation to obtain the unknown reaction torque, T_{R1}, also. Hence, on taking only the last two equations of 7.102, we get

$$\frac{\rho J l}{6}\begin{bmatrix} 4 & 1 \\ 1 & 2 \end{bmatrix}\begin{Bmatrix} \ddot{\varphi}_{z2} \\ \ddot{\varphi}_{z3} \end{Bmatrix} + \frac{GJ}{l}\begin{bmatrix} 2 & -1 \\ -1 & 1 \end{bmatrix}\begin{Bmatrix} \varphi_{z2} \\ \varphi_{z3} \end{Bmatrix} = \begin{Bmatrix} T_{E2}^{(1)} + T_{E2}^{(2)} \\ T_{E3}^{(2)} \end{Bmatrix} \tag{7.103}$$

Equation 7.103 can be solved if external torques are specified; this analysis is known as *forced vibrations.* If external torque is zero, the problem is called *free vibration.*

7.2.5 Free Torsional Vibrations

For free vibrations, the solution of Equation 7.103 can be assumed to be of the following form:

$$\begin{Bmatrix} \varphi_{z2}(t) \\ \varphi_{z3}(t) \end{Bmatrix} = \begin{Bmatrix} \Phi_{z2} \\ \Phi_{z3} \end{Bmatrix} e^{j\omega_{nf}t} \quad \text{so that} \quad \begin{Bmatrix} \ddot{\varphi}_{z2} \\ \ddot{\varphi}_{z3} \end{Bmatrix} = -\omega_{nf}^2 \begin{Bmatrix} \Phi_{z2} \\ \Phi_{z3} \end{Bmatrix} e^{j\omega_{nf}t} \tag{7.104}$$

where Φ is the amplitude of torsional vibrations, ω_{nf} is the torsional natural frequency, and $j = \sqrt{-1}$. On substituting Equation 7.104 into the homogeneous part of Equation 7.103, we get

$$\left(-\omega_{nf}^2 \frac{\rho J l}{6} \begin{bmatrix} 4 & 1 \\ 1 & 2 \end{bmatrix} + \frac{GJ}{l} \begin{bmatrix} 2 & -1 \\ -1 & 1 \end{bmatrix} \right) \begin{Bmatrix} \Phi_{z2} \\ \Phi_{z3} \end{Bmatrix} = \begin{Bmatrix} 0 \\ 0 \end{Bmatrix} \tag{7.105}$$

Equation 7.105 is a standard eigenvalue problem, and for a nontrivial solution, the following determinant should be zero, i.e.,

$$\begin{vmatrix} \left(\dfrac{2GJ}{l} - \omega_{nf}^2 \dfrac{2\rho J l}{3} \right) & \left(-\dfrac{GJ}{l} - \omega_{nf}^2 \dfrac{\rho J l}{6} \right) \\ \left(-\dfrac{GJ}{l} - \omega_{nf}^2 \dfrac{\rho J l}{6} \right) & \left(\dfrac{GJ}{l} - \omega_{nf}^2 \dfrac{\rho J l}{3} \right) \end{vmatrix} = 0 \tag{7.106}$$

On taking the determinant of Equation 7.106, we get a polynomial in terms of ω_{nf}. It is called the *characteristics equation* or the *frequency equation*, and has the following form:

$$\omega_{nf}^4 - \frac{60G}{7\rho l^2}\omega_{nf}^2 + \frac{36G^2}{7\rho^2 l^4} = 0 \tag{7.107}$$

Equation 7.107 is a quadratic polynomial in terms of the square of ω_{nf}, and can be solved as

$$\omega_{nf}^2 = \frac{30G}{7\rho l^2} \pm \sqrt{\left(\frac{30G}{7\rho l^2}\right)^2 - \frac{36G^2}{7\rho^2 l^4}} \quad \text{or} \quad \omega_{nf}^2 = \frac{6G}{\rho l^2} \pm \frac{5.367G}{\rho l^2} = \frac{0.6492G}{\rho l^2} \quad \text{and} \quad \frac{7.923G}{\rho l^2}$$

Hence, the torsional natural frequencies are (on substituting $l = 0.5L$)

$$\omega_{nf_1} = 0.8057\sqrt{\frac{G}{\rho l^2}} = 1.6114\sqrt{\frac{G}{\rho L^2}} \quad \text{and} \quad \omega_{nf_2} = 2.8146\sqrt{\frac{G}{\rho l^2}} = 5.6292\sqrt{\frac{G}{\rho L^2}} \tag{7.108}$$

which are more than that obtained by the analytical method discussed in the previous section (refer Table 7.1) with $\omega_{nf_1} = 1.57\sqrt{G/(\rho L^2)}$ and $\omega_{nf_2} = 4.71\sqrt{G/(\rho L^2)}$. The finite element always gives upper bounds of natural frequencies because it overpredicts the stiffness, which is in turn due to the underestimate of displacements. With the increased number of elements, it is expected that it should gradually converge to the analytically calculated natural frequencies.

For a single element, on equating the first diagonal element of the matrix in Equation 7.106 to zero, we get $\omega_{nf_1} = 3.464\sqrt{G/(\rho L^2)}$, which is still higher than that given by Equation 7.108 for two elements.

The procedure of the FEM has been introduced in the present section with the help of a simple cantilever rod. Natural frequencies and mode shapes could also be obtained by using the solution of the eigen eigenvalue problem. Equation 7.105 has the following standard form of the eigenvalue problem:

$$(\mathbf{A}-\lambda\mathbf{I})\boldsymbol{\Phi}=\mathbf{0} \tag{7.109}$$

with

$$\mathbf{A}=\mathbf{K}^{-1}\mathbf{M}=\left(\frac{GJ}{l}\begin{bmatrix}2 & -1\\ -1 & 1\end{bmatrix}\right)^{-1}\frac{\rho Jl}{6}\begin{bmatrix}4 & 1\\ 1 & 2\end{bmatrix}$$

$$=\frac{\rho l^2}{6G}\begin{bmatrix}1 & 1\\ 1 & 2\end{bmatrix}\begin{bmatrix}4 & 1\\ 1 & 2\end{bmatrix}=\frac{\rho L^2}{24G}\begin{bmatrix}5 & 3\\ 6 & 5\end{bmatrix}$$

and

$$\lambda=\frac{1}{\omega_{nf}^2}$$

Eigenvalues and eigenvectors of the matrix, **A**, are given by

$$\lambda=\begin{Bmatrix}\lambda_1\\ \lambda_2\end{Bmatrix}=\begin{Bmatrix}1/\omega_{nf1}^2\\ 1/\omega_{nf2}^2\end{Bmatrix}=\begin{Bmatrix}9.2426/24\\ 0.7574/24\end{Bmatrix}\frac{\rho L^2}{G}\Rightarrow\begin{Bmatrix}\omega_{nf1}\\ \omega_{nf2}\end{Bmatrix}=\begin{Bmatrix}1.6114\\ 5.6292\end{Bmatrix}\sqrt{\frac{G}{\rho L^2}}$$

and

$$\boldsymbol{\Phi}=\left[\begin{Bmatrix}0.5774\\ 0.8165\end{Bmatrix}_1\quad\begin{Bmatrix}-0.5774\\ 0.8165\end{Bmatrix}_2\right]$$

which gives the same natural frequencies as obtained in Equation 7.108 along with the mode shapes in the form of eigenvectors that could be normalized according to the convenience requirement. On also including the boundary condition at the fixed end, the mode shape vector takes the following form:

$$\boldsymbol{\Phi}=\left[\begin{Bmatrix}0\\ 0.5774\\ 0.8165\end{Bmatrix}_1\quad\begin{Bmatrix}0\\ -0.5774\\ 0.8165\end{Bmatrix}_2\right]$$

In the above modal matrix, the first column belongs to the first mode and the second to the second mode. It can be observed that, for the first mode, there would not be any node (i.e., a zero rotational displacement, except at the fixed end of the shaft); however, for the second mode, there would be a node because two values have the opposite sign. The mode shape could be used to obtain displacements at intermediate locations.

Now, through more examples, the application of FEM on torsional vibration analyses is illustrated.

Example 7.4

Determine natural frequencies and mode shapes for a rotor system as shown in Figure 7.14. Neglect the mass of the shaft and assume the disks are thin.

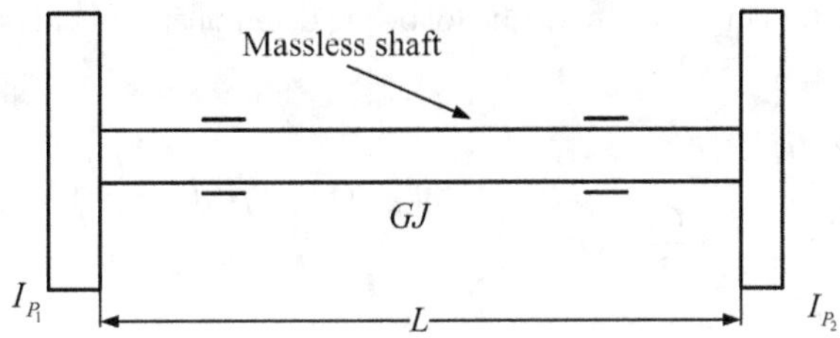

FIGURE 7.14 A two-disk rotor system.

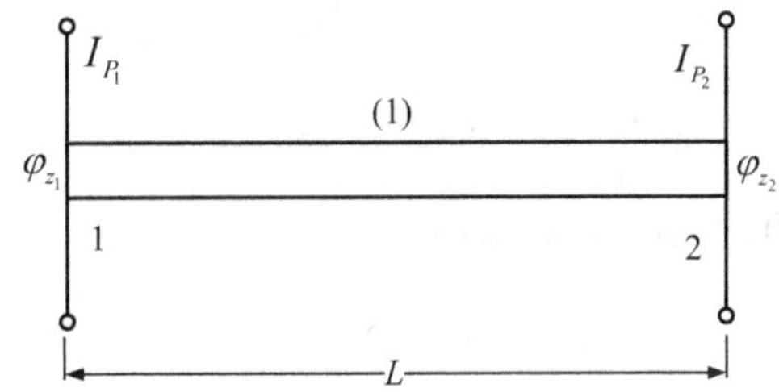

FIGURE 7.15 A two-disk rotor system discretized as a single element.

Solution:

Because the present problem contains only lumped masses, let us discretize the shaft into only one element as shown in Figure 7.15. Noting Equation 7.98, the finite element equation of the one-element rotor system can be written as

$$\begin{bmatrix} I_{p1} & 0 \\ 0 & I_{p2} \end{bmatrix} \begin{Bmatrix} \ddot{\varphi}_{z1} \\ \ddot{\varphi}_{z2} \end{Bmatrix} + \frac{GJ}{L} \begin{bmatrix} 1 & -1 \\ -1 & 1 \end{bmatrix} \begin{Bmatrix} \varphi_{z1} \\ \varphi_{z2} \end{Bmatrix} = \begin{Bmatrix} -T_{R_1} \\ T_{R_2} \end{Bmatrix} \tag{a}$$

It should be noted that the mass of the shaft is neglected; hence, the consistent mass matrix is not present, and only lumped masses due to rigid disks as diagonal elements in the mass matrix are present since these are at the node itself.

Because for the present problem both nodes are free, the following (natural) boundary conditions apply:

$$T_{R_1} = T_{R_2} = 0 \tag{b}$$

For the simple harmonic motion, we have the following relation:

$$\begin{Bmatrix} \ddot{\varphi}_{z1} \\ \ddot{\varphi}_{z2} \end{Bmatrix} = -\omega_{nf}^2 \begin{Bmatrix} \varphi_{z1} \\ \varphi_{z2} \end{Bmatrix} \tag{c}$$

where ω_{nf} is the torsional natural frequency. On substituting Equations (b) and (c) into Equation (a), we get

$$\begin{bmatrix} \left(\frac{GJ}{L} - \omega_{nf}^2 I_{p1}\right) & -\frac{GJ}{L} \\ -\frac{GJ}{L} & \left(\frac{GJ}{L} - \omega_{nf}^2 I_{p2}\right) \end{bmatrix} \begin{Bmatrix} \varphi_{z1} \\ \varphi_{z2} \end{Bmatrix} = \begin{Bmatrix} 0 \\ 0 \end{Bmatrix} \tag{d}$$

For the nontrivial solution of Equation (d), the following determinant should be 0:

$$\begin{vmatrix} \left(\dfrac{GJ}{L} - \omega_{nf}^2 I_{P1}\right) & -\dfrac{GJ}{L} \\ -\dfrac{GJ}{L} & \left(\dfrac{GJ}{L} - \omega_{nf}^2 I_{P2}\right) \end{vmatrix} = 0 \Rightarrow \omega_{nf}^2 \left\{ \omega_{nf}^2 I_{P1} I_{P2} - \frac{GJ}{L}\left(I_{P1} + I_{P2}\right) \right\} = 0 \tag{e}$$

which gives a frequency equation. It gives the torsional natural frequencies as

$$\omega_{nf_1} = 0 \quad \text{and} \quad \omega_{nf_2} = \sqrt{\frac{GJ}{L} \frac{\left(I_{P1} + I_{P2}\right)}{I_{P1} I_{P2}}} \tag{f}$$

which are exactly the same as obtained by the closed form solution with the torsional stiffness of the shaft $k_t = GJ/L$.

On substituting $\omega_{nf_1} = 0$ into Equation (d), we get

$$\varphi_{z1} = \varphi_{z2} \tag{e}$$

Hence, the zero natural frequency is corresponding to the rigid body mode of the shaft: In this case, both rigid disks have the same rotational displacements and the shaft does not experience any twisting moment, which is of little practical importance. On substituting ω_{nf_2} into Equation (d), we get

$$\left(\frac{GJ}{L} - \frac{GJ}{L}\frac{\left(I_{P1} + I_{P2}\right)}{I_{P1} I_{P2}} I_{P1}\right)\varphi_{z1} - \left(\frac{GJ}{L}\right)\varphi_{z2} = 0 \quad \text{which gives} \quad \frac{\varphi_{z1}}{\varphi_{z2}} = -\frac{I_{P2}}{I_{P1}} \tag{f}$$

Hence, mode shapes are also exactly the same as solution obtained by the analytical method in the closed form. Because the mass of the shaft is not considered, even when more elements are considered, the same results are expected and it is left to the reader as an exercise. However, if the distributed mass of the shaft is also considered, then with more elements the solution is expected to be better because the modeling of the shaft is refined on increasing the number of elements. This is illustrated in subsequent examples.

Example 7.5

A cantilever continuous rod has a rigid disk at the free end and also is supported by a torsional spring of stiffness k_t at the free end as shown in Figure 7.16. Use the following parameters: the polar mass moment of inertia of the disk $I_p = 0.02$ kg-m², torsional stiffness of the spring at the free end $k_t = 100$ Nm/rad, the length of the rod $L = 0.4$ m, the diameter of the rod $d = 0.015$ m, the mass density of the rod material $\rho = 7800\,\text{kg/m}^3$, and the modulus of rigidity of the rod material $G = 0.8 \times 10^{11}\,\text{N/m}^2$. Obtain the first two torsional natural frequencies of the system.

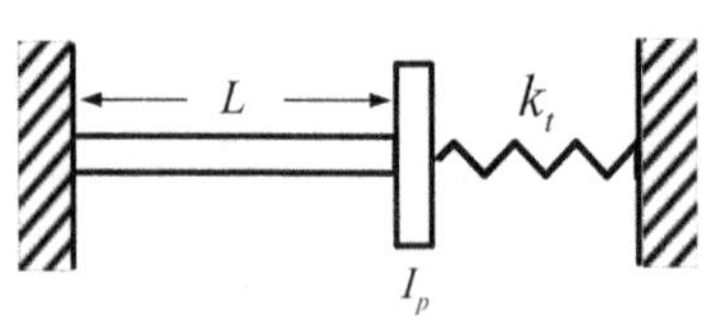

FIGURE 7.16 A cantilever rod with a rigid disk and a torsional spring at the free end.

Solution:

We have the following data for the rotor system when the shaft is discretized into three elements as shown in Figure 7.17. The shaft is divided into two equal length elements.

$$I_p = 0.02\ \text{kg-m}^2,\quad k_t = 100\ \text{N-m/rad},\quad L = 0.4\ \text{m},\quad d = 0.015\ \text{m},\quad G = 0.8\times 10^{11}\ \text{N/m}^2$$

$$J = \frac{\pi}{32}d^4 = 4.97\times 10^{-9}\ \text{m}^4,\quad l = \frac{L}{2} = \frac{0.4}{2} = 0.2\ \text{m},\quad \frac{GJ}{l} = 1988.04\ \text{N-m}$$

$$\frac{\rho J l}{6} = 1.292\times 10^{-6}\ \text{kg-m}^2$$

The governing equation of the first element can be written as

$$1.292\times 10^{-6}\begin{bmatrix} 2 & 1 \\ 1 & 2 \end{bmatrix}\begin{Bmatrix} \ddot{\varphi}_{z1} \\ \ddot{\varphi}_{z2} \end{Bmatrix} + 1988.04\begin{bmatrix} 1 & -1 \\ -1 & 1 \end{bmatrix}\begin{Bmatrix} \varphi_{z1} \\ \varphi_{z2} \end{Bmatrix} = \begin{Bmatrix} -T_{R1} \\ T_{R2} \end{Bmatrix} \qquad \text{(a)}$$

Let us consider that there is no disk and no spring attached to the shaft at node 3. Then, the second element equation becomes

$$1.292\times 10^{-6}\begin{bmatrix} 2 & 1 \\ 1 & 2 \end{bmatrix}\begin{Bmatrix} \ddot{\varphi}_{z2} \\ \ddot{\varphi}_{z3} \end{Bmatrix} + 1988.04\begin{bmatrix} 1 & -1 \\ -1 & 1 \end{bmatrix}\begin{Bmatrix} \varphi_{z2} \\ \varphi_{z3} \end{Bmatrix} = \begin{Bmatrix} -T_{R2} \\ T_{R3} \end{Bmatrix} \qquad \text{(b)}$$

However, the third element contains a rigid disk at node 3 and it contributes toward only the inertia $I_p\ddot{\varphi}_{z3}$. Also, a torsional spring is attached between nodes 3 and 4; hence, we can write (from free body diagrams of the disk and spring, respectively, in Figure 7.17d through 7.16e)

$$-T_{R3} - k_t(\varphi_{z3} - \varphi_{z4}) = I_p\ddot{\varphi}_{z3} \quad \text{and} \quad -T_{R4} + k_t(\varphi_{z4} - \varphi_{z3}) = 0 \qquad \text{(c)}$$

which can be alternatively written in the following matrix form:

$$0.02\begin{bmatrix} 1 & 0 \\ 0 & 0 \end{bmatrix}\begin{Bmatrix} \ddot{\varphi}_{z3} \\ \ddot{\varphi}_{z4} \end{Bmatrix} + 100\begin{bmatrix} 1 & -1 \\ -1 & 1 \end{bmatrix}\begin{Bmatrix} \varphi_{z3} \\ \varphi_{z4} \end{Bmatrix} = \begin{Bmatrix} -T_{R3} \\ T_{R4} \end{Bmatrix} \qquad \text{(d)}$$

It should be noted that Equations (c) and (d) are one and the same; only the form of the presentation is different. Now, on assembling Equations (a), (b), and (d), we get

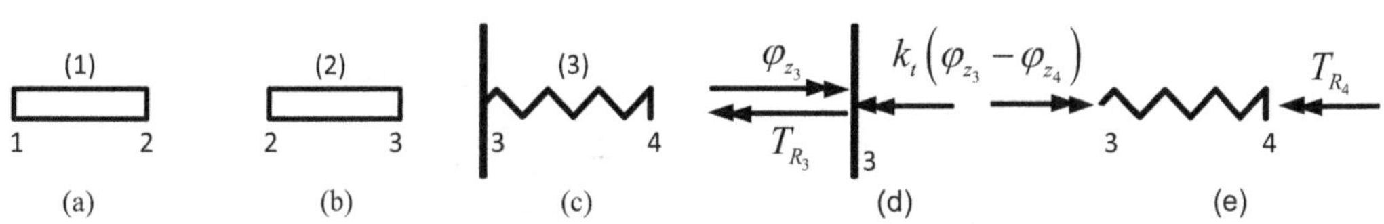

FIGURE 7.17 Discretization of the rotor into three elements and free body diagrams of disk and spring: (a) shaft element 1; (b) shaft element 2; (c) disk and spring as element 3; (d) free body diagram of disk; (e) free body diagram of spring.

$$1.29\times10^{-6}\begin{bmatrix} 2 & 1 & 0 & 0 \\ 1 & (2+2) & 1 & 0 \\ 0 & 1 & \left(2+\dfrac{0.02}{1.29\times10^{-6}}\right) & 0 \\ 0 & 0 & 0 & 0 \end{bmatrix}\begin{Bmatrix} \ddot{\varphi}_{z1} \\ \ddot{\varphi}_{z2} \\ \ddot{\varphi}_{z3} \\ \ddot{\varphi}_{z4} \end{Bmatrix}$$
$$+1.988.04\begin{bmatrix} 1 & -1 & & \\ -1 & (1+1) & -1 & \\ & -1 & \left(1+\dfrac{100}{1988.04}\right) & -1.05 \\ & & -1.05 & 1.05 \end{bmatrix}\begin{Bmatrix} \varphi_{z1} \\ \varphi_{z2} \\ \varphi_{z3} \\ \varphi_{z4} \end{Bmatrix}=\begin{Bmatrix} -T_{R_1} \\ 0 \\ 0 \\ T_{R_4} \end{Bmatrix} \tag{e}$$

On simplification, we get

$$1.29\times10^{-6}\begin{bmatrix} 2 & 1 & 0 & 0 \\ 1 & 4 & 1 & 0 \\ 0 & 1 & 15{,}480 & 0 \\ 0 & 0 & 0 & 0 \end{bmatrix}\begin{Bmatrix} \ddot{\varphi}_{z1} \\ \ddot{\varphi}_{z2} \\ \ddot{\varphi}_{z3} \\ \ddot{\varphi}_{z4} \end{Bmatrix}$$
$$+1988.04\begin{bmatrix} 1 & -1 & 0 & 0 \\ -1 & 2 & -1 & 0 \\ 0 & -1 & 1.05 & -1.05 \\ 0 & 0 & -1.05 & 1.05 \end{bmatrix}\begin{Bmatrix} \varphi_{z1} \\ \varphi_{z2} \\ \varphi_{z3} \\ \varphi_{z4} \end{Bmatrix}=\begin{Bmatrix} -T_{R_1} \\ 0 \\ 0 \\ T_{R_4} \end{Bmatrix} \tag{f}$$

On application of boundary conditions at nodes 1 and 4, that is $\varphi_{z1}=\ddot{\varphi}_{z1}=\varphi_{z4}=\ddot{\varphi}_{z4}=0$, the above equation will reduce to (on considering middle two equations, i.e., after eliminating the first and second rows and columns, which contain unknown reaction torques)

$$1.29\times10^{-6}\begin{bmatrix} 4 & 1 \\ 1 & 15{,}480 \end{bmatrix}\begin{Bmatrix} \ddot{\varphi}_{z2} \\ \ddot{\varphi}_{z3} \end{Bmatrix}+1988.04\begin{bmatrix} 2 & -1 \\ -1 & 1.05 \end{bmatrix}\begin{Bmatrix} \varphi_{z2} \\ \varphi_{z3} \end{Bmatrix}=\begin{Bmatrix} 0 \\ 0 \end{Bmatrix} \tag{g}$$

This is a homogeneous equation. For simple harmonic motion, we can write the above equation as

$$\begin{bmatrix} 3976.08-5.168\times10^{-6}\omega_{nf}^2 & -1988.04-1.292\times10^{-6}\omega_{nf}^2 \\ -1988.04-1.292\times10^{-6}\omega_{nf}^2 & 2087.44-0.02\omega_{nf}^2 \end{bmatrix}\begin{Bmatrix} \varphi_{z2} \\ \varphi_{z3} \end{Bmatrix}=\begin{Bmatrix} 0 \\ 0 \end{Bmatrix} \tag{h}$$

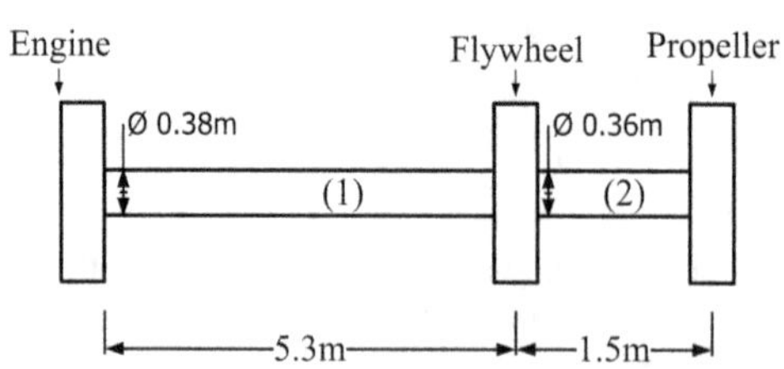

FIGURE 7.18 A three-disk rotor model.

For a nontrivial solution, the determinant of the above matrix should be equal to zero, i.e.,

$$\left(3976.08-5.168\times10^{-6}\omega_{nf}^2\right)\left(2087.44-0.02\omega_{nf}^2\right)-\left(-1988.04-1.292\times10^{-6}\omega_{nf}^2\right)^2=0 \tag{i}$$

which gives a polynomial of the following form:

$$\omega_{nf}^4-7.695\times10^8\omega_{nf}^2+4.2047\times10^{13}=0 \tag{j}$$

Hence, the natural frequencies become

$$\omega_{nf_1}=233.76\text{ rad/s}\left(37.20\text{ Hz}\right);\quad \omega_{nf_2}=27{,}730\text{ rad/s}\left(4414.7\text{ Hz}\right) \tag{k}$$

For comparison, by considering the shaft as massless and with no spring at the free end, the natural frequency is given as (for this simple case, only one natural frequency exists)

$$\omega_{nf_1}=\sqrt{\frac{GJ}{LI_d}}=\sqrt{\frac{0.8\times10^{11}\times4.97\times10^{-9}}{0.4\times0.02}}=222.93\text{ rad/s}=35.48\text{ Hz} \tag{l}$$

which is less than the fundamental natural frequency obtained by the FEM (i.e., 37.20 Hz) for the actual rotor system. This is expected. (i) The additional spring at the free end provides more stiffness, so the natural frequency is more in the actual case compared with the simplified rotor system with a massless shaft and without a spring at the free end, and (ii) FEM always give an upper bound of natural frequencies; however, we have not presented any closed form solution to the actual problem. Hence, the comparison is not of the same system. However, the polar mass moment of inertia of the shaft has very little effect as can be seen here.

The polar mass moment of inertia of the shaft has an effect of decreasing the natural frequency, when it is considered in the analysis. For a continuous cantilever shaft with a disk at the free end, the closed form solution is given as (see Example 7.3)

$$\omega_{nf_1}=0.8605\sqrt{\frac{GJ}{LI_d}}=0.8605\sqrt{\frac{0.8\times10^{11}\times4.97\times10^{-9}}{0.4\times0.02}}=191.83\text{ rad/s}=30.53\text{ Hz} \tag{m}$$

and

$$\omega_{nf_2}=3.4256\sqrt{\frac{GJ}{LI_d}}=3.4256\sqrt{\frac{0.8\times10^{11}\times4.97\times10^{-9}}{0.4\times0.02}}=763.68\text{ rad/s}=121.54\text{ Hz} \tag{n}$$

On comparing Equations (l) and (m), it can be seen that while considering the shaft inertia there is a decrease in the first natural frequency. On comparing Equations (k) and (n), it can be observed that the second mode natural frequency has a large difference for the cases with and without a spring at the free end. These comparisons have been made to highlight the effectiveness of the FEM and the possible trend of natural frequencies due to various complexities in the system (i.e., additional spring, additional rigid disk, and addition of shaft inertia). It should be noted that in the finite element analysis, only two elements of the shaft (total three with the spring as an element) were considered; if we increase the number of elements, the natural frequencies of the actual system are expected to be refined.

Example 7.6

A marine reciprocating engine, flywheel, and propeller are approximately equivalent to the following three-rotor system. The engine has a crank 50 cm long and a connecting rod 250 cm long. The engine revolving parts are equivalent to 50 kg at the crank radius, and the piston and pin masses are 41 kg. The connecting rod mass is 52 kg, and its center of gravity is 26 cm from the crank pin center.

The mass of the flywheel is 200 kg with a radius of gyration of 25 cm. The propeller has a polar mass moment of inertia of 6 kg-m². The equivalent shaft between the engine masses and the flywheel is 38 cm in diameter and 5.3 m long, and that between the flywheel and the propeller is 36 cm in diameter and 1.5 m long. Find the torsional natural frequencies and mode shapes of the system by the FEM.

Solution:

Using the TMM, the same problem was solved in the previous chapter (see Example 6.14). Hence, from the previous chapter, we have the following data for the three-disk rotor system corresponding to the reciprocating engine, flywheel, and propeller as shown in Figure 7.18.

The equivalent polar moment of inertia of the engine is $I_{P_e} = 29.95$ kg-m^2. The polar mass moment of inertia of the flywheel is $I_{P_f} = 12.5$ kg-m^2. For the propeller, the polar mass moment of inertia is given as $I_{P_p} = 6$ kg-m^2. The torsional stiffnesses of shaft segments (1) and (2) are given as $k_{t_1} = 3\times 10^7$ N-m/rad and $k_{t_2} = 8.67\times 10^7$ N-m/rad. Elements (1) and (2) are shown in Figure 7.19.

The elemental equation for element (1) is

$$\begin{bmatrix} 29.95 & 0 \\ 0 & 0 \end{bmatrix}\begin{Bmatrix} \ddot{\varphi}_{z1} \\ \ddot{\varphi}_{z2} \end{Bmatrix} + 3\times 10^7\begin{bmatrix} 1 & -1 \\ -1 & 1 \end{bmatrix}\begin{Bmatrix} \varphi_{z1} \\ \varphi_{z2} \end{Bmatrix} = \begin{Bmatrix} -T_1 \\ T_2 \end{Bmatrix} \quad \text{(a)}$$

Similarly, on considering element (2), which consists of two disks, one at either end of the shaft, we have

$$\begin{bmatrix} 12.5 & 0 \\ 0 & 6 \end{bmatrix}\begin{Bmatrix} \ddot{\varphi}_{z2} \\ \ddot{\varphi}_{z3} \end{Bmatrix} + 8.67\times 10^7\begin{bmatrix} 1 & -1 \\ -1 & 1 \end{bmatrix}\begin{Bmatrix} \varphi_{z2} \\ \varphi_{z3} \end{Bmatrix} = \begin{Bmatrix} -T_2 \\ T_3 \end{Bmatrix} \quad \text{(b)}$$

On assembling Equations (a) and (b), we get

$$\begin{bmatrix} 29.95 & 0 & 0 \\ 0 & 12.5 & 0 \\ 0 & 0 & 6 \end{bmatrix}\begin{Bmatrix} \ddot{\varphi}_{z1} \\ \ddot{\varphi}_{z2} \\ \ddot{\varphi}_{z3} \end{Bmatrix} + 10^7\begin{bmatrix} 3 & -3 & 0 \\ -3 & 3+8.67 & -8.67 \\ 0 & -8.67 & 8.67 \end{bmatrix}\begin{Bmatrix} \varphi_{z1} \\ \varphi_{z2} \\ \varphi_{z3} \end{Bmatrix} = \begin{Bmatrix} -T_1 \\ 0 \\ T_3 \end{Bmatrix} \quad \text{(c)}$$

For the present case, the boundary conditions are $T_1 = T_3 = 0$, and for simple harmonic motion during the free vibration with frequency ω_{nf}, we have

$$-\omega_{nf}^2\begin{bmatrix} 29.95 & 0 & 0 \\ 0 & 12.5 & 0 \\ 0 & 0 & 6 \end{bmatrix}\begin{Bmatrix} \varphi_{z1} \\ \varphi_{z2} \\ \varphi_{z3} \end{Bmatrix} + 10^7\begin{bmatrix} 3 & -3 & 0 \\ -3 & 11.67 & -8.67 \\ 0 & -8.67 & 8.67 \end{bmatrix}\begin{Bmatrix} \varphi_{z1} \\ \varphi_{z2} \\ \varphi_{z3} \end{Bmatrix} = \begin{Bmatrix} 0 \\ 0 \\ 0 \end{Bmatrix} \quad \text{(d)}$$

or

$$\begin{bmatrix} 3\times 10^7 - 29.95\omega_{nf}^2 & -3\times 10^7 & 0 \\ -3\times 10^7 & 11.67\times 10^7 - 12.5\omega_{nf}^2 & -8.67\times 10^7 \\ 0 & -8.67\times 10^7 & 8.67\times 10^7 - 6\omega_{nf}^2 \end{bmatrix}\begin{Bmatrix} \varphi_{z1} \\ \varphi_{z2} \\ \varphi_{z3} \end{Bmatrix} = \begin{Bmatrix} 0 \\ 0 \\ 0 \end{Bmatrix} \quad \text{(e)}$$

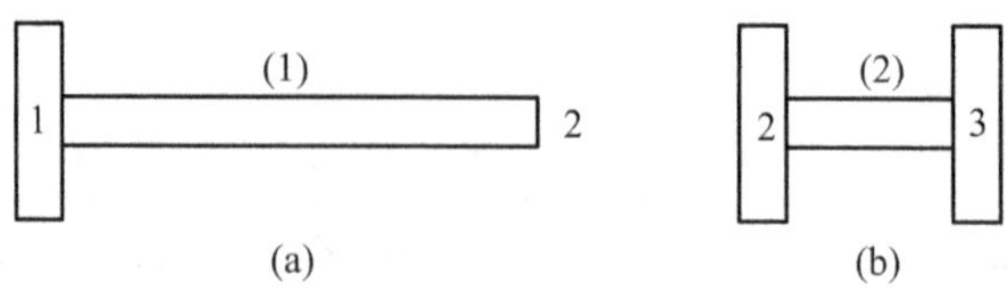

FIGURE 7.19 (a) The first element and (b) the second element.

On setting the determinant of the above matrix to zero, from the characteristic polynomial, we get the natural frequencies of the system as

$$\omega_{nf1} = 0, \quad \omega_{nf2} = 1.58 \times 10^3 \text{ rad/s}, \quad \omega_{nf3} = 4.72 \times 10^3 \text{ rad/s} \tag{f}$$

in which one of the natural frequencies is zero because of the free–free end conditions. The same can also be obtained more conveniently by using eigenvalue analysis with the help of a computer program. For this, let us premultiply Equation (d) by the inverse of the mass matrix **M** to get

$$\left(\mathbf{D} - \omega_{nf}^2 \mathbf{I}\right)\varphi_z = \mathbf{0} \tag{g}$$

with

$$\mathbf{D} = \mathbf{M}^{-1}\mathbf{K} = \begin{bmatrix} 1.0\times10^6 & -1.0\times10^6 & 0 \\ -2.4\times10^6 & 9.34\times10^6 & -6.94\times10^6 \\ 0 & -14.45\times10^6 & 14.45\times10^6 \end{bmatrix} \tag{h}$$

where **I** is the identity matrix. The eigenvalue of matrix **D** is the square of the natural frequencies and the columns of the eigenmatrix are the corresponding mode shapes. These are given as

$$\text{Eigenvalues} = \begin{Bmatrix} \omega_{nf1}^2 \\ \omega_{nf2}^2 \\ \omega_{nf3}^2 \end{Bmatrix} = \begin{Bmatrix} 0 \\ 2.593\times10^6 \\ 2.261\times10^7 \end{Bmatrix}, \Rightarrow \begin{Bmatrix} \omega_{nf1} \\ \omega_{nf2} \\ \omega_{nf3} \end{Bmatrix} = \begin{Bmatrix} 0 \\ 1610.31 \\ 4754.32 \end{Bmatrix} \text{rad/s}$$

and

$$\text{Eigenvector (or modal) matrix} = \begin{bmatrix} \begin{Bmatrix} -0.577 \\ -0.577 \\ -0.577 \end{Bmatrix}_1 & \begin{Bmatrix} 0.387 \\ -0.587 \\ -0.711 \end{Bmatrix}_2 & \begin{Bmatrix} 0.023 \\ -0.477 \\ 0.879 \end{Bmatrix}_3 \end{bmatrix}$$

It should be noted that the natural frequencies obtained are close to what we obtained in Example 6.14. From the eigenvector (or modal) matrix, the mode shapes could be plotted after normalization, e.g., by assigning the first-row elements as 1 and correspondingly scaling the other two rows scaled as follows:

$$\text{Normalized eigenvector matrix} = \begin{bmatrix} \begin{Bmatrix} 1 \\ 1 \\ 1 \end{Bmatrix}_1 & \begin{Bmatrix} 1 \\ -1.517 \\ -1.837 \end{Bmatrix}_2 & \begin{Bmatrix} 1 \\ -21.636 \\ 39.954 \end{Bmatrix}_3 \end{bmatrix}$$

Observe that the mode corresponding to the first natural frequency is a rigid body mode (all nodes have the same rotational displacements in the first eigenvector), the one corresponding to the second natural frequency has one node of the shaft between the engine and the flywheel (corresponding to a sign change in the second eigenvector), and the one corresponding to the third mode has two nodes at either side of the flywheel (corresponding to two sign changes in the third eigenvector).

Example 7.7

For a continuous shaft of 3 m length and 0.03 m diameter, obtain the torsional natural frequencies up to the fifth mode and plot the corresponding mode shapes for the free–free boundary

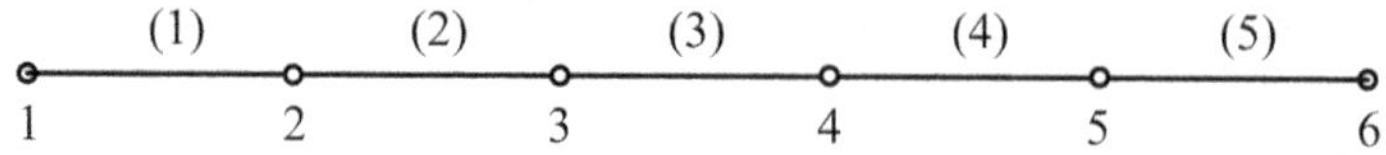

FIGURE 7.20 A continuous rod discretized into five elements with six nodes.

conditions. Perform a convergence study by increasing the number of elements (i.e., 5, 10, 20, 50, 100, and 500), compare the natural frequencies with the closed form analytical solutions, and discuss the results. The following properties of the shaft should be taken: $\rho = 7800\,\text{kg/m}^3$ and $G = 0.8 \times 10^{11}\,\text{N/m}^2$.

Solution:

The shaft is discretized into five equal numbers of elements as shown in Figure 7.20 for illustration; however, the torsional natural frequencies are then presented with a higher number of elements also.

The properties of the elements are

$$l = L/5 = 3/5 = 0.6\ \text{m}, \quad J = \frac{\pi}{32}(0.03)^4 = 7.95 \times 10^{-8}\ \text{m}^4$$

$$\frac{\rho J l}{6} = \frac{7800 \times 7.95 \times 10^{-8} \times 0.6}{6} = 6.20 \times 10^{-5}\ \text{kg-m}^2, \quad \frac{GJ}{l} = \frac{0.8 \times 10^{11} \times 7.95 \times 10^{-8}}{0.6} = 1.06 \times 10^4\ \text{N-m}$$

The elemental equation of the jth element for torsional vibrations is given as

$$6.20 \times 10^{-5} \begin{bmatrix} 2 & 1 \\ 1 & 2 \end{bmatrix} \begin{Bmatrix} \ddot{\varphi}_{zj} \\ \ddot{\varphi}_{zj+1} \end{Bmatrix} + 1.06 \times 10^4 \begin{bmatrix} 1 & -1 \\ -1 & 1 \end{bmatrix} \begin{Bmatrix} \varphi_{zj} \\ \varphi_{zj+1} \end{Bmatrix} = \begin{Bmatrix} -T_j \\ T_{j+1} \end{Bmatrix}, \quad j = 1,2,\ldots,5 \tag{a}$$

All elements have the same size and properties; for example, for the third element, the above equation has the following form:

$$6.20 \times 10^{-5} \begin{bmatrix} 2 & 1 \\ 1 & 2 \end{bmatrix} \begin{Bmatrix} \ddot{\varphi}_{z3} \\ \ddot{\varphi}_{z4} \end{Bmatrix} + 1.06 \times 10^4 \begin{bmatrix} 1 & -1 \\ -1 & 1 \end{bmatrix} \begin{Bmatrix} \varphi_{z3} \\ \varphi_{z4} \end{Bmatrix} = \begin{Bmatrix} -T_3 \\ T_4 \end{Bmatrix} \tag{b}$$

On assembling the elemental equations of all five elements, we get

$$6.20 \times 10^{-5} \begin{bmatrix} 2 & 1 & 0 & 0 & 0 & 0 \\ 1 & 4 & 1 & 0 & 0 & 0 \\ 0 & 1 & 4 & 1 & 0 & 0 \\ 0 & 0 & 1 & 4 & 1 & 0 \\ 0 & 0 & 0 & 1 & 4 & 1 \\ 0 & 0 & 0 & 0 & 1 & 2 \end{bmatrix} \begin{Bmatrix} \ddot{\varphi}_{z1} \\ \ddot{\varphi}_{z2} \\ \ddot{\varphi}_{z3} \\ \ddot{\varphi}_{z4} \\ \ddot{\varphi}_{z5} \\ \ddot{\varphi}_{z6} \end{Bmatrix}$$

$$+1.06 \times 10^4 \begin{bmatrix} 1 & -1 & 0 & 0 & 0 & 0 \\ -1 & 2 & -1 & 0 & 0 & 0 \\ 0 & -1 & 2 & -1 & 0 & 0 \\ 0 & 0 & -1 & 2 & -1 & 0 \\ 0 & 0 & 0 & -1 & 2 & -1 \\ 0 & 0 & 0 & 0 & -1 & 1 \end{bmatrix} \begin{Bmatrix} \varphi_{z1} \\ \varphi_{z2} \\ \varphi_{z3} \\ \varphi_{z4} \\ \varphi_{z5} \\ \varphi_{z6} \end{Bmatrix} = \begin{Bmatrix} -T_1 \\ 0 \\ 0 \\ 0 \\ 0 \\ T_6 \end{Bmatrix} \tag{c}$$

which has the standard form

$$\mathbf{I}_p\ddot{\boldsymbol{\varphi}}_z + \boldsymbol{K}_t\boldsymbol{\varphi}_z = \boldsymbol{T}_R \tag{d}$$

The boundary conditions are the free–free ends of the rod; hence, we have $T_1 = T_6 = 0$ in Equation (c). Hence, after application of the boundary conditions, the size of the matrices also remains the same, i.e., 6×6. For simple harmonic oscillation with frequency ω_{nf}, Equation (d) takes the following standard eigenvalue problem form:

$$\left(\mathbf{K}_t - \omega_{nf}^2\mathbf{I}_p\right)\boldsymbol{\varphi}_z = \mathbf{0} \tag{e}$$

or

$$\left(\mathbf{D} - \omega_{nf}^2\mathbf{I}\right)\boldsymbol{\varphi}_z = \mathbf{0} \tag{f}$$

with

$$\mathbf{D} = \mathbf{I}_p^{-1}\mathbf{K}_t \tag{g}$$

On obtaining the eigenvalue of the matrix **D**, we get the following natural frequencies (for five elements) after taking the square root of the eigenvalues (all are in rad/s):

$$\omega_{nf_1} = 0;\quad \omega_{nf_2} = 3409.1;\quad \omega_{nf_3} = 7152.2;\quad \omega_{nf_4} = 11{,}503.0;\quad \omega_{nf_5} = 16{,}114.0;\quad \omega_{nf_6} = 18{,}490.0$$

Mode shapes can be obtained from the corresponding eigenvectors from the eigenmatrix. Figure 7.21 shows mode shapes up to the fifth mode. It should be noted that flexible modes have nodes (e.g., second mode has one node, third node has two nodes, and so on).

Here, one may ask that, in the finite element formulation of the torsional vibration, the shape function (interpolation function) we are using is linear. So, the mode shape should be also linear. How did the mode shapes become smooth curves? The answer to this is that a linear function is between two nodes; if the number of nodes is very high, then it will represent better displacement variation, resulting in smooth curves.

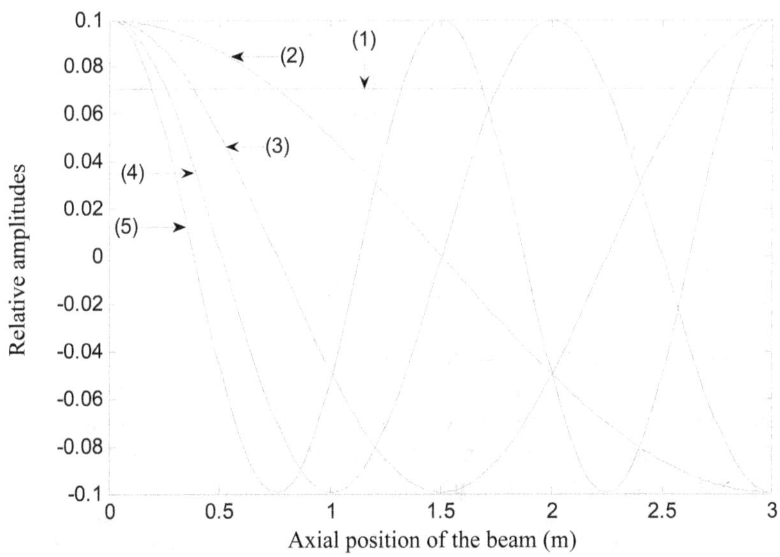

FIGURE 7.21 Mode shapes: (1) first mode, (2) second mode, (3) third mode, (4) fourth mode, and (5) fifth mode.

TABLE 7.2
Convergence of Torsional Natural Frequencies (in rad/s) with the Number of Elements

S.N.	Natural Frequency	Number of Rod Elements (5)	(10)	(20)	(50)	(100)	(500)	Exact Value
1	ω_{nf_1}	0	0	0	0	0	0	0
2	ω_{nf_2}	3409.1	3367.5	3357.2	3354.3	3353.9	3353.7	3353.7
3		7152.2	6818.2	6735.0	6711.8	6708.5	6707.4	6707.4
4	ω_{nf_4}	11,503.0	10,436.0	10,154.0	10,076.0	10,065.0	10,061.0	10,061.0
5	ω_{nf_5}	16,114.0	14,304.0	13,636.0	13,450.0	13,424.0	13,415.0	13,415.0
6	ω_{nf_6}	18,490.0	18,490.0	17,202.0	16,838.0	16,786.0	16,769.0	16,769.0

For comparison with the exact solution (refer to Table 7.1), we have

$$\omega_{nf} = \frac{n\pi}{L}\sqrt{\frac{G}{\rho}} \tag{h}$$

with $L = 3$ m, $G = 8.0 \times 10^{11}$ N/m^2, and $\rho = 7850$ kg/m. We have the following exact natural frequencies of the rotor system (all are in rad/s):

$$\omega_{nf_1} = 0;\quad \omega_{nf_2} = 3353.7;\quad \omega_{nf_3} = 6707.4;\quad \omega_{nf_4} = 10,061.0;\quad \omega_{nf_5} = 13,415;\quad \omega_{nf_6} = 16,769.0$$

On comparison with the FEM solution, we see that lower natural frequencies are quite close to the exact values; however, for higher modes, the natural frequencies by the FEM are quite high compared with the exact values. To get more accuracy for higher modes by the FEM, more elements need to be considered. Table 7.2 shows such a convergence study up to 500 elements.

7.3 DEVELOPMENT OF THE FINITE ELEMENT FOR A SIMPLE GEAR PAIR

Figure 7.22 shows a gear-pair element. The counterclockwise direction is taken as the positive direction for all rotational displacements as per the practice and convenience in finite element formulations. The gear ratio, n, (which is an inverse of the speed ratio) is defined as

$$n = \frac{\omega_1}{\omega_2} \tag{7.110}$$

where ω_1 and ω_2 are the rotational speeds of the driver (pinion or gear 1) and driven (gear 2) shafts. The sense of rotation has been indicated in the figure.

For no slip and no backlash conditions and for perfectly rigid gear teeth (it is assumed that gear teeth are not providing any torsional flexibility; however, in the real case bending of the tooth and the deformation at the contact zone due to the Hertzian contact stress or elasto-hydrodynamic lubrication might give rise to an appreciable amount of effective torsional flexibility, especially for the case when the power transmission is very high), from Figure 7.22, we have

$$\varphi_{z_{g2}} = -\frac{1}{n}\varphi_{z_{g1}} \tag{7.111}$$

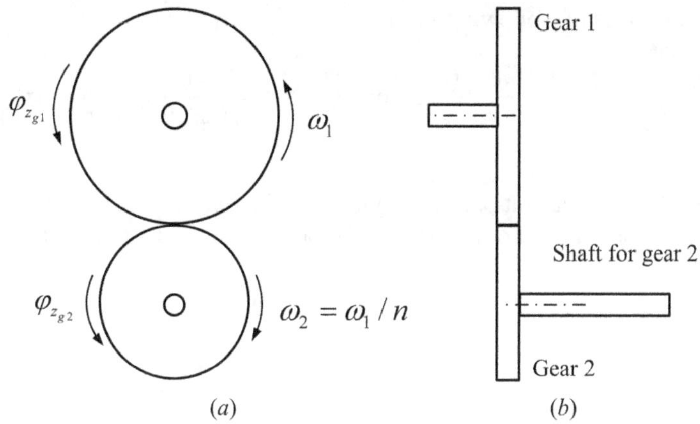

FIGURE 7.22 A simple gear-pair element: (a) gear-pair element; (b) gear-pair element with shaft 2.

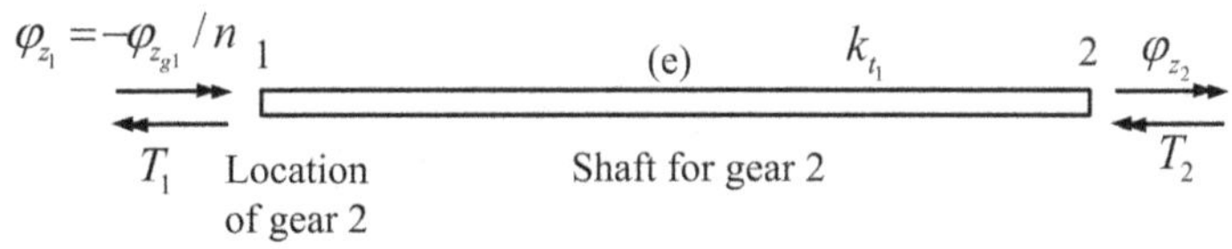

FIGURE 7.23 A gear-pair shaft element with gear 2 at node 1.

where $\varphi_{z_{g1}}$ and $\varphi_{z_{g2}}$ are the rotational displacements of gears 1 and 2, respectively. The negative sign is due to the way these have been chosen in Figure 7.22. Because $\varphi_{z_{g2}}$ is defined in terms of $\varphi_{z_{g1}}$, we can eliminate $\varphi_{z_{g2}}$ from the state vector of the system. It should be noted that we are not obtaining an equivalent gear system such as obtained earlier in Chapter 6 (Kahraman et al., 1992).

7.3.1 The Consistent Mass and Stiffness Matrices of a Gear-Pair Element

To consider the inertia of the shaft also in the gear-pair element, the earlier formulation of the consistent mass and stiffness matrices and torque vectors in the torsional vibration of a rod needs to be modified. Because the governing equations of the rod are the same, the only change is in the gear-pair shaft element boundary condition (refer to Figure 7.23). So, there will be changes in the shape function derivation. Let the gear-pair shaft element nodes 1 and 2 have φ_{z1} and φ_{z2} angular displacements, respectively. Here, the gear is attached at node 1, so its angular displacement is related to the pinion as $\varphi_{z1} = -\varphi_{z_{g1}}/n$.

At any general gear-pair shaft element (*e*), the following geometric boundary conditions exist (refer to Figure 7.23):

$$\text{At first node:} \quad z = 0; \quad \varphi_z^{(e)}(0,t) = -\varphi_{z_{g1}}(t)/n \tag{7.112}$$

and

$$\text{At second node:} \quad z = l; \quad \varphi_z^{(e)}(l,t) = \varphi_{z2}(t) \tag{7.113}$$

On application of Equations 7.112 and 7.113 into Equation 7.61, the constants of the interpolation function (i.e., the shape function) can be obtained as

$$a = -\frac{\varphi_{z_{g1}}(t)}{n} \quad \text{and} \quad b = \frac{\varphi_{z2}(t) + n\varphi_{z_{g1}}(t)}{nl} \tag{7.114}$$

Hence, Equation 7.61 takes the following form:

$$\varphi_z^{(e)}(z,t) = -\frac{\varphi_{z_{g1}}(t)}{n} + \frac{\varphi_{z2}(t) + n\varphi_{z_{g1}}(t)}{nl} z \quad \text{or} \quad \varphi_z^{(e)}(z,t) = \left(\frac{-1}{n}\right)\left(1 - \frac{z}{l}\right)\varphi_{z_{g1}}(t) + \left(\frac{z}{l}\right)\varphi_{z2}(t) \tag{7.115}$$

It should be noted that the gear ratio, n, is playing a major role in the shape function. In view of Equation 7.115, the shape function of the form of Equation 7.56 can be written as

$$\varphi_z^{(e)}(z,t) = \mathbf{N}_g(z)\varphi_z^{(ne)}(t) \tag{7.116}$$

with

$$\mathbf{N}_g(z) = \lfloor N_{g1}(z) \quad N_{g2}(z) \rfloor; \quad \varphi_z^{(ne)}(t) = \begin{Bmatrix} \varphi_{z_{g1}}(t) \\ \varphi_{z2}(t) \end{Bmatrix} \tag{7.117}$$

and

$$N_{g1}(z) = \left(\frac{-1}{n}\right)\left(1 - \frac{z}{l}\right); \quad N_{g2}(z) = \left(\frac{z}{l}\right) \tag{7.118}$$

where N_{g1} and N_{g2} are interpolation functions corresponding to the first and second nodes, respectively, of the gear-pair shaft element (refer to Figure 7.23). These have the following characteristics, e.g., $N_{g1} = -1/n$ at $z=0$, $N_{g2}=0$ at $z=l$; and $N_{g1}=0$ at $z=0$, $N_{g2}=1$ at $z=l$. Hence, the consistent mass matrix takes the following form (refer to Equation 7.84):

$$\mathbf{M}^{(e)} = \int_0^l \rho J \mathbf{N}_g^T \mathbf{N}_g dz = \int_0^l \rho J \begin{bmatrix} N_{1g}^2 & N_{1g}N_{2g} \\ N_{2g}N_{1g} & N_{2g}^2 \end{bmatrix} dz = \frac{\rho J l}{6}\begin{bmatrix} 2/n^2 & -1/n \\ -1/n & 2 \end{bmatrix} \tag{7.119}$$

Please note the change in the sign for off-diagonal terms of the mass matrix apart from the appearance of the gear ratio in the matrix. Similarly, the consistent stiffness matrix takes the following form:

$$\mathbf{K}^{(e)} = \int_0^l GJ \mathbf{N}_{g,z}^T \mathbf{N}_{g,z} dz = \int_0^l GJ \begin{bmatrix} N_{1g,z}^2 & N_{1g,z}N_{2g,z} \\ N_{2g,z}N_{1g,z} & N_{2g,z}^2 \end{bmatrix} dz = \frac{GJ}{l}\begin{bmatrix} 1/n^2 & 1/n \\ 1/n & 1 \end{bmatrix} \tag{7.120}$$

Here also, similar terms of the gear ratio appear with change in sign of off-diagonal terms. The reaction and external torque vectors take the following form (here also observe the change in sign and the gear ratio in the terms at node 1):

$$\mathbf{T}_R^{(ne)} = GJ \begin{Bmatrix} N_{1g} \\ N_{2g} \end{Bmatrix} \varphi_{z,z}^{(e)}\Big|_0^l = \begin{Bmatrix} GJ\varphi_{z,z}^{(e)} N_{1g}\big|_{z=l} - GJ\varphi_{z,z}^{(e)} N_{1g}\big|_{z=0} \\ GJ\varphi_{z,z}^{(e)} N_{2g}\big|_{z=l} - GJ\varphi_{z,z}^{(e)} N_{2g}\big|_{z=0} \end{Bmatrix}$$

$$= \begin{Bmatrix} 0 + GJ\varphi_{z_{g1},z}^{(e)} / n \\ GJ\varphi_{z2,z}^{(e)} - 0 \end{Bmatrix} = \begin{Bmatrix} T_{R_{g1}} / n \\ T_{R_{g2}} \end{Bmatrix}^{(ne)} \tag{7.121}$$

and

$$\mathbf{T}_E^{(ne)} = \int_0^l \mathbf{N}_g^T \left[T^{(e)}(z,t) + T_0^{(e)}(t)\delta^*\left(z - z_0^{(e)}\right) \right] dz \tag{7.122}$$

7.3.2 Lumped Mass Matrix of the Gear Pair

The equivalent inertia torque of the gear pair with respect to reference shaft 1 is given as (refer to Chapter 6 for more details regarding the equivalent polar mass moments of inertia from disks in two shaft systems, which are rotating at different speeds)

$$\left\{ I_{Pg1} + \frac{I_{Pg2}}{n^2} \right\} \ddot{\varphi}_{z\,g1} \tag{7.123}$$

where I_{Pg1} and I_{Pg2} are the polar mass moments of inertia of actual gears 1 and 2, respectively. This can be seen from the kinetic energy of the gear pair also, as

$$T = \frac{1}{2} I_{Pg1}\dot{\varphi}_{zg1}^2 + \frac{1}{2} I_{Pg2}\dot{\varphi}_{zg2}^2 = \frac{1}{2} I_{Pg1}\dot{\varphi}_{zg1}^2 + \frac{1}{2} I_{Pg2}\left(-\frac{\dot{\varphi}_{zg1}}{n}\right)^2 = \frac{1}{2}\left(I_{Pg1} + \frac{I_{Pg2}}{n^2}\right)\dot{\varphi}_{zg1}^2 \tag{7.124}$$

Expression 7.123 could be written in the mass matrix form, as

$$\begin{bmatrix} \left(I_{Pg1} + \dfrac{I_{Pg2}}{n^2}\right) & 0 \\ 0 & 0 \end{bmatrix} \begin{Bmatrix} \ddot{\varphi}_{zg1} \\ \ddot{\varphi}_{z2} \end{Bmatrix} \tag{7.125}$$

It should be worth mentioning that φ_{z2} is the actual rotational displacement of the right-hand end of the shaft that is attached to gear 2. Hence, no scaling is required of the mode shapes obtained from the above formulation.

The following examples illustrate the use of the gear-pair element for geared and branched systems by using the FEM.

Example 7.8

For a simple gear-rotor system as shown in Figure 7.24, find the torsional natural frequencies. Shaft A has a diameter of 5 cm and a length of 0.75 m, and shaft *B* has a diameter of 4 cm and a

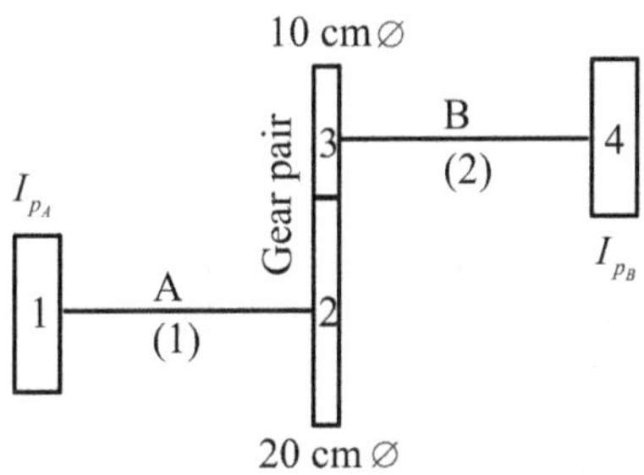

FIGURE 7.24 Two disks with a geared system.

length of 1.0 m. Take the modulus of rigidity of the shaft G equal to 0.8×10^{11} N/m^2, and the polar mass moments of inertia of disks and gears are $I_{P_A} = 24\ \text{Nm}^2$, $I_{P_B} = 10\ \text{Nm}^2$, $I_{P_{gA}} = 5\ \text{Nm}^2$, and $I_{P_{gB}} = 3\ \text{Nm}^2$.

Solution:

Now this problem is solved by using the FEM for the illustration of the method for geared rotor systems. The pinion and the gear have an appreciable amount of the polar mass moment of inertia. Let us divide the geared rotor system into two elements (Figure 7.24). Denote the node number of the disk on branch A as 1 and the gear as 2. For branch B, the node number of the gear is 3 and for the disk it is 4 (Figure 7.25).

The following data can be obtained for the present rotor system:

$$J_A = \frac{\pi}{32} d_A^4 = 6.136\times10^{-7}\ \text{m}^4;\quad J_B = \frac{\pi}{32} d_B^4 = 2.51\times10^{-7}\ \text{m}^4$$

and

$$k_1 = k_A = \frac{GJ_A}{l_A} = \frac{0.8\times10^{11}\times6.136\times10^{-7}}{0.75} = 6.545\times10^4\ \text{Nm/rad};\quad k_2 = k_B = 2.011\times10^4\ \text{Nm/rad}$$

Now elemental equations are written one by one for each element, and the same would then be assembled to form the global equation of motion of the geared rotor system.

Element (1): The elemental equation can be written as (neglecting the polar mass moment of inertia of the shaft)

$$\frac{1}{9.81}\begin{bmatrix} 24 & 0 \\ 0 & 5 \end{bmatrix}\begin{Bmatrix} \ddot{\varphi}_{z1} \\ \ddot{\varphi}_{z2} \end{Bmatrix} + 10^4\begin{bmatrix} 6.545 & -6.545 \\ -6.545 & 6.545 \end{bmatrix}\begin{Bmatrix} \varphi_{z1} \\ \varphi_{z2} \end{Bmatrix} = \begin{Bmatrix} -T_1 \\ T_2 \end{Bmatrix} \tag{a}$$

Element (2): The elemental equation can be written as

$$\frac{1}{9.81}\begin{bmatrix} 3 & 0 \\ 0 & 10 \end{bmatrix}\begin{Bmatrix} \ddot{\varphi}_{z3} \\ \ddot{\varphi}_{z4} \end{Bmatrix} + 10^4\begin{bmatrix} 2.011 & -2.011 \\ -2.011 & 2.011 \end{bmatrix}\begin{Bmatrix} \varphi_{z3} \\ \varphi_{z4} \end{Bmatrix} = \begin{Bmatrix} -T_3 \\ T_4 \end{Bmatrix} \tag{b}$$

Now, from the above equation, φ_{z_3} has to be replaced by φ_{z_2} because we have $\varphi_{z3} = -\varphi_{z2}/n$ where n is the gear ratio ($n=2$). Hence, it takes the form (refer to Equations 7.120 and 7.121)

$$\frac{1}{9.81}\begin{bmatrix} 3/2^2 & 0 \\ 0 & 10 \end{bmatrix}\begin{Bmatrix} \ddot{\varphi}_{z2} \\ \ddot{\varphi}_{z4} \end{Bmatrix} + 10^4\begin{bmatrix} 2.011/2^2 & 2.011/2 \\ 2.011/2 & 2.011 \end{bmatrix}\begin{Bmatrix} \varphi_{z2} \\ \varphi_{z4} \end{Bmatrix} = \begin{Bmatrix} T_3/2 \\ T_4 \end{Bmatrix} \tag{c}$$

Now Equations (a) and (c) can be assembled to get the global governing equation:

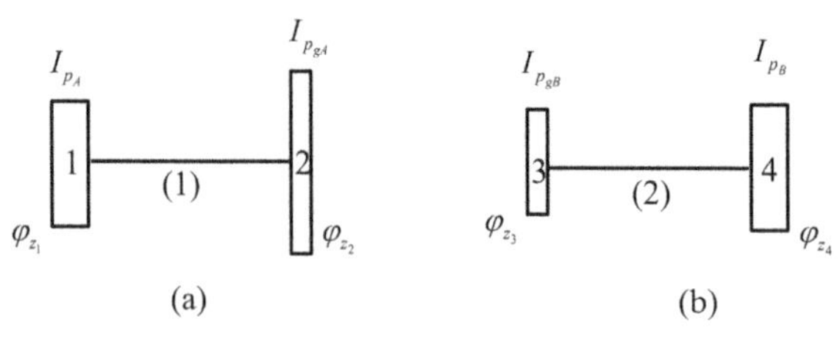

FIGURE 7.25 Elements with nodal variables: (a) the first element; (b) the second element.

$$\frac{1}{9.81}\begin{bmatrix} 24 & 0 & 0 \\ 0 & 5+0.75 & 0 \\ 0 & 0 & 10 \end{bmatrix}\begin{Bmatrix} \ddot{\varphi}_{z1} \\ \ddot{\varphi}_{z2} \\ \ddot{\varphi}_{z4} \end{Bmatrix}+10^4\begin{bmatrix} 6.545 & -6.545 & 0 \\ -6.545 & 6.545+0.503 & 1.006 \\ 0 & 1.006 & 2.011 \end{bmatrix}$$

$$\begin{Bmatrix} \varphi_{z1} \\ \varphi_{z2} \\ \varphi_{z4} \end{Bmatrix}=\begin{Bmatrix} -T_1 \\ T_2+(T_3/2) \\ T_4 \end{Bmatrix} \tag{d}$$

At the gear pair, we have $T_2+(T_3/2)=0$; disks at the ends are free; hence, $T_1=T_4=0$. Thus, on application of boundary conditions, Equation (d) takes the following form for free vibrations:

$$\left(-\omega_{nf}^2\frac{1}{9.81}\begin{bmatrix} 24 & 0 & 0 \\ 0 & 5.75 & 0 \\ 0 & 0 & 10 \end{bmatrix}+10^4\begin{bmatrix} 6.545 & -6.545 & 0 \\ -6.545 & 7.048 & 1.006 \\ 0 & 1.006 & 2.011 \end{bmatrix}\right)\begin{Bmatrix} \varphi_{z1} \\ \varphi_{z2} \\ \varphi_{z4} \end{Bmatrix}=\begin{Bmatrix} 0 \\ 0 \\ 0 \end{Bmatrix} \tag{e}$$

from which we can write

$$\mathbf{D}=\mathbf{M}^{-1}\mathbf{K}=9.81\begin{bmatrix} 24 & 0 & 0 \\ 0 & 5.75 & 0 \\ 0 & 0 & 10 \end{bmatrix}^{-1}\times 10^4\times\begin{bmatrix} 6.545 & -6.545 & 0 \\ -6.545 & 7.048 & 1.006 \\ 0 & 1.006 & 2.011 \end{bmatrix} \tag{f}$$

From the eigenvalue analysis, the natural frequencies and mode shapes can be obtained from the matrix **D** and are given as

$$\omega_{nf_1}=0, \qquad \varphi_z^{(1)}=\lfloor\, 1 \quad 1 \quad -0.5 \,\rfloor$$

$$\omega_{nf_2}=142.13\ \text{rad/s}, \qquad \varphi_z^{(2)}=\lfloor\, 0.1929 \quad 0.0472 \quad 0.9801 \,\rfloor$$

$$\omega_{nf_3}=382.77\ \text{rad/s}, \qquad \varphi_z^{(3)}=\lfloor\, -0.2174 \quad 0.9731 \quad 0.0757 \,\rfloor$$

Here superscript on φ_z represents the mode number. On adding gear displacement (with the help of gear ratio) and normalizing the mode shapes (with respect to the maximum displacement equating to unity) are given as

$$\varphi_z^{(1)}=\lfloor\, 1 \quad 1 \quad -0.5 \quad -0.5 \,\rfloor;\quad \varphi_z^{(2)}=\lfloor\, 0.1968 \quad 0.0482 \quad -0.0241 \quad 1.0000 \,\rfloor;$$

$$\varphi_z^{(3)}=\lfloor\, -0.2234 \quad 1.0000 \quad -0.5000 \quad 0.0778 \,\rfloor$$

The eigenvector gives a relative rotational displacement at various node locations. For the rigid body mode, it can be seen that nodes 1 and 2 have rotational displacement of 1 unit each, whereas node 4 has a rotational displacement of 0.5 units, because of the gear ratio effect. For obtaining rotational displacement at node 3, we need to use the gear ratio to angular displacement at node 2, which gives rotational displacement at node 3 as 0.5 units. The same has to be followed for other modes. The results can be compared with Example 6.12 by the TMM, and they are quite close.

Example 7.9

Obtain the torsional natural frequencies of a branched rotor system as shown in Figure 7.26. Take the polar mass moment of inertia of rotors and gears as $I_{PA}=0.01$ kg-m^2, $I_{PE}=I_{PB}=0.005$ kg-m^2, and $I_{PF}=I_{PC}=I_{PD}=0.006$ kg-m^2. Take the gear ratios between various gear pairs as $n_{BC}=3$ and $n_{BD}=4$. The shaft lengths are $l_{AB}=l_{CE}=l_{DF}=0.25$ m, and the diameters are $d_{AB}=0.03$ m, $d_{CE}=0.02$ m, and $d_{DF}=0.02$ m. Take the shaft modulus of rigidity $G=0.8\times 10^{11}$ N/m^2.

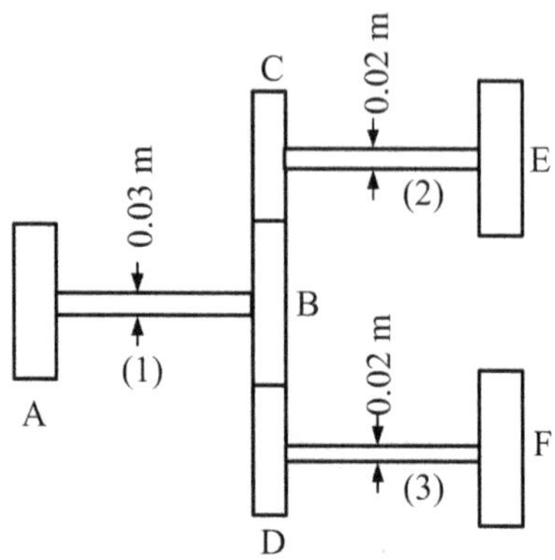

FIGURE 7.26 A branched system with gears.

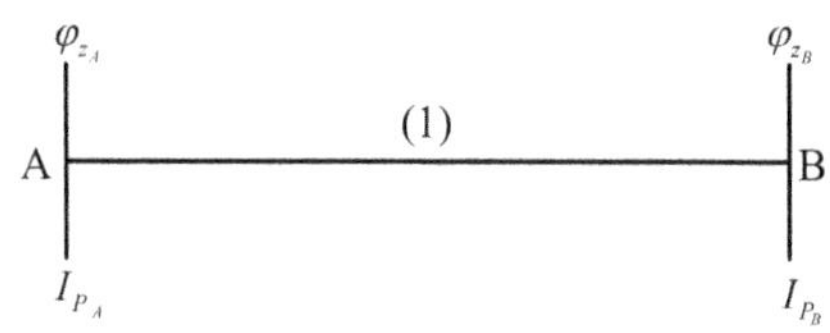

FIGURE 7.27 Element (1).

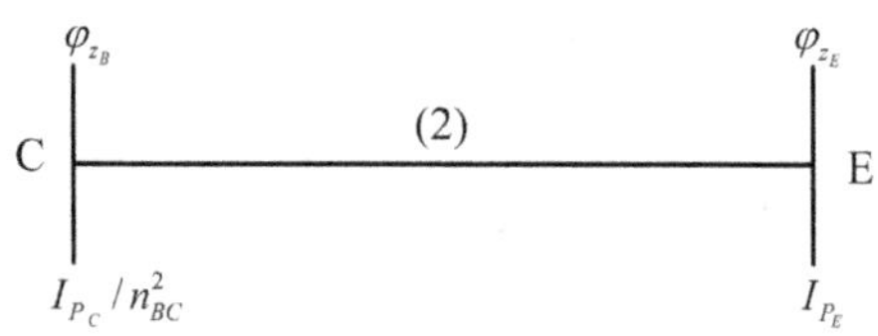

FIGURE 7.28 Element (2).

Solution:

The branched system is divided into three elements. Figure 7.26 shows various element numbers of the branched rotor system. Now various element equations can be obtained as follows:

Element (1): Figure 7.27 shows element (1) with nodal variables at nodes A and B. Because this shaft element is the input shaft, there is no change in the inertia terms as well as in the stiffness terms.

The equations of motion for element (1) can be written as (neglecting the inertia of the shaft)

$$\begin{bmatrix} I_{PA} & 0 \\ 0 & I_{PB} \end{bmatrix} \begin{Bmatrix} \ddot{\varphi}_{zA} \\ \ddot{\varphi}_{zB} \end{Bmatrix} + \begin{bmatrix} k_{t_1} & -k_{t_1} \\ -k_{t_1} & k_{t_1} \end{bmatrix} \begin{Bmatrix} \varphi_{zA} \\ \varphi_{zB} \end{Bmatrix} = \begin{Bmatrix} -T_A \\ T_B \end{Bmatrix} \tag{a}$$

Element (2): Figure 7.28 shows element (2) with nodal variables at nodes C and E. The actual nodal variable at gear C is related to the nodal variable of gear B; hence, the nodal variable of gear B is used. This affects both the mass and stiffness matrices as well as the internal torque vector.

The equations of motion for element (2) can be written as

$$\begin{bmatrix} I_{PC}/n_{BC}^2 & 0 \\ 0 & I_{PE} \end{bmatrix} \begin{Bmatrix} \ddot{\varphi}_{zB} \\ \ddot{\varphi}_{zE} \end{Bmatrix} + \begin{bmatrix} k_{t_2}/n_{BC}^2 & k_{t_2}/n_{BC} \\ k_{t_2}/n_{BC} & k_{t_2} \end{bmatrix} \begin{Bmatrix} \varphi_{zB} \\ \varphi_{zE} \end{Bmatrix} = \begin{Bmatrix} T_C/n_{BC} \\ T_E \end{Bmatrix} \tag{b}$$

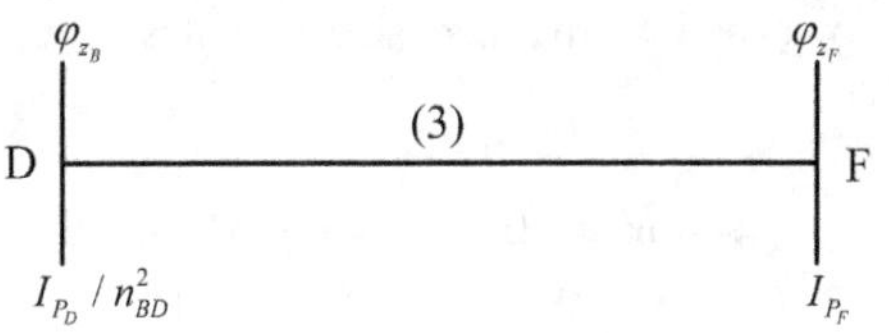

FIGURE 7.29 Element (3).

Element (3): Figure 7.29 shows element (3) with nodal variables. Here also instead of the actual rotational displacement of gear *D*, the rotational displacement of gear *B* is used.

For element (3), we can write the equation of motion as

$$\begin{bmatrix} I_{PD}/n_{BD}^2 & 0 \\ 0 & I_{PF} \end{bmatrix} \begin{Bmatrix} \ddot{\varphi}_{zB} \\ \ddot{\varphi}_{zF} \end{Bmatrix} + \begin{bmatrix} k_{t_3}/n_{BD}^2 & k_{t_3}/n_{BD} \\ k_{t_3}/n_{BD} & k_{t_3} \end{bmatrix} \begin{Bmatrix} \varphi_{zB} \\ \varphi_{zF} \end{Bmatrix} = \begin{Bmatrix} T_D/n_{BD} \\ T_F \end{Bmatrix} \tag{c}$$

In the present problem, we have discretized the system into three elements. However, because at the branched point, the rotational displacements of three gears (i.e., *B*, *C*, *D*) are related, this leads to a finite element system model of only a four-DOF system. Hence, on assembling Equations (a) through (c), we get

$$\begin{bmatrix} I_{PA} & 0 & 0 & 0 \\ 0 & \left\{I_{PB} + \left(I_{pc}/n_{BC}^2\right) + \left(I_{PD}/n_{BD}^2\right)\right\} & 0 & 0 \\ 0 & 0 & I_{PE} & 0 \\ 0 & 0 & 0 & I_{PF} \end{bmatrix} \begin{Bmatrix} \ddot{\varphi}_{zA} \\ \ddot{\varphi}_{zB} \\ \ddot{\varphi}_{zE} \\ \ddot{\varphi}_{zF} \end{Bmatrix}$$

$$+ \begin{bmatrix} k_{t_1} & -k_{t_1} & 0 & 0 \\ -k_{t_1} & \left\{k_{t_1} + \left(k_{t_2}/n_{BC}^2\right) + \left(k_{t_3}/n_{BD}^2\right)\right\} & k_{t_2}/n_{BC} & k_{t_3}/n_{BD} \\ 0 & k_{t_2}/n_{BC} & k_{t_2} & 0 \\ 0 & k_{t_3}/n_{BC} & 0 & k_{t_3} \end{bmatrix} \begin{Bmatrix} \varphi_{zA} \\ \varphi_{zB} \\ \varphi_{zE} \\ \varphi_{zF} \end{Bmatrix} = \begin{Bmatrix} -T_A \\ 0 \\ T_E \\ T_F \end{Bmatrix} \tag{d}$$

It should be noted that in the assembled form, the second row in the torque column is zero because we have the following condition at the branch point:

$$T_B + \frac{T_C}{n_{BC}} + \frac{T_D}{n_{BD}} = 0 \tag{e}$$

From the free boundary conditions at shaft ends *A*, *E*, and *F*, we have

$$T_A = T_E = T_F = 0 \tag{f}$$

Hence, on application of boundary conditions in Equation (d), all the terms in the right-hand side vector are zero. Hence, Equation (d) forms an eigenvalue problem with a matrix size of 4×4, for a nontrivial solution of it, we get the following condition:

$$\left|\mathbf{K} - \omega_{nf}^2 \mathbf{M}\right| = 0 \tag{g}$$

Where **K** and **M** are the assembled stiffness and mass matrices, respectively. Equation (g) gives the frequency equation as a polynomial, and the roots of the polynomial give torsional natural frequencies of the system. However, for a matrix of size 4×4, it is quite cumbersome. Alternatively, Equation (g) could be written as

$$\left|\mathbf{D} - \omega_{nf}^2 \mathbf{I}\right| = 0 \quad \text{with} \quad \mathbf{D} = \mathbf{M}^{-1}\mathbf{K} \tag{h}$$

in which the eigenvalue of matrix **D** is related to natural frequencies (square root of the eigenvalue is the natural frequency) and the eigenvector is related to the mode shape. Hence, this method requires eigenvalue analysis and is preferred for large matrices. It should be noted that the stiffness matrix, **K**, is a singular matrix because we have free–free boundary conditions. For the present case, we have the following data:

$$I_{PA} = 0.01\ \text{kg-m}^2;\quad I_{PE} = 0.005\ \text{kg-m}^2;\quad I_{PF} = 0.006\ \text{kg-m}^2$$

$$I_{PB} = 0.005\ \text{kg-m}^2;\quad I_{PC} = 0.006\ \text{kg-m}^2;\quad I_{PD} = 0.006\ \text{kg-m}^2$$

$$n_{BC} = 3;\quad n_{BD} = 4;$$

$$k_{tAB} = \frac{GJ_{AB}}{l_{AB}} = 2.54\times 10^4\ \text{N-m/rad};\quad k_{tCE} = k_{tDF} = 0.50\times 10^4\ \text{N-m/rad}$$

The assembled stiffness and mass matrices have the following form:

$$\mathbf{K} = \begin{bmatrix} 25.4 & -25.4 & 0 & 0 \\ -25.4 & 26.32 & 1.68 & 1.26 \\ 0 & 1.68 & 5.02 & 0 \\ 0 & 1.26 & 0 & 5.02 \end{bmatrix} \times 10^3\ \text{N-m/rad}$$

and

$$\mathbf{M} = \begin{bmatrix} 0.01 & 0 & 0 & 0 \\ 0 & 0.006 & 0 & 0 \\ 0 & 0 & 0.005 & 0 \\ 0 & 0 & 0 & 0.006 \end{bmatrix} \text{kg-m}^2$$

Using Equation (g) or (h), the natural frequencies can be obtained as

$$\omega_{nf_1} = 0\ \text{rad/s};\quad \omega_{nf_2} = 922.22\ \text{rad/s}$$

$$\omega_{nf_3} = 1015.70\ \text{rad/s};\quad \omega_{nf_4} = 2614.50\ \text{rad/s}$$

In the former method, the relative amplitudes of rotational displacements (i.e., mode shapes) are obtained by substituting these natural frequencies one at a time in Equation (d). In the latter method, mode shapes (eigenvectors) are obtained by the eigenvalue analysis of the matrix $\mathbf{D} = \mathbf{M}^{-1}\mathbf{K}$. The eigenvalue gives the square of natural frequencies, and the eigenvectors give the mode shapes as described previously. The mode shapes are shown in Figure 7.30 for all four modes. It should be noted that relative rotational displacements obtained from Equation (d) are actual displacements, and no further scaling by the gear ratio is required as we did for the equivalent geared system. The rotational displacements of disks *C* and *D* are related (i.e., these are not independent) to the rotational displacement of disk *B* with respective gear ratios. Thus, these rotational displacements have been obtained with these relations for the completeness of mode shapes.

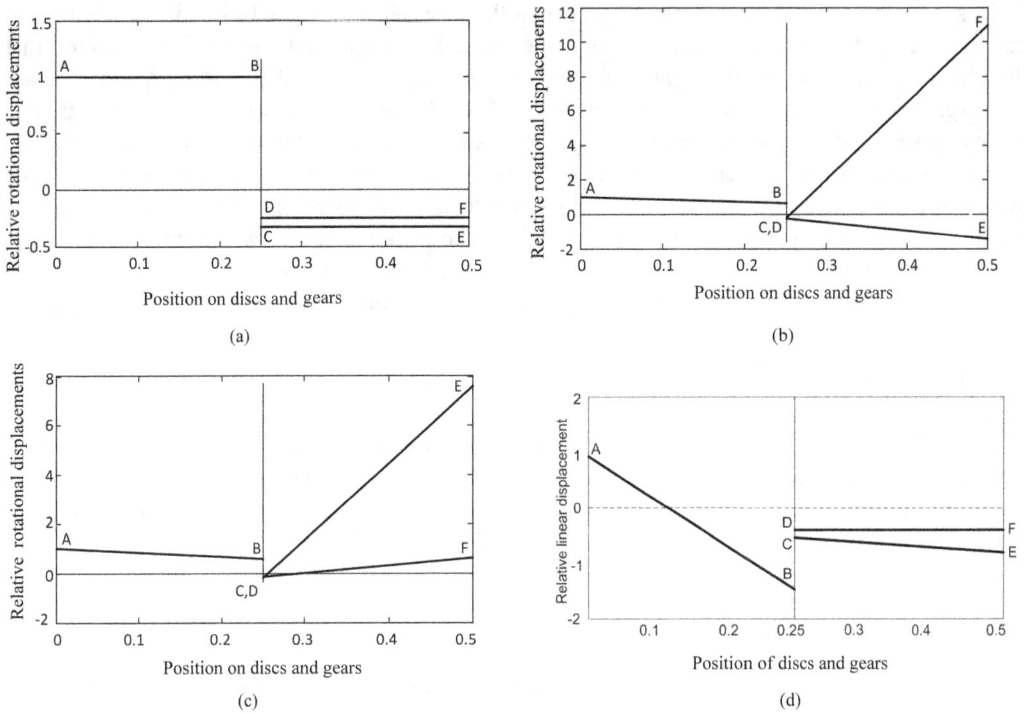

FIGURE 7.30 (a) Mode shape of the geared system for $\omega_{nf_1} = 0$. (b) Mode shape of the geared system for $\omega_{nf_2} = 922.22$ rad/s. (c) Mode shape of the geared system for $\omega_{nf_3} = 1015.70$ rad/s. (d) Mode shape of the geared system for $\omega_{nf_4} = 2614.50$ rad/s.

7.4 CONCLUDING REMARKS

To summarize, in the present chapter, torsional vibrations of rods by the continuous system approach are initially presented. Hamilton's principle and Newton's second law are used to obtain the governing equations, and associated geometric (i.e., related to the rotational displacement) and natural (i.e., related to the torque) boundary conditions. The form of the governing equation is similar to the wave equation. The closed form solutions for simple boundary conditions are presented by the method of separation of variables, and the corresponding natural frequencies (eigen frequencies) and mode shapes (eigen functions) are tabulated and plotted, respectively. Because for more complex rotor-support systems, obtaining the closed form solution is very difficult and sometimes it is impractical, the need for an approximate method (e.g., FEM) is felt. The finite element analysis for the torsional vibration is presented by the Galerkin and Rayleigh–Ritz methods. The analysis presented for the torsional case is also of the similar form as that of the axial vibrations of rods. In that case, the axial displacement replaces the rotational displacement and the axial force replaces the torque. The axial stiffness of the rod is given as $k_a = EA/l$ (where A is the cross-sectional area of the rod), whereas the torsional stiffness is given as $k_t = GJ/l$. With these analogies, all the analyses of the torsional vibration can be used for axial vibrations also. The finite element analysis is extended to rotors with branched systems also by developing elemental equations for the gear-pair element. However, the flexibility of the gear teeth is not considered and they are assumed to be rigid. The epicyclic gear train torsional analysis (Botman, 1976; Yang et al., 2024) is not considered in which the rotation of the carrier gives rise to a system with periodic coefficients that needs to solve by means of Floquet's theory, which is beyond the scope of the book. Now in subsequent chapters transverse vibration of the multi-DOF rotor system will be analyzed.

In some cases, a coupling is present between the transverse and axial directions, when a rod has initial constant tension or compression. Due to the initial constant tension, it is expected that the effective transverse stiffness increases, and due to this, the natural frequency also increases. However, for the initial constant compression of the rod, the transverse stiffness decreases, and correspondingly, the transverse natural frequency also decreases. In the limit, when this compressive load is equal to or greater than the buckling load, the stiffness and so natural frequency of the system become zero. Another issue we have not touched in the present chapter is how to tackle the increasing number of matrices while using the FEM. There are *condensation schemes* (or *reduction schemes*) that allow the unwanted DOFs to be eliminated from the assembled system matrices, and this will be discussed during the application of the FEM in transverse vibrations.

EXERCISE PROBLEMS

Exercise 7.1 Obtain the following equations of motion by using Newton's second law of motion of a continuous rod subjected to a uniform torque $T(z, t)$ for the pure torsional vibration. The notations have the following nomenclature: GJ is the modulus of rigidity, L is the length, ρ is the mass density, and J is the polar area moment of inertia of the rod.

$$\rho J\ddot{\varphi}_z - GJ\varphi_{z,zz} - T_{z_e}(z,t) = 0$$

Exercise 7.2 A cantilever rod has a thin rigid disk (of mass M with r as the radius of gyration) at the free end, and it is also supported by a torsional spring of stiffness k_t at the free end (see Figure 7.31). Let us take GJ as the modulus of rigidity, L as the length, $I_p(z)=\rho J$ as the polar mass moment of inertia per unit length, ρ as the mass density, and J as the polar area moment of inertia of the rod. Write down equations of motion and boundary conditions for torsional vibrations. Obtain the expression of frequency equations and eigenfunctions (mode shapes) in the closed form.

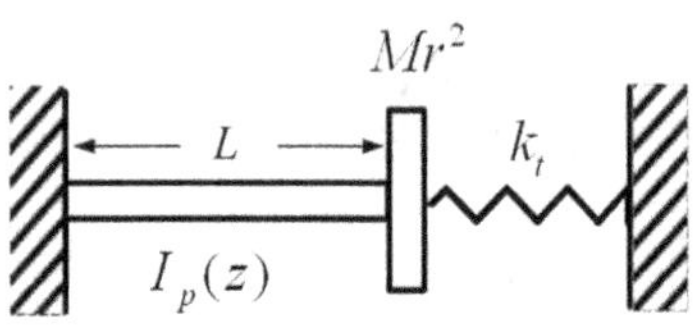

FIGURE 7.31 A cantilever rod with a rigid disk and a torsional spring at the free end.

Exercise 7.3 Consider a rod with E as the Young's modulus, A as the area of the cross-section, L as the length, $m(z)=\rho A$ as the mass per unit length, and $u_z(z)$ as the axial displacement at a position z. Write down equations of motion and boundary conditions for *axial vibrations*. Obtain the expressions of frequency equations and eigenfunctions (mode shapes) in closed form for the following boundary conditions: (1) fixed–free, (2) free–free, and (3) fixed–fixed.

Exercise 7.4 Formulate the equation of motion and the eigenvalue problem for torsional vibrations of a nonuniform rod that is clamped at one end and free at the other end. Consider only the torsional vibrations. Let $GJ(z)=2GJ_o\,\{1-(z/L)\}$ and $I_p(z)=2I_{po}\{1-(z/L)\}$, where J_o and I_{po} are constant quantities. Obtain expressions of natural frequencies and eigenfunctions (mode shapes).

Exercise 7.5 Obtain the natural frequencies and mode shapes of the rotor system shown in Figure 7.32 for the following parameters: $I_{p1} = 0.02$ kg-m^2, $I_{p2} = 0.08$ kg-m^2, $l=1$ m, $d=0.01$ m,

$\rho=7800\,kg/m^3$, and $G=0.8\times10^{11}\,N/m^2$. Use the FEM (at least two elements) and compare the results by considering the problem with and without the mass of the shaft.

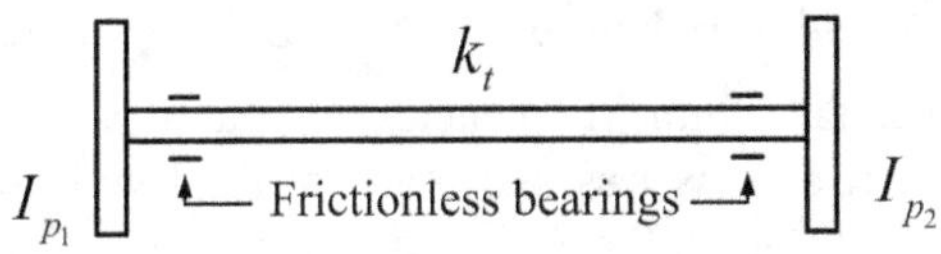

FIGURE 7.32 Two-disk rotor system.

Exercise 7.6 All the exercise problems provided for the TMM in the previous chapter can be attempted with the FEM. Compare your results obtained by the TMM and FEM, and interpret the observations made. Does the TMM underestimate and the FEM overestimate natural frequencies?

Exercise 7.7 For a continuous shaft of 3 m length and 0.03 m diameter, obtain the torsional natural frequencies up to the fifth mode and plot the corresponding mode shapes for the following boundary conditions: (i) fixed–free and (ii) fixed–fixed. Use the FEM. Compare natural frequencies with closed form analytical solutions, perform a convergence study by increasing the number of elements (i.e., 5, 10, 50, and 100), and discuss the result. The following properties of the shaft should be taken: $\rho=7800\,kg/m^3$ and $G=0.8\times10^{11}\,N/m^2$.

Exercise 7.8 For a continuous shaft of 3 m length and 0.03 m diameter, obtain the torsional natural frequencies up to the fifth mode and plot the corresponding mode shapes for the following boundary conditions: (i) a cantilevered shaft with a disk at the free end; (ii) a fixed–fixed shaft with a disk at the midspan. The polar mass moment of inertia of the disk $I_p=0.02$ kg-m^2. Use the FEM. Compare natural frequencies with closed form analytical solutions. The following properties of the shaft should be taken: $\rho=7800\,kg/m^3$ and $G=0.8\times10^{11}\,N/m^2$.

Exercise 7.9 Obtain the torsional natural frequencies and mode shapes of an epicyclic gear train as shown in Figure 7.33. Find also the locations of nodal points on the shaft. The gear mounted on shaft *B* is a planetary gear, and the gear on shaft *A* is a sun gear. Consider the polar mass moment of inertia of the shaft, the arm, and gears as negligible. Shaft *A* has a 5 cm diameter and 0.75 m length, and shaft *B* has a 4 cm diameter and 1.0 m length. Take the modulus of rigidity of the shaft *G* equal to $0.8\times10^{11}\,N/m^2$; the polar mass moments of inertia of the disks are $I_A=24$ Nm^2 and $I_B=10$ Nm^2. The angular speeds of shaft A and the arm are 600 rpm counter-clockwise (CCW) and 300 rpm clockwise (CW), respectively. Use the FEM.

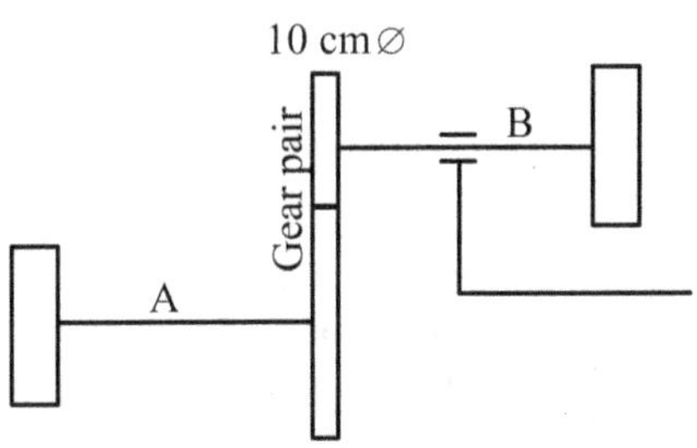

FIGURE 7.33 An epicyclic geared system.

Exercise 7.10 A rod has two rigid disks at either end and is supported by two torsional springs of stiffness $k_{t_1} = k_{t_2} = k_t$ at each end of the rod. Use the following parameters: the polar mass moments of inertia of the disks are $I_{p1} = I_{p2} = 0.002$ kg-m^2, torsional stiffness of the spring at free end k_t=200 Nm/rad, length of the rod l=0.5 m, diameter of the rod d=0.01 m, mass density of the rod material ρ=7800 kg/m^3, and modulus of rigidity of the rod material $G=0.8\times10^{11}$ N/m^2. Obtain the natural frequencies of the system by considering the rod as only two elements.

Exercise 7.11 Find the torsional natural frequencies of a cantilever rotor system as shown in Figure 7.34. Consider the shaft as massless and made of steel with a modulus of rigidity of 0.8×10^{11} N/m^2 and a mass density of 7800 kg/m^3. A disk is mounted at the free end of the shaft with a polar mass moment of inertia 0.01 kg-m^2. In the diagram, all dimensions are in cm. Use the FEM and compare results with the TMM.

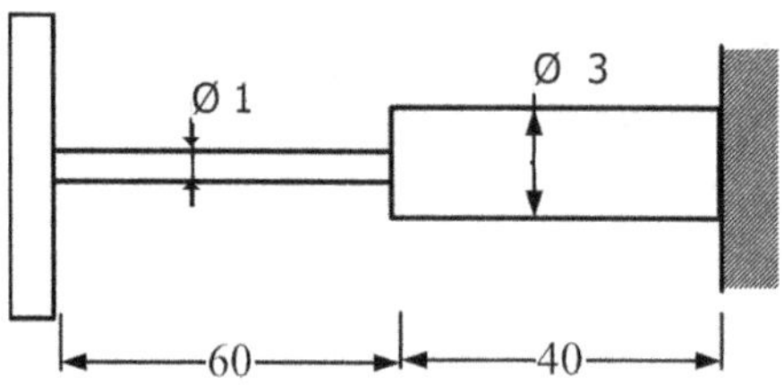

FIGURE 7.34 A cantilever rotor with a stepped shaft.

Exercise 7.12 Find the torsional natural frequencies and mode shapes of the rotor system shown in Figure 7.35. B_1 and B_2 are frictionless bearings, which provide free–free end conditions; and D_1, D_2, D_3, and D_4 are rigid disks. The shaft is made of steel with modulus of rigidity $G=0.8\times10^{11}$ N/m^2 and a uniform diameter d=20 mm. The various shaft lengths are as follows: B_1D_1=150 mm, D_1D_2=50 mm, D_2D_3=50 mm, D_3D_4=50 mm, and D_4B_2=150 mm. The mass of disks are m_1=4 kg, m_2=5 kg, m_3=6 kg, and m_4=7 kg. Consider the shaft as massless. Consider the disks as thin and take the diameter of the disks as D_1=8 cm, D_2=10 cm, D_3=12 cm, and D_4=14 cm.

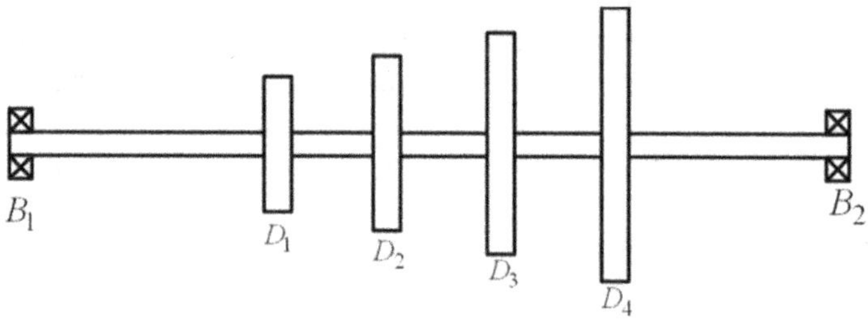

FIGURE 7.35 A multidisc rotor system.

Exercise 7.13 Obtain the torsional natural frequency of an overhung rotor system as shown in Figure 7.36. Take the polar mass moment of inertia of the disk as I_p=0.04 kg-m^2. The massless shaft has the following properties: Lengths are a=0.3 m, b=0.7 m, the uniform diameter is 10 mm, and the modulus of rigidity $G=0.8\times10^{11}$ N/m^2. Bearing A is flexible and provides a torsional restoring torque with its torsional stiffness equal to 5% of the torsional stiffness of the shaft segment having length a. Consider bearing B as a fixed bearing. Use the FEM.

Exercise 7.14 A cantilever rotor has a massless flexible shaft (of diameter d=20 mm and length L=40 cm) and carries a thin disk (of polar mass moment of inertia 0.02 kg-m^2) at the free end. The shaft has an axial through-hole of diameter 3 mm at a distance of 5 mm from the center of

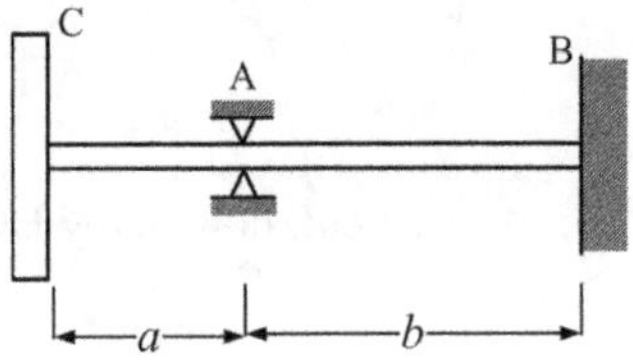

FIGURE 7.36 An overhung rotor.

the shaft and parallel to the shaft axis. Analyze the free torsional vibration of this rotor using (i) the FEM and (ii) the analytical method (closed form solutions). Obtain the torsional natural frequency of the system. Take $G = 0.8 \times 10^{11}\,\mathrm{N/m^2}$.

Exercise 7.15 For a tapered shaft element of length l with diameters d_1 and d_2 at node 1 and node 2, respectively. Obtain a simplified expression of the mass coefficient (m_{11}) and the stiffness coefficient (k_{12}) of respective matrices using the FEM. Herein, G is the modulus of rigidity and ρ is the density of the shaft material. Derivation may be started from the weak formulation. Refer to Appendix 2.1 for tapered shaft geometric parameters.

Exercise 7.16 Choose a single answer:

i. For the torsional vibration of a uniform rod with the distributed inertia and stiffness properties, the fundamental natural frequency expression for fixed–fixed end conditions is given as

A. $\dfrac{3\pi}{2L}\sqrt{\dfrac{G}{\rho}}$

B. $\dfrac{3\pi}{2L}\sqrt{\dfrac{E}{\rho}}$

C. $\dfrac{\pi}{2L}\sqrt{\dfrac{G}{\rho}}$

D. $\dfrac{\pi}{2L}\sqrt{\dfrac{E}{\rho}}$

ii. The FEM is
A. an approximate method
B. an exact method
C. either exact or approximate method and it depends on the number of elements
D. neither exact nor approximate method

iii. During the discretization of the shaft into a number of elements in the FEM,
A. the element length must be equal
B. the element length may be unequal
C. the element length must be unequal
D. the element length should be of decreasing order

iv. The degree of polynomial of the shape function, in the finite element formulation of the torsional vibration case, is
A. one
B. two

C. three
D. four

v. For the torsional vibration in a finite element, how many DOFs are considered?
A. one
B. two
C. three
D. four

vi. In non-weak finite element formulation of torsional vibration, the requirement of completeness conditions in the shape function for the polynomial degree would be
A. one
B. two
C. three
D. four

vii. In finite element formulation, the compatibility conditions are
A. between two nodes of the same element
B. between any two nodes of different elements
C. between two common nodes of different elements
D. among all nodes of two different elements

viii. In the finite element formulation, the consistent elemental mass matrix will have
A. only diagonal terms as nonzero
B. only diagonal terms as zero
C. only off-diagonal terms as nonzero
D. only off-diagonal terms as zero
E. all diagonal and off-diagonal are nonzero

ix. In the finite element formulation for the torsional vibration, if one end of the shaft has torsional spring of stiffness k_t, then the corresponding boundary condition will be (φ_z is the torsional displacement at the end of the shaft)
A. $T_R = -k_t\varphi_z$
B. $T_R = k_t\varphi_z$
C. $T_R = 0$
D. $\varphi_z = 0$

x. In the FEM with the consistent mass matrix, generally on increasing the number of elements, a particular natural frequency value
A. asymptotically decreases to a fixed value
B. asymptotically increases to a fixed value
C. asymptotically decreases and then increases to a fixed value
D. remain fixed

xi. In the FEM with the consistent mass matrix, on increasing the number of elements, the number of natural frequencies obtained
A. increases
B. decreases
C. remain the same
D. it depends on the choice of element

xii. In the finite element formulation of a gear-pair element, if the bearing which is supporting very close to the gear (driven gear) has a torsional stiffness k_t, then in the stiffness matrix of the gear-pair element, its contribution will be

A. $\begin{bmatrix} k_t & -k_t \\ -k_t & k_t \end{bmatrix}$

B. $\begin{bmatrix} k_t/n^2 & k_t/n \\ k_t/n & k_t \end{bmatrix}$

C. $\begin{bmatrix} k_t/n^2 & -k_t/n \\ -k_t/n & k_t \end{bmatrix}$

D. $\begin{bmatrix} k_t/n^2 & 0 \\ 0 & 0 \end{bmatrix}$

xiii. For a geared rotor system for the torsional vibration, the finite element formulation with a continuous inertia consideration of the shaft will give the mass matrix of the shaft as (n is the gear ratio, ρ is the mass density of the shaft, J is the second polar moment of area, and l is the length of the element)

A. $\frac{\rho Jl}{6}\begin{bmatrix} 2/n^2 & -1/n \\ -1/n & 2 \end{bmatrix}$

B. $\frac{\rho Jl}{6}\begin{bmatrix} 2/n^2 & 1/n \\ 1/n & 2 \end{bmatrix}$

C. $\frac{\rho Jl}{6}\begin{bmatrix} 2 & -1 \\ -1 & 2 \end{bmatrix}$

D. $\frac{\rho Jl}{6}\begin{bmatrix} 2 & 1 \\ 1 & 2 \end{bmatrix}$

xiv. For an eigenvalue defined as $(\mathbf{K}^{-1}\mathbf{M} - \lambda\mathbf{I})\mathbf{X} = \mathbf{0}$ where $\mathbf{K}$ is the stiffness matrix, $\mathbf{M}$ is the mass matrix, $\mathbf{I}$ is the identity matrix, $\mathbf{0}$ is the null vector, and λ is the eigenvalue, the natural frequency of the system will be equal to

A. square root of the eigenvalue
B. square root of the reciprocal of the eigenvalue
C. the eigenvalue
D. the reciprocal of the eigenvalue

xv. For a rotor system, the following are the mass and stiffness matrices: $\mathbf{M} = \begin{bmatrix} 3 & 1 \\ 1 & 6 \end{bmatrix}$ and $\mathbf{K} = \begin{bmatrix} 6 & -1 \\ -1 & 3 \end{bmatrix}$; the natural frequencies of the system in rad/s will be

A. 0.6542, 1.5286

B. 0.4279, 2.3368

C. 0.6534, 3.2546

D. 0.2456, 1.6825

xvi. In analysis of the free vibration using the FEM, a fixed–fixed shaft will require minimum how many element(s) to be considered to get at least one natural frequency of the system?
   A. 1
   B. 2
   C. 3
   D. 4

xvii. For a geared rotor system, the gear shaft has a stiffness matrix (without gear effect) as $\mathbf{K} = \begin{bmatrix} 900 & -150 \\ -150 & 600 \end{bmatrix}$, if the gear ratio is 3, then the corresponding stiffness matrix with gear effect will be

   A. $\mathbf{K} = \begin{bmatrix} 100 & 50 \\ 50 & 600 \end{bmatrix}$

   B. $\mathbf{K} = \begin{bmatrix} 100 & -50 \\ -50 & 600 \end{bmatrix}$

   C. $\mathbf{K} = \begin{bmatrix} 300 & 50 \\ 50 & 200 \end{bmatrix}$

   D. $\mathbf{K} = \begin{bmatrix} 300 & -50 \\ -50 & 200 \end{bmatrix}$

ANSWERS TO SELECTED EXERCISE PROBLEMS

7.1 Not applicable
7.2

$$\omega_{nf_i} = \alpha_i L \sqrt{\frac{G}{\rho L^2}}$$

with

$\alpha_1 L = 0.25;\ \alpha_2 L = 3.17;\ \alpha_3 L = 6.30$

7.3 Not applicable
7.4 Not applicable
7.5 0.00 rad/s, 70.06 rad/s and 11094.50 rad/s
7.6 1681.65 rad/s, 5160.66 rad/s, 8987.21 rad/s, 13340.22 rad/s, 18028.20 rad/s and 21632.72 rad/s
7.7 3392.15 rad/s, 7016.46 rad/s, 11094.00 rad/s, 15689.29 rad/s and 20126.14 rad/s
7.8 i 320.66 rad/s, 3423.80 rad/s, 7032.75 rad/s, 11104.74 rad/s, 15696.04 rad/s and 20128.61 rad/s
ii 641.32 rad/s, 7016.46 rad/s, 7081.13 rad/s, 15689.29 rad/s and 15716.23 rad/s
7.9 0.00 rad/s, 158.95 rad/s and 5824.80 rad/s (epicylic gear ratio 2/7)
7.10 313.83 rad/s, 1612.45 rad/s and 22314.62 rad/s
7.11 $\omega_{nf_1} = 360.0$ rad/s; $\omega_{nf_2} = 13,797.5$ rad/s;
7.12 0.00 rad/s, 1372.06 rad/s, 2451.10 rad/s, 3751.06 rad/s, 36994.88 rad/s and 37059.86 rad/s
7.13 46.03 rad/s and 12314.83 rad/s
7.14 $\omega_{nf_1} = 393.8$ rad/s (with a hole away from the center); $\omega_{nf_2} = 396.1$ rad/s (with a hole at the center line).

7.15 $$\mathbf{M}=\frac{\pi\rho l}{240}\begin{bmatrix} 2\left(d_1^2+3d_1d_2+6d_2^2\right) & \left(3d_1^2+4d_1d_2+3d_2^2\right) \\ \left(3d_1^2+4d_1d_2+3d_2^2\right) & 2\left(d_1^2+3d_1d_2+6d_2^2\right) \end{bmatrix}$$

and

$$\mathbf{K}=\frac{\pi G}{160l}\begin{bmatrix} \left(d_1^4+d_1^3d_2+d_1^2d_2^2+d_1d_2^3+d_2^4\right) & -\left(d_1^4+d_1^3d_2+d_1^2d_2^2+d_1d_2^3+d_2^4\right) \\ -\left(d_1^4+d_1^3d_2+d_1^2d_2^2+d_1d_2^3+d_2^4\right) & \left(d_1^4+d_1^3d_2+d_1^2d_2^2+d_1d_2^3+d_2^4\right) \end{bmatrix}$$

7.16 (i) C (ii) A (iii) B (iv) A (v) B (vi) B
(vii) C (viii) E (ix) A (x) A (xi) A (xii) D

REFERENCES

Bathe, K.J., 1982, *Finite Element Procedures in Engineering Analysis*. Englewood Cliffs, NJ: Prentice-Hall.

Botman, M., 1976, Epicyclic gear vibrations, *ASME Journal of Engineering for Industry*, *98*, 811–815.

Cook, R.D., Malkus, D.S., Plesha, M.E., and Witt, R.J., 2002, *Concepts and Applications of Finite Element Analysis*, 4th ed. New York, NY: John Wiley & Sons.

Dixit, U.S., 2009, *Finite Element Methods for Engineers*. New Delhi: Cengage Learning.

Huebner, K.H., Dewhirst, D.L., Smith, D.E., and Byrom, T.G., 2001, *The Finite Element Method for Engineers*, 4th ed. New York, NY: Wiley-Interscience Publication, John Wiley & Sons, Inc.

Hughes, T.J.T., 1986, *The Finite Element Method*. Englewood Cliffs, NJ: Prentice-Hall.

Kahraman, A., Nevzat Ozguven, H., Houser, D.R., and Zakrajsek, J.J., 1992, Dynamic analysis of geared rotors by finite elements. *Transactions of the ASME, Journal of Mechanical Design*, *114*, 507.

Kreyszig, E, 2006, *Advanced Engineering Mathematics*. Hoboken, NJ: John Wiley & Sons.

Meirovitch, L., 1986, *Elements of Vibration Analysis*. New York, NY: McGraw Hill Book Co.

Rao, J.S., 1992, *Advanced Theory of Vibration*. New Delhi: Wiley Eastern Limited.

Reddy, J.N., 1993, *An Introduction to the Finite Element Method*. New York, NY: McGraw-Hill.

Thomson, W.T. and Dahleh, M.D., 1998, *Theory of Vibration with Applications*, 5th ed. New Delhi: Pearson Education Inc.

Yang, Y., Hu, N., Li, Y., Cheng, Z., and Shen, G., 2024, Dynamic modeling and analysis of planetary gear system for tooth fault diagnosis, *Mechanical Systems and Signal Processing*, *207*, 1–33, 110946.

Zienkiewicz, O.C. and Taylor, R.L., 1989, *The Finite Element Method*, 3rd ed. New York, NY: McGraw-Hill.

8 Transverse Vibrations of Rotor Systems by the Influence Coefficients and Transfer Matrix Methods

Transverse vibrations have been considered previously in great detail for several single-mass rotor systems. The thin disk and long rigid rotors were considered with various complexities at supports, for example, the rigid disk mounted on a flexible massless shaft with rigid bearings (e.g., the simply supported, cantilever, and overhung in Chapter 2), flexible bearings (anisotropic and cross-coupled stiffness and damping properties), and flexible foundations (Chapter 4). Higher order effects, i.e., the gyroscopic moment on rotors for the simple and most general cases of motion, were also described in detail (Chapter 5). However, in the actual case, as we have seen in the previous two chapters for torsional vibrations, the rotor system can have several masses (e.g., turbine blades, propellers, flywheels, gears, etc.) or distributed mass and stiffness properties, and multiple supports, and other such components like coupling, seals, etc. While dealing with torsional vibrations, we did consider multi-degree-of-freedom (DOF) rotor systems and continuous rotor systems. Mainly five methods were dealt with, that is, the Newton's second law of motion (or the D'Alembert principle), Lagrange's equations, extended Hamilton's principle, the transfer matrix method (TMM), and the finite element method. We will be extending the idea of these methods from torsional vibrations to transverse vibrations along with some additional methods, which are suitable for the analysis of multi-DOF rotor system transverse vibrations. In the present chapter, we will consider the analysis of multi-DOF rotor systems by the influence coefficient method, the TMM, and Dunkerley's approximate method. The main focus of these methods is to estimate the rotor system natural frequencies, mode shapes, and forced responses. The relative merit and demerit of these methods are discussed. The continuous rotor system with an analytical approach and discretized multi-DOF rotor systems through the finite element method without and with gyroscopic effects will be treated in Chapters 9 and 10, respectively. In Chapter 1 and 2 of Volume II, we will consider internal damping of shaft material and flexible bearing in the finite element analysis. Conventional methods of vibrations like the modal analysis, Rayleigh–Ritz method, weighted sum approach, collocation method, mechanical impedance (or receptance) method, and dynamic stiffness method are not covered exclusively in the present textbook because this information is readily available elsewhere (Meirovitch, 1986; Thomson and Dahleh, 1998). However, the basic concepts of these will be used directly whenever it is required with proper referencing.

8.1 INFLUENCE COEFFICIENT METHOD

In transverse vibrations due to coupling of the translational and rotational (tilting due to bending) displacements, the analysis becomes more complex compared with torsional vibrations. A force in a shaft can produce translational and rotational displacements; similarly, a moment can produce rotational and translational displacements. *Influence coefficients* can be used to relate these parameters (the force, the moment, and the translational and rotational displacements) relatively easily. In the present section, the influence coefficient method is used to calculate natural frequencies and forced responses of rotating machines. For up to three-DOF rotor systems, the hand calculation is feasible; however, for more than three DOFs, the help of computer routines (e.g., MATLAB, Scilab,

DOI: 10.1201/9781003661559-8

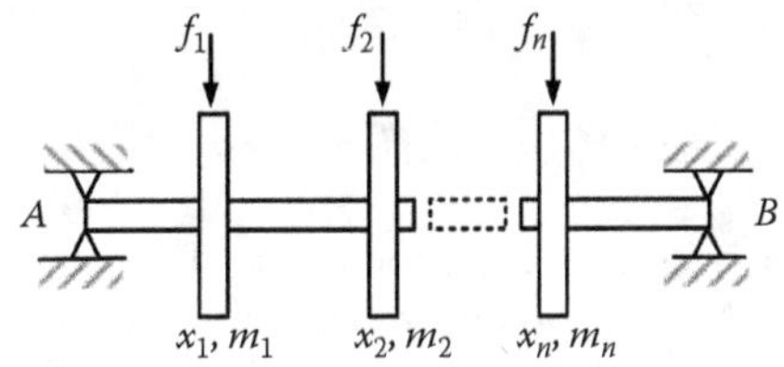

FIGURE 8.1 A multi-DOF rotor system mounted on rigid bearings.

Python, etc.) is necessary. The method is described for multi-DOFs, i.e., n number of disks mounted on a flexible massless shaft (Figure 8.1) and supported by rigid bearings, which can be extended for multi-DOF rotor systems with flexible multi-supports also.

8.1.1 Static Case

Let $f_1, f_2, \ldots, f_n$ be static forces on disks 1, 2, …, n, respectively, and $x_1, x_2, \ldots, x_n$ are the corresponding shaft deflections at disks. The reference for the measurement of the shaft displacement is from the bearing axis (i.e., the undeformed position of the shaft) and the gravity effect is neglected (i.e., static deflections are small). The system under consideration is a linear system. If a force f is applied to the disk of mass m_1, then deflection of disk 1 is proportional to f, i.e.,

$$x_1 \propto f \quad \text{or} \quad x_1 = \alpha_{11} f \tag{8.1}$$

where α_{11} is a proportionality constant, which depends upon the elastic properties of the shaft and support conditions (e.g., simply supported, fixed supports, free supports, multi-supports, etc.). It should be noted that we will have deflections at other disk locations (i.e., 2, 3, …, n) as well due to force at disk 1 due to the elastic coupling. If the same force f is applied to the disk of mass m_2 instead of mass m_1, then the deflection of disk 1 will still be proportional to the force, i.e.,

$$x_1 \propto f \quad \text{or} \quad x_1 = \alpha_{12} f \tag{8.2}$$

where α_{12} is another proportionality constant (the first subscript represents the displacement position, and the second subscript represents the force location). In general, for a symmetrical shaft, we have $\alpha_{12} \neq \alpha_{11}$. Similarly, if force f is applied to the disk of mass m_n, then the deflection at disk 1 will be

$$x_1 = \alpha_{1n} f \tag{8.3}$$

where α_{1n} is another proportionality constant. If forces $f_1, f_2, \ldots, f_n$ are applied at the locations of respective masses simultaneously, then the total deflection at disk 1 will be summation of all the displacements obtained above by the use of *superposition theorem*, as

$$x_1 = \alpha_{11} f_1 + \alpha_{12} f_2 + \cdots + \alpha_{1n} f_n \tag{8.4}$$

In the equation, it is assumed that displacements are small so that a linear relation exists between the forces applied and corresponding displacement produced. Similarly, we can write displacements at other disk locations as

$$x_2 = \alpha_{21} f_1 + \alpha_{22} f_2 + \cdots + \alpha_{2n} f_n \tag{8.5}$$

and

$$x_n = \alpha_{n1} f_1 + \alpha_{n2} f_2 + \cdots + \alpha_{nn} f_n \tag{8.6}$$

Here, α_{2j}, α_{3j}, …, α_{nj}, where $j=1, 2, \ldots, n$ are another set of proportionality constants and can be defined as described for α_{1j} ($j=1, 2, \ldots, n$) above. Hence, in general, α_{ij} is defined as a displacement at station i due to a unit external force at station j, while keeping all other external forces zero. From reciprocal relations (Srinath, 2009), we have in general for a linear elastic body $\alpha_{ij}=\alpha_{ji}$. Equations 8.4 through 8.6 can be combined in matrix form as

$$\begin{Bmatrix} x_1 \\ x_2 \\ \vdots \\ x_n \end{Bmatrix} = \begin{bmatrix} \alpha_{11} & \alpha_{12} & \cdots & \alpha_{1n} \\ \alpha_{21} & \alpha_{22} & \cdots & \alpha_{2n} \\ \vdots & \vdots & \ddots & \vdots \\ \alpha_{n1} & \alpha_{n2} & \cdots & \alpha_{nn} \end{bmatrix} \begin{Bmatrix} f_1 \\ f_2 \\ \vdots \\ f_n \end{Bmatrix} \tag{8.7}$$

It should be noted that due to the transverse force actually both the translational and rotational displacements take place, i.e., an elastic coupling exists between the translational and rotational displacements. We have already seen such coupling in Chapter 2 due to the bending of the shaft. In Chapter 4, we saw coupling between horizontal and vertical plane transverse translational motions (x and y) (and also between orthogonal transverse rotational motions, i.e., φ_x and φ_y) due to the dynamic properties of *fluid-film bearings*. In Chapter 5, we saw coupling between horizontal and vertical plane transverse rotational motions (φ_x and φ_y) due to *gyroscopic couples*.

Similarly, a moment gives rotational displacement as well as translational displacement. The method can be extended to account for the rotational displacement (tilting), φ_y, of the disk, and for the application of the point moment, M, at various disk locations along the shaft. Then, Equation 8.7 takes the following form:

$$\eta = \alpha \mathbf{f} \tag{8.8}$$

with

$$\eta = \begin{Bmatrix} x_1 \\ \vdots \\ x_n \\ \varphi_{y1} \\ \vdots \\ \varphi_{yn} \end{Bmatrix}; \quad \mathbf{f} = \begin{Bmatrix} f_1 \\ \vdots \\ f_n \\ M_1 \\ \vdots \\ M_n \end{Bmatrix} \quad \text{and} \quad \alpha = \begin{bmatrix} \alpha_{11} & \alpha_{12} & \cdots & \alpha_{1(2n)} \\ \alpha_{21} & \alpha_{22} & \cdots & \alpha_{2(2n)} \\ \vdots & \vdots & \ddots & \vdots \\ \alpha_{(2n)1} & \alpha_{(2n)2} & \cdots & \alpha_{(2n)(2n)} \end{bmatrix} \tag{8.9}$$

which gives

$$\mathbf{f} = \alpha^{-1} \eta \tag{8.10}$$

where α_{ij} with $i,j=1, 2, \ldots, 2n$ are influence coefficients. The first subscript defines the translational (or rotational) displacement location, and the second subscript defines the force (or moment) location. The analysis so far has referred only to static loads applied to the shaft. When the displacement of the disk changes rapidly with time, for dynamic loads, apart from the elastic force of shaft, the inertia force of disk also will come into the picture. It should be noted that the present analysis considers only a single-plane deformation; however, similar analysis is valid for another orthogonal plane. However, due to the symmetrical nature of the shaft geometry elastic coupling in two orthogonal planes will not exist.

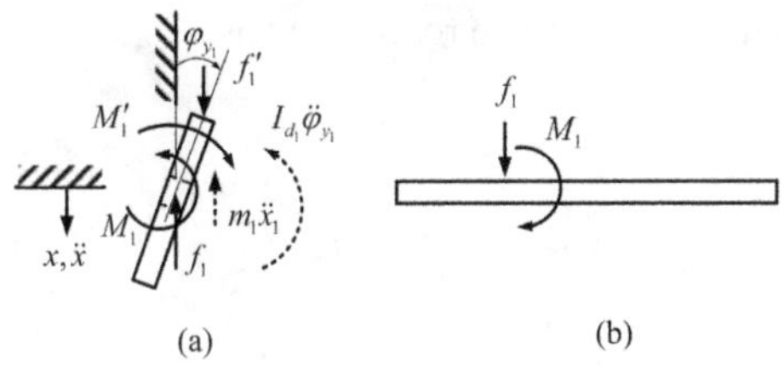

FIGURE 8.2 Free body diagrams of (a) a disk and (b) the shaft for a general motion.

8.1.2 Dynamic Case

In Figure 8.2, free body diagrams of a disk and the shaft are shown. Let $f_1'(t)$ and $M_1'(t)$ be the external dynamic force and moment (herein prime does not represent derivative) on the disk m_1, whereas $f_1(t)$ and $M_1(t)$ are the dynamic reaction force and moment transmitted to the shaft, respectively (which is equal and opposite to the reaction force and moment of the shaft on the disk). From the force and moment balance of disk 1, we have

$$f_1'(t) - f_1(t) = m_1\ddot{x}_1 \quad \text{and} \quad M_1'(t) - M_1(t) = I_{d_1}\ddot{\varphi}_{y1} \tag{8.11}$$

where I_d is the diametral mass moment of inertia of the disk. Due to the small angular tilt of the disk, the effect of f_1' on moment equations is not considered.

Similarly at other disk locations, we can write

$$f_2'(t) - f_2(t) = m_2\ddot{x}_2 \quad \text{and} \quad M_2'(t) - M_2(t) = I_{d_2}\ddot{\varphi}_{y2} \tag{8.12}$$

and

$$f_n'(t) - f_n(t) = m_n\ddot{x}_n \quad \text{and} \quad M_n'(t) - M_n(t) = I_{d_n}\ddot{\varphi}_{yn} \tag{8.13}$$

Substituting for $f_1(t), f_2(t), \ldots, f_n(t)$ and $M_1(t), M_2(t), \ldots M_n(t)$ from Equations 8.11 through 8.13 and remembering that for the simple harmonic motion of disks $\ddot{x} = -\omega^2 x$ and $\ddot{\varphi}_y = -\omega^2\varphi_y$ (for the free vibration $\omega = \omega_{nf}$, where ω_{nf} is the natural frequency of the rotor system, and for the forced vibration, ω is the excitation frequency), Equation 8.8 gives

$$\eta(t) = \alpha\begin{Bmatrix} f_1' + \omega^2 m_1 x_1 \\ \vdots \\ f_n' + \omega^2 m_n x_n \\ M_1' + \omega^2 I_{d_1}\varphi_{y1} \\ \vdots \\ M_n' + \omega^2 I_{d_n}\varphi_{yn} \end{Bmatrix} \tag{8.14}$$

which can be expanded as

$$\eta(t) = \alpha\mathbf{f}' + \omega^2\alpha\begin{Bmatrix} m_1 x_1 \\ \vdots \\ m_n x_n \\ I_{d_1}\varphi_{y1} \\ \vdots \\ I_{d_n}\varphi_{yn} \end{Bmatrix} \quad \text{with} \quad \mathbf{f}'(t) = \begin{Bmatrix} f_1' \\ \vdots \\ f_n' \\ M_1' \\ \vdots \\ M_n' \end{Bmatrix} \tag{8.15}$$

In view of Equation 8.9, Equation 8.15 can be rearranged as

$$\eta(t) = \alpha \mathbf{f}'(t)$$

$$+\omega^2 \begin{bmatrix} \alpha_{11}m_1 & \cdots & \alpha_{1n}m_n & \alpha_{1(n+1)}I_{d1} & \cdots & \alpha_{1(2n)}I_{d_n} \\ \vdots & \ddots & \vdots & \vdots & \ddots & \vdots \\ \alpha_{n1}m_1 & \cdots & \alpha_{nn}m_n & \alpha_{n(n+1)}I_{d1} & \cdots & \alpha_{n(2n)}I_{d_n} \\ \alpha_{(n+1)1}m_1 & \cdots & \alpha_{(n+1)n}m_n & \alpha_{(n+1)(n+1)}I_{d1} & \cdots & \alpha_{(n+1)(2n)}I_{d_n} \\ \vdots & \ddots & \vdots & \vdots & \ddots & \vdots \\ \alpha_{(2n)1}m_1 & \cdots & \alpha_{(2n)n}m_n & \alpha_{(2n)(n+1)}I_{d1} & \cdots & \alpha_{(2n)(2n)}I_{d_n} \end{bmatrix} \eta(t) \tag{8.16}$$

which can be written in more compact form as

$$\left(\mathbf{A} - \frac{1}{\omega^2}\mathbf{I}\right)\eta(t) = -\frac{1}{\omega^2}\alpha\mathbf{f}'(t) \tag{8.17}$$

with

$$\mathbf{A} = \begin{bmatrix} \alpha_{11}m_1 & \cdots & \alpha_{1n}m_n & \alpha_{1(n+1)}I_{d1} & \cdots & \alpha_{1(2n)}I_{d_n} \\ \vdots & \ddots & \vdots & \vdots & \ddots & \vdots \\ \alpha_{n1}m_1 & \cdots & \alpha_{nn}m_n & \alpha_{n(n+1)}I_{d1} & \cdots & \alpha_{n(2n)}I_{d_n} \\ \alpha_{(n+1)1}m_1 & \cdots & \alpha_{(n+1)n}m_n & \alpha_{(n+1)(n+1)}I_{d1} & \cdots & \alpha_{(n+1)(2n)}I_{d_n} \\ \vdots & \ddots & \vdots & \vdots & \ddots & \vdots \\ \alpha_{(2n)1}m_1 & \cdots & \alpha_{(2n)n}m_n & \alpha_{(2n)(n+1)}I_{d1} & \cdots & \alpha_{(2n)(2n)}I_{d_n} \end{bmatrix} \tag{8.18}$$

Disk displacements x and φ_y can be calculated for known applied loads (e.g., the unbalance forces and moments) as

$$\eta(t) = \mathbf{R}\mathbf{f}'(t) \tag{8.19}$$

with

$$\mathbf{R} = -\frac{1}{\omega^2}\left(\mathbf{A} - \frac{1}{\omega^2}\mathbf{I}\right)^{-1}\alpha \tag{8.20}$$

where **R** represents the *receptance matrix,* and for the present case, it contains only real elements. For free vibrations, the right-hand side of Equation 8.17 is zero, and $\omega = \omega_{nf}$, i.e.,

$$\left(\mathbf{A} - \frac{1}{\omega_{nf}^2}\mathbf{I}\right) = \mathbf{0} \tag{8.21}$$

which is only satisfied when

$$\left|\mathbf{A} - \frac{1}{\omega_{nf}^2}\mathbf{I}\right| = 0 \tag{8.22}$$

The frequency equation and system natural frequencies can be calculated from this. Alternatively, through eigenvalue analysis of matrix **A**, system natural frequencies and mode shapes can be obtained directly.

To include damping effect in the formulation, damping forces in terms of damping coefficients c_{ij} $(i, j, = 1, 2, N)$ multiplied by the corresponding velocities in Equations 8.11 through 8.13 could be introduced. Regrading estimation of damping coefficients, the Rayleigh's damping model can be used and is described in Section 9.6. In general, the receptance matrix, **R**, may contain complex elements when damping forces also act upon the shaft (or when motion in both planes is considered for unbalance response calculation), in which case applied forces and disk displacements will not all be in phase with one another. Hence, a more general form of Equation 8.19 would have both the real and imaginary parts of $\boldsymbol{\eta}$, $\mathbf{f}'$, and **R**. In such case, the real and imaginary parts of each equation need to be separated, and then these can be assembled again into real matrix form, which will be double the size of complex matrices. Some of the steps are described below:

$$\eta_r + j\eta_i = (\mathbf{R}_r + j\mathbf{R}_i)(\mathbf{f}'_r + j\mathbf{f}'_i)$$

where r and i refer to the real and imaginary parts, respectively. The above equation can be expanded as

$$\eta_r + j\eta_i = (\mathbf{R}_r\mathbf{f}'_r - \mathbf{R}_i\mathbf{f}'_i) + j(\mathbf{R}_i\mathbf{f}'_r + \mathbf{R}_r\mathbf{f}'_i)$$

Now on equating the real and imaginary parts on both sides of equations, we get

$$\eta_r = \mathbf{R}_r\mathbf{f}'_r - \mathbf{R}_i\mathbf{f}'_i \quad \text{and} \quad \eta_i = \mathbf{R}_i\mathbf{f}'_r + \mathbf{R}_r\mathbf{f}'_i$$

The above equations can be combined again as

$$\begin{Bmatrix} \eta_r \\ \eta_i \end{Bmatrix} = \begin{bmatrix} \mathbf{R}_r & -\mathbf{R}_i \\ \mathbf{R}_i & \mathbf{R}_r \end{bmatrix} \begin{Bmatrix} \mathbf{f}'_r \\ \mathbf{f}'_i \end{Bmatrix} \tag{8.23}$$

It can be observed that now the size of the matrix and vectors are double that in Equation 8.19. It should again be noted that the present analysis is only for one-plane motion, and for two-plane motion, the size of the matrix would be double the size of the matrix in Equation 8.23. Now, through simple numerical examples (for two or more DOFs), some of the basic concepts of the present method are illustrated.

Example 8.1

Obtain transverse natural frequencies of a rotor system as shown in Figure 8.3. Take the mass of the disk as $m = 10$ kg and the diametral mass moment of inertia of the disk as $I_d = 0.02$ kg-m². The disk is placed at 0.25 m from the right support. The shaft has a diameter of 10 mm and a span length of 1 m. The shaft is assumed to be massless. Take the Young's modulus of the shaft as $E = 2.1 \times 10^{11}$ N/m². Consider single-plane motion only. Compare the results for the case when the disk is at the midspan.

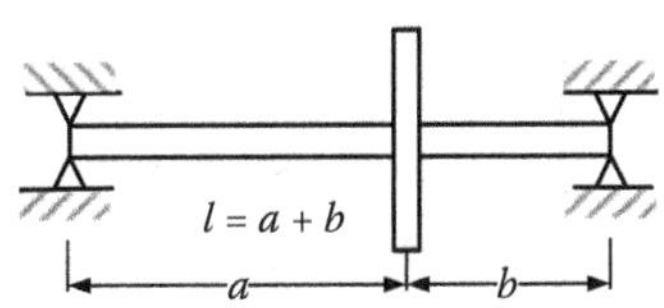

FIGURE 8.3 A rotor system.

Solution:

Influence coefficients for a simply supported shaft are defined as (Appendix 8.1)

$$\begin{Bmatrix} y \\ \varphi_x \end{Bmatrix} = \begin{bmatrix} \alpha_{11} & \alpha_{12} \\ \alpha_{21} & \alpha_{22} \end{bmatrix} \begin{Bmatrix} f_y \\ M_{yz} \end{Bmatrix}$$

with

$$\alpha_{11} = \frac{a^2b^2}{3EIl}; \quad \alpha_{12} = -\frac{3a^2l - 2a^3 - al^2}{3EIl}$$

and

$$\alpha_{21} = \frac{ab(b-a)}{3EIl}; \quad \alpha_{22} = -\frac{3al - 3a^2 - l^2}{3EIl}$$

It should be noted that subscript 1 represents a force or a translational displacement, and subscript 2 represents a moment or a rotational displacement. To obtain the natural frequencies of the rotor system with a single disk, from Equation 8.21, we have

$$\left(\mathbf{A} - \frac{1}{\omega_{nf}^2}\mathbf{I}\right)\eta = \mathbf{0} \Rightarrow \begin{bmatrix} \left(\alpha_{11}m - \dfrac{1}{\omega_{nf}^2}\right) & \alpha_{12}I_d \\ \alpha_{21}m & \left(\alpha_{22}I_d - \dfrac{1}{\omega_{nf}^2}\right) \end{bmatrix} \begin{Bmatrix} y \\ \varphi_x \end{Bmatrix} = \begin{Bmatrix} 0 \\ 0 \end{Bmatrix} \tag{a}$$

The determinant of the above matrix would give the frequency equation as

$$mI_d\omega_{nf}^4\left(\alpha_{11}\alpha_{22} - \alpha_{12}^2\right) - \omega_{nf}^2\left(\alpha_{11}m + \alpha_{22}I_d\right) + 1 = 0 \tag{b}$$

For the present problem, we have

$$\alpha_{11} = 1.137 \times 10^{-4} \text{ m/N}; \quad \alpha_{12} = -3.03 \times 10^{-4} \text{ m/N-m};$$

$$\alpha_{21} = -3.03 \times 10^{-4} \text{ rad/N}; \quad \alpha_{22} = 1.41 \times 10^{-3} \text{ rad/N-m}$$

Equation (a) becomes,

$$\omega_{nf}^4 - 8.505 \times 10^4 \omega_{nf}^2 + 7.3 \times 10^7 = 0 \tag{c}$$

which gives two natural frequencies of the system, as

$$\omega_{nf_1} = 29.44 \text{ rad/s} \quad \text{and} \quad \omega_{nf_2} = 289.15 \text{ rad/s}$$

It should be noted that the translational and rotational motions are coupled for the present transverse vibrations because the disk is offset from its midspan; however, when the disk is at the midspan, then the translational and rotational motions will be decoupled (i.e., $\alpha_{12} = \alpha_{21} = 0$). Natural frequencies for such a case for pure translational motion would be ($a = b = 0.5$ m, $\alpha_{11} = 2.0220 \times 10^{-4}$ m/N, and $\alpha_{22} = 0.80882 \times 10^{-4}$ rad/N-m)

$$\omega_{nf_1} = \sqrt{\frac{1}{m\alpha_{11}}} = \sqrt{\frac{1}{10 \times 2.021 \times 10^{-4}}} = 22.24 \text{ rad/s} \tag{d}$$

Similarly, for pure translation motion of the disk, we have

$$\omega_{nf2} = \sqrt{\frac{1}{I_d\alpha_{22}}} = \sqrt{\frac{1}{0.02\times 0.8084\times 10^{-3}}} = 248.63 \text{ rad/s} \quad \text{(e)}$$

These expressions (d) and (e) also, of course, can be obtained from frequency Equation (a) for $\alpha_{12}=\alpha_{21}=0$.

Example 8.2

Find the transverse natural frequency of a stepped shaft rotor system as shown in Figure 8.4. Consider the shaft as massless and made of steel with Young's modulus $E=2.1\times 10^{11}\,\text{N/m}^2$. The disk could be considered as a point mass of 10 kg. The circular shaft is simply supported at the ends. In the figure, all dimensions are in cm.

Solution:

To simplify the analysis, let us consider that the rotational displacement of the disk is negligibly small. From Equation 8.21, we have

$$\left(\mathbf{A}-\frac{1}{\omega_{nf}^2}\mathbf{I}\right)\eta=\mathbf{0}\Rightarrow\left(\alpha_{11}m-\frac{1}{\omega_{nf}^2}\right)y=0$$

from which the natural frequency is given as

$$\omega_{nf} = \sqrt{\frac{1}{m\alpha_{11}}}$$

Hence, now the next step would be to obtain the influence coefficient, α_{11}. Using the *energy method,* this influence coefficient is obtained as follows (refer to Timoshenko and Young (1968) for more details about the method).

For a load F, when it acts at the disk, reaction forces at bearings can be obtained as (Figure 8.5)

$$\sum M_A=0 \quad\Rightarrow\quad F_B\times 1-F\times 0.6=0 \quad\Rightarrow\quad F_B=0.6F$$

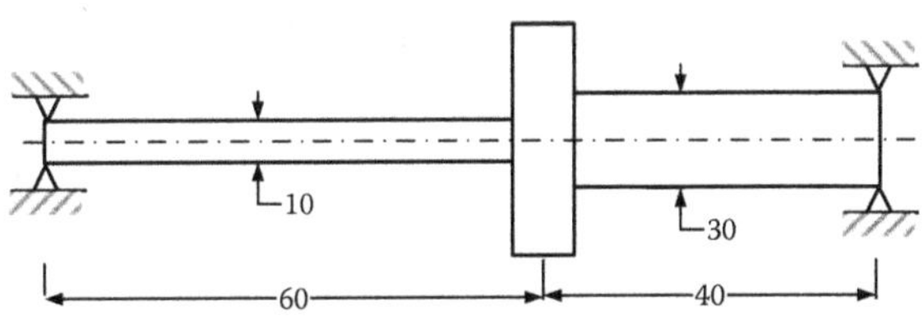

FIGURE 8.4 A simply supported stepped shaft.

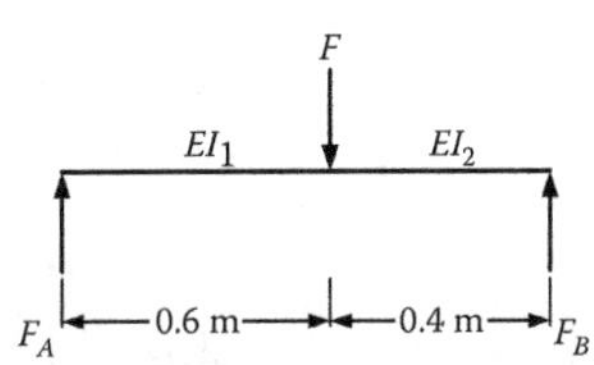

FIGURE 8.5 A free body diagram of the rotor system.

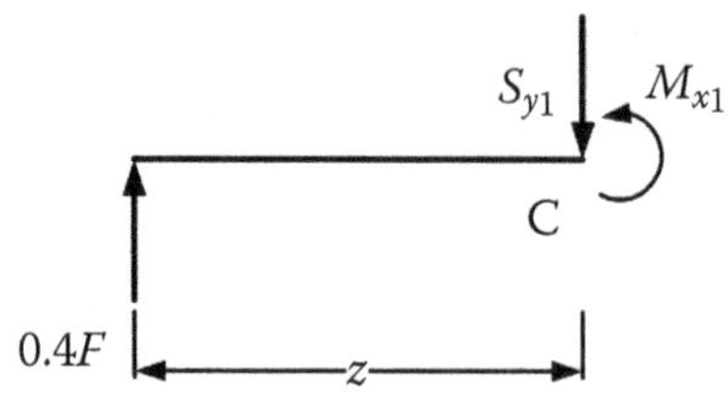

FIGURE 8.6 Free body diagram of the shaft section $0 \le z \le 0.6$ m.

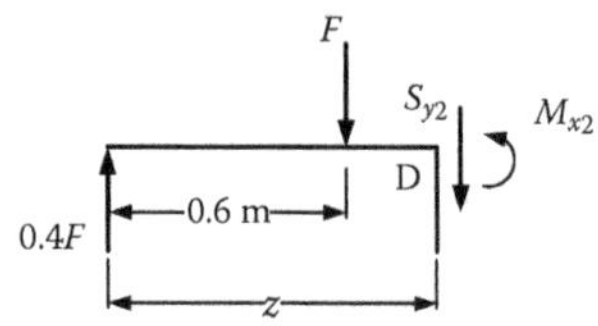

FIGURE 8.7 Free body diagram of the shaft section 0.6 m $\le z \le$ 1.0 m.

and

$$\sum F = 0 \quad \Rightarrow \quad F_A + F_B = F \quad \Rightarrow \quad F_A = 0.4F$$

From Figure 8.6, the bending moment in the shaft section $0 \le z \le 0.6$m can be obtained as

$$\sum M_C = 0 \Rightarrow M_{x_1} - 0.4Fz = 0 \Rightarrow M_{x_1} = 0.4Fz \tag{a}$$

From Figure 8.7, the bending moment in the shaft $0.6 \le z \le 1.0$ m can be obtained as

$$\sum M_D = 0 \Rightarrow M_{x_2} + F(z - 0.6) - 0.4Fz = 0 \Rightarrow M_{x_2} = 0.6F(1 - z) \tag{b}$$

The strain energy stored in the shaft from bending moments can be obtained as

$$U = \int_0^{0.6} \frac{M_{x_1}^2 dz}{2EI_1} + \int_{0.6}^{1.0} \frac{M_{x_2}^2 dz}{2EI_2}$$

From *Castigliano's theorem,* the translational displacement can be obtained as

$$\delta = \frac{\partial U}{\partial F} = \int_0^{0.6} \frac{M_{x_1} \dfrac{\partial M_{x_1}}{\partial F}}{EI_1} dz + \int_{0.6}^{1.0} \frac{M_{x_2} \dfrac{\partial M_{x_2}}{\partial F}}{EI_2} dz \tag{c}$$

On substituting Equations (a) and (b) into Equation (c), we get

$$\delta = \int_0^{0.6} \frac{(0.4Fz)(0.4z)}{EI_1} dz + \int_{0.6}^{1.0} \frac{\{0.6F(1-z)\}\{0.6(1-z)\}}{EI_2} dz$$

$$= \int_0^{0.6} \frac{0.16Fz^2}{EI_1} dz + \int_{0.6}^{1.0} \frac{0.36F(z^2 - 2z + 1)}{EI_2} dz = \frac{0.01152F}{EI_1} + \frac{0.00768F}{EI_2}$$

The stiffness of the beam is given as

$$k = \frac{1}{\alpha_{11}} = \frac{F}{\delta} = \left(\frac{0.01152}{EI_1} + \frac{0.00768}{EI_2} \right)^{-1}$$

From above, in fact it can be observed that the two-shaft segment can be thought of as connected parallel to each other at the disk location.

With $E = 2.1 \times 10^{11}\ \text{N/m}^2$; $I_1 = \frac{\pi}{64} 0.1^4 = 4.9087 \times 10^{-6}\ \text{m}^4$; $I_2 = \frac{\pi}{64} 0.3^4 = 3.9761 \times 10^{-4}\ \text{m}^4$, we have

$$k = \frac{1}{\alpha_{11}} = 8.8752 \times 10^7\ \text{N/m} \tag{d}$$

Hence, the natural frequency is given as

$$\omega_{nf} = \sqrt{\frac{1}{m\alpha_{11}}} = \sqrt{\frac{8.8752 \times 10^7}{10}} = 2979.12\ \text{rad/s}$$

The influence coefficient can be also obtained by the *singular function approach,* and for more details, readers are referred to Timoshenko and Young (1968).

Example 8.3

Find the transverse natural frequencies and mode shapes of the rotor system shown in Figure 8.8. B is a fixed end, and D_1 and D_2 are rigid disks. The shaft is made of steel with Young's modulus E=2.1×10^{11} N/m^2 and a uniform diameter of d=10 mm. Shaft lengths are BD_2=50 mm and D_1D_2=75 mm. The masses of the disks are m_1=2 kg and m_2=5 kg. Consider the shaft as massless and neglect the diametral mass moment of inertia of the disks.

Solution:

For simplicity of the analysis, the shaft is considered as massless and disk masses are considered as point masses (i.e., the diametral mass moments of inertia of the disks are neglected, which means rotational displacements are ignored). The first step would be to obtain the influence coefficients corresponding to two-disk locations acted on by concentrated forces. Basically, we need to derive translational displacements at two-disk locations due to forces F_1 and F_2 acting at these locations as shown in Figure 8.9. These deflection relations are often available in a tabular form in standard

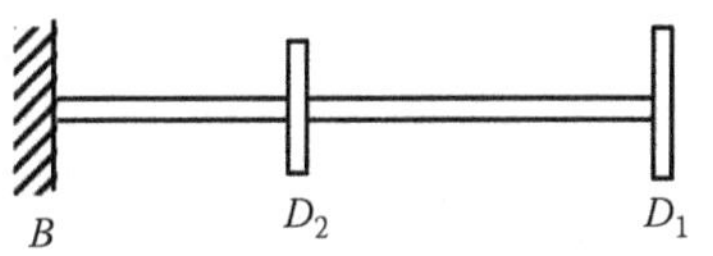

FIGURE 8.8 A two-disk cantilever shaft system.

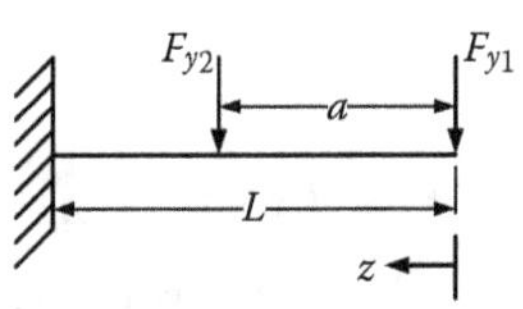

FIGURE 8.9 A shaft with two concentrated forces.

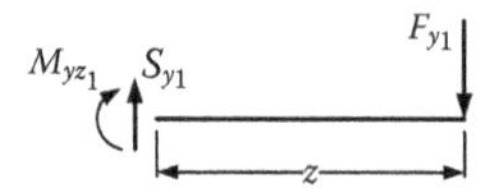

FIGURE 8.10 The free body diagram of a shaft segment for $0 \le z \le a$.

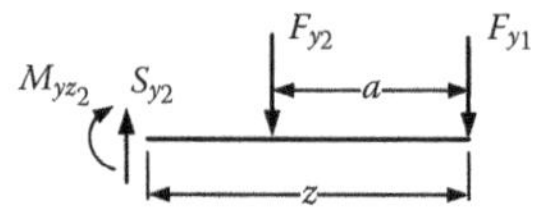

FIGURE 8.11 The free body diagram of a shaft segment for $a \le z \le L$.

handbooks (e.g., Young and Budynas, 2002). However, for the present problem, the calculation of influence coefficients is explained by the energy method, the singularity function, and a tabular method to clarify the basic concept of the procedure for completeness in the case where a standard handbook is not available or for a particular rotor system configuration where the influence coefficients are not tabulated.

Energy method: In this method, we need to obtain the strain energy due to bending moments in the shaft. Bending moments at different segments of the shaft can be obtained as

i. Shaft segment for $0 \le z \le a$

From Figure 8.10, the bending moment can be written as

$$M_{yz_1} = -F_{y_1} z \tag{a}$$

ii. Shaft segment for $a \le z \le L$

From Figure 8.11, the bending moment can be written as

$$M_{yz_2} = -F_{y_1} z - F_{y_2}(z-a) \tag{b}$$

The total strain energy is given by

$$U = \int_0^a \frac{M_{yz_1}^2 dz}{2EI} + \int_a^L \frac{M_{yz_2}^2 dz}{2EI} \tag{c}$$

Using the Castigliano theorem, the translational deflection at station 1 can be written as

$$y_1 = \frac{\partial U}{\partial F_{y_1}} = \int_0^a \frac{M_{yz_1} \frac{\partial M_{yz_1}}{\partial F_{y_1}} dz}{EI} + \int_a^L \frac{M_{yz_2} \frac{\partial M_{yz_2}}{\partial F_{y_1}} dz}{EI} \tag{d}$$

On substituting Equations (a) and (b) into Equation (d), we get

$$y_1 = \frac{\partial U}{\partial F_{y_1}} = \int_0^a \frac{(-F_{y_1} z)(-z) dz}{EI} + \int_a^L \frac{\{-F_{y_1} z - F_{y_2}(z-a)\}(-z) dz}{EI}$$

$$= \frac{F_{y_1}}{EI} \frac{z^3}{3}\bigg|_0^a + \frac{F_{y_1}}{EI} \frac{z^3}{3}\bigg|_a^L + \frac{F_{y_2}}{EI}\left(\frac{z^3}{3} - \frac{az^2}{2}\right)\bigg|_a^L$$

$$= \frac{F_{y_1}}{3EI}\left(a^3 + L^3 - a^3\right) + \frac{F_{y_2}}{EI}\left\{\frac{\left(L^3 - a^3\right)}{3} - \frac{a\left(L^2 - a^2\right)}{2}\right\}$$

which finally takes the following form:

$$y_1 = \left(\frac{L^3}{3EI}\right)F_{y1} + \left\{\frac{(L^3 - a^3)}{3EI} - \frac{a(L^2 - a^2)}{2EI}\right\}F_{y2} \tag{e}$$

Equation (e) has the following form:

$$y_1 = \alpha_{y1f1}F_{y1} + \alpha_{y1f2}F_{y2} \tag{f}$$

Hence, for $a = 0.075$ m, $L = 0.125$ m, and $EI = 103.1$ N-m^2, we have

$$\alpha_{y1f1} = 6.316 \times 10^{-6} \text{ m/N} \quad \text{and} \quad \alpha_{y1f2} = 1.314 \times 10^{-6} \text{ m/N} \tag{g}$$

The deflection at station 2 can be obtained as

$$y_2 = \frac{\partial U}{\partial F_{y2}} = \int_0^a \frac{M_{yz1}\frac{\partial M_{yz1}}{\partial F_{y2}}dz}{EI} + \int_a^L \frac{M_{yz2}\frac{\partial M_{yz2}}{\partial F_{y2}}dz}{EI} \tag{h}$$

On substituting Equations (a) and (b) into Equation (h), we get

$$y_2 = \frac{\partial U}{\partial F_{y2}} = \int_0^a \frac{(-F_{y1}z)(0)dz}{EI} + \int_a^L \frac{\{-F_{y1}z - F_{y2}(z-a)\}(-z+a)dz}{EI}$$

$$= 0 + \frac{F_{y1}}{EI}\left(\frac{z^3}{3} - \frac{az^2}{2}\right)\Bigg|_a^L + \frac{F_{y2}}{EI}\left(\frac{z^3}{3} - \frac{2az^2}{2} + a^2 z\right)\Bigg|_a^L$$

which finally takes the form

$$y_2 = \left(\frac{(L^3 - a^3)}{3EI} - \frac{a(L^2 - a^2)}{2EI}\right)F_{y1} + \left\{\frac{(L^3 - a^3)}{3EI} - \frac{a(L^2 - a^2)}{EI} + \frac{a^2(L-a)}{EI}\right\}F_{y2} \tag{i}$$

Equation (i) has the following form:

$$y_2 = \alpha_{y2f1}F_{y1} + \alpha_{y2f2}F_{y2} \tag{j}$$

Hence, for $a = 0.075$ m, $L = 0.125$ m, and $EI = 103.1$ N-m^2, we have

$$\alpha_{y2f1} = 1.314 \times 10^{-7} \text{ m / N} \quad \text{and} \quad \alpha_{y2f2} = 0.404 \times 10^{-6} \text{ m / N} \tag{k}$$

Method of the singularity function: Now the influence coefficients are obtained by the singularity function method for illustration. The singularity function (< >) is defined as

$$\langle f(z)\rangle = f(z) \quad \text{if} \quad f(z) > 0$$
$$= 0 \quad \text{if} \quad f(z) < 0$$

The bending moment at any cross-section along the shaft span can be written as

$$EIy'' = F_{y1}\langle z\rangle + F_{y2}\langle z - 0.075\rangle \tag{m}$$

which implies that for $0 \le z \le a$, the contribution from the F_{y_2} would not be there since $(z-0.075)$ would be a negative quantity, so it becomes zero. However, for $a \le z \le L$, contributions from both F_{y_1} and F_{y_2} would be there. On integrating Equation (m) twice, we get the following expressions:

$$EIy' = F_{y1}\frac{1}{2}\langle z\rangle^2 + F_{y2}\frac{1}{2}\langle z-0.075\rangle^2 + c_1 \tag{n}$$

and

$$EIy = F_{y1}\frac{\langle z\rangle^3}{6} + \frac{F_{y2}}{6}\langle z-0.075\rangle^3 + c_1 z + c_2 \tag{o}$$

where the integration constants c_1 and c_2 are obtained by the boundary conditions of the problem, and are given as

i. For $y'\big|_{z=0.125} = 0$ from Equation (n), we have

$$F_{y1}\frac{\langle 0.125\rangle^2}{2} + F_{y2}\frac{\langle 0.05\rangle^2}{2} + c_1 = 0$$

which gives

$$c_1 = -0.0078F_{y_1} - 0.0013F_{y_2} \tag{p}$$

ii. For $y\big|_{z=0.125} = 0$ from Equation (o), we have

$$F_{y1}\frac{\langle 0.125\rangle^3}{6} + F_{y2}\frac{\langle 0.05\rangle^3}{6} + 0.125\left(-0.0078F_{y_1} - 0.0013F_{y_2}\right) + c_2 = 0$$

which gives

$$c_2 = 6.51\times10^{-4}F_{y_1} + 1.354\times10^{-4}F_{y2} \tag{q}$$

Finally, Equation (o) becomes

$$EIy = F_{y1}\frac{\langle z\rangle^3}{6} + \frac{F_{y2}}{6}\langle z-0.075\rangle^3 + \left(-0.0078F_{y_1} - 0.0013F_{y_2}\right)z$$
$$+ 6.51\times10^{-4}F_{y_1} + 1.354\times10^{-4}F_{y2} \tag{r}$$

On evaluating the deflection at station 1 for $z=0$ (i.e., at the free end), we have

$$y\big|_{z=0} = \frac{1}{EI}\left\{6.51\times10^{-4}F_{y_1} + 1.354\times10^{-4}F_{y2}\right\}$$

which has the following form

$$y_1 = \alpha_{y_1 f_1}F_{y_1} + \alpha_{y_1 f_2}F_{y2}$$

with

$$\alpha_{y_1 f_1} = 6.316\times10^{-6}\ \text{m/N} \quad \text{and} \quad \alpha_{y_1 f_2} = 1.314\times10^{-6}\ \text{m/N}$$

For the deflection at station 2 for $z=0.075$ m, we get

$$y\big|_{z=0.075} = \frac{1}{EI}\left\{\begin{matrix} \left(\dfrac{0.075^3}{6} - 0.075\times 0.0078 + 6.51\times 10^{-4}\right)F_{y1} \\ +\left(-0.0013\times 0.075 + 1.354\times 10^{-4}\right)F_{y2} \end{matrix}\right\} \tag{s}$$

which has the following form

$$y_2 = \alpha_{y_2 f_1} F_{y1} + \alpha_{y_2 f_2} F_{y2}$$

with

$$\alpha_{y_2 f_1} = 1.314\times 10^{-6}\ \text{m/N} \quad \text{and} \quad \alpha_{y_2 f_2} = 0.404\times 10^{-6}\ \text{m/N}$$

which is the same as obtained by the energy method.

Tabular method from deflection relations: For this method, the z-axis is shifted to the fixed end. For a force f_y at a distance a from the fixed end of a cantilever beam, the transverse deflection can be written as (refer to Appendix 8.1)

$$y = \frac{f_y z^2}{6EI}(3a - z) \quad \text{for} \quad 0 < z < a \quad \text{and} \quad y = \frac{f_y a^2}{6EI}(3z - a) \quad \text{for} \quad a < z < L$$

where L is the total span of the beam and z is measured from the fixed end. Please note that for influence coefficient calculation one force at a time is considered. Let l_1 and l_2 be distances from the fixed end to the force f_{y2} and from the force f_{y2} to the force f_{y1}, respectively. Then, influence coefficients can be defined as (please note that a refers to the force location, and z, the displacement location)

$$\alpha_{11} = \frac{y_1}{f_{y1}}\bigg|_{\substack{a=l_1+l_2 \\ z=l_1+l_2}} = \frac{(l_1+l_2)^2}{6EI}\{3(l_1+l_2)-(l_1+l_2)\} = \frac{(l_1+l_2)^3}{3EI}$$

$$\alpha_{22} = \frac{y_2}{f_{y2}}\bigg|_{\substack{a=l_1 \\ z=l_1}} = \frac{l_1^2}{6EI}(3l_1 - l_1) = \frac{l_1^3}{3EI}$$

and

$$\alpha_{21} = \frac{y_2}{f_{y1}}\bigg|_{\substack{a=l_1+l_2 \\ z=l_1}} = \alpha_{12} = \frac{y_1}{f_{y2}}\bigg|_{\substack{a=l_1 \\ z=l_1+l_2}} = \frac{l_1^2}{6EI}\{3(l_1+l_2)-l_1\}$$

Table 8.1 shows the systematic calculation of influence coefficients for the above expressions. Influence coefficients have been obtained by substituting l_1=0.05 m and l_2=0.075 m in the above expressions. For the present case, in fact the last column is of interest and it can be seen that these values of influence coefficients match with the previously obtained influence coefficients.

For free vibrations, from Equation 8.21, we have

$$\begin{bmatrix} \left(\alpha_{y_1 f_1} m_1 - \dfrac{1}{\omega_{nf}^2}\right) & \alpha_{y_1 f_2} m_2 \\ \alpha_{y_2 f_1} m_1 & \left(\alpha_{y_2 f_2} m_2 - \dfrac{1}{\omega_{nf}^2}\right) \end{bmatrix} \begin{Bmatrix} y_1 \\ y_2 \end{Bmatrix} = \begin{Bmatrix} 0 \\ 0 \end{Bmatrix} \tag{t}$$

For a nontrivial solution, this gives

TABLE 8.1
Calculation of Influence Coefficients from Deflection Relations

S.N.	Displacement/Force Locations[a]	a(m)	z(m)	a_{ii} (m/N)	$a_{ij} = a_{ji}$ (m/N)
1	(1, 1)	0.125	0.125	6.316×10^{-6}	—
2	(2, 2)	0.05	0.05	0.404×10^{-6}	—
3	(1, 2)	0.05	0.125	—	1.314×10^{-6}

[a] When numbers in the brackets are the same, then displacement and force are at the same location (e.g., (1, 1) means the force and displacement considered are at location 1). When the numbers in the brackets are different, then the displacement and force are at the corresponding location (e.g., (1, 2) means the displacement at location 1 and force at location 2 are considered).

$$\begin{vmatrix} \left(\alpha_{y_1 f_1} m_1 - \dfrac{1}{\omega_{nf}^2}\right) & \alpha_{y_1 f_2} m_2 \\ \alpha_{y_2 f_1} m_1 & \left(\alpha_{y_2 f_2} m_2 - \dfrac{1}{\omega_{nf}^2}\right) \end{vmatrix} = 0 \tag{u}$$

which gives the frequency equation as

$$m_1 m_2 \omega_{nf}^4 \left(\alpha_{y_1 f_1}\alpha_{y_2 f_2} - \alpha_{y_1 f_2}^2\right) - \omega_{nf}^2\left(\alpha_{y_1 f_1} m_1 + \alpha_{y_2 f_2} m_2\right) + 1 = 0 \tag{v}$$

For the present problem, we have

$$\alpha_{y_1 f_1} = 6.316\times10^{-6}\ \text{m / N},\quad \alpha_{y_1 f_2} = \alpha_{y_2 f_1} = 1.314\times10^{-6}\ \text{m / N} \quad \text{and} \quad \alpha_{y_2 f_2} = 0.404\times10^{-6}\ \text{m / N}$$

$$m_1 = 2\ \text{kg} \quad \text{and} \quad m_2 = 5\ \text{kg}$$

Equation (u) becomes

$$\omega_{nf}^4 - 1.772\times10^6 \omega_{nf}^2 + 1.209\times10^{11} = 0$$

which gives two natural frequencies of the system for pure translational motion, as

$$\omega_{nf_1} = 266.67\ \text{rad/s} \quad \text{and} \quad \omega_{nf_2} = 1304.0\ \text{rad/s}$$

Mode shapes corresponding to these natural frequencies are plotted in Figure 8.12, which can be obtained from Equation (t) corresponding to the natural frequencies obtained. While plotting mode shapes, the maximum displacement corresponding to a particular mode has been chosen as unity and other displacements have been normalized accordingly. Corresponding to the first mode, both disks have an *in-phase* motion, whereas, for the second mode, two disks have an *antiphase* motion. It should be noted that for the case when the disks have appreciable diametral mass moments of inertia, then along with the forces at stations 1 and 2, the moments also need to be considered while deriving influence coefficients. In that case, we will have 16 influence coefficients, i.e., the size of the influence coefficient matrix would be 4×4; however, the symmetry conditions of influence coefficients will still prevail. Correspondingly, we would have four natural frequencies and the corresponding mode shapes.

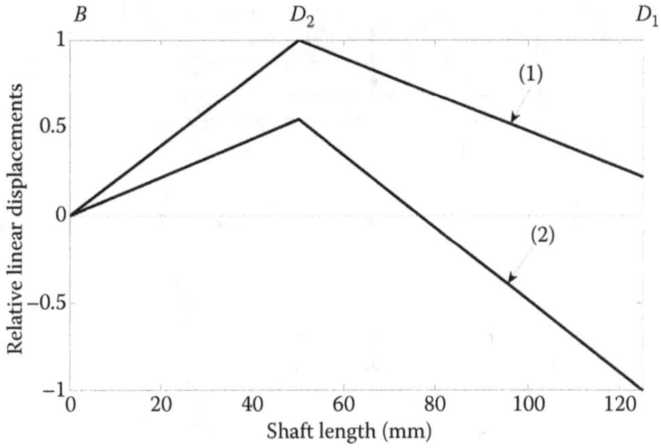

FIGURE 8.12 Mode shapes: (1) first mode shape, and (2) second mode shape.

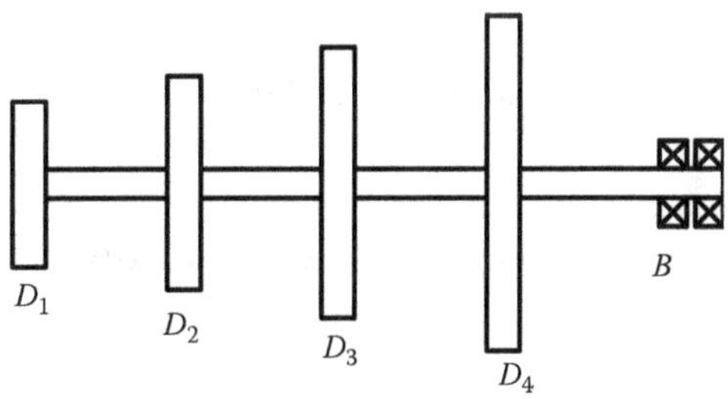

FIGURE 8.13 A multidisc cantilever rotor.

Example 8.4

Find the transverse natural frequencies and mode shapes of a rotor system as shown in Figure 8.13. B is a fixed bearing, which provides a fixed support boundary condition, and D_1, D_2, D_3, and D_4 are rigid disks. The shaft is made of steel with Young's modulus $E = 2.1 \times 10^{11}\,\text{N/m}^2$ and uniform diameter $d = 20\,\text{mm}$. The various shaft lengths are as follows: $D_1D_2 = 50\,\text{mm}$, $D_2D_3 = 50\,\text{mm}$, $D_3D_4 = 50\,\text{mm}$, and $D_4B_2 = 150\,\text{mm}$. The masses of the disks are $m_1 = 4\,\text{kg}$, $m_2 = 5\,\text{kg}$, $m_3 = 6\,\text{kg}$, and $m_4 = 7\,\text{kg}$. Consider the shaft as massless and the disks as point masses, i.e., neglect the diametral and polar mass moments of inertia of all disks.

Solution:

The first step would be to obtain influence coefficients. We have the translational deflection, y, and force relations from the strength of material for a cantilever shaft with a concentrated force at any point (Figure 8.14) as (see Appendix 8.1)

$$y = \frac{f_y z^2}{6EI}(3a - z) \quad \text{for} \quad 0 < z < a \quad \text{and} \quad y = \frac{f_y a^2}{6EI}(3z - a) \quad \text{for} \quad a < z < L \tag{a}$$

Let us take stations 1, 2, 3, and 4 at each of the disk locations. Figure 8.15 shows a cantilever shaft with forces at stations 1 and 2. Station 1 is at the free end, and it has an intermediate station 2 between station 1 and the fixed support. Lengths L, l_1, and l_2 are shown in the figure, and for this case, we have $L = l_1 + l_2$. Between two stations, we have the following influence coefficients (which relates the force to the translational transverse displacement from Equation (a)):

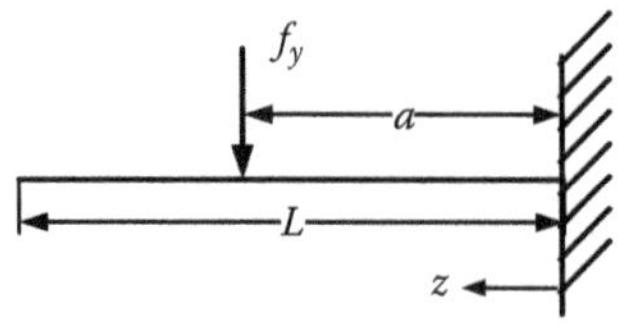

FIGURE 8.14 A cantilever shaft with a concentrated force at any point.

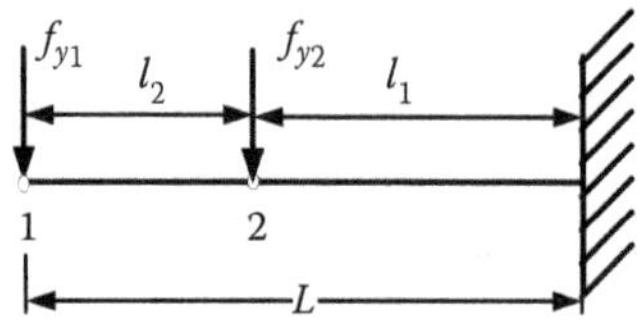

FIGURE 8.15 A cantilever shaft with a force at the free end and another at the intermediate point.

TABLE 8.2
A Calculation Procedure of Influence Coefficients in a Multidisc Cantilever Shaft

	Displacement/ Force Locations[a]	$l_1 (m)$	$l_2 (m)$	$\alpha_{ii} (m/N)$	$\alpha_{ij} = \alpha_{ji} (m/N)$
1	(1, 1)	0.30	0.00	5.4567×10^{-6}	—
2	(2, 2)	0.25	0.00	3.1578×10^{-6}	—
3	(3, 3)	0.20	0.00	1.6168×10^{-6}	—
4	(4, 4)	0.15	0.00	6.8209×10^{-7}	—
5	(1, 2)	0.25	0.05	—	4.1052×10^{-6}
6	(1, 3)	0.20	0.10	—	2.8294×10^{-6}
7	(1, 4)	0.15	0.15	—	1.7052×10^{-6}
8	(2, 3)	0.20	0.05	—	2.2231×10^{-6}
9	(2, 4)	0.15	0.10	—	1.3642×10^{-6}
10	(3, 4)	0.15	0.05	—	1.0231×10^{-6}

[a] Numbers in parentheses represent the displacement and force station numbers, respectively.

$$\alpha_{11} = \left.\frac{y_1}{f_{y1}}\right|_{\substack{a=l_1+l_2\\ z=l_1+l_2}} = \frac{(l_1+l_2)^2}{6EI}\{3(l_1+l_2)-(l_1+l_2)\} = \frac{(l_1+l_2)^3}{3EI}$$

$$\alpha_{22} = \left.\frac{y_2}{f_{y2}}\right|_{\substack{a=l_1\\ z=l_1}} = \frac{l_1^2}{6EI}(3l_1-l_1) = \frac{l_1^3}{3EI}; \tag{b}$$

$$\alpha_{21} = \left.\frac{y_2}{f_{y1}}\right|_{\substack{a=l_1+l_2\\ z=l_1}} = \alpha_{12} = \left.\frac{y_1}{f_{y2}}\right|_{\substack{a=l_1\\ z=l_1+l_2}} = \frac{l_1^2}{6EI}\{3(l_1+l_2)-l_1\}$$

Similar relations would also be valid between stations (1, 3) and (1, 4), in which values of l_1 and l_2 would be taken according to the node location (refer to Table 8.2 for details).

Figure 8.16 shows a cantilever shaft with a force at station 2, and it has a station 3 between the force and the fixed end. Between the two stations, we have the following influence coefficients:

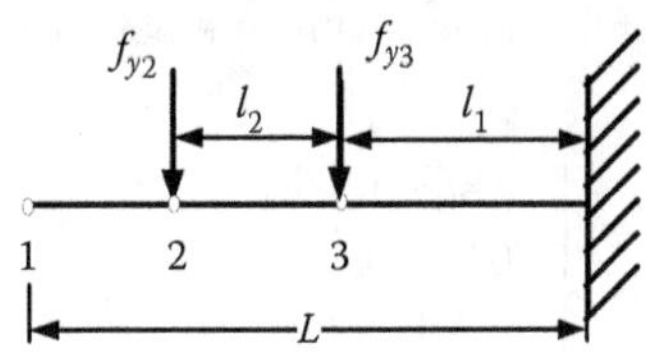

FIGURE 8.16 A cantilever shaft with two forces at intermediate stations.

$$\alpha_{22}\Big|_{z=l_1+l_2}^{a=l_1+l_2} = \frac{(l_1+l_2)^3}{3EI}; \quad \alpha_{33}\Big|_{z=l_1}^{a=l_1} = \frac{l_1^3}{3EI};$$

$$\alpha_{32}\Big|_{z=l_1}^{a=l_1+l_2} = \alpha_{23}\Big|_{z=l_1+l_2}^{a=l_1} = \frac{a^2}{6EI}(2l_1+3l_2) \tag{c}$$

These relations could also be used between stations (2, 4) and (3, 4).

It should be noted that, in Figure 8.16, the shaft segment from force f_{y2} to the free end would act as a rigid shaft. This will not contribute to the deformation of the shaft at station 2, and it will act as if it is not present at all (this is true only for the massless shaft assumption).

For example, when disks 1 and 2 are present, we have $l_1=0.25$ m and $l_2=0.05$ m. Hence, the influence coefficients from Equation (a) take the following values: $\alpha_{11}=5.4567\times10^{-6}$ m/N, $\alpha_{12}=\alpha_{21}=4.1052\times10^{-6}$ m/N, and $\alpha_{22}=3.1578\times10^{-6}$ m/N. Similarly, when disks 2 and 3 are present, we have $l_1=0.20$ m and $l_2=0.05$ m. Hence, the influence coefficients from Equation (b) take the following values: $\alpha_{22}=3.1578\times10^{-6}$ m/N, $\alpha_{23}=\alpha_{32}=2.2231\times10^{-6}$ m/N, and $\alpha_{33}=1.6168\times10^{-6}$ m/N. Along similar lines, other influence coefficients can be calculated; this is summarized in Table 8.2

Now for free vibrations, the governing equation has the following form:

$$\left(\begin{bmatrix} \alpha_{y_1f_1}m_1 & \alpha_{y_1f_2}m_2 & \alpha_{y_1f_3}m_3 & \alpha_{y_1f_4}m_4 \\ & \alpha_{y_2f_2}m_2 & \alpha_{y_2f_3}m_3 & \alpha_{y_2f_4}m_4 \\ & & \alpha_{y_3f_3}m_3 & \alpha_{y_3f_4}m_4 \\ sym & & & \alpha_{y_4f_4}m_4 \end{bmatrix} - \frac{1}{\omega_{nf}^2}\begin{bmatrix} 1 & 0 & 0 & 0 \\ 0 & 1 & 0 & 0 \\ 0 & 0 & 1 & 0 \\ 0 & 0 & 0 & 1 \end{bmatrix}\right)\begin{Bmatrix} y_1 \\ y_2 \\ y_3 \\ y_4 \end{Bmatrix} = \begin{Bmatrix} 0 \\ 0 \\ 0 \\ 0 \end{Bmatrix} \tag{d}$$

which can be written for the data of the present problem as

$$\left(\mathbf{A} - \frac{1}{\omega_{nf}^2}\mathbf{I}\right)\eta = \mathbf{0} \tag{e}$$

with

$$\mathbf{A} = \begin{bmatrix} \alpha_{y_1f_1}m_1 & \alpha_{y_1f_2}m_2 & \alpha_{y_1f_3}m_3 & \alpha_{y_1f_4}m_4 \\ & \alpha_{y_2f_2}m_2 & \alpha_{y_2f_3}m_3 & \alpha_{y_2f_4}m_4 \\ & & \alpha_{y_3f_3}m_3 & \alpha_{y_3f_4}m_4 \\ sym & & & \alpha_{y_4f_4}m_4 \end{bmatrix}$$

$$= 10^{-4}\begin{bmatrix} 0.2183 & 0.2053 & 0.1698 & 0.1194 \\ 0.1642 & 0.1579 & 0.1334 & 0.0955 \\ 0.1132 & 0.1112 & 0.0970 & 0.0716 \\ 0.0682 & 0.0682 & 0.0614 & 0.0477 \end{bmatrix}$$

The eigenvalue and eigenvector of the above matrix are obtained as

$$\left[1/\omega_{nf}^2\right]=\begin{Bmatrix}1/\omega_{nf1}^2\\1/\omega_{nf2}^2\\1/\omega_{nf3}^2\\1/\omega_{nf4}^2\end{Bmatrix}=10^{-4}\begin{Bmatrix}0.5076\\0.0121\\0.0010\\0.0002\end{Bmatrix}$$

and

$$\mathrm{H}=\left[\begin{Bmatrix}Y_1\\Y_2\\Y_3\\Y_4\end{Bmatrix}_{\omega_{nf1}}\quad\begin{Bmatrix}Y_1\\Y_2\\Y_3\\Y_4\end{Bmatrix}_{\omega_{nf2}}\quad\begin{Bmatrix}Y_1\\Y_2\\Y_3\\Y_4\end{Bmatrix}_{\omega_{nf3}}\quad\begin{Bmatrix}Y_1\\Y_2\\Y_3\\Y_4\end{Bmatrix}_{\omega_{nf4}}\right]$$

$$=\begin{bmatrix}0.7082 & 0.6740 & 0.5085 & -0.2763\\0.5434 & 0.0317 & -0.5467 & 0.6740\\0.3836 & -0.4402 & -0.3758 & -0.6371\\0.2368 & -0.5925 & 0.5489 & 0.2519\end{bmatrix}$$

Hence, the transverse natural frequencies are

$$\omega_{nf_1}=140.36\text{ rad/s};\quad \omega_{nf_2}=909.09\text{ rad/s};\quad \omega_{nf_3}=3162.28\text{ rad/s};\quad \omega_{nf_4}=7071.07\text{ rad/s}$$

The corresponding eigenvectors (mode shapes) can be plotted. It should be noted that in the eigenvector matrix (or the modal matrix), **H**, the first column components have no sign change, which corresponds to the fundamental mode; however, in subsequent columns, i.e., 2, 3, and 4, there are sign changes, and for each sign change, there would be a node in that particular mode (e.g., for the second mode there is a node between station 2 and 3 (single node), for the third mode there are nodes between stations 1 and 2 and 3 and 4 (two nodes), and finally for the fourth mode there are nodes between stations 1 and 2, 2 and 3, and 3 and 4 (three nodes)).

8.2 TRANSFER MATRIX METHOD

In the influence coefficient method, governing equations are derived by considering the equilibrium of the whole system. Difficulties with such an approach are that for a large system, it becomes very complex and we need to evaluate influence coefficients for each case separately depending on boundary conditions and load distributions. In the TMM, which is also called the Myklestad and Prohl method (Myklestad, 1944; Prohl, 1945), the shaft is divided into a number of imaginary smaller beam elements and governing equations are derived for each of these elements in order to determine the overall system behavior. The TMM has an advantage over the influence coefficient method in that the size of matrices being handled does not increase with the number of DOFs. It demands less computer memory, and associated *ill-conditioning problems* (i.e., nearly singular matrices) are fewer. Also, this method is relatively simple and straightforward in application. For the present case, disks are considered as point masses. In the case when the mass of the shaft is appreciable, then it is divided into a number of smaller masses concentrated (or lumped) at junctions (or stations) of beam segments so that concentrated masses and the shaft may be modeled as shown in Figure 8.17. The station number can be assigned wherever there is change in the state vector (i.e., the translational and rotational (tilting) displacements, shear forces, and bending moments), for example at the step change in shaft diameter, and at disks, couplings, and support positions. Let stations be numbered from 0 to $(n+1)$ from the left end of the shaft toward the right end of the shaft as shown in the figure.

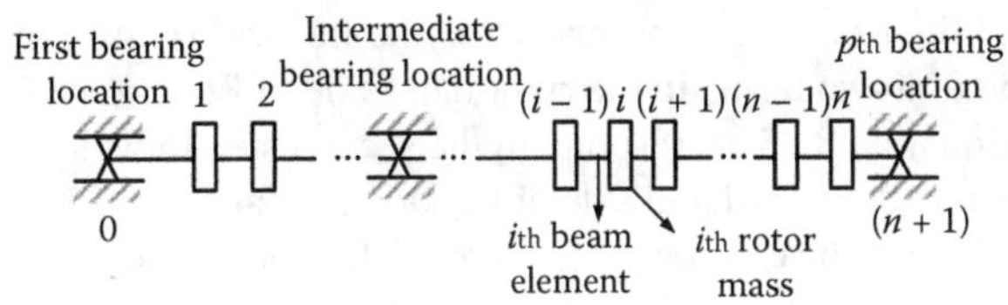

FIGURE 8.17 Modeling a real rotor with discrete elements.

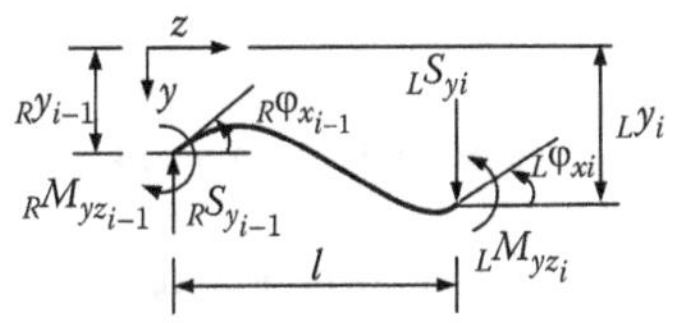

FIGURE 8.18 A free body diagram of the ith shaft segment.

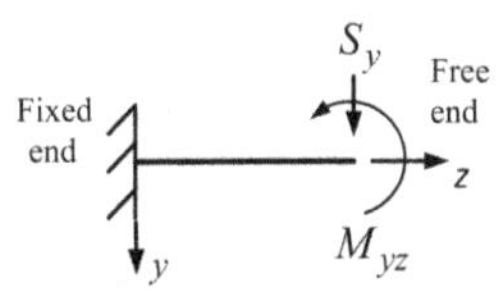

FIGURE 8.19 A cantilever beam.

8.2.1 Field Matrix

Figure 8.18 shows the free body diagram of the ith shaft segment, which is between stations $(i-1)$ and i. Let S_y be the shear force in the y direction, M_{yz} be the bending moment in the y–z plane, y be the transverse translational displacement in the y direction, and φ_x be the transverse rotational displacement in the y–z plane. The back-subscript R refers to the *right* of a particular mass, and L refers to the *left* of a particular mass. Directions for the shear force and the bending moment are chosen according to the positive sign conversion of the strength of material (Timoshenko and Young, 1968).

The displacement (translational or rotational) at the ith station will be equal to the sum of the relative displacement between the ith and $(i-1)$th stations and the absolute displacement of the $(i-1)$th station. As a first step for obtaining the relative displacement, the $(i-1)$th station can be considered to be a fixed end as shown in Figure 8.19. Hence, the displacement and the slope at the free end are related to the applied moment and the shear force at the free end by considering the beam as though it were a cantilever. Then as a second step, the displacement and the slope of the fixed end are given by considering the beam as rigid and displacements of the other end are obtained. On the assumption of small displacements, hence, finally the above two steps can be superposed to get the total displacement and slope of the ith shaft segment at the left of the ith station (i.e., at the ith mass), as

$$_{L}y_i = {}_{R}y_{i-1} - {}_{R}\varphi_{x_{i-1}} l - \frac{_{L}M_{yzi} l^2}{2EI} + \frac{_{L}S_{yi} l^3}{3EI} \tag{8.24}$$

and

$$_{L}\varphi_{x_i} = {}_{R}\varphi_{x_{i-1}} + \frac{_{L}M_{yzi} l}{EI} - \frac{_{L}S_{yi} l^2}{2EI} \tag{8.25}$$

where l is the length of the shaft segment, and EI is the flexural rigidity of the shaft segment. It should be noted that the first two terms in the right-hand side of Equation 8.24 and the first term in the right-hand side of Equation 8.25 correspond to the second step in which the shaft is considered rigid. The last two terms on the right-hand side of Equations 8.24 and 8.25 are from the first step in which the shaft is considered to be cantilevered. From the free body diagram (Figure 8.18), the shear force and the bending moment at either ends of the ith shaft segment are related as

$$ {}_L S_{yi} = {}_R S_{yi-1} \tag{8.26}$$

and

$$ {}_L M_{yzi} = {}_R M_{yzi-1} + {}_L S_{yi} l \tag{8.27}$$

On substituting for ${}_L S_{yi}$ and ${}_L M_{yzi}$ from Equations 8.26 and 8.27 into Equations 8.24 and 8.25, these equations can be rewritten as

$$ {}_L y_i = {}_R y_{i-1} - {}_R \varphi_{x_{i-1}} l - \frac{\left({}_R M_{yzi-1} + {}_R S_{yi-1} l\right) l^2}{2EI} + \frac{{}_R S_{yi-1} l^3}{3EI} \tag{8.28}$$

$$ {}_L \varphi_{x_i} = {}_R \varphi_{x_{i-1}} + \frac{\left({}_R M_{yzi-1} + {}_R S_{yi-1} l\right) l}{EI} - \frac{{}_R S_{yi-1} l^2}{2EI} \tag{8.29}$$

$$ {}_L S_{yi} = {}_R S_{yi-1} \tag{8.30}$$

and

$$ {}_L M_{yzi} = {}_R M_{yzi-1} + {}_R S_{yi-1} l \tag{8.31}$$

These equations can be rearranged and expressed in matrix form as

$$ {}_L \mathbf{S}_i = \mathbf{F}_i \, {}_R \mathbf{S}_{i-1} \tag{8.32}$$

with

$$ {}_L \mathbf{S}_i = {}_L\begin{Bmatrix} -y \\ \varphi_x \\ M_{yz} \\ S_y \end{Bmatrix}_i ; \quad \mathbf{F}_i = \begin{bmatrix} 1 & l & l^2/2EI & l^3/6EI \\ 0 & 1 & l/EI & l^2/2EI \\ 0 & 0 & 1 & l \\ 0 & 0 & 0 & 1 \end{bmatrix}_i ; \quad {}_R \mathbf{S}_{i-1} = {}_R\begin{Bmatrix} -y \\ \varphi_x \\ M_{yz} \\ S_y \end{Bmatrix}_{i-1} \tag{8.33}$$

where $\mathbf{F}_i$ is the *field matrix* for the ith shaft segment, and $\mathbf{S}_i$ is the *state vector* at the ith mass. The field matrix, $\mathbf{F}_i$, transforms the state vector from the left of a shaft segment to the right of the shaft segment. The damping is not considered in the present analysis, and because of this, the field matrix remains real. Equation 8.32 is for the motion in the vertical plane. A similar set of equations can be written for the motion in the horizontal plane. Hence, on considering motion in both planes, we have

$$ {}_L \mathbf{S}_{v_i} = \mathbf{F}_i \, {}_R \mathbf{S}_{v_{i-1}} \quad \text{and} \quad {}_L \mathbf{S}_{h_i} = \mathbf{F}_i \, {}_R \mathbf{S}_{h_{i-1}} \tag{8.34}$$

where v and h represent the vertical and horizontal planes, respectively. The field matrix is the same due to the symmetric shaft (depending upon the sign convention signs of angular displacement and moments may change). In general, state vectors will have both real and imaginary components due to coupled two-plane motion while considering gyroscopic moments and damping in the system. The effect of gyroscopic moments will be considered subsequently. Equation 8.34 is expanded to give a more general form as

$$ {}_L\mathbf{S}_i^* = \mathbf{F}_i^* \, {}_R\mathbf{S}_{i-1}^* \tag{8.35}$$

with

$$ {}_L\mathbf{S}_i^* = {}_L\begin{Bmatrix} \mathbf{S}_{h_r} \\ \mathbf{S}_{h_j} \\ \mathbf{S}_{v_r} \\ \mathbf{S}_{v_j} \\ 1 \end{Bmatrix}_i ; \quad \mathbf{F}_i^* = \begin{bmatrix} \mathbf{F} & \mathbf{0}_1 & \mathbf{0}_1 & \mathbf{0}_1 & \mathbf{0}_3 \\ \mathbf{0}_1 & \mathbf{F} & \mathbf{0}_1 & \mathbf{0}_1 & \mathbf{0}_3 \\ \mathbf{0}_1 & \mathbf{0}_1 & \mathbf{F} & \mathbf{0}_1 & \mathbf{0}_3 \\ \mathbf{0}_1 & \mathbf{0}_1 & \mathbf{0}_1 & \mathbf{F} & \mathbf{0}_3 \\ \mathbf{0}_2 & \mathbf{0}_2 & \mathbf{0}_2 & \mathbf{0}_2 & 1 \end{bmatrix}_i ; \quad {}_R\mathbf{S}_{i-1}^* = {}_R\begin{Bmatrix} \mathbf{S}_{h_r} \\ \mathbf{S}_{h_j} \\ \mathbf{S}_{v_r} \\ \mathbf{S}_{v_j} \\ 1 \end{Bmatrix}_{i-1} \tag{8.36}$$

$$ \mathbf{0}_1 = \begin{bmatrix} 0 & 0 & 0 & 0 \\ 0 & 0 & 0 & 0 \\ 0 & 0 & 0 & 0 \\ 0 & 0 & 0 & 0 \end{bmatrix}; \quad \mathbf{0}_2 = \begin{bmatrix} 0 & 0 & 0 & 0 \end{bmatrix}; \quad \mathbf{0}_3 = \begin{Bmatrix} 0 \\ 0 \\ 0 \\ 0 \end{Bmatrix} $$

where $\mathbf{S}^*$ is the *modified state vector* of size 17×1 and $\mathbf{F}^*$ is the *modified field matrix* of size 17×17. Subscripts h and v refer to the horizontal and vertical directions, respectively; r and j refer to the real and imaginary parts, respectively. The last equation of an identity has been added to facilitate the inclusion of the unbalance in the analysis, as will be made clear later (it was used for torsional vibrations also in Chapter 6). It should be noted that in the simplest case for single-plane motion and without damping (i.e., imaginary components will be zero) the size of the modified field matrix $\mathbf{F}_i^*$ will be 5×5 and is given as

$$ {}_L\mathbf{S}_i^* = {}_L\begin{Bmatrix} -y \\ \varphi_x \\ M_{yz} \\ S_y \\ 1 \end{Bmatrix}_i ; \quad \mathbf{F}_i^* = \begin{bmatrix} 1 & l & l^2/2EI & l^3/6EI & 0 \\ 0 & 1 & l/EI & l^2/2EI & 0 \\ 0 & 0 & 1 & l & 0 \\ 0 & 0 & 0 & 1 & 0 \\ 0 & 0 & 0 & 0 & 1 \end{bmatrix}_i ; \quad {}_R\mathbf{S}_{i-1}^* = {}_R\begin{Bmatrix} -y \\ \varphi_x \\ M_{yz} \\ S_y \\ 1 \end{Bmatrix}_{i-1} \tag{8.37}$$

which has a similar form as Equation 8.33 except for the inclusion of an identity equation in the last row.

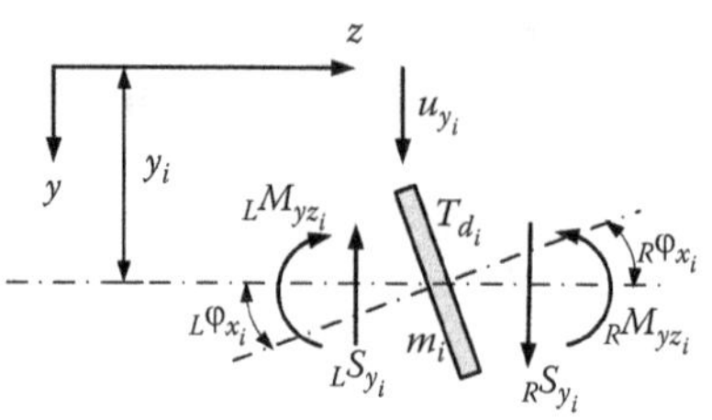

FIGURE 8.20 Forces, moments, and the translational and rotational displacements on a thin disk.

8.2.2 Point Matrix

Figure 8.20 shows the free body diagram of the disk of mass m_i, where u_{y_i} is the unbalance force at the ith location. The relationship between forces and displacements at the thin disk is given by its equations of motion

$$_RS_{yi} - {_LS_{yi}} + u_{yi} = m_i\ddot{y}_i$$

For simple harmonic motion, it gives

$$_RS_{yi} - {_LS_{yi}} + u_{yi} = -m_i\omega^2 y_i \tag{8.38}$$

where ω is the spin speed of the shaft for the forced vibration analysis. It should be noted that a synchronous whirl is considered here. It should be kept in mind that for free vibrations, the unbalance force, u_y, would be zero and $\omega = \omega_{nf}$.

Similarly, for the moment and the rotational displacement, we have the following relationship:

$$_RM_{yz_i} - {_LM_{yz_i}} = I_{d_i}\ddot{\varphi}_{x_i} \quad \text{or} \quad _RM_{yz_i} - {_LM_{yz_i}} = -\omega^2 I_{d_i}\varphi_{x_i} \tag{8.39}$$

where I_d is the diametral mass moment of inertia of the disk. Moreover, because the translational and rotational (slope) displacements on each side of the mass are equal, we can rewrite all governing equations for a thin disk as

$$-_Ry_i = -_Ly_i \tag{8.40}$$

$$_R\varphi_{x_i} = {_L\varphi_{x_i}} \tag{8.41}$$

$$_RM_{yz_i} = -\omega^2 I_{d_i}\, {_L\varphi_{x_i}} + {_LM_{yz_i}} \tag{8.42}$$

and

$$_RS_{yi} = -m_i\omega^2\, {_Ly_i} + {_LS_{yi}} - u_{yi} \tag{8.43}$$

On combining these equations in matrix form, we have

$$_R\mathbf{S}_i = \mathbf{P}_i\, {_L\mathbf{S}_i} + \mathbf{u}_i \tag{8.44}$$

with

$$ {}_R\mathbf{S}_i = {}_R\begin{Bmatrix} -y \\ \varphi_x \\ M_{yz} \\ S_y \end{Bmatrix}_i ; \quad \mathbf{P}_i = \begin{bmatrix} 1 & 0 & 0 & 0 \\ 0 & 1 & 0 & 0 \\ 0 & -\omega^2 I_d & 1 & 0 \\ m\omega^2 & 0 & 0 & 1 \end{bmatrix}_i ; \quad {}_L\mathbf{S}_i = {}_L\begin{Bmatrix} -y \\ \varphi_x \\ M_{yz} \\ S_y \end{Bmatrix}_i ; \quad \mathbf{u}_i = \begin{Bmatrix} 0 \\ 0 \\ 0 \\ -u_y \end{Bmatrix}_i \tag{8.45} $$

where $\mathbf{P}_i$ is the *point matrix* and $\mathbf{u}_i$ is the unbalance vector of the *i*th disk mass. The point matrix, $\mathbf{P}_i$, transforms the state vector from the left of a disk to the right of the disk. Equation 8.44 can be written for both the horizontal and vertical planes, as

$$ {}_R\mathbf{S}_{v_i} = \mathbf{P}_i {}_L\mathbf{S}_{v_i} + \mathbf{u}_{v_i} \quad \text{and} \quad {}_R\mathbf{S}_{h_i} = \mathbf{P}_i {}_L\mathbf{S}_{h_i} + \mathbf{u}_{h_i} \tag{8.46} $$

In general, because the unbalance will also have both real and imaginary components, Equation 8.44 can be expanded to a more general form, as

$$ {}_R\mathbf{S}_i^* = \mathbf{P}_i^* {}_L\mathbf{S}_i^* \tag{8.47} $$

with

$$ {}_R\mathbf{S}_i^* = {}_R\begin{Bmatrix} \mathbf{S}_{h_r} \\ \mathbf{S}_{h_j} \\ \mathbf{S}_{v_r} \\ \mathbf{S}_{v_j} \\ 1 \end{Bmatrix}_i ; \quad \mathbf{P}_i^* = \begin{bmatrix} \mathbf{P} & \mathbf{0}_1 & \mathbf{0}_1 & \mathbf{0}_1 & \mathbf{u}_{h_r} \\ \mathbf{0}_1 & \mathbf{P} & \mathbf{0}_1 & \mathbf{0}_1 & \mathbf{u}_{h_j} \\ \mathbf{0}_1 & \mathbf{0}_1 & \mathbf{P} & \mathbf{0}_1 & \mathbf{u}_{v_r} \\ \mathbf{0}_1 & \mathbf{0}_1 & \mathbf{0}_1 & \mathbf{P} & \mathbf{u}_{v_j} \\ \mathbf{0}_2 & \mathbf{0}_2 & \mathbf{0}_2 & \mathbf{0}_2 & 1 \end{bmatrix}_i ; \quad {}_L\mathbf{S}_i^* = {}_L\begin{Bmatrix} \mathbf{S}_{h_r} \\ \mathbf{S}_{h_j} \\ \mathbf{S}_{v_r} \\ \mathbf{S}_{v_j} \\ 1 \end{Bmatrix}_i \tag{8.48} $$

$$ \mathbf{0}_1 = \begin{bmatrix} 0 & 0 & 0 & 0 \\ 0 & 0 & 0 & 0 \\ 0 & 0 & 0 & 0 \\ 0 & 0 & 0 & 0 \end{bmatrix} ; \quad \mathbf{0}_2 = \begin{bmatrix} 0 & 0 & 0 & 0 \end{bmatrix} $$

where $\mathbf{P}_i^*$ is the *modified point matrix* and the size of the matrix is 17×17. By putting 1 in the last row of the state vector, unbalance force components have been accommodated in the last column of the modified point matrix itself. It should be noted that in the simplest case for single-plane motion (e.g., for the vertical plane) the size of the modified point matrix, $\mathbf{P}_i^*$, is 5×5 and the state vector, $\mathbf{S}^*$, is 5×1; these are given as

$$ \mathbf{P}_i^* = \begin{bmatrix} 1 & 0 & 0 & 0 & 0 \\ 0 & 1 & 0 & 0 & 0 \\ 0 & -\omega^2 I_d & 1 & 0 & 0 \\ m\omega^2 & 0 & 0 & 1 & -u_y \\ 0 & 0 & 0 & 0 & 1 \end{bmatrix}_i \quad \text{and} \quad \mathbf{S}^* = \begin{Bmatrix} -y \\ \varphi_x \\ M_{yz} \\ S_y \\ 1 \end{Bmatrix} \tag{8.49} $$

8.2.3 Overall Transfer Matrix

Once we have the point and field matrices, we can use them to form the overall transfer matrix to relate the state vector at one extreme end station (i.e., the left) to the other extreme end (i.e., the right), provided we do not have intermediate supports, which will be dealt with subsequently. Equations 8.35 and 8.47 can be combined as

$$ {}_R\mathbf{S}_i^* = \mathbf{P}_i^* {}_L\mathbf{S}_i^* = \mathbf{P}_i^*\mathbf{F}_i^* {}_R\mathbf{S}_{i-1}^* = \mathbf{U}_i {}_R\mathbf{S}_{i-1}^* \tag{8.50} $$

with

$$ \mathbf{U}_i^* = \mathbf{P}_i^*\mathbf{F}_i^* $$

where $\mathbf{U}_i^*$ is the modified transfer matrix for the *i*th segment, which transforms the state vector from the right of the $(i-1)$th station to the right of *i*th station. The transfer matrix for all $(n+1)$ stations (i.e., from the 0th to *n*th as shown in Figure 8.21) in the system may be obtained in a similar manner as Equation 8.50, as follows:

$$ \begin{aligned} {}_R\mathbf{S}_1^* &= \mathbf{U}_1^* {}_R\mathbf{S}_0^* \\ {}_R\mathbf{S}_2^* &= \mathbf{U}_2^* {}_R\mathbf{S}_1^* = \mathbf{U}_2^*\mathbf{U}_1^* {}_R\mathbf{S}_0^* \\ {}_R\mathbf{S}_3^* &= \mathbf{U}_3^* {}_R\mathbf{S}_2^* = \mathbf{U}_3^*\mathbf{U}_2^*\mathbf{U}_1^* {}_R\mathbf{S}_0^* \\ &\vdots \\ {}_R\mathbf{S}_n^* &= \mathbf{U}_n^* {}_R\mathbf{S}_{n-1}^* = \mathbf{U}_n^*\mathbf{U}_{n-1}^* \ldots \mathbf{U}_3^*\mathbf{U}_2^*\mathbf{U}_1^* {}_R\mathbf{S}_0^* = \mathbf{T}^* {}_R\mathbf{S}_0^* \end{aligned} \tag{8.51} $$

where $\mathbf{T}^*$ is the *overall modified transfer matrix* for the complete rotor system and the size of the matrix is still 17×17 in the most general form.

Now the TMM is illustrated for a very simple case when there is no coupling between the vertical and horizontal planes (e.g., from gyroscopic effects or from bearings) and no damping in the system (i.e., in a single-plane displacements and forces are in phase with each other); in this case, the overall transfer matrix $\mathbf{T}$ will take a size of 5×5. Equation 8.51 can be written in the expanded form, as

$$ {}_R\begin{Bmatrix} -y \\ \varphi_x \\ M_{yz} \\ S_y \\ 1 \end{Bmatrix}_n = \begin{bmatrix} t_{1,1} & t_{1,2} & t_{1,3} & t_{1,4} & t_{1,5} \\ t_{2,1} & t_{2,2} & t_{2,3} & t_{2,4} & t_{2,5} \\ t_{3,1} & t_{3,2} & t_{3,3} & t_{3,4} & t_{3,5} \\ t_{4,1} & t_{4,2} & t_{4,3} & t_{4,4} & t_{4,5} \\ 0 & 0 & 0 & 0 & 1 \end{bmatrix} \begin{Bmatrix} -y \\ \varphi_x \\ M_{yz} \\ S_y \\ 1 \end{Bmatrix}_0 \tag{8.52} $$

To determine the system characteristics, it is first necessary to define the system boundary conditions, which describe the system support conditions.

For illustration of the TMM, let us consider a simply supported multi-DOF rotor system as shown in Figure 8.21, for which translational displacements and moments are zero at the supports

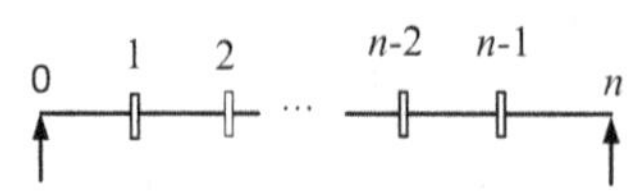

FIGURE 8.21 A simply supported multi-DOF rotor system.

(i.e., $y_0 = y_n = 0$ and $M_{yz0} = M_{yzn} = 0$). These states are corresponding to the first and third rows in the state vector $\{S\}$ and the above equation takes the following form:

$$\begin{Bmatrix} 0 \\ \varphi_x \\ 0 \\ S_y \\ 1 \end{Bmatrix}_n = \begin{bmatrix} t_{1,1} & t_{1,2} & t_{1,3} & t_{1,4} & t_{1,5} \\ t_{2,1} & t_{2,2} & t_{2,3} & t_{2,4} & t_{2,5} \\ t_{3,1} & t_{3,2} & t_{3,3} & t_{3,4} & t_{3,5} \\ t_{4,1} & t_{4,2} & t_{4,3} & t_{4,4} & t_{4,5} \\ 0 & 0 & 0 & 0 & 1 \end{bmatrix} \begin{Bmatrix} 0 \\ \varphi_x \\ 0 \\ S_y \\ 1 \end{Bmatrix}_0 \tag{8.53}$$

On considering only the first and third set of equations from the matrix Equation 8.53, because they have zero (known quantity) on the left-hand side vector, we get

$$\begin{bmatrix} t_{1,2} & t_{1,4} \\ t_{3,2} & t_{3,4} \end{bmatrix} \begin{Bmatrix} \varphi_x \\ S_y \end{Bmatrix}_0 = \begin{Bmatrix} -t_{1,5} \\ -t_{3,5} \end{Bmatrix} \tag{8.54}$$

The remaining two equations, that is, the second and fourth sets of equations from the matrix Equation 8.53, can be written as

$$\begin{Bmatrix} \varphi_x \\ S_y \end{Bmatrix}_n = \begin{bmatrix} t_{2,2} & t_{2,4} \\ t_{4,2} & t_{4,4} \end{bmatrix} \begin{Bmatrix} \varphi_x \\ S_y \end{Bmatrix}_0 + \begin{Bmatrix} t_{2,5} \\ t_{4,5} \end{Bmatrix} \tag{8.55}$$

Now, depending upon the free or forced vibration analysis, Equations 8.54 and 8.55 will be used to obtain natural frequencies and mode shapes or forced responses, respectively. In the following section, these have been described in detail.

8.2.4 Free Vibrations

For free vibrations, the frequency of vibration, ω, in Equation 8.45 will be the natural frequency, ω_{nf}, of the system. For the present illustration of the simply supported shaft, Equation 8.54 becomes a homogeneous equation because the right-hand side terms (i.e., terms of t's with the second subscript as 5) are related to unbalance forces, which are zero in free vibrations. Hence, we have

$$\begin{bmatrix} t_{1,2} & t_{1,4} \\ t_{3,2} & t_{3,4} \end{bmatrix} \begin{Bmatrix} \varphi_x \\ S_y \end{Bmatrix}_0 = \begin{Bmatrix} 0 \\ 0 \end{Bmatrix} \tag{8.56}$$

For a nontrivial solution, the determinant of the above matrix will be zero, hence

$$f(\omega_{nf}) = t_{1,2}t_{3,4} - t_{3,2}t_{1,4} = 0 \tag{8.57}$$

Equation 8.57 is a frequency equation in the form of a polynomial, and it contains natural frequencies as unknown roots of the frequency equation. In the case of simple systems, the frequency can be obtained in an explicit form. By hand calculations, it is feasible to obtain roots of a polynomial only up to the third degree with the help of closed form solutions. However, for complex systems, any convenient root-searching techniques of numerical analysis (e.g., the incremental method, the bisection method, and the Newton–Raphson method) could be used. In such cases, there is no need to multiply the field and point matrices in the symbolic form. These matrices should be multiplied in the numerical form by choosing a suitable guess value for the natural frequency. Once multiplications of overall transfer matrix are done, the condition of Equation 8.57 may be checked to see whether it is satisfactory within certain acceptable limits after the final multiplications of matrices. Based on the residue of Equation 8.57 or its derivatives, the next guess value of the natural frequency may be decided and the process may be repeated until the residue is reduced to the desired level of accuracy. Because Equation 8.57 has, in general, several roots, the procedure may be repeated to obtain the remaining roots. Care should be exercised in obtaining all the roots (natural frequencies) without stepping over any of them. Chapter 6 on torsional vibrations using the TMM may be referred to for more details of the root-searching algorithm.

To obtain mode shapes, from Equation 8.56, we have

$$S_{y0} = -\frac{t_{1,2}}{t_{1,4}}\varphi_{x0} = -\frac{t_{3,2}}{t_{3,4}}\varphi_{x0} \tag{8.58}$$

Let us take a reference value of the rotational displacement at station 0 as $\varphi_{x0} = 1$. On substituting the second term (the third term can also be used) of Equation 8.58 into Equation 8.55 for free vibrations, we get

$$\begin{Bmatrix} \varphi_x \\ S_y \end{Bmatrix}_n = -\begin{bmatrix} t_{2,2} & t_{2,4} \\ t_{4,2} & t_{4,4} \end{bmatrix}\begin{Bmatrix} 1 \\ t_{1,2}/t_{1,4} \end{Bmatrix} \tag{8.59}$$

Now the state vector at the nth station is known completely. By back substitution of the state vector at the nth station into Equation 8.51, we can get the state vectors at other stations. It should be noted that the t's are a function of the natural frequency; hence, from the procedure described above, we will get a mode shape corresponding to a particular natural frequency. The procedure can be repeated for each of the system natural frequencies to get the corresponding mode shapes. In general, for a system of n degrees of freedom, we will have n natural frequencies and that many number of mode shapes. Moreover, the mode shape for a particular natural frequency has the unique relative amplitudes of state variables. For other kinds of simple boundary conditions, Table 8.3 provides frequency equations and equations to obtain mode shapes.

8.2.5 Forced Vibrations

It should be noted that the extreme right-hand side vector of Equations 8.54 and 8.55 is unbalance forcing terms (i.e., terms of t's with the second subscript as 5). Other terms of t's contain ω, which is the spin speed of the shaft. Equation 8.54 can be used to obtain the state vector at station 0. Subsequently, state vectors at the nth and other locations can be obtained by Equations 8.55 and 8.51, respectively. These state vectors contain both the rotational and translational displacements, and the reaction force and the moment. Hence, unbalance responses and reaction forces are known, and can be obtained for various speeds one at a time. Equations 8.54 and 8.55 are tabulated in Table 8.4 along with similar equations for other standard boundary conditions for a ready reference.

8.2.6 Gyroscopic Effects

If gyroscopic effects are allowed in Equation 8.47, the modified point matrix, $\mathbf{P}^*$, will get some extra terms; however, the modified field matrix, $\mathbf{F}^*$, will not be affected. For free vibrations, the formulation has a more general form since equations will contain both the natural whirl frequency, ν, (while considering gyroscopic effects, we tried to distinguish the whirl natural frequency, ν, symbol by in place of ω_{nf}; while the former is *speed dependent* $\nu(\omega)$, the latter is speed independent so that $\nu(\omega=0)=\omega_{nf}$) and the spin speed, ω, of the shaft. This leads to the speed dependency of the natural whirl frequency $\nu(\omega)$.

TABLE 8.3
Equations for the Calculation of Natural Frequencies and Mode Shapes

S.N.	Boundary Conditions	Station Numbers	Equations to Get Natural Frequencies	Equations to Get Mode Shapes
1	Simply supported (pinned–roller support)	0: pinned end, n: roller support	$\begin{bmatrix} t_{1,2} & t_{1,4} \\ t_{3,2} & t_{3,4} \end{bmatrix} \begin{Bmatrix} \varphi_x \\ S_y \end{Bmatrix}_0 = \begin{Bmatrix} 0 \\ 0 \end{Bmatrix}$	${}_R\begin{Bmatrix} \varphi_x \\ S_y \end{Bmatrix}_n = \begin{bmatrix} t_{2,2} & t_{2,4} \\ t_{4,2} & t_{4,4} \end{bmatrix} \begin{Bmatrix} \varphi_x \\ S_y \end{Bmatrix}_0$
2	Cantilever (fixed–free)	0: fixed end, n: free end	$\begin{bmatrix} t_{3,3} & t_{3,4} \\ t_{4,3} & t_{4,4} \end{bmatrix} \begin{Bmatrix} M_{yz} \\ S_y \end{Bmatrix}_0 = \begin{Bmatrix} 0 \\ 0 \end{Bmatrix}$	${}_R\begin{Bmatrix} -y \\ \varphi_x \end{Bmatrix}_n = \begin{bmatrix} t_{1,3} & t_{1,4} \\ t_{2,3} & t_{2,4} \end{bmatrix} \begin{Bmatrix} M_{yz} \\ S_y \end{Bmatrix}_0$
3	Fixed–fixed	0, n: fixed ends	$\begin{bmatrix} t_{1,3} & t_{1,4} \\ t_{2,3} & t_{2,4} \end{bmatrix} \begin{Bmatrix} M_{yz} \\ S_y \end{Bmatrix}_0 = \begin{Bmatrix} 0 \\ 0 \end{Bmatrix}$	${}_R\begin{Bmatrix} M_{yz} \\ S_y \end{Bmatrix}_n = \begin{bmatrix} t_{3,3} & t_{3,4} \\ t_{4,3} & t_{4,4} \end{bmatrix} \begin{Bmatrix} M_{yz} \\ S_y \end{Bmatrix}_0$
4	Free–free	0, n: free ends	$\begin{bmatrix} t_{3,1} & t_{3,2} \\ t_{4,1} & t_{4,2} \end{bmatrix} \begin{Bmatrix} -y \\ \varphi_z \end{Bmatrix}_0 = \begin{Bmatrix} 0 \\ 0 \end{Bmatrix}$	$\begin{Bmatrix} -y \\ \varphi_z \end{Bmatrix}_n = \begin{bmatrix} t_{1,1} & t_{1,2} \\ t_{2,1} & t_{2,2} \end{bmatrix} \begin{Bmatrix} -y \\ \varphi_z \end{Bmatrix}_0$
5	Fixed–pinned	0: fixed end, n: pinned end	$\begin{bmatrix} t_{1,3} & t_{1,4} \\ t_{3,3} & t_{3,4} \end{bmatrix} \begin{Bmatrix} M_{yz} \\ S_y \end{Bmatrix}_0 = \begin{Bmatrix} 0 \\ 0 \end{Bmatrix}$	$\begin{Bmatrix} \varphi_x \\ S_y \end{Bmatrix}_n = \begin{bmatrix} t_{2,3} & t_{2,4} \\ t_{4,3} & t_{4,4} \end{bmatrix} \begin{Bmatrix} M_{yz} \\ S_y \end{Bmatrix}_0$

TABLE 8.4
Governing Equations for Forced Vibration with Different Boundary Conditions

S.N.	Boundary Conditions	Station Numbers	Equations to Get State Vectors	
1	Simply supported (pinned–roller supports)	0: pinned end, n: roller support	$\begin{bmatrix} t_{1,2} & t_{1,4} \\ t_{3,2} & t_{3,4} \end{bmatrix} \begin{Bmatrix} \varphi_x \\ S_y \end{Bmatrix}_0 = \begin{Bmatrix} -t_{1,5} \\ -t_{3,5} \end{Bmatrix}$	${}_R\begin{Bmatrix} \varphi_x \\ S_y \end{Bmatrix}_n = \begin{bmatrix} t_{2,2} & t_{2,4} \\ t_{4,2} & t_{4,4} \end{bmatrix} \begin{Bmatrix} \varphi_x \\ S_y \end{Bmatrix}_0 + \begin{Bmatrix} t_{2,5} \\ t_{4,5} \end{Bmatrix}$
2	Cantilever (fixed–free)	0: fixed end, n: free end	$\begin{bmatrix} t_{3,3} & t_{3,4} \\ t_{4,3} & t_{4,4} \end{bmatrix} \begin{Bmatrix} M_{yz} \\ s_y \end{Bmatrix}_0 = \begin{Bmatrix} -t_{3,5} \\ -t_{4,5} \end{Bmatrix}$	${}_R\begin{Bmatrix} -y \\ \varphi_x \end{Bmatrix}_n = \begin{bmatrix} t_{1,3} & t_{1,4} \\ t_{2,3} & t_{2,4} \end{bmatrix} \begin{Bmatrix} M_{yz} \\ S_y \end{Bmatrix}_0$
3	Fixed–fixed	0, n: fixed ends	$\begin{bmatrix} t_{1,3} & t_{1,4} \\ t_{2,3} & t_{2,4} \end{bmatrix} \begin{Bmatrix} M_{yz} \\ S_y \end{Bmatrix}_0 = \begin{Bmatrix} 0 \\ 0 \end{Bmatrix}$	${}_R\begin{Bmatrix} M_{yz} \\ S_y \end{Bmatrix}_n = \begin{bmatrix} t_{3,3} & t_{3,4} \\ t_{4,3} & t_{4,4} \end{bmatrix} \begin{Bmatrix} M_{yz} \\ S_y \end{Bmatrix}_0$
4	Free–free	0, n: free ends	$\begin{bmatrix} t_{3,1} & t_{3,2} \\ t_{4,1} & t_{4,2} \end{bmatrix} \begin{Bmatrix} -y \\ \varphi_z \end{Bmatrix}_0 = \begin{Bmatrix} 0 \\ 0 \end{Bmatrix}$	$\begin{Bmatrix} -y \\ \varphi_z \end{Bmatrix}_n = \begin{bmatrix} t_{1,1} & t_{1,2} \\ t_{2,1} & t_{2,2} \end{bmatrix} \begin{Bmatrix} -y \\ \varphi_z \end{Bmatrix}_0$
5	Fixed–pinned	0: fixed end, n: pinned end	$\begin{bmatrix} t_{1,3} & t_{1,4} \\ t_{3,3} & t_{3,4} \end{bmatrix} \begin{Bmatrix} M_{yz} \\ S_y \end{Bmatrix}_0 = \begin{Bmatrix} 0 \\ 0 \end{Bmatrix}$	$\begin{Bmatrix} \varphi_x \\ S_y \end{Bmatrix}_n = \begin{bmatrix} t_{2,3} & t_{2,4} \\ t_{4,3} & t_{4,4} \end{bmatrix} \begin{Bmatrix} M_{yz} \\ S_y \end{Bmatrix}_0$

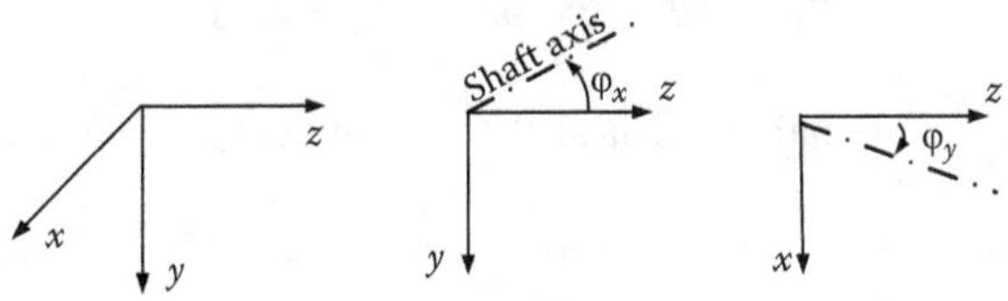

FIGURE 8.22 Coordinate axes and positive conventions for rotational displacements: (a) rectangular three-dimensional axis system; (b) tilting of the shaft axis in the y–z plane; (c) tilting of the shaft axis in the z–x plane.

Equilibrium equations for the moment in the y–z (vertical) and z–x (horizontal) planes, also considering gyroscopic moments, are given as (noting Equation 8.39 and for gyroscopic terms refer to Chapter 5)

$$ {}_RM_{yz_i} - {}_LM_{yz_i} = I_{d_i}\ddot{\varphi}_{x_i} + I_{p_i}\omega\dot{\varphi}_{y_i} \tag{8.60} $$

and

$$ {}_RM_{zx_i} - {}_LM_{zx_i} = I_{d_i}\ddot{\varphi}_{y_i} - I_{p_i}\omega\dot{\varphi}_{x_i} \tag{8.61} $$

where I_p is the polar mass moment of inertia of the disk. In the modified point matrix, the following rows and columns element will be affected: (i) Equations 8.60 and 8.61 are equations of moments, and in the state vector **S**, the 3rd, 7th, 11th, and 15th rows are for moment equations. Hence, only these rows in the modified matrix will be affected. Now in Equations 8.60 and 8.61, additional terms for φ_y and φ_x slopes (see Figure 8.22) appear because of gyroscopic effects. In the modified state vector $\mathbf{S}^*$, the rotational displacements are at the 2nd, 6th, 10th, and 14th rows. Hence, in the modified point matrix $\mathbf{P}^*$, columns 2, 6, 10, and 14 will be affected. For more clarity of the above explanation, let us express rotational displacements and moments as

$$ \varphi = \left(\Phi_r + j\Phi_j\right)e^{j\nu t};\quad M = \left(M_r + jM_j\right)e^{j\nu t} \tag{8.62} $$

where ν is the natural whirl frequency with gyroscopic effects, and subscripts r and j represent the real and imaginary parts of a complex quantity, respectively. On substituting Equation 8.62 into Equation 8.60, we get

$$ \left({}_RM_{yz_r} + j{}_RM_{yz_j}\right) - \left({}_LM_{yz_r} + j{}_LM_{yz_j}\right) = -I_{d_i}\nu^2\left(\Phi_{x_r} + j\Phi_{x_j}\right) + jI_{p_i}\omega\nu\left(\Phi_{y_r} + j\Phi_{y_j}\right) \tag{8.63} $$

Separating the real and imaginary parts from Equation 8.63, we get

$$ {}_RM_{yz_r} - {}_LM_{yz_r} = -I_{d_i}\nu^2\Phi_{x_r} - I_{p_i}\omega\nu\Phi_{y_j} \tag{8.64} $$

and

$$ {}_RM_{yz_j} - {}_LM_{yz_j} = -I_{d_i}\nu^2\Phi_{x_j} + I_{p_i}\omega\nu\Phi_{y_r} \tag{8.65} $$

It should be noted that Equation 8.64 corresponds to the 11th row in Equation 8.47 and the gyroscopic effects term (i.e., $-I_{p_i}\omega\nu$ which is the coefficient of φ_{y_j} that is the sixth row of the modified state vector $\mathbf{S}^*$) in Equation 8.64 will correspond to the sixth column in Equation 8.47. Similarly, from Equation 8.65, it can be seen that gyroscopic effects will introduce another term $I_{p_i}\omega$ at the 15th row and 2nd column corresponding to M_{yz_j} and φ_{y_r}, respectively. Hence, we have

$$P_{11,6} = -I_{p_i}\omega\nu \quad \text{and} \quad P_{15,2} = I_{p_i}\omega\nu \tag{8.66}$$

where $P_{i,j}$ represents the additional element at the ith row and jth column of the modified point matrix with gyroscopic effects.

Similarly, using Equation 8.65, we get the additional element of the modified point matrix as

$$P_{3,14} = -P_{7,10} = I_p\omega\nu \tag{8.67}$$

Hence, the modified point matrix with gyroscopic effect would relate state vectors on either side of the ith disk as follows (refer to Equation 8.47):

$${}_R\mathbf{S}_i^* = \mathbf{P}_i^* \, {}_L\mathbf{S}_i^* \tag{8.68}$$

with

$${}_R\mathbf{S}_i^* = {}_R\begin{Bmatrix} \mathbf{S}_{h_r} \\ \mathbf{S}_{h_j} \\ \mathbf{S}_{v_r} \\ \mathbf{S}_{v_j} \\ 1 \end{Bmatrix}_i ; \quad P_i^* = \begin{bmatrix} \mathbf{P} & \mathbf{0}_1 & \mathbf{0}_1 & \mathbf{G}_1 & \mathbf{u}_{h_r} \\ \mathbf{0}_1 & \mathbf{P} & \mathbf{G}_2 & \mathbf{0}_1 & \mathbf{u}_{h_j} \\ \mathbf{0}_1 & \mathbf{G}_3 & \mathbf{P} & \mathbf{0}_1 & \mathbf{u}_{v_r} \\ \mathbf{G}_4 & \mathbf{0}_1 & \mathbf{0}_1 & \mathbf{P} & \mathbf{u}_{v_j} \\ \mathbf{0}_2 & \mathbf{0}_2 & \mathbf{0}_2 & \mathbf{0}_2 & 1 \end{bmatrix}_i ; \quad {}_L\mathbf{S}_i^* = {}_L\begin{Bmatrix} \mathbf{S}_{h_r} \\ \mathbf{S}_{h_j} \\ \mathbf{S}_{v_r} \\ \mathbf{S}_{v_j} \\ 1 \end{Bmatrix}_i$$

$$\mathbf{S}_h = \begin{Bmatrix} -x \\ \varphi_y \\ M_{zx} \\ S_x \end{Bmatrix}; \quad \mathbf{P} = \begin{bmatrix} 1 & 0 & 0 & 0 \\ 0 & 1 & 0 & 0 \\ 0 & -\omega^2 I_d & 1 & 0 \\ m\omega^2 & 0 & 0 & 1 \end{bmatrix}; \quad \mathbf{S}_v = \begin{Bmatrix} -y \\ \varphi_x \\ M_{yz} \\ S_y \end{Bmatrix}; \quad \mathbf{u}_h = \begin{Bmatrix} 0 \\ 0 \\ 0 \\ -u_x \end{Bmatrix};$$

$$\mathbf{u}_v = \begin{Bmatrix} 0 \\ 0 \\ 0 \\ -u_y \end{Bmatrix}$$

and

$$\mathbf{G}_1 = \mathbf{G}_4 = \begin{bmatrix} 0 & 0 & 0 & 0 \\ 0 & 0 & 0 & 0 \\ 0 & I_p\omega\nu & 0 & 0 \\ 0 & 0 & 0 & 0 \end{bmatrix}; \quad \mathbf{G}_2 = \mathbf{G}_3 = \begin{bmatrix} 0 & 0 & 0 & 0 \\ 0 & 0 & 0 & 0 \\ 0 & -I_p\omega\nu & 0 & 0 \\ 0 & 0 & 0 & 0 \end{bmatrix};$$

$$\mathbf{0}_1 = \begin{bmatrix} 0 & 0 & 0 & 0 \\ 0 & 0 & 0 & 0 \\ 0 & 0 & 0 & 0 \\ 0 & 0 & 0 & 0 \end{bmatrix}; \quad \mathbf{0}_2 = \begin{bmatrix} 0 & 0 & 0 & 0 \end{bmatrix}; \quad \mathbf{0}_3 = \begin{Bmatrix} 0 \\ 0 \\ 0 \\ 0 \end{Bmatrix}$$

Now through numerical examples, the TMM will be illustrated for simple as well as complex rotor systems, especially with gyroscopic effects and intermediate supports.

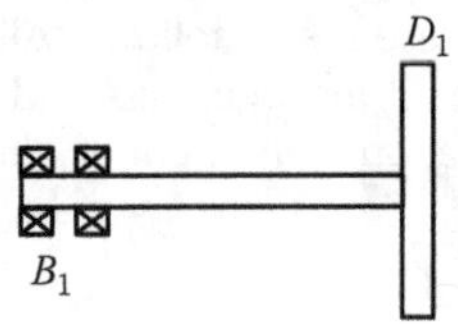

FIGURE 8.23 A cantilever rotor.

Example 8.5

Obtain the unbalance response and transverse critical speeds of a cantilever rotor system as shown in Figure 8.23. End B_1 of the shaft is fixed. The length of the shaft is 0.2 m, and the diameter is 0.01 m. The disk is thin and has 1 kg of mass with a radius of 3.0 cm. Neglect the mass of the shaft and the gyroscopic effect. Take an unbalance of 3 g at a radius of 2 cm. Choose the shaft speed suitably so as to cover two critical speeds in the unbalance response. Take $E = 2.1 \times 10^{11}\ \mathrm{N/m^2}$.

Solution:

Let station 0 be the fixed end and station 1 be the free end. Consider only single-plane motion. The overall transformation can be written as

$$_R\mathbf{S}_1^* = \mathbf{T}^*\mathbf{S}_0^* \tag{a}$$

with

$$\mathbf{T}^* = \mathbf{P}_1^*\mathbf{F}_1^* \tag{b}$$

$$\mathbf{P}_1^* = \begin{bmatrix} 1 & 0 & 0 & 0 & 0 \\ 0 & 1 & 0 & 0 & 0 \\ 0 & -I_d\omega^2 & 1 & 0 & 0 \\ m\omega^2 & 0 & 0 & 1 & -u_y \\ 0 & 0 & 0 & 0 & 1 \end{bmatrix}; \quad \mathbf{F}_1^* = \begin{bmatrix} 1 & l & l^2/2EI & l^3/6EI & 0 \\ 0 & 1 & l/EI & l^2/2EI & 0 \\ 0 & 0 & 1 & l & 0 \\ 0 & 0 & 0 & 1 & 0 \\ 0 & 0 & 0 & 0 & 1 \end{bmatrix} \tag{c}$$

$$\mathbf{S}^* = \begin{Bmatrix} -y \\ \varphi_x \\ M_{yz} \\ S_y \\ 1 \end{Bmatrix} \tag{d}$$

$$\mathbf{P}_1^*\mathbf{F}_1^* = \begin{bmatrix} 1 & 0 & 0 & 0 & 0 \\ 0 & 1 & 0 & 0 & 0 \\ 0 & -I_d\omega^2 & 1 & 0 & 0 \\ m\omega^2 & 0 & 0 & 1 & -u_y \\ 0 & 0 & 0 & 0 & 1 \end{bmatrix} \begin{bmatrix} 1 & l & l^2/2EI & l^3/6EI & 0 \\ 0 & 1 & l/EI & l^2/2EI & 0 \\ 0 & 0 & 1 & l & 0 \\ 0 & 0 & 0 & 1 & 0 \\ 0 & 0 & 0 & 0 & 1 \end{bmatrix}$$

$$= \begin{bmatrix} 1 & l & l^2/2EI & l^3/6EI & 0 \\ 0 & 1 & l/EI & l^2/2EI & 0 \\ 0 & -\omega^2 I_d & 1-\left(\omega^2 I_d l/EI\right) & l-\left(\omega^2 I_d l^2/2EI\right) & 0 \\ m\omega^2 & m\omega^2 l & m\omega^2 l^2/2EI & 1+\left(m\omega^2 l^3/6EI\right) & -u_y \\ 0 & 0 & 0 & 0 & 1 \end{bmatrix} \tag{e}$$

Boundary conditions for the present case are as follows: All the translational and rotational displacements at station 0 are zero, and all the moments and shear forces are zero at the right of station 1. Hence, the state vector at station 0 and right of station 1 will have the following form:

$$\mathbf{S}_0^* = \left\{ 0 \quad 0 \quad M_{yz} \quad S_y \quad 1 \right\}^T \tag{f}$$

and

$$\mathbf{S}_1^* = \left\{ y \quad \varphi_x \quad 0 \quad 0 \quad 1 \right\}^T \tag{g}$$

From free vibrations, the unbalance, u_y, is zero. Hence, the third and fourth rows will give an eigenvalue problem of the following form:

$$\begin{bmatrix} 1-\left(\omega_{nf}^2 I_d l \big/ EI\right) & l-\left(\omega_{nf}^2 I_d l^2 \big/ 2EI\right) \\ m\omega_{nf}^2 l^2 \big/ 2EI & 1+\left(m\omega_{nf}^2 l^3 \big/ 6EI\right) \end{bmatrix} \begin{Bmatrix} M_{yz} \\ S_y \end{Bmatrix}_0 = \begin{Bmatrix} 0 \\ 0 \end{Bmatrix} \tag{h}$$

which gives a polynomial of the following form:

$$\omega_{nf}^4 - 12EI\frac{\frac{1}{3}ml^2 + I_d}{ml^3 I_d}\omega_{nf}^2 + \frac{12E^2I^2}{ml^4 I_d} = 0 \tag{i}$$

For the present problem, we have the following data:

$$m = 1 \text{ kg}, \quad r = 0.03 \text{ m}, \quad I_d = \frac{1}{4}mr^2 = 2.25\times10^{-4} \text{ kg-m}^2, \quad l = 0.2 \text{ m}$$

$$E = 2.1\times10^{11} \text{ N/m}^2, \quad d = 0.01 \text{ m}, \quad \text{and} \quad I = \pi d^4/64 = 4.91\times10^{-10} \text{ m}^4$$

Hence from Equation (i), we get

$$\omega_{nf}^4 - \left(9.3176\times10^6\right)\omega_{nf}^2 + \left(3.5421\times10^{11}\right) = 0$$

which gives the natural frequencies as

$$\omega_{nf_1} = 195.4 \text{ rad/s} \quad \omega_{nf_2} = 3046.2 \text{ rad/s}$$

Now for the unbalance response from Equations (a), (e), (g), and (f), we have

$$\begin{bmatrix} 1 & l & l^2 \big/ 2EI & l^3 \big/ 6EI & 0 \\ 0 & 1 & l \big/ EI & l^2 \big/ 2EI & 0 \\ 0 & -\omega^2 I_d & 1-\left(\omega^2 I_d l \big/ EI\right) & l-\left(\omega^2 I_d l^2 \big/ 2EI\right) & 0 \\ m\omega^2 & m\omega^2 l & m\omega^2 l^2 \big/ 2EI & 1+\left(m\omega^2 l^3 \big/ 6EI\right) & -u_y \\ 0 & 0 & 0 & 0 & 1 \end{bmatrix} \begin{Bmatrix} 0 \\ 0 \\ M_{yz} \\ S_y \\ 1 \end{Bmatrix}_0 = \begin{Bmatrix} -y \\ \varphi_x \\ 0 \\ 0 \\ 1 \end{Bmatrix}_1 \tag{j}$$

The first two rows and the last three rows give

$$\begin{bmatrix} l^2/2EI & l^3/6EI \\ l/EI & l^2/2EI \end{bmatrix} \begin{Bmatrix} M_{yz} \\ S_y \end{Bmatrix} = \begin{Bmatrix} -y \\ \varphi_x \end{Bmatrix} \tag{k}$$

and

$$\begin{bmatrix} 1-\left(\omega^2 I_d l/EI\right) & l-\left(\omega^2 I_d l^2/2EI\right) & 0 \\ m\omega^2 l^2/2EI & 1+\left(m\omega^2 l^3/6EI\right) & -u_y \\ 0 & 0 & 1 \end{bmatrix} \begin{Bmatrix} M_{yz} \\ S_y \\ 1 \end{Bmatrix} = \begin{Bmatrix} 0 \\ 0 \\ 1 \end{Bmatrix} \tag{l}$$

The form of Equations (k) and (l) is similar to that given in Table 8.4. Equation (k) can be written as

$$\begin{bmatrix} 1-\left(\omega^2 I_d l/EI\right) & l-\left(\omega^2 I_d l^2/2EI\right) \\ m\omega^2 l^2/2EI & 1+\left(m\omega^2 l^3/6EI\right) \end{bmatrix} \begin{Bmatrix} M_{yz} \\ S_y \end{Bmatrix} = \begin{Bmatrix} 0 \\ u_y \end{Bmatrix} \tag{m}$$

or

$$\begin{Bmatrix} M_{yz} \\ S_y \end{Bmatrix} = \begin{bmatrix} 1-\left(\omega^2 I_d l/EI\right) & l-\left(\omega^2 I_d l^2/2EI\right) \\ m\omega^2 l^2/2EI & 1+\left(m\omega^2 l^3/6EI\right) \end{bmatrix}^{-1} \begin{Bmatrix} 0 \\ u_y \end{Bmatrix} \tag{n}$$

Equation (n) can be used to get the bending moment and the shear force due to the unbalance force. Hence, on substituting Equation (n) into Equation (k), we get

$$\begin{Bmatrix} -y \\ \varphi_x \end{Bmatrix} = \begin{bmatrix} l^2/2EI & l^3/6EI \\ l/EI & l^2/2EI \end{bmatrix} \begin{bmatrix} 1-\left(\omega^2 I_d l/EI\right) & l-\left(\omega^2 I_d l^2/2EI\right) \\ m\omega^2 l^2/2EI & 1+\left(m\omega^2 l^3/6EI\right) \end{bmatrix}^{-1} \begin{Bmatrix} 0 \\ u_y \end{Bmatrix} \tag{o}$$

Equation (o) can be solved for a particular spin speed to get the unbalance response (y and φ_x). Then, in a similar way, the spin speed can be varied to get the variation of the unbalance response with the spin speed. A plot of y and φ_x with respect to the spin speed of the rotor is given in Figure 8.24a and b. The resonant condition can be seen as large amplitudes of vibration, and it indicates critical speeds. Due to the absence of damping in the rotor system at the resonance, the vibration amplitude looks very high; however, in the real rotor system, always sufficient damping is present to make the vibration amplitude during crossing of the resonance a reasonable value. A similar plot can be obtained from Equation (n) for M_{yz} and S_y and is shown in Figure 8.24c and d. It can be observed from all four plots that critical speeds are the same as natural frequencies obtained by free vibration analysis ($\omega_{nf_1} = 195.4$ rad / s and $\omega_{nf_2} = 3046.2$ rad/s). In the plots of the shear force and the bending moment at the support, antiresonances can be seen in between the two critical speeds. This indicates that two modes of vibrations have canceling effects on the shear force and the bending moment. The frequency at which the antiresonance occurs for the shear force is not the same as that of the bending moment. It indicates that it is not a system characteristic and its location may change for different unbalance force magnitudes and positions; however, the critical speed has fixed frequencies since it is a system characteristic.

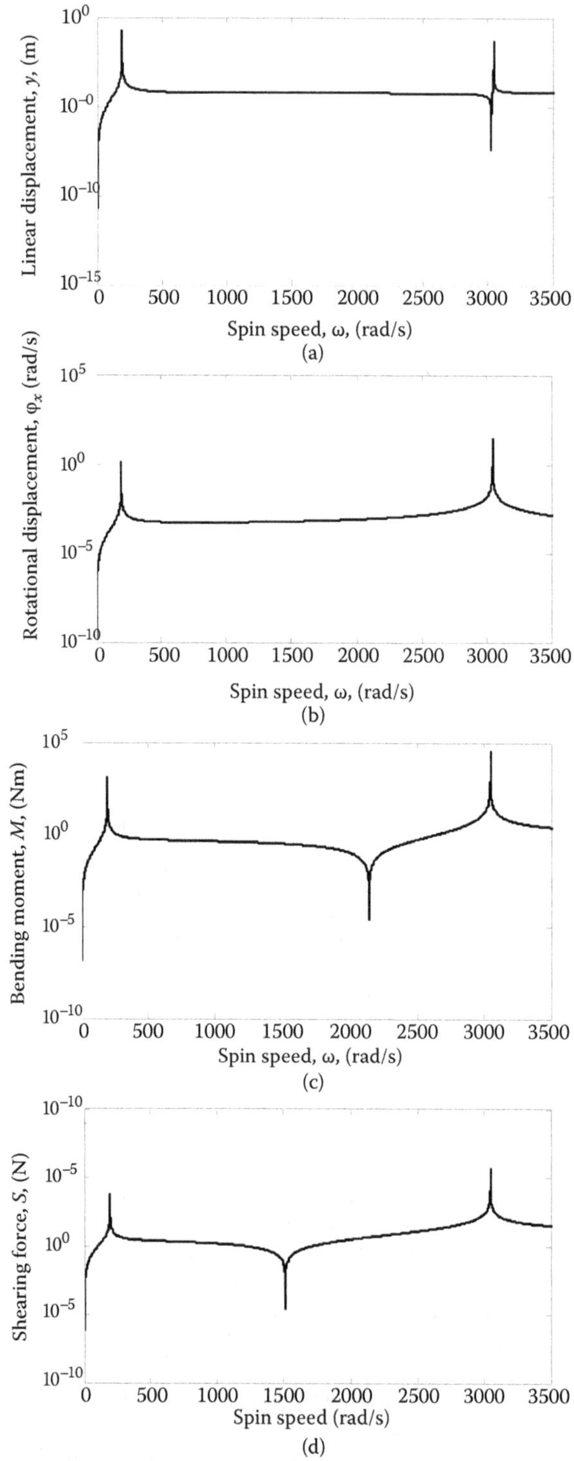

FIGURE 8.24 (a) Variation of the unbalance translational response with the spin speed. (b) Variation of the unbalance rotational response with the spin speed. (c) Variation of the support bending moment with the spin speed. (d) Variation of the support shearing force with the spin speed.

Example 8.6

Obtain the transverse natural frequencies and corresponding mode shapes of the rotor system shown in Figure 8.25. Take the mass of the disk as, m=10 kg, and the diametral mass moment of inertia as I_d=0.02 kg-m², and the disk is placed at 0.25 m from the right support. The shaft has a diameter of 10 mm and a span length of 1 m. The shaft material has Young's modulus E=2.1×10¹¹ N/m²; consider the shaft as massless. Neglect gyroscopic effects and take one-plane motion only. Compare the results with the influence coefficient method of Example 8.1.

Solution:

Station numbers 0, 1, and 2 can be assigned at the left support, at the disk, and at the right support, respectively. Then, the overall transformation of state vectors is given as

$$ {}_R\mathbf{S}_2 = \mathbf{F}_2\mathbf{P}_1\mathbf{F}_1\mathbf{S}_0 \tag{a}$$

which can be expanded as

$$ {}_R\mathbf{S}_2 = \begin{bmatrix} 1 & l & 0.5\beta l & \frac{1}{6}\beta l^2 \\ 0 & 1 & \beta & 0.5\beta l \\ 0 & 0 & 1 & l \\ 0 & 0 & 0 & 1 \end{bmatrix}_2 \begin{bmatrix} 1 & 0 & 0 & 0 \\ 0 & 1 & 0 & 0 \\ 0 & -\omega_{nf}^2 I_d & 1 & 0 \\ m\omega_{nf}^2 & 0 & 0 & 1 \end{bmatrix}_1 \begin{bmatrix} 1 & l & 0.5\beta l & \frac{1}{6}\beta l^2 \\ 0 & 1 & \beta & 0.5\beta l \\ 0 & 0 & 1 & l \\ 0 & 0 & 0 & 1 \end{bmatrix}_1 \mathbf{S}_0 \tag{b}$$

with

$$\beta_1 = l_1 / EI \quad \text{and} \quad \beta_2 = l_2 / EI$$

Subscripts 1 and 2 outside the matrices belong to the following parameters: $\beta, l,$ and I_d. On multiplication of matrices, we get

$$ {}_R\mathbf{S}_2 = \begin{bmatrix} 1+\frac{1}{6}m_1\beta_2 l_2^2\omega_{nf}^2 & l_2 - 0.5 I_{d_1}\beta_2 l_2\omega_{nf}^2 & 0.5\beta_2 l_2 & \frac{1}{6}\beta_2 l_2^2 \\ 0.5 m_1\beta_2 l_2\omega_{nf}^2 & 1 - I_{d_1}\beta_2\omega_{nf}^2 & \beta_2 & 0.5\beta_2 l_2 \\ m_1 l_2\omega_{nf}^2 & -I_{d_1}\omega_{nf}^2 & 1 & l_2 \\ m_1\omega_{nf}^2 & 0 & 0 & 1 \end{bmatrix}_1 \begin{bmatrix} 1 & l & 0.5\beta l & \frac{1}{6}\beta l^2 \\ 0 & 1 & \beta & 0.5\beta l \\ 0 & 0 & 1 & l \\ 0 & 0 & 0 & 1 \end{bmatrix}_1 \mathbf{S}_0$$

which finally takes the following form:

$$ {}_R\begin{Bmatrix} -y \\ \varphi_x \\ M_{yz} \\ S_y \end{Bmatrix}_2 = \begin{bmatrix} t_{11} & t_{12} & t_{13} & t_{14} \\ t_{21} & t_{22} & t_{23} & t_{24} \\ t_{31} & t_{32} & t_{33} & t_{34} \\ t_{41} & t_{42} & t_{43} & t_{44} \end{bmatrix} \begin{Bmatrix} -y \\ \varphi_x \\ M_{yz} \\ S_y \end{Bmatrix}_0 \tag{c}$$

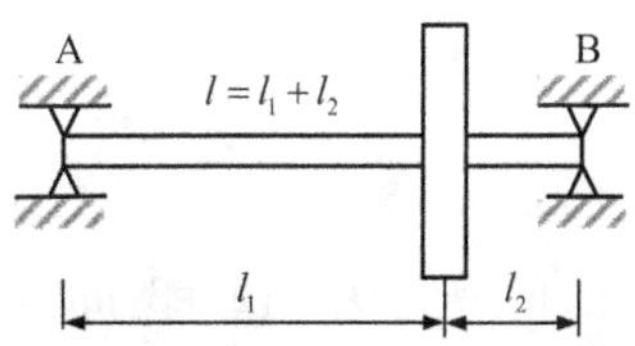

FIGURE 8.25 A simply supported rotor system.

with

$$t_{11}=1+\frac{1}{6}m_1\beta_1 l_2^2\omega_{nf}^2;\quad t_{12}=l_1\left(1+\frac{1}{6}m_1\beta_2 l_2^2\omega_{nf}^2\right)+\left(l_2-0.5I_{d_1}\beta_2 l_2\omega_{nf}^2\right)$$

$$t_{13}=05\beta_1 l_1\left(1+\frac{1}{6}m_1\beta_2 l_2^2\omega_{nf}^2\right)+\beta_1\left(l_2-0.5I_{d_1}\beta_2 l_2\omega_{nf}^2\right)+0.5\beta_2 l_2$$

$$t_{14}=\frac{1}{6}\beta_1 l_1^2\left(1+\frac{1}{6}m_1\beta_2 l_2^2\omega_{nf}^2\right)+0.5\beta_1 l_1\left(l_2-0.5I_{d_1}\beta_2 l_2\omega_{nf}^2\right)+l_1\left(0.5\beta_2 l_2\right)+\frac{1}{6}\beta_2 l_2^2$$

$$t_{21}=0.5m_1\beta_2 l_2\omega_{nf}^2;\quad t_{22}=l_2\left(0.5m_1\beta_2 l_2\omega_{nf}^2\right)+\left(1-I_{d_1}\beta_2\omega_{nf}^2\right)$$

$$t_{23}=0.5\beta_1 l_1\left(0.5m_1\beta_2 l_2\omega_{nf}^2\right)+\beta_1\left(1-I_{d_1}\beta_2\omega_{nf}^2\right)+\beta_2$$

$$t_{24}=\frac{1}{6}\beta_1 l_1^2\left(0.5m_1\beta_2 l_2\omega_{nf}^2\right)+0.5\beta_1 l_1\left(1-I_{d_1}\beta_2\omega_{nf}^2\right)+l_1\left(\beta_2\right)+0.5\beta_2 l_2$$

$$t_{31}=m_1 l_2\omega_{nf}^2;\quad t_{32}=l_1\left(m_1 l_2\omega_{nf}^2\right)+\left(-I_{d_1}\omega_{nf}^2\right)$$

$$t_{33}=0.5\beta_1 l_1\left(m_1 l_2\omega_{nf}^2\right)-\beta_1\left(-I_{d_1}\omega_{nf}^2\right)+1$$

$$t_{34}=\frac{1}{6}\beta_1 l_1^2\left(m_1 l_2\omega_{nf}^2\right)+0.5\beta_1 l_1\left(I_{d_1}\omega_{nf}^2\right)+l_1+l_2;\quad t_{41}=m_1\omega_{nf}^2 \tag{d}$$

$$t_{42}=l_1\left(m_1\omega_{nf}^2\right);\quad t_{43}=0.5\beta_1 l_1\left(m_1\omega_{nf}^2\right);\quad t_{44}=\frac{1}{6}\beta_1 l_1^2\left(m_1\omega_{nf}^2\right)+1$$

The following boundary conditions are applied for the present case (Figure 8.25):

$$y_0=M_0={}_Ry_2={}_RM_2=0 \tag{e}$$

On application of the boundary conditions in Equation (c), the following set of equations is obtained:

$$\begin{Bmatrix}0\\0\end{Bmatrix}=\begin{bmatrix}t_{12} & t_{14}\\ t_{32} & t_{34}\end{bmatrix}\begin{Bmatrix}\varphi_x\\ S_y\end{Bmatrix}_0 \quad\text{and}\quad {}_R\begin{Bmatrix}\varphi_x\\ S_y\end{Bmatrix}_2=\begin{bmatrix}t_{22} & t_{24}\\ t_{42} & t_{44}\end{bmatrix}\begin{Bmatrix}\varphi_x\\ S_y\end{Bmatrix}_0 \tag{f}$$

From the first set of Equation (e), the frequency equation takes the following form:

$$t_{12}t_{34}-t_{14}t_{32}=0 \tag{g}$$

On substituting Equation (e) into Equation (g), we get

$$\left\{l_1\left(1+\frac{1}{6}m_1\beta_2 l_2^2\omega_{nf}^2\right)+\left(l_2-0.5I_{d_1}\beta_2 l_2\omega_{nf}^2\right)\right\}\left\{\frac{1}{6}\beta_1 l_1^2\left(m_1 l_2\omega_{nf}^2\right)+0.5\beta_1 l_1\left(-I_{d_1}\omega_{nf}^2\right)+l_1+l_2\right\}$$
$$-\left\{\frac{1}{6}\beta_1 l_1^2\left(1+\frac{1}{6}m_1\beta_2 l_2^2\omega_{nf}^2\right)+0.5\beta_1 l_1\left(l_2-0.5I_{d_1}\beta_2 l_2\omega_{nf}^2\right)+l_1\left(0.5\beta_2 l_2\right)+\frac{1}{6}\beta_2 l_2^2\right\}$$
$$\left\{l_1\left(m_1 l_2\omega_{nf}^2\right)+\left(-I_{d_1}\omega_{nf}^2\right)\right\}=0$$

which simplifies to

$$\left(m_1 I_{d_1}\beta_1\beta_2 l_1^2 l_2^2\right)\omega_{nf}^4-3\left\{\left(l_1^2+\beta_1+l_1^2\beta_2\right)I_{d_1}+(\beta_1+\beta_2)l_1^2 l_2^2 m_1\right\}\omega_{nf}^2+9(l_1+l_2)^2=0 \tag{h}$$

After substituting numerical values in Equation (h), we get the natural frequencies of the rotor system as

$$\omega_{nf_1} = 29.45 \text{ rad/s} \quad \text{and} \quad \omega_{nf_2} = 289.23 \text{ rad/s}$$

which are very close to that obtained in Example 8.1.

The relative translational and rotational displacements, and the bearing and shaft reaction forces:

i. For $\omega_{nf_1} = 29.45\,\text{rad/s}$: Let us assume $\varphi_{zo} = 1\,\text{rad}$; from Equation (f), we have

$$S_{yo} = -\frac{t_{12}(\omega_{nf})}{t_{14}(\omega_{nf})} = -6.12\times 10^2 \text{ N}$$

$${}_R\varphi_{x_2} = t_{22}\varphi_{xo} + t_{24}S_{y_o} = t_{22} - \frac{t_{24}t_{12}}{t_{14}} = -1.39 \text{ rad}$$

$${}_R S_{y_2} = t_{42}\varphi_{xo} + t_{44}S_{y_o} = t_{42} - \frac{t_{44}t_{12}}{t_{14}} = 1.93\times 10^3 \text{ N}$$

Now, the state vectors at stations 0 and 2 are completely known. Hence, the state vectors at the left and right of station 1 can be obtained as

$$_L\begin{Bmatrix} -y \\ \varphi_x \\ M_{yz} \\ S_y \end{Bmatrix}_1 = \begin{bmatrix} 1 & l & 0.5\beta l & \frac{1}{6}\beta l^2 \\ 0 & 1 & \beta & 0.5\beta l \\ 0 & 0 & 1 & l \\ 0 & 0 & 0 & 1 \end{bmatrix}_1 \begin{Bmatrix} -y \\ \varphi_x \\ M_{yz} \\ S_y \end{Bmatrix}_0 = \begin{bmatrix} 1 & l & 0.5\beta l & \frac{1}{6}\beta l^2 \\ 0 & 1 & \beta & 0.5\beta l \\ 0 & 0 & 1 & l \\ 0 & 0 & 0 & 1 \end{bmatrix}_1 \begin{Bmatrix} 0 \\ 1 \\ 0 \\ -t_{12}/t_{14} \end{Bmatrix}_2$$

$$= \begin{Bmatrix} 0.299 \\ -0.805 \\ -496.197 \\ -661.596 \end{Bmatrix}$$

and

$$_R\begin{Bmatrix} -y \\ \varphi_x \\ M_{yz} \\ S_y \end{Bmatrix}_1 = \begin{bmatrix} 1 & 0 & 0 & 0 \\ 0 & 1 & 0 & 0 \\ 0 & -\omega_{nf}^2 I_d & 1 & 0 \\ m\omega_{nf}^2 & 0 & 0 & 1 \end{bmatrix}_1 {}_L\begin{Bmatrix} -y \\ \varphi_x \\ M_{yz} \\ S_y \end{Bmatrix}_1$$

$$= \begin{bmatrix} 1 & 0 & 0 & 0 \\ 0 & 1 & 0 & 0 \\ 0 & -\omega_{nf}^2 I_d & 1 & 0 \\ m\omega_{nf}^2 & 0 & 0 & 1 \end{bmatrix}_1 \begin{Bmatrix} 0.299 \\ -0.805 \\ -496.197 \\ -661.596 \end{Bmatrix} = \begin{Bmatrix} 0.299 \\ -0.805 \\ -482.234 \\ 1928.937 \end{Bmatrix}$$

It should be noted that for relative amplitudes of displacements, reaction forces and moments have no quantitative significance. In fact, estimation of unbalance responses would give the exact value of loads at various stations of the shaft.

ii. For $\omega_{nf_2} = 289.23$ rad/s: Let us again assume $\varphi_{zo} = 1$ rad. For this case also, along the same lines as for the previous case, we get the following results:

$$Q_0 = -1.116\times10^3\ \text{N};\quad {}_R\varphi_{x2} = 1.088\ \text{rad};\quad {}_RS_{y2} = -1.033\times10^4\ \text{N};$$

$$\begin{Bmatrix} -y \\ \varphi_x \\ M_{yz} \\ S_y \end{Bmatrix}_{L1} = \begin{bmatrix} -0.011 \\ -2.044 \\ -836.783 \\ -1115.7 \end{bmatrix}_1 \quad\text{and}\quad \begin{Bmatrix} -y \\ \varphi_x \\ M_{yz} \\ S_y \end{Bmatrix}_{R1} = \begin{bmatrix} -0.011 \\ -2.044 \\ 2580 \\ -1033 \end{bmatrix}_1$$

Example 8.7

Find the transverse natural frequencies and mode shapes of the two-disk rotor system shown in Figure 8.26. B is a fixed end, and D_1 and D_2 are rigid disks. The masses of the disks are $m_1=5$ kg and $m_2=2$ kg, and the diametral mass moments of inertia are $I_{d_1} = 0.03$ kg-m^2 and $I_{d_2} = 0.01$ kg-m^2. The shaft is made of steel with modulus of elasticity $E=2.1\ (10)^{11}$ N/m^2, and of uniform diameter $d=10$ mm. The shaft lengths are $BD_1=50$ mm, and $D_1D_2=75$ mm. Consider the shaft as massless.

Solution:

Example 8.7 by the TMM is the same as Example 8.3 by the influence coefficient method and Exercise 9.2 by the finite element method (FEM). Figure 8.27 shows the station numbers with 0 at the fixed end, and 1 and 2 at the subsequent disks.

We have the following data:

$$I = \frac{\pi}{64}d^4 = \frac{\pi}{64}0.01^4 = 4.91\times10^{-10}\ \text{m}^4 \quad \text{and} \quad EI = 103.1\ \text{N-m}^2$$

The transformation of the state between stations 0 and 1 can be written as

$$_R\mathbf{S}_1 = \mathbf{P}_1\mathbf{F}_1\mathbf{S}_0 = \mathbf{U}_1\mathbf{S}_0 \tag{a}$$

with

$$\mathbf{P}_1 = \begin{bmatrix} 1 & 0 & 0 & 0 \\ 0 & 1 & 0 & 0 \\ 0 & -I_d\omega_{nf}^2 & 1 & 0 \\ m\omega_{nf}^2 & 0 & 0 & 1 \end{bmatrix}_1 = \begin{bmatrix} 1 & 0 & 0 & 0 \\ 0 & 1 & 0 & 0 \\ 0 & -0.03\omega_{nf}^2 & 1 & 0 \\ 5\omega_{nf}^2 & 0 & 0 & 1 \end{bmatrix} \tag{b}$$

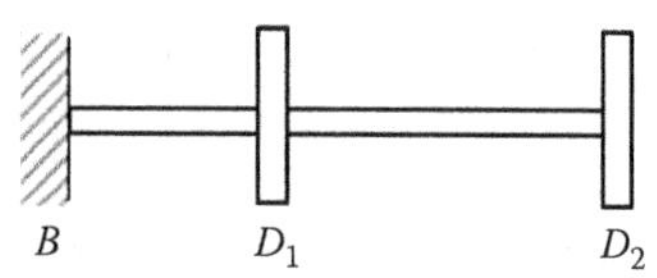

FIGURE 8.26 A cantilever two-disk system.

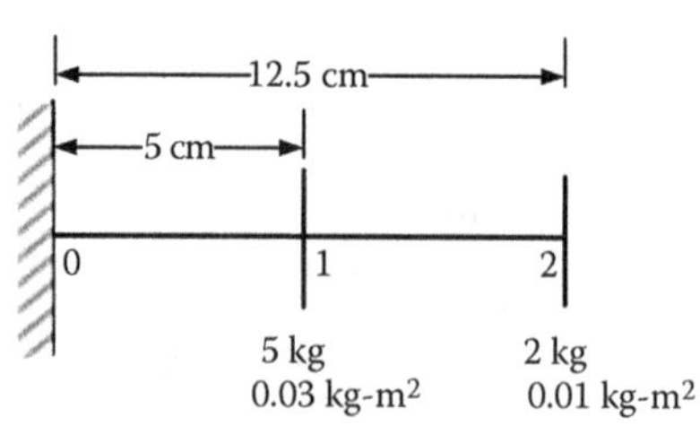

FIGURE 8.27 The cantilever rotor with the shaft and disk properties.

$$\mathbf{F}_1 = \begin{bmatrix} 1 & l & l^2/2EI & l^3/6EI \\ 0 & 1 & l/EI & l^2/2EI \\ 0 & 0 & 1 & l \\ 0 & 0 & 0 & 1 \end{bmatrix}_1 = \begin{bmatrix} 1 & 0.05 & 1.21\times10^{-5} & 2.02\times10^{-7} \\ 0 & 1 & 4.85\times10^{-4} & 1.21\times10^{-5} \\ 0 & 0 & 1 & 0.05 \\ 0 & 0 & 0 & 1 \end{bmatrix} \tag{c}$$

$$\mathbf{U}_1 = \mathbf{P}_1\mathbf{F}_1 = \begin{bmatrix} 1 & 0.05 & 1.21\times10^{-5} & 2.02\times10^{-7} \\ 0 & 1 & 4.85\times10^{-4} & 1.21\times10^{-5} \\ 0 & -0.03\omega_{nf}^2 & 1-\left(1.46\times10^{-5}\right)\omega_{nf}^2 & 0.05-\left(3.63\times10^{-7}\right)\omega_{nf}^2 \\ 5\omega_{nf}^2 & 0.25\omega_{nf}^2 & \left(6.05\times10^{-5}\right)\omega_{nf}^2 & 1+\left(1.01\times10^{-6}\right)\omega_{nf}^2 \end{bmatrix} \tag{d}$$

Similarly, the transformation of the state between stations 1 and 2 can be written as

$${}_R\mathbf{S}_2 = \mathbf{P}_2\mathbf{F}_2\,{}_R\mathbf{S}_1 = \mathbf{U}_2\,{}_R\mathbf{S}_1 \tag{e}$$

with

$$\mathbf{P}_2 = \begin{bmatrix} 1 & 0 & 0 & 0 \\ 0 & 1 & 0 & 0 \\ 0 & -0.01\omega_{nf}^2 & 1 & 0 \\ 2\omega_{nf}^2 & 0 & 0 & 1 \end{bmatrix}, \quad \mathbf{F}_2 = \begin{bmatrix} 1 & 0.075 & 2.73\times10^{-5} & 6.82\times10^{-7} \\ 0 & 1 & 7.27\times10^{-4} & 2.73\times10^{-5} \\ 0 & 0 & 1 & 0.075 \\ 0 & 0 & 0 & 1 \end{bmatrix} \tag{f}$$

$$\mathbf{U}_2 = \mathbf{P}_2\mathbf{F}_2 = \begin{bmatrix} 1 & 0.075 & 2.73\times10^{-5} & 6.82\times10^{-7} \\ 0 & 1 & 7.27\times10^{-4} & 2.73\times10^{-5} \\ 0 & -0.01\omega_{nf}^2 & 1-\left(7.27\times10^{-6}\right)\omega_{nf}^2 & 0.075-\left(2.73\times10^{-7}\right)\omega_{nf}^2 \\ 2\omega_{nf}^2 & 0.15\omega_{nf}^2 & \left(5.46\times10^{-5}\right)\omega_{nf}^2 & 1+\left(1.36\times10^{-6}\right)\omega_{nf}^2 \end{bmatrix} \tag{g}$$

On substituting Equation (a) into Equation (e), we get the overall transformation of the state vector from station 0 to 2, as

$${}_R\mathbf{S}_2 = \mathbf{U}_2\mathbf{U}_1\mathbf{S}_0 = \mathbf{T}\mathbf{S}_0 \tag{h}$$

with

$$\mathbf{T} = \mathbf{U}_2\mathbf{U}_1 = \mathbf{P}_2\mathbf{F}_2\mathbf{P}_1\mathbf{F}_1 = \begin{bmatrix} t_{11} & t_{12} & t_{13} & t_{14} \\ t_{21} & t_{22} & t_{23} & t_{24} \\ t_{31} & t_{32} & t_{33} & t_{34} \\ t_{41} & t_{42} & t_{43} & t_{44} \end{bmatrix} \tag{i}$$

where

$$t_{11} = 1+0.341\times10^{-5}\omega_{nf}^2;\quad t_{12} = 0.125-0.648\times10^{-6}\omega_{nf}^2;\quad t_{13} = 0.758\times10^{-4}+0.357\times10^{-9}\omega_{nf}^2$$

$$t_{14} = 0.316\times10^{-5}-0.922\times10^{-11}\omega_{nf}^2;\quad t_{21} = 0.136\times10^{-3}\omega_{nf}^2;\quad t_{22} = 1-0.150\times10^{-4}\,\omega_{nf}^2$$

$$t_{23} = 0.121\times10^{-2}-0.896\times10^{-8}\omega_{nf}^2;\quad t_{24} = 0.758\times10^{-4}-0.236\times10^{-9}\omega_{nf}^2$$

$$t_{31} = 0.5\left(0.750\times10^{-1} - 0.273\times10^{-6}\omega_{nf}^2\right)\omega_{nf}^2$$

$$t_{32} = -0.100\times10^{-1}\omega_{nf}^2 - 0.300\times10^{-1}\left(1.0 - 0.727\times10^{-5}\omega_{nf}^2\right)\omega_{nf}^2 + 0.250\left(0.750\times10^{-1} - 0.273\times10^{-6}\omega_{nf}^2\right)\omega_{nf}^2$$

$$t_{33} = -0.485\times10^{-5}\omega_{nf}^2 + \left(1.0 - 0.727\times10^{-5}\omega_{nf}^2\right)\left(1.0 - 0.146\times10^{-4}\omega_{nf}^2\right) + 0.605\times10^{-4}\left(0.750\times10^{-1} - 0.273\times10^{-6}\omega_{nf}^2\right)\omega_{nf}^2$$

$$t_{34} = -0.121\times10^{-6}\omega_{nf}^2 + \left(1.0 - 0.727\times10^{-5}\omega_{nf}^2\right)\left(0.500\times10^{-1} - 0.363\times10^{-6}\omega_{nf}^2\right) + \left(0.750\times10^{-1} - 0.273\times10^{-6}\omega_{nf}^2\right)\left(0.101\times10^{-5}\omega_{nf}^2 + 1\right)$$

$$t_{41} = 2.0\omega_{nf}^2 + 0.5\left(0.136\times10^{-5}\omega_{nf}^2 + 1.0\right)\omega_{nf}^2$$

$$t_{42} = 0.250\omega_{nf}^2 - 0.164\times10^{-5}\omega_{nf}^4 + 0.250\left(0.136\times10^{-5}\omega_{nf}^2 + 1.0\right)\omega_{nf}^2$$

$$t_{43} = 0.970\times10^{-4}\omega_{nf}^2 + 0.546\times10^{-4}\omega_{nf}^2\left(1.0 - 0.146\times10^{-4}\omega_{nf}^2\right) + 0.605\times10^{-4}\left(1.0 + 0.136\times10^{-5}\omega_{nf}^2\right)\omega_{nf}^2$$

$$t_{44} = 0.222\times10^{-5}\omega_{nf}^2 + 0.546\times10^{-4}\omega_{nf}^2\left(0.5\times10^{-1} - 0.363\times10^{-6}\omega_{nf}^2\right) + \left(1 + 0.136\times10^{-5}\omega_{nf}^2\right)\left(1.0 + 0.101\times10^{-5}\omega_{nf}^2\right)$$

The expanded form of Equation (i) has the following form:

$$\left\{\begin{matrix} -y \\ \varphi_x \\ M_{yz} \\ S_y \end{matrix}\right\}_2^R = \begin{bmatrix} t_{11} & t_{12} & t_{13} & t_{14} \\ t_{21} & t_{22} & t_{23} & t_{24} \\ t_{31} & t_{32} & t_{33} & t_{34} \\ t_{41} & t_{42} & t_{43} & t_{44} \end{bmatrix} \left\{\begin{matrix} -y \\ \varphi_x \\ M_{yz} \\ S_y \end{matrix}\right\}_0 \quad \text{(j)}$$

The boundary conditions are as follows: (1) y and φ_x at station 0 are zero, and (2) the free end shear force, S_y, and bending moment, M_{yz}, are zero. On substituting these boundary conditions in Equation (j), we get

$$\left\{\begin{matrix} -y \\ \varphi_x \\ 0 \\ 0 \end{matrix}\right\}_2^R = \begin{bmatrix} t_{11} & t_{12} & t_{13} & t_{14} \\ t_{21} & t_{22} & t_{23} & t_{24} \\ t_{31} & t_{32} & t_{33} & t_{34} \\ t_{41} & t_{42} & t_{43} & t_{44} \end{bmatrix} \left\{\begin{matrix} 0 \\ 0 \\ M_{yz} \\ S_y \end{matrix}\right\}_0 \quad \text{(k)}$$

From the last two expressions of Equation (k), we have

$$\begin{bmatrix} t_{33} & t_{34} \\ t_{43} & t_{44} \end{bmatrix} \begin{Bmatrix} M_{yz} \\ S_y \end{Bmatrix}_0 = \begin{Bmatrix} 0 \\ 0 \end{Bmatrix} \tag{l}$$

Hence, on taking the determinant of Equation (l) equal to zero, the frequency equation is

$$t_{33}(\omega_{nf})t_{44}(\omega_{nf}) - t_{34}(\omega_{nf})t_{43}(\omega_{nf}) = 0$$

which is

$$(3.62\times10^{-23})\omega_{nf}^8 - (1.83\times10^{-16})\omega_{nf}^6 + (2.2\times10^{-10})\omega_{nf}^4 - (4.14\times10^{-5})\omega_{nf}^2 + 1 = 0$$

From this, the transverse natural frequencies can be obtained by the root-searching numerical method described earlier by expressing the frequency equation as a function of the following form:

$$\begin{aligned} f(\omega_{nf}) &= t_{33}(\omega_{nf})t_{44}(\omega_{nf}) - t_{34}(\omega_{nf})t_{43}(\omega_{nf}) \\ &= (3.62\times10^{-23})\omega_{nf}^8 - (1.83\times10^{-16})\omega_{nf}^6 + (2.2\times10^{-10})\omega_{nf}^4 - (4.14\times10^{-5})\omega_{nf}^2 + 1 \end{aligned} \tag{n}$$

Alternatively, through commercial software the roots of equations can be directly obtained. The roots of this function are natural frequencies and are obtained as

$$\omega_{nf_1} = 168.27\,\text{rad/s};\quad \omega_{nf_2} = 442.14\,\text{rad/s};\quad \omega_{nf_3} = 1223.26\,\text{rad/s}\quad \text{and}\quad \omega_{nf_4} = 1826.21\,\text{rad/s}$$

Since herein both translational and rotational displacements are considered, we have four natural frequencies as against two in Example 8.3. For obtaining the mode shape, from the first two equations of (*k*), we have

$$_R\begin{Bmatrix} -y \\ \varphi_x \end{Bmatrix}_2 = \begin{bmatrix} t_{13} & t_{14} \\ t_{23} & t_{24} \end{bmatrix} \begin{Bmatrix} M_{yz} \\ S_y \end{Bmatrix}_0 \tag{o}$$

For a given natural frequency, the first equation of the matrix Equation (l) gives

$$_0M_{yz} = -\frac{t_{34}}{t_{33}}\,{}_0S_y \tag{p}$$

By choosing $_Ry_2 = 1$ as a reference value of the displacement, the first equation of the matrix Equation (m) gives

$$t_{13}M_{yz} + t_{14}S_y = -1 \tag{q}$$

On substituting Equation (p) into Equation (q), the shear force at station 0 can be obtained as

$$-t_{13}\frac{t_{34}}{t_{33}}\,{}_0S_y + t_{14}\,{}_0S_y = 1 \quad \text{or} \quad {}_0S_y = \frac{t_{33}}{t_{14}t_{33} - t_{13}t_{34}} \tag{r}$$

The bending moment at station 0 can be obtained now from Equation (p). Now, we have obtained the state vector at station 0 completely, and using the transformation matrices, the state vectors at all other station can be obtained and are given as

i. *For the first natural frequency*: The state vectors are given as

$$\begin{Bmatrix} -y \\ \varphi_x \\ M_{yz} \\ S_y \end{Bmatrix}_0 = \begin{Bmatrix} 0 \\ 0 \\ -1.91\times 10^4 \\ 8.75\times 10^4 \end{Bmatrix}; \quad {}_R\begin{Bmatrix} -y \\ \varphi_x \\ M_{yz} \\ S_y \end{Bmatrix}_1 = \begin{Bmatrix} -0.212 \\ -8.215 \\ -7.69\times 10^3 \\ 5.64\times 10^4 \end{Bmatrix};$$

$${}_R\begin{Bmatrix} -y \\ \varphi_x \\ M_{yz} \\ S_y \end{Bmatrix}_2 = \begin{Bmatrix} -1.001 \\ -12.265 \\ 0 \\ 0 \end{Bmatrix}$$

From the above state vectors, we have

$$\begin{Bmatrix} -y_0 \\ -y_1 \\ -y_2 \end{Bmatrix} = \begin{Bmatrix} 0 \\ -0.212 \\ -1.001 \end{Bmatrix} \text{ and } \begin{Bmatrix} \varphi_{x0} \\ \varphi_{x1} \\ \varphi_{x2} \end{Bmatrix} = \begin{Bmatrix} 0 \\ -8.215 \\ -12.265 \end{Bmatrix} \equiv \begin{Bmatrix} 0 \\ -0.6698 \\ -1 \end{Bmatrix}$$

Figure 8.28 shows the mode shape for y and φ_x corresponding to the first natural frequency.

ii. *For the second natural frequency*: The state vectors are given as

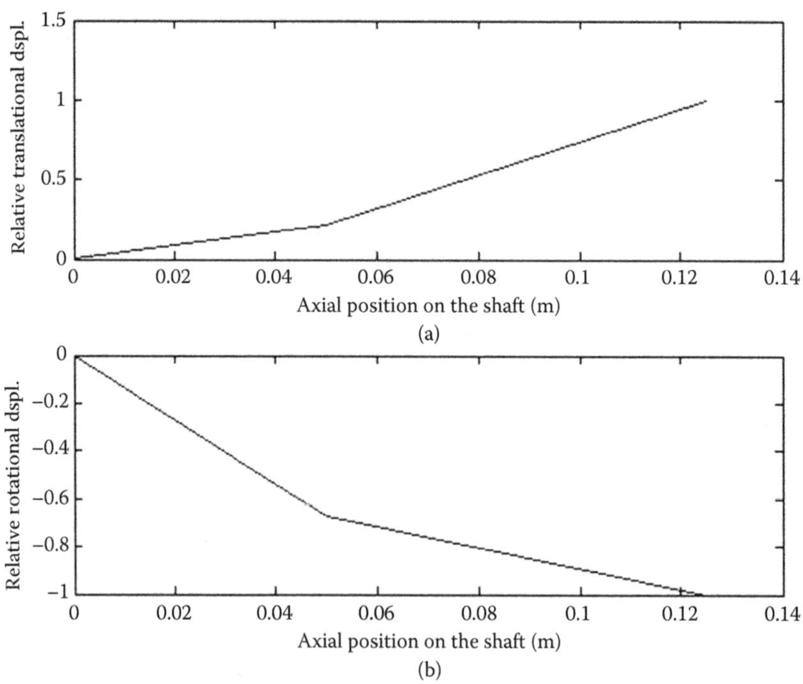

FIGURE 8.28 The mode shapes corresponding to the first natural frequency: (a) for the translational displacement and (b) for the rotational displacement.

$$\begin{Bmatrix} -y \\ \varphi_x \\ M_{yz} \\ S_y \end{Bmatrix}_0 = \begin{Bmatrix} 0 \\ 0 \\ 29.32\times10^4 \\ -209.14\times10^4 \end{Bmatrix}; \quad \begin{Bmatrix} -y \\ \varphi_x \\ M_{yz} \\ S_y \end{Bmatrix}_1^R = \begin{Bmatrix} 3.508 \\ 101.613 \\ -45.41\times10^4 \\ 39.1\times10^4 \end{Bmatrix}; \quad \begin{Bmatrix} -y \\ \varphi_x \\ M_{yz} \\ S_y \end{Bmatrix}_2^R = \begin{Bmatrix} -1 \\ -217.81 \\ 0 \\ 0 \end{Bmatrix}$$

So we have

$$\begin{Bmatrix} -y_0 \\ -y_1 \\ -y_2 \end{Bmatrix} = \begin{Bmatrix} 0 \\ 3.508 \\ -1 \end{Bmatrix}, \text{ and } \begin{Bmatrix} \varphi_{x0} \\ \varphi_{x1} \\ \varphi_{x2} \end{Bmatrix} = \begin{Bmatrix} 0 \\ 101.613 \\ -217.81 \end{Bmatrix} \equiv \begin{Bmatrix} 0 \\ 1 \\ -2.144 \end{Bmatrix}$$

Figure 8.29 shows the mode shape for y and φ_z corresponding to the second natural frequency.

iii. For the third natural frequency: The state vectors are given as

$$\begin{Bmatrix} -y \\ \varphi_x \\ M_{yz} \\ S_y \end{Bmatrix}_0 = \begin{Bmatrix} 0 \\ 0 \\ -16.41\times10^4 \\ 718.54\times10^4 \end{Bmatrix}; \quad \begin{Bmatrix} -y \\ \varphi_x \\ M_{yz} \\ S_y \end{Bmatrix}_1^R = \begin{Bmatrix} -1.463 \\ 35.078 \\ -15.41\times10^4 \\ 299\times10^4 \end{Bmatrix}; \quad \begin{Bmatrix} -y \\ \varphi_x \\ M_{yz} \\ S_y \end{Bmatrix}_2^R = \begin{Bmatrix} -1 \\ 4.677 \\ 0 \\ 0 \end{Bmatrix}$$

From the above state vectors, we have

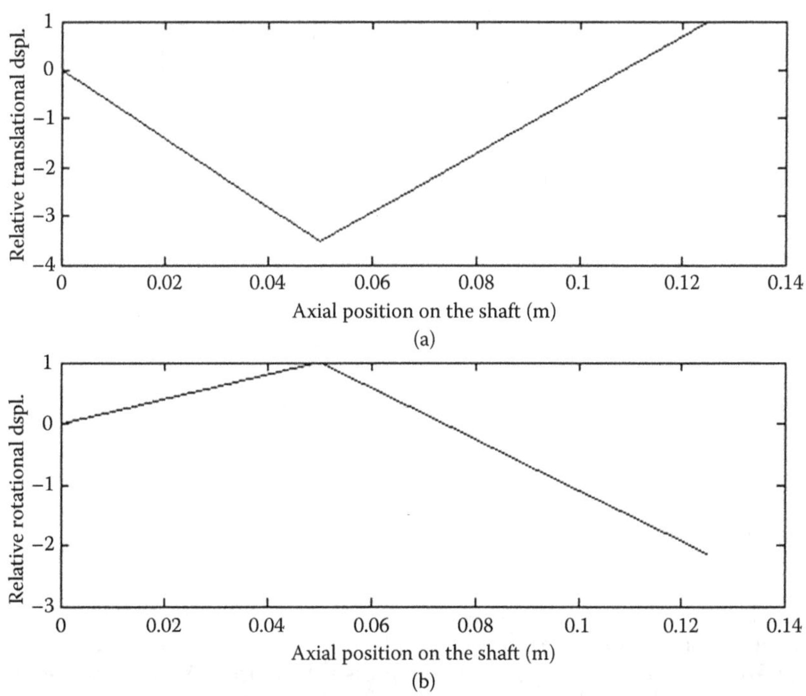

FIGURE 8.29 The mode shapes corresponding to the second natural frequency: (a) for the translational displacement and (b) for the rotational displacement.

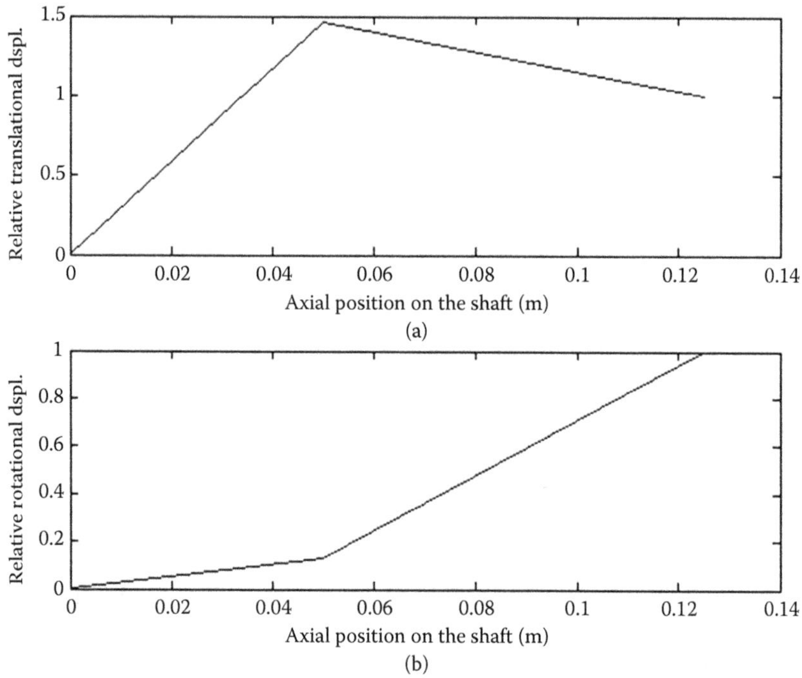

FIGURE 8.30 The mode shapes corresponding to the third natural frequency: (a) for the translational displacement and (b) for the rotational displacement.

$$\begin{Bmatrix} -y_0 \\ -y_1 \\ -y_2 \end{Bmatrix} = \begin{Bmatrix} 0 \\ -1.463 \\ -1 \end{Bmatrix} \quad \text{and} \quad \begin{Bmatrix} \varphi_{x0} \\ \varphi_{x1} \\ \varphi_{x2} \end{Bmatrix} = \begin{Bmatrix} 0 \\ 35.078 \\ 4.677 \end{Bmatrix} \equiv \begin{Bmatrix} 0 \\ 0.133 \\ 1 \end{Bmatrix}$$

Figure 8.30 shows the mode shapes for y and φ_z corresponding to the third natural frequency.

iv. For the fourth natural frequency: The state vectors are given as

$$\begin{Bmatrix} -y \\ \varphi_x \\ M_{yz} \\ S_y \end{Bmatrix}_0 = \begin{Bmatrix} 0 \\ 0 \\ 20.83\times 10^4 \\ -834.24\times 10^4 \end{Bmatrix}; \quad {}_R\begin{Bmatrix} -y \\ \varphi_x \\ M_{yz} \\ S_y \end{Bmatrix}_1 = \begin{Bmatrix} -3.61 \\ 128.45 \\ -42.4\times 10^4 \\ 667\times 10^4 \end{Bmatrix}; \quad {}_R\begin{Bmatrix} -y \\ \varphi_x \\ M_{yz} \\ S_y \end{Bmatrix}_2 = \begin{Bmatrix} -1 \\ 2.283 \\ 0 \\ 0 \end{Bmatrix}$$

From the above state vectors, we have

$$\begin{Bmatrix} -y_0 \\ -y_1 \\ -y_2 \end{Bmatrix} = \begin{Bmatrix} 0 \\ -3.61 \\ -1 \end{Bmatrix} \quad \text{and} \quad \begin{Bmatrix} \varphi_{x0} \\ \varphi_{x1} \\ \varphi_{x2} \end{Bmatrix} = \begin{Bmatrix} 0 \\ 128.45 \\ 2.283 \end{Bmatrix} \equiv \begin{Bmatrix} 0 \\ 1 \\ 0.018 \end{Bmatrix}$$

Figure 8.31 shows the mode shapes for y and φ_z corresponding to the fourth natural frequency. On careful observation of mode shapes, it can be seen that for mode 1 (Figure 8.28a) and mode 2 (Figure 8.29a), there is an increase in the node (zero-crossing) in translational displacement mode shapes. However, in mode 3 (Figure 8.30b) and mode 4 (Figure 8.31b), there is a change of the slope in rotational displacement mode shapes.

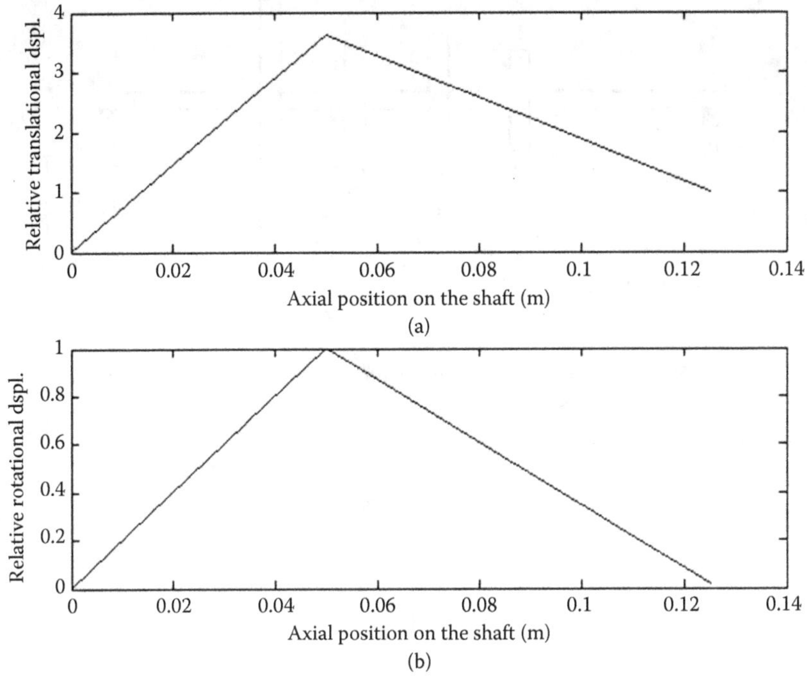

FIGURE 8.31 The mode shapes corresponding to the fourth natural frequency: (a) for the translational displacement and (b) for the rotational displacement.

Example 8.8

Find the transverse natural frequencies and mode shapes of the rotor system shown in Figure 8.32. B_1 and B_2 are bearings, which provide simply supported end conditions, and D_1, D_2, D_3, and D_4 are rigid disks. The shaft is made of steel with Young's modulus $E=2.1\times10^{11}$ N/m^2 and uniform diameter $d=20$ mm. The various shaft lengths are as follows: $B_1D_1=150$ mm, $D_1D_2=50$ mm, $D_2D_3=50$ mm, $D_3D_4=50$ mm, and $D_4B_2=150$ mm. The masses of the disks are $m_1=4$ kg, $m_2=5$ kg, $m_3=6$ kg, and $m_4=7$ kg. Consider the shaft as massless. Consider the disks as thin and take the diameter of the disks as $d_1=8\,\text{cm}$, $d_2=10\,\text{cm}$, $d_3=12\,\text{cm}$, and $d_4=14\,\text{cm}$; neglect the gyroscopic effects.

Solution:

The disks have the following data:

$$m_1=4\text{ kg},\quad m_2=5\text{ kg},\quad m_3=6\text{ kg},\quad m_4=7\text{ kg}$$

$$d_1=0.08\text{ m},\quad d_2=0.1\text{ m},\quad d_3=0.12\text{ m},\quad d_4=0.14\text{ m}$$

$$I_{d_1}=\frac{1}{4}m_1r_1^2=\frac{1}{4}\times4\times0.04^2=0.0016\text{ kg-m}^2,\quad I_{d_2}=\frac{1}{4}\times5\times0.05^2=0.003125\text{ kg-m}^2$$

$$I_{d3}=\frac{1}{4}\times6\times0.06^2=0.0054\text{ kg-m}^2,\quad I_{d4}=\frac{1}{4}\times7\times0.07^2=0.008576\text{ kg-m}^2$$

The shaft has $EI=1649.34$ N-m^2 and the following dimensions according to station numbers:

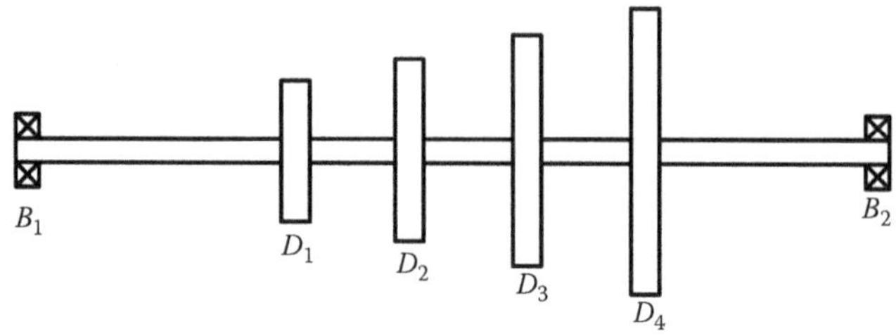

FIGURE 8.32 A multidisc rotor system.

$$l_1 = 150 \text{ mm}, \quad l_2 = 50 \text{ mm}, \quad l_3 = 50 \text{ mm}, \quad l_4 = 50 \text{ mm}, \quad l_5 = 150 \text{ mm}$$

Now, the overall transformation of the state vector can be written as

$$_R\mathbf{S}_5 = \mathbf{T}\mathbf{S}_0 \tag{a}$$

with

$$\mathbf{T} = \mathbf{F}_5\mathbf{P}_4\mathbf{F}_4\mathbf{P}_3\mathbf{F}_3\mathbf{P}_2\mathbf{F}_2\mathbf{P}_1\mathbf{F}_1 \tag{b}$$

$$\mathbf{P}_i = \begin{bmatrix} 1 & 0 & 0 & 0 \\ 0 & 1 & 0 & 0 \\ 0 & -\omega_{nf}^2 I_{d_i} & 1 & 0 \\ m_i\omega_{nf}^2 & 0 & 0 & 1 \end{bmatrix}; \quad \mathbf{F}_i = \begin{bmatrix} 1 & l & l^2/2EI & l^3/6EI \\ 0 & 1 & l/EI & l^2/2EI \\ 0 & 0 & 1 & l \\ 0 & 0 & 0 & 1 \end{bmatrix}_i; \quad \mathbf{S} = \begin{Bmatrix} -y \\ \varphi_x \\ M_{yz} \\ S_y \end{Bmatrix} \tag{c}$$

From Table 8.3, we have the eigenvalue problem for the simply supported boundary conditions as

$$\begin{bmatrix} t_{1,2} & t_{1,4} \\ t_{3,2} & t_{3,4} \end{bmatrix} \begin{Bmatrix} \varphi_x \\ S_y \end{Bmatrix}_0 = \begin{Bmatrix} 0 \\ 0 \end{Bmatrix} \tag{d}$$

which gives the frequency equation as

$$f(\omega_{nf}) = \left\{ t_{1,2}(\omega_{nf}) t_{3,4}(\omega_{nf}) - t_{1,4}(\omega_{nf}) t_{3,2}(\omega_{nf}) \right\} = 0 \tag{e}$$

On solving the roots of the above function by the root-searching method, we obtain the following natural frequencies:

$$\omega_{nf_1} = 215 \text{ rad/s}, \quad \omega_{nf_2} = 1076 \text{ rad/s}, \quad \omega_{nf_3} = 2470 \text{ rad/s}, \quad \omega_{nf_4} = 3847 \text{ rad/s}$$
$$\omega_{nf_5} = 6398 \text{ rad/s}, \quad \omega_{nf_6} = 8885 \text{ rad/s}, \quad \omega_{nf_7} = 10{,}776 \text{ rad/s}, \quad \omega_{nf_8} = 13{,}282 \text{ rad/s}$$

From Table 8.3, the eigenvector can be obtained from the following equation:

$$_R\begin{Bmatrix} \varphi_x \\ S_y \end{Bmatrix}_5 = \begin{bmatrix} t_{2,2} & t_{2,4} \\ t_{4,2} & t_{4,4} \end{bmatrix} \begin{Bmatrix} \varphi_x \\ S_y \end{Bmatrix}_0 \tag{f}$$

For example, for a particular natural frequency, ω_{nf_1}, from Equations (d) and (f), we have

$$\begin{bmatrix} t_{1,2}(\omega_{nf1}) & t_{1,4}(\omega_{nf1}) \\ t_{3,2}(\omega_{nf1}) & t_{3,4}(\omega_{nf1}) \end{bmatrix} \begin{Bmatrix} \varphi_x \\ S_y \end{Bmatrix}_0 = \begin{Bmatrix} 0 \\ 0 \end{Bmatrix} \quad \text{and}$$

$$_R\begin{Bmatrix} \varphi_x \\ S_y \end{Bmatrix}_5 = \begin{bmatrix} t_{2,2}(\omega_{nf1}) & t_{2,4}(\omega_{nf1}) \\ t_{4,2}(\omega_{nf1}) & t_{4,4}(\omega_{nf1}) \end{bmatrix} \begin{Bmatrix} \varphi_x \\ S_y \end{Bmatrix}_0 \tag{g}$$

Now on choosing $\varphi_x = 1$ as a reference value, from Equation (g) and noting the boundary condition and value of ω_{nf1}, we get the state vector at the zeroth and fifth stations as

$$\mathbf{S}_0 = \begin{Bmatrix} -y \\ \varphi_x \\ M_{yz} \\ S_y \end{Bmatrix}_0 = \begin{Bmatrix} 0 \\ 1 \\ 0 \\ -66{,}905.27 \end{Bmatrix} \quad \text{and} \quad {}_R\mathbf{S}_5 = \begin{Bmatrix} 0 \\ -1.033338 \\ 0 \\ 72{,}938.64 \end{Bmatrix}$$

At other stations, the state vectors can also be obtained as

$${}_R\mathbf{S}_1 = \mathbf{P}_1\mathbf{F}_1\mathbf{S}_0 = \begin{Bmatrix} 0.127182 \\ 0.543645 \\ -10{,}076.00 \\ -43{,}389.27 \end{Bmatrix}; \quad {}_R\mathbf{S}_2 = \mathbf{P}_2\mathbf{F}_2\,{}_R\mathbf{S}_1 = \begin{Bmatrix} 0.146180 \\ 0.205305 \\ -12{,}275.12 \\ -9603.40 \end{Bmatrix}$$

and

$${}_R\mathbf{S}_3 = \mathbf{P}_2\mathbf{F}_2\,{}_R\mathbf{S}_2 = \begin{Bmatrix} 0.147021 \\ -0.174094 \\ -12{,}711.83 \\ 31{,}172.87 \end{Bmatrix}; \quad {}_R\mathbf{S}_4 = \mathbf{P}_3\mathbf{F}_3\,{}_R\mathbf{S}_3 = \begin{Bmatrix} 0.129076 \\ -0.535830 \\ -10{,}940.80 \\ 72{,}938.64 \end{Bmatrix}$$

These state vectors for the fundamental natural frequency are summarized in Table 8.5. The mode shapes for the translational and rotational displacements are extracted in the table and are drawn as shown in Figure 8.33a and b, respectively. Along the same lines, other state vectors corresponding to various natural frequencies can be obtained and are given in Tables 8.6 through 8.12; the corresponding mode shapes from the second to the fifth modes are shown in Figure 8.33c and d.

TABLE 8.5
Relative Values of State Vectors at Various Stations for the Fundamental Natural Frequency

	State Vectors at the First Natural Frequency, 215 rad/s					
	$\mathbf{S}_0$	$\mathbf{S}_1$	$\mathbf{S}_2$	$\mathbf{S}_3$	$\mathbf{S}_4$	$\mathbf{S}_5$
y	0	0.127182	0.146180	0.147021	0.129076	0
φ_x	1.00	0.543645	0.205305	−0.174094	−0.535830	−1.033338
M_{yz}	0	−10,076.00	−12,275.12	−12,711.83	−10,940.80	0
S_y	−66,905.27	−43,389.27	−9603.40	31,172.87	72,938.64	72,938.64

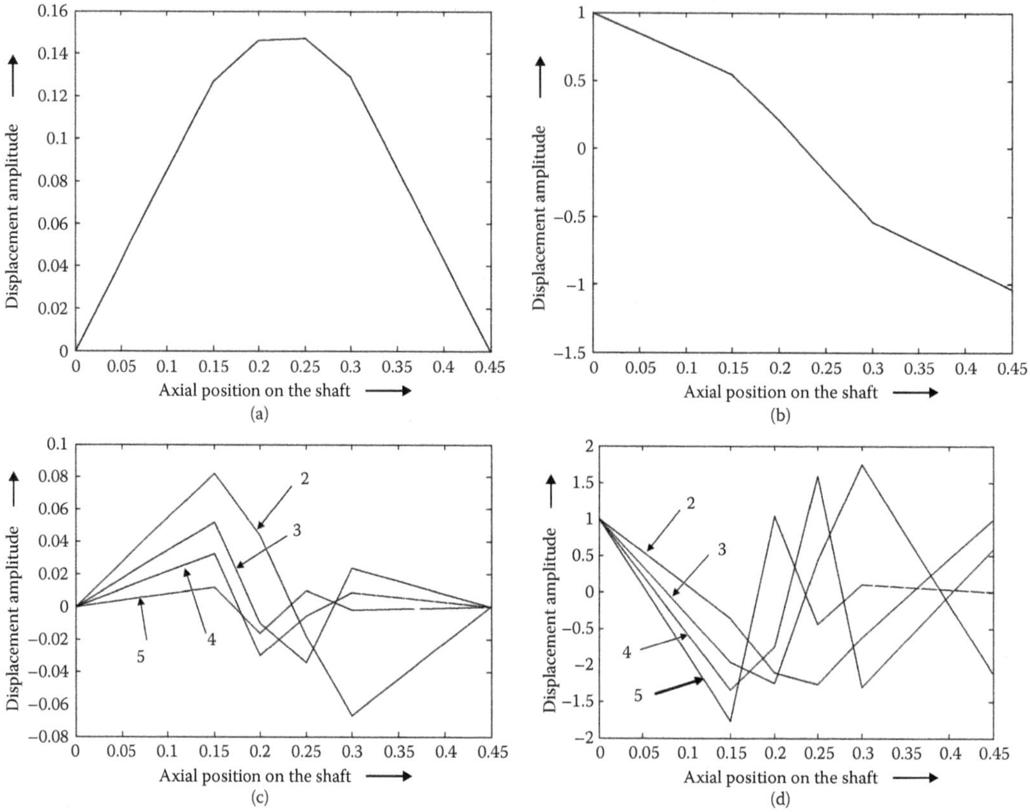

FIGURE 8.33 (a) The fundamental mode shape of translational displacements, (b) the fundamental mode shape of rotational (slope) displacements, (c) higher mode shapes (second to fifth) of translational displacements, and (d) higher mode shapes (second–fifth) of rotational (slope) displacements.

Example 8.9

Obtain the bending critical speeds of a rotor system as shown in Figure 8.34. Take the mass of the disk, $m = 5$ kg, and its diametral mass moment of inertia, $I_d = 0.02$ kg-m^2. The lengths of the shaft segments are $a = 0.3$ m and $b = 0.7$ m, and the diameter of the shaft is 0.01 m. Neglect the gyroscopic effects. $E = 2.1 \times 10^{11}$ N/m^2.

Solution:

Figure 8.35 shows the station numbering and free body diagrams of various segments and supports. Additional difficulty in the present problem in the intermediate support, which requires additional consideration in the transfer matrix analysis. It is due to an unknown support reaction, which comes in the formulation.

For shaft segment (1) as shown in Figure 8.35b, the state vector can be related as

$$ {}_L\mathbf{S}_1 = \mathbf{F}_1\mathbf{P}_1\mathbf{S}_0 = \mathbf{U}_1\mathbf{S}_0 \tag{a} $$

with

$$ \mathbf{U}_1 = \mathbf{F}_1\mathbf{P}_1 $$

TABLE 8.6
Relative Values of State Vectors at Various Stations for the Second Natural Frequency

State Vectors at the Second Natural Frequency, 1076 rad/s						
	S_0	S_1	S_2	S_3	S_4	S_5
y	0	0.08201	0.04415	−0.01806	−0.06734	0
φ_x	1.00	−0.35964	−1.10908	−1.26954	−0.62340	0.98042
M_{yz}	0	−29,234.01	−16,196.41	13,547.38	35,270.03	0
S_y	−19,9334.91	18,0497.74	43,6132.02	31,0671.00	−235,133.53	−235,133.53

TABLE 8.7
Relative Values of State Vectors at Various Stations for the Third Natural Frequency

State Vectors at the Third Natural Frequency, 2470 rad/s						
	S_0	S_1	S_2	S_3	S_4	S_5
y	0	0.05172	−0.009865	−0.03457	0.02368	0
φ_x	1.00	−0.96553	−1.25193	0.43362	1.7472	−1.11239
M_{yz}	0	−33,799.42	38,772.86	58,144.09	−62,887.40	0
S_y	−28,8162.68	974,074.79	673,141.85	−592,459.95	419,249.34	419,249.34

TABLE 8.8
Relative Values of State Vectors at Various Stations for the Fourth Natural Frequency

State Vectors at the Fourth Natural Frequency, 3847 rad/s						
	S_0	S_1	S_2	S_3	S_4	S_5
y	0	0.0323	−0.03034	−0.0055	0.008635	0
φ_x	1.00	−1.3525	−0.7590	1.5790	−1.30565	0.57196
M_{yz}	0	−19,708.29	93,968.80	−65,909.34	41,291.063	0
S_y	−344,903.01	1,571,465.29	−673,757.87	−1,169,875.76	−275,273.75	−275,273.75

TABLE 8.9
Relative Values of State Vectors at Various Stations for the Fifth Natural Frequency

State Vectors at the Fifth Natural Frequency, 6398 rad/s						
	S_0	S_1	S_2	S_3	S_4	S_5
y	0	0.01159	−0.01634	0.01017	−0.002194	0
φ_x	1.00	−1.7681	1.0278	−0.43533	0.103774	−0.010208
M_{yz}	0	54,931.80	−1947.42	1647.72	−2506.62	0
S_y	−405,838.58	1,491,974.8	−1,852,691.57	645,435.50	16,710.77	16,710.77

TABLE 8.10
Relative Values of State Vectors at Various Stations for the Sixth Natural Frequency

State Vectors at the Sixth Natural Frequency, 8885 rad/s

	S_0	S_1	S_2	S_3	S_4	S_5
y	0	0.6159	2.78	8.48	−9.555	0
φ_x	1.00	10.31	125.4	0.0000004289	19.06	1.01
M_{yz}	0	−1,09,8548.67	−2,225,8104.4	−140,562,945.6	−4,168,712.59	0
S_y	1,366,381.12	195,881,901.58	1,290,696,326.5	5,308,048,574.2	27,791,417.25	27,791,417.25

TABLE 8.11
Relative Values of State Vectors at Various Stations for the Seventh Natural Frequency

State Vectors at the Seventh Natural Frequency, 10776 rad/s

	S_0	S_1	S_2	S_3	S_4	S_5
y	0	0.01552	0.2102	−0.2306	0.038953	0
φ_x	1.00	−1.6894	11.200	3.7243	−0.976793	0.035605
M_{yz}	0	254,747.85	−3,468,821.24	640,227.57	22,263.86	0
S_y	-394,294.80	6,818,067.54	128,888,160.07	-31,812,064.91	−148,425.72	−148,425.72

TABLE 8.12
Relative Values of State Vectors at Various Stations for the Eighth Natural Frequency

State Vectors at the Eighth Natural Frequency, 13282 rad/s

	S_0	S_1	S_2	S_3	S_4	S_5
y	0	−0.03468	0.02929	−0.001166	−0.0007505	0
φ_x	1.0	−2.69371	−1.06703	0.05671	−0.1437	8.9769
M_{yz}	0	679,094.86	16,463.86	3645.74	200,574.14	0
S_y	−541,528.33	−25,017,431.96	824,200.49	−410,329.07	−1,337,160.94	−1,337,160.94

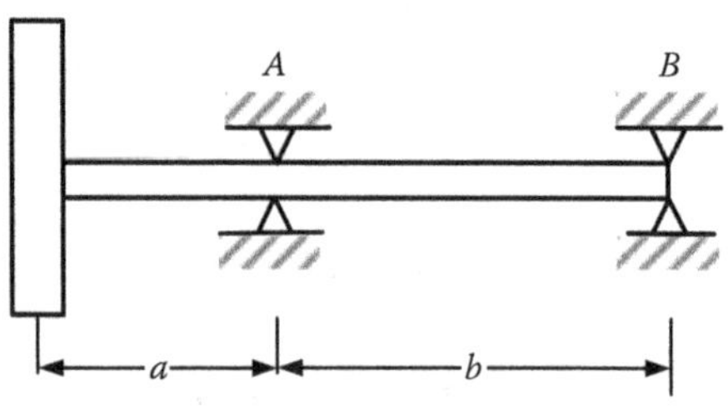

FIGURE 8.34 An overhang rotor system.

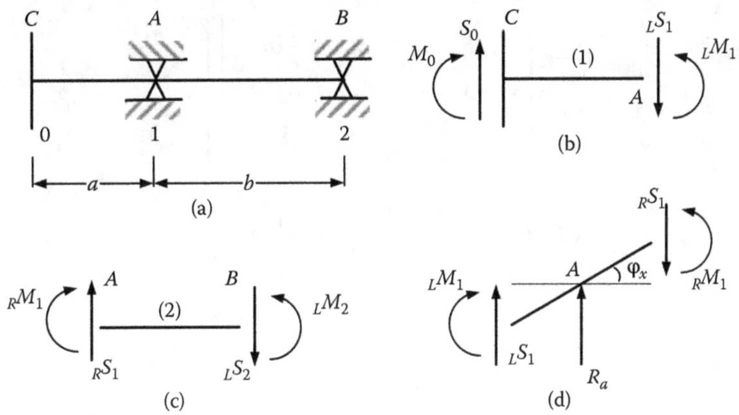

FIGURE 8.35 Free body diagrams of rotor segments.

For the second shaft segment, the state vector can be written as

$$_L\mathbf{S}_1 = \mathbf{F}_2\,{}_R\mathbf{S}_1 = \mathrm{U}_2\,{}_R\mathbf{S}_1 \tag{b}$$

with

$$\mathbf{U}_2 = \mathbf{F}_2$$

The boundary conditions for the overhang rotor, as shown in Figure 8.35a, can be written as

$$\text{At node 0:}\quad S_0 = M_0 = 0 \tag{c}$$

$$\text{At node 1:}\quad y_1 = 0;\quad {}_LM_1 = {}_RM_1 \neq 0 \quad \text{and} \quad {}_LS_1 - {}_RS_1 + R_a = 0 \tag{d}$$

where R_a is *the support reaction* at bearing A.

$$\text{At node 2:}\quad y_2 = 0 \quad \text{and} \quad {}_RM_2 = {}_LM_2 = 0 \tag{e}$$

From the first shaft segment, i.e., Equation (a), on application of end conditions of Equation (c), we have

$$_L\begin{Bmatrix} 0 \\ \varphi \\ M \\ S \end{Bmatrix}_1 = \begin{bmatrix} u_{11} & u_{12} & u_{13} & u_{14} \\ u_{21} & u_{22} & u_{23} & u_{24} \\ u_{31} & u_{32} & u_{33} & u_{34} \\ u_{41} & u_{42} & u_{43} & u_{44} \end{bmatrix}_1 \begin{Bmatrix} -y \\ \varphi \\ 0 \\ 0 \end{Bmatrix}_0 \tag{f}$$

with

$$\mathbf{U}_1 = \begin{bmatrix} u_{11} & u_{12} & u_{13} & u_{14} \\ u_{21} & u_{22} & u_{23} & u_{24} \\ u_{31} & u_{32} & u_{33} & u_{34} \\ u_{41} & u_{42} & u_{43} & u_{44} \end{bmatrix} = \begin{bmatrix} 1 & l & 0.5\alpha l & \frac{1}{6}\alpha l^2 \\ 0 & 1 & \alpha & 0.5\alpha l \\ 0 & 0 & 1 & l \\ 0 & 0 & 0 & 1 \end{bmatrix}_1 \begin{bmatrix} 1 & 0 & 0 & 0 \\ 0 & 1 & 0 & 0 \\ 0 & -\omega_{nf}^2 I_d & 1 & 0 \\ m\omega_{nf}^2 & 0 & 0 & 1 \end{bmatrix}_1$$

$$= \begin{bmatrix} 1+\frac{1}{6}m\omega_{nf}^2\alpha l^2 & l-0.5I_d\omega_{nf}^2 l\alpha & 0.5\alpha l & \frac{1}{6}\alpha l^2 \\ 0.5m\omega_{nf}^2\alpha l & 1-I_d\omega_{nf}^2\alpha & \alpha & 0.5\alpha l \\ m\omega_{nf}^2 l & -I_d\omega_{nf}^2 & 1 & l \\ m\omega_{nf}^2 & 0 & 0 & 1 \end{bmatrix}_1 \tag{g}$$

and

$$\alpha = \frac{l}{EI} \tag{h}$$

From Equation (f), the first set of equations gives

$$0 = -u_{11}y_0 + u_{12}\varphi_0 \Rightarrow \varphi_0 = \frac{u_{11}}{u_{12}}y_0 \tag{i}$$

On using Equation (i) in Equation (f), we get the state vectors on the left of station 1 as

$$\begin{aligned} {}_L y_1 &= 0 \\ {}_L\varphi_1 &= -u_{21}y_0 + u_{22}\frac{u_{11}}{u_{12}}y_0 = \left(-u_{21} + \frac{u_{22}u_{11}}{u_{12}}\right)y_0 \\ {}_L M_1 &= -u_{31}y_0 + u_{32}\frac{u_{11}}{u_{12}}y_0 = \left(-u_{31} + \frac{u_{32}u_{11}}{u_{12}}\right)y_0 \\ {}_L S_1 &= -u_{41}y_0 + u_{42}\frac{u_{11}}{u_{12}}y_0 = \left(-u_{41} + \frac{u_{42}u_{11}}{u_{12}}\right)y_0 \end{aligned} \tag{j}$$

Now noting Equations (d) and (j), we can obtain the state vectors on the right of station 1 as

$$\begin{aligned} {}_R y_1 &= {}_L y_1 = 0 \\ {}_R\varphi_1 &= {}_L\varphi_1 = \left(-u_{21} + \frac{u_{22}u_{11}}{u_{12}}\right)y_0 \\ {}_R M_1 &= {}_L M_1 = \left(-u_{31} + \frac{u_{32}u_{11}}{u_{12}}\right)y_0 \\ {}_R S_1 &= {}_L S_1 + R_a = \left(-u_{41} + \frac{u_{42}u_{11}}{u_{12}}\right)y_0 + R_a \end{aligned} \tag{k}$$

Equation (k) can be written as

$$\begin{Bmatrix} y \\ \varphi \\ M \\ S \end{Bmatrix}_1^R = \begin{bmatrix} 0 & 0 \\ \bar{u}_{21} & 0 \\ \bar{u}_{31} & 0 \\ \bar{u}_{41} & 1 \end{bmatrix} \begin{Bmatrix} -y_0 \\ R_a \end{Bmatrix} \tag{l}$$

with

$$\bar{u}_{21} = u_{21} - \frac{u_{22}u_{11}}{u_{12}}; \quad \bar{u}_{31} = u_{31} - \frac{u_{32}u_{11}}{u_{12}} \quad \text{and} \quad \bar{u}_{41} = u_{41} - \frac{u_{42}u_{11}}{u_{12}}$$

Equation (l) can be written in a standard form as

$$_R\mathbf{S}_1 = \bar{\mathbf{U}}_1\bar{\mathbf{S}}_1 \tag{m}$$

with

$$\bar{\mathbf{U}}_1 = \begin{bmatrix} 0 & 0 & 0 & 0 \\ \bar{u}_{21} & 0 & 0 & 0 \\ \bar{u}_{31} & 0 & 0 & 0 \\ \bar{u}_{41} & 0 & 0 & 1 \end{bmatrix}; \quad \text{and} \quad \bar{\mathbf{S}}_1 = \begin{Bmatrix} -y_0 \\ 0 \\ 0 \\ R_a \end{Bmatrix}$$

For the second shaft segment, i.e., Equation (b) and noting Equation (m), we have

$$_L\mathbf{S}_2 = \mathbf{U}_2\bar{\mathbf{U}}_1\bar{\mathbf{S}}_1 = \begin{bmatrix} 1 & l & 0.5\alpha l & \frac{1}{6}l^2\alpha \\ 0 & 1 & \alpha & 0.5\alpha l \\ 0 & 0 & 1 & l \\ 0 & 0 & 0 & 1 \end{bmatrix}_2 \begin{bmatrix} 0 & 0 & 0 & 0 \\ \bar{u}_{21} & 0 & 0 & 0 \\ \bar{u}_{31} & 0 & 0 & 0 \\ \bar{u}_{41} & 0 & 0 & 1 \end{bmatrix} \begin{Bmatrix} -y_0 \\ 0 \\ 0 \\ R_a \end{Bmatrix} \tag{n}$$

On simplifying Equation (n), we get

$$_L\begin{Bmatrix} y \\ \varphi_x \\ M_{yz} \\ S_y \end{Bmatrix}_2 = \begin{bmatrix} \left(l\bar{u}_{21} + 0.5\alpha l\bar{u}_{31} + \frac{1}{6}\alpha l^2\bar{u}_{41}\right) & 0 & 0 & \frac{1}{6}\alpha l^2 \\ \left(\bar{u}_{21} + \alpha\bar{u}_{31} + 0.5\alpha l\bar{u}_{41}\right) & 0 & 0 & 0.5\alpha l \\ \left(\bar{u}_{31} + l\bar{u}_{41}\right) & 0 & 0 & l \\ \left(\bar{u}_{41}\right) & 0 & 0 & 1 \end{bmatrix}_2 \begin{Bmatrix} -y_0 \\ 0 \\ 0 \\ R_a \end{Bmatrix} \tag{o}$$

On expanding Equation (o) and noting the boundary conditions (e), we have

$$\begin{aligned} _Ry_2 = {_Ly_2} = 0 &= \left(l_2\bar{u}_{21} + 0.5\alpha_2 l_2\bar{u}_{31} + \frac{1}{6}\alpha_2 l_2^2\bar{u}_{41}\right)(-y_0) + \frac{1}{6}\alpha_2 l_2^2 R_a \\ _R\varphi_2 = {_L\varphi_2} &= \left(\bar{u}_{21} + \alpha_2\bar{u}_{31} + 0.5\alpha_2 l_2\bar{u}_{41}\right)(-y_0) + 0.5\alpha_2 l_2 R_a \\ _RM_2 = {_LM_2} = 0 &= \left(\bar{u}_{31} + l_2\bar{u}_{41}\right)(-y_0) + l_2 R_a \\ _RS_2 = {_LS_2} = R_b &= \bar{u}_{41}(-y_0) + R_a \end{aligned} \tag{p}$$

From the first equation of Equation (p), we have

$$R_a = \frac{y_0\left(l_2\bar{u}_{21} + 0.5\alpha_2 l_2\bar{u}_{31} + \frac{1}{6}\alpha_2 l_2^2\bar{u}_{41}\right)}{\frac{1}{6}\alpha_2 l_2^2} \tag{q}$$

On substituting Equation (q) in the third equation of the set of Equations (p), and noting that $y_0 \neq 0$, we get

$$-\left(\bar{u}_{31}+l_2\bar{u}_{41}\right)+\frac{l_2\left(l_2\bar{u}_{21}+0.5\alpha_2 l_2\bar{u}_{31}+\frac{1}{6}\alpha_2 l_2^2\bar{u}_{41}\right)}{\frac{1}{6}\alpha_2 l_2^2}=0 \tag{r}$$

On simplification of Equation (r), we get

$$\bar{u}_{21}+\frac{1}{3}\alpha_2\bar{u}_{31}=0 \tag{s}$$

which is the frequency equation. On substituting variables defined in Equation (l) into Equation (s), we have

$$\left(u_{21}-\frac{u_{22}u_{11}}{u_{12}}\right)+\frac{1}{3}\alpha_2\left(u_{31}-\frac{u_{32}u_{11}}{u_{12}}\right)=0 \tag{t}$$

In view of the variables defined in Equations (g) and (h), the frequency Equation (t) becomes

$$m\omega_{nf}^2\frac{l_1^2}{2EI}-\frac{\left(1+\omega_{nf}^2 m\frac{l_1^3}{6EI}\right)\left(1-\omega_{nf}^2 I_d\frac{l_1}{EI}\right)}{\left(l_1-\omega_{nf}^2 I_d\frac{l_1^2}{2EI}\right)}+\frac{l_2}{3EI}\left(m\omega_{nf}^2 l_1\right)-\frac{l_2}{3EI}\frac{\left(-\omega_{nf}^2 I_d\right)\left(1+\omega_{nf}^2 m\frac{l_1^3}{6EI}\right)}{\left(l_1-\omega_{nf}^2 I_d\frac{l_1^2}{2EI}\right)}=0$$

which can be simplified as

$$mI_d l_1^3\left(3l_1+4l_2\right)\omega_{nf}^4-6EI\left(2ml_1^3+6I_d l_1+2I_d l_2+2ml_1^2 l_2\right)\omega_{nf}^2+36(EI)^2=0 \tag{u}$$

For the present problem, we have

$$I_d=0.02\ \text{m}^4;\quad l_1=0.3\ \text{m};\quad l_2=0.7\ \text{m};\quad m=5\ \text{kg};\quad d=0.01\ \text{m}$$

$$I=\frac{\pi}{64}(0.01)^4=4.909\times10^{-10}\ \text{m}^4;\quad EI=1.031\times10^6\ \text{Nm}^2$$

On substituting in Equation (u), we get

$$\omega_{nf}^4-5.97\times10^4\omega_{nf}^2+3.83\times10^7=0$$

which gives

$$\omega_{nf_1}=25.47\ \text{rad/s}\quad\text{and}\quad\omega_{nf_2}=243.00\ \text{rad/s}$$

The mode shapes can be obtained by using transfer matrices between the various intermediate stations derived in Equations (q), (p), (o), (m), (b), and (a). For reference, $y_0=1$ may be chosen. This is left to readers as an exercise.

An alternative method for the above example is explained now. A transformation matrix to transform the state vector from the left of a support to the right of the support can be developed as follows (refer to Equation (d)):

$$_R\begin{Bmatrix}-y\\ \varphi\\ M\\ S\\ 1\end{Bmatrix}_1=\begin{bmatrix}1&0&0&0&0\\0&1&0&0&0\\0&0&1&0&0\\0&0&0&1&R_A\\0&0&0&0&1\end{bmatrix}_L\begin{Bmatrix}-y\\ \varphi\\ M\\ S\\ 1\end{Bmatrix}_1\quad\text{or}\quad {}_R\mathbf{S}_1^*=\mathbf{U}_{spA}^*\,{}_L\mathbf{S}_1^* \tag{v}$$

where now all vectors and matrices are modified to accommodate the reaction force from the support. Modifying all other transformation in Equations (a) and (b), we get the modified overall transformation as

$$ {}_L\mathbf{S}_2^* = \mathbf{U}_2^*\mathbf{U}_{spA}^*\mathbf{U}_1^*\mathbf{S}_0^* = \mathbf{T}^*\mathbf{S}_0^* \tag{w}$$

with

$$\mathbf{U}_{spA}^*\mathbf{U}_1^* = \begin{bmatrix} 1 & 0 & 0 & 0 & 0 \\ 0 & 1 & 0 & 0 & 0 \\ 0 & 0 & 1 & 0 & 0 \\ 0 & 0 & 0 & 1 & R_A \\ 0 & 0 & 0 & 0 & 1 \end{bmatrix} \begin{bmatrix} u_{11} & u_{12} & u_{13} & u_{14} & 0 \\ u_{11} & u_{22} & u_{23} & u_{24} & 0 \\ u_{31} & u_{32} & u_{33} & u_{34} & 0 \\ u_{41} & u_{42} & u_{43} & u_{44} & 0 \\ 0 & 0 & 0 & 0 & 1 \end{bmatrix} = \begin{bmatrix} u_{11} & u_{12} & u_{13} & u_{14} & 0 \\ u_{11} & u_{22} & u_{23} & u_{24} & 0 \\ u_{31} & u_{32} & u_{33} & u_{34} & 0 \\ u_{41} & u_{42} & u_{43} & u_{44} & R_A \\ 0 & 0 & 0 & 0 & 1 \end{bmatrix}$$

$$\mathbf{T}^* = \mathbf{U}_2^*\mathbf{U}_{spA}^*\mathbf{U}_1^*$$

$$= \begin{bmatrix} \bar{u}_{11} & \bar{u}_{12} & \bar{u}_{13} & \bar{u}_{14} & 0 \\ \bar{u}_{11} & \bar{u}_{22} & \bar{u}_{23} & \bar{u}_{24} & 0 \\ \bar{u}_{31} & \bar{u}_{32} & \bar{u}_{33} & \bar{u}_{34} & 0 \\ \bar{u}_{41} & \bar{u}_{42} & \bar{u}_{43} & \bar{u}_{44} & 0 \\ 0 & 0 & 0 & 0 & 1 \end{bmatrix} \begin{bmatrix} u_{11} & u_{12} & u_{13} & u_{14} & 0 \\ u_{11} & u_{22} & u_{23} & u_{24} & 0 \\ u_{31} & u_{32} & u_{33} & u_{34} & 0 \\ u_{41} & u_{42} & u_{43} & u_{44} & R_A \\ 0 & 0 & 0 & 0 & 1 \end{bmatrix} = \begin{bmatrix} c_{11} & c_{12} & c_{13} & c_{14} & \bar{u}_{14}R_A \\ c_{11} & c_{22} & c_{23} & c_{24} & \bar{u}_{24}R_A \\ c_{31} & c_{32} & c_{33} & c_{34} & \bar{u}_{34}R_A \\ c_{41} & c_{42} & c_{43} & c_{44} & \bar{u}_{44}R_A \\ 0 & 0 & 0 & 0 & 1 \end{bmatrix}$$

which can be expanded as

$$\left\{ \begin{matrix} -y \\ \varphi \\ M \\ S \\ 1 \end{matrix} \right\}_{L\,2} = \begin{bmatrix} t_{11} & t_{12} & t_{13} & t_{14} & t_{15}R_A \\ t_{21} & t_{22} & t_{23} & t_{24} & t_{25}R_A \\ t_{31} & t_{32} & t_{33} & t_{34} & t_{35}R_A \\ t_{41} & t_{42} & t_{43} & t_{44} & R_A \\ 0 & 0 & 0 & 0 & 1 \end{bmatrix} \left\{ \begin{matrix} -y \\ \varphi \\ M \\ S \\ 1 \end{matrix} \right\}_0 \tag{x}$$

The boundary conditions given in Equations (c) and (e) can be applied in Equation (x) to get

$$\left\{ \begin{matrix} 0 \\ \varphi \\ 0 \\ S \\ 1 \end{matrix} \right\}_{L\,2} = \begin{bmatrix} t_{11} & t_{12} & t_{13} & t_{14} & t_{15}R_A \\ t_{21} & t_{22} & t_{23} & t_{24} & t_{25}R_A \\ t_{31} & t_{32} & t_{33} & t_{34} & t_{35}R_A \\ t_{41} & t_{42} & t_{43} & t_{44} & R_A \\ 0 & 0 & 0 & 0 & 1 \end{bmatrix} \left\{ \begin{matrix} -y \\ \varphi \\ 0 \\ 0 \\ 1 \end{matrix} \right\}_0 \tag{y}$$

which gives

$$\left\{ \begin{matrix} 0 \\ 0 \end{matrix} \right\} = \begin{bmatrix} t_{11} & t_{12} \\ t_{31} & t_{32} \end{bmatrix} \left\{ \begin{matrix} -y \\ \varphi \end{matrix} \right\} + \left\{ \begin{matrix} t_{15} \\ t_{35} \end{matrix} \right\} R_A \quad \text{and} \quad {}_L\left\{ \begin{matrix} \varphi \\ S \end{matrix} \right\}_2 = \begin{bmatrix} t_{21} & t_{22} \\ t_{41} & t_{42} \end{bmatrix} \left\{ \begin{matrix} -y \\ \varphi \end{matrix} \right\}_0 + \left\{ \begin{matrix} t_{25} \\ t_{45} \end{matrix} \right\} R_A \tag{z}$$

The first expression of Equation (z) can be rearranged as

$$\left\{ \begin{matrix} 0 \\ 0 \end{matrix} \right\} = \begin{bmatrix} t_{11} & t_{12} & t_{15} \\ t_{31} & t_{32} & t_{35} \end{bmatrix} \left\{ \begin{matrix} -y \\ \varphi \\ R_A \end{matrix} \right\} \tag{a1}$$

or

$$\begin{Bmatrix} 0 \\ 0 \end{Bmatrix} = \begin{bmatrix} t_{11} & t_{31} \\ t_{12} & t_{32} \\ t_{15} & t_{35} \end{bmatrix} \begin{bmatrix} t_{11} & t_{12} & t_{15} \\ t_{31} & t_{32} & t_{35} \end{bmatrix} \begin{Bmatrix} -y \\ \varphi \\ R_A \end{Bmatrix}$$

$$= \begin{bmatrix} t_{11}^2 + t_{31}^2 & t_{11}t_{12} + t_{31}t_{32} & t_{11}t_{15} + t_{31}t_{35} \\ t_{11}t_{12} + t_{31}t_{32} & t_{12}^2 + t_{32}^2 & t_{15}t_{12} + t_{35}t_{32} \\ t_{11}t_{15} + t_{31}t_{35} & t_{15}t_{12} + t_{35}t_{32} & t_{15}^2 + t_{35}^2 \end{bmatrix} \begin{Bmatrix} -y_0 \\ \varphi_0 \\ R_A \end{Bmatrix} \tag{b1}$$

For the nontrivial solution of Equation (b1), we should have

$$\begin{vmatrix} t_{11}^2 + t_{31}^2 & t_{11}t_{12} + t_{31}t_{32} & t_{11}t_{15} + t_{31}t_{35} \\ t_{11}t_{12} + t_{31}t_{32} & t_{12}^2 + t_{32}^2 & t_{15}t_{12} + t_{35}t_{32} \\ t_{11}t_{15} + t_{31}t_{35} & t_{15}t_{12} + t_{35}t_{32} & t_{15}^2 + t_{35}^2 \end{vmatrix} = 0$$

It can be verified that Equations (b1) and (u) will be identical. It should be noted that the transformation between the right and left side of the support B can be written as

$$_R\begin{Bmatrix} -y \\ \varphi \\ M \\ S \\ 1 \end{Bmatrix}_2 = \begin{bmatrix} 1 & 0 & 0 & 0 & 0 \\ 0 & 1 & 0 & 0 & 0 \\ 0 & 0 & 1 & 0 & 0 \\ 0 & 0 & 0 & 1 & R_B \\ 0 & 0 & 0 & 0 & 1 \end{bmatrix} {}_L\begin{Bmatrix} -y \\ \varphi \\ M \\ S \\ 1 \end{Bmatrix}_2 \quad \text{or} \quad {}_R\mathbf{S}_2^* = \mathbf{U}_{spB}^* \, {}_L\mathbf{S}_2^* \tag{c1}$$

where subscript *sp* represents support. It will be used for getting the state vector on the right of support B.

Example 8.10

Obtain the variation of the transverse natural frequency with the shaft speed (i.e., obtain the Campbell diagram) of the cantilever rotor system shown in Figure 8.36. From this Campbell diagram, obtain the critical speeds. The end B_1 of the shaft has fixed end conditions. The length of the shaft is 0.2 m, and the diameter is 0.01 m. The disk is thin and has 1 kg of mass; the radius of the disk is 3.0 cm. Consider the gyroscopic effects of the disk; however, neglect the mass of the shaft. Take a range of the shaft speed such that it covers at least two critical speeds in the Campbell diagram. Use the TMM.

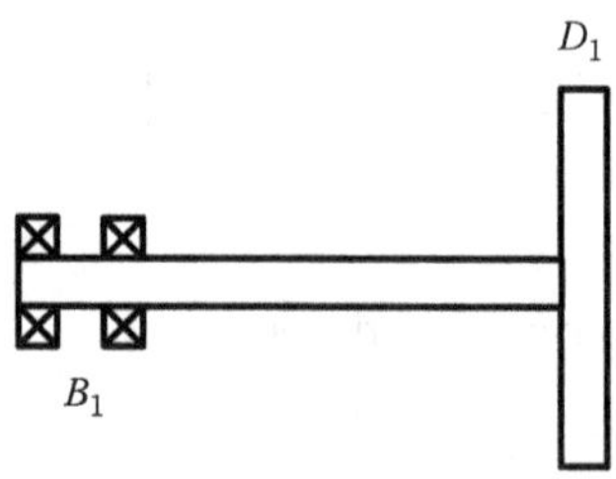

FIGURE 8.36 A cantilever rotor.

Solution:

Because of the gyroscopic effect, the motion in the two orthogonal planes will be coupled. This requires consideration of the modified field and point transfer matrices as given by Equations 8.35 and 8.47. In addition, the gyroscopic effect will introduce the following new terms in the point matrix, $P^*_{3,14} = -P^*_{7,10} = -P^*_{11,6} = P^*_{15,2} = I_p\omega\nu$, where the subscripts represent the row and column, respectively, in the modified point matrix, and ω is the spin speed of the shaft. Let station 0 be the fixed end and station 1 be the free end. For free vibrations, the overall transformation can be written as

$$_R\mathbf{S}^*_1 = \mathbf{T}^*\mathbf{S}^*_0 \tag{a}$$

with

$$\mathbf{T}^* = \mathbf{P}^*_1\mathbf{F}^*_1 \tag{b}$$

$$\mathbf{P}^*_1 = \begin{bmatrix} \mathbf{P} & \mathbf{0}_1 & \mathbf{0}_1 & \mathbf{G}_p & \mathbf{0}_3 \\ \mathbf{0}_1 & \mathbf{P} & \mathbf{G}_q & \mathbf{0}_1 & \mathbf{0}_3 \\ \mathbf{0}_1 & \mathbf{G}_p & \mathbf{P} & \mathbf{0}_1 & \mathbf{0}_3 \\ \mathbf{G}_q & \mathbf{0}_1 & \mathbf{0}_1 & \mathbf{P} & \mathbf{0}_3 \\ \mathbf{0}_2 & \mathbf{0}_2 & \mathbf{0}_2 & \mathbf{0}_2 & 1 \end{bmatrix}; \quad \mathbf{F}^*_1 = \begin{bmatrix} \mathbf{F} & \mathbf{0}_1 & \mathbf{0}_1 & \mathbf{0}_1 & \mathbf{0}_3 \\ \mathbf{0}_1 & \mathbf{F} & \mathbf{0}_1 & \mathbf{0}_1 & \mathbf{0}_3 \\ \mathbf{0}_1 & \mathbf{0}_1 & \mathbf{F} & \mathbf{0}_1 & \mathbf{0}_3 \\ \mathbf{0}_1 & \mathbf{0}_1 & \mathbf{0}_1 & \mathbf{F} & \mathbf{0}_3 \\ \mathbf{0}_2 & \mathbf{0}_2 & \mathbf{0}_2 & \mathbf{0}_2 & 1 \end{bmatrix}_1 \tag{c}$$

$$\mathbf{P} = \begin{bmatrix} 1 & 0 & 0 & 0 \\ 0 & 1 & 0 & 0 \\ 0 & -\nu^2 I_d & 1 & 0 \\ m\nu^2 & 0 & 0 & 1 \end{bmatrix} \qquad \mathbf{F} = \begin{bmatrix} 1 & l & l^2/2EI & l^3/6EI \\ 0 & 1 & l/EI & l^2/2EI \\ 0 & 0 & 1 & l \\ 0 & 0 & 0 & 1 \end{bmatrix} \tag{d}$$

$$\mathbf{G}_p = \begin{bmatrix} 0 & 0 & 0 & 0 \\ 0 & 0 & 0 & 0 \\ 0 & I_p\omega\nu & 0 & 0 \\ 0 & 0 & 0 & 0 \end{bmatrix}; \quad \mathbf{G}_q = \begin{bmatrix} 0 & 0 & 0 & 0 \\ 0 & 0 & 0 & 0 \\ 0 & -I_p\omega\nu & 0 & 0 \\ 0 & 0 & 0 & 0 \end{bmatrix}$$

$$\mathbf{0}_1 = \begin{bmatrix} 0 & 0 & 0 & 0 \\ 0 & 0 & 0 & 0 \\ 0 & 0 & 0 & 0 \\ 0 & 0 & 0 & 0 \end{bmatrix}; \quad \mathbf{0}_2 = \begin{bmatrix} 0 & 0 & 0 & 0 \end{bmatrix}; \quad \mathbf{0}_3 = \begin{Bmatrix} 0 \\ 0 \\ 0 \\ 0 \end{Bmatrix} \tag{e}$$

$$\mathbf{S}^* = \begin{Bmatrix} \mathbf{S}_{h_r} \\ \mathbf{S}_{h_j} \\ \mathbf{S}_{v_r} \\ \mathbf{S}_{v_j} \\ 1 \end{Bmatrix}; \quad \mathbf{S}_h = \begin{Bmatrix} -x \\ \varphi_y \\ M_{xz} \\ S_x \end{Bmatrix}; \quad \mathbf{S}_v = \begin{Bmatrix} -y \\ \varphi_x \\ M_{yz} \\ S_y \end{Bmatrix}; \quad \mathbf{u} = \begin{Bmatrix} 0 \\ 0 \\ 0 \\ 0 \end{Bmatrix} \tag{f}$$

$$\mathbf{T}^* = \mathbf{P}^*\mathbf{F}^* = \begin{bmatrix} \mathbf{PF} & \mathbf{0}_1 & \mathbf{0}_1 & \mathbf{G}_p\ \mathbf{F} & \mathbf{0}_3 \\ \mathbf{0}_1 & \mathbf{PF} & \mathbf{G}_q\ \mathbf{F} & \mathbf{0}_1 & \mathbf{0}_3 \\ \mathbf{0}_1 & \mathbf{G}_p\ \mathbf{F} & \mathbf{PF} & \mathbf{0}_1 & \mathbf{0}_3 \\ \mathbf{G}_q\ \mathbf{F} & \mathbf{0}_1 & \mathbf{0}_1 & \mathbf{PF} & \mathbf{0}_3 \\ \mathbf{0}_2 & \mathbf{0}_2 & \mathbf{0}_2 & \mathbf{0}_2 & 1 \end{bmatrix}_1 \tag{g}$$

$$\mathbf{PF} = \begin{bmatrix} 1 & 0 & 0 & 0 \\ 0 & 1 & 0 & 0 \\ 0 & -\nu^2 I_d & 1 & 0 \\ m\nu^2 & 0 & 0 & 1 \end{bmatrix} \begin{bmatrix} 1 & l & l^2/2EI & l^3/6EI \\ 0 & 1 & l/EI & l^2/2EI \\ 0 & 0 & 1 & l \\ 0 & 0 & 0 & 1 \end{bmatrix}$$

$$= \begin{bmatrix} 1 & l & l^2/2EI & l^3/6EI \\ 0 & 1 & l/EI & l^2/2EI \\ 0 & -\nu^2 I_d & 1-\left(\nu^2 I_d l/EI\right) & l-\left(\nu^2 I_d l^2/2EI\right) \\ m\nu^2 & m\nu^2 l & m\nu^2 l^2/2EI & 1+\left(m\nu^2 l^3/6EI\right) \end{bmatrix} \tag{h}$$

$$\mathbf{G}_p\mathbf{F} = \begin{bmatrix} 0 & 0 & 0 & 0 \\ 0 & 0 & 0 & 0 \\ 0 & I_p\omega\nu & 0 & 0 \\ 0 & 0 & 0 & 0 \end{bmatrix} \begin{bmatrix} 1 & l & l^2/2EI & l^3/6EI \\ 0 & 1 & l/EI & l^2/2EI \\ 0 & 0 & 1 & l \\ 0 & 0 & 0 & 1 \end{bmatrix} = \begin{bmatrix} 0 & 0 & 0 & 0 \\ 0 & 0 & 0 & 0 \\ 0 & I_p\omega\nu & I_p\omega\nu l/EI & I_p\omega\nu l^2/2EI \\ 0 & 0 & 0 & 0 \end{bmatrix} \tag{i}$$

$$\mathbf{G}_q\mathbf{F} = \begin{bmatrix} 0 & 0 & 0 & 0 \\ 0 & 0 & 0 & 0 \\ 0 & -I_p\omega\nu & 0 & 0 \\ 0 & 0 & 0 & 0 \end{bmatrix} \begin{bmatrix} 1 & l & l^2/2EI & l^3/6EI \\ 0 & 1 & l/EI & l^2/2EI \\ 0 & 0 & 1 & l \\ 0 & 0 & 0 & 1 \end{bmatrix}$$

$$= \begin{bmatrix} 0 & 0 & 0 & 0 \\ 0 & 0 & 0 & 0 \\ 0 & -I_p\omega\nu & -I_p\omega\nu l/EI & -I_p\omega\nu l^2/2EI \\ 0 & 0 & 0 & 0 \end{bmatrix} \tag{j}$$

The boundary conditions for the present case are such that all the translational and rotational displacements at station 0 are zero, and all the moments and shear forces are zero at the right of station 1. Hence, the state vectors at station 0 and 1 have the following form:

$$\mathbf{S}_0^* = \left\{ 0 \quad 0 \quad M_{xz_r} \quad S_{x_r} \quad 0 \quad 0 \quad M_{xz_i} \quad S_{x_i} \quad 0 \quad 0 \quad M_{yz_r} \quad S_{y_r} \quad 0 \quad 0 \quad M_{yz_i} \quad S_{y_i} \quad 1 \right\}^T$$

and

$${}_R\mathbf{S}_0^* = \left\{ -x_r \quad \varphi_{y_r} \quad 0 \quad 0 \quad -x_i \quad \varphi_{y_r} \quad 0 \quad 0 \quad -y_r \quad \varphi_{x_r} \quad 0 \quad 0 \quad -y_i \quad \varphi_{x_i} \quad 0 \quad 0 \quad 1 \right\}^T$$

Following rows 3, 4, 7, 8, 11, 12, 15, and 16 will give the eigenvalue problem of the following form:

$$\begin{bmatrix} 1-\left(\nu^2 I_d l/EI\right) & l-\left(\nu^2 I_d l^2/2EI\right) & 0 & 0 & 0 & 0 & I_p\omega\nu l/EI \\ m\nu^2 l^2/2EI & 1+\left(m\nu^2 l^3/6EI\right) & 0 & 0 & 0 & 0 & 0 \\ 0 & 0 & 1-\left(\nu^2 I_d l/EI\right) & l-\left(\nu^2 I_d l^2/2EI\right) & -I_p\omega\nu l/EI & -I_p\omega\nu l^2/2EI & 0 \\ 0 & 0 & m\nu^2 l^2/2EI & 1+\left(m\nu^2 l^3/6EI\right) & 0 & 0 & 0 \\ 0 & 0 & -I_p\omega\nu l/EI & -I_p\omega\nu l^2/2EI & 1-\left(\nu^2 I_d l/EI\right) & l-\left(\nu^2 I_d l^2/2EI\right) & 0 \\ 0 & 0 & 0 & 0 & m\nu^2 l^2/2EI & 1+\left(m\nu^2 l^3/6EI\right) & 0 \\ I_p\omega\nu l/EI & I_p\omega\nu l^2/2EI & 0 & 0 & 0 & 0 & l-\left(\nu^2 I_d l^2/2EI\right) \\ 0 & 0 & 0 & 0 & 0 & 0 & 1+\left(m\nu^2 l^3/6EI\right) \end{bmatrix}$$

$$\times \begin{Bmatrix} M_{xz_r} \\ S_{xz_r} \\ M_{xz_i} \\ S_{xz_i} \\ M_{yz_r} \\ Syz_r \\ M_{yz_i} \\ S_{yz_i} \end{Bmatrix} = \begin{Bmatrix} 0 \\ 0 \\ 0 \\ 0 \\ 0 \\ 0 \\ 0 \\ 0 \end{Bmatrix} \qquad \text{(k)}$$

We need to find v for which the determinant of the above matrix is zero, i.e.,

$$|f(\nu,\omega)| = 0 \tag{I}$$

It should be noted that the characteristic equation (or frequency equation) function $f(\nu, \omega)$ is a function of whirl frequency, ν, and the spin speed of the shaft, ω. Hence, whirl frequencies have to be obtained for a particular spin speed at a time. For different operating speeds, solutions will help in plotting the Campbell diagram. Figure 8.37 shows a flowchart of the solution algorithm to get the Campbell diagram from the above frequency equation.

Figure 8.38 has been obtained for a particular spin speed of the rotor at 1000 rad/s. This plot gives four whirl frequencies corresponding to one spin speed of the rotor. With a decrease in the spin speed, the first two roots (and similarly, the third and fourth roots) come closer (Figure 8.39 for ω=500 rad/s). It should be noted that for ω=0, in fact we would get only two roots (Figure 8.40 for ω=0 rad/s). This is due to the absence of gyroscopic couple when the spin speed of the rotor is zero. On the other hand, when the spin speed is increased, separation between the first two pairs (and similarly, for the third and fourth roots) would be faster (Figure 8.41 for ω=1500 rad/s) due to an increase in the gyroscopic effect. Similar, four whirl frequencies can be obtained for different spin speeds one at a time. These whirl frequencies then could be plotted as a Campbell diagram as shown in Figure 8.42.

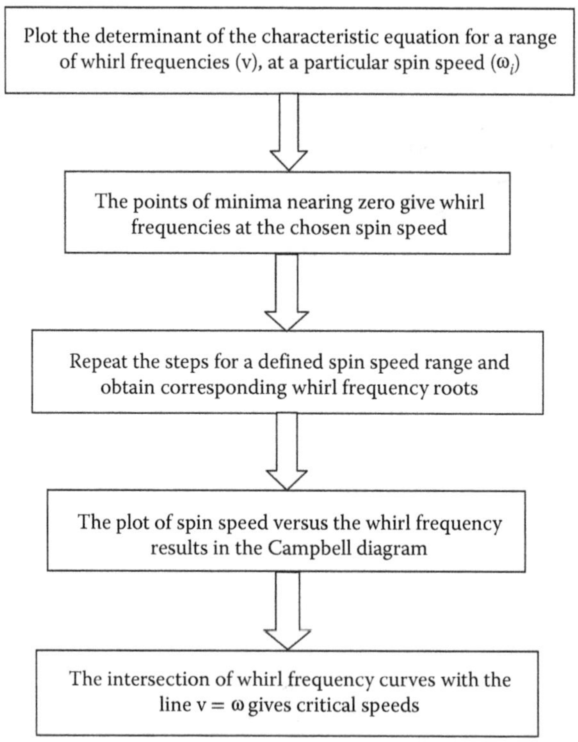

FIGURE 8.37 Flowchart for a procedure to obtain the Campbell plot from the frequency equation.

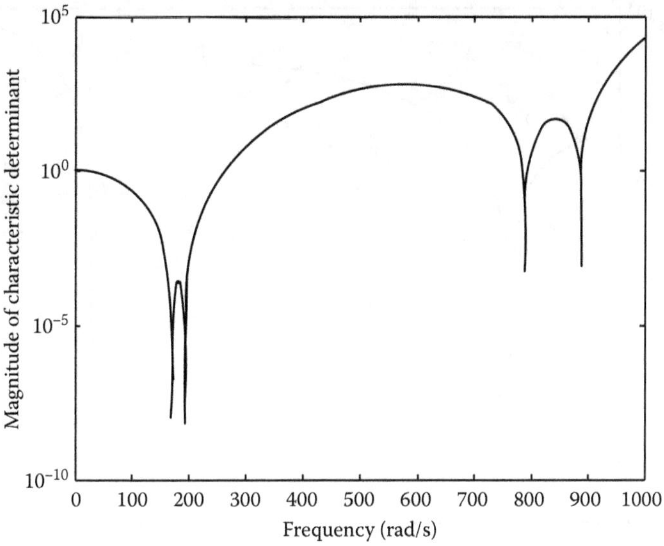

FIGURE 8.38 A plot for variation of the characteristic equation as a function of the whirl frequency with 1000 rad/s as the spin speed of the rotor.

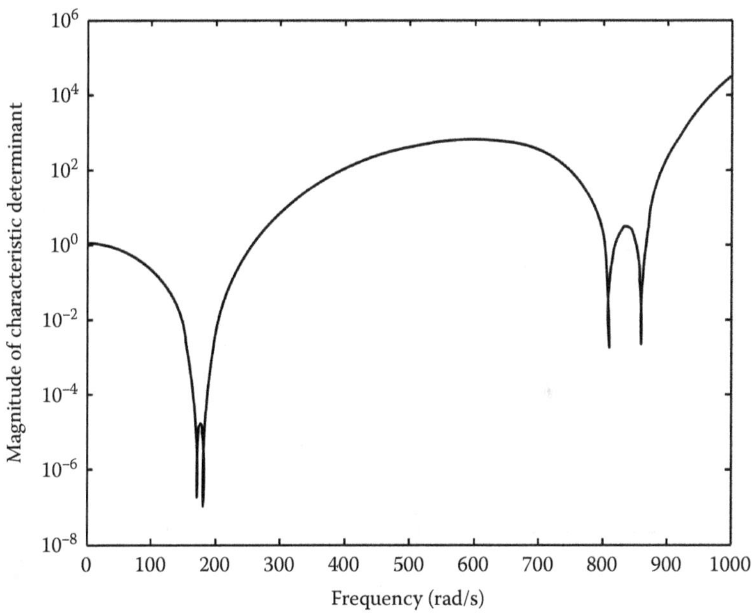

FIGURE 8.39 A plot for variation of the characteristic equation as a function of the whirl frequency with 500 rad/s as the spin speed of the rotor.

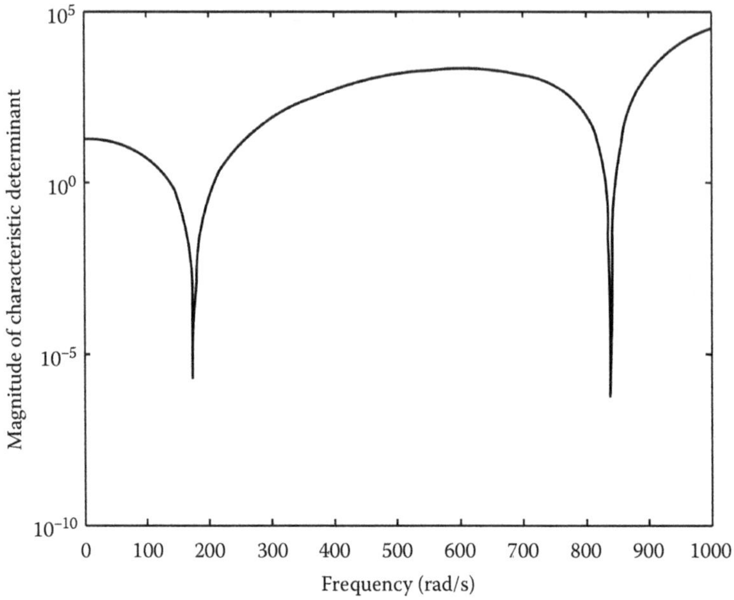

FIGURE 8.40 A plot for variation of the characteristic equation as a function of the whirl frequency with 0 rad/s as the spin speed of the rotor.

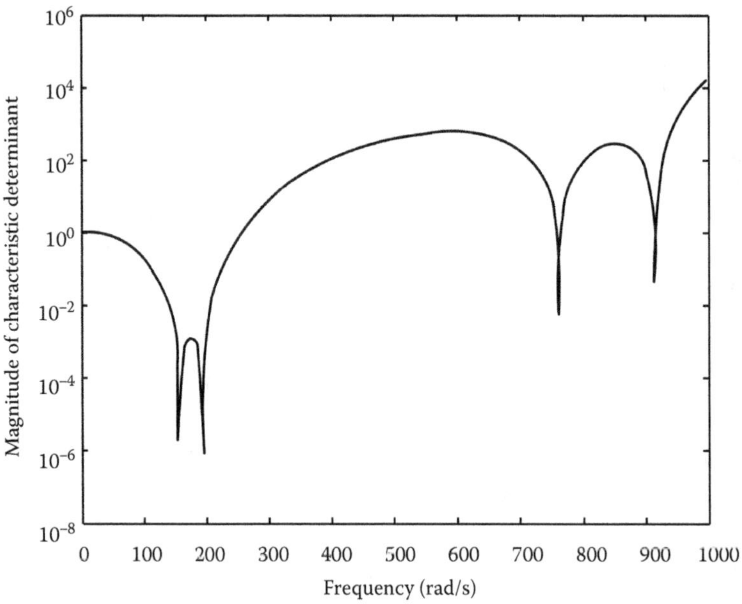

FIGURE 8.41 A plot for variation of the characteristic equation as a function of the whirl frequency with 1500 rad/s as the spin speed of the rotor.

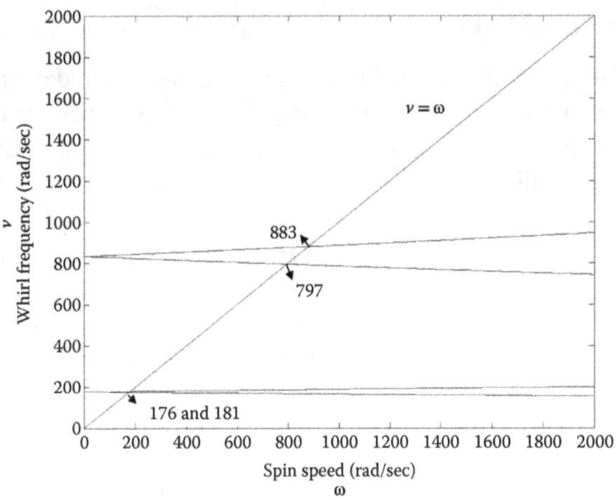

FIGURE 8.42 The Campbell diagram for the rotor system.

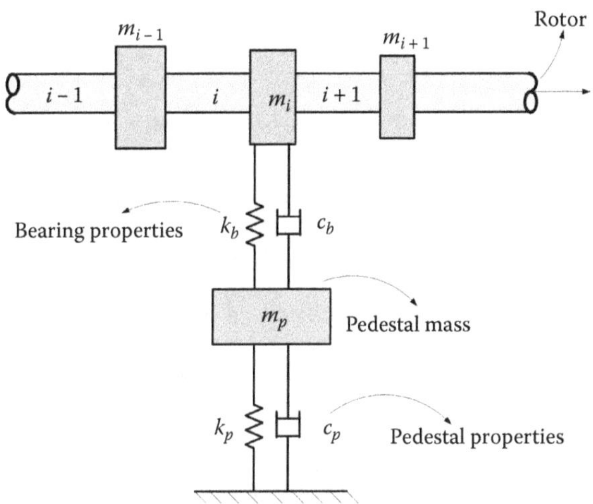

FIGURE 8.43 A shaft with an intermediate bearing and its pedestal.

If one wants to build a linear model of a multi-segment rotor that is mounted on multiple bearings, each of them is of stiffness coefficient k_b and damping coefficient c_b. Bearing pedestals are mounted to the ground through k_p and $c_{p,}$ and the pedestals have a mass of m_p each. Brief steps are given to use the TMM that will help to evaluate the linear response of the rotor and intermediate pedestals. We do not have a "chain" layout but a "tree" layout of the system as shown in Figure 8.43.

1. On taking the free body diagram of mass m_i (Figure 8.43), additional forces will come in the shear force balance equation corresponding to bearing forces (both x_i and x_p will appear). Herein, x_p is the pedestal displacement.
2. To get the transfer matrix, one needs to replace x_p with x_i. We can get this relation from the free body diagram of pedestal mass, m_p, and by taking the force balance.
3. On replacing x_p in terms of x_i from Step 2 in Step 1, we will obtain force relations that can be used to modify the point (mass) matrix.

Please note that one can write the damping force as $j\omega c x_p$ and the inertia force as $-\omega^2 m_p x_p$.

8.3 DUNKERLEY'S FORMULA

Dunkerley's formula (Dunkerley, 1894) can be used to calculate the fundamental transverse natural frequency without the help of the numerical methods. This method gives a very rough estimation of the natural frequency. From the influence coefficient method, the natural frequency of the system is obtained by the following conditions:

$$|\mathbf{A}| = 0 \tag{8.69}$$

For a two degrees of freedom system, as shown in Figure 8.44a, by considering only the translational displacements, the above equation will be of the following form:

$$\begin{vmatrix} \left(m_1\alpha_{11} - \dfrac{1}{\omega_{nf}^2}\right) & m_2\alpha_{12} \\ m_1\alpha_{21} & \left(m_2\alpha_{22} - \dfrac{1}{\omega_{nf}^2}\right) \end{vmatrix} = 0 \Rightarrow \left(m_1\alpha_{11} - \frac{1}{\omega_{nf}^2}\right)\left(m_2\alpha_{22} - \frac{1}{\omega_{nf}^2}\right) - m_1 m_2 \alpha_{12}\alpha_{21} = 0$$

which can be rearranged as

$$\left(\frac{1}{\omega_{nf}^2}\right)^2 - \left(m_1\alpha_{11} + m_2\alpha_{22}\right)\left(\frac{1}{\omega_{nf}^2}\right) + m_1 m_2\left(\alpha_{11}\alpha_{22} - \alpha_{12}\alpha_{21}\right) = 0 \tag{8.70}$$

but for a polynomial whose first coefficient is unity, the second coefficient is equal to minus the sum of the roots of the equation (Scarborough, 1966):

$$\frac{1}{\omega_{nf1}^2} + \frac{1}{\omega_{nf2}^2} = \left(m_1\alpha_{11} + m_2\alpha_{22}\right) \tag{8.71}$$

Now define

$$\frac{1}{\omega_{11}^2} = m_1\alpha_{11} \quad \text{and} \quad \frac{1}{\omega_{22}^2} = m_2\alpha_{22} \tag{8.72}$$

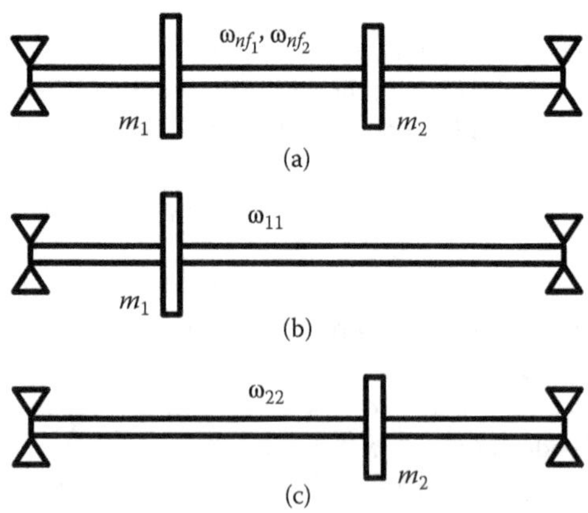

FIGURE 8.44 Two-DOF and single-DOF rotor systems: (a) two-DOF rotor system; (b) single-DOF rotor system with only left disk; (c) single-DOF rotor system with only right disk.

where ω_{11} and ω_{22} are the natural frequencies of the single-DOF system, respectively, when mass m_1 alone is present (Figure 8.44b) and when mass m_2 alone is present (Figure 8.44c). Hence, we have

$$\frac{1}{\omega_{nf1}^2}+\frac{1}{\omega_{nf2}^2}=\frac{1}{\omega_{11}^2}+\frac{1}{\omega_{22}^2} \tag{8.73}$$

It can be proved that in general for multi-DOF systems, we can write

$$\frac{1}{\omega_{nf1}^2}+\frac{1}{\omega_{nf2}^2}+\cdots+\frac{1}{\omega_{nfN}^2}=\frac{1}{\omega_{11}^2}+\frac{1}{\omega_{22}^2}+\cdots+\frac{1}{\omega_{NN}^2} \tag{8.74}$$

where N is the total DOFs of the rotor system. In most cases, the fundamental frequency ω_{nf1} will be much lower than all other higher natural frequencies, so the above equation may be written as

$$\frac{1}{\omega_{nf1}^2}+\varepsilon=\sum_{i=1}^{N}\frac{1}{\omega_{ii}^2} \tag{8.75}$$

where ε is a small positive quantity. The above equation can be rearranged as

$$\omega_{nf1}^2=\frac{1}{\left(-\varepsilon+\sum_{i=1}^{N}\frac{1}{\omega_{ii}^2}\right)} \tag{8.76}$$

On neglecting ε, we can find the fundamental natural frequency of the rotor system as

$$\omega_{nf1}^2=\frac{1}{\left(\sum_{i=1}^{N}\frac{1}{\omega_{ii}^2}\right)} \tag{8.77}$$

It should be noted that because ε is a positive quantity, we will get a lower bound of the fundamental frequency from the present method as opposed to the FEM in which we always get higher bounds of natural frequencies.

$$\frac{1}{\omega_{nf1}^2}\cong\frac{1}{\omega_{11}^2}+\frac{1}{\omega_{22}^2}+\cdots+\frac{1}{\omega_{NN}^2} \tag{8.78}$$

Dunkerley first suggested this. Equation 8.78 always gives a value for the fundamental frequency that is slightly lower than the true value, by virtue of the approximation involved. There are other approximate methods available that can be used to obtain bounds of natural frequencies of the rotor system, for example *Rayleigh's quotient* (Meirovitch, 1986).

Example 8.11

Find the transverse natural frequencies and mode shapes of the rotor system shown in Figure 8.45. B is a fixed end, and D_1 and D_2 are rigid disks. The shaft is made of steel with Young's modulus $E=2.1\times10^{11}$ N/m² and a uniform diameter $d=10$ mm. The shaft lengths are $BD_2=50$ mm, and $D_1D_2=75$ mm. The masses of the disks are $m_1=2$ kg and $m_2=5$ kg. Consider the shaft as massless, and neglect the diametral mass moments of inertia of the disks.

Solution:

For Figure 8.46, we have

$$EI=103.1\ \text{N-m}^2,\quad l_1=0.125\ \text{m},\quad l_2=0.05\ \text{m}$$

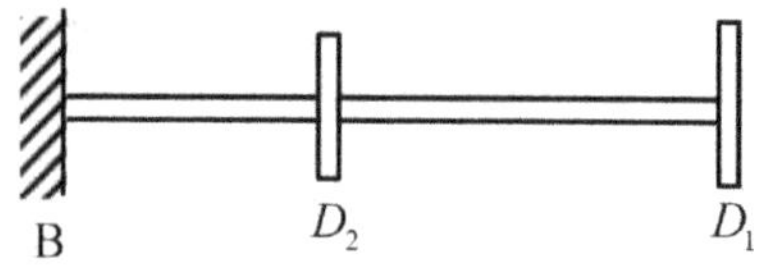

FIGURE 8.45 Two-disk cantilever shaft system.

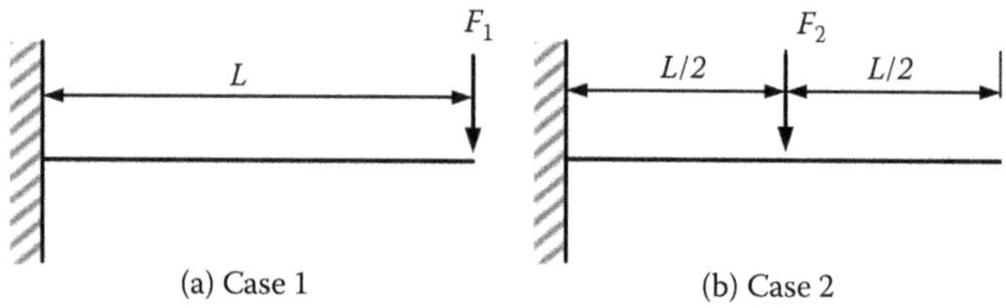

FIGURE 8.46 Cantilever rotor systems with a single disk: (a) Case 1; (b) Case 2.

$$k_{11} = \frac{3EI}{l_1^3} = \frac{3 \times 103.1}{0.125^3} = 0.158 \times 10^6 \text{ N/m}, \quad k_{22} = \frac{3EI}{l_2^3} = \frac{3 \times 103.1}{0.05^3} = 2.47 \times 10^6 \text{ N/m}$$

$$\omega_{11}^2 = \frac{k_{11}}{m_1} = \frac{0.158 \times 10^6}{2} = 0.079 \times 10^6, \quad \omega_{22}^2 = \frac{k_{22}}{m_2} = \frac{2.47 \times 10^6}{5} = 0.494 \times 10^6$$

The system natural frequency from Dunkerley's formula is given as

$$\frac{1}{\omega_{nf_1}^2} = \frac{1}{\omega_{11}^2} + \frac{1}{\omega_{22}^2} = \left(\frac{1}{0.079} + \frac{1}{0.494}\right) \times 10^{-6} = 14.68 \times 10^{-6}$$

or

$$\omega_{nf_1} = 260.10 \text{ rad/s}$$

Using the TMM, the value of the fundamental natural frequency was $\omega_{nf_1} = 266.67$ rad/s. Hence, Dunkerley's formula gives a reasonably good estimate of the fundamental natural frequency, and it gives the lower bound.

Example 8.12

Find the fundamental transverse natural frequency of the rotor system shown in Figure 8.47. B_1 and B_2 are bearings that provide a simply supported end condition, and D_1, D_2, D_3, and D_4 are rigid disks. The shaft is made of steel with Young's modulus $E=2.1 \times 10^{11}$ N/m² and uniform diameter $d=20$ mm. The various shaft lengths are as follows: $B_1D_1=150$ mm, $D_1D_2=50$ mm, $D_2D_3=50$ mm, $D_3D_4=50$ mm, and $D_4D_2=150$ mm. The masses of the disks are $m_1=4$ kg, $m_2=5$ kg, $m_3=6$ kg, and $m_4=7$ kg. Consider the shaft as massless. Consider the disks as point masses.

Solution:

The influence coefficient for a simply supported shaft with a disk is given as

$$\alpha = \frac{a^2b^2}{3EIl} \quad \text{with} \quad l = a + b = 0.45 \text{ m} \tag{a}$$

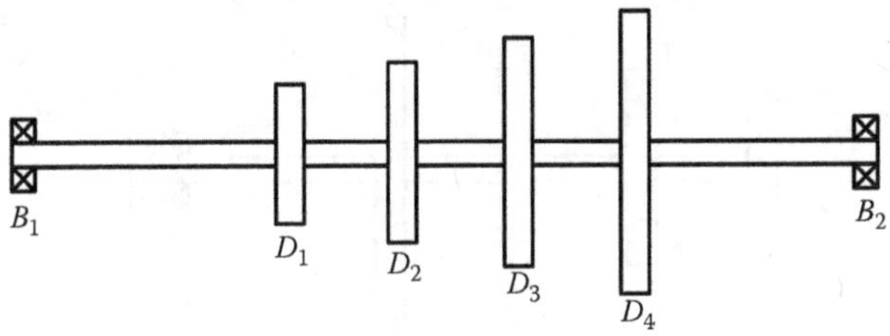

FIGURE 8.47 A multidisc rotor system.

TABLE 8.13
Calculation Procedure of the Fundamental Natural Frequency Using Dunkerley's Formula

S.N.	a_i (m)	b_i (m)	Influence Coefficient, α_{ii} (m/N)	Mass of the Disk, m_i (kg)	For a Single-DOF, $1/\omega_{ii}^2 = m_i\alpha_{y_i f_i}$	The Fundamental Natural Frequency $\omega_{nf_1} = \sqrt{1/\left(\sum 1/\omega_{ii}^2\right)}, (rad/s)$
1	0.150	0.300	9.0946×10^{-7}	4	3.6378×10^{-6}	211.50
2	0.200	0.250	1.1228×10^{-6}	5	5.6139×10^{-6}	
3	0.250	0.200	1.1228×10^{-6}	6	6.7367×10^{-6}	
4	0.300	0.150	9.0946×10^{-7}	7	6.3662×10^{-6}	
					$\Sigma=2.2355\times10^{-5}$	

where l is the span of the shaft, and a and b are the disk position from the left and right bearings. The natural frequency of a single-DOF rotor system can be obtained as

$$\omega_{ii}^2 = \frac{1}{m_i\alpha_{y_i f_i}} \tag{b}$$

We have $d=0.02$ m, $l=0.45$ m, and $EI=1649.34$ N-m². Hence, $EIl=742.20$ N-m³. Table 8.13 summarizes the calculation of the fundamental natural frequency From TMM method in Example 8.8, the fundamental natural frequency of the rotor system is 215 rad/s, which is close to 211.5 rad/s from the present method.

Example 8.13

Find the transverse natural frequencies and mode shapes of the rotor system shown in Figure 8.48. B is a fixed bearing that provides a fixed support end condition, and D_1, D_2, D_3, and D_4 are rigid disks. The shaft is made of steel with modulus of rigidity $E=2.1\times10^{11}$ N/m² and uniform diameter $d=20$ mm. The various shaft lengths are as follows: $D_1D_2=50$ mm, $D_2D_3=50$ mm, $D_3D_4=50$ mm, and $D_4B_2=150$ mm. The masses of the disks are $m_1=4$ kg, $m_2=5$ kg, $m_3=6$ kg, and $m_4=7$ kg. Consider the shaft as massless. Consider the disk as point masses, i.e., neglect the diametral and polar mass moments of inertia of all disks.

Solution:

The influence coefficient for a cantilever shaft with a disk at the free end is given as

$$\alpha_{yf} = \frac{L^3}{3EI} \tag{a}$$

where L is the span of the shaft. The natural frequency of a single-DOF rotor system can be obtained as

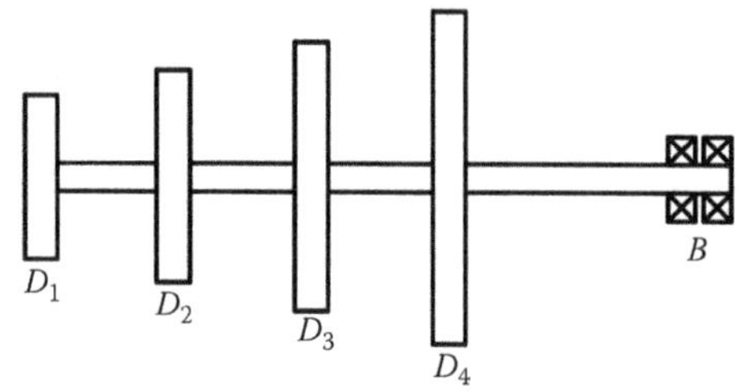

FIGURE 8.48 A multidisc cantilever rotor.

TABLE 8.14
Calculation Procedure of the Fundamental Natural Frequency Using Dunkerley's Formula

S.N.	L_i (m)	Influence Coefficient, α_{ii} (m/N)	Mass of the Disk, m_i (kg)	For a Single-DOF, $1/\omega_{ii}^2 = m_i\alpha_{yi\,fi}$	The Fundamental Natural Frequency $\omega_{nfi} = \sqrt{1/\left(\sum 1/\omega_{ii}^2\right)}$, (rad / s)
1	0.30	5.4567×10^{-6}	4	2.1827×10^{-5}	92.08
2	0.25	3.1578×10^{-6}	5	1.5789×10^{-5}	
3	0.20	8.0841×10^{-6}	6	4.8504×10^{-5}	
4	0.15	4.5473×10^{-6}	7	3.1831×10^{-5}	
				$\Sigma=1.1795\times10^{-4}$	

$$\omega_{ii}^2 = \frac{1}{m_i\alpha_{yi\,fi}}, \quad i = 1,2,3,4 \tag{b}$$

We have $d=0.02$ m and $EI=1649.34$ N-m^2. Table 8.14 summarizes the calculation of the fundamental natural frequency.

8.4 CONCLUDING REMARKS

To summarize, in the present chapter, we studied methods of calculation of natural frequencies and forced responses. Two main methods, namely the influence coefficient and TMMs, are given detailed treatment. Durkerley's formula for the approximate estimation of the fundamental frequency is presented. All three methods are illustrated with simple examples keeping the calculation complexity to a minimum, while retaining various basic features of the solution method for a variety of cases. The application of these methods for larger systems is then straightforward; however, it requires the help of a computer programming. The influence coefficient method is simple in application; however, it requires calculations of influence coefficients with the help of load–deflection relations, which are different for different systems. Moreover, when the DOF of the system increases the matrix size increases, so higher computational time is required for large rotor systems to solve the eigenvalue problem. The TMM is quite systematic and effective even for multi-DOF systems. It does not require any other prior calculation of the system matrix element as in the influence coefficient method. The overall size of the matrix remains the same, and it does not increase with the DOF of the system. However, for the calculation of natural frequencies, this method requires a root-searching numerical method, which is time consuming, and there is a risk of missing some roots. Moreover, with the TMM for the case of intermediate support condition, analysis becomes cumbersome and still distributed properties of shaft in TMM could not be considered because of complexity involved. The dynamic matrix method (not described here) is similar to the TMM in that it relates to the state vectors at different stations; however, while assembling the component equations, the size of the matrix no longer remains the same, and it increases with the DOF of the system. In fact, the finite element method is the improvised version of this method, as we will see in the next chapter. Sustainable design of a rotor for transverse vibration focuses on minimizing material waste and energy loss while keeping operational speeds away from destructive critical speeds. The primary engineering goal is to shift the rotor's natural frequencies away from its operating frequency using eco-friendly materials and optimized geometries.

EXERCISE PROBLEMS

Use both the influence coefficient and TMMs for the entire problem unless otherwise stated in the problem. Dunkerley's formula can be used to obtain a rough estimate of the fundamental natural frequency.

Exercise 8.1 Obtain the transverse natural frequencies of the rotor shown in Figure 8.49. The rotor is assumed to be fixed at one end and free at the other. Take the mass of the disk as $m=2$ kg and its diametral mass moment of inertia as $I_d=0.05$ kg-m^2. The shaft is assumed to be massless, and its length and diameter are 0.2 and 0.01 m, respectively. Take the Young's modulus $E=2.1\times10^{11}$ N/m^2 for the shaft material.

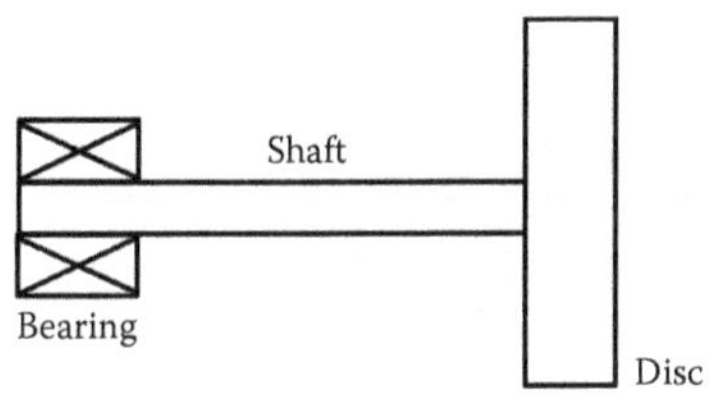

FIGURE 8.49 A cantilever rotor system.

Exercise 8.2 Find the transverse natural frequencies and mode shapes of the rotor system shown in Figure 8.50 by the influence coefficient method. Take $EI=2$ MNm^2 for the shaft, and the diametral mass moment of inertia of the disk is negligible.

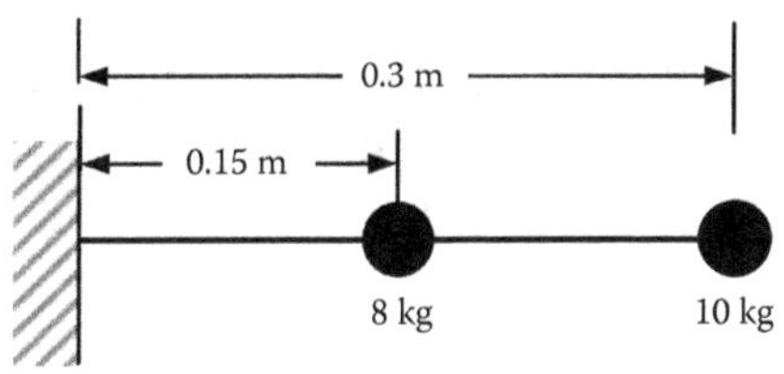

FIGURE 8.50 Two-mass cantilever rotor system.

Exercise 8.3 For Exercise 8.2, when the left and right disks have, respectively, diametral mass moments of inertia $I_{d_1}=0.05$ kg-m^2 and $I_{d_2}=0.06$ kg-m^2, obtain the transverse natural frequencies and mode shapes of the rotor system.

Exercise 8.4 Obtain the transverse natural frequency of the rotor system shown in Figure 8.51. The mass of the disk m is 5 kg, and the diametral mass moment of inertia, I_d, is 0.02 kg-m^2. The shaft lengths are $a=0.3$ m and $b=0.7$ m. The diameter of the shaft is 10 mm. Take the Young's modulus $E=2.1\times10^{11}$ N/m^2 for the shaft material.

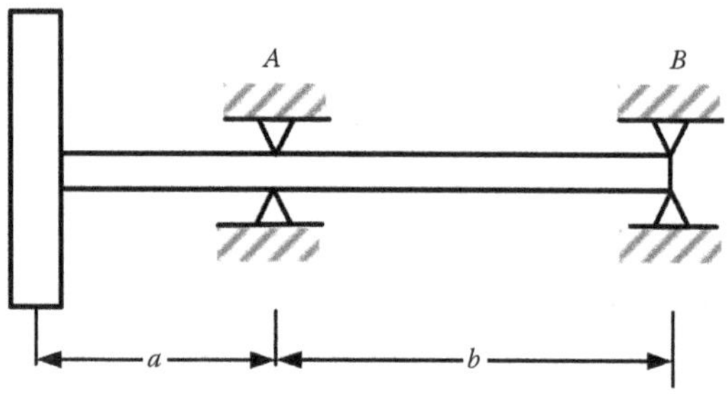

FIGURE 8.51 An overhung rotor system.

Exercise 8.5 Obtain the transverse natural frequencies of the rotor system shown in Figure 8.52. The mass of the disk is $m = 5$ kg, and the diametral mass moment of inertia is I_d=0.02 kg-m^2. The shaft lengths are a=0.3 m and b=0.7 m, the diameter of the shaft is 10 mm and the modulus of elasticity of the shaft is E=2.1×10^{11} N/m^2. Consider two different cases, i.e., when bearing A is (i) a simple support and (ii) a flexible support, which provides a bending stiffness equal to two times the bending stiffness of a cantilever shaft having length a. Bearing B is a fixed bearing. Refer to Appendix 2.1.

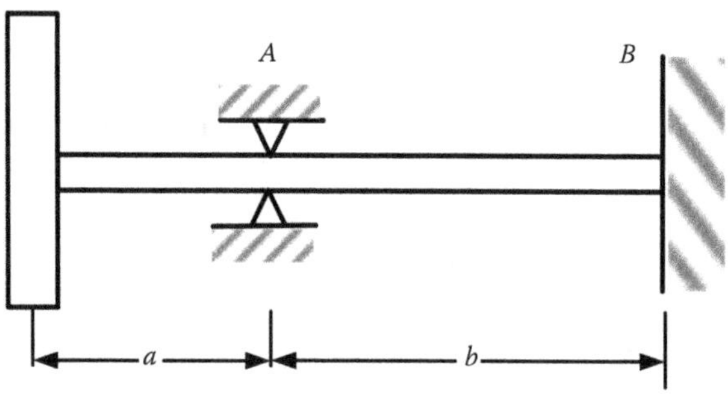

FIGURE 8.52 An overhung rotor system.

Exercise 8.6 Find the transverse natural frequencies and mode shapes of the rotor system shown in Figure 8.53. B_1 and B_2 are bearings, which provide a simply supported end condition, and D_1 and D_2 are rigid disks. The shaft is made of steel with Young's modulus E=2.1×10^{11} N/m^2 and uniform diameter d=10 mm. The various shaft lengths are as follows: B_1D_1=50 mm, D_1D_2=75 mm, and D_2B_2=50 mm. The masses of disks are m_1=4 kg and m_2=6 kg. Consider the shaft as massless. Consider the following cases: (i) Neglect the diametral mass moments of inertia of both disks and (ii) take $I_{d_1} = 0.05$ kg-m^2 and $I_{d_2} = 0.06$ kg-m^2.

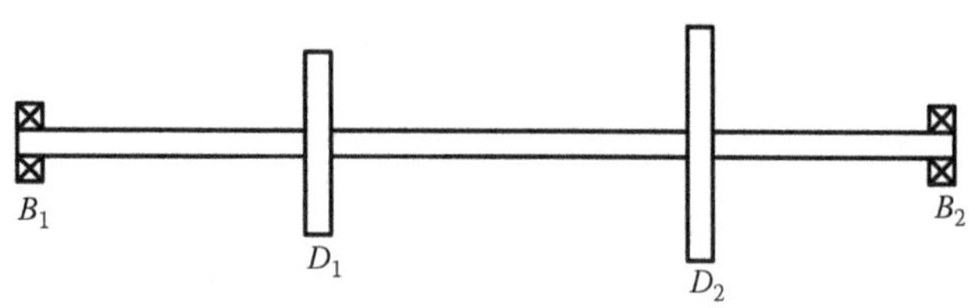

FIGURE 8.53 A simply supported shaft with two disks.

Exercise 8.7 Find the transverse natural frequencies and mode shapes of the rotor system shown in Figure 8.54. B_1 and B_2 are bearings that provide a simply supported end conditions, and D_1, D_2, D_3, D_4, and D_5 are rigid disks. The shaft is made of steel with Young's

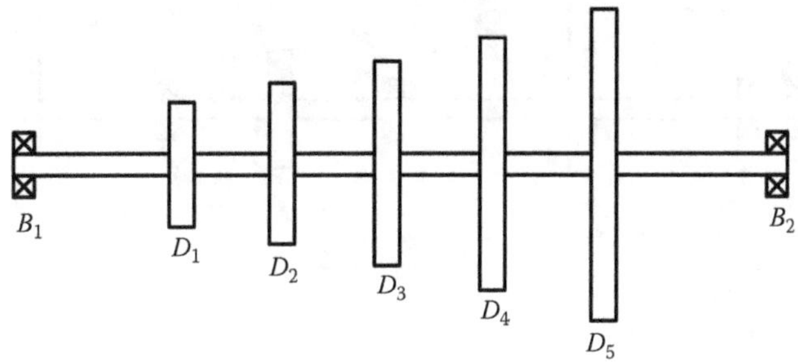

FIGURE 8.54 A multidisc rotor system with simply supported end conditions.

modulus $E=2.1\times10^{11}\,\text{N/m}^2$ and uniform diameter $d=20\,\text{mm}$. The various shaft lengths are as follows: $B_1D_1=150\,\text{mm}$, $D_1D_2=50\,\text{mm}$, $D_2D_3=50\,\text{mm}$, $D_3D_4=50\,\text{mm}$, $D_4D_5=50\,\text{mm}$, and $D_5B_2=150\,\text{mm}$. The masses of the disks are $m_1=4\,\text{kg}$, $m_2=5\,\text{kg}$, $m_3=6\,\text{kg}$, $m_4=7\,\text{kg}$, and $m_5=8\,\text{kg}$. Consider the shaft as massless. Consider the following cases: (i) Consider the disk as point masses, i.e., neglect the diametral and polar mass moments of inertia of all disks, and (ii) consider the disks as thin and take the diameter of disks as $d_1 = 8$ cm, $d_2 = 10$ cm, $d_3 = 12$ cm, $d_4 = 14$ cm, and $d_5 = 16$ cm; however, neglect the gyroscopic effects.

For a Jeffcott rotor with an offset disk, the following influence coefficients are valid (Refer to Table 8.17 for deflection relations in terms of z):

$$\begin{Bmatrix} x \\ \varphi_y \end{Bmatrix} = \begin{bmatrix} \alpha_{11} & \alpha_{12} \\ \alpha_{21} & \alpha_{22} \end{bmatrix} \begin{Bmatrix} f_x \\ M_{zx} \end{Bmatrix}$$

with

$$\begin{bmatrix} \alpha_{11} & \alpha_{12} \\ \alpha_{21} & \alpha_{22} \end{bmatrix} = \begin{bmatrix} \dfrac{a^2b^2}{3EIl} & \dfrac{-\left(3a^2l-2a^3-al^2\right)}{3EIl} \\ \dfrac{ab(b-a)}{3EIl} & \dfrac{-\left(3al-3a^2-l^2\right)}{3EIl} \end{bmatrix}$$

where l is the span length, and a and b are the distances of the disk from the left and right supports, respectively.

Exercise 8.8 Find all the transverse natural frequencies and draw corresponding mode shapes of the rotor system shown in Figure 8.55. B_1 is a fixed support (i.e., with transverse translational and rotational (slope) displacements equal to zero), and B_2 and B_3 are bearings with a simply support condition (i.e., with transverse translational displacement equal to zero). Shaft segments, of each diameter 0.01 m, have the following dimensions: $B_1D_1=50\,\text{mm}$, $D_1B_2=50\,\text{mm}$, $B_2D_2=25\,\text{mm}$, $D_2B_3=25\,\text{mm}$, and $B_3D_3=30\,\text{mm}$. The shaft is made of steel with Young's modulus $E=2.1\times10^{11}\,\text{N/m}^2$. The masses of disks are $m_1=1\,\text{kg}$, $m_2=1.5\,\text{kg}$, and $m_3=0.75\,\text{kg}$. Consider two cases: (i) The shaft is massless and the disks are rigid lumped masses; (ii) the disks are thin and the diameter of the disks are $d_1 = 12$ cm, $d_2 = 6$ cm, and $d_3 = 12$ cm (neglect the gyroscopic effects). Compare the order of magnitude of the torsional (with $G=0.8\times10^{11}\,\text{N/m}^2$) and the bending natural frequencies for this so obtained for the same system.

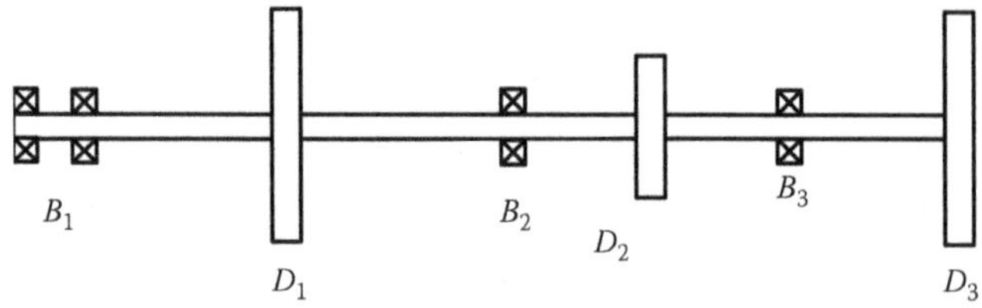

FIGURE 8.55 Three-support rotor system.

Exercise 8.9 Find the unbalance response and critical speeds of the rotor system shown in Figure 8.56. B is a bearing with fixed end conditions, and D_1 and D_2 are rigid disks. The shaft is made of steel with Young's modulus $E=2.1\times10^{11}\,\text{N/m}^2$ and uniform diameter $d=10\,\text{mm}$. The various shaft lengths are as follows: $BD_2=50\,\text{mm}$ and $D_2D_1=75\,\text{mm}$. The diametral mass moments of inertia of the disks are $I_{d_1}=0.04\ \text{kg-m}^2$ and $I_{d_2}=0.1\ \text{kg-m}^2$. Take an unbalance of 2 g at a radius of 5 cm at a convenient location and orientation. Neglect the mass of the shaft.

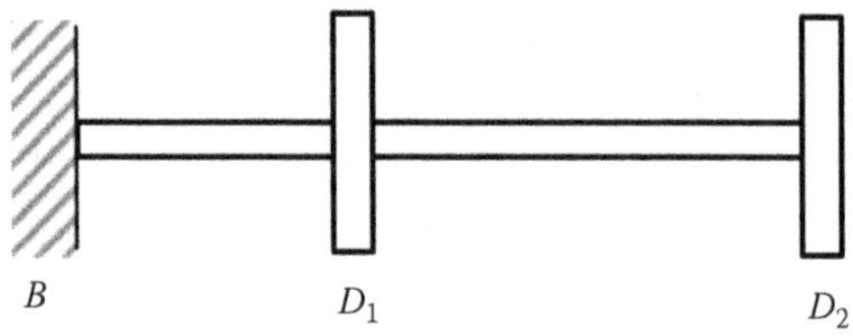

FIGURE 8.56 A two-disk cantilever rotor.

Exercise 8.10 Obtain the transverse natural frequencies of the cantilever rotor system shown in Figure 8.57. The end B_1 of the shaft has a fixed end condition, and the other end is free. The length of the shaft is 0.4 m, and the diameter is 0.1 m. The disk is thin and has 1 kg mass, 0.04 kg-m² polar mass moment of inertia, and 0.02 kg-m² diametral mass moment of inertia. Neglect the mass of the shaft, and consider the gyroscopic effects. Take the shaft speed as 10,000 rpm. Obtain the Campbell diagram and show the critical speed on it.

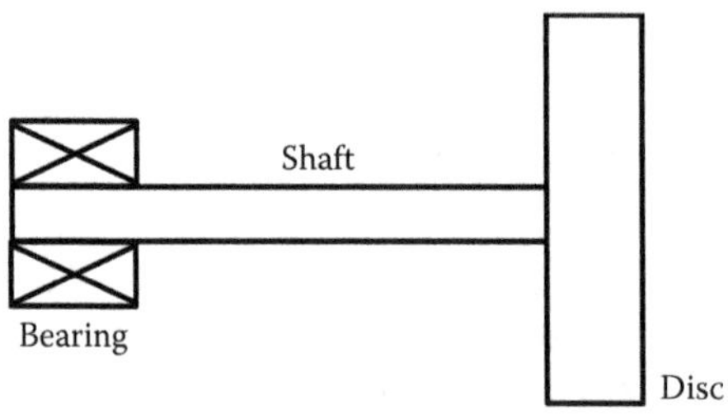

FIGURE 8.57 A cantilever rotor system.

Exercise 8.11 Obtain the transverse natural frequencies of the rotor system shown in Figure 8.58 for the following parameters: (i) Disks are point masses with $m_1=5\,\text{kg}$, $m_2=8\,\text{kg}$, $l=1\,\text{m}$, $d=0.02\,\text{m}$, $\rho=7800\,\text{kg/m}^3$, and $E=2.1\times10^{11}\,\text{N/m}^2$, and (ii) the disks are thin with radii of 10 and 15 cm with the other parameters the same as case (i).

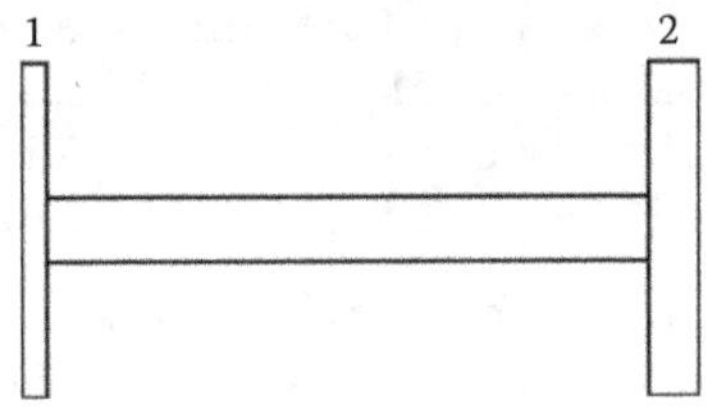

FIGURE 8.58 Two-disk rotor system.

Exercise 8.12 Find the transverse natural frequencies and mode shapes of the rotor system shown in Figure 8.59. B_1 and B_2 are fixed supports, and D_1 and D_2 are rigid disks. The shaft is made of steel with Young's modulus $E = 2.1 \times 10^{11}\,\text{N/m}^2$, and has a uniform diameter of $d = 10\,\text{mm}$. The different shaft lengths are as follows: $B_1D_1 = 50\,\text{mm}$, $D_1D_2 = 75\,\text{mm}$, and $D_2B_2 = 50\,\text{mm}$. The masses of the thin disks are $m_1 = 2\,\text{kg}$ and $m_2 = 3\,\text{kg}$, and the radii are $r_1 = 5\,\text{cm}$ and $r_2 = 8\,\text{cm}$. Consider the shaft as massless.

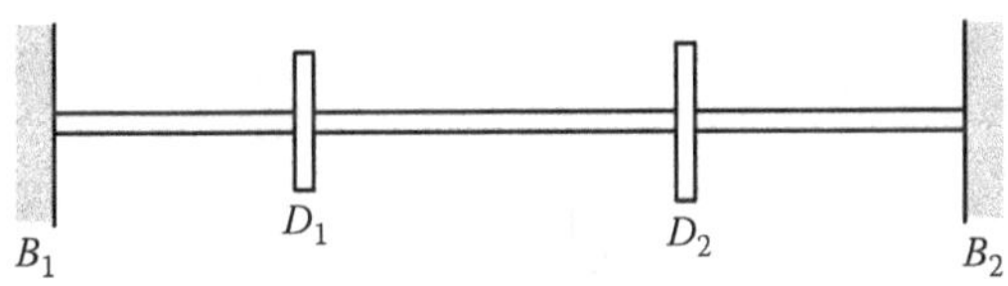

FIGURE 8.59 Two-disk rotor system with rigid end supports.

Exercise 8.13 Find the fundamental transverse natural frequency of the rotor system shown in Figure 8.60 using Durkerley's formula. Take $EI = 2$ MN-m^2 for the shaft, and the mass moment of inertia of the disk is negligible.

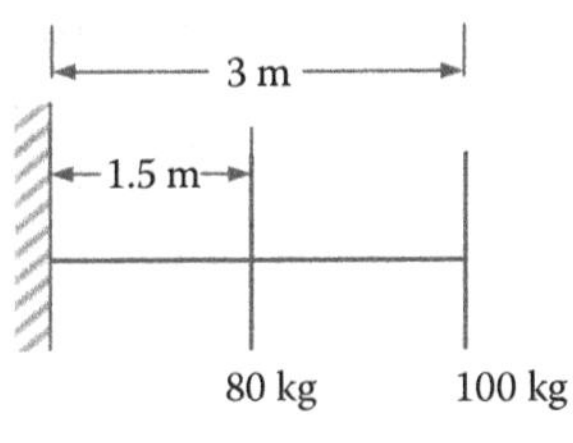

FIGURE 8.60 A cantilever rotor system.

Exercise 8.14 Obtain the transverse natural frequencies of a rotor system as shown in Figure 8.61. The masses of the disks are $m_1 = 5\,\text{kg}$, and $m_2 = 8\,\text{kg}$. The shaft length is $l = 0.3\,\text{m}$, the diameter of the shaft is 15 mm, and the modulus of elasticity of the shaft is $E = 2.1 \times 10^{11}\,\text{N/m}^2$. Both the bearings are roller supports.

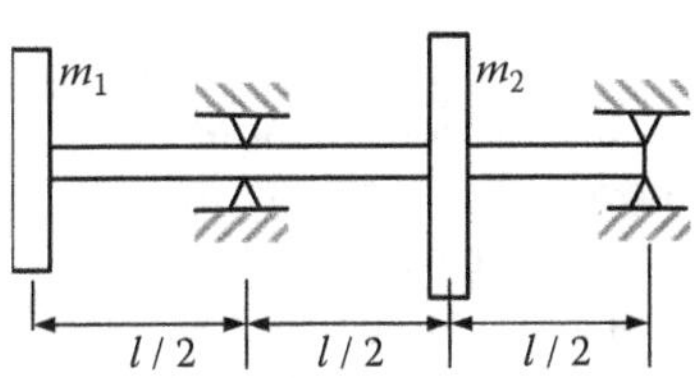

FIGURE 8.61 An overhung rotor system.

Exercise 8.15 Obtain the transverse natural frequencies of the rotor system shown in Figure 8.61. The masses of the disks are $m_1=5$ kg and $m_2=8$ kg, and the radii are $r_1=5$ cm and $r_2=8$ cm. The shaft length is $l=0.3$ m, the diameter of the shaft is 15 mm, and the modulus of elasticity of the shaft is $E=2.1\times 10^{11}$ N/m^2. Both the bearings are roller supports.

Exercise 8.16 Find the transverse natural frequencies and mode shapes of the rotor system shown in Figure 8.62. B is a fixed bearing that provides a fixed support end condition, and D_1, D_2, D_3, D_4, and D_5 are rigid disks. The shaft is made of steel with Young's modulus $E=2.1\times 10^{11}$ N/m^2 and uniform diameter $d=20$ mm. The various shaft lengths are as follows: $D_1D_2=50$ mm, $D_2D_3=50$ mm, $D_3D_4=50$ mm, $D_4D_5=50$ mm, and $D_5B_2=50$ mm. The masses of the disks are $m_1=4$ kg, $m_2=5$ kg, $m_3=6$ kg, $m_4=7$ kg, and $m_5=8$ kg. Consider the shaft as massless. Consider two cases: (i) Consider the disks as point masses, i.e., neglect the diametral and polar mass moments of inertia of all disks; (ii) consider the disks as thin and take the diameter of the disks as $d_1 = 12$ cm, $d_2 = 6$ cm, $d_3 = 12$ cm, $d_4 = 14$ cm, and $d_5 = 16$ cm (neglect the gyroscopic effects).

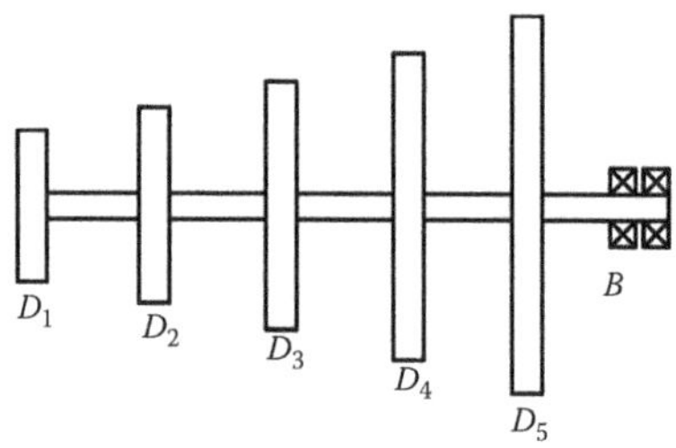

FIGURE 8.62 A multidisc cantilever rotor.

Exercise 8.17 Consider the rotor system shown in Figure 8.63 for obtaining the transverse natural frequency. Two flexible massless shafts are connected by a coupling (i.e., joined by a pin). A thin disk, of mass 3 kg and diametral mass moment of inertia 0.01 kg-m^2, is attached to one of the shafts (assume toward the left side of the shaft), and it is not interfering with the relative motion between the two shafts. The other shaft ends have fixed conditions. Take the length of each of the shafts as 0.5 m and the diameter as 0.05 m. Young's modulus $E=2.1\times 10^{11}$ N/m^2. Use the TMM.

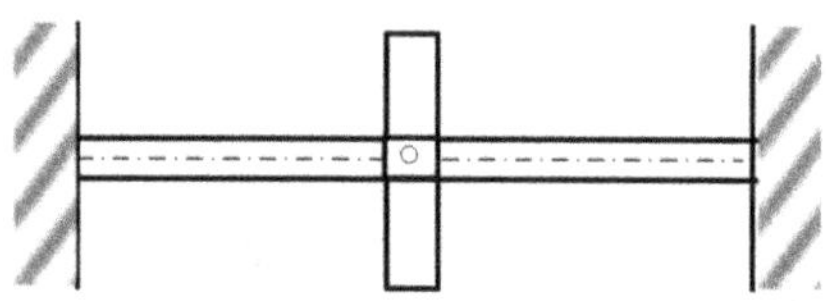

FIGURE 8.63 A rotor system with a coupling.

Exercise 8.18 Consider the rotor system shown in Figure 8.63 for obtaining the transverse natural frequency. Two flexible massless shafts are connected by a coupling (i.e., joined by a pin). A thin disk, of mass 3 kg and diametral mass moment of inertia 0.01 kg-m^2, is attached to one of the shafts (assume it is toward the left side of the shaft) and it is not interfering with the relative motion between the two shafts. The other ends of the shafts have fixed conditions. Take the length of the shafts as 0.6 m (left side) and 0.4 m (right side), and the diameter as 0.05 m. Young's modulus $E=2.1\times 10^{11}$ N/m^2. *Use the influence coefficient method.*

Exercise 8.19 Obtain the transverse natural frequency of the coaxial-shaft rotor system, which is modeled as shown in Figure 8.64. The disks have masses of 3 and 2 kg, respectively, on shaft A and shaft *B*. Shafts *A* and *B* are, respectively, 2 and 1.5 cm in diameter with a length of 40 cm each. Neglect the inertia of shafts. The bearing between two coaxial shafts provides an effective transverse stiffness of 100 MN/m. Take $E=2.1\times10^{11}$ N/m^2. Use the TMM.

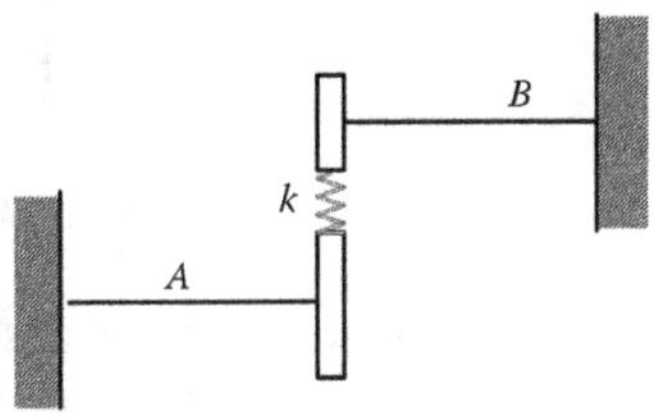

FIGURE 8.64 A two-cantilever rotor system connected by a flexible coupling.

Exercise 8.20 Using the *influence coefficient method*, find the transverse natural frequencies of the rotor system shown in Figure 8.65a. B_1 and B_2 are bearings that provide a simply supported end condition, and D_1 and D_2 are rigid point disks. The shaft is made of steel with Young's modulus $E=2.1\times10^{11}$ N/m^2 and uniform diameter $d=10$ mm. The various shaft lengths are as follows: $B_1D_1=50$ mm, $D_1D_2=75$ mm, and $D_2B_2=50$ mm. The masses of the disks are $m_1=4$ kg and $m_2=6$ kg. Consider the shaft as massless, and neglect the diametral mass moments of inertia of both disks. Corresponding to Figure 8.65b, the formula for the translational transverse displacement is given below.

$$y=\frac{Pbz}{6lEI}\left(l^2-z^2-b^2\right) \qquad \text{for} \quad (0<z<a)$$

$$y=\frac{Pb}{6lEI}\left\{\frac{l}{b}(z-a)^3+\left(l^2-b^2\right)z-z^3\right\} \qquad \text{for} \quad (a<z<l)$$

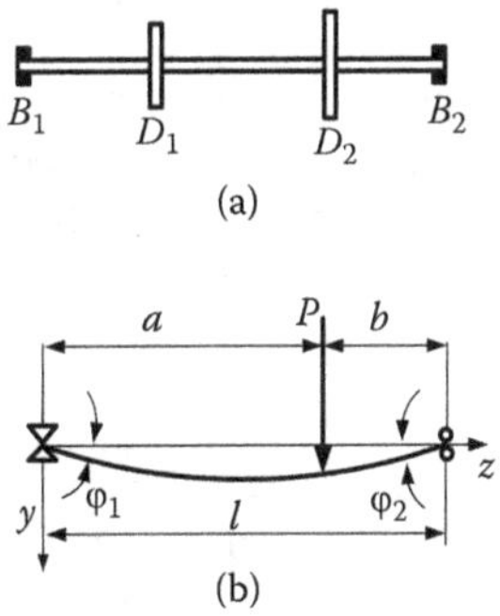

FIGURE 8.65 (a) A simply supported shaft with two disks. (b) A simply supported shaft with a concentrated load *P* at any point.

Exercise 8.21 Obtain the natural frequencies of the rotor system shown in Figure 8.66. Assume the rotor has free–free boundary conditions. Take the shaft material Young's modulus as $E=2.1\times10^{11}$ N/m^2 and density $\rho=7800$ kg/m^3. The mass of the rotor is 15 kg, and the diametral mass moments of inertia of the disk is 0.2 kg-m^2. Take the shaft diameter as 0.02 m and the length as 0.80 m. Use the TMM.

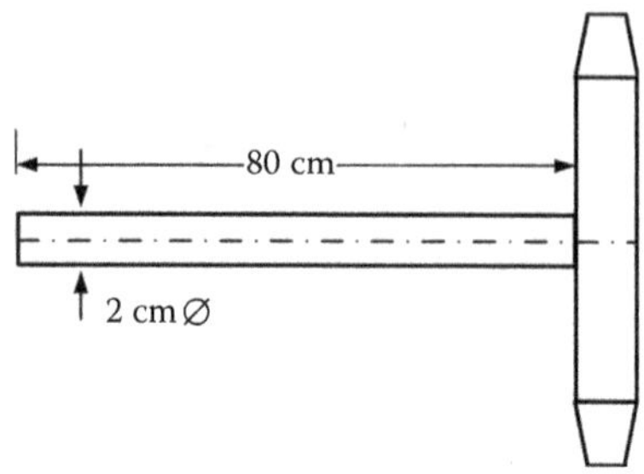

FIGURE 8.66 A rotor with free–free boundary conditions.

Exercise 8.22 In the TMM for transverse vibration if the gyroscopic effect of the disk is also considered, then with the help of a free body diagram write down the changes in the governing equation (relations for the translational and rotational displacements, and force and moment equations) for the field and point matrices and derive new field and point matrices in the expanded form.

Exercise 8.23 Obtain the transverse natural frequency of the rotor-bearing system shown in Figure 8.67 by the FEM (Chapter 9) or the TMM. Both bearings have fixed end conditions. At the disk location, the rotor is experiencing translational and rotational (tilting) stiffness, k_y and k_{φ_x} (torsional spring is not shown in the figure for brevity). Take $L = 1$ m, $a = 0.3$ m, $d = 0.15$ m, $k_y = 1$ MN/m, $k_{\varphi_x} = 100$ kN-m/rad, $m = 5$ kg, $I_d = 0.02$ kg-m², $E = 2.1 \times 10^{11}$ N/m², and $\rho = 7800$ kg/m³. Consider one-plane motion only.

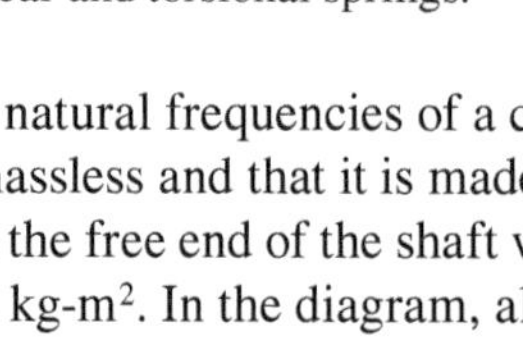

FIGURE 8.67 A rotor system with linear and torsional springs.

Exercise 8.24 Find the transverse natural frequencies of a cantilever rotor system as shown in Figure 8.68. Consider the shaft as massless and that it is made of steel with Young's modulus of 2.1×10^{11} N/m². A disk is mounted at the free end of the shaft with a mass of 10 kg, and the diametral mass moment of inertia is 0.04 kg-m². In the diagram, all dimensions are in cm.

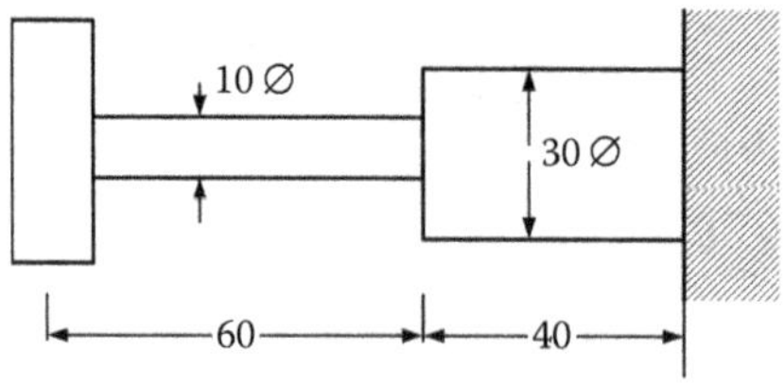

FIGURE 8.68 A stepped shaft with cantilever end conditions.

Exercise 8.25 Obtain the transverse natural frequencies of the rotor system shown in Figure 8.69. The rotor system consists of a uniform shaft symmetrically supported by three bearings and symmetrically mounted with two disks. Bearings prevent translational transverse displacements. The total length of the shaft is 1 m, and the diameter is 10 mm. The Young's modulus of the shaft material $E=2.1\times10^{11}$ N/m^2. The disks each have mass 2 kg, and the inertia of the shaft can be neglected.

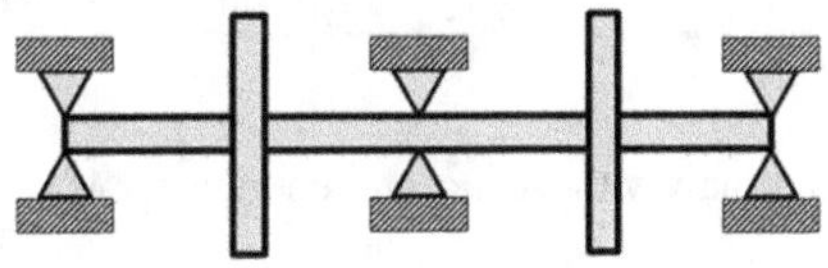

FIGURE 8.69 A multi-support rotor system.

Exercise 8.26 Obtain the *transverse natural frequencies* of the rotor system as shown in Figure 8.70. Use only the TMM or the FEM (Chapter 9). The diameter of the shaft is 1 cm, and its length is 90 cm with fixed–fixed end conditions. Consider the mass of the shaft. The disk is placed 30 cm from one of the bearing ends. The disk is thin with a mass of 3 kg and a diameter of 10 cm. Springs support the disk; each spring's stiffness is $k=10$ kN/m. Consider only single-plane motion and neglect the gyroscopic effect. Take for the shaft material properties $E=2.1\times10^{11}$ N/m^2 and $\rho=7800$ kg/m^3.

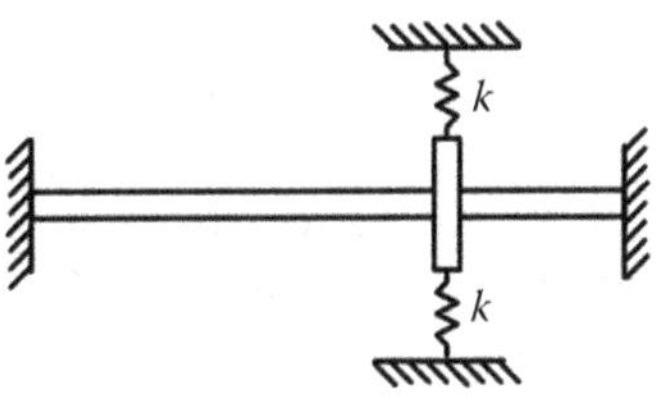

FIGURE 8.70 A rotor system with springs.

Exercise 8.27 Obtain the transverse natural frequencies of the rotor system shown in Figure 8.71. It has simply supported conditions and identical thin disks ($I_d=0.01$ kg-m^2), and these are mounted very close to the support itself so that they have pure tilting motion (transverse rotational displacements). It is assumed that the supports do not interfere with the motion of the disks. Use *the influence coefficient method* only. The relations of transverse rotational displacements due to the moment at one end are given in Table 8.15 for Figure 8.72. Take $E=2.1\times10^{11}$ N/m^2. The length of the shaft is 0.6 m, and its diameter is 0.01 m. Neglect the mass of the shaft and gyroscopic effects of disks.

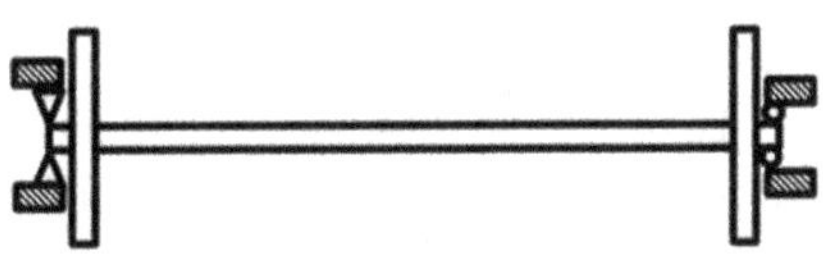

FIGURE 8.71 A rotor system with disks near supports.

TABLE 8.15
A Simply Supported Shaft with a Concentrated Moment *M* at One End Point

$\varphi_1 = \dfrac{Ml}{6EI}$	$y = \dfrac{Ml}{6EI}\left(1 - \dfrac{z^2}{l^2}\right)$
$\varphi_2 = \dfrac{Ml}{3EI}$	

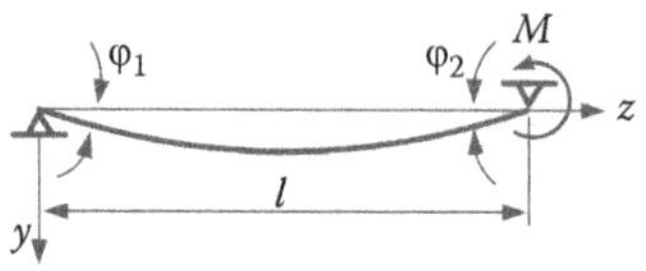

FIGURE 8.72 A simply supported shaft with a concentrated moment M at one end point.

Exercise 8.28 Obtain the transverse natural frequencies of the rotor system shown in Figure 8.73 for free–free boundary conditions. Use *the TMM* only. Take $E=2.1\times10^{11}\,\text{N/m}^2$. All the disks have the same mass of 3 kg. Between two adjacent disks, the shaft has a length of 0.4 m and a diameter of 0.01 m. Neglect the mass of the shaft and the diametral mass moment of inertia of the disk. Neglect gyroscopic effects.

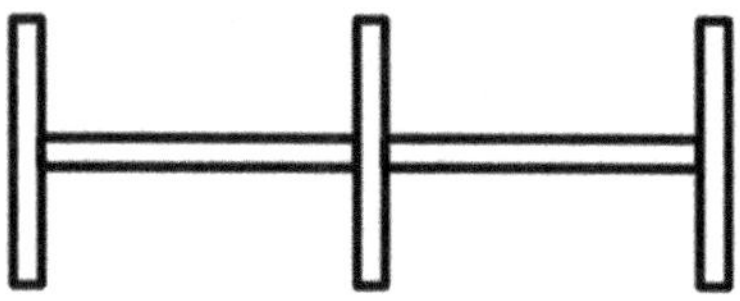

FIGURE 8.73 A three-disk rotor system with free end conditions.

Exercise 8.29 Use the TMM to obtain the natural frequencies of a free–free rotor system. Obtain the transverse natural frequencies of the rotor system for the following parameters: (i) Disks are point masses with $m_1=5\,\text{kg}$, $m_2=8\,\text{kg}$, $l=1\,\text{m}$, $d=0.02\,\text{m}$, $\rho=7800\,\text{kg/m}^3$, and $E=2.1\times10^{11}\,\text{N/m}^2$ and (ii) the disks are thin with radii of 10 and 15 cm with the other parameters the same as case (i). Use the TMM.

Exercise 8.30 A motor shaft drives another (second) shaft through a pulley and belt arrangement. The motor shaft is modeled as a cantilever shaft (0.01 m diameter and length of 0.08 m) with a rigid disk (i.e., a pulley of mass 0.5 kg) at the free end. The belt in tension is modeled as a spring (1 MN/m) and the second shaft system as simply supported (0.01 m diameter and length of 0.30 m) with a pulley (0.75 kg) at its midspan. Neglect the mass of the shafts; however, consider its flexibility. Take the material of the shaft $E=2.1\times10^{11}\,\text{N/m}^2$. Obtain the transverse natural frequencies of the complete rotor system.

Exercise 8.31 A rotor system, as shown in Figure 8.74, has simply supported conditions, and two thin disks ($m=2\,\text{kg}$ and $I_d=0.01$ kg-m^2) are one mounted at the midspan (which has pure translational displacement) and another very close to the left support itself (which has pure tilting motion, i.e., transverse rotational displacement). It is assumed that the support does not interfere the motion of the end disk. Relations of the transverse translational/rotational displacement due to a force at the midspan and relations of transverse translational/rotational displacement due to moment at one end are given in Table 8.16. Take $E=2.1\times10^{11}\,\text{N/m}^2$. The length of the shaft is 0.6 m, and its diameter is 0.01 m. Neglect the mass of the shaft and gyroscopic effect. Obtain the critical speed of the system for the transverse vibration. Use the influence coefficient method only (refer to Figures 8.76–8.77).

TABLE 8.16
A Simply Supported Shaft with a Moment at One End and a Force at the Midspan

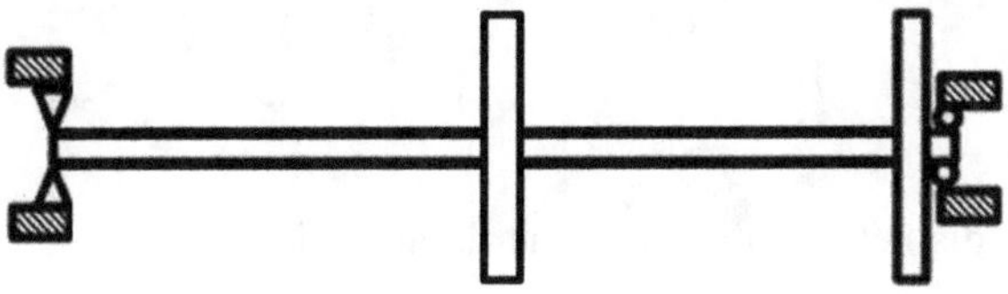

FIGURE 8.74 A simply supported shaft with a disk at the midspan and another at the left end.

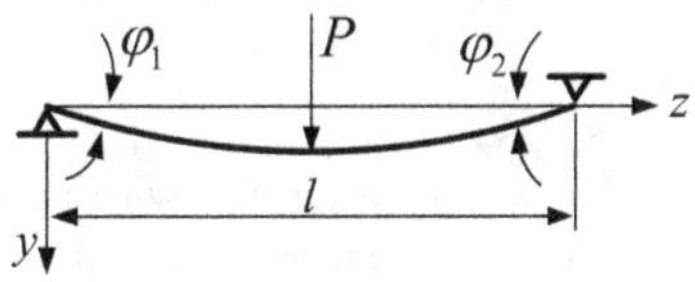

FIGURE 8.75 Simply supported shaft with a force at the midspan.

$$y = \frac{Mz}{6EIl}\left(1 - \frac{z^2}{l^2}\right); \quad 0 < z < l$$

$$y = \frac{Pz}{12EI}\left(\frac{3l^2}{4} - z^2\right); \quad 0 < z < l/2$$

$\varphi = dy/dz$

Exercise 8.32 Multiple choice questions: Select a single answer only.

i. Dunkerley's formula gives the lower bound of the fundamental frequency.
   A. true
   B. false

ii. There is an approximation involved in the formulation of the influence coefficient method for a multi-DOF rotor system.
   A. true
   B. false

iii. There is an approximation involved in the solution procedure by the influence coefficient method for a multi-DOF rotor system.
   A. true
   B. false

iv. There is an approximation involved in the solution procedure by the TMM for a multi-DOF rotor system.
   A. true
   B. false

v. There is an approximation involved in the formulation of the TMM for a multi-DOF rotor system.
   A. true
   B. false

vi. For an N-disk rotor with negligible diametral mass moments of inertia, the number of transverse natural frequencies would be
   A. N
   B. $2N$
   C. $4N$
   D. N^2

vii. For an N-disk rotor with appreciable diametral mass moment of inertia without gyroscopic effects, the number of transverse natural frequencies would be
   A. N
   B. $2N$
   C. $4N$
   D. N^2

viii. For an N-disk rotor with appreciable diametral moment of inertia and with gyroscopic effects, the number of transverse natural frequencies would be
   A. N
   B. $2N$
   C. $4N$
   D. N^2

ix. A point matrix in the TMM relates
   A. state vectors at either side of a disk
   B. state vectors at either side of a shaft
   C. state vectors from a shaft to another shaft
   D. state vectors at a disk to another disk

x. A field matrix in the TMM relates
   A. state vectors at either side of a disk
   B. state vectors at either side of a shaft
   C. state vectors from a shaft to another shaft
   D. state vectors at a disk to another disk

xi. A transfer matrix relates
   A. state vectors from a shaft to another shaft
   B. state vectors at a disk to another disk
   C. state vectors from one end of the rotor system to the other end
   D. either (A) or (B)

xii. An overall transfer matrix relates
   A. state vectors from a shaft to another shaft
   B. state vectors at a disk to another disk
   C. state vectors from one end of the rotor system to the other end
   D. either (A) or (B)

xiii. On analyzing free transverse vibration of a Jeffcott rotor (a disk at the midspan) using the influence coefficient method for a pure tilting motion, the natural frequency will be (herein, I_d is the diametral mass moment of inertia, m is the mass, EI is the flexural rigidity, and l is the shaft span)
   A. $\sqrt{\frac{12EI}{I_d l}}$
   B. $\sqrt{\frac{3EI}{I_d l}}$
   C. $\sqrt{\frac{48EI}{ml^3}}$
   D. $\sqrt{\frac{3EI}{ml^3}}$

xiv. For a cantilever beam with a point moment at free end, the influence coefficient relating to the linear displacement with the applied moment will be (EI is the flexural rigidity and l is the shaft span)

A. $\dfrac{l^2}{2EI}$

B. $\dfrac{l^2}{3EI}$

C. $\dfrac{l^2}{12EI}$

D. $\dfrac{l^2}{48EI}$

APPENDIX 8.1: LOAD–DEFLECTION RELATIONS FOR VARIOUS BOUNDARY CONDITIONS OF THE SHAFT

Table 8.17 shows the load–deflection relations for various boundary conditions of the shaft.

ANSWERS TO SELECTED EXERCISE PROBLEMS

8.1 $\omega_{nf1} = 84.6$ rad/s; $\omega_{nf2} = 333.8$ rad/s.

8.2 $\omega_{nf1} = 4967.1$ rad/s; $\omega_{nf2} = 30{,}249.1$ rad/s.

8.3 $\omega_{nf1} = 4132.2$ rad/s; $\omega_{nf2} = 19042.2$ rad/s; $\omega_{nf3} = 40{,}672.5$ rad/s; $\omega_{nf4} = 55{,}845.1$ rad/s.

8.4 $\omega_{nf1} = 25.5$ rad/s; $\omega_{nf2} = 243.0$ rad/s.

8.5 Part I: $\omega_{nf1} = 28.0$ rad/s; $\omega_{nf2} = 245.5$ rad/s.
Part II: $\omega_{nf1} = 24.2$ rad/s; $\omega_{nf2} = 175.6$ rad/s.

8.6 Part I: $\omega_{nf1} = 389.3$ rad/s; $\omega_{nf2} = 1269.1$ rad/s.
Part II: $\omega_{nf1} = 200.1$ rad/s; $\omega_{nf2} = 489.4$ rad/s; $\omega_{nf3} = 781.9$ rad/s; $\omega_{nf4} = 1337.9$ rad/s.

8.7 $\omega_{nf1} = 158.2$ rad/s; $\omega_{nf2} = 749.7$ rad/s; $\omega_{nf3} = 1709.2$ rad/s; $\omega_{nf4} = 2679.9$ rad/s.
$\omega_{nf5} = 3818.8$ rad/s; $\omega_{nf6} = 6207.0$ rad/s; $\omega_{nf7} = 7540.1$ rad/s; $\omega_{nf8} = 9172.1$ rad/s.

8.8 $\omega_{nf1} = 1495.7$ rad/s; $\omega_{nf2} = 3632.7$ rad/s; $\omega_{nf3} = 4310.8$ rad/s; $\omega_{nf4} = 5925.5$ rad/s,
$\omega_{nf5} = 8463.0$ rad/s; $\omega_{nf6} = 9564.4$ rad/s; $\omega_{nf7} = 174{,}200.7$ rad/s; $\omega_{nf8} = 349{,}429.1$ rad/s.

8.9 $\omega_{nf1} = 60.4$ rad/s; $\omega_{nf2} = 135.6$ rad/s; $\omega_{nf3} = 991.7$ rad/s; $\omega_{nf4} = 1765.1$ rad/s.

8.10 Not applicable

8.11 $\omega_{nf1} = 0$ rad/s; $\omega_{nf2} = 0$ rad/s; $\omega_{nf3} = 319.6$ rad/s; $\omega_{nf4} = 760.7$ rad/s.
(Results are without shaft inertia and by FEM/TMM)
$\omega_{nf1} = 0$ rad/s; $\omega_{nf2} = 0$ rad/s; $\omega_{nf3} = 225.23$ rad/s; $\omega_{nf4} = 577.07$ rad/s.
(Results are with shaft inertia and by FEM)

8.12 $\omega_{nf1} = 953.7$ rad/s; $\omega_{nf2} = 2143.1$ rad/s; $\omega_{nf3} = 2565.5$ rad/s; $\omega_{nf4} = 3530.6$ rad/s.

8.13 $\omega_{nf1} = 45.0$ rad/s.

8.14 $\omega_{nf1} = 166.4$ rad/s; $\omega_{nf2} = 455.4$ rad/s.

8.15 $\omega_{nf1} = 162.6$ rad/s; $\omega_{nf2} = 446.1$ rad/s; $\omega_{nf3} = 1309.2$ rad/s; $\omega_{nf4} = 2006.3$ rad/s,
$\omega_{nf5} = 17{,}819.3$ rad/s; $\omega_{nf6} = 17{,}906.9$ rad/s.

8.16 $\omega_{nf1} = 192.3$ rad/s; $\omega_{nf2} = 979.9$ rad/s; $\omega_{nf3} = 2151.9$ rad/s; $\omega_{nf4} = 3334.2$ rad/s,
$\omega_{nf5} = 5062.5$ rad/s; $\omega_{nf6} = 5451.5$ rad/s; $\omega_{nf7} = 7806.6$ rad/s; $\omega_{nf8} = 9268.9$ rad/s.

8.17 $\omega_{nf1} = 916.5$ rad/s; $\omega_{nf2} = 5316.9$ rad/s; $\omega_{nf3} = 9030.2$ rad/s.

8.18 $\omega_{nf1} = 1116.0$ rad/s; $\omega_{nf2} = 8216.3$ rad/s.

8.19 $\omega_{nf1} = 142.7$ rad/s; $\omega_{nf2} = 9129.7$ rad/s.

8.20 $\omega_{nf1} = 389.3$ rad/s; $\omega_{nf2} = 1269.1$ rad/s.

TABLE 8.17
Load–Deflection Relations for Various Boundary Conditions of the Shaft

S.N.	Boundary Conditions	Rotational Displacement	Translational Displacement at Any Location of the Shaft	Maximum Translational Displacement (δ^{max})
1	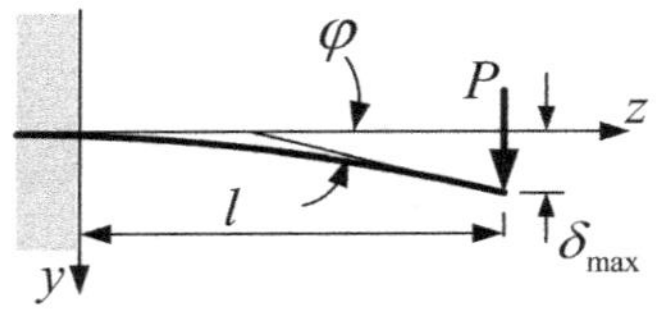 **FIGURE 8.76(A)** A cantilever shaft with a concentrated load P at the free end.	$\varphi\vert_{z=l}=\frac{pl^2}{2EI}$	$y=\frac{Pz^2}{6EI}(3l-z)$	$y\vert_{z=1}=\frac{Pl^3}{3EI}$
2	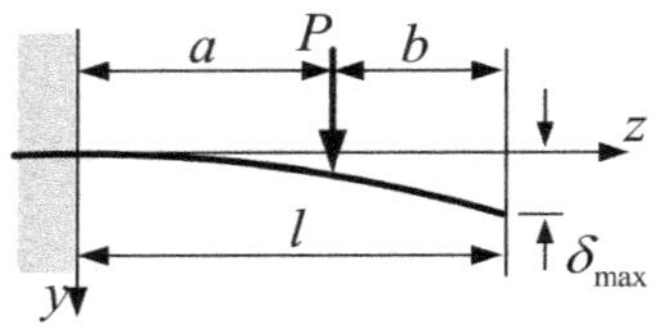 **FIGURE 8.76(B)** A cantilever shaft with a concentrated load P at $z=a$.	$\varphi\vert_{z=l}=\frac{pa^2}{2EI}$	$y=\frac{Pz^2}{6EI}(3l-z)$ for $0<z<a$ $y\vert_{z=1}=\frac{Pa^2}{6EI}(3l-a)$ for $a<z<1$	$y=\frac{Pa^2}{6EI}(3z-a)$
3	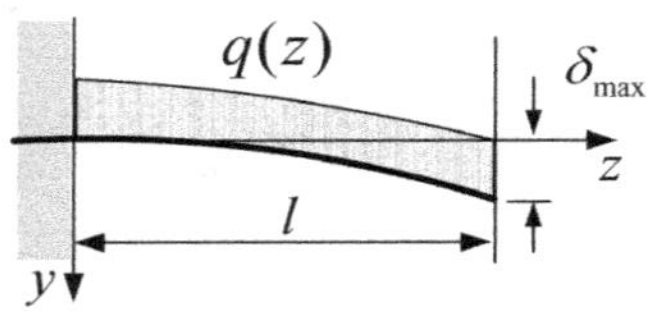 **FIGURE 8.76(C)** A cantilever shaft with a uniformly distributed load $q(z)=q_0$ over the whole span.	$\varphi\vert_{z=l}=\frac{q_0l^3}{2EI}$	$y=\frac{q_0z^2}{24EI}(z^2+6l^2-4lz)$	$y\vert_{z=1}=\frac{q_0l^4}{8EI}$

(Continued)

TABLE 8.17 (*Continued*)
Load–Deflection Relations for Various Boundary Conditions of the Shaft

S.N.	Boundary Conditions	Rotational Displacement	Translational Displacement at Any Location of the Shaft	Maximum Translational Displacement (δ^{max})
4	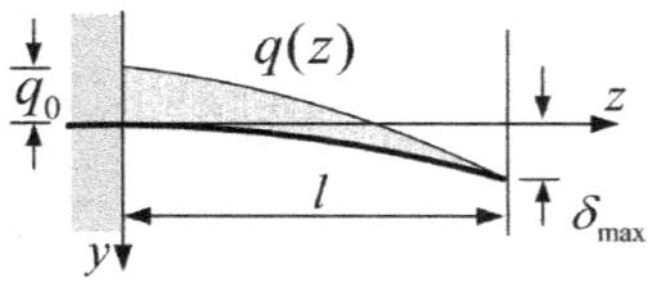 **FIGURE 8.76(D)** A cantilever shaft with a linearly varying distributed load $q(z)$ over the whole span.	$\varphi\big\|_{z=l} = \frac{q_0 l^3}{24EI}$	$y = \frac{q_0 z^2}{120lEI}\left(10l^2 - 10l^2 z + 5lz^2 - z^3\right)$	$y\big\|_{z=1} = \frac{q_0 l^4}{30EI}$
5	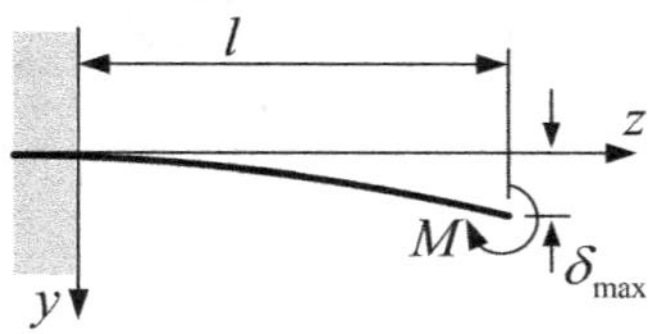 **FIGURE 8.76(E)** A cantilever shaft with a concentrated moment M at the free end.	$\varphi\big\|_{z=l} = \frac{Ml}{EI}$	$y = \frac{Mz^2}{2EI}$	$y\big\|_{z=1} = \frac{Ml^2}{2EI}$
6	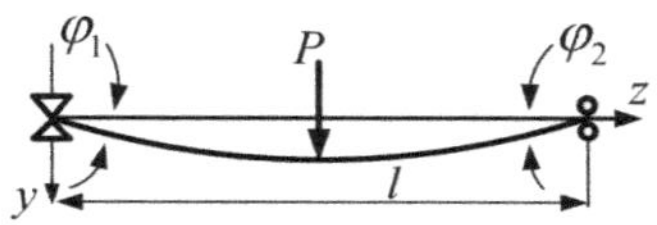 **FIGURE 8.76(F)** A simply supported shaft with a concentrated load P at the midspan.	$\varphi_1 = \varphi_2 = \frac{pl^2}{16EI}$	$y = \frac{Pz}{12EI}\left(\frac{3l^2}{4} - z^2\right)$ for $0 < z < \frac{1}{2}$	$y\big\|_{z=0.5l} = \frac{Pl^2}{48EI}$

(*Continued*)

TABLE 8.17 (*Continued*)

Load–Deflection Relations for Various Boundary Conditions of the Shaft

S.N.	Boundary Conditions	Rotational Displacement	Translational Displacement at Any Location of the Shaft	Maximum Translational Displacement (δ^{max})
7	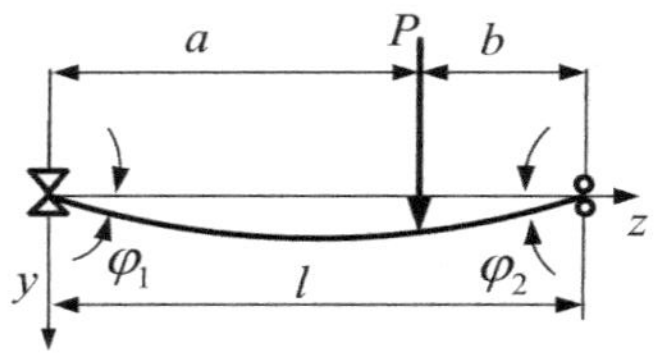 **FIGURE 8.76(G)** A simply supported shaft with a concentrated load *P* at any point.	$\varphi_1 = \frac{Pb\left(l^2 - b^2\right)}{6lEI}$ $\varphi_2 = \frac{Pab(2l - b)}{6lEI}$	$y = \frac{Pbz}{6lEI}\left(l^2 - z^2 - b^2\right)$ for $[0 < z < a]$ $y = \frac{Pb}{6lEI}\left[\frac{l}{b}(z-a)^3 + \left(l^2 - b^2\right)z - z^3\right]$ for $[a < z < l]$	$y\big\rvert_{z=\sqrt{\frac{l^2-b^2}{3}}} = \frac{Pb\left(l^2 - b^2\right)^{3/2}}{9\sqrt{3}lEI}$ and at the center, if $a > b$ $y\big\rvert_{z=0.5l} = \frac{Pb}{48EI}\left(3l^2 - 4b^2\right)$
8	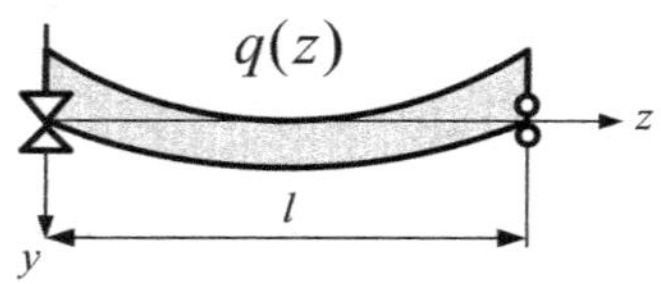 **FIGURE 8.76(H)** A simply supported shaft with a uniformly distributed load $q(z) = q_0$ over the whole span.	$\varphi_1 = \varphi_2 = \frac{ql^3}{24EI}$	$y = \frac{qz}{24EI}\left(l^3 - 2lz^2 + z^3\right)$	$y\big\rvert_{z=0.5l} = \frac{5ql^4}{384EI}$
9	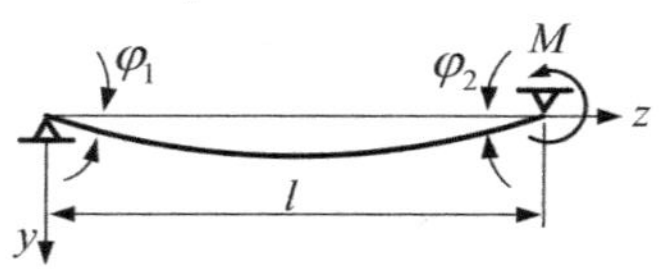 **FIGURE 8.76(I)** A simply supported shaft with a concentrated moment *M* at one end point.	$\varphi_1 = \frac{Ml}{6EI}$ $\varphi_2 = \frac{Ml}{3EI}$	$y = \frac{Mlz}{6EI}\left(1 - \frac{z^2}{l^2}\right)$	$y\big\rvert_{z=l/\sqrt{3}} = \frac{5Ml^2}{9\sqrt{3}EI}$ and at the center $y\big\rvert_{z=0.5l} = \frac{Ml^2}{16EI}$

(*Continued*)

TABLE 8.17 (*Continued*)
Load–Deflection Relations for Various Boundary Conditions of the Shaft

S.N.	Boundary Conditions	Rotational Displacement	Translational Displacement at Any Location of the Shaft	Maximum Translational Displacement (δ^{max})
10	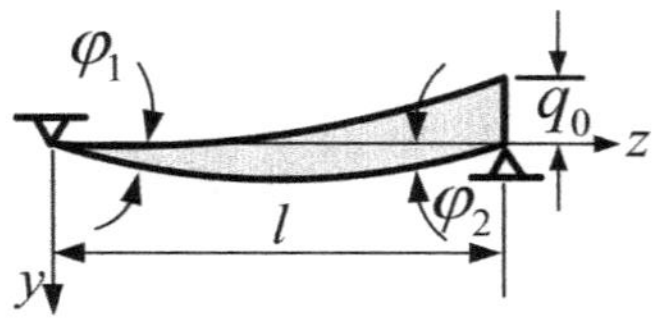 **FIGURE 8.76(J)** A simply supported shaft with a linearly varying load.	$\varphi_1 = \frac{7q_0l^3}{360EI}$ $\varphi_2 = \frac{q_0l^3}{45EI}$	$y = \frac{q_0z}{360lEI}\left(7l^4 - 10l^2z^2 + 3z^4\right)$	$y\big\|_{z=0.519l} = 0.00652\frac{q_0l^4}{EI}$ and at the center $y\big\|_{z=0.5l} = 0.00651\frac{q_0l^4}{EI}$

8.21 $\omega_{nf1} = 0.0$ rad/s; $\omega_{nf2} = 0.0$ rad/s; $\omega_{nf3} = 245.2$ rad/s; $\omega_{nf4} = 1456.4$ rad/s.
8.22 Not applicable
8.23 $\omega_{nf1} = 18{,}074.4$ rad/s; $\omega_{nf2} = 71{,}720.4$ rad/s.
8.24 $\omega_{nf1} = 1156.7$ rad/s; $\omega_{nf2} = 13{,}043.6$ rad/s.
8.25 $\omega_{nf1} = 140.7$ rad/s; $\omega_{nf2} = 212.7$ rad/s.
8.26 $\omega_{nf1} = 139.6$ rad/s; $\omega_{nf2} = 1050.7$ rad/s.
8.27 $\omega_{nf1} = 185.4$ rad/s; $\omega_{nf2} = 321.1$ rad/s.
8.28 $\omega_{nf1} = 0.0$ rad/s; $\omega_{nf2} = 0.0$ rad/s; $\omega_{nf3} = 69.5$ rad/s
8.29 Part I: $\omega_{nf1} = 0.0$ rad/s; $\omega_{nf2} = 0.0$ rad/s; $\omega_{nf3} = 306.1$ rad/s; $\omega_{nf4} = 1331.7$ rad/s
Part II: $\omega_{nf1} = 0.0$ rad/s; $\omega_{nf2} = 0.0$ rad/s; $\omega_{nf3} = 637.1$ rad/s; $\omega_{nf4} = 1515.9$ rad/s.
8.30 $\omega_{nf1} = 631.8$ rad/s; $\omega_{nf2} = 2039.0$ rad/s.
8.31 $\omega_{nf1} = 100.38$ rad/s; $\omega_{nf2} = 365.94$ rad/s.
8.32

(i) A	(ii) A	(iii) A	(iv) A	(v) B
(vi) A	(vii) B	(viii) C	(ix) A	(x) B
(xi) B	(xii) C	(xiii) A	(xiv) A	

REFERENCES

Dunkerley, S., 1894, On the whirling and vibration of shafts, *Philosophical Transactions of the Royal Society of London*, Series A, *185*, Part I, 279.
Meirovitch, L., 1986, *Elements of Vibration Analysis*. New York, NY: McGraw Hill Book.
Myklestad, N.O., 1944, A new method of calculating natural modes of uncoupled bending vibrations, *Journal of Aeronautical Science*, *11*(2), 153–162.
Prohl, M.A., 1945, A general method for calculating the critical speeds of flexible rotors, *Journal of Applied Mechanics*, *12*(*3*), 142–148.
Scarborough, J.B., 1966, *Numerical Mathematical Analysis*, 6th ed. Baltimore, MD: Johns Hopkins Press.
Srinath, L.S., 2009, *Advanced Mechanics of Solid*, 3rd ed. New Delhi, India: Tata McGraw-Hill Education.
Thomson, W.T. and Dahleh, M.D., 1998, *Theory of Vibration with Applications*, 5th ed. New Delhi: Pearson Education Inc.
Timoshenko, S.P. and Young, D.H., 1968, *Elements of Strength of Materials*, An East-West edition, 5th ed. New Delhi: D. Van Nostrand.
Young, W.C. and Budynas, R.G., 2002, *Roark's Formulas for Stress and Strain*, *General Engineering Series*, New York, NY: McGraw-Hill International Edition.

9 Transverse Vibrations of Rotor Systems by the Continuous and Finite Element Methods

In Chapter 8, we studied transverse vibrations of multi-degree-of-freedom (DOF) rotor systems by considering the shaft as massless (or lumped mass) and flexible by using the influence coefficient and transfer matrix methods. While dealing with torsional vibrations, we learned that rotors could be modeled accurately, if treated as a continuous system (i.e., distributed inertia and elastic properties). The present chapter considers transverse vibrations of the shaft as having distributed mass and stiffness properties. Equations of motion are obtained by using Newton's second law of motion and Hamilton's principle. These equations are partial differential equations as we have seen in the case of torsional vibration. For simple boundary conditions, closed-form solutions are obtained by the method of separation of variables. However, the continuous system approach becomes impracticable for more complex boundary conditions. Hence, we need to resort to some approximate methods, for example, the finite element method (FEM). For the finite element (FE) analysis of rotor systems, the Euler–Bernoulli beam model is considered initially for developing the elemental mass and stiffness matrices. Both the free and forced vibrations are analyzed for a variety of conditions at supports using the FEM. In the FEM, the overall size of matrices increases with the number of DOFs of the system. It is often required to reduce the size of the actual matrices to be solved to save computational time. Such cases can be handled using static and dynamic reductions (or condensations). These reductions introduce some amount of approximation in solutions. Subsequent chapters will discuss higher order effects such as rotary inertia, shear deformation, and gyroscopic moment in the dynamic analysis of rotor systems using the FEM.

9.1 GOVERNING EQUATIONS IN CONTINUOUS SYSTEMS

In the present section, one-dimensional governing equations with boundary conditions (i.e., the boundary-value problem) of a shaft will be formulated for transverse vibrations by using Newton's law of motion and Hamilton's principle. Extended Hamilton's principle requires various energies, for example, the strain and kinetic energies, and the work done by external forces. Consider a circular shaft of length, L, with a cross-sectional area, A, as shown in Figure 9.1. It is acted upon by

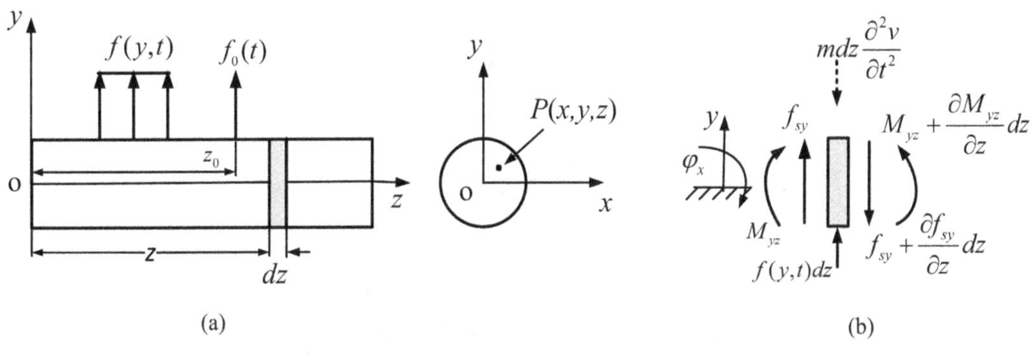

FIGURE 9.1 (a) Euler–Bernoulli beam with loadings. (b) Free body diagram of a shaft infinitesimal segment.

DOI: 10.1201/9781003661559-9

an external distributed force, $f(z, t)$, over a length of the shaft, and by a concentrated force, $f_0(t)$, at $z=z_0$ in the direction of the y-axis. Thus, the shaft is subjected to lateral (transverse or bending) vibrations in the vertical direction (i.e., y direction). For the Euler–Bernoulli beam, motions in the vertical and horizontal planes are uncoupled, that is, the vertical force produces only vertical displacements (translational and rotational) and the horizontal force produces only horizontal displacements. Therefore, the analyses could be done separately in the two orthogonal planes and would remain identical. The analysis presented here is for the vertical plane y–z and that for the horizontal plane z–x will remain the same (however, there will be change in signs of angular displacements and moments, refer to Appendix 9.1).

Let us consider a cross-sectional plane at a distance, z, as shown in Figures 9.1 and 9.2. Let v be the transverse displacement in the y direction of the neutral axis at the distance z. For small displacements, all the points of the chosen plane have the same transverse displacement. According to the Euler–Bernoulli beam theory, a plane cross section before bending remains plane even after the bending. Thus, the plane has a rotation about the x-axis and is given by a slope, $\varphi_x = -\dfrac{\partial v}{\partial z}$, of the elastic curve (i.e., the neutral axis) as shown in Figure 9.2. To define the position of the chosen plane, two displacement variables are required, v and φ_x. These displacements are assumed to be small. Let us choose a point P with (x, y, z) as its coordinates in the chosen plane at a distance z.

9.1.1 Displacement Fields

The displacement field of the point P, under the assumption of small displacements, can be obtained as given by Equations 9.1 and 9.2 (see Figure 9.3):

$$v_x = 0; \quad v_y = v(z,t); \quad v_z = y\varphi_x = -y\frac{\partial v(z,t)}{\partial z} = -yv' \tag{9.1}$$

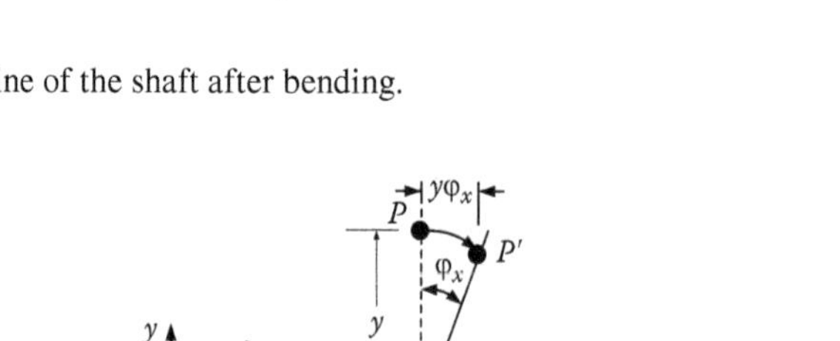

FIGURE 9.2 Elastic line of the shaft after bending.

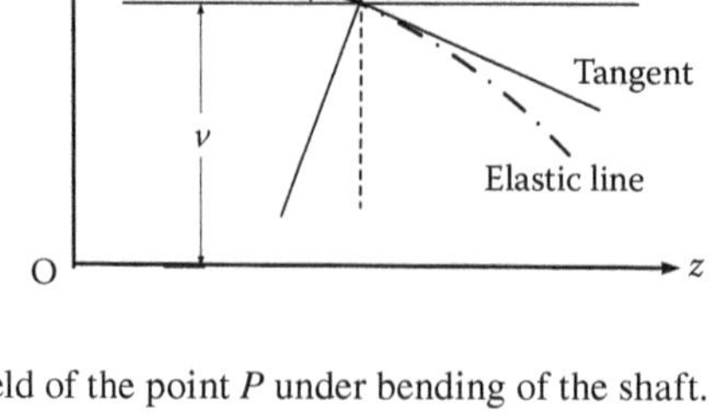

FIGURE 9.3 Displacement field of the point P under bending of the shaft.

with

$$\varphi_x = -\frac{\partial v}{\partial z} \tag{9.2}$$

where v is the displacement component of the point P, and the subscripts (i.e., x, y, z) represent the directions. The prime (′) represents the derivative with respect to the spatial coordinate z, once.

9.1.2 Strain and Stress Fields

Correspondingly, the strain and stress fields can be given by Equations 9.3 through 9.5:

$$\varepsilon_{zz} = \frac{\partial v_z}{\partial z} = -yv''(z,t); \quad \varepsilon_{yz} = \left(\frac{\partial v_y}{\partial z} + \frac{\partial v_z}{\partial y}\right) = \left(\frac{\partial v}{\partial z} + \frac{\partial(-yv')}{\partial y}\right) = (v' - v') = 0$$
$$\varepsilon_{xx} = \varepsilon_{yy} = \varepsilon_{xy} = \varepsilon_{zx} = 0 \tag{9.3}$$

and

$$\sigma_{zz} = -Eyv''(z,t); \quad \sigma_{xx} = \sigma_{yy} = \tau_{xy} = \tau_{yz} = \tau_{zx} = 0 \tag{9.4}$$

9.1.3 Energy Expressions

The strain energy is given by Equation 9.5:

$$U = \frac{1}{2}\int_V \left[\sigma_{xx}\varepsilon_{xx} + \sigma_{yy}\varepsilon_{yy} + \sigma_{zz}\varepsilon_{zz} + \tau_{xy}\varepsilon_{xy} + \tau_{yz}\varepsilon_{yz} + \tau_{zx}\varepsilon_{zx}\right] dV \tag{9.5}$$

On substituting the strain and stress fields from Equations 9.3 and 9.4 into Equation 9.5, we get Equation 9.6 as follows:

$$U = \frac{1}{2}\int_V \sigma_{zz}\varepsilon_{zz} dV = \frac{1}{2}\int_0^L\int_A Ey^2 v''^2 dA dz = \frac{1}{2}\int_0^L E\left(\int_A y^2 dA\right) v''^2 dz = \frac{1}{2}\int_0^L EI_{xx}\left\{v''(z,t)\right\}^2 dz \tag{9.6}$$

with

$$I_{xx} = \int_A y^2 dA$$

where I_{xx} is the area moment of inertia of the cross section of the shaft about the axis of bending, that is, the x-axis.

The kinetic energy is given by Equation 9.7:

$$T = \frac{1}{2}\int_0^L\int_A \rho\left(\dot{v}_x^2 + \dot{v}_y^2 + \dot{v}_z^2\right) dA dz \tag{9.7}$$

Neglecting the axial component of the velocity, $\dot{v}_z \approx 0$, the velocity field can be written as follows:

$$\dot{v}_x = 0, \quad \dot{v}_y = \dot{v}(z,t) \quad \text{and} \quad \dot{v}_z \approx 0 \tag{9.8}$$

In view of Equation 9.8, Equation 9.7 gives Equation 9.9:

$$T = \frac{1}{2}\int_0^L\int_A \rho\left(\frac{\partial v}{\partial t}\right)^2 dAdz = \frac{1}{2}\int_0^L \rho A\{\dot{v}(z,t)\}^2 dz \tag{9.9}$$

where ρ is the mass density of the shaft material, A is the area of the cross section of the shaft, and dot (.) represents the partial derivative with respect to the time t. It should be kept in mind that in the above kinetic energy expression only the translational inertia of the shaft is considered. The rotational inertia is ignored in Euler–Bernoulli beam theory. However, the rotary inertia (i.e., Rayleigh's beam theory) will be considered in Chapter 10 along with the gyroscopic moment and shear deformation effects (i.e., the Timoshenko beam theory). The work done (nonconservative) by external forces (the distributed as well as the concentrated) is as follows:

$$W_{nc} = \int_0^L f(z,t)vdz + \int_0^L f_0(t)\delta^*(z - z_0)vdz \tag{9.10}$$

where $f(z,t)$ is the distributed transverse dynamic force in some portion of the shaft length and $f_0(t)$ is the concentrated transverse dynamic load at the axial location z_0. The symbol (δ^*) is the *Dirac delta function* defined as follows:

$$\delta^*(z - z_0) = 1 \quad \text{for} \quad z = z_0, \quad \text{and} \quad \delta^*(z - z_0) = 0 \quad \text{otherwise} \tag{9.11}$$

It has the following property:

$$\int_{-\infty}^{\infty} g(z)\delta^*(z - z_0)dz = g(z_0) \tag{9.12}$$

Use of the Dirac delta function in Equation 9.10 enables the representation of the point (or the concentrated) dynamic force as a distributed force description in the form of an integral.

9.1.4 Equations of Motion

Hamilton's principle for a *conservative dynamical system* states that the variation of the integral of the Lagrangian function $L = (T - U)$ over a given interval of time is stationary (see Chapter 7 for more details). Hence, the elemental equation of motion and boundary conditions can be obtained as follows:

$$\int_{t_1}^{t_2} [\delta(T - U) + \delta W_{nc}]dt = 0 \tag{9.13}$$

where the above form is called the *extended Hamilton's principle*, which is valid for *nonconservative dynamical systems* also. Herein, δ is a variational operator. On substituting Equations 9.6, 9.9, and 9.10 into Equation 9.13, we get

$$\delta\int_{t_1}^{t_2}\int_0^L\left\{\frac{1}{2}\rho A\left(\frac{\partial v(z,t)}{\partial t}\right)^2 - \frac{1}{2}EI_{xx}\left(\frac{\partial^2 v(z,t)}{\partial z^2}\right)^2 + f(z,t)v + f_0(t)\delta^*(z - z_0)v\right\}dzdt = 0 \tag{9.14}$$

Taking the variational operator inside the integral of the above equation, we get Equation 9.15 as follows:

$$\int_{t_1}^{t_2}\int_0^L \left\{ \begin{array}{l} \rho A\left(\dfrac{\partial v(z,t)}{\partial t}\right)\delta\left(\dfrac{\partial v(z,t)}{\partial t}\right) - EI_{xx}\left(\dfrac{\partial^2 v(z,t)}{\partial z^2}\right)\delta\left(\dfrac{\partial^2 v(z,t)}{\partial z^2}\right) \\ +f(z,t)(\delta v) + f_0(t)\delta^*(z-z_0)(\delta v) \end{array} \right\} dz dt = 0 \tag{9.15}$$

To obtain the weak formulation of the above equation, integration by parts of the first two terms with respect to t and z, respectively, results in

$$\begin{aligned} &\int_0^L \left\{ \rho A\left(\frac{\partial v(z,t)}{\partial t}\right)\delta v(z,t)\Big|_{t_1}^{t_2} - \int_{t_1}^{t_2} \rho A\left(\frac{\partial^2 v(z,t)}{\partial t^2}\right)\delta v(z,t)dt \right\} dz \\ &+\int_{t_1}^{t_2} \left\{ \begin{array}{l} -EI_{xx}\left(\dfrac{\partial^2 v(z,t)}{\partial z^2}\right)\delta\left(\dfrac{\partial v(z,t)}{\partial z}\right)\Bigg|_0^L + \dfrac{\partial}{\partial z}\left(EI_{xx}\dfrac{\partial^2 v(z,t)}{\partial z^2}\right)\delta v(z,t)\Bigg|_0^L \\ -\displaystyle\int_0^L \dfrac{\partial^2}{\partial z^2}\left(EI_{xx}\dfrac{\partial^2 v(z,t)}{\partial z^2}\right)\delta v(z,t)dz \end{array} \right\} dt \\ &+\int_{t_1}^{t_2}\int_0^L \left\{ f(z,t) + f_0(t)\delta^*(z-z_0) \right\}\delta v dz dt = 0 \end{aligned} \tag{9.16}$$

The variation of the displacement, v, at two instances t_1 and t_2 is zero; hence, the first term is zero. On rearranging Equation 9.16, we get

$$\begin{aligned} &\int_{t_1}^{t_2}\left[\int_0^L \left\{ \frac{\partial^2}{\partial z^2}\left(EI_{xx}\frac{\partial^2 v}{\partial z^2}\right) + \rho A\frac{\partial^2 v}{\partial t^2} - f - f_0\delta^*(z-z_0) \right\}\delta v dz \right. \\ &\left. + EI_{xx}\frac{\partial^2 v}{\partial z^2}\delta\left(\frac{\partial v}{\partial z}\right)\Bigg|_0^L - \frac{\partial}{\partial z}\left(EI_{xx}\frac{\partial^2 v}{\partial z^2}\right)\delta v\Bigg|_0^L \right] dt = 0 \end{aligned} \tag{9.17}$$

Since δv is the variation of displacement, which is arbitrary, we will have the differential equation of motion from the coefficient of δv within the double integral of the above equation, as follows:

$$EI_{xx}v''''(z,t) + \rho A\ddot{v}(z,t) = f(z,t) + f_0(t)\delta^*(z-z_0) \tag{9.18}$$

Similarly, the last two terms in Equation 9.17 should be individually equated to zero to get the boundary conditions as given below:

$$EI_{xx}v''(z,t)\delta v'(z,t)\big|_0^L = 0 \quad \text{and} \quad EI_{xx}v'''(z,t)\delta v(z,t)\big|_0^L = 0 \tag{9.19}$$

where depending upon the boundary conditions, variations δv and $\delta v'$ or their coefficients v'' and v''' can be zero. Equations 9.18 and 9.19 form a *boundary-value problem.* It should be noted that Equation 9.18 is a linear partial differential equation. The independent variables are spatial, z, and temporal, t, in nature.

9.1.5 Alternative Method

In this section, the equation of motion will be derived using Newton's second law of motion.

Consider a free body diagram of an infinitesimal shaft segment as shown in Figure 9.1b. M_{yz} is the bending moment in *the* y–z plane and f_{sy} is the shear force in the y direction (shear force and bending moment sign convention is based on the strength of the material). The moment balance about the right plane of the shaft segment gives us Equation 9.20 as follows:

$$M_{yz} - \left(M_{yz} + \frac{\partial M_{yz}}{\partial z} dz \right) - \left(m dz \frac{\partial^2 v}{\partial t^2} \right) \frac{dz}{2} + f_{sy} dz - f(z,t) dz(dz) = 0 \tag{9.20}$$

with $m = \rho A$ is the mass per unit length of the shaft. External forces produce small terms such as $(dz)^2$ in the above equation, and therefore, they cannot be considered. Diametral mass moment of inertia and associated rotatory inertia is also ignored, which is generally considered in Rayleigh's beam theory. On simplifying the above equation and neglecting higher degree terms of dz, we get

$$\frac{\partial M_{yz}}{\partial z} - f_{sy} = 0 \quad \text{or} \quad EI \frac{\partial^3 v}{\partial z^3} - f_{sy} = 0 \tag{9.21}$$

with

$$M_{yz} = EI \frac{\partial^2 v}{\partial z^2}$$

Differentiating the above equation with respect to z, we get

$$EI \frac{\partial^4 v}{\partial z^4} - \frac{\partial f_{sy}}{\partial z} = 0 \tag{9.22}$$

Next we take the force equilibrium of the shaft segment to get

$$f(z,t) dz + f_0(t) \delta^*(z - z_0) dz + f_{sy} - \left(f_{sy} + \frac{f_{sy}}{\partial z} dz \right) = m dz \frac{\partial^2 v}{\partial t^2} \tag{9.23}$$

which can be simplified to

$$m \frac{\partial^2 v}{\partial t^2} + \frac{\partial f_{sy}}{\partial z} = f(z,t) + f_0(t) \delta^*(z - z_0) \tag{9.24}$$

On substituting Equation 9.22 in Equation 9.24, we get

$$m \frac{\partial^2 v}{\partial t^2} + EI \frac{\partial^4 v}{\partial z^4} = f(z,t) + f_0(t) \delta^*(z - z_0) \tag{9.25}$$

which is the same as derived using Hamilton's principle.

The governing equation can be solved relatively easy for both free and forced vibrations in the closed form for simple boundary conditions, for example, the simply supported, free–free, fixed–free, and fixed–fixed support conditions.

It should be noted that terms $EI_{xx}v''$ and $EI_{xx}v'''$ represent the *bending moment* and the *shear force*, respectively, whereas v' and v are the transverse rotational and translational displacements, respectively. Table 9.1 summarizes the support conditions for which these field variables become zero. It can be observed that the bending moment and the transverse rotational displacement are never zero simultaneously; similarly, the shear force and the transverse translational displacement never become zero simultaneously. For free vibration analysis purposes, the closed-form solution procedures are presented in subsequent sections for a few boundary conditions.

TABLE 9.1
Field Variables for Different Boundary Conditions

S.N.	Boundary Conditions	$EI_{xx}v''$	v'	$EI_{xx}v'''$	v
1	Simply supported	0	–	–	0
2	Freely supported	0	–	0	–
3	Fixed supported	–	0	–	0

9.2 NATURAL FREQUENCIES AND MODE SHAPES

Natural frequencies (or eigenvalues) and mode shapes (or eigenfunctions) in the closed form are obtained by considering the homogeneous (for the free vibration) beam governing Equation 9.18. We consider the undamped bending vibrations of the shaft with a uniform sectional property. For free vibration, let the external force $f(z, t) = f_0(t) = 0$. Based on the synchronous (or harmonic) motion assumption for the free vibration, the response takes the following form:

$$v(z,t) = \chi(z)\eta(t) \quad \text{so that} \quad v''''(z,t) = \chi''''(z)\eta(t) \quad \text{and} \quad \ddot{v}(z,t) = \chi(z)\ddot{\eta}(t) \tag{9.26}$$

where $\chi(z)$ is the *eigenfunction* (i.e., the *mode shape*), and for the free vibration, $\eta(t)$ has the following form:

$$\eta(t) = A\cos\omega_{nf}t + B\sin\omega_{nf}t \quad \text{so that} \quad \ddot{\eta}(t) = -\omega_{nf}^2\eta(t) \tag{9.27}$$

in which ω_{nf} is the *natural frequency*, and A and B are the integral constants to be determined from the initial conditions of the problem. Equation 9.27 gives the condition of simple harmonic motion, which ensures synchronous motion. In synchronous motion, all particles of the shaft have a fixed ratio even at different times. On substituting Equations 9.26 and 9.27 in Equation 9.18 without force terms, we get

$$EI_{xx}\chi''''(z)\eta(t) + \rho A\chi(z)\ddot{\eta}(t) = 0 \tag{9.28}$$

or

$$\left\{EI_{xx}\chi''''(z) - \rho A\omega_{nf}^2\chi(z)\right\}\eta(t) = 0 \tag{9.29}$$

Since $\eta(t) \neq 0$, we have

$$\frac{d^4\chi(z)}{dz^4} - \beta^4\chi(z) = 0 \tag{9.30}$$

with

$$\beta^4 = \frac{\rho A\omega_{nf}^2}{EI_{xx}} \tag{9.31}$$

The general solution of Equation 9.30, which is now an ordinary differential equation of the fourth order, is given as follows:

$$\chi(z) = C_1\sin\beta z + C_2\cos\beta z + C_3\sinh\beta z + C_4\cosh\beta z \tag{9.32}$$

where C_1, C_2, C_3, and C_4 are integration constants to be evaluated from boundary conditions.

Let us consider a shaft with the *simply supported end conditions*. Boundary conditions at shaft ends imply the following conditions on the eigen function:

$$\chi(0)=\chi(L)=0 \quad \text{and} \quad \chi''(0)=\chi''(L)=0 \tag{9.33}$$

On taking the derivative of Equation 9.32 twice with respect to z, we get

$$\chi''(z)=-C_1\beta^2\sin\beta z - C_2\beta^2\cos\beta z + C_3\beta^2\sinh\beta z + C_4\beta^2\cosh\beta z \tag{9.34}$$

Applying the four boundary conditions of Equation 9.33 in Equations 9.32 and 9.34, we get

$$\chi(0)=C_1\times 0 + C_2 + C_3\times 0 + C_4 = 0$$

$$\chi(L)=C_1\sin\beta L + C_2\cos\beta L + C_3\sinh\beta L + C_4\cosh\beta L = 0$$

$$\chi''(0)=-C_1\beta^2\times 0 - C_2\beta^2 + C_3\beta^2\times 0 + C_4\beta^2 = 0$$

and

$$\chi''(L)=-C_1\beta^2\sin\beta L - C_2\beta^2\cos\beta L + C_3\beta^2\sinh\beta L + C_4\beta^2\cosh\beta L = 0 \tag{9.35}$$

which could be written in matrix form as

$$\begin{bmatrix} 0 & 1 & 0 & 1 \\ \sin\beta L & \cos\beta L & \sinh\beta L & \cosh\beta L \\ 0 & -\beta^2 & 0 & \beta^2 \\ -\beta^2\sin\beta L & -\beta^2\cos\beta L & \beta^2\sinh\beta L & \beta^2\cosh\beta L \end{bmatrix} \begin{Bmatrix} C_1 \\ C_2 \\ C_3 \\ C_4 \end{Bmatrix} = \begin{Bmatrix} 0 \\ 0 \\ 0 \\ 0 \end{Bmatrix} \tag{9.36}$$

From the first and third sets of equations in the matrix Equation 9.36 (i.e., $C_2+C_4=0$ and $C_2-C_4=0$), we get $C_2=C_4=0$. Hence, on elimination of the first and third rows, and the second and fourth columns of the above equation, we get

$$\begin{bmatrix} 0 & 1 & 0 & 1 \\ \sin\beta L & \cos\beta L & \sinh\beta L & \cosh\beta L \\ 0 & -\beta^2 & 0 & \beta^2 \\ -\beta^2\sin\beta L & -\beta^2\cos\beta L & \beta^2\sinh\beta L & \beta^2\cosh\beta L \end{bmatrix} \begin{Bmatrix} C_1 \\ 0 \\ C_3 \\ 0 \end{Bmatrix} = \begin{Bmatrix} 0 \\ 0 \\ 0 \\ 0 \end{Bmatrix} \tag{9.37}$$

Finally, the above equation reduces to

$$\begin{bmatrix} \sin\beta L & \sinh\beta L \\ -\beta^2\sin\beta L & \beta^2\sinh\beta L \end{bmatrix} \begin{Bmatrix} C_1 \\ C_3 \end{Bmatrix} = \begin{Bmatrix} 0 \\ 0 \end{Bmatrix} \tag{9.38}$$

The determinant of the homogeneous matrix Equation 9.38 gives

$$\beta\sin\beta L\sinh\beta L = 0 \tag{9.39}$$

In the above equation, all three terms have a common solution, that is, $\beta=0$. The first and third terms (i.e., β and $\sinh\beta L$) have no other solutions; however, the second term (i.e., $\sin \beta L$) has an infinite number of solutions. Excluding the trivial solution of $\beta=0$ (i.e., for $\beta\neq 0$ we have $\sinh\beta L \neq 0$), we obtain the frequency equation as

$$\sin\beta L = 0 \quad \text{with} \quad \beta \neq 0 \tag{9.40}$$

which will be satisfied for

$$\beta_n L = n\pi, \quad n = 1,2,\ldots,\infty \tag{9.41}$$

In view of Equation 9.41 and from Equation 9.31, the natural frequency in nth mode is given by

$$\omega_{nf_n} = (n\pi)^2 \sqrt{\frac{EI_{xx}}{\rho A L^4}} \tag{9.42}$$

Thus, we have an interesting relation between the fundamental mode and higher modes in the case of simply supported end conditions considered here, that is $\omega_{nf_n} = n^2\omega_{nf_1}$.

Now to get the mode shape (or eigen function), from the first set of Equation 9.38, we have

$$C_1 \sin\beta L + C_3 \sinh\beta L = 0 \tag{9.43}$$

Since $\sin\beta L = 0$ even when $\beta L \neq 0$, (e.g., $\beta L = \pi$), $C_1\neq 0$. However, since we have $\sinh\beta L \neq 0$ when $\beta L \neq 0$, $C_3=0$. Thus, now we have $C_2=C_3=C_4=0$, and in view of Equation 9.41 and from Equation 9.32, we have

$$\chi_n(z) = C_1 \sin\beta z = C_1 \sin\frac{n\pi}{L} z \tag{9.44}$$

The procedure will remain the same for a shaft with other boundary conditions. Frequency equations and natural frequencies for the fixed–fixed, fixed–free, free–free, and fixed–pinned shaft boundary conditions are given in Table 9.2. It should be noted that the frequency equations for fixed–fixed and free–free boundary conditions are the same. For free–free boundary conditions, two "zero" frequencies correspond to the translational and rotational rigid body modes, respectively. In the rigid body modes, the shaft would have rigid body up and down motion in the former case, whereas it would have rigid body transverse rotation about the center of gravity in the latter case. For the fixed–fixed case, two "zero" frequencies correspond to the translational and rotational rigid body modes and are not feasible. This is due to the boundary condition that needs to be satisfied while solving the differential equation. For the fixed–fixed case, both ends must have zero transverse translational and rotational displacements, and for the rigid body motion, all the particles of the shaft should have the same displacements (either translational or rotational). Therefore, the rigid body modes are only possible with zero translational and zero rotational displacements, which means the only possible motion corresponding to two zero natural frequencies from the frequency equations of the fixed–fixed case is that of zero displacements, that is, no motion at all. Hence, these zero natural frequencies have no practical significance for fixed–fixed boundary conditions of the shaft. This could be proved by directly solving differential Equation 9.30 with zero natural frequencies subject to the boundary conditions of the fixed–fixed case, and it is left for the reader as an exercise.

Orthogonality conditions: The most important property of normal modes (or eigen function or characteristic function) is the orthogonality. It is this property that makes possible the uncoupling

TABLE 9.2
Frequency Equations, Natural Frequencies, and Mode Shapes for Different Boundary Conditions

S.N.	Boundary Conditions	Frequency Equation	Natural Frequency, ω_{nf}	Mode Shape Function
1	Simply supported	$\sin\beta L = 0$ with $\beta^4 = \frac{\rho A\omega_{nf}^2}{EI}$	$(n\pi)^2\sqrt{\frac{EI_{xx}}{\rho AL^4}}$ $n = 1,2,3,\ldots$	$\sin\beta_n z$
2	Fixed–fixed supports	$\cos\beta L\cosh\beta L = 1$	(Approximate, however, the same as case 4, except 0) $\left\{\frac{(2n+1)\pi}{2}\right\}^2\sqrt{\frac{EI_{xx}}{\rho AL^4}}$ $n = 1,2,3,\ldots$	$\cosh\beta_n z - \cos\beta_n z - \frac{\cosh\beta_n L - \cos\beta_n L}{\sinh\beta_n L - \sin\beta_n L}(\sinh\beta_n z - \sin\beta_n z)$
3	Fixed–free (cantilevered) supports	$\cos\beta L\cosh\beta L = -1$	$3.52\sqrt{\frac{EI_{xx}}{\rho AL^4}}$ for n = 1; $\left\{\frac{(2n-1)\pi}{2}\right\}^2\sqrt{\frac{EI_{xx}}{\rho AL^4}}$ $n = 2,3,4,\ldots$	$\cosh\beta_n z - \cos\beta_n z - \frac{\cosh\beta_n L + \cos\beta_n L}{\sinh\beta_n L + \sin\beta_n L}(\sinh\beta_n z - \sin\beta_n z)$
4	Free–free supports	$\cos\beta L\cosh\beta L = 1$ (Same as case 4)	$(n\pi)^2\sqrt{\frac{EI_{xx}}{\rho AL^4}}$ $n = 0,0,1.506,2.500,\ldots$	$\cosh\beta_n z + \cos\beta_n z - \frac{\cosh\beta_n L - \cos\beta_n L}{\sinh\beta_n L - \sin\beta_n L}(\sinh\beta_n z + \sin\beta_n z)$
5	Fixed–pinned supports	$\tan\beta L - \tanh\beta L = 0$	$\left(\frac{1}{4}n\pi\right)^2\sqrt{\frac{EI_{xx}}{\rho AL^4}}$ $n = 5,9,13,\ldots$	$\cosh\beta_n z - \cos\beta_n z - \frac{\cosh\beta_n L + \cos\beta_n L}{\sinh\beta_n L + \sin\beta_n L}(\sinh\beta_n z - \sin\beta_n z)$

of the equations of motion in different modes of vibration. For two different frequencies $\omega_m \neq \omega_n$, normal modes must satisfy the following condition:

$$\int_0^L \rho A \chi_m(z)\chi_n(z)dz = 0 \quad \text{for} \quad m \neq n \tag{9.45}$$

For $m=n$

$$\int_0^L \rho A \chi_m^2(z)dz = M_m \tag{9.46}$$

in which M_m is called the "generalized mass" in the mth mode. Along similar lines, we have

$$\int_0^L EI_{xx}\chi_m''(z)\chi_n''(z)dz = 0 \quad \text{for} \quad m \neq n \tag{9.47}$$

For $m=n$

$$\int_0^L EI_{xx}\chi_m''^2(z)dz = K_m \tag{9.48}$$

in which K_m is called the *generalized stiffness* in the mth mode. These properties are useful for modal analysis in forced vibrations. These orthogonality conditions can be verified for eigenfunctions corresponding to a simply supported shaft, that is $\chi_m = c_m \sin(m\pi z/L)$.

9.3 FORCED VIBRATIONS

The orthogonality property will now be used to obtain uncoupled equations of motion for the continuous system corresponding to the natural mode of vibration, which is easier to handle. Having obtained natural frequencies and mode shapes, the transverse displacement of the shaft for forced vibrations can be written as follows:

$$v(z,t) = \sum_{i=1}^{\infty} \chi_i(z)\eta_i(t) \tag{9.49}$$

in which $\eta_i(t)$ is the generalized coordinate. The above expression is called the *expansion theorem.* Physical interpretation of this is that the response will consist of contributions from all possible modes, and these contributions depend upon initial conditions (for free vibrations) or type of forcing (for forced vibrations). Theoretically, infinite numbers of modes for continuous systems are possible. However, in practice, contribution of higher modes toward the response is negligible. Hence, only the first few modes are considered in computation. On substituting Equation 9.49 in Equation 9.18, multiplying both sides by χ_k, and integrating in the domain of the shaft, we get

$$\begin{aligned} &\int_0^L \left\{ EI_{xx} \sum_{i=1}^{\infty} \chi_i''''(z)\chi_k(z)\eta_i(t) + \rho A \sum_{i=1}^{\infty} \chi_i(z)\chi_k(z)\ddot{\eta}_i(t) \right\} dz \\ &= \int_0^L \left\{ f(z,t) + f_0(t)\delta^*(z-z_0) \right\} \chi_k(z)dz \end{aligned} \tag{9.50}$$

Applying the orthogonality condition of normal modes to Equations 9.45 through 9.48, the partial differential equation of motion is discretized into uncoupled, ordinary differential equations of motion as follows:

$$\ddot{\eta}_i(t) + \omega_{nfi}^2 \eta_i(t) = f_i(t), \quad i = 1,2,\ldots,\infty \tag{9.51}$$

with

$$\omega_{nfi} = \sqrt{\frac{K_i}{M_i}}$$

where ω_{nfi} is the ith natural frequency and f_i is *the generalized force* in the ith mode given as follows:

$$f_i(t) = \frac{1}{M_i}\int_0^L \left\{ f(z,t) + f_0(t)\delta^*(z - z_0) \right\} \chi_i(z)\,dz \tag{9.52}$$

The solution of Equation 9.51, for a general force, can be obtained by the *Duhamel or convolution integral* (Meirovitch, 1986) as given below:

$$\eta_i(t) = \frac{1}{\omega_{nfi}}\int_0^t f_i(\tau)\sin\omega_{nfi}(t-\tau)\,d\tau + \eta_i(0)\cos\omega_{nfi}t + \frac{\dot{\eta}_i(0)}{\omega_{nfi}}\sin\omega_{nfi}t \tag{9.53}$$

where $\eta_i(0)$ and $\dot{\eta}_i(t)$ are the initial conditions and τ is a dummy variable. An explicit form of the steady-state force response can be obtained through a simple example as follows.

Example 9.1

Consider a simply supported shaft with a uniform cross section subjected to a sinusoidal force, $F_0 \sin\omega t$, applied to a section $z_0 = 0.5L$, that is, at the midspan. Obtain the steady-state forced response of the shaft system.

Solution:

From Equation 9.52, the generalized force is given by

$$\begin{aligned} f_i(t) &= \frac{1}{M_i}\int_0^L \delta^*(z - 0.5L) F_0 \sin\omega t \sin\left(\frac{i\pi z}{L}\right) dz \\ &= \frac{1}{M_i}\sin\left(\frac{i\pi}{2}\right) F_0 \sin\omega t = \frac{1}{M_i}(-1)^{(i-1)/2} \\ &\quad F_0 \sin\omega t, \quad i = 1,3,5,\ldots \end{aligned} \tag{9.54}$$

The generalized mass, from Equation 9.46, is given as

$$M_i = \int_0^L \rho A \sin^2\left(\frac{i\pi z}{L}\right) dz = \frac{\rho A L}{2} \tag{9.55}$$

In view of Equation 9.55, Equation 9.54 gives

$$f_i(t) = (-1)^{(i-1)/2} \frac{2F_0}{\rho AL} \sin \omega t, \quad i = 1,3,5,\ldots \tag{9.56}$$

For the zero initial displacement and velocity, the response from Equation 9.53 can be written as follows:

$$\begin{aligned} \eta_i(t) &= (-1)^{(i-1)/2} \frac{2F_0}{\rho AL} \frac{1}{\omega_{nfi}} \int_0^t \sin \omega t \sin \omega_{nfi}(t-\tau) d\tau \\ &= (-1)^{(i-1)/2} \frac{2F_0}{\rho AL} \frac{1}{\omega_{nfi}} \frac{\left(\sin \omega t - \dfrac{\omega}{\omega_{nfi}} \sin \omega_{nfi} t\right)}{\omega_{nfi} \left\{1 - \left(\dfrac{\omega}{\omega_{nfi}}\right)^2\right\}} \quad i = 1,3,5,\ldots \end{aligned} \tag{9.57}$$

The steady-state transverse displacement of the shaft from Equation 9.49, in view of Equation 9.57, can be written as follows (dropping the second term within the bracket (i.e., $\sin \omega_{nfi} t$), which corresponds to the transient part, and in the actual case dies out after some time in the presence of damping):

$$v(z,t) = \sum_{i=1}^{\infty} \chi_i(z) \eta_i(t) = \frac{2F_0}{\rho AL} \left[\sum_{i=1,3,5,\ldots}^{\infty} (-1)^{(i-1)/2} \frac{\sin(i\pi z/L)}{\omega_{nfi}^2 \left\{1 - (\omega/\omega_{nfi})^2\right\}} \right] \sin \omega t \tag{9.58}$$

It is apparent that all even modes do not contribute to the steady-state forced response for the present problem. This is due to the force applied at the midspan, and all even modes have a node at the midspan. It is also of interest to compare the contribution of various modes. This comparison can be done on the basis of maximum modal displacement, disregarding the manner in which these displacements combine. The amplitude will indicate the relative importance of the modes. Furthermore, for all modes (except even modes), the modal contribution is simply in proportion to $1/\omega_{nfi}^2$. Therefore, in higher modes, the factor $1/\omega_{nfi}^2$ becomes smaller, and hence its contribution can be ignored in the superimposition of modes.

The above analysis is based on the continuous approach and becomes impractical for real rotor systems because of the complex boundary conditions. For such cases, the FEM is quite versatile and practical, which discretizes the continuous rotor system to a discrete rotor system. However, the advantage of a closed-form solution for simple cases is that it could be used for checking and performing the convergence study of the computer code developed based on the FEM, and subsequently the developed code could be used for more complex problems.

9.4 BRIEF REVIEW ON APPLICATION OF THE FEM IN ROTOR-BEARING SYSTEMS

In the past several decades, various methods have been developed to analyze the dynamic behavior of multi-DOF rotor-bearing systems, such as the influence coefficient, transfer matrix, dynamic stiffness, mechanical impedance, and FEMs. Of these methods, the FEM is one which is particularly well suited for modeling the large scale and complex rotor-bearing-foundation systems. As compared to the torsional and axial vibrations, researchers attempted more complex transverse vibration analyses of rotor-bearing systems in the beginning. Even the studies on the coupled axial, torsional, and transverse vibrations of rotor systems have been performed with other complexities such as rotary inertia, shear deformation, gyroscopic couple, and internal and external damping.

The Euler–Bernoulli beam theory (refer to Section 9.1) accounts for the primary effect of bending in beams, which is due to *pure bending*. In this theory, any plane cross section of the beam before bending is assumed to remain plane after bending, normal to the elastic (neutral) axis. Therefore, a beam cross section has not only the translation but also the rotation. Rayleigh's beam theory (refer to Section 11.8 in volume II) accounts for the higher effect of the energy arising out of this cross-sectional rotation, which he called the *rotary inertia*. Subsequently, Timoshenko accounted for the *shear strain energy* in the beam due to the shear force. The Timoshenko beam (refer to Chapter 10) usually refers to a beam in which both the rotary inertia and shear deformation effects are taken into account along with the primary effect of pure bending. Higher effects of rotary inertia and shear deformation are predominant in the transverse vibration of a beam having a large cross section (i.e., for thick beams; for a circular shaft, we have the condition for a thick beam as $r/L > 0.1$, where r is the radius of the shaft and L is the length of the shaft). If a Timoshenko shaft spins, then *gyroscopic effects* also have an important role along with the rotary inertia and shear effects.

Historically, Ruhl (1970) and Ruhl and Booker (1972) were among the first researchers to utilize the FEM to study the stability and unbalance response of turbo-rotor systems. In their FE formulations, only the elastic bending and translational kinetic energies were included. However, many effects such as rotary inertia, shear deformations, gyroscopic moments, and internal and external damping were all neglected in their FE analyses. These higher order effects can be very important for some configurations, as discussed by Dimentberg (1961) and Tondl (1965). McVaugh and Nelson (1976) generalized Ruhl's work by utilizing the Rayleigh beam model to devise an FE formulation including the effects of rotary inertia, gyroscopic moments, and axial load to simulate a flexible rotor system supported on the linear stiffness and (viscous) damping bearings. In order to facilitate the computations of natural whirl frequencies and unbalance responses, element equations were transformed into a rotating frame of reference for the case of isotropic bearings (i.e., cross-coupled coefficients and direct coefficients are equal for both stiffness and damping). Also to save computational time, the Guyan (static) reduction procedure (1965) was adopted to reduce the size of system matrices. Nelson and Zorzi (1977, 1980) extended the work of McVaugh and Nelson by the inclusion of both internal viscous and hysteretic damping in the same FE model and with axial load, respectively. At about same time, Rouch and Kao (1979) and Nelson (1980) utilized the Timoshenko beam theory for establishing shape functions. Based on these shape functions, the FE matrices of governing equations were derived. In these system FE matrices, a shear parameter was included in the shape functions to take into account the effect of transverse deformations. Comparisons were made between the FE analysis and the classical closed-form Timoshenko beam theory analysis for nonrotating and rotating shafts.

Ozguven and Ozkan (1984) and Edney et al. (1990) presented combined effects of the shear deformation and internal damping to analyze the natural whirl frequencies and unbalance responses of rotor-bearing systems. Chen and Ku (1991) developed a C^0 class (i.e., a compatibility requirement of only field variables instead of C^1, in which field variable compatibility is required up to the first derivative) Timoshenko beam FE model to study the combined effects of shear deformations and internal damping on the forward and backward whirl frequencies and onset speeds of the instability threshold of a flexible rotor system supported on the linear stiffness and damping bearings. Cleghorn and Tabarrok (1992) developed an FE model for the free lateral (transverse) vibration analysis of linearly tapered Timoshenko beams. The derived mass matrix was approximate, and the stiffness matrix was exact. Hou et al. (1996) developed an FE model of the Timoshenko beam to analyze free vibrations of nonuniform beams mounted on variable two-parameter foundations. The characteristic of this model was that the cross-sectional area, moment of inertia, and shear modulus were all assumed to vary in polynomial forms, implying that the beam element can deal with commonly seen nonuniform beams having different cross sections, such as the rectangular, circular, tubular, and even complex thin-walled sections, as well as foundations consisting of beams which vary in a general way. This beam element model enabled the user to handle the vibration analysis of more general beams in structures. Chen and Peng (1997) studied the stability of the rotating shaft with dissimilar stiffness and discussed influences of the stiffness ratio and axial compressive loads. An FE model of a Timoshenko beam was adopted to approximate the shaft, and effects of gyroscopic moments and torsional rigidities were taken into account. Results showed that with the existence of the dissimilar stiffness unstable zones occurred. Critical speeds were decreased, and instability regions were enlarged when the stiffness ratio was increased. The increase of the stiffness ratio consequently made the rotating shaft unstable. When the axial compressive load increased, critical speeds decreased, and zones of instability enlarged.

Mohiuddin and Khulief (1994, 1999) presented an FE formulation of a rotor-bearing system. The model coupled the bending and torsional motions of the rotating shaft and was derived using the Lagrangian approach. The model accounts for gyroscopic effects as well as the inertia coupling between the bending and torsional vibrations. A reduced order model was obtained using model truncation. Model transformations involved the complete mode shapes of a general rotor system with gyroscopic effects and anisotropic bearings. Nelson (1998) gave a comprehensive review of some of the modeling and analysis procedures developed for understanding and simulation of the characteristics of rotor dynamic systems. Ku (1998) analyzed the instability of the rotor system considering internal damping. Yardimoglu (2006) developed an FE model for transverse vibration tapered Timoshenko beams with a rectangular cross section. The displacement function was obtained from the coupled displacement field approach by considering the tapering functions of breath and depth of the beam, which reduced the number of nodal variables. The polynomial coefficients of the transverse displacement and cross-sectional rotation are coupled by considering the differential equations of equilibrium. Now the trend in the field is the utilization of these FE models for response prediction and model updating with the help of experimental measurement data. Although three-dimensional FE models are gradually becoming popular in industries, computation time is still a challenge. FE analysis simulates isolated physics under hypothetical load conditions during the design phase. The digital twins continuously replicate a physical asset's live condition by streaming real-time IoT sensor data. :But at present the industries do not discard FEA; they embed reduced-order FEA models into the Digital Twin architecture for real-time stress monitoring.

The above review gives an idea that there is a vast amount of literature available on the FE analysis of rotor systems. However, for a beginner, it is difficult to follow this literature. The aim of the subsequent section is to give a lucid presentation of the FE formulation of the simplest rotor model, that is, for the Euler–Bernoulli beam model. The Timoshenko beam model with a gyroscopic couple will be covered in Chapter 10.

9.5 FE FORMULATION OF EULER–BERNOULLI BEAM

Transversely vibrating shafts are most frequently modeled using the Euler–Bernoulli beam hypothesis. This is because of its simplicity and well-established and accurate prediction of real motion in the case of relatively thinner shafts. We considered the Euler–Bernoulli beam theory in the preceding sections, and equations of motion were derived using the continuous system approach using Newton's second law of motion and Hamilton's principle. Now the FE formulation and associated consistent mass and stiffness matrices and consistent force vector are obtained by using the Galerkin method (Cook et al., 1989). For the present case, the whirling motion of the beam in the vertical and horizontal directions is uncoupled. The analysis in the vertical plane (i.e., y–z plane) is presented, which is identical to the horizontal plane (i.e., z–x plane). Only difference comes in terms of change in signs in angular displacements and moments during the formulation (refer to Appendix 9.1). For obtaining bending natural whirl frequencies, the eigenvalue problem formulation is presented. Numerical examples are presented for obtaining system natural whirl frequencies, mode shapes, and unbalance responses.

Boundary conditions of a problem are not required while developing the FE formulation for a shaft. In fact, we would require them after the assembly of elements for a complete rotor system. Let

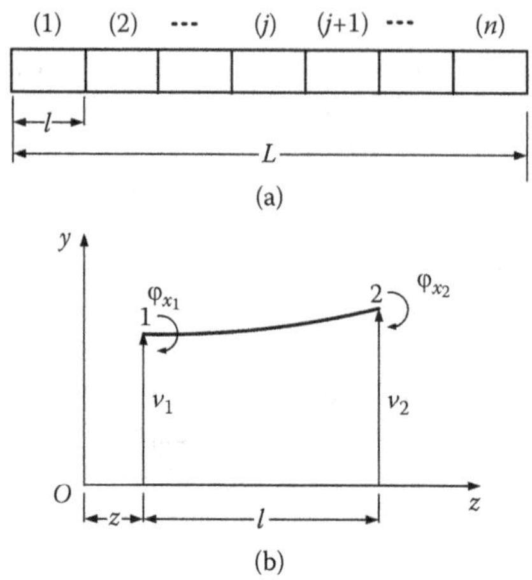

FIGURE 9.4 (a) Shaft discretized into n elements. (b) Typical beam element in the y–z plane.

us discretize a given shaft into several FEs (Figure 9.4a), and consider an element at a distance z in a global coordinate system in the plane y–z as shown in Figure 9.4. The length of the element is l. The element has two nodes 1 and 2. Let v be the nodal translational displacement and φ_x be the nodal rotational displacement at any intermediate point of the shaft element; these displacements with subscripts 1 and 2 belong to the respective nodes. The translational and rotational displacements are acting in the positive axis directions at the nodal points and are shown in Figure 9.4b.

In the FE model, continuous displacement variables (v, φ_x) are approximated in terms of discretized displacements at nodes of an element $\left(v_1, v_2, \varphi_{x1}, \varphi_{x2}\right)$. Therefore, the displacement can be expressed within the element by using the appropriate shape functions as follows:

$$v^{(e)}(z,t) = \mathbf{N}(z)\,\eta^{(ne)}(t) \tag{9.59}$$

where $\mathbf{N}(z)$ is the row vector of the shape function, $\eta^{(ne)}(t)$ is the nodal displacement vector, and superscripts (e) and (ne) represent an *element* and *nodes of the element*, respectively. Let r be the number of nodal displacements on the element; the number of shape functions is then also equal to r. The explicit form of the shape functions and nodal displacement vector will be provided in Section 9.5.2.

9.5.1 FE Formulation in a Weak Form

On substituting the approximate elemental solution of Equation 9.59 in the equation of motion (Equation 9.18), the residue is given by

$$R^{(e)} = EI_{xx}v^{(e)''''} + \rho A\ddot{v}^{(e)} - f(z,t) - f_0(t)\delta^*(z - z_0) \tag{9.60}$$

The residue can be minimized over the element as follows:

$$\int_0^l N_i(z)R^{(e)}dz = 0; \quad i = 1,2,\ldots,r \tag{9.61}$$

where N_i is the weight function and l is the element length. The Galerkin method is used to minimize the residue wherein weight functions are assumed to be the same as shape functions. On substituting the residue in the above equation, we get

$$\int_0^l N_i(z)\left\{\rho A\frac{\partial^2 v^{(e)}}{\partial t^2}+\frac{\partial^2}{\partial z^2}\left(EI_{xx}\frac{\partial^2 v^{(e)}}{\partial z^2}\right)-f(z,t)-f_0(t)\delta^*(z-z_0)\right\}dz=0;\quad i=1,2,\ldots,r \tag{9.62}$$

On rearranging Equation 9.62, we get

$$\int_0^l \rho AN_i\frac{\partial^2 v^{(e)}}{\partial t^2}dz+\underbrace{\int_0^l \underset{I}{N_i}\frac{\partial^2}{\partial z^2}\left(EI_{xx}\frac{\partial^2 v^{(e)}}{\partial z^2}\right)dz}_{II}$$
$$-\int_0^l N_i\left\{f(z,t)+f_0(t)\delta^*(z-z_0)\right\}dz=0;\quad i=1,2,\ldots,r \tag{9.63}$$

On integrating (so as to equalize the order of the spatial derivative within the integral to two) the second term of Equation 9.63 by parts twice and keeping the other integrals the same, we have

$$\int_0^l \rho AN_i\frac{\partial^2 v^{(e)}}{\partial t^2}dz+N_i\frac{\partial}{\partial z}\left(EI_{xx}\frac{\partial^2 v^{(e)}}{\partial z^2}\right)\Bigg|_0^l-N_i'\left(EI_{xx}\frac{\partial^2 v^{(e)}}{\partial z^2}\right)\Bigg|_0^l$$
$$+\int_0^l N_i''EI_{xx}\frac{\partial^2 v^{(e)}}{\partial z^2}dz-\int_0^l N_i\left\{f(z,t)+f_0(t)\delta^*(z-z_0)\right\}dz=0;\quad i=1,2,\ldots,r \tag{9.64}$$

where prime (′) represents the partial derivative with respect to z. The above equation is the weak FE formulation of the governing equation. The completeness and compatibility requirement of field variables (i.e., the translational and rotational displacements) can be obtained from the above equation as follows. The highest order derivative with respect to z in Equation 9.64 is third, so the completeness of v, φ_x, φ_x', and φ_x'' is required (i.e., the derivative of the displacement up to the same order; note that $\varphi_x''=-v'''$). Hence, a polynomial of third degree for the translational displacement will be able to satisfy the above condition. The completeness conditions need to satisfy within elements and at boundaries. The highest order derivative with respect to z in only integral terms of Equation 9.64 is second, so compatibility up to v and φ_x is required at the common boundary of two elements (i.e., the derivative of the displacement up to one order less; note that $\varphi_x=-v'$).

9.5.2 Derivations of Shape Functions

Expressing the transverse displacement of the element as a function of nodal DOFs, v_1, φ_{x1}, v_2, and φ_{x2}, we have

$$v^{(e)}(z,t)=\mathbf{N}(z)\,\eta^{(ne)}(t) \tag{9.65}$$

with

$$\eta^{(ne)}(t)=\begin{bmatrix} v_1 & \varphi_{x1} & v_2 & \varphi_{x2}\end{bmatrix}^T;\quad \mathbf{N}(z)=\begin{bmatrix} N_1(z) & N_2(z) & N_3(z) & N_4(z)\end{bmatrix}$$

It should be noted that for two-noded elements with two DOFs per node, we have $r=4$ for the total DOFs of the element. The translational and rotational displacements (i.e., v_1, φ_{x1}, v_2, and φ_{x2}) at nodes 1 and 2 of the element are specified, and these give four boundary conditions for the element from which four constants in the shape function can be determined uniquely. Let us assume the transverse displacement $v(z)$ to be a cubic polynomial, which has four unknown coefficients, to predict the displacements within the element as follows:

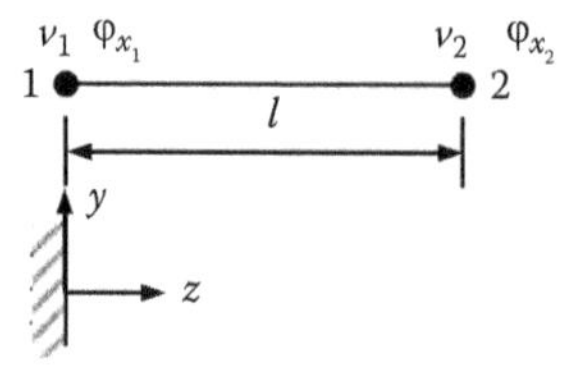

FIGURE 9.5 Shaft element.

$$v^{(e)}(z,t) = a_1 z^3 + a_2 z^2 + a_3 z + a_4 = \begin{bmatrix} z^3 & z^2 & z & 1 \end{bmatrix} \begin{Bmatrix} a_1 \\ a_2 \\ a_3 \\ a_4 \end{Bmatrix} \tag{9.66}$$

so that

$$\varphi_x^{(e)}(z,t) = -\frac{\partial v^{(e)}(z,t)}{\partial z} = -\left(3a_1 z^2 + 2a_2 z + a_3\right) = -\begin{bmatrix} 3z^2 & 2z & 1 & 0 \end{bmatrix} \begin{Bmatrix} a_1 \\ a_2 \\ a_3 \\ a_4 \end{Bmatrix} \tag{9.67}$$

where a_1, a_2, a_3, and a_4 are constants to be determined from the boundary conditions of the element as shown in Figure 9.5. Equation 9.66 is expected to satisfy the governing differential Equation 9.18 of the beam. In addition, the cubic displacement shape function satisfies the continuity condition of both the translational and rotational displacements at the nodes, which will also be shown in Section 9.5.3.

Each node has two DOFs and these are specified as follows:

$$z = 0, \quad v(0) = v_1; \quad z = l, \quad v(0) = v_2; \quad z = 0, \quad \varphi_x(l) = \varphi_{x1}; \quad z = l, \quad \varphi_x(l) = \varphi_{x2} \tag{9.68}$$

All four unknown constants can be determined from these four specified quantities (i.e., boundary conditions).

On application of boundary conditions of the element at the two nodes into Equations 9.66 and 9.67, we have

$$v(0,t) = v_1(t) = a_4; \quad \varphi_x(0,t) = \varphi_{x1}(t) = -a_3$$

and

$$v(l,t) = v_2(t) = a_1 l^3 + a_2 l^2 + a_3 l + a_4; \quad \varphi_x(l,t) = \varphi_{x2}(t) = -\left(3a_1 l^2 + 2a_2 l + a_3\right) \tag{9.69}$$

The above equations could be written in matrix form, as given by

$$\mathbf{Aa}(t) = \eta^{(ne)}(t) \tag{9.70}$$

with

$$\mathbf{A} = \begin{bmatrix} 0 & 0 & 0 & 1 \\ 0 & 0 & -1 & 0 \\ l^3 & l^2 & l & 1 \\ -3l^3 & -2l & -1 & 0 \end{bmatrix}; \quad \mathbf{a}(t) = \begin{Bmatrix} a_1 \\ a_2 \\ a_3 \\ a_4 \end{Bmatrix}; \quad \eta^{(ne)}(t) = \begin{Bmatrix} v_1 \\ \varphi_{x1} \\ v_2 \\ \varphi_{x2} \end{Bmatrix} \tag{9.71}$$

which can be solved for unknown constants as

$$\mathbf{a}(t) = \mathbf{A}^{-1}\ \eta^{(ne)}(t) \tag{9.72}$$

with

$$\mathbf{A}^{-1} = \begin{bmatrix} 2/l^3 & -1/l^2 & -2/l^3 & -1/l^2 \\ -3/l^2 & 2/l & 3/l^2 & 1/l \\ 0 & -1 & 0 & 0 \\ 1 & 0 & 0 & 0 \end{bmatrix}$$

On substituting Equation 9.72 into Equation 9.66, we get

$$\begin{aligned} v^{(e)}(z,t) &= \lfloor\ z^3 \quad z^2 \quad z \quad 1\ \rfloor \mathbf{A}^{-1}\eta^{(ne)}(t) \\ &= \lfloor\ N_1(z) \quad N_2(z) \quad N_3(z) \quad N_4(z)\ \rfloor \begin{Bmatrix} v_1 \\ \varphi_{x1} \\ v_2 \\ \varphi_{x2} \end{Bmatrix}^{(ne)} = \mathbf{N}(z)\eta^{(ne)}(t) \end{aligned} \tag{9.73}$$

with

$$\begin{aligned} \mathbf{N}(z) &= \lfloor\ z^3 \quad z^2 \quad z \quad 1\ \rfloor \mathbf{A}^{-1} \\ &= \left\lfloor \left(1-3\frac{z^2}{l^2}+2\frac{z^3}{l^3}\right) \quad -\left(z-2\frac{z^2}{l}+\frac{z^3}{l^2}\right) \quad \left(3\frac{z^2}{l^2}-2\frac{z^3}{l^3}\right) \quad -\left(-\frac{z^2}{l}+\frac{z^3}{l^2}\right) \right\rfloor \end{aligned} \tag{9.74}$$

and on substituting Equation 9.72 into Equation 9.67, we get

$$\begin{aligned} \varphi_x^{(e)}(z,t) &= -\lfloor\ 3z^2 \quad 2z \quad 1 \quad 0\ \rfloor \mathbf{A}^{-1}\eta^{(ne)}(t) \\ &= \lfloor\ N_1'(z) \quad N_2'(z) \quad N_3'(z) \quad N_4'(z)\ \rfloor \begin{Bmatrix} v_1 \\ \varphi_{x1} \\ v_2 \\ \varphi_{x2} \end{Bmatrix}^{(ne)} = \mathbf{N}'(z)\eta^{(ne)}(t) \end{aligned} \tag{9.75}$$

with

$$\begin{aligned} \mathbf{N}'(z) &= -\lfloor\ 3z^2 \quad 2z \quad 1 \quad 0\ \rfloor \mathbf{A}^{-1} \\ &= \left\lfloor -\left(-6\frac{z}{l^2}+6\frac{z^2}{l^3}\right) \quad \left(1-4\frac{z}{l}+3\frac{z^2}{l^2}\right) \quad -\left(6\frac{z}{l^2}-6\frac{z^2}{l^3}\right) \quad \left(-2\frac{z}{l}+3\frac{z^2}{l^2}\right) \right\rfloor \end{aligned} \tag{9.76}$$

It should be noted that

$$N_1 = 1, \text{ and } N_2 = N_3 = N_4 = 0 \text{ for } z = 0, \quad N_3 = 1, \quad \text{and} \quad N_1 = N_2 = N_4 = 0 \text{ for } z = l$$

and

$$N_2' = 1, \text{ and } N_1' = N_3' = N_4' = 0 \text{ for } z = 0, \quad N_4' = 1, \quad \text{and} \quad N_1' = N_2' = N_3' = 0 \text{ for } z = l$$

It should be noted that N_1 and N_3 have a unit value at $z = 0$ and at for $z = l$, respectively. Within the element, the variation of both is cubic in z. Similarly, N_2' and N_4' have a unit value at$z = 0$ and for $z = l$, respectively. Within the element, the variation of both is quadratic in z. On taking one more derivative of Equation 9.76 with respect to z, we get

$$\mathbf{N}''(z) = \left\lfloor -\left(-6\frac{1}{l^2} + 12\frac{z}{l^3}\right) \quad \left(-4\frac{1}{l} + 6\frac{z}{l^2}\right) \quad -\left(6\frac{1}{l^2} - 12\frac{z}{l^3}\right) \quad \left(-\frac{2}{l} + 6\frac{z}{l^2}\right) \right\rfloor$$

If we define ξ_1 and ξ_2 as natural coordinates, such that

$$\xi_1 = 1 - \frac{z}{l} \quad \text{and} \quad \xi_2 = \frac{z}{l} \tag{9.77}$$

Equations 9.74 and 9.76 will take the following simple form:

$$\mathbf{N}(\xi_1, \xi_2) = \left\lfloor \left(3\xi_1^3 - 2\xi_2^2\right) \quad -l\xi_1^2\xi_2 \quad \left(3\xi_1^2 - 2\xi_2^3\right) \quad l\xi_1\xi_2^2 \right\rfloor \tag{9.78}$$

and

$$\begin{aligned} \frac{\partial \mathbf{N}(\xi_1, \xi_2)}{\partial \xi_1} &= \left\lfloor 9\xi_1 \quad -2l\xi_1\xi_2 \quad 6\xi_1 \quad l\xi_2^2 \right\rfloor; \\ \frac{\partial \mathbf{N}(\xi_1, \xi_2)}{\partial \xi_2} &= \left\lfloor -4\xi_2 \quad -l\xi_1^2 \quad -6\xi_2^2 \quad 2l\xi_1\xi_2 \right\rfloor \end{aligned} \tag{9.79}$$

The above form of the shape functions helps in evaluating the integral relatively easily, which is required frequently at the time of calculation of the elemental mass and stiffness matrices.

9.5.3 Satisfaction of the Compatibility and Completeness Conditions

In the previous section, we stated the requirements of the compatibility and completeness conditions for the chosen shape functions. In the present section, we will see whether the chosen cubic polynomial satisfies these conditions.

Compatibility conditions: Consider two neighboring elements j and $(j+1)$ for checking interelement compatibility as shown in Figure 9.6.

From Equations 9.73 and 9.75, the translational and rotational displacements for the jth element can be written as follows:

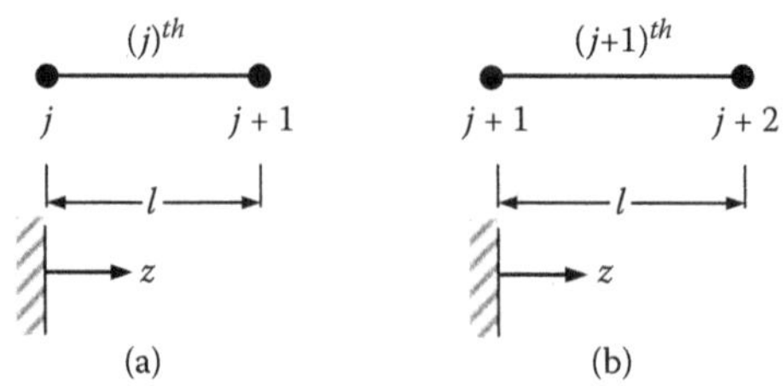

FIGURE 9.6 Two neighboring elements (a) jth element and (b) $(j+1)$th element.

$$v^{(j)}(z,t)=\left\lfloor \left(1-3\frac{z^2}{l^2}+2\frac{z^3}{l^3}\right)\ -\left(z-2\frac{z^2}{l}+\frac{z^3}{l^2}\right)\left(3\frac{z^2}{l^2}-2\frac{z^3}{l^3}\right)\ -\left(-\frac{z^2}{l}+\frac{z^3}{l^2}\right)\right\rfloor\begin{Bmatrix} v_j(t)\\ \varphi_{x_j}(t)\\ v_{j+1}(t)\\ \varphi_{x_{j+1}}(t)\end{Bmatrix} \tag{9.80}$$

and

$$\varphi_x^{(j)}(z,t)=\left\lfloor -\left(-6\frac{z}{l^2}+6\frac{z^2}{l^3}\right)\left(1-4\frac{z}{l}+3\frac{z^2}{l^2}\right)\ -\left(6\frac{z}{l^2}-6\frac{z^2}{l^3}\right)\left(-2\frac{z}{l}+3\frac{z^2}{l^2}\right)\right\rfloor\begin{Bmatrix} v_j(t)\\ \varphi_{x_j}(t)\\ v_{j+1}(t)\\ \varphi_{x_{j+1}}(t)\end{Bmatrix} \tag{9.81}$$

Now the translational and rotational displacements at node $(j+1)$ of the jth element (i.e., at $z=l$), which is a common node between the jth and $(j+1)$th elements, can be written from the above equations, as given below:

$$v^{(j)}\Big|_{z=l}=\lfloor\ 0\quad 0\quad 1\quad 0\ \rfloor\begin{Bmatrix} v_j\\ \varphi_{x_j}\\ v_{j+1}\\ \varphi_{x_{j+1}}\end{Bmatrix}=v_{j+1} \tag{9.82}$$

and

$$\varphi_x^{(j)}\Big|_{z=l}=\lfloor\ 0\quad 0\quad 0\quad 1\ \rfloor\begin{Bmatrix} v_j\\ \varphi_{x_j}\\ v_{j+1}\\ \varphi_{x_{j+1}}\end{Bmatrix}=\varphi_{x_{j+1}} \tag{9.83}$$

Similarly, for the $(j+1)$th element, we have

$$v^{(j+1)}(z,t)=\left\lfloor \left(1-3\frac{z^2}{l^2}+2\frac{z^3}{l^3}\right)\ -\left(z-2\frac{z^2}{l}+\frac{z^3}{l^2}\right)\left(3\frac{z^2}{l^2}-2\frac{z^3}{l^3}\right)\ -\left(-\frac{z^2}{l}+\frac{z^3}{l^2}\right)\right\rfloor\begin{Bmatrix} v_{j+1}(t)\\ \varphi_{x_{j+1}}(t)\\ v_{j+2}(t)\\ \varphi_{x_{j+2}}(t)\end{Bmatrix} \tag{9.84}$$

and

$$\varphi_x^{(j+1)}(z,t)=\left\lfloor -\left(-6\frac{z}{l^2}+6\frac{z^2}{l^3}\right)\left(1-4\frac{z}{l}+3\frac{z^2}{l^2}\right)\ -\left(6\frac{z}{l^2}-6\frac{z^2}{l^3}\right)\left(-2\frac{z}{l}+3\frac{z^2}{l^2}\right)\right\rfloor\begin{Bmatrix} v_{j+1}(t)\\ \varphi_{x_{j+1}}(t)\\ v_{j+2}(t)\\ \varphi_{x_{j+2}}(t)\end{Bmatrix} \tag{9.85}$$

Now the translational and rotational displacements at node $(j+1)$ of the $(j+1)$th element (i.e., at $z=0$) can be written, from the above equation, as follows:

$$v^{(j+1)}\Big|_{z=0} = \lfloor 1 \quad 0 \quad 0 \quad 0 \rfloor \begin{Bmatrix} v_{j+1} \\ \varphi_{x_{j+1}} \\ v_{j+2} \\ \varphi_{x_{j+2}} \end{Bmatrix} = v_{j+1} \tag{9.86}$$

and

$$\varphi_x^{j+1}\Big|_{z=0} = \lfloor 0 \quad 1 \quad 0 \quad 0 \rfloor \begin{Bmatrix} v_{j+1} \\ \varphi_{x_{j+1}} \\ v_{j+2} \\ \varphi_{x_{j+2}} \end{Bmatrix} = \varphi_{x_{j+1}} \tag{9.87}$$

Hence, the shape function is compatible for v and φ_x since we obtained the same value for the translational and rotational displacements at node $(j+1)$ while considering the jth and $(j+1)$th elements independently (i.e., from Equations 9.82, 9.83, 9.86, and 9.87).

Completeness conditions: For a cubic shape function, the completeness up to the second derivative covers the requirements of the present case, as it does not vanish on taking the derivative up to the second order.

9.5.4 FE Governing Equations

While formulating the weak form of the elemental equation by Galerkin's method in Section 9.5.1, we obtained sets of equations in the form of Equation 9.64. Now assembling such equations for all values of $i=1, 2, 3, 4$ (i.e., $r=4$), we get

$$\int_0^l \rho A \mathbf{N} \ddot{v}^{(e)} dz + \int_0^l EI_{xx} \mathbf{N}_{,zz} v_{,zz}^{(e)} dz = \mathbf{S}_y + \mathbf{M}_{yz} + \int_0^l \mathbf{N}\left\{f(z,t) + f_0(t)\delta^*(z - z_0)\right\} dz \tag{9.88}$$

with

$$\mathbf{N} = \begin{Bmatrix} N_1 \\ N_2 \\ N_3 \\ N_4 \end{Bmatrix}; \quad \mathbf{N}_{,zz} = \begin{Bmatrix} N_{1,zz} \\ N_{2,zz} \\ N_{3,zz} \\ N_{4,zz} \end{Bmatrix}$$

$$\mathbf{S}_y = -\begin{Bmatrix} N_1\left(EI_{xx} v_{,zz}^{(e)}\right)_{,z}\Big|_0^l \\ N_2\left(EI_{xx} v_{,zz}^{(e)}\right)_{,z}\Big|_0^l \\ N_3\left(EI_{xx} v_{,zz}^{(e)}\right)_{,z}\Big|_0^l \\ N_4\left(EI_{xx} v_{,zz}^{(e)}\right)_{,z}\Big|_0^l \end{Bmatrix} = -\begin{Bmatrix} 0 - \left(EI_{xx} v_{,zz}^{(e)}\right)_{,z}\Big|_{z=0} \\ 0 - 0 \\ \left(EI_{xx} v_{,zz}^{(e)}\right)_{,z}\Big|_{z=l} - 0 \\ 0 - 0 \end{Bmatrix} = -\begin{Bmatrix} S_{y1} \\ 0 \\ -S_{y2} \\ 0 \end{Bmatrix}$$

$$\mathbf{M}_{yz} = \left\{ \begin{array}{c} N_{1,z}\left(EI_{xx}v_{,zz}^{(e)}\right)\Big|_0^l \\ N_{2,z}\left(EI_{xx}v_{,zz}^{(e)}\right)\Big|_0^l \\ N_{3,z}\left(EI_{xx}v_{,zz}^{(e)}\right)\Big|_0^l \\ N_{4,z}\left(EI_{xx}v_{,zz}^{(e)}\right)\Big|_0^l \end{array} \right\} = \left\{ \begin{array}{c} 0-0 \\ 0-\left(EI_{xx}v_{,zz}^{(e)}\right)\Big|_{z=0} \\ 0-0 \\ \left(EI_{xx}v_{,zz}^{(e)}\right)\Big|_{z=l}-0 \end{array} \right\} = \left\{ \begin{array}{c} 0 \\ -M_{yz1} \\ 0 \\ M_{yz2} \end{array} \right\}$$

and defining

$$S_{y1} = -f_{sy1} = -\left(EI_{xx}v_{,zz}^{(e)}\right)_{,z}\Big|_{z=0}; \quad M_{yz1} = \left(EI_{xx}v_{,zz}^{(e)}\right)\Big|_{z=0};$$

$$S_{y2} = -f_{sy2} = -\left(EI_{xx}v_{,zz}^{(e)}\right)_{,z}\Big|_{z=l} \quad \text{and} \quad M_{yz2} = \left(EI_{xx}v_{,zz}^{(e)}\right)\Big|_{z=l}$$

where $f_{sy}=-S_y$ and M_{yz} represent the shear force (see Figures 9.1) and the bending moment, respectively; $\{N\}$ is a vector of shape functions. It should be noted again that

$$N_1 = 0 \quad \text{for} \quad z = l \quad \text{and} \quad N_1 = 1 \quad \text{for} \quad z = 0$$
$$N_2 = 1 \quad \text{for} \quad z = l \quad \text{and} \quad N_2 = 0 \quad \text{for} \quad z = 0$$

Similarly, we have

$$N_{1,z} = 0 \quad \text{for} \quad z = l \quad \text{and} \quad N_{1,z} = 1 \quad \text{for} \quad z = 0$$
$$N_{2,z} = 1 \quad \text{for} \quad z = l \quad \text{and} \quad N_{2,z} = 0 \quad \text{for} \quad z = 0$$

On substituting Equation 9.73 into Equation 9.88, we get

$$\begin{aligned} &\int_0^l \rho A \mathbf{N}^T(z)\,\mathbf{N}(z)\,dz\ddot{\eta}^{(ne)}(t) + \int_0^l EI_{xx}\mathbf{N}_{,zz}^T(z)\mathbf{N}_{,zz}dz\eta^{(ne)}(t) \\ &\quad = \mathbf{f}_R^{(ne)}(t) + \int_0^l \mathbf{N}^T(z)\left\{f(z,t)+f_0(t)\delta^*(z-z_0)\right\}dz \end{aligned} \tag{9.89}$$

with

$$\left(\mathbf{f}_R^{(ne)}(t)\right)^T = \lfloor\ -S_{y1} \quad -M_{yz1} \quad S_{y2} \quad M_{yz2}\ \rfloor$$

Equation 9.89 can be written as follows:

$$\mathbf{M}^{(e)}\ddot{\eta}^{(ne)}(t) + \mathbf{K}^{(e)}\eta^{(ne)}(t) = \mathbf{f}_{ext}(t)^{(ne)} + \mathbf{f}_R(t)^{(ne)} \tag{9.90}$$

with

$$\eta^{(ne)}(t) = \left\{ v_1 \quad \varphi_{x1} \quad v_2 \quad \varphi_{x2} \right\}^T; \quad \mathbf{f}_{ext}^{(ne)}(t) = \int_0^l \mathbf{N}^T(z)\left\{f(z,t)+f_0(t)\delta^*(z-z_0)\right\}dz$$

where $\mathbf{M}^{(e)}$ is the *consistent mass matrix* of the element, $\mathbf{K}^{(e)}$ is the stiffness matrix of the element, $\eta^{(ne)}(t)$ is the displacement vector, $\mathbf{f}_R^{(ne)}(t)$ is the reactive or internal nodal force vector, and $\mathbf{f}_{ext}^{(ne)}(t)$ is the consistent external nodal force vector. The consistent stiffness matrix is given as follows:

$$\mathbf{K}^{(e)} = \int_0^l EI_{xx}\{\mathbf{N}''(z)\}^T \mathbf{N}''(z)dx = \int_0^l EI_{xx}\begin{Bmatrix} N_2'' \\ N_2'' \\ N_3'' \\ N_4'' \end{Bmatrix}\lfloor N_1'' \quad N_2'' \quad N_3'' \quad N_4'' \rfloor dz$$

$$= \int_0^l EI_{xx}\begin{bmatrix} (N_1'')^2 & N_1''N_2'' & N_1''N_3'' & N_1''N_4'' \\ N_2''N_1'' & (N_2'')^2 & N_2''N_3'' & N_2''N_4'' \\ N_3''N_1'' & N_3''N_2'' & (N_3'')^2 & N_3''N_4'' \\ N_4''N_1'' & N_4''N_2'' & N_4''N_3'' & (N_4'')^2 \end{bmatrix} dz \tag{9.91}$$

Here, the prime represents the derivative with respect to the spatial coordinate, z. On substituting Equation 9.74 into Equation 9.91, we get

$$\mathbf{K}^{(e)} = \frac{EI_{xx}}{l^3}\begin{bmatrix} 12 & -6l & -12 & -6l \\ & 4l^2 & 6l & 2l^2 \\ & & 12 & 6l \\ sym & & & 4l^2 \end{bmatrix} \tag{9.92}$$

Elements of the stiffness matrix can be written as given below:

$$k_{ij} = \int_0^l EI_{xx}N_i''N_j''\, dz \quad i, j = 1,2,3,4 \tag{9.93}$$

For example,

$$k_{12} = \int_l^0 EI_{xx}N_1''N_2''dz = -\int_l^0 EI_{xx}\left(-\frac{6}{l^2}+12\frac{z}{l^3}\right)\left(-\frac{4}{l}+6\frac{z}{l^2}\right)dz$$

$$= -\int_l^0 EI_{xx}\left\{\frac{24}{l^3}-(16+48)\frac{z}{l^4}+72\frac{z^2}{l^5}\right\}dz = -\frac{EI_{xx}}{l^3}(6l) = -\frac{6EI_{xx}}{l^2}$$

The consistent mass matrix is given by

$$\mathbf{M}^{(e)} = \int_0^l \rho A\mathbf{N}^T\mathbf{N}dz$$

$$= \int_0^l \rho A\begin{bmatrix} N_1^2 & N_1N_2 & N_1N_3 & N_1N_4 \\ N_2N_1 & N_2^2 & N_2N_3 & N_2N_4 \\ N_3N_1 & N_3N_2 & N_3^2 & N_3N_4 \\ N_4N_1 & N_4N_2 & N_4N_3 & N_4^2 \end{bmatrix}dz = \frac{\rho Al}{420}\begin{bmatrix} 156 & -22l & 54 & 13l \\ & 4l^2 & -13l & -3l^2 \\ & & 156 & 22l \\ sym & & & 4l^2 \end{bmatrix} \tag{9.94}$$

Elements of the mass matrix can be written as follows:

$$m_{ij} = \int_0^l \rho A N_i N_j dz \quad i, j = 1,2,3,4 \tag{9.95}$$

For example,

$$m_{12} = \int_0^l \rho A N_1 N_2 dz = -\int_0^l \rho A\left(1-3\frac{z^2}{l^2}+2\frac{z^3}{l^3}\right)\left(z-2\frac{z^2}{l}+\frac{z^3}{l^2}\right)dz = -\frac{\rho A l}{420}(22l) = -\frac{11\rho A l^2}{210}$$

It should be noted that both the mass and stiffness matrices are symmetric. For the present formulation, we neglected the rotary inertia while deriving the equation of motion of the continuous beam. The rotary inertia terms appear in the above elemental mass matrix (corresponding to even rows and even columns) are due to consideration of the element as a whole and its equivalent effects at nodes. Damping has not been considered yet, and Rayleigh's (or proportionate) damping will be discussed in Section 9.6.

9.5.5 Consistent Load Vector

In the previous section, we derived FE equations wherein the consistent mass and stiffness matrices were also obtained with the help of chosen shape functions. In the present section, we will obtain the consistent force vector for a variety of loading conditions.

Uniformly distributed force: Let $f(x, t)=f(t)$ be the uniformly distributed force per unit length (see Figure 9.7).

Noting Equations 9.90 and 9.74, the consistent force vector is given as follows:

$$\mathbf{f}_{ext}^{(ne)} = \int_0^l f(t)\mathbf{N}^T(z)dz = \int_0^l f(t)\begin{Bmatrix} N_1(z) \\ N_2(z) \\ N_3(z) \\ N_4(z) \end{Bmatrix} dz$$

$$= \int_0^l f(t)\begin{Bmatrix} \left(1-3\frac{z^2}{l^2}+2\frac{z^3}{l^3}\right) \\ -\left(z-2\frac{z^2}{l}+\frac{z^3}{l^2}\right) \\ \left(3\frac{z^2}{l^2}-2\frac{z^3}{l^3}\right) \\ -\left(-\frac{z^2}{l}+\frac{z^3}{l^2}\right) \end{Bmatrix} dz = \begin{Bmatrix} \frac{1}{2}f(t)l \\ \frac{-1}{12}f(t)l^2 \\ \frac{1}{2}f(t)l \\ \frac{1}{12}f(t)l^2 \end{Bmatrix} \tag{9.96}$$

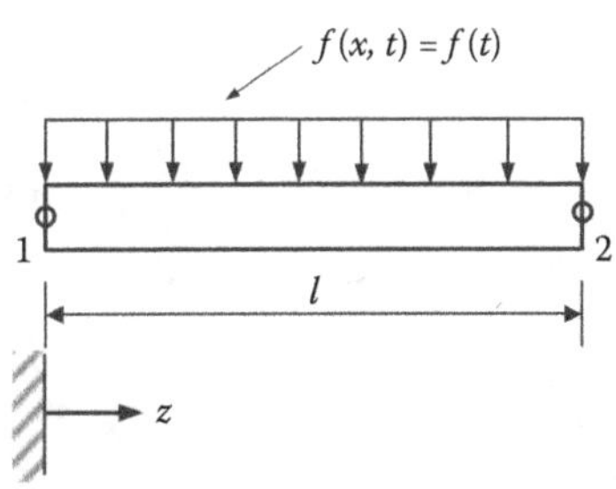

FIGURE 9.7 Beam element with uniformly distributed force.

It should be noted that because of the uniformly distributed force over the element, both forces and moments are present at both nodes of the element. In fact, the total force over the element $f(t)l$ is the same as the total force at the nodes of the element, that is, on adding the first and third rows of the consistent force vector. In addition to these forces, we have moments at the second and fourth rows. While following the lumped system approach, one would have distributed the total force equally at the nodes without any moment terms.

A single concentrated force: A single force acting at $z = z_0$ could be written as follows:

$$f(z,t) = f_0(t)\delta^*(z - z_0) \tag{9.97}$$

where $\delta^*()$ is the *Dirac delta function*.

The consistent force vector can be written as follows:

$$\mathbf{f}_{ext}^{(ne)}(t) = \int_0^l f_0(t)\delta^*(z - z_0)\mathbf{N}^T(z)\,dz = f_0(t)\mathbf{N}^T(z_0)$$

For cases shown in Figure 9.8, the *consistent external force* vector takes the following form:

$$\text{(i) For } z_0 = 0, \quad \mathbf{f}_{ext}^{(ne)}(t) = f_0(t)\left\{ 1 \quad 0 \quad 0 \quad 0 \right\}^T \tag{9.98}$$

$$\text{(ii) For } z_0 = l, \quad \mathbf{f}_{ext}^{(ne)}(t) = f_0(t)\left\{ 0 \quad 0 \quad 1 \quad 0 \right\}^T \tag{9.99}$$

$$\text{(iii) For } z_0 = \frac{1}{2}l, \quad \mathbf{f}_{ext}^{(ne)}(t) = f_0(t)\left\{ \frac{1}{2} \quad \frac{-1}{8}l \quad \frac{1}{2} \quad \frac{1}{8}l \right\}^T \tag{9.100}$$

The first and third rows of the force vector represent the equivalent force at the nodes of the external force applied to the element at various locations as shown in Figure 9.8. Similarly, the second and fourth rows represent the equivalent moment at the nodes due to external force. For cases (i) and (ii), since the external force is at node 1 or 2, respectively, the corresponding equivalent force remains the same. However, for case (iii), the external force is equally distributed at nodes 1 and 2, and in addition to that, it also produces moments at both nodes. Hence, a load inside the element will produce equivalent forces and moments at both nodes. This is also true when an external moment is applied instead of an external force.

A linearly varying force: Let the linearly varying force (Figure 9.9) over an element be expressed as follows:

$$f^{(e)}(z,t) = a + bz \tag{9.101}$$

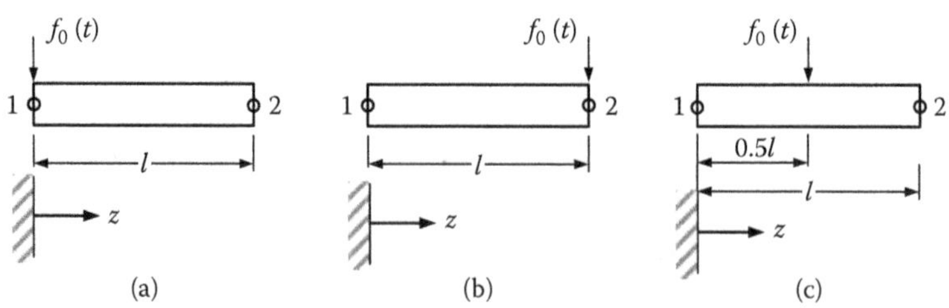

FIGURE 9.8 Beam element with a concentrated force: (a) $z_0=0$, (b) $z_0=l$, and (c) $z_0=0.5\,l$.

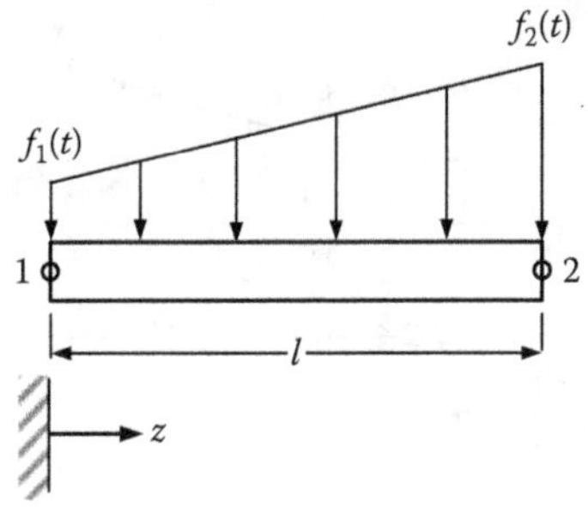

FIGURE 9.9 Linearly varying force.

where a and b are unknown constants to be determined from the force distribution conditions, for example, the amplitudes of forces at ends.

For such a linear force distribution over the element, at least two nodes are required on an element and corresponding force values $f_1(t)$ and $f_2(t)$ must be specified, that is,

$$\text{at} \quad z = 0, \quad f^{(e)}(0,t) = f_1(t) \Rightarrow f_1(t) = a$$

and

$$\text{at} \quad z = l, \quad f^{(e)}(l,t) = f_2(t) \Rightarrow f_2(t) = a + bl$$

which can be solved to give

$$a = f_1 \quad \text{and} \quad b = \frac{f_2 - f_1}{l}$$

Then, the assumed form of the element force becomes

$$f^{(e)}(z,t) = f_1(t) + \frac{f_2(t) - f_1(t)}{l} z = \left[\left(1 - \frac{z}{l}\right) \quad \frac{z}{l} \right] \begin{Bmatrix} f_1(t) \\ f_2(t) \end{Bmatrix} = \mathbf{N}_f(z)\, \mathbf{f}^{(ne)}(t) \tag{9.102}$$

where $\mathbf{N}_f(z)$ is *the shape function vector for the force*. Hence, the consistent load vector would be

$$\mathbf{f}_{ext}^{(ne)}(t) = \int_0^l \mathbf{N}^T(z)\, \mathbf{N}_f(z) dz\, \mathbf{f}^{(ne)}(t)$$

$$= \begin{bmatrix} \frac{7}{20} l & \frac{3}{20} l \\ \frac{-1}{20} l^2 & \frac{-1}{30} l^2 \\ \frac{3}{20} l & \frac{7}{20} l \\ \frac{1}{30} l^2 & \frac{1}{20} l^2 \end{bmatrix} \begin{Bmatrix} f_1(t) \\ f_2(t) \end{Bmatrix} = \begin{Bmatrix} \left(\frac{7}{20} f_1(t) + \frac{3}{20} f_2(t) \right) l \\ -\left(\frac{1}{20} f_1(t) + \frac{1}{30} f_2(t) \right) l^2 \\ \left(\frac{3}{20} f_1(t) + \frac{7}{20} f_2(t) \right) l \\ \left(\frac{1}{30} f_1(t) + \frac{1}{20} f_2(t) \right) l^2 \end{Bmatrix} \tag{9.103}$$

A quadratic variation of the force: In a quadratic polynomial, one needs to determine three constants. Hence, at least three nodes are required as shown in Figure 9.10, with node 3 as an *internal*

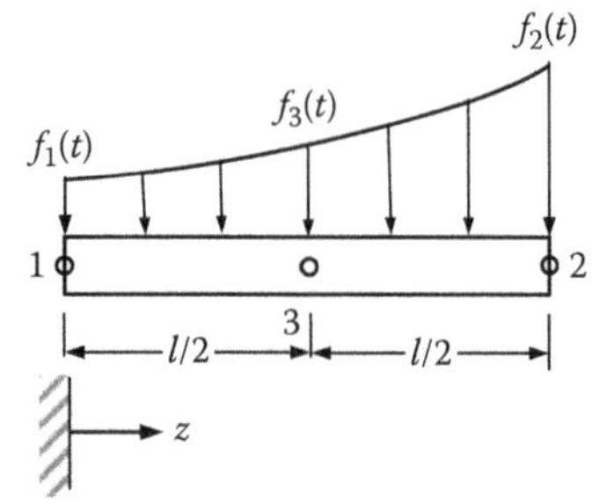

FIGURE 9.10 Beam element with a quadratic variation of the force.

node. At these nodes (i.e., at $z=0$, $0.5\,l$, and l), the corresponding force values $f_1(t)$, $f_2(t)$, and $f_3(t)$ must be specified, so that the element force can be written as follows:

$$f^{(e)}(z,t)=a+bz+cz^2=\begin{bmatrix} N_{f1}(z) & N_{f2}(z) & N_{f3}(z) \end{bmatrix}\begin{Bmatrix} f_1(t) \\ f_2(t) \\ f_3(t) \end{Bmatrix}=\mathbf{N}_f(z)\mathbf{f}^{(ne)}(t) \quad (9.104)$$

with

$$\mathbf{N}_f(z)=\begin{bmatrix} \left\{\left(1-\frac{z}{l}\right)\left(1-\frac{2z}{l}\right)\right\} & \left\{\frac{z}{l}\left(\frac{2z}{l}-1\right)\right\} & \left\{\frac{4z}{l}\left(1-\frac{z}{l}\right)\right\} \end{bmatrix}$$

and

$$\mathbf{f}^{(ne)}(t)=\begin{Bmatrix} f_1(t) \\ f_2(t) \\ f_3(t) \end{Bmatrix}$$

Hence, the nodal external force vector is defined as:

$$\mathbf{f}_{ext}^{(ne)}(t)=\int_0^l \mathbf{N}^T(z)\mathbf{N}_f(z)dz\mathbf{f}^{(ne)}(t) \quad (9.105)$$

Obtaining the explicit expression of the same is left for the reader as an exercise. For other forms of complex force distribution function, a numerical integration could be done.

9.5.6 System Equations of Motion

In previous sections, we obtained the elemental mass and stiffness matrices, and the internal and external force vectors. The objective of the present section is to demonstrate how to get system mass and stiffness matrices, force vectors in terms of the elemental mass and stiffness matrices, and force vectors as given in Equation 9.90. To illustrate the same, let us take a two-element system with a rigid disk at node 2 as shown in Figure 9.11a. Nodal displacement variables are shown in Figure 9.11b at each of the three nodes. Figure 9.11c shows two elements with the corresponding internal shear forces and bending moments, wherein the superscript specifies the element number and the subscript specifies the node number. Here, the disk is assumed to be at node 2 of element 1

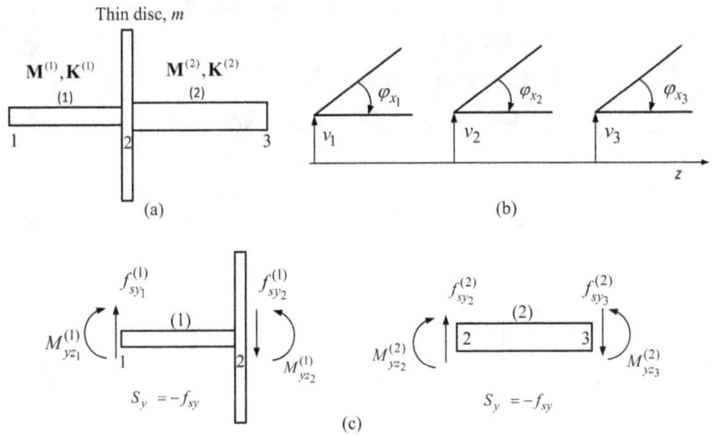

FIGURE 9.11 Two-element shaft with a rigid thin disk: (a) discretization of system, (b) nodal generalized coordinates, and (c) nodal forces and moments in elements.

(alternatively, it can be taken at node 1 of element 2). The elemental differential equations for elements 1 and 2, respectively, can be written as follows:

$$\mathbf{M}^{(1)}\ddot{\boldsymbol{\eta}}^{(1)}(t)+\mathbf{K}^{(1)}\boldsymbol{\eta}^{(1)}(t)=\mathbf{f}_R^{(1)}(t) \tag{9.106}$$

and

$$\mathbf{M}^{(2)}\ddot{\boldsymbol{\eta}}^{(2)}(t)+\mathbf{K}^{(2)}\boldsymbol{\eta}^{(2)}(t)=\mathbf{f}_R^{(2)}(t) \tag{9.107}$$

where the superscript represents the element number. External forces are not considered here.

The mass and stiffness matrices for each of the elements are obtained as in Equation 9.90. The corresponding elemental mass and stiffness matrices are added to get the global mass and stiffness matrices by considering connectivity between various nodes. The elemental stiffness and mass matrices of Equations 9.106 and 9.107 can be stated as follows (note that the point mass m is assumed to be attached to element 1 at node 2):

$$\mathbf{K}^{(1)}=\begin{bmatrix} k_{11}^{(1)} & k_{12}^{(1)} & k_{13}^{(1)} & k_{14}^{(1)} \\ k_{21}^{(1)} & k_{22}^{(1)} & k_{23}^{(1)} & k_{24}^{(1)} \\ k_{31}^{(1)} & k_{32}^{(1)} & k_{33}^{(1)} & k_{34}^{(1)} \\ k_{41}^{(1)} & k_{42}^{(1)} & k_{43}^{(1)} & k_{44}^{(1)} \end{bmatrix}, \quad \mathbf{K}^{(2)}=\begin{bmatrix} k_{33}^{(2)} & k_{34}^{(2)} & k_{35}^{(2)} & k_{36}^{(2)} \\ k_{43}^{(2)} & k_{44}^{(2)} & k_{45}^{(2)} & k_{46}^{(2)} \\ k_{53}^{(2)} & k_{54}^{(2)} & k_{55}^{(2)} & k_{56}^{(2)} \\ k_{63}^{(2)} & k_{64}^{(2)} & k_{65}^{(2)} & k_{66}^{(2)} \end{bmatrix}$$

$$\mathbf{M}^{(1)}=\begin{bmatrix} m_{11}^{(1)} & m_{12}^{(1)} & m_{13}^{(1)} & m_{14}^{(1)} \\ m_{21}^{(1)} & m_{22}^{(1)} & m_{23}^{(1)} & m_{24}^{(1)} \\ m_{31}^{(1)} & m_{32}^{(1)} & \left(m_{33}^{(1)}+m\right) & m_{34}^{(1)} \\ m_{41}^{(1)} & m_{42}^{(1)} & m_{43}^{(1)} & m_{44}^{(1)} \end{bmatrix}, \quad \mathbf{M}^{(2)}=\begin{bmatrix} m_{33}^{(2)} & m_{34}^{(2)} & m_{35}^{(2)} & m_{36}^{(2)} \\ m_{43}^{(2)} & m_{44}^{(2)} & m_{45}^{(2)} & m_{46}^{(2)} \\ m_{53}^{(2)} & m_{54}^{(2)} & m_{55}^{(2)} & m_{56}^{(2)} \\ m_{63}^{(2)} & m_{64}^{(2)} & m_{65}^{(2)} & m_{66}^{(2)} \end{bmatrix}$$

The nodal displacement and internal force vectors are defined as follows:

TABLE 9.3
Connectivity Matrix

Element Number	Local	Nodes
1	1	2
2	3	4

$$\eta^{(1)}(t)=\begin{Bmatrix} v_1 \\ \varphi_{x1} \\ v_2 \\ \varphi_{x2} \end{Bmatrix};\quad \eta^{(2)}(t)=\begin{Bmatrix} v_2 \\ \varphi_{x2} \\ v_3 \\ \varphi_{x3} \end{Bmatrix};\quad \mathbf{f}_R^{(1)}(t)=\begin{Bmatrix} -S_{y1}^{(1)} \\ -M_{yz1}^{(1)} \\ S_{y2}^{(1)} \\ M_{yz2}^{(1)} \end{Bmatrix};\quad \mathbf{f}_R^{(2)}(t)=\begin{Bmatrix} -S_{y2}^{(2)} \\ -M_{yz2}^{(2)} \\ S_{y3}^{(2)} \\ M_{yz3}^{(2)} \end{Bmatrix}$$

The relationship between the elemental (local) nodal and the system (or global) displacements is established through the connectivity matrix given in Table 9.3.

From Equation 9.106 and the connectivity matrix (Table 9.3), the system model can be given as follows:

$$\mathbf{M}^{(s)}\ddot{\eta}^{(s)}(t)+\mathbf{K}^{(s)}\eta^{(s)}(t)=\mathbf{f}_R^{(s)}(t) \tag{9.108}$$

with

$$\mathbf{M}^{(s)}=\begin{bmatrix} m_{11}^{(1)} & m_{12}^{(1)} & m_{13}^{(1)} & m_{14}^{(1)} & 0 & 0 \\ m_{21}^{(1)} & m_{22}^{(1)} & m_{23}^{(1)} & m_{24}^{(1)} & 0 & 0 \\ m_{31}^{(1)} & m_{32}^{(1)} & \left(m_{33}^{(1)}+m_{33}^{(2)}+m\right) & \left(m_{34}^{(1)}+m_{34}^{(2)}\right) & m_{35}^{(2)} & m_{36}^{(2)} \\ m_{41}^{(1)} & m_{42}^{(1)} & \left(m_{43}^{(1)}+m_{43}^{(2)}\right) & \left(m_{44}^{(1)}+m_{44}^{(2)}\right) & m_{45}^{(2)} & m_{46}^{(2)} \\ 0 & 0 & m_{53}^{(2)} & m_{54}^{(2)} & m_{55}^{(2)} & m_{56}^{(2)} \\ 0 & 0 & m_{63}^{(2)} & m_{64}^{(2)} & m_{65}^{(2)} & m_{66}^{(2)} \end{bmatrix}$$

$$\mathbf{K}^{(s)}=\begin{bmatrix} k_{11}^{(1)} & k_{12}^{(1)} & k_{13}^{(1)} & k_{14}^{(1)} & 0 & 0 \\ k_{21}^{(1)} & k_{22}^{(1)} & k_{23}^{(1)} & k_{24}^{(1)} & 0 & 0 \\ k_{31}^{(1)} & k_{32}^{(1)} & \left(k_{33}^{(1)}+k_{33}^{(2)}\right) & \left(k_{34}^{(1)}+k_{34}^{(2)}\right) & k_{35}^{(2)} & k_{36}^{(2)} \\ k_{41}^{(1)} & k_{42}^{(1)} & \left(k_{43}^{(1)}+k_{43}^{(2)}\right) & \left(k_{44}^{(1)}+k_{44}^{(2)}\right) & k_{45}^{(2)} & k_{46}^{(2)} \\ 0 & 0 & k_{53}^{(2)} & k_{54}^{(2)} & k_{55}^{(2)} & k_{56}^{(2)} \\ 0 & 0 & k_{63}^{(2)} & k_{64}^{(2)} & k_{65}^{(2)} & k_{66}^{(2)} \end{bmatrix}$$

$$\eta^{(s)}(t)=\begin{Bmatrix} v_1 \\ \varphi_{x1} \\ v_2 \\ \varphi_{x2} \\ v_3 \\ \varphi_{x3} \end{Bmatrix} \quad \text{and} \quad \mathbf{f}_R^{(s)}(t)=\begin{Bmatrix} -S_{y1}^{(1)} \\ -M_{yz1}^{(1)} \\ S_{y2}^{(1)}-S_{y2}^{(2)} \\ M_{yz2}^{(1)}-M_{yz2}^{(2)} \\ S_{y3}^{(2)} \\ M_{yz3}^{(2)} \end{Bmatrix}=\begin{Bmatrix} -S_{y1}^{(1)} \\ -M_{yz1}^{(1)} \\ 0 \\ 0 \\ S_{y3}^{(2)} \\ M_{yz3}^{(2)} \end{Bmatrix}$$

where $\mathbf{M}^{(s)}$ and $\mathbf{K}^{(s)}$ are called the assembled mass and stiffness matrices, respectively, and $\eta^{(s)}(t)$ and $\mathbf{f}_R^{(s)}(t)$ are the assembled displacement and force vectors, respectively. Superscript s represents the *system* or *assembled* matrices and vectors.

9.5.7 Eigenvalue Problem

For obtaining natural frequencies, the reduced system of equations is obtained by applying boundary conditions to Equation 9.108 and eliminating rows and columns corresponding to applied boundary conditions. This sets the corresponding natural frequencies equal to zero, which is associated with the rigid body translation and rotation. The remaining natural frequencies are obtained by setting the determinant of the reduced system of equations equal to zero. The reduced system of equations after applying boundary conditions becomes

$$\bar{\mathbf{M}}^{(s)}\ddot{\bar{\eta}}(t)+\bar{\mathbf{K}}^{(s)}\bar{\eta}(t)=\mathbf{0}^{(s)} \tag{9.109}$$

For example, if both ends of the shaft are fixed for the previous two-element single disk example, then we have the following boundary conditions:

$$v_1=\varphi_{x1}=v_3=\varphi_{x3}=0 \tag{9.110}$$

Hence, elimination of the first, second, fifth, and sixth rows and columns in the assembled mass and stiffness matrices yields the reduced mass and stiffness matrices, and the reduced nodal and internal force vectors as follows:

$$\bar{\mathbf{M}}^{(R)}=\begin{bmatrix}\left(m_{33}^{(1)}+m_{33}^{(2)}+m\right) & \left(m_{34}^{(1)}+m_{34}^{(2)}\right)\\ \left(m_{43}^{(1)}+m_{43}^{(2)}\right) & \left(m_{44}^{(1)}+m_{44}^{(2)}\right)\end{bmatrix};$$
$$\bar{\mathbf{K}}^{(R)}=\begin{bmatrix}\left(k_{33}^{(1)}+k_{33}^{(2)}\right) & \left(k_{34}^{(1)}+k_{34}^{(2)}\right)\\ \left(k_{43}^{(1)}+k_{43}^{(2)}\right) & \left(k_{44}^{(1)}+k_{44}^{(2)}\right)\end{bmatrix} \tag{9.111}$$

and

$$\bar{\eta}^{(R)}(t)=\begin{Bmatrix}v_2\\ \varphi_{x2}\end{Bmatrix};\quad \mathbf{f}_R^{(R)}(t)=\begin{Bmatrix}0\\0\end{Bmatrix} \tag{9.112}$$

For a simple harmonic motion during free vibration, we have $\bar{\eta}^{(R)}=\eta_0 e^{j\omega_{nf}t}$ where ω_{nf} is the natural frequency of the system. Hence, from Equation 9.109, the associated eigenvalue problem becomes

$$\bar{\mathbf{K}}^{(R)}\eta_0=\omega_{nf}^2\bar{\mathbf{M}}^{(R)}\eta_0 \tag{9.113}$$

Equation 9.113 can be expressed in the form of

$$\left(\bar{\mathbf{M}}^{(R)}-\lambda\bar{\mathbf{K}}^{(R)}\right)\eta_0=\mathbf{0}\quad\text{with}\quad\lambda=1/\omega_{nf}^2 \tag{9.114}$$

Premultiplying Equation 9.114 by inverse of $\bar{\mathbf{K}}^{(R)}$, the eigenvalue problem becomes

$$(\mathbf{A}-\lambda\mathbf{I})\eta_0=\mathbf{0} \tag{9.115}$$

with

$$\mathbf{A}=\left(\bar{\mathbf{K}}^{(R)}\right)^{-1}\bar{\mathbf{M}}^{(R)}$$

Equation 9.115 possesses a nontrivial solution, if and only if the determinant of the coefficients η_0 vanishes. Thus, the characteristic equation can be written as given below:

$$\Delta(\lambda)=\det(\mathbf{A}-\lambda\mathbf{I})=|\mathbf{A}-\lambda\mathbf{I}|=0 \tag{9.116}$$

where $\Delta(\lambda)$ is a *frequency equation* in the form of a polynomial of degree n in λ_r, where n is the number of DOFs (also the degree of the characteristic polynomial). Equation 9.116 possesses n real and positive roots, and λ_r is related to system natural frequencies by $\lambda_r=1/\omega_{nf_r}^2$ with $r=1, 2, \ldots, n$. If η_r represents the eigenvector (mode shape) corresponding to the eigenvalue λ_r, then n solutions of the eigenvalue problem (Equation 9.115) can be written as

$$(\mathbf{A}-\lambda_r\mathbf{I})\eta_r=\mathbf{0}; \quad r=1,2,\ldots,n \tag{9.117}$$

In case the stiffness matrix, $\bar{\mathbf{K}}^s$, is a singular matrix (e.g., free–free conditions), then Equation 9.113 takes the following form:

$$(\tilde{\lambda}\mathbf{I}-\mathbf{D})\eta_0=\mathbf{0} \tag{9.118}$$

with

$$\tilde{\lambda}=\omega_{nf}^2; \quad \mathbf{D}=\left(\bar{\mathbf{M}}^{(R)}\right)^{-1}\bar{\mathbf{K}}^{(R)} \tag{9.119}$$

Thus, the characteristic equation can be written as

$$\Delta(\tilde{\lambda})=\det(\mathbf{D}-\tilde{\lambda}\mathbf{I})=|\mathbf{D}-\tilde{\lambda}\mathbf{I}|=0 \tag{9.120}$$

where $\Delta(\tilde{\lambda})$ is now the *frequency equation* in the form of a polynomial of degree n in $\tilde{\lambda}_r$, where n is the number of DOFs. Equation 9.120 possesses n real and positive roots, and $\tilde{\lambda}_r$ is related to the system natural frequencies by $\tilde{\lambda}_r=1/\omega_{nf_r}^2$ with $r=1, 2, \ldots, n$. If η_r represents the eigenvector (mode shape) corresponding to the eigenvalue $\tilde{\lambda}_r$, then n solutions of the eigenvalue problem (Equation 9.120) can be written as

$$(\mathbf{D}-\tilde{\lambda}_r\mathbf{I})\eta_r=\mathbf{0} \quad \text{with} \quad r=1,2,\ldots,n \tag{9.121}$$

In fact, for multi-DOFs, there is no need to obtain the frequency equation in explicit form, and only the eigenvalues and eigenvectors of matrix **A** or **D** need to be obtained. These eigenvalues and eigenvectors can be used to obtain natural frequencies and mode shapes. These procedures will be illustrated through the following examples.

Example 9.2

Find out the transverse natural frequencies and mode shapes of a rotor system as shown in Figure 9.12. The following data are given: diameter of the shaft $d=10\,\text{mm}$, density of the shaft material $\rho=7800\,\text{kg/m}^3$, Young's modulus of the shaft material $E=2.1\times10^{11}\,\text{N/m}^2$, and length of

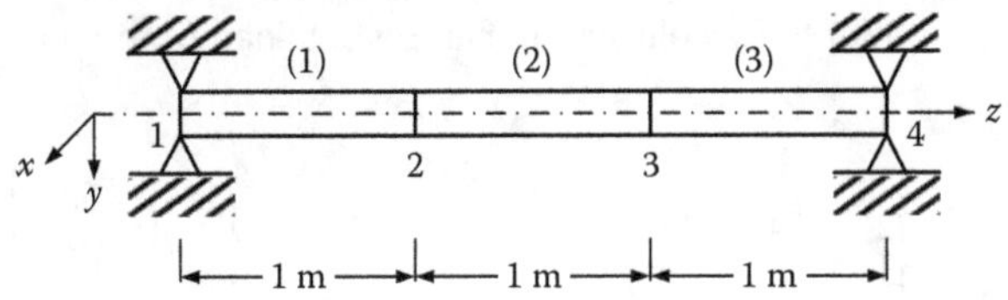

FIGURE 9.12 Simply supported shaft.

the shaft $L = 3$ m. Perform a convergence study to show how the natural frequencies converge to those obtained from the closed-form solution of continuous system analysis with the increasing number of elements.

Solution:

Initially, let us divide the shaft into three elements as shown in Figure 9.12. This will help in illustrating the FE procedure, and extraction of natural frequencies and mode shapes from the eigenvalue analysis with a relatively smaller matrix size. From the data given, we have

$$A = 7.854\times10^{-5}\ \text{m}^2;\ I = 4.91\times10^{-10}\ \text{m}^4;\rho Al/420 = 1.46\times10^{-3}\ \text{kg};\ EI/l^3 = 103.11\,\text{N/m}$$

In view of Equation 9.90, the FE equation for element (1) can be written as follows:

$$1.46\times10^{-3}\begin{bmatrix} 156 & -22 & 54 & 13 \\ & 4 & -13 & -3 \\ & & 156 & 22 \\ sym & & & 4 \end{bmatrix}\begin{Bmatrix} \ddot{v}_1 \\ \ddot{\varphi}_{x1} \\ \ddot{v}_2 \\ \ddot{\varphi}_{x2} \end{Bmatrix} + 103.11\begin{bmatrix} 12 & -6 & -12 & -6 \\ & 4 & 6 & 2 \\ & & 12 & 6 \\ sym & & & 4 \end{bmatrix}\begin{Bmatrix} v_1 \\ \varphi_{x1} \\ v_2 \\ \varphi_{x2} \end{Bmatrix} = \begin{Bmatrix} -S_{y1} \\ -M_{yz1} \\ S_{y2} \\ M_{yz2} \end{Bmatrix} \tag{a}$$

Similarly, the FE equation for elements (2) and (3) can be written as

$$1.46\times10^{-3}\begin{bmatrix} 156 & -22 & 54 & -13 \\ & 4 & -13 & -3 \\ & & 156 & 22 \\ sym & & & 4 \end{bmatrix}\begin{Bmatrix} \ddot{v}_2 \\ \ddot{\varphi}_{x2} \\ \ddot{v}_3 \\ \ddot{\varphi}_{x3} \end{Bmatrix} + 103.11\begin{bmatrix} 12 & -6 & -12 & -6 \\ & 4 & 6 & 2 \\ & & 12 & 6 \\ sym & & & 4 \end{bmatrix}\begin{Bmatrix} v_2 \\ \varphi_{x2} \\ v_3 \\ \varphi_{x3} \end{Bmatrix} = \begin{Bmatrix} -S_{y2} \\ -M_{yz2} \\ S_{y3} \\ M_{yz3} \end{Bmatrix} \tag{b}$$

and

$$1.46\times10^{-3}\begin{bmatrix} 156 & -22 & 54 & 13 \\ & 4 & -13 & -3 \\ & & 156 & 22 \\ sym & & & 4 \end{bmatrix}\begin{Bmatrix} \ddot{v}_3 \\ \ddot{\varphi}_{x3} \\ \ddot{v}_4 \\ \ddot{\varphi}_{x4} \end{Bmatrix} + 103.11\begin{bmatrix} 12 & -6 & -12 & -6 \\ & 4 & 6 & 2 \\ & & 12 & 6 \\ sym & & & 4 \end{bmatrix}\begin{Bmatrix} v_3 \\ \varphi_{x3} \\ v_4 \\ \varphi_{x4} \end{Bmatrix} = \begin{Bmatrix} -S_{y3} \\ -M_{yz3} \\ S_{y4} \\ M_{yz4} \end{Bmatrix} \tag{c}$$

The assembled FE equation of the whole system can be obtained from Equations (a) through (c) as follows:

$$1.46\times10^{-3}\begin{bmatrix} 156 & -22 & 54 & 13 & 0 & 0 & 0 & 0 \\ & 4 & -13 & -3 & 0 & 0 & 0 & 0 \\ & & 312 & 0 & 54 & 13 & 0 & 0 \\ & & & 8 & -13 & -3 & 0 & 0 \\ & & & & 312 & 0 & 54 & 13 \\ & & & & & 8 & -13 & -3 \\ & & & & & & 156 & 22 \\ sym & & & & & & & 4 \end{bmatrix}\begin{Bmatrix} \ddot{v}_1 \\ \ddot{\varphi}_{x1} \\ \ddot{v}_2 \\ \ddot{\varphi}_{x2} \\ \ddot{v}_3 \\ \ddot{\varphi}_{x3} \\ \ddot{v}_4 \\ \ddot{\varphi}_{x4} \end{Bmatrix}$$

$$+103.11\begin{bmatrix} 12 & -6 & -12 & -6 & 0 & 0 & 0 & 0 \\ & 4 & 6 & 2 & 0 & 0 & 0 & 0 \\ & & 24 & 0 & -12 & -6 & 0 & 0 \\ & & & 8 & 6 & 2 & 0 & 0 \\ & & & & 24 & 0 & -12 & -6 \\ & & & & & 8 & 6 & 2 \\ & & & & & & 12 & 6 \\ sym & & & & & & & 4 \end{bmatrix}\begin{Bmatrix} v_1 \\ \varphi_{x1} \\ v_2 \\ \varphi_{x2} \\ v_3 \\ \varphi_{x3} \\ v_4 \\ \varphi_{x4} \end{Bmatrix}=\begin{Bmatrix} -S_{y1} \\ -M_{yz1}=0 \\ 0 \\ 0 \\ 0 \\ 0 \\ S_{y4} \\ M_{yz4}=0 \end{Bmatrix} \quad \text{(d)}$$

Now the boundary conditions for simply supported shaft ends are $v_1=0,\ v_4=0,\ M_{yz1}=0$, and $M_{yz4}=0$. On applying the boundary conditions in Equation (d), we observe that at the right-hand side the internal force vectors, except the first and seventh rows, are all zero (or specified). The second and sixth rows of Equation (d) can be later used for finding the eigenvectors. Reducing the assembled equations after eliminating the first and seventh rows and columns in Equation (d), we get

$$1.46\times10^{-3}\begin{bmatrix} 4 & -13 & -3 & 0 & 0 & 0 \\ & 312 & 0 & 54 & 13 & 0 \\ & & 8 & -13 & -3 & 0 \\ & & & 312 & 0 & 13 \\ & & & & 8 & -3 \\ sym & & & & & 4 \end{bmatrix}\begin{Bmatrix} \ddot{\varphi}_{x1} \\ \ddot{v}_2 \\ \ddot{\varphi}_{x2} \\ \ddot{v}_3 \\ \ddot{\varphi}_{x3} \\ \ddot{\varphi}_{x4} \end{Bmatrix}$$

$$+103.11\begin{bmatrix} 4 & 6 & 2 & 0 & 0 & 0 \\ & 24 & 0 & -12 & -6 & 0 \\ & & 8 & 6 & 2 & 0 \\ & & & 24 & 0 & -6 \\ & & & & 8 & 2 \\ sym & & & & & 4 \end{bmatrix}\begin{Bmatrix} \varphi_{x1} \\ v_2 \\ \varphi_{x2} \\ v_3 \\ \varphi_{x3} \\ \varphi_{x4} \end{Bmatrix}=\begin{Bmatrix} 0 \\ 0 \\ 0 \\ 0 \\ 0 \\ 0 \end{Bmatrix} \quad \text{(e)}$$

For the simple harmonic motion during free vibration, we have $\ddot{\eta}=-\omega_{nf}^2\eta$; hence, Equation (e) could be written as follows:

$$\left(103.11\begin{bmatrix} 4 & -6 & 2 & 0 & 0 & 0 \\ & 24 & 0 & -12 & -6 & 0 \\ & & 8 & 6 & 2 & 0 \\ & & & 24 & 0 & -6 \\ & & & & 8 & 2 \\ sym & & & & & 4 \end{bmatrix}\right.$$

$$\left. -1.46\times10^{-3}\omega_{nf}^2\begin{bmatrix} 4 & -13 & -3 & 0 & 0 & 0 \\ & 312 & 0 & 54 & 13 & 0 \\ & & 8 & -13 & -3 & 0 \\ & & & 312 & 0 & 13 \\ & & & & 8 & -3 \\ sym & & & & & 4 \end{bmatrix}\right)\begin{Bmatrix} \varphi_{x1} \\ v_2 \\ \varphi_{x2} \\ v_3 \\ \varphi_{x3} \\ \varphi_{x4} \end{Bmatrix} = \begin{Bmatrix} 0 \\ 0 \\ 0 \\ 0 \\ 0 \\ 0 \end{Bmatrix} \tag{f}$$

Equation (f) also has the following form:

$$\left(\mathbf{K} - \omega_{nf}^2\mathbf{M}\right)\eta_0 = \mathbf{0} \tag{g}$$

which has an eigenvalue problem in the following form:

$$\left(\mathbf{A} - \lambda\mathbf{I}\right)\eta_0 = \mathbf{0} \quad \text{with} \quad \mathbf{A} = \mathbf{M}^{-1}\mathbf{K} \quad \text{and} \quad \lambda = \omega_{nf}^2 \tag{h}$$

Natural frequencies can be obtained from eigenvalues of matrix **A** by the relations given in Equation (h), that is, by taking the square root of the eigenvalues. The natural frequencies obtained for three elements for the simply supported beam are given in Table 9.4. Exact natural frequencies from analytical closed-form equations have also been tabulated, and we see that except for the first two natural frequencies, the others are deviating a lot from the exact values. With the FEM, natural frequencies for different numbers of elements can be obtained along similar lines and have been tabulated to study the convergence trend. It can be seen that with 10 elements itself the convergence has occurred with quite good accuracy up to the third mode and with reasonable accuracy up to the fifth mode. However, with 50 elements, all have converged quite close to the exact values. If the natural frequencies for higher modes are required, the number of elements required for convergence may be still larger.

For each natural frequency, there will be a corresponding mode shape. Table 9.5 shows typical eigenvectors corresponding to the first and second natural frequencies (which can also be obtained from the eigenvalue problem formulated for three elements in Equation (f) after substituting ω_{nfi}

TABLE 9.4
Convergence Study of Natural Frequencies (rad/s)

	Natural Frequencies by the FEM (Number of Shaft Elements)				
Mode No.	**(3)**	**(6)**	**(10)**	**(50)**	**Natural Frequencies by the Analytical Method**
I	14.237	14.226	14.225	14.225	14.225
II	57.574	56.947	56.907	56.901	56.901
III	142.100	128.532	128.095	128.027	128.027
IV	264.223	230.294	227.980	227.604	227.603
V	472.774	365.071	357.034	355.633	355.631

and ω_{nf_2}, respectively, one at a time). The sequence of eigenvector components will remain the same as that of $\boldsymbol{\eta}_0$ in Equation (f). Hence, the relative translational and rotational displacements corresponding to the simply supported shaft for the first and second natural frequencies can be extracted from Table 9.5 as shown in Tables 9.6 and 9.7, respectively. It should be noted that since only three elements have been taken for illustration, only four points are obtained in the mode shapes. The mode shape values can be interpolated by the shape function (cubic polynomial) of the beam element in between node points. For n elements, we will have $(n+1)$ nodal values.

Figure 9.13 shows mode shapes (i.e., for translational and rotational displacements) corresponding to the first five modes with 50 elements. It should be noted that mode shapes are relative translational (or rotational) displacements; hence, with different normalization procedures, absolute displacements may change, but relative displacements in a particular mode will remain the same. The interpolation function from Equations 9.74 and 9.75 can be used to interpolate displacements between nodes for getting smooth mode shapes, since these have compatible properties of v and

TABLE 9.5
Eigenvectors Corresponding to the First and Second Natural Frequencies with Three Elements

DOF	First Column	Second Column
φ_{x1}	−1.0000	−1.0000
v_2	−0.8270	−0.4148
φ_{x2}	−0.5000	0.5000
v_3	−0.8270	0.4148
φ_{x3}	0.5000	0.5000
φ_{x4}	1.0000	−1.0000

TABLE 9.6
Relative Translational Displacements at Various Nodes with Three Elements

DOF	Translational Displacements (I-Mode)	Translational Displacements (II-Mode)
v_1	0.0000	0.0000
v_2	−0.8270	−0.4148
v_3	−0.8270	0.4148
v_4	0.0000	0.0000

TABLE 9.7
Relative Rotational Displacements at Various Nodes for Three Elements

DOF	Rotational Displacements (I-Mode)	Rotational Displacements (II-Mode)
φ_{x1}	−1.0000	−1.0000
φ_{x2}	−0.5000	0.5000
φ_{x3}	0.5000	0.5000
φ_{x4}	1.0000	−1.0000

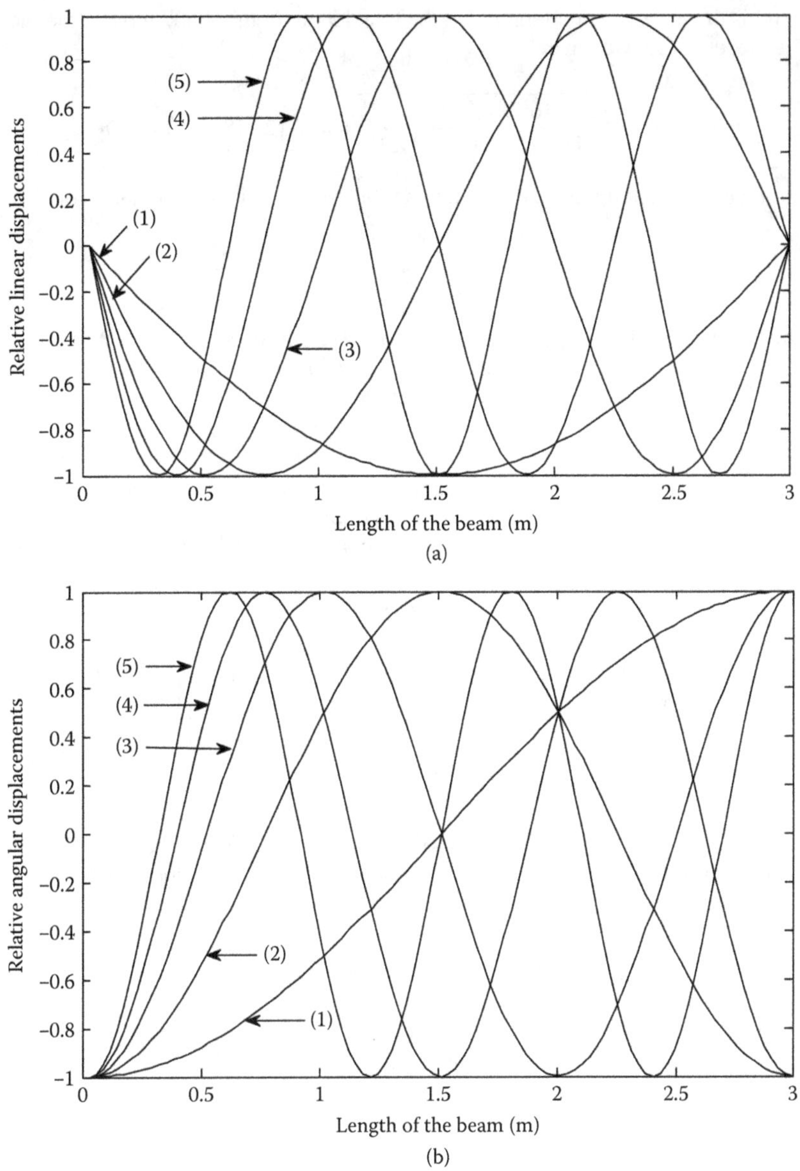

FIGURE 9.13 (a) Normalized translational displacement mode shapes: (1) first mode shape, (2) second mode shape, (3) third mode shape, (4) fourth mode shape, and (5) fifth mode shape. (b) Normalized rotational displacement mode shapes: (1) first mode shape, (2) second mode shape, (3) third mode shape, (4) fourth mode shape, and (5) fifth mode shape.

φ_x at common nodes between two neighboring elements. Figure 9.13a shows that the simply supported shaft has $(n-1)$ nodes (i.e., the location of the zero displacement), where n is the number of modes.

Analytical solution: Natural frequencies for the simply supported continuous shaft using the analytical method (i.e., the closed-form expression) are given as (see Table 9.2) follows:

$$\omega_{nf_n} = n^2\pi^2\sqrt{\frac{EI}{mL^4}} \text{ rad/s}, \quad n = 1,2,3,\ldots \tag{i}$$

where $m = \rho A$ is the mass per unit length, L is the total length of the shaft, and n is the mode number. For the present case, we have

$$\omega_{nf} = n^2\pi^2\sqrt{\frac{2.1\times10^{11}\times4.9087\times10^{-10}}{7800\times7.854\times10^{-5}\times3^4}} = n^2\pi^2(2.078)^{1/2} = 14.225n^2 \quad \text{(j)}$$

Corresponding to different modes (i.e., n), the natural frequency is tabulated in Table 9.4.

Example 9.3

Find out the transverse natural frequencies of the rotor system shown in Figure 9.14. The following rotor data are given: for the rigid disk $m = 10\,\text{kg}$, $I_d = 0.02\ \text{kg-m}^2$; for the shaft $d = 0.01\ \text{m}$, $L = 1\ \text{m}$, and $E = 2.1\times10^{11}\ \text{N/m}^2$. Consider the shaft as massless.

Solution:

This simple example has been chosen to illustrate the inclusion of a disk in the governing equation and the application of simply supported boundary conditions. Single-plane motion is considered since there is no coupling of motion in two orthogonal planes. Dividing the rotor in four equal elements (i.e., $l = L/4 = 0.25$ m) as shown in Figure 9.15 gives the following governing equations:

The elemental equation for element (1) can be written as follows:

$$\begin{bmatrix} 0 & 0 & 0 & 0 \\ & 0 & 0 & 0 \\ & & 0 & 0 \\ sym & & & 0 \end{bmatrix}\begin{Bmatrix} \ddot{v}_1 \\ \ddot{\varphi}_{x1} \\ \ddot{v}_2 \\ \ddot{\varphi}_{x2} \end{Bmatrix} + \frac{EI}{l^3}\begin{bmatrix} 12 & -6l & -12 & -6l \\ & 4l^2 & 6l & 2l^2 \\ & & 12 & 6l \\ sym & & & 4l^2 \end{bmatrix}\begin{Bmatrix} v_1 \\ \varphi_{x1} \\ v_2 \\ \varphi_{x2} \end{Bmatrix} = \begin{Bmatrix} -S_{y1} \\ -M_{yz1} \\ S_{y2} \\ M_{yz2} \end{Bmatrix} \quad \text{(a)}$$

Similarly, for the other elements, elemental equations can be written by changing the corresponding nodal variables in Equation (a). If the disk is assumed to be in element (3), the mass matrix for element (3) becomes

$$\mathbf{M}\ddot{\eta} = \begin{bmatrix} 0 & 0 & 0 & 0 \\ 0 & 0 & 0 & 0 \\ 0 & 0 & m & 0 \\ 0 & 0 & 0 & I_d \end{bmatrix}\begin{Bmatrix} \ddot{v}_3 \\ \ddot{\varphi}_{x3} \\ \ddot{v}_4 \\ \ddot{\varphi}_{x4} \end{Bmatrix} \quad \text{(b)}$$

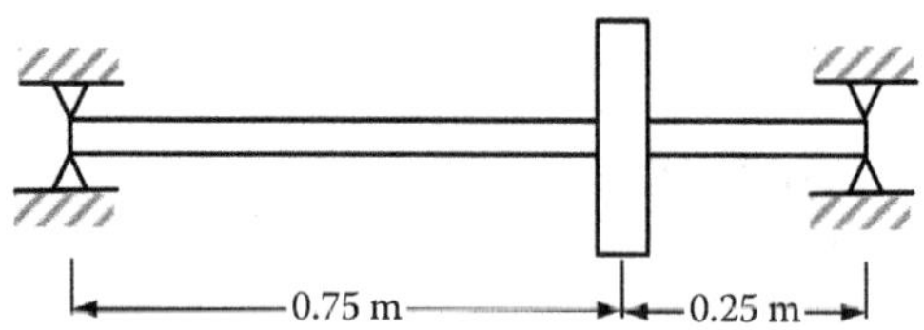

FIGURE 9.14 Simply supported shaft with an offset disk.

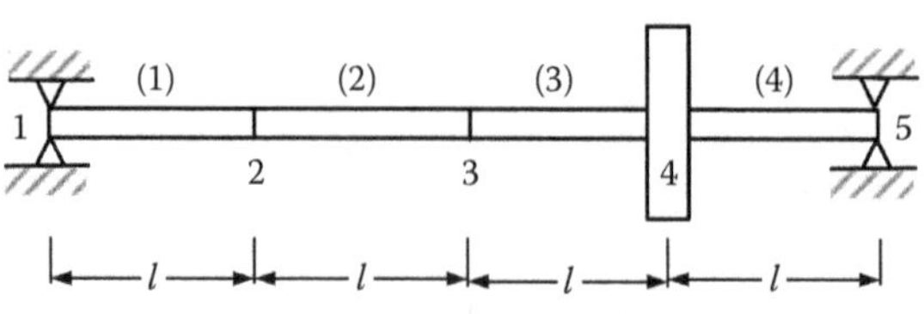

FIGURE 9.15 FE discretization of the rotor system.

On assembling the elemental equations for all elements, we get

$$\begin{bmatrix} 0 & 0 & 0 & 0 & 0 & 0 & 0 & 0 & 0 & 0 \\ & 0 & 0 & 0 & 0 & 0 & 0 & 0 & 0 & 0 \\ & & 0 & 0 & 0 & 0 & 0 & 0 & 0 & 0 \\ & & & 0 & 0 & 0 & 0 & 0 & 0 & 0 \\ & & & & 0 & 0 & 0 & 0 & 0 & 0 \\ & & & & & 0 & 0 & 0 & 0 & 0 \\ & & & & & & m & 0 & 0 & 0 \\ & & & & & & & I_d & 0 & 0 \\ & & & & & & & & 0 & 0 \\ sym & & & & & & & & & 0 \end{bmatrix} \begin{Bmatrix} \ddot{v}_1 \\ \ddot{\varphi}_{x1} \\ \ddot{v}_2 \\ \ddot{\varphi}_{x2} \\ \ddot{v}_3 \\ \ddot{\varphi}_{x3} \\ \ddot{v}_4 \\ \ddot{\varphi}_{x4} \\ \ddot{v}_5 \\ \ddot{\varphi}_{x5} \end{Bmatrix} \tag{c}$$

$$+\frac{EI}{l^3}\begin{bmatrix} 12 & -6l & -12 & -6l & 0 & 0 & 0 & 0 & 0 & 0 \\ & 4l^3 & 4l^3 & 2l^2 & 0 & 0 & 0 & 0 & 0 & 0 \\ & & 24 & 0 & -12 & -6l & 0 & 0 & 0 & 0 \\ & & & 8l^2 & 6l & 2l^2 & 0 & 0 & 0 & 0 \\ & & & & 24 & 0 & -12 & -6l & 0 & 0 \\ & & & & & 8l^2 & 6l & 2l^2 & 0 & 0 \\ & & & & & & 24 & 0 & -12 & -6l \\ & & & & & & & 8l^2 & 6l & 2l^2 \\ & & & & & & & & 12 & 6l \\ sym & & & & & & & & & 4l^2 \end{bmatrix} \begin{Bmatrix} v_1 \\ \varphi_{x1} \\ v_2 \\ \varphi_{x2} \\ v_3 \\ \varphi_{x3} \\ v_4 \\ \varphi_{x4} \\ v_5 \\ \varphi_{x5} \end{Bmatrix} = \begin{Bmatrix} -S_{y1} \\ -M_{yz1} = 0 \\ 0 \\ 0 \\ 0 \\ 0 \\ 0 \\ 0 \\ S_{y5} \\ M_{yz5} = 0 \end{Bmatrix}$$

The boundary conditions for the simply supported shaft are $v_1 = v_5 = 0 = \ddot{v}_1 = \ddot{v}_5$ and $M_{yz1} = M_{yz5} = 0$. Application of the boundary conditions in Equation (c) and elimination of the first and ninth rows and columns gives

$$\begin{bmatrix} 0 & 0 & 0 & 0 & 0 & 0 & 0 & 0 \\ & 0 & 0 & 0 & 0 & 0 & 0 & 0 \\ & & 0 & 0 & 0 & 0 & 0 & 0 \\ & & & 0 & 0 & 0 & 0 & 0 \\ & & & & 0 & 0 & 0 & 0 \\ & & & & & m & 0 & 0 \\ & & & & & & I_d & 0 \\ \text{sym} & & & & & & & 0 \end{bmatrix} \begin{Bmatrix} \ddot{\varphi}_{x1} \\ \ddot{v}_2 \\ \ddot{\varphi}_{x2} \\ \ddot{v}_3 \\ \ddot{\varphi}_{x3} \\ \ddot{v}_4 \\ \ddot{\varphi}_{x4} \\ \ddot{\varphi}_{x5} \end{Bmatrix} + \tag{d}$$

$$\frac{EI}{l^3}\begin{bmatrix} 4l^3 & 6l & 2l^2 & 0 & 0 & 0 & 0 & 0 \\ & 24 & 0 & -12 & -6l & 0 & 0 & 0 \\ & & 8l^2 & 6l & 2l^2 & 0 & 0 & 0 \\ & & & 24 & 0 & -12 & -6l & 0 \\ & & & & 8l^2 & 6l & 2l^2 & 0 \\ & & & & & 24 & 0 & -6l \\ & & & & & & 8l^2 & 2l^2 \\ \text{sym} & & & & & & & 4l^2 \end{bmatrix} \begin{Bmatrix} \varphi_{x1} \\ v_2 \\ \varphi_{x2} \\ v_3 \\ \varphi_{x3} \\ v_4 \\ \varphi_{x4} \\ \varphi_{x5} \end{Bmatrix} = \begin{Bmatrix} 0 \\ 0 \\ 0 \\ 0 \\ 0 \\ 0 \\ 0 \\ 0 \end{Bmatrix}$$

For free vibration, on substituting $\ddot{\eta} = -\omega_{nf}^2 \eta$ into Equation (d), we get an eigenvalue problem of the following form:

$$\left(-\omega_{nf}^2 \mathbf{M} + \mathbf{K}\right)\eta = \mathbf{0} \tag{e}$$

Substituting $m=10$ kg, $I_d=0.02$ kg-m^2, $l=0.25$ m, and $EI/l^3 = 6593.92$ N/m in Equation (e), it can be solved to give the following natural frequencies of the rotor system:

$$\omega_{nf1}^2 = 867.18 \Rightarrow \omega_{nf1} = 29.45 \text{ rad/s} \quad \text{and} \quad \omega_{nf2}^2 = 83{,}652.13 \Rightarrow \omega_{nf2} = 289.23 \text{ rad/s}$$

Comparing it with the influence coefficient method ($\omega_{nf1} = 22.24$ rad/s; $\omega_{nf2} = 248.70$ rad/s) suggests that by the FEM we got the upper bounds of natural frequencies. With an increase in the number of elements, it is expected that natural frequencies will come closer to that of the influence coefficient method, as seen in the previous example. The advantage of the FEM is that now we have the flexibility of including the effect of the shaft mass in determining the natural frequencies without much effort. We need to consider the consistent mass matrix in shaft elements chosen along with the disk mass. The corresponding mode shapes also can be plotted with the help of eigen vectors. This is left to the reader as an exercise.

Example 9.4

Calculate the transverse natural frequencies of an overhung rotor system as shown in Figure 9.16. The mass of the disk is $m = 5$ kg and its diametral mass moment of inertia $I_d=0.02$ kg-m^2. The length of the shaft segments is $a=0.3$ m and $b=0.7$ m, and the diameter of the shaft is 10 mm. Neglect the gyroscopic effects. $\rho=7800$ kg/m^3 and $E=2.1\times10^{11}$ N/m^2.

Solution:

The main characteristic of the present rotor system is that it has an intermediate support. The present example describes how to handle such intermediate supports even for multi-support systems, which is very common in actual rotor systems. For illustration, the shaft can be divided into two elements with three nodes as shown in Figure 9.17.

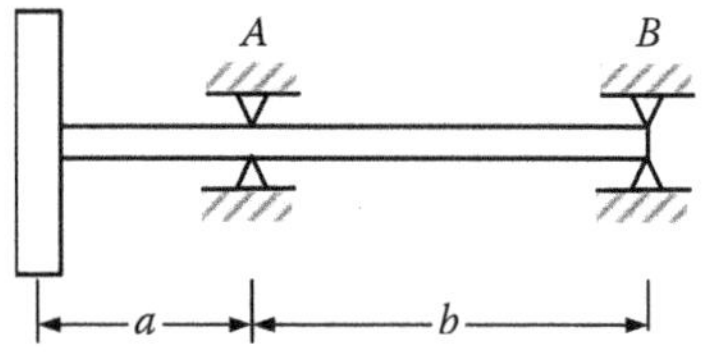

FIGURE 9.16 Overhung rotor system.

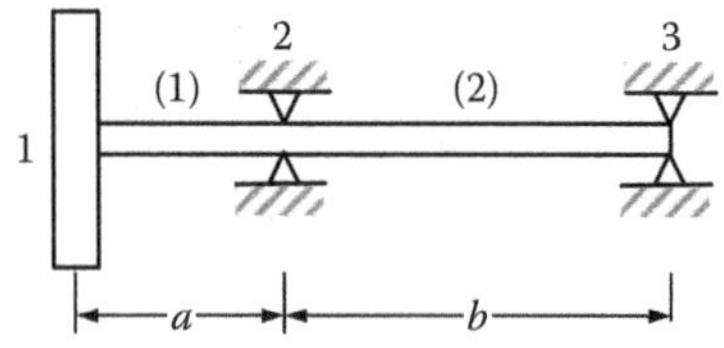

FIGURE 9.17 Discretization of the shaft into two elements.

The elemental consistent mass and stiffness matrices of a shaft in vertical plane (*y-z* plane) are given as follows:

$$\mathbf{M}^{(e)}=\frac{\rho Al}{420}\begin{bmatrix}156 & -22l & 54 & 13l\\ & 4l^2 & -13l & -3l^2\\ & & 156 & 22l\\ sym & & & 4l^2\end{bmatrix},\quad \mathbf{K}^{(e)}=\frac{EI}{l^3}\begin{bmatrix}12 & -6l & -12 & -6l\\ & 4l^2 & 6l & 2l^2\\ & & 12 & 6l\\ sym & & & 4l^2\end{bmatrix} \qquad \text{(a)}$$

Some of the common properties of elements are $A=\frac{\pi}{4}d^2=\frac{\pi}{4}0.01^2=7.854\times10^{-5}\ \text{m}^2$, $I=\frac{\pi}{64}d^4=\frac{\pi}{64}0.01^4=4.91\times10^{-10}\ \text{m}^4$, ρ=7800 kg/m³, and E=2.1×10¹¹ N/m².

i. *Element 1*: The following properties are taken for element 1: $l=0.3$ m, $\frac{\rho Al}{420}=4.376\times10^{-4}$ kg, and $EI/l^3=3.818\times10^3$ N/m. At node 1, we need to consider the mass of the disk, $m=5$ kg, and its diametral mass moment of inertia, I_d=0.02 kg-m². The governing equation for element 1 would be

$$4.376\times10^{-4}\begin{bmatrix}156+\dfrac{5}{4.376\times10^{-4}} & -6.6 & 54 & 3.9\\ & 0.36+\dfrac{0.02}{4.376\times10^{-4}} & -3.9 & -0.27\\ & & 156 & 6.6\\ sym & & & 0.36\end{bmatrix}\begin{Bmatrix}\ddot{v}_1\\ \ddot{\varphi}_{x1}\\ \ddot{v}_2\\ \ddot{\varphi}_{x2}\end{Bmatrix} \qquad \text{(b)}$$

$$+3.818\times10^3\begin{bmatrix}12 & -1.8 & -12 & -1.8\\ & 0.36 & 1.8 & 0.18\\ & & 12 & 1.8\\ sym & & & 0.36\end{bmatrix}\begin{Bmatrix}v_1\\ \varphi_{x1}\\ v_2\\ \varphi_{x2}\end{Bmatrix}=\begin{Bmatrix}-S_{y1}\\ -M_{yz1}\\ S_{y2}\\ M_{yz2}\end{Bmatrix}$$

ii. *Element 2*: The following properties are taken for element 2: $l=0.7$ m, $\frac{\rho Al}{420}=1.0\times10^{-3}$ kg, and $EI_{yy}/l^3=0.3005\times10^3$ N/m. The governing equation for element 2 would be

$$1.0\times10^{-3}\begin{bmatrix}156.0 & -15.4 & 54.0 & 9.1\\ & 1.96 & -9.1 & -1.47\\ & & 156.0 & 15.4\\ sym & & & 1.96\end{bmatrix}\begin{Bmatrix}\ddot{v}_2\\ \ddot{\varphi}_{x2}\\ \ddot{v}_3\\ \ddot{\varphi}_{x3}\end{Bmatrix} \qquad \text{(c)}$$

$$+0.3\times10^3\begin{bmatrix}12.0 & -4.2 & -12.0 & -4.2\\ & 1.96 & 4.2 & 0.98\\ & & 12 & 4.2\\ sym & & & 1.96\end{bmatrix}\begin{Bmatrix}v_2\\ \varphi_{x2}\\ v_3\\ \varphi_{x3}\end{Bmatrix}=\begin{Bmatrix}-S_{y2}\\ -M_{yz2}\\ S_{y3}\\ M_{yz3}\end{Bmatrix}+\begin{Bmatrix}R_2\\ 0\\ 0\\ 0\end{Bmatrix}$$

It should be noted that R_2 has been added in the extreme right-hand side vector of Equation (c) as an external force at node 2, and it corresponds to the reaction of the intermediate support. The reaction force R_2 exists along with the shaft internal shear force $-S_{y2}$ as given in the first vector in the right-hand side of Equation (c). On combining Equations (b) and (c), we get

$$\begin{bmatrix} 5.0683 & -0.0029 & 0.0236 & 0.0017 & 0 & 0 \\ & 0.0202 & -0.0017 & -0.0001 & 0 & 0 \\ & & 0.2275 & -0.0128 & 0.0551 & 0.0093 \\ & & & 0.0022 & -0.0093 & -0.0015 \\ & & & & 0.1593 & 0.0157 \\ sym & & & & & 0.0020 \end{bmatrix} \begin{Bmatrix} \ddot{v}_1 \\ \ddot{\varphi}_{x1} \\ \ddot{v}_2 \\ \ddot{\varphi}_{x2} \\ \ddot{v}_3 \\ \ddot{\varphi}_{x3} \end{Bmatrix}$$

$$+10^4 \times \begin{bmatrix} 4.5815 & -0.6872 & -4.5815 & -0.6872 & 0 & 0 \\ & 0.1374 & 0.6872 & 0.0687 & 0 & 0 \\ & & 4.9421 & 0.5610 & -0.3606 & -0.1262 \\ & & & 0.1963 & 0.1262 & 0.0295 \\ & & & & 0.3606 & 0.1262 \\ sym & & & & & 0.0589 \end{bmatrix} \tag{d}$$

$$\begin{Bmatrix} v_1 \\ \varphi_{x1} \\ v_2 \\ \varphi_{x2} \\ v_3 \\ \varphi_{x3} \end{Bmatrix} = \begin{Bmatrix} -S_{y1} \\ -M_{yz1} \\ 0 \\ 0 \\ S_{y3} \\ M_{yz3} \end{Bmatrix} + \begin{Bmatrix} 0 \\ 0 \\ R_2 \\ 0 \\ 0 \\ 0 \end{Bmatrix}$$

It can be seen from Equation (d) that at node 2, the shaft internal reaction forces and moments get canceled on assembling of elemental equations. However, the reaction force, R_2, from the intermediate support, which is unknown, will exist in the assembled Equation (d). Now the assembled equation is ready for the application of boundary conditions.

The boundary conditions of the present problem are given as follows:

$$\text{At node 1}: S_{y1} = M_{yz1} = 0; \quad \text{At node 2}: v_2 = 0; \quad \text{At node 3}: v_3 = M_{yz3} = 0 \tag{e}$$

It should be noted that at node 2 the bending moment will not be zero as in the case of node 3 because node 2 is an intermediate support and node 3 is a simple support. On application of boundary conditions, forces R_2 and S_{y3} will remain nonzero on the right-hand side of Equation (e). These are present in the third and fifth rows. Hence, by eliminating the corresponding third and fifth rows and columns, Equation (e) reduces to

$$\begin{bmatrix} 5.0683 & -0.0029 & 0.0017 & 0 \\ & 0.0202 & -0.0001 & 0 \\ & & 0.0022 & -0.0015 \\ sym & & & 0.0020 \end{bmatrix} \begin{Bmatrix} \ddot{v}_1 \\ \ddot{\varphi}_{x1} \\ \ddot{\varphi}_{x2} \\ \ddot{\varphi}_{x3} \end{Bmatrix}$$

$$+10^4 \begin{bmatrix} 4.5815 & -0.6872 & -0.6872 & 0 \\ & 0.1374 & 0.0687 & 0 \\ & & 0.1963 & 0.0295 \\ sym & & & 0.0589 \end{bmatrix} \begin{Bmatrix} v_1 \\ \varphi_{x1} \\ \varphi_{x2} \\ \varphi_{x3} \end{Bmatrix} = \begin{Bmatrix} 0 \\ 0 \\ 0 \\ 0 \end{Bmatrix} \tag{f}$$

For a simple harmonic motion during free vibrations, Equation (f) can be written as a standard eigenvalue problem as follows:

$$\left(\mathbf{D} - \omega_{nf}^2 \mathbf{I}\right)\eta_0 = \mathbf{0} \tag{g}$$

with

$$\mathbf{M}=\begin{bmatrix} 5.0683 & -0.0029 & 0.0017 & 0 \\ & 0.0202 & -0.0001 & 0 \\ & & 0.0022 & -0.0015 \\ sym & & & 0.0020 \end{bmatrix}, \quad \mathbf{K}=10^4\begin{bmatrix} 4.5815 & -0.6872 & -0.6872 & 0 \\ & 0.1374 & 0.0687 & 0 \\ & & 0.1963 & 0.0295 \\ sym & & & 0.0589 \end{bmatrix} \tag{h}$$

$$\eta_0=\begin{Bmatrix} v_1 \\ \varphi_{x1} \\ \varphi_{x2} \\ \varphi_{x3} \end{Bmatrix}, \quad \mathbf{D}=\mathbf{M}^{-1}\mathbf{K} \tag{i}$$

where **I** is the identity matrix of size 4×4. The square root of the eigenvalue of the matrix **D** gives natural frequencies as follows:

$$\omega_{nf1}=25.29\text{ rad/s},\quad \omega_{nf2}=234.87\text{ rad/s},\quad \omega_{nf3}=444.89\text{ rad/s},\quad \omega_{nf4}=1667.90\text{ rad/s}$$

For more refined elements (20 elements), the above four natural frequencies are found to converge to

$$\omega_{nf1}=25.29\text{ rad/s},\quad \omega_{nf2}=233.46\text{ rad/s},\quad \omega_{nf3}=364.18\text{ rad/s},\quad \omega_{nf4}=1167.90\text{ rad/s}$$

and the eigenvectors are mode shapes (the translational and rotational displacements) corresponding to these natural frequencies and are plotted in Figure 9.18. The normalization is such that the maximum amplitude in each mode is assigned as unity. The third and fifth equations of the expression in Equation (d), which was eliminated earlier, give reaction forces at supports 2 and 3. Although for free vibration it has no practical significance, it will be useful for forced vibrations.

Example 9.5

Find at least three transverse natural frequencies and draw the corresponding mode shapes of the rotor system as shown in Figure 9.19. The shaft support at B_1 is the fixed support (with zero transverse translational and rotational (slope) displacements), and at B_2 and B_3 are intermediate roller supports. The shaft is made of steel with Young's modulus $E=2.1\times10^{11}$ N/m^2 and mass density $\rho=7800$ kg/m^3. The masses of thin disks D_1, D_2, and D_3, respectively, are $m_1=1$ kg, $m_2=1.5$ kg, $m_3=0.75$ kg, and the radius of gyration, r_g, of all the disks is $5/\sqrt{2}$ cm. The lengths of various shaft segments are as follows: B_1D_1=50 mm, D_1B_2=50 mm, B_2D_2=25 mm, D_2B_3=25 mm, and B_3D_3=30 mm. The shaft has a uniform diameter of 10 mm. Neglect the gyroscopic effects.

Solution:

The shaft is discretized into five elements for the illustration of the problem as shown in Figure 9.20. While choosing nodes care should be taken so that every important component, support, disk, etc., has a node and its contribution is easily handled in the FE formulation. Choosing a large number of nodes is expected to improve the determination of natural frequencies.

For element length l (which is not constant for different elements of the present problem), the mass and stiffness matrices are given as follows:

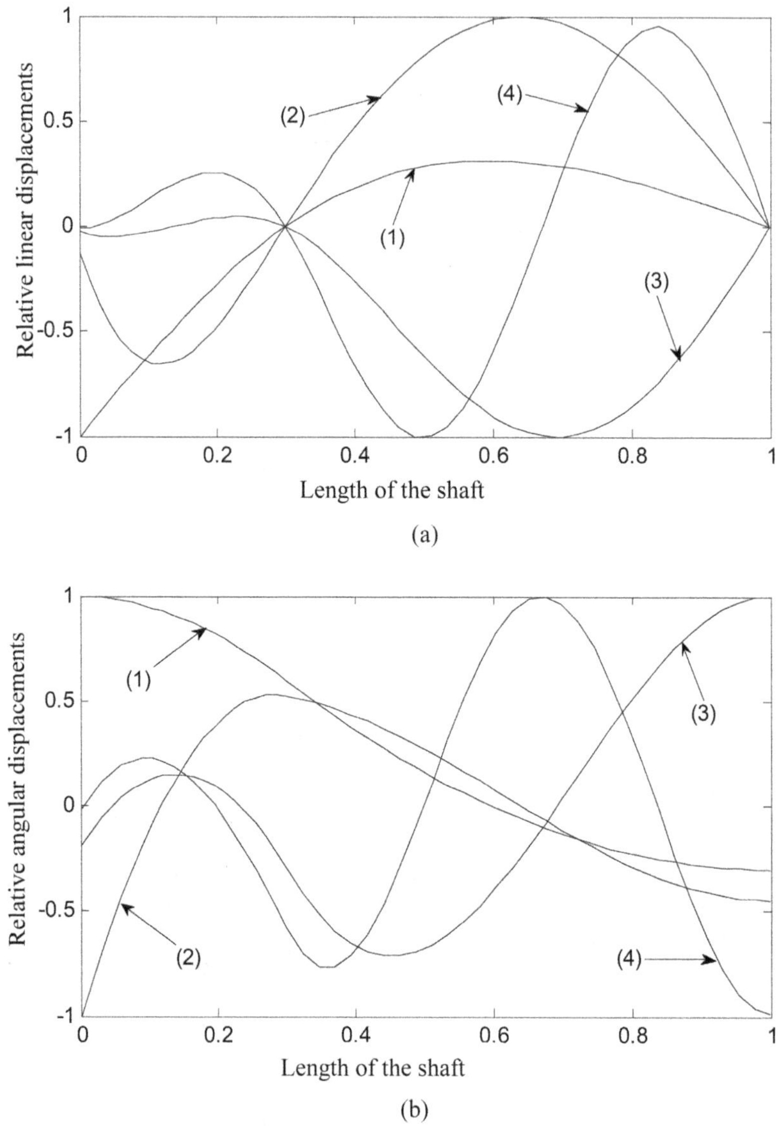

FIGURE 9.18 (a) Normalized translational displacement mode shapes: (1) first mode shape, (2) second mode shape, (3) third mode shape, and (4) fourth mode shape. (b) Normalized rotational displacement mode shapes: (1) first mode shape, (2) second mode shape, (3) third mode shape, and (4) fourth mode shape.

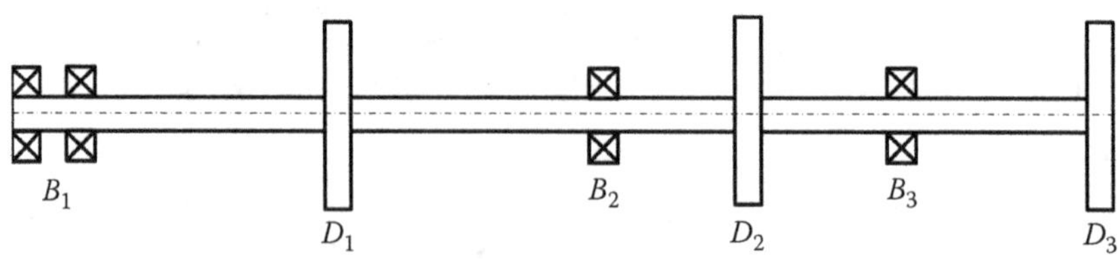

FIGURE 9.19 Rotor system with multidisc and multi-intermediate support.

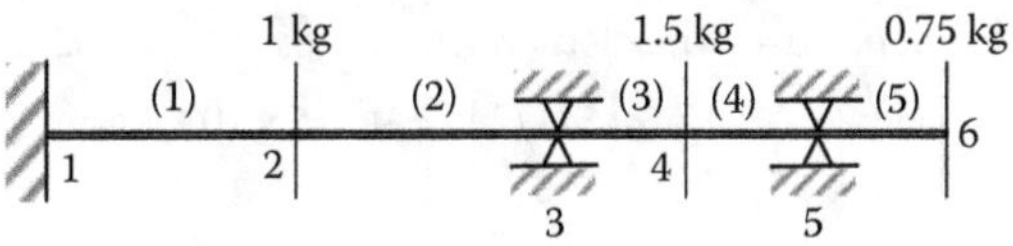

FIGURE 9.20 Discretization of the rotor system into elements.

$$\mathbf{M}^{(e)}=\frac{\rho Al}{420}\begin{bmatrix}156 & -22l & 54 & 13l\\ -22l & 4l^2 & -13l & -3l^2\\ 54 & -13l & 156 & 22l\\ 13l & -3l^2 & 22l & 4l^2\end{bmatrix};\quad \mathbf{K}^{(e)}=\frac{EI}{l^3}\begin{bmatrix}12 & -6l & -12 & -6l\\ -6l & 4l^2 & 6l & 2l^2\\ -12 & 6l & 12 & 6l\\ -6l & 2l^2 & 6l & 4l^2\end{bmatrix} \tag{a}$$

We have the following properties of the shaft: $E = 2.1\times10^{11}$ N/m^2 and $\rho = 7800$ kg/m^3. In addition to this, various elements have different properties, which are given as follows:

i. *Element 1 (with a disk at node 2)*: The element properties are $l = 0.05\,\text{m}$, $I=\frac{\pi}{64}0.01^4 = 4.9\times10^{-10}\ \text{m}^4$, $m = 1\,\text{kg}$, and $I_d=\frac{1}{2}mr_g^2=\frac{1}{2}\times1\times\left(5/\sqrt{2}\right)^2 = 6.25\times10^{-4}\ \text{kg-m}^2$. From Equation (a), we get

$$\mathbf{M}^{(1)}=10^{-4}\begin{bmatrix}113.77 & -0.80 & 39.38 & 0.47\\ -0.80 & 0.0073 & -0.47 & -0.0055\\ 39.38 & -0.47 & (113.77+10{,}000) & 0.80\\ 0.47 & -0.0055 & 0.80 & (0.0073+6.25)\end{bmatrix}$$

and

$$\mathbf{K}^{(1)}=10^{6}\begin{bmatrix}9.88 & -0.247 & -9.88 & -0.247\\ -0.247 & 0.0082 & 0.247 & 0.0041\\ -9.88 & 0.247 & 9.88 & 0.247\\ -0.247 & 0.004 & 0.247 & 0.0082\end{bmatrix} \tag{b}$$

ii. *Element 2 (with no disks at nodes 2 and 3)*: The element properties are $l = 0.05$ m and $I = 4.9\times10^{-10}$ m^4. From Equation (a), we get

$$\mathbf{M}^{(2)}=10^{-4}\begin{bmatrix}113.77 & -0.80 & 39.38 & 0.47\\ -0.80 & 0.0073 & -0.47 & -0.0055\\ 39.38 & -0.47 & 113.77 & 0.80\\ 0.47 & -0.0055 & 0.80 & 0.0073\end{bmatrix}$$

and

$$\mathbf{K}^{(2)}=10^{6}\begin{bmatrix}9.88 & -0.247 & -9.88 & -0.247\\ -0.247 & 0.0082 & 0.247 & 0.0041\\ -9.88 & 0.247 & 9.88 & 0.247\\ -0.247 & 0.004 & 0.247 & 0.0082\end{bmatrix} \tag{c}$$

iii. *Element 3 (with a disk at node 4)*: The element properties are $l = 0.025\,\text{m}$, $I = 4.9\times10^{-10}\ \text{m}^4$, $m = 1\,\text{kg}$, and $I_d = \frac{1}{2}mr_g^2 = \frac{1}{2}\times1.5\times\left(5/\sqrt{2}\right)^2 = 9.375\times10^{-4}\ \text{kg-m}^2$. From Equation (a), we get

$$\mathbf{M}^{(3)} = 10^{-4}\begin{bmatrix} 56.8 & -0.201 & 19.7 & 0.12 \\ -0.201 & 0.001 & -0.12 & -0.001 \\ 19.69 & -0.12 & (56.8+15{,}000) & 0.201 \\ 0.12 & -0.001 & 0.201 & (0.001+9.375) \end{bmatrix}$$

and

$$\mathbf{K}^{(3)} = 10^{6}\begin{bmatrix} 79 & -0.99 & -79 & -0.99 \\ -0.99 & 0.016 & 0.99 & 0.008 \\ -79 & 0.99 & 79 & 0.99 \\ -0.99 & 0.008 & 0.99 & 0.016 \end{bmatrix} \tag{d}$$

iv. *Element 4 (with no disks at nodes 4 and 5)*: The element properties are $l = 0.025$ m and $I = 4.9\times10^{-10}\ \text{m}^4$. From Equation (a), we get

$$\mathbf{M}^{(4)} = 10^{-4}\begin{bmatrix} 56.8 & -0.201 & 19.7 & 0.12 \\ -0.201 & 0.001 & -0.12 & -0.001 \\ 19.69 & -0.12 & 56.8 & 0.201 \\ 0.12 & -0.001 & 0.201 & 0.001 \end{bmatrix};\ \mathbf{K}^{(4)} = 10^{6}\begin{bmatrix} 79 & -0.99 & -79 & -0.99 \\ -0.99 & 0.016 & 0.99 & 0.008 \\ -79 & 0.99 & 79 & 0.99 \\ -0.99 & 0.008 & 0.99 & 0.016 \end{bmatrix} \tag{e}$$

v. *Element 5 (with a disk at node 6)*: The element properties are $l = 0.03\,\text{m}$, $I = 4.9\times10^{-10}\ \text{m}^4$, $m = 0.75$ kg, and $I_d = \frac{1}{2}mr_g^2 = \frac{1}{2}\times0.75\times\left(5/\sqrt{2}\right)^2 = 4.6875\times10^{-4}\ \text{kg-m}^2$. From Equation (a), we get

$$\mathbf{M}^{(5)} = 10^{-4}\begin{bmatrix} 68.28 & -0.29 & 23.6 & 0.17 \\ -0.29 & 0.002 & -0.17 & -0.001 \\ 23.6 & -0.17 & (68.28+7500) & 0.29 \\ 0.17 & -0.001 & 0.29 & (0.002+4.6875) \end{bmatrix}$$

and

$$\mathbf{K}^{(5)} = 10^{6}\begin{bmatrix} 45.7 & -0.69 & -45.7 & -0.69 \\ -0.69 & 0.014 & 0.69 & 0.007 \\ -45.7 & 0.69 & 45.7 & 0.69 \\ -0.69 & 0.007 & 0.69 & 0.014 \end{bmatrix} \tag{f}$$

Noting Equations (b) through (f), the assembled form of the governing equation of the rotor system has the following form:

$$
10^{-4}\begin{bmatrix}
113.8 & -0.8 & 39.4 & 0.47 & 0 & 0 & 0 & 0 & 0 & 0 & 0 & 0\\
-0.8 & 0.01 & -0.47 & -0.01 & 0 & 0 & 0 & 0 & 0 & 0 & 0 & 0\\
39.4 & -0.47 & -10,227.54 & 0 & 39.4 & 0.47 & 0 & 0 & 0 & 0 & 0 & 0\\
0.47 & -0.01 & 0 & 6.27 & -0.47 & -0.01 & 0 & 0 & 0 & 0 & 0 & 0\\
0 & 0 & 39.4 & -0.47 & 170.6 & 0.6 & 19.7 & 0.12 & 0 & 0 & 0 & 0\\
0 & 0 & 0.47 & -0.01 & 0.6 & 0.011 & -0.12 & -0.001 & 0 & 0 & 0 & 0\\
0 & 0 & 0 & 0 & 19.69 & -0.12 & 15,113.77 & 0 & 19.7 & 0.12 & 0 & 0\\
0 & 0 & 0 & 0 & 0.12 & -0.001 & 0 & 9.395 & -0.12 & -0.001 & 0 & 0\\
0 & 0 & 0 & 0 & 0 & 0 & 19.69 & -0.12 & 125.1 & -0.089 & 23.6 & 0.17\\
0 & 0 & 0 & 0 & 0 & 0 & 0.12 & -0.001 & -0.089 & 0.003 & -0.17 & -0.001\\
0 & 0 & 0 & 0 & 0 & 0 & 0 & 0 & 23.6 & -0.17 & 7568.26 & 0.29\\
0 & 0 & 0 & 0 & 0 & 0 & 0 & 0 & 0.17 & -0.001 & 0.29 & 4.7075
\end{bmatrix}
$$

$$
\begin{Bmatrix}\ddot{v}_1\\ \ddot{\varphi}_{x_1}\\ \ddot{v}_2\\ \ddot{\varphi}_{x_2}\\ \ddot{v}_3\\ \ddot{\varphi}_{x_3}\\ \ddot{v}_4\\ \ddot{\varphi}_{x_4}\\ \ddot{v}_5\\ \ddot{\varphi}_{x_5}\\ \ddot{v}_6\\ \ddot{\varphi}_{x_6}\end{Bmatrix}
+10^{6}\begin{bmatrix}
9.88 & -0.247 & -9.88 & -0.247 & 0 & 0 & 0 & 0 & 0 & 0 & 0 & 0\\
-0.247 & 0.0082 & 0.247 & 0.004 & 0 & 0 & 0 & 0 & 0 & 0 & 0 & 0\\
-9.88 & 0.247 & 19.76 & 0 & -9.88 & -0.247 & 0 & 0 & 0 & 0 & 0 & 0\\
-0.247 & 0.004 & 0 & 0.0164 & 0.247 & 0.004 & 0 & 0 & 0 & 0 & 0 & 0\\
0 & 0 & -9.88 & 0.247 & 88.88 & -0.743 & -79 & -0.99 & 0 & 0 & 0 & 0\\
0 & 0 & -0.247 & 0.004 & -0.743 & 0.0242 & 0.99 & 0.008 & 0 & 0 & 0 & 0\\
0 & 0 & 0 & 0 & -79 & 0.99 & 158 & 0 & -79 & -0.99 & 0 & 0\\
0 & 0 & 0 & 0 & -0.99 & 0.008 & 0 & 0.032 & 0.99 & 0.008 & 0 & 0\\
0 & 0 & 0 & 0 & 0 & 0 & -79 & 0.99 & 124.7 & 0.3 & -45.7 & -0.69\\
0 & 0 & 0 & 0 & 0 & 0 & -0.99 & -0.99 & 0.3 & 0.03 & 0.69 & 0.007\\
0 & 0 & 0 & 0 & 0 & 0 & 0 & 0 & -45.7 & 0.69 & 45.7 & 0.69\\
0 & 0 & 0 & 0 & 0 & 0 & 0 & 0 & -0.69 & 0.007 & 0.69 & 0.014
\end{bmatrix}
$$

$$
\begin{Bmatrix}v_1\\ \varphi_{x_1}\\ v_2\\ \varphi_{x_2}\\ v_3\\ \varphi_{x_3}\\ v_4\\ \varphi_{x_4}\\ v_5\\ \varphi_{x_5}\\ v_6\\ \varphi_{x_6}\end{Bmatrix}
=\begin{Bmatrix}-S_{y_1}\\ -M_{yz_1}\\ 0\\ 0\\ 0\\ 0\\ 0\\ 0\\ 0\\ 0\\ S_{y_6}\\ M_{yz_6}\end{Bmatrix}
+\begin{Bmatrix}0\\ 0\\ 0\\ 0\\ R_3\\ 0\\ 0\\ 0\\ R_5\\ 0\\ 0\\ 0\end{Bmatrix}
\qquad \text{(g)}
$$

where R_3 and R_5 are unknown support reactions at nodes 3 and 5, respectively. The boundary conditions of the present problem are as follows (Figure 9.20):

$$\text{Node 1}: v_1 = \varphi_{x_1} = 0; \quad \text{Node 3}: v_3 = 0; \quad \text{Node 5}: v_5 = 0; \quad \text{Node 6}: S_{y6} = M_{yz6} = 0 \tag{h}$$

On application of the boundary conditions given in Equation (h), governing Equation (g) takes the following reduced form after eliminating the first, second, fifth, and ninth rows and columns:

$$10^{-4}\begin{bmatrix} 1.02\times10^4 & 0 & 0.47 & 0 & 0 & 0 & 0 & 0 \\ 0 & 6.27 & -0.01 & 0 & 0 & 0 & 0 & 0 \\ 0.47 & -0.01 & 0.011 & -0.12 & -0.001 & 0 & 0 & 0 \\ 0 & 0 & -0.12 & 1.51\times10^4 & 0 & 0.12 & 0 & 0 \\ 0 & 0 & -0.001 & 0 & 9.395 & -0.001 & 0 & 0 \\ 0 & 0 & 0 & 0.12 & -0.001 & 0.003 & -0.17 & -0.001 \\ 0 & 0 & 0 & 0 & 0 & -0.17 & 7568.26 & 0.29 \\ 0 & 0 & 0 & 0 & 0 & -0.001 & 0.29 & 4.708 \end{bmatrix}\begin{Bmatrix} \ddot{v}_2 \\ \ddot{\varphi}_{x2} \\ \ddot{\varphi}_{x3} \\ \ddot{v}_4 \\ \ddot{\varphi}_{x4} \\ \ddot{\varphi}_{x5} \\ \ddot{v}_6 \\ \ddot{\varphi}_{x6} \end{Bmatrix}$$

$$+10^{6}\begin{bmatrix} 19.76 & 0 & -0.247 & 0 & 0 & 0 & 0 & 0 \\ 0 & 0.0164 & 0.004 & 0 & 0 & 0 & 0 & 0 \\ -0.247 & 0.004 & 0.0242 & 0.99 & 0.008 & 0 & 0 & 0 \\ 0 & 0 & 0.99 & 158 & 0 & -0.99 & 0 & 0 \\ 0 & 0 & 0.008 & 0 & 0.032 & 0.008 & 0 & 0 \\ 0 & 0 & 0 & -0.99 & 0.008 & 0.03 & 0.69 & 0.007 \\ 0 & 0 & 0 & 0 & 0 & 0.69 & 45.7 & 0.69 \\ 0 & 0 & 0 & 0 & 0 & 0.007 & 0.69 & 0.014 \end{bmatrix}\begin{Bmatrix} v_2 \\ \varphi_{x2} \\ \varphi_{x3} \\ v_4 \\ \varphi_{x4} \\ \varphi_{x5} \\ v_6 \\ \varphi_{x6} \end{Bmatrix} = \begin{Bmatrix} 0 \\ 0 \\ 0 \\ 0 \\ 0 \\ 0 \\ 0 \\ 0 \end{Bmatrix} \tag{i}$$

which gives the following form:

$$\mathbf{M}\ddot{\eta} + \mathbf{K}\eta = \mathbf{0} \tag{j}$$

For the simple harmonic motion during free vibrations, Equation (j) can be written as a standard eigenvalue problem as follows:

$$\left(\mathbf{D} - \omega_{nf}^2 \mathbf{I}\right)\eta_0 = \mathbf{0} \tag{k}$$

with

$$\mathbf{D} = \mathbf{M}^{-1}\mathbf{K} \tag{l}$$

The square root of the eigenvalue of matrix **D** gives the first four natural frequencies as follows:

$$\omega_{nf_1} = 1662.5 \text{ rad/s}, \quad \omega_{nf_2} = 3723.2 \text{ rad/s}, \quad \omega_{nf_3} = 4996.5 \text{ rad/s}, \quad \omega_{nf_4} = 5316.1 \text{ rad/s}$$

Eigenvectors are mode shapes corresponding to these natural frequencies. Mode shapes for the first three natural modes are plotted in Figure 9.21. It can be observed that for higher modes the overhang disk has a relatively smaller amplitude of vibrations as compared to other disk locations.

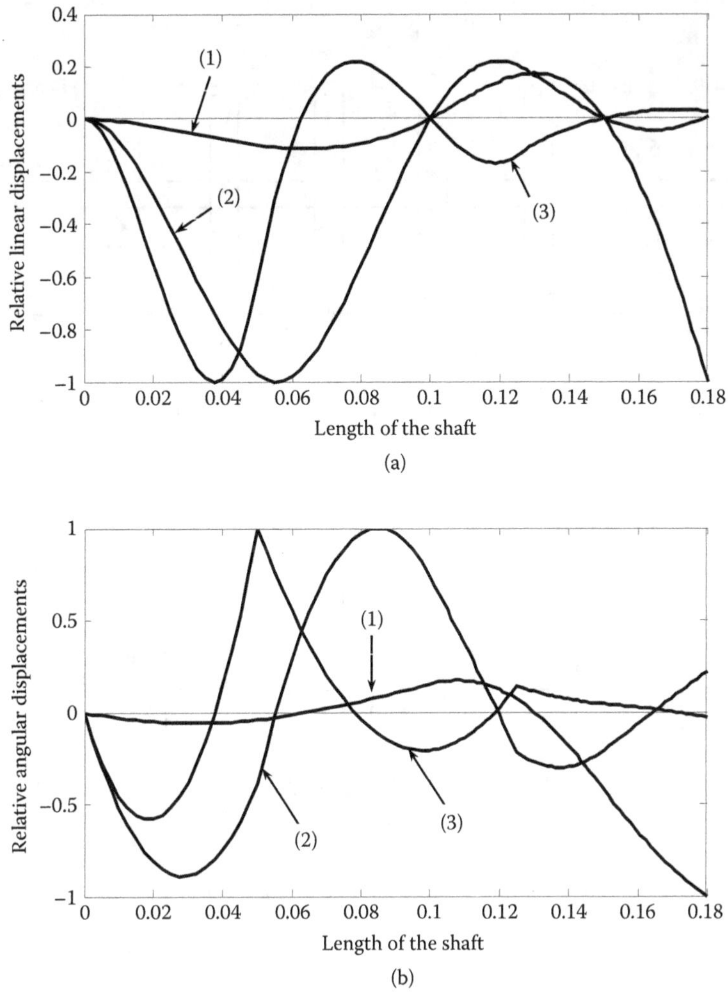

FIGURE 9.21 (a) Normalized translational displacement mode shapes: (1) first mode shape, (2) second mode shape, and (3) third mode shape. (b) Normalized rotational displacement mode shapes: (1) first mode shape, (2) second mode shape, and (3) third mode shape.

Example 9.6

A typical simply supported multidisc rotor system as shown in Figure 9.22 is analyzed to show the application of the present FE model. The physical properties of the rotor-bearing system are given as follows: diameter of the shaft=0.1 m, length of the shaft=3.5 m, Young's modulus of the shaft material=2.08×10^{11} N/m^2, mass density of the shaft material=7830 kg/m^3, number of rigid disks=4, and mass of each rigid disk=60.3 kg. The shaft segments between two neighboring disks, between the bearing and neighboring disks, and the overhang portion are 0.5 m length each.

Solution:

The rotor is modeled as seven elements and fourteen elements as shown in Figures 9.22 and 9.23, respectively. In the case of a seven-element member, (i) identical rigid bearings are located at nodes 2 and 7, and (ii) rigid disks are located at nodes 3, 4, 5 and 6. In the case of the 14-element

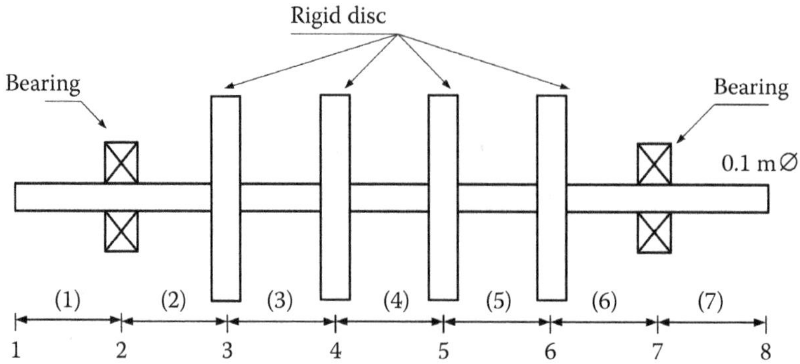

FIGURE 9.22 Rotor-bearing system with rigid disks (seven elements of 0.5 m each).

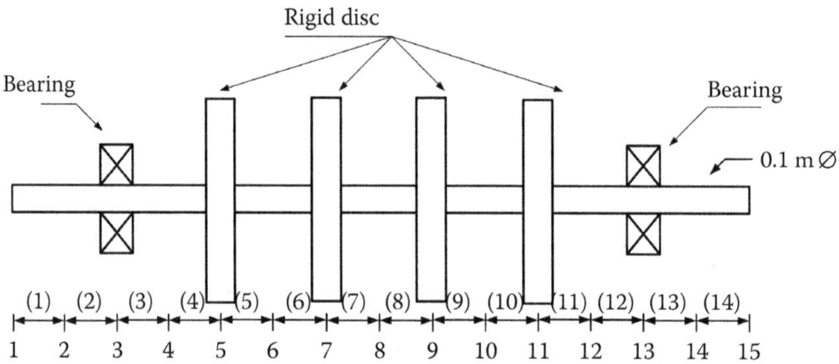

FIGURE 9.23 Rotor-bearing system with rigid disks (14 elements of 0.25 m each).

TABLE 9.8
Natural Whirl Frequencies of the Rotor-Bearing System with Rigid Disks

	Transverse Natural Frequencies (rad/s)	
Mode No.	**For Seven Elements**	**For 14 Elements**
1	116.1756	115.8136
2	438.4482	438.2795
3	861.5609	860.7477
4	1209.4922	1207.0327

member, (i) identical rigid bearings are located at nodes 3 and 13, and (ii) rigid disks are located at nodes 5, 7, 9, and 11. The shaft has a uniform cross section. Rigid disks are considered as point masses, and the mass of the shaft is also considered. The first four transverse natural frequencies are obtained for the chosen models and are given in Table 9.8. The results show that not much improvement in natural frequencies occurred in the 14-element model as compared to the seven-element model; this suggests that convergence has already occurred with the seven-element model. The mode shapes are shown in Figure 9.24.

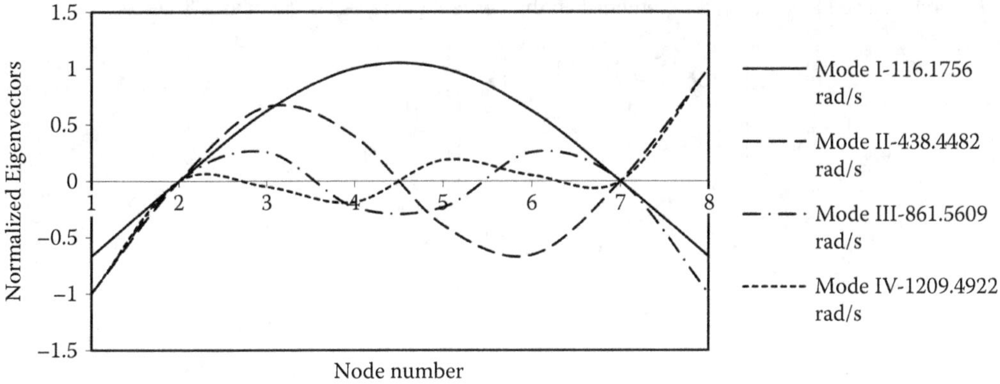

FIGURE 9.24 Mode shapes for the rotor-bearing system with rigid disks.

Example 9.7

Perform forced vibration analysis of the rotor system shown in Figure 9.25. The shaft has a diameter $d = 10\,\text{mm}$, and the disk has a mass $m_{disc} = 1.5\,\text{kg}$ (ignore the diametral mass moment of inertia of the disk). The unbalance in the disk is given as $m_{unb} = 0.005\,\text{kg}$, $r_{unb} = 0.05\,\text{m}$, and $\phi_{unb} = 30^\circ$ phase at time $t = 0$. Plot the unbalance response (i.e., the amplitude and the phase) variation with respect to the speed of the rotor at the disk location. Choose the upper limit of the speed of the rotor such that plots should depict the first two critical speeds.

Solution:

The shaft element is taken such that the disk is always at a node point. Figure 9.25 shows the discretized rotor system with three elements. It is assumed the disk is on element 2. The phase is measured with respect to some physical reference on the shaft. Let us assume that the reference mark is aligned with the y-axis at time $t = 0$. The angular speed, ω, is in the counterclockwise direction as seen in Figure 9.26. In the present case, both plane motions would be considered simultaneously.

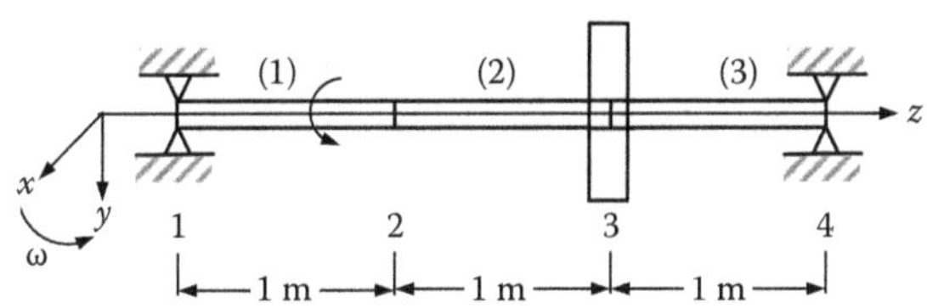

FIGURE 9.25 Single disk rotor system.

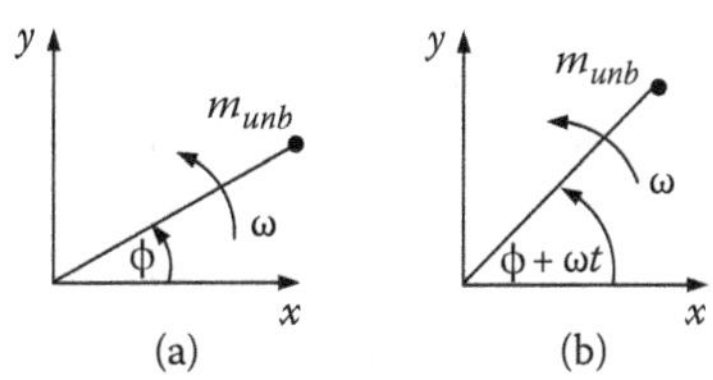

FIGURE 9.26 Unbalance on the disk (a) at $t = 0$ and (b) at time t.

Element 1: The FE equation for element 1 in the vertical plane y–z is given as follows:

$$1.46\times10^{-3}\begin{bmatrix} 156 & -22 & 54 & 13 \\ & 4 & -13 & -3 \\ & & 156 & 22 \\ sym & & & 4 \end{bmatrix}\begin{Bmatrix} \ddot{v}_1 \\ \ddot{\varphi}_{x1} \\ \ddot{v}_2 \\ \ddot{\varphi}_{x2} \end{Bmatrix} + 103.11\begin{bmatrix} 12 & -6 & -12 & -6 \\ & 4 & 6 & 2 \\ & & 12 & 6 \\ sym & & & 4 \end{bmatrix}\begin{Bmatrix} v_1 \\ \varphi_{x1} \\ v_2 \\ \varphi_{x2} \end{Bmatrix} = \begin{Bmatrix} 0 \\ 0 \\ 0 \\ 0 \end{Bmatrix} + \begin{Bmatrix} -S_{y1} \\ -M_{yz1} \\ S_{y2} \\ M_{yz2} \end{Bmatrix} \tag{a}$$

Similarly, the FE equation for element 1 in the horizontal plane (z–x) is given as follows:

$$1.46\times10^{-3}\begin{bmatrix} 156 & 22 & 54 & -13 \\ & 4 & 13 & -3 \\ & & 156 & -22 \\ sym & & & 4 \end{bmatrix}\begin{Bmatrix} \ddot{u}_1 \\ \ddot{\varphi}_{y1} \\ \ddot{u}_2 \\ \ddot{\varphi}_{y2} \end{Bmatrix} + 103.11\begin{bmatrix} 12 & 6 & -12 & 6 \\ & 4 & -6 & 2 \\ & & 12 & -6 \\ sym & & & 4 \end{bmatrix}\begin{Bmatrix} u_1 \\ \varphi_{y1} \\ u_2 \\ \varphi_{y2} \end{Bmatrix} = \begin{Bmatrix} 0 \\ 0 \\ 0 \\ 0 \end{Bmatrix} + \begin{Bmatrix} -S_{x1} \\ -M_{zx1} \\ S_{x2} \\ M_{zx2} \end{Bmatrix} \tag{b}$$

On combining elemental Equations (a) and (b) for the vertical and horizontal planes, we finally get the FE equation for element (1) as follows:

$$1.40\times10^{-3}\begin{bmatrix} 156 & 22 & 0 & 0 & 54 & -13 & 0 & 0 \\ & 4 & 0 & 0 & 13 & -3 & 0 & 0 \\ & & 156 & -22 & 0 & 0 & 54 & 13 \\ & & & 4 & 0 & 0 & -13 & -3 \\ & & & & 156 & -22 & 0 & 0 \\ & & & & & 4 & 0 & 0 \\ & & & & & & 156 & 22 \\ sym & & & & & & & 4 \end{bmatrix}\begin{Bmatrix} \ddot{u}_1 \\ \ddot{\varphi}_{y1} \\ \ddot{v}_1 \\ \ddot{\varphi}_{x1} \\ \ddot{u}_2 \\ \ddot{\varphi}_{y2} \\ \ddot{v}_2 \\ \ddot{\varphi}_{x2} \end{Bmatrix} + 103.11\begin{bmatrix} 12 & 6 & 0 & 0 & -12 & 6 & 0 & 0 \\ & 4 & 0 & 0 & -6 & 2 & 0 & 0 \\ & & 12 & -6 & 0 & 0 & -12 & -6 \\ & & & 4 & 0 & 0 & 6 & 2 \\ & & & & 12 & -6 & 0 & 0 \\ & & & & & 4 & 0 & 0 \\ & & & & & & 12 & 6 \\ sym & & & & & & & 4 \end{bmatrix}\begin{Bmatrix} u_1 \\ \varphi_{y1} \\ v_1 \\ \varphi_{x1} \\ u_2 \\ \varphi_{y2} \\ v_2 \\ \varphi_{x2} \end{Bmatrix} = \begin{Bmatrix} 0 \\ 0 \\ 0 \\ 0 \\ 0 \\ 0 \\ 0 \\ 0 \end{Bmatrix} + \begin{Bmatrix} -S_{x1} \\ -M_{yz1} \\ -S_{y1} \\ -M_{zx1} \\ S_{x2} \\ M_{yz2} \\ S_{y2} \\ M_{zx2} \end{Bmatrix} \tag{c}$$

Element 2: The FE equation for element 2 (with nodes 2 and 3) can be written similarly to element 1 with the only exception that a disk is present at node 3 and it also has unbalance at node 3 (i.e., in the fifth and seventh rows of the external force vector).

$$10^{-3}\begin{bmatrix} 156 & 22 & 0 & 0 & 54 & -13 & 0 & 0 \\ & 4 & 0 & 0 & 13 & -3 & 0 & 0 \\ & & 156 & -22 & 0 & 0 & 54 & 13 \\ & & & 4 & 0 & 0 & -13 & -3 \\ & & & & \left(156+1.5/1.46\times10^{-3}\right) & -22 & 0 & 0 \\ & & & & & 4 & 0 & 0 \\ & & & & & & \left(156+1.5/1.46\times10^{-3}\right) & 22 \\ sym & & & & & & & 4 \end{bmatrix}\begin{Bmatrix} \ddot{u}_2 \\ \ddot{\varphi}_{y2} \\ \ddot{v}_2 \\ \ddot{\varphi}_{x2} \\ \ddot{u}_3 \\ \ddot{\varphi}_{y3} \\ \ddot{v}_3 \\ \ddot{\varphi}_{x3} \end{Bmatrix}$$

$$+103.11\begin{bmatrix} 12 & 6 & 0 & 0 & -12 & 6 & 0 & 0 \\ & 4 & 0 & 0 & -6 & 2 & 0 & 0 \\ & & 12 & -6 & 0 & 0 & -12 & -6 \\ & & & 4 & 0 & 0 & 6 & 2 \\ & & & & 12 & -6 & 0 & 0 \\ & & & & & 4 & 0 & 0 \\ & & & & & & 12 & 6 \\ sym & & & & & & & 4 \end{bmatrix}\begin{Bmatrix} u_2 \\ \varphi_{y2} \\ v_2 \\ \varphi_{x2} \\ u_3 \\ \varphi_{y3} \\ v_3 \\ \varphi_{x3} \end{Bmatrix} = \begin{Bmatrix} 0 \\ 0 \\ 0 \\ 0 \\ 2.5\times10^{-6}\left[0.866+j0.5\right] \\ 0 \\ (-j)2.5\times10^{-6}\left[0.866+j0.5\right] \\ 0 \end{Bmatrix} e^{j\omega t} + \begin{Bmatrix} -S_{x2} \\ -M_{yz2} \\ -S_{y2} \\ -M_{zx2} \\ S_{x3} \\ M_{yz3} \\ S_{y3} \\ M_{zx3} \end{Bmatrix} \qquad \text{(d)}$$

In Equation (d) since the fifth and seventh rows are equations of motion corresponding to displacements u_3 and v_3, the lumped mass at node 3 of the disk contributed inertia terms $m_{disc}\ddot{u}_3$ and $m_{disc}\ddot{v}_3$, which came in as diagonal terms in the mass matrix of Equation (d). The unbalance force is given as $m_b r_b \omega^2 e^{j\phi} e^{j\omega t}$ in the horizontal direction $f_{x3}(t)$, and the unbalance force in the vertical direction $f_{y3}(t)$ can be obtained for shaft rotations as in Figure 9.27a as $f_{y3}(t) = -jf_{x3}(t)$ and in Figure 9.27b as $f_{y3}(t) = jf_{x3}(t)$.

Element 3: The FE equation for element 3 is given as follows:

$$1.46\times10^{-3}\begin{bmatrix} 156 & 22 & 0 & 0 & 54 & -13 & 0 & 0 \\ & 4 & 0 & 0 & 13 & -3 & 0 & 0 \\ & & 156 & -22 & 0 & 0 & 54 & 13 \\ & & & 4 & 0 & 0 & -13 & -3 \\ & & & & 156 & -22 & 0 & 0 \\ & & & & & 4 & 0 & 0 \\ & & & & & & 156 & 22 \\ sym & & & & & & & 4 \end{bmatrix}\begin{Bmatrix} \ddot{u}_3 \\ \ddot{\varphi}_{y3} \\ \ddot{v}_3 \\ \ddot{\varphi}_{x3} \\ \ddot{u}_4 \\ \ddot{\varphi}_{y4} \\ \ddot{v}_4 \\ \ddot{\varphi}_{x4} \end{Bmatrix} \qquad \text{(e)}$$

$$+103.11\begin{bmatrix} 12 & 6 & 0 & 0 & -12 & 6 & 0 & 0 \\ & 4 & 0 & 0 & -6 & 2 & 0 & 0 \\ & & 12 & -6 & 0 & 0 & -12 & -6 \\ & & & 4 & 0 & 0 & 6 & 2 \\ & & & & 12 & -6 & 0 & 0 \\ & & & & & 4 & 0 & 0 \\ & & & & & & 12 & 6 \\ sym & & & & & & & 4 \end{bmatrix}\begin{Bmatrix} u_3 \\ \varphi_{y3} \\ v_3 \\ \varphi_{x3} \\ u_4 \\ \varphi_{y4} \\ v_4 \\ \varphi_{x4} \end{Bmatrix} = \begin{Bmatrix} 0 \\ 0 \\ 0 \\ 0 \\ 0 \\ 0 \\ 0 \\ 0 \end{Bmatrix} + \begin{Bmatrix} -S_{x3} \\ -M_{yz3} \\ -S_{y3} \\ -M_{zx3} \\ S_{x4} \\ M_{yz4} \\ S_{y4} \\ M_{zx4} \end{Bmatrix}$$

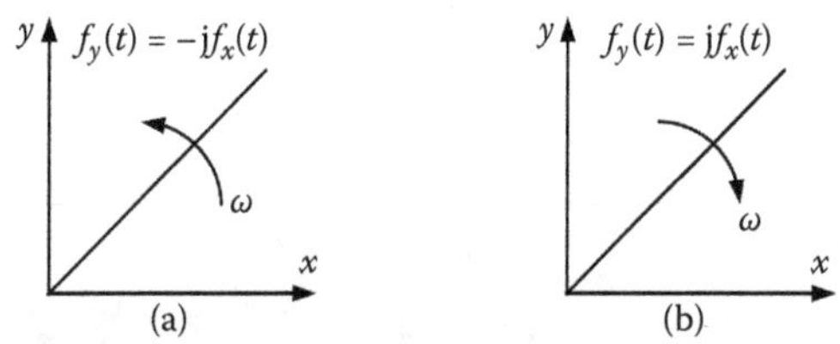

FIGURE 9.27 Unbalance forces in the horizontal and vertical directions for the direction of rotor rotation: (a) counterclockwise and (b) clockwise.

Assembled equation: On assembling all of the elemental equations (c) through (e), we get

$$
1.46\times10^{-3}\begin{bmatrix}
156 & 22 & 0 & 0 & 54 & -13 & 0 & 0 & 0 & 0 & 0 & 0 & 0 & 0 & 0 & 0\\
 & 4 & 0 & 0 & 13 & -3 & 0 & 0 & 0 & 0 & 0 & 0 & 0 & 0 & 0 & 0\\
 & & 156 & -22 & 0 & 0 & 54 & 13 & 0 & 0 & 0 & 0 & 0 & 0 & 0 & 0\\
 & & & 4 & 0 & 0 & -13 & -3 & 0 & 0 & 0 & 0 & 0 & 0 & 0 & 0\\
 & & & & (156+156) & (-22+22) & 0 & 0 & 54 & 13 & 0 & 0 & 0 & 0 & 0 & 0\\
 & & & & & (4+4) & 0 & 0 & 13 & 3 & 0 & 0 & 0 & 0 & 0 & 0\\
 & & & & & & (156+156) & (22-22) & 0 & 0 & 54 & 13 & 0 & 0 & 0 & 0\\
 & & & & & & & (4+4) & 0 & 0 & -13 & -3 & 0 & 0 & 0 & 0\\
 & & & & & & & & (1183.4+156) & (-22+22) & 0 & 0 & 54 & -13 & 0 & 0\\
 & & & & & & & & & (4+4) & 0 & 0 & 13 & -3 & 0 & 0\\
 & & & & & & & & & & (1183.4+156) & (22-22) & 0 & 0 & 54 & 13\\
 & & & & & & & & & & & (4+4) & 0 & 0 & 13 & -3\\
 & & & & & & & & & & & & 156 & -22 & 0 & 0\\
 & & & & & & & & & & & & & 4 & 0 & 0\\
 & & & & & & & & & & & & & & 156 & 22\\
sym & & & & & & & & & & & & & & & 4
\end{bmatrix}
\begin{Bmatrix}
\ddot{u}_1\\ \ddot{\varphi}_{y1}\\ \ddot{v}_1\\ \ddot{\varphi}_{x1}\\ \ddot{u}_2\\ \ddot{\varphi}_{y2}\\ \ddot{v}_2\\ \ddot{\varphi}_{x2}\\ \ddot{w}_3\\ \ddot{\varphi}_{y3}\\ \ddot{v}_3\\ \ddot{\varphi}_{x3}\\ \ddot{w}_4\\ \ddot{\varphi}_{y4}\\ \ddot{v}_4\\ \ddot{\varphi}_{x4}
\end{Bmatrix}
$$

$$
+103.11\begin{bmatrix}
12 & 6 & 0 & 0 & -12 & 6 & 0 & 0 & 0 & 0 & 0 & 0 & 0 & 0 & 0 & 0 \\
 & 4 & 0 & 0 & -6 & 2 & 0 & 0 & 0 & 0 & 0 & 0 & 0 & 0 & 0 & 0 \\
 & & 12 & -6 & 0 & 0 & -12 & -6 & 0 & 0 & 0 & 0 & 0 & 0 & 0 & 0 \\
 & & & 4 & 0 & 0 & 6 & 2 & 0 & 0 & 0 & 0 & 0 & 0 & 0 & 0 \\
 & & & & (12+12) & (-6+6) & 0 & 0 & -12 & 6 & 0 & 0 & 0 & 0 & 0 & 0 \\
 & & & & & (4+4) & 0 & 0 & 6 & 2 & 0 & 0 & 0 & 0 & 0 & 0 \\
 & & & & & & (12+12) & (6-6) & 0 & 0 & -12 & -6 & 0 & 0 & 0 & 0 \\
 & & & & & & & (4+4) & 0 & 0 & 6 & 2 & 0 & 0 & 0 & 0 \\
 & & & & & & & & (12+12) & (-6+6) & 0 & 0 & -12 & 6 & 0 & 0 \\
 & & & & & & & & & (4+4) & 0 & 0 & -6 & 2 & 0 & 0 \\
 & & & & & & & & & & (12+12) & (6-6) & 0 & 0 & -12 & 6 \\
 & & & & & & & & & & & (4+4) & 0 & 0 & 6 & 2 \\
 & & & & & & & & & & & & 12 & -6 & 0 & 0 \\
 & & & & & & & & & & & & & 4 & 0 & 0 \\
 & & & & & & & & & & & & & & 12 & 6 \\
sym & & & & & & & & & & & & & & & 4
\end{bmatrix}
\begin{Bmatrix} u_1 \\ \varphi_{y_1} \\ v_1 \\ \varphi_{x_1} \\ u_2 \\ \varphi_{y_2} \\ v_2 \\ \varphi_{x_2} \\ u_3 \\ \varphi_{y_3} \\ v_3 \\ \varphi_{x_3} \\ u_4 \\ \varphi_{y_4} \\ v_4 \\ \varphi_{x_4} \end{Bmatrix}
$$

$$
\begin{Bmatrix} 0 \\ 0 \\ 0 \\ 0 \\ 0 \\ 0 \\ 0 \\ 0 \\ 2.5\times10^{-6}\omega^2(0.866+j0.5) \\ 0 \\ 2.5\times10^{-6}\omega^2(0.5+j0.866) \\ 0 \\ 0 \\ 0 \\ 0 \\ 0 \end{Bmatrix} e^{j\omega t} + \begin{Bmatrix} -S_{x_1} \\ -M_{yz_1} \\ -S_{y_1} \\ -M_{zx_1} \\ 0 \\ 0 \\ 0 \\ 0 \\ 0 \\ 0 \\ 0 \\ 0 \\ S_{x_4} \\ M_{yz_4} \\ S_{y_4} \\ M_{zx_4} \end{Bmatrix} \tag{f}
$$

Boundary conditions: The boundary conditions for a simply supported shaft are as given below:

$$u_1 = 0, \quad v_1 = 0, \quad u_4 = 0, \quad \text{and} \quad v_4 = 0 \tag{g}$$

and

$$M_{yz1} = 0, \quad M_{zx1} = 0, \quad M_{yz4} = 0, \quad \text{and} \quad M_{zx4} = 0 \tag{h}$$

Reduced system equations: On application of Equation (h) into Equation (f), rows 2, 4 to 12, 14, and 16 (total 12) are known or specified in the right-hand side vector. The remaining four rows (1st, 3rd, 13th, and 15th) can be removed since these contain shear forces as additional unknowns on the right-hand side. Displacement information from Equation (g) gives the 1st, 3rd, 13th, and 15th columns of the mass and stiffness matrices multiplied by zero. Therefore, these columns can also be eliminated since they will not contribute any terms in the retained 12 equations. Shear forces at bearing locations are nothing but forces being transmitted through bearings and can be obtained once displacements are known from the four eliminated equations. Equation (f), after applying boundary conditions, reduces to

$$1.46\times10^{-3}\begin{bmatrix} 4 & 0 & 13 & -3 & 0 & 0 & 0 & 0 & 0 & 0 & 0 & 0 \\ & 4 & 0 & 0 & -13 & -3 & 0 & 0 & 0 & 0 & 0 & 0 \\ & & 312 & 0 & 0 & 0 & 54 & -13 & 0 & 0 & 0 & 0 \\ & & & 8 & 0 & 0 & 13 & -3 & 0 & 0 & 0 & 0 \\ & & & & 312 & 0 & 0 & 0 & 54 & -13 & 0 & 0 \\ & & & & & 8 & 0 & 0 & -13 & -3 & 0 & 0 \\ & & & & & & 1339.4 & 0 & 0 & 0 & -13 & 0 \\ & & & & & & & 8 & 0 & 0 & -3 & 0 \\ & & & & & & & & 1339.4 & 0 & 0 & 13 \\ & & & & & & & & & 8 & 0 & -3 \\ & & & & & & & & & & 4 & 0 \\ sym & & & & & & & & & & & 4 \end{bmatrix}\begin{Bmatrix} \ddot{\varphi}_{y1} \\ \ddot{\varphi}_{x1} \\ \ddot{u}_2 \\ \ddot{\varphi}_{y2} \\ \ddot{v}_2 \\ \ddot{\varphi}_{x2} \\ \ddot{u}_3 \\ \ddot{\varphi}_{y3} \\ \ddot{v}_3 \\ \ddot{\varphi}_{x3} \\ \ddot{\varphi}_{y4} \\ \ddot{\varphi}_{x4} \end{Bmatrix}$$

$$+103.11\begin{bmatrix} 4 & 0 & -6 & 2 & 0 & 0 & 0 & 0 & 0 & 0 & 0 & 0 \\ & 4 & 0 & 0 & 6 & 2 & 0 & 0 & 0 & 0 & 0 & 0 \\ & & 24 & 0 & 0 & 0 & -12 & 6 & 0 & 0 & 0 & 0 \\ & & & 8 & 0 & 0 & -6 & 2 & 0 & 0 & 0 & 0 \\ & & & & 24 & 0 & 0 & 0 & -12 & -6 & 0 & 0 \\ & & & & & 8 & 0 & 0 & 6 & 2 & 0 & 0 \\ & & & & & & 24 & 0 & 0 & 0 & 6 & 0 \\ & & & & & & & 8 & 0 & 0 & 2 & 0 \\ & & & & & & & & 24 & 0 & 0 & -6 \\ & & & & & & & & & 8 & 0 & 2 \\ & & & & & & & & & & 4 & 0 \\ sym & & & & & & & & & & & 4 \end{bmatrix}\begin{Bmatrix} \varphi_{y1} \\ \varphi_{x1} \\ u_2 \\ \varphi_{y2} \\ v_2 \\ \varphi_{x2} \\ u_3 \\ \varphi_{y3} \\ v_3 \\ \varphi_{x3} \\ \varphi_{y4} \\ \varphi_{x4} \end{Bmatrix} = \begin{Bmatrix} 0 \\ 0 \\ 0 \\ 0 \\ 0 \\ 0 \\ 2.5\times10^{-6}\omega^2(0.866+j0.5) \\ 0 \\ 2.5\times10^{-6}\omega^2(0.5-j0.866) \\ 0 \\ 0 \\ 0 \end{Bmatrix} e^{j\omega t} \tag{i}$$

Forced responses: Since excitation is a simple harmonic $\mathbf{f}(t) = \mathbf{F}e^{j\omega t}$, $\eta(t) = \bar{\eta}e^{j\omega t}$ and $\ddot{\eta} = -\omega^2\eta e^{j\omega t} = -\omega^2\bar{\eta}$ where ω is the rotor spin speed, and $\bar{\eta}$ is the complex displacement vector and it contains a complex quantity, that is, the magnitude and phase information of various displacement components. Hence, Equation (i) can be written as follows:

$$-\omega^2\mathbf{M}\bar{\eta}(\omega) + \mathbf{K}\bar{\eta}(\omega) = \mathbf{F}(\omega) \tag{j}$$

which can be solved as follows:

$$\bar{\eta}(\omega) = \left(\mathbf{K} - \omega^2\mathbf{M}\right)^{-1} \mathbf{F}(\omega) \tag{k}$$

It should be noted that matrices **M** and **K** contain real quantities, whereas force vector **F** and the corresponding displacement vector $\bar{\eta}$ contain complex terms. The response (amplitudes and phases) can be plotted with respect to shaft speeds. Figure 9.28 shows such typical plots. Critical speeds can be observed at two places where the amplitude of the unbalance response has spikes, whereas in the phase plot it is accompanied by a change in phase on the order of 180° in the

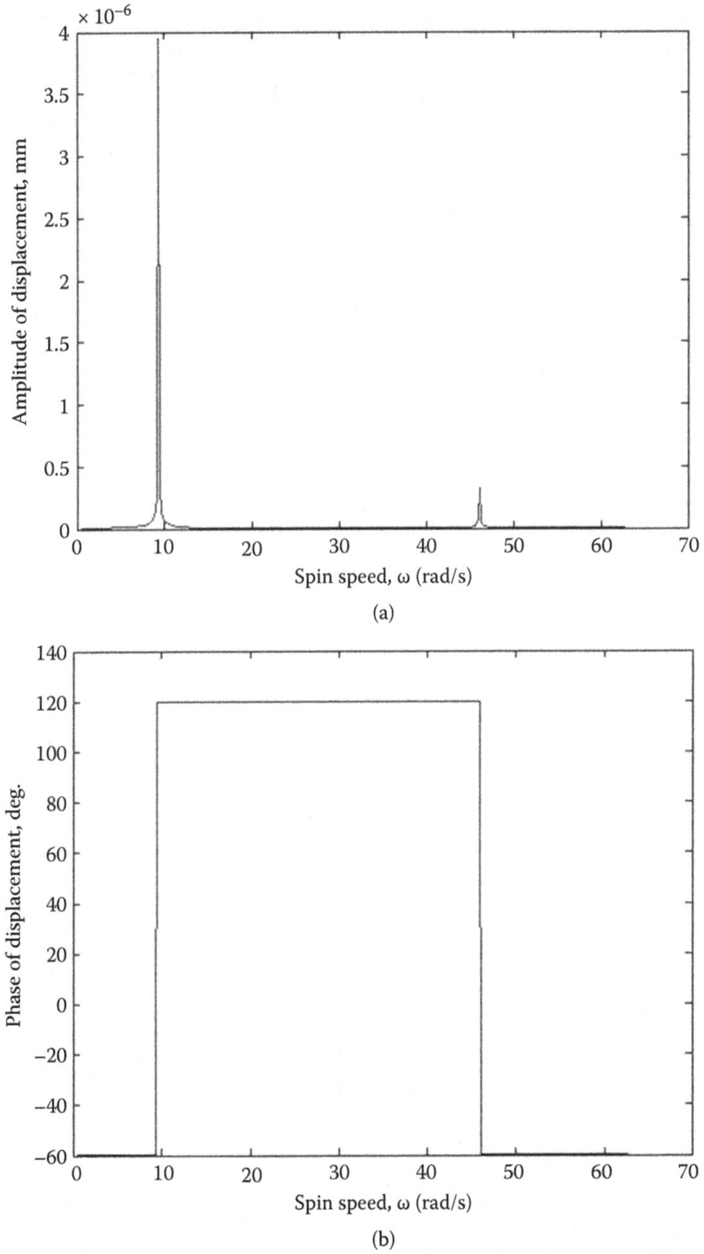

FIGURE 9.28 Forced response variation with respect to the spin speed of the shaft: (a) amplitude of the response and (b) phase of the response.

absence of damping. We have done two-plane analysis for the present case; however, there is no coupling in two planes as such, and it would still be possible to obtain the unbalance response by considering only single-plane motion (in the vertical or horizontal plane).

Example 9.8

For the turbine-generator rotor system shown in Figure 9.29, obtain all the transverse natural frequencies and mode shapes. Turbine stages are treated as rigid thin disks with $m_1=30$ kg, $r_1=5$ cm, $m_2=33.33$ kg, $r_2=7$ cm, $m_3=34.09$ kg, $r_3=9$ cm, $m_4=34.80$ kg, $r_4=11$ cm, $m_5=35.71$ kg, and $r_5=13$ cm (m is the mass and r is the radius). The couplings are treated as rigid thin disks with $m_6=m_7=5$ kg and $r_6=r_7=4$ cm. The generator is treated as rigid thin disks: $m_8=m_9=m_{10}=m_{11}=m_{12}=m_{13}=20$ kg with $r_8=r_9=r_{10}=r_{11}=r_{12}=r_{13}=4$ cm. Each shaft segment between the disks, between a disk and bearing, and between a bearing and coupling has 0.5 m length. The diameters of the turbine and generator shafts are 0.1 and 0.15 m, respectively. Take $E=2.1\times10^{11}$ N/m^2. Neglect the diametral and polar mass moment of inertia of various disks and treat the coupling as a pin joint.

Solution:

The rotor system shown in Figure 9.29 is divided into 15 elements as shown in Figure 9.30. The coupling is considered as a lumped mass (i.e., $m=m_6+m_7=10$ kg), and at coupling, the displacements from the seventh and eight elements are equal; however, the slopes are different because it was modeled as a pin joint. Bearings are assumed to be rigid. The polar mass moment of inertia (i.e., the gyroscopic effect) and diametral mass moment of inertia (i.e., the rotary inertia effect) of the disks are neglected. The length of each element is taken as 0.5 m as shown in Figure 9.30. *FEM formulation:* The coupling is considered as a lumped mass (i.e., $m=m_6+m_7=10$ kg) and is assumed to have negligible stiffness. The polar and diametral mass moments of inertia of the disks are considered in the analysis.

The elemental equation for the first element is given by (considering single-plane motion in the horizontal plane z-x and note the changes in signs of some of elements related to angular displacements, refer to Appendix 9.1)

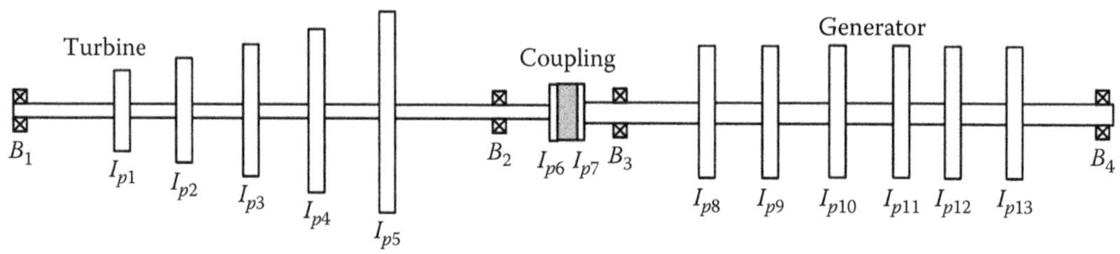

FIGURE 9.29 Turbine-generator model.

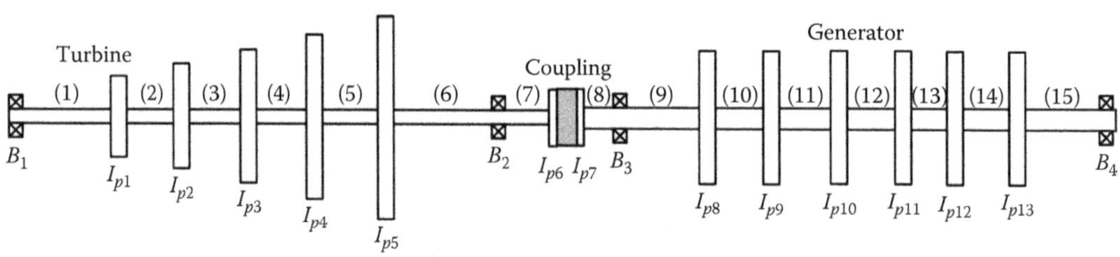

FIGURE 9.30 Rotor system divided into 15 elements.

$$\frac{\rho A_1 l}{420}\begin{bmatrix} 156 & 22l & 54 & -13l \\ 22l & 4l^2 & 13l & -3l^2 \\ 54 & 13l & 156+\left(420m_1/\rho A_1 l\right) & -22l \\ -13l & -3l^2 & -22l & 4l^2 \end{bmatrix}\begin{Bmatrix} \ddot{u}_1 \\ \ddot{\varphi}_{y1} \\ \ddot{u}_2 \\ \ddot{\varphi}_{y2} \end{Bmatrix}$$

$$+\frac{EI_1}{l^3}\begin{bmatrix} 12 & 6l & -12 & 6l \\ 6l & 4l^2 & -6l & 2l^2 \\ -12l & -6l & 12 & -6l \\ 6l & 2l^2 & -6l & 4l^2 \end{bmatrix}\begin{Bmatrix} u_1 \\ \varphi_{y1} \\ u_2 \\ \ddot{\varphi}_{y2} \end{Bmatrix}=\begin{Bmatrix} -S_1 \\ -M_1 \\ S_2 \\ M_2 \end{Bmatrix} \tag{a}$$

where l is the length of the shaft element, u and φ_y are the translational and rotational displacements at a node of an element, respectively, and m_1 represents the disk mass at node 2 and is assumed to act at the right node of element 1.

For the second element, the elemental equation is given as follows:

$$\frac{\rho A_2 l}{420}\begin{bmatrix} 156 & 22l & 54 & -13l \\ 22l & 4l^2 & 13l & -3l^2 \\ 54 & 13l & 156+\left(420m_2/\rho A_2 l\right) & -22l \\ -13l & -3l^2 & -22l & 4l^2 \end{bmatrix}\begin{Bmatrix} \ddot{u}_2 \\ \ddot{\varphi}_{y2} \\ \ddot{u}_3 \\ \ddot{\varphi}_{y3} \end{Bmatrix}$$

$$+\frac{EI_2}{l^3}\begin{bmatrix} 12 & 6l & -12 & 6l \\ 6l & 4l^2 & -6l & 2l^2 \\ -12l & -6l & 12 & -6l \\ 6l & 2l^2 & -6l & 4l^2 \end{bmatrix}\begin{Bmatrix} u_2 \\ \varphi_{y2} \\ u_3 \\ \varphi_{y3} \end{Bmatrix}=\begin{Bmatrix} -S_2 \\ -M_2 \\ S_3 \\ M_3 \end{Bmatrix}$$

For the third element, the elemental equation is given as follows:

$$\frac{\rho A_3 l}{420}\begin{bmatrix} 156 & 22l & 54 & -13l \\ 22l & 4l^2 & 13l & -3l^2 \\ 54 & 13l & 156+\left(420m_3/\rho A_3 l\right) & -22l \\ -13l & -3l^2 & -22l & 4l^2 \end{bmatrix}\begin{Bmatrix} \ddot{u}_3 \\ \ddot{\varphi}_{y3} \\ \ddot{u}_4 \\ \ddot{\varphi}_{y4} \end{Bmatrix}$$

$$+\frac{EI_3}{l^3}\begin{bmatrix} 12 & 6l & -12 & 6l \\ 6l & 4l^2 & -6l & 2l^2 \\ -12l & -6l & 12 & -6l \\ 6l & 2l^2 & -6l & 4l^2 \end{bmatrix}\begin{Bmatrix} u_3 \\ \varphi_{y3} \\ u_4 \\ \varphi_{y4} \end{Bmatrix}=\begin{Bmatrix} -S_3 \\ -M_3 \\ S_4 \\ M_4 \end{Bmatrix}$$

For the fourth element, the elemental equation is given as follows:

$$\frac{\rho A_4 l}{420}\begin{bmatrix} 156 & 22l & 54 & -13l \\ 22l & 4l^2 & 13l & -3l^2 \\ 54 & 13l & 156+\left(420m_4/\rho A_4 l\right) & -22l \\ -13l & -3l^2 & -22l & 4l^2 \end{bmatrix}\begin{Bmatrix} \ddot{u}_4 \\ \ddot{\varphi}_{y4} \\ \ddot{u}_5 \\ \ddot{\varphi}_{y5} \end{Bmatrix}$$

$$+\frac{EI_4}{l^3}\begin{bmatrix} 12 & 6l & -12 & 6l \\ 6l & 4l^2 & -6l & 2l^2 \\ -12l & -6l & 12 & -6l \\ 6l & 2l^2 & -6l & 4l^2 \end{bmatrix}\begin{Bmatrix} u_4 \\ \varphi_{y4} \\ u_5 \\ \varphi_{y5} \end{Bmatrix}=\begin{Bmatrix} -S_4 \\ -M_4 \\ S_5 \\ M_5 \end{Bmatrix}$$

For the fifth element, the elemental equation is given as follows:

$$\frac{\rho A_5 l}{420}\begin{bmatrix} 156 & 22l & 54 & -13l \\ 22l & 4l^2 & 13l & -3l^2 \\ 54 & 13l & 156+(420m_5/\rho A_5 l) & -22l \\ -13l & -3l^2 & -22l & 4l^2 \end{bmatrix}\begin{Bmatrix} \ddot{u}_5 \\ \ddot{\varphi}_{y5} \\ \ddot{u}_6 \\ \ddot{\varphi}_{y6} \end{Bmatrix}$$

$$+\frac{EI_5}{l^3}\begin{bmatrix} 12 & 6l & -12 & 6l \\ 6l & 4l^2 & -6l & 2l^2 \\ -12l & -6l & 12 & -6l \\ 6l & 2l^2 & -6l & 4l^2 \end{bmatrix}\begin{Bmatrix} u_5 \\ \varphi_{y5} \\ u_6 \\ \varphi_{y6} \end{Bmatrix}=\begin{Bmatrix} -S_5 \\ -M_5 \\ S_6 \\ M_6 \end{Bmatrix}$$

For the sixth element, the elemental equation is given as follows:

$$\frac{\rho A_6 l}{420}\begin{bmatrix} 156 & 22l & 54 & -13l \\ 22l & 4l^2 & 13l & -3l^2 \\ 54 & 13l & 156 & -22l \\ -13l & -3l^2 & -22l & 4l^2 \end{bmatrix}\begin{Bmatrix} \ddot{u}_6 \\ \ddot{\varphi}_{y6} \\ \ddot{u}_7 \\ \ddot{\varphi}_{y7} \end{Bmatrix}$$

$$+\frac{EI_6}{l^3}\begin{bmatrix} 12 & 6l & -12 & 6l \\ 6l & 4l^2 & -6l & 2l^2 \\ -12l & -6l & 12 & -6l \\ 6l & 2l^2 & -6l & 4l^2 \end{bmatrix}\begin{Bmatrix} u_6 \\ \varphi_{y6} \\ u_7 \\ \varphi_{y7} \end{Bmatrix}=\begin{Bmatrix} -S_6 \\ -M_6 \\ S_7 \\ M_7 \end{Bmatrix}+\begin{Bmatrix} 0 \\ 0 \\ R_7 \\ 0 \end{Bmatrix}$$

For the seventh element, the elemental equation is given as follows:

$$\frac{\rho A_7 l}{420}\begin{bmatrix} 156 & 22l & 54 & -13l \\ 22l & 4l^2 & 13l & -3l^2 \\ 54 & 13l & 156+\{420(m_6+m_7)/\rho A_7 l\} & -22l \\ -13l & -3l^2 & -22l & 4l^2 \end{bmatrix}\begin{Bmatrix} \ddot{u}_7 \\ \ddot{\varphi}_{y7} \\ \ddot{u}_8^{(7)} \\ \ddot{\varphi}_{y8}^{(7)} \end{Bmatrix}$$

$$+\frac{EI_7}{l^3}\begin{bmatrix} 12 & 6l & -12 & 6l \\ 6l & 4l^2 & -6l & 2l^2 \\ -12l & -6l & 12 & -6l \\ 6l & 2l^2 & -6l & 4l^2 \end{bmatrix}\begin{Bmatrix} u_7 \\ \varphi_{y7} \\ u_8^{(7)} \\ \varphi_{y8}^{(7)} \end{Bmatrix}=\begin{Bmatrix} -S_7 \\ -M_7 \\ S_8 \\ M_8 \end{Bmatrix}$$

It should be noted that the eighth node of the seventh element is the coupling location (for the present case, it is modeled as a pin joint). The superscript represents the element number of the translational and rotational displacements. It has been introduced to have a distinct slope at either end of node 8 for the seventh and eighth elements. It will be clearer during the application of boundary conditions, since $u_8^{(7)} = u_8^{(8)}$, but $\varphi_{y8}^{(7)} \neq \varphi_{y8}^{(8)}$. Now considering the generator shaft, for the eighth element, the elemental equation is given as follows:

$$\frac{\rho A_8 l}{420}\begin{bmatrix} 156 & 22l & 54 & -13l \\ 22l & 4l^2 & 13l & -3l^2 \\ 54 & 13l & 156 & -22l \\ -13l & -3l^2 & -22l & 4l^2 \end{bmatrix}\begin{Bmatrix} \ddot{u}_8^{(8)} \\ \ddot{\varphi}_{y8}^{(8)} \\ \ddot{u}_9 \\ \ddot{\varphi}_{y9} \end{Bmatrix}$$

$$+\frac{EI_8}{l^3}\begin{bmatrix} 12 & 6l & -12 & 6l \\ 6l & 4l^2 & -6l & 2l^2 \\ -12l & -6l & 12 & -6l \\ 6l & 2l^2 & -6l & 4l^2 \end{bmatrix}\begin{Bmatrix} u_8^{(8)} \\ \varphi_{y8}^{(8)} \\ u_9 \\ \varphi_{y9} \end{Bmatrix} = \begin{Bmatrix} -S_8 \\ -M_8 \\ S_9 \\ M_9 \end{Bmatrix} + \begin{Bmatrix} 0 \\ 0 \\ R_9 \\ 0 \end{Bmatrix}$$

For the ninth element, the elemental equation is given as follows:

$$\frac{\rho A_9 l}{420}\begin{bmatrix} 156 & 22l & 54 & -13l \\ 22l & 4l^2 & 13l & -3l^2 \\ 54 & 13l & 156+420\left(m_9 / \rho A_9 l\right) & -22l \\ -13l & -3l^2 & -22l & 4l^2 \end{bmatrix}\begin{Bmatrix} \ddot{u}_9 \\ \ddot{\varphi}_{y9} \\ \ddot{u}_{10} \\ \ddot{\varphi}_{y10} \end{Bmatrix}$$

$$+\frac{EI_9}{l^3}\begin{bmatrix} 12 & 6l & -12 & 6l \\ 6l & 4l^2 & -6l & 2l^2 \\ -12l & -6l & 12 & -6l \\ 6l & 2l^2 & -6l & 4l^2 \end{bmatrix}\begin{Bmatrix} u_9 \\ \varphi_{y9} \\ u_{10} \\ \varphi_{y10} \end{Bmatrix} = \begin{Bmatrix} -S_9 \\ -M_9 \\ S_{10} \\ M_{10} \end{Bmatrix}$$

Similar elemental equations are written for the generator shaft. In the above elemental equations

$$\begin{aligned} A_1 &= A_2 \cdots = A_7 = \pi d_1^2 / 4 = \pi \times 0.3^2/4 = 0.0706\ \text{m}^2 \\ A_8 &= A_9 = \cdots = A_{15} = \pi d_8^2 / 4 = \pi \times 0.1^2/4 = 0.196349\ \text{m}^2 \\ I_1 &= I_2 = \cdots = I_7 = \pi d_1^4 / 64 = \pi \times 0.3^4/64 = 3.9761 \times 10^{-4}\ \text{m}^4 \\ I_8 &= I_9 = \cdots = I_{15} = \pi d_8^4 / 64 = \pi \times 0.5^4/64 = 3.067 \times 10^{-3}\ \text{m}^4 \end{aligned} \tag{b}$$

where d_1 and d_8 are the diameters of the turbine and generator shafts, respectively. After adding all of the elemental equations, the global matrix is given by

$$\mathbf{M}_{33\times33}\ddot{\eta}(t)_{33\times1} + \mathbf{K}_{33\times33}\eta(t)_{33\times1} = \mathbf{f}(t)_{33\times1} \tag{c}$$

The right-side force vector in Equation (c) is given by

$$\boldsymbol{\eta}_{33\times 1}=\begin{Bmatrix} u_1 \\ \varphi_{y1} \\ u_2 \\ \varphi_{y2} \\ \vdots \\ \varphi_{y8}^{(7)} \\ \varphi_{y8}^{(8)} \\ \vdots \\ u_{16} \\ \varphi_{y16} \end{Bmatrix} \quad \mathbf{f}(t)_{33\times 1}=\begin{Bmatrix} -S_1 \\ -M_1 \\ 0 \\ \vdots \\ 0 \\ 0 \\ 0 \\ 0 \\ \vdots \\ S_{16} \\ M_{16} \end{Bmatrix}+\begin{Bmatrix} 0 \\ 0 \\ 0 \\ \vdots \\ R_7 \\ 0 \\ R_9 \\ 0 \\ \vdots \\ 0 \\ 0 \end{Bmatrix} \tag{d}$$

It should be noted that $\varphi_{y8}^{(7)}$ and $\varphi_{y8}^{(8)}$ appear in the displacement vector since these are independent variables due to the pin joint (i.e., the coupling model) at node 8. Here, R_7 and R_9 are intermediate reactions at the supports (i.e., at the seventh and ninth nodes). Boundary conditions for simply supported nodes 1 and 16, and at intermediate nodes 7 and 9, are given as

$$u_1=0;\quad M_1=0;\quad u_{16}=0;\quad M_{16}=0;\quad u_7=0;\quad u_9=0;\quad u_8^{(7)}=u_8^{(8)} \text{ and } \varphi_{y8}^{(7)}\neq\varphi_{y8}^{(8)} \tag{e}$$

After applying boundary conditions, and eliminating the first (due to S_1), seventh (due to R_7), ninth (due to R_9), and thirty-second (due to S_{16}) rows and columns, we have governing equations with the force vector having zero entries, as follows:

$$\mathbf{M}_{29\times 29}\ddot{\boldsymbol{\eta}}_{29\times 1}+\mathbf{K}_{29\times 29}\boldsymbol{\eta}_{29\times 1}=\mathbf{0} \tag{f}$$

For free vibrations, we have the following eigenvalue problem:

$$\left(-\omega^2\mathbf{M}+\mathbf{K}\right)\bar{\boldsymbol{\eta}}=\mathbf{0} \tag{g}$$

Eigenvectors are found and again the variables deleted while applying boundary conditions are replaced with zeros to get the relative displacement of the shaft at various nodes. This eigenvector contains both translational and rotational displacements. So, these eigenvectors are separated to get the translational and rotational displacements. Then the relative displacement versus spin speed of the shaft is drawn. The square root of eigenvalues gives the natural frequencies for an undamped system.

Results of the free vibration analysis: Table 9.9 gives the first 12 natural frequencies of the system. Mode shapes have not been plotted for brevity.

Forced vibration analysis: For the forced vibration analysis, the unbalance mass in the first and last disks is considered. The initial unbalance mass is taken as 0.005 kg, and its radial and angular

TABLE 9.9
First 12 Transverse Natural Frequencies of the Turbine-Generator System

S.N.	Natural Frequency (rad/s)	S.N.	Natural Frequency (rad/s)
1	509.4	7	8277.9
2	515.8	8	10,683.0
3	1645.3	9	13,009.0
4	2061.9	10	17,562.0
5	4183.7	11	18,913.0
6	7005.2	12	22,524.0

positions are 0.05 m with the 30° phase, respectively, with some reference of the shaft. The force vector is given by

$$\mathbf{F} = \begin{Bmatrix} 0 & 0 & m_{t1} r_{t1} \omega^2 e^{j\theta_{t1}} & 0 & 0 & 0 & \cdots & m_{t13} r_{t13} \omega^2 e^{j\theta_{t13}} & 0 & 0 & 0 \end{Bmatrix}^T \tag{h}$$

To find the forced vibration responses, we have

$$\bar{\eta} = \mathbf{D}^{-1}\,\mathbf{F} \quad \text{with} \quad \mathbf{D} = \mathbf{K} - \omega^2 \mathbf{M} \tag{i}$$

where $\bar{\eta}$ contains both the transverse translational and rotational displacements containing magnitude and phase information as follows:

$$\text{Magnitude of displacement } \bar{\eta}_k = \sqrt{\left(\bar{\eta}_k^r\right)^2 + \left(\bar{\eta}_k^i\right)^2}; \quad k = 1, 2, \ldots, N \tag{j}$$

and

$$\text{Phase of displacement } \bar{\eta}_k = \tan^{-1}\left(\frac{\bar{\eta}_k^i}{\bar{\eta}_k^r}\right); \quad k = 1, 2, \ldots, N \tag{k}$$

Responses at nodes 4 and 13 are generated by taking the spin speed of the shaft one at a time over a speed range. For clarity, the same response is drawn by splitting into two ranges. Forced response displacement amplitudes and phases with respect to the spin speed of the rotor are shown in Figure 9.31a through d at the fourth node of the turbine and in Figure 9.32a through d at the fourteenth node of the generator shaft.

From the forced response phase plots, it is clear that there is a sudden change of 180° (shown in radians) at critical speeds and there is a corresponding peak in displacement plots. Analysis in the vertical plane, *y-z*, will also be giving similar critical speeds due to the symmetry of the rotor.

9.6 PROPORTIONAL DAMPING

The equation of motion for the shaft element with damping for free vibrations can be written as follows:

$$\mathbf{M}^{(e)}\ddot{\eta}^{(e)}(t) + \mathbf{C}^{(e)}\dot{\eta}^{(e)}(t) + \mathbf{K}^{(e)}\eta^{(e)}(t) = \mathbf{0} \tag{9.122}$$

where $\mathbf{M}^{(e)}$ is the element mass matrix, $\mathbf{C}^{(e)}$ is the element damping matrix, $\mathbf{K}^{(e)}$ is the element stiffness matrix, and $\eta^{(e)}(t)$ is the element vector of nodal DOFs. The damping matrix $\mathbf{C}^{(e)}$ of the beam element can be expressed in terms of the element mass and stiffness matrices as given below (Clough and Penzien, 1993):

$$\mathbf{C}^{(e)} = a_0 \mathbf{M}^{(e)} + a_1 \mathbf{K}^{(e)} \tag{9.123}$$

where a_0 and a_1 are *Rayleigh damping factors*. The relation between the damping ratio ζ and the natural frequency ω_{nf} in terms of a_0 and a_1 is given as follows:

$$\zeta_n = \frac{a_0}{2\omega_{nf_n}} + \frac{a_1 \omega_{nf_n}}{2} \tag{9.124}$$

where Rayleigh damping factors, a_0 and a_1, can be evaluated by the solution of a pair of simultaneous equations, if damping ratios ζ_m and ζ_n are associated with two specific natural frequencies ω_{nf_m} and ω_{nf_n} are known. Writing Equation 9.124 for each of these two cases and expressing them in matrix form leads to

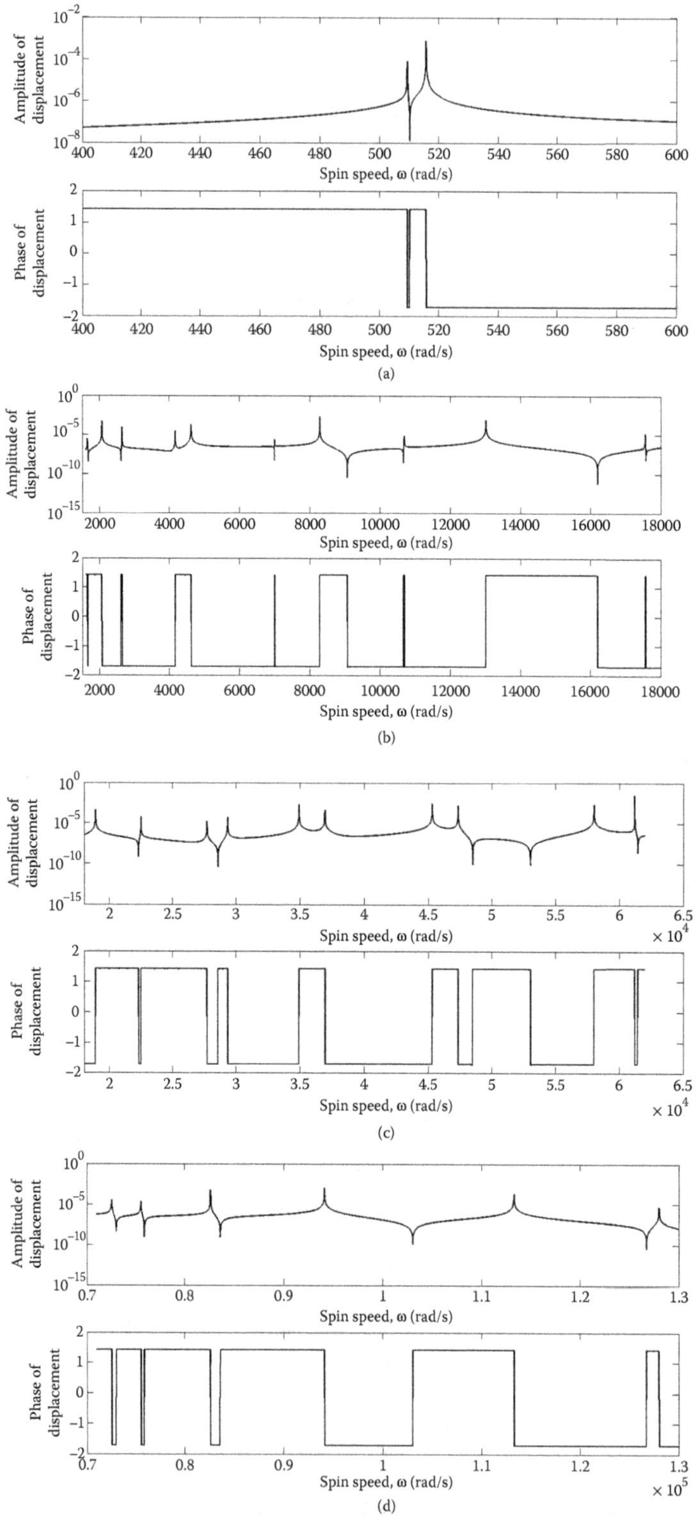

FIGURE 9.31 (a) Amplitude of displacement (in m) and phase (in rad) variation with the spin speed in the range of 400–600 rad/s, (b) 1600–18,000 rad/s, (c) 18,000–65,000 rad/s, and (d) 70,000–130,000 rad/s.

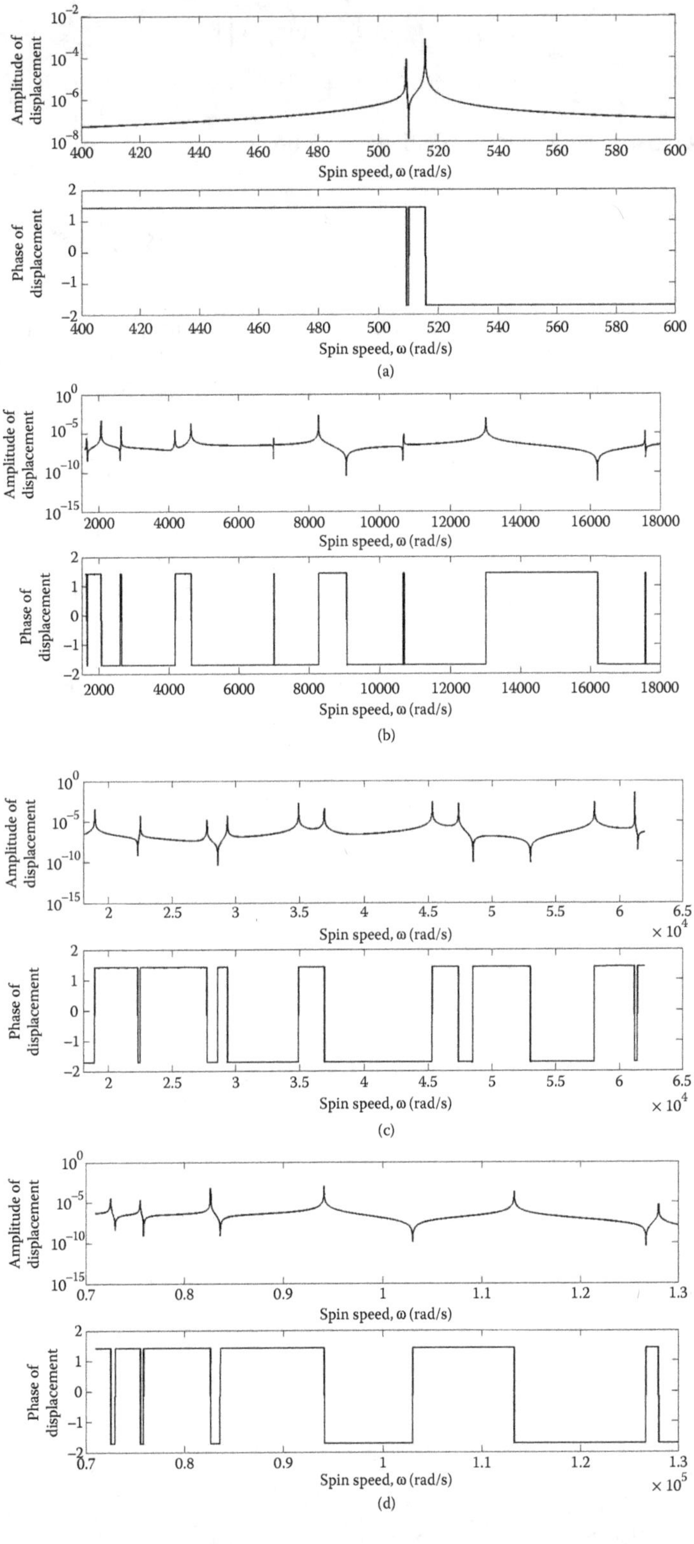

FIGURE 9.32 The amplitude of displacement (in m) and phase (in rad) variation with the spin speed in the range of (a) 400–700 rad/s, (b) 1600–18,000 rad/s, (c) 18,000–65,000 rad/s, and (d) 70,000–130,000 rad/s.

$$\begin{Bmatrix} \zeta_m \\ \zeta_n \end{Bmatrix} = \frac{1}{2}\begin{bmatrix} \omega_{nf_m}^{-1} & \omega_{nf_m} \\ \omega_{nf_n}^{-1} & \omega_{nf_n} \end{bmatrix}\begin{Bmatrix} a_0 \\ a_1 \end{Bmatrix} \tag{9.125}$$

From Equation 9.125, a_0 and a_1 can be calculated as follows:

$$\begin{Bmatrix} a_0 \\ a_1 \end{Bmatrix} = 2\frac{\omega_{nf_m}\omega_{nf_n}}{\omega_{nf_n}^2 - \omega_{nf_m}^2}\begin{bmatrix} \omega_{nf_n} & -\omega_{nf_m} \\ -\omega_{nf_n}^{-1} & \omega_{nf_m}^{-1} \end{bmatrix}\begin{Bmatrix} \zeta_m \\ \zeta_n \end{Bmatrix} \tag{9.126}$$

where ω_{nf_m} and ω_{nf_n} are the system fundamental and highest natural frequencies of interest. For most engineering structures, the damping ratio is on the order of 0.01. However, it is advisable to experimentally estimate the modal damping ratios for relevant modes from the real system.

Example 9.9

For Example 9.2, obtain Rayleigh's damping coefficients, considering the first two natural frequencies, and for damping ratios of 0.01 and 0.015 for the first and second natural modes, respectively. Obtain the damping matrix for the first element by using these damping coefficients.

Solution:

The first and second natural frequencies of the rotor system of Example 9.2 are given by

$$\omega_{nf_1} = 14.225 \text{ rad/s} \quad \text{and} \quad \omega_{nf_2} = 56.901 \text{ rad/s}$$

and the corresponding damping ratios are given as follows:

$$\zeta_1 = 0.01 \quad \text{and} \quad \zeta_2 = 0.015$$

Now Rayleigh's damping coefficients are given as follows:

$$\begin{Bmatrix} a_0 \\ a_1 \end{Bmatrix} = 2\frac{\omega_{nf_1}\omega_{nf_2}}{\omega_{nf_2}^2 - \omega_{nf_1}^2}\begin{bmatrix} \omega_{nf_2} & -\omega_{nf_1} \\ -\omega_{nf_2}^{-1} & \omega_{nf_1}^{-1} \end{bmatrix}\begin{Bmatrix} \zeta_1 \\ \zeta_2 \end{Bmatrix}$$

$$= 2\frac{14.225 \times 56.901}{56.901^2 - 14.225^2}\begin{bmatrix} 56.901 & -14.225 \\ -1/56.901 & 1/14.225 \end{bmatrix}\begin{Bmatrix} 0.01 \\ 0.015 \end{Bmatrix} = \begin{Bmatrix} 0.1897 \\ 0.0005 \end{Bmatrix}$$

Hence, the damping matrix for the first element is (in the z–x plane)

$$\mathbf{C}^{(e)} = a_0\mathbf{M}^{(e)} + a_1\mathbf{K}^{(e)}$$

$$= a_0 \times 1.46 \times 10^{-3}\begin{bmatrix} 156 & 22 & 54 & -13 \\ & 4 & 13 & -3 \\ & & 156 & -22 \\ sym & & & 4 \end{bmatrix} + a_1 \times 103.11\begin{bmatrix} 12 & 6 & -12 & 6 \\ & 4 & -6 & 2 \\ & & 12 & -6 \\ sym & & & 4 \end{bmatrix}$$

$$= \begin{bmatrix} 0.623 & 0.296 & -0.565 & 0.286 \\ 0.296 & 0.194 & -0.286 & 0.096 \\ -0.565 & -0.286 & 0.623 & -0.296 \\ 0.286 & 0.096 & -0.296 & 0.1944 \end{bmatrix}$$

It should be noted that the damping matrix also remains symmetric. Determination of natural frequencies (i.e., the eigenvalue problem formulation in the state space form) with damping in the system will be dealt in Chapter 2 of Volume II, when we will consider rotors mounted on fluid-film bearings by using the FEM.

9.7 STATIC AND DYNAMIC REDUCTIONS

The static (or Guyan) and dynamic reduction methods are generally used to reduce certain DOFs in FE equations. Similarly, we have expansion schemes to overcome the limitation of the number of measurements in practical rotors. The response is measured at only a limited number of locations and over a limited frequency range; hence, only a relatively small number of mode shape vectors with a reduced number of elements are measured. Comparing measured experimental data with numerically generated data is difficult due to incomplete information. One way to compare the data is to reduce the number of DOFs in the analytical model or expand the number of DOFs from the limited measured DOFs. The most popular and the simplest method is the static reduction introduced by Guyan (1965), which is only accurate at a low frequency range. The static method can be modified to reproduce the exact response of a structure at any frequency. This is called the dynamic reduction method and is an extension to Guyan's method. Paz (1984) used the dynamic reduction method in an iterative algorithm to save computational effort in calculating the eigensystem of a structure. O'Callahan (1989) introduced a technique known as the improved reduction system, which is an improvement of the static reduction method, and provides a perturbation to the transformation from the static case by including the inertia term as pseudostatic forces. Friswell and Mottershead (1995) gave an iterative improved reduction system method that converges to the same transformation as the system equivalent reduction expansion process (O'Callahan et al., 1989). The system equivalent reduction expansion process uses the computed eigenvectors to produce the transformation between the master and slave coordinates. Kane and Torby (1991) gave a brief review on reduction methods and compared capabilities and requirement of various methods. Dharmaraju et al. (2005) and Tiwari and Dharmaraju (2006) developed high-frequency and hybrid condensation schemes, respectively, well suited to crack parameter identification. Karthikeyan et al. (2007) extended the hybrid condensation scheme for a rotor system with damping. Prasad and Tiwari (2019) included a known gyroscopic matrix also in the reduction scheme, which is suitable for high-speed rotors.

9.7.1 Static (Guyan) Reduction

The method essentially consists of elimination of certain DOFs. The DOFs eliminated in this process are called slaves, and those retained for the analysis are called masters. Generally retained coordinates (*masters*) coincide with lumped disks, bearing locations, unbalance or balancing plane locations, and any other external force locations. Discarded coordinates (slaves) correspond to points in the model that are noncritical or of secondary interest such as rotational DOFs, intermediate locations on the shaft, etc. In identification of system parameters, since it is not possible to measure responses at all DOFs in a practical situation, this is overcome by applying static reduction to governing system equations. The state vector η, the force vector $\mathbf{f}$, the mass matrix $\mathbf{M}$, and the stiffness matrix $\mathbf{K}$ are partitioned into subvectors and matrices relating to the master DOFs, which are to be retained, and the slave DOFs, which are to eliminated. If no force is applied at the slave DOFs, a general governing FE equation can be partitioned as

$$\mathbf{M}\ddot{\eta}(t)+\mathbf{K}\eta(t)=\mathbf{f}(t) \tag{9.127}$$

with

$$\mathbf{M}=\begin{bmatrix}\mathbf{m}_{mm} & \mathbf{m}_{ms}\\ \mathbf{m}_{sm} & \mathbf{m}_{ss}\end{bmatrix};\quad \mathbf{K}=\begin{bmatrix}\mathbf{k}_{mm} & \mathbf{k}_{ms}\\ \mathbf{k}_{sm} & \mathbf{k}_{ss}\end{bmatrix};\quad \mathbf{f}=\begin{Bmatrix}\mathbf{f}_m\\ \mathbf{0}\end{Bmatrix};\quad \eta=\begin{Bmatrix}\eta_m\\ \eta_s\end{Bmatrix} \tag{9.128}$$

where subscripts m and s represent the master and the slave, respectively. As the name suggests static reduction (or condensation) is based on the assumption that the excitation frequencies are such that the inertia terms are negligibly small as compared to the stiffness term. This might happen with the engineering approximation when the excitation frequency range is relatively low. However, it has been reported that error involved in such an approximation in the estimation of natural frequencies and unbalance responses are acceptably low even in the high-frequency range. Neglecting inertia terms for the second set of equations in Equation 9.127 gives

$$\mathbf{k}_{sm}\eta_m+\mathbf{k}_{ss}\eta_s=\mathbf{0}\Rightarrow\eta_s=-\mathbf{k}_{ss}^{-1}\mathbf{k}_{sm}\eta_m \tag{9.129}$$

and from the identity, we have

$$\eta_m=\mathbf{I}\eta_m \tag{9.130}$$

On combining Equations 9.129 and 9.130, we obtain the static transformation as

$$\begin{Bmatrix}\eta_m\\ \eta_s\end{Bmatrix}=\begin{bmatrix}\mathbf{I}\\ -\mathbf{k}_{ss}^{-1}\,\mathbf{k}_{sm}\end{bmatrix}\eta_m=\mathbf{T}^s\eta_m\quad\text{with}\quad \mathbf{T}^s=\begin{bmatrix}\mathbf{I}\\ -\mathbf{k}_{ss}^{-1}\,\mathbf{k}_{sm}\end{bmatrix} \tag{9.131}$$

where the superscript s represents the *static* reduction, and $\mathbf{T}^s$ denotes the static transformation between the full state vector $\boldsymbol{\eta}$ and the reduced master coordinate vector $\boldsymbol{\eta}_m$. On substituting Equation 9.131 into Equation 9.127and premultiplying both sides of the equation by $\left(\mathbf{T}^s\right)^T$, the resulting equation after rearranging becomes

$$\mathbf{M}^s\ddot{\eta}_m+\mathbf{K}^s\eta_m=\mathbf{f}^s \tag{9.132}$$

with

$$\mathbf{M}^s=(\mathbf{T}^s)^T\begin{bmatrix}\mathbf{m}_{mm} & \mathbf{m}_{ms}\\ \mathbf{m}_{sm} & \mathbf{m}_{ss}\end{bmatrix}\mathbf{T}^s;\quad \mathbf{K}^s=(\mathbf{T}^s)^T\begin{bmatrix}\mathbf{k}_{mm} & \mathbf{k}_{ms}\\ \mathbf{k}_{sm} & \mathbf{k}_{ss}\end{bmatrix}\mathbf{T}^s;\quad \mathbf{f}^s=(\mathbf{T}^s)^T\begin{Bmatrix}\mathbf{f}\\ \mathbf{0}\end{Bmatrix} \tag{9.133}$$

Equation 9.132 is the reduced form of the system governing equation and due to this the overall size of the matrices to be handled greatly reduces and in turn so does the computational time. The same transformation as in Equation 9.131 is valid with the damping and gyroscopic matrices as well, and the reduced damping and gyroscopic matrices take the following form:

$$\mathbf{C}^s=(\mathbf{T}^s)^T\begin{bmatrix}\mathbf{c}_{mm} & \mathbf{c}_{ms}\\ \mathbf{c}_{sm} & \mathbf{c}_{ss}\end{bmatrix}\mathbf{T}^s\quad\text{and}\quad \mathbf{G}^s=(\mathbf{T}^s)^T\begin{bmatrix}\mathbf{G}_{mm} & \mathbf{G}_{ms}\\ \mathbf{G}_{sm} & \mathbf{G}_{ss}\end{bmatrix}\mathbf{T}^s \tag{9.134}$$

While developing the static reduction transformation, all dynamic effects—that is, the mass, damping, and gyroscopic effects—are ignored. However, in dynamic reduction, the gyroscopic matrix (since it is generally known) can be retained, which is not very well reported in the literature (Prasad and Tiwari, 2019).

Example 9.10

For Example 9.2, apply static condensation and eliminate all of the rotational DOFs from the assembled equation of motion for three elements. Subsequently, obtain the transverse natural frequencies of the rotor system.

Solution:

The reduced form of the FE equation in the example has the following form (Equation e):

$$1.46\times10^{-3}\begin{bmatrix} 4 & -13 & -3 & 0 & 0 & 0 \\ -13 & 312 & 0 & 54 & 13 & 0 \\ -3 & 0 & 8 & -13 & -3 & 0 \\ 0 & 54 & -13 & 312 & 0 & 13 \\ 0 & 13 & -3 & 0 & 8 & -3 \\ 0 & 0 & 0 & 13 & -3 & 4 \end{bmatrix}\begin{Bmatrix} \ddot{\varphi}_{x1} \\ \ddot{v}_2 \\ \ddot{\varphi}_{x2} \\ \ddot{v}_3 \\ \ddot{\varphi}_{x3} \\ \ddot{\varphi}_{x4} \end{Bmatrix} \tag{a}$$

$$+103.11\begin{bmatrix} 4 & 6 & 2 & 0 & 0 & 0 \\ 6 & 24 & 0 & -12 & -6 & 0 \\ 2 & 0 & 8 & 6 & 2 & 0 \\ 0 & -12 & 6 & 24 & 0 & -6 \\ 0 & -6 & 2 & 0 & 8 & 2 \\ 0 & 0 & 0 & -6 & 2 & 4 \end{bmatrix}\begin{Bmatrix} \varphi_{x1} \\ v_2 \\ \varphi_{x2} \\ v_3 \\ \varphi_{x3} \\ \varphi_{x4} \end{Bmatrix} = \begin{Bmatrix} 0 \\ 0 \\ 0 \\ 0 \\ 0 \\ 0 \end{Bmatrix}$$

The first step is to separate the master (translational displacements) and slave (rotational displacements) DOFs, as

$$1.46\times10^{-3}\begin{bmatrix} 312 & 54 & -13 & 0 & 13 & 0 \\ 54 & 312 & 0 & -13 & 0 & 13 \\ -13 & 0 & 4 & -3 & 0 & 0 \\ 0 & -13 & -3 & 8 & -3 & 0 \\ 13 & 0 & 0 & -3 & 8 & -3 \\ 0 & 13 & 0 & 0 & -3 & 4 \end{bmatrix}\begin{Bmatrix} \ddot{v}_2 \\ \ddot{v}_3 \\ \ddot{\varphi}_{x1} \\ \ddot{\varphi}_{x2} \\ \ddot{\varphi}_{x3} \\ \ddot{\varphi}_{x4} \end{Bmatrix} \tag{b}$$

$$+103.11\begin{bmatrix} 24 & -12 & 6 & 0 & -6 & 0 \\ -12 & 24 & 0 & 6 & 0 & -6 \\ 6 & 0 & 4 & 2 & 0 & 0 \\ 0 & 6 & 2 & 8 & 2 & 0 \\ -6 & 0 & 0 & 2 & 8 & 2 \\ 0 & -6 & 0 & 0 & 2 & 4 \end{bmatrix}\begin{Bmatrix} v_2 \\ v_3 \\ \varphi_{x1} \\ \varphi_{x2} \\ \varphi_{x3} \\ \varphi_{x4} \end{Bmatrix} = \begin{Bmatrix} 0 \\ 0 \\ 0 \\ 0 \\ 0 \\ 0 \end{Bmatrix}$$

In the above equations, the following rearrangement of columns has been made in the mass and stiffness matrices: 2C to 1C, 4C to 2C, 1C to 3C, and 3C to 4C. Also, the displacement vector has the following rearrangement of rows: 2R to 1R, 4R to 2R, 1R to 3R, and 3R to 4R, where C and R represent the column and row, respectively. *Often one makes a mistake by not rearranging row-wise and column-wise. It is essential to make row-wise rearrangements to segregate forces and moments along with translational and rotational displacements (column-wise rearrangement).* With these rearrangements of column and row wise, we get symmetric matrices.

Now we have the following submatrices and subvectors as defined in Equation 9.129:

$$\mathbf{k}_{mm} = 103.11\begin{bmatrix} 24 & -12 \\ -12 & 24 \end{bmatrix}; \quad \mathbf{k}_{ss} = 103.11\begin{bmatrix} 4 & 2 & 0 & 0 \\ 2 & 8 & 2 & 0 \\ 0 & 2 & 8 & 2 \\ 0 & 0 & 2 & 4 \end{bmatrix}; \quad \mathbf{k}_{ms} = 103.11\begin{bmatrix} 6 & 0 & -6 & 0 \\ 0 & 6 & 0 & -6 \end{bmatrix}$$

$$\mathbf{k}_{sm} = 103.11\begin{bmatrix} 6 & 0 \\ 0 & 6 \\ -6 & 0 \\ 0 & -6 \end{bmatrix}; \quad \eta_m = \begin{Bmatrix} v_2 \\ v_3 \end{Bmatrix}; \quad \eta_s = \begin{Bmatrix} \varphi_{x1} \\ \varphi_{x2} \\ \varphi_{x3} \\ \varphi_{x4} \end{Bmatrix} \tag{c}$$

The static condensation transformation matrix is obtained as

$$\begin{Bmatrix} \eta_m \\ \eta_s \end{Bmatrix} = \begin{bmatrix} \mathbf{I} \\ -\mathbf{k}_{ss}^{-1}\mathbf{k}_{sm} \end{bmatrix}\eta_m = \mathbf{T}^s\eta_m \tag{d}$$

Hence, we have

$$-\mathbf{k}_{ss}^{-1}\mathbf{k}_{sm} = -\left(103.11\begin{bmatrix} 4 & 2 & 0 & 0 \\ 2 & 8 & 2 & 0 \\ 0 & 2 & 8 & 2 \\ 0 & 0 & 2 & 4 \end{bmatrix}\right)^{-1} \times 103.11\begin{bmatrix} 6 & 0 \\ 0 & 6 \\ -6 & 0 \\ 0 & -6 \end{bmatrix} = \begin{bmatrix} -1.6 & 0.4 \\ 0.2 & -0.8 \\ 0.8 & -0.2 \\ -0.4 & 1.6 \end{bmatrix} \tag{e}$$

Hence, we have the following transformations:

$$\begin{Bmatrix} v_3 \\ \varphi_{x2} \\ v_2 \\ \varphi_{x1} \\ \varphi_{x3} \\ \varphi_{x4} \end{Bmatrix} = \begin{bmatrix} 1.000 & 0.000 \\ 0.000 & 1.000 \\ -1.6 & 0.4 \\ 0.2 & -0.8 \\ 0.8 & -0.2 \\ -0.4 & 1.6 \end{bmatrix}\begin{Bmatrix} v_3 \\ \varphi_{x2} \end{Bmatrix} \tag{g}$$

Hence, the reduced form of the matrix becomes

$$\mathbf{M}^s\ddot{\eta}_m + \mathbf{K}^s\eta_m = \mathbf{0} \tag{h}$$

with

$$\mathbf{M}^s = (\mathbf{T}^s)^T\begin{bmatrix} \mathbf{m}_{mm} & \mathbf{m}_{ms} \\ \mathbf{m}_{sm} & \mathbf{m}_{ss} \end{bmatrix}\mathbf{T}^s = \begin{bmatrix} 1.000 & 0.000 \\ 0.000 & 1.000 \\ -1.6 & 0.4 \\ 0.2 & -0.8 \\ 0.8 & -0.2 \\ -0.4 & 1.6 \end{bmatrix}^T$$

$$\times\left(1.46\times10^{-3}\begin{bmatrix} 312 & 54 & -13 & 0 & 13 & 0 \\ 54 & 312 & 0 & -13 & 0 & 13 \\ -13 & 0 & 4 & -3 & 0 & 0 \\ 0 & -13 & -3 & 8 & -3 & 0 \\ 13 & 0 & 0 & -3 & 8 & -3 \\ 0 & 13 & 0 & 0 & -3 & 4 \end{bmatrix}\right)\begin{bmatrix} 1.000 & 0.000 \\ 0.000 & 1.000 \\ -1.6 & 0.4 \\ 0.2 & -0.8 \\ 0.8 & -0.2 \\ -0.4 & 1.6 \end{bmatrix} = \begin{bmatrix} 0.5747 & 0.0359 \\ 0.0359 & 0.5747 \end{bmatrix}$$

$$\mathbf{K}^s = (\mathbf{T}^s)^T \begin{bmatrix} \mathbf{k}_{mm} & \mathbf{k}_{ms} \\ \mathbf{k}_{sm} & \mathbf{k}_{ss} \end{bmatrix} \mathbf{T}^s$$

$$= \begin{bmatrix} 1.000 & 0.000 \\ 0.000 & 1.000 \\ -1.6 & 0.4 \\ 0.2 & -0.8 \\ 0.8 & -0.2 \\ -0.4 & 1.6 \end{bmatrix}^T \left(103.11 \begin{bmatrix} 24 & -12 & 6 & 0 & -6 & 0 \\ -12 & 24 & 0 & 6 & 0 & -6 \\ 6 & 0 & 4 & 2 & 0 & 0 \\ 0 & 6 & 2 & 8 & 2 & 0 \\ -6 & 0 & 0 & 2 & 8 & 2 \\ 0 & -6 & 0 & 0 & 2 & 4 \end{bmatrix} \right) \begin{bmatrix} 1.000 & 0.000 \\ 0.000 & 1.000 \\ -1.6 & 0.4 \\ 0.2 & -0.8 \\ 0.8 & -0.2 \\ -0.4 & 1.6 \end{bmatrix}$$

$$= \begin{bmatrix} 986.856 & -866.124 \\ -866.124 & 989.856 \end{bmatrix}$$

Hence, we have the following eigenvalue problem:

$$-\omega_{nf}^2 \begin{bmatrix} 0.5747 & 0.0359 \\ 0.0359 & 0.5747 \end{bmatrix} \begin{Bmatrix} v_2 \\ v_3 \end{Bmatrix} + \begin{bmatrix} 989.856 & -866.124 \\ -866.124 & 989.856 \end{bmatrix} \begin{Bmatrix} v_2 \\ v_3 \end{Bmatrix} = \begin{Bmatrix} 0 \\ 0 \end{Bmatrix} \tag{i}$$

which gives the natural frequencies as

$$\omega_{nf_1} = 14.24 \text{ rad/s}; \quad \omega_{nf_2} = 58.72 \text{ rad/s}$$

On comparison with Example 9.2 ($\omega_{nf_1} = 14.24$ rad/s; $\omega_{nf_2} = 57.57$ rad/s), with the same number of elements, it could be seen that not much difference in the determined fundamental natural frequency even with the condensations of some of DOFs. However, now the size of the matrix has decreased drastically from 6×6 to 2×2. For a greater number of elements, it is expected that other higher natural frequency estimations would be improved.

Example 9.11

Apply static condensation for Example 9.6 by taking all the rotational DOFs except at the disks as slave DOFs and compare the accuracy of the natural frequencies with the no condensation case.

Solution:

The *static condensation* is employed to reduce the order (or the size) of the eigenvalue problem, when dealing with 14-element model, which reduces the expense of computing eigenvalues and eigenvectors. All rotational DOFs (11) except at disks are selected as slaves DOFs. After the condensation scheme, the order of eigenvalue problem becomes 17 ($2\times14-11=17$). Table 9.10a shows the reordering of filed variables during static condensation. Table 9.10b shows that, with the condensation scheme, the first two natural whirl frequencies are in excellent agreement, but, for the third and fourth natural whirl frequencies, agreement is not so good. It is observed that error increases with the mode number and no definite trend is observed (i.e., at lower modes, obtained natural frequencies are more, and at higher mode, less than without reduction). For large DOF systems, the condensation of a large number of DOFs may introduce more errors.

TABLE 9.10A
During Static Condensation Reordering of Field Variables

Node Numbers	Original Field Variable Order	Field Variables	Master/Slave DOFs	Reshuffled Order Number Assigned	Reshuffled Field Variables and Its Order: Field Variable	Order Number and M/S
1	1	y_1	M	1	y_1	M-1
	2	φ_{x1}	S	18	y_2	M-2
2	3	y_2	M	2	y_4	M-3
	4	φ_{x2}	S	19	y_5	M-4
3	5	$y_3 = 0$	- (Bearing 1)	—	y_6	M-5
	6	φ_{x3}	S	20	y_7	M-6
4	7	y_4	M	3	y_8	M-7
	8	φ_{x4}	S	21	y_9	M-8
5	9	y_5	M	4	y_{10}	M-9
	10	φ_{x5}	M (disk 1)	14	y_{11}	M-10
6	11	y_6	M	5	y_{12}	M-11
	12	φ_{x6}	S	22	y_{14}	M-12
7	13	y_7	M	6	y_{15}	M-13
	14	φ_{x7}	M (disk 2)	15	φ_{x5}	M-14
8	15	y_8	M	7	φ_{x7}	M-15
	16	φ_{x8}	S	23	φ_{x9}	M-16
9	17	y_9	M	8	φ_{x11}	M-17
	18	φ_{x9}	M (disk 3)	16	φ_{x1}	S-18
10	19	y_{10}	M	9	φ_{x2}	S-19
	20	φ_{x10}	S	24	φ_{x3}	S-20
11	21	y_{11}	M	10	φ_{x4}	S-21
	22	φ_{x11}	M (disk 4)	17	φ_{x6}	S-22
12	23	y_{12}	M	11	φ_{x8}	S-23
	24	φ_{x12}	S	25	φ_{x10}	S-24
13	25	$y_{13} = 0$	- (Bearing 2)	—	φ_{x12}	S-25
	26	φ_{x13}	S	26	φ_{x13}	S-26
14	27	y_{14}	M	12	φ_{x14}	S-27
	28	φ_{x14}	S	27	φ_{x15}	S-28
15	29	y_{15}	M	13	—	—
	30	φ_{x15}	S	28	—	—

TABLE 9.10B
Natural Whirl Frequencies

Mode No.	Natural Whirl Frequencies (rad/s): $\omega_{nf}^{(7)}$[a]	$\omega_{nf}^{(14)}$	$\omega_{nf}^{(14)}$ Static Reduction (% Error)
1	116.17	116.17	116.17 (1.6063×10^{-5})
2	438.40	438.32	438.33 (0.0010)
3	861.50	860.73	860.82 (0.0115)
4	1209.45	1207.01	1207.44 (0.0355)

[a] Values in the parenthesis denote the number of elements considered.

9.7.2 Dynamic Reduction

Equation 9.127 can be expanded into two equations in the frequency domain as follows:

$$-\omega^2\mathbf{m}_{mm}\eta_m - \omega^2\mathbf{m}_{ms}\eta_s + \mathbf{k}_{mm}\eta_m + \mathbf{k}_{ms}\eta_s = \mathbf{F}_m \tag{9.135}$$

and

$$-\omega^2\mathbf{m}_{sm}\eta_m - \omega^2\mathbf{m}_{ss}\eta_s + \mathbf{k}_{sm}\eta_m + \mathbf{k}_{ss}\eta_s = \mathbf{0} \tag{9.136}$$

where ω is the frequency (e.g., for the unbalance excitation, the spin speed of the rotor, or for free vibration, a natural frequency), and η and **F** are the complex displacement and force vectors, respectively. Equation 9.136 can be rearranged as

$$\eta_s = -\left(\mathbf{k}_{ss} - \omega^2\mathbf{m}_{ss}\right)^{-1}\left(\mathbf{k}_{sm} - \omega^2\mathbf{m}_{sm}\right)\eta_m \tag{9.137}$$

with an identity equation

$$\eta_m = \mathbf{I}\eta_m \tag{9.138}$$

Equations 9.137 and 9.138 can be combined as follows:

$$\begin{Bmatrix} \eta_m \\ \eta_s \end{Bmatrix} = \mathbf{T}^d\eta_m \tag{9.139}$$

with

$$\mathbf{T}^d = \begin{bmatrix} \mathbf{I} \\ \cdots\cdots\cdots\cdots\cdots\cdots \\ -\left(\mathbf{k}_{ss} - \omega^2\mathbf{m}_{ss}\right)^{-1}\left(\mathbf{k}_{sm} - \omega^2\mathbf{m}_{sm}\right) \end{bmatrix} \tag{9.140}$$

where $\mathbf{T}^d$ is the transformation matrix for the dynamic condensation. In the above transformation apart from the mass and stiffness matrices, the frequency, ω, also appears. It is generally called the central frequency, and it is chosen as the geometrical mean of frequencies of interest. It should be noted that for $\omega=0$ Equation 9.140 reduces to Equation 9.131; that is, the static condensation is a special form of more general dynamic condensation. If the frequency is fixed in dynamic condensation, then one can also use it in the time domain, as the transformation is then fixed. One has to decide the frequency where the transformation is exact. This can be the major excitation frequency or perhaps in the middle of the frequency range of interest.

On substituting Equation 9.139 into the frequency domain equation of motion (Equation 9.127), the resulting equation becomes

$$\left(-\omega^2\mathbf{M}^d + \mathbf{K}^d\right)\eta_m = \mathbf{F}^d \tag{9.141}$$

with

$$\mathbf{M}^d = \left(\mathbf{T}^d\right)^T\mathbf{M}\mathbf{T}^d;\quad \mathbf{K}^d = \left(\mathbf{T}^d\right)^T\mathbf{K}\mathbf{T}^d;\quad \mathbf{F}^d = \left(\mathbf{T}^d\right)^T\mathbf{F} \tag{9.142}$$

where $\mathbf{M}^d$ and $\mathbf{K}^d$ are the reduced mass and stiffness matrices, respectively, and η_m and $\mathbf{F}_m$ are the reduced response and force vectors, respectively. The procedure of elimination of slave DOFs as discussed above is generally called the *dynamic reduction* scheme (Paz, 1984). The damping and gyroscopic matrices are also reduced by the transformation matrix defined by Equation 9.140 as follows:

$$\mathbf{C}^d = \left(\mathbf{T}^d\right)^T \mathbf{C}\mathbf{T}^d \quad \text{and} \quad \mathbf{G}^d = \left(\mathbf{T}^d\right)^T \mathbf{G}\mathbf{T}^d \tag{9.143}$$

However, the assumption involved with the reduction defined in Equation 9.143 is that matrices **C** and **G** are negligibly small as compared to matrices **M** and **K**. There are several investigations still under way on condensation schemes. Tiwari and Dharmaraju (2006) developed a high-frequency condensation scheme, in which the inertia term is so dominant that other terms like stiffness and damping forces are negligible. For such cases, the transformation matrix can be obtained by dropping stiffness terms from Equation 9.140; this can be written as follows:

$$\mathbf{T}^{hf} = \begin{bmatrix} \mathbf{I} \\ -\mathbf{m}_{ss}^{-1}\mathbf{m}_{sm} \end{bmatrix} \tag{9.144}$$

where superscript *hf* represents high frequency. In developing dynamic condensation, even the gyroscopic matrix can be included, as generally it is known.

Example 9.12

For Example 9.2, condense all rotational DOFs by using the dynamic condensation technique. Consider only three elements for the illustration of the method.

Solution:

The dynamic condensation transformation matrix is defined as

$$\begin{Bmatrix} \eta_m \\ \eta_s \end{Bmatrix} = \mathbf{T}^d = \begin{bmatrix} \mathbf{I} \\ -\left(\mathbf{k}_{ss} - \omega^2\mathbf{m}_{ss}\right)^{-1}\left(\mathbf{k}_{sm} - \omega^2\mathbf{m}_{sm}\right) \end{bmatrix}\eta_m \tag{a}$$

From the previous example, we have

$$\mathbf{k}_{mm} = 103.11\begin{bmatrix} 24 & -12 \\ -12 & 24 \end{bmatrix}; \quad \mathbf{k}_{ss} = 103.11\begin{bmatrix} 4 & 2 & 0 & 0 \\ 2 & 8 & 2 & 0 \\ 0 & 2 & 8 & 2 \\ 0 & 0 & 2 & 4 \end{bmatrix}; \quad \mathbf{k}_{ms} = 103.11\begin{bmatrix} 6 & 0 & -6 & 0 \\ 0 & -6 & 0 & -6 \end{bmatrix} \tag{b}$$

$$\mathbf{k}_{sm} = 103.11\begin{bmatrix} 6 & 0 \\ 0 & 6 \\ -6 & 0 \\ 0 & -6 \end{bmatrix}; \eta_m = \begin{Bmatrix} v_2 \\ v_3 \end{Bmatrix}; \quad \eta_s = \begin{Bmatrix} \varphi_{x1} \\ \varphi_{x2} \\ \varphi_{x3} \\ \varphi_{x4} \end{Bmatrix} \tag{c}$$

$$\mathbf{m}_{mm} = 1.46\times10^{-3}\begin{bmatrix} 312 & 54 \\ 54 & 312 \end{bmatrix}; \quad \mathbf{m}_{ss} = 1.46\times10^{-3}\begin{bmatrix} 4 & -3 & 0 & 0 \\ -3 & 8 & -3 & 0 \\ 0 & -3 & 8 & -3 \\ 0 & 0 & -3 & 4 \end{bmatrix} \tag{d}$$

$$\mathbf{m}_{ms} = 1.46 \times 10^{-3} \begin{bmatrix} -13 & 0 & 13 & 0 \\ 0 & -13 & 0 & 13 \end{bmatrix}; \quad \mathbf{m}_{sm} = 1.46 \times 10^{-3} \begin{bmatrix} -13 & 0 \\ 0 & -13 \\ 13 & 0 \\ 0 & 13 \end{bmatrix} \tag{e}$$

For the dynamic condensation, we need the frequency at which we need to transform equations. From the static condensation, we have some knowledge of the natural frequencies of interest. Let us take 37.5 rad/s as the average value of the frequency.

$$\left(\mathbf{k}_{ss} - \omega^2 \mathbf{m}_{ss}\right)^{-1} = \left(103.11 \begin{bmatrix} 4 & 2 & 0 & 0 \\ 2 & 8 & 2 & 0 \\ 0 & 2 & 8 & 2 \\ 0 & 0 & 2 & 4 \end{bmatrix} - 37.5^2 \times 1.46 \times 10^{-3} \begin{bmatrix} 4 & -3 & 0 & 0 \\ -3 & 8 & -3 & 0 \\ 0 & -3 & 8 & -3 \\ 0 & 0 & -3 & 4 \end{bmatrix} \right)^{-1}$$

$$= \begin{bmatrix} 0.0029 & -0.0008 & 0.0003 & -0.0001 \\ -0.0008 & 0.0016 & -0.0005 & 0.0003 \\ 0.0003 & -0.0005 & 0.0016 & -0.0008 \\ -0.0001 & 0.0003 & -0.0008 & 0.0029 \end{bmatrix} \tag{f}$$

and

$$\left(\mathbf{k}_{sm} - \omega^2 \mathbf{m}_{sm}\right) = 103.11 \begin{bmatrix} 6 & 0 \\ 0 & 6 \\ -6 & 0 \\ 0 & -6 \end{bmatrix} - 37.5^2 \times 1.46 \times 10^{-3} \begin{bmatrix} -13 & 0 \\ 0 & -13 \\ 13 & 0 \\ 0 & 13 \end{bmatrix}$$

$$= 10^3 \begin{bmatrix} 0.6454 & 0 \\ 0 & 0.6454 \\ -0.6454 & 0 \\ 0 & -0.6454 \end{bmatrix} \tag{g}$$

So, we have

$$-\left(\mathbf{k}_{ss} - \omega^2 \mathbf{m}_{ss}\right)^{-1} \left(\mathbf{k}_{sm} - \omega^2 \mathbf{m}_{sm}\right) = \begin{bmatrix} -1.7148 & 0.4505 \\ 0.2252 & -0.8574 \\ 0.8574 & -0.2252 \\ -0.4505 & 1.7148 \end{bmatrix} \tag{h}$$

Hence, the dynamic transformation is as follows:

$$\begin{Bmatrix} v_2 \\ v_3 \\ \varphi_{x1} \\ \varphi_{x2} \\ \varphi_{x3} \\ \varphi_{x4} \end{Bmatrix} = \begin{bmatrix} 1 & 0 \\ 0 & 1 \\ -1.7148 & 0.4505 \\ 0.2252 & -0.8574 \\ 0.8574 & -0.2252 \\ -0.4505 & 1.7148 \end{bmatrix} \begin{Bmatrix} v_2 \\ v_3 \end{Bmatrix} \tag{i}$$

Hence, the reduced form of the matrix becomes

$$\mathbf{M}^s \ddot{\eta}_m + \mathbf{K}^s \eta_m = \mathbf{0} \tag{j}$$

with

$$\mathbf{M}^d = \left(\mathbf{T}^d\right)^T \begin{bmatrix} \mathbf{m}_{mm} & \mathbf{m}_{ms} \\ \mathbf{m}_{sm} & \mathbf{m}_{ss} \end{bmatrix} \mathbf{T}^d$$

$$= \begin{bmatrix} 1 & 0 \\ 0 & 1 \\ -1.7148 & 0.4505 \\ 0.2252 & -0.8574 \\ 0.8574 & -0.2252 \\ -0.4505 & 1.7148 \end{bmatrix}^T \left(1.46\times 10^{-3} \begin{bmatrix} 312 & 54 & -13 & 0 & 13 & 0 \\ 54 & 312 & 0 & -13 & 0 & 13 \\ -13 & 0 & 4 & -3 & 0 & 0 \\ 0 & -13 & -3 & 8 & -3 & 0 \\ -13 & 0 & 0 & -3 & 8 & -3 \\ 0 & 13 & 0 & 0 & -3 & 4 \end{bmatrix} \right)$$

$$\begin{bmatrix} 1 & 0 \\ 0 & 1 \\ -1.7148 & 0.4505 \\ 0.2252 & -0.8574 \\ 0.8574 & -0.2252 \\ -0.4505 & 1.7148 \end{bmatrix}$$

or

$$\mathbf{M}^d = \begin{bmatrix} 0.5858 & 0.0293 \\ 0.0293 & 0.5858 \end{bmatrix}$$

and

$$\mathbf{K}^d = \left(\mathbf{T}^d\right)^T \begin{bmatrix} \mathbf{k}_{mm} & \mathbf{k}_{ms} \\ \mathbf{k}_{sm} & \mathbf{k}_{ss} \end{bmatrix} \mathbf{T}^d$$

$$= 103.11 \begin{bmatrix} 1 & 0 \\ 0 & 1 \\ -1.7148 & 0.4505 \\ 0.2252 & -0.8574 \\ 0.8574 & -0.2252 \\ -0.4505 & 1.7148 \end{bmatrix}^T \begin{bmatrix} 24 & -12 & 6 & 0 & -6 & 0 \\ -12 & 24 & 0 & 6 & 0 & -6 \\ 6 & 0 & 4 & 2 & 0 & 0 \\ 0 & 6 & 2 & 8 & 2 & 0 \\ -6 & 0 & 0 & 2 & 8 & 2 \\ 0 & -6 & 0 & 0 & 2 & 4 \end{bmatrix} \begin{bmatrix} 1 & 0 \\ 0 & 1 \\ -1.7148 & 0.4505 \\ 0.2252 & -0.8574 \\ 0.8574 & -0.2252 \\ -0.4505 & 1.7148 \end{bmatrix}$$

or

$$\mathbf{K}^d = 10^3 \begin{bmatrix} 0.9978 & -0.8709 \\ -0.8709 & 0.9978 \end{bmatrix}$$

Hence, we have the following eigenvalue problem:

$$-\omega_{nf}^2 \begin{bmatrix} 0.5858 & 0.0293 \\ 0.0293 & 0.5858 \end{bmatrix} \begin{Bmatrix} v_2 \\ v_3 \end{Bmatrix} + 10^3 \begin{bmatrix} 0.9978 & -0.8709 \\ -0.8709 & 0.9978 \end{bmatrix} \begin{Bmatrix} v_2 \\ v_3 \end{Bmatrix} = \begin{Bmatrix} 0 \\ 0 \end{Bmatrix} \tag{k}$$

for which the eigenvalues and eigenvectors as follows:

$$\begin{Bmatrix} \omega_{nf1}^2 \\ \omega_{nf2}^2 \end{Bmatrix} = 10^3 \begin{Bmatrix} 3.3582 \\ 0.2064 \end{Bmatrix} \quad \text{and} \quad \left[\begin{Bmatrix} v_2 \\ v_3 \end{Bmatrix}_{\omega_{nf1}} \quad \begin{Bmatrix} v_2 \\ v_3 \end{Bmatrix}_{\omega_{nf2}} \right] = \begin{bmatrix} 0.7071 & 0.7071 \\ -0.7071 & -0.7071 \end{bmatrix} \tag{l}$$

which gives the natural frequencies as follows:

$$\omega_{nf_1} = 14.37 \text{ rad/s}; \quad \omega_{nf_2} = 57.95 \text{ rad/s}$$

Comparing Examples 9.2 and 9.10 with the same number of elements, we see that there is not much difference in the obtained natural frequencies even with the condensations of some of the DOFs. However, now the size of the matrix has decreased drastically from 6×6 to 2×2. The difference between the condensed and non-condensed cases may be appreciable in very large DOF systems. The present example, in fact, illustrates the method but not the potential of the dynamic condensation as such.

9.8 CONCLUDING REMARKS

In the present chapter, we considered the transverse vibration of a continuous system for some simple boundary conditions. Closed-form expressions of natural frequencies and mode shapes (eigenfunctions) are summarized for various simple boundary conditions. The main focus of the present chapter is to analyze the free and forced vibrations by using the FEM. The elemental equation of the Euler–Bernoulli beam model is derived using Galerkin's method. Applications of various simple boundary conditions such as the simply supported, cantilevered, overhung, and intermediate supports have been illustrated through examples. The purpose of the examples is to have clarity of the intermediate steps involved in the formulation of the elemental and assembled governing equations, application of boundary conditions, extraction of modal parameters, and forced response information. Proportional damping or Rayleigh's damping is also described. The static and dynamic condensation schemes are described in detail, which are used either to reduce the size of matrices to be handled for the final solution of the eigenvalue problem or to obtain the forced response. The forced response of simple rotor systems is obtained for the most common type of excitation, that is, unbalances. However, it can be easily extended for other periodic forces by using Fourier analysis, as any periodic force can be written using the sine and cosine series form. Fourier series will be dealt with in more detail in Chapter 6 of Volume II. Response of rotor system for aperiodic force can be obtained using convolution integral (Duhamel equation) and for detail readers can refer to an excellence book by Thomson and Dahleh (1998). In the next chapter, we will continue with the FE analysis of rotors by including some higher effects such as gyroscopic effects, rotary inertia, and shear deformation, which are predominant at high spin speeds for thick rotating shafts.

EXERCISES

Note: Assume the density 7800 kg/m^3 for all exercise problems.

Exercise 9.1 Find the transverse natural frequencies of the rotor system shown in Figure 9.33. The mass of the disk is m=5 kg, and the diametral mass moment of inertia is I_d=0.02 kg-m^2. The lengths of the shaft are a=0.3 m and b=0.7 m. The diameter of the shaft is 10 mm. Bearing

A has a roller support, and Bearing B has the fixed support condition. Neglect the mass of the shaft. Use the FEM.

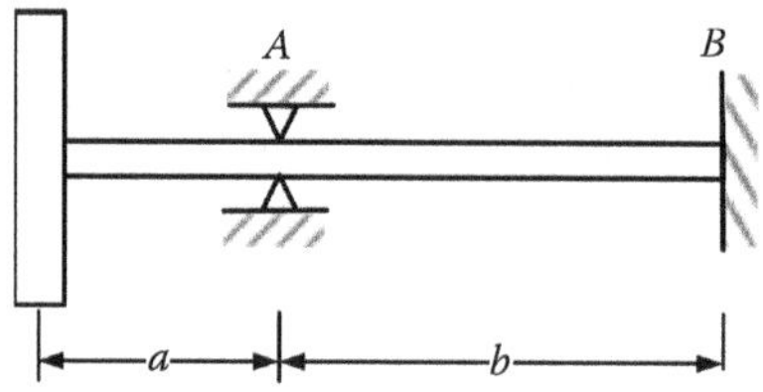

FIGURE 9.33 Overhung rotor system with one end fixed and an intermediate support.

Exercise 9.2 Find the transverse natural frequencies of the rotor system shown in Figure 9.34. $EI = 2$ MN-m^2 for the shaft, and the diametral mass moment of inertia of the disk is negligible. The shaft is made of steel with 2.1×10^{11} N/m^2 as Young's modulus, E, and 7800 kg/m^3 as the mass density, ρ.

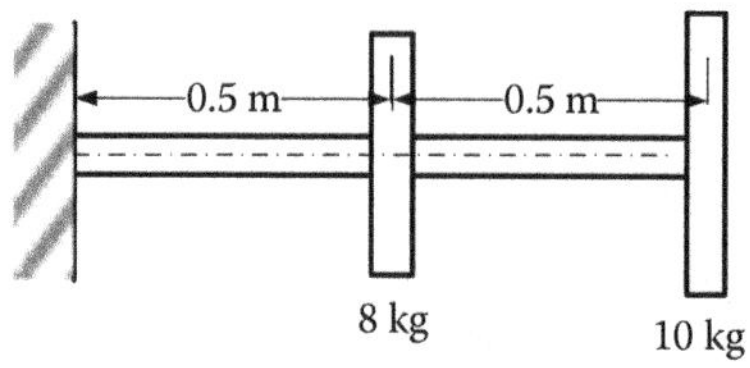

FIGURE 9.34 Cantilever rotor with two disks.

Exercise 9.3 Find the transverse natural frequency of the rotor system shown in Figure 9.35. Consider the shaft as massless and made of steel with 2.1×10^{11} N/m^2 as Young's modulus, E, and 7800 kg/m^3 as the mass density, ρ. The disk has 10 kg mass. The shaft is simply supported at the ends (all dimensions in the figure are in cm).

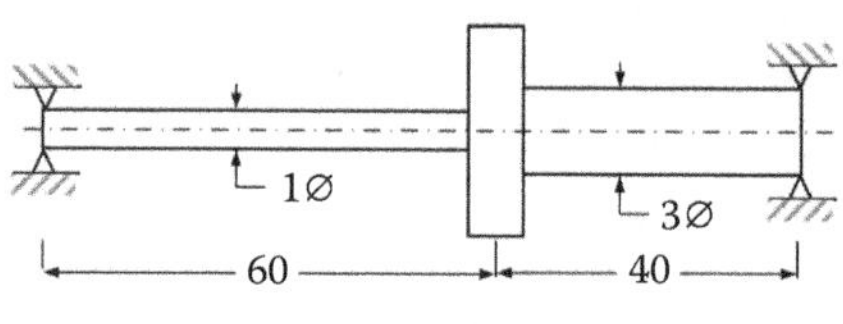

FIGURE 9.35 Stepped shaft rotor system.

Exercise 9.4 For Exercises 9.1 through 9.3, plot the translational and rotational displacements (with both amplitude and phase) of disks with respect to the spin speed of the rotor (minimum spin frequency of the rotor is 0.1 rad/s, and maximum is at least 5 rad/s above the second critical speed of the rotor system). Assume imbalances of 20 gm-mm at one of the disks with the 30° phase with some shaft reference point. Check whether the critical speeds are in agreement with that obtained by free vibration analysis.

Exercise 9.5 Find the bending natural frequency of the rotor shown in Figure 9.36. The rotor is assumed to have a fixed support at one end. The mass of the disk $m = 2$ kg and the diametral mass moment of inertia $I_d = 0.05$ kg-m^2. The shaft is assumed to be massless, and its length and diameter are 0.2 and 0.01 m, respectively. Young's modulus $E = 2.1 \times 10^{11}$ N/m^2 for the shaft material.

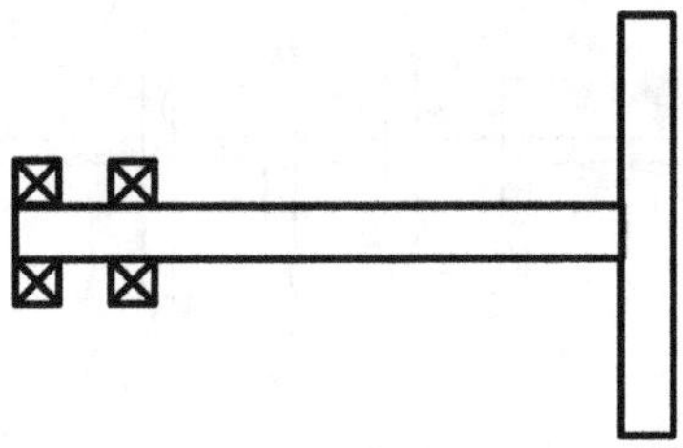

FIGURE 9.36 Cantilever rotor system.

Exercise 9.6 Find the transverse natural frequencies and mode shapes of the rotor system for Exercise 9.2, when the left and right disks have diametral mass moments of inertia $I_{d1}=0.05$ kg-m^2 and $I_{d2}=0.06$ kg-m^2, respectively.

Exercise 9.7 Find the transverse natural frequencies of the rotor system shown in Figure 8.34. The mass of the disk is $m=5$ kg, and the diametral mass moment of inertia is $I_d=0.02$ kg-m^2. The shaft lengths are $a=0.3$ m and $b=0.7$ m, the diameter of the shaft is 10 mm, and the modulus of elasticity of the shaft is $E=2.1\times10^{11}$ N/m^2. Consider two different cases, that is, when bearing A is (i) a simple support and (ii) a flexible support, which provides a bending stiffness equal to 5% of a cantilevered shaft segment having length a. Bearing B is simply supported.

Exercise 9.8 Find the bending critical speeds and mode shapes of the rotor system shown in Figure 9.37. B_1 and B_2 are bearings, which provide simply supported end conditions, and D_1 and D_2 are rigid disks. The shaft is made of steel with modulus of rigidity $E=2.1\times10^{11}$ N/m^2, the mass density, $\rho=7800$ kg/m^3, and uniform diameter $d=10$ mm. The various shaft lengths are as follows: $B_1D_1=50$ mm, $D_1D_2=75$ mm, and $D_2B_2=50$ mm. The masses of the disks are $m_{d1}=4$ kg and $m_{d2}=6$ kg. Consider the shaft as massless. Consider the following cases: (i) Neglect the diametral mass moment of inertia of both disks and (ii) take $I_{d1}=0.05$ kg-m^2 and $I_{d2}=0.06$ kg-m^2.

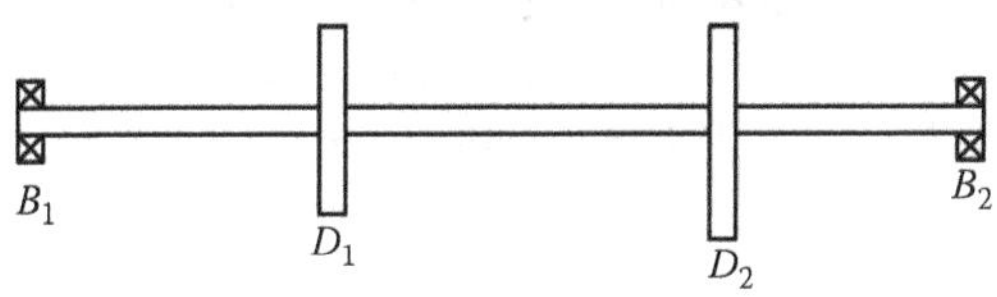

FIGURE 9.37 Simply supported rotor with two disks.

Exercise 9.9 Find the bending critical speeds and mode shapes of the rotor system shown in Figure 9.38. B_1 and B_2 are bearings, which provide simply supported end conditions, and D_1, D_2, D_3, and D_4 are rigid disks. The shaft is made of steel with modulus of rigidity $E=2.1\times10^{11}$ N/m^2 and uniform diameter $d=10$ mm. The various shaft lengths are as follows: $B_1D_1=150$ mm, $D_1D_2=50$ mm, $D_2D_3=50$ mm, $D_3D_4=50$ mm, and $D_4B_2=150$ mm. The masses of the disks are $m_{d1}=4$ kg, $m_{d2}=5$ kg, $m_{d3}=6$ kg, and $m_{d4}=7$ kg. Consider the shaft as massless. Consider the following cases: (i) Neglect the diametral mass moment of inertia of all disks and (ii) take $I_{d1}=0.05$ kg-m^2, $I_{d2}=0.06$ kg-m^2, $I_{d3}=0.07$ kg-m^2, and $I_{d4}=0.08$ kg-m^2.

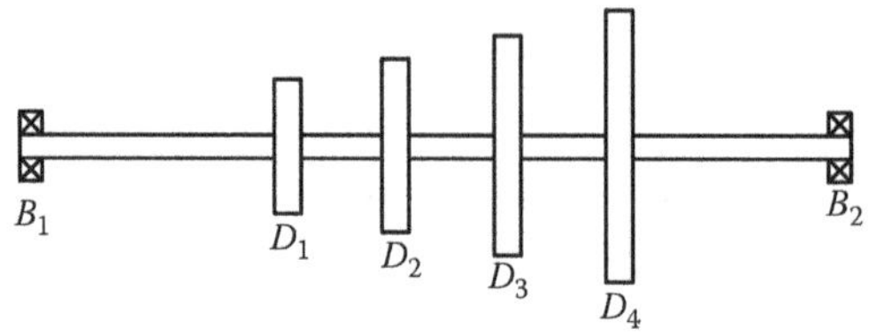

FIGURE 9.38 Multidisc simply supported rotor system.

Exercise 9.10 Find the transverse natural frequencies and mode shapes of the rotor system shown in Figure 9.39. B is a bearing with a fixed end condition, and D_1 (near the bearing) and D_2 are rigid disks. The shaft is made of steel with Young's modulus $E=2.1\times10^{11}$ N/m² and uniform diameter $d=10$ mm. The various shaft lengths are as follows: $BD_1=50$ mm and $D_1D_2=75$ mm. The diametral mass moments of inertia of the disks are $I_{d1}=0.04$ kg-m² (near the bearing) and $I_{d2}=0.1$ kg-m². Neglect the mass of the shaft.

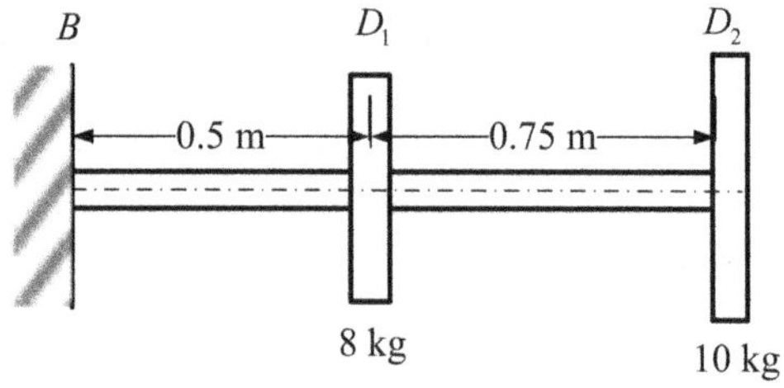

FIGURE 9.39 Two-disk cantilever rotor.

Exercise 9.11 Consider the rotor system shown in Figure 9.40. Two flexible shafts are connected by a coupling (i.e., pin joint). The other ends of the shafts have fixed conditions. Formulate system governing equations using the FEM for free transverse vibrations considering two elements. Obtain the eigenvalue problem for the boundary condition mentioned. Take the length of each of the shaft as 0.5 m and the diameter as 0.05 m. The shaft has Young's modulus $E=2.1\times10^{11}$ N/m² and mass density $\rho=7800$ kg/m³. State the assumptions made.

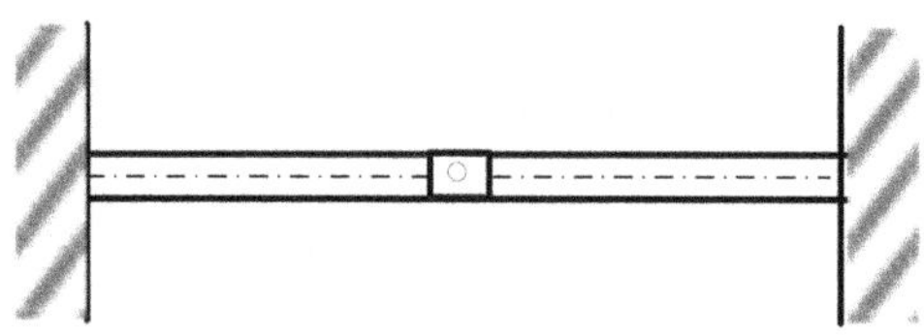

FIGURE 9.40 Two shafts connected by pin.

Exercise 9.12 Find the transverse natural frequencies of an overhung rotor system as shown in Figure 9.41. Take the masses of disks as $m_1=5$ kg and $m_2=3$ kg, and its diametral mass moments of inertia are $I_{d1}=0.02$ kg-m² and $I_{d2}=0.015$ kg-m², respectively. The lengths of the shaft segments are $a=0.3$ m and $b=c=0.7$ m, and the diameter of the shaft is 10 mm. Neglect the gyroscopic effects. Take $\rho=7800$ kg/m³, and $E=2.1\times10^{11}$ N/m² for the shaft material.

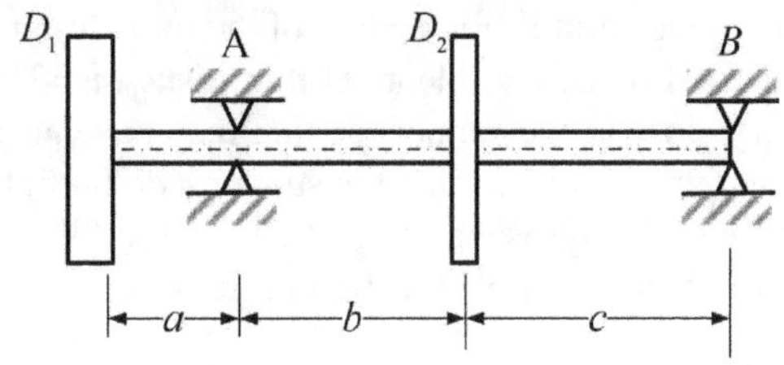

FIGURE 9.41 Overhung rotor system.

Exercise 9.13 The rotor in Figure 9.42 is a typical twin-spool aeroengine rotor model, comprising lower pressure and high pressure spools rotating coaxially at different speeds with a speed ratio (high pressure/lower pressure) of 1.5. Table 9.11 summarizes all the geometrical and physical parameters of the rotor-bearing system. Obtain the critical speeds of such a rotor system with the help of a Campbell diagram.

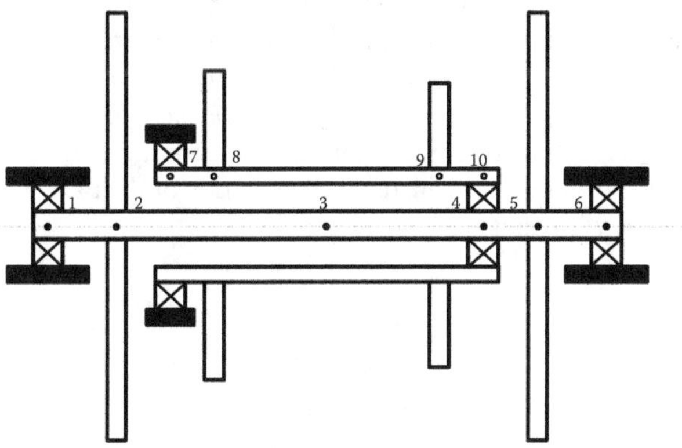

FIGURE 9.42 Twin-spool aeroengine rotor model with a lower pressure rotor (elements 1–6) and high pressure rotor (elements 7–10).

TABLE 9.11
The Geometrical and Physical Properties of the Twin-Spool Aeroengine Rotor Model

				Bearings		Disk	
Node	Axial Distance (mm)	Inner Diameter (mm)	Outer Diameter (mm)	Stiffness (MN/m) $k_{xx}=k_{yy}$	Damping (kNs/m) $c_{xx}=c_{yy}$	Mass (kg)	I_p (kg-mm²)
1	0.0	30.4	0.0	2795	5.260	—	—
2	76.2	30.4	0.0	—	—	4.904	0.02712
3	254.0	30.4	0.0	—	—	—	—
4	406.4	30.4	0.0	8.7598	—	—	—
5	457.2	30.4	0.0	—	—	4.203	0.02034
6	508.0	30.4	0.0	17.519	3.507	—	—
7	152.4	50.8	38.1	17.519	3.507	—	—
8	203.2	50.8	38.1	—	—	3.327	0.01469
9	355.6	50.8	38.1	—	—	2.227	0.09720
10	406.4	50.8	38.1	8.7598	—	—	—

Exercise 9.14 Find the transverse natural frequencies of the rotor system shown in Figure 9.43. It has simply supported end conditions, and identical thin disks ($m=2$ kg and $I_d=0.01$ kg-m^2). These disks are mounted very close to the support so that they have pure tilting motion (transverse rotational displacements). It is assumed that the supports do not interfere in the motion of the disks. Use the FEM and neglect gyroscopic effects. $E=2.1\times10^{11}$ N/m^2 and $\rho=7800$ kg/m^3. The length of the shaft is 0.6 m, and its diameter is 0.01 m.

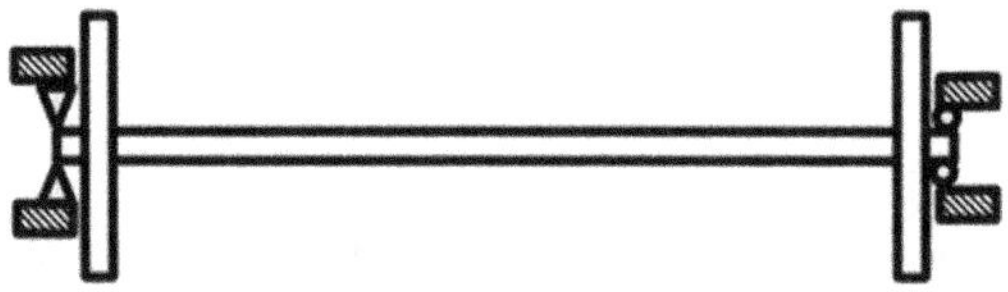

FIGURE 9.43 Rotor system with disks near supports.

Exercise 9.15 In a propeller as shown in Figure 9.44, the axle has the axial and transverse vibrations and because of that the blades (two in numbers) have only transverse vibrations. Find the natural frequencies of the combined system. Consider only the translational DOFs of masses. Prefer the FEM. Neglect the mass of the axle and blades, and assume lumped (point) masses as shown in the figure. Let $m_1=5$ kg and $m_2=m_3=2$ kg. The axle has a length of 0.15 m and diameter of 0.02 m. The blade has a length of 0.35 m and diameter of 0.01 m. $E=2.1\times10^{11}$ N/m^2 for the axle and the shaft material. Give all the steps during the generation of the FE governing equations for the axial and translational vibrations.

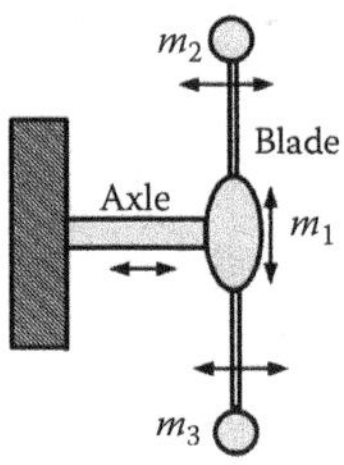

FIGURE 9.44 Aircraft propeller.

Exercise 9.16 A shaft has two heavy disks mounted on it. Find the lowest critical speed of the original system considering the transverse natural frequencies of each disk separately as 100 Hz and 200 Hz (use Dunkerley's formula of Chapter 8).

Exercise 9.17 Find the transverse natural frequencies of a rotor system with two coaxial shafts, which is modeled as shown in Figure 9.45. The disks have masses of 3 kg and 2 kg, respectively, on shaft A and shaft B. Shafts A and B are 2 and 1.5 cm in diameter, respectively, with a length of 40 cm each. Neglect the inertia of shafts. The bearing between two coaxial shafts provides an effective transverse (linear) stiffness of 100 MN/m. $E=2.1\times10^{11}$ N/m^2.

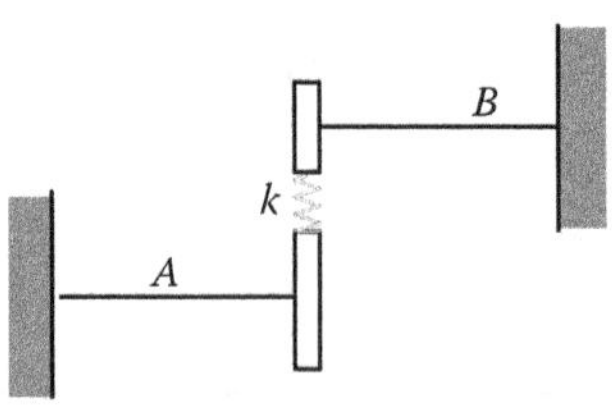

FIGURE 9.45 Coaxial shaft rotor model.

Exercise 9.18 Find the equations of motion for transverse vibration of the rotor system shown in Figure 9.46 using the FEM. Consider single-plane motion. Write elemental equations, assembled equations, boundary conditions, imposition of boundary conditions, the reduced form of equations of motion, and the eigenvalue problem. Let d be the diameter of the shaft, m the mass of the disk, I_d the diametral mass moment of inertia of the disk, ρ the mass density of the shaft material, and E the Young's modulus of the shaft material. Bearings have k_1 and k_2 direct stiffness coefficients, and c_1 and c_2 direct damping coefficients. Consider inertia of the shaft in the formulation.

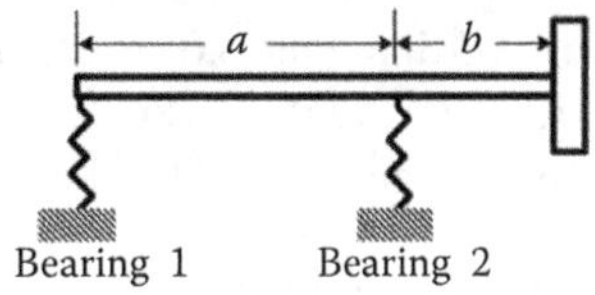

FIGURE 9.46 Flexible rotor on flexible bearings.

Exercise 9.19 A cantilever rotor with a massless flexible shaft (of diameter $d=20\,\text{mm}$ and length $L=40\,\text{cm}$) carries a thin disk (of mass 5 kg and diametral mass moment of inertia 0.01 kg-m^2) at the free end. The disk has an initial tilt of 3° with respect to the shaft axis at an angular postilion of 120° with respect to some shaft reference, and radially eccentric by 200 μm and at an angle of 30° with respect to the same reference position on the shaft. Consider the gyroscopic effect of the disk (refer to Chapter 10). It is to be analyzed for forced transverse vibration using the FEM. Write elemental equations, assembled equations, boundary conditions, imposition of boundary conditions, the reduced form of the equations of motion, and the forced vibration vector. $\omega=6000\,\text{rpm}$, and $E=2.1\times10^{11}\,\text{N/m}^2$. Obtain the Campbell diagram and show critical speeds.

Exercise 9.20 Find *the expanded* form (step-by-step elemental form) of the governing equations of motion for transverse vibration and the final form of the eigenvalue problem for the rotor-bearing system shown in Figure 9.47 using the FEM. Consider single-plane motion only (y–z plane only). $a=0.6\,\text{m}$, $b=0.4\,\text{m}$, shaft diameter $d=0.1\,\text{m}$, disk mass $m=1\,\text{kg}$, $I_d=0.1$ kg-m^2, bearing stiffness $k=1$ kN/m, shaft material $\rho=7800\,\text{kg/m}^3$, and $E=2.1\times10^{11}\,\text{N/m}^2$.

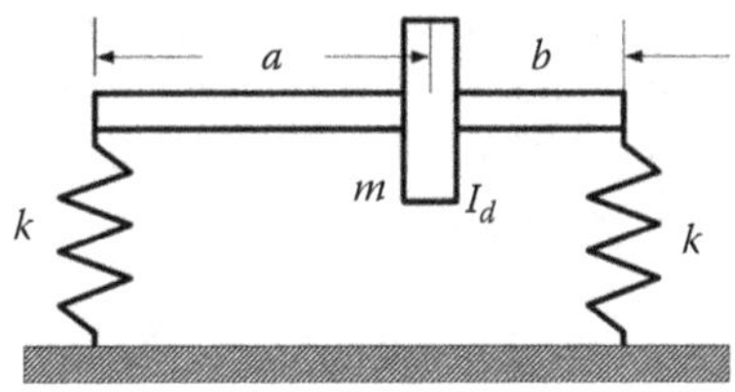

FIGURE 9.47 Rotor on flexible bearings.

Exercise 9.21 Obtain transverse natural frequencies of the rotor system shown in Figure 9.48 using the FEM. The length of the shaft in left side of the disk is 0.6 m with a total length of 1.0 m, and its diameter is 0.01 m. The mass of the disk is 3 kg, and the diametrical moment of inertia of the disk is 0.01 kg-m^2. The stiffness of the bending (tilting) spring, k_{bsp}, (it resists the transverse angular displacement) is equal to half that of a cantilever shaft with the same diameter and a length of 0.6 m. Take for shaft $E=2.1\times10^{11}\,\text{N/m}^2$ and $\rho=7800\,\text{kg/m}^3$.

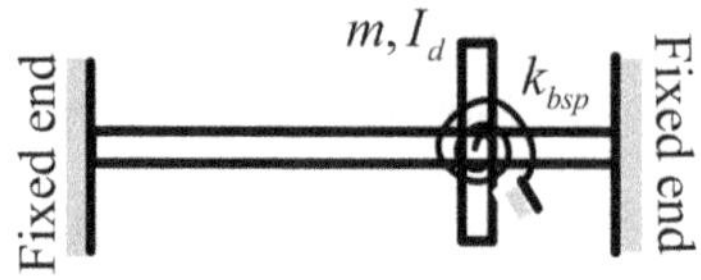

FIGURE 9.48 A fixed–fixed shaft with a disk and bending (tilting) spring.

Exercise 9.22 A shaft of length 1 m and diameter 0.02 m is supported at ends as a fixed–free (cantilever) support. Take $E=2.1 \times 10^{11}$ N/m^2 and $\rho=7800$ kg/m^3. Obtain the first two critical speeds of the rotor system by the FEM for transverse vibration.

Exercise 9.23 Obtain the transverse natural frequency of the rotor-bearing system as shown in Figure 9.49 by the FEM. Both bearings have fixed end conditions. At the disk location, the rotor is experiencing linear and tilting stiffness, k_y and k_{φ_x}. Take $L=1$ m, $a=0.3$ m, $d=0.15$ m, $k_y=1$ MN/m, $k_{\varphi_x}=100$ kN-m/rad, $m=5$ kg, $I_d=0.02$ kg-m^2, $E=2.1\times10^{11}$ N/m^2, and $\rho=7800$ kg/m^3, Consider one plane motion only.

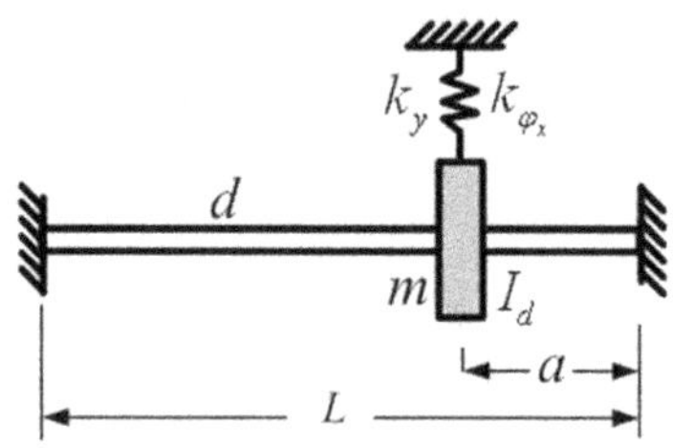

FIGURE 9.49 A rotor system with a spring support.

Exercise 9.24 Consider a rotor system as shown in Figure 9.50. One end of the shaft has a fixed support; at the midspan, it is roller supported, and at the disk end, it is spring supported. The spring provides resistance to both transverse translational and rotational (tilting) displacements. Consider the vertical plane motion only. Formulate system governing equations using the FEM for free transverse vibrations considering only two elements. Provide boundary conditions at each node. Obtain the eigen value problem for the boundary condition mentioned. Obtain the natural frequencies of the rotor system. Neglect the gyroscopic effect. Take the length of each of the shaft as 0.5 m and the diameter as 0.02 m. The shaft has the Young's modulus of $E=2.1\times10^{11}$ N/m^2, and the mass density of $\rho=7800$ kg/m^3. Spring has a linear stiffness of 100 kN/m and a rotational (tilting) stiffness of 100 N-m/rad. The disk has a mass of 2 kg and a diametral mass moment of inertia of 0.015 kg-m^2. If required, the inertia of the shaft may be ignored.

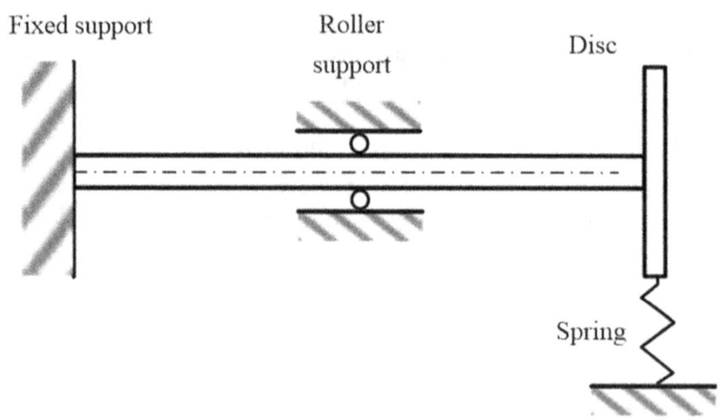

FIGURE 9.50 A fixed shaft with roller supported at the midspan and spring supported at the disk end.

Exercise 9.25 Multiple choice questions, select only the best single answer:

i. In the expansion theorem while performing the modal analysis of a rotor system
   A. the response is expressed as a summation of eigenfunctions with constant coefficients
   B. the eccentricity is expressed as a summation of eigenfunctions with constant coefficients
   C. the force is expressed as a summation of eigenfunctions with constant coefficients
   D. bearing force is expressed as a function of the stiffness and damping coefficients

ii. In Rayleigh's proportional damping, the following is true (where α and β are constants, **M** and **K** are mass and stiffness matrices, respectively, of a rotor system):
   A. $\alpha\mathbf{M}+\beta\mathbf{K}^2$
   B. $\alpha\mathbf{M}^2+\beta\mathbf{K}$
   C. $\alpha\mathbf{M}+\beta\mathbf{K}$
   D. $\alpha\mathbf{M}-\beta\mathbf{K}$

iii. The proportional damping is generally expressed as (where α, β, and γ are constants, **M**, **K**, and **G** are mass, stiffness, and gyroscopic matrices, respectively, of a rotor system):
   A. $\alpha\mathbf{M}+\beta\mathbf{G}$
   B. $\alpha\mathbf{M}+\beta\mathbf{K}$
   C. $\alpha\mathbf{K}+\beta\mathbf{G}$
   D. $\alpha\mathbf{M}+\beta\mathbf{G}+\gamma\mathbf{K}$

iv. The static condensation is a technique used in FEMs
   A. to reduce the overall DOFs of the system
   B. to increase the overall DOFs of the system
   C. to simplify the system equation without disturbing the DOFs of the system
   D. to decouple the system equation without disturbing the DOFs of the system

v. While developing the static condensation scheme
   A. the stiffness and damping matrices are ignored
   B. the mass and damping matrices are ignored
   C. the stiffness and mass matrices are ignored
   D. the force vector is ignored

vi. With general damping, the dynamic system would have eigenvalues that are
   A. complex
   B. real
   C. pure imaginary
   D. mixture of real and imaginary

vii. With general damping, the dynamic system would have eigenvectors that are
   A. complex
   B. real
   C. pure imaginary
   D. mixture of real and imaginary

viii. With proportional damping, the dynamic system would have eigenvectors that are
A. complex with a purely real or imaginary part
B. real
C. pure imaginary
D. mixture of real and imaginary

ix. A shaft would have the following boundary condition applicable at an intermediate support:
A. the slope of the shaft changes
B. the shear force remains the same
C. the bending moment remains the same
D. the displacement changes

x. In transverse vibrations, the antiresonance occurs due to
A. special resonance conditions
B. instability in the system
C. cancellation of two neighboring modes
D. free vibration phenomenon

xi. For equations of motion $\begin{bmatrix} 1 & 0 \\ 0 & 0.01 \end{bmatrix} \begin{Bmatrix} \ddot{x} \\ \ddot{\varphi}_y \end{Bmatrix} + \begin{bmatrix} 500 & 50 \\ 50 & 5 \end{bmatrix} \begin{Bmatrix} x \\ \varphi_y \end{Bmatrix} = \begin{Bmatrix} 0 \\ 0 \end{Bmatrix}$ after eliminating the rotational DOF by using the static reduction method, the reduced mass and stiffness terms (in corresponding) SI units would be
A. 1.0001, 500.05
B. 0.9999, 490.05
C. 1.0001, 490.05
D. 0.9999, 500.05

xii. In FE formulation for the transverse vibration, one end of the flexible shaft has a linear spring of stiffness k in the transverse direction, the corresponding boundary condition will be (herein, S_y is the shear force, k in the stiffness of linear spring, y is the linear displacement of the shaft, φ_x is the angular displace of the shaft, and l is the length of the shaft element)
A. $S_y = -ky$
B. $S_y = -k(0.5\varphi_x l)$
C. $S_y = -k(y + 0.5\varphi_x l)$
D. $S_y = -k(y - 0.5\varphi_x l)$

xiii. In FE formulation of the transverse vibration of the shaft, the derivative of the shape function N_4 with respect to spatial coordinate gives a value at $z=0$, as
A. 0
B. 1
C. –1
D. 2

xiv. For FE formulation of the transverse vibration of beam, the stiffness matrix term $EI\int_0^l N_2''(z)N_4''(z)dz$ will have one of the following forms

A. $\frac{2EI}{l}$

B. $-\frac{6EI}{l^2}$

C. $\frac{6EI}{l^2}$

D. $-\frac{12EI}{l^3}$

xv. In the transverse vibration of a rotor system for an intermediate pinned support, the following boundary conditions prevails

A. the displacement and bending moment are zero
B. only the displacement is zero
C. the bending moment is zero but not the shear force
D. the bending moment is not zero but the slope is zero

xvi. For a rotor system, the y-axis is chosen positive in the vertical downward direction and x-axis in the horizontal right direction, if the disk plane is in the x-y plane and the rotor is rotating count-clockwise direction, then the unbalance forces in the x and y directions will be related as

A. $f_y = jf_x$
B. $f_y = -jf_x$
C. $f_y = j^4 f_x$
D. $f_y = j^2 f_x$

xvii. Obtain Rayleigh's damping coefficients, considering the first two natural frequencies (30 and 120 rad/s), and for damping ratios of 0.01 and 0.015 for the first and second natural modes, respectively.

A. 0.40 and 0.0002
B. 0.30 and 0.0003
C. 0.14 and 0.0015
D. 0.45 and 0.0003

xviii. For transverse vibration of a shaft with the fixed–pinned end condition, if only a single element is considered, then the stiffness matrix after the application of boundary conditions will reduce to

$$\mathbf{K} = \begin{bmatrix} 120 & -60 & -120 & -60 \\ -60 & 40 & 60 & 20 \\ -120 & 60 & 120 & 60 \\ -60 & 20 & 60 & 40 \end{bmatrix} \quad \text{with} \quad \eta = \begin{Bmatrix} v_1 \\ \varphi_{x1} \\ v_2 \\ \varphi_{x2} \end{Bmatrix}$$

A. $\mathbf{K}_{red} = [40]$

B. $\mathbf{K}_{red} = \begin{bmatrix} 120 & 60 \\ 60 & 40 \end{bmatrix}$

C. $\mathbf{K}_{red} = \begin{bmatrix} 120 & -60 \\ -60 & 40 \end{bmatrix}$

D. $\mathbf{K}_{red} = [120]$

xix. If two cantilever shafts, facing each other, are joined by a pin at their free ends and if each cantilever shaft is modeled as a single element for transverse vibration in a single plane, then the reduced form of equations of motion after the application of boundary conditions will have the stiffness matrix of the following size

A. 1×1
B. 2×2
C. 3×3
D. 4×4

xx. If a simply supported shaft has two disks very close to its supports so that its motion is not restricted in the transverse tilt and if the system is modeled by a single element, then the reduced form of the stiffness matrix after the application of boundary conditions will have the following matrix size

A. 1×1
B. 2×2
C. 3×3
D. 4×4

xxi. If two cantilever shafts, facing each other, are joined by a linear spring at their free ends and if each cantilever shaft is modeled as a single element for transverse vibration, then the reduced form of equations of motion after the application of boundary conditions will have the stiffness matrix of the following size

A. 6×6
B. 2×2
C. 3×3
D. 4×4

Appendix 9.1 Comparison of Transverse Vibration Continuous and FE Analyses in the y-z Plane with That of the z-x Plane

Description of Equations	Equations in the *y–z* Plane	Equations in the *z–x* Plane
Displacement field	$v_x = 0;\ v_y = v(z,t);\ v_z = y\varphi_x(z,t) = -yv'(z,t)$	$u_x = u(z,t);\ u_y = 0;\ u_z = -x\varphi_y(z,t) = -xu'(z,t)$
Angular displacement	$\varphi_x = -\dfrac{\partial v}{\partial z} = -v'$	$\varphi_y = \dfrac{\partial u}{\partial z} = u'$
Bending moment and shear force (refer to Figure 11.26)	$M_{yz} = EI_{xx}v^{(e)}_{,zz}$ and $f_{sy} = \left(EI_{xx}v_{,zz}\right)_{,z}$	$M_{zx} = EI_{yy}u^{(e)}_{,zz}$ and $f_{sx} = \left(EI_{yy}u_{,zz}\right)_{,z}$
Equation of motion	$EI_{xx}v''''(z,t) + \rho A\ddot{v}(z,t) = f_y(z,t) + f_{y0}(t)\delta^*(z-z_0)$	$EI_{yy}u''''(z,t) + \rho A\ddot{u}(z,t) = f_x(z,t) + f_{x0}(t)\delta^*(z-z_0)$
Boundary conditions	$EI_{xx}v''(z,t)\delta v'(z,t)\Big\rvert_0^L = 0$ and $EI_{xx}v'''(z,t)\delta v(z,t)\Big\rvert_0^L = 0$	$EI_{yy}u''(z,t)\delta u'(z,t)\Big\rvert_0^L = 0$ and $EI_{yy}u'''(z,t)\delta u(z,t)\Big\rvert_0^L = 0$
Shape functions	$\mathbf{N}(z) = \left[\left(1-3\dfrac{z^2}{l^2}+2\dfrac{z^3}{l^3}\right)\ \ -\left(z-2\dfrac{z^2}{l}+\dfrac{z^3}{l^2}\right)\ \ \left(3\dfrac{z^2}{l^2}-2\dfrac{z^3}{l^3}\right)\ \ -\left(-\dfrac{z^2}{l}+\dfrac{z^3}{l^2}\right)\right]$	$\mathbf{N}(z) = \left[\left(1-3\dfrac{z^2}{l^2}+2\dfrac{z^3}{l^3}\right)\ \ \left(z-2\dfrac{z^2}{l}+\dfrac{z^3}{l^2}\right)\ \ \left(3\dfrac{z^2}{l^2}-2\dfrac{z^3}{l^3}\right)\ \ \left(-\dfrac{z^2}{l}+\dfrac{z^3}{l^2}\right)\right]$
Elemental mass matrix	$\mathbf{M}^{(e)} = \dfrac{\rho Al}{420}\begin{bmatrix} 156 & -22l & 54 & 13l \\ & 4l^2 & -13l & -3l^2 \\ & & 156 & 22l \\ sym & & & 4l^2 \end{bmatrix}$	$\mathbf{M}^{(e)} = \dfrac{\rho Al}{420}\begin{bmatrix} 156 & 22l & 54 & -13l \\ & 4l^2 & 13l & -3l^2 \\ & & 156 & -22l \\ sym & & & 4l^2 \end{bmatrix}$
Elemental stiffness matrix	$\mathbf{K}^{(e)} = \dfrac{EI_{xx}}{l^3}\begin{bmatrix} 12 & -6l & -12 & -6l \\ & 4l^2 & 6l & 2l^2 \\ & & 12 & 6l \\ sym & & & 4l^2 \end{bmatrix}$	$\mathbf{K}^{(e)} = \dfrac{EI_{xx}}{l^3}\begin{bmatrix} 12 & 6l & -12 & 6l \\ & 4l^2 & -6l & 2l^2 \\ & & 12 & -6l \\ sym & & & 4l^2 \end{bmatrix}$
Elemental displacement vector	$\eta^{(ne)}(t) = \{\ v_1\ \ \varphi_{x1}\ \ v_2\ \ \varphi_{x2}\ \}^T$	$\eta^{(ne)}(t) = \{\ u_1\ \ \varphi_{y1}\ \ u_2\ \ \varphi_{y2}\ \}^T$
Elemental reaction force vector	$\left\{\mathbf{f}^{(ne)}_{yR}(t)\right\}^T = \left[\ -S_{y1}\ \ -M_{yz1}\ \ S_{y2}\ \ M_{yz2}\ \right]$ with $S_y = -f_{sy} = -\left(EI_{xx}v_{,zz}\right)_{,z}$	$\left\{\mathbf{f}^{(ne)}_{xR}(t)\right\}^T = \left[\ -S_{x1}\ \ -M_{zx1}\ \ S_{x2}\ \ M_{zx2}\ \right]$ with $S_x = -f_{sx} = -\left(EI_{yy}u_{,zz}\right)_{,z}$

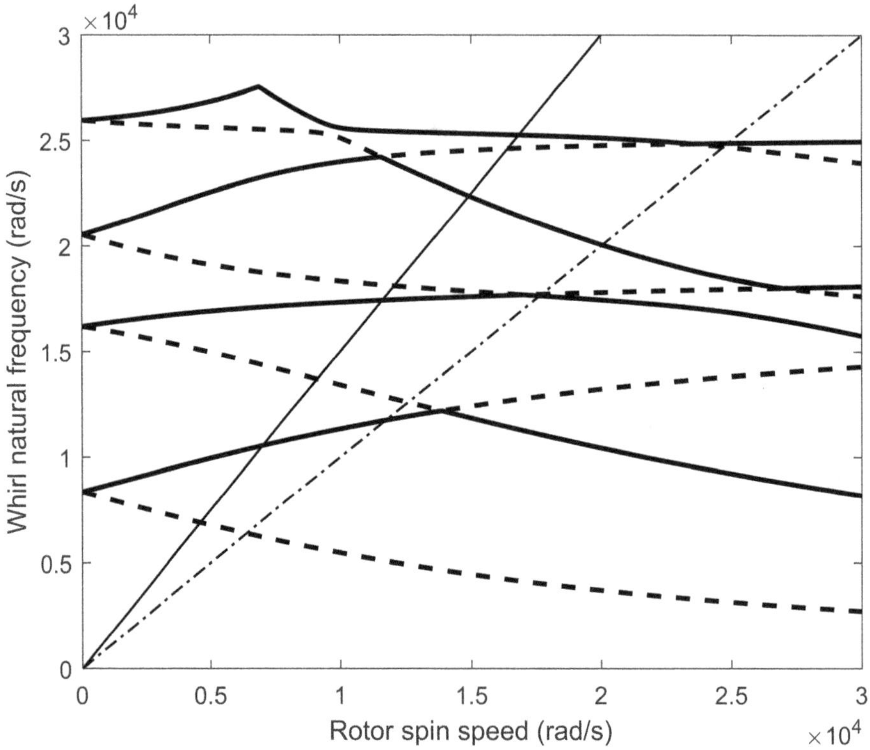

FIGURE 9.51 Campbell diagram showing critical speeds and curve veering phenomenon.

ANSWERS TO SELECTED EXERCISE PROBLEMS

(Results shows for minimum number of elements and with shaft mass unless otherwise stated)

9.1 $\omega_{nf_1} = 27.8$ rad/s; $\omega_{nf_2} = 241.2$ rad/s; $\omega_{nf_3} = 966.4$ rad/s;

9.2 $\omega_{nf_1} = 437.4$ rad/s; $\omega_{nf_2} = 2699.3$ rad/s; $\omega_{nf_3} = 9927.9$ rad/s; $\omega_{nf_4} = 25{,}526.0$ rad/s.

9.3 $\omega_{nf_1} = 28.4$ rad/s; $\omega_{nf_2} = 736.4$ rad/s; $\omega_{nf_3} = 2675.7$ rad/s; $\omega_{nf_4} = 10{,}691.8$ rad/s.

9.4 $\omega_{nf_1} = 27.8$ rad/s; $\omega_{nf_2} = 241.2$ rad/s; $\omega_{nf_3} = 966.4$ rad/s

9.5 $\omega_{nf_1} = 84.4$ rad/s; $\omega_{nf_2} = 330.6$ rad/s; $\omega_{nf_3} = 7413.5$ rad/s; $\omega_{nf_4} = 26{,}629.0$ rad/s.

9.6 With shaft mass (two elements)
$\omega_{nf_1} = 436.1$ rad/s; $\omega_{nf_2} = 2643.5$ rad/s; $\omega_{nf_3} = 8427.8$ rad/s; $\omega_{nf_4} = 16{,}446.3$ rad/s.
Without shaft mass
$\omega_{nf_1} = 738.8$ rad/s; $\omega_{nf_2} = 5113.8$ rad/s; $\omega_{nf_3} = 15{,}662.4$ rad/s; $\omega_{nf_4} = 26{,}492.6$ rad/s.

9.7 Part I: $\omega_{nf_1} = 25.3$ rad/s; $\omega_{nf_2} = 234.9$ rad/s; $\omega_{nf_3} = 444.9$ rad/s; $\omega_{nf_4} = 1667.9$ rad/s.
Part II: $\omega_{nf_1} = 24.0$ rad/s; $\omega_{nf_2} = 146.0$ rad/s; $\omega_{nf_3} = 297.7$ rad/s; $\omega_{nf_4} = 792.2$ rad/s.

9.8 For $D_1 D_2 = 0.75$.
With FEM with shaft mass, natural frequencies are (three elements)
$\omega_{nf_1} = 199.9$ rad/s; $\omega_{nf_2} = 489.4$ rad/s; $\omega_{nf_3} = 779.8$ rad/s; $\omega_{nf_4} = 1334.0$ rad/s.
With FEM without shaft mass, natural frequencies are (three elements)
$\omega_{nf_1} = 200.1$ rad/s; $\omega_{nf_2} = 489.4$ rad/s; $\omega_{nf_3} = 781.9$ rad/s; $\omega_{nf_4} = 1337.9$ rad/s

9.10 $\omega_{nf_1} = 3.9$ rad/s; $\omega_{nf_2} = 28.7$ rad/s; $\omega_{nf_3} = 74.0$ rad/s; $\omega_{nf_4} = 182.0$ rad/s.

9.11 $\omega_{nf_1} = 916.5$ rad/s; $\omega_{nf_2} = 5316.9$ rad/s; $\omega_{nf_3} = 9030.2$ rad/s.

9.12 $b = c = 0.7$ m (number of elements 3)

$\omega_{nf1} = 16.2$ rad/s; $\omega_{nf2} = 37.7$ rad/s; $\omega_{nf3} = 204.8$ rad/s; $\omega_{nf4} = 248.6$ rad/s; $\omega_{nf5} = 600.6$ rad/s; $\omega_{nf6} = 1020.9$ rad/s.

9.13 First backward 6400 rpm and first forward 11,800 rpm, second backward 12,600 rpm and second forward 17,600 rpm (refer Figure 9.51).

9.14 $\omega_{nf1} = 167.79$ rad/s; $\omega_{nf2} = 316.12$ rad/s

9.15 $\omega_{nf1} = 59.3$ rad/s; $\omega_{nf2} = 112.5$ rad/s; $\omega_{nf3} = 2.187.1$ rad/s; $\omega_{nf4} = 2196.7$ rad/s; $\omega_{nf5} = 9167.9$ rad/s.

9.16 $\omega_{nf1} = 89.4$ Hz (562.0 rad/s)

9.17 $\omega_{nf1} = 4.4$ rad/s; $\omega_{nf2} = 79.3$ rad/s; $\omega_{nf3} = 106.6$ rad/s; $\omega_{nf4} = 285.9$ rad/s

9.18 Not applicable

9.19 Not applicable

9.20 $\omega_{nf1} = 5.7$ rad/s; $\omega_{nf2} = 9.8$ rad/s; $\omega_{nf3} = 2866.6$ rad/s; $\omega_{nf4} = 8279.9$ rad/s

9.21 $\omega_{nf1} = 83.06$ rad/s; $\omega_{nf2} = 396.03$ rad/s

9.22 $\omega_{nf1} = 91.65$ rad/s; $\omega_{nf2} = 903.02$ rad/s

9.23 $\omega_{nf1} = 18{,}074.4$ rad/s; $\omega_{nf2} = 71{,}720.4$ rad/s.

9.24 $\omega_{nf1} = 224.95$ rad/s; $\omega_{nf2} = 768.77$ rad/s; $\omega_{nf3} = 2321.30$ rad/s.

9.25

(i) A	(ii) C	(iii) B	(iv) A	(v) B	(vi) A
(vii) A	(viii) A	(ix) C	(x) C	(xi) C	(xii) A
(xiii) A	(xiv) A	(xv) B	(xvi) A	(xvii) A	(xviii) A
(xix) C	(xx) B	(xxi) D			

REFERENCES

Chen, L.W. and Ku, D.M., 1991, Finite element analysis of natural whirl speeds of rotating shafts, *Computers & Structures*, *40*, 741–747.

Chen, L.W. and Peng, W.K., 1997, Stability analyses of a Timoshenko shaft with dissimilar lateral moments of inertia, *Journal of Sound and Vibration*, *207(1)*, 33–46.

Cleghorn, W.L. and Tabarrok, B. 1992, Finite element formulation of a tapered Timoshenko beam for free lateral vibration analysis, *Journal of Sound and Vibration*, *152*(*3*), 461–470.

Clough, R.W. and Penzien, J., 1993, *Dynamics of Structures*. New York, NY: McGraw-Hill.

Cook, D., Malkus, D.W., and Plesha, M.E., 1989, *Concepts and Applications of Finite Element Analysis*, 3rd ed. New York, NY: John Wiley & Sons.

Dharmaraju, N., Tiwari, R., and Talukdar, S., 2005, Development of a novel hybrid reduction scheme for identification of an open crack model in a beam, *Mechanical Systems and Signal Processing*, *19*, 633–657.

Dimentberg, F.M., 1961, *Flexural Vibrations of Rotating Shafts*. London: Butterworths.

Edney, S.L., Fox, C.H.J., and Williams, E.J., 1990, Tapered Timoshenko finite elements for rotor dynamics analysis, *Journal of Sound and Vibration*, *137*, 463–481.

Friswell, M.I. and Mottershead, J.E., 1995, *Finite Element Model Updating in Structural Dynamics*. New York, NY: Kluwer Academic Publishers.

Guyan, R.J., 1965, Reduction of stiffness and mass matrices, *AIAA Journal*, *3*(*2*), 280.

Hou, Y.C., Tseng, C.H., and Ling, S.F. 1996, A new high-order non-uniform Timoshenko beam finite element on variable two-parameter foundations for vibration analysis, *Journal of Sound and Vibration*, *191*(*1*), 91–106.

Kane, K. and Torby, B.J., 1991, The extended modal reduction method applied to rotor dynamic problems, *Transactions of American Society of Mechanical Engineering, Journal of Vibration and Acoustics*, *113*(*1*), 79–84.

Karthikeyan, M., Tiwari, R., and Talukdar, S., 2007, Development of a technique to locate and quantify a beam crack based on modal parameters, *Transactions of ASME, Journal of Vibration and Acoustics*, *129*(*3*), 390–395.

Ku, D.M., 1998, Finite element analysis of whirl speeds of rotor-bearing systems with internal damping, *Transactions of the ASME, Journal Mechanical Design*, *102*, 599–619.

McVaugh, J.M. and Nelson, H.D., 1976, The dynamics of rotor bearing systems using finite elements, *Transactions of the ASME, Journal of Engineering for Industry*, *98*, 593–600.

Meirovitch, L., 1986, *Elements of Vibration Analysis*, 2nd ed. New York, NY: McGraw Hill.

Mohiuddin, M.A. and Khulief, Y.A., 1994, Modal characteristics of rotors using a conical shaft finite element, *Computer Methods in Applied Mechanics & Engineering*, *115*, 125–144.

Mohiuddin, M.A. and Khulief, Y.A., 1999, Coupled bending torsional vibration of rotors using finite elements, *Journal of Sound and Vibration*, *223*(2), 297–316.

Nelson, H.D., 1980, A finite rotating shaft element using Timoshenko beam theory, *Transactions of the ASME, Journal of Mechanical Design*, *102*, 793–803.

Nelson, H.D., 1998, Rotordynamic modelling and analysis procedures: A review, *JSME International Journal, Series C*, *41*(*1*), 1–12.

Nelson, H.D. and Zorzi, E.S., 1977, Finite element simulation of rotor bearing systems with internal damping, *Transactions of the ASME, Journal of Engineering for Power*, *99*, 71–76.

Nelson, H.D. and Zorzi, E.S., 1980, The dynamics of rotor bearing systems with axial torque: A finite element approach, *Transactions of the ASME, Journal of Mechanical Engineering Design*, *102*, 158–161.

O'Callahan, J., 1989, A Procedure for an Improved Reduced System (IRS) Model, *Seventh International Modal Analysis Conference*, Las Vegas, NV, January 17–21.

O'Callahan, J., Avitabile, P., and Riemer, R., 1989, System Equivalent Reduction Expansion Process, *7th International Modal Analysis Conference*, Las Vegas, January, 29–37.

Ozguven, H.N. and Ozkan, Z.L., 1984, Whirl speeds and unbalance response of multibearing rotors using finite elements, *Transactions of the ASME, Journal of Vibration, Acoustics, Stress, and Reliability in Design*, *106*, 72–79.

Paz, M., 1984, Dynamic condensation, *AIAA Journal*, *22*(*5*), 724–727.

Prasad, V. and Tiwari, R., 2019, Identification of speed-dependent active magnetic bearing parameters and rotor balancing in high-speed rotor systems, *Journal of Dynamic Systems, Measurement, and Control*, *141*(*4*), 041013.

Rouch, K.E. and Kao, J.S., 1979, A tapered beam finite element for rotor dynamics analysis, *Journal of Sound and Vibration*, *66*, 119–140.

Ruhl, R.L., 1970, Dynamics of Distributed Parameter Rotor Systems: Transfer Matrix & Finite Element Techniques, Ph.D. Dissertation, Cornell University, Ithaca, NY.

Ruhl, R.L. and Booker, J.F., 1972, A finite element model for distributed parameter turborotor systems, *Transactions of the ASME, Journal of Engineering for Industry*, *94*(*1*), 126–132.

Thomson, W.T. and Dahleh, M.D., 1998, *Theory of Vibration with Applications*, 5th Ed. New Delhi, India:Pearson Education, Inc.

Tiwari, R. and Dharmaraju, N., 2006, Development of a condensation scheme for transverse rotational degrees of freedom elimination in identification of beam crack parameters, *Mechanical Systems and Signal Processing*, *20*(*8*), 2148–2170.

Tondl, A., 1965, *Some Problems of Rotor Dynamics*. London: Chapman and Hall, pp. 100–102.

Yardimoglu, B., 2006, Vibration analysis of rotating tapered Timoshenko beams by a new finite element model, *Shock and Vibration*, *13*, 117–126.

10 Transverse Vibrations of Rotor Systems with Higher-Order Effects by the Continuous and Finite Element Methods

In Chapter 5, gyroscopic effects on a rotor with a single disk and massless elastic shaft were discussed in great detail by using quasi-static and dynamic analyses with the help of the analytical approach. Multi-degree-of-freedom (DOF) systems with the gyroscopic effect of thin disks were described while discussing the transfer matrix method in Chapter 8. There we did not consider gyroscopic effects due to thick flexible shafts (i.e., a shaft with distributed mass and stiffness properties). In the previous chapter, we dealt with the transverse vibration analysis of a simple rotor with the analytical (i.e., the continuous) and numerical (finite element) approaches without considering the important higher effects, such as shear deformation, rotary inertia, and gyroscopic effects. In the present chapter, the finite element analysis will be extended initially to a single disk of a simple rotor (i.e., the Euler–Bernoulli beam with distributed properties). Then, it will be extended for more general rotor systems (i.e., the Timoshenko beam model) with the help of a powerful analysis tool of the finite element method (FEM). The Timoshenko beam theory will be used for the development of a governing equation of the continuous system analysis. The finite element formulation will be developed for the spinning Timoshenko beam, which includes higher effects like rotary inertia, shear deformation, and gyroscopic effects. The eigenvalue problem will be developed through the state-space form of the governing equation. With the help of examples, the extraction of modal parameters from the special eigenvalue problem will be explained. The standard Campbell diagram for various cases is obtained, which shows the natural whirl frequency variation with the spin speed of the shaft for asynchronous whirl. This diagram can be used to obtain critical speeds of any complex rotor system.

10.1 GYROSCOPIC EFFECT IN ROTOR SYSTEMS WITH A SINGLE RIGID DISK

In this section, a finite element analysis for a cantilever rotor with gyroscopic effects, as shown in Figure 10.1, will be illustrated. However, the procedure can be extended to other boundary conditions (simple rotors, rotors with an intermediate support, two-spool rotors, etc.). The shaft has been modeled with consistent mass and stiffness matrices. However, only the gyroscopic couple due to the disk is considered, and its effect due to the shaft inertia is neglected at present.

Since we will be considering the gyroscopic couple of the disk alone, we need to consider both plane motions simultaneously. Let us model the rotor as a single element with two nodes.

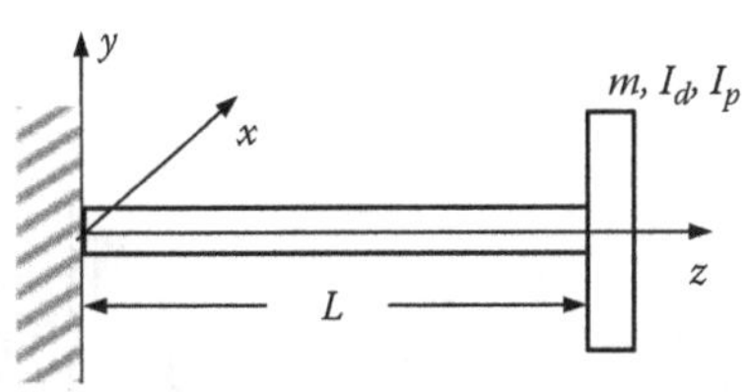

FIGURE 10.1 A cantilever rotor system.

DOI: 10.1201/9781003661559-10

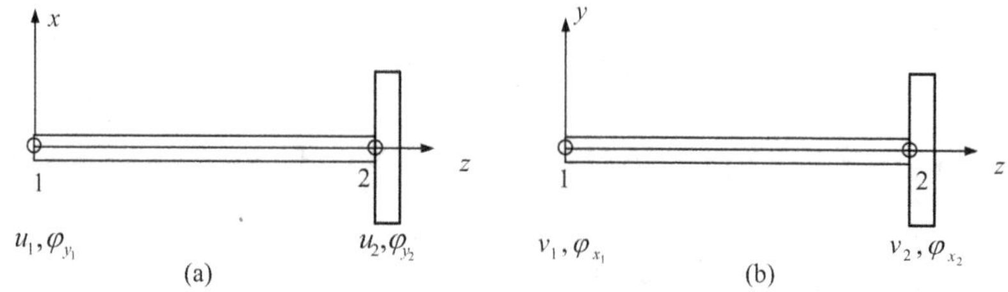

FIGURE 10.2 A rotor element: (a) element in the z–x plane and (b) element in the y–z plane.

Elemental equations of motion in the z–x plane as shown in Figure 10.2a (without gyroscopic effects) can be written as

$$\bar{m}\begin{bmatrix} 156 & 22l & 54 & -13l \\ & 4l^2 & -13l & -3l^2 \\ & & 156+\dfrac{m}{\bar{m}} & -22l \\ sym & & & 4l^2+\dfrac{I_d}{\bar{m}} \end{bmatrix}\begin{Bmatrix} \ddot{u}_1 \\ \ddot{\varphi}_{y1} \\ \ddot{u}_2 \\ \ddot{\varphi}_{y2} \end{Bmatrix} + \frac{EI}{l^3}\begin{bmatrix} 12 & 6l & -12 & 6l \\ & 4l^2 & -6l & 2l^2 \\ & & 12 & -6l \\ sym & & & 4l^2 \end{bmatrix}\begin{Bmatrix} u_1 \\ \varphi_{y1} \\ u_2 \\ \varphi_{y2} \end{Bmatrix} = \begin{Bmatrix} -S_{x1} \\ -M_{xz1} \\ S_{x2} \\ M_{xz2} \end{Bmatrix} \tag{10.1}$$

with $\bar{m} = \dfrac{\rho A l}{420}$.

The elemental equations of motion in the y–z plane as shown in Figure 10.2b (without gyroscopic effects) can be written as follows (terms corresponding to angular displacements and related to moments will have sign changes; refer to Appendix 9.1):

$$\bar{m}\begin{bmatrix} 156 & -22l & 54 & 13l \\ & 4l^2 & 13l & -3l^2 \\ & & 156+\dfrac{m}{\bar{m}} & 22l \\ sym & & & 4l^2+\dfrac{I_d}{\bar{m}} \end{bmatrix}\begin{Bmatrix} \ddot{v}_1 \\ \ddot{\varphi}_{x1} \\ \ddot{v}_2 \\ \ddot{\varphi}_{x2} \end{Bmatrix} + \frac{EI}{l^3}\begin{bmatrix} 12 & -6l & -12 & -6l \\ & 4l^2 & 6l & 2l^2 \\ & & 12 & 6l \\ sym & & & 4l^2 \end{bmatrix}\begin{Bmatrix} v_1 \\ \varphi_{x1} \\ v_2 \\ \varphi_{x2} \end{Bmatrix} = \begin{Bmatrix} -S_{y1} \\ -M_{yz1} \\ S_{y2} \\ M_{yz2} \end{Bmatrix} \tag{10.2}$$

In Chapter 5, the gyroscopic couple effect of the disk alone was described and the gyroscopic matrix is given as (please note that the disk is at node 2)

$$-\omega \mathbf{G}\eta_2 = -\omega\begin{bmatrix} 0 & 0 & 0 & 0 \\ 0 & 0 & 0 & I_p \\ 0 & 0 & 0 & 0 \\ 0 & -I_p & 0 & 0 \end{bmatrix}\begin{Bmatrix} \dot{u}_2 \\ \dot{\varphi}_{y2} \\ \dot{v}_2 \\ \dot{\varphi}_{x2} \end{Bmatrix}. \tag{10.3}$$

It could be seen that the gyroscopic effect leads to coupling of motions in the y–z and z–x planes. It can be expanded in the following form to accommodate both the nodes of the shaft element (1), as

$$-\omega \mathbf{G} \eta^{(1)} = -\omega \begin{bmatrix} 0 & 0 & 0 & 0 & 0 & 0 & 0 & 0 \\ 0 & 0 & 0 & 0 & 0 & 0 & 0 & 0 \\ 0 & 0 & 0 & 0 & 0 & 0 & 0 & 0 \\ 0 & 0 & 0 & 0 & 0 & 0 & 0 & 0 \\ 0 & 0 & 0 & 0 & 0 & 0 & 0 & 0 \\ 0 & 0 & 0 & 0 & 0 & 0 & 0 & I_p \\ 0 & 0 & 0 & 0 & 0 & 0 & 0 & 0 \\ 0 & 0 & 0 & 0 & 0 & -I_p & 0 & 0 \end{bmatrix} \begin{Bmatrix} \dot{u}_1 \\ \dot{\varphi}_{y1} \\ \dot{v}_1 \\ \dot{\varphi}_{x1} \\ \dot{u}_2 \\ \dot{\varphi}_{y2} \\ \dot{v}_2 \\ \dot{\varphi}_{x2} \end{Bmatrix} \tag{10.4}$$

In Equations 10.1 and 10.2, now the gyroscopic couple effect (Equation 10.4) will be added and equations of motion of the element, in both planes, can be written as

$$\bar{m} \begin{bmatrix} 156 & 22l & 0 & 0 & 54 & -13l & 0 & 0 \\ & 4l & 0 & 0 & 13l & -3l^2 & 0 & 0 \\ & & 156 & -22l & 0 & 0 & 54 & 13l \\ & & & 4l^2 & 0 & 0 & -13l & -3l^2 \\ & & & & \left(156+\frac{m}{m}\right) & -22l & 0 & 0 \\ & & & & & \left(4l^2+\frac{I_d}{\bar{m}}\right) & 0 & 0 \\ & & & & & & \left(156+\frac{m}{\bar{m}}\right) & 22l \\ sym & & & & & & & \left(4l^2+\frac{I_d}{\bar{m}}\right) \end{bmatrix} \begin{Bmatrix} \ddot{u}_1 \\ \ddot{\varphi}_{y1} \\ \ddot{v}_1 \\ \ddot{\varphi}_{x1} \\ \ddot{u}_2 \\ \ddot{\varphi}_{y2} \\ \ddot{v}_2 \\ \ddot{\varphi}_{x2} \end{Bmatrix}$$

$$-\omega \begin{bmatrix} 0 & 0 & 0 & 0 & 0 & 0 & 0 & 0 \\ 0 & 0 & 0 & 0 & 0 & 0 & 0 & 0 \\ 0 & 0 & 0 & 0 & 0 & 0 & 0 & 0 \\ 0 & 0 & 0 & 0 & 0 & 0 & 0 & 0 \\ 0 & 0 & 0 & 0 & 0 & 0 & 0 & 0 \\ 0 & 0 & 0 & 0 & 0 & 0 & 0 & I_p \\ 0 & 0 & 0 & 0 & 0 & 0 & 0 & 0 \\ 0 & 0 & 0 & 0 & 0 & -I_p & 0 & 0 \end{bmatrix} \begin{Bmatrix} \dot{u}_1 \\ \dot{\varphi}_{y1} \\ \dot{v}_1 \\ \dot{\varphi}_{x1} \\ \dot{u}_2 \\ \dot{\varphi}_{y2} \\ \dot{v}_2 \\ \dot{\varphi}_{x2} \end{Bmatrix}$$

$$+\frac{EI}{l^3} \begin{bmatrix} 12 & 6l & 0 & 0 & -12 & 6l & 0 & 0 \\ & 4l^2 & 0 & 0 & -6l & 2l^2 & 0 & 0 \\ & & 12 & -6l & 0 & 0 & -12 & -6l \\ & & & 4l^2 & 0 & 0 & 6l & 2l^2 \\ & & & & 12 & -6l & 0 & 0 \\ & & & & & 4l^2 & 0 & 0 \\ & & & & & & 12 & 6l \\ sym & & & & & & & 4l^2 \end{bmatrix} \begin{Bmatrix} u_1 \\ \varphi_{y1} \\ v_1 \\ \varphi_{x1} \\ u_2 \\ \varphi_{y1} \\ v_2 \\ \varphi_{x2} \end{Bmatrix} = \begin{Bmatrix} 0 \\ 0 \\ 0 \\ 0 \\ 0 \\ 0 \\ 0 \\ 0 \end{Bmatrix} + \begin{Bmatrix} -S_{x1} \\ -M_{xz1} \\ -S_{y1} \\ -M_{yz1} \\ S_{x2} \\ M_{xz2} \\ S_{y2} \\ M_{yz2} \end{Bmatrix} \tag{10.5}$$

The above formulation could alternatively be obtained by a more formal approach by using the additional kinetic energy term of the disk developed in Chapter 5 due to gyroscopic moments in the finite element formulations of Euler–Bernoulli beams as described in Chapter 9, which will lead to similar results. This will be illustrated while developing the finite element formulation with the Timoshenko beam model (in which the shear deformation, rotary inertia, and gyroscopic effects of shafts will also be considered) in subsequent sections. For the present illustration of a cantilever rotor, the boundary conditions are given as

$$u_1 = v_1 = \varphi_{y1} = \varphi_{x1} = 0 \quad \text{and} \quad M_{xz2} = S_{x2} = M_{yz2} = S_{y2} = 0 \tag{10.6}$$

After application of the boundary conditions (Equation 10.6) in Equation 10.5, on elimination of the first four rows and columns, the equations of motion reduce to

$$\bar{m}\begin{bmatrix} 156+\frac{m}{\bar{m}} & -22l & 0 & 0 \\ & 4+\frac{I_d}{\bar{m}} & 0 & 0 \\ & & 156+\frac{m}{\bar{m}} & 22l \\ sym & & & 4\frac{I_d}{\bar{m}} \end{bmatrix}\begin{Bmatrix} \ddot{u}_2 \\ \ddot{\varphi}_{y2} \\ \ddot{v}_2 \\ \ddot{\varphi}_{x2} \end{Bmatrix} - \omega\begin{bmatrix} 0 & 0 & 0 & 0 \\ 0 & 0 & 0 & I_p \\ 0 & 0 & 0 & 0 \\ 0 & -I_p & 0 & 0 \end{bmatrix}\begin{Bmatrix} \dot{u}_2 \\ \dot{\varphi}_{y2} \\ \dot{v}_2 \\ \dot{\varphi}_{x2} \end{Bmatrix}$$

$$+\frac{EI}{l^3}\begin{bmatrix} 12 & -6l & 0 & 0 \\ & 4l^2 & 0 & 0 \\ & & 12 & 6l \\ sym & & & 4l^2 \end{bmatrix}\begin{Bmatrix} u_2 \\ \varphi_{y2} \\ v_2 \\ \varphi_{x2} \end{Bmatrix} = \begin{Bmatrix} 0 \\ 0 \\ 0 \\ 0 \end{Bmatrix} \tag{10.7}$$

which is an equation of motion of the following standard form:

$$\mathbf{M}\ddot{\eta}^R - \omega\mathbf{G}\dot{\eta}^R + \mathbf{K}\eta^R = \mathbf{0} \tag{10.8}$$

where η^R is the reduced displacement vector after the application of boundary conditions. The solution of the above equation is described in Chapter 5 using the characteristic equation. However, herein now a more powerful method based on an eigenvalue problem will be described. For the asynchronous whirl, the eigenvalue problem of the form of Equation 10.8 is not a standard one. For obtaining the standard form of the eigenvalue, Equation 10.8 must be expressed in state-space form. To illustrate the transformation, let us consider the following simple example.

Example 10.1

Obtain the state-space form of the following standard governing equation for vibrations of a spring-mass-damper system. Then, obtain the following frequency equation:

$$m\ddot{x} + c\dot{x} + kx = 0 \tag{a}$$

Solution:

Let us express the velocity as

$$\dot{x} = V \tag{b}$$

From Equation (a), we get

$$\dot{V} = -\frac{c}{m}V - \frac{k}{m}x \tag{c}$$

Equations (b) and (c) can be combined as

$$\begin{Bmatrix} \dot{x} \\ \dot{V} \end{Bmatrix} = \begin{bmatrix} 0 & 1 \\ -k/m & -c/m \end{bmatrix} \begin{Bmatrix} x \\ V \end{Bmatrix} \tag{d}$$

which can be written in the standard state-space form as

$$\dot{\mathbf{h}} = \mathbf{D}\mathbf{h} \tag{e}$$

with

$$\dot{\mathbf{h}} = \begin{Bmatrix} \dot{x} \\ \dot{V} \end{Bmatrix}; \quad \mathbf{D} = \begin{bmatrix} 0 & 1 \\ -k/m & -c/m \end{bmatrix}; \quad \mathbf{h} = \begin{Bmatrix} x \\ V \end{Bmatrix}$$

The solution of the above equation can be taken as

$$\mathbf{h} = \mathbf{h}_0 e^{\lambda t} \tag{f}$$

so that

$$\dot{\mathbf{h}} = \lambda \mathbf{h}_0 e^{\lambda t}$$

where $\mathbf{h}_0$ is the complex amplitude and λ is the eigenvalue. On substituting Equation (f) into Equation (e), we get

$$\lambda \mathbf{h} = \mathbf{D}\mathbf{h} \tag{g}$$

or

$$(\mathbf{D} - \lambda \mathbf{I})\mathbf{h} = \mathbf{0}$$

where **I** is the identity matrix. This is a standard form of eigenvalue problem, in which the matrix **D** will be used to obtain eigenvalues and eigenvectors. The frequency equation can be obtained as

$$|\mathbf{D} - \lambda \mathbf{I}| = 0 \Rightarrow \left| \begin{bmatrix} 0 & 1 \\ -k/m & -c/m \end{bmatrix} - \lambda \begin{bmatrix} 1 & 0 \\ 0 & 1 \end{bmatrix} \right| = 0 \Rightarrow \begin{vmatrix} -\lambda & 1 \\ -k/m & (-c/m - \lambda) \end{vmatrix} = 0$$

or

$$\left(\frac{c}{m}\lambda + \lambda^2\right) + \left(\frac{k}{m}\right) = 0 \Rightarrow \lambda^2 + \frac{c}{m}\lambda + \frac{k}{m} = 0 \tag{h}$$

which is a familiar frequency equation for a single-DOF spring-mass-damper system (refer to Chapter 2). In fact, the matrix **D** can be directly used for obtaining eigenvalues and eigenvectors in multi-DOF systems.

Now coming back to the conversion of Equation 10.8 to the state-space form, let

$$\dot{\eta} = \mathbf{V} \tag{10.9}$$

so that Equation 10.8 takes the following form:

$$\dot{\mathbf{V}} = \omega\mathbf{M}^{-1}\mathbf{G}\mathbf{V} - \mathbf{M}^{-1}\mathbf{K}\eta \tag{10.10}$$

Combining the above two equations, we have

$$\dot{\mathbf{h}} = \mathbf{D}\mathbf{h} \tag{10.11}$$

with

$$\mathbf{D} = \begin{bmatrix} \mathbf{0} & \mathbf{I} \\ -\mathbf{M}^{-1}\mathbf{K} & \omega\mathbf{M}^{-1}\mathbf{G} \end{bmatrix}; \quad \mathbf{h} = \begin{Bmatrix} \eta \\ \mathbf{V} \end{Bmatrix} \tag{10.12}$$

Let us assume the solution of Equation 10.11 has the following form:

$$\mathbf{h} = \mathbf{h}_0 e^{\lambda t} \tag{10.13}$$

where λ is the eigenvalue. On substituting Equation 10.13 into Equation 10.11, we get

$$\lambda\mathbf{h}_0 = \mathbf{D}\mathbf{h}_0 \quad \text{or} \quad (\mathbf{D} - \lambda\mathbf{I})\mathbf{h}_0 = \mathbf{0} \tag{10.14}$$

which is a standard eigenvalue problem. Eigenvalues of Equation 10.14 appear as pure imaginary conjugate pairs with magnitudes equal to natural whirl frequencies. This is because no damping is considered in the system at present, which is related to the real part of the eigenvalue. For the case when the gyroscopic effect is not present, the eigenvalue will be of the form $\pm j\alpha$ and $\pm j\alpha$. However, with the gyroscopic effect, it takes the form $\pm j\alpha^+$ and $\pm j\alpha^-$, where $\alpha^+ > \alpha > \alpha^-$ and they correspond to the forward and backward whirls, respectively. A detailed illustration of the solution of the eigenvalue problem will be given in the next example, when the FEM will be applied to rotors with gyroscopic effects.

Example 10.2

Obtain the forward and backward synchronous transverse critical speeds for a general motion of a rotor as shown in Figure 10.3. The rotor is assumed to have a fixed support at one end. Take the mass of the thin disk as $m=1$ kg with a radius of 3.0 cm. The shaft length, L, and diameter, d, are 0.2 and 0.01 m, respectively. Take the shaft's Young's modulus $E=2.1\times10^{11}$ N/m². Using the FEM and considering the mass of the shaft (without rotary inertia and gyroscopic effects) with material density $\rho=7800$ kg/m³, obtain the first two forward and backward synchronous bending critical speeds by drawing the Campbell diagram.

Solution:

On taking a single element (this is for illustration, but more elements up to a certain number are expected to give better results) for the present problem, we have the following data:

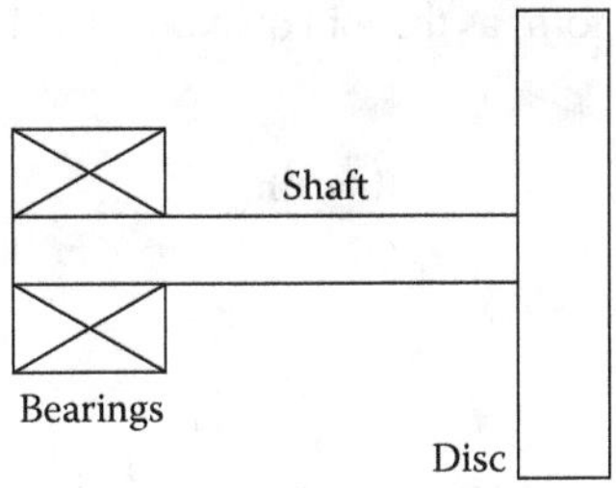

FIGURE 10.3 A cantilever rotor.

$$m = 1\text{ kg};\quad I_p = \frac{1}{2}mr^2 = 4.50\times10^{-4}\text{ kg-m}^2;$$

$$I_d = \frac{1}{4}mr^2 = 2.25\times10^{-4}\text{ kg-m}^2$$

$$E = 2.1\times10^{11}\text{ N/m}^2,\quad \rho = 7800\text{ kg/m}^3;$$

$$l = L = 0.2\text{ m};\quad d = 0.15\text{ m};$$

$$A = \frac{\pi}{4}d^2 = \frac{\pi}{4}0.01^2 = 7.854\times10^{-5}\text{ m}^2;$$

$$I = \frac{\pi}{64}d^4 = \frac{\pi}{64}0.01^4 = 4.9087\times10^{-10}\text{ m}^4$$

$$\frac{\rho A l}{420} = \frac{7800\times7.854\times10^{-5}\times0.2}{420} = 2.9172\times10^{-4}\text{ kg}$$

$$\frac{EI}{l^3} = \frac{2.1\times10^{11}\times4.9087\times10^{-10}}{0.2^3} = 1.2885\times10^4\text{ N/m}$$

From Equation 10.7, we have

$$10^{-2}\begin{bmatrix} 104.5508 & -0.1284 & 0 & 0 \\ & 0.0272 & 0 & 0 \\ & & 104.5508 & 0.1284 \\ sym & & & 0.0272 \end{bmatrix}\begin{Bmatrix} \ddot{u}_2 \\ \ddot{\varphi}_{y2} \\ \ddot{v}_2 \\ \ddot{\varphi}_{x2} \end{Bmatrix}$$

$$-\omega\times10^{-3}\begin{bmatrix} 0 & 0 & 0 & 0 \\ 0 & 0 & 0 & 0.4500 \\ 0 & 0 & 0 & 0 \\ 0 & -0.4500 & 0 & 0 \end{bmatrix}\begin{Bmatrix} \dot{u}_2 \\ \dot{\varphi}_{y2} \\ \dot{v}_2 \\ \dot{\varphi}_{x2} \end{Bmatrix} \qquad \text{(a)}$$

$$+10^5\begin{bmatrix} 1.5463 & -0.1546 & 0 & 0 \\ & 0.0206 & 0 & 0 \\ & & 1.5463 & 0.1546 \\ sym & & & 0.0206 \end{bmatrix}\begin{Bmatrix} u_2 \\ \varphi_{y2} \\ v_2 \\ \varphi_{x2} \end{Bmatrix} = \begin{Bmatrix} 0 \\ 0 \\ 0 \\ 0 \end{Bmatrix}$$

Hence, Equation (a) has the same form as that of Equation 10.8. The state-space form can be written as

$$\dot{\mathbf{h}} = \mathbf{D}\mathbf{h} \tag{b}$$

with

$$\mathbf{D} = \begin{bmatrix} \mathbf{0} & \mathbf{I} \\ -\mathbf{M}^{-1}\mathbf{K} & \omega\mathbf{M}^{-1}\mathbf{G} \end{bmatrix}; \quad \mathbf{h} = \begin{Bmatrix} \eta \\ \dot{\eta} \end{Bmatrix}$$

$$\eta = \begin{Bmatrix} u_2 \\ \varphi_{y2} \\ v_2 \\ \varphi_{x2} \end{Bmatrix}; \quad \dot{\eta} = \begin{Bmatrix} \dot{u}_2 \\ \dot{\varphi}_{y2} \\ \dot{v}_2 \\ \dot{\varphi}_{x2} \end{Bmatrix} \tag{c}$$

$$\mathbf{M} = 10^{-2} \begin{bmatrix} 104.5508 & -0.1284 & 0 & 0 \\ & 0.0272 & 0 & 0 \\ & & 104.5508 & 0.1284 \\ sym & & & 0.0272 \end{bmatrix} \tag{d}$$

$$\mathbf{G} = 10^{-3} \begin{bmatrix} 0 & 0 & 0 & 0 \\ 0 & 0 & 0 & 0.4500 \\ 0 & 0 & 0 & 0 \\ 0 & -0.4500 & 0 & 0 \end{bmatrix}$$

$$\mathbf{K} = 10^{5} \begin{bmatrix} 1.5463 & -0.1546 & 0 & 0 \\ & 0.0206 & 0 & 0 \\ & & 1.5463 & 0.1546 \\ sym & & & 0.0206 \end{bmatrix} \tag{e}$$

To compare eigenvalues and eigenvectors without and with gyroscopic effects, the results from both analyses are provided. Table 10.1 lists the eigenvalues and eigenvectors corresponding to one element eigen analysis with 1 rad/s (~10 rpm, i.e., very slow speed for which the gyroscopic effect is expected to be negligibly small) of rotor speed. Similarly, these are provided in Table 10.2 for 1500 rad/s. In eigenvalues, the real part is related to the damping and the imaginary part is related to the natural whirl frequency of the system. It can be observed that the imaginary part of eigenvalues in serial numbers 1–4 is all the same; 1 and 2 are complex conjugates. Thus, the whirl frequency from serial number 1 is positive and the whirl natural frequency from serial number 2 is exactly the same as serial number 1 but with a negative sign. The negative frequency has no physical significance; hence, it can be omitted for this case. Without the gyroscopic effect, eigenvalues from serial numbers 1 and 3 are the same, and eigenvalues from serial numbers 2 and 4 are complex conjugates of 1 and 3, respectively. Hence, there is only one natural whirl frequency (i.e., 192.68 rad/s, see Table 10.1) from the first four eigenvalues. Similarly, from serial numbers 5–8, there is another natural whirl frequency (i.e., 2757.55 rad/s, see Table 10.1). Hence, without gyroscopic effect case with one element, we could get only two natural frequencies (refer to Table 10.1). Eigenvectors are also appearing in complex conjugates (see Table 10.1). It should be noted that with zero speed the dynamic matrix $[D]$ becomes singular, since all the diagonal elements are zero with $\omega = 0$. Hence, for this case, it will be worthwhile to write the eigenvalue problem as

$$\left(\mathbf{M}^{-1}\mathbf{K} - \lambda\mathbf{I}\right)\eta = \mathbf{0} \tag{f}$$

TABLE 10.1
Eigenvalue and Eigenvectors without a Gyroscopic Couple (i.e., at a Speed of 1 rad/s)

S.N.	Eigenvalue	Eigenvectors
1	$-0.0000+0.1927\times10^3j$	$4.8421\times10^{-4}+0.0000j$
		$0.0036+0.0000j$
		$0.0000-4.8421\times10^{-4}j$
		$0.0000-0.0036j$
		$0.0000+0.0933j$
		$0.0000-0.7009j$
		$0.0933+0.0000j$
		$0.7009+0.0000j$
2	$-0.0000-0.1927\times10^3j$	$4.8421\times10^{-4}+0.0000j$
		$0.0036+0.0000j$
		$0.0000+4.8421\times10^{-4}j$
		$0.0000+0.0036j$
		$0.0000-0.0933j$
		$0.0000\pm0.7009j$
		$0.0933+0.0000j$
		$0.7009+0.0000j$
3	$0.0000+0.1926\times10^3j$	$-4.8423\times10^{-4}+0.0000j$
		$-0.0036+0.0000j$
		$0.0000-4.8423\times10^{-4}j$
		$0.0000-0.0036j$
		$0.0000-0.0933j$
		$0.0000-0.7009j$
		$0.0933+0.0000j$
		$0.7009+0.0000j$
4	$0.0000-0.1926\times10^3j$	$-4.8423\times10^{-4}+0.0000j$
		$-0.0036+0.0000j$
		$0.0000+4.8423\times10^{-4}j$
		$0.0000+0.0036j$
		$0.0000+0.0933j$
		$0.0000+0.7009j$
		$0.0933+0.0000j$
		$0.7009+0.0000j$
5	$-0.0000+2.7567\times10^3j$	$0.0000+1.8793\times10^{-7}j$
		$0.0000-2.5650\ 10^{-4}j$
		$1.8793\times10^{-7}+0.0000j$
		$-2.5650\times10^{-4}-0.0000j$
		$-5.1807\times10^{-4}-0.0000j$
		$0.7071+0.0000j$
		$-0.0000+5.1807\times10^{-4}j$
		$0.0000-0.7071j$
6	$-0.0000-2.7567\times10^3j$	$0.0000-1.8793\times10^{-7}j$
		$0.0000+2.5650\times10^{-4}j$
		$1.8793\times10^{-7}+0.0000j$
		$-2.5650\times10^{-4}-0.0000j$
		$-5.1807\times10^{-4}-0.0000j$

(Continued)

TABLE 10.1 (*Continued*)
Eigenvalue and Eigenvectors without a Gyroscopic Couple (i.e., at a Speed of 1 rad/s)

S.N.	Eigenvalue	Eigenvectors
		$0.7071+0.0000j$
		$-0.0000-5.1807\times 10^{-4}j$
		$0.0000+0.7071j$
7	$0.0000+2.7584\times 10^{3}j$	$0.0000+1.8721\times 10^{-7}j$
		$0.0000-2.5635\times 10^{-4}j$
		$1.8721\times 10^{-7}+0.0000j$
		$-2.5635\times 10^{-4}-0.0000j$
		$-5.1639\times 10^{-4}-0.0000j$
		$0.7071+0.0000j$
		$-0.0000+5.1639\times 10^{-4}j$
		$0.0000-0.7071j$
8	$0.0000-2.7584\times 10^{3}j$	$0.0000-1.8721\times 10^{-7}j$
		$0.0000+2.5635\times 10^{-4}j$
		$1.8721\times 10^{-7}+0.0000j$
		$-2.5635\times 10^{-4}-0.0000j$
		$-5.1639\times 10^{-4}-0.0000j$
		$0.7071+0.0000j$
		$-0.0000-5.1639\times 10^{-4}j$
		$0.0000+0.7071j$

TABLE 10.2
Eigenvalue and Eigenvectors with Gyroscopic Effects at 1500 rad/s

S.N.	Eigenvalue	Eigenvectors
1	$-0.0000+0.1742\times 10^{3}j$	$-0.0000-0.0005j$
		$-0.0000-0.0040j$
		$0.0005-0.0000j$
		$0.0040-0.0000j$
		$0.0881+0.0000j$
		$0.7016+0.0000j$
		$0.0000+0.0881j$
		$0.0000+0.7016j$
2	$-0.0000-0.1742\times 10^{3}j$	$-0.0000+0.0005j$
		$-0.0000+0.0040j$
		$0.0005+0.0000j$
		$0.0040+0.0000j$
		$0.0881-0.0000j$
		$0.7016+0.0000j$
		$0.0000-0.0881j$
		$0.0000-0.7016j$
3	$0.0000+0.2105\times 10^{3}j$	$-0.0005-0.0000j$
		$-0.0033-0.0000j$

(*Continued*)

TABLE 10.2 (*Continued*)
Eigenvalue and Eigenvectors with Gyroscopic Effects at 1500 rad/s

S.N.	Eigenvalue	Eigenvectors
		0.0000−0.0005j
		0.0000−0.0033j
		−0.0000+0.0996j
		−0.0000+0.7000j
		−0.0996−0.0000j
		−0.7000+0.0000j
4	0.0000−0.2105×10^3j	−0.0005−0.0000j
		−0.0033−0.0000j
		0.0000−0.0005j
		0.0000−0.0033j
		−0.0000+0.0996j
		−0.0000+0.7000j
		−0.0996−0.0000j
		−0.7000+0.0000j
5	−0.0000+1.8034×10^3j	0.0000+1.3635×10^{-6}j
		−0.0000−0.0004j
		−1.3635×10^{-6}+0.0000j
		0.0004−0.0000j
		−0.0025−0.0000j
		0.7071+0.0000j
		−0.0000−0.0025j
		0.0000+0.7071j
6	−0.0000−1.8034×10^3j	−0.0000−1.3635×10^{-6}j
		−0.0000+0.0004j
		−1.3635×10^{-6}−0.0000j
		0.0004+0.0000j
		−0.0025+0.0000j
		0.7071+0.0000j
		−0.0000+0.0025j
		0.0000−0.7071j
7	0.0000+4.2662×10^3j	6.9367×10^{-8}+0.0000j
		0.0002+0.0000j
		0.0000−6.9367×10^{-8}j
		−0.0000−0.0002j
		−0.0000+0.0003j
		−0.0000+0.7071j
		0.0003+0.0000j
		0.7071+0.0000j
8	0.0000−4.2662×10^3j	6.9367×10^{-8}−0.0000j
		0.0002−0.0000j
		0.0000+6.9367×10^{-8}j
		−0.0000+0.0002j
		−0.0000−0.0003j
		−0.0000−0.7071j
		0.0003−0.0000j
		0.7071+0.0000j

which is a familiar form discussed in detail in Chapters 7 and 9 for, respectively, torsional and transverse vibrations.

The eigenvalue and eigenvector for matrix **D**, for example, at 1500 rad/s, are provided in Table 10.2. It can be observed that these eigenvalues are also purely imaginary and occur in complex conjugates (e.g., serial numbers 1 and 2, 3 and 4, 5 and 6, and 7 and 8). Serial numbers 1–4 belong to the first natural mode (as in the case of without gyroscopic effects these were all the same; however, now 1 and 3 are not the same). Eigenvalues 2 and 4 are complex conjugates of eigenvalues 1 and 3, respectively; hence, only eigenvalues 1 and 3 need to be considered (for the same reason as given for the case without gyroscopic effects). Serial numbers 1 (174.20 rad/s) and 3 (210.50 rad/s) belong to the backward and forward whirls, respectively (which are below and above the first natural frequency without a gyroscopic couple, i.e., 192.68 rad/s). Similarly, serial numbers 5 (1803.4 rad/s) and 7 (4266.2 rad/s) belong to the backward and forward whirls, respectively (which are below and above the second natural frequency without a gyroscopic couple, i.e., 2757.55 rad/s); and serial numbers 6 and 8 are exactly the same as 5 and 7 with a negative sign (real parts in all eigenvalues are zero).

Eigenvectors corresponding to these eigenvalues are also listed in Table 10.2. It can be observed that they are appearing as complex numbers. However, on close observation, either the real part is zero or the imaginary part is zero. For example, for eigenvalue 1, the real parts of the first and second eigenvector components are zero, whereas the imaginary parts of the third and fourth eigenvector components are zero. When the real part is zero, the phase is 90°, and when the imaginary part is zero, the phase is 0° since a complex number is represented as $(\cos\theta + j\sin\theta)$.

Table 10.3 lists critical speeds up to the third mode with and without a gyroscopic couple. It should be noted that the critical speed without a gyroscopic couple is always between the corresponding mode forward and backward whirl critical speeds, which indicate the splitting of the natural frequency in the case of gyroscopic effect (refer to Chapters 5 and 8 for more in-depth discussions on gyroscopic effects and the Campbell diagram). It can be observed in Figure 10.4 that the split in the whirl natural frequency is distinct at higher speeds and at higher mode numbers, which has been plotted by eigenvalues (natural whirl frequency) obtained by varying the spin speed of the rotor. Figure 10.5a and b shows mode shapes (absolute value of eigenvectors) for the translational and rotational displacements, respectively, which has been obtained by eigenvectors for the rotor system with 20 elements (refer to Chapter 9 for a detailed procedure).

TABLE 10.3
Critical Speeds with Different Numbers of Elements

	Critical Speed without Gyroscopic Effects (rad/s)		Critical Speed without Gyroscopic Effects (rad/s)			
			For 1 Element		For 20 Elements	
Mode No.	1 Element	20 Elements	Backward Whirl	Forward Whirl	Backward Whirl	Forward Whirl
1	192.65	192.65	190.32	195.02	190.32	195.02
2	2757.55	2724.22	1709.47	[a]	1706.29	6773.36
3	[b]	7993.37	[b]	[b]	7543.02	19,858.0

[a] Critical speed could not be obtained because the estimated natural whirl frequency with only one element never intersected the $\omega_{nf} = \omega$ line. This is because the FEM always overestimates the natural frequencies and error in the second natural frequency is expected to be more with a single element; hence, variation of the natural whirl frequency was of a diverging nature.

[b] With a single element, it is possible to get critical speeds only up to the second mode.

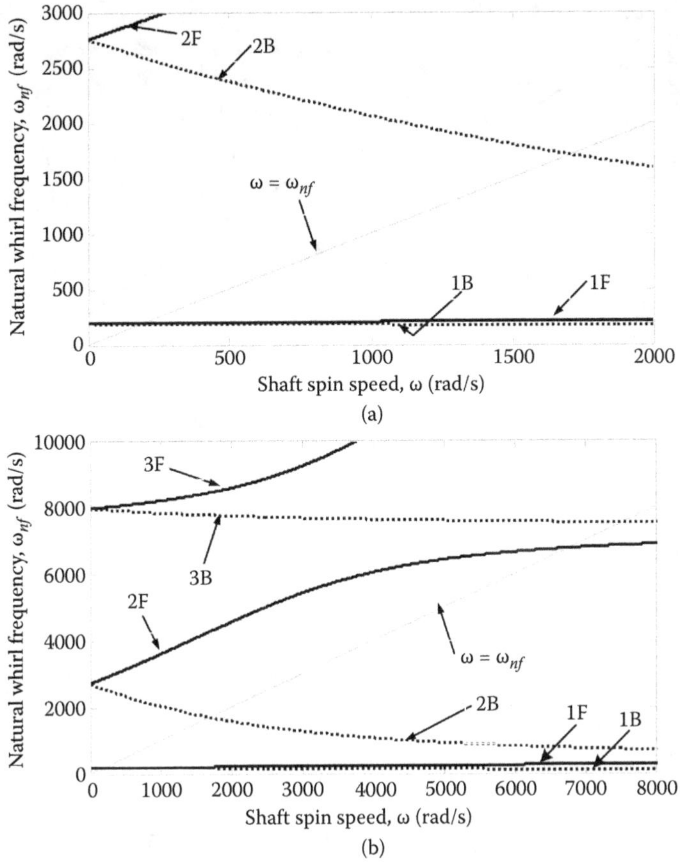

FIGURE 10.4 (a) Normalized translational displacement mode shapes at $\omega = 100$ rad/s (forward): (1) first mode shape, (2) second mode shape; (b) normalized rotational displacement mode shapes at $\omega = 100$ rad/s (forward): (1) first mode shape, and (2) second mode shape.

10.2 TIMOSHENKO BEAM THEORY

In a previous section, we discussed the gyroscopic effect of a disk alone in a slender flexible beam. The cross-sectional dimensions of the beam were small in comparison with the length. In this case, the transverse rotation of a beam element is equal to the slope of the elastic line of the shaft. When we investigate higher frequency modes, an infinitesimal element would have an appreciable amount of *rotary inertia*. When the beam is thick in that case, the cross-sectional dimensions of the beam are comparable with the length and the *shear effect* becomes predominant. The Euler–Bernoulli beam theory assumes that plane cross sections remain plane and perpendicular to the longitudinal axis after bending. This assumption implies that all shear strains are zero. When the normality assumption is not used—i.e., the plane sections remain plane but not necessarily to the longitudinal (neutral) axis after the deformation—the transverse shear strain, ε_{yz}, is not zero. Therefore, the rotation of a transverse normal plane about the x-axis is not equal to $-dv/dz$. The beam theory based on these relaxed assumptions is called a shear deformation beam theory, commonly known as the Timoshenko beam theory.

Consider a short stubby beam as shown in Figure 10.6a before deformation. Let L be the length of the shaft, A be the cross-sectional area, E be the Young's modulus, and ρ be the mass density of

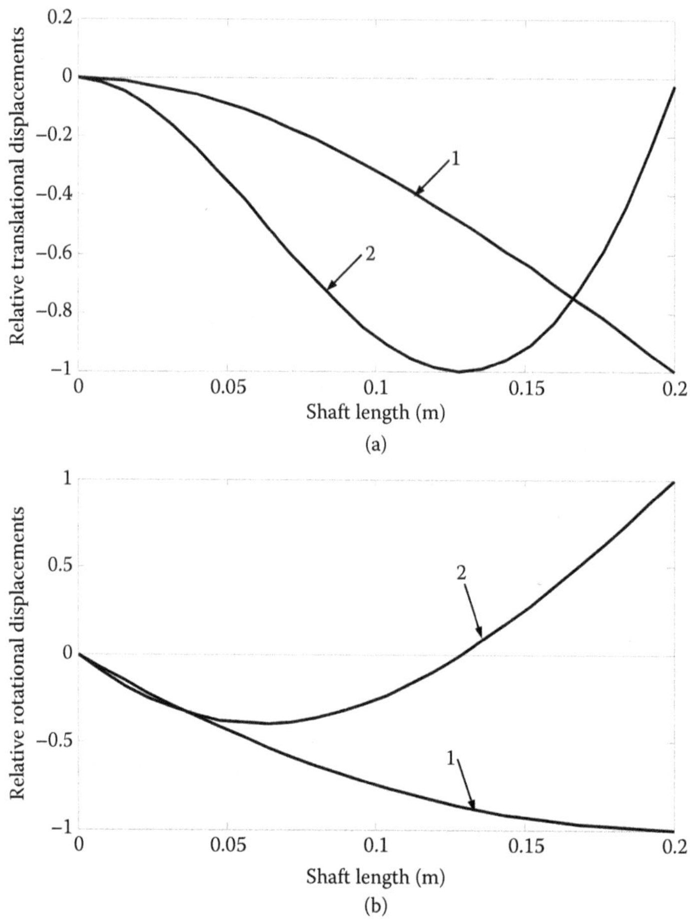

FIGURE 10.5 (a) Campbell diagram (with 1 element). (b) Campbell diagram (with 20 elements).

the shaft material. Assume that the deformation of the beam is purely due to the shear and that a vertical element before deformation remains vertical after deformation, and moves by distance v_s (subscript s represents the pure shear) in the transverse y direction as shown in Figure 10.6b. For the pure shear case, there will also be no coupling in deformations in the two transverse directions (i.e., in the x and y directions).

Since there is no coupling of motion in two transverse directions and on neglecting displacement in the axial direction, the displacement field is given by

$$v_x = 0; \quad v_y = v_s(z,t); \quad v_z = 0 \tag{10.15}$$

where v_x, v_y, and v_z are displacement components in the x, y, and z directions, respectively. Line elements tangential to the elastic line of the beam undergo a rotation $\beta(z,t)$ in plane y-z due to the shear as shown in Figure 10.6c. Engineering strains from the displacement fields of Equation 10.15 are

$$\varepsilon_{yz} = \frac{\partial v_y}{\partial z} + \frac{\partial v_z}{\partial y} = v_s' \tag{10.16}$$

and

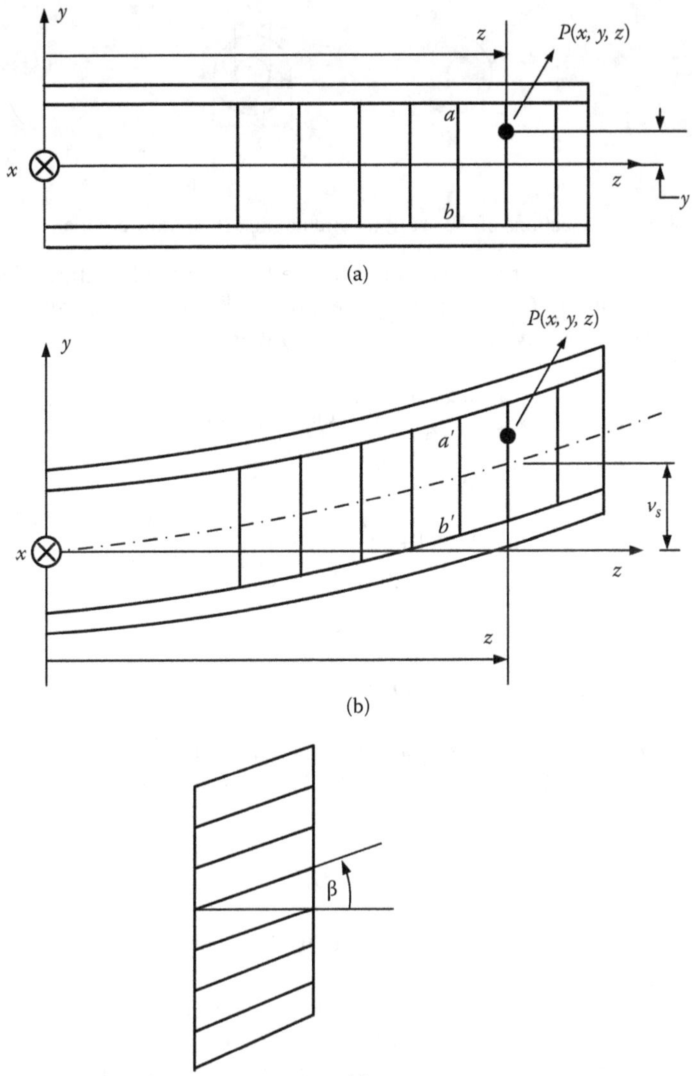

FIGURE 10.6 (a) A beam before deflection. (b) The beam after the shear deflection. (c) Rotation due to shear.

$$\varepsilon_{xx} = 0; \quad \varepsilon_{yy} = 0; \quad \varepsilon_{zz} = 0; \quad \varepsilon_{xy} = 0; \quad \varepsilon_{zx} = 0 \tag{10.17}$$

where prime (′) represents the derivative with respect to z.

The stress field consists of only one component:

$$\tau_{yz} = G\varepsilon_{yz} = Gv_s' \tag{10.18}$$

where G is the shear modulus of the beam material. Equation 10.18 shows that the shear stress across the cross section of the beam is uniform and independent of the coordinate, y. However, if we introduce explicit dependency of the stress field on another independent variable, y, then analysis becomes unnecessarily complicated. In the actual case, this is not true and therefore a shear correction factor k_{sc} is applied to Equation 10.18, rather than making the theory more complicated by having one more independent parameter y in the analysis.

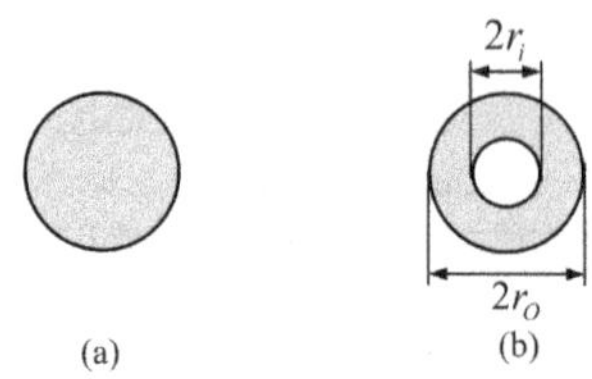

FIGURE 10.7 Typical cross sections of the shaft: (a) circular and (b) hollow circular.

Several definitions of the shear correction factors are found in the literature. The present investigation uses the one given by Cowper (1966) (mentioned also by Shames and Dym (2005), and Hutchinson (2001)). These are given for some cases as (Figure 10.7)

$$k_{sc} = \frac{6(1+\nu)}{7+6\nu} \quad \text{for a circular cylinder} \tag{10.19}$$

and

$$k_{sc} = \frac{6(1+\nu)(1+\bar{r})^2}{(7+6\nu)(1+\bar{r})^2+(20+12\nu)\bar{r}^2} \quad \text{for a hollow cylinder} \tag{10.20}$$

with

$$\bar{r} = r_i / r_o$$

where ν is the Poisson's ratio, and r_i and r_o are the inner and outer radii of the cross section of the hollow shaft, respectively, as shown in Figure 10.7b. Thus, considering the shear correction factor, Equation 10.18 can be written as

$$\tau_{yz} = k_{sc} G v_s' \tag{10.21}$$

In general, the beam is not subjected to pure shear only. The significant component of the deformation arises due to the bending moment, as in the case of Euler and Rayleigh (as compared to the Timoshenko beam model, which considers only the rotary inertia effect without shear deformation effects) beams. Combining the effect of shear considered above with the Euler beam theory (Chapter 9), the following displacement field for a point $P(x, y, z)$ as shown in Figure 10.8 is assumed:

$$v_x = 0; \quad v_y = v(z,t) \quad \text{and} \quad v_z = y\varphi_x(z,t) \tag{10.22}$$

with

$$v = v_s + v_b; \quad v' = v_s' + v_b' = -\beta - \varphi_x \tag{10.23}$$

The total slope of the beam, v', consists of two parts, one due to the pure bending, which is $-\varphi_x(z,t)$ and the other due to only the shear, which is $-\beta(z,t)$. The axial displacement v_z of a point at a distance y from the center (or elastic) line is only due to the bending slope and the shear force makes no contribution to this. So, there are two independent variables in this problem. One independent variable is the total displacement of the point, $v(z,t)$, and the other is the slope due to the bending, $\varphi_x(z,t)$. The engineering strain field from Equation 10.22 is

$$\varepsilon_{zz} = \frac{\partial v_z}{\partial z} = \frac{\partial(y\varphi_x)}{\partial z} = y\frac{\partial \varphi_x}{\partial z} = y\varphi_x' \tag{10.24}$$

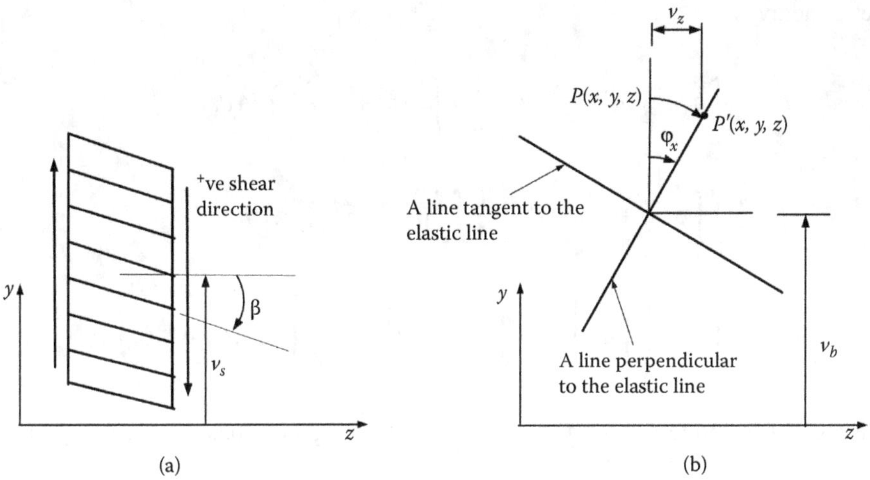

FIGURE 10.8 A displacement field of the Timoshenko beam: (a) shear and (b) bending (Euler).

$$\varepsilon_{yz} = \frac{\partial v_z}{\partial y} + \frac{\partial v_y}{\partial z} = \frac{\partial(y\varphi_x(z,t))}{\partial y} + \frac{\partial(v(z,t))}{\partial z} = \varphi_x(z,t)\frac{\partial(y)}{\partial y} + v'(z,t) = (v' + \varphi_x) \quad (10.25)$$

and

$$\varepsilon_{xx} = 0; \quad \varepsilon_{yy} = 0; \quad \varepsilon_{xy} = 0; \quad \varepsilon_{zx} = 0 \quad (10.26)$$

The corresponding stress field with the shear correction factor is given as

$$\sigma_{zz} = Ey\varphi_x'; \quad \tau_{yz} = k_{sc}G(v' + \varphi_x) \quad (10.27)$$

and

$$\sigma_{yy} = 0; \quad \sigma_{xx} = 0; \quad \tau_{xy} = \tau_{zx} = 0 \quad (10.28)$$

The strain (conservative) energy is

$$\begin{aligned} U &= \frac{1}{2}\int_V (\sigma_{xx}\varepsilon_{xx} + \sigma_{yy}\varepsilon_{yy} + \sigma_{zz}\varepsilon_{zz} + \tau_{xy}\varepsilon_{xy} + \tau_{yz}\varepsilon_{yz} + \tau_{zx}\varepsilon_{zx})\,dV \\ &= \frac{1}{2}\int_0^L\int_A (\sigma_{zz}\varepsilon_{zz} + \tau_{yz}\varepsilon_{yz})\,dA\,dx \\ &= \frac{1}{2}\int_0^L\int_A \left[(Ey\varphi_x')(y\varphi_x') + \left\{k_{sc}G(v'+\varphi_x)(v'+\varphi_x)\right\}\right]dA\,dx \\ &= \frac{1}{2}\int_0^L \left\{E(\varphi_x')^2\left(\int_A y^2\,dA\right) + k_{sc}G(v'+\varphi_x)^2\left(\int_A dA\right)\right\}dx \\ &= \frac{1}{2}\int_0^L \left\{EI_{xx}\varphi_x'^2 + k_{sc}AG(v'+\varphi_x)^2\right\}dz \text{ with } I_{xx} = \int_A y^2\,dA \quad \text{and} \quad A = \int_A dA \end{aligned} \quad (10.29)$$

The kinetic energy is

$$\begin{aligned} T &= \frac{1}{2}\int_V \rho\left(\dot{u}_x^2+\dot{u}_y^2+\dot{u}_z^2\right)dV = \frac{1}{2}\int_0^L\int_A \rho\left(\dot{u}_x^2+\dot{u}_y^2+\dot{u}_z^2\right)dA\,dz \\ &= \frac{1}{2}\int_0^L\int_A \rho\left(\dot{u}_y^2+\dot{u}_z^2\right)dA\,dz = \frac{1}{2}\int_0^L\int_A \rho\left(\dot{v}^2+y^2\dot{\varphi}_x^2\right)dA\,dz \\ &= \frac{1}{2}\int_0^L\left\{\rho\dot{v}^2\left(\int_A dA\right)+\rho\dot{\varphi}_x^2\left(\int_A y^2\,dA\right)\right\}dz = \frac{1}{2}\int_0^L\left\{\rho A\dot{v}^2+\rho I_{xx}\dot{\varphi}_x^2\right\}dz \end{aligned} \tag{10.30}$$

If $f_y(z,t)$ is the distributed force for transverse loads in the y-z plane, then the work done (nonconservative) by external forces can be written as

$$W_{nc} = \int_0^L f_y(z,t)\,v\,dz \tag{10.31}$$

The elemental equation of motion and boundary conditions can be obtained from the extended Hamilton's principle, as follows:

$$\int_{t_1}^{t_2}\left[\delta(T-U)+\delta W_{nc}\right]dt = 0 \tag{10.32}$$

Substituting Equations 10.29 through 10.31 into Equation 10.32, we get

$$\int_{t_1}^{t_2}\left[\delta\begin{pmatrix}\dfrac{1}{2}\displaystyle\int_0^L\left\{\rho A\dot{v}^2+\rho I_{xx}\dot{\varphi}_x^2\right\}dz \\ -\dfrac{1}{2}\displaystyle\int_0^L\left\{EI_{xx}\varphi_x'^2+k_{sc}AG\left(v'+\varphi_x\right)^2\right\}dz\end{pmatrix}+\delta\left(\int_0^L f_y(z,t)\,v\,dz\right)\right]dt = 0 \tag{10.33}$$

On operating the variation operator, from Equation 10.33, we get

$$\int_{t_1}^{t_2}\left[\begin{pmatrix}\displaystyle\int_0^L\left\{\rho A\dot{v}\delta(\dot{v})+\rho I_{xx}\dot{\varphi}_x\delta(\dot{\varphi}_x)\right\}dz \\ -\displaystyle\int_0^L\left\{EI_{xx}\varphi'\delta(\varphi_x')+k_{sc}AG\left(v'+\varphi_x\right)\delta\left(v'+\varphi_x\right)\right\}dz\end{pmatrix} \\ +\left(\displaystyle\int_0^L f_y(z,t)\,\delta v\,dz\right)\right]dt = 0 \tag{10.34}$$

On changing the order of variation and differentiation in Equation 10.34, we get

$$\int_{t_1}^{t_2}\left[\left(\int_0^L\left\{\rho A\dot{v}\frac{\partial}{\partial t}(\delta v)+\rho I_{xx}\dot{\varphi}_x\frac{\partial}{\partial t}(\delta\varphi_x)\right\}dz\right)\right]dt$$
$$-\int_{t_1}^{t_2}\left[\int_0^L\left\{EI_{xx}\varphi_x'\frac{\partial}{\partial z}(\delta\varphi_x)+k_{sc}AG(v'+\varphi_x)\left(\frac{\partial}{\partial z}\delta v+\delta\varphi_x\right)\right\}dz\right]dt \tag{10.35}$$
$$+\int_{t_1}^{t_2}\left[\int_0^L f_y(z,t)\delta v dz\right]dt=0$$

On performing integration by parts of terms, which has both the differential and variational operators (i.e., the first four terms), in Equation 10.35, we get

$$\int_0^L\{\rho A\dot{v}(\delta v)\}dz\Big|_{t_1}^{t_2}-\int_{t_1}^{t_2}\left[\int_0^L\{\rho A\ddot{v}(\delta v)\}dz\right]dt+\int_0^L\{\rho I_{xx}\dot{\varphi}_x(\delta\varphi_x)\}dz\Big|_{t_1}^{t_2}$$
$$-\int_{t_1}^{t_2}\left[\int_0^L\{\rho I_{xx}\ddot{\varphi}_x(\delta\varphi_x)\}dz\right]dt-\int_{t_1}^{t_2}\left[EI_{xx}\varphi_x'(\delta\varphi_x)\right]dt\Big|_0^L$$
$$+\int_{t_1}^{t_2}\left[\int_0^L\{EI_{xx}\varphi_x''(\delta\varphi_x)\}dz\right]dt-\int_{t_1}^{t_2}\left[\{k_{sc}AG(v'+\varphi_x)(\delta v)\}\right]dt\Big|_0^L \tag{10.36}$$
$$+\int_{t_1}^{t_2}\left[\int_0^L k_{sc}AG(v''+\varphi_x')(\delta v)dz\right]dt-\int_{t_1}^{t_2}\left[\int_0^L\{k_{sc}AG(v'+\varphi_x)(\delta\varphi_x)\}dz\right]dt$$
$$+\int_{t_1}^{t_2}\left[\int_0^L f_y(z,t)\delta v dz\right]dt=0$$

The first and third terms of Equation 10.36 will vanish, since variations are defined to be zero at t_1 and t_2. The remaining terms can be rearranged in the following form:

$$-\int_{t_i}^{t_2}\left[\int_0^L\{\rho A\ddot{v}-k_{sc}AG(v''+\varphi_x')-f_y(z,t)\}\delta v dz\right]dt$$
$$-\int_{t_1}^{t_2}\left[\int_0^L\{\rho I_{xx}\ddot{\varphi}_x-EI_{xx}\varphi_x''+k_{sc}AG(v'+\varphi_x)\}dz\delta\varphi_x\right]dt \tag{10.37}$$
$$-\int_{t_1}^{t_2}[EI_{xx}\varphi_x'(\delta\varphi_x)]dt\Big|_0^L-\int_{t_1}^{t_2}\left[\{k_{sc}AG(v'+\varphi_x)(\delta v)\}\right]dt\Big|_0^L=0$$

Variations δv and $\delta\varphi_x$ in the spatial domain are arbitrary; this yields the differential equations of motion as

$$\rho A\ddot{v} - k_{sc}AG\left(v'' + \varphi_x'\right) = f_y\left(z,t\right) \tag{10.38}$$

and

$$\rho I_{xx}\ddot{\varphi}_x - EI_{xx}\varphi_x'' - k_{sc}AG\left(v' + \varphi_x\right) = 0 \tag{10.39}$$

and the boundary conditions as

$$k_{sc}AG\left(v' + \varphi_x\right)\delta v\Big|_0^L = 0 \quad \text{and} \quad EI_{xx}\varphi_x'\delta\varphi_x\Big|_0^L = 0 \tag{10.40}$$

The equations of motion (Equations 10.38 and 10.39) can be combined to a single equation. On rearranging Equation 10.38 for free vibrations (i.e., $f_y(z, t)=0$), we get

$$\varphi_x' = -v'' + \frac{\rho}{k_{sc}G}\ddot{v} \tag{10.41}$$

which gives

$$\varphi_x'' = -v''' + \frac{\rho}{k_{sc}G}\ddot{v}' \tag{10.42}$$

$$\ddot{\varphi}_x' = -\ddot{v}'' + \frac{\rho}{k_{sc}G}\ddddot{v}$$

and

$$\varphi_x''' = -v'''' + \frac{\rho}{k_{sc}G}\ddot{v}''$$

On differentiating Equation 10.39 with respect to z, we get

$$EI_{xx}\varphi_x''' - \rho I_{xx}\ddot{\varphi}_x' - k_{sc}AG\left(v'' + \varphi_x'\right) = 0 \tag{10.43}$$

On substituting Equations 10.41 and 10.42 into Equation 10.43, we get

$$EI_{xx}\left(-v'''' + \frac{\rho}{k_{sc}G}\ddot{v}''\right) - \rho I_{xx}\left(-\ddot{v}'' + \frac{\rho}{k_{sc}G}\ddddot{v}\right) - k_{sc}AG\left\{v'' + \left(-v'' + \frac{\rho}{k_{sc}G}\ddot{v}\right)\right\} = 0 \tag{10.44}$$

which gives the Timoshenko beam governing equation for free vibrations as

$$EI_{xx}\frac{\partial^4 v}{\partial z^4} + \rho A\frac{\partial^2 v}{\partial t^2} - \rho I_{xx}\left(1 + \frac{E}{k_{sc}G}\right)\frac{\partial^4 v}{\partial t^2\,\partial z^2} + \frac{\rho^2 I_{xx}}{k_{sc}G}\frac{\partial^4 v}{\partial t^4} = 0 \tag{10.45}$$

It should be noted that due to the rotary inertia and the shear deformation in the Timoshenko beam, two extra terms (the third and fourth terms on the left-hand side of the above equation) appear in the equation of motion as compared to the Euler–Bernoulli beam, in which only the first two terms on the left-hand side appear. The third and fourth terms containing k_{sc} are related to the shear effect, whereas those the third term without the shear effect are related to the rotary inertia (i.e., $\rho I_{xx}\ddot{v}''$). It should be noted that a similar equation of motion could be developed for the z–x plane of the following form:

$$EI_{yy}\frac{\partial^4 u}{\partial z^4}+\rho A\frac{\partial^2 u}{\partial t^2}-\rho I_{yy}\left(1+\frac{E}{k_{sc}G}\right)\frac{\partial^4 u}{\partial t^2\,\partial z^2}+\frac{\rho^2 I_{yy}}{k_{sc}G}\frac{\partial^4 u}{\partial t^4}=0 \tag{10.46}$$

where u is the translational transverse displacement in the x-axis direction, and I_{yy} is the moment of inertia of the cross section about the y-axis. At present, we have not considered gyroscopic effects, which comes into effect when the shaft spins.

10.2.1 Closed Form Solution

Equation 10.45 is solved for a simply supported end condition of a shaft. For obtaining the closed form solution of Equation 10.45 for the specified boundary conditions, it can be simplified by using a general solution of the following form:

$$v(z,t)=A_n\sin\left(\frac{nz\pi}{L}\right)\sin\omega_{nf}t \tag{10.47}$$

where n is the mode number, A_n is the arbitrary constant, and ω_{nf} is the natural frequency of the shaft. On substituting Equation 10.47 into Equation 10.45, we get

$$-\frac{\rho^2 L^4}{Ek_{sc}G}\omega_{nf}^4+\left[\frac{\rho AL^4}{EI_{xx}}+\frac{\rho L^2}{E}\left(1+\frac{E}{k_{sc}G}\right)(n\pi)^2\right]\omega_{nf}^2-(n\pi)^4=0 \tag{10.48}$$

On solving Equation 10.48 for ω_{nf}, we get

$$\omega_{nf}^2=\frac{-b+\sqrt{b^2-4ac}}{2a} \tag{10.49}$$

with

$$b=\left\{\frac{\rho AL^4}{EI_{xx}}+\frac{\rho l^2}{E}\left(1+\frac{E}{k_{sc}G}\right)(n\pi)^2\right\};\quad a=-\frac{\rho^2 L^4}{Ek_{sc}G};\quad c=-(n\pi)^4 \tag{10.50}$$

The above exact solution will be used for comparing natural frequencies for a simply supported end condition of a shaft with the FEM as a benchmark solution for the convergence study. Along similar lines for other boundary conditions, expressions of natural frequencies could be attempted.

10.3 FINITE ELEMENT FORMULATIONS OF THE TIMOSHENKO BEAM

For the finite element analysis, we need to discretize the beam into a number of elements as shown in Figure 10.9. Consider a finite element of the shaft of length l in an elemental coordinate system x–y–z. Since the two orthogonal transverse plane motions are uncoupled, the deformation of the element is initially considered in the y–z plane with two nodes, 1 and 2, as shown in Figure 10.10. The motion in the z–x plane could be analyzed along similar lines. Let v be the nodal translational

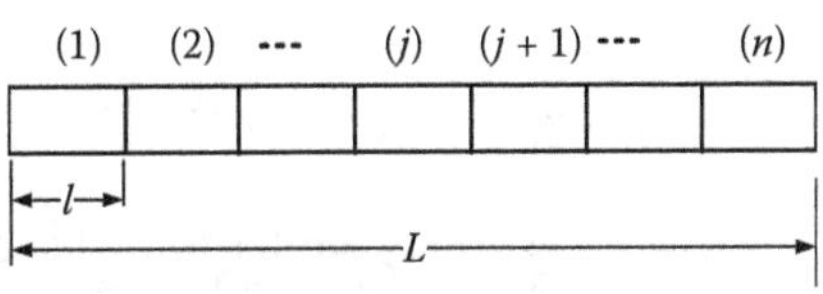

FIGURE 10.9 Discretization of a beam into a number of elements.

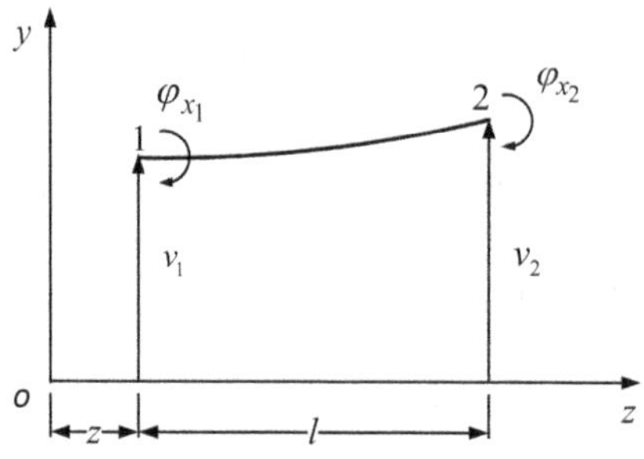

FIGURE 10.10 A typical beam element in the y–z plane.

transverse displacement of the shaft element. Let $v' = dv/dz$ be the total slope of the beam, which consists of two parts, one due to the bending, which is $-\varphi_x$, and the other due to the shear, which is $-\beta$ and is given by $-(\beta+\varphi_x)$. From the displacement field of Equation 10.22, we know that the axial displacement of a point $P(x, y, z)$ at a distance y from the center (or elastic) line is only due to the bending slope and that the shear force has no contribution to this. In the previous section, we observed that there are two independent variables in this problem (i.e., v and φ_x).

From Equations 10.38 and 10.39 without considering the work done by external forces, the equations of motion are

$$\rho A\ddot{v} - k_{sc}AG\left(v' + \varphi_x\right)' = 0 \tag{10.51}$$

and

$$EI_{xx}\varphi_x'' - \rho I_{xx}\ddot{\varphi}_x + k_{sc}AG\left(v' + \varphi_x\right) = 0 \tag{10.52}$$

In the finite element model, the continuous displacement field can be approximated in terms of discretized generalized displacements of element nodes. In the present study, each element in a single plane (e.g., y–z) has two nodes and each node has two generalized displacements (one translational and the other rotational). Therefore, displacements can be obtained within the element by using appropriate shape functions to be derived in the subsequent subsection.

10.3.1 Weak Formulations of the Timoshenko Beam Element for the Static Case

Let the static case be considered first by dropping time derivative terms. The equations of motion (Equations 10.51 and 10.52) could be written as below while the boundary conditions remain the same:

$$k_{sc}AG\left(v' + \varphi_x\right)' = 0 \tag{10.53}$$

and

$$EI_{xx}\varphi_x'' - k_{sc}AG\left(v' + \varphi_x\right) = 0 \tag{10.54}$$

On assuming the approximate solution of the following form, we have

$$v^{(e)}(z) = \mathbf{S}(z)\ \mathbf{v}^{(ne)}(t) \tag{10.55}$$

and

$$\varphi_x^{(e)}(z) = \mathbf{T}(z)\,\mathbf{v}^{(ne)}(t) \tag{10.56}$$

where $\mathbf{S}(z)$ and $\mathbf{T}(z)$ are shapes function vectors for the translational and rotational displacements, respectively, and $\mathbf{v}^{(ne)}(t)$ is the nodal displacement vector, which contains nodal displacements and unknown interpolation coefficients (refer Equation 10.69). On substituting the approximate solutions of Equations 10.55 and 10.56 in the equations of motion (Equations 10.53 and 10.54), the residue of each equation is given by

$$R_1^{(e)} = k_{sc}AG\left(v^{(e)\prime} + \varphi_x^{(e)}\right)' \tag{10.57}$$

and

$$R_2^{(e)} = EI\varphi_x^{(e)\prime\prime} - k_{sc}AG\left(v^{(e)\prime} + \varphi_x^{(e)}\right) \tag{10.58}$$

Employing the Galerkin principle, one has

$$\int_0^l \mathbf{S}(z)\,R_1^{(e)}dz = \mathbf{0} \tag{10.59}$$

and

$$\int_0^l \mathbf{T}(z)\,R_2^{(e)}dz = \mathbf{0} \tag{10.60}$$

where $\mathbf{0}$ is the null vector of the same size $\mathbf{S}$ or $\mathbf{T}$. Using Equations 10.57 and 10.58 in Equations 10.59 and 10.60, we get

$$\int_0^l \mathbf{S}(z)\,k_{sc}AG\left(v^{(e)\prime} + \varphi_x^{(e)}\right)'\,dz = \mathbf{0} \tag{10.61}$$

and

$$\int_0^l \mathbf{T}(z)\left\{EI_{xx}\varphi_x^{(e)\prime\prime} - k_{sc}AG\left(v^{(e)\prime} + \varphi_x^{(e)}\right)\right\}dz = \mathbf{0} \tag{10.62}$$

On performing integration by parts, we get

$$\mathbf{S}(z)\,k_{sc}AG\left(v^{(e)\prime} + \varphi_x^{(e)}\right)\Big|_0^l - \int_0^l \mathbf{S}'(z)\,k_{sc}AG\left(v^{(e)\prime} + \varphi_x^{(e)}\right)dz = \mathbf{0} \tag{10.63}$$

and

$$\mathbf{T}(z)\,EI_{xx}\varphi_x^{(e)\prime}\Big|_0^l - \int_0^l \mathbf{T}'(z)\,EI_{xx}\varphi_x^{(e)\prime}dz - \int_0^l \mathbf{T}(z)\,k_{sc}AGv^{(e)\prime}dz$$
$$-\int_0^l \mathbf{T}(z)\,k_{sc}AG\varphi_x^{(e)}dz = \mathbf{0} \tag{10.64}$$

On looking into the completeness and compatibility requirements, we need the compatibility requirement up to $v^{(e)}$ and $\varphi_x^{(e)}$ (i.e., up to the first derivative of displacement, v) and the completeness requirement up to $v^{(e)}$ and $\varphi_x^{(e)}$ (i.e., up to the value of the variables themselves). Hence, according to these requirements, a linear interpolation function would serve the purpose. The variables $v^{(e)}$ and $\varphi_x^{(e)}$ do not have the same physical units; they can be interpolated, in general, with different degrees of interpolation. As we know for thin beams (i.e., Euler–Bernoulli beam), we have $\varphi_x^{(e)} - v^{(e)\prime}$; hence, a linear polynomial for $v^{(e)}$ implies a constant for $\varphi_x^{(e)}$. This will make the bending potential energy $\int_0^l \frac{1}{2} EI \left(\varphi_x^{(e)\prime}\right)^2 dz$ equal to zero. This numerical problem is known as *shear locking* (since it originates from the shear effect). To overcome this, several alternative methods have been developed in the literature (Reddy, 2003).

10.3.2 Derivation of Shape Functions

Now on assuming that $v^{(e)}$ is a cubic polynomial, $\varphi_x^{(e)}$ should be of the same order as $v^{(e)\prime}$, that is, a quadratic. These are the exact shape functions in the static analysis. The inter-element compatibility requires that v_1, φ_{x1}, v_2, and φ_{x2} must be continuous. Therefore, $v^{(e)}$ and $\varphi_x^{(e)}$ can be assumed as

$$v^{(e)}(\xi,\eta) = v_1\xi + v_2\eta + a\xi\eta + b\left(\xi^2\eta - \xi\eta^2\right) \tag{10.65}$$

and

$$\varphi_x^{(e)}(\xi,\eta) = \varphi_{x1}\xi + \varphi_{x2}\eta + c\xi\eta \tag{10.66}$$

with

$$\xi = 1 - \frac{z}{l} \tag{10.67}$$

and

$$\eta = \frac{z}{l}$$

where a, b, and c are unknown interpolation coefficients, and ξ and η are called natural coordinates. All four boundary conditions of the element in Figure 10.10 are satisfied as

$$\begin{aligned} v^{(e)}\Big|_{z=0} &= v_1 \\ v^{(e)}\Big|_{z=l} &= v_2 \\ \varphi_x\Big|_{z=0} &= \varphi_{x1} \\ \varphi\Big|_{z=l} &= \varphi_{x2} \end{aligned} \tag{10.68}$$

because $\xi\eta$ and $\left(\xi^2\eta - \xi\eta^2\right)$ vanish at boundaries of the element. Therefore a, b, and c are assumed to be internal DOFs, which will be eliminated, subsequently. Because of these DOFs, we are able to make the shape function of the required polynomial degree (i.e., a cubic for $v^{(e)}$ and a quadratic for $\varphi_x^{(e)}$) without violating the desired boundary conditions. Observe that without internal DOFs, the assumed shape function would be linear for both $v^{(e)}$ and $\varphi_x^{(e)}$. The presence of $\xi\eta$ and $\left(\xi^2\eta - \xi\eta^2\right)$ is necessary to complete the polynomial of the required degree to avoid shear locking. Equations 10.65 and 10.66 can be written as

$$v^{(e)}(\xi,\eta) = \lfloor \xi \quad 0 \quad \eta \quad 0 \quad \xi\eta \quad (\xi^2\eta - \xi\eta^2) \quad 0 \rfloor \lfloor v_1 \quad \varphi_{x1} \quad v_2 \quad \varphi_{x2} \quad a \quad b \quad c \rfloor^T = \mathbf{S}\mathbf{v}^{(ne)} \tag{10.69}$$

and

$$\varphi^{(e)}(\xi,\eta) = \lfloor 0 \quad \xi \quad 0 \quad \eta \quad 0 \quad 0 \quad \xi\eta \rfloor \lfloor v_1 \quad \varphi_{x1} \quad v_2 \quad \varphi_{x2} \quad a \quad b \quad c \rfloor^T = \mathbf{T}\mathbf{v}^{(ne)} \tag{10.70}$$

with

$$\mathbf{S}|_{z=0} = \lfloor 1 \quad 0 \quad 0 \quad 0 \quad 0 \quad 0 \quad 0 \rfloor \text{ and } \mathbf{S}|_{z=l} = \lfloor 0 \quad 0 \quad 1 \quad 0 \quad 0 \quad 0 \quad 0 \rfloor \tag{10.71}$$

and

$$\mathbf{T}|_{z=0} = \lfloor 0 \quad 1 \quad 0 \quad 0 \quad 0 \quad 0 \quad 0 \rfloor \text{ and } \mathbf{T}|_{z=l} = \lfloor 0 \quad 0 \quad 0 \quad 1 \quad 0 \quad 0 \quad 0 \rfloor \tag{10.72}$$

where $\mathbf{u}^{(ne)}(t)$ is the nodal displacement vector. On substituting Equations 10.69 through 10.72 in Equations 10.63 and 10.64, we get

$$\int_0^l k_{sc}AG\{\mathbf{S}'(z)\}^T \mathbf{S}'(z)dz\, \mathbf{v}^{(ne)}(t) + \int_0^l k_{sc}AG\{\mathbf{S}'(z)\}^T \mathbf{T}(z)dz\mathbf{v}^{(ne)} = \begin{Bmatrix} -k_{sc}AG(v'+\varphi_x)|_{z=0} \\ 0 \\ k_{sc}AG(v'+\varphi_x)|_{z=l} \\ 0 \\ 0 \\ 0 \\ 0 \end{Bmatrix} \tag{10.73}$$

and

$$\int_0^l EI_{xx}\{\mathbf{T}'(z)\}^T \mathbf{T}(z)dz\, \mathbf{v}^{(ne)}(t) + \int_0^l k_{sc}AG\{\mathbf{T}(z)\}^T \mathbf{S}'(z)dz\, \mathbf{v}^{(ne)}(t) + \int_0^l k_{sc}AG\{\mathbf{T}(z)\}^T \mathbf{T}(z)dz\, \mathbf{v}^{(ne)}(t) = \begin{Bmatrix} 0 \\ EI_{xx}\varphi_x'|_{z=0} \\ 0 \\ -EI_{xx}\varphi_x'|_{z=l} \\ 0 \\ 0 \\ 0 \end{Bmatrix} \tag{10.74}$$

On combining Equations 10.73 and 10.74, the stiffness relationship can be obtained as

$$\mathbf{K}_{7\times7}\mathbf{v}^{(ne)}_{7\times1}(t) = \begin{Bmatrix} \mathbf{f}^{(ne)}(t) \\ \mathbf{0}_{3\times1} \end{Bmatrix}_{7\times1} \tag{10.75}$$

with

$$\mathbf{K} = \mathbf{K}_1 + \mathbf{K}_2 + \mathbf{K}_3 + \mathbf{K}_4 + \mathbf{K}_5 \tag{10.76}$$

$$\mathbf{K}_1 = \int_0^l k_{sc} AG \{\mathbf{S}'(z)\}^T \mathbf{S}'(z)\,dz$$

$$= k_{sc}AG \begin{bmatrix} 1/l & 0 & -1/l & 0 & 0 & 0 & 0 \\ 0 & 0 & 0 & 0 & 0 & 0 & 0 \\ -1/l & 0 & 1/l & 0 & 0 & 0 & 0 \\ 0 & 0 & 0 & 0 & 0 & 0 & 0 \\ 0 & 0 & 0 & 0 & 1/(3l) & 0 & 0 \\ 0 & 0 & 0 & 0 & 0 & 1/(5l) & 0 \\ 0 & 0 & 0 & 0 & 0 & 0 & 0 \end{bmatrix} \tag{10.77}$$

$$\mathbf{K}_2 = \int_0^l k_{sc} AG \{\mathbf{S}'(z)\}^T \mathbf{T}dz$$

$$= k_{sc}AG \begin{bmatrix} 0 & 1/2 & 0 & 1/2 & 0 & 0 & -1/6 \\ 0 & 0 & 0 & 0 & 0 & 0 & 0 \\ 0 & -1/2 & 0 & -1/2 & 0 & 0 & 1/6 \\ 0 & 0 & 0 & 0 & 0 & 0 & 0 \\ 0 & -1/6 & 0 & 1/6 & 0 & 0 & 0 \\ 0 & 0 & 0 & 0 & 0 & 0 & -1/30 \\ 0 & 0 & 0 & 0 & 0 & 0 & 0 \end{bmatrix} \tag{10.78}$$

$$\mathbf{K}_3 = \int_0^l EI_{xx} \{\mathbf{T}'(z)\}^T \mathbf{T}'dz$$

$$= EI_{xx} \begin{bmatrix} 0 & 0 & 0 & 0 & 0 & 0 & 0 \\ 0 & 1/l & 0 & -1/l & 0 & 0 & 0 \\ 0 & 0 & 0 & 0 & 0 & 0 & 0 \\ 0 & -1/l & 0 & 1/l & 0 & 0 & 0 \\ 0 & 0 & 0 & 0 & 0 & 0 & 0 \\ 0 & 0 & 0 & 0 & 0 & 0 & 0 \\ 0 & 0 & 0 & 0 & 0 & 0 & 1/3l \end{bmatrix} \tag{10.79}$$

$$\mathbf{K}_4 = \int_0^l k_{sc} AG \{\mathbf{T}(z)\}^T \mathbf{S}(z) dz$$

$$= k_{sc}AG \begin{bmatrix} 0 & 0 & 0 & 0 & 0 & 0 & 0 \\ 1/2 & 0 & -1/2 & 0 & -1/6 & 0 & 0 \\ 0 & 0 & 0 & 0 & 0 & 0 & 0 \\ 1/2 & 0 & -1/2 & 0 & 1/6 & 0 & 0 \\ 0 & 0 & 0 & 0 & 0 & 0 & 0 \\ 0 & 0 & 0 & 0 & 0 & 0 & 0 \\ -1/6 & 0 & 1/6 & 0 & 0 & -1/30 & 0 \end{bmatrix} \tag{10.80}$$

$$\mathbf{K}_5 = \int_0^l k_{sc} AG \{\mathbf{T}(z)\}^T \mathbf{T}(z) dz$$

$$= k_{sc}AG \begin{bmatrix} 0 & 0 & 0 & 0 & 0 & 0 & 0 \\ 0 & l/3 & 0 & l/6 & 0 & 0 & -l/12 \\ 0 & 0 & 0 & 0 & 0 & 0 & 0 \\ 0 & l/6 & 0 & l/3 & 0 & 0 & -l/12 \\ 0 & 0 & 0 & 0 & 0 & 0 & 0 \\ 0 & 0 & 0 & 0 & 0 & 0 & 0 \\ 0 & -l/12 & 0 & -l/12 & 0 & 0 & l/30 \end{bmatrix} \tag{10.81}$$

and

$$\mathbf{K} = \begin{bmatrix} \frac{k}{l} & \frac{k}{2} & \frac{-k}{l} & \frac{k}{2} & 0 & 0 & \frac{-k}{6} \\ \frac{k}{2} & \left(\frac{kl}{3}+\frac{EI}{l}\right) & \frac{-k}{2} & \left(\frac{kl}{6}-\frac{EI}{l}\right) & \frac{-k}{6} & 0 & \frac{-kl}{12} \\ \frac{-k}{l} & \frac{-k}{2} & \frac{k}{l} & \frac{-k}{2} & 0 & 0 & \frac{k}{6} \\ \frac{k}{2} & \left(\frac{kl}{6}-\frac{EI}{l}\right) & \frac{-k}{2} & \left(\frac{kl}{3}+\frac{EI}{l}\right) & \frac{k}{6} & 0 & \frac{-kl}{12} \\ 0 & \frac{-k}{6} & 0 & \frac{k}{6} & \frac{k}{3l} & 0 & 0 \\ 0 & 0 & 0 & 0 & 0 & \frac{k}{5l} & \frac{-k}{30} \\ \frac{-k}{6} & \frac{-kl}{12} & \frac{k}{6} & \frac{-kl}{12} & 0 & \frac{-k}{30} & \left(\frac{kl}{30}+\frac{EI}{3l}\right) \end{bmatrix} \tag{10.82}$$

where

$$k = k_{sc}AG \tag{10.83}$$

Equation 10.75 has a vector $\mathbf{u}^{(ne)}$, which contains three internal DOFs that need to be eliminated at this stage with the help of the static condensation scheme (refer to Chapter 9). Defining the external DOFs (four in number) of the element as masters and internal DOFs (three in number) of the element as slaves, the stiffness matrix in Equation 10.75 is subdivided as

$$\begin{bmatrix} (\mathbf{k}_{11})_{44} & (\mathbf{k}_{12})_{43} \\ (\mathbf{k}_{21})_{34} & (\mathbf{k}_{22})_{33} \end{bmatrix}_{77} \mathbf{v}^{(ne)}(t) = \begin{Bmatrix} \mathbf{f}^{(ne)}(t) \\ \mathbf{0} \end{Bmatrix} \tag{10.84}$$

with

$$\mathbf{k}_{11} = \frac{2EI_{xx}}{\Phi l^3}\begin{bmatrix} 6 & & & sym \\ 3l & (0.5\Phi+2)l^2 & & \\ -6 & -3l & 6 & \\ 3l & (-0.5\Phi+1)l^2 & -3l & (0.5\Phi+2)l^2 \end{bmatrix}$$

$$\mathbf{k}_{22} = \frac{k_{sc}AG}{30l}\begin{bmatrix} 10 & 0 & 0 \\ 0 & 6 & -l \\ 0 & -l & \left(1+\frac{5\Phi}{6}\right)l^2 \end{bmatrix}$$

$$\mathbf{k}_{12}^T = \mathbf{k}_{21} = \frac{k_{sc}AG}{12}\begin{bmatrix} 0 & -2 & 0 & 2 \\ 0 & 0 & 0 & 0 \\ -2 & -l & 2 & -l \end{bmatrix}$$

$$\mathbf{u}^{(ne)}(t) = \begin{Bmatrix} \eta^{(ne)}(t) \\ \begin{Bmatrix} a(t) \\ b(t) \\ c(t) \end{Bmatrix} \end{Bmatrix}$$

$$\Phi = \frac{12EI_{xx}}{k_{sc}GAl^2}$$

$$\eta^{(ne)}(t) = \lfloor v_1 \quad \varphi_{x1} \quad v_2 \quad \varphi_{x2} \rfloor^T$$

and

$$\mathbf{f}^{(ne)}(t) = \lfloor -S_{y1} \quad -M_{yz1} \quad S_{y2} \quad M_{yz2} \rfloor^T$$

$$S_{y1} = k_{sc}AG\left(v' + \varphi_x\right)\Big|_{z=0}$$

$$M_{yz1} = EI_{xx}\varphi'_x\Big|_{z=0}$$

$$S_{y2} = k_{sc}AG\left(v' + \varphi_x\right)\Big|_{z=l}$$

$$M_{yz2} = EI_{xx}\varphi'_x\Big|_{z=l}$$

where Φ is called the shear parameter, S_y is the shear force, and M_{yz} is the bending moment. Equation 10.84 can be written as

$$\mathbf{k}_{11}\eta^{(ne)}(t) + \mathbf{k}_{12}\begin{Bmatrix} a(t) \\ b(t) \\ c(t) \end{Bmatrix} = \mathbf{f}^{(ne)}(t) \quad \text{and} \quad \mathbf{k}_{21}\eta^{(ne)}(t) + \mathbf{k}_{22}\begin{Bmatrix} a(t) \\ b(t) \\ c(t) \end{Bmatrix} = \mathbf{0} \tag{10.85}$$

The second set of Equation 10.85 gives

$$\begin{Bmatrix} a(t) \\ b(t) \\ c(t) \end{Bmatrix} = -\mathbf{k}_{22}^{-1}\mathbf{k}_{21}\lfloor v_1 \quad \varphi_{x1} \quad v_2 \quad \varphi_{x2} \rfloor^T = \begin{bmatrix} 0 & \dfrac{l}{2} & 0 & -\dfrac{l}{2} \\ \mu & \dfrac{\mu l}{2} & -\mu & \dfrac{\mu l}{2} \\ \dfrac{6\mu}{l} & 3\mu & -\dfrac{6\mu}{l} & 3\mu \end{bmatrix}\begin{Bmatrix} v_1 \\ \varphi_{x1} \\ v_2 \\ \varphi_{x2} \end{Bmatrix} \tag{10.86}$$

with

$$\mu = 1/(1+\Phi) \tag{10.87}$$

On substituting Equation 10.86 into the first set of Equation 10.85, we get

$$\mathbf{k}_{11}\eta^{(ne)}(t) + \mathbf{k}_{12}\left\{-\mathbf{k}_{22}^{-1}\mathbf{k}_{21}\eta^{(ne)}(t)\right\} = \mathbf{f}^{(ne)} \tag{10.88}$$

which can be written as

$$\mathbf{k}\eta^{(ne)}(t) = \mathbf{f}^{(ne)}(t) \tag{10.89}$$

with

$$\mathbf{k} = \mathbf{k}_{11} - \mathbf{k}_{12}\mathbf{k}_{22}^{-1}\mathbf{k}_{21} \tag{10.90}$$

where **k** is the condensed stiffness matrix. Now internal DOFs can be eliminated from the shape functions also. On substituting *a*, *b*, and *c* obtained from Equation 10.86 into Equation 10.69, we get

$$
\begin{aligned}
v^{(e)}(\xi,\eta) &= \begin{bmatrix} \xi & 0 & \eta & 0 & \xi\eta & (\xi^2\eta-\xi\eta^2) & 0 \end{bmatrix} \\
&\quad \times \begin{bmatrix} v_1 & \varphi_{x1} & v_2 & \varphi_{x2} & a & b & c \end{bmatrix}^T \\
&= \begin{bmatrix} \xi & 0 & \eta & 0 \end{bmatrix} \begin{Bmatrix} v_1 \\ \varphi_{x1} \\ v_2 \\ \varphi_{x2} \end{Bmatrix} + \begin{bmatrix} \xi\eta & (\xi^2\eta-\xi\eta^2) & 0 \end{bmatrix} \begin{Bmatrix} a \\ b \\ c \end{Bmatrix} \\
&= \begin{bmatrix} \xi & 0 & \eta & 0 \end{bmatrix} \begin{Bmatrix} v_1 \\ \varphi_{x1} \\ v_2 \\ \varphi_{x2} \end{Bmatrix} \\
&\quad + \begin{bmatrix} \xi\eta & (\xi^2\eta-\xi\eta^2) & 0 \end{bmatrix} \begin{bmatrix} 0 & \frac{l}{2} & 0 & -\frac{l}{2} \\ \mu & \frac{\mu l}{2} & -\mu & \frac{\mu l}{2} \\ \frac{6\mu}{l} & 3\mu & -\frac{6\mu}{l} & 3\mu \end{bmatrix} \begin{Bmatrix} v_1 \\ \varphi_{x1} \\ v_2 \\ \varphi_{x2} \end{Bmatrix}
\end{aligned} \tag{10.91}
$$

which can be simplified as

$$
\begin{aligned}
v^{(e)}(z,t) &= N_1(z)v_1(t) - N_2(z)\varphi_{x_1}(t) + N_3(z)\,v_2(t) - N_4(z)\varphi_{x_2}(t) \\
&= \mathbf{N}(z)\,\eta^{(ne)}(t)
\end{aligned} \tag{10.92}
$$

with

$$N_i(z) = \frac{1}{1+\Phi}\left[\alpha_i(z) + \Phi\beta_i(z)\right]$$

$$\Phi = \frac{12EI_{xx}}{k_{sc}GAl^2}$$

$$i = 1,2,3,4$$

$$\alpha_1 = 1 - 3\eta^2 + 2\eta^3$$

$$\beta_1 = 1 - \eta$$

$$\alpha_2 = \left(\eta - 2\eta^2 + \eta^3\right)l$$

$$\beta_2 = \left(\eta - \eta^2\right)l/2$$

$$\alpha_3 = 3\eta^2 - 2\eta^3$$

$$\beta_3 = \eta$$

$$\alpha_4 = \left(-\eta^2 + \eta^3\right) l$$

$$\beta_4 = \left(-\eta + \eta^2\right) l/2$$

$$\eta = \frac{z}{l}$$

and

$$\eta^{(ne)}(t) = \lfloor \; v_1 \quad \varphi_{x1} \quad v_2 \quad \varphi_{x2} \; \rfloor^T$$

where $\mathbf{d}^{(ne)}$ is the *nodal displacement vector* and $\mathbf{N}$ is the *translational shape function vector.* The $\alpha_i(z)$ functions are associated with the bending deformation and the $\beta_i(z)$ functions are due to the shear deformation of a Timoshenko beam. Similarly, on substituting *a*, *b*, and *c* obtained from Equation 10.86 in Equation 10.70, we get

$$\begin{aligned} \varphi_x^{(e)}(z,t) &= -M_1(z)\, v_1(t) + M_2(z)\varphi_{x_1}(t) - M_3(z)\, v_2(t) + M_4(z)\varphi_{x_2}(t) \\ &= \mathbf{M}(z)\, \eta^{(ne)}(t) \end{aligned} \tag{10.93}$$

with

$$M_i(z) = \frac{1}{1+\Phi}\left[\varepsilon_i(z) + \Phi\delta_i(z)\right]$$

$$\Phi = \frac{12EI_{xx}}{k_{sc}GAl^2}$$

$$i = 1, 2, 3, 4$$

$$\varepsilon_1 = \left(6\eta^2 - 6\eta\right)/l$$

$$\delta_1 = 1$$

$$\varepsilon_2 = 1 - 4\eta + 3\eta^2$$

$$\delta_2 = 1 - \eta$$

$$\varepsilon_3 = \left(6\eta - 6\eta^2\right)/l$$

$$\delta_3 = 0$$

$$\varepsilon_4 = 3\eta^2 - 2\eta$$

$$\delta_4 = \eta$$

$$\eta = \frac{z}{l}$$

where **M** is the *rotational shape function vector.* Functions, $\varepsilon_i(z)$, are associated with the bending deformation and functions $\delta_i(z)$ are due to the shear deformation of a Timoshenko beam.

10.3.3 Weak Formulation of the Timoshenko Beam Element for the Dynamic Case

For the dynamic case, the shape functions derived in the previous section will still be valid and could take the following form:

$$\begin{aligned} v^{(e)}(z,t) &= N_1(z)v_1(t) + N_2(z)\varphi_{x1}(t) + N_3(z)v_2(t) + N_4(z)\varphi_{x2}(t) \\ &= \mathbf{N}(z)\,\eta^{(ne)}(t) \end{aligned} \tag{10.94}$$

and

$$\begin{aligned} \varphi_x^{(e)}(z,t) &= -M_1(z)v_1(t) + M_2(z)\varphi_{x1}(t) - M_3(z)v_2(t) + M_4(z)\varphi_{x2}(t) \\ &= \mathbf{M}(z)\,\eta^{(ne)}(t) \end{aligned} \tag{10.95}$$

On substituting Equations 10.94 and 10.95 in the equations of motion (Equations 10.51 and 10.52), residues of each equation of motion are given by

$$R_1^{(e)} = \rho A\ddot{v}^{(e)} - k_{sc}AG\left(v''^{(e)} + \varphi_x'^{(e)}\right) \tag{10.96}$$

and

$$R_2^{(e)} = EI_{xx}\varphi_x^{(e)''} - \rho I_{xx}\ddot{\varphi}_x^{(e)} - k_{sc}AG\left(v^{(e)'} + \varphi_x^{(e)}\right) \tag{10.97}$$

Using the Galerkin method to minimize the residue, one has

$$\int_0^l \mathbf{N}(z)\,R_1^{(e)}dz = \mathbf{0} \tag{10.98}$$

and

$$\int_0^l \mathbf{M}(z)\,R_2^{(e)}dz = \mathbf{0} \tag{10.99}$$

On substituting of Equations 10.96 and 10.97 in Equations 10.98 and 10.99, using integration by parts the weak formulation can be obtained as

$$\int_0^l \rho A \mathbf{N}^T(z)\mathbf{N}(z)dz\;\ddot{\eta}^{(ne)}(t)+\int_0^l k_{sc}AG\{\mathbf{N}'(z)\}^T\mathbf{N}'(z)dz\;\eta^{(ne)}(t)$$

$$+\int_0^l k_{sc}AG\{\mathbf{N}'(z)\}^T\mathbf{M}(z)dz\;\eta^{(ne)}(t) \tag{10.100}$$

$$=\begin{Bmatrix} -k_{sc}AG\left(v^{(e)\prime}+\varphi_x^{(e)}\right)\Big|_{z=0} \\ 0 \\ k_{sc}AG\left(v^{(e)\prime}+\varphi_x^{(e)}\right)\Big|_{z=l} \\ 0 \end{Bmatrix}$$

and

$$\int_0^l \rho I_{xx}\mathbf{M}^T(z)\mathbf{M}(z)dz\ddot{\eta}^{(ne)}(t)+\int_0^l EI_{xx}\{\mathbf{M}'(z)\}^T\mathbf{M}'(z)dz\;\eta^{(ne)}(t)$$

$$+\int_0^l k_{sc}AG\mathbf{M}^T(z)\mathbf{N}'(z)dz\;\eta^{(ne)}(t) \tag{10.101}$$

$$+\int_0^l k_{sc}AG\mathbf{M}^T(z)\mathbf{M}(z)dz\eta^{(ne)}(t)=\begin{Bmatrix} 0 \\ -EI_{xx}\varphi_x^{(e)'}\Big|_{z=0} \\ 0 \\ EI_{xx}\varphi_{xx}^{(e)'}\Big|_{z=l} \end{Bmatrix}$$

Equations 10.100 and 10.101 can be written as

$$\mathbf{M}_1\ddot{\eta}^{(ne)}(t)+\mathbf{K}_1\eta^{(ne)}(t)+\mathbf{K}_2\eta^{(ne)}(t)=\begin{Bmatrix} -k_{sc}AG\left(v'+\varphi_x\right)\big|_{z=0} \\ 0 \\ k_{sc}AG\left(v'+\varphi_x\right)\big|_{z=l} \\ 0 \end{Bmatrix} \tag{10.102}$$

and

$$\mathbf{M}_2\ddot{\eta}^{(ne)}(t)+\mathbf{K}_3^{(ne)}\;\eta(t)+\mathbf{K}_4^{(ne)}\;\eta(t)+\mathbf{K}_5^{(ne)}\;\eta(t)=\begin{Bmatrix} 0 \\ EI\varphi_x'\big|_{z=0} \\ 0 \\ -EI\varphi_x'\big|_{z=l} \end{Bmatrix} \tag{10.103}$$

with

$$\mathbf{M}_1 = \int_0^l \rho A \mathbf{N}^T(z)\mathbf{N}(z)dz = \frac{\rho A l}{840(1+\Phi)^2}$$

$$\begin{bmatrix} 4\left(70\Phi^2+147\Phi+78\right) & & & sym \\ \left(35\Phi^2+77\Phi+44\right)l & \left(7\Phi^2+14\Phi+8\right)l^2 & & \\ 4\left(35\Phi^2+63\Phi+27\right) & \left(35\Phi^2+63\Phi+26\right)l & \left(70\Phi^2+147\Phi+78\right) & \\ -\left(35\Phi^2+63\Phi+26\right)l & -\left(7\Phi^2+14\Phi+6\right)l^2 & -\left(35\Phi^2+77\Phi+44\right)l & \left(70\Phi^2+14\Phi+8\right)l^2 \end{bmatrix}$$

$$\mathbf{M}_2 = \int_0^l \rho I_{xx} \mathbf{M}^T(z)\mathbf{M}(z)dz$$

$$= \frac{\rho I_{xx}}{30(1+\Phi)^2 l}\begin{bmatrix} 36 & & & sym \\ -3(5\Phi-1)l & \left(10\Phi^2+5\Phi+4\right)l^2 & & \\ -36 & 3(5\Phi-1)l & 36 & \\ -3(5\Phi-1)l & \left(5\Phi^2-5\Phi-1\right)l^2 & 3(5\Phi-1)l & \left(10\Phi^2+5\Phi+4\right)l^2 \end{bmatrix}$$

$$\mathbf{K}_1 = \int_0^l k_{sc}AG\{\mathbf{N}'(z)\}^T \mathbf{N}'(z)dz = \frac{k_{sc}AG}{60(1+\Phi)^2 l}$$

$$\begin{bmatrix} 12\left(5\Phi^2+10\Phi+6\right) & & & sym \\ 6l & \left(5\Phi^2+10\Phi+8\right)l^2 & & \\ -12\left(5\Phi^2+10\Phi+6\right) & -6l & 12\left(5\Phi^2+10\Phi+6\right) & \\ 6l & -\left(5\Phi^2+10\Phi+2\right)l^2 & -6l & \left(5\Phi^2+10\Phi+8\right)l^2 \end{bmatrix}$$

$$\mathbf{K}_2 = \int_0^l k_{sc}AG\{\mathbf{N}'(z)\}^T \mathbf{M}(z)dz$$

$$= \frac{k_{sc}AG}{60(1+\Phi)^2 l}\begin{bmatrix} 12(5\Phi+6) & & & \text{sym} \\ 6l & \left(5\Phi^2+10\Phi+8\right)l^2 & & \\ -12(5\Phi+6) & 6\left(5\Phi^2+5\Phi-1\right)l & 12(5\Phi+6) & \\ 6l & -\left(5\Phi^2+10\Phi+2\right)l^2 & -6l & \left(5\Phi^2+10\Phi+8\right)l^2 \end{bmatrix}$$

$$\mathbf{K}_3 = \int_0^l EI_{xx}\left\{\mathbf{M}'(z)\right\}^T \mathbf{M}'(z)dz = \frac{EI_{xx}}{(1+\Phi)^2 l^3}\begin{bmatrix} 12 & & & sym \\ 6l & (\Phi^2+2\Phi+4)l^2 & & \\ -12 & -6l & 12 & \\ 6l & -(\Phi^2+2\Phi-2)l^2 & -6l & (\Phi^2+2\Phi+4)l^2 \end{bmatrix}$$

$$\mathbf{K}_4 = \int_0^l k_{sc}AG\mathbf{M}^T(z)\mathbf{N}'(z)dz = \frac{k_{sc}AG}{60(1+\Phi)^2 l}$$

$$\begin{bmatrix} 12(5\Phi+6) & & & sym \\ -6(5\Phi^2+5\Phi-1)l & (5\Phi^2+10\Phi+8)l^2 & & \\ -12(5\Phi+6) & -6l & 12(5\Phi+6) & \\ -6(5\Phi^2+5\Phi-1)l & -(5\Phi^2+10\Phi+2)l^2 & 6(5\Phi^2+5\Phi-1)l & (5\Phi^2+10\Phi+8)l^2 \end{bmatrix}$$

$$\mathbf{K}_5 = \int_0^l k_{sc}AG\mathbf{M}^T(z)\mathbf{M}(z)dz$$

$$= \frac{k_{sc}AG}{30(1+\Phi)^2 l}\begin{bmatrix} 36 & & & sym \\ -3(5\Phi-1)l & (10\Phi^2+5\Phi+4)l^2 & & \\ -36 & 3(5\Phi-1)l & 36 & \\ -3(5\Phi-1)l & (5\Phi^2-5\Phi-1)l^2 & 3(5\Phi-1)l & (10\Phi^2+5\Phi+4)l^2 \end{bmatrix}$$

and

$$\Phi = \frac{12EI_{xx}}{k_{sc}GAl^2}$$

On combining Equations 10.102 and 10.103, we get

$$\mathbf{M}^{(e)}\ddot{\eta}^{(ne)}(t) + \mathbf{K}^{(e)(ne)}\eta(t) = \mathbf{f}^{(ne)}(t) \tag{10.104}$$

with

$$\mathbf{M}^{(e)} = \mathbf{M}_1 + \mathbf{M}_2 = \mathbf{M}_T + \mathbf{M}_R$$

$$\mathbf{M}_T = \mathbf{M}_{T0} + \Phi\mathbf{M}_{T1} + \Phi^2\mathbf{M}_{T2}$$

$$\mathbf{M}_R = \mathbf{M}_{R0} + \Phi\mathbf{M}_{R1} + \Phi^2\mathbf{M}_{R2}$$

and

$$\mathbf{K}^{(e)} = \mathbf{K}_1 + \mathbf{K}_2 + \mathbf{K}_3 + \mathbf{K}_4 + \mathbf{K}_5 = \mathbf{K}_0^{(e)} + \Phi\mathbf{K}_1^{(e)}$$

where $\mathbf{f}^{(ne)}$ is a generalized force vector, and the mass matrix $\mathbf{M}$ consists of the translational mass matrix $\mathbf{M}_T$ and the rotational mass matrix $\mathbf{M}_R$. Details of the mass matrix and stiffness matrices are given in Appendix 10.1. Now, through examples, the effects of rotary inertia and shear deformation will be illustrated.

Example 10.3

Obtain the natural frequency parameters defined by a nondimensional form as $\bar{\omega}_{nf}\left(\bar{\omega}_{nf}^4 = \rho AL^4\omega_n^2/EI\right)$ of a uniform, nonrotating simply supported Timoshenko beam for the first four modes. The slenderness parameter R (=$r/2L$) is to be varied from 0.02 (thin beam) to 0.10 (thick beam). Herein, r is the radius of the shaft and L is the length of the shaft. Show the comparison for different numbers of elements and with the closed form analytical formula.

Solution:

Since in previous chapters the usual finite element procedures have been dealt with in detail (i.e., regarding the elemental equations, assembly procedures, application of boundary conditions, eigenvalue problem formulations, etc.), here those details for the present problem are omitted. The uniform, nonrotating simply supported Timoshenko beam for obtaining natural frequencies is first discretized into five elements. From the eigenvalue formulation of the present problem, the first four lowest natural frequencies are obtained and compared with the published work (Ku, 1998) and with the classical closed form solution given by Shames and Dym (2005). Ku (1998) considered C^0 continuity (i.e., compatibility up to translational displacements) for the Timoshenko beam model, whereas the model represented here considers C^1 continuity (i.e., the compatibility up to the translational and rotational displacements). Comparisons of nondimensional natural frequencies for the first four modes are presented in a tabular form in Tables 10.4 through 10.7, respectively. In the published work of Ku (1998), only the first two natural frequencies are available. The present study extracts the natural frequencies in the first four modes. These have been compared with classical closed form solutions. These studies are also conducted by discretizing the beam into three and seven elements to make convergence comparisons; there are summaries in Tables 10.4 through 10.7.

Due to the shear effect, deflection of the beam increases, which reduces the effective stiffness. So, the nondimensionalized natural frequency decreases as the diameter of the beam is increased (i.e., with the slenderness ratio). This trend can be observed in all four modes. The accuracy of the present solution is well within 5% for all cases. By the FEM, we overpredict with coarser discretization and we can see that as the number of elements increases from 3 to 7 for the first two modes and from 3 to 15 for the third and fourth modes, the natural frequency

TABLE 10.4
Natural Frequency Parameters $\bar{\omega}_{nf}$ of a Uniform, Nonrotating Simply Supported Timoshenko Beam for the First Mode

	Present Work			Ku (1998)			Shames and
R	[a](3)	(5)	(7)	(3)	(5)	(7)	Dym (2005)
0.02	3.1316	3.1300	3.1294	3.13460	3.1297	3.1293	3.1325
0.04	3.0992	3.0964	3.0956	3.11039	3.0942	3.0938	3.1061
0.06	3.0497	3.0449	3.0439	3.07206	3.0403	3.0399	3.0644
0.08	2.9881	2.9810	2.9792	3.01909	2.9736	2.9733	3.0103
0.10	2.9190	2.9098	2.9073	2.96344	2.8995	2.8992	2.9472

[a] Number in the brackets indicates the number of elements considered.

TABLE 10.5
Natural Frequency Parameters $\bar{\omega}_{nf}$ of a Uniform, Nonrotating Simply Supported Timoshenko Beam for the Second Mode

	Present Work			Ku (1998)			Shames and
R	(3)	(5)	(7)	(3)	(5)	(7)	Dym (2005)
0.02	6.2467	6.2033	6.1957	6.2875	6.2018	6.1913	6.2122
0.04	6.0555	5.9864	5.9702	6.0335	5.9589	5.9497	6.0210
0.06	5.8041	5.7082	5.6831	5.7170	5.6540	5.6462	5.7573
0.08	5.5373	5.4198	5.3873	5.3973	5.3439	5.3374	5.4671
0.10	5.2779	5.1454	5.1076	5.0995	5.0481	5.0481	5.1792

TABLE 10.6
Natural Frequency Parameters $\bar{\omega}_{nf}$ of a Uniform, Nonrotating Simply Supported Timoshenko Beam for the Third Mode

	Present Work					Shames and
R	(3)	(5)	(7)	(10)	(15)	Dym (2005)
0.02	9.8418	9.2159	9.1661	9.1444	9.1344	9.1934
0.04	9.6009	8.6629	8.5740	8.5285	8.5049	8.6364
0.06	9.2564	8.0541	7.9361	7.8726	7.8389	7.9839
0.08	8.8619	7.4939	7.3590	7.2849	7.2450	7.3659
0.10	8.1258	7.0046	6.8613	6.7813	6.7381	6.8137

TABLE 10.7
Natural Frequency Parameters $\bar{\omega}_{nf}$ of a Uniform, Nonrotating Simply Supported Timoshenko Beam for the Fourth Mode

	Present Work					Shames and
R	(3)	(5)	(7)	(10)	(15)	Dym (2005)
0.02	13.2640	12.2161	12.0504	11.9727	11.9361	12.0421
0.04	12.5987	11.2187	10.9781	10.8396	10.7659	10.9357
0.06	11.6571	10.2279	9.9587	9.78718	9.6931	9.8244
0.08	10.3753	9.3770	9.1074	8.9223	8.8190	8.8755
0.10	8.4569	7.8377	7.7423	7.6888	7.6595	8.0456

parameter is decreasing. Convergence (not much improvement) can especially be observed in Tables 10.6 and 10.7 between elements 10–15. The difference in the closed form solution of Shames and Dym (2005) and the FEM results could be due to the approximate nature of the eigenfunction (i.e., $\sin\frac{n\pi z}{L}$ and chosen in the closed form solution). It will be interesting to compare natural frequencies with Euler–Bernoulli beam theory. This will give idea about for what slenderness ratio, which beam theory is sufficient to consider without compromising on the accuracy.

Example 10.4

A typical simply supported rotor disk system as shown in Figure 10.11 is to be analyzed for obtaining the whirl natural frequencies to show the application of the FEM. The physical properties of the rotor system are given as following: The diameter of shaft is 0.1 m, the length of shaft is 3.5 m, the Young's modulus of material of the shaft is 2.08×10^{11} N/m^2, the mass density of the shaft material is 7830 kg/m^3, the Poisson's ratio is 0.3, the number of rigid disks is 4, and the mass of each rigid disk is 60.3 kg.

Solution:

The rotor is discretized into the 7 and 14 elements, respectively, as shown in Figures 10.11 and 10.12. In the case of seven elements, two identical rigid bearings are located at nodes 2 and 7, and four rigid disks are located at nodes 3, 4, 5, and 6. In the case of the 14-element member, two identical rigid bearings are located at nodes 3 and 13, and rigid disks are located at nodes 5, 7, 9, and 11. The shaft is assumed to be uniform along the span.

Rigid disks are considered as point masses and these point masses are added to the mass matrix corresponding to the locations of rigid disk translational accelerations. The assembled mass and stiffness matrices are obtained by the usual method, and boundary conditions are applied to get the reduced form of the dynamic matrix. Natural whirl frequencies are obtained by solving the eigenvalue problem (just to mention that the shaft is not rotating for the present case). They are obtained for the 7- and 14-element model, and are given in Table 10.8. Results show that good convergence has already occurred with the seven-element model.

For the present case, R is less than 0.02 so the effect of shear is less but as the rotary inertia is also considered, the natural whirl frequencies of the rotor-bearing system decrease as compared to the Euler–Bernoulli case. This decrease is 0.3%, 1.8%, 2%, and 3% for the first, second, third,

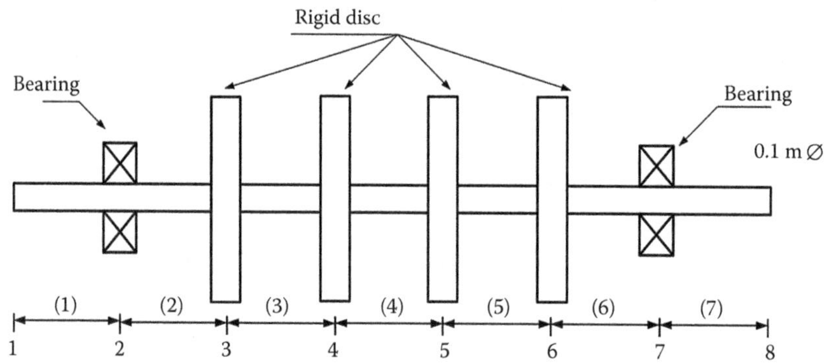

FIGURE 10.11 Rotor-bearing system with rigid disks (seven elements of 0.50-m length each).

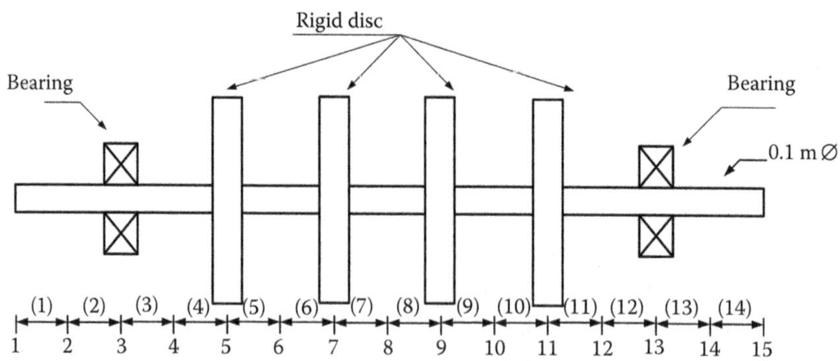

FIGURE 10.12 Rotor-bearing system with rigid disks (14 elements of 0.25-m length each).

TABLE 10.8
Natural Whirl Frequencies of the Rotor-Bearing System with Rigid Disks (for the Nonrotating Timoshenko Beam Case)

	Natural Whirl Frequencies without Considering Shear and Rotary Inertia Effects (rad/s)		Natural Whirl Frequencies with Shear and Rotary Inertia Effects (rad/s)	
Mode No.	For Seven Elements	For 14 Elements	For Seven Elements	For 14 Elements
1	116.1756	115.8136	115.6111	115.3816
2	438.4482	438.2795	431.0719	430.8983
3	861.5609	860.7477	839.1099	838.1357
4	1209.4922	1207.0327	1168.2006	1165.2664

and fourth modes, respectively. The difference is more at higher modes as the rotary inertia becomes predominant at higher frequencies. It should be noted that in the present problem, the rotor is assumed to be nonrotating and because of this gyroscopic couple is absent. The case of a rotating shaft with rotary inertia, shear deformation, and gyroscopic couple is considered in the next section.

10.4 WHIRLING OF TIMOSHENKO SHAFTS

In the present section, the analysis of whirling of a spinning Timoshenko shaft is presented, which includes higher effects like the rotary inertia, the shear effect, and the gyroscopic effect (Nelson, 1980, 1998). A finite element formulation including these effects is presented by using the consistent mass matrix approach. Because of the gyroscopic effect, two perpendicular transverse motions are now coupled and most importantly the natural frequencies of the rotor system depend upon the angular speed of the shaft. This leads to the forward and backward whirls phenomena, which we have already described in detail for the single mass rotor system in Chapter 5 and in Chapter 8 with the transfer matrix method. The finite element formulation in accordance with the real coordinate system with the gyroscopic effect yields the gyroscopic matrix as skew symmetric and all other matrices as symmetric. The eigenvalue problem (without damping) gives eigenvalues as pure imaginary and eigenvectors as complex. The finite element analysis of the Timoshenko rotating beam compared with results obtained by Weaver, Timoshenko, and Young (1990) and Eshleman and Eubanks (1969); they determined critical speeds from approximate relations.

10.4.1 Equations of Motion of a Spinning Timoshenko Shaft

The analysis of the previous section could be extended for the present case with inclusion of gyroscopic effects. However, this would lead to the coupling of motions in two orthogonal planes. The translation of the cross section centerline (elastic line) neglecting the axial motion is given by two displacements $u(z, t)$ and $v(x, t)$ in the x and y direction, respectively, which consists of contribution due to both the bending deformation and contribution due to the shear deformation:

$$u(z,t) = u_b(z,t) + u_s(z,t) \quad \text{and} \quad v(z,t) = v_b(z,t) + v_s(z,t) \tag{10.105}$$

The rotation of the cross section is described by the rotation angles $\varphi_x(z,t)$ and $\varphi_y(z,t)$ about the x and y axes, respectively, which are associated with bending deformation of the element as

$$\varphi_x(z,t) = -\frac{\partial v_b(z,t)}{\partial z} \quad \text{and} \quad \varphi_y(z,t) = \frac{\partial u_b(z,t)}{\partial z} \tag{10.106}$$

The different signs in expressions due to the direction of the slope and the angular displacement in the y-z and z-x planes are different. On differentiating Equation 10.105, we get

$$u'(z,t)=\frac{\partial u_b(z,t)}{\partial z}+u_s'(z,t) \quad \text{and} \quad v'(z,t)=\frac{\partial v_b(z,t)}{\partial z}+v_s'(z,t) \tag{10.107}$$

On noting Equation 10.106, Equation 10.107 takes the form

$$u'(z,t)=\varphi_y(z,t)+u_s'(z,t) \quad \text{and} \quad v'(z,t)=-\varphi_x(z,t)+v_s'(z,t) \tag{10.108}$$

which can be written as

$$u_s'(z,t)=u'(z,t)-\varphi_y(z,t) \quad \text{and} \quad v_s'(z,t)=v'(z,t)+\varphi_x(z,t) \tag{10.109}$$

where u_s' and v_s' are shear strains. For the uniform differential shaft element located at z, the potential energy of an element of length can be expressed as

$$U=\frac{1}{2}\int_0^l\left[EI\left\{(\varphi_y')^2+(\varphi_x')^2\right\}+k_{sc}AG\left\{u_s'^2+v_s'^2\right\}\right]dz \tag{10.110}$$

where the first term is the elastic bending and the second term is the shear deformation energy. On noting Equation 10.109, we get

$$U=\frac{1}{2}\int_0^l\left[EI\left\{(\varphi_y')^2+(\varphi_x')^2\right\}+k_{sc}AG\left\{(u'-\varphi_y)^2+(v'+\varphi_x)^2\right\}\right]dz \tag{10.111}$$

The kinetic energy of a shaft element rotating at a constant speed, ω, including the translational and rotational forms, is given by

$$T=\frac{1}{2}\int_0^l\left\{\rho A\left(\dot{u}^2+\dot{v}^2\right)+\rho I\left(\dot{\varphi}_y^2+\dot{\varphi}_x^2\right)+I_P\omega\left(\dot{\varphi}_x\varphi_y-\dot{\varphi}_y\varphi_x\right)\right\}dz+\frac{1}{2}\int_0^l I_P\omega^2 dz \tag{10.112}$$

In Equation 10.112, various kinetic energy terms are contributed as follows: The first term is due to transverse translational motions, the second term is due to transverse rotational motions, the third term is due to gyroscopic couples, and the last term is due to spinning of the rotor. In Chapter 5, derivation of terms related to gyroscopic moments have been described in detail, while discussing the energy method. Angular momentum in the y–z plane and z–x plane are opposite in direction and have the same magnitude. The quantity $\varphi_x\varphi_y$ is a constant quantity in the conservative system, so it gives

$$\frac{d(\varphi_x\varphi_y)}{dt}=0; \quad \text{so that } \dot{\varphi}_x\varphi_y=-\varphi_x\dot{\varphi}_y \tag{10.113}$$

Noting Equation 10.113, Equation 10.112 becomes

$$T=\frac{1}{2}\int_0^l\left\{\rho A\left(\dot{u}^2+\dot{v}^2\right)+\rho I\left(\dot{\varphi}_y^2+\dot{\varphi}_x^2\right)-2I_P\omega\dot{\varphi}_y\varphi_x\right\}dz+\frac{1}{2}\int_0^l I_P\omega^2 dz \tag{10.114}$$

If $f_x(z,t)$ and $f_y(z,t)$ are distributed transverse forces in the x and y directions, respectively, then the work done by external forces is

$$W_{nc}=\int_0^l\left(f_x u+f_y v\right)dz \tag{10.115}$$

From extended Hamilton's principle, we have

$$\int_{t_1}^{t_2} \{\delta(T-U)+\delta W_{nc}\}\,dt = 0 \tag{10.116}$$

On substituting Equations 10.111, 10.114, and 10.115 into Equation 10.116, we get

$$\begin{aligned}
&\int_{t_1}^{t_2}\left[\delta\left(\frac{1}{2}\int_0^l \left\{\rho A\left(\dot{u}^2+\dot{v}^2\right)+\rho I\left(\dot{\varphi}_y^2+\dot{\varphi}_x^2\right)-2I_P\omega\dot{\varphi}_y\varphi_x+I_P\omega^2\right\}dz\right)\right]dt \\
&-\int_{t_1}^{t_2}\left[\delta\left(\frac{1}{2}\int_0^l \left[EI\left\{(\varphi_y')^2+(\varphi_x')^2\right\}+k_{sc}AG\left\{(u'-\varphi_y)^2+(v'+\varphi_x)^2\right\}\right]dz\right)\right]dt \\
&+\int_{t_1}^{t_2}\left[\delta\left\{\int_0^l (f_x u+f_y v)\,dz\right\}\right]dt = 0
\end{aligned} \tag{10.117}$$

On operating the variation operator in Equation 10.117, we get

$$\begin{aligned}
&\int_{t_1}^{t_2}\left[\int_0^l \left\{\rho A\dot{u}\delta(\dot{u})+\rho A\dot{v}\delta(\dot{v})+\rho I\dot{\varphi}_x\delta(\dot{\varphi}_x)+\rho I\dot{\varphi}_y\delta(\dot{\varphi}_y)-2I_P\omega\delta(\dot{\varphi}_y)\varphi_x-2I_P\omega\dot{\varphi}_y\delta(\varphi_x)\right\}dz\right]dt \\
&-\int_{t_1}^{t_2}\left[\int_0^l \left\{EI\varphi_y'\delta(\varphi_y')+EI\varphi_x'\delta(\varphi_x')+k_{sc}AG(u'-\varphi_y)\delta(u'-\varphi_y)+k_{sc}AG(v'+\varphi_x)\delta(v'+\varphi_x)\right\}dz\right]dt \\
&+\int_{t_1}^{t_2}\int_0^l \{f_x\delta u+f_y\delta v\}\,dz\,dt = 0
\end{aligned} \tag{10.118}$$

On changing the order of variation and differentiation in Equation 10.118, we get

$$\begin{aligned}
&\int_{t_1}^{t_2}\left[\int_t^0 \left\{\begin{matrix}\rho A\dot{u}\dfrac{d}{dt}(\delta u)+\rho A\dot{v}\dfrac{d}{dt}(\delta v)+\rho I\,\dot{\varphi}_y\dfrac{d}{dt}(\delta\varphi_y)+\rho I\dot{\varphi}_x\dfrac{d}{dt}(\delta\varphi_x) \\ -I_p\omega\dfrac{d}{dt}(\delta\varphi_y)\varphi_x-I_p\omega\dot{\varphi}_y\delta(\varphi_x)\end{matrix}\right\}dz\right]dt \\
&-\int_{t_1}^{t_2}\left[\int_0^l \left\{\begin{matrix}EI\varphi_y'\dfrac{d}{dz}(\delta\varphi_y)+EI\varphi_x'\dfrac{d}{dz}(\delta\varphi_x)+k_{sc}AG(u'-\varphi_y)\left(\dfrac{d}{dz}(\delta u)-\delta\varphi_y\right) \\ +k_{sc}AG(v'+\varphi_x)\left(\dfrac{d}{dz}(\delta v)+\delta\varphi x\right)\end{matrix}\right\}dz\right]dt \\
&+\int_{t_1}^{t_2}\int_0^l \{f_x\delta u+f_y\delta v\}\,dz\,dt = 0
\end{aligned} \tag{10.119}$$

On performing integration by parts of terms, which has both differential and variational operators, in Equation 10.119, we get

$$
\begin{aligned}
&\int_0^l \left\{\rho A\dot{u}(\delta u)\right\}dz\Big|_{t_1}^{t_2} - \int_{t_1}^{t_2}\left[\int_0^l \left\{\rho A\ddot{u}(\delta u)\right\}dz\right]dt + \int_0^l \left\{\rho A\dot{v}(\delta v)\right\}dz\Big|_{t_1}^{t_2} - \int_{t_1}^{t_2}\left[\int_0^l \left\{\rho A\ddot{v}(\delta v)\right\}dz\right]dt \\
&+\int_0^l \left\{\rho I\dot{\varphi}_y(\delta\varphi_y)\right\}dz\Big|_{t_1}^{t_2} - \int_{t_1}^{t_2}\left[\int_0^l \left\{\rho I\ddot{\varphi}_y(\delta\varphi_y)\right\}dz\right]dt + \int_0^l \left\{\rho I\dot{\varphi}_x(\delta\varphi_x)\right\}dz\Big|_{t_1}^{t_2} - \int_{t_1}^{t_2}\left[\int_0^l \left\{\rho I\ddot{\varphi}_x(\delta\varphi_x)\right\}dz\right]dt \\
&-\int_0^l I_P\omega\varphi_x\delta(\varphi_y)dz\Big|_{t_1}^{t_2} + \int_{t_1}^{t_2}\left[\int_0^l \left\{I_P\omega\varphi_x\delta(\varphi_y) - I_P\omega\dot{\varphi}_y\delta(\varphi_x)\right\}dz\right]dt \\
&-\int_{t_1}^{t_2}\left[EI\varphi_y'(\delta\varphi_y)\right]dt\Big|_0^l + \int_{t_1}^{t_2}\left[\int_0^l \left\{EI\varphi_y''(\delta\varphi_y)\right\}dz\right]dt - \int_{t_1}^{t_2}\left\{EI\varphi_x'(\delta\varphi_x)\right\}dt\Big|_0^l + \int_{t_1}^{t_2}\left[\int_0^L \left\{EI\varphi_x''(\delta\varphi_x)\right\}dz\right]dt \\
&-\int_{t_1}^{t_2}\left\{k_{sc}AG(u'-\varphi_y)(\delta u)\right\}dt\Big|_0^l + \int_{t_1}^{t_2}\left[\int_0^l \left\{k_{sc}AG(u''-\varphi_y')(\delta u)\right\}dz\right]dt + \int_{t_1}^{t_2}\left[\int_0^l \left\{k_{sc}AG(u'-\varphi_y)(\delta\varphi_y)\right\}dz\right]dt \\
&-\int_{t_1}^{t_2}\left\{k_{sc}AG(v'+\varphi_x)(\delta v)\right\}dt\Big|_0^l + \int_{t_1}^{t_2}\left[\int_0^l \left\{k_{sc}AG(v''+\varphi_x')(\delta v)\right\}dz\right]dt - \int_{t_1}^{t_2}\left[\int_0^l \left\{k_{sc}AG(v'+\varphi_x)(\delta\varphi_x)\right\}dz\right]dt \\
&+\int_{t_1}^{t_2}\left[\int_0^l (f_x\delta u + f_y\delta v)dz\right]dt = 0
\end{aligned}
\tag{10.120}
$$

The first, third, fifth, seventh, and ninth terms of Equation 10.120 will vanish, since variations are defined to be zero at t_1 and t_2. The remaining terms can be rearranged in the following form:

$$
\begin{aligned}
&-\int_{t_1}^{t_2}\left[\int_0^l \left\{\rho A\ddot{u} - k_{sc}AG(u''-\varphi_y') - f_x\right\}\delta u dz\right]dt - \int_{t_1}^{t_2}\left[\int_0^l \{\rho A\ddot{v} + k_{sc}AG(v''+\varphi_x') - f_y)\}\delta v dz\right]dt \\
&-\int_{t_1}^{t_2}\left[\int_0^l \left\{\rho I\ddot{\varphi}_y - EI\varphi_y'' - k_{sc}AG(u'-\varphi_y) - I_P\omega\dot{\varphi}_x\right\}dz\delta\varphi_y\right]dt \\
&-\int_{t_1}^{t_2}\left[\int_0^l \left\{\rho I\ddot{\varphi}_x - EI\varphi_x'' - k_{sc}AG(v'+\varphi_x) + I_P\omega\dot{\varphi}_y\right\}dz\delta\varphi_x\right]dt \\
&-\int_{t_1}^{t_2}\left[EI\varphi_y'(\delta\varphi_y)\right]dt\Big|_0^l - \int_{t_1}^{t_2}\left[EI\varphi_x'(\delta\varphi_x)\right]dt\Big|_0^l - \int_{t_1}^{t_2}\left[\left\{k_{sc}AG(u'-\varphi_y)(\delta u)\right\}\right]dt\Big|_0^l \\
&-\int_{t_1}^{t_2}\left[\left\{k_{sc}AG(v'+\varphi_x)(\delta v)\right\}\right]dt\Big|_0^l = 0
\end{aligned}
\tag{10.121}
$$

Variations δu, $\delta\varphi_y$, δv, and $\delta\varphi_x$ in the spatial domain are arbitrary so their coefficients must be zero; this yields the differential equations of motion as

$$\rho A\ddot{u} - k_{sc}AG\left(u'' - \varphi_y'\right) = f_x; \quad EI\varphi_y'' + k_{sc}AG\left(u' - \varphi_y\right) - \rho I\ddot{\varphi}_y + I_P\omega\dot{\varphi}_x = 0$$
$$\rho A\ddot{v} - k_{sc}AG\left(v'' + \varphi_x'\right) = f_y; \quad EI\varphi_x'' + k_{sc}AG\left(v' + \varphi_x\right) - \rho I\ddot{\varphi}_x - I_P\omega\dot{\varphi}_y = 0 \tag{10.122}$$

and the boundary (geometrical and natural) conditions are

$$k_{sc}AG\left(u' - \varphi_y\right)\delta u\Big|_0^l = 0, \quad EI\varphi_x'\delta\varphi_x\Big|_0^l = 0 \tag{10.123}$$

and

$$k_{sc}AG\left(v' + \varphi_x\right)\delta v\Big|_0^l = 0, \quad EI\varphi_y'\delta\varphi_y\Big|_0^l = 0 \tag{10.124}$$

It should be noted that now all four equations of motion (Equation 10.122) are coupled (basically two orthogonal plane motions are coupled) due to gyroscopic couple terms and need to be solved, simultaneously. In the next subsection, no closed form solution would be attempted but the finite element formulation of governing equations is presented.

10.4.2 Finite Element Formulation

Now the finite element formulation for governing Equation 10.122 will be developed by using Galerkin's method. As compared to the previous section on finite element formulations, here two-plane governing equations need to be considered, simultaneously. In the finite element model, the continuous displacement field can be approximated in terms of generalized displacements of the element nodes. In the present finite element model (see Figures 10.13 and 10.14), each element has two nodes and each node has four generalized displacements (two translational and the other two rotational).

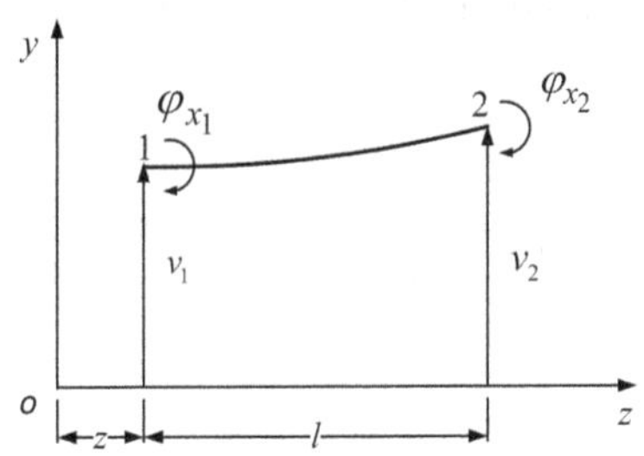

FIGURE 10.13 A typical beam element in the y–z plane.

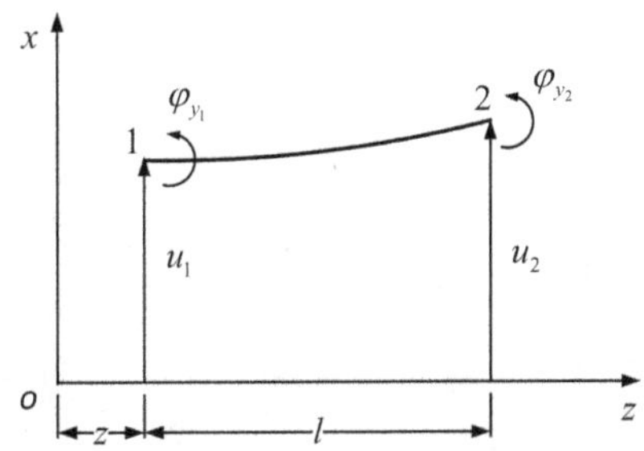

FIGURE 10.14 A typical rotor element in the z–x plane.

The translational displacement within the element can be obtained by using the appropriate shape functions and are defined as

$$u(z,t) = \mathbf{N}_u(z)\ \eta^{(ne)}(t) \quad \text{and} \quad v(z,t) = \mathbf{N}_v(z)\ \eta^{(ne)}(t) \tag{10.125}$$

with

$$\eta^{(ne)}(t) = \lfloor u_1 \quad v_1 \quad \varphi_{x1} \quad \varphi_{y1} \quad u_2 \quad v_2 \quad \varphi_{x2} \quad \varphi_{y2} \rfloor^T$$

and

$$\mathbf{N}_u = \lfloor N_1 \quad 0 \quad 0 \quad N_2 \quad N_3 \quad 0 \quad 0 \quad N_4 \rfloor$$

$$\mathbf{N}_v = \lfloor 0 \quad N_1 \quad -N_2 \quad 0 \quad 0 \quad N_3 \quad -N_4 \quad 0 \rfloor$$

In matrix form, this could be combined as

$$\begin{Bmatrix} u(z,t) \\ v(z,t) \end{Bmatrix} = \mathbf{N}_t(z)\ \eta(t)^{(ne)} \tag{10.126}$$

with

$$\mathbf{N}_t(z) = \begin{bmatrix} \mathbf{N}_u(z) \\ \mathbf{N}_v(z) \end{bmatrix} = \begin{bmatrix} N_1 & 0 & 0 & N_2 & N_3 & 0 & 0 & N_4 \\ 0 & N_1 & -N_2 & 0 & 0 & N_3 & -N_4 & 0 \end{bmatrix}$$

where $\mathbf{N}_t(z)$ is called the translational shape function matrix. Also, N_i ($i=1, 2, 3, 4$) are the same as that derived in the previous section, which is given in Equation 10.92, and $\eta^{(ne)}(t)$ is the nodal displacement vector. It should be carefully noted that in the previous section one-plane motion was considered, and accordingly, in Equation 10.92, the order of stacking of displacements is different, and can be chosen as per convenience; however, the mass and stiffness matrices might take different forms that should be taken care of. Now, the angular displacement within the element can be obtained by

$$\varphi_x(z,t) = \mathbf{M}_{\varphi_x}(z)\ \eta^{(ne)}(t) \quad \text{and} \quad \varphi_y(z,t) = \mathbf{M}_{\varphi_y}(z)\ \eta^{(ne)}(t) \tag{10.127}$$

and

$$\mathbf{M}_{\varphi_x}(z) = -\lfloor 0 \quad M_1 \quad -M_2 \quad 0 \quad 0 \quad M_3 \quad -M_4 \quad 0 \rfloor$$

$$\mathbf{M}_{\varphi_y}(z) = \lfloor M_1 \quad 0 \quad 0 \quad M_2 \quad M_3 \quad 0 \quad 0 \quad M_4 \rfloor$$

In matrix form we have

$$\begin{Bmatrix} \varphi_x(z,t) \\ \varphi_y(z,t) \end{Bmatrix} = \mathbf{M}_t(z)\ \eta^{(ne)}(t) \tag{10.128}$$

with

$$\mathbf{M}_t(z) = \begin{bmatrix} \mathbf{M}_{\varphi_x}(z) \\ \mathbf{M}_{\varphi_y}(z) \end{bmatrix} = \begin{bmatrix} 0 & -M_1 & M_2 & 0 & 0 & -M_3 & M_4 & 0 \\ M_1 & 0 & 0 & M_2 & M_3 & 0 & 0 & M_4 \end{bmatrix}$$

Herein, $\mathbf{M}_t(z)$ is called the *rotational shape function matrix.* $M_{i,}$ $i=1, 2, 3, 4$ is given in Equation 10.93.

10.4.3 Weak Form Finite Element Formulations

Since in the previous section the weak form finite element formulation without gyroscopic couple effect has already been performed in great detail, for the present case the same is described briefly. Neglecting the work done by external forces and substituting Equations 10.125 and 10.127 in the equations of motion (Equation 10.122), residues can be given as

$$\begin{aligned} R_1^{(e)} &= \rho A\ddot{u}^{(e)} - k_{sc}AG\left(u^{(e)\prime} - \varphi_y^{(e)}\right)' \\ R_2^{(e)} &= \rho A\ddot{v}^{(e)} - k_{sc}AG\left(v^{(e)\prime} + \varphi_x^{(e)}\right)' \\ R_3^{(e)} &= EI\varphi_x''^{(e)} + k_{sc}AG\left(v'^{(e)} + \varphi_x^{(e)}\right) - \rho I\ddot{\varphi}_x^{(e)} - I_P\omega\dot{\varphi}_y^{(e)} \\ R_4^{(e)} &= EI\varphi_y''^{(e)} + k_{sc}AG\left(u'^{(e)} - \varphi_y^{(e)}\right) - \rho I\ddot{\varphi}_y^{(e)} + I_P\omega\dot{\varphi}_x^{(e)} \end{aligned} \tag{10.129}$$

The Galerkin method is used to minimize the residue. So, applying the weight function equivalent to the shape functions, residues can be minimized as

$$\begin{aligned} &\int_0^l \mathbf{N}_u R_1^{(e)} dz = \mathbf{0} \\ &\int_0^l \mathbf{N}_v R_2^{(e)} dz = \mathbf{0} \\ &\int_0^l \mathbf{M}_{\varphi_x} R_3^{(e)} dz = \mathbf{0}; \quad \int_0^l \mathbf{M}_{\varphi_y} R_4^{(e)} dz = \mathbf{0} \end{aligned} \tag{10.130}$$

Using Equations 10.129 and 10.130, weak formulations can be obtained and the equation of motion for the finite element takes the form

$$\mathbf{M}^{(e)}\ddot{\boldsymbol{\eta}}^{(ne)}(t) - \omega\mathbf{G}^{(e)}\dot{\boldsymbol{\eta}}^{(ne)}(t) + \mathbf{K}^{(e)}\boldsymbol{\eta}^{(ne)}(t) = \mathbf{f}^{(ne)}(t) \tag{10.131}$$

with

$$\mathbf{M}^{(e)} = \mathbf{M}_T + \mathbf{M}_R$$

$$\mathbf{K}^{(e)} = \mathbf{K}_0 + \Phi\mathbf{K}_1$$

$$\mathbf{G}^{(e)} = \mathbf{G}_0 + \Phi\mathbf{G}_1 + \Phi^2\mathbf{G}_2$$

where $\mathbf{M}^{(e)}$ is the elemental mass matrix, $\mathbf{K}^{(e)}$ is the elemental stiffness matrix, and $\mathbf{G}^{(e)}$ is the elemental gyroscopic matrix. The mass matrix $\mathbf{M}^{(e)}$ consists of the translational mass matrix $\mathbf{M}_T$ and the rotational mass matrix $\mathbf{M}_R$, which are given as

$$\mathbf{M}_T = \mathbf{M}_{T0} + \Phi\mathbf{M}_{T1} + \Phi^2\mathbf{M}_{T2}; \quad \mathbf{M}_R = \mathbf{M}_{R0} + \Phi\mathbf{M}_{R1} + \Phi^2\mathbf{M}_{R2} \tag{10.132}$$

Details of the mass, stiffness, and gyroscopic matrices are given in Appendix 10.2.

10.4.4 Rigid Disk Element

Rigid disk equations of motion have already been described in detail in Chapter 5. The rigid disk equation of motion can be written as

$$\mathbf{M}^d\ddot{\eta}^d(t) - \omega\mathbf{G}^d\dot{\eta}^d(t) + \mathbf{0}\eta(t) = \mathbf{f}^d(t) \tag{10.133}$$

where $\mathbf{M}^d$ is the disk mass matrix, $\mathbf{G}^d$ is the gyroscopic matrix, $\mathbf{0}$ is a null stiffness matrix, and $\mathbf{f}^d(t)$ is the vector of generalized external forces (refer to Appendix 10.2).

10.4.5 System Equations of Motion

Equations of motion of the complete system can be obtained by assembling the contribution of each elemental equation of motion, i.e., Equations 10.131 and 10.133 (refer to Chapter 9 for details of the assembly procedure). The system equation of motion becomes

$$\mathbf{M}\ddot{\eta} - \omega\mathbf{G}\dot{\eta} + \mathbf{K}\eta = \mathbf{f}(t) \tag{10.134}$$

where $\mathbf{M}$ is the assembled mass matrix, $\mathbf{G}$ is the assembled gyroscopic matrix, $\mathbf{K}$ is the assembled stiffness matrix, $\mathbf{f}(t)$ is the assembled external force vector, and $\eta(t)$ is the assembled vector of nodal coordinates of the whole system.

10.4.6 Eigenvalue Problem

For the sake of computation of eigenvalues and eigenvectors, the homogeneous part of Equation 10.134 is written in the first-order state vector form as

$$\begin{bmatrix} \mathbf{0} & \mathbf{M} \\ \mathbf{M} & -\omega\mathbf{G} \end{bmatrix}\dot{\mathbf{h}} + \begin{bmatrix} -\mathbf{M} & \mathbf{0} \\ \mathbf{0} & \mathbf{K} \end{bmatrix}\mathbf{h} = \mathbf{0} \tag{10.135}$$

In compact form this is

$$\mathbf{M}^*\dot{\mathbf{h}} + \mathbf{K}^*\mathbf{h} = \mathbf{0} \tag{10.136}$$

with

$$\mathbf{h} = \begin{Bmatrix} \dot{\eta} \\ \eta \end{Bmatrix}$$

$$\mathbf{M}^* = \begin{bmatrix} \mathbf{0} & \mathbf{M} \\ \mathbf{M} & -\omega\mathbf{G} \end{bmatrix}$$

$$\mathbf{K}^* = \begin{bmatrix} -\mathbf{M} & \mathbf{0} \\ \mathbf{0} & \mathbf{K} \end{bmatrix}$$

The solution is assumed to be of the form

$$\mathbf{h} = \mathbf{h}_0\, e^{\lambda t} \tag{10.137}$$

This gives

$$\dot{\mathbf{h}} = \lambda \mathbf{h}_0 e^{\lambda t} \tag{10.138}$$

On substituting Equations 10.137 and 10.138 into Equation 10.136, we get

$$\lambda \mathbf{M}^* \mathbf{h}_0 e^{\lambda t} + \mathbf{K}^* \mathbf{h}_0 e^{\lambda t} = \mathbf{0} \tag{10.139}$$

Therefore, the associated eigenvalue problem becomes

$$(\mathbf{D} + \lambda \mathbf{I})\mathbf{h}_0 = \mathbf{0} \quad \text{with} \quad \mathbf{D} = \mathbf{M}^{*^{-1}}\mathbf{K}^* \tag{10.140}$$

The eigenvalues of Equation 10.140 appear as pure imaginary conjugate pairs with magnitude equal to the natural whirl frequencies. When the gyroscopic effect is dropped, eigenvalues appear as repeated pairs of pure imaginary roots due to uncoupled motions in orthogonal planes. When the gyroscopic effect is considered, it leads to the splitting of eigenvalues into forward and backward whirls (i.e., with the increase in spin speed, increasing natural whirl frequencies represent forward whirls, whereas decreasing natural whirl frequencies represent backward whirls). Because of the presence of gyroscopic terms, the natural whirl frequencies are a function of the rotational or spin speed of the shaft. To determine the critical speeds, a Campbell diagram can be drawn to find the intersection points of natural whirl frequencies with one per revolution excitation (i.e., for synchronous whirl, the $\lambda = \omega$ line) and higher multiples of per revolution excitations ($\lambda = n\omega$ line, with $n=2$, 3, …) (refer to Chapter 5).

Example 10.5

Obtain the forward and backward critical speeds of a uniform cross section of a simply supported spinning Timoshenko shaft. Consider gyroscopic effects also. Show the results up to the second modes with different number of elements and for different values of the slenderness ratio.

Solution:

To evaluate the natural whirl frequencies of a uniform cross section, the simply supported rotating (spinning) Timoshenko beam model in which gyroscopic effects also play a predominant role is discretized into five different elements. The comparison is carried out for the forward and backward critical speeds, $\lambda=\pm 1$, and the results are compared to the values published by Eshleman and Eubanks (1969). Critical speeds are nondimensionalized by using the natural frequency parameter, $\bar{\omega}_{nf}\left(\bar{\omega}_{nf}^2 = \rho A L^4 \omega_{nf}^2 / EI\right)$. The results are also obtained for ten elements. These results are listed in Tables 10.9 and 10.10 for the first and second critical speeds, respectively. The effect of shear continues to increase as the slenderness parameter R ($R = r/2L$, where r is the radius of beam) is changed from 0.02 to 0.10. The results agree with the values published by Eshleman and Eubanks (1969).

TABLE 10.9
First Nondimensionalized Critical Speeds of a Simply Supported Rotating Timoshenko Beam

	Backward Whirl ($\lambda = -1$)			Forward Whirl ($\lambda = 1$)		
	Present Work		Eshleman and Eubanks (1969)	Present Work		Eshleman and Eubanks (1969)
R	(5)	(10)		(5)	(10)	
0.02	3.1240	3.1234	3.1251	3.1360	3.1354	3.1373
0.04	3.0745	3.0739	3.0780	3.1187	3.1180	3.1245
0.06	3.0028	3.0018	3.0067	3.0899	3.0887	3.1037
0.08	2.9183	2.9166	2.9193	3.0494	3.0473	3.0757
0.10	2.8291	2.8269	2.8234	2.9985	2.9954	3.0416

TABLE 10.10
Second Nondimensionalized Critical Speeds of a Simply Supported Rotating Timoshenko Beam

	Backward Whirl ($\lambda = -1$)			Forward Whirl ($\lambda = 1$)		
	Present Work		Eshleman and Eubanks (1969)	Present Work		Eshleman and Eubanks (1969)
R	(5)	(10)		(5)	(10)	
0.02	6.1597	6.1522	6.1560	6.2487	6.2408	6.2489
0.04	5.8591	5.8440	5.8387	6.1278	6.1084	6.1515
0.06	5.5132	5.4914	5.4493	5.9280	5.8945	6.0045
0.08	5.1858	5.1582	5.0670	5.6734	5.6279	5.8250
0.10	4.8937	4.8621	4.7231	5.3964	5.3258	5.6287

Example 10.6

Obtain the forward and backward critical speeds of the rotor system of Example 10.4 by considering the rotary and gyroscopic effects of the shaft along with the disks. In the present example, the effects of spinning of the rotor have been considered, and hence obtain the Campbell diagram. Each disk is having polar mass moment of inertia and diametral mass moment of inertia values 1.1 and 0.7 kg-m^2, respectively.

Solution:

A typical rotor disk system as shown in Figure 10.12 is analyzed to show the application of the present finite element model. A typical rotor-bearing system as shown in Figures 10.11 and 10.12 is analyzed to show the application of the finite element model for the present case (i.e., the rotary inertia, shear, and gyroscopic effects). The analysis is performed with different numbers of elements (7 and 14) to establish the convergence. The physical properties of the shaft are the same as in Example 10.4. The natural whirl frequencies as a function of spin speed are obtained by solving the eigenvalue problem at different spin speeds, sequentially. The critical speed map, called the Campbell diagram, is plotted for the first four natural whirl frequencies as a function of the spin speed. Critical speeds are obtained as intersection points of natural whirl frequencies with the synchronous whirl. Variation of natural whirl frequencies of the rotor system as a function of the spin speed including the rotary inertia, shear effect, and gyroscopic effects, is given in Table 10.11 for the 14-element model. The forward and backward whirl natural frequencies for the first four modes for each spin speed is given in Table 10.11. Figure 10.15 shows the Campbell diagram for the 7-element model, and Figure 10.16 the shows the Campbell diagram for the 14-element model. Critical speeds obtained from the Campbell diagram are given in Table 10.12 for the 7- and 14-element models.

TABLE 10.11
Variation of Natural Whirl Frequencies of the Rotor System as a Function of Spin Speed Including Rotary Inertia and Shear Effect (14-Element Model with Rigid Disks)

Spin Speed (rad/s)	Natural Whirl Frequencies (rad/s)		Spin Speed (rad/s)	Natural Whirl Frequencies (rad/s)	
	Backward	Forward		Backward	Forward
0	115.3816	115.3816	1256.64	108.0046	123.0490
	430.8983	430.8983		405.7617	455.3722
	838.1357	838.1357		798.4256	874.7258
	1165.2664	1165.2664		1117.5498	1206.8125
418.88	112.8870	117.9086	1675.52	105.6212	125.6569
	422.5738	439.1482		397.3076	463.3240
	825.2409	850.6818		784.5246	886.2290
	1150.0536	1179.7865		1100.2929	1219.3679
837.76	110.4272	120.4653	2094.4	103.2791	128.2859
	414.1895	447.3102		388.8451	471.1565
	812.0016	862.8781		770.3143	897.3932
	1134.1449	1193.6279		1082.4155	1231.3258
1047.2	109.2111	121.7540	2513.27	100.9801	130.9332
	409.9800	451.3544		380.3932	478.8624
	805.2550	868.8453		755.8155	908.2256
	1125.9320	1200.3007		1063.9768	1242.7197

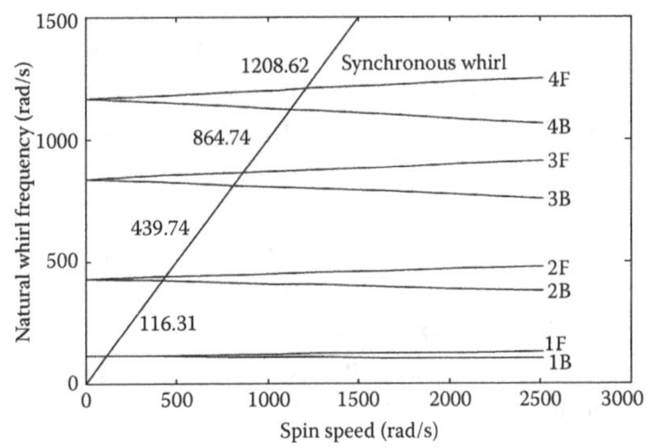

FIGURE 10.15 Campbell diagram for the rotor disk system (seven-element model).

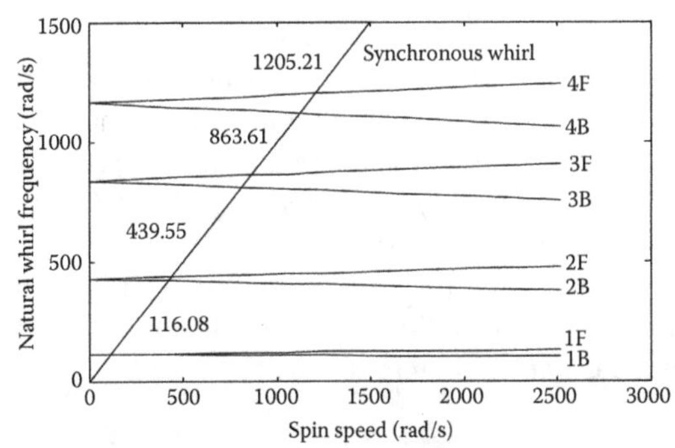

FIGURE 10.16 Campbell diagram for the rotor-bearing system (14-element model).

TABLE 10.12
Critical Speeds of the Rotor-Bearing System with Rigid Disks Including Shear and Rotary Inertia

Mode No.	Critical Speeds without Gyroscopic Effects (rad/s)	Natural Whirl Frequencies with Gyroscopic effect (rad/s)			
		For Seven Elements		For 14 Elements	
		Backward	Forward	Backward	Forward
1	115.3816	114.9266	116.3128	114.6986	116.0819
2	430.8983	422.6644	439.7359	422.5013	439.5510
3	838.1357	813.6305	864.7369	812.7908	863.6147
4	1165.2664	1125.3705	1208.6196	1122.9023	1205.2135

10.5 CONCLUDING REMARKS

In the present chapter, finite element analyses of a simple rotor system, including the gyroscopic effect, are considered. Initially, elemental equations are derived for a simple rotor system with rigid disk and flexible massless shaft by considering the effect of a gyroscopic couple. The gyroscopic couple introduces complexity in four ways: First, both orthogonal motions need to be considered simultaneously for the analysis; second, the gyroscopic effect makes the natural whirl frequency dependent upon the spin speed of the rotor; third, the gyroscopic effect introduces a velocity term in the equation of motion; and fourth, the eigenvalue problem must be formulated in the state-space form. The FEM is applied to the solution of the governing partial differential equation of the Timoshenko beam with rotary and gyroscopic effects. Elemental matrices are presented for the mass, stiffness, and gyroscopic effect. With numerical illustration, standard analysis by using the Campbell diagram is presented to obtain the variations of whirl natural frequencies with the spin speed of the shaft. Critical speeds are obtained from such Campbell diagrams. The external and internal damping effects are not considered in this chapter. The external damping due to bearings will be considered in Chapter 2 of volume II, wherein the Campbell diagram will contain logarithmic decrement also (which gives the instability condition of rotor systems). The internal damping in simple rotor systems will be dealt in Chapter 1 of volume II, which gives instability to rotor systems. For the finite element formulation with the internal damping along with other higher effects, readers can also refer to paper by Ku (1998). Sustainable futuristic rotor design focuses on optimizing aerodynamics, reducing fuel burn, and utilizing lightweight composite materials. Key strategies involve shape-adaptive morphing blades, eco-mode hybrid propulsion integration, and advanced open-rotor architectures.

EXERCISE PROBLEMS

Exercise 10.1 Obtain the forward and backward synchronous bending (transverse) critical speeds for a general motion of a rotor as shown in Figure 10.17. The rotor is assumed to have a fixed support at one end. Take the mass of the disk $m=2\,\text{kg}$, the polar mass moment of inertia $I_p=0.01$ kg-m^2, and the diametral mass moment of inertia $I_d=0.005$ kg-m^2. The shaft is assumed to be massless and its length and diameter are 0.2 and 0.1 m, respectively. Take Young's modulus of the shaft as $E=2.1\times10^{11}\,\text{N/m}^2$. Using the FEM and considering the mass of the shaft with material density $\rho=7800\,\text{kg/m}^3$, obtain the first two forward and backward synchronous bending critical speeds by drawing the Campbell diagram.

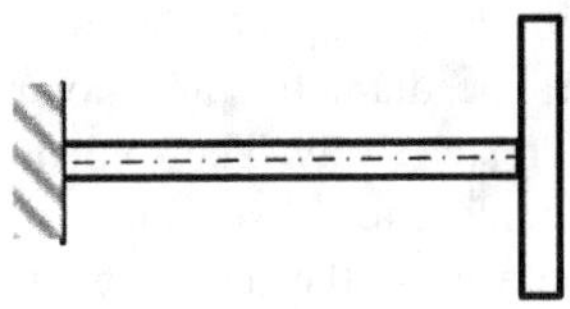

FIGURE 10.17 A cantilever rotor.

Exercise 10.2 Obtain the transverse forward and backward synchronous critical speeds for a general motion of a rotor system as shown in Figure 10.18. Take the mass of the disk as $m=10$ kg, the diametral mass moment of inertia as $I_d=0.02$ kg-m², and the polar mass moment of inertia as $I_p=0.04$ kg-m². The disk is placed at 0.25 m from the right support. The shaft has a diameter of 10 mm, total span length of 1 m, and Young's modulus of 2.1×10^{11} N/m². Consider the mass of the shaft with material density $\rho=7800$ kg/m³ and take into account the gyroscopic effects of the disk.

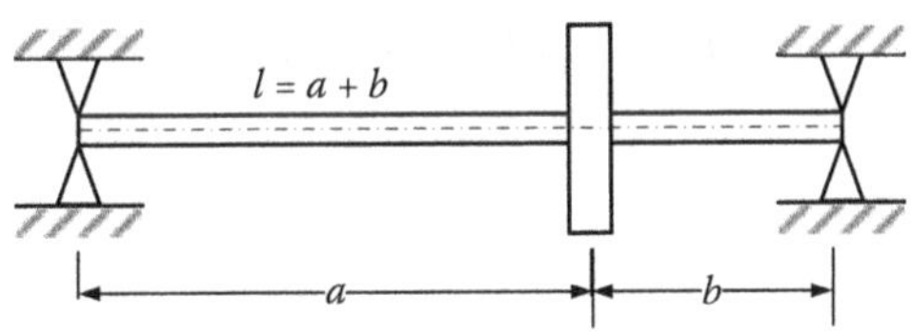

FIGURE 10.18 A rotor system.

Exercise 10.3 Formulate the standard eigenvalue problem for the following rotor system equations of motion:

$$\mathbf{M}\ddot{\boldsymbol{\eta}} - \omega\mathbf{G}\dot{\boldsymbol{\eta}} + \mathbf{K}\boldsymbol{\eta} = \mathbf{f}$$

where **M**, **G**, and **K** are the mass, gyroscopic, and stiffness matrices, respectively; $\boldsymbol{\eta}$ and **f** are the response and force vectors, respectively; and ω is the spin speed of the rotor. Discuss the characteristics of the eigenvalues and interpret them physically for the above case.

Exercise 10.4 Find the bending critical speeds of the rotor system shown in Figure 10.19 by using the finite element analysis. Consider the shaft Young's modulus as $E=2.1\times10^{11}$ N/m². The mass of the disk, m, is 10 kg, and the mass moments of inertia of the disk about its principal axes are $I_p=0.02$ kg-m² and $I_d=0.01$ kg-m². The tilting of the disk from the vertical axis is 2° due to a mounting error. Discuss whether the gyroscopic effect will be present or not. Obtain the critical speeds accordingly. Consider the mass of the shaft with material density $\rho=7800$ kg/m³.

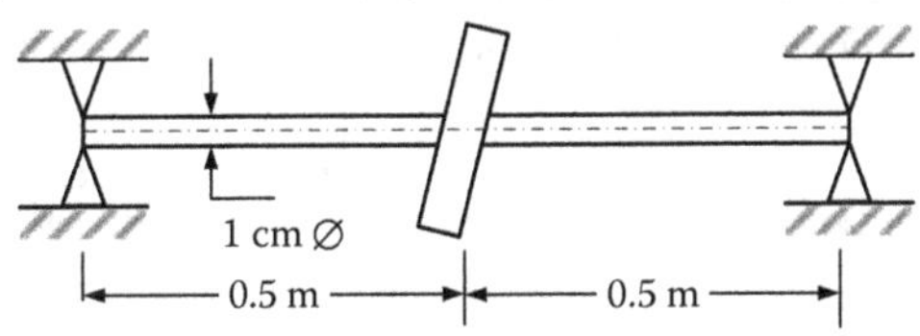

FIGURE 10.19 A simply supported rotor with a tilted rigid disk.

Exercise 10.5 A shaft with diameter $d=0.0155$ m, $EI=600$ N-m² of total length $l=0.6$ m is fixed at one end and a roller support at 0.3 m from the fixed end. At a quarter length from the simply supported bearing, the shaft carries a thin disk of mass $m=6$ kg and of diameter moment of inertia, $I_d=0.03$ kg-m². Find (a) the nonrotational natural frequency and (b) the whirl frequency.

Exercise 10.6 The rotor of a turbine is 13.6 kg in mass and is mounted at the midspan of a shaft with bearings 0.4064 m apart. The diametral and polar mass moment of inertia of the disk are 0.01 kg-m^2 and 0.01 kg-m^2, respectively. The rotor is known to have an unbalance of 0.2879 kg-cm and 120° phase with respect to a fixed mark of the shaft. Obtain the Campbell diagram and unbalance response. Determine the forces exerted on the bearings at a speed of 6000 rpm if the diameter of the steel shaft is 2.54 cm. Assume the shaft to be simply supported at the bearings. Take E=200 GN-m^{-2}. Consider the inertia of the shaft.

Exercise 10.7 A shaft of total length l=1 m, supported on end bearings, carries two disks at the quarter-length points. The disks have mass m=1 kg, the polar mass moment of inertia I_p=0.02 kg-m^2, the diametral mass moment of inertia I_d=0.01 kg-m^2, the shaft diameter, d, is 0.01 m, and $E = 2.1 \times 10^{11}$ N/m^2. Obtain the Campbell diagram up to fourth mode and discuss the results.

Exercise 10.8 Derive the equations of motion and boundary conditions of a rotor system for transverse (bending) vibrations by using Hamilton's principle for the following strain energy, U, and kinetic energy, T, terms:

$$U = \frac{1}{2}\int_0^l EI\left\{\varphi_x'^2 + \varphi_y'^2\right\} dz$$

and

$$T = \frac{1}{2}\int_0^l \rho A\left\{\dot{u}_x^2 + \dot{u}_y^2\right\} dz + \frac{1}{2}\int_0^l I_d\left\{\dot{\varphi}_x^2 + \dot{\varphi}_y^2\right\} dz + \int_0^l \omega^2 I_p dz - \int_0^l \omega I_p \dot{\varphi}_x \dot{\varphi}_y dz$$

with

$$\varphi_x = -\frac{du_y}{dz}; \quad \varphi_y = \frac{du_x}{dz}$$

where u_x and u_y are the rotor translational displacements in the horizontal and vertical directions, respectively; I_d and I_p are the diametral and polar mass moments of inertia of the shaft, respectively; EI is the modulus of rigidity of the shaft; A is the cross-sectional area of the shaft; l is the length of the shaft; ρ is the mass density of the shaft material; ω is the shaft spin speed; and *dot* and *prime* represent the time, t, and spatial (in the shaft axial direction, z) derivatives, respectively, given the physical significance of contributions of each term in the strain and kinetic energy expressions.

For E=2.1×10^{11} N/m^2, ρ=7800 k/m^3, d=0.1 m, and l=1.0 m, obtain the Campbell diagram and critical speeds.

Exercise 10.9 Develop the finite element formulation for the following equations of motion of a rotor by using the Galerkin method (there is no need to obtain elements of the mass matrix, stiffness matrix, etc.; instead keep them in the integral form):

$$EI\frac{\partial^4 u_x}{\partial z^4} - mr^2\left(\frac{\partial^4 u_x}{\partial z^2 \partial t^2} + 2\omega\frac{\partial^3 u_y}{\partial z^2 \partial t}\right) + m\frac{\partial^2 u_x}{\partial t^2} = 0$$

and

$$EI\frac{\partial^4 u_y}{\partial z^4} - mr^2\left(\frac{\partial^4 u_y}{\partial z^2 \partial t^2} - 2\omega\frac{\partial^3 u_x}{\partial z^2 \partial t}\right) + m\frac{\partial^2 u_y}{\partial t^2} = 0$$

where m (= ρA) is the mass per unit length of the shaft material; r is the radius of gyration of the shaft; u_x and u_y are the rotor translational displacements in the horizontal and vertical directions, respectively; ω is the shaft rotational frequency; EI is the modulus of rigidity of the shaft; and A

is the cross-sectional area of the shaft (refer to Chapter 1 of Volume II for the derivation of the above equations of motion). Comment about the shape function including, whether there would be any changes as compared to the Euler–Bernoulli beam model.

For $E=2.1\times10^{11}\,\text{N/m}^2$, $\rho=7800$ k/m³, $d=0.1$ m, and $l=1.0$ m, obtain the Campbell diagram and critical speeds.

Exercise 10.10 Consider the rotor system shown in Figure 10.20. The shaft has distributed mass and stiffness properties. Bearings B_1 and B_2 are simply supported bearings. The span of the shaft is 3 m, the diameter of the shaft is 10 mm, the mass density of the shaft material is 7800 kg/m³, and Young's modulus of the shaft material is $2.1\times10^{11}\,\text{N/m}^2$. Proceed with dividing the beam into three elements and writing the elemental and global equations, and application of the boundary conditions. Determine the bending critical speeds of the rotor system by using the developed finite element system equations.

For $E=2.1\times10^{11}\,\text{N/m}^2$, $\nu=0.1325$, $\rho=7800$ k/m³, $d=0.1$ m, and $L=1.0$ m, obtain the Campbell diagram and critical speeds.

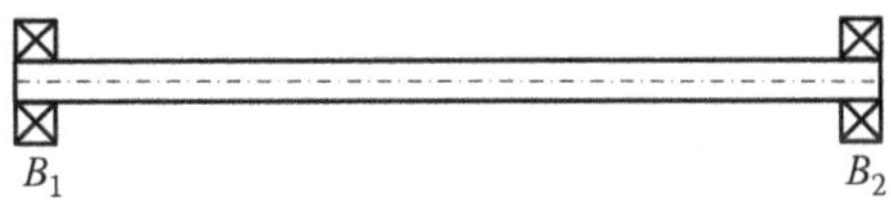

FIGURE 10.20 A uniform continuous simply supported shaft.

Exercise 10.11 Obtain four bending critical speeds of the rotor system as shown in Figure 10.21. Take the mass of the disk as $m=5$ kg, and the diametral mass moment of inertia as $I_d=0.02$ kg-m². Take shaft length $a=0.3$ m and $b=0.7$ m. The diameter of the shaft is 3 cm. Use the Timoshenko beam model and consider gyroscopic effects. Take $E=2.1\times10^{11}\,\text{N/m}^2$, $\nu=0.1325$, and $\rho=7800$ k/m³.

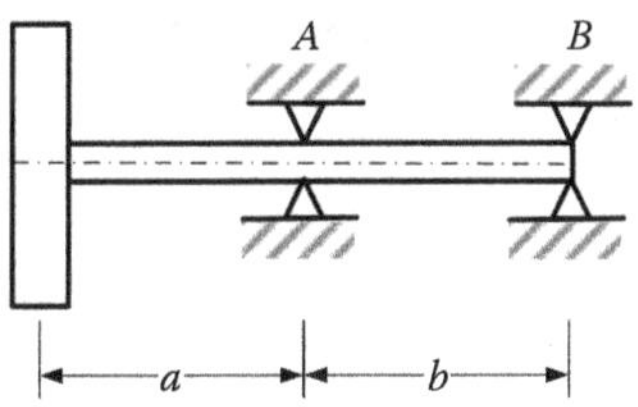

FIGURE 10.21 An overhung rotor system.

Exercise 10.12 Find four bearing critical speeds of the rotor system shown in Figure 10.22 using finite element analysis. The following rotor data are given: $m=10$ kg, $I_d=0.02$ kg-m², $d=0.3$ m, $L=1$ m, $E=2.1\times10^{11}\,\text{N/m}^2$, $\nu=0.1325$, and $\rho=7800$ k/m³. Use the Timoshenko beam and consider the gyroscopic effects.

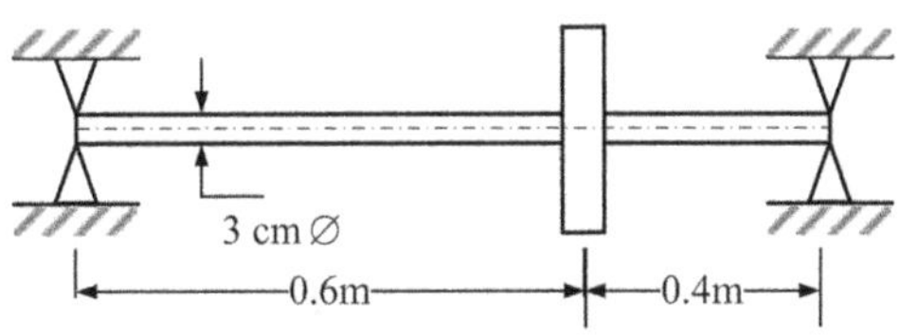

FIGURE 10.22 A uniform continuous shaft with a rigid disk.

Exercise 10.13 For a pure angular (transverse) motion of the rotor system shown in Figure 10.23, obtain the critical speeds. Choose the FEM or transfer matrix method (TMM) appropriately (*do not use the quasi-static method*). Neglect the inertia of the shaft; however, consider the rotary inertia and gyroscopic effects of the disk. Take a=0.5, shaft diameter d=0.1 m, thin disk diametral mass moment of inertia I_d=0.1 kg-m^2, and E=2.1×10^{11} N/m^2.

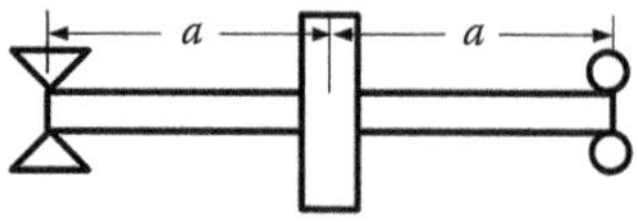

FIGURE 10.23 A Jeffcott rotor.

Exercise 10.14 Choose a single answer from the multiple choice questions.

i. For the eigenvalue problem of the rotor system with a gyroscopic couple and without damping, the eigenvalues are expected to be
   A. complex
   B. pure real
   C. pure imaginary
   D. zero

ii. Using the FEM with fewer numbers of elements, the natural frequency of the rotor system as compared to the exact one is expected to be
   A. always more
   B. always less
   C. equal
   D. either less or more

iii. While using the FEM in a rotor system with gyroscopic effects (with damping in bearings), the eigenvalue problem gives eigenvalues as
   A. pure real
   B. complex
   C. real as well as complex
   D. pure imaginary

iv. A rotor model with the Timoshenko beam theory includes
   A. bending effect only
   B. bending, shear, rotary, and gyroscopic effects
   C. bending, shear, and gyroscopic effects
   D. bending, shear, and rotary effects

v. In the Timoshenko beam theory for a nonrotating beam, the following effect is not included:
   A. gyroscopic
   B. shear deformation
   C. rotary inertia
   D. bending

APPENDIX 10.1

10.1 TIMOSHENKO BEAM MODEL

Matrices are corresponding to the z-x plane and the y-z plane; the sign changes will take place of terms of angular displacements and the corresponding moments as discussed in Section 10.1 and in Chapter 9.

10.1.1 Translational Mass Matrix

For stacking of the displacement vector, η, refer to Equation 10.92.

$$\mathbf{M}_T = \mathbf{M}_{T0} + \Phi\mathbf{M}_{T1} + \Phi^2\mathbf{M}_{T2}$$

where

$$\mathbf{M}_{T0} = \frac{\rho A l}{420(1+\Phi)^2}\begin{bmatrix} 156 & & & sym \\ 22l & 4l^2 & & \\ 54 & 13l & 156 & \\ -13l & -3l^2 & -22l & 4l^2 \end{bmatrix}$$

$$\mathbf{M}_{T1} = \frac{\rho A l}{420(1+\Phi)^2}\begin{bmatrix} 294 & & & sym \\ 38.5l & 7l^2 & & \\ 126 & 31.5l & 294 & \\ -31.5l & -7l^2 & -38.5l & 7l^2 \end{bmatrix}$$

$$\mathbf{M}_{T2} = \frac{\rho A l}{420(1+\Phi)^2}\begin{bmatrix} 140 & & & sym \\ 17.5l & 3.5l^2 & & \\ 70 & 17.5 & 140 & \\ -17.5 & -3.5l^2 & -17.5l & 3.5l^2 \end{bmatrix}$$

10.1.2 Rotational Mass Matrix

$$\mathbf{M}_R = \mathbf{M}_{R0} + \Phi\mathbf{M}_{R1} + \Phi^2\mathbf{M}_{R2}$$

where

$$\mathbf{M}_{R0} = \frac{\rho I}{(1+\Phi)^2}\begin{bmatrix} \frac{6}{5l} & & & sym \\ \frac{1}{10} & \frac{2l}{15} & & \\ \frac{-6}{5l} & \frac{-1}{10} & \frac{6}{5l} & \\ \frac{1}{10} & \frac{-l}{30} & \frac{-1}{10} & \frac{l}{15} \end{bmatrix}$$

$$\mathbf{M}_{R1} = \frac{\rho I}{(1+\Phi)^2}\begin{bmatrix} 0 & & & sym \\ \frac{-1}{2} & \frac{l}{6} & & \\ 0 & \frac{1}{2} & 0 & \\ \frac{-1}{2} & \frac{-l}{6} & \frac{1}{2} & \frac{l}{6} \end{bmatrix}$$

$$\mathbf{M}_{R2} = \frac{\rho I}{(1+\Phi)^2}\begin{bmatrix} 0 & & & sym \\ 0 & \frac{l}{3} & & \\ 0 & 0 & 0 & \\ 0 & \frac{l}{6} & \frac{1}{2} & \frac{l}{6} \end{bmatrix}$$

10.1.3 Stiffness Matrix

$$\mathbf{K}^{(e)} = \mathbf{K}_0^{(e)} + \Phi\mathbf{K}_1^{(e)}$$

where

$$\mathbf{K}_0^{(e)} = \frac{EI}{(1+\Phi)l^3}\begin{bmatrix} 12 & & & sym \\ 6l & 4l^2 & & \\ -12 & -6l & 12 & \\ 6l & 2l^2 & -6l & 4l^2 \end{bmatrix}$$

$$\mathbf{K}_1^{(e)} = \frac{EI}{(1+\Phi)l^3}\begin{bmatrix} 0 & & & sym \\ 0 & l^2 & & \\ 0 & 0 & 0 & \\ 0 & -l^2 & 0 & l^2 \end{bmatrix}$$

APPENDIX 10.2

10.2 SPINNING TIMOSHENKO SHAFT MODEL

10.2.1 Translational Mass Matrix

For the sequence of the displacement vector, η, refer to Equation 10.125.

$$\mathbf{M}_T^{(e)} = \mathbf{M}_{T0}^{(e)} + \Phi\mathbf{M}_{T1}^{(e)} + \Phi^2\mathbf{M}_{T2}^{(e)}$$

$$\mathbf{M}_{T0}^{(e)} = \frac{\rho A l}{420(1+\Phi)^2}\begin{bmatrix} 156 & & & & & & & sym \\ 0 & 156 & & & & & & \\ 0 & -22l & 4l^2 & & & & & \\ 22l & 0 & 0 & 4l^2 & & & & \\ 54 & 0 & 0 & 13l & 156 & & & \\ 0 & 54 & -13l & 0 & 0 & 156 & & \\ 0 & 13l & -3l^2 & 0 & 0 & 22l & 4l^2 & \\ -13l & 0 & 0 & -3l^2 & -22l & 0 & 0 & 4l^2 \end{bmatrix}$$

$$\mathbf{M}_{T1}^{(e)} = \frac{\rho A l}{420(1+\Phi)^2}\begin{bmatrix} 294 & & & & & & & sym \\ 0 & 294 & & & & & & \\ 0 & -38.5l & 7l^2 & & & & & \\ 38.5l & 0 & 0 & 7l^2 & & & & \\ 126 & 0 & 0 & 31.5l & 294 & & & \\ 0 & 126 & -31.5l & 0 & 0 & 294 & & \\ 0 & 31.5l & -7l^2 & 0 & 0 & 38.5l & 7l^2 & \\ -31.5l & 0 & 0 & -7l^2 & -38.5l & 0 & 0 & 7l^2 \end{bmatrix}$$

$$\mathbf{M}_{T2}^{(e)} = \frac{\rho A l}{420(1+\Phi)^2}\begin{bmatrix} 140 & & & & & & & sym \\ 0 & 140 & & & & & & \\ 0 & -17.5l & 3.5l^2 & & & & & \\ 17.5l & 0 & 0 & 3.5l^2 & & & & \\ 70 & 0 & 0 & 17.5l & 140 & & & \\ 0 & 70 & -17.5l & 0 & 0 & 140 & & \\ 0 & 17.5l & -3.5l^2 & 0 & 0 & 17.5l & 3.5l^2 & \\ -17.5l & 0 & 0 & -3.5l^2 & -17.5l & 0 & 0 & 3.5l^2 \end{bmatrix}$$

10.2.2 Rotational Mass Matrix

$$\mathbf{M}_{R}^{(e)} = \mathbf{M}_{R0}^{(e)} + \Phi\mathbf{M}_{R1}^{(e)} + \Phi^2\mathbf{M}_{R2}^{(e)}$$

$$\mathbf{M}_{R0}^{(e)} = \frac{\rho I}{(1+\Phi)^2}\begin{bmatrix} \frac{6}{5l} & & & & & & & sym \\ 0 & \frac{6}{5l} & & & & & & \\ 0 & -\frac{1}{10} & \frac{2l}{15} & & & & & \\ \frac{1}{10} & 0 & 0 & \frac{2l}{15} & & & & \\ -\frac{6}{5l} & 0 & 0 & -\frac{1}{10} & \frac{6}{5l} & & & \\ 0 & -\frac{6}{5l} & \frac{1}{10} & 0 & 0 & \frac{6}{5l} & & \\ 0 & -\frac{1}{10} & -\frac{l}{30} & 0 & 0 & \frac{1}{10} & \frac{2l}{15} & \\ \frac{1}{10} & 0 & 0 & -\frac{l}{30} & -\frac{1}{10} & 0 & 0 & \frac{2l}{15} \end{bmatrix}$$

$$\mathbf{M}_{R1}^{(e)}=\frac{\rho I}{(1+\Phi)^2}\begin{bmatrix} 0 & & & & & & & sym \\ 0 & 0 & & & & & & \\ 0 & \frac{1}{2} & \frac{l}{6} & & & & & \\ -\frac{1}{2} & 0 & 0 & \frac{l}{6} & & & & \\ 0 & 0 & 0 & \frac{1}{2} & 0 & & & \\ 0 & 0 & -\frac{1}{2} & 0 & 0 & 0 & & \\ 0 & \frac{1}{2} & -\frac{l}{6} & 0 & 0 & -\frac{1}{2} & \frac{l}{6} & \\ -\frac{1}{2} & 0 & 0 & -\frac{l}{6} & \frac{1}{2} & 0 & 0 & \frac{l}{6} \end{bmatrix}$$

$$\mathbf{M}_{R2}^{(e)}=\frac{\rho I}{(1+\Phi)^2}\begin{bmatrix} 0 & & & & & & & sym \\ 0 & 0 & & & & & & \\ 0 & 0 & \frac{l}{3} & & & & & \\ 0 & 0 & 0 & \frac{l}{3} & & & & \\ 0 & 0 & 0 & 0 & 0 & & & \\ 0 & 0 & 0 & 0 & 0 & 0 & & \\ 0 & 0 & \frac{l}{6} & 0 & 0 & 0 & \frac{l}{3} & \\ 0 & 0 & 0 & \frac{l}{6} & 0 & 0 & 0 & \frac{l}{3} \end{bmatrix}$$

10.2.3 Stiffness Matrix

$$\mathbf{K}^{(e)}=\mathbf{K}_0^{(e)}+\Phi\mathbf{K}_1^{(e)}$$

$$\mathbf{K}_0^{(e)}=\frac{EI}{(1+\Phi)l^3}\begin{bmatrix} 12 & & & & & & & sym \\ 0 & 12 & & & & & & \\ 0 & -6l & 4l^2 & & & & & \\ 6l & 0 & 0 & 4l^2 & & & & \\ -12 & 0 & 0 & -6l & 12 & & & \\ 0 & -12 & 6l & 0 & 0 & 12 & & \\ 0 & -6l & 2l^2 & 0 & 0 & 6l & 4l^2 & \\ 6l & 0 & 0 & 2l^2 & -6l & 0 & 0 & 4l^2 \end{bmatrix}$$

$$\mathbf{K}_1^{(e)} = \frac{EI}{(1+\Phi)l^3}\begin{bmatrix} 0 & & & & & & & sym \\ 0 & 0 & & & & & & \\ 0 & 0 & l^2 & & & & & \\ 0 & 0 & 0 & l^2 & & & & \\ 0 & 0 & 0 & 0 & 0 & & & \\ 0 & 0 & 0 & 0 & 0 & 0 & & \\ 0 & 0 & -l^2 & 0 & 0 & 0 & l^2 & \\ 0 & 0 & 0 & -l^2 & 0 & 0 & 0 & l^2 \end{bmatrix}$$

10.2.4 Gyroscopic Matrix

$$\mathbf{G}^{(e)} = \mathbf{G}_0^{(e)} + \Phi\mathbf{G}_1^{(e)} + \Phi^2\mathbf{G}_2^{(e)}$$

with (r is the radius of the shaft and A is the cross section area of the shaft so that $I=Ar^2/4$)

$$\mathbf{G}_0^{(e)} = \frac{\rho A r^2}{60(1+\Phi)^2 l}\begin{bmatrix} 0 & & & & & & & skew\text{-}sym \\ 36 & 0 & & & & & & \\ -3l & 0 & 0 & & & & & \\ 0 & -3l & 4l^2 & 0 & & & & \\ 0 & 36 & -3l & 0 & 0 & & & \\ -36 & 0 & 0 & -3l & 36 & 0 & & \\ -3l & 0 & 0 & l^2 & 3l & 0 & 0 & \\ 0 & -3l & -l^2 & 0 & 0 & 3l & 4l^2 & 0 \end{bmatrix}$$

$$\mathbf{G}_1^{(e)} = \frac{\rho A r^2}{60(1+\Phi)^2 l}\begin{bmatrix} 0 & & & & & & & skew\text{-}sym \\ 0 & 0 & & & & & & \\ 15l & 0 & 0 & & & & & \\ 0 & 15l & 5l^2 & 0 & & & & \\ 0 & 0 & 15l & 0 & 0 & & & \\ 0 & 0 & 0 & 15l & 0 & 0 & & \\ 15l & 0 & 0 & 5l^2 & 0 & 0 & 0 & \\ 0 & 15l & -5l^2 & 0 & -15l & 5l^2 & 5l^2 & 0 \end{bmatrix}$$

$$\mathbf{G}_2^{(e)} = \frac{\rho A r^2}{60(1+\Phi)^2 l}\begin{bmatrix} 0 & & & & & & & skew\text{-}sym \\ 0 & 0 & & & & & & \\ 0 & 0 & 0 & & & & & \\ 0 & 0 & 10l^2 & 0 & & & & \\ 0 & 0 & 0 & 0 & 0 & & & \\ 0 & 0 & 0 & 0 & 0 & 0 & & \\ 0 & 0 & 0 & -5l^2 & 0 & 0 & 0 & \\ 0 & 0 & 5l^2 & 0 & 0 & 0 & 10l^2 & 0 \end{bmatrix}$$

10.2.5 Rigid Disk Element Matrices

$$\mathbf{M}^d = \begin{bmatrix} m_d & 0 & 0 & 0 \\ 0 & m_d & 0 & 0 \\ 0 & 0 & I_d & 0 \\ 0 & 0 & 0 & I_d \end{bmatrix}$$

$$\mathbf{G}^d = \begin{bmatrix} 0 & 0 & 0 & 0 \\ 0 & 0 & 0 & 0 \\ 0 & 0 & 0 & -I_p \\ 0 & 0 & I_p & 0 \end{bmatrix}$$

$$\eta^d = \left\{ u_x \quad u_y \quad \varphi_y \quad \varphi_x \right\}^T$$

ANSWERS TO SELECTED EXERCISE PROBLEMS

10.1 Refer Table 10.13

TABLE 10.13
Critical Speeds with Different Numbers of Elements

	Critical Speed without Gyroscopic Effects ($I_p = 0$ and $I_d \neq 0$) (rad/s)		Critical Speed with Gyroscopic Effects (rad/s)			
			For One Element		For Ten Elements	
Mode no.	One Element	Ten Elements	Backward whirl	Forward whirl	Backward whirl	Forward whirl
1	129.04	129.04	114.29	148.08	114.29	148.08
2	680.95	680.91	445.33	[a]	445.32	[a]
3	[b]	7321.96	[b]	[b]	7265.55	7303.07

[a] Critical speed could not be obtained because estimated natural whirl frequency with only one element never intersected the $\omega_{nf} = \omega$ line. This is due to the fact that the FEM always gives over estimation of the natural frequencies and error in the second natural frequency is expected to be more with a single element; hence, variation of the natural whirl frequency was diverging nature.

[b] With a single element, it is possible to get critical speeds only up to the second mode.

10.2 The first pair of backward and forward critical speeds are 28.38 rad/s and 29.17 rad/s. Similarly, the second pair is 162.95 rad/s and 335.95 rad/s. For the third mode only the backward whirl mode is feasible at 373.27 rad/s.

10.3 Not applicable

10.4 The first pair of backward and forward critical speeds are 21.92 rad/s and 21.92 rad/s. Similarly, the second pair is 193.73 rad/s and 943.82 rad/s. For the third mode again, the gyroscopic effect is absent with the third pair of critical speeds are 1070.24 rad/s and 1070.24 rad/s.

10.5 The first pair of backward and forward critical speeds for four elements are 69.59 rad/s and 82.58 rad/s. Similarly, the second pair is 314.18 rad/s and 3480.87 rad/s. For the third mode again only the backward critical speed is feasible at 3506.16 rad/s.

10.6 The first pair of backward and forward critical speeds are 461.92 rad/s and 461.92 rad/s. Similarly, the second pair is 2012.84 rad/s and 15,545.98 rad/s. For the third mode again, the gyroscopic effect is absent with the third pair of critical speeds are 16,553.62 rad/s and 16,553.62 rad/s.

10.7 Critical speeds: 55.79 rad/s (B1), 63.97 rad/s (F1), 185.62 rad/s (B2), 185.62 rad/s (F2), 261.07 rad/s (B3.).

10.8 Critical speeds for the backward and forward modes, respectively, are (1.2686×10^3, 1.2842×10^3) rad/s for the first mode, (4.9415×10^3, 5.1855×10^3) rad/s for the second mode, (1.0668×10^4, 1.1856×10^4) rad/s for the third mode, (1.7993×10^4, 2.1577×10^4) rad/s for the fourth mode.

10.9 Refer Table 10.14

TABLE 10.14
Comparison of FEM-Based Critical Speeds (in rad/s) for Different Numbers of Elements with Closed Form Solution

Mode Number	Forward Critical Speed (Two Elements)	Backward Critical Speed (Two Elements)	Forward Critical Speed (Ten Elements)	Backward Critical Speed (Ten Elements)	Forward Critical Speed (20 Elements)	Backward Critical Speed (20 Elements)	Forward Critical Speed (Closed Form)	Backward Critical Speed (Closed Form)
1	1287.31	1283.35	1282.26	1278.31	1282.25	1278.30	1284.24	1268.59
2	5719.85	5648.79	5153.52	5090.33	5153.01	5089.82	5185.46	4941.46
3	14,502.33	14,081.30	11,692.02	11,371.85	11,686.17	11,366.16	11,856.25	10,668.25
4	26,759.19	25,389.35	21,044.15	20,029.96	21,011.54	19,999.02	21,576.77	17,993.05

10.10 Refer Table 10.15

TABLE 10.15
Comparison of FEM-Based Critical Speeds (in rad/s) for Different Numbers of Elements with Closed Form Solution

Mode Number	Forward Critical Speed (Two Elements: Timoshenko Spinning Shaft)	Backward Critical Speed (Two Elements: Timoshenko Spinning Shaft)	Forward Critical Speed (Four Elements: Timoshenko Spinning Shaft)	Backward Critical Speed (Four Elements: Timoshenko Spinning Shaft)	Forward Critical Speed (20 Elements: Rayleigh Spinning Shaft)	Backward Critical Speed (20 Elements: Rayleigh Spinning Shaft)	Critical Speed (Closed Form for the Non-Spinning Shaft)
1	1254.56	1250.77	1234.10	1230.36	1282.25	1278.30	1264.99
2	5541.69	5474.52	4908.78	4851.50	5153.01	5089.82	4891.61
3	13,804.20	13,436.38	11,063.17	10,788.32	11,686.17	11,366.16	10,465.30
4	25,383.64	24,195.56	21,457.71	20,503.55	21,011.54	19,999.02	17,503.31

10.11 Refer Table 10.16

TABLE 10.16
Comparison of FEM-Based Critical Speeds (in rad/s) for Different Numbers of Elements

Mode Number	Forward Critical Speed (Two Elements: Timoshenko Spinning Shaft)	Backward Critical Speed (Two Elements: Timoshenko Spinning Shaft)	Forward Critical Speed (Four Elements: Timoshenko Spinning Shaft)	Backward Critical Speed (Four Elements: Timoshenko Spinning Shaft)
1	226.84	205.64	227.09	205.93
2	1289.93	1078.54	1062.88	974.96
3	4917.61	1552.64	3924.56	1408.70
4	—	5071.27	7399.20	3983.34

10.12 Refer Table 10.17

TABLE 10.17
Comparison of FEM-Based Critical Speeds (in rad/s) for Different Numbers of Elements

Mode Number	Forward Critical Speed (Two Elements: Timoshenko Spinning Shaft)	Backward Critical Speed (Two Elements: Timoshenko Spinning Shaft)	Forward Critical Speed (Four Elements: Timoshenko Spinning Shaft)	Backward Critical Speed (Four Elements: Timoshenko Spinning Shaft)
1	184.06	183.61	184.25	183.80
2	1755.19	1066.62	1435.78	1035.96
3	4503.91	2586.14	3472.57	2060.09
4	—	5268.81	6153.48	4045.65

10.13 The first pair of backward and forward critical speeds are 1113.58 rad/s and 1113.58 rad/s. Similarly, the second pair of critical speeds are 4401.07 rad/s and 6278.47 rad/s for the backward and forward whirl, respectively. For the third mode again, the gyroscopic effect is absent with the third pair of critical speeds are 13,025.94 rad/s and 13,025.94 rad/s. In the fourth mode only, the backward whirl critical speed is present at 14,092.07 rad/s.

10.14

(i) C (ii) A (iii) B (iv) B (v) A

REFERENCES

Cowper, G.R. 1966, The shear coefficient in Timoshenko's beam theory, *ASME Journal of Applied Mechanics, 33(2)*, 335–340.

Eshleman, R.L. and Eubanks, R.A., 1969, On the critical speeds of a continuous rotor, *ASME Journal of Engineering for Industry, 91(4B)*, 1180–1188.

Hutchinson, J.R. (2001), Shear coefficients for Timoshenko beam theory, *Journal of Applied Mechanics, 68(1)*, 87–92.

Ku, D.-M., 1998, Finite element analysis of whirl speeds for rotor-bearing systems with internal damping, *Mechanical Systems and Signal Processing, 12(5)*, 599–610.

Nelson, H.D., 1980, A finite rotating shaft element using Timoshenko beam theory, *Transactions of the ASME, Journal of Mechanical Design, 102*, 793–803.

Nelson, H.D., 1998, Rotordynamic modelling and analysis procedures: A review. *JSME International Journal, Series C, 41(1)*, 1–12.

Reddy, J.N., 2003, *An Introduction to the Finite Element Method*. New Delhi: McGraw Hill.

Shames, I.H. and Dym, C.L., 2005, *Energy and Finite Element Methods in Structural Mechanics*. New Delhi: New Age International.

Weaver, W., Timoshenko, S.P., and Young, D.H., 1990, *Vibration Problems in Engineering*. New York: John Wiley & Sons.

Index

Q

R

S

T

U

W

For Product Safety Concerns and Information please contact our EU
representative GPSR@taylorandfrancis.com
Taylor & Francis Verlag GmbH, Kaufingerstraße 24, 80331 München, Germany

www.ingramcontent.com/pod-product-compliance
Lightning Source LLC
LaVergne TN
LVHW080621120926
847902LV00005B/58

* 9 7 8 1 0 4 1 1 0 3 5 5 4 *